HYDROLOGY AND WATER RESOURCES OF INDIA

Water Science and Technology Library

VOLUME 57

The titles published in this series are listed at the end of this volume.

HYDROLOGY AND WATER RESOURCES OF INDIA

by

SHARAD K. JAIN
National Institute of Hydrology,
Roorkee, India

PUSHPENDRA K. AGARWAL
National Institute of Hydrology,
Roorkee, India

and

VIJAY P. SINGH
Department of Biological and Agricultural Engineering,
Texas A & M University,
College Station, Texas, U.S.A.

Springer

A C.I.P. Catalogue record for this book is available from the Library of Congress.

ISBN-10 1-4020-5179-4 (HB)
ISBN-13 978-1-4020-5179-1 (HB)
ISBN-10 1-4020-5180-8 (e-book)
ISBN-13 978-1-4020-5180-7 (e-book)

Published by Springer,
P.O. Box 17, 3300 AA Dordrecht, The Netherlands.

www.springer.com

Cover Illustration: Map of India

Printed on acid-free paper

Dedicated to

SKJ: My family, with respect and affection

PKA: My late father

VPS: Anita, Vinay and Arti

Saraswati saryuh sindhurmibhirma ho
maheervasa yantu vakshinih |
Devirapo matarah sudaytnavo
ghravatpayo madhumanno archat ||
Yajur Veda (X,64.9)

Water from wells, rivers, rain, and
from any other source on earth
should be used wisely as it is gift of nature,
for well-being of all.

PREFACE

India is truly a country of continental scale in every respect – topography, climate, culture, and of course water resources. Currently, India is one of the fastest growing economies of the world. Needless to say, the water sector has a crucial role to play in the social well-being and the quality of life.

The need for a book describing India's hydrological practices and water resources, problems in their exploitation, and prospects for their solution has long been felt. A book of somewhat similar nature was published more than three decades ago. During this intervening time a lot of changes have occurred in India not only in the practice of hydrological and water resources technology but also in the environment, society, landscape and land use, agricultural practices, and even climate. Thus, this book attempts to provide a more up-to-date account of water resources of India. It is aimed at students, practitioners, mangers, planners, administrators, and anyone who may be interested in the development and management of water resources in India.

The book is organized in four sections. The first section *Introduction* has one chapter, *Physical Environment of India*, which gives a broad overview of the country. The second section, *Hydrology and Hydrometeorology*, is comprised of five chapters dealing with the major components of the hydrologic cycle for India. *Water Budget and Population of India* are discussed in Chapter 2, followed by a discussion of *Rainfall and Analysis of Rainfall Data* in Chapter 3; *Evaporation and Other Meteorological Data* in Chapter 4; *Runoff and Streamflow* in Chapter 5; and *Groundwater* in Chapter 6. These chapters describe instruments and measurement techniques, and the techniques of analysis that are used in India in order to measure and analyze the components of the hydrologic cycle. Maps, figures, and tables are extensively used to illustrate the quantities of concern and their spatial and temporal variations.

The third section, comprising 10 chapters, focuses on *River Basins of India.* Beginning with an overview of the various river basins and their water resources in Chapter 7 entitled '*River Basins of India*', it goes on to discuss individual basins or a group of basins. Chapter 8 discusses *Ganga Basin*; Chapter 9 *Brahmaputra Basin*; Chapter 10 *Indus Basin*; Chapter 11 *Narmada Basin*; Chapter 12 *Tapi, Sabarmati, and Mahi Basins*; Chapter 13 *Mahanadi, Subernarekha, and Brahamani Basins*; Chapter 14 *Krishna and Godavari Basins*; Chapter 15 *Cauvery and Pennar Basins*; and Chapter 16 *Other Basins and Islands.*

The fourth section entitled *Water Uses, Projects, Problems and Governance* is the final section of the book and is comprised of 8 chapters. Chapter 17 describes the *Major Uses of Water* in India. It discusses at length the water uses for irrigation, hydropower, municipal and industrial uses, environment flow, recreation, navigation, etc. *Problems Related to Water Resources* form the subject matter of Chapter 18, followed by Chapter 19 on *Reservoirs and Lakes* in India, and Chapter 20 on *Water Quality Related Aspects.* A practitioner should also be conversant with legal aspects and this perspective is presented in Chapter 21 entitled *Constitutional Provisions, Inter-state Water Disputes and Treaties. Inter Basin Water Transfer* is becoming a major initiative in India these days in order to overcome problems due to spatial and temporal mismatch between water availability and demands. A detailed description of the present proposals being debated in India and other related aspects are presented in Chapter 22. A brief discussion of the *Institutions in the Field of Hydrology and Water Resources* in India forms the subject matter of the Chapter 23. The final Chapter 24 is devoted to *Water Governance* in India.

Water Resources of India is a vast subject and despite this voluminous book, several aspects could not be covered here. We hope that the book will be helpful to those who have interest in the water resources of India. For this book, data was compiled from a very large number of sources and at times data from different sources were quite different. We have attempted to present a consistent and the most likely status of things. It is, however, likely that there are some mistakes in the book and we request the alert readers to bring these to our notice.

Sharad Jain
Pushpendra Agarwal
Vijay Singh

April 2006
Roorkee/College Station

ACKNOWLEDGEMENTS

A large number of colleagues helped us at various stages of preparation of the manuscript. We had long consultations with them and many of our colleagues shared the data and figures that they had. Mr. C. P. Kumar and Dr. Sanjay Jain provided a lot of information, relevant books, photographs, addresses of web-pages, etc. Dr. Sanjay Jain also provided the photo for book cover and his help is thankfully acknowledged. Useful data were also provided by many of our colleagues from National Institute of Hydrology, including Dr. Pratap Singh, Dr. M. K. Goel, Dr. P. K. Bhunya, Dr. K. S. Ramasastri, Dr. K. D. Sharma, and Mr. A. K. Lohani. Dr. Nagesh Kumar of Indian Institute of Science, Bangalore, reviewed a chapter and gave data for some basins. We are thankful to Mr. Hemant Chowdhary, Dr. S. K. Mishra, Prof. G. C. Mishra, Dr. M. L. Kansal, Dr. Vijay Kumar, Mr. S. S. Kanwar, Mr. Pankaj Garg, Mr. A. R. Senthil Kumar, and Mr. Yatveer Singh for the useful information provided by them.

The staff of NIH library, particularly Mohd. Furqanullah, were always very cooperative. Mr. N. K. Varshney prepared excellent drawings at short notice, and Mr. T. P. Panicker and Mr. Rajesh Agarwal provided help with editing.

Completion of a book of this nature requires work till late at night and on weekends over a long time span. Our families not only had to put up with the inconvenience on numerous occasions but also helped a lot and kept us free from other responsibilities. Their support is gratefully acknowledged.

SKJ, PKA, VPS

TABLE OF CONTENTS

LIST OF FIGURES

LIST OF TABLES

Chapter 14. Krishna and Godavari Basins 641

SECTION 1

INTRODUCTION

CHAPTER 1

PHYSICAL ENVIRONMENT OF INDIA

Earth is also known as a blue planet because more than three-fourths of the surface area is covered by water. According to estimates, the total amount of water on earth is 1.4 million cubic km and if this is uniformly spread around the earth as a layer, the thickness of this layer will be nearly 3 km. However, about 97% of earth's water is contained in oceans and sea and fresh water is only 2.7% of the total available water. Out of this, nearly three quarter (75.2%) is frozen in polar regions and about 22.6% is buried as ground water. A considerable part of ground water, known as static water, is too deep underground for easy exploitation. A small proportion of the remaining water is available in rivers, lakes, soil, etc. Thus only a very small fraction of earth's water is utilizable by the mankind.

If the water use data are viewed on a global scale, more than $3,240\,km^3$ of fresh water are used annually. There is considerable variation in sector-wise water use throughout the world. Agriculture is the primary user in Asia, while municipal and industrial sectors are major users in most of Europe and North America. Overall, agriculture is the main consumer of water and accounts for 69% of the use followed by 23% for industries and 8% for domestic use. The use of water for various purposes has significantly risen during the last century and the trend continues. This has led to focus the attention on sustainable use of water resources and maintenance of environmental quality. The first and necessary step in this endeavour is inventory of water resources of a region.

The objective of the present book is to explain hydrology and water resources of India. To that end, it is helpful to get familiarized with the physical environment of India.

1.1. TOPOGRAPHY

India occupies the south-central peninsula of the Asian continent. Besides the mainland, there are two groups of islands, namely Lakshadweep in the Arabian Sea and Andaman & Nicobar Islands in the Bay of Bengal. The mainland of India lies between 8°4′ N and 37°6′ N latitude and 68°7′ E and 97°25′ E longitude. The Andaman and Nicobar islands lie to the south east of the mainland and Lakshadweep to the southwest. A world map showing the location of India is given in Figure 1.

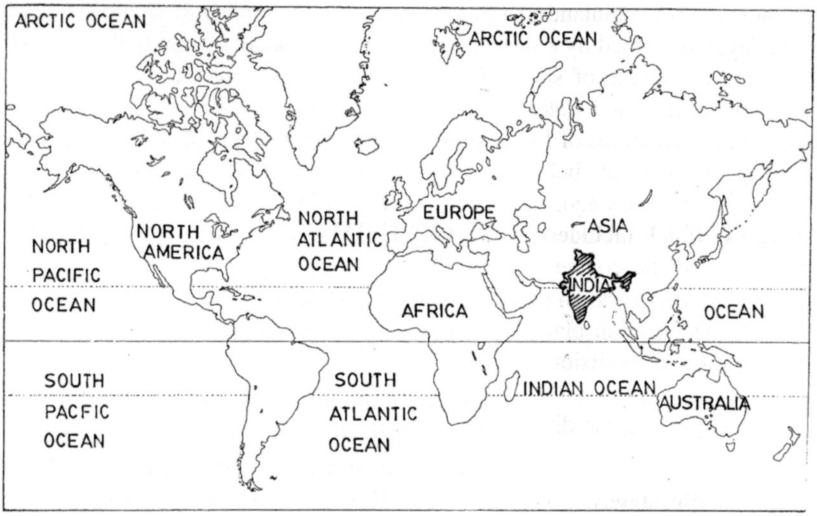

Figure 1. Location of India in the world

Note that the country completely lies in the northern hemisphere. India has a long coastline, extending to about 7,500 km which permits numerous ocean routes to the rest of the world for trade and travel.

India is endowed with almost all the important topographical features, such as high mountains, extensive plateaus, and wide plains traversed by mighty rivers. The country is bounded by Himalayas in the North and has a large peninsular region tapering towards the Indian Ocean. The Himalayas in the north are the major mountain ranges of the world. The other prominent mountains of India include the Aravallis, the Vindhyachals, the Satpuras, the Eastern Ghats, and the Western Ghats. The mountains are the primary source of rivers which derive their flow from rainfall and snow and glacier melt. The plateaus are another striking feature of topography in India and they range in elevation from 300 to 900 m.

With a geographical area of 3,287,263 sq. km, India is the seventh largest country in the world. The countries whose area is larger than India are: Russia, Canada, China, United States of America, Brazil and Australia. India occupies about nearly 2.42% of the land area of the earth. The latitudinal and longitudinal extent of India is almost of the same magnitude in degrees, about 30°. The distance between the extreme north to south tip is about 3,200 km while the east-west extent is 3,000 km. The Indian Standard Time (IST) corresponds to the Meridian of 82°30′ E. Thus, IST is 5 hours 30 minutes ahead of G.M.T. Due to vast extent in the longitudinal direction, the difference between local times of the two extreme points in the east and the west is about two hours. The Tropic of Cancer divides the country into two nearly equal parts. The northern part mainly consists of the Himalayas and the plains of Indus and Ganga. The southern part is of triangular shape.

To the north of the mainland of India lies the mountain chain of Himalayas. The word Himalayas is formed by *him* (snow) with *alaya* (abode), meaning thereby that Himalayas are the abode of snow. The presence of Himalayas has an important bearing on the climate and water resources of the country.

The current land features of India have evolved over a period of millions of years due to tectonic movements below the crust of the earth as well as external natural agents. Millions of years ago, India was a part of a great continent known as the Gondwanaland which included the continent of Australia, Africa, South America and Antarctica. In the course of time, this continent got split into a number of plates and the Indo-Australian plate began to slowly drift towards North. This plate collided with the Eurasian plate millions of years ago and was pushed beneath it. Due to the impact of collision force, the sedimentary rocks of the area folded to form Himalayas.

The major physiographic divisions of India are described in the following.

1.1.1. The Himalayas

The Himalayas mountain region extends all along the northern border of the country. From the eastern border of West Pakistan to the frontiers of Burma, the Himalayas run for a distance of about 2,400 km in the east-west direction in the shape of an arc. The width of the mountains varies from 400 km in the west to 150 km in the east. Himalayas cover an area of about $500,000\,\text{km}^2$. The average elevation of the eastern half portion is greater than that of the western half.

Mountainous area in India occupies 96.06 M-ha divided into: Himalayas 51.43 M-ha, Vindhya region 9.27 M-ha, Eastern Ghats 18.02 M-ha, Western Ghats 7.74 M-ha, and Satpura Ranges 6.60 M-ha. The area of Himalayas above various contour heights is given in Table 1.

The Himalayas, which are geologically young mountains, consist of three main series running parallel to each other. The northern most range is known as the Great Himalayas or the Himadri. In this range, the average altitude is about 6,000 m and this range contains a number of high peaks including the Mount Everest (height $= 8,848$ m) which is the highest peak in the world. This peak is located in Nepal where it is known as Sagarmatha. The second highest peak of Hamalayas is Kanchenjunga (8,598 m) which is the highest peak in India. Other prominent peaks are Nanga Parbat (8,126 m) and Nanda Devi (7,817 m).

Table 1. Area of Himalayas above various contour heights

Contour height	Area (million km^2)	Percent
1,500	4.6	100
2,100	3.90	85
3,000	3.20	69
3,600 to 4,500	2.10	45
5,400 to 6,000	0.56	12

The second range, the Middle Himalayas or Himachal, lies to the south of Himadri. Here, the altitude is between 3,700 to 4,500 m and the width is about 50 km. Popular hill stations of North India, such as Darjeeling, Dharamshala, Dalhousie, Shimla, and Mussorrie are located in this range.

The southern most range is the outer Himalayas or the Shivaliks. Here the altitude ranges from 900 to 1,100 m and the width varies between 10 and 15 km. In between the Himachal and Shivalik ranges are plateaus and flat bottom valleys of thick gravel and alluvium. Locally, these are known as *duns*, e.g., Dehradun.

In the east-west direction, Himalayas can be divided into four parts. The Western Himalayas cover the state of Jammu & Kashmir and a part of Himachal Pradesh. The part between Satluj River and the Kali River is known as Kumaon Himalayas and the Nepal Himalayas lie between the Kali River and the Tista River. The area between the Tista and the Brahmaputra Rivers is known as the Assam Himalayas. The major rivers of north India that originate from Himalayas are the Indus, Satluj, Ganga, and the Brahmaputra. Being snowfed, these rivers are perennial and carry large amount of water and sediments. The origin of the Indus, the Satluj and the Brahmaputra is near the Kailash Mansarovar region.

Besides the snow covered peaks, glaciers and pristine rivers, Himalayas are known for beautiful valleys the world over. The valleys of Kashmir, Kulu, Kangra, and Khasi and Garo hills are known for their scenic beauty and salubrious climate attracting millions of tourists every year.

1.1.2. The Northern Plains

The Himalayas are young mountains and the rivers originating from them carry large volumes of sediments. As these rivers enter plains, their velocity and hence the sediment carrying capacity reduces which forces them to dump the sediment. The north Indian plains of Indus and Ganga were formed by the alluvium that was carried by the rivers originating from Himalayas. This has led to the formation of vast northern plains of thick and fertile alluvium in north India. This coupled with favorable climate and adequate water supply has made this region highly fertile. No wonder that this region was the birthplace of the most ancient and enduring civilization on earth, known as the Indus valley civilization.

The northern plains extend from the mouths of the Indus in the west to the mouths of the Ganga-Brahmaputra in the east, a distance of about 3,200 km. The width of the plains varies between 300 km to 150 km.

The western part of these plains has five rivers – the Indus and its tributaries, the Chenab, the Ravi, the Beas and the Satluj. Locally, the land between two rivers is known as the Doab (*Do* means two and *ab* means water). [Readers may recall the equivalent Greek word Mesopotamia, which also means the land between two rivers. The famous Mesopotamian civilization arose in the valley of Euphrates and Tigris, modern Iraq, about 5,000 BC.] By the same analogy, the plain formed by five rivers in north India is known as Punjab (Punj means five). After the partition of India in 1947, a large part of Punjab is now in Pakistan.

The plains of Ganga valley can be divided into four parts based on relief.A narrow belt of land near the foot of the Shivaliks is known as Bhabar. Its width is about 8–16 km and it is covered with pebbles. A wet and marshy belt known as the Terai lies south of Bhabar. This belt supports forests and wildlife. The older alluvium of the plain is known as Bhangar. The continuous deposition of alluvium leads to the formation of terrace-like features on the flood plains which are known as Khadar.

1.1.3. The Peninsular Region

This is a triangular shaped region whose vertex is near Kanyakumari at the southern tip and base is near the line joining Calcutta to Saurashtra in Gujarat. In this region, the elevation varies between 300 and 1,800 m. The peninsular region may be further subdivided into two parts: the central highlands and the Deccan Plateau. The northern part of the peninsular region which consists of low mountain ranges and igneous rocks forms the central islands. The north-western part of this area is dotted with very old mountains, known as the Aravallis. The southern boundary of this highland is formed by Vindhyachal mountain range. Further west to Aravallis lies the Thar desert. The Malwa plateau lies between the Aravallis and the Vindhyachal range. The area east to Malwa plateau is known as Bundelkhand. The valley of Narmada River lies to the south of Bundelkhand. To the east of Narmada valley is the Chottanagpur plateau whose major part comes under the newly formed Jharkhand state.

The southern part of the peninsular region is known as the Deccan plateau which extends southward from the Satpura range. The western boundary of this plateau is formed by a mountain chain known as Western Ghats. This is a low mountain range near the western coast of India. In the Maharashtra state, these mountains are known as Sahyadri, in Tamil Nadu they are known as Nilgiri, and in Kerala they are known as Cardamom hills. The hills near the eastern boundary of the plateau are known as Eastern Ghats. The general slope of the peninsular plateau is towards east as evidenced by the flow direction of major rivers. But the north part of the plateau slopes towards west. The Major mountain ranges in India are given in Figure 2.

1.1.4. The Coastal Plains

The coastal plains lie between the Western Ghats and the Arabian Sea on the west coast and the Eastern Ghats and the Bay of Bengal on the east coast. These are narrow strips of land which are desiccated by a number of rivers. The northern part of the west coast plains is known as the Konkan Coast and the southern part as the Malabar Coast. The western coast has a number of big seaports, such as Mumbai and Cochin. The eastern coastal plains are wider than the western plains and the alluvial deposits are thicker. A number of rivers, such as the Mahanadi, the Godavari, the Krishna, and the Cauvery, have the deltas on the eastern coast.

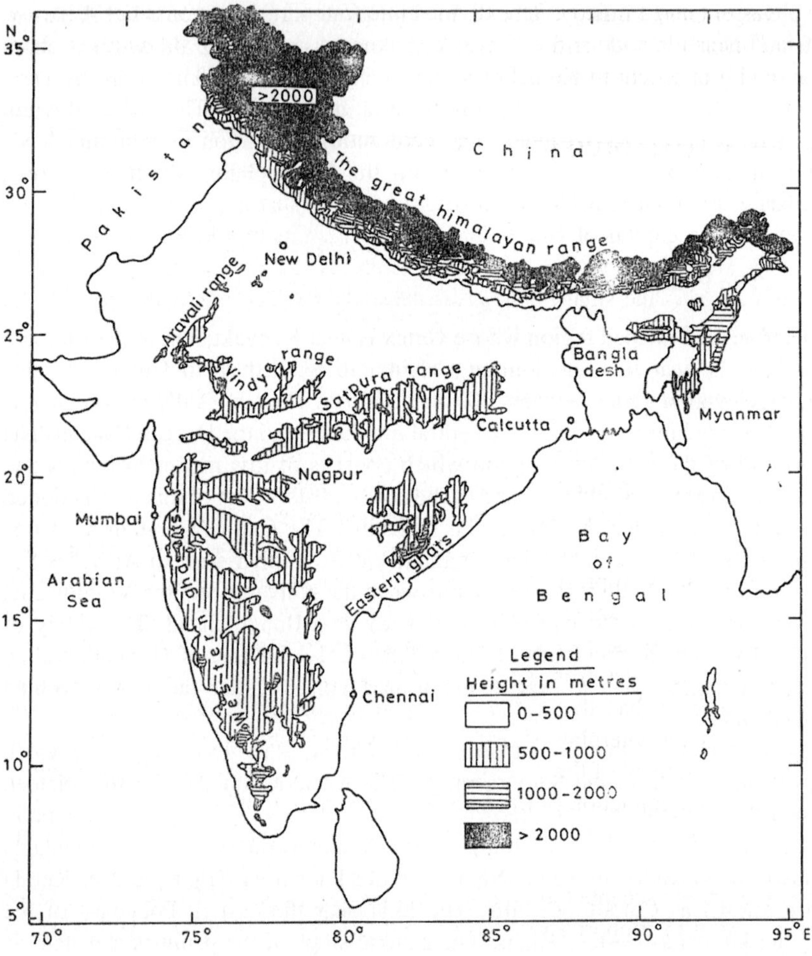

Figure 2. Major mountain ranges in India

1.1.5. The Islands

The Andaman and Nicobar islands consist of a total of 572 picturesque islands in the South Eastern part of the Bay of Bengal. They lie along an arc in long and narrow broken chain, approximately North-South over a distance of nearly 800 km. Most of these islands are of small size, less than 50 km in length.

The other group of islands known as the Lakshadweep group is in the Arabian Sea. This tiniest Union Territory of India with an area of 32 sq. km is an archipelago mainly consisting of ten inhabited islands and 17 uninhabited islands. The inhabited islands are Kavaratti, Agatti, Amini, Kadmat, Kiltan, Chetlat, Bitra,

Andrott, Kalpeni and Minicoy. The islands are located in the Arabian Sea between 8° to 12°13″ North latitude and 71° to 74° East longitude, 220 to 440 km away from the coastal city of Kochi in Kerala.

1.2. POLITICAL DIVISIONS OF INDIA

The Republic of India is a Union consisting of 28 states and 7 union territories. New Delhi is the capital of India. The area of each state and union territory is presented in Table 2. In terms of area, Rajasthan is the largest state followed by Maharashtra; Goa is the smallest state in terms of area. The political map of India is shown in Figure 3.

1.3. CLIMATE

The presence of mighty mountains in the north, extensive plateaus and the ocean in the south has an important influence on the climate of India. Geographic and physiographic characteristics have greatly influenced the climatic characteristics of the country. India is a country with extremes of climates. The Indian sub-continent experiences tropical monsoon climate in general due to the Himalayas which function as an effective meteorological barrier. Here temperature varies from more than 47° C at some places in summer to below −40° C at many places in Himalayas; rainfall varies from almost negligible to about 11,000 mm per annum. Cherrapunji in India has the distinction of receiving the highest rainfall in the world, caused by the interplay of the vigorous sweep of monsoon currents and the funnel shaped alignment of the adjoining ranges. This wide range of climatic conditions working in conjunction with a range of topographic and soil/rock properties produces a complex and interesting pattern of water resources distribution over the country.

According to an accepted definition, a place is classified as arid if it lacks sufficient rainfall to support agriculture. The moisture availability index (MAI) is defined as:

$$\text{MAI} = 75\% \text{ of probable rainfall (P75)}/\text{ET}_0 \tag{1}$$

where ET_0 is the reference crop evapotranspiration. MMI is also widely used to classify climate. Based on MMI, most places in India are arid during Oct. to May.

1.3.1. Temperature

Temperature and rainfall are two important variables that characterize climate. In India, the hottest months are April, May and June. With the onset of monsoons, the maximum temperature drops significantly, although the weather can be quite humid at times. Temperatures start falling from October onwards and are at the lowest during December-January. With the passage of vernal equinox in March, cool weather gives way to hot summer.

Table 2. Area and population of Indian states and union territories

S. N.	States/Union Territory	Capital	Area in sq. km	Rank in terms of area
		States		
1	Andhra Pradesh	Hyderabad	275,068	4
2	Arunachal Pradesh	Itanagar	83,743	14
3	Assam	Dispur	78,438	16
4	Bihar	Patna	94,163	12
5	Chattisgarh	Raipur	135,194	10
6	Goa	Panaji	3,702	28
7	Gujarat	Gandhinagar	196,024	7
8	Haryana	Chandigarh	44,212	20
9	Himachal Pradesh	Shimla	55,673	17
10	Jammu & Kashmir	Srinagar	222,236	6
11	Jharkhand	Ranchi	79,714	15
12	Karnataka	Bangalore	191,791	8
13	Kerala	Thiruvananthapuram	38,863	21
14	Madhya Pradesh	Bhopal	308,252	2
15	Maharashtra	Mumbai	307,690	3
16	Manipur	Imphal	22,327	23
17	Meghalaya	Shillong	22,429	22
18	Mizoram	Aizawl	21,081	24
19	Nagaland	Kohima	16,579	25
20	Orissa	Bhubaneshwar	155,707	9
21	Punjab	Chandigarh	50,382	19
22	Rajasthan	Jaipur	342,239	1
23	Sikkim	Gangtok	7,096	27
24	Tamil Nadu	Chennai	130,058	11
25	Tripura	Agartala	10,486	26
26	Uttar Pradesh	Lucknow	241,068	5
27	Uttaranchal	Dehradun	53,343	18
28	West Bengal	Kolkata	88,752	13
		Union territories		
1	Andaman & Nicobar Islands	Port Blair	8,249	1
2	Chandigarh	Chandigarh	114	5
3	Dadra & Nagar Haveli	Silvassa	491	4
4	Daman & Diu	Daman	112	6
5	Lakshadweep	Kavaratti	32	7
6	Pondicherry	Pondicherry	492	3
7	National Capital Territory	Delhi	1,483	2
	Total		3,287,263	

Eastern parts of Andhra Pradesh, Tamil Nadu and some parts of Gujarat and Orissa are the warmest parts of the country. Jammu & Kashmir, Himachal Pradesh, Uttaranchal, and Arunachal Pradesh are the coolest states. Over the central parts of India, the maximum recorded temperatures often exceed 45° C while along the west coast, the maximum temperatures fall in the range between 35°–40° C. Smaller values of maximum temperatures of around 25° C are recorded in parts of Himachal

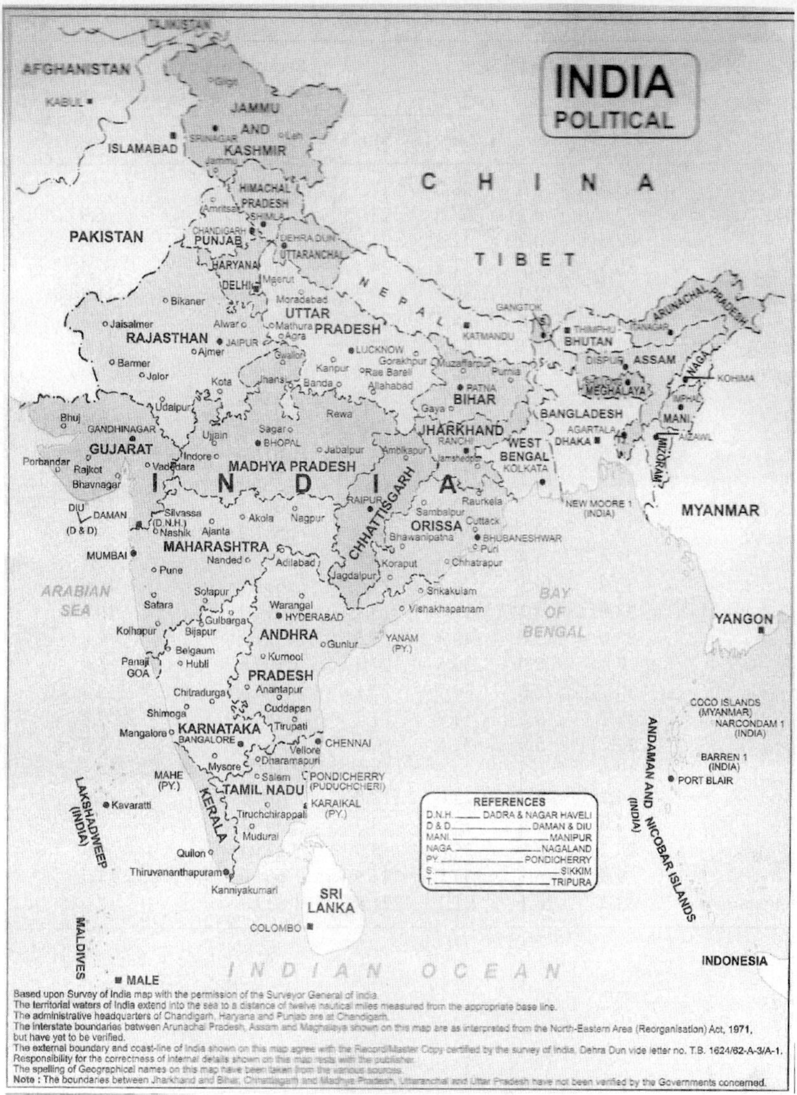

Figure 3. Political map of India

Pradesh and Jammu & Kashmir. Low temperature dropping to the vicinity of −40° C have been recorded in the northern most parts of India at Kargil.

Table 3 gives the maximum and the minimum temperatures in the months of January and July, rainfall, and mean wind speed for important cities of India. The maximum and minimum summer and winter temperatures for various states and union territories in India is given in Table 4.

Table 3. Selected climatic variables at important cities of India

Station	Temperature (°C) January Max.	January Min.	July Max.	July Min.	Annual rainfall (mm)	Annual mean wind speed km/hour
Aizawl	23.4	8.0	28.2	17.3	2,263.8	7.3
Ahemadabad	32.4	9.3	37.5	23.7	823.1	6.9
Amritsar	22.6	0.2	41.4	21.9	649.1	8.6
Balasore	30.5	10.3	34.2	24.1	1,688.6	7.6
Bangalore	29.6	12.3	30.2	18.1	923.7	11.5
Bhopal	29.6	5.5	35.3	21.3	1,208.9	8.3
Chennai	30.3	18.1	37.8	23.2	1,215.3	11.8
Cherrapunji	18.9	4.7	26.0	16.9	11,418.7	8.9
Darjeeling	13.9	0.0	21.9	12.7	2,758.4	3.1
Dehradun	23.1	2.7	34.7	20.7	2,313.7	3.2
Guwahati	26.5	8.2	34.9	24.1	1,637.2	3.2
Hyderabad	31.9	10.7	34.0	21.0	764.4	12.6
Jaipur	26.8	2.7	39.5	22.9	648.1	12.4
Kolkata	29.9	9.9	34.6	24.5	1,581.8	4.9
Lucknow	27.4	4.8	38.3	23.9	992.4	3.1
Margaon	32.9	19.1	29.7	22.2	2,611.7	13.6
Mumbai	33.0	16.1	31.8	23.1	2,099.2	10.5
New Delhi	27.0	3.0	40.3	23.6	714.2	9.8
Patna	27.1	7.0	36.5	24.4	1,109.8	6.8
Port Blair	30.7	22.0	30.7	22.1	3,180.5	13.5
Raipur	31.0	9.5	34.9	21.7	1,388.2	7.8
Ranchi	27.7	5.9	32.5	20.9	1,462.7	5.0
Shillong	19.0	0.3	26.6	16.3	2,415.3	4.1
Shimla	14.3	−3.3	25.0	12.8	1,480.6	3.6
Sibsagar	25.6	6.9	35.6	23.3	2,504.3	4.4
Silchar	28.2	8.8	35.6	23.9	3,225.4	1.8
Srinagar	9.4	−6.7	35.5	14.5	664.0	4.0
Thiruvananthapuram	33.3	20.1	31.2	21.7	1,839.3	7.9

1.3.2. Precipitation

The annual precipitation over the country including snowfall is about 4,000 cubic kilometers which is equivalent to about 120 cm of rainfall depth. This amount is the largest anywhere in the world for a country of a comparable size. The annual rainfall however fluctuates widely. Over the Khasi and Jaintia Hills it is 1,100 cm, while towards the north, in the Brahmaputra valley, the rainfall drops to 200 cm. Cherrapunji, which has an elevation of 1,330 m records a rainfall of the order of 1,142 cm in a year and as much as 104 cm in a day. Out of this, seasonal rainfall is of the order of 3,000 cubic kilometers. In India, rainfall is received through South-West and North-East monsoons, cyclonic depressions, and western disturbances.

Table 4. Maximum and minimum summer and winter temperatures for various states and union territories in India

States/ Union territories	Summer		Winter	
	Max. °C	Min. °C	Max. °C	Min. °C
	States			
Andhra Pradesh	41	20	32	13
Arunachal Pradesh				
Assam	35	18	26	17
Bihar	47	20	28	04
Chhattisgarh				
Goa	32	21		
Gujarat	41	27	29	14
Haryana	46	35	14	02
Himachal Pradesh	33	14	15	00
Jammu & Kashmir				
Jharkhand				
Karnataka	35	26	25	14
Kerala	35	21		
Madhya Pradesh	48	22	23	04
Maharashtra	39	22	34	12
Manipur	29	14	25	07
Meghalaya	25	15	16	04
Mizoram	29	18	24	11
Nagaland	18	29	24	11
Orissa	49	27	16	05
Punjab	45	35	14	00
Rajasthan	47	17	32	07
Sikkim				
Tamil Nadu	43	18		
Tripura	35	24	27	13
Uttar Pradesh	45	11	32	02
Uttaranchal				
West Bengal	40	24	26	07
	Union territories			
Andaman & Nicobar Islands	33	22	31	22
Chandigarh	43	35	14	7
Dadra & Nagar Haveli	41	27	29	14
Daman & Diu	32	21		
Lakshadweep				
Pondicherry	43	18		
National Capital Territory of Delhi	45	35	20	07

Most of the rainfall takes place under the influence of South-West monsoon between June and September, except in Tamil Nadu where it occurs under the influence of North-East monsoon during October-November. Rainfall in India shows great variations, unequal seasonal distribution, unequal geographical distribution, and frequent departure from the normal.

In the parlance of the India Meteorological Department (IMD), a day with a rainfall of 2.5 mm or more is known as a rainy day. The mean annual number of rainy days over India varies from less than 20 over the northwestern parts (West Rajasthan and Kutchh region of Gujarat), to more than 180 in the north-east parts. In the southern parts of the West Coast also, the annual number of rainy days is quite high, about 140 days. The annual number of rainy days is around 40–60 over central parts of India. From the observed spatial pattern, the mean intensity of rainfall is found to vary between 10 and 40 mm per rainy day. In the extreme northern parts, the lowest value is below 10 mm/day. The intensity is near about 10 mm/rainy day over north-western India. Along the West Coast as well as in some parts of north-eastern India, the highest value is about 40 mm/day.

The average annual rainfall over the country is shown in Figure 4. As seen from the figure, there exists a considerable amount of areal variation in the annual

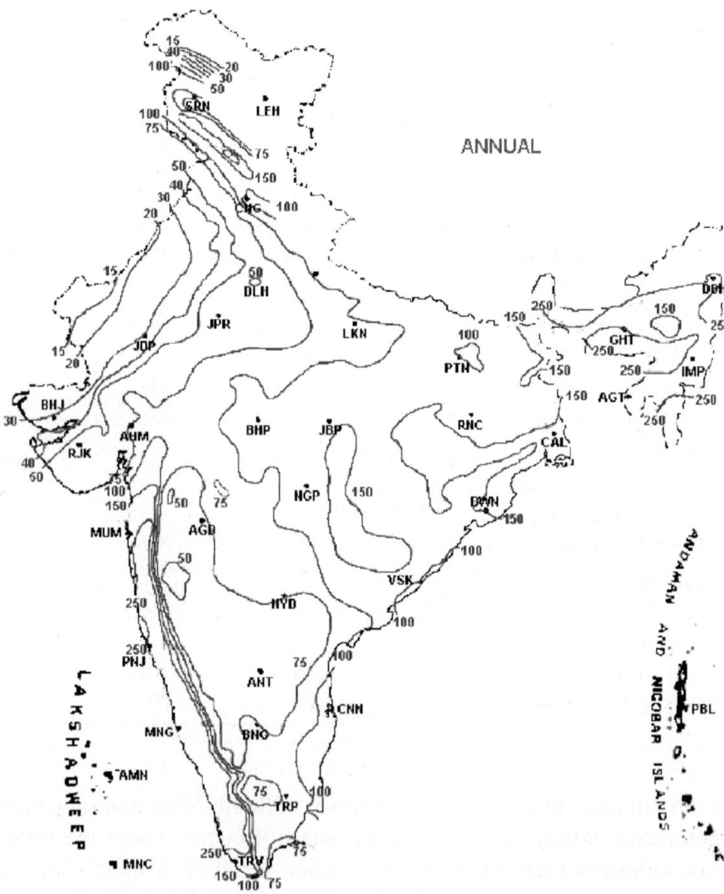

Figure 4. Average annual rainfall over India

rainfall in India. The annual rainfall varies from about 100 mm in western deserts to about 1,100 mm in north-eastern parts of the country. The wettest states of India are Meghalaya and Arunachal Pradesh, Assam, sub-Himalayas, West Bengal, Andaman & Nicobar Islands, Konkan, Coastal Karnataka and Kerala also receive rainfall in the range 250–400 cm per year. More than half of the precipitation takes place in about 15 days and that too in about 100 hours in a year. The number of rainy days varies from about 5 in western deserts to 150 in the north-east. Such a wide variation makes the task of water resources engineers very challenging indeed.

On the west coast of India, heavy rainfall occurs along the slopes of the Western Ghats on the windward side, rapidly decreasing on the leeward side. On the east coast of India, rainfall is the highest near the coast and decreases inland. In the northern plains, rainfall decreases from 150 cm in West Bengal to nearly 10 cm in extreme west Rajasthan.

In the Himalayas, annual rainfall in the eastern Himalayas is more than that in the western side, being more than 200 cm in eastern Himalayas and only about 70 cm in the western Himalayas. It is also higher in the lower foothills and rapidly decreases in the upper ranges.

1.3.3. Monsoons

The word ‘monsoon’ is an Arabic word which means winds changing directions. It is commonly used to denote the seasonal reversal of the wind direction along the shores of the Indian Ocean. In Indian sub-continent, the winds generally blow from the southwest during half of the year and from the northeast during the other half. The reversal of direction is due to the effects of differential heating as the Himalayan plateau heats up during the summer, causing the air to rise and be replaced by the warm, moist air from over the Indian Ocean. Towards the end of May when the weather is the hottest in India, the south-east trade winds from the south Indian Ocean cross the equator and after deflecting, due to the rotation of the earth, extend rapidly into the north Indian Ocean, the Bay of Bengal and the Arabian Sea as South-Westerlies. These deflected trade winds first enter the south Bay of Bengal early in May and later get established over both the sea areas. This westerly current which extends from the Arabic coast to the China Sea across India is called the Indian south-west monsoon.

The SW or the summer monsoon is the single most noteworthy feature of the Indian climate. This monsoon starts from the equatorial belt and hits the Indian subcontinent in two distinct currents. These are known as the Bay of Bengal branch that sets in the North-eastern part of the country and the Arabian Sea branch that hits at the southern part of the peninsula. The first branch moves westwards, and the second northwards, and they together cover the whole country. Andaman & Nicobar Islands receive rains from the Bay of Bengal branch which normally sets in by 20th May. Monsoon reaches Kerala by June 1 and advances along the Konkan coast in early June. Normally, monsoon sets over the entire country by the middle of July.

During the monsoon season, sky is generally cloudy and there are frequent spells of intense rainfall. Figure 5 shows the dates on onset of monsoon over the country.

As the summer monsoon heralds the major rainy season, its arrival is eagerly awaited by all sections of Indian society, the leaders, the planners, the decision makers, the farmers, the traders, and so on, with great expectations. Monsoon rains give relief from sweltering heat and trigger the beginning of sowing operations on a large scale.

The Bay of Bengal is a source of cyclonic systems of low pressure called 'monsoon depressions' during this season. They form in the northern part of the Bay of Bengal with an average frequency of about 2 to 3 per month. Further they move in a northward or north-westward direction and bring well-distributed rainfall over the central and northern parts of the country. The distribution of rainfall over northern and central India is critically influenced by the path taken by these depressions. During the latter half of the month of September, the SW monsoon current becomes feeble and begins withdrawing from the north-western parts of India. It withdraws from almost all parts of the country by the end of September, and is slowly replaced by a northerly continental airflow. The retreating monsoon winds cause occasional showers along the east coast of Tamil Nadu, but decrease towards the interior.

During the monsoon season, it does not rain every day. There are some periods known as 'breaks in the monsoon', when rainfall activity is absent and weather is uncomfortably hot and humid. The usual duration of breaks is 3–7 days and they are frequent in August. Prolonged breaks may cause severe damages to the

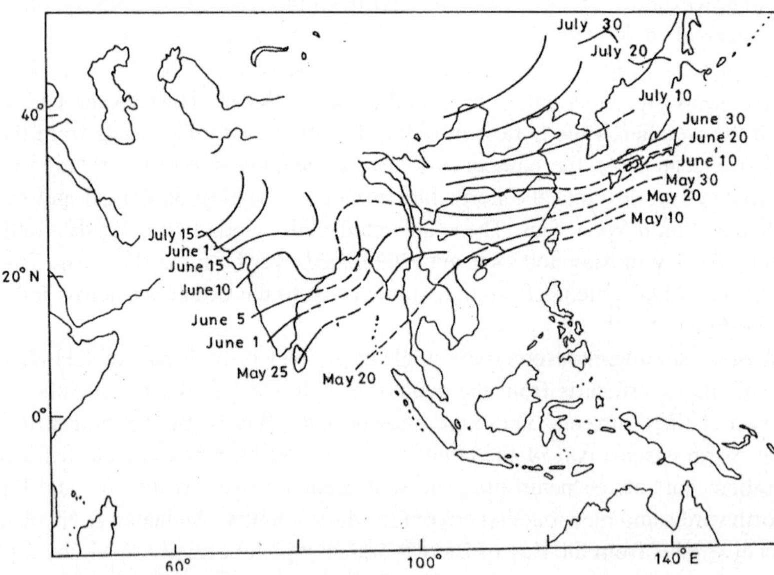

Figure 5. Normal dates of onset of monsoon

crops. Global features like *El Nino*, northern hemispheric temperatures and snow cover influence the year-to-year variability of monsoon rains. Heavy rainfall in the mountainous catchments during monsoon causes floods over the plains. The withdrawal of monsoon takes place in the month of September. Figure 6 shows the normal dates of withdrawal of monsoon.

Although the SW monsoon covers the entire country for four monsoon months, its actual stay at a specific place depends on the dates of its onset and withdrawal. Over Western Rajasthan, monsoon may last for about 70 days while the duration may be more than 120 days over the south-western peninsula.

The post-monsoon or north-east (NE) monsoon season is known as a transitional season. During this transitional season the north-easterly airflow becomes established over the subcontinent and produces the winter or NE monsoon rains over the southern tip of the country. During this season tropical cyclones form in the Bay of Bengal and result in heavy rainfall along their path. Due to these storms, rainfall

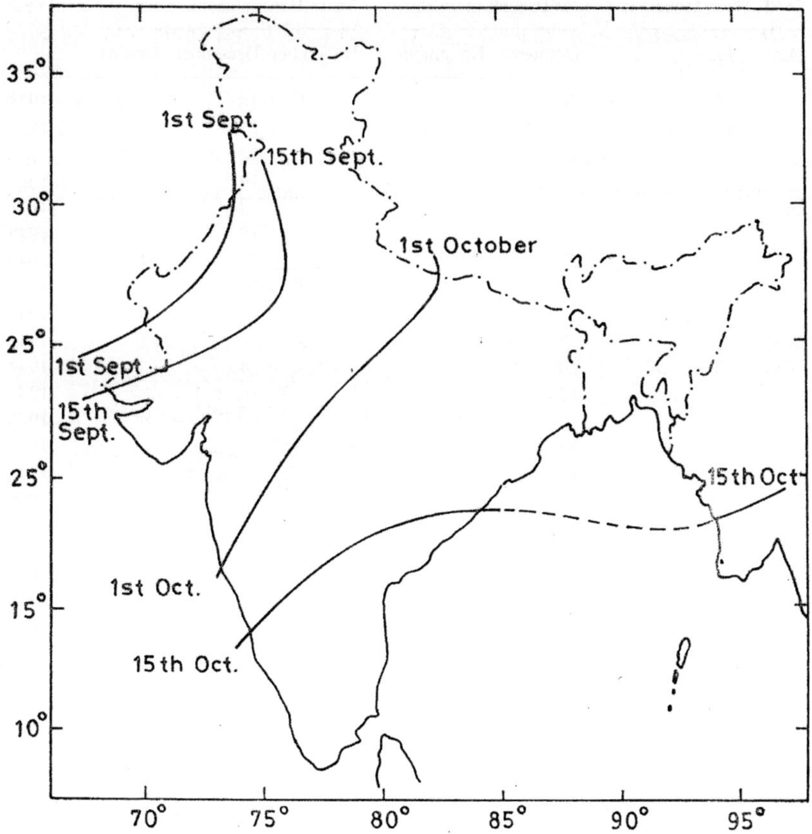

Figure 6. Normal dates of withdrawal of monsoon

has been received mainly in many parts of Tamil Nadu and some parts of Andhra Pradesh and Karnataka. Due to high-velocity winds and tidal waves of these storms, widespread damages also occur in the coastal regions.

During the monsoon season the day temperatures all over the country begin falling sharply. Over north-western India the mean temperatures fall from about 38° C in October, to 28° C in November. Decrease in humidity levels and clear skies also found over most parts of north and central India after mid-October.

As the south-west monsoon withdraws, a north-easterly flow of air begins. This air picks up moisture from the low-pressure areas in the Bay of Bengal and hits the coastal areas of Orissa and Tamil Nadu to cause rainfall. During this period, severe tropical cyclones are also formed in the Bay of Bengal and the Arabian Sea. These cyclones are responsible for intense rainfall in the coastal areas. Table 5 gives the normal dates of onset and withdrawal of monsoons.

Table 5. Normal dates of onset and withdrawal of monsoons

Year	South West Monsoon		North East monsoon		North East monsoon			Year
	May	June	October	November	November	December	January	
1934		8	24			18		1934
1935		12	13			28		1935
1936	19		22			22		1936
1937		4	27			19		1937
1938	26		27			9		1938
1939		5	14			17		1939
1940		14	27				9	1940
1941	23		21			19		1941
1942		10		1			5	1942
1943	29		5				13	1943
1944		3	20			24		1944
1945		5	12			10		1945
1946	29		16				8	1946
1947		3	19				4	1947
1948		11	16				3	1948
1949	23		20			1		1949
1950	27		15			16		1950
1951	31			6		2		1951
1952	20		15			15		1952
1953		7	14				13	1953
1954	31		15				11	1954
1955	29		20			10		1955
1956	21			26		28		1956
1957		1	16			·	3	1957
1958		14	29			19		1958
1959	31		21			4		1959
1960	14		21				18	1960
1961	18		24			15		1961
1962	17		9				13	1962
1963	31		20			10		1963
1964		6	29			27		1964

1965	26		15		12		1965
1966		1	6		22		1966
1967		9	15		20		1967
1968		8	15		21		1968
1969	17		14			3	1969
1970	26		14			12	1970
1971		27	19		19		1971
1972		18	21		27		1972
1973		4	17		31		1973
1974	26			5	30		1974
1975	31		25		20		1975
1976	31		15		26		1976
1977	29		10		6		1977
1978	29		21			1	1978
1979		11	22		15		1979
1980		1	10		6		1980
1981	31		23		24		1981
1982	29		18		19		1982
1983		13	24			18	1983
1984	30			3		7	1984
1985	28		29			17	1985
1986	4		26		30		1986
1987		2	20		26		1987
1988	27			3		6	1988
1989		3	29			10	1989
1990	28		18		27		1990
1991		2	19		23		1991
1992		5		2	9		1992
1993	28		13			1	1993
1994	28		18		24		1994
1995		8	23		13		1995
1996		3	10		19		1996
1997		8	13		23		1997
1998		2	28			4	1998
1999	25		4			12	1999
2000		1		5		2	2000
2001	23		16				2001

The water vapor carried by the monsoon during June-September is about 1,110 M ha-m. About 300 M ha-m (= 27% of the total) of this moisture precipitates in the form of rainfall. During the remaining eight months of the year, there is a substantial amount of moisture over the country. The precipitation during this eight-month period is of the order of 100 M ha-m. In South India covering Tamil Nadu, Andhra Pradesh, and Kerala States, north-eastern monsoon also significantly contributes to precipitation.

In the beginning of December, weather disturbances originating in the extra-tropical region enter India from Afghanistan and Pakistan. These are known as western disturbances and cause moderate to heavy rain and snowfall in the Jammu & Kashmir and Himachal Pradesh States. Light to moderate rainfall is also experienced in the northern plains.

1.3.4. Seasons in India

Based on the temperature and precipitation variations, there are four distinct seasons in India.

(i) The Cold Weather Season

The cold weather season starts in early December and is over by the end of February. January and February are the coldest months. In these months the temperature remains cool and dry. The temperature varies between 10° C–15° C in the northern India as well as about 25° C in the southern India. Due to the western disturbances, heavy rainfall on the coast of Tamil Nadu occurs in this period. The north-west part of India also receives some rainfall or snowfall in this season. In other parts of the country, the months of January and February are cloudless and rainless and the weather remains pleasant.

(ii) The Hot Weather Season

The hot season starts in the month of March and lasts till mid June. Weather is very hot during this season due to vertical sunshine over India. The highest day temperature reaches close to 50° C in some places. Pre-monsoon showers are found to occur in Chottanagpur, Kerala, and Western Ghats due to low-pressure moist winds from Arabian Sea. The northern plains remain dry and hot winds called *loo* blow during the day. Sometimes dust storms occur in Punjab, Haryana and Uttar Pradesh followed by light rain and cool breeze, thereby lowering the temperature to a great extent.

Thunderstorms associated with light rain are frequently formed in this season. Hails also fall at some places in this season and at times these cause damage to crops. Over the northern plains a few dust storms, locally known as *andhis*, also occur. Strong winds accompany these storms which carry substantial quantity of dust whose erosion is aided by the dry weather. Winds carry large quantity of loose material such as papers, dry leaves, polythene bags etc. This material settles down soon only if the winds are followed by showers. Over the eastern and north-eastern regions in the states of Bihar, West Bengal and Assam, violent thunderstorms with strong winds and rain lasting for short durations also occur.

(iii) The Advancing Monsoon Season

As already described above, this season runs from mid June to September. Heavy rainfall in the whole of India occurs due to monsoon winds starting from Bay of Bengal and Arabian Sea. Sometimes one can observe rainfall almost non-stop for several days at various intensities. The weather becomes hot and sultry when there is a break in monsoon.

(iv) The Retreating Monsoon Season

This season runs from October to November. In this season sky is usually clear and humidity is low. Monsoon starts retreating in late September to early October. The months of October and early November form a period of transition from hot rainy season to dry winter season.

The period of transition when low-pressure area is transferred from north-west of India to Bay of Bengal is marked by the formation of cyclones. Sometimes these cyclones attain such a great speed that they become devastating. Normally these cyclones affect coastal areas of Andhra Pradesh, Orissa, and Bangladesh. Great loss of life and property due to these havocs took place in Bangladesh in 1970, Andhra Pradesh in 1977 and 1997. The recent disaster in 1999 in Orissa is unforgettable, when more than 10,000 people were killed.

1.3.5. Number of Rainy Days

A day on which 2.5 mm or more of rain falls is counted as a rainy day. The pattern of rainy days generally follows the annual rainfall pattern described earlier.The Western Ghats, Assam, portions of sub-Himalayan West Bengal and some higher elevations of the Himalayas up to Punjab have more than one hundred rainy days. In extreme west Rajasthan the number of rainy days is less than ten. As in the case of rainfall distribution, the number of rainy days in the northern plains decreases from east to west. In the peninsula, between 40 and 50 rainy days occur over the semi-arid regions extending from mid Maharashtra to Tamil Nadu. Over Orissa, Madhya Pradesh and adjoining Andhra Pradesh, there are between 50 to 75 rainy days. On the west coast, the number of rainy days is as high as 137 at a few places.

The rainfall regions may be divided into four parts as mentioned below:

(i) *Heavy Rainfall Regions*
These rainfall regions receive more than 200 centimeter of rainfall annually. Assam, Arunachal Pradesh, Meghalaya, Sikkim, West Bengal, and southern slopes of Eastern Himalayas fall under this category. Western Coastal regions and Western Ghats also receive heavy rainfall from the Arabian Sea monsoon. The heaviest rainfall in the world (more than 1,142 cm) is found at Mawsynram (near Cherrapunji) in Meghalaya. It should be noted that Mawsynram is a new station established near Cherrapunji. It was found that the rainfall at Mawsynram is higher than that at Cherrapunji.

(ii) *Areas of Moderate Rainfall*
The areas where the annual rainfall is between 100–200 cm come under this category. Bengal, Bihar, Eastern U.P. and Sub-mountain regions of Punjab, Orissa, Madhya Pradesh, and the east coast of Tamil Nadu receive moderate rainfall.

(iii) *Regions of Low Rainfall*
These areas receive an annual rainfall between 50–100 cm. Most of the Deccan region, Gujarat, Eastern Rajasthan, Western U.P., Haryana and Northern Punjab are the areas of low rainfall.

(iv) *Areas of Scanty Rainfall*
These areas receive less than 50 cm annual rainfall. Scanty rainfall is found mostly in Kutch and Western Rajasthan. Due to Aravalli hills, Western Rajasthan remains

dry. Scanty rainfall is also found in Southern Haryana, South-West Punjab, Ladakh in Kashmir, and Lahul-Spiti in Himachal Pradesh.

1.3.6. Snow and Glaciers

Precipitation occurring in the Himalayas at heights of 2,450 m and above usually gets solidified as snow. All the great peaks of Himalayan mountains are, therefore, covered with snow. The mountainous area covered by snow is about 80% of the total area of Himalayas. In Himalayas, the western part gets more snow than the eastern and gets it earlier. The snowline, the lowest line on a mountain at which snow exists throughout the year, is about 5,490 m at the equator and 610 m in Greenland. In temperate zones, it is about 3,050 m. In winters, the snow line can descend to altitudes as low as 2,500 m.

As successive snowfalls occur, pressure on the lower layers increases and the snow becomes a granular ice. When the weight of ice increases, the depth of ice being more than 76 m, it begins to move and is known as a glacier. The rate of movement varies from 0.3 m to several metres a year. Table 6 shows snow covered areas of Himalayas lying between 70° and 100° E and south of 40° N in different months during 1979–82.

(i) Indian Himalayan Glaciers

Glaciers can be conceived as natural reservoirs which store precipitation in the winter season and gradually release it as melt water in summers, thereby augmenting flows into the rivers. Himalayas are the home to a multitude of glaciers and are the largest reservoir of snow and ice outside the polar regions. Snowmelt runoff and glacier melt runoff makes these rivers perennial in nature. In the rivers originating from Himalayas, significant snow and glacier melt runoff contribution begins in April when seasonal snow cover starts ablating. This contribution continues till October/ November depending upon the climate conditions.

The principal glaciers in Himalayas can be divided in four groups:

a. Punjab Himalaya Group,
b. Garhwal Himalaya Group,
c. Nepal Himalaya Group, and
d. Assam Himalaya Group.

Table 6. Snow covered areas of Himalayas lying between 70° and 100 °E and south of 40 °N during different months (after Dhanju, 1983)

Month	Snow covered area (10^6 km^2)	Remarks
September, 1979	0.93	Minimum snow cover
January, 1980	2.11	Maximum snow cover
September, 1980	0.74	Minimum snow cover
March, 1981	3.67	Maximum snow cover
September, 1981	0.55	Minimum snow cover
January, 1982	2.44	Maximum snow cover

(a) *Punjab Himalaya Group of Glaciers.* Under the Punjab Himalaya Group, the major glaciers are: Rakhiot Glacier, Kolhai Glacier, Neh-Nar Glacier, Sarbal Glacier, Kangriz Glacier, Brahma Glacier, Drung Drung Glacier, Mulkila Group Glaciers, Barashigri Glacier, Dibi Bokri Glacier, Gara Glacier and Gorgarang Glacier.
(b) *Garhwal Himalaya Group of Glaciers.* Garhwal Himalaya Group of Glaciers includes Gangotri Glacier, Santopath Glacier, Kedarnath Glacier, Milam Glacier, Pindari Glacier, Shankulapa Glacier, and Poting Glacier.
(c) *Nepal Himalaya Group of Glaciers.* Yaling Glacier, Chong Kumadan Glacier, Rundun Glacier, Glaciers adjoining to Dhaulagiri and Annapurna Peaks, Kang Shung Glacier, Rupal Glacier, Khumbu Glacier, Glaciers adjoining to Makalu Peak, and Zemu Glaciers fall under this group.
(d) *Assam Himalaya Group of Glaciers.* Glaciers adjoining Kanchenjunga peak, Sanlung Glacier and Glaciers adjoining Gyara Pari peak falls under the Assam Himalaya Group of glaciers.

(ii) Principal Glacier fed river systems of Himalayas
Since the snow glacial melt is a significant part of the flow in Himalayan rivers, estimation of the melt rate of these bodies and the total volume of water expected in the melt season is of vital use for water resources planning and management. Due to paucity of detailed hydrological studies on Himalayan glaciers, database of Indian glaciers is poor.

A number of important major rivers, namely, Indus, Ganga and Brahmaputra and their tributaries originate from Glaciers of Himalayas. The important features of principal glacier fed river systems of Himalayas are shown in Table 7.

Snow and glaciers melt during hot months (March to June) and yield large summer flows in the Himalayan rivers, such as the Indus, the Ganga, the Brahmaputra, and their tributaries. Table 7 contains data of principal glacier-fed river systems of Himalayas.

Detailed investigations have been conducted only on a few glaciers. In the following, summary information about selected glaciers is given.

Gangotri Glacier: Gangotri Glacier is one of the largest glaciers in the Himalayas. The glacier is located in the Uttarkashi District of Uttaranchal State (U.A.) falling in the Garhwal Himalayan Region. The location of Gangotri Glacier is shown in Figure 7. The snout of the Gangotri Glacier is known as 'Gomukh' and a proglacial melt water stream, known as Bhagirathi River, emerges out from Gomukh at an elevation of 4,000 m. To reach at the Glacier, one has to trek for about 18 km from the Gangotri town. Rough estimate based on the topography of the area and some field observations suggest that the depth of the glacier is about 200 m. Gangotri Glacier contains a large number of crevasses spread all over the ablation zone. These crevasses are well exposed when seasonal snow accumulated in the ablation zone is depleted.

Gangotri glacier system is a cluster of many glaciers comprising of main Gangotri glacier (length: 30.20 km; width: 0.20–2.35 km; area: 86.32 km^2) as trunk part of

Table 7. Principal Glacier fed river system of Himalayas

Major river system	Name of river/tributary	Catchment area (sq. km) covered by		Percentage glaciation
		Mountains	Glaciers	
Indus	Indus	268,482	8,790	3.3
	Jhelum	33,670	170	5.0
	Chenab	27,195	2,944	10.0
	Ravi	8,029	206	2.5
	Sutlej	47,915	1,295	2.7
	Beas	14,504	638	4.4
	Yamuna	11,655	125	1.1
	Ganga	23,051	2,312	10.0
Ganga	Ramganga	6,734	3	0.04
	Kali	16,317	997	6.01
	Karnali	53,354	1,543	2.9
	Gandak	37,814	1,845	4.9
	Kosi	61,901	1,318	2.1
	Tista	12,432	495	4.0
	Raidak	26,418	195	0.7
Brahmaputra	Manas	31,080	528	1.7
	Subansiri	18,130	725	4.0
	Brahmaputra	256,928	1,080	0.4
	Dibang	12,950	90	0.7
	Lohit	20,720	425	2.01
Gross		1,001,294	25,724	2.6

Source: NIH (1991)).

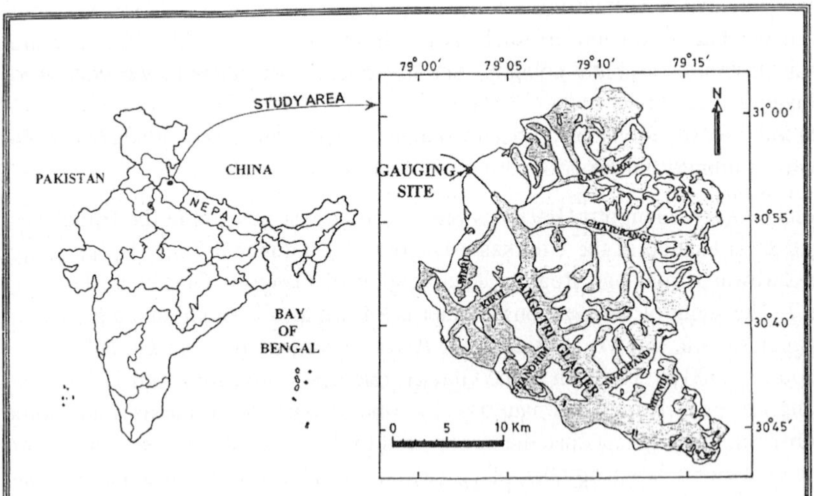

Figure 1: Location map of the study area.

Figure 7. Location map of the Gangotri Glacier

the system. It flows in the northwest direction. The major glacier tributaries of the Gangotri Glacier system are Raktvarn Glacier (55.30 km^2), Chaturangi Glacier (67.70 km^2), Kirti Glacier (33.14 km^2), Swachand Glacier (16.71 km^2), Ghanohim Glacier (12.97 km^2), Meru Glacier (6.11 km^2), Maindi Glacier (4.76 km^2) and few other having glacierised area of about 3.08 km^2. The elevation area of the Gangotri Glacier varies from 4,000–7,000 m. The Gangotri Glacier area also has several high peaks around it, notably the majestic snow clad Shivling peak ($\approx 6{,}500$ m).

The Gangotri Glacier lies in the central crystalline zone. The Himalayan crystallines along the Bhagirathi valley are composed of Pelitic and semi-pelitic meta-sediments with acid and basic intrusive. The area is situated north of Main Central Thrust (MCT) which separates the metamorphics from the underlying very low grades of unmetamorphosed sedimentary sequence of the lesser Himalaya. Mica schists are the predominant rock found over the MCT. Further, northeast of Gangotri these schists are intruded by hard and massive granite known as Gangotri granite. From Gangotri town towards Gomukh, the Gangotri granite gradually changes into fine grained, well-foliated, garnetiferrous gneiss and augen gneiss intruded by fine-grained aplitic veins.

The Bhagirathi river valley is a broad U-shaped with high sidewalls, which is a characteristic of its glacial origin. The lower part of the ablation zone of the glacier is covered with thick supraglacial moraine and shows development of few lakes. Due to the recession, the location of Gomukh has also moved upstream with time. Morainic material present between Chirwasa and Gomukh in the form of tillite hillocks are considered as evidences of the extent of Gangotri Glacier. It is NW-SE trending valley within the granitic terrain. The prominent geomorphic landforms formed by the glacial environment are different levels of lateral, and recessional moraines, U-shaped glacial troughs, terraces and outwash plains.

Thick vegetation is found from Gangotri town to Deo Gad. Deo Gad, located about 5 km upstream to the Gangotri town, is a tributary to the Bhagirathi River. The vegetation gradually reduces beyond Deo Gad. There is little vegetation around Chirwasa, a place between Deo Gad and Gomukh, but beyond Chirwasa, the vegetation is significantly reduced up to Gomukh. As such there is very little vegetation between Chirwasa and Gomukh. Common flora found in the area are Himalayan cedar *(Botanical name: Cedrus deodara)* which dominate on the slopes along with few Spruce *(Picea smitbiana),* Silver fir *(Abies pindrow)* and Blue pine *(Pinus wallichiana)*. The Indian birch *(Betula utilis)* is the only tree found in and around the Bhojwasa. The important faunas include leopard, musk deer and varau deer. Many different species of birds also found in the area.

A preliminary analysis of limited rainfall data showed that the average summer season rainfall (May-October) was about 260 mm. Mean monthly temperatures for May, June, July, August, September and October were 8.8, 10.3, 11.7, 10.8, 7.7 and 5.3° c, respectively, suggesting that July is the warmest month. Average daily maximum and minimum temperatures over the summer season was 14.6° C and 3.9° C, respectively. Day-time wind speed is about 4 times stronger than the night-time wind speed. Mean daily sunshine hours were 5.6 hours. Monthly pan evaporation was 150.6, 113.4, 106.9, 85.5, 92.0 and 97.6 mm for the month from May to October respectively. Meteorological conditions represent dry weather conditions in the study area.

The mean daily discharge ranges between 5 and 194 m^3/s. The mean monthly discharge near the snout for May, June, July, August, September and October was 27.3, 74.1, 121.8, 105.7, 57.0 and 19.7 m3/s, respectively; the maximum discharge occurs in July. The months of July and August contribute about 57% to the total summer discharge. The strong storage characteristics of the Gangotri Glacier are reflected by the comparable magnitude of runoff observed during day-time and night-time. Mean monthly suspended sediment concentration for May, June, July, August, September and October during the study period were 1,942, 2,063, 3,658, 2,551, 646 and 160 mg/l, respectively. Mean monthly suspended sediment load for corresponding months was found to be 6,002, 14,113, 39,371, 24,075, 4,371 and 267 tones, respectively.

A detailed description of the glacier system as well as observed data for the same are given in NIH (2003).

The Dokriani Glacier: The Dokriani Glacier is a valley type glacier located in Garhwal region of Himalayas. This glacier lies between latitudes 31°49′ to 31°52′ N and longitudes 78°47′ to 78°51′ E. The glacier originates in the vicinity of Janoli (6,633 m) and Draupadi ka Danda (5,716 m) peaks. It is situated about 30 km ENE of Bhukki village. A small stream, known as Din Gad, originates from the Dokriani Glacier. It follows a narrow valley and meets Bhagirathi River at Bhukki. The total drainage area of this glacier is about 23 km^2 out of which about 10.3 km^2 is glaciated (the remaining part has rock outcrops, slopes, etc.). The elevation of glacier varies from about 3,950–5,800 m and its length is about 5.5 km whereas its width varies from 0.1–2.0 km from snout to the accumulation zone. The middle part of the glacier is highly fractured and consists of crevasses, moulins, glacier table, and ground moraines. The crevasses are mainly transverse type which is wide and long. Sometimes longitudinal crevasses are also seen along the sides of the glacier.

The snout of glacier is situated at an elevation of about 4,000 m and is covered by huge boulders and debris. The lower portion of the glacier is almost completely covered by debris. The material of these moraines has been derived from the side of valley. This glacier is bounded by two large lateral moraines which are about 200 m in height. Besides these, several other lateral moraines are present at different altitudes. These different levels of moraines are indicative of the past extension of the glacier.

The Kolhai Glacier (Jammu & Kashmir): This glacier is situated in the Lidder Valley in Jammu & Kashmir State and extends from 75°19′ N to 75°22′ N latitude and 34°8′ E to 34°12′ E longitude. The elevation of the glacier ranges from 3,650 m to 4,800 m. On the basis of orientation and its location in adjoining valleys around the Kolhai peak, the entire glacier can be divided into four parts. The average width of 3 parts of the glacier is approximately 1 km and that of the other part is about 0.75 km. The length of the glacier varies from 2.5 km to 6 km. The glaciated area has been assessed to be around 24 km^2.

Deep and wide crevasses exist all along the snout and ablation zone of the glacier. Some of these crevasses are quite large – at a few places, the cracks in the ice

Table 8. Physical characteristics of the glaciers with snout ages and average flow rates

Glacier	Location	Altitude (m)	Length of glacier (km)	^{32}Si Model age (yr.)	Post flow rate (m/yr)	Modern flow rate (m/yr)
Nehnar (Kashmir)	34°09′N, 75°31′N	3,920–4,925	3.4	500	6	> 12
Chhota Shigri (H.P.)	32°15′N, 77°31′N	4,050–5,000	9.0	250	28	23
Gara (H.P.)	31°30′N, 78°26′N	4,710–5,600	6.0	200	20	60
Gorgarang (H.P.)	31°26′N, 78°24′N	4,765–5,360	3.5	160	18	NA
Zemu (Sikkim)	27°43′N, 88°17′N	4,260–6,000	26.0	120	200	NA
Chang me Khangpu (Sikkim)	27°58′N, 88°42′N	4,850–5,800	5.8	100	40	13

are more than 10 m wide. A small but deep lake exists on the southern side of the glacier. This lake is known as Doodhnag Lake. It is oval shaped with elevation about 3,750 m and surface area about 2 km^2.

The Chhota Shigri Glacier: The Chhota Shigri Glacier lies on the northern slope of the main ridge of the Pir Panjal Range in the east of the Rohtang Pass (H.P.). The high, steep ridges and mountain terrain provide an ideal condition for the development of this glacier. The Chhota Shigri Glacier is located at 32°15′ N and 77°31′ E, covering about 10 km^2 area. The total drainage area of Chhota Shigri Glacier stream is approximately 45 km^2. There is very high gradient from accumulation to ablation area and snout. The glacier melt drains out in a single confined stream and meets the Chandra River. Lateral moraines are present all along the body of the glacier up to the accumulation zone.

(iii) Radio Isotope Study of glaciers

Isotopic techniques are increasing being applied to study diverse glaciological problems on selected Himalayan glaciers. Problems related to ice dynamic (movement of glaciers ice, accumulation ratio of ice) based on natural and artificial radio isotopes like ^{32}Si, ^{210}Pb, ^{197}Cs etc. and climatic variations in the Himalayan environment based on stable isotopes have been studied. These isotopes provide ages (residence time of snow/ice in the glacier) of the glacier ice all along the glacier and provide time index to study various processes and glaciological parameters of different time scales. The physical characteristics of the glaciers with snow ages and average flow rates have been given in Table 8.

1.4. METEOROLOGICAL SUB DIVISIONS

The entire country has been divided into twenty two meteorological sub-divisions as detailed in Table 9. These sub-division are used in computations, such as weather forecasting, and estimation of mean rainfall. The Table also gives the information about the average annual and seasonal rainfall in the various meteorological sub-divisions.

Table 9. Meteorological sub-divisions of India

SN	Sub divisions	Area (km^2)	Average annual rainfall (mm)	Average seasonal rainfall (June-Sep.) (mm)	Average seasonal rainfall/average annual rainfall (%)
1	Andaman and Nicobar Islands	8,249	2,964	1,600	54
2	Arunachal Pradesh	82,578	2,997	2,085	70
3	Assam & Meghalaya	101,012	2,497	1,824	73
4	Nagaland, Manipur, Mizoram and Tripura	70,447	2,314	2,092	90
5	Sub Himalayan West Bengal and Sikkim	28,924	2,779	2,172	78
6	Gangetic West Bengal	66,228	1,429	1,079	76
7	Orissa	155,782	1,484	1,143	77
8	Bihar				
	Jharkhand (Bihar Plateau)	79,638	1,371	1,125	82
	Bihar Plains	94,238	1,204	1,023	85
9	Uttar Pradesh				
	Uttar Pradesh East	146,509	1,014	893	85
	Uttar Pradesh West	96,732	836	726	88
	Uttaranchal (Hills of West Uttar Pradesh)	51,122	1,750	1,409	81
10	Haryana, Chandigarh and Delhi	45,821	556	463	83
11	Punjab	50,362	611	467	76
12	Himachal Pradesh	55,673	1,518	993	65
13	Jammu & Kashmir	222,236	997	458	46
	Rajasthan				
14	Rajasthan West	195,086	310	275	89
	Rajasthan East	147,128	700	647	92
	Madhya Pradesh				
15	Madhya Pradesh West	229,550	1,043	945	91
	Madhya Pradesh East	213,291	1,398	1,227	88
	Gujarat				
16	Gujarat Region	86,597	967	920	95
	Saurashtra & Kachchh	109,990	515	479	93
	Maharashtra				
	Konkan & Goa	34,095	2,881	2,705	94
17	Madhya Maharashtra	115,306	940	788	94
	Marathawada	64,525	794	660	93
	Vidarbha	97,537	1,102	960	97
	Andhra Pradesh				
18	Coastal Andhra Pradesh	93,045	1,008	572	57
	Telangana	114,726	931	759	82
	Rayalseema	69,043	676	367	54
19	Tamil Nadu and Pondicherry	130,549	1,007	301	30
	Karnataka				
20	Coastal Karnataka	18,717	3,292	2,886	88
	Interior Karnataka North	79,895	685	447	65
	Interior Karnataka South	93,161	1,271	868	68
21	Kerala	38,864	2,978	1,998	67
22	Lakshadweep	32	1,496	943	63

Source: CWC (2002).

1.5. SOILS OF INDIA

The soils on which the society depends so much have evolved over thousands of years. From the agricultural point of view, soil may be defined as the material comprising weathered rock minerals, which, together with organic matter, water and air, provides a medium for the growth of plants. This medium is the basic source of all human and animal food.

Indian soils are generally classified into four major types: (i) the Indo-Gangetic alluvium soils; (ii) the black cotton or regur soils; (iii) the red soils lying on metamorphic rocks; and (iv) the laterite soils. Figure 8 contains a soil map of India.

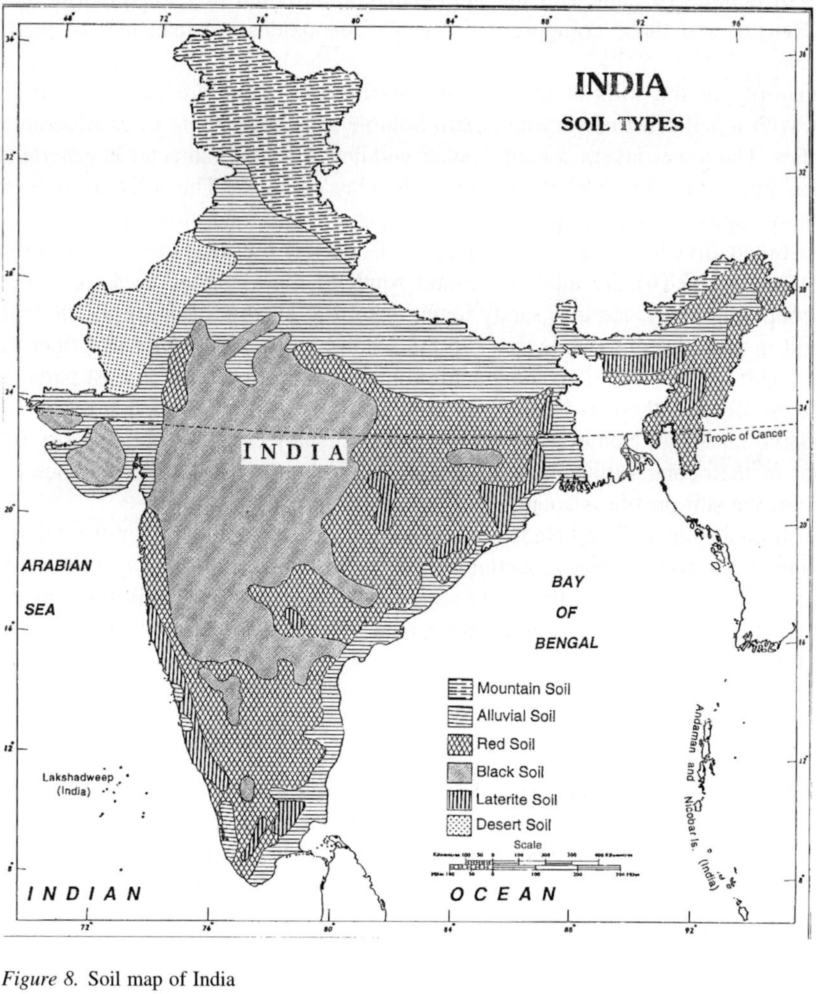

Figure 8. Soil map of India

1.5.1. Indo-Gangetic Alluvium

The Indo-Gangetic alluvium is by far the largest and most important of the soil groups of India. The soils of this group cover about 777,000 square kilometres. They are distributed mainly in the northern, north-western and north-eastern parts, including the Punjab, Haryana, Uttar Pradesh, Bihar, Bengal, and parts of Assam and Orissa.

Although there are local variations in properties, the basic characteristics of these soils depend on the fact that these have been deposited by the numerous tributaries of the Indus, the Ganga and the Brahmaputra River systems. While draining the Himalayas, these rivers bring down the products formed by the weathering of rocks.

According to their age, these soils are classified as old alluvium and new alluvium. At times, layers of hard pans are observed in the soil profile at various levels. Layers of pebbles or kankar in the Indo-Gangetic alluvium of Uttar Pradesh and West Bengal, and those composed of impure iron oxides are instances of these hard pans.

A majority of the soils of the Punjab and Haryana plains are loams or sandy loams, with a soil crust of varying depth. Soluble salts are present in considerable quantities. The lower layers contain kankar nodules. The soil character is generally alkaline due to the presence of sodium in the clay complex. The soils are rich in phosphorus and potash, but are deficient in organic matter and nitrogen.

Two broad divisions are distinguished in Bihar: (i) the alluvium found north of the Ganga and (ii) the alluvium found south of the Ganga. The soils in the first group are clayey, loam to sandy loam, neutral to alkaline, rich in potash, and deficient in phosphorous, while those in the second group are heavier and finer in texture, with higher potassium oxide and of acidic pH in the southernmost parts.

In West Bengal, there is hardly any regularity in the manner of deposition of river-borne materials. Some of the early deposits display considerable diversity on account of their long subjugation to climatic and other controls. The presence of arsenic in the soil profile is a cause of concern here.

The alluvial soils of Tamil Nadu are transported soils, found mainly in the deltaic areas and on the coastal line. A section of the profile shows alternate layers of sand and silt. The composition of the strata varies with the nature of the silt brought by rivers which, in turn, varies with the catchment areas and the tracts through which rivers flow.

1.5.2. Black Cotton Soils

The typical soil of the Deccan Trap is the regur or black cotton soil. It is common in Maharashtra, western parts of Madhya Pradesh, Karnataka, and some parts of Tamil Nadu. Generally, black soil areas have high fertility, although some areas in the uplands display low productivity. The soils on the slopes and the uplands are thin and somewhat sandy, but those in the plains are darker, deeper and richer. Moreover, these are constantly enriched by deposits washed down from the hills.

Black soils are highly argillaceous, fine-grained, and dark with a high proportion of calcium and magnesium carbonates. They are very tenacious of moisture, and are exceedingly sticky when wet. Owing to considerable contraction on drying, large and deep cracks are formed in dry season. The soils contain much iron and fairly high quantities of lime, magnesia, alumina, and potash. However, they are poor in phosphorus, nitrogen and organic matter. In all regur soils, in general, there is a layer rich in kankar nodules formed by the segregation of calcium carbonate at some depth below the surface and above the weathered rocks.

In Maharashtra, the soils derived from the Deccan Trap occupy quite a large area. On the uplands and on the slopes, the soils are light coloured, thin and poor. On the lowlands and in the valleys, relatively clayey black soils are found. Along the Ghats, the soils are very coarse and gravelly. The soil is often some 6 metres deep in the valleys of the Tapi, the Narmada, the Godavari and the Krishna. The subsoil contains a good deal of lime. Outside the Deccan Trap area, the black cotton soil predominates in the Surat and Broach districts of Gujarat. In Madhya Pradesh, two distinct kinds of black soils are found: (i) deep heavy black soil covering the Narmada valley, and (ii) shallow black soil. The cotton-growing areas generally have deep and heavy black soils, although soils of lighter texture are also found. The black soils of Karnataka are fairly heavy, with a high salt concentration. They are generally rich in lime and magnesia.

1.5.3. Red Soils

Red soils extend practically over the whole Archaean basement of Peninsular India, from Bundelkhand to the extreme south, covering 2,072,000 square kilometres, embracing south Bengal, Orissa, parts of Madhya Pradesh, eastern Andhra Pradesh, Karnataka and a major part of Tamil Nadu. These soils also occur in Santhal Parganas in Bihar, and in the Mirzapur, Jhansi and Hamirpur districts of Uttar Pradesh. They were produced as a result of meteoric weathering of ancient crystalline and metamorphic rocks.

As the name suggests, the colour of these soils is generally red, grading sometimes into brown, chocolate, yellow, grey and even black. The redness is due more to a general diffusion than to a high proportion of iron content. Red soils can be divided into two broad subgroups: (i) red loams of argillaceous character with a cloddy structure and possessing a few concretions, and (ii) red earths with loose and friable topsoil. The light sandy red and yellow soils found in the Mahanadi basin are of alluvial origin. The formations in the north and on the west coast of Kerala consist mostly of the sands deposited from the sea.

In the upland regions the soils are poor, thin, gravelly and light-coloured, while in the plains and valleys these are of much more fertile, deep, dark varieties. They are generally poor in nitrogen, phosphorus, and humus. Compared with regur, they are poor in potash and iron oxide, and are also uniformly low in phosphorus.

More than two-thirds of the cultivated area in Tamil Nadu is covered by red soils. They are in-situ formations, produced from the rock below under the influence of climatic conditions. The rocks are acidic, consisting of mica or red granites. The soils are shallow and open in texture. They have a low exchange capacity and are deficient in organic matter and plant nutrients. The predominant type in the eastern tract of Karnataka is the red soil overlying granite. It is rich in potash, iron and alumina. The acidic soils in the south of Bihar are red soils. In West Bengal, the red soils are the transported soils from the hills of the Chhota Nagpur plateau.

1.5.4. Laterites

Laterite is a soil type peculiar to India and some other tropical countries, characterized by the intermittent occurrence of moist climate. In formation it varies from compact to vesicular rock composed essentially of a mixture of hydrated oxides of aluminium and iron, with small quantities of manganese oxides, titania, etc. It is produced by the atmospheric weathering of several types of rocks. Laterites may break and be carried to lower levels by streams. When redeposited, they become a compact mass by the segregate action of hydrates. Thus there are high-level laterites resting on rocks, and low-level laterites formed in the usual way of detrital deposits.

Laterites are especially well developed on the summits of the hills of Karnataka, Kerala, Madhya Pradesh, coastal region of Orissa, south Maharashtra, Malabar and part of Assam. All lateritic soils are generally very poor in lime and magnesia, and deficient in nitrogen. Occasionally, the Phosphorus content may be high, but there is deficiency of K_2O.

In Tamil Nadu, there are both high-level and low-level laterites which are formed from a variety of rock materials under certain climatic and weather conditions. They are both in-situ and sedimentary formations, and are found along the coastal region where rainfall is heavy and the climate humid. The laterites at lower elevations grow rice, whereas those at higher elevations grow tea, cinchona, rubber, and coffee. The soils are rich in nutrients and contain 10 to 20 per cent organic matter.

In Kerala, in between the broad sea belt consisting of sandy soil and sandy loams and the eastern regions comprising forest and plantation soils, the mainland contains residual laterite. It is poor in total and available P_2O_5, available K_2O and CaO. The laterite soils of Karnataka are comparable with the laterites found in Malabar, the Nilgiris, etc. The soils have very low lime content on account of severe leaching and erosion. In West Bengal, the area between the Damodar and the Bhagirathi is interspersed with some basaltic and granitic hills with a laterite capping. The laterites of Orissa are largely found capping hills and plateaus, occasionally in considerable thickness. Two types of laterites have been distinguished: (i) the laterite murrum, and (ii) the laterite rock. These are also found occurring together.

1.6. LAND USE AND LAND MANAGEMENT

The land is used for agriculture, for growing forests, for grazing animals, for mining, for installing industries and for construction of houses, roads, railways, etc. For sustainable development and prosperity of any country, the proper and wise use of

Table 10. Land use classification of India

S. N.	Classification	Area (M-ha)
1	Total geographical area	328.73
2	Total cultivated area	184.38
3	Total gross cropped area	188.15
4	Net sown area	142.82
5	Gross irrigated area	70.64
6	Net irrigated area	53.00
7	Area under forest	68.39
8	Uncultivable area	
(i)	Area put to non-agricultural uses	22.51
(ii)	Barren land	18.77
(iii)	Permanent pastures and other grazing lands	11.24
(iv)	Cultivable waste	14.21
9	Fallow land	
(i)	Current fallow	13.53
(ii)	Other fallow land	9.77

the land is required. The land use depends on the kind of land, its depth, fertility, water retention capacity, available mineral contents, and means of transportation, etc. The use of land for agriculture depends on soil type, irrigation facilities, and climate.

In India, about 51.09% of the land is under cultivation, 21.81% under forest and 3.92% under pasture. Built up areas and uncultivated land occupy about 12.34% (Kundra, 1999). About 5.17% of the total land is uncultivated waste, which can be converted into agricultural land. The other types of land comprises up o 4.67%. The land use classification of India is given in Table 10. The land use classification of the country has also been presented in Figure 9.

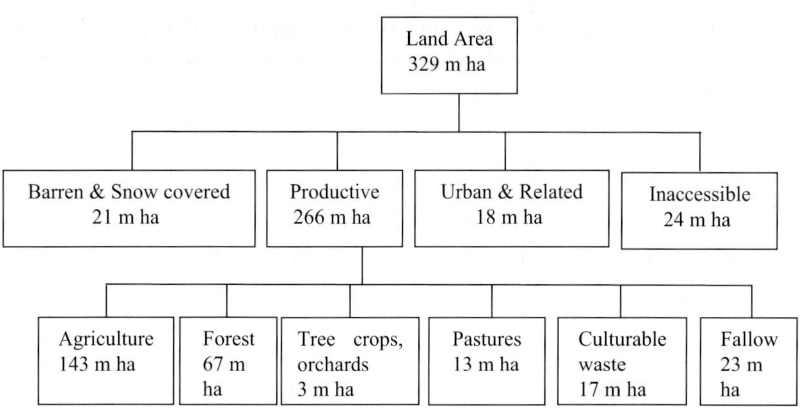

Figure 9. Land use classification of India

In India, the subject of land is under the exclusive jurisdiction of States. There is no national legislation which restricts transfer of agricultural land to other uses. However, State Governments have enacted legislation on the subject, which provides restriction on use of land for non-agricultural purposes. Statutory environmental clearance under Environment (Protection) Act is required for the following types of agricultural development projects and human settlement projects:

- Agricultural Development Projects: Major irrigation projects with command area of 10,000 ha and more.
- Human Settlement Projects: Located in the Coastal Regulation Zone.

If the land involved is forestland, prior clearance is to be obtained under Forest (Conservation) Act to use forests for non-forest purposes. In case some activity involves National Parks or Sanctuaries and if it is not beneficial to the wildlife, it cannot be taken up in those areas. Regarding buildings in the coastal regulation zone, there are restrictions on height, plinth area, withdrawal of groundwater, disposal of waste, etc. Construction of buildings is prohibited in the sensitive areas within Coastal Regulation Zone. Further, land type and local surface and ground water conditions are increasingly being taken into account in selecting sites and techniques for waste disposal. This has been found to have a positive impact on land use.

Land inventories are being generated on the basis of various characteristics of the soil such as soil type, slope characteristics, climatic and hydrological data, vegetation cover, land capability, land irrigability etc. These are updated regularly. The organizations/agencies are involved in maintaining and updating the land inventories are: National Bureau of Soil Survey and Land Use Planning (Indian Council for Agricultural Research); All India Soil and Land Use Survey Organization (Ministry of Agriculture); National Remote Sensing Agency; Indian Space Research Organization; Department of Land Resources (Ministry of Rural development); Ministry of Environment and Forests; and Survey of India. The National Land Use and Conservation Board (NLCB) in the Ministry of Agriculture is the national level policy planning, coordinating and monitoring agency for issues concerning land resources management.

Land and water are mutually reinforcing resource systems. However since the land use pattern has perceptible influence on hydrological characteristic and the soil erosion factors, there is an urgent need to have an integrated water-land management.

1.7. FORESTS

India is rich in bio-diversity – the country is one of the 12 mega-diversity nations. The country has 7 % of the world's biodiversity and supports 16 major vegetation types. About 200 million people in India depend on forests for their livelihoods – directly or indirectly. Forests are important in environmental and economic sustainability, provide numerous goods and services, and maintain life-support systems.

The Indian forests ranges from evergreen tropical rain forests in the Andaman and Nicobar Islands, the Western Ghats, and the north-eastern states, to dry alpine scrub that are found in the Himalayas. Interspersed between these two extremes are semi-evergreen rain forests, deciduous monsoon forest, thorn forests, subtropical pine forests, and montane zone forests.

The Forest Survey of India (FSI) assesses the forest cover of the country using satellite imageries and ground verification and publishes The State of Forests Report (*www.fsiorg.net*) is published after every two year. India's total forest and tree cover for the year 2003 has been assessed by FSI to be 778, 229 km^2. This consists of 51,285 sq. km of very dense forest, 339,279 sq. km of moderately dense forest, 287,769 sq. km of open forest, and 99,896 sq. km of tree cover. This implies that forest cover occupied 20.64 % of country's geographic area, tree cover occupied 3.04% area, yielding total 23.68 % of the covered area. Madhya Pradesh accounts for the largest forest cover of the country at 76,429 sq. km followed by Arunachal Pradesh (68,019 sq. km), and Orissa (48,366 sq. km). Table 11 gives statewise forest coverage of India.

Realizing the role of forests in controlling soil erosion, moderation of floods, recharging of ground aquifers, as habitat for wildlife, conservation of bio-diversity and gene pool, etc., several programmes have been launched.

The forest coverage of the country has shown a net increase of 33,896 sq. km during the period from 1997 to 1999 (FSI, 1999). The area of dense forest has increased by 10,098 sq. km; that of mangroves by 44 sq. km. While some states have registered a net gain in the forest area, others show a declining trend. For instance, in Andhra Pradesh, the net increase of 939 sq. km in forest cover has occurred in 5 years (1993–98), whereas in Mizoram, the loss of 437 sq. km has occurred in 4 years (1994–98). Other states where gains have been recorded are: Arunachal Pradesh, Gujarat, Haryana, Himachal Pradesh, Karnataka, Madhya Pradesh, Maharashtra Orissa, Punjab, Rajasthan Tripura, U.P., West Bengal and Delhi. The states where forest cover has declined are Assam, Bihar, Goa, Mizoram, Kerala, Manipur, Meghalaya, Nagaland, Sikkim, and Andaman and Nicobar Islands. Forests in the northeast of India spread over 14000 sq. km of the Himalayas foothills north of the Brahmaputra River has been described as 'extraordinary' and 'a jewel in the crown of Indian forests' by the WWF. The targets about forests are to achieve 25 % forest cover by 2007 and 33 per cent by 2012 but this appears to be difficult.

The reasons for gains and losses are varied and many. The inclusion of established large block plantations in the forest cover, improvement in the density of forest due to joint forest management and natural regeneration of mangroves have been instrumental in the increasing of forest area. Negative growths are attributed to shifting cultivation, encroachment and tree felling.

The forests of India provide the fuel and fodder for rural people, raw material for industries, a habitat for plants and animals, a sink for CO_2 emissions, and a protective cover for soils. The Forest (Conservation) Act, 1980, amended in 1988, stipulates a massive afforestation programme involving public to establish reserves and re-vegetate degraded lands through joint forest management. In 1979, large-scale afforestation programmes began with the establishment of Forest Development Corporations in the states and launching of Social Forestry Projects. The annual

Table 11. Statewise forest coverage in India

SN	States/Union Territories	Forest cover (km^2)	Tree cover (km^2)	Total (km^2)
1	Andhra Pradesh	44,419	12,120	56,539
2	Arunachal Pradesh	68,019	363	68,382
3	Assam	27,826	935	28,761
4	Bihar	5,558	1,620	7,178
5	Chhattisgarh	55998	6,723	62,721
6	Goa	2,156	136	2,292
7	Gujarat	14,946	10,586	25,532
8	Haryana	1,517	1,415	2,932
9	Himachal Pradesh	14,353	491	14,844
10	Jammu & Kashmir	21,267	3,826	25,093
11	Jharkhand	22,716	5,012	22,728
12	Karnataka	36,449	5,371	41,820
13	Kerala	15,577	1,903	17,480
14	Madhya Pradesh	76,429	7,250	83,679
15	Maharashtra	46,865	9,320	56,185
16	Manipur	17,219	136	17,355
17	Meghalaya	16,839	352	17,191
18	Mizoram	18,430	130	18,560
19	Nagaland	13,609	217	13,826
20	Orissa	48,366	6,381	54,747
21	Punjab	1,580	1,608	3,188
22	Rajasthan	15,826	8,638	24,464
23	Sikkim	3,262	22	3,284
24	Tamil Nadu	22,643	4,991	27,634
25	Tripura	8,093	116	8,209
26	Uttar Pradesh	14,118	7,715	21,833
27	Uttaranchal	24,465	571	25,036
28	West Bengal	12,343	1,731	14,074
		Union Territories		
1	Andaman & Nicobar	6,964	33	6,997
2	Chandigarh	15	8	23
3	Dadra & Nagar Haveli	225	35	260
4	Daman & Diu	8	6	14
5	Delhi (NCT)	170	98	268
6	Pondicherry	40	35	75
7	Lakshadweep Islands	23	2	25
	Total	678,333	99,896	778,229

Source: www.fsiorg.net

planting rates were about $10,000\,km^2$ (1980–1985), $17,800\,km^2$ (1985–1990) and about $15,000\,km^2$ after 1991 (NATCOM, 2003). These steps have definitely helped India to conserve its forests and put some check on the diversion of forest land to non-forest uses. But despite such measures, the average growing stock in India is $74\,m^3/ha$, much lower than the global average of $110\,m^3/ha$.

Due to fuel wood and timber extraction, livestock grazing, and fires, forests in India face continues threat of degradation. The projected climate change is likely to further enhance the stresses on the forest ecosystems and will adversely affect their ability to provide good and services to the society. However, the good news is that due to various conservation measures, the forest cover of India has been increasing steadily over the years. It is noteworthy that this increase is despite the diversion of about $43,200\,\text{km}^2$ of forestland for non-forest purposes such as agriculture ($26,200\,\text{km}^2$), and developmental activities, urbanization, industrialization, and road construction. Between the years 1994 and 2000, a net increase of $42,141\,\text{km}^2$ in the forest cover has been recorded. An increase of $46,690\,\text{km}^2$ in dense forest ($> 40\%$ tree canopy cover) is mainly due to the enhancement of many open forest areas to the dense forest category. World's Forests Report (1999) had acknowledged that India was the only developing country in the world where the forest cover was actually increasing.

Uncontrolled grazing by livestock in forest areas is one of the most important reasons for forest degradation in India. Basically, this destroys the seedlings and young recruits. Estimates show that about 78% of India's forests are affected by grazing. Continued illegal felling many times aided by connivance between officers and mafia, extraction of fuel-wood and non-timber products, invasion by weeds, and forest fires (some of which may not be natural) are the other factors responsible for degradation of forests. Another reason for the degradation of forests is shifting cultivation or jhumming (slash, burn, cultivate and abandon) which is mostly practiced in the northeastern states. According to various estimates, the North-East region accounts for a major share of the 2.75 million hectares of land in India that is under jhumming. Half a million families are already dependent on it and the pressure on land and ecology will increase manifold in the coming years. Fire has affected about 53% of forests in India, out of which 9% are frequent incidences of fires.

1.7.1. Classification of Forests

For administrative purpose, the forest area of India has been classified as Reserve Forests, Protected Forests, and Unclassified Forests. According to the latest assessment for 2001 the area under reserve forest accounts for about 42 M-ha. Some experts have divided the variety of Indian forests into 16 major types.

The forests of India may be divided into major vegetation regions as described in the following.

(i) Tropical Rain Forests

Being a warm and wet region throughout the year, trees available in the forests of this region do not have a distinct season of shedding leaves. These forests are known as evergreen forests and are found in the areas where rainfall in the dry season is more than 200 centimetres. Hence, these forests are also known as typical rain-forests and are found in the rainy slopes of the Western Ghats, Plains of West Bengal and Orissa and North Eastern India. Trees grow very vigorously in these areas and the height of the trees can be up to 60 meters and above. The number of species is too large and too mixed to exploit each one of them commercially. Some of the useful commercial trees found in these forests are Ebony, Mahogany and Rosewood.

(ii) Tropical Deciduous Forests

Because of the formation of natural cover almost all over India, especially in the regions having rainfall between 75–200 centimetres, these forests are monsoon forests par excellence. Economically these forests are of great importance. These are called deciduous because they shed leaves for about 6–8 weeks in summer. Being less fire resistant, these forests require a lot of care. Tropical deciduous forests may be divided into two groups: (i) moist and (ii) dry deciduous.

(a) Moist Deciduous Forests. Moist Deciduous Forests are found on the eastern slopes of the Western Ghats. Teak is an important species found in this region. This type of forests is also found in the north-eastern part of the peninsula, i.e., around Chhotanagpur plateau covering east Madhya Pradesh, Chattisgarh, West Orissa and South Bihar. These forests are common along the Shiwalik hills in north India.

(b) Dry Deciduous Forests. Dry Deciduous Forests Sal is the most important tree found in dry deciduous forests. It has been noticed that the dry deciduous forests are replacing the moist deciduous forests.

(iii) The Thorn and Scrub Forests

These forests are found in the areas where rainfall is under 75 centimetres. These forests are found in the north-western part of the country, from Saurashtra in the west to Punjab in the north. In the eastern part of the country, these forests are also found in Malwa Plateau of northern Madhya Pradesh and Bundelkhand Plateau of South west Uttar Pradesh. Some of the useful trees found in this type of forests are Kikar, Babul, Khair and Date Palms. These forests gradually fade away into scrubs and thorny bushes, which constitute typical desert vegetation.

(iv) Tidal Forests

The tidal forests are generally found along the coasts and the rivers. The noteworthy characteristic of tidal forests is that they can survive in both fresh and salty water. Mangrove trees cover large areas of tidal forests. Sundari is a well-known tidal tree. The Sundarban forests in Ganga-Brahmaputra delta have got this name due to dominance of Sundari tree.

On the basis of topography and climate, the forests of India have been broadly classified into the following categories:

- Tropical forests;
- Montane sub-tropical forests;
- Montane temperate forests;
- Sub-alpine forests; and
- Alpine forests.

The dominant forest types are the tropical dry deciduous forest (38%) and tropical moist deciduous forest (32%). The other important forest types are tropical evergreen, tropical thorn, sub-tropical pine and alpine forest.

Following a different classification, the forest cover can be divided into three types, namely: dense forest, open forest and mangroves. There are 4.73 M-ha of scrubland in addition to the reported forest cover.

Dense forest

If the canopy density is 40 per cent and above, the forests are termed as dense forest.

Open forest

If the canopy density is between 10–40 per cent, the forests are termed as Open forest.

Indian Mangroves

Mangroves are salt-tolerant forest ecosystems found mainly in the tropical and sub-tropical inter-tidal regions in estuaries and coasts. They are reservoirs of a large number of plant and animal species. Mangroves exhibit remarkable capacity for salt tolerance, stabilize the shoreline, and act as a barrier against encroachments by the sea.

India is home to some of the best mangroves in the world and these occur all along the Indian coastline in sheltered estuary, tidal creeks, backwaters, salt marshes and mud flats covering a total area of 4,827 sq km. Mangroves are mainly distributed along the east coast of the country and to a lesser extent, along the west coast. The Sunderbans, covering an area of about $10,000\,km^2$ along the Ganges-Brahmaputra delta, constitute the largest mangrove wetland in the world. The Mahanadi mangrove in Orissa, the Godavari and Krishna mangroves in Andhra Pradesh, the Pichavaram and Muthupet mangroves in the Cauvery delta of Tamil Nadu, the mangroves in the Gulf of Kutchh in Gujarat, and those in the Andaman and Nicobar islands are the other important mangroves in India.

For conservation and management of mangroves, 15 mangroves have been identified for intensive conservation and management. These are: Northern Andaman and Nicobar, Sunderbans (West Bengal), Bhitarkanika (Orissa), Coringa, Godavari Delta and Krishna Estuary (Andhra Pradesh), Mahanadi Delta (Orissa), Pitchavaram and Point Calimer (Tamil Nadu), Goa, Gulf of Kutch (Gujarat), Coondapur (Karnataka), Achra/ Ratnagiri (Maharashtra) and Vembanad (Kerala).

Unfortunately, the mangroves of the country except those of the Andaman and Nicobar, are considerably degraded and shrunk in area. Shrinkage of the mangrove areas have been due to large development of agriculture in the deltas of the major rivers, the reclamation of the coastal wetlands for human settlement and the exploitation of mangroves for products such as fuel. The mangrove cover of the country has been estimated to have reduced by 35 percent during 1985 to 1995.

Mangroves are also under threat due to climate change related phenomenon such as sea-level rise, storm surges, and precipitation and temperature changes. Sea-level rise would submerge additional mangrove areas. Natural disasters such as tsunami waves of 2004 also cause damages to mangroves.

Coral reefs are shallow-water tropical marine ecosystems. These are characterized by high biomass production and rich diversity of flora and fauna. Four coral areas have been identified for conservation and management. These are: Gulf of Mannar, Andaman and Nicobar Islands, Lakshadweep Islands, and Gulf of Kuchch.

1.7.2. Additional Zones of Vegetation in Mountainous Region

(i) The Eastern Himalayan Region
This region includes hills in upper Assam, Sikkim, Bhutan and Nepal. Rainfall is heavy in the outer ranges and there are thick forests of Sal trees. This area is famous as a tea-growing area. Tea lovers all over the world consider Darjeeling tea to be the best tea in the world. Cultivation of rice is done in some places. Forests are the main source of wealth in this sub-region.

(ii) The Western and Kumaon Himalayan Region
This region includes Kumaon, Garhwal, Himachal Pradesh and Jammu and Kashmir. As compared with the eastern Himalayas, the climate here is dry. Northern parts receive more of winter rainfall and the climate is almost of the Mediterranean type. Horticulture occupies a prominent place in the agricultural economy of this region, the major produce being apples, cherries, apricots, peaches, pears, and plums. Apples from Kashmir and Himachal Pradesh are exported to many countries. Other cultivated crops are potato, wheat, maize and rice. Thick forests of Deodar and Pine are found in this region. This area also houses the famous 'valley of flowers'.

1.7.3. Indian Forestry Policy

India is one of the few countries which have a forest policy since 1894. The main plank of the revised Forestry Policy of 1988 is protection, conservation and development of forest. Its aims are:

- maintenance of environmental stability through preservation and restoration of ecological balance;
- conservation of natural heritage;
- check soil erosion and denudation in catchments area of rivers, lakes and reservoirs;
- check extension of sand dunes in desert areas and along coastal tracts;
- increase in forest tree cover through afforestation and social forestry programmes;
- take steps to meet requirements of fuel wood, fodder, minor forest produce and soil timber of rural and tribal populations; and
- increase in productivity of forest to meet the national needs.

Under the Forest (Conservation) Act, 1980, prior approval of the central government is necessary to divert forest lands for non-forest purposes.

Participatory Forest Management is an effective means of regenerating degraded forests and it has found widespread use in India. This approach promotes active participation and involvement of the people in forest conservation and development, including the development of micro-level plans and their implementation. In 1990, the guidelines were issued highlighting the need and the procedure for the involvement of village communities and voluntary agencies in the protection and development of degraded forests. Majority of the States have created mechanisms for public participation in the management of degraded forests. Approximately 10 M ha of forest area is being maintained through Village Forest Protection Committees.

1.7.4. Importance of Forests in India

Some of the major life-support systems of economic and environmental importance of forests are as follows:

(i) Biodiversity
A wide variety of flora and fauna consisting of more than 5,150 species of plants, 16,214 species of insects, 44 mammals, 42 birds, 164 reptiles, 121 amphibians and 435 fish, is found in the Indian forests (NATCOM, 2003). However, many plant and animal species are under various degrees of threat recently due to heavy pressures on natural resources. To conserve biodiversity of the country, about 14.8 M-ha area of forests comprising 80 National Parks and 441 Wildlife Sanctuaries have been converted into protected area.

(ii) Biomass supply
The forests meet nearly 40% of the energy needs and 30% of the fodder needs of India. Many tribals of India live in close proximity of forests and largely depend on them for their fuel requirement. It is estimated that approximately 270 Mt of fuelwood, 280 Mt of fodder, over 12 million m^3 of timber, and several non-timber forest products (such as fruits, nuts, edible flowers, medicinal herbs, rattan and bamboo, honey and gum) are annually harvested from forests.

(iii) Livelihoods to forest dependent communities
About 15,000 plant species are found in India; about 20% of these species yield non-timber products. Millions of forest dwellers and agricultural communities depend on forests for a range of non-timber forest products. Moreover, all forest related activities are labour intensive and lead to rural employment generation.

(iv) Gross Domestic Product
The forest sector provides a range of goods and services whose value is pegged at Rs. 26,000 crores annually. Nearly half of it is from fuelwood.

Concerns have been raised about the impact of deforestation in Himalayas on floods in the Ganga basin. Results of some studies have sown that the deforestation of the Himalayas is not likely to have a significant effect on the extent of the floods in the plains and the delta below.

1.8. AGRICULTURE

Traditionally, India is as agriculture based country. The agriculture sector has a vital place in the economic development of India as it contributes 23% of GDP and employs about 64% of the workforce. Of course, in the recent times, the share of agriculture in the GDP has been continuously declining and it is expected that this trend will continue in near future. Even then, agriculture will have important bearings on India's economy, as it helps to feed a growing population, employs a large labour force, and provides raw material to agro-based industries. India is one of the few developing countries that has the potential to produce crops in almost all land class types namely: dry, semidry, moist, and (sub)humid.

In the 1960s, remarkable improvement in grain yield was realized through the 'green revolution' and later with improved agricultural practices and inputs. These practices include improved mechanized farming, increased net area under irrigation (31 M-ha in 1970–1971; 53 M-ha in 1994–1995; and 57 M-ha in 1998–1999) and net sown area (119 M-ha in 1950–1951 that has increased and is almost stabilized at 143 M-ha over the past decade). Fertilizer consumption, which was 2.6 Mt in 1970–1971, grew to 16.6 Mt in 2000–2001. Food grain production grew at an annual rate of 3% from 1984–85 to 1994–95 and since the 1950s, India's food grain production has increased by four times. Randhawa (1980) has compiled an exhaustive history of agriculture in India.

Three cropping systems are followed in India depending upon rainfall and other climatic conditions. At places where rainfall is in the range 375 to 625 mm, mono-cropping or single crop is grown. Where rainfall is in the range 650 to 750 mm, farmers go for intercropping. When the annual rainfall is 750 mm or more, two or more crops are harvested.

1.8.1. Agriculture Seasons in India

India has mainly two agricultural seasons, namely, the Kharif and the Rabi and the crops grown in these seasons are known as Kharif and Rabi crops, respectively.

(i) The Kharif Season

This season begins with the arrival of monsoon in June or in early July. The crops are sown in June-July and harvested in September-October. The Kharif crops include rice, millets, maize, groundnuts, jute and cotton. Pulses are also grown during this season.

(ii) The Rabi Season

This season follows Kharif. Crops are sown in November and are harvested in April-May. The Rabi season production largely depends upon subsoil moisture. The major crops are wheat, gram, oil seeds like mustard and rapeseed.

Besides these two main seasons, farmers in irrigated areas are able to reap a third harvest during May to July. This season is known as hot weather or Jayad. Moong and Urad are popular crops of this season. Watermelon and Cucumber are also grown in this season.

In addition to food crops, India also produces a large number of non-food or cash crops like, Sugarcane, Tea, Coffee, Cotton, Tobacco, Rubber, and Spices, etc. Sugarcane is a perennial crop that occupies land year around. Table 12 gives the crop area of major crops for 1993–94 and percentage share of the various crops.

In India, wet agriculture is practiced in areas of high rainfall or irrigated areas. With sufficient availability of moisture and suitable soil, farmers may be able to

Table 12. All India Cropping Pattern for 1993–94

Crop	Area (1000 Hectares)	Percentage share to gross cropped area
Rice	42,624	22.90
Wheat	25,135	13.50
Jowar	12,768	6.80
Maize	6,079	3.30
Bajra	9,899	5.30
Rabi	2,045	1.10
Barley	809	0.40
Other Cereals	1,943	1.10
Coarse Cereals	33,543	18.00
Total Cereals	101,302	54.40
Gram	6,265	3.40
Tur	3,469	1.80
Other Pulses	13,643	7.30
Total Pulses	23,377	12.50
Total Foodgrains	124,679	66.90
Sugarcane	3,851	2.10
Condiments and Species	2,876	1.50
Total Fruits	2,908	1.60
Vegetables	4,266	2.30
Oilseeds	28,260	15.20
Fibers	8,479	4.50
Tobacco	430	0.20
Other Crops	10,671	5.70
Gross Cropped Area	186,420	100.00

reap three crops. Dryland agriculture refers to those areas where rainfall is low and irrigation inadequate. In these areas, conservation of moisture is important and crops grown include Jowar, Bajra, and pulses which need less moisture.

India is famous all over the world for high quality tea, such as the Darjeeling tea. Tea plants thrive in hot and humid climate. Tea is mainly grown in West Bengal, Assam, Tamil Nadu, and Kerala. The crop requires 150 to 300 cm of annual rainfall and 20–30° C temperature. Coffee is a related crop that also requires similar climatic conditions albeit lesser rainfall. It is grown mainly in Karnataka, Kerala, and Tamil Nadu and large production of the crop has made India a leading coffee exporter. Table 13 shows the production of the principal crops according to 1991–92.

1.8.2. Agro-Climatic Zones

India presents a large number of complex agro-climatic situations. The Planning Commission after examining the earlier studies at the regionalization of the agricultural economy has recommended that agricultural planning be done on the basis

Table 13. The production of the principal crops according to 1991–92

Crops	Production in lakh ton
Bajra	46.44
Cotton	98.36
Jowar	83.57
Jute	88.51
Maize	70.83
Oil Seeds	182.77
Pulses	120.51
Rice	736.64
Sugarcane	2492.56
Wheat	550.87

of agro-climatic regions. For resource development, the country has been broadly divided into fifteen agricultural regions based on agro-climatic features, particularly soil type, climate including temperature and rainfall and its variation and water resources availability. This classification forms the basis for agricultural planning in the country. These 15 regions are given in Table 14 (Chowdhary et al., 1993). This zone based planning aims at the scientific management of regional resources without adversely affecting the natural resources and the environment.

Another agro-climatic classification divides the country in 20 zones as described in Table 15. These zones are shown in Figure 10.

Table 14. Agro-climatic Classification of India

SN	Name of the agro-climatic zone	Area covers under the agro-climatic zone	Rainfall in the zone	Soil type in the zone	Main crop in the zone
1	Northern most zone	Jammu & Kashmir	50–75 cm	Podsolic type	Maize & Paddy
2	Eastern Himalayan zone	Areas of Eastern Himalayas	About 250 cm	Alluvial and deep black soils	Paddy
3	Western Himalayan zone	Hilly region of Western Himalayas (Outside Ladakh) and its foothills	Much diversity from Eastern Himalayan zone	Brown hilly and alluvial soil	Paddy
4	Lower Assam hills	Tripura, Mizoram, Manipur	About 250 cm	–	Paddy

5	Sutlej Yamuna alluvial plains	Haryana, Punjab, adjoining areas of Jammu & Kashmir & Uttar Pradesh	Low rainfall zone	Alluvial and calcarious soil	Paddy, Cotton
6	The arid west zone	Areas of Thar desert, Western Rajasthan and adjoining areas	Low rainfall (20–30 cm annually)	Sandy soil	Jowar, Bajra
7	The Central semi-arid zone	Eastern Rajasthan, West Madhya Pradesh and adjoining areas of Uttar Pradesh and Gujarat	Over 75 cm	Medium black to red-black soil	Sorghum, Jowar, Maize
8	Lower Ganga basin	Ganga basin in Uttar Pradesh, hills & plain of east Madhya Pradesh	100–150 cm	Medium black to red-black soil	Paddy
9	Eastern Ganga and Mahanadi basin	Gangetic plains in Buhar and West Bengal, The Bhihar plateau and Mahanadi Basin	Moderate rainfall	Rich alluvial intercepted by red, yellow and laterite soils	Paddy & Jute
10	Gujarat & neigh-bourhood	Parts of Gujarat & Maharashtra	Low rainfall (50–70 cms)	Sandy loam to loamy soil	Groundnut, jowar, bajra & cotton
11	The northern and central Western Ghats	Coastal areas of Maharashtra & Karnataka	Heavy rainfall	Acidic soil (pH $\leq$ 3)	Paddy
12	Central highlands	Leeward side of Western Ghats	Drought prone area	Medium black in north and red sandy in south	Jowar, Cotton, Ground nut,
13	Godavari basin	Andhra Pradesh	High rainfall	Sandy coastal alluvial to red soil	Paddy, groundnut, jowar
14	Lower Western Ghats	Kerala, adjoining hills of Tamil Nadu and Karnataka	High rainfall	–	Topioca and Paddy
15	Semi-arid Tamil Nadu tract	Areas of Tamil Nadu	High rainfall	Red sandy to deltaic alluvial soils	Jowar, groundnut

Source: Chowdhary et al. (1993).

Table 15. Agro-climatic zones of India

Zone No.	Name	Area (km^2)	Annual average precipitation (mm)
1	Western Himalayas	142,844	638
2	Western Plain	338,612	612
3	Karnataka Plateau	47,968	1,018
4	Northern Plain and Central Highland	327,968	946
5	Central Highlands, Gujarat Plain	181,696	924
6	Deccan Plateau	314,404	1,080
7	Deccan Plateau and Eastern Ghats	166,908	918
8	Eastern Ghats and Tamil Nadu uplands	186,400	1,163
9	Northern Plain	116,412	1,189
10	Central Highland	237,672	1,258
11	Moderately to Gently sloping	148,968	1,346
12	Eastern Plateau and Eastern Ghats	280,968	1,355
13	Eastern plain	114,176	1,359
14	Western Himalayas	190,016	1,083
15	Assam and Bengal Plain	123,020	2,248
16	Eastern Himalayas	100,372	2,546
17	North Eastern Hills	104,620	2,914
18	Eastern Central Plain	64,292	1,148
19	Western Ghats and Coastal Plains	122,960	1,763
20	Islands of Andaman Nicobar and Laksha Dweep groups	8,281	2,964 and 1,496

Agriculture scenario in India

Due to the Green Revolution, food grain production in India has dramatically increased from 50 Mt in 1951 to 212 Mt in 2002. Also, the mean cereal productivity has increased from 500 kg/ha to almost 1800 kg/ha. Between 1950 and 2000, annual cereal production per capita rose from 121.5 kg to 191.0 kg. These increases in productivity have been largely due to area expansion, large-scale cultivation of new high yielding semi-dwarf varieties, increased application of irrigation, fertilizers and biocides, and support by government policies. But even with this rise, there is much scope for increase in land productivity.

Presently, India has 190 M-ha gross sown area, 142 M-ha net sown area, and 40% of the area is being irrigated. Similar revolutions have also been found in the production of milk (operation flood), fish, eggs, sugar, and a few other crops. Consequently, India is now the largest producer of milk, fruits, cashew nuts, coconuts and tea, the second largest producer of wheat, vegetables, sugar and fish, and the third largest producer of rice in the world. As a consequence, the per capita availability of food grains has risen in the country from 350 g in 1951, to about 500 g per day, availability of milk from less than 125 g to 210 g per day; all this despite the increase in population from 350 million to 1.04 billion. This growth in agricultural production has also led to comfortable position of surplus food stocks with the

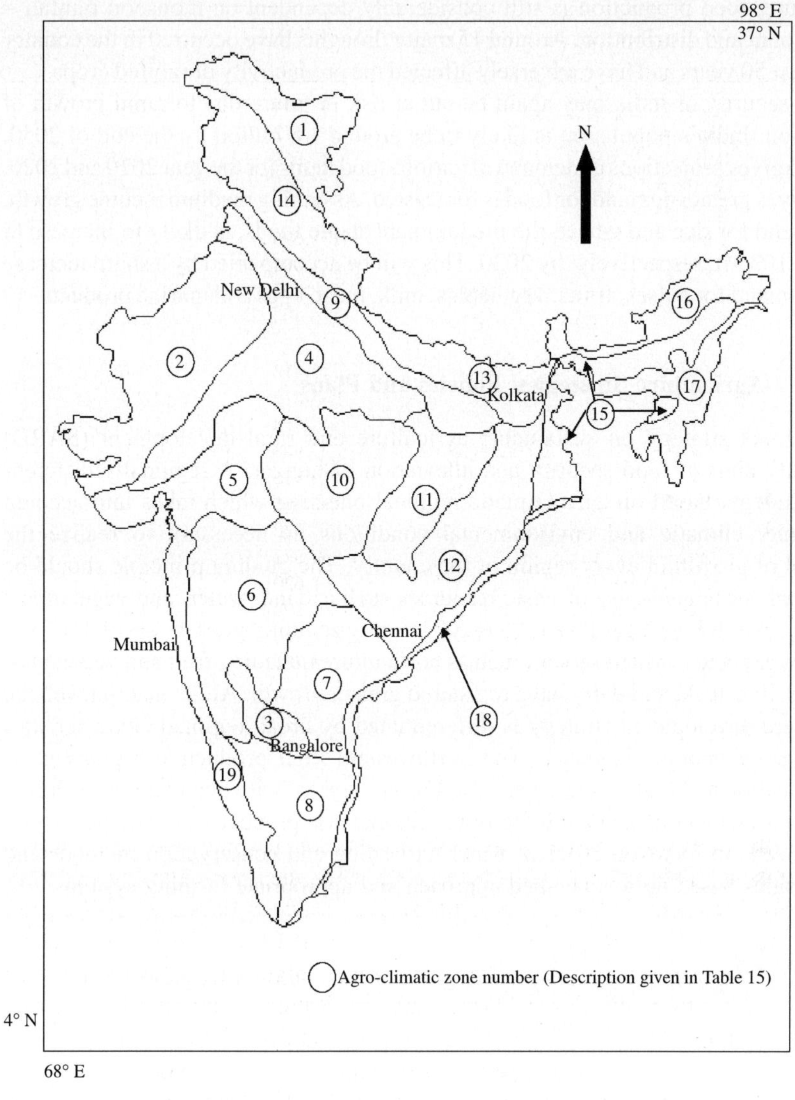

Figure 10. Agro Ecological Region of India

government. Of course, it is a serious concern that some sections of population remain malnourished. Due to the availability of buffer stocks, the nation could manage the severe droughts of 1987, 1999–2000, and 2002–2003 without large import of food and did not face problems of food security.

It is to be noted that the benefits of the green revolution in India were largely confined to the irrigated areas. Despite laudable progress in irrigation and agriculture, food production is still considerably dependent on monsoon rainfall – both amount and distribution. Around 15 major droughts have occurred in the country in the past 50 years and have adversely affected the productivity of rainfed crops.

Food security of India may again be put at risk in future due to rapid growth of population. India's population is likely to be around 1.6 billion by the end of 2050. Table 16 gives projections of demand of various food items for the year 2010 and 2020. Naturally, a greater demand for food is forecasted. Assuming medium income growth, the demand for rice and wheat, the predominant staple foods, is likely to increase to 122 and 103 Mt, respectively, by 2020. This will be accompanied by a sharp increase in the demand for pulses, fruits, vegetables, milk, meat, eggs and marine products.

1.8.3. Agriculture Strategies, Policies and Plans

The national strategy on sustainable agriculture and rural development (SARD) essentially aims at food security and alleviation of hunger. A regionally differentiated strategy, based on agro climatic regional planning which takes into account agronomic, climatic and environmental conditions, is necessary to realize the potential of growth in every region of the country. The guiding principle should be ecological, sustainable use of basic resources such as land, water, and vegetation.

For the Ninth Five Year Plan (1997–2002), an agricultural growth rate of 4.5% per annum is expected. Allied sectors such as horticulture including fruit and vegetables, fisheries, livestock, and dairy have registered greater growth. At the macro level, the agriculture development strategy is differentiated by broad regional characteristics of an agro-economic character. The northwestern high productive regions grow diversification and high value crops. The Eastern region, with abundant water has to show improvement of irrigation facilities and increase productivity. The peninsular region needs to focus on efficient water harvesting and conservation methods and technologies based on a watershed approach and appropriate farming systems.

Table 16. Food Demand assuming a 5% GDP growth at constant prices

Items	Production (Mt)	Demand (Mt)	
	1999–2000	2010	2020
Rice	85.4	103.6	122.1
Wheat	71.0	85.8	102.8
Coarse grains	29.9	34.9	40.9
Total cereals	184.7	224.3	265.8
Pulses	16.1	21.4	27.8
Fruits	41.1	56.3	77.0
Vegetables	84.5	112.7	149.7
Milk	75.3	103.7	142.7
Meat and eggs	3.7	5.4	7.8

Source: NATCOM (2003).

A multi-prong approach is being employed to address problems of land degradation:

- Prevention of soil loss from the catchments,
- Integrated approach to catchment treatment,
- Land use to match land capability,
- Reduction of run-off from the catchments to reduce peak flow into the river system,
- Execution of watershed development programme,
- Generation of data on land suitability, soil and capability for regulating land use,
- Capacity building in soil and water conservation.

Being predominantly an agricultural country, agriculture is the backbone of the economy of India. About 70% of the Indian population is engaged in agriculture. Agriculture provides food as well as raw materials for industries. Several agricultural products like tea, coffee, cotton, oilseeds etc. are exported to other countries. National information on sustainable agriculture is disseminated to decision-makers, advisory organizations and farmers via the Internet: *http://goidirectory.nic.in/*.

1.9. FLORA OF INDIA

Plants of a particular region, which have several species, are referred to as flora. About 49,000 species of plants have already been found in India, which represents the widest range for any country of the world. Out of 49,000 species of flora, about 5,000 species are found exclusively in India including flowering and non-flowering plants. The assemblage of plant species living in association with each other in a given environmental framework is known as vegetation. Climatic conditions, natural vegetation, and soil are closely interrelated with each other. The original vegetative cover of India consisted of forest, grassland and scrubs.

Surveys of the floral and faunal resources in the country are carried out by the Botanical Survey of India and the Zoological Survey of India. The National Institute of Oceanography and other specialized institutions and universities further strengthen the taxonomic database. The diversity of the country's biological resources is yet to be fully surveyed. Approximately 65% of the total geographical area has been surveyed to date. Based on this, over 47,000 species of plants and 81,000 species of animals have been recorded. The biological diversity of the country is so rich that it may play a very important and crucial role in the future survival of mankind if it is conserved and used with the utmost care. Today, two hot spots in biological diversity have been identified in the country, namely, the Eastern Himalayan region and the Western Ghats.

1.10. FAUNA OF INDIA

Fauna refers to species of animals. With an extreme wide variety of flora, fauna in India are equally rich and varied. About 81,000 species of fauna are found in India. In the fresh and marine water of the country, about 2,500 species of fish are available. Likewise, more than 1,200 species of birds are also found in the country. In addition, there are reptiles, amphibians, mammals, and small insects and worms.

A wide variety of mammals, e.g., Elephant, Lion, Tiger, Camel, Wild assess, Nilgai, Gazal, Deer, etc., are found in the country. Elephant is found in the jungles of Assam, Uttaranchal, Kerala, and Karnataka. Camels and wild asses are found in extremely hot and arid deserts. Among the animals of prey, the Indian lion distinguishes itself as the only species found anywhere in the world. Its natural habitat is the Gir forests of Saurashtra in Gujarat. Lion is considered to be the most majestic animal in Indian jungles and is known as the king of forests. In addition, tiger is one of the most powerful species in Indian jungles. Tiger has been designated as the national animal of India. The famous Bengal tiger (which is also the mascot of the Louisiana State University, USA) is found in the tidal forests in the Sundarbans. The other important animals belonging to the cat family are leopards, clouded leopards, and snow leopards. The snow leopard is found in the upper reaches of the Himalayas along with Lesser Panda. Several species of interesting animals, including wild sheep, mountain goats, shrew etc., are found in the Himalayan ranges. India has several species of monkeys. Langoor is the most common amongst them. Red-faced monkeys roam around in populated areas of north India.

There are more than 1,200 species of birds found in the country. These include peacock, pheasants, geese, ducks, munahs, parakeets, pigeons, crane, hornbills and sunbirds, and Indian cuckoo. Peacock is the national bird of India.

1.10.1. Biodiversity

Establishment of a Protected Areas Network, under the Wildlife (Protection) Act, 1972, comprising of Biosphere Reserves, National Parks and Sanctuaries, both terrestrial and aquatic, has been a positive step towards conservation of animal genetic resources. This Network today comprises of 10 Biosphere Reserves, 89 National Parks, 504 Sanctuaries, along with dedicated conservation programmes such as Project Tiger, Crocodile Rehabilitation and project Elephant.

Eight biodiversity rich areas of the country have been designated as Biosphere Reserves applying the United Nations Educational, Scientific and Cultural Organization's Man and the Biosphere (UNESCO MAB) criteria. India is one of the 12 mega biodiversity centres in the world. The country is divided into 10 biogeographic regions: Trans-Himalayan, Himalayan, Indian Desert, Semi-Arid, Western Ghats, Deccan Peninsula, Gangetic Plains, North-East India, Islands, and Coasts.

Biosphere reserves

Biosphere reserves are multi-purpose protected areas to preserve the genetic diversity in representative eco-systems. The major objectives of biosphere reserves are: (i) to conserve diversity and integrity of plants, animals and micro-organisms; (ii) to promote research on ecological conservation and other environmental aspects.

Approximately 5.3% of the total geographical area of country has been earmarked for extensive in situ conservation of habitats and eco-systems through a protection area network of 80 National Parks and 44 Wildlife Sanctuaries. The Central and State Governments together run and manage 33 Botanical Gardens. In addition, universities have their own Botanical Gardens.

Special efforts are being made in India to preserve the wild life of birds and animals. To preserve tigers, project tiger has been implemented which has been a great success. Steps have also been taken for the protection of biological diversity of our land. Under this scheme so far eleven biosphere reserves have been set up namely: (i) Nilgiri biosphere reserve in Karnataka; (ii) Nanda Devi biosphere reserve in Uttar Pradesh; (iii) Nokrek biosphere reserve in Meghalaya; (iv) Great Nicobar in Tamil Nadu; (v) Gulf of Mannar in Tamil Nadu; (vi) Manas biosphere reserve in Assam; (vii) Sunderbans biosphere reserve in West Bengal; (viii) Similipal biosphere reserve in Orissa; (ix) Dibru Saikhowa; (x) Dehong Deband and (xi) Pachmarhi. In these biosphere reserves, plant and animal species are being protected so that this natural heritage can be transmitted to future generation in all its natural vigour and glory. Today in the country there are 84 national parks and 447 wild life sanctuaries covering an area of 150,000 sq. km. Some important wildlife sanctuaries are Jim Corbett National Park, Kaziranga National Park in Assam, Rajaji national park, Manas, Bandipur, and Sariska. Some important bird sanctuaries are Bharatpur Bird Sanctuary, Sultanpur, and Vedanthangal. Well known biosphere reserves are at Nanda Devi, Nilgiri, Rann of Kutch, and Valley of Flowers.

1.11. WETLANDS

The natural resources of a country constitute its greatest wealth and India is fortunate in this respect. Its rich flora and fauna have made the World Bank declare India as one of the twelve 'megadiversity' countries of the world. Many people consider wetlands as waste areas and hence destroy or use them for other development activities. Obviously, this is a short-sighted view.

Wetlands are areas of land where the water level remains near or above the surface of the ground for most of the year. The wetlands cover about 6 per cent of the earth's land surface. There are several kinds of wetlands, such as marshes, swamps, lagoons, bogs, fens, and mangroves. They are home to some of the richest and most diverse and fragile of natural resources. As they support a variety of plant and animal life, biologically they are one of the most productive systems in the world. Figure 11 shows a view of Loktak Lake, Manipur, India.

The richness and usefulness of the wetlands was first brought to the notice of the world by a convention on wetlands held in 1971 in Ramsar, an Iranian city situated on the shores of the Caspian Sea. The Ramsar Convention on Wetlands is an inter-governmental treaty with 135 contracting parties. To mark the date of the signing of the convention on wetlands, 2nd February of each year is observed as World Wetlands Day (WWD). There are 1,235 wetland sites totaling 106.6 million hectares, designated for inclusion in the Ramsar List of Wetlands of International

Figure 11. Loktak Lake, Manipur, India

Importance. India too is a signatory to the Ramsar Convention. As a part of conservation strategy a data book called Montreaux Record is kept of all those wetlands which require international help for conservation.

Wetlands perform many useful functions. They help check floods, prevent coastal erosion and mitigate the effects of natural disasters, such as cyclones and tidal waves. They also store water for long periods. Many wading birds and water fowl like ergets, herons and ibises nest in wetlands. These areas also provide food and shelter for mammals, such as mink, otters and swamp antelopes like sitatunga. They also act as natural filters and help remove a whole range of pollutants from water, such as viruses from sewage works or heavy metals from industrial plants. In short, the terrestrial, marine and atmospheric ecosystems interact in complex, dynamic and often unknown ways.

The natural wetland ecosystems of India can be classified in three main groups: a) marine ecosystems, e.g., coral reefs, b) coastal ecosystems, e.g., mangroves, and c) inland freshwater ecosystems, e.g., rivers, lakes and marshes. A wetland inventory of India was prepared by the Space Application Centre (SAC) for the years 1991–1992 using satellite imageries. Total 27,403 wetlands occupying an area of $75,819\,km^2$ have been listed in this inventory. Of these, 53% wetlands fall under the category of coastal wetlands and the rest are inland wetlands.

1.11.1. Indian Wetlands

Wetlands are widely regarded as natural filters. In India a total area of 40,494 sq. km is classified as wetlands. This includes 14.05 sq. km natural wetlands and balance is man-made. Thus wetlands area consists only 1.21 per cent of the total land surface. Most of the wetlands in India are directly or indirectly linked with major river systems. Nearly 28% of the area of 93 wetlands is under total protection.

India has a wealth of wetland eco-systems distributed in different geographical regions, for example, cold arid zone of Ladakh in the North, the wet humid climate of Imphal in the East, the warm arid zone of Rajasthan, and the wet and humid zone of Southern Peninsula. About 20 wetlands have been identified for conservation and management.

A directory on wetlands in India has been published which gives information on location, area and ecological categorization of wetlands in different parts of the country. India is a signatory to the Convention on Wetlands of international importance, especially as Waterfowl Habitat (Ramsar Convention) and six Indian Wetlands, viz., Keoladeo National Park, Bharatpur and Sambar (Rajasthan), Chilka (Orissa), Loktak (Manipur), Wullar (Jammu & Kashmir), and Harike (Punjab) have been designated under this Convention.

The Ministry of Environment and Forests is the nodal agency for the conservation, regeneration and protection of natural resources. Wetlands, which have an important role in the environmental development programmes, did not receive enough attention until recently. Efforts to conserve wetlands in India began in 1987 and the main focus of efforts is on biological methods of conservation rather than adopting engineering options. A national wetland mapping project has also been initiated for an integrated approach on conservation.

Recently, eleven wetlands in India have been categorized for seeking international assistance to save them from distress (Vasantha, 2003). These include Point Calimere in Tamil Nadu, Astamudi, Sasthamkolta Lake and Vembanad wetlands in Kerala, Kolleru Lake in Andhra Pradesh, Bhitrakanika mangroves in Orissa, Pong Dam Lake in Himachal Pradesh, East Calcutta wetlands in West Bengal, Bhoj wetlands in Madhya Pradesh, Tsomoriri in Jammu and Kashmir, and Deepor Beel fresh water Lake in Assam.

Due to its importance in the socio-economic and cultural life, Loktak Lake (surface area = 26 sq. km) is considered to be the lifeline of Manipur. This lake is the largest natural freshwater lake in north-eastern part of India. The southern portion of Loktak Lake forms the Keibul Lamjao National Park, which is the only floating wildlife sanctuary of the world. This park is a mass of floating islands, locally called phumdis, and is the habitat of endangered mammal, the brow-antlered deer (cervas eldi eldi) locally known as Sangai. The Loktak Development Authority has been created to restore and develop resources of this lake.

Chilika is the largest brackish water lagoon that sprawls along the east coast of India in the Mahanadi delta. Its water-spread area varies between 906 and 1,105 sq. km. Identified as one of the hotspots of biodiversity of India, it is the largest wintering ground for migratory water fowl in the country. The Chilika Development Authority was created to conserve and restore the Chilika lagoon eco-system with its rich biodiversity and aquatic resources for the benefit of all stakeholders.

Ramsar wetlands in India

Significance of wetlands was globally recognized as exclusive habitats for waterfowl at a convention held at Ramsar in February, 1971. With Increase in understanding it is now considered that wetlands have considerable contribution towards national economies by providing fish and wildlife habitats, water quality improvement, flood protection, natural products, hydrological control, recreation and aesthetics, etc. Ramsar Convention is one of the most important international treaties for conservation of nature and natural resources. It has identified a list of specific parameters

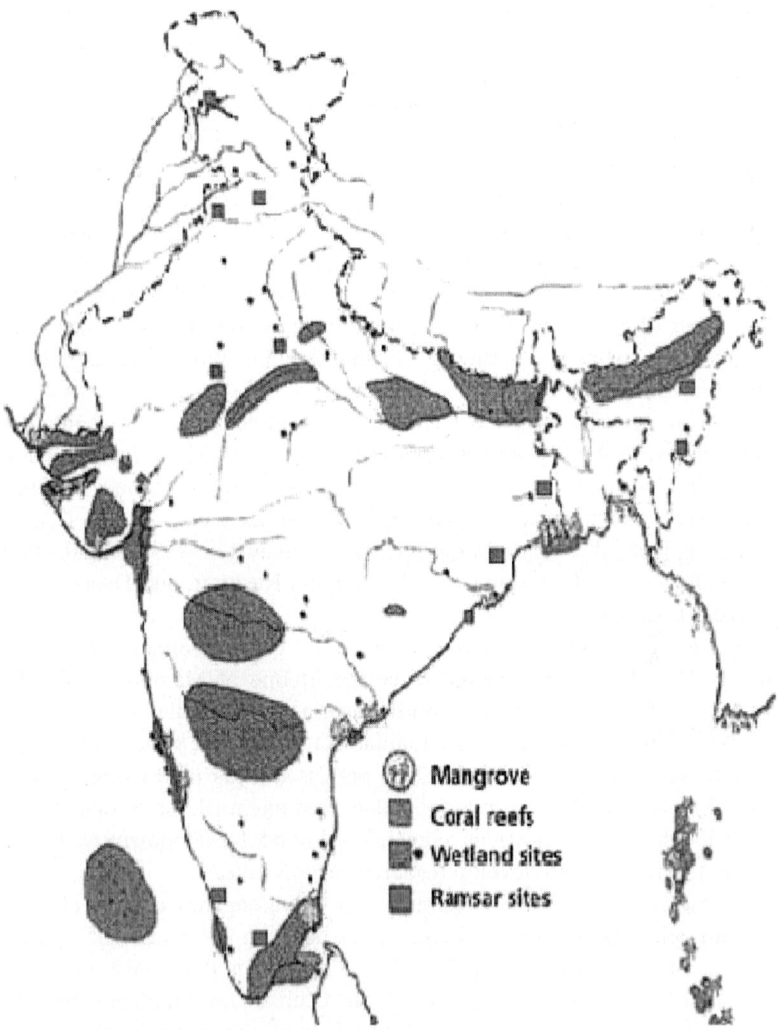

Figure 12. Figure showing important wetlands of India

& based on these parameters a list of wetlands of international importance is being maintained. These wetlands meet at least one of the criteria points of this Ramsar Convention. The location of the important wetlands is shown in Figure 12. The area of each wetland is presented in Table 17.

1.11.2. Inland or Freshwater Wetlands

This category includes natural lakes, swamps, marshes, man-made reservoirs, and tanks etc. In the inventory prepared by the Space Application Centre (SAC), the number of relevant natural inland wetlands is 14,657; the total wetland area of India is 7.58 M-ha. Of this 5.3 M-ha is covered by natural wetlands whereas 2.26 M-ha is occupied by man-made wetlands (NATCOM, 2003). Some of the man-made wetlands such as the one at Bharatpur in Rajasthan are exceptionally rich in bird species. However, many inland wetlands of India have been reclaimed for urban settlement, agriculture, construction of roads, and have severely degraded by pollution from a variety of sources. According to the studies made by the Wildlife Institute of India, 70-80 percent of freshwater marshes and lakes in the Gangetic floodplains have been lost over the past five decades mostly due to reasons stated previously. Table 18 summarizes the number and areas of inland wetlands.

Table 17. Ramsar Wetlands in India

S. N.	Name of Wetland	State	Area (ha)
1	Ashtamudi Wetland	Kerala	61,400
2	Bhitarkanika Mangroves	Orissa	65,000
3	Bhoj Wetland	Madhya Pradesh	3,201
4	Chilika Lake	Orissa	116,500
5	Deepor Beel	Assam	4,000
6	East Calcutta Wetlands	West Bengal	12,500
7	Harike Lake	Punjab	4,100
8	Kanjli	Punjab	183
9	Keoladeo National Park	Rajasthan	2,873
10	Kolleru Lake	Andhra Pradesh	90,100
11	Loktak Lake	Manipur	26,600
12	Point Calimere Wildlife and Bird	Tamil Nadu	38,500
13	Pong Dam Lake	Himachal Pradesh	15,662
14	Ropar	Punjab	1,365
15	Sambhar Lake	Rajasthan	24,000
16	Sasthamkotta Lake	Kerala	373
17	Tsomoriri	Jammu & Kashmir	12,000
18	Vembanad-Kol Wetland	Kerala	151,250
19	Wular Lake	Jammu & Kashmir	18,900

Source: http://www.water-mgmt.com/en/wetlands.htm

Table 18. Number and areas of inland wetlands

Inland Wetland Category	Number	Area (sq. km)
	Natural	
Lakes/Ponds	4, 646	6, 795
Ox-bow lakes	3, 197	1, 511
Waterlogged (Seasonal)	4, 921	2, 857
Playas	79	1, 185
Swamps/Marshes	1, 814	1, 978
	Man-made	
Reservoirs	2, 208	14, 820
Tanks	5, 549	5, 583
Waterlogged	892	773
Abandoned Quarries (water)	105	58
Ash ponds/Cooling ponds	33	29
Total Inland Wetlands	23, 444	35, 589

1.11.3. Coastal Wetlands

As stated earlier, the Indian coastline including the islands is nearly 7,500 km long. In the inventory prepared by SAC, there are 3,959 coastal wetlands which have been classified under 13 major wetland types as shown in Table 19. These wetlands cover a geographical area of 40, 230 km^2.

1.12. LAKES

The hydrology of lake has attracted the attention of Indian Hydrologists after 1980's. Lakes play a significant role in shaping the socio-cultural, socio-economic, ecological and hydrological balance of a region. Lakes are storage elements of a

Table 19. Coastal wetlands of India

SN	Types of Coastal Wetlands	Area (sq. km)
1	Tidal Mudflats	23, 621
2	Mangroves	4, 871
3	Estuaries	1, 540
4	Lagoons	1, 564
5	Sand Beaches	4, 210
6	Marshes	1, 698
7	Other Vegetated Wetlands	1, 391
8	Coral Reefs	841
9	Creeks	192
10	Backwaters	171
11	Rocky Coasts	177
12	Salt-Pans	655
13	Aquaculture Ponds	769

local or regional hydrologic system. They alter the quantity and quality regime of the water flowing through the system. Although, there are some differences in the geometrical, hydraulic and biological characteristics of natural lakes and reservoirs, similar concepts, principle, and models are used for planning, and development of lakes and reservoirs. Moreover, lakes and reservoirs are in most cases being used for similar purposes. The recorded significant differences between natural lakes and man made inland water reservoirs are:

- Reservoirs have both larger drainage areas and larger surface areas associated with them than natural lakes;
- Reservoirs have both larger mean and maximum depths than natural lakes;
- Reservoirs have shorter hydraulic detention times than natural lakes;
- Reservoirs have lower total phosphorus and chlorophyll concentrations than natural lakes- despite the fact that reservoirs in general receive higher total phosphorus and nitrogen loadings:
- Reservoirs also exhibit longitudinal density and concentration gradients which are phenomena not usually observed in natural lakes.

In many respects, lakes, can be considered as "little ocean". Like oceans they receive inflow from rivers, exhibit vertical stratification, undergo cycling and sedimentation, lose water through evaporation and so forth. In freshwater lakes, water also leaves via surface or subsurface outlet. Due to these characteristics, a lake is considered as a portion of a drainage system where water is retained for considerably larger periods than in normal river channels.

Lakes in general, represent additional storage capacity of hydrologic systems. Natural or artificial changes in the storage either on quantity or quality of alter not only the streamflow regime but also the water balance in the region. These effects may be of particular significance in arid and semi-arid regions.

Quantitatively, lakes constitute only about 0.1% of the total water at the earth's surface. They have received proportionately greater attention because of their importance to humans. In many places, lakes are used as sources of drinking water, as a receptacle for sewage and agricultural runoff, for recreation, and for industrial purposes. Because of their generally small size, they can be severely altered by human activities.

Natural and artificial lakes are a major source of water in many regions in India. Excessive siltation, withdrawals from contributing streams upstream, and changing land use in the watersheds has contributed to depletion or shrinking of these water bodies. The water quality in lakes is affected by run-off loaded with fertilizers, insecticides, pesticides coupled with discharges from industries as well as human settlements. Other threats to lakes include encroachments and weed infestation. When the lake system begins to deteriorate, the birds and fishes begin to die or move, fauna is spoiled, and obnoxious weeds proliferate.

Lakes either smaller or larger in size, depending upon their depth of water available and purpose of uses, are known in different names in different parts like, Jheels, Bheels, Marshes, and Tanks etc. However, purposes for which they are being used for are being provided services, and problems faced by them are identical in nature. A national inventory of wetlands was prepared in 1980s. In India, 1,193 wetlands, covering an area and the remaining of 3,904,543 ha were recorded. Out of these, 572 were natural wetlands, 542 were man-made habitats. Some 938 wetlands were freshwater, 134 brackish and 19 coastal.

Major interventions for improving the lake systems in the country include watershed management, dredging operations, treatment of effluents before discharge into the lakes and disposal of solid wastes away from the lakes. A National Lake Conservation Programme has been launched to revitalize natural lakes. In the first phase, the programme aims to cover nine lakes: Hussain Sagar (AP), Sukhuna (Chandigarh), Dal and Nagin (J & K), Bhoj (MP), Powai (Maharashtra), Udiapur Lake (Rajasthan), Kodaikanal and Ooty (TN), Nainital (UA), and Rabindra Sagar (WB). Further details about lakes are given in Chapter 19.

1.13. HYDROLOGY IN ANCIENT INDIA

Natural entities and forces, such as Sun, Earth, Rivers, Ocean, Wind, Water, etc. have been worshipped in India as Gods since time immemorial. Perhaps it is not a sheer coincidence that the King of these Gods is Indra, the God of Rain. Clearly, ancient Indians were aware of the importance of rain and other hydrologic variables for the society. The ancient Indian literature contains numerous references to hydrology and a reading of it suggests that those people knew the basic concepts of hydrological processes and measurements. Important concepts of modern hydrology are scattered in various verses of Vedas, Puranas, Meghmala, Mahabharat, Mayurchitraka, Vrhat Sanhita and other ancient Indian works. Biswas (1969) has rightly remarked that the growth of modern science in Europe would have been hardly possible without the background of pioneering contributions from India, China and Arabian countries, well up to the 12th century A.D. This section contains a brief compilation of information pertaining to hydrology in ancient Indian literature. A detailed treatment on this subject is available in NIH (1990).

In Vedic age, Indians had developed the concept that water gets divided into minute particles due to the effect of sunrays and wind. In various places in Puranas, it is stated that water cannot be created or destroyed, only its state is changed in various phases of the hydrologic cycle. Evaporation, condensation, cloud formation, precipitation and its measurement were well understood in India in Vedic times. Effects of Yajna, forests, reservoirs, etc. on rainfall; classification of clouds, their colour, rainfall capacity, etc.; forecasting of rainfall on the basis of natural phenomena, such as colour of sky, clouds, wind direction, lightening, and the activities of animals; all these were well developed in India even before the 10th century B.C. Contrivance to measure rainfall was developed during the time of Kautilya (4th century B.C.) which had the same principle as that of modern hydrology except that the weight measure of Drona was adopted instead of modern depth measurement of rainfall. The knowledge of monsoon winds and height of clouds along with the division of atmosphere was well developed in the Vedic age. These people also developed technique of knowing the slope of an area by means of a flowing river.

Variation in the height of water table with place, hot and cold springs, ground water utilization by means of wells, well construction methods and equipment are fully described in 54th chapter of Vrhat Sanhita named as 'Dakargala'. Sunrays, wind, humidity, vegetation, etc., as the major causes of evapotranspiration were well realized.

It is very interesting to learn that Varahamihira in as early as 550 A.D. presented a simple method for obtaining potable water from a contaminated source of water. Various plant materials along with solar heating, aeration, quenching of water with fire heated stones, gold, silver, iron or sand were suggested for this purpose. The change in the quality of water with the months of year and suitability of water from different sources for various uses were described.

Efficient water use, lining of canals, construction of dams, tanks, essential requirements for the construction of good tanks, bank protection methods, spillways and other minor aspects were given due consideration in ancient times in India. Well organized water pricing system was prevalent during the time of Kautilya. Various references are available in the Vedas emphasizing the importance of efficient water use so as to reduce the intensity of water scarcity and drought, etc.

The Vedic texts which are more than 3,000 years old contain valuable references on hydrologic cycle. The most important concepts, on which the modern science of hydrology is founded, are scattered in Vedas in various verses which are in the form of hymns and prayers addressed to various deities. Likewise, other Sanskrit literature has valuable discourses regarding hydrology.

Atisthanteenam viveshnanam kashthanam madhyaey nihitam shareeram,
Bratrasya nidyam vi varantyapo deerghatam aashaydindrashatruha.

Verse I, 32, 10 says that the water is never stationary, but it continuously gets evaporated and due to smallness of particles we ca not see the upgoing water particles. In the Varahamihira's Vraht Sanhita (550 A.D.), three Chapters are devoted to hydrometeorology comprising Pregnancy of clouds (Chapter 21), Pregnancy of air (Chapter 22) and quantity of rainfall (Chapter 23). Slokas 1 and 2 of Dakargelam (Chapter 54 of Vraht Samhita) state the importance of science of ground water exploration which helps man to ascertain the existence of water. These are as follows:

Dharmyam yashashyam va vadabhaytoham dakargalam yen jaloplabdhiha
Punsam yathagdeshu shirastathaiva chhitavapi pronnatnimnasanstha.
Ekayna vardayna rasayna chambhyashchyutam namasto vasudha vishayshanta
Nana rastvam bahuvarnatam cha gatam pareekshyam chhititulyamayva.

The water veins beneath the earth are like vein's in the human body, some higher and some lower. The water falling from sky assumes various colours and tastes from differences in the nature of the earth.

In Linga Purana a full fledged chapter (I, 36) has been devoted to the science of hydrology. It scientifically explains evaporation, condensation, rainfall with suitable examples and says that the water cannot be destroyed, only its state is changed:

Dandhaymanayshu charachayshu godhoombhootastvabha nishkramantee
Ya ya oordhva mastraynayrita vai tastastvabhamyagnivayucha.
Ato dhoomagnivatanam sanyogstavamuchyatay
Vareeni varshteetyabhrambhrasyeshah sahastradrik.

i.e. "after being heated by sun, water contained in most of the materials on earth gets converted to smoke (vapour) and ascends to sky with the air and subsequently gets converted to cloud. Thus the combination of smoke, fire and air is the cause of cloud formation. These clouds cause rainfall under the guidance of Lord Indra, having thousand eyes.

Vayu (51. 14-15-16) states like this:

Aadityapateetam suryaganeha somam sankramatay jalam
Nadeebhirvayuyuktabhirlokadhanam pravartatay.
Yatsomatstravatay surya tadbhayshvavatishthatay
Megha vayunighatain visrajant jalam bhuvi.
Evamutikshapyatay chaiva patatay cham punarjalam
Na nashmu udkasyasti tadev parivartatay.

i.e. the water evaporated by sun ascends to atmosphere through the capillarity of air, and there gets cooled and condensed. After formation of clouds it rains by the force of air. Thus, water is not lost in all these processes but gets converted from one form to other continuously.

Verses of Rigveda (I, 27.6; I, 32.8):

Vibhaktasi chitrabhano sindhoroorma upak aa
Sagho dashushay chharasi.
Nadam na bhinnamuya shayanam mano ruhana atim yantyapah
Yashchidwatro mahina paryatishthattasamhih patsutah shirbbhoova.

This verse explains that all water that goes to the sky with wind by the heat of Sun rays gets converted to clouds and then again after the penetration by sunrays it rains and gets stored into rivers, ponds, ocean, etc. Two verses (V, 54, 2 & V, 55, 5) explain the cloud-bearing winds as the cause of rainfall, viz.:

Pra vo marootaststavisha udnyavo vayovridho ashwayujah parijayah
San vighuta dadhati vashati tritah swarntyapoivana parijayah.
Udeeryatha marootah samudrato youam vrishtim varshyatha pureeshidam
Na vo dastra up dasyanti dhanayvah shubam yatamanu ratha avratsat.

"O cloud-bearing winds, your troops are rich in water, they are strengtheners of life, and are your strong bonds, they shed water and augment food, and are harnessed with waves that wander far and spread everywhere. Combined with lightning, the triple-group (of wind, cloud and lightning) roars aloud, and the water falls upon the earth."

By the time of Kautilya (4th century B.C.), Indians had developed the method and instrumental devices for measuring rainfall. This raingauge was known as Varshaman. Kautilya describes its construction in these words "In front of the store house, a bowel (Kunda) with its mouth as wide as an aratni (nearly 18 inches) shall be set up as rain gauge," (Arthasastra, Book II, Chapter V). Kautilya was acquainted with the distribution of rainfall in various areas. He furnishes a very accurate scientific description of the same with statistics. The quantity of rain that falls in the country of Jangala (desert countries) is 16 dronas (4 Adak = 1 drona and one adak equals nearly 7 lb, 11 oz), half as much in moist countries (the countries which are fit for agriculture), 13.5 dronas in the countries of Asmakas (Maharashtra); 23 dronas in Avanti, and an immense quantity in the western countries, the border of the Himalayas and the countries where water channels are used in agriculture. From this it is evident that the spirit of the methodology of the measurement of rainfall given by Kautilya is the same as we have today, the only difference is that he expresses it in weight measures while we use a linear measure nowadays (Arth, Chapter XXIV, Book II, P. 130).

Further discussing the geographical details of rainfall, he observes "when one third of the requisite quantity of rain falls both during the commencement and closing months of the rainy season, and two third in the middle, then the rainfall is considered very even."

Discussing the classification of clouds and interrelationship of rainfall and agriculture, the celebrated author adds "there are the clouds which continuously rain for seven days; eighty are they that pour minute drops; and sixty are they that appear with the sunshine". When rain, free from wind and unmingled with sunshine falls so as to render three turns of ploughing possible, then the reaping of good harvest is certain.

The Vrhat Sanhita and Mayuracitraka by Varahamihira are two very important treatises which are replete with climatological and meteorological information, although they abound in astrological guesses, they contain sufficient scientific facts also. The Vrhat Samhita has three chapters (21st, 22nd, and 23rd) on climatology and meteorology.

The Jains have made considerable contribution in the field of meteorology. The 'Prajnapana' and 'Avasyaka Curnis' provide outstanding studies of the various types of winds. This tradition must have been far older than these treatises. The 'Prajnapana' makes reference of snowfall and hailstorm. The 'Trilokasara' of Nemichandra says that there are seven types of periodic clouds. They rain for seven days each in the rainy season. Then there are twelve species of white clouds. They also bring rain for seven days each. Thus the season of rainfall extends over 133 days in all.

Buddhists too, at least before 400 B.C., have attempted at a very scientific classification of clouds and four species mentioned by them can be compared with the most important four species enumerated in modern meteorology. So much of subtle observation at such an early date is an achievement of the finest order.

Verses (184. 15–17) of Mahabharata state that the plants drink water through their roots. The mechanism of water uptake by plants is explained by the example of water rise through a pipe. It is said that the water uptake process is facilitated by the conjunction of air. This clearly reveals the knowledge of capillary action of soil in the movement of water up and down the plant.

Two Mantras of Atharvaveda say that if the water source is on mountains, then the river formed will be perennial and will flow with high speed (AV.I., 15.3) and (AV.11, 3.1). Similarly Verse (II, 3.1) reveals the same fact saying that the rivers originating from snowclad mountains will keep on flowing in summers also.

The Vishnu Purana (II, 5.3) classified the soils of subterranean region in seven categories, (i) Black (2) White or Yellowish (3) Blue or Red (4) Yellow (5) Gravelly (6) Hilly or boulder and (7) Golden hued.

Regarding the occurrence of ground water, it says: "If there is a termite mound nearby to the east of a Jambu tree, plenty of sweet water, yielding for a long time occurs at a depth of two Purushas, at a distance of three hastas (cubit) to the south of the tree (Vr.S.54.9). Similarly, an Arjuna tree with a termite mound to the north shows water at a depth of 3.5 Purushas at a distance of 3 hastas to the west".

Yajurveda also contains some knowledge about evaporation along with transpiration. Verse 28.43 states that the vegetation attracts water from earth and evaporates it to the atmosphere due to heat, wind, etc. to form clouds.

SECTION 2

HYDROLOGY AND HYDROMETEOROLOGY

CHAPTER 2

WATER BUDGET AND POPULATION OF INDIA

The hydrologic cycle can be quantitatively represented by an equation which represents the continuity of mass. Many forms of this expression, called the water balance or water budget equation, are possible by sub-dividing, consolidating, or eliminating some of the terms, depending upon the purpose of computation. The water balance approach enables a quantitative evaluation of water resources and changes therein. The study of the water balance of lakes, river basins, unsaturated zone, and ground-water basins forms a basis for planning of projects and rational use of water resources in time and space (e.g., inter-basin transfer, stream flow regulation, irrigation scheduling, etc.).

Current information on the water balance of river and lake basins for short time intervals (season, week, and day) is used for operational management. For instance, the mean annual rainfall of 117 cm over the country is equivalent to a volume of water of about $3{,}840 \times 10^9\,m^3$. About 49 percent of annual rainfall is converted into surface runoff and ground water and the remaining 51 percent is lost to the atmosphere by evaporation and transpiration.

2.1. THE WATER BALANCE EQUATION

Water balance equation evolves from the principle of conservation of mass, often referred to as the continuity equation. According to this equation, for a given volume of space and during any period of time, the difference between total input and output equals the change of water storage within the volume. Therefore, the use of a water-balance technique implies the availability of both measured storage and fluxes. However, by judicious selection of the control volume and period of time for which the balance equation will be applied, some measurements may be eliminated (Ferguson and Znamensky, 1981).

Consider the water balance equation for a water body. The input may comprise precipitation (P) which may be rainfall, snow, or both; surface inflow (Q_{SI}); and ground water inflow (Q_{GI}). The outflow part of the equation includes evaporation from the catchment surface (E), surface water outflow (Q_{SO}), and ground water outflow (Q_{GO}). When inflow exceeds outflow, the total water stored in the basin

(ΔS) increases; an inflow less than the outflow results in decreased storage. As the components of water balance equation are subject to measurement errors, the equation includes a discrepancy term (ε). Consequently the general form of the water balance equation may be written as:

$$P + Q_{SI} + Q_{GI} - E - Q_{SO} - Q_{GO} - \Delta S - \varepsilon = 0 \quad (1)$$

The components of a water balance equation may be expressed as a mean depth of water for the control volume (mm), or as volume of water (m^3).

The hydrologic budget or water balance of a drainage basin is a mathematical statement of its hydrologic cycle. It is expressed by equating the difference between inflow, I, and outflow, O, of a drainage basin to the rate of change of storage within the basin, ΔS, for a specified period of time, Δt. When the basin is considered as a black-box system or as a reservoir, its hydrologic budget can be expressed as

$$\frac{\Delta S}{\Delta t} = \overline{I} - \overline{O} \quad (2a)$$

$$\frac{S_2 - S_1}{\Delta t} = \frac{I_2 + I_1}{2} - \frac{O_2 + O_1}{2} \quad (2b)$$

where $\overline{I}$ and $\overline{O}$ are, respectively, the average inflow and the average outflow for the time interval Δt, which is assumed to be small to justify averaging of inflow and outflow. Subscripts 1 and 2 correspond to the values of the variables at the start and the end of the time interval $\Delta t = t_2 - t_1$. If I and O vary continuously with time t, then eq. (2.2) can be written as

$$\frac{dS(t)}{dt} = I(t) - O(t) \quad (3)$$

Implied in eq. (2) or (3) is that I, O, and S do not vary in space or are spatially lumped. Eq. (3) is also referred to as the spatially lumped continuity equation, or sometimes as the water budget.

All hydrologic analyses of drainage basins must satisfy eq. (3), or else the analysis is incomplete and is therefore not reasonable. The appearance of this equation is deceiving in its simplicity. For most hydrologic problems, more than one variable is unknown and, therefore, eq. (3) cannot be solved without additional information. Without an extra relation between S(t) and O(t) with or without I(t), O(t) cannot be evaluated. Furthermore, I and O are not known as continuous explicit functions of time. The difficulty of this equation lies in evaluating the component variables.

In eq. (3), I and O are expressed as rates having the dimensions of $L^3/T/L^2$ ($= L/T$) or L^3/T. By integrating, eq. (3) can also be written with quantities expressed in volumetric units as

$$\int dS(t) = \int [I(t) - O(t)]dt$$

$$S(t) - S(0) = \int_0^t I(t)dt - \int_0^t O(t)dt \tag{4}$$

$$= V_I(t) - V_O(t)$$

in which S(0) is the initial storage or storage at $t = 0, V_I(t)$ and $V_O(t)$ are volumes of inflow and outflow at time t having the dimensions of $L^3/L^2(=L)$ or L^3. Eq. (3) or its variant in eq. (2b) or (4) is the fundamental governing equation for hydrologic analysis and synthesis.

For a drainage basin, the inflow may be comprised of rainfall, snowfall, hail, and other forms of precipitation. Surface runoff, subsurface runoff, groundwater runoff, evaporation, transpiration, and infiltration may constitute outflow. The components of storage may include surface storage (over the ground, including storage in channels and reservoirs, depression and detention storage), subsurface storage (within the rootzone), groundwater storage (within the aquifers), and interception (over vegetation, buildings, etc.). Eq. (3) can be rewritten by including all these components as

$$\frac{d}{dt}(S_s + S_m + S_g + S_i) = I_r + I_{sn} - O_{sr} - O_{sb} - O_g - e - e_t - f \tag{5}$$

where S with a subscript denotes a component of storage with the subscript s for surface storage, m for soil-moisture storage, g for groundwater storage, and i for interception storage; I_r is rainfall intensity, I_{sn} is rate of snowfall, O_{sr} is surface runoff, O_{sb} is subsurface runoff, O_g is groundwater runoff, e is rate of evaporation, e_t is rate of transpiration, and f is infiltration rate. Eq. (5) can be written, analogous to eq. (4), in volumetric units as

$$S_s(t) - S_s(O) + S_m(t) - S_m(O) + S_g(t) - S_g(O) + S_i(t) - S_i(O)$$
$$= P - V_Q - E - T - F \tag{6}$$

in which $P = \int_0^t (I_r + I_{sn})dt$ = volume (or depth) of precipitation, $V_Q = \int_0^t (O_{sr} + O_{sb} + O_g)dt$ = volume (or amount) of runoff, $E = \int_0^t edt$ = amount of evaporation, $T = \int_0^t edt$ = transpiration (volume of water transpired), t and $F = \int_0^t fdt$ = cumulative infiltration. The term V_Q provides the estimate of basin yield. The terms E, T, F, and S constitute what are frequently known as hydrologic abstractions.

Eq. (1) may be simplified or made more complex, depending upon several factors. These are the available data, the purpose of computation, the type and size of water body (river basin, an administrative district, lake or reservoir, etc.) and its hydrographic and hydrologic features. In large river basins, Q_{GI} and Q_{GO} are small compared with other terms and may be ignored. Usually, there is no surface water inflow into a river basin with a distinct watershed divide (assuming no artificial diversions from other basins), and therefore, Q_{SI} may not be included in the water balance equation. Thus, for a river basin eq. (1) can be simplified as:

$$P - E - Q_{SO} - \Delta S - \varepsilon = 0 \tag{7}$$

Depending on the specific problem, the terms of eq. (7) may be further subdivided. For example, in compiling the water balance for short time intervals, the change in the total water storage (ΔS) in a river basin may be subdivided into several parts. These could typically be the changes of moisture storage in the soil (ΔM), in aquifers (ΔG), in lakes and reservoirs (ΔL), in river channels (ΔS_C), in glaciers (ΔS_G), and in snow cover (ΔS_S). Thus, ΔS can be expressed as:

$$\Delta S = \Delta M + \Delta G + \Delta L + \Delta S_C + \Delta S_G + \Delta S_S \tag{8}$$

Special features of the water balance equation for different time intervals.

The computation of the mean annual water balance is one of the simplest problems, since it is possible to disregard the changes in water storage in the basin (ΔS), which are difficult to determine. Over a long period, positive and negative variations in water storage for individual years tend to balance out, and their net value at the end of a long period may be assumed to be zero.

The reverse situation occurs when computing the water balance for short time intervals. The shorter the time interval, the more precise are the requirements for measurement or computation of the water balance components and the more subdivided should be the values of S and other elements. This results in a complex water balance equation which is difficult to close within acceptable errors.

The accuracy of water balance computations generally increases with an increase in the river basin's area. This is due to the fact that the smaller is the basin area, the more complicated is its water balance equation. The reason is that it is difficult to estimate some components of the balance, such as ground water exchange with adjacent basins, water storage in lakes, reservoirs, swamps, and glaciers, the dynamics of the water balance of forests, and irrigated and drained land. The effect of these factors gradually decreases with an increase in the river basin area.

The complexity of the computation of the water balance of lakes, reservoirs, ground water basins and mountain-glacier basins tend to increase with increase in area. This is due to the increased difficulty of accurately measuring and computing the numerous important water balance components of large water bodies, such as lateral inflow and variations in water storage in large lakes and reservoirs, precipitation on their water surface, etc.

2.1.1. Closing of the Water Balance Equation

To close the water balance equation, it is essential to measure or compute all the water balance elements using independent methods, wherever possible. Measurements or estimation of water balance elements usually involves errors due to shortcomings in the techniques used. The water balance equation, therefore, usually does not balance and the discrepancy (ε) is included as a residual term of the water balance equation. A low value of ε does not necessarily mean that all the components have been estimated accurately, because the errors may cancel each other.

If the value of a water balance component cannot be obtained by direct measurement or computation, it may be evaluated as a residual term in the equation. In this case, the term includes an unknown error which may even be larger than the value of the component. Similar considerations apply when the measured values of one component are used to estimate the values of another component. The estimated value will include errors due to the imperfections of the formula and in the measured component.

2.2. WORLD WATER INVENTORY

The total quantity of water in the world is around 1,357.5 million cubic km (M km^3). Over 97% of the world's water is in the oceans and seas and is too salty for most productive uses. Two third of the remaining water is found in the form of ice, caps, glaciers, permafrost, swamps and deep aquifers. About 108,000 km^3 precipitates annually on the earth surface. About 60% (61,000 km^3) evaporates directly back into the atmosphere, leaving 47,000 km^3 flowing towards the sea. If this quantity were evenly distributed, it would have amounted to approximately 9,000 m^3 per person per year in the year 1990.

Thus an amount of only 35 M km^3 fresh water is available on the earth. Out of this about 21.6 M km^3 is Antarctica ice and 10.5 M km^3 is fresh ground water. The distribution of fresh water on the earth is given in Table 1.

The total amount of water received on the earth in the form of precipitation is about 0.42 M km^3. Out of which 0.32 M km^3 is received on the oceans and seas the rest 0.10 M km^3 is on the earth. The water carried by the rivers and springs to the sea each year is about 0.038 M km^3. However, only 4% of this total river flow is used for irrigation and the rest flows to the sea unusable.

The water balance of the oceans is shown in Table 2. It is found from the table that there is considerable water transfer between the oceans and the evaporation and precipitation values vary from ocean to ocean.

Shiklomanov (2000) estimated the mean global renewable water resources at 42,750 km^3/year. The water balance of the continental land mass is shown on Table 3. From the table one can see that Africa is the driest continent in the world where only 20% of the precipitation is going as runoff, whereas the continents of North America and Europe have the highest runoff rates. In absolute terms, Asia has the largest water resources.

Every year the rivers of the world discharge about 40,000 km^3 of water into the sea. Nearly half of it (20,000 km^3) is discharged into the Atlantic ocean where four

Table 1. World Water Reserves

Item	Area 10^6 km^2	Volume 1000 km^3	Depth of run-off (m)	Percent of total water	Percent of fresh water
World ocean	361.3	1,338,000	3700	96.5	–
Ground water					
Gravitational, capillary	134.8	23,400[1]	174	1.7	–
Fresh	134.8	10,530	78	0.76	30.1
Soil moisture	82.0	16.5	0.2	0.001	0.05
Antarctica ice	13.98	21,600	1546	1.56	61.7
Other ice and snow	2.25	2,464	1848	0.179	7.04
Ground ice in permafrost	21.0	300.0	14	0.022	0.86
Lakes					
Fresh	1.2	91	73.6	0.007	0.26
Saline	0.8	85.4	103.8	0.006	–
Marshes	2.7	11.47	4.28	0.0008	0.03
Rivers	148.8	2.12	0.014	0.0002	0.006
Biological water	510.0	1.12	0.002	0.0001	0.003
Atmospheric water	510.0	12.9	0.025	0.001	0.04
Total water	510.0	1,385,984.6	2718	100	–
Fresh water	148.8	35,029.21	235	2.53	100

Ignoring ground water reserves in Antarctica, estimated at 2 million km^3.
Source: UNESCO (1978).

large rivers (Amazon, Congo, Orinoko and Parana) fall. The total annual average flow of the total rivers of the world is 1.2 million m^3/s. The average discharge of the Amazon, which is the largest river of the world, is 200,000 m^3/s. The India's largest river Brahmaputra and the second largest river Ganga have an average discharge of 16,200 m^3/s and 15,600 m^3/s respectively.

2.3. WATER BALANCE FOR INDIA

The average annual precipitation received in India is 4,000 km^3, out of which 700 km^3 is immediately lost to the atmosphere, 2,150 km^3 soaks into the ground

Table 2. Annual water balance of oceans

Ocean	Area (M km^2)	Precipitation (mm)	Inflow through rivers (km^3)	Evaporation (mm)	Water Exchange with other oceans (mm)
Atlantic	92.373	780	20,000	1,040	–60
Arctic	14.09	240	4,400	120	350
Indian	73.92	1,010	4,800	1,380	–300
Pacific	179.68	1,210	10,800	1,140	130

Table 3. Annual Water Balance of Continents

Continent	Area (10^6 km^2)	Precipitation (mm)	Total Runoff (mm)	Average water resources (km^3)	Runoff as % of Precipitation	Evaporation
Africa	30.3	686	139	4,050	20	547
Asia	44.5	726	293	13,510	40	433
Australia & Oceania	8.6	736	226	2,400	30	510
Europe	9.9	734	319	2,900	43	415
North America	24.2	670	287	7,850	43	383
South America	17.8	1,648	583	12,030	35	1,065

and 1,150 km^3 flows as surface runoff. The total water resources in the country have been estimated as 1,953 km^3. Nearly 62% or 1,202 km^3 of the total water resources is available in the Ganga-Brahmaputra-Meghna basin. The remaining 23 basins have 751 km^3 of the total water resources.

The annual water availability in terms of utilizable water resources in India is 1,122 km^3. Besides this, the quantity of 123 km^3 to 169 km^3 additional return flow will also be available from increased use from irrigation, domestic and industrial purposes by the year 2050. The per capita availability of utilizable water, which was about 3,000 m^3 in the year 1951, has been reduced to 1,100 m^3 in 1998 and is expected to be 687 m^3 by the year 2050 (see Table 4).

The following discussion gives various components of water balance for India. Note that the magnitudes of various terms go on changing as new projects come up and as the pattern of water use changes. Further, differences are expected in the figures given by different investigators.

2.3.1. Atmospheric Water Balance

The atmospheric water balance equation for India can be written by equating inflow with outflow plus the change in storage (see Figure 1):

$$V_I + E_T + V_{AI} = P + V_0 + V_{AE} + \varepsilon \tag{9}$$

where V_I represents the inflow of water vapour to the Indian atmosphere from land routes and sea routes, E_T is the total evap otranspiration, V_{AI} is the initial

Table 4. Per capita availability of water

Year	1951	1991	2010	2025	2050
Population (10^6)	361	846.3	1,157	1,333	1,581
Average Water Resources (m^3/person/year)	3,008	128.3	938	814	687

Source: WG (1999a).

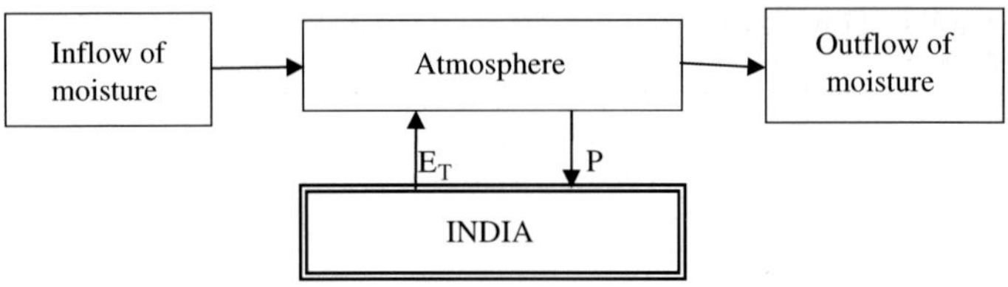

Figure 1. Atmospheric Water Balance

water vapour present in the atmosphere, P is the total precipitation, V_0 is the outgoing water vapour, V_{AE} is the water vapour present at the end of period under consideration.

Components of climatic water balance for India in terms of water surplus and deficit show wide regional variations. Western Ghats have considerable surplus of water as are the Mahanadi and Godavari rivers. The rest of the Peninsula and leeward side of Western Ghats does not have much surplus water, except in monsoon months.

On the basis of observations, the average annual values for inflow of water vapour from the Arabian Sea has been estimated as 770 million ha-m and from the Bay of Bengal as 370 million ha-m of water. Note that 1 ha-m $=$ 1 hectare $\times$ 1 m $=$ 100 m $\times$ 100 m $\times$ 1 m $=$ 10,000 m^3. This unit is frequently used in India. The inflow of water vapour from land routes during non-monsoon months has been estimated as 300 million ha-m of water which also includes winter rains in South India. This has been arrived at by considering non-monsoon rainfall of 100 million ha-m and the possibility of 33 percent of water vapour precipitating as rain during non-monsoon months. The water vapour present at the beginning and end of the year has been assumed to be the same and it has been estimated from available data of Indian Meteorological Department (IMD) as equivalent to 40 mm of precipitable water over the total area of the country. This comes to 13 million ha-m of water. The average annual precipitation, including rain as well as snow, has been estimated as 400 million ha-m. The total evapotranspiration has been estimated on the basis of detailed estimates of water use balance at 278 million ha-m.

Substituting the estimated values of V_I, E_T, V_{AI}, P and V_{AE} in the equation of atmospheric water balance, the outgoing water vapour V_o has been estimated as 1,264 million ha-m which includes the discrepancy term ε. The results of atmospheric water balance are given in Table 5.

2.3.2. Hydrologic Water Balance

The equation for hydrologic water balance of the country for average annual conditions can be written as:

$$P + I = Q_S + E_T + Q_g + \Delta S + \varepsilon \tag{10}$$

Table 5. Atmospheric Water Balance of India (All components in million ha-m)

Inflow Term	Value	Outflow Term	Value
V_{AI}	13	P	400
V_I	1,440	V_o	1,318
E_T	278	V_{AE}	13
Total	1,731	Total	1,731

where P is the total precipitation, E_T is total evapotranspiration, I is the total inflow as surface water (I_s) and ground water (I_g), Q_s is the outflow as surface water to oceans and other countries, (Q_g) is the ground outflow, and ΔS represents the change in soil moisture storage. A diagrammatic representation of the hydrologic water balance for India is shown in Figure 2.

The surface water inflow (I_s) in the country is supplied by northern rivers, such as the Brahmaputra, the Teesta, the Kosi, the Kamla, the Bagmati, the Gandak, the Ghaghra, the Gomati, and other small streams. This flow has been estimated as 20 million ha-m by the Irrigation Commission (1972). The ground water inflow has been assumed as 20 percent of surface water inflow, i.e., 4 million ha-m. The surface water outflow Q_s was estimated as 150 million ha-m by Irrigation Commission (1972) for 18 river basins of the country. Considering the latest runoff and water utilization data, it is estimated that the surface water outflow from the country is about 126 million ha-m.

The ground water outflow has been assumed to be about 13 percent of the surface water outflow on the basis of a general pattern and thus Q_g has been estimated as 20 million ha-m. The values for total precipitation P and total evapotranspiration E_T were estimated for atmospheric water balance also. The change in storage has been assumed as zero for average conditions. The results of annual hydrologic water balance are given in Table 6.

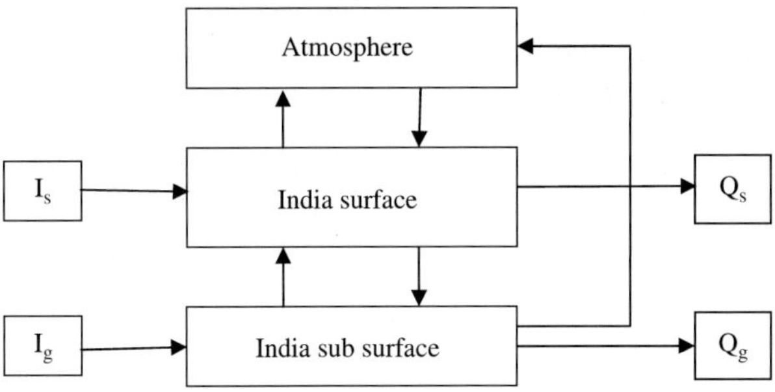

Figure 2. Hydrologic Water Balance

Table 6. Hydrologic Water Balance of India (All components in million ha-m per year)

Inflow Term	Value	Outflow Term	Value
P	400	E_T	278
I_s	20	Q_s	126
I_g	4	Q_g	20
Total	424	Total	424

2.3.3. Water Use Balance

The equation of water use balance for an irrigated land for average annual conditions can be written as:

$$E_{TI} = (R_S + D_S - N_S)\alpha + (T_p - N_g)\beta + P_E \tag{11}$$

where E_{TI} is evapotranspiration from irrigated areas, R_S is diversion from river sources, D_S is supply from detention storage, T_p is withdrawal from ground water zone, P_E is the effective precipitation which is consumptively used by irrigated crops, and N_S is the non-irrigation use of surface water and N_g for ground water. The term α is the consumptive use efficiency for surface water and β is that for ground water. On the basis of experiments and field experience, α can be taken as 33 percent for unlined canals. For ground water after accounting for losses in water courses and field application efficiency, β has been estimated as 50 percent. A diagrammatic representation of the water use balance equation is shown in Figure 3.

The component E_{TI} was estimated by the National Commission on Agriculture (1976). It was assumed that the consumptive use of irrigated crops is equal to the irrigation water requirement. Singh (1995) estimated the component E_{TI} as

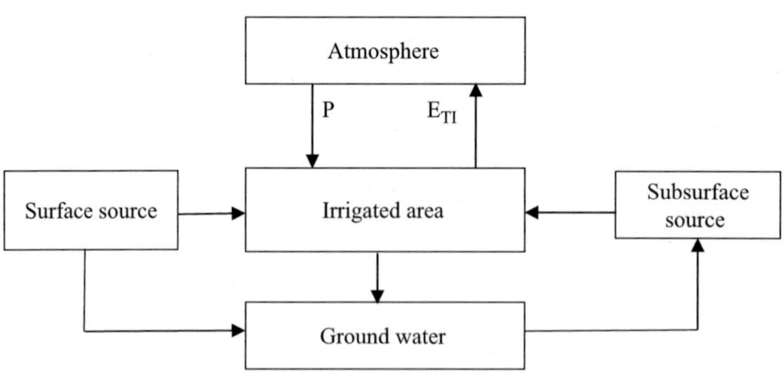

Figure 3. Water Use Balance for India

26 million ha-m. Nag and Kathpalia (1975) had estimated non-irrigation use for industrial, domestic and other purposes as 3 million ha-m. However, this use over the last 30 years has significantly increased and the current estimates place it at 12 million ha-m. The estimated non-irrigation use from surface water sources (N_S) is 8 million ha-m and from ground water sources (N_g) is 4 million ha-m. The water diverted from rivers (R_S) has been estimated as 50 million ha-m. The live storage capacity created in the country up to 1997 has been estimated as 17.7 million ha-m. Assuming evaporation losses at 15% of live storage, the losses come out to be 2.655 million ha-m and the balance 15.045 million ha-m is release for irrigation and non-irrigation uses (D_S).

The National Commission on Agriculture (1976) estimated 6.10 million open wells in the country in 1971. During the fourth plan, there were 21,000 state tube wells having electric pumps and 782,000 private wells were used for irrigation with diesel or electric pumps. Assuming reasonable running time of 2,000 to 3,000 hours per annum for state tube wells and 500 to 1,000 hours per annum for private wells; and their average discharges as 135,000 litres/hour and 30,000 litres/hour, respectively, the total draft from ground water (T_p) was estimated as 13 million ha-m per annum. By the end of 1990s, the number of dug wells has increased to 10.51 million, shallow tube wells to 6.74 million, and public tube wells to 90,000. The amount of water utilized to irrigate about 74 million ha is estimated at 80 million ha-m.

2.3.4. Water Use Balance for Other Areas

The equation for water use balance considering all areas can be expressed as:

$$E_T = E_{TI} + E_L + E_F + E_{UC} + E_W + E_{Tg} + E_R + \varepsilon \tag{12}$$

where E_L represents the immediate evaporation from land surfaces, E_F represents the transpiration from forests and vegetation, E_{UC} represents the evapotranspiration from un-irrigated crops, E_W represents the evaporation from water bodies, E_{Tg} represents the evapotranspiration from ground water, and E_R represents the evaporation from remaining areas. All these components can be expressed in million ha-m of water.

The component E_{TI} has been estimated in the water use balance of irrigated areas as 26 million ha-m. The component E_L has been estimated as 128 million ha-m of water by considering meteorological data. This consists of 70 million ha-m that is immediate evaporation and another 58 million ha-m evaporation from soil. It has been estimated that about 64 million ha of land is under forest and assuming that from this area, evapotranspiration occurs at a rate of 0.85 m per year, the component E_F comes to be 55 million ha-m. The net area under un-irrigated crops has been estimated as 87 million ha and assuming that 43 cm of effective rainfall is used consumptively, the component E_{UC} comes out to be 37 million ha-m. The evaporation from water bodies E_W occurs at the potential rate throughout the year and it has been estimated as 7 million ha-m. For waterlogged area, estimates vary from 2.5 to 4.0 million ha and from this area, evapotranspiration occurs at a potential rate. Loss under this head are estimated at 6 million ha-m.

Addition of all these components gives a total of 259 million ha-m. E_T representing the total evapotranspiration was estimated under hydrological water balance as 278 million ha-m. The remaining component E_R representing evaporation from the remaining areas works out to be 19 million ha-m which includes the discrepancy term ε.

2.3.5. Ground Water Balance

The equation for ground water balance for average year may be written as:

$$R_R + R_F + R_I + I_g = T_p + E_{Tg} + S_E + S_g + Q_g + \eta \qquad (13)$$

where R_R is the recharge due to rainfall, R_F is the flood flow recharge when the river is in high stage, R_I is the recharge due to irrigation, I_g is the groundwater inflow from outside, T_p is the withdrawal by wells, E_{Tg} is the evapotranspiration from waterlogged areas, S_E is the effluent seepage, S_g is the net change in ground water storage, and Q_g is the ground water outflow. The various components are shown in Figure 4.

The component R_F for all rivers has been estimated as 8 million ha-m. The recharge due to irrigation (R_I) can be obtained from the water use data as the difference between water from surface and ground water sources and the consumptive water use for irrigation. This component has been estimated as 18 million ha-m. The component I_g has been taken as 4 million ha-m and Q_g as 20 million ha-m in hydrological water balance. Considering baseflows of rivers, the effluent seepage S_E has been estimated as 45 million ha-m. In the absence of any data the change in ground water storage S_g has been assumed as zero. This should

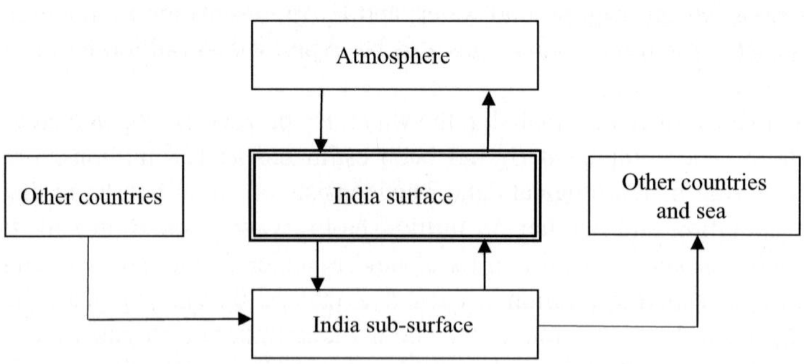

Figure 4. Ground water balance

Table 7. Ground water balance of India (All components in million ha-m)

Inflow Term	Value	Outflow Term	Value
R_R	61	T_P	13
R_F	58	E_{Tg}	13
R_I	18	S_E	45
I_g	4	Q_g	20
Total	91	Total	91

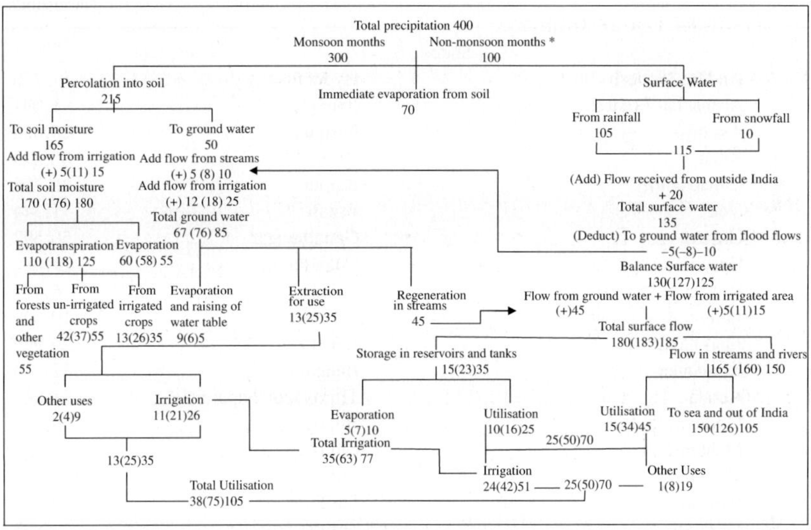

Figure 5. India –approximate distribution of average annual water resources as in 1974, 2000, 2025 AD (million ha-m). The numbers in brackets are those for the year 2000

be reasonable for long-term data. By knowing all the components of ground water balance the remaining component R_R representing the recharge due to rainfall has been evaluated as 61 million ha-m. This also includes the discrepancy term ε. The results of the ground water balance are given in Table 7.

The overall water balance for India showing all the components is shown in Figure 5.

2.3.6. Hydrologic Water Balance

The equation for hydrologic water balance of the country for average annual conditions can be written as:

$$P + I = Q_S + E_T + Q_g + \Delta S + \varepsilon \tag{14}$$

2.4. POPULATION AND NUTRITION REQUIREMENTS

The demographic data forms an important input in planning in general and this is also true for the water sectors. The demand for the municipal water use directly depends upon population. In India, the major water use is for agriculture and these demands depend upon population and nutritional requirements. Furthermore, the population size is directly or indirectly related to demands for electric energy,

Table 8. Population of Indian states and union territories (2001 census)

S. N.	States/Union Territory	Capital	Population (thousands)
	States		
1	Andhra Pradesh	Hyderabad	75,728
2	Arunachal Pradesh	Itanagar	1,091
3	Assam	Dispur	26,638
4	Bihar	Patna	82,879
5	Chhatisgarh	Raipur	20,796
6	Goa	Panaji	1,344
7	Gujarat	Gandhinagar	50,597
8	Haryana	Chandigarh	21,083
9	Himachal Pradesh	Shimla	6,077
10	Jammu & Kashmir	Srinagar	10,069
11	Jharkhand	Ranchi	26,909
12	Karnataka	Bangalore	52,734
13	Kerala	Thiruvananthapuram	31,839
14	Madhya Pradesh	Bhopal	60,385
15	Maharashtra	Mumbai	95,752
16	Manipur	Imphal	2,389
17	Meghalaya	Shillong	2,306
18	Mizoram	Aizawl	891
19	Nagaland	Kohima	1,989
20	Orissa	Bhubaneshwar	36,707
21	Punjab	Chandigarh	24,289
22	Rajasthan	Jaipur	56,473
23	Sikkim	Gangtok	540
24	Tamil Nadu	Chennai	62,111
25	Tripura	Agartala	3,191
26	Uttar Pradesh	Lucknow	166,053
27	Uttaranchal	Dehradun	8,480
28	West Bengal	Kolkata	80,221
	Union territories		
1	Andaman & Nicobar Islands	Port Blair	356
2	Chandigarh	Chandigarh	901
3	Dadra & Nagar Haveli	Silvassa	221
4	Daman & Diu	Daman	158
5	Lakshadweep	Kavaratti	61
6	Pondicherry	Pondicherry	974
7	National Capital Territory of Delhi	Delhi	13,782
Total			1,027,015

maintenance of environment, recreation, etc. Due to these reasons, the information about population in India and its projections as well as nutritional requirements is being provided here.

2.4.1. Population of India

After China, India is the second most populous country in the world. In India, census is carried out every 10 years in the year ending with 1, such as 1991, 2001, and so on. The population of India based on the last census (2001) was 102.7015 crore. Among the states, U.P. with a population of 16.6053 crore had the largest population while Sikkim was the state with the smallest population. Only five countries of the world namely, China, India, United States, Indonesia, and Brazil have more people than does U.P. The population of the various states and union territories is given in Table 8.

To give a historical perspective of the population growth in India, the data of population at various times during the last century is given in Table 9. During the last three decades of the 20th century, there has been a very high growth of population with the average annual growth often exceeding 2%. India has seen a rapid decline in crude, death rate in the last century as a result of improved medical facilities and sanitation.

From the point of water resources development, urbanization is of particular interest since it influences demand of water for municipal uses. For urban areas, the per capita demand for domestic use is much larger than for rural areas. The relevant norms are discussed later. The criteria for classification of urban areas in India and the number of such areas in 1991 is given in Table 10.

Table 11 lists urban population for census operations carried during the last century and the percentage of urban population for class-I & class-II cities. Table 8

Table 9. Population statistics of India, 1901–1991

Census year	Total population (million)	Average annual growth rate	Crude birth rate (per 1,000)	Crude death rate (per 1,000)
1901	238.4	0.30	–	–
1911	252.0	0.56	49.2	42.6
1921	251.2	−0.03	48.1	47.2
1931	278.9	1.06	46.2	36.3
1941	318.5	1.34	45.2	31.2
1951	361.0	1.26	39.9	27.4
1961	439.1	1.98	40.9	22.8
1971	548.2	2.20	40.0	17.8
1981	683.3	2.24	37.8	15.4
1991	846.3	2.14	32.5	11.4
2001	1,027.0			

Source: Working group (1999) water requirements.

Table 10. Classification of Indian urban areas

Class of urban area	Population size	Number in 1991
Class I	> 100,000	300
Class II	50,000–99,999	345
Class III	20,000–49,999	945
Class IV	10,000–19,999	1,167
Class V	5,000–9,999	740
Class VI	< 5,000	197
All India		3,678*

Source: Working Group (1999) Water Management.

Table 11. Total urban population and share of class I & II cities

Census year	Urban population	Urban population of class – I cities	Percent of urban population of class – I cities	Urban population of class – II cities	Percent of urban population of class – II cities
1901	25,616,051	6,586,347	25.7	2,892,094	11.3
1911	25,580,199	6,955,756	27.2	2,682,565	10.5
1921	27,691,306	8,142,241	29.4	2,873,264	10.4
1931	32,976,018	10,090,279	30.6	3,941,183	11.9
1941	43,558,665	16,519,922	37.9	4,970,555	11.4
1951	61,629,646	27,308,404	44.3	6,130,907	9.9
1961	77,562,000	39,380,309	50.7	8,535,706	11.0
1971	106,966,534	60,123,375	56.2	12,029,893	11.2
1981	156,188,507	94,292,998	60.3	18,191,847	11.6
1991	217,611,000	13,880,200	63.7	23,309,000	10.7

Source: WG (1999a).

extends this information by providing projections of urbanization up to the year 2050 by which the population of India is expected to stabilize. Of special importance is the rapid rise of urbanization in India which shows the stress on infrastructure development, including water supply and sanitation.

The projection of population as per the UN medium variant and the urban population for the various years is given in Table 12.

India's population is growing with such a high rate that it has nearly doubled in the last 35 years. The population of the country has already crossed the landmark of 1 billion. Today, after China, India is the second most populous country in the world and if the present trend of population increasing continues, India will be the most populous country of the world by the middle of the current century. The population growth is putting additional pressure on natural resources as well as complicating the development works. This is a cause of worry for planners and developers. Arguably, the control of population should be the top priority of the nation.

Table 12. Projections of India's population

Year	Population as per medium variant (million)	Percent urban population for medium variant
2010	1,189	33.8
2020	1,327	37.2
2030	1,455	41.2
2040	1,564	45.3
2050	1,640	74*

Source: WG (1999a).

2.4.2. Population Projections by Different Sources

The long-term planning has to take into account the population growth. A number of individuals and agencies have estimated the likely population of India by the year 2025 and 2050. According to the estimates adopted by National Commission for Integrated Water Resources Development (NCIWRD, 1999), by the year 2025 the population is expected to be 1,333 million in high growth scenario and 1,286 million in low growth scenario. For the year 2050, the high rate of population growth is likely to result in about 1,581 million people while the low growth projections place the number at nearly 1,346 million.

To control the rapid population growth rate, the several efforts are being made by the government. Due to the continuous efforts, a visible fall has been observed in the rate of population growth. In the past, the average number of children born to an Indian couple was around six and the birth rate was around 40–45 per 1,000 (up to 1971). Due to various efforts of the government, the birth rate has considerably reduced and the average number of children born to an Indian woman during her reproductive life has decreased from 6 to 3.3.

While the average birth rate in the country is declining, its spatial distribution is uneven. Some states, such as Kerala, Tamil Nadu, and Goa, have achieved a goal of an average of two children per family, whereas in Bihar, Madhya Pradesh, Rajasthan, and Uttar Pradesh, the average number of children per family still is between four to six.

Projections of population by several agencies and individuals are given in Table 13.

Table 14 shows the projected urban and rural population of India for various years up to year 2050, by which time the population is likely to stabilize. NCIWRDP (1999) has accepted the higher limit of population as 1,581 million and the lower limit as 1,345.9 million for the year 2050. The recent trends in the Indian society show that most couples in urban areas go for two children while the families in rural areas are also of smaller size these days. Expectedly, this trend will accelerate. Hopefully then, the population of the country shall stabilize before 2050 at a lower level.

Table 13. Population projections by different sources

S. N.	Reference	Population (in million) in the year		
		2010	2025	2050
1	UN 1994 Revised			
A	Low variant	1,156.6	1,286.3	1,345.9
B	Middle variant	1,189.0	1,392.0	1,640.0
C	High variant	1,221.7	1,501.5	1,980.0
2	Registrar general of India (1996)	1,162.0		
3	Visaria & Visaria (standard) 1996	1,146.0	133.0	1,581.0

Source: WG (1999a).

Table 14. Urban and rural population projections for India

Scenario	Year 2010		Year 2025		Year 2050	
	Urban	Rural	Urban	Rural	Urban	Rural
UN medium variant	402	787	630	762	1,007	633

Source: WG (1999b).

2.4.3. Nutrition Requirements

The Indian economy has traditionally been agriculture-based. At the time of independence, it was crucial to develop irrigation to increase agricultural production to make the country self-sufficient in food grains and for poverty alleviation. Accordingly, irrigation sector was assigned a very high priority in the 5-year plans. Giant schemes, such as the Bhakra Nangal, Hirakud, Damodar Valley, Nagarjunasagar, Rajasthan Canal project, etc., were taken up to increase irrigation potential and maximize agricultural production.

In 1990–91, India's production of food grain was 176.4 million tones, including 162.1 million tones of cereals and 14.3 million tones of pulses. In the last two decades, it has been found that the requirement of cereals in urban areas is about 135 kg/capita/year which is stationary, whereas the requirements of cereals in rural areas have shown a declining trend from 185 to 175/capita/year. The consumption of milk, eggs and live stock products have shown a rapid growth rate.

Keeping in view the level of consumption, the estimated food grain requirements per capita per year has been estimated to be 198 kg for the year 2010, 218 kg for the year 2025 and 284 kg for the year 2050 (See Table 15). After considering factors, such as feed requirement, losses in storage and transport, seed requirement, and buffer stock, the projected food grain and feed demand for 2025 would be 320 million tons (high demand scenario) and 308 million tons (low demand scenario). The requirement for the year 2050 would be 494 million tons (high demand scenario) and 420 million tons (low demand scenario).

Table 15. Food demand estimates

Item	Unit/year	1993–96	2010	2025	2050
Assumptions:					
Population – Total	Million	939.6	1,168	1,370.1	1,622.9
Population – Urban	Million	247.4	370.4	504.5	752.8
Population – Rural	Million	692.2	797.6	865.6	870.1
A. Growth rate of aggregate expenditure	Percentage		4	4	4
i. Food grain	Million tons	154.16	214.79	276.92	395.44
ii. Milk & Milk products	Million tons	48.97	94.16	160.68	357.15
ii. Edible oils	Million tons	4.33	7.13	10.94	21.76
iv. Meat & Fish	Million tons	3.69	6.57	10.7	23.02
v. Sugar & Gur	Million tons	9.25	15.61	24.15	46.28
vi. Fruits & vegetable	Million tons	38.58	67.44	110.62	249.46
Total of foodgrain & feed	Million tons	158	221	288	419
Food grain & feed per head	Million tons	168	189	210	258
B. Growth rate of aggregate expenditure	Percentage		4.5	4.5	4.5
i. Food grain	Million tons		219.35	287.26	434.58
ii. Milk & Milk products	Million tons		101.38	181.47	445
ii. Edible oils	Million tons		7.49	11.99	26.42
iv. Meat & Fish	Million tons		6.99	11.91	28.4
v. Sugar & Gur	Million tons		16.55	26.74	56.21
vi. Fruits & vegetable	Million tons		71.55	123.33	311.08
Total of foodgrain & feed	Million tons		226	299	464
Food grain & feed per head	Million tons		194	218	286
C. Growth rate of aggregate expenditure	Percentage		5	5	5
i. Food grain	Million tons		223.8	298.42	485.39
ii. Milk & Milk products	Million tons		108.84	204.32	558.61
ii. Edible oils	Million tons		7.86	13.16	32.47
iv. Meat & Fish	Million tons		7.41	13.25	35.36
v. Sugar & Gur	Million tons		17.51	29.58	69.04
vi. Fruits & vegetable	Million tons		75.82	137.94	390.99
Total of foodgrain & feed	Million tons		231	312	522
Food grain & feed per head	Million tons		198	228	321

Source: WG (1999a).

Estimates of food grain requirements for various years have been given by NCIWRD (1999) for various growth scenarios, viz., 4, 4.5, and 5%. The commission accepted the scenario of a 4.5% growth rate of aggregate expenditure and computed future foodgrain requirements.

Furthermore, assuming feed requirements, the losses in storage and transportation, seed requirement, and carryover for years of monsoon failure at 12.5%, the final demands are given in Table 16.

Table 16. Projected foodgrain and feed demand for India

Demand	Year		
	2010	2025	2050
High Demands (million tones)	224	291	449
Low Demands (million tones)	222	280	382
With addition of seed, feed, wastage, etc.			
High Demands (million tones)	247	320	494
Low Demands (million tones)	245	308	420

Source: WG (1999a).

2.5. WATER POVERTY INDEX

We live in a water-challenged world where freshwater crisis is deepening with time, because each year about 80 million additional people stake their claims to the earth's limited water resources. Sadly, nearly all the projected three billion people who world be added over the next half of the century will be born in countries that are already experiencing water shortages. By 2050, India is projected to add 613 million people and China another 211 million.

Even with today's six billion people, the world has a huge water deficit. Using data on overpumping occurring in China, India, Saudi Arabia, North Africa and the U.S., Postel (1999) calculates an annual depletion of aquifers to be 160 billion cubic metres or 160 billion tons. Assuming that it takes 1,000 tons of water to produce one ton of grain, this 160 billion ton water deficit is equal to 160 million tons of grain. At an average world grain consumption of nearly 300 kg per person per year, this would feed 480 million people. Stated differently, 480 million of the world's six billion people are being fed with grain produced with the unsustainable use of water.

Data show that nearly 70 percent of the water consumed worldwide is used for irrigation. In the increasingly intense competition for water among sectors, agriculture almost always loses. The 1,000 tons of water used in India to produce one ton of wheat worth about (Rs. 8,000) can also be used to expand industrial output by (Rs. 400,000) or 50 times.

In addition to population growth, urbanization and industrialization also expand the demand for water. As villagers, who are traditionally reliant on the village wells, move to urban life, their residential water use can easily triple. Industrialization takes even more water than urbanization. Rising affluence generates additional demand for water.

Two options immediately come to mind to control water demand: stabilize population or raise water productivity. Another management strategy is to raise the price of water to reflect its cost and minimize wastage. Shifting to more water-efficient technologies and crops offer a huge potential to raise water productivity. These shifts will accelerate if the price of water closely reflects its true value.

2.5.1. Water Poverty Index for India

The Water Poverty Index (WPI) provides a framework to understand how water can best be managed to meet the needs of people. This is a holistic approach to water resource evaluation, in keeping with the Sustainable Livelihoods Approach used to evaluate development progress. The approach to calculate WPI is based on the formulation of a framework, which incorporates a wide range of variables. A team of researchers from the UK's Centre for Ecology and Hydrology and experts from the World Water Council have developed WPI. The WPI demonstrates that it is not the amount of water resources available that determine poverty levels in a country, but the effectiveness of how these resources are used. The scores of the index range on a scale of 1 to 100, with the total being generated as a weighted additive value of five components which are:

Resource

This is a measure of ground and surface water availability, adjusted for quality and reliability.

Access

This indicates the effective access to water that people have for their survival.

Use

This captures some measure of how water is used, including sectoral shares.

Capacity

This variable represents human and financial capacity to manage the system.

Environment

This tries to capture an evaluation of ecological integrity related to water.

The WPI assigns a value of 20 points as the best score for each of its five categories. A country that completely meets the criteria in all five categories would have a score of 100. The highest-ranking country, Finland, has a WPI of 78 points, while Haiti has a WPI of just 35.

The WPI reveals that some of the world's richest nations, such as the United States and Japan, fare poorly in water ranking, while some developing countries score in the top ten. According to the WPI, the top 10 water-rich nations in the world in descending order are: Finland, Canada, Iceland, Norway, Guyana, Suriname, Austria, Ireland, Sweden and Switzerland. The 10 countries lowest on the WPI are all in the developing world namely: Haiti, Niger, Ethiopia, Eritrea, Malawi, Djibouti, Chad, Benin, Rwanda, and Burundi. The WPI demonstrates a strong connection between 'water poverty' and 'income poverty.'

For India, the scores on various parameters have been provided by Lawrence et al. (2003) as follows: Resource – 6.8; Access – 11.0; Capacity – 12.1; Use = 13.8; Environment = 9.5; and WPI = 53.2. Since the scores are low on almost all counts, a holistic and concerted long-term strategy is necessary. Furthermore, scores on parameters, such as resource, are likely to dwindle rapidly in view of rapidly increasing population and industrial growth.

CHAPTER 3

RAINFALL AND ANALYSIS OF RAINFALL DATA

In India, rainfall is mainly dependent on the southwest and northeast monsoons, western disturbances, shallow cyclonic depressions, and violent local storms. Most of the rainfall in India takes place under the influence of southwest monsoons between June and September. An exception is Tamil Nadu where rainfall is under the influence of northeast monsoons during October to November.

Modern metrological observations in India began in 1793 with the establishment of the first Indian meteorological observatory in Chennai (the then Madras). When the India Meteorological Department (IMD) was set-up in 1875, a network of nearly 90 weather observatories was created.

The rainfall in India shows great variations of all types: seasonal (temporal) and geographical. It generally exceeds 1,000 mm in areas to the East of Longitude 78 °E. It extends to 2,500 mm along almost the entire West Coast and Western Ghats and over most of Assam and Sub-Himalayan West Bengal. In the western part of the nation, rainfall diminishes rapidly from 500 mm to less than 150 mm in the extreme west. The peninsular part has large areas of rainfall below 600 mm with pockets of even 500 mm.

India has a long history of rainfall measurement. The Calcutta observatory possesses the longest rainfall records in India going back to 1784. At the time of independence there were about 2,750 rainfall stations in India. India Meteorological Department (IMD) has prepared rainfall maps with rainfall data of 2,700 raingauge stations during the period 1901–1950. Maps of annual rainfall and coefficient of variation have also been prepared (IMD, 1981).

3.1. MEASUREMENT OF RAINFALL

The amount, intensity and areal distribution of precipitation are essential in many hydrological studies. The total amount of precipitation, which reaches the ground in a stated period is expressed as the depth to which it would cover in liquid form on a horizontal projection of the earth's surface. Similarly, snowfall is also expressed as the snow water equivalent of fresh snow in a stated time period covering an even horizontal surface (WMO, 1983).

The units of precipitation are linear and daily amounts of precipitation should be read to the nearest 0.1 mm. Weekly, fortnightly and monthly amounts should, however, be read to the nearest 1 mm at least.

All observations of rain are taken in India at 08:30 hours IST to ensure standardization and inter-comparison of rainfall from different rain gauge stations. The rainfall measured at 08:30 hours on any particular date is entered against that date and it is understood that the rainfall so registered has been received in 24 hours preceding 08:30 hours of the day of observation.

3.1.1. Precipitation Gauges

A rain gauge basically consists of a collector for delineating the area of the rain and a funnel leading to a storage device. The precipitation thus collected is measured by transferring the contents to a graduated measuring jar, which goes along with the particular rain gauge. Different types of gauges are used to measure rainfall and snowfall. Since the size, shape and exposure affect the precipitation caught by a gauge, it is desirable to use a standard gauge. The Symon's raingauge has been adopted as the standard in India. This raingauge is shown in Figure 1.

The Symon's raingauge consists of a funnel which has an accurately turned and beveled gunmetal rim 127 mm in diameter, cylindrical body, a receiver with a narrow neck and a splayed base which is fixed in the ground. This gauge is now being replaced by a fiber glass reinforced plastic (FRP) rain gauge. The collector

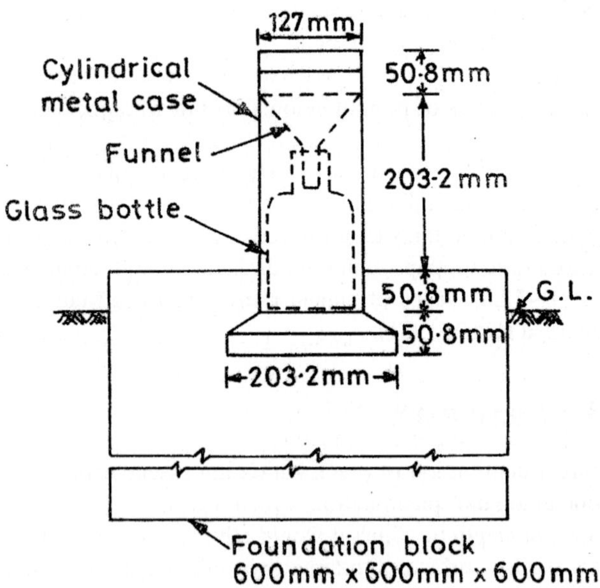

Figure 1. Symon's raingauge

area of the FRP gauge is $200\,cm^2$. In heavy rainfall areas, a gauge with a smaller orifice (usually $100\,cm^2$) area is preferred to allow collection of less volume (more depth) of rainfall.

The raingauge shall be fixed on a masonry concrete foundation $60 \times 60 \times 60\,cm^3$ sunk into the ground. The gauge should be cemented into the platform in such a way that the rim of the gauge is approximately 30 cm above the ground level. This height is necessary to prevent the splashing of water into the gauge. The Indian Standard IS: 4,986–1983 emphasizes the following points while installing a gauge:

i) The gauge should be placed on the level ground. The gauge should not be placed on the slope or terrace and never on a wall or roof.
ii) The distance between the raingauge and the nearest object should be generally four times the height of the nearest object/obstruction but in no case the distance should be less than twice the height of the object/obstruction.
iii) Great care should be taken for mountainous and coastal stations such that the gauges are not unduly exposed to the sweep of the wind. A belt of trees or a wall on the side of the prevailing wind at a distance as mentioned in (ii) above is recommended as an efficient wind shelter.
iv) In the hills where it is difficult to find a level space, the site for the gauge shall be chosen with a minimum level area of $6\,m \times 6\,m$ where it is best shielded from high winds and where the wind does not cause eddies.

3.1.2. Recording Gauges

Three types of recording precipitation gauges are in general use. They are the float type, the tilting or tipping bucket type and the weighing type. In India, the float and siphon type-recording rain gauge, usually referred to as Self Recording Rain Gauge (SRRG), is most commonly used as part of the measurement network. The tipping and weighing type are used by some organizations at their experimental stations. This raingauge is shown in Figure 2.

The float and siphon recording raingauge consists of a collector and rainfall recording mechanism mounted on a base. The recording mechanism consists of a float chamber and a siphon chamber. The siphon is fixed to the float chamber with provision for adjusting to enable correct siphoning. The lower end of the float chamber is provided with a hexagonal cap, which can be removed for cleaning the float chamber. A wire gauge filter is provided which fits into the inlet tube to check the entry of other material. The recording pen is mounted on the stem of the float. The time is marked by a horizontally rotating drum driven by a clock. The clock mechanism may be mechanical or quartz type.

The gauge generally has a capacity of 10 mm of rainfall for each siphoning and has a collector having a rim area of $325\,cm^2$ (203.4 mm diameter). To record the precipitation at stations where heavy rainfall with high intensity is experienced, it is preferable to have a raingauge with a collector having a rim area of

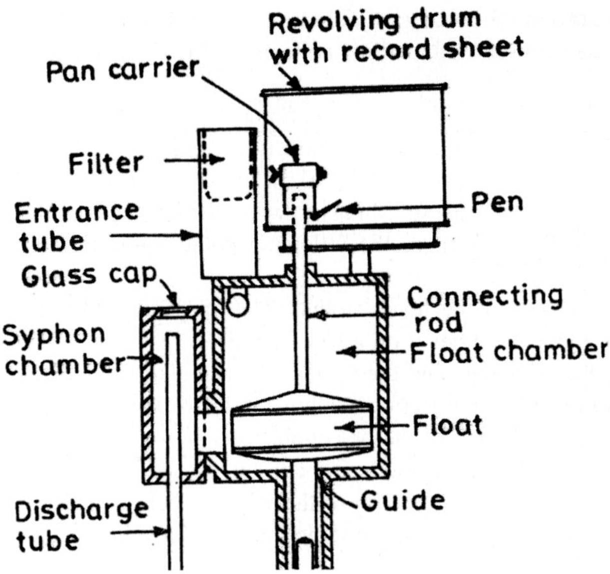

Figure 2. Float and siphon type-recording raingauge

130 cm^2 (128.6 mm diameter). With this collector, the raingauge would have a capacity of 25 mm of rainfall for each siphoning.

The float and siphon recording raingauge gives a continuous record of rainfall. The instrument can record the total amount of rain with the times of start and cessation. The recording raingauge is generally used in conjunction with a non-recording raingauge installed close by for use as a standard by means of which the readings of the recording raingauge could be checked and if necessary adjusted.

3.1.3. Installation and Use of Recording Raingauge

The recording raingauge should be placed at a distance of at least 2 m from the non-recording raingauge and preferably 3 m away. An appropriate chart is wrapped on the clock drum while ensuring that corresponding horizontal lines on the overlapping portions are coincident and that the bottom of the chart is as near the flange at the bottom as possible.

The clock is wound and the pen is set after putting enough ink. Water is poured gradually into the inlet tube until the water begins to be siphoned off. After siphoning is complete the pen should rest against the zero line on the chart. The chart is changed at the same time every day, usually between 08:00 and 09:00 IST. For more details on the installation and use of the recording raingauge BIS code IS: 8,389–1983 may be referred to.

3.1.4. Design of Raingauge Networks

Since precipitation is one of the most important inputs in hydrological analysis, it is necessary to ensure that the network is adequate to obtain representative observations of precipitation. Scientific planning of the network design is necessary so that rainfall is estimated with desired accuracy at the minimum cost. However, the design of networks has been a rather neglected subject and most networks are inadequate. Even in areas with relatively large numbers of long established stations, there are usually gaps and deficiencies.

A precipitation network is a system for collection of data with due consideration to the needs as well as the economy. The problem of network design is to determine: i) how many stations are required? ii) where they are to be located? and, iii) how long they are to be operated?

The optimum density of a precipitation network, i.e., the number of gauges per unit area depends on the purposes to which the observed data are to be put. The Bureau of Indian Standards (BIS) ISI: 4,987–1994 recommended that one raingauge for 520 km^2 shall be sufficient. However, if the catchment lies in the path of low-pressure systems, which cause precipitation in the area during their movement, their network should be denser particularly in the upstream. In not too elevated regions with an average elevation of one kilometer above sea level, the network density shall be one raingauge in 260 km^2 to 390 km^2. In predominantly hilly areas, where very heavy rainfall is experienced, the network density should be at least one raingauge for every 130 km^2. If, however, the catchment is in the rain shadow region of high mountains, such a dense network may not be necessary. Table 1 gives the desired and actual network density in some basins in Himalayan region.

Clearly, the network is not dense enough and many more stations need to be installed to adequately measure the rainfall and its variation. Another aspect of concern is that at higher altitudes, the network density is still low.

Installation of SRRGs has to be commensurate with the requirements. The BIS standard recommends that at least 10% of the raingauge stations should be equipped with SRRGs.

Recognizing the need for increasing the network of raingauges for hydrologic purposes, IMD in collaboration with Central Water and Power Commission

Table 1. Raingauge Stations in Himalayan Region

Altitude range (m)	Desired network density	Actual density			
		Total	Ganga	Brahmaputra	Indus
<1, 500	370	86	43	10	33
1,500–2,100	460	57	28	9	20
2,100–3,000	470	35	18	13	4
3,000–5,000	730	28	17	8	3

reviewed the network of ordinary raingauges in the country in 1956–57. They recommended one raingauge for every 500 km^2 area. The raingauge network at present comprises about 5,000 stations whose data are archived by IMD. This gives one raingauge in about every 600 km^2 and the average distance between adjacent raingauge stations is from 20 to 30 km. In addition, about 4,650 raingauges are being maintained by the Railways, Forest, Agricultural and Irrigation Departments, etc. About 522 self-recording raingauge are being maintained by Railways, Forest, Agriculture and Irrigation Departments. The hourly, daily and monthly rainfall data of IMD stations are available in the National Data Center, IMD, Pune, on computer media. Daily rainfall records for about 3,000 stations spread over the country are available for a period of over 95 years.

3.1.5. Optimum Network Design

The main objective of rainfall network is to provide estimates of rainfall during a particular interval at a specified point or over an area. The optimum network should, therefore, be such that by interpolation between values of different stations, it should be possible to determine the required characteristics of the precipitation data with sufficient accuracy. The BIS Standard and the IMD (1972) have recommended a statistical procedure for estimating the optimum number of raingauges required for a network of raingauges over a basin.

$$N = \left[\frac{C_v}{P}\right]^2 \tag{1}$$

where N = optimum number of raingauge stations, C_v = coefficient of variation of the rainfall values of the existing raingauge stations, and P = desired degree of percentage error in the estimate of the basin mean rainfall. The percentage error P is generally taken as 10% for working out the optimum number. Obviously a decrease in the percentage error would mean an increase in the number of gauges required.

3.2. MEASUREMENT OF SNOW

Solid precipitation is far more difficult to measure than liquid precipitation. Compared to rainfall, snowfall over an area is more uniform, but its accumulation and retention on the ground are highly heterogeneous because of the buoyancy of snow. Snow measurement consists of the actual depth of snow unaffected by wind and water content of such a depth. Except for special applications, the water content of snow is more useful than the actual depth of snow. The data on snow depth and the average density are used to estimate the water equivalent of snow. In the earlier days, the necessity for undertaking snow and glacier studies arose for forecasting runoff from the Himalayan rivers in the critical summer months of April, May, and June.

The depth of snowfall is the amount of fresh snow deposited over a given period (generally one day); the water equivalent of fresh snow is the amount of liquid precipitation represented by that snowfall. The depth of fresh snowfall is generally measured by depth probes (graduated rulers) by inserting them vertically in the snow. The measurements are made on a snowboard whose surface is kept free of snow before snowfall. A snowboard (at least 40 cm*40 cm) is a piece of plywood or light weight metal which provides a reference level for measurement. Snowboards should be located where the effect of wind is the minimum.

A non-recording snow gauge consists of open receptacles with vertical sides, usually in the form of cylinders. For small snowfalls, non-recording rain gauges are used for rain as well as snow measurement and often give satisfactory results. In the case of snowfall measurement, the funnel receiver is removed. The amount of snow collected in the cylindrical gauge is melted either by wrapping the snow gauge with a warm cloth or by adding an adequate and accurately measured quantity of warm water. Under conditions of heavy snowfall, the snow accumulated over the collector is carefully pressed into the gauge, melted and measured.

Various sizes of the gauge orifice and height are used in different countries. The non-recording snow gauge adopted for use in India consists of a simple cylindrical collector having a diameter of 23 cm. Now these gauges are being replaced by snow gauges with 500 cm^2 collectors. The gauge is normally exposed at a height of about 2 m above the ground, and mounted on a metal stand. A standard snow gauge being used in India is shown in Figure 3.

The storage gauges are used to measure total seasonal water equivalent of snowfall in remote areas where daily observations are impractical. Such gauges have capacities to store large snowfall during periods as long as a year.

Like recording rain gauges, there are also snow gauges for continuous recording of snow water equivalent. These are used in locations where continuous records are required and where gauges cannot be regularly attended by the observer. Weighing type snow gauges are most commonly used to measure both rain and snow. The capacity of weighing type recording snow gauges ranges from 300–600 mm water equivalent and the time resolution capability may vary from 5 minutes to several hours. Two most common weighing recording type gauges are Fischer and Porter and Universal snow gauges.

A snow pillow is a device to measure the water equivalent of snow. Pressure pillows are filled with antifreeze liquid and a pipe connects the pillow to a stilling well. The liquid is pressed into the stilling well by the weight of falling snow and this liquid level in the well is related to the water equivalent of snow accumulated over the pillow. The variation in load on pillow may be recorded by an automatic water level recorder. A snow pillow gives a point value of water equivalent of the snow cover accumulated on its surface. Snow pillows are of various shapes and sizes, and different materials are used to construct them.

Snowfall in the Western Himalayas is being monitored by IMD, Central Water Commission (CWC), Bhakra Beas Management Board (BBMB), and Snow and Avalanche Study Establishment (SASE). IMD measures snowfall by using snow

Figure 3. Standard snow gauge being used in India

gauge and snow depth is measured by snow stakes. CWC has a network of 33 snow gauges in the Chenab basin and 9 stations in Yamuna basin with a snow pillow installed at Jubbal. BBMB has a network of 21 snow gauge stations in Satluj basin above Rampur where snow gauges and snow stakes are used to monitor snow. SASE has a good network of snow gauges at very high elevations in the Himalayas, extending from Kashmir to the hills of Uttaranchal. At some of the locations, weighing type precipitation gauges or snow stakes are used; snow pillows are being used at a few locations.

3.2.1. Recording Snow Gauges

Like recording rain gauges, there are also snow gauges for continuous recording of snow water equivalent. Recording snow gauges are designed for use in locations where continuous records are required and for locations where gauges cannot be regularly attended by the observer. The following snow recording gauges are used.

Weighing type snow gauge

The gauges most commonly used to measure both rain and snow are the weighing type snow gauges, in which a pen arm continuously records the weight of accumulated snowfall on a clock-turned drum. The continuous recording may be either by means of a spring mechanism or with a system of balance weight. The spring after accommodation of precipitation compresses and activates the recording mechanism. The mechanical displacement of the spring can be converted to a digital output signal, which can be recorded in situ or telemetered by a suitable device to the desired destination. This type of gauge normally has no provision for emptying itself but by a system of livers, it is possible to make the pen traverse the chart any number of times. The capacity of weighing type recording snow gauges ranges from 300–600 mm water equivalent and the time resolution capability may vary from 5 minutes to several hours. The main advantage of this type of instrument is its capability to record snow, hail and mixture of snow and rain. It does not require the solid precipitation to be melted before it can be recorded because it functions on the weighing principle.

The two most common weighing recording type snow gauges are Fischer and Porter, and Universal snow gauges. The Fischer and Porter gauge has an orifice diameter of 20.3 cm, and a collector capacity of 630 mm water equivalent. The Universal snow gauge also has an orifice of the dimension of 20.3 cm. However, it has a standard maximum capacity of about 300 mm water equivalent, including antifreeze. A weighing type snow gauge is shown in Figure 4.

Electrically heated tipping bucket snow gauge

In principle, a tipping bucket snow gauge is similar to the tipping bucket rain gauge, except that an additional device is employed to melt the solid precipitation. Usually, to heat the funnel and orifice, a nichrome wire is embedded in the double wall case. The wire is densely packed at the top of the gauge in order to concentrate the heat at the funnel. A second heater is mounted under the gauge to provide heat to the tipping bucket mechanism and to prevent freezing of water collected in the bucket. For most snow conditions, satisfactory melting is achieved when thermostat is set at 5 °C.

A bucket will tip when it accumulates a standard amount, usually 0.25 mm, of precipitation. The tipping of one bucket will cause an empty bucket to come into position to be filled. Each tip is recorded through an event recording mechanism.

3.2.2. Snow Pillow

High cost of obtaining samples by having personnel travel to snow areas and the need for more rapid acquisition of information have led to the development of sensors that can be read remotely and automatically. One of these sensors is snow pillow. A snow pillow is a device to measure the water equivalent of snow. Pressure pillows are filled with antifreeze liquid and a pipe connects the pillow to stilling well. The antifreeze solution used in the storage type of snow gauges can be used

Figure 4. Weighing type snow gauge

in the pillow. The liquid is pressed into the stilling well by the weight of falling snow and this liquid level in the well is directly related to the water equivalent of snow accumulated over the pillow. The variation in load on pillow may be recorded by means of an automatic water level recorder. On site and/or telemetry data acquisition system can be installed to provide continuous measurements of the water equivalent through the use of a standpipe and float actuated charts or digital recorders. Alternatively, the pressure changes in the pillow can be monitored by a transducer, whose electrical output can be interfaced with the telemetry system. The snow pillow assumes that snow acts as a perfect fluid and unusual conditions can affect the accuracy of measurements.

A snow pillow gives a point value of the average water equivalent of the snow cover accumulated on its surface. It is a fair indicator of the actual water equivalent of snow in many heavy snow areas, especially during the primary snow accumulation period. Sometimes, pressure plates are also used rather than original butyl rubber pillow. The size of the pressure plate or snow pillow necessary for best accuracy depends upon the amount of snow cover normally expected. Greater depths of snow require large pillow sizes.

Snow pillows are designed in various shapes and sizes, and different materials are used to construct them. Now-a-days, the pillows are fabricated from butyl rubber, neoprene rubber, sheet metal or stainless steel. A 3.66 m diameter snow pillow is enough for most snow cover depths.

3.2.3. Relationship between Snow Covered Area and Runoff

In many studies, regression analysis has been carried out to correlate snow-cover area and runoff. Efforts have also been made to correlate winter snowfall and snowmelt runoff. The snow cover area can be easily assessed from satellite imageries. Dey and Goswami (1983) presented the results of studies involving the utilization of satellite snow cover observations for seasonal streamflow estimates in Western Himalayas. A regression model relating seasonal flow from April through July 1974 to early April snow-cover explained 73% of variance of measured flows in Indus River. Remotely sensed snow-cover area data provides the best available input in empirical snowmelt prediction techniques for remote Himalayan basins. The study has also indicated high correlation of concurrent flows, in adjoining Himalayan basins like Indus and Kabul.

A regression model of snow-cover area versus runoff was evaluated by Dey and Goswami (1984), using data of Satluj, Indus, and Chenab rivers. The mean seasonal snowmelt runoff (April to June) in Indus, Satluj and Chenab rivers are 4,027, 735, and 1,508 cumec, respectively, for catchment areas of 162,100, 38,000 and 26, 155 km^2. The following relationship between snow-cover area and seasonal runoff were established:

$$Y = 0.06493\ X - 0.363325 \qquad \text{for Satluj River} \tag{2}$$

$$Y = 0.472\ X + 4.73895 \qquad \text{for Indus River} \tag{3}$$

where Y = Seasonal run-off (April–July) in $10^9\,m^3$ and X = Average percent of snow-cover of the basin.

A conceptual snowmelt runoff model (SNOWMOD) was developed by Singh and Jain (2003) has been extensively tested using the data of Himalayan rivers.

The importance of permanent snow-covered area in any study of snowmelt in Himalayan basins was realized long ago. Studies were carried out in 1980s for glacierized mountains (upper Indus in Pakistan) and a model was developed for annual variation of run-off and it's forecasting. The approach is based on identification of a number of glaciological and climatological factors other than snow-covered area. Neglecting rainfall, runoff and ground water discharge and also losses, the total melt water runoff has been assumed to be the sum of three components: (i) complete melting of a glacier snow pack, (ii) complete melting of glacier ablation snow cover, and (iii) glacier ice melt from a contributing fraction of area, The useful information is provided about characteristics of high mountain

Table 2. The Snow-cover and runoff characteristics

S.N.	River basin	Area (km^2)	Mean run-off (mm) Apr-Aug.	Snow-cover (%)	Ice cover mean estimates (%)
1	Hunza	13, 000	763	88	38
2	Gilgit	26, 000	578	86	27
3	Indus	16, 0000	303	83	11
4	Shyok	33, 000	292	93	9
5	Jhelum	25, 000	644	74	2

basins in the Himalayan region based upon 1975–1978 data. The details of the results are given in Table 2.

Limited studies on snowmelt processes and simulation of snowmelt runoff have been completed for Himalayan basins. This is mainly because of poor network in the Himalayan catchments. In the recent years, the condition of the network in several Indian snowbound catchments has improved.

3.3. STORAGE OF RAINFALL DATA

India has a long tradition of systematic observations, dating back centuries in different fields, including meteorology, geology, agriculture, sea level and land survey. Specialized institutes are carrying out these observations since the early 19th century. Changes/improvements have been made in observational networks according to needs; networks have evolved and have been modernized although more efforts in this direction are needed. With rapid advances in satellite technology in India, space-based systems have contributed considerably to enhancing observational capabilities.

In India, rainfall data collected by state organizations is generally stored only in the form of printed records. Systematic computerization of the data has begun in some states only recently. As IMD maintains a large network of observatories, it collects a huge volume of data. The computerization, quality control and archival of this data in electronic media is being done by the National Data Centre of IMD at Pune.

3.3.1. Daily Rainfall Data

During the initial period of storing data on computer compatible media, IMD was storing data on punched paper cards. The daily rainfall data were punched in a 31-card format until 1970. From 1971 onwards, IMD switched over to a 24-card format. In the 31-card format, the data of 12 months for each day were punched on each card together with station related information, year and date. In the 24-card format, each month's rainfall data were punched on 2 cards, 16 days data on the first card and 15 days data and monthly total on the second card.

3.3.2. Hourly Rainfall Data

Data of hourly rainfall recorded at the self-recording raingauges maintained by either India Meteorological Department or other organizations were rarely published in printed form. IMD however, used to store the data of self-recording raingauge on punched cards and magnetic tapes. Besides station code, the first card would contain year, month, date and card number data of hourly rainfall corresponding to 1st to 16th hour. The second card, besides station code and other details, contains data of hourly rainfall, time of maximum rainfall occurrence and total rainy duration in the day, given in hours and minutes.

3.3.3. Computerization of Data Under the Hydrology Project

In the 1990s, an ambitious project, known as the Hydrology Project Phase I was launched in eight states of peninsular India. Among the objectives of the project was computerization of hydrological data in these states. Huge amounts of historical data have been computerized in the form of a Hydrological Information System (HIS). More details of this project and its achievements are given in a later chapter.

3.4. ESTIMATION OF MISSING DATA

While working with rainfall data, one often comes across missing data situations. Data for the period of missing rainfall data could be filled using estimated value. The length of the period up to which the data could be filled is dependent on individual judgment. Rainfall for the missing period is estimated either by using the normal ratio method or the distance power method.

3.4.1. Normal Ratio Method

In the normal ratio method, the rainfall R_A at station A is estimated as a function of the normal monthly or annual rainfall of station A and those of the neighbouring stations for the period of missing data at station A:

$$R_A = \frac{\sum_{i=1}^{n} \frac{NR_A}{NR_i} - R_i}{n} \qquad (4)$$

where R_A is the estimated rainfall at station A, R_i is the rainfall at surrounding station i, NR_A is the normal monthly or seasonal rainfall at station A, NR_i is the normal monthly or seasonal rainfall at station i, and n is the number of surrounding stations whose data are used for estimation.

3.4.2. Adjustment of Data

Adjustment of data has two principal objectives. The first is to make the record homogeneous with a given environment and the second is to eliminate or reduce extraneous influences by correcting for changes in gauge location or exposure. Adjustment for these errors is made by 'Double Mass Analysis'. This is a graphical method to identify and adjust inconsistencies in a station's data by comparing with the trend of reference stations' data. As the name itself implies, a double mass curve's both axis are accumulated precipitation values. Usually, the accumulated seasonal or annual precipitation values of reference station(s) is taken as abscissa and those of the station under test as ordinate.

A change in the regime of the raingauge, such as a change in exposure, change in location is revealed by a change in the slope of the straight line fit. The older records are adjusted by multiplying the precipitation values by the ratio of the slope of the later period to the slope of the earlier period.

3.5. DISAGGREGATION OF DAILY DATA

The network of recording raingauges in India being sparse in comparison to that of daily (non-recording) rain gauges, many times it is necessary to convert the daily rainfall into shorter period intervals. The information of short interval rainfall is used together with information of daily rainfall from nearby non-recording (daily) gauges.

Mass curve is a graphical display of accumulated rainfall versus time. Mass curve of accumulated rainfall at (non-recording) daily stations and recording stations (SRRG) are prepared by plotting the accumulated rainfall values against time for the storm duration under analysis. A comparison of the mass curves of the recording raingauge stations with those of the non-recording stations would help in deciding which recording raingauge or group of gauges could be considered as representative of which of the non-recording raingauge for purposes of distributing daily rainfall into hourly rainfall.

The hourly rainfall data at each of the two SRRG stations is plotted on a graph to prepare the mass curve of hourly rainfall. The daily rainfall data at each of the four stations is cumulated and plotted on a graph. The points are joined to form the mass curve of daily rainfall. The mass curves of daily rainfall are compared with those of hourly rainfall to determine which of the daily rain gauge stations are represented by which of the SRRG stations.

To convert daily rainfall into hourly rainfall, the hourly rainfall from 08:00 hours to 08:00 hours for consecutive days is accumulated and the rainfall during each hour is expressed as a ratio of the total rainfall during 24 hours (08:00 to 08:00). These ratios are used to distribute the daily rainfall for the corresponding duration at those non-recording raingauge stations, which are represented by the SRRG.

3.5.1. Aggregation of Precipitation Data in Time

For the estimation of design flood using the unit hydrograph approach short increments of design storm would be required. In India, in view of the rather limited length of hourly rainfall data and the sparse network of SRRG stations, it is not possible to derive the PMP estimates of less than 1 observational day duration directly. As such, recourse is taken to determine the short interval design rainfall increments by applying a time distribution based on the hourly rainfall data as described above. Also, for developing intensity – duration – frequency (IDF) relationships, determination of maximum rainfall intensities for different durations by aggregation would be required.

The steps involved in deriving the time distribution are indicated below:

Step 1: Select all such storm spells whose 24 hour totals have exceeded at least 150 mm and 48 hour totals 200 mm.

Step 2: Compute the maximum hourly rainfall totals for 1, 2, 3, 6, 9, 12, 15, 18, 21, 24, …48 hours using only consecutive hourly rainfall data.

Step 3: Express the maximum rainfall totals computed in step 2 as percentage to the total rainfall amount of the 24-hour duration or 48-hour duration.

Step 4: Repeat the procedure in step 2 and step 3 for all selected storm spells.

Step 5: Plot the percentages of different durations from each spell on a graph paper and draw smooth curves. Separate graphs are plotted for 24 hour and 48 hours as needed.

3.5.2. Estimation of Mean Areal Precipitation

Precipitation observations from gauges are point measurements. But the precipitation process exhibits appreciable spatial variation over relatively short distances. An accurate assessment of mean areal precipitation is a pre-requisite and basic input in hydrological analysis. Numerous methods of computing areal rainfall from point raingauge measurements have been proposed. The most commonly used methods are the Arithmetic average method, the Thiessen polygon method, and the Isohyetal method. The choice of a method is dependent on the quality and nature of data, the importance of use and required precision, the availability of time and skills of the analyst. Some commonly used methods are describe below.

The simplest technique to compute the average precipitation depth over a catchment area is arithmetic average of the values at gauges within the area for the period of concern. If the gauges are relatively uniformly distributed over the catchment and the values do not greatly differ, this technique yields reliable results.

3.5.3. Thiessen Polygon Method

The Thiessen Polygon method is used with non-uniform station spacing and gives weights to station's precipitation data according to the area, which is closer to that station than to any other station. This area is found by drawing the perpendicular bisectors of the lines joining the nearby station so that the polygons are formed around stations. The polygons thus formed around each station are the boundaries of the effective area assumed to be controlled by station. The area governed by each station is planimetered and expressed as a percentage of the total area, the weighted average precipitation for the basin is computed by multiplying each station precipitation amount by its assigned percentage of area and totaling.

The weighted average precipitation is given as:

$$P = \frac{\sum_{i=1}^{n} P_i W_i}{\sum_{i=1}^{n} W_i} \tag{5}$$

where P is the average catchment precipitation, P_i is the precipitation at stations 1 to n, and W_i the weights of respective stations. The advantage of this method is that stations outside the catchment may also be used to assign weights of marginal stations within the catchment. A drawback is that it assumes that precipitation between two stations varies linearly and does not make allowance for variation due to orography. Also, whenever a set of stations are added to or removed from the network, a new set of polygons has to be drawn. If a few observations are missing, it is convenient to estimate the missing data than to construct a new set of polygons.

Besides the above, there are many other methods, such as kriging, that are used for estimation of the mean areal precipitation. For further details of processing of precipitation data, reference may be made to Singh (1992), WMO (1994), Singh and Singh (2001), and Jain and Singh (2003).

3.5.4. Relationship between Point and Areal Rainfall

Rainfall varies temporally as well as spatially. An isohyet is a contour line showing the locus of equal rainfall depth. An individual storm may have a spatial distribution or pattern in the form of concentric isohyets, leading to the formation of a storm eye that depicts the center of the storm. In general, storm patterns are not static, rather they move gradually in the direction of prevailing winds. For regional rainfall mapping, isohyets are commonly referred to as isopluvials. These maps show contours of equal rainfall depth, applicable for a range of durations, frequencies, and catchment sizes. For large catchments, intense storms (or thunderstorms) may cover only a part of the whole area.

The maximum intensity of rainfall occurs at the center of the storm and the intensity gradually decreases radially away from the center. A large number of formulae have been developed which relate storm point rainfall to storm areal rainfall. For a rainfall of a given duration, the average depth decreases with increase in area exponentially as:

$$\bar{p} = p_0 \exp(-KA^n) \tag{6}$$

Table 3. Values of coefficients K and n for storms of different durations

Duration	K	n	Remark
1 day	0.0008526	0.6614	
2 days	0.0009877	0.6306	
3 days	0.001745	0.5961	
1 day	0.004472	0.599	For Brahmaputra basin
2 days	0.009152	0.50683	

where $\bar{p}$ is the average depth (cm) over an area A (km^2), p_0 is the highest amount of rainfall (cm) at the storm center, and K and n are constants for a given region. On the basis of 42 severe most storms in North India, the values of K and n were obtained by Dhar & Bhattacharya (1975) for storms of different durations as shown in Table 3.

Since it is unlikely that the storm center falls over a raingauge station, the exact determination of p_0 is rare. Hence, while analyzing large area storms, the highest rainfall is taken as the average depth of central zone of 25 km^2. IMD has suggested the following relation to determine the reduction factor for rainfall over areas up to 300 km^2 for durations varying from 30 minutes to 24 hrs:

$$P = \exp(-A^{1/3}/8T^{0.5}) \tag{7}$$

where P is the areal to point rainfall ratio, A is the basin area in sq. miles, and T is the storm duration in hours.

3.6. RAINFALL OF VARIOUS DURATIONS

Here we discuss some features of rainfalls of various durations for the country. Apart from being of statistical interest, these data are of considerable interest in planning and management of water resources projects, ecological planning, environmental systems management, agriculture and irrigation planning, forestry, public health engineering, and so on.

3.6.1. Annual Rainfall

Table 4 shows annual normal rainfall for some basins of India. Among basins, rainfall is the maximum in the Brahmaputra basin which also has the highest number of rainy days. The Luni basin receives the least rainfall and has only 20 rainy days.

Table 5 gives the coefficient of skewness of observed rainfall at these stations. Also shown in the table are the numbers of years of data used. The skewness is highest in the dry regions of the country.

Some statistical properties of annual extreme rainfall series at selected stations in various meteorological sub-divisions are given in Table 6. The annual rainfall over

Table 4. Annual normals of rainfall (cm) for major river basins of India

River basin	Annual normal	
	Rainfall (cm)	No. of rainy days
Brahmaputra	268.7	112
Cauvery	119.8	65
Ganga	115.8	52
Godavari	116.1	59
Indus	88.7	46
Luni	40.6	20
Mahanadi	142.2	69
Mahi & Sabarmati	84.0	37
Narmada	123.1	50
Tapi	79.7	46

Source: Rakhecha and Soman (1992).

Table 5. Coefficient of skewness of annual extreme series at selected stations

S. N.	Station (State/Sub-division)	No. of years of record	Skewness
1.	Pasighat Arunachal Pradesh	50	0.75
2.	Guwahati (Assam)	61	1.21
3.	Sekoni (Assam)	58	1.39
4.	Deemapur (Nagaland)	57	0.88
5.	Berhampore (West Bengal)	70	1.28
6.	Sambalpur (Orissa)	69	0.92
7.	Deogarh (Bihar)	62	1.30
8.	Darbhanga (Bihar)	68	1.00
9.	Kundra (west U.P.)	69	0.91
10.	Okhimmath (west U.P.)	68	1.28
11.	Hydergarh (east U.P.)	68	0.65
12.	Panipat (Haryana)	68	1.06
13.	Dasuyia (Punjab)	70	1.66
14.	Hamirpur (H.P.)	69	1.60
15.	Kangra (H.P.)	62	0.59
16.	Kothai (H.P.)	70	1.27
17.	Baramula (J&K)	66	0.64
18.	Shri Ranbir Singh Pura (J&K)	65	1.08
19.	Barmer (west Rajasthan)	69	1.93
20.	Bikaner (west Rajasthan)	69	1.00
21.	Pali (east Rajasthan)	70	0.56
22.	Phalodi (east Rajasthan)	69	2.30
23.	Kota (east Rajasthan)	69	0.85
24.	Sagar (M.P.)	70	1.08
25.	Shahpur (M.P.)	66	1.27
26.	Raipur (M.P.)	70	1.67
27.	Janakpur (M.P.)	30	0.79
28.	Bhuj (Gujarat)	70	2.86
29.	Tharad (Gujarat)	69	1.60

30.	Dahanu (Maharashtra)	69	1.13
31.	Brahmpuri (Maharashtra)	69	0.67
32.	Nanded (Maharashtra)	70	1.69
33.	Chintalapudi (A.P.)	69	1.02
34.	Borlan (A.P.)	46	1.15
35.	Kodanur (A.P.)	45	0.81
36.	Madurai (Tamil Nadu)	70	1.28
37.	Mangalore (Karnataka)	70	1.64
38.	Virajpet (Karnataka)	70	1.28
39.	Bijapore (Karnataka)	70	1.28
40.	Adur (Kerala)	70	0.74

Source: Mukeerjee et al. (1991).

India has been shown in Figure 4 of Chapter 1. Figure 5 shows the co efficient of variation of monsoon rainfall in India.

3.6.2. Monthly Rainfall

The average monthly rainfall for selected stations in India is given in Table 7. As explained earlier, around 80% of annual rainfall is received in four monsoon months at most places. Of interest here is high rainfall in winter months at stations in Tamil Nadu.

Table 8 lists the normal monthly and annual rainfall at selected places in Himalayas. Data given in this table are useful to get an idea of the spatial and temporal variation in rainfall in this region.

3.6.3. Seasonal Rainfall

The winter season rainfall (January–February) of India has been shown in Figure 6. The pre-monsoon (March-May), monsoon (June-September) and post-monsoon (October-December) rainfall of India have been presented in Figure 7, 8 and 9 respectively.

Further, the percent departure in all India summer monsoon rainfall for the period 1871 to 1990 has been plotted in Figure 10. In the years 1877, 1899, 1918, and 1972, the deficit in rainfall was more than 20% while in 1917 and 1961, the percent departure was close to 20%.

Table 9 gives seasonal normal values of rainfall for selected river basins in India. Note that the duration of seasons in the table is not equal. Further, Table 10 gives the mean seasonal rainfall and annual rainfall for different meteorological sub-divisions of the country.

On the basis of rainfall distribution and other meteorological parameters, the country has been divided by IMD into different meteorologically homogeneous subdivisions, as described in Chapter 1. Using rainfall data for the years 1891–1970,

Table 6. Statistical estimates of annual extreme series at some stations

S. N	Station	Mean	Standard deviation	Max. recorded	Return periods along with their standard errors [SE(x_T)]							
					x_5	S.E.	x_{50}	S.E.	x_{100}	S.E.	x_{1000}	S.E.
1.	Pasighat	213.5	82.50	467.9	284.2	18.52	458.9	36.17	509.7	41.57	677.5	59.68
2.	Guwahati	102.5	36.20	232.9	127.7	6.33	193.7	12.37	212.9	14.21	276.3	20.41
3.	Sekoni	99.3	29.00	198.4	117.5	4.82	166.4	9.41	180.7	10.82	227.7	15.53
4.	Deemapur	97.7	33.80	197.1	119.7	5.93	179.5	11.59	196.8	13.32	254.2	19.12
5.	Berhampore	108.9	44.32	286.3	137.5	6.89	214.5	13.47	236.8	15.48	310.8	22.22
6.	Sambalpur	148.7	56.38	312.4	187.2	9.23	289.5	18.04	319.3	20.73	417.6	29.76
7.	Deogarh	100.5	46.63	264.2	102.4	7.07	176.6	13.80	198.1	15.86	269.4	22.76
8.	Darbhanga	127.5	46.70	266.7	158.5	7.55	241.5	14.74	295.6	16.94	345.3	24.31
9.	Okhimath	104.9	35.00	208.3	125.1	5.13	181.6	10.02	198.0	11.52	252.3	16.54
10.	Kundra	114.1	58.39	315.0	159.3	10.38	274.3	20.27	307.7	23.29	418.2	33.44
11.	Hydergarh	112.5	40.45	233.2	77.1	5.97	142.7	11.66	161.8	13.40	224.9	19.23
12.	Panipat	96.5	45.10	254.0	127.1	7.38	208.3	14.42	231.9	16.57	310.0	23.79
13.	Dasuyia	95.1	36.30	245.9	118.8	5.61	181.4	10.95	199.5	12.58	259.7	18.07
14.	Hamirpur	116.3	50.00	351.0	148.7	7.73	234.4	15.10	259.3	17.35	341.6	24.91
15.	Kangra	156.6	44.03	306.3	191.5	8.51	280.9	16.62	306.9	19.10	392.7	27.41
16.	Kothai	72.3	27.65	182.9	90.3	4.33	138.6	8.45	152.6	9.71	199.0	13.95
17.	Baramula	58.97	17.70	107.4	72.1	3.16	106.5	6.18	116.5	7.11	149.5	10.21
18.	Sri R.S. Pura	103.5	34.30	207.0	128.4	5.88	191.3	11.48	209.7	13.2	270.4	18.9
19.	Barmer	66.95	46.90	285.8	91.5	6.28	161.1	12.28	181.4	14.1	248.3	20.26
20.	Bikaner	61.96	32.50	165.6	83.8	5.21	141.9	10.18	158.9	11.70	214.8	16.79

21.	Pali	85.35	37.09	206.6	115.22	6.81	191.2	13.30	213.3	15.28	286.3	21.94
22.	Phalodi	55.05	35.44	225.0	71.9	4.57	122.6	8.93	137.4	10.27	186.1	14.74
23.	Kota	103.0	42.00	249.2	134.4	7.32	215.5	14.30	239.1	15.4	317.0	23.59
24.	Sagar	135.2	59.64	381.5	167.3	8.43	261.4	16.47	288.8	18.93	379.2	27.18
25.	Shahpur	111.6	42.20	289.6	87.0	6.29	155.2	12.29	175.0	14.13	240.6	20.28
26.	Raipur	129.1	51.60	370.3	95.0	7.44	178.0	14.53	202.1	16.69	281.9	23.96
27.	Janakpur	98.0	25.67	165.1	78.0	5.89	121.1	11.51	133.6	13.23	174.9	18.99
28.	Bhuj	88.3	74.94	467.0	119.4	8.26	211.5	16.12	238.3	18.53	326.8	26.60
29.	Tharad	116.7	78.96	370.3	158.7	10.95	280.1	21.39	315.4	24.59	432.0	35.30
30.	Dahanu	203.6	77.50	481.0	257.4	12.78	399.0	24.96	440.2	28.68	576.2	41.17
32.	Brahmpuri	148.9	61.25	323.6	104.7	8.89	203.3	17.36	231.9	19.95	326.5	28.64
32.	Nanded	93.4	38.01	254.0	115.1	5.38	175.2	10.51	192.6	12.08	250.4	17.35
33.	Chintalapudi	86.6	26.36	160.8	103.4	4.10	148.9	8.01	162.1	9.12	205.8	13.23
34.	Borlan	114.9	47.67	243.8	144.2	9.01	225.8	17.60	249.5	20.22	327.8	29.04
35.	Kodanur	79.4	53.90	298.2	60.4	10.16	143.9	19.85	169.2	22.81	248.5	32.75
36.	Madurai	90.2	4.86	229.0	116.5	6.00	183.6	11.73	203.1	13.48	267.5	19.36
37.	Mangalore	159.0	41.55	360.9	189.0	6.91	266.1	13.50	288.6	15.51	362.6	22.27
38.	Virajpet	156.3	60.35	366.5	193.8	9.12	295.6	17.80	325.2	20.46	422.9	29.37
39.	Bijapur	68.7	26.68	181.1	88.2	4.48	138.2	8.76	152.8	10.07	200.9	14.45
40.	Adur	113.9	34.00	223.5	138.1	5.73	202.0	11.18	220.6	15.07	282.0	18.45

Source: Mukeerjee et al. (1991).

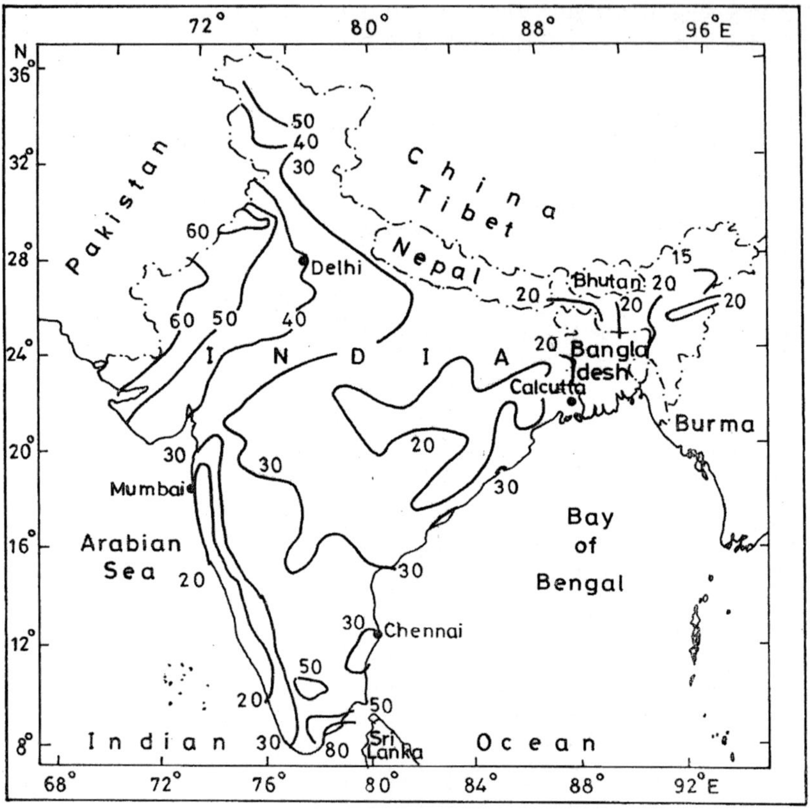

Figure 5. Coefficient of variation of monsoon rainfall

Dhar and Rakhecha (1979) estimated the monthly and annual depth of rainfall on different subdivisions for each of the years. Table 10 gives the mean seasonal and annual average rainfall depth obtained on the basis of 80 years of data for different subdivisions. The amount of summer and winter monsoon rainfall expressed as a percentage of annual rainfall and the coefficient of variability (CV) of annual rainfall have also been included in the table. Among the different subdivisions of the country the mean annual rainfall is highest in the Coastal Karnataka and lowest in western Rajasthan. Low CV values are obtained over high rainfall regions and high CV over low rainfall regions. The amount of seasonal rainfall during four months of June to September is more than 75% of the annual rainfall in almost all the subdivisions, excepting Assam, Jammu and Kashmir and some subdivisions of peninsular India, lowest being for Tamil Nadu. The average annual rainfall for the country on the basis of 80-year data has been estimated as 117 cm. The CV was found to be 9.5% thereby indicating stable nature of Indian rainfall. Monthly, seasonal and annual rainfall values for major river basins of the country using 50 years data are given in Table 11.

Table 7. Average monthly rainfall for selected stations in India

City	Elevation	Jan	Feb	Mar	Apr	May	Jun	Jul	Aug	Sep	Oct	Nov	Dec
Mount Abu	1,219 m	7.0	6.0	4.0	3.0	24.0	118.0	558.0	629.0	238.0	19.0	6.0	13.0
Agra	169 m	13.0	12.5	8.0	5.0	10.0	62.0	217.0	206.0	116.0	18.0	4.0	8.0
Ahmedabad	56 m	0.5	2.0	1.0	1.0	10.0	93.0	310.0	204.0	106.0	9.0	4.0	1.0
Amritsar	91 m	29.5	31.5	24.0	15.0	15.0	48.0	161.0	155.0	71.0	10.0	3.0	13.0
Aurangabad	579 m	7.0	4.0	5.0	6.0	18.0	188.0	171.0	122.0	170.0	45.5	24.	7.0
Bangalore	921 m	6.0	7.0	10.0	41.0	106.0	73.0	100.4	126.5	169.0	149.0	68.0	11.0
Bhopal	501 m	6.0	4.0	9.0	6.0	14.5	173.5	501.0	278.0	265.0	42.0	26.0	5.0
Bodhgaya	113 m	18.0	30.0	12.5	6.0	24.0	164.0	335.5	349.0	190.5	49.0	12.5	5.0
Bombay	11 m	4.0	2.0	1.0	4.0	16.5	484.0	165.5	340.0	264.0	64.5	13.5	2.0
Calcutta	64 m	9.0	30.0	35.5	44.5	139.5	271.5	125.0	328.0	253.0	114.0	21.0	5.0
Chandigarh	61 m	52.0	9.0	26.5	10.0	11.0	71.0	269.0	253.0	188.0	52.0	8.0	24.0
Cherrapunji	131 m	19.0	54.0	185.0	666.0	1281.0	2694.0	2447.0	1780.0	1101.0	493.0	60.0	12.5
Cochin	Sea level	23.0	20.0	51.0	125.5	297.0	723.0	592.0	353.0	195.0	340.0	178.0	41.0
Darjeeling	2,134 m	13.5	30.0	48.0	105.0	245.0	614.0	835.0	575.0	480.0	239.0	21.0	7.0
Dehradun	682 m	59.0	63.0	32.0	16.5	37.0	217.0	668.0	731.0	270.0	82.0	9.0	26.0
Guwahati	55 m	10.0	36.0	50.5	145.0	236.0	312.0	312.0	261.0	167.0	71.0	14.0	4.0
Goa	1.022 m	11.0	9.0	4.0	4.0	7.0	80.0	233.3	248.0	228.0	11.0	4.0	8.7
Haridwar	290 m	66.0	30.0	27.0	7.0	42.0	124.0	539.0	36.0	223.0	136.0	3.5	38.0
Indore	564 m	6.0	4.0	2.0	3.0	13.0	147.0	282.0	207.0	164.0	31.0	15.5	7.0
Jaipur	431 m	11.0	8.0	9.0	4.0	14.5	57.0	197.0	205.0	82.0	12.0	4.0	8.0
Jaisalmer	225 m	2.1	1.2	2.6	1.5	5.2	6.8	89.5	85.8	13.9	1.3	4.9	2.2
Jodhpur	221 m	4.0	6.0	3.0	3.0	10.0	36.0	101.0	123.0	61.0	8.0	3.0	3.0
Khajuraho	280 m	4.5	4.4	3.2	7.2	1.3	88.9	348.4	537.9	115.4	106.7	1.8	3.4
Kargil	3,048 m	45.2	51.6	86.3	31.2	23.6	10.3	5.6	10.2	7.6	4.8	4.3	25.8
Leh	3,170 m	11.8	8.6	11.9	6.5	6.5	4.3	15.7	19.5	12.2	7.1	2.9	8.0
Lucknow	120 m	19.0	19.9	9.0	6.0	20.0	113.0	305.0	292.3	188.0	32.5	6.0	8.0
Chennai	26	36.0	10.0	7.0	15.0	26.0	47.0	91.0	116.0	119.0	306.0	355.0	138.0

(Continued)

Table 7. (Continued)

City	Elevation	Jan	Feb	Mar	Apr	May	Jun	Jul	Aug	Sep	Oct	Nov	Dec
Madurai	107 m	20.0	13.5	18.0	55.0	70.0	40.0	49.5	104.0	119.0	180.0	145.0	51.0
Mussorie	2,006 m	69	87.0	57.0	34.0	42.0	223.0	697.5	694.0	253.0	29.0	4.0	35.0
Mysore	767 m	4.0	6.0	13.0	58.0	42.0	62.5	68.0	84.0	126.0	149.0	70.0	10.0
Nagpur	311 m	9.0	16.5	15.5	15.5	19.0	224	371.0	290.0	203.5	55.0	20.0	12.0
Nainital	1,939 m	70.0	75.0	53.0	21.9	76.0	363.0	72.0	692.0	332.0	59.0	11.0	25.0
New Delhi	239 m	25.0	21.0	13.0	8.0	13.0	77.0	179.0	184.0	123.0	10.0	2.5	11.0
Ootacamund	2,286 m	33.0	13.0	32.5	77.0	160.0	167.0	212.0	144.0	148.0	203.0	161.0	44.0
Patna	53 m	15.0	19.0	11.0	7.0	36.0	181.0	294.0	332.0	218.0	58.0	9.3	6.0
Pune	559 m	1.5	1.5	1.5	1.5	27.0	114.0	167.0	90.0	134.0	90.0	27.0	4.0
Puri	6 m	10.0	23.0	13.0	15.0	73.0	189.0	262.0	277.0	230.0	183.0	81.0	6.7
Shillong	1,496 m	13.0	27.0	50.0	129.0	261.0	347.0	347.0	317.0	299.5	171.0	41.0	7.4
Shimla	2,205 m	66.0	74.0	60.0	46.0	64.0	153.0	414.0	428.0	424.0	30.0	13.0	33.5
Srinagar	1,768 m	74.0	72.0	92.0	93.0	16.5	36.0	59.3	61.5	39.5	30.0	11.0	33.5
Tiruchirapalli	78 m	25.0	12.5	9.0	46.0	84.0	41.0	34.0	97.0	120.9	272.0	148.0	71.0
Trivandrum	61 m	20.0	19.0	39.0	116.0	224.0	334.0	197.0	120.6	114.5	272.5	177.0	63.5
Udaipur	577 m	5.0	4.0	2.5	3.0	18.0	77.0	217.0	178.0	98.0	14.5	3.0	3.0
Varanasi	81 m	2.0	2.0	2.5	5.0	12.0	15.0	12.0	4.0	4.5	1.0	1.0	—

Table 8. Normal monthly and annual rainfall (mm) at selected places in Himalayas

Station	J	F	M	A	M	J	J	A	S	O	N	D	Annual
Uttar Pradesh and Uttaranchal													
Mussorie	79.3	89.7	66.8	37.1	54.9	208.3	782.3	800.3	329.4	41.1	12.5	36.1	2, 537.8
Dehra Dun	57.9	66.8	37.9	19.6	35.8	184.4	655.6	713.0	304.5	41.9	7.6	24.9	2, 149.0
Rajpur	69.3	65.8	43.2	24.4	39.4	246.1	968.0	1058.7	392.4	43.9	9.1	25.7	2, 386.9
Nanital	69.9	73.1	52.6	38.1	84.1	390.9	769.4	750.1	362.7	61.0	12.9	25.4	2, 690.2
Mukteswar	56.9	62.0	48.5	36.3	56.4	176.0	316.0	306.3	201.7	43.4	8.9	24.6	1, 337.0
Almora	42.9	49.3	42.2	27.9	48.3	143.8	264.9	234.2	130.3	32.3	7.4	21.3	1, 044.8
Pithoragarh	44.2	56.4	40.4	28.2	72.6	182.9	299.7	287.3	149.3	33.3	6.9	21.8	1, 223.0
Ranikhet	54.1	62.0	46.2	31.0	50.0	144.0	331.5	344.4	165.9	33.5	7.4	22.9	1, 292.9
Pawi	60.7	66.8	55.1	32.0	51.6	132.3	326.1	359.9	148.3	33.8	8.1	28.2	1, 302.9
Srinagar	55.1	56.1	36.8	22.1	42.7	118.4	244.1	223.3	97.3	24.9	5.6	21.8	948.2
Joshimath	65.8	92.7	98.5	54.4	35.1	88.9	176.3	184.7	108.5	28.2	12.2	27.4	972.7
Jammu and Kashmir													
Dras	103.9	107.9	146.8	102.8	56.4	16.5	13.2	17.0	15.2	19.6	16.5	57.9	673.0
Kargil	36.8	38.3	59.7	42.4	24.9	6.6	6.9	9.7	9.7	5.8	3.1	20.6	264.5
Leh	10.9	7.6	8.9	6.1	5.3	4.8	12.7	16.5	9.1	3.1	1.8	5.8	92.6
Himachal Pradesh													
Kotgarh	69.9	81.3	86.9	61.5	67.6	105.2	239.3	210.3	107.7	24.9	12.5	35.8	1, 102.9
Kotkhai	67.1	82.5	67.3	48.3	51.1	92.5	212.1	182.6	105.4	21.6	9.9	31.0	971.4
West Bengal													
Darjeeling	10.9	31.7	54.1	113.0	231.4	597.1	792.2	643.4	445.5	142.2	24.6	6.3	3, 092.4
Kalimpong	9.9	23.6	29.2	81.3	144.8	409.2	612.9	504.4	312.2	104.4	15.5	6.6	2, 254.0

Source: Chandra (1988).

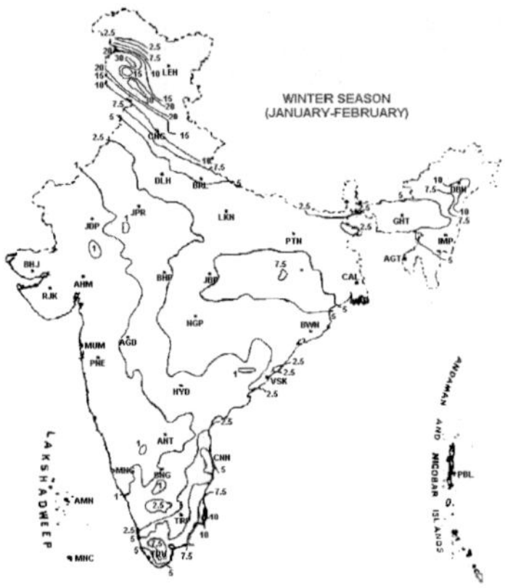

Figure 6. Map of Winter Season Rainfall of India (January–February)

Although India is endowed with abundant renewable water resources, the availability in time is restricted to a period of 4 months of the southwest monsoon season. Above 80–90% of the flow in the rivers is confined to the southwest monsoon season. For the rest of the year, except for snow fed rivers there is inadequate flow in the rivers. Further, the distribution of rainfall varies widely from one part of the country to another. These factors limit the scope for easy management of water resources. One way to make an efficient use of the available water resources is by the conservation of the monsoon flow through storage reservoirs.

Using long period rainfall data, several workers have estimated the mean depth of rainfall for various regions and river basins. IMD (1970a; 1971; 1977) has calculated the mean monthly and annual values of rainfall for different basins.

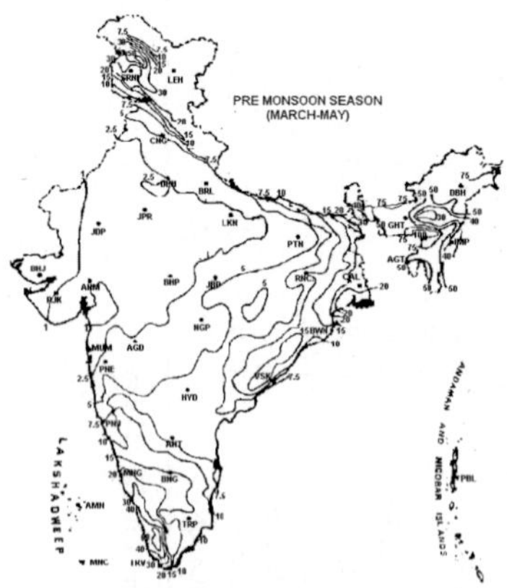

Figure 7. Map of Pre-Monsoon Season Rainfall of India

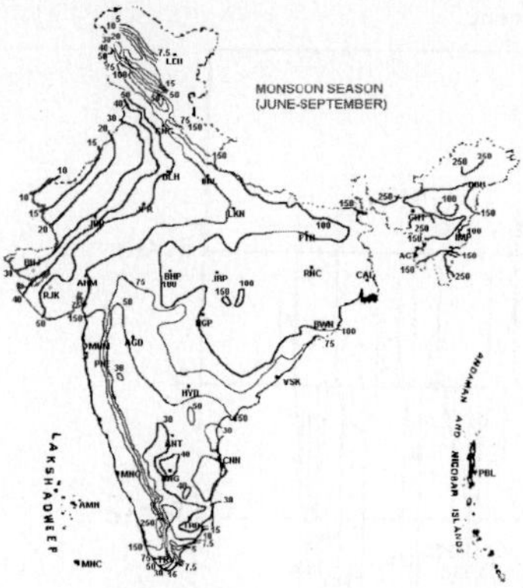

Figure 8. Map of Monsoon Season Rainfall of India

Using rainfall data of stations for the period 1901–1950 IMD (1981) has published a comprehensive Rainfall Atlas of India, which contains 98 maps on different aspects of rainfall distribution.

3.6.4. High Rainfall

Cherrapunji is located in the northeastern part of our country in the Khasi hills on the Shillong plateau. It is known all over the world for a heavy rainfall – an average of about 1,143 cm in a year, second only to 1,168 cm recorded in Hawaii. In the year 1860–61, Cherrapunji had the highest recorded total rainfall in a twelve month period – 2,647 cm. Also, a record rain of 930 cm fell in a month in July 1861.

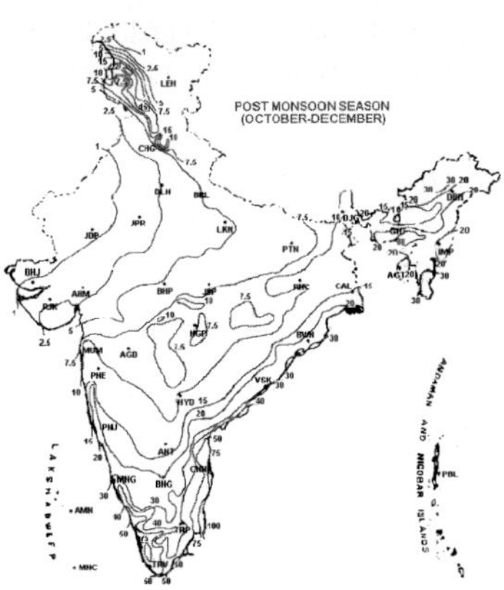

Figure 9. Map of Post Monsoon Season Rainfall of India. (October-December)

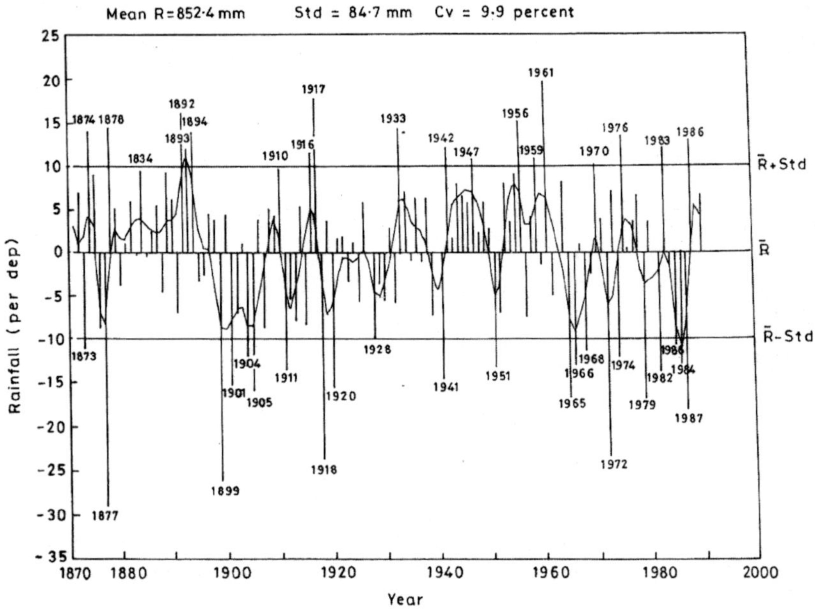

Figure 10. All India summer monsoon rainfall 1871–1990

The observed maximum rainfall events of the world for selected durations are given in Table 12. It is seen from this table that for durations of 15 days and higher,

Table 9. Table showing seasonal normals of rainfall (cm) for major river basins of India

River basin	Jan–Feb		Mar–May		Jun–Sep		Oct–Dec	
	Rainfall 1(cm)	No. of rainy days	Rainfall 1(cm)	No. of rainy days	Rainfall 1(cm)	No. of rainy days	Rainfall 1(cm)	No. of rainy days
Indus	12.6	8	13.8	9	56.4	25	5.9	4
Brahmaputra	5.4	5	56.6	31	188.7	67	18.0	9
Ganga	4.0	3	7.8	5	97.3	41	6.7	3
Narmada	2.9	1	2.7	5	110.8	36	6.7	8
Godavari	3.6	2	5.3	5	95.9	47	12.3	6
Mahanadi	4.0	3	7.3	6	120.5	54	10.4	6
Tapi	1.6	1	2.0	2	68.9	39	7.2	4
Cauvery	3.0	2	18.3	11	64.0	33	34.5	19
Mahi & Sabarmati	0.5	0	1.1	1	80.3	35	2.1	1
Luni	0.9	1	1.4	1	37.3	17	0.9	1

Source: Rakhecha & Soman (1992).

Table 10. Mean seasonal and annual rainfall (cm) for different sub-divisions of India using data of period 1891 to 1970

Sub-divisions	Jan–Feb	Mar–May	Jun–Sep	Oct–Dec	Annual	% of annual		CV
						Jun-Sep	Oct-Dec	
North Assam	5.3	57.7	149.7	15.4	228.1	66	7	11
South Assam	5.5	73.4	180.8	23.6	283.3	64	8	11
Sub-Himalayan west Bengal	2.8	42.5	219.1	15.5	279.9	78	5	15
Gangetic west Bengal	3.6	17.4	107.8	13.1	141.9	76	9	15
Orissa	3.8	12.7	112.5	19.6	148.6	76	13	14
Bihar Plateau	4.6	8.9	110.4	9.7	133.6	83	7	13
Bihar Plains	3.2	7.2	101.6	7.	119.2	85	6	16
East Uttar Pradesh	3.3	3.0	89.4	5.9	101.6	88	6	20
West Uttar Pradesh	5.3	4.3	88.3	5.1	103.0	86	5	19
Haryana, Chandigarh & Delhi	3.6	2.9	45.7	2.5	54.7	83	5	28
Punjab	5.9	4.8	47.9	3.4	62.0	77	5	34
Himachal Pradesh	16.5	15.2	134.5	8.3	174.5	77	5	21
Jammu & Kashmir	17.2	22.6	50.6	9.3	99.7	51	9	22
West Rajasthan	0.9	1.5	27.0	0.9	30.3	89	3	40
East Rajasthan	1.4	1.8	61.5	2.4	67.1	92	3	23
West Madhya Pradesh	2.2	2.1	94.0	5.0	103.3	91	5	20
East Madhya Pradesh	3.8	4.8	119.3	7.3	135.2	88	5	15
Gujarat	0.4	0.9	92.2	3.1	96.6	95	0.3	29
Saurashtra & Kutch	0.3	0.9	48.9	2.1	52.2	94	4	37
Konkan	0.3	3.4	276.5	13.4	293.6	94	5	18
Madhya Maharashtra	0.5	4.2	85.3	10.4	100.4	85	10	23
Marathwada	1.2	3.5	70.1	7.9	82.7	85	9	25
Vidarbha	2.5	3.6	93.8	7.2	107.1	87	7	19
Coastal Andhra Pradesh	1.9	8.4	57.7	31.5	99.5	58	32	18
Telangana	1.5	5.6	74.5	9.5	91.1	82	10	21
Rayalseema	1.2	7.5	38.2	20.8	67.7	56	31	20
Tamil Nadu	4.6	14.3	34.6	47.1	100.6	34	47	14
Coastal Karnataka	0.3	16.7	289.7	26.1	332.8	87	8	15
North Interior Karnataka	0.5	9.5	46.7	14.6	71.3	65	20	19
South Interior Karnataka	0.8	16.8	74.4	22.5	114.5	65	20	20
Kerala	0.3	40.8	200.5	50.3	294.9	68	17	14
India as a whole	3.9	12.4	89.0	11.9	117.2	76	10	10

Source: Rakhecha & Soman (1992).

Table 11. Sub-division wise monthly, seasonal, and annual rainfall (cm) normal (based on data 1901–50)

SN	Sub-Division	Jan	Feb	Mar	Apr	May	Jun	Jul	Aug	Sep	Oct	Nov	Dec	Jan–Feb	Mar–May	Jun–Sep	Oct–Dec	Annual
1	Bay Islands	4.8	3.3	3.0	6.8	34.9	52.7	37.9	40.2	46.8	28.6	23.2	17.1	8.1	44.7	177.7	68.9	299.4
2	Assam	1.8	3.8	8.1	21.3	33.8	48.4	44.7	39.6	31.7	14.4	3.1	0.9	5.6	63.2	164.4	18.4	251.6
3	Sub-Himalayan West Bengal	1.0	2.0	3.6	11.8	32.9	67.5	68.3	57.5	50.4	15.8	1.5	0.4	3.0	48.3	243.6	17.7	312.6
4	Gangetic West Bengal	1.3	2.6	2.7	4.3	10.8	24.1	31.8	31.9	20.7	10.9	2.2	0.3	3.9	17.8	108.4	13.4	143.5
5	Orissa	1.4	2.6	2.1	3.5	7.1	21.3	35.2	33.6	23.7	13.2	4.0	0.6	4.0	12.7	113.7	17.8	148.2
6	Bihar Plateau	2.1	3.1	2.0	1.9	5.1	19.6	35.8	35.4	21.8	8.3	1.6	0.5	5.2	9.0	112.6	10.4	137.2
7	Bihar Plains	1.4	2.0	1.1	1.6	4.8	17.2	31.1	31.4	2.3	5.9	0.9	0.3	3.4	7.4	102.3	7.2	120.3
8	Uttar Pradesh, East	1.6	1.9	0.9	0.7	1.5	10.1	30.2	29.6	19.1	4.2	0.6	0.6	3.5	3.0	88.9	5.4	100.8
9	Uttar Pradesh, West,	2.4	2.5	1.4	0.9	1.5	9.4	28.9	28.8	16.8	2.6	0.4	1.0	4.9	3.7	83.9	3.9	96.4
10	Punjab (Incl. Delhi & Haryana)	2.7	2.6	2.1	1.2	1.2	4.8	17.9	17.2	10.1	1.1	0.3	1.3	5.3	4.5	50.0	2.7	62.5
11	Jammu & Kashmir	9.6	9.6	10.8	7.8	5.2	5.5	16.7	17.5	7.5	2.6	1.5	5.1	19.2	23.8	47.2	9.3	99.5
12	Rajasthan, East	0.9	0.6	0.5	0.3	0.9	6.8	24.3	22.8	11.2	1.2	0.4	0.5	1.5	1.6	65.2	2.1	70.4
13	Rajasthan, West	0.5	0.6	0.4	0.3	0.8	2.9	9.9	11.0	4.0	0.4	0.1	0.3	1.1	1.5	27.7	0.8	31.1
14	Madhya Pradesh, West	1.3	1.0	0.7	0.4	0.9	12.7	35.0	29.4	17.6	2.9	1.7	0.7	2.3	2.1	94.7	5.3	10.4

15	Madhya Pradesh, East	1.8	2.6	1.7	1.6	1.7	18.9	42.9	39.5	21.5	5.9	1.5	0.5	4.4	5.0	122.9	7.9	140.2
16	Gujarat Region	0.2	0.2	0.1	0.3	0.7	13.0	40.0	24.1	15.8	2.3	0.8	0.1	0.4	1.1	93.0	3.2	97.7
17	Saurashtra & Kutch	0.2	0.2	0.1	0.3	0.6	7.4	20.6	10.4	6.6	1.3	0.4	0.1	0.4	1.0	45.0	1.8	48.2
18	Konkan	0.3	0.1	0.1	0.6	2.9	61.7	107.6	63.0	37.3	10.3	2.9	0.3	0.4	3.6	269.6	13.6	287.2
19	Madhya Maharashtra	0.5	0.2	0.3	1.1	2.4	14.8	28.1	17.9	16.1	6.8	3.1	0.6	0.7	3.8	77.0	10.5	92.1
20	Marathwada	0.6	0.6	0.7	0.8	1.6	13.8	17.1	14.0	19.6	4.8	2.9	0.8	1.2	3.1	64.5	8.6	77.4
21	Vidarbha	1.2	1.9	1.1	1.1	1.2	17.5	34.3	25.8	18.1	4.8	2.1	0.7	3.1	3.5	95.7	7.7	110.0
22	Coastal Andhra Pradesh	1.0	1.4	1.2	2.5	5.1	10.4	15.3	14.8	16.6	18.9	11.7	2.1	2.4	8.8	57.0	32.6	100.8
23	Telangana	0.5	1.5	1.2	2.2	2.3	13.3	24.1	19.0	19.3	6.4	2.5	0.4	2.0	5.7	75.7	9.3	92.7
24	Rayalaseema	0.9	0.6	0.7	2.0	5.0	5.8	8.1	9.6	13.4	11.2	8.3	2.2	1.5	7.7	36.9	21.7	67.8
25	Tamil Nadu	3.7	1.6	2.3	5.1	7.3	6.1	6.9	9.6	10.6	19.3	19.6	8.7	5.3	14.7	33.2	47.6	100.8
26	Coastal Mysore	0.3	0.2	0.5	3.0	10.9	84.4	110.0	63.4	28.4	17.5	6.6	1.3	0.5	14.4	286.2	25.4	326.5
27	Interior Mysore, North	0.3	0.3	0.7	2.8	5.1	8.8	12.7	9.8	12.7	9.5	3.9	0.9	0.6	8.5	44.1	14.3	67.5
28	Interior Mysore, South	0.5	0.5	1.1	4.9	10.3	17.7	31.1	20.9	14.5	15.0	6.7	1.3	1.0	16.3	84.2	23.0	124.5
29	Kerala	1.9	2.0	4.6	11.5	24.5	66.7	67.8	41.7	24.0	30.6	19.1	5.1	3.9	40.6	200.3	54.8	299.6
30	Arabian Sea Islands	3.2	1.2	1.3	3.8	15.2	34.5	27.5	21.3	15.4	16.0	11.5	6.3	4.4	20.3	98.7	33.8	157.2

Normals for Himachal Pradesh are not available.

Table 12. Observed extreme rainfall events of the world

Location	Date	Rainfall depth (cm)	Duration
Plumb Point, Jamaica	May 12, 1916	19.8	15 min
Belouve, Reunion	Feb. 28, 1964	108.7	9 hrs
Belouve, Reunion	Feb. 28–29, 1964	134.0	12 hrs
Cilaos, Reunion	Mar. 15–16, 1952	187.0	1 day
Cilaos, Reunion	Mar. 15–17, 1952	250.0	2 days
Cilaos, Reunion	Mar. 15–18, 1952	324.0	3 days
Cherrapunji, India	Sept. 12–15, 1974	372.1	4 days
Cilaos, Reunion	Mar. 13–18, 1952	385.4	5 days
Cherrapunji, India	June 24–30, 1931	479.8	15 days
Cherrapunji, India	July 1861	930.0	31 days
Cherrapunji, India	May–July 1861	1, 637.0	3 months
Cherrapunji, India	Apr.–Sept. 1861	2, 245.0	6 months
Cherrapunji, India	Aug. 1860–July 1861	2, 646.0	1 year
Cherrapunji, India	1860–1861	4, 077.0	2 years

Source: Singh (1982).

Cherrapunji, India, has the world record of the highest rainfall. This station received 4,077 cm of rainfall during 1860–61.

Based on the rainfall records available for different raingauge stations in the world (Subramanya, 1984), a list of world's greatest recorded rainfall of various durations can be assembled. While plotting this data on a log-log paper an enveloping straight line drawn to the plotted points obeys the following equation:

$$P_m = 42.16 D^{0.475} \tag{8}$$

where P_m is the extreme rainfall depth in cm, and D is the duration in hours. The results of the above equations are useful for the PMP estimation.

Large parts of India receive heavy rainfalls associated with certain meteorological situations. Rainfall by a depression can be as high as 30 to 40 cm a day. Information regarding the greatest rainfall for a particular duration that has occurred or is likely to occur at a place is of great practical utility in the design of hydraulic structures. It enables the designer in evaluating the largest quantity of water that will be caught over an area from which he can derive the amount of water that will run off, depending upon the land characteristics.

Several workers have studied rainfall records to determine the magnitude of the largest rainfall that had occurred over different stations. Using daily rainfall data of about 300 stations from 1875 to 1982, Rakhecha (1992) prepared a generalized map of the highest point rainfall for 24 hours duration in India. Figures 11, 12, 13 give the highest 1-day, 2-day, and 3-day rainfall for India. These maps give information about the maximum rainfall that can be expected over small areas in any part of the country.

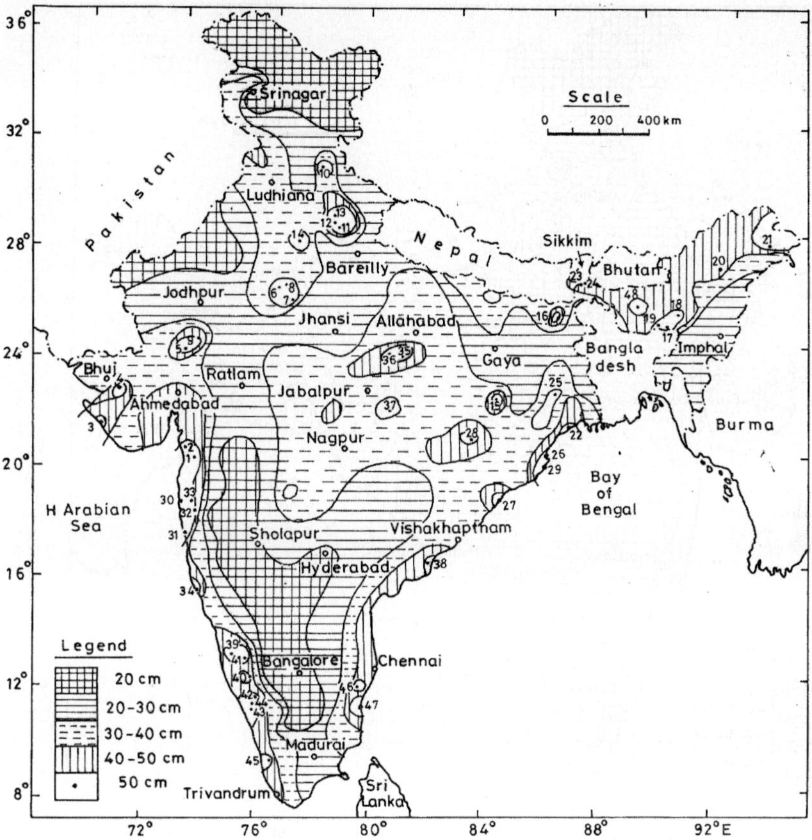

Figure 11. Highest 1-day rainfall for India

3.6.5. Severe Rainstorms Over India

Normally, in India severe rainstorms are associated with meteorological systems, such as active low pressure areas, depressions and cyclonic storms. Generally these systems originate from the neighboring seas of the Bay of Bengal and the Arabian Sea and after crossing the respective coastal areas these move inland. Their lifespan is, on average, 3 to 6 days over land. Associated with these disturbances heavy to very heavy rainfall occurs over a wide area of the country through with which these disturbances travel. Generally the duration of a rainstorm over a given area of region associated with these disturbances is of the order of about 2 to 3 days.

During the period of 110 years from 1880 to 1990, about 97 severe rainstorms of 2 to 3 days duration affected the country. Their distribution over different states of the country is given in Table 13. Figure 14 shows the tracks of cyclonic storms from 1891–1960 for the July month.

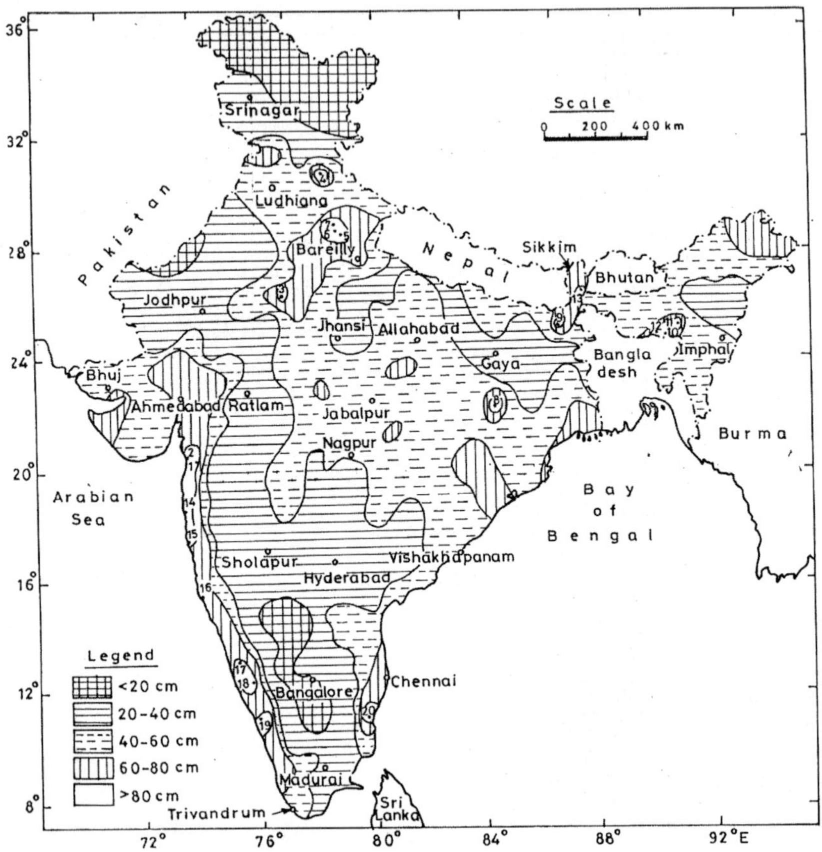

Figure 12. Highest 2-day rainfall for India

The break up of meteorological disturbances which caused these 97 severe rainstorms is given in Table 14.

Figure 15 gives the tracks of cyclones in the month of October and Figure 16 gives similar map for the month of November.

3.6.6. Precipitation Distribution in Mountainous Areas

Mountains have a strong impact on spatial distribution of precipitation. Knowledge of precipitation distribution is a basic and important requirement for planning and managing water resources, simulation of runoff and preparation of precipitation maps of the basin/region. In large mountainous basins, weather systems interact with topography and result in highly non-uniform precipitation. The uplift of moisture-laden air currents striking against a mountainous barrier provides good rainfall on the windward side. Gradients in the amount and intensity of precipitation depend

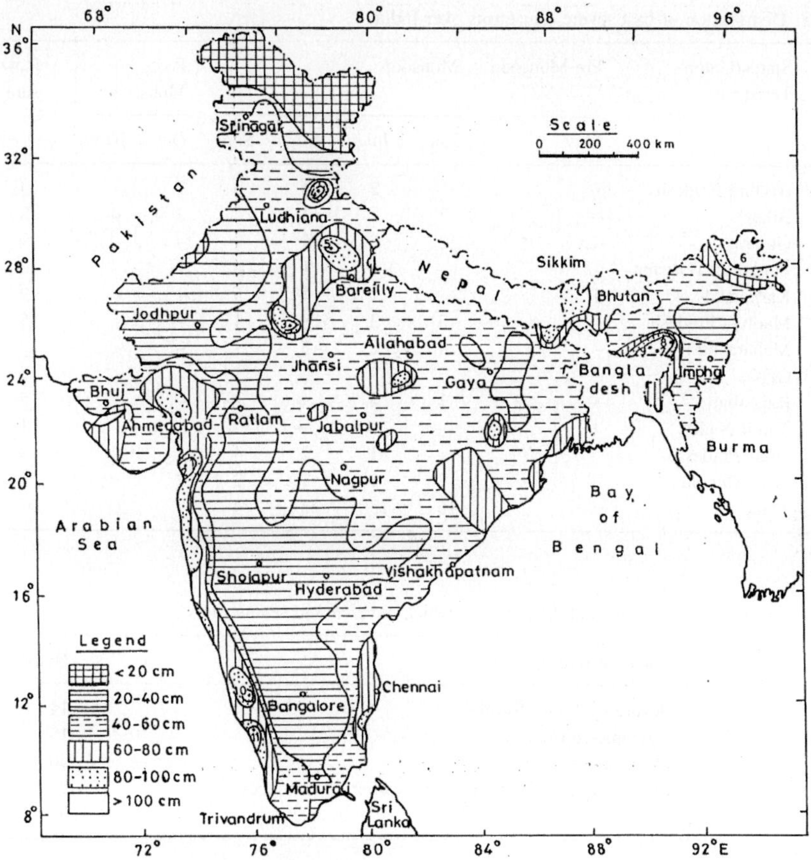

Figure 13. Highest 3-day rainfall for India

upon several factors, such as topography, strength of moisture bearing wind, its moisture content and orientation of the mountain range with respect to the prevailing wind direction. Depending on the relief of a mountain, there may not be continuous increase in precipitation with altitude: above a particular altitude, it may begin to decrease.

A summary of important precipitation distribution studies carried out in Himalayas is given in Table 15. The precipitation distribution for Himalayas is poorly known as compared with many other mountains of the world. Singh et al. (1995) and Singh and Kumar (1997) studied the distribution of precipitation with altitude for some important basins in the western Himalayan region, encompassing the outer, middle and greater Himalayan ranges.

These studies show that rainfall increases linearly with elevation for basins in the outer Himalayan range. In some cases, more number of rainy days are responsible

Table 13. Distribution of past severe rainstorms over India

SN	States/Union Territory	Pre-Monsoon	Monsoon				Post-Monsoon		Row sum
		May	Jun	Jul	Aug	Sep	Oct	Nov	
1	Andhra Pradesh	1	–	2	1	3	2	–	9
2	Bihar	–	2	1	1	2	2	–	8
3	Gujarat	–	2	6	1	–	–	–	9
4	Punjab & Haryana	–	–	–	1	4	1	–	6
5	Karnataka	1	–	1	1	5	–	1	9
6	Madhya Pradesh	–	3	3	8	1	–	–	15
7	Maharashtra	–	4	2	5	1	–	–	12
8	Orissa	–	3	2	2	1	–	–	8
9	Rajasthan	–	1	4	–	–	–	–	5
10	Tamil Nadu	1	–	–	–	–	–	–	1
11	Uttar Pradesh	–	–	1	2	4	1	–	8
12	West Bengal	–	2	1	1	2	2	–	8
	Total	3	16	23	23	24	7	1	97

Table 14. Details of meteorological disturbances which have caused severe rainstorms

SN	Cause of rainstorms	Number
1	Severe cyclonic storms	14
2	Cyclonic storms	19
3	Depressions	40
4	Low pressure areas and other weather disturbances	24
	Total	97

for higher rainfall at higher altitudes, while in some cases higher rain intensity is responsible for increasing rainfall with altitude in the outer Himalayan range.

The middle Himalayan range of the Beas basin has exceptionally heavy rainfall on the windward side and much less (less than half) rain on the leeward side. Rainfall gradients are 106 mm/100 m on windward and 13 mm/100 m on leeward of this range. Dharamshala, located on the windward side, experiences exceptionally heavy rainfall (average 1,972 mm) during the monsoon season due to the orographic effect. A sudden rise in the altitude of the middle Himalayan range near Dharamshala behaves as a giant mountain barrier and increases rainfall very significantly on the windward side of this range.

Different trends of rainfall variation with elevation are observed in different seasons in the middle Himalayan range with a linear increase in annual rainfall. Seasonal analysis indicates that rainfall varies linearly with altitude for the post-monsoon and monsoon seasons in the middle Himalayan range; annual rainfall also follows a similar distribution. But, for winter and pre-monsoon seasons, rainfall first increases and then decreases after a certain elevation. A second-order polynomial fits

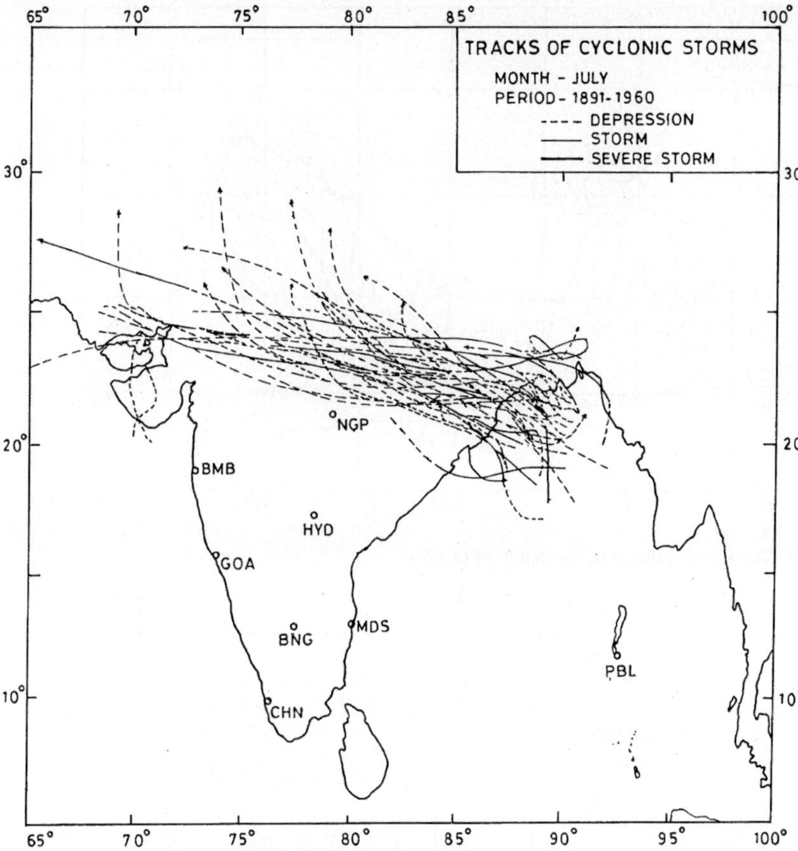

Figure 14. Tracks of cyclonic storms from 1891–1960

reasonably well for these two seasons. There is little rain in the greater Himalayan range because most of the moisture is precipitated over the outer and middle Himalayan ranges.

Rainfall follows an exponentially decreasing trend with altitude in the greater Himalayan range. The average annual rainfall decreases from outer Himalayas to greater Himalayas in the Satluj basin. In greater Himalayas, it is about one-sixth of outer Himalayas rainfall. The maximum rainfall occurs in the middle Himalayan range in the Beas basin. Spatial correlation is higher in the outer Himalayas range than in other ranges, which may be because of lesser relief in the outer Himalayan range than in other ranges.

Snowfall increases linearly with elevation in greater Himalayas. Snowfall gradients for the Spiti and Baspa sub-basins are 43 mm/100 m and 10 mm/100 m, respectively. The ratio of snowfall to annual precipitation varies linearly with altitude. All station records show more than 60% snow contribution to annual

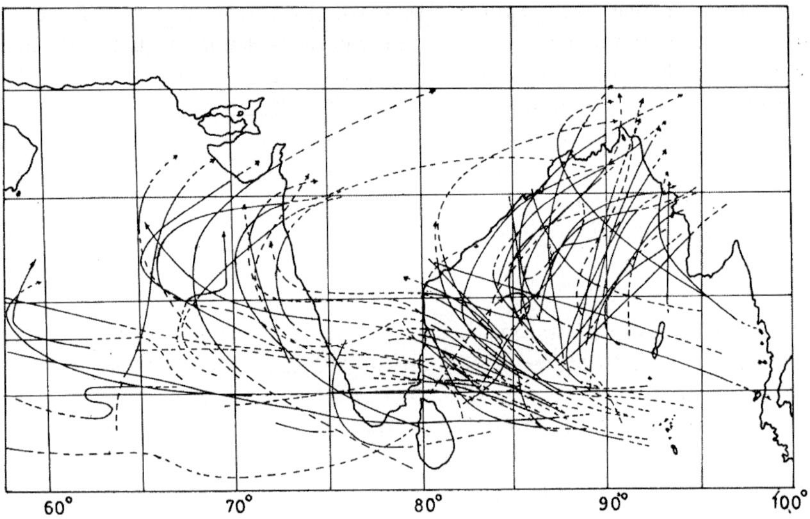

Figure 15. Tracks of cyclones in the month of October

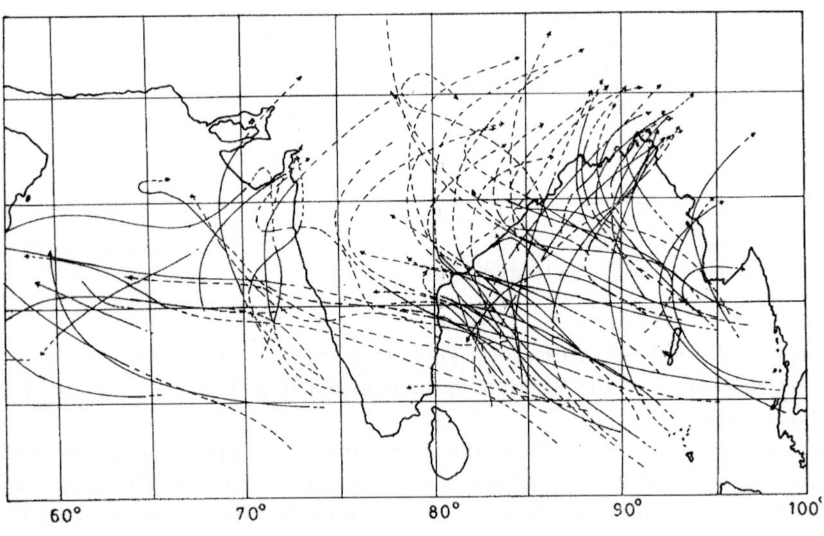

Figure 16. Tracks of cyclones in the month of November

precipitation. An extrapolation of the relationship indicates that snow and rain contribute equally at about 2,000 m, and all the precipitation occurs as snow above 5,000 m.

An evaluation of seasonal contribution to the annual rainfall suggests that monsoon rainfall contributes most to the annual rainfall, 45–71% in the Satluj basin, and 39–71% in the Beas basin, over all Himalayan ranges, as shown in Table 16 and 17. Minimum rainfall occurs in the post-monsoon season in the outer and middle Himalayas because of the lower moisture content in this season. But, in the greater Himalayan range, minimum rainfall is in winter because most of the precipitation falls as snow over this range. The contribution of pre-monsoon rainfall to annual rainfall increases from outer Himalayas to greater Himalayas and becomes significant to annual rainfall in the greater Himalayan range. Winter rainfall is also significant in the middle Himalayan range for both basins.

Spatial correlation function

Spatial correlation of rainfall appears to be higher in outer Himalayas than in other ranges. It may be because of lesser relief in the outer Himalayan ranges compared with middle and greater Himalayan ranges. Spatial correlation of rainfall is better on the leeward side than on the windward side in the middle Himalayan range. A study of spatial correlation for annual rainfall and snowfall for both Satluj and Beas basins revealed that spatial correlation functions followed an exponential form:

$$r(d) = r(0)\exp(-d/d_o) \tag{9}$$

Table 15. Summary of some important world-wide precipitation distribution studies

Authors/ reference	Study area/basin	Type of precipitation	Relationship with altitude/gradients	Other specific conclusions
Dhar and Rakhecha (1981)	Central Himalayas (Nepal Himalayas)	Rain	polynomial of 4th degree	Maximum rainfall occurs at 2,000–2,400 m
Higuchi et al. (1982)	Nepal Himalayas	Rain	–	Rainfall decreased with altitude in the range from 2,800–4,500 m
Singh et al. (1994; 1995)	Chenab basins, western Himalayas	Rain, snow	Second order polynomial for annual rainfall on windward of outer Himalayas, windward and lee-ward of middle Himalayas. Linear increase on leeward of outer Himalayas and exponential decrease in the greater Himalayas. Linear increase in snow with altitude.	Spill over effect was noticed in the outer Himalayas. Maximum rainfall was in the outer Himalayas

(Continued)

Table 15. (Continued)

Authors/ reference	Study area/basin	Type of precipitation	Relationship with altitude/gradients	Other specific conclusions
Singh and Kumar (1997)	Satluj and Beas basins, western Himalayas	Rain, snow	Rainfall increase linearly in outer Himalayas. In the middle Himalayas second order polynomial trends for winter and pre-monsoon, and linear increase for post monsoon and monsoon seasons. Rainfall exponentially decreases in greater Himalayas.	

Source: Adapted from Singh and Singh (2001).

Table 16. Seasonal distribution of average rainfall in different ranges of Himalayas in the Satluj basin

Himalayan range	Side	Rainfall (mm)				
		Winter	Pre-monsoon	monsoon	Post-monsoon	Annual
Outer	Windward	155 (10.9%)	146 (10.3%)	1,010 (71.3%)	105 (7.4%)	1,416
	Leeward	172 (14.3%)	201 (16.7%)	725 (60.3%)	106 (8.8%)	1,203
	Average	164 (12.5%)	174 (13.3%)	868 (66.2%)	106 (8.0%)	1,312
Middle	Leeward	209 (28.0%)	128 (17.2%)	336 (45.0%)	73 (9.8%)	746
Greater	Leeward	6 (3.0%)	75 (37.5%)	105 (52.5%)	14 (7.0%)	200

Source: Singh and Kumar (1997).

Table 17. Seasonal distribution of average rainfall in different ranges of Himalayas in the Beas basin

Himalayan range	Side	Rainfall (mm)				
		Winter	Pre-monsoon	Monsoon	Post-monsoon	Annual
Outer	Leeward	179 (12.8%)	148 (10.6%)	985 (70.6%)	83 (5.9%)	1,395
Middle	Windward	346 (13.6%)	323 (12.7%)	1,740 (68.3%)	140 (5.5%)	2,549
	Leeward	327 (31.1%)	210 (19.9%)	413 (39.2%)	102 (9.7%)	1,052
	Average	337 (18.7%)	267 (14.8%)	1,077 (59.8%)	121 (6.7%)	1,801

Source: Singh and Kumar (1997).

Table 18. Spatial correlation function for different Himalayan ranges of Satluj and Beas basins

Basin	Himalayan range	Type of precip.	Spatial correlation function	Distance for r = 0.75
Spiti sub-basin	Greater (Leeward)	Snow	$r(d) = 1.200\exp(-0.0277\,d)$	17 km
Baspa sub-basin	Greater (Leeward)	Snow	$r(d) = 1.039\exp(-0.0280\,d)$	12 km
Beas	Outer (Leeward)	Rain	$r(d) = 0.877\exp(-0.0035\,d)$	43 km
	Middle (Leeward)	Rain	$r(d) = 1.231\exp(-0.0441\,d)$	5 km
	Middle (Leeward)	Rain	$r(d) = 0.851\exp(-0.0037\,d)$	34 km

Source: Singh and Kumar (1997).

where r(d) is the spatial correlation coefficient, d is the distance between two stations and d_o is the correlation radius (the distance at which the correlation reduces e times) and r(0) is the correlation function at zero distance). Correlation coefficient (r) for rainfall and snowfall series as determined for different ranges. Spatial correlation functions obtained for different ranges and types of precipitation, and distance for which spatial correlation coefficient is greater than 0.75 are given in Table 18.

3.6.7. Variation of Precipitation with Elevation

Dhar and Bhattacharya (1977) studied the variation of precipitation with elevation in central Himalayas and a relationship between precipitation and elevation was obtained using 15 to 20 years data of more than 50 stations in this region. It was found that there are two zones of maximum precipitation: one near the foot of the Himalayas and other at an elevation of 2.0 to 2.4 km. For higher elevations beyond 2.4 km, precipitation was found to decrease sharply as one move across central Himalayas.

Statistically significant relationships between annual rainfall and elevation have been derived for some parts of the lesser Himalayas. However, such relationships have only a local predictive value and hold for a certain range in altitude. Above a critical elevation (2,500–3,500 m), a reduction in precipitation is observed. North of the Greater Himalayas, precipitation decreases rapidly to less than 500 mm annually. Equally low amounts of precipitation have been reported for Tibetan Plateau. The annual precipitation at Lhasa has been reported to be 350 mm. A similar overall pattern was observed for the Kumaon and Garhwal Himalayas where annual precipitation of 2,000 mm was reported.

Upadhyay and Bahadur (1982) carried out a study of the variation of precipitation in Himalayas. The Himalayas Mountain system was conceived to be constituted of three parallel longitudinal ranges:

1. The outer Himalayas or Shiwalik ranges with a height from 1,000–1,300 m and width from 10 to 50 km.
2. The lesser or middle Himalayas with a height ranging from 2,000–3,300 m and width between 60 to 80 km.
3. The greater Himalayas with a height of 6,100 m and average width of about 200 km.

Data of rainfall from seven sub-regions in Western Himalayas having homogeneous topographic aspects were used to study the variation of precipitation with altitude. Table 19 shows the annual precipitation in different sub-zones. It has been seen that the precipitation gradient decreases or even becomes negative when a considerable increase of wind speed occurs with increasing elevation; this partly explains the decrease of precipitation after a certain elevation in the Himalayas. This elevation is noticed to be generally around 2,000 m.

Based on the study, it was found that participation is influenced by altitude in three ways:

Table 19. Variation of precipitation with altitude

Station	Latitude (N)	Longitude (E)	Height (m)	Annual precipitation (cm)
Kangra	32°06′	76°15′	733	196.6
Palampur	32°7′	76°32′	1,250	263.7
Dharamsala	32°13′	76°19′	1,387	300.9
Doon Valley				
Dehradun	30°19′	78°02′	679	207.5
Raipur	30°18′	78°05′	750	209.7
Rajpur	30°24′	78°05′	914	300.7
Mussoorie	30°27′	78°05′	2,042	247.0
Almora	29°36	79°40′	1,572	105.4
Ranikhet	29°38′	79°26′	1,810	133.7
Mukteshwar	29°28′	79°39′	2,311	132.5
Kumaon Region				
Haldwani	29°13′	79°31′	440	199.5
Kathgodam	29°17′	79°32′	513	209.2
Nainital	29°3′	79°27′	1,934	253.9
Joshimath	30°16′	79°15′	769	142.3
Ukhimath	30°30′	79°15′	1,220	201.1
Birangkhal	30°15′	69°15′	1,530	122.8
Joshimath	30°33′	79°35′	1,840	95.4
Kulu	31°57′	77°7′	1,215	100.6
Benjar	31°38′	77°20′	1,524	110.6
Kathoi	31°1′8′	77°32′	1,608	101.2
Keylong	32°35′	77°4′	3,166	61.4
Ksauli	30°53′	76°58′	1,844	163.7
Kotgarh	31°18′	77°29′	1,949	115.3
Shimla	31°06′10	77°10′	2,202	159.0

1. The amount of precipitation increases with altitude up to a certain level and decreases thereafter. The level of maximum varies greatly from place to place depending on local topography. It is generally observed to be between altitudes of 1,500 to 2,500 m.
2. Average variability of precipitation generally increases with elevation.
3. At higher altitudes, the period of maximum precipitation is generally earlier than that on foothills.

Rao (1990) has provided data of the variation of precipitation with elevation in the Beas catchment which is given in Table 20.

Table 20. Variation of Precipitation with Elevation in Beas Catchment

Station	Elevation (m)	Mean annual precipitation (cm)	Standard deviation	Coeff. of variation %
Dharamsala	1,211	366	184	54
Palampur	1,217	256	96	38
Hamirpur	786	142	55	40
Kangra	701	196	53	27
Kulu	1,236	100	30	30
Mandi	752	157	25	16
Jogindernagar	1,221	223	56	25
Sundernagar	1,193	159	27	17
Banjar	1,522	109	28	26

3.7. RAIN STORM ANALYSIS

Hydrometeorological analysis of severe rainstorms is important for planning and designing water resources projects. In the absence of actual records of stream flow, rainfall data are useful for assessing the flood potentialities as well as the design flood of a project. The information required is the depth of rain which falls in different durations of time and over specified area. A rainstorm has three hydrologic dimensions and their relationships are of vital importance for assessing flood potential of a region. These are Depth-Area-Duration (DAD) curves.

Rainfall data are employed to estimate the design storm for use with a suitable rainfall runoff model to estimate design flood. Rainstorm analysis is a first step in the design storm estimation procedure. The design storm (rainfall) is a magnitude of rainfall and its distribution, which is developed for the design of hydraulic structures. It has three components, namely, the rainfall amount, the areal distribution of rainfall, and the time distribution of rainfall. The duration of design rainfall is determined, considering the size of the drainage basin, duration of the flood, and the type of structures.

3.7.1. Types of Design Storms

Design storms are derived on the basis of (i) statistical analysis and (ii) physical approach. Two types of storms are considered for design purposes. They are the Probable Maximum Precipitation (PMP) and the Standard Project Storm (SPS). The PMP is defined as "the theoretically greatest depth of precipitation for a given duration that is physically possible over a given station or a specified area" (WMO, 1986). Estimates of PMP for different durations over an area are required to calculate the probable maximum flood (PMF) hydrograph. The use of PMP or SPS depends upon the degree of safety required for the structure. The PMP is an estimate of the greatest depth of rainfall for a given duration that is physically possible over a drainage basin. The PMF hydrograph is used in the design of spillways of large dams. According to prevailing norms, the major reservoirs with a capacity of 60 million m^3 and more should be designed for PMF derived from PMP. SPS is defined as that rainstorm which is reasonably capable of occurring in the region of the problem basin. It is generally the most severe rainstorm which has occurred in the region of the basin.

Design storm

Abnormally high floods with very low probability of occurrence are caused by extreme storms. Consequently, flood flow estimates have to be based on data of extreme rainfall events. Extreme rainfall events are few and to enable consideration of not only the storms which have occurred over the catchment but also those which could have occurred, methods based on transposition of storms, transposition of depth-area-duration relationships and maximization of storms have been developed.

For maximization of storms, methods of moisture maximization and spatial maximization are employed. Moisture maximization attempts to estimate the maximum possible moisture content of the air mass at a given location, temperature and humidity and purports to represent the upper estimate of the possible depth of precipitation. Usually rain storm efficiency is less than 25% and all the moisture flowing into the storm region is not precipitated. Actual precipitation is always much smaller. This gave rise to the concept of "maximum possible precipitation" but it needs to be highlighted that data available on extremes are few and the process causing such events is still poorly understood. The term probable maximum precipitation (PMP) has come into wide use but it does not mean that precipitation exceeding PMP cannot take place. Cost of protection from the maximum possible flood (if such a flood existed) would be too high to justify the project and hence a lower level of protection is generally accepted. Therefore, terms PMP and probable maximum flood (PMF) are now generally understood to cover this lower level of protection. Of course, this lower level may vary from project to project and country to country.

Rationalization of PMP procedures

The assumptions made or implied in storm transposition and moisture maximization procedures may be valid for extra tropical regions where precipitation is chiefly due to fronts (warm moist air rising over cold dry air). In a semi-arid country, such as India where severe rainfall storms are invariably associated with low-pressure systems such as depressions and cyclones in combination with other rain causing phenomena, validity of the assumption of pseudo-adiabatic atmosphere is often questioned.

The storm transposition and maximization method has thus serious limitations in its application to catchments in India. An alternative method to arrive at PMP estimates is a statistical procedure developed by Hershfield and Wilson (1960) and modified by Dhar et al. (1983) to suit the prevailing conditions in India. WMO (1986) have noted that the procedure is used mostly for making quick estimates for basins of no more than 1,000 km^2 but has been used for much larger areas. Indian Institute of Tropical Meteorology (IITM, 1989) has published the atlas giving isolines of 1-day PMP values estimated by this statistical procedure. By using depth-area relationships based on Indian data, the point PMP values can be adjusted to various sizes of catchment areas.

Dhar & Nandargi (1989) have documented severe rainfall storms of 1-day, 2-day and 3-day durations and presented Depth-Area envelope curves for 13 plain regions of the country. These can be converted into percent depth-area curves for use with atlas to obtain areal 1-day, 2-day and 3-day PMP values. The methods by Dhar et al. (1983) with IITM atlas gives reasonable PMP values comparable to those obtained by the traditional meteorological method for catchments of area not more than 10,000 km^2. Most of the dams in the country have catchment areas less than 10,000 km^2. In-depth storm and flood studies are necessary for dams with large catchment areas, such as Narmada Sagar.

Temporal distribution of design storm rainfall

Procedure for arriving at temporal distribution of severe rainfall storms has been reviewed by Rao (1990) who concluded that the effect of chronological distribution of PMP on PMF hydrograph peak and shape would be large and hence needs serious consideration. To arrive at a realistic chronological distribution of PMP, the following principles enunciated in WMO (1986) for estimation of PMP may be followed:

1. The sequence should be in accordance with the storm characteristics of the area. A study of self-recording rain gauge data of severe storms indicated that heavy rainfall spells generally vary from 9 to 18 hrs with a clear break from the next spell. Patterns of chronological distribution with "one bell" for each day of the storm is thus realistic and may be adopted for catchments in India.
2. The sequence should be such that the maximum summation of increments for any given duration may be less than or equal to but not more than the PMP for the same duration.
3. Increments of rainfall should be areal and not point values. Temporal distribution provided by IMD as percentages of 24-hour point rainfall obtained from maximum rainfall depth duration data by fitting the mean curve (or envelope if

conservative values are desirable) may be corrected for the area by multiplying with an appropriate correction factor.

On the basis of data from 12 dense networks, Rao (1990) concluded that it is possible to represent the relationship between areal-point rainfall ratio (P), area A (sq. km) and duration T (hours) as:

$$P = \exp(-A^{0.333}/11^{*}T^{0.5}) \tag{10}$$

Rao (1990) recommended that eq. (10) may be adopted for applying the areal correction to the temporal distribution pattern of point rainfall or adopt the simpler chronological distribution pattern of “one rectangle” instead of “one bell” for each day of PMS.

3.7.2. Storm Selection and Analysis

The first step in the rainstorm analysis is a thorough understanding of the meteorology of large storms in the region of analysis. This requires a detailed study of the synoptic and dynamic features of all major storms. A judicious selection of a few storms, which are representative of the whole catchment, needs to be made. Before deciding the appropriate storm depth for design purposes, either basin centered or storm centered storm analysis is carried out.

By reviewing relevant records, such as daily or multi-hourly rainfall data, depression/ storm tracks, recent and historical storm details, and flood data, a preliminary selection of storm is made. Based on these, a list of all rain periods is made. The next step is to fix an appropriate threshold value for storm selection after considering the following:

- If the catchment lies in a semi-arid region a lower depth of rainfall is adopted as a threshold while a higher depth is adopted for humid regions. A suitable threshold would be 5 cm in humid regions and 2.5 cm in semi-arid regions.
- For smaller catchments, a higher threshold and for larger catchments, a smaller threshold is used. An average catchment depth of 10 cm for catchment areas up to 5,000 km^2, 5 to 8 cm for catchment areas between 5,000 and 10,000 km^2, and 5 cm for catchment areas greater than 10,000 km^2 may be appropriate.

For the storms periods listed, the average rainfall using data of stations located within and around the basin is determined and compared with the threshold value. All the storms whose daily depth equals or exceeds the threshold are considered for further analysis.

For the short listed storms, isohyetal maps are prepared. The scale generally used for the isohyetal maps is 1 cm = 10 km. A common practice is to prepare total storm depth maps of 1 day, 2 days, and 3 days, depending on the storm duration and requirement of design storm duration. Where the movement of storm over elongated catchments is involved, daily isohyetal maps may have to be prepared.

The design storm studies by Rakhecha & Soman (1992) revealed that in the most severe 3-day rainstorm, 50% of basin rainfall is obtained on the heaviest of the 3 days, 30% on the second heaviest day, and only 20% on the third day. Further, 3-day maximum basin rain depths varies from 20–30% in many basins, thereby showing that more than one fifths of the annual rainfall over a basin can occur in the course of three days. The spillway design storm of a river basin does not depend upon the mean annual rainfall, as a low annual rainfall basin could have high design storm and vice versa.

3.7.3. Estimation of PMP

Two main approaches are followed to estimate PMP. The first is the physical approach in which PMP is derived by analysis of a number of major rainstorms in and near the basin and then transposing the severest storm pattern over the basin to obtain the maximum rain depths for different durations. The transposed rain depths are then adjusted for moisture maximization. The areal rainfall values are adjusted to the highest amount of moisture index corresponding to a maximum persisting dew point temperature. Rakhecha et al. (1990) have given generalized maps of highest 24-hour persisting dew point temperatures for individual months in June to September for India. The PMP estimation by the physical method is described in IMD (1972).

The second approach is a statistical procedure in which PMP at a particular location is determined from frequency analysis of the annual maximum rainfall. Normally, the Gumbel or the log-normal distribution is used. A statistical method for estimating PMP for small areas was developed by Hershfield (1961, 1965) by using Chow's general frequency equation in a modified form:

$$X_e = \overline{X} + k_m \sigma \tag{11}$$

$$k_m = (X_1 - \overline{X}_{n-1})/\sigma_{n-1} \tag{12}$$

where X_e is the PMP rainfall for a given station for specific duration; k_m is the frequency factor and X_1; $\overline{X}_n$ and σ are, respectively, the highest, mean and standard deviation for a series of n annual maximum rainfall values of a given duration; and $\overline{X}_{n-1}$, σ_{n-1} are, respectively, the mean and standard deviation for this series excluding the highest value from the series. Dhar and Kulkarni (1974) applied the Hershfield method to about 1,000 stations having long period rainfall data in the plains of North India and prepared a generalized map of PMP for 1-day duration. The generalized one-day PMP distribution for India is shown in Figure 17.

The Indian Institute of Tropical Meteorology (IITM, 1989) has published an atlas containing generalized charts of PMP estimates based on the Hershfield method for 1-day duration for different states of India. The 1-day PMP for stations in different states of India varies between 30 and 110 cm.

For design purposes, hydrologists also need PMP values for rainstorms of two to three days as synoptic scale disturbances, which cause heavy rainfall over India during monsoons lasting for 2 to 3 days. Rakhecha (1992) carried out a study on the estimation of PMP for 2-day duration for stations in the Indian peninsula

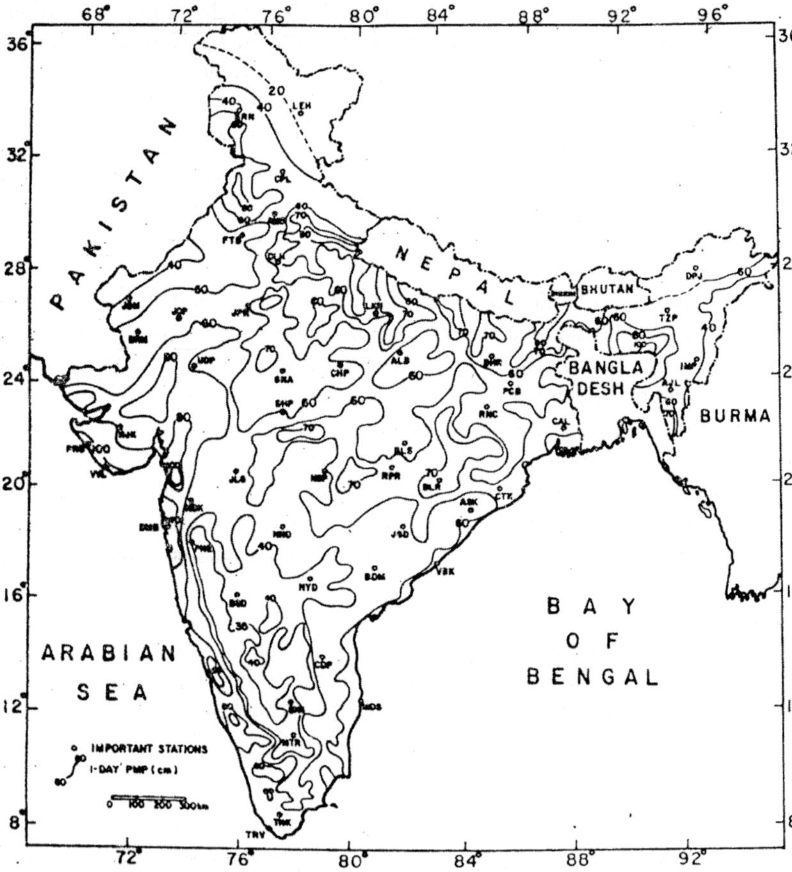

Figure 17. Generalized one day probable maximum precipitation map

lying between 8° to 20°N. A generalized map showing the spatial distribution of 2-day PMP is shown in Figure 18. The 2-day PMP estimates over the Indian peninsula varied from 40 to 95 cm and the average ratio of 2-day PMP to the highest observed 2-day rainfall was 1.76.

3.7.4. Depth-Duration and Depth-Area-Duration Analysis

Data on volumes of precipitation during severe storms is important to examine and study storms suitable for design. Such information is generally presented in the form of maximum average depth of storm precipitation over various areas, such as 100 km^2, 500 km^2, etc. and is known as Depth-Area-Duration (DAD) data. The DAD data is presented in tabular as well as graphical form. WMO (1994) recommends Depth-Duration (D-D) and DAD for storm analysis.

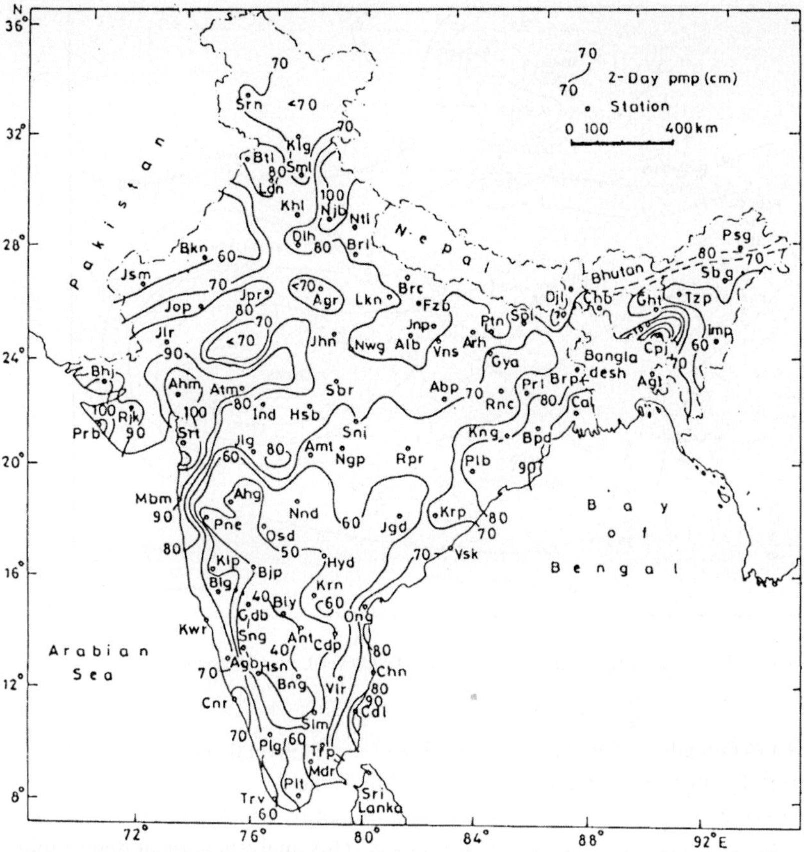

Figure 18. Generalized two-day PMP map

Design storm depths for different durations are obtained by DD or DAD analysis of severe rainstorms in and around a river basin on the basis of 70–80 years of rainfall data. Estimates of PMP for the catchments are obtained by storm transposition and moisture maximization techniques. Frequencies of maximum rainfall of different durations are also worked out by extreme value techniques. The DD analysis is carried out using the boundary of the catchment up to a desired site to estimate storm depths over the catchment for both in-situ and transposed storms. The average depths of precipitation for different durations, 1 day, 2 days, and 3 days, etc. are computed in the same way as in the case of the isohyetal method. These depths for different durations constitute the DD data.

Where a good network of recording raingauges is available, the DD analysis could be performed for shorter duration (1-hr, 3-hr, 6-hr, 12-hr, etc.) rainfalls. Alternatively, analysis can be performed by distributing rainfall observed at non-recording stations using the data of nearby representative SRRG station. The storm

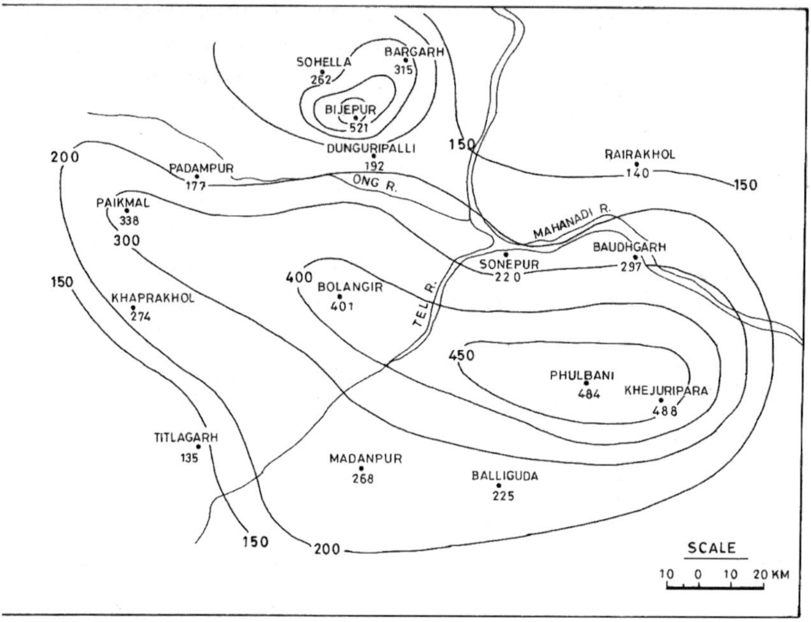

Figure 19. One day storm isohyetal map of 30 August 1982 over Lower Mahanadi

centered DAD analysis for the total one-day storm isohyetal map of 30th August 1982 storm (shown in Figure 19) is presented in Table 21.

The DAD analysis is carried out using data of storms which have occurred over different regions and for principal flood seasons. This analysis aims at determining the maximum depth of precipitation for various size areas during storm periods of 1-day, 2-days, 3-days durations. Such values, when determined for each of the transposable storms, provide the requisite data for estimation of the design storm.

The steps of DAD analysis are given below:

Table 21. Depth-Area-Duration analysis computation for one-day storm (30.08.1982)

Isohyetal value (point)	Area planimetered in cm^2	Area in km^2	Net Area in km^2	Average depth (mm)	Volume in mm * km^2		Average rainfall depth
					Increment	Cumulative	
488	–	–	–	–	–	–	488.0
450	13.4	1,340	1,340	469	628,460	628,460	469.0
400	38.1	3,810	2,470	425	1,049,750	1,678,210	440.5
300	91.2	9,120	5,310	350	1,858,500	3,536,710	387.8
200	177.1	17,710	8,590	250	2,147,500	5,684,210	321.0

1. The selected storm is assigned a definite beginning and end using rainfall records of stations in the region. Typically, the storm duration is from a period of no rain to the next period of no rain. This ensures that the storm totals from all stations are for the same time interval. Analysis is performed for the period of the principal burst.
2. For each day of the total storm period, separate isohyetal maps for each duration, i.e., maximum values for 1-day, 2-days, 3-days, etc. are drawn. Since the greatest rainfall of one day may not occur simultaneously for areas of all sizes under consideration, it may be necessary to carry analysis for two alternative days for establishing the maximum one-day rainfall. Similarly, a two-day duration map is prepared.
3. In storm centered DAD analysis, the last closed isohyet is taken as boundary and in case of catchment centred analysis, the catchment boundary is considered.
4. The isohyetal maps are divided into zones to represent principal rainfall centres. Starting with the central isohyet in each zone, the net area encompassed between isohyets is determined.
5. The average isohyetal value is multiplied by the area to compute the volume. The volumes are accumulated for every successive isohyetal range and then divided by the total area encompassed by a particular isohyet to determine the maximum depth of the storm for the corresponding area.
6. Analyses are repeated for other durations also. The incremental maps could be prepared for shorter durations, such as 3, 6, 9, 12 and 18 hours also where adequate data of recording stations are available.

For further details on DAD analysis, WMO (1969) may be referred to. A step-by-step procedure for analysis is described in IMD (1972).

The total storm depth-area curve for each duration is constructed by plotting the average depth against the accumulated area. The starting point on the curve is the central storm precipitation (highest value). Considering the sparse network of precipitation stations in India, the probability of recording the highest point precipitation is small. In the study of most large area storms, it is reasonable to assume that the highest station precipitation represents the average depth over an appreciable area rather than the maximum point precipitation. The depth area curves are plotted for different durations for each of the major rain storms. Conventionally a logarithmic scale is used for area and the linear scale for precipitation depths. A typical set of DAD curves for a storm is shown in Figure 20.

Abbi et al. (1970a) made a detailed study of 85 heavy rainstorms over North Bengal and showed that the 11–13 June 1950 rainstorm contributed the highest average depth of rainfall for 1–3 day durations. Dhar et al. (1974, 1982) analyzed all the severe recorded rainstorms of 1-day, 2-day and 3-day durations over different regions of North Indian plains and over the central parts of Indian peninsula.

The results of DAD analysis of severe rainstorms of different river basins/regions were published in the proceedings of Workshops on "Unusual storm events and their relevance to dam safety" by CBIP (1989, 1990). The DAD values for 1-, 2-, and 3-day durations for the most predominant rainstorms obtained by various

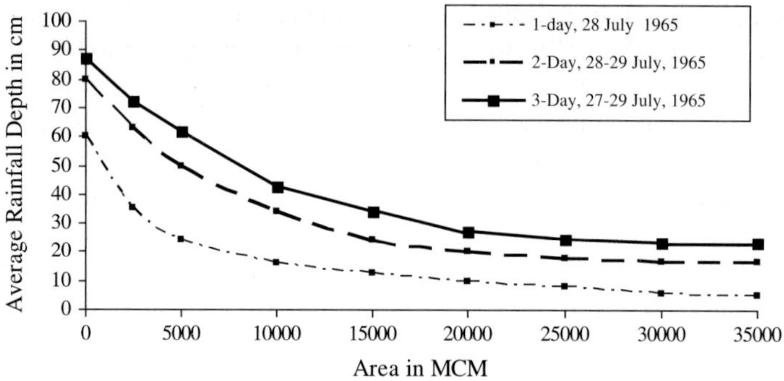

Figure 20. A typical set of DAD curves for a storm

workers are given in Table 22. The maximum point rainfall for these 13 rainstorms varied from 24 to 99 cm for 1-day, from 44 to 127 cm for 2-days, and from 60 to 145 cm for 3-days durations. The main centers of these storms are shown in Figure 21.

3.7.5. Frequency of Maximum Rainfall of Different Durations

An important problem in hydrometeorology is to determine the maximum values of rainfall of different durations that may be expected to occur with given frequency. The return period is the average interval of time after which rainfall of a specified amount will be equaled or exceeded. Thus, 1-day rainfall with a 10-year return period is that magnitude of daily rainfall which can be expected to be equaled or exceeded, on average, ten times in 100 years.

Several investigators have carried out a return period analysis of maximum one-day rainfall for many stations having long period rainfall data. Dhar and Kulkarni (1971a) prepared nomograms for different sub-divisions of North India for estimating maximum 1-day point rainfall for 2 to 25 year return periods. Ayyar and Prasad (1971) studied maximum 1-hour rainfall of about 100 SRRG stations having rainfall data of 20 years and prepared generalized maps of 1-hour rainfall of return periods from 2 to 50 years. Caution should be exercised in determining rainfall of specified return periods from generalized charts because local conditions may cause differences in rainfall characteristics. If records are available, it is always better to analyze the in-situ data.

3.8. RAINFALL INTENSITY – DURATION – FREQUENCY STUDIES

A rainfall event or storm denotes a period of time having significant rainfall that is preceded and followed by a period of nil or insignificant rainfall. Rainfall duration is the time elapsed from the beginning to the end of the rainfall event. It is generally

Table 22. DAD values (cm) of severe rainstorms over India

Severe rainstorm date	Affected region	Duration (day)	Area in km^2					
			10	100	1×10^3	5×10^3	10×10^3	20×10^3
17–18 Sept. 1880	Uttar Pradesh	1	82	82	78	63	52	40
		2	104	103	99	87	77	62
20–22 Sept. 1990	Bengal	1	44	43	41	36	33	28
		2	73	72	67	58	52	44
		3	83	82	78	69	62	52
19–21 Sept. 1926	Madhya Pradesh	1	36	36	35	33	30	26
		2	65	65	63	57	53	47
		3	83	82	81	76	71	62
26–28 July, 1927	Gujarat	1	54	53	48	39	33	29
		2	100	94	79	63	56	49
		3	129	126	114	94	83	71
1–3 July, 1930	Maharashtra	1	36	36	31	24	22	19
		2	71	70	58	40	33	28
		3	77	76	66	47	39	35
1–3 July, 1941	South Gujarat	1	99	97	85	65	54	43
		2	127	126	118	97	83	66
		3	145	143	134	117	105	86
17–19 May, 1943	Tamil Nadu	1	42	41	37	29	25	21
		2	72	72	69	55	46	37
		3	95	95	91	73	61	49
3–5 Oct. 1955	Punjab	1	50	47	45	40	35	29
		2	72	70	64	56	51	44
		3	72	71	67	59	53	47
1–3 Oct. 1961	Bihar	1	37	37	36	32	28	23
		2	55	54	53	49	44	35
		3	58	57	57	44	50	42
28–30 Sept. 1964	Karnataka	1	24	23	23	22	21	19
		2	44	43	32	27	25	22
		3	62	61	51	38	34	30
13–15 July, 1965	Andhra Pradesh	1	51	49	39	25	20	16
		2	54	52	41	27	23	20
		3	60	57	45	30	27	23
18–20 July, 1981	Rajasthan	1	56	56	54	45	37	27
		2	84	83	76	62	52	40
		3	97	95	85	71	61	48
28–30 Aug. 1982	Orissa	1	52	52	51	45	38	30
		2	70	70	69	65	59	50
		3	88	88	84	74	66	55

Source: Rakhecha & Soman (1992).

measured in hours but it may be of the order of several minutes for small catchments and days for large catchments.

The rainfall depth and duration vary widely depending on the geographic location, climate, and time of the year. Rainfall intensity also varies widely in time and

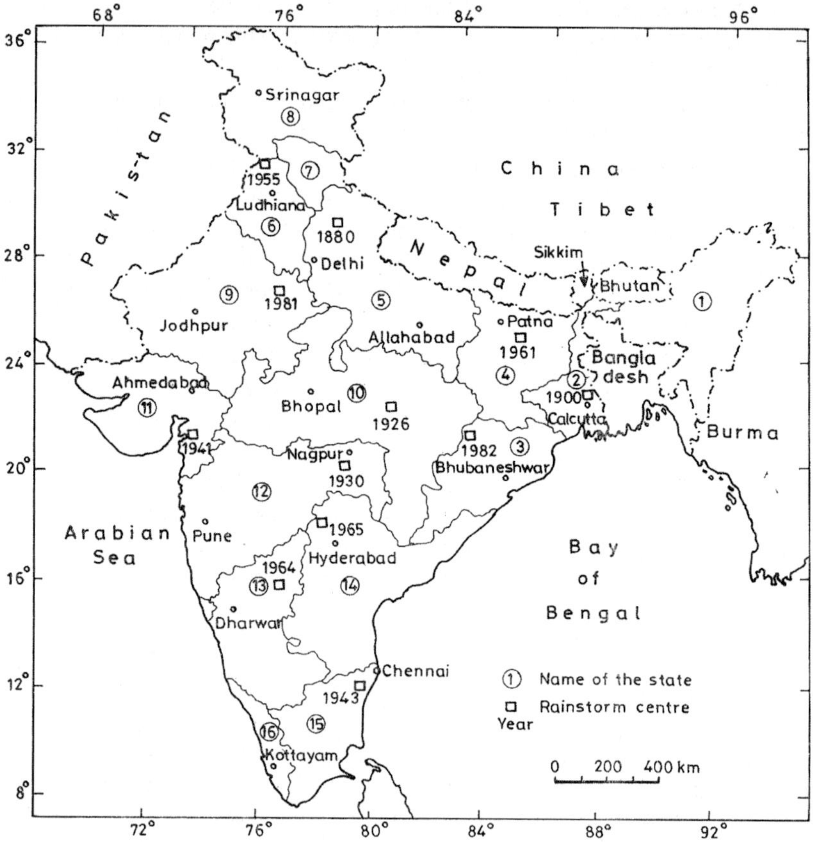

Figure 21. Main Centers of Rainstorms

space. Typically, rainfall intensities range from 0.1 to 30 mm/hr, but at times these can be as large as 300 mm/hr in extreme cases.

Rainfall frequency refers to the average elapsed time between two rainfall events of the same depth and duration. The actual elapsed time, however, varies widely and is, therefore, interpreted statistically. For example, if at a certain location, a 100-mm rainfall event lasting for 6 hours occurs on average once in 50 years, the 100-mm, 6-hour rainfall frequency for this location will be 1 in 50 years or 0.2. The inverse of the rainfall frequency is its return period or the recurrence interval. Generally, larger rainfall depths are associated with longer return periods and vice versa.

A long historical record of rainfall is needed to ascertain the statistical properties of the annual maximum rainfall. Due to the lack of long rainfall records, extrapolations are usually necessary to estimate rainfall depths associated with long return periods. Such extrapolations involve a certain amount of risk. When the risk involves human life, the concepts of rainfall frequency and return period are no longer considered adequate for design purposes. Instead, a reasonable maximization of the meteorological factors associated with extreme rainfall is used.

The rainfall intensities for events of shorter duration (1 hour or less) can be expressed in terms of the average values. However, for longer events instantaneous values of rainfall intensity are more important, especially for determination of flood peaks. Temporal distribution of rainfall depicts the variation of depth (or intensity) with time. It can be expressed in either discrete or continuous form. The discrete form is referred to as a hyetograph that represents a histogram of rainfall depth (or intensity) as ordinates. The continuous form is the temporal rainfall distribution, a function describing the rate of rainfall accumulation with time.

The intensity of storms decreases with increase in the storm duration. Further, a storm of any given duration will have a larger intensity if its return period is large. In other words, for a storm of given duration, storms of higher intensity in that duration are rarer than storms of smaller intensity. In many design problems related to watershed management, such as runoff disposal and erosion control, it is necessary to know the rainfall intensities of different durations and different return periods. The interdependency between intensity i (cm/h), duration D (hour) and return period T (years) is commonly expressed in a general form as

$$i = \frac{kT^n}{(b+t)^m} \tag{13}$$

where k, b, m, and n are constants for a given catchment. Typical values of these for a few places in India based on the reported studies of the Central Soil and Water Conservation Research and Training Institute, Dehradun, are given in Table 23.

Extreme point rainfall values of different durations and for different return periods have been evaluated by IMD and the isopluvial maps covering the entire country

Table 23. Typical values of constants in equation 13

Place	k	n	b	m
Bhopal	6.93	0.189	0.50	0.878
Nagpur	11.45	0.156	1.25	1.032
Chandigarh	5.82	0.160	0.40	0.750
Bellary	6.16	0.694	0.50	0.972
Raipur	4.68	0.139	0.15	0.928
New Delhi	5.2	0.157	0.5	1.107
Nagpur	11.5	0.156	0.25	0.990
Bhuj	3.8	0.199	0.25	0.990
Kolkota	5.9	0.115	0.15	0.924
Bangalore	6.3	0.126	0.50	1.128

Source: Mishra and Soman (2003) and Rakhecha et al. (1992).

have been prepared. These are available for rainfall durations of 15 min., 30 min., 45 min., 1 h, 3 h, 6 h, 9 h, 12 h, 15 h and for return periods of 2, 5, 10, 25, 50 and 100 years.

The relationship between intensity (I), frequency (F) and duration (D) at any location can be determined from an analysis of records obtained from self-recording raingauges. The magnitudes of rains of various durations, such as 5 min., 15 min., 30 min., 60 min., and so on, are determined and these data for selected durations can then be used to determine the magnitudes of rains of various frequencies. For example, the 10-year frequency of 30 min rainfall can be determined from a record of annual maximum 30 minutes values. The average rainfall intensity when plotted against duration for a particular frequency provides an intensity-duration curve which is an important tool for the prediction of runoff.

Ram Babu et al. (1979) have developed rainfall intensity-duration-frequency equations and nomographs for 42 stations as well as for different zones of the country, namely north, central, eastern, western and southern. These equations and nomograms enable a quick determination of rainfall intensity for any desired duration and frequency. The rainfall depth increases with storm duration and is described mathematically as

$$P = at^n \tag{14}$$

where P is the rainfall depth, t is the storm duration, a is a coefficient, and n is an exponent, and a positive number less than 1. Typically, n varies from 0.2 to 0.5. By analyzing storm data, equation (14) can be used to predict the storm depth as a function of storm duration. The applicability of such an equation is, however, limited to the regional or local conditions for which it was derived.

For extreme rainfall events (world's greatest observed rainfall events) given in Table 12, the resulting enveloping curve can be described as given in Figure 22.

$$P = 4.8685\, t^{0.4888} \tag{15}$$

where P is in centimeters and t is in minutes. For eq. (15), the coefficient of determination is 0.996. A less accurate, but adequate, predictive relation (Mishra and Singh, 2003) is:

$$P = \frac{a\,t}{b+t} \tag{16}$$

where $a = 4.54 \times 10^3$ cm and $b = 2.39 \times 10^5$ minutes. Equation (16) fits the data with a coefficient of determination of 0.96. A comparison of the predicted rainfall values from equations (15) and (16) shows that low rainfall depths are less accurately predicted by equation (16). However, eq. (15) over-predicts high rainfall depths. Equation (15) remains unchanged if the highest two rainfall points are ignored whereas the constants of equation (16) change and the coefficient of

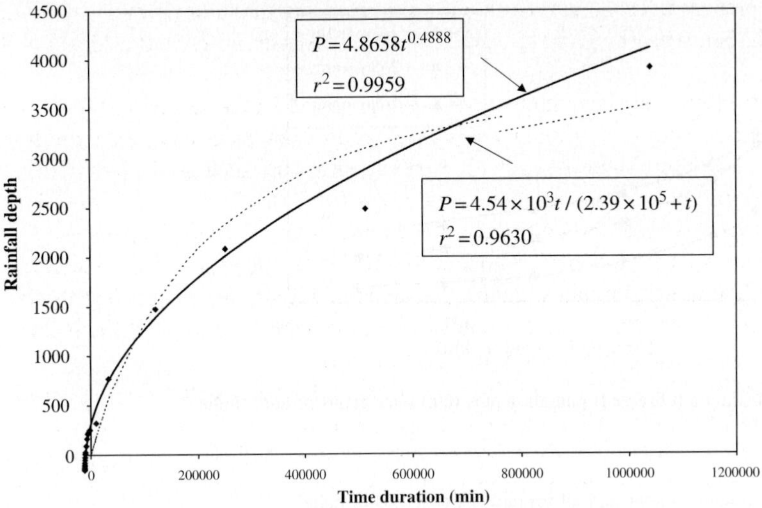

Figure 22. Rainfall depth-duration relation for the world's greatest observed rainfall events

determination improves. The former is somewhat insensitive to very high rainfall values, whereas the latter is sensitive. It is because the trend of equation (15) is governed by low rainfall values, in contrast with equation (16). Thus, extrapolation of equation (15) may result in over-prediction of high values and the trend of equation (16) passing through the middle of the highest two rainfall values exhibits a mean trend. The advantage of equation (16) is that it is dimensionally homogeneous.

Maximum observed rainfall depths over the north-central Indian plains were computed by Subramanya (1984). These depths were computed from observed data of two most severe recorded rainstorms (storm of 17–18 September 1880 over north-west Uttar Pradesh and storm of 28–30 July 1927 over north Gujarat). The first storm was more severe but of shorter duration and smaller areal extent while the storm over Gujarat covered larger areas and was of longer duration.

Table 24 gives the maximum 24-hour rainfall magnitudes ever observed in India. Figures 23 gives the maximum (observed) rainfall depths (cm) over plains of north India. It is interesting to note that on January 07, 1952, Foc in the Reunion Islands (east of Madagaskar, Indian Ocean) recorded more than 1,800 mm of rainfall. As far as Asia is concerned, the record of maximum rainfall was set in Paishih, Taiwan in 1962 when 1,250 mm rain fell. In terms of intensity, the most intense event was observed in Barot, West Indies, where 38.1 mm of rainfall took place in a single minute on November 26, 1970 (The Times of India, New Delhi, July 29, 2005). At Cherrapunji, the rainiest month was July 1861 when 9,300 mm of rain was recorded while the rainiest year was from August 1960 to July 1861 when 26,461 mm of rain fell.

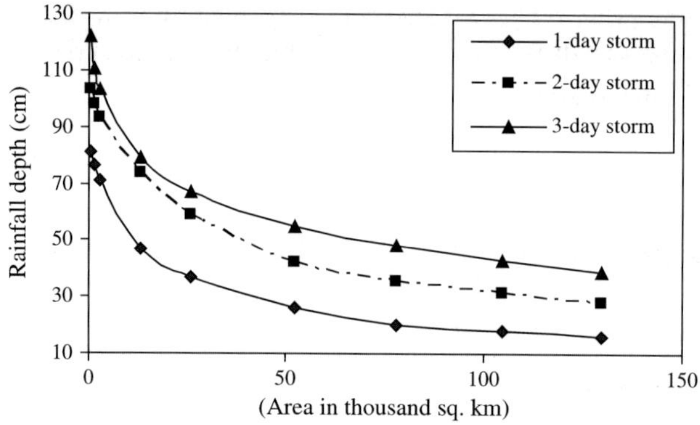

Figure 23. Maximum (observed) rainfall depths (cm) over plains of north India

Table 24. Maximum 24-hour rainfall magnitudes observed in India

Place	24-hour rainfall (mm)	Date of occurrence
Aminidivi (Lakshadweep)	1,168	May 06, 2004
Cherrapunji (Meghalaya)	1,036	June 14, 1876
Cherrapunji (Meghalaya)	998	July 12, 1910
Kasauli (Himachal)	996	June 18, 1899
Mousinram (Meghalaya)	989	July 10, 1952
Dharampur (Gujarat)	987	July 02, 1941
Cherrapunji (Meghalaya)	985	Sep 13, 1974
Mumbai (Maharashtra)	984	July 26, 2005

3.8.1. Storm Intensity and Duration

Storm intensity and duration are inversely related. Division of equation (10) by t yields the average (uniform) rainfall intensity, i_o, as follows:

$$i_o = at^{n-1} \tag{17}$$

or

$$i_o = \frac{a}{t^m} \tag{18}$$

where $m = 1 - n$. Since n is less than 1, m is also less than 1. Similarly, i_o can be derived from equation (16) as:

$$i_o = \frac{a}{b+t} \tag{19}$$

where a and b are constants which can be determined by regression analysis. A general form of the intensity-duration relation combining the features of equations (14) and (18) is frequently given as

$$i_o = \frac{a}{(b+t)^m} \tag{20}$$

For $b = 0$, equation (20) reduces to equation (14) and for $m = 1$, the former reduces to equation (19).

3.8.2. Depth-Duration-Frequency

It is often necessary for midsize catchments to determine several depth-duration curves, each for a different frequency or return period. Depth-duration-frequency (DDF) curves are plotted with duration on the abscissa, depth on the ordinate, and frequency (or alternatively, return period) as a third variable. The return period is attached with constant 'a' in equation (14) as

$$a = kT^n \tag{21}$$

where k is the coefficient, n is the exponent, and T is the return period. Coupling of equation (21) with equation (14) leads to a relation for the depth-duration-frequency as:

$$P = k(Tt)^n \tag{22}$$

Similarly, a relation of the form of equation (12) can be obtained as:

$$P = \frac{kT^n t}{b+t} \tag{23}$$

A general form of equation (19), similar to equation (16), can be given as:

$$P = \frac{kT^n t}{(b+t)^m} \tag{24}$$

The DDF curves can be constructed using frequency analysis of recorded rainfall data. The steps are as follows:

1. Select the duration of rainfall, such as 5, 10, 20, 30, 60, or 100 min.
2. For the duration selected, compute the annual maximum rainfall depth for each year of record. If there are 30 years of data, then there will be 30 annual maximum depths associated with the selected duration.
3. Fit an appropriate frequency curve to these values. The methods of frequency analysis are described elsewhere (Singh, 1992).
4. Obtain from the fitted frequency curve the values of rainfall depth for the selected periods such as 5, 10, 20, 30, 50, 80, and 100 years. Tabulate these values.

5. Repeat steps (ii) and (iii) for different durations.
6. Rearrange the data as rainfall depth against duration for various return periods or frequencies.
7. For a selected frequency, plot on log-log graph paper rainfall depth values on the ordinate versus duration on the abscissa, and fit a smooth curve or connect the points linearly. Do the same for other frequencies. These plots will be a family of DDF curves.

Typical depth-duration-frequency curves are plotted in Figure 24. It is seen from this figure that as the return period or duration increases, the rainfall depth increases and vice versa.

3.8.3. Intensity-Duration-Frequency

Similar to the depth-duration-frequency curves, intensity-duration-frequency (IDF) curves are plotted with duration on the abscissa, intensity on the ordinate, and frequency (or alternatively, return period) as a third variable. Coupling of equation (21) with equation (22) leads to an intensity-duration-frequency relation as:

$$i_o = \frac{kT^n}{(b+t)^m} \tag{25}$$

For a given catchment, constants k, b, m, and n are evaluated from measured data or local experience. Typical values of these constants for a few places in India are given in Table 25. These values are based on the reported studies of the Central Soil and Water Conservation Research and Training Institute, Dehradun, India. The values of Table 23 correspond to rainfall intensity, i_o, measured in cm/h; duration, t, in hr; and return period, T, in years.

Extreme point rainfall values of different durations and for different return periods have been evaluated by the Indian Meteorological Department and isopluvial maps

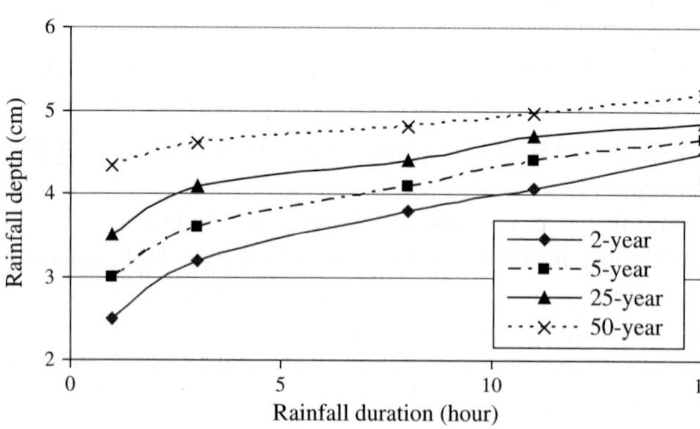

Figure 24. Typical variation of the maximum rainfall depth of various durations with return period

Table 25. Representative values of constants in eq. (25)

Place	k	n	b	m
Bellary	6.16	0.694	0.50	0.972
Bhopal	6.93	0.189	0.50	0.878
Chandigarh	5.82	0.160	0.40	0.750
Nagpur	11.45	0.156	1.25	1.032
Raipur	4.68	0.139	0.15	0.928

covering the entire country have been prepared. These are available for rainfall durations of 15 min, 30 min, 45 min, 1 h, 3 h, 6 h, 9 h, 12 h, 15 h and for return periods of 2, 5, 10, 25, 50, and 100 years.

Example 3.1 Determine 1-hr and 6-hr maximum rainfall depths and corresponding intensities having return periods of 2, 10, 20, 50, and 100 years that are likely to occur in Chandigarh (India).

Solution: Here, t = 1 hr and 6 hrs. For Chandigarh, the values of the constants of equation (25) are: k = 5.82, n = 0.160, b = 0.40, and m = 0.75. As an example, the rainfall intensity for t = 1 hr and T = 2 years is computed from equation (25) as:

$$i_o = \frac{kT^n}{(b+t)^m} = \frac{5.82 \times 2^{0.160}}{(0.40+1)^{0.75}} = 5.05\,\text{cm/hr} \tag{26}$$

The corresponding maximum rainfall depth is: $5.05 \times 1 = 5.05$ cm. Other computations are given in Table 26.

Construction of IDF curves is similar to that of DDF curves, except for that the former utilizes the maximum rainfall intensity and latter the maximum rainfall depth. From a given maximum rainfall depth and its duration, the maximum rainfall intensity can be computed by dividing the maximum rainfall depth by the associated duration. The resulting plots will be a family of IDF curves.

3.8.4. Rainfall Depth-Area-Duration

Generally, the greater the catchment area, the smaller is the spatially averaged storm depth. This variation of storm depth with catchment area has led to the concept of

Table 26. Computation of rainfall depths and intensities for various return periods (Example 3.1)

Rainfall	Duration (hr)	Return period				
		2	10	20	50	100
Intensity (cm/hr)	1	5.05	6.54	7.30	8.46	9.45
	6	1.62	2.09	2.34	2.70	3.02
Depth (cm)	1	5.05	6.54	7.30	8.46	9.45
	6	9.72	12.54	14.04	16.20	18.12

point depth, defined as the storm depth associated with a given point area. A point area is the smallest area below which the variation of storm depth with catchment area is assumed to be insignificant. It is taken of the order of 25 km^2 (Ponce, 1989). The point depth applies for all areas less than the point area. For larger areas, a reduction in the point depth is necessary to account for the decrease of the storm depth with catchment area. This depth reduction is accomplished with a depth-area reduction chart, a function relating the catchment area (abscissa) to the point depth percentage (ordinate). The storm duration is a curve parameter in a depth-area reduction chart.

The depth-area-duration (DAD) analysis is another way of portraying the reduction of storm depth with area, with duration as a third variable. To construct a DAD chart, two methods are used: (1) the mass-curve method and (2) the incremental isohyetal method. Each method portrays the time division of rainfall differently. The mass-curve method involves construction of mass curves for individual stations, an average mass curve for the entire area, and the construction of one isohyetal map for the total storm rainfall. When rainfall measurements are available at different times of the day, construction of mass curves is fairly easy.

The incremental isohyetal method involves construction of a number of isohyetal maps. If the isohyetal pattern is simple and rainfall readings are available at a uniform time each day, then this method is relatively simple. In many cases, especially in developing countries, rainfall is recorded and reported once a day, and the standard durations are, therefore, integral multiples of a day, such as 2 days, 3 days, and so on; and hence, the incremental isohyetal method may be preferable.

3.9. LONG RANGE FORECAST OF MONSOON

India was the first country to start a systematic development of long range forecast techniques to estimate monsoon rainfall. Such developments started more than a century ago in the 1880s. As noted by Gowariker et al. (1991), Blanford noticed a negative association of excessive winter and spring snowfalls in Himalayas with the southwest monsoon over India and issued monsoon forecast from 1882 to 1885 based on the Himalayan snow cover.

Gowariker et al. (1991) developed a power regression model for long range forecast of monsoon rainfall. This model utilizes signals from 16 parameters which are globally significant in regional and global forcings for the subsequent monsoon. To determine the exact form of the power regression model, long-term data were used. Various model parameters were determined by using step-by-step iteration together with method of least squares. The model given by Gowariker et al. (1991) is:

$$\frac{R+\alpha_0}{\beta_0} = C_0 + \sum_{i=1}^{16} C_i \left(\frac{X_i + \alpha_i}{\beta_i}\right)^{p_i} \tag{27}$$

where R is the monsoon rainfall over India as a percentage of the long period average, Xs are variable of the model, αs, βs, ps, and Cs are model parameters.

In the first stage of model development, the best suited order of parameters was determined. The model of eq. (27) was developed by arranging the variables in order of their decreasing correlation with the monsoon rainfall. The variables of the model are described in Table 27.

The values of parameters of the power regression model are given in Table 28.

The authors have reported that the forecast rainfall lie roughly within ±4% of the actual values and the forecasts have been able to explain 94% variance of the observed rainfall.

3.10. RELATION BETWEEN RAINFALL EXTREMES AND EL NINO CHARACTERISTICS

Along the coast of Peru-Ecuador in South America there is an ocean current called Peru or Humboldt current. The El Nino (The Child) is identified with warming of this ocean current. In some years, the current may extend southwards along the coast of Peru to Latitude 12 °S, killing plankton and fish in coastal waters. The occurrence of El Nino events is on the basis of disruption of fishery, hydrological data, sea surface temperature and rainfall along and near the Peru-Ecuador coast. Intensities can be defined based on the positive sea-surface temperature anomalies along the coast, viz. Strong in excess of 3 °C; Moderate, 2.0 °–3.0 °C; Weak, 1.0 °–2.0 °C. El Nino is the oceanic component of general El Nino/Southern Oscillation (ENSO) phenomenon, while the Southern Oscillation, generally represented by Tahiti minus Darwin atmospheric difference (T-D), is its atmospheric component. Anomalies in the two may not always occur simultaneously.

Table 27. Variables of long range monsoon forecast model

Subscript of X	Corresponding variable
1	50 hPa wind pattern (winter)
2	Eurasian snow cover (December)
3	500 hPa ridge (April)
4	Central India temperature (May)
5	10 hPa zonal wind pattern (January)
6	East coast of India temperature (March)
7	Northern hemisphere surface pressure anomaly (January to April)
8	Argentina pressure (April)
9	Northern hemisphere temperature (January & February)
10	(Spring) Southern Oscillation Index
11	El Nino (Previous year)
12	North India temperature (March)
13	Indian ocean equatorial pressure (January to May)
14	El-Nino (Same year)
15	Himalayan snow cover (January to March)
16	Darwin pressure (Spring)

Table 28. Parameters of long range monsoon forecast model

i	α_i	β_i	p_i	C_i
0	0	1		0.1303086×10^{2}
1	50	10	1.4	$-0.5824688 \times 10^{-1}$
2	0	10	−2.0	0.3026653×10^{2}
3	0	10	−4.0	-0.1733019×10^{1}
4	0	10	4.0	0.6958737×10^{0}
5	50	10	−0.4	-0.1956036×10^{2}
6	0	10	4.0	0.4374519×10^{0}
7	0	10	4.0	0.4687574×10^{-2}
8	0	100	−3.9	-0.1662928×10^{6}
9	50	10	−4.0	-0.4617612×10^{5}
10	50	10	4.0	0.5844691×10^{-1}
11	50	10	4.0	$-0.4456572 \times 10^{-2}$
12	0	10	4.0	-0.1497441×10^{0}
13	0	1,000	3.3	-0.1183124×10^{4}
14	50	10	4.0	$-0.1344925 \times 10^{-1}$
15	50	10	−4.0	0.1940398×10^{5}
16	0	1,000	3.3	0.1257985×10^{4}

In literature, the term ENSO is used for the general phenomenon of El Nino/Southern Oscillation. While analyzing the influence of these on droughts and floods in India, Kane (2004) used the notations given in Table 29.

Various combinations of these were seen in Indian data. In particular, some prominent events were ENSOW, i.e., El Nino (EN) existed, SOI minima (SO) also existed and, eastern equatorial pacific SST was warmer (W). The SOI and SST plots (12 monthly running means) were used to check whether the SOI minima or SST maxima occurred in the middle of the calendar year (May-August). If so, the events were termed as ENSOW-U, i.e., Unambiguous ENSOW. If the SOI minima or SST maxima occurred in the earlier or later part of the year (not in the middle), the events were termed as ENSOW-A, i.e. Ambiguous ENSOW. The year having neither an EN nor SO nor W nor C were termed as non-events. Symbols S (Strong), M (Moderate, W (Weak) indicates strength of the El Ninos involved. Figure 25

Table 29. Terms defined by Kane (2004) for ENSO phenomena

EN	Presence of El Nino at Puerto Chicama (8°S, 80°W, Peru Coast) and Nino 1+2 region (0–10°S, 90–80°W).
SO	Presence of minimum in the Southern Oscillation Index (SOI). SOI is atmospheric pressure difference Tahiti between Darwin (T-D), i.e., maximum in (D-T).
W	Presence of maximum (positive anomalies) in the sea surface temperature (SST) in the eastern equatorial pacific (Nino 3 region (5°N – 5°S, 150°W – 90°W) and/or Wright SST). Wright SO index is based on pressure at a number of stations.
C	Presence of minimum (negative anomalies) in the sea surface temperature (SST) in the eastern equatorial pacific (Nino 3 region and/or Wright SST). These are La Ninas.

shows the departure of monsoon rainfall amounts and Wright's Southern Oscillation Index (June-August) from 100 year mean.

3.10.1. Droughts in India

The years when drought occurred in India during 1925–1994 are listed in Table 30, with the designations of years used in Kane (1997a, b; 1998).

Thus among the 22 droughts (10 mild, d; 12 severe, D), 8 were associated with Unambiguous ENSOW and 2 with Ambiguous ENSOW. This indicated that ENSOW (especially ENSOW-U) is a combination favourable for droughts (Kane, 2004). However, it is neither necessary nor sufficient. In half the number of events, droughts occurred in other categories of years. It may also be noted that the interactions between El Nino/La Nina and the climate anomalies (droughts, floods etc.) are non-linear in nature. As such, the distinction d or D or f, F may not be very meaningful.

In addition to the above, Indian summer monsoon rainfall data is also available since 1871 and the Wright SST index is also available since then. These are presented in Table 31.

Out of the above 17 drought events (9 severe and 8 mild), only 7 were associated with ENSOW-U and 10 were associated with other types of events. This also confirms favourable but not exclusiveness of ENSOW type events for causing droughts.

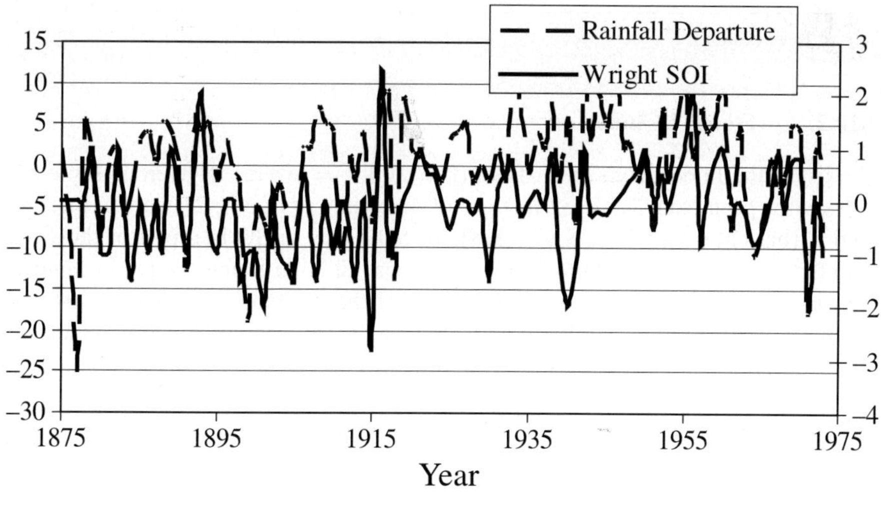

Figure 25. Departure of rainfall amounts and Wright's Southern Oscillation Index (June–August) from 100 year mean

Table 30. Occurrence of droughts in India

Year	Drought	Associated event	Year	Drought	Associated event
1925	d	ENSOW-A (Jan–Dec)	1966	D	Non-event
1928	D	C (La Nina)	1968	D	W
1930	d	ENSOW-U (Mar–Dec)	1972	D	ENSOW-U (Feb–Dec)
1932	d	EN (Feb–Jun)	1974	D	SO
1939	d	EN (Jan)	1979	D	SOW
1941	D	ENSOW-U (Jan–Jun)	1982	D	ENSOW-U (Oct–Dec)
1951	D	ENSOW-U (Mar–Dec)	1985	d	Non-event
1952	d	Non-event	1986	D	W
1957	d	ENSOW-U (Feb–Dec)	1987	D	ENSOW-U (Jan–Nov)
1962	d	Non-event	1991	d	ENSOW-A (Jul–Sep)
1965	D	ENSOW-U (Mar–Dec)	1992	d	ENSOW-A (Jan–Aug)

Source: Kane (2004).

Table 31. Occurrence of droughts in India (additional data).

Year	Drought	Associated event	Year	Drought	Associated event
1873	D	ENSO (W not seen)	1905	D	ENSOW-U (Jan–Dec)
1876	d	Non-event	1907	d	ENC- (W not seen)
1877	D	ENSOW-U (Feb–Dec)	1911	D	ENSOW-U (Aug–Dec)
1888	d	ENSOW-U (Jan–Dec)	1912	d	ENSO (Jan–Feb)
1891	d	ENSO (Apr–Aug)	1913	d	SOW (No data for W)
1899	D	ENSOW-U (Jul–Dec)	1915	d	Non-event
1901	D	Non-event	1918	D	ENSOW-U (No data for W)
1902	d	ENSOW-U (Jan–Dec)	1920	D	W (Jun–Oct)
1904	D	SOW(Jul–Dec)			

Source: Kane (2004).

3.10.2. Floods in India

The list of the floods from 1871–1996 (f = mild; F = severe) is given in Table 32.

It can be seen from Table 32 that out of 43 events of floods, 21 were associated with C (basically La Nina), 8 with ENC or SOC (EN or SO in a part of the year, C in the rest), 5 with ENSOW-A (mostly II year El Nino events, active in January but disappearing soon after, leaving the rest of the years as C), 5 non-events and 4 other types (ENW, SO, SOW). Thus, La Nina events seem to be better associated with floods than El Nino events are with droughts probably because La Nina events run smoothly and last longer.

The All-India area-weighted mean summer monsoon rainfall, based on a homogeneous rainfall data set of 306 rain-gauges in India, developed by the Indian Institute of Tropical Meteorology, is widely considered to be a reliable index of summer monsoon activity over the Indian region. The alternating sequences of multi-decadal periods having relatively more frequent droughts and floods is a characteristic of the

Table 32. Occurrence of floods in India

1872	F	C	1914	F	ENSOW-A	1955	f	C
1874	F	ENC	1916	F	C	1956	F	C
1875	F	C	1917	F	ENC	1959	F	SO
1878	F	ENSOW-A	1926	F	ENSOW-A	1961	F	Non-event
1879	F	C	1933	F	C	1964	F	C
1882	F	C	1934	F	C	1970	F	C
1884	F	ENW	1936	F	SOC	1973	f	ENC
1887	F	ENC	1938	F	C	1975	F	C
1889	F	ENC	1942	F	C	1978	f	Non-event
1890	f	C	1944	f	SOW	1983	F	ENSOW-A
1892	F	C	1945	F	Non-event	1988	F	C
1893	F	C	1946	F	SOC	1990	f	non-event
1894	F	C	1947	F	Non-event	1994	F	SOW
1908	f	C	1949	F	SOC			
1910	F	C	1953	F	ENSOW-A			

Source: Kane (2004).

well-known epochal pattern of monsoon variability. Analysis associates the rainfall anomalies with El Niño and La Niña occurrences. It was found that a majority of the El Niño events are associated with drought years over India, while the La Niña events are associated with flood years.

CHAPTER 4

EVAPORATION AND OTHER METEOROLOGICAL DATA

This chapter provides a discussion about evaporation and some meteorological variables that are important in hydrological studies.

4.1. EVAPORATION

Evaporation is the process whereby liquid water is converted to vapor by the transfer of water molecules to the atmosphere. This process is accompanied by a transfer of energy. Evaporation tends to reduce water temperature. Latent heat of vaporization is the heat required (in calories) to convert 1 gram of water at 1 atmosphere (atm) pressure to vapor. It requires 539 cal of heat to evaporate 1 gram of water. About 70 M ha-m of water is estimated to evaporate from water bodies and land surface in the country. Obviously, evaporation is an important component of the hydrologic cycle. This discussion is focused on evaporation from a free water surface. Evaporation also occurs beneath the soil surface as heat energy penetrates and vaporizes the moisture contained on and between the soil grains.

Water also evaporates directly from the solid state by the process of sublimation. A significant amount of water vapor enters the atmosphere by sublimation from snow and ice. However, this contribution is far less than evaporation from the liquid water in the oceans, lakes, and rivers. Evaporation is a function of climatic variables, such as incoming solar radiation, air and water temperature, sunshine hours, wind speed, and saturation vapor pressure deficit. Pan evaporation provides an estimate of open water evaporation.

The amount of water vapor that can be held in the air is a function of temperature. Warm air can hold far more water vapor than cold air. This relation provides the basis for understanding not only ET but also precipitation. As the vapor pressure of the air becomes saturated and equals the vapor pressure of the water, evaporation becomes zero. By the same token, evaporation is a maximum when the vapor pressure of the air is low, thus producing the maximum gradient between the vapor pressure of the air and that of water.

4.1.1. Factors Affecting Evaporation

The most important factors affecting evaporation are: (a) vapor pressure difference, (b) temperature, (c) barometric pressure, (d) humidity, (e) wind, (f) water quality, and (g) water depth and soil type.

Vapor Pressure and Humidity: Vapor pressure and humidity are interdependent. According to Dalton's law, the vapor-pressure gradient determines the rate of evaporation. The greater the vapor pressure gradient, the larger is evaporation. As the gradient tends to zero, evaporation also tends to zero. But this is complicated by the temperature of the air and the water. If the water has lower temperature than the air, evaporation will be low because the vapor pressure of cold water is low. When air is warmer than water, evaporation continues until the vapor pressure of air equals that of water.

Temperature

The movement of water molecules increases with temperature. The kinetic energy of water molecules increases as temperature rises and this permits molecules to escape from liquid water to the air more rapidly. For this reason, the warmer the water, the higher is the evaporation. Of course, temperature and other factors are interrelated, and it is not possible to isolate the effect of each factor on evaporation.

Atmospheric Pressure

Evaporation is greater at lower atmospheric pressures than at higher atmospheric pressures because fewer air molecules are present in less dense air, there is less likelihood of escaping water molecules colliding with air molecules. Atmospheric pressure decreases with altitude and, therefore, evaporation is higher at higher altitudes. This effect is somewhat offset by decreasing temperatures with increasing altitude.

Wind

A thin film of saturated vapor exists over a water surface under certain conditions. If this film were left undisturbed, it would act as an insolating buffer between the water surface and the unsaturated air above and evaporation would slow down. This condition seldom exists in nature because wind tends to remove the saturated film and re-expose the water surface to unsaturated air. Evaporation increases with wind velocity until the vapor-pressure gradient reaches some nearly constant relation. At this point, a further increase in wind velocity will not increase evaporation appreciably. The effect of increased wind velocity on evaporation is believed to be related to the size of the water body. Wind removes water vapor from small bodies of water rather quickly, whereas on large water bodies, longer time is required to accomplish the same effect.

Water Quality

The vapor pressure of water is reduced when solids are dissolved in water. Pure water has a higher vapor pressure than salt water. A salt content of 1% slows the rate of evaporation by about 1%. Since the ocean generally has a salt content of a little over 3%, sea-water evaporation is lower than the freshwater rate by about 3%.

Water Depth and Soil Type
Shallow-water bodies are more rapidly heated than the deep-water bodies and are, therefore, evaporated more rapidly. Intercepted and ponded precipitation is evaporated rapidly after rain-fall when insolation provides renewed heating. Soils have an important role in evaporation: dark soils absorb incoming radiation and convert it to heat more effectively than lighter-colored soils.

Lake Effect
Water absorbs and stores heat energy. A larger water body stores greater energy than a smaller one. Lakes and ponds have varying effects on evaporation, depending on the depth of the water body. Evaporation rates are higher for shallower water bodies. Further, vapor pressure of the warm lake is much higher than the cold air thus providing favorable conditions for evaporation. Shallow lakes and ponds are heated by the sun to the bottom, raising the water temperature quickly, and increasing evaporation with rising temperature. Such water bodies, including ponding on a drainage basin, are rapidly evaporated because they have little heat storage.

Evaporation is small during a runoff event, for the vapor pressure gradient is low. In general, a runoff event from a convective storm lasts only a few hours, and this short time, coupled with a low vapor-pressure gradient, would result in very small evaporation. Although a cyclonic storm has a longer duration, evaporation occurring over its duration and the duration of runoff is still small and can be neglected. Evaporation from storage reservoirs in arid areas can be large and is a direct loss for the project purpose. Evaporation from soils in the drainage basin is important because this antecedent moisture deficit must be satisfied before any significant runoff occurs.

4.1.2. Methods to Estimate Evaporation

Evaporation from water bodies can be determined by (1) the water budget, (2) the energy budget, (3) mass transfer methods, (4) combination methods, and (5) evaporation formulas. In the first method, water balance equation is employed to estimate evaporation. The energy-budget method for determining evaporation is similar to the water budget method except that the energy budget deals with the conservation of energy rather than water. The energy available for evaporation is obtained by considering the incoming energy, the outgoing energy and the energy stored in the water body for a given time interval.

In the simplest mass-transfer method, evaporation rate is estimated as a function of wind speed and the saturation vapor-pressure deficit. Several empirical expressions have been proposed, some simple and some fairly complex. The combination methods combine the energy-budget and mass-transfer methods. The most popular combination method for computing evaporation from free water surfaces is that developed by Penman (1948). This method combines fundamental physical principles and empirical concepts based on meteorological observations. Singh (1982) has described this method in detail.

4.1.3. Transpiration

Transpiration is the process by which plants utilize water for their metabolism and growth. Plants remove water from the soil through their root system and transpire it to the atmosphere through stomata in their leaves. These stomata actively transpire water vapour to the atmosphere during daylight but close after darkness begins, whereupon transpiration ceases.

Transpiration is affected by physiological and environmental factors. Stomata open and close in response to environmental conditions such as light and dark, heat and cold, and so on. Stomata allow carbon dioxide to enter the plant in the process of photosynthesis. Important physiological factors include (a) density and behaviour of stomata, (b) extent and character of protective coverings, (c) leaf structure, and (d) plant diseases. All these factors are influenced by the types of plants and the density of those plants. With adequate soil moisture, the greater the plant density, the greater the amount of moisture lost by transpiration.

Vapour pressure gradient, temperature, solar radiation, wind, and available soil moisture are the most important factors affecting transpiration. The vapour-pressure gradient is a measure of the energy required to move the water from the leaf to the air. Plants transpire little when the vapour-pressure gradient is low as during a rainfall event. On the other hand, plants transpire rapidly when warm, dry air surrounds them and plenty of soil moisture is available.

The rate of transpiration is doubled for approximately each 10 °C rise in temperature. Because a leaf is dark colored, it absorbs incoming solar radiation effectively and its temperature becomes higher than the surrounding air. This condition exists even though evaporation cooling from transpiration lowers the temperature of the leaf.

Absorption of solar energy by a leaf raises its temperature and its aqueous vapour pressure. Thus, transpiration increases with increasing insolation. Wind usually increases transpiration by removing the film of moisture-laden air next to the leaf and consequently increasing the vapour-pressure gradient. Gentle winds have been found to be more effective in increasing transpiration than strong winds. Transpiration is greatly affected by the amount of soil moisture. As the plant uses moisture from the soil, the capillary forces holding moisture in the soil become stronger and it is more difficult for the plant roots to remove moisture. Transpiration ceases when the soil moisture approaches the permanent wilting point.

4.1.4. Measurement of Transpiration

Methods for measuring transpiration are related to the size and nature of the plants. Small plants are placed in a small closed container and the amount of moisture transpired is measured. A drying agent is usually placed in the container to absorb the transpired water vapour. This moisture can be measured by weighing the drying agent before and after the test. A phytometer is a large vessel, filled with soil, in which plants are rooted. The soil surface is sealed to prevent evaporation. So the only moisture escape is by transpiration. The lost moisture can be determined by weighing the plant and the container before and after the test. This method yields good results as long as natural environmental conditions are maintained (Singh 1982).

4.1.5. Evapotranspiration

Evapotranspiration (ET) and consumptive use include both the transpiration by vegetation, and evaporation from water surfaces, soil, snow, ice, and vegetation. Consumptive use differs from ET only in that it includes the water used to make plant tissue. For all practical purposes, these two terms are synonymous. ET and consumptive use convert water to vapour which cannot be used again. Some water consumed by animals is converted to vapour in the metabolic process and, therefore, is not available for further use. The units of ET and consumptive use are cm of depth for a specified period. Annual mean potential evapotranspiration over India is shown in Figure 1.

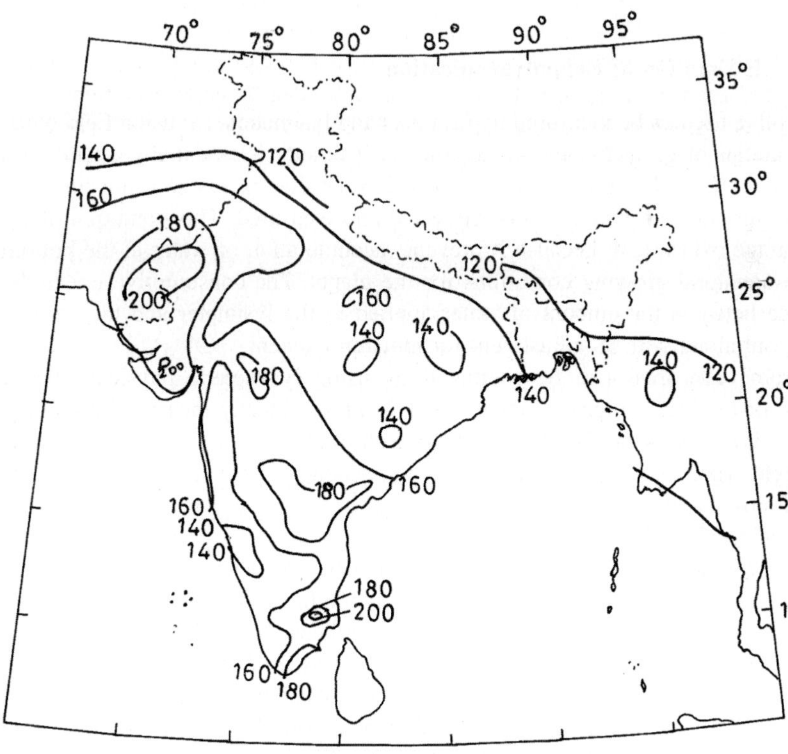

Figure 1. Annual mean potential evapotranspiration (cm) over India

Despite the widespread application of the ET concept, there has been considerable ambiguity in the use of various terms such as potential ET and reference crop ET. To overcome this, the Food and Agricultural Organization (FAO) of the United Nations has introduced uniformity and standardization in interpretation and use of various terms such as potential ET and reference crop ET in the report FAO-56 (Allen et al., 1998). FAO-56 discourages the use of the term potential ET due to ambiguities in its definition. Moreover, FAO recommended that a hypothetical reference surface "closely resembling an extensive surface of green grass of uniform height, actively growing, completely shading the ground and with adequate water" be adopted as reference surface. In the FAO approach, the surface characteristics that influence ET are quantified in an unambiguous fashion.

ET from a reference surface, not short of water, is called the reference crop ET or reference ET and is denoted as ET_0. The reference surface is a hypothetical grass reference crop with specific characteristics. Further, crop ET under standard conditions (ET_c) refers to the ET from excellently managed, disease-free, large, well-watered fields that achieve full production under the given climatic conditions. Further, due to suboptimal crop management and environmental constraints that affect crop growth and limit evapotranspiration (ET_c) under non-standard conditions generally requires a correction.

4.1.6. Estimation of Evapotranspiration

Consumptive use can be measured by (a) tanks and lysimeters, (b) using field plots, and (c) studies of groundwater fluctuations. A lysimeter is essentially a tank with a pervious bottom. The bottom arrangement is such that excess soil moisture will drain through the soil, which can be collected and measured. This condition offers an advantage over a tank because it prevents accumulation of water at the bottom to cause unnatural growing conditions for the plants. The consumptive use is the difference between the amount of water applied to the lysimeter and the amount draining out along with an adjustment for moisture content.

ET from crop areas can be estimated by using lysimeters, physically-based methods, or by using empirical equations. ET can be directly measured through a lysimeter. Lysimeters can be divided in three classes:
(*a*) Weight type,
(*b*) Hydraulic based, and
(*c*) Volumetric based.
A weighing type lysimeter uses mechanical balance to determine the change in the water content of the control volume. The hydraulics-based equipment employs hydrostatic principles of weighing. In the volumetric based lysimeters, ET is measured by the amount of water added or removed from the control volume to keep constant water content. The site selected for installation of lysimeters should be typical of the surrounding area with respect to soil characteristics, slope, and vegetative cover. Although lysimeters are difficult to construct and maintain, if properly maintained, these can provide quite accurate data.

Specially designed field plots are also used to determine ET under field conditions. Surface runoff from these plots is collected and measured; water input in the form of precipitation or irrigation is measured. Deep percolation is captured by underground drains. To determine ET, losses as runoff or deep percolation are subtracted from the water input and a correction is made for soil moisture.

Measurement of ET with a lysimeter is time-consuming and needs careful planning. Installation and maintenance of a lysimeter requires skilled manpower, instruments, and finances. Due to these reasons, indirect methods based on climatological data are frequently used for estimation of ET_0. This requires measurements of meteorological variables which influence evaporation. Empirical equations have been developed to predict consumptive use for different crops and different conditions. These relations have helped by replacing difficult and expensive measurements. In India, gravimetric/volumetric lysimeters have been installed at nearly 40 stations for ET measurements. Besides, there are some installations maintained by academic institutes.

To estimate ET from a well-watered agricultural crop, reference ET from a standard surface (ET_0) is first estimated. This value is multiplied by an empirical crop coefficient to obtain the ET from the crop (ET_c). The crop coefficient accounts for the difference between the standard surface and the crop. Reference ET is expressed in the units of depth/time, e.g., mm/day. It is a climatic parameter expressing the evaporative power of the atmosphere at the given space and time coordinates. Crop and soil features are not involved in computing it.

Numerous reference ET equations have been developed and are being used depending upon the availability of weather data. These equations range in sophistication from empirical solar radiation- or temperature-based equations to complex resistance-based equations. Blaney and Criddle (1962) developed an empirical relation between ET, mean air temperature, and mean percentage of day-time hours. This relation has been used extensively. The underlying assumption of this procedure is that the heating of the air and evaporation share the heat budget in a fixed proportion. As a result, ET varies directly with the sum of the products of mean monthly air temperature and monthly percentage of daytime hours with an actively growing crop with sufficient soil moisture. Both meteorological and crop effects are included in the Blaney-Criddle method. By accounting for rainfall in the month or season, one can compute the crop irrigation water requirement for a given irrigation efficiency.

Thornthwaite (1948) derived an equation to be used especially in the central U.S. for limited water conditions. This equation produces monthly estimates of ET as a function of mean monthly temperature, and two coefficients which depend on the location. The combination approach of Penman (1948) links evaporation dynamics with the flux of net radiation and aerodynamic transport characteristics of a natural surface. Based on the observations that latent heat transfer in plant stems is influenced not only by these abiotic factors, Monteith (1965) introduced a surface conductance term that accounted for the response of leaf stomata to its hydrologic environment. This modified form of the Penman equation is widely known as the Penman-Monteith (PM) equation.

The PM equation is physically based because it attempts to incorporate the physiological and aerodynamic characteristics of the reference surface. While the use of the modified Penman method (Doorenbos and Pruit, 1977) was earlier recommended by FAO, recent studies have suggested that this method overestimates ET. FAO has now recommended the use of the PM method to compute reference ET from a grass surface and has specified a grass reference ET equation. Recent studies have shown that the reference ET computed using the PM equation yields estimates that are close to observed reference ET values.

FAO-56 Penman-Monteith Method

As described in the Irrigation and Drainage Paper 56 (Allen et al. 1998), the FAO has adopted the Penman-Monteith equation (named here FAO56-PM) as the standard technique to compute reference ET. The FAO56-PM can be used for hourly or daily time steps. For hourly time steps, the equation is stated as (Allen et al. 1998):

$$ET_0 = \frac{0.408\Delta(R_n - G) + \gamma \frac{37}{T_{hr} + 273} u_2 \left[e^0(T_{hr}) - e_a\right]}{\Delta + \gamma(1 + 0.34u_2)} \tag{1}$$

where ET_0 is grass reference ET in mm/hour, R_n is the net radiation at the grass surface in MJ per m^2 per hour, G is the soil heat flux density in MJ per m^2 per hour, T is the mean hourly air temperature in °C, u_2 is the mean hourly wind speed at 2 m height in m/s, $e^0(T_{hr})$ is the saturation vapour pressure in kPa at air temperature T_{hr}, e_a is the actual hourly vapour pressure in kPa, Δ is the slope of vapour pressure versus temperature curve in kPa per °C, and γ is the psychrometric constant in kPa per °C. Allen et al. (1998) have described the procedure and steps for the application of the Penman-Monteith equation for various time step sizes.

4.1.7. Evaporimeters

The U. S. Weather Bureau Class A pan (modified), is used to measure evaporation in India as per I.S. 6939–1992. It is a circular pan of 1.22 m diameter and 0.255 m depth, made out of 1 mm copper sheet painted with white paint. It rests on a white painted wooden stand (Figure 2) and water is filled in it. Water is added daily to keep the levels within a limited range.

Measuring cylinder is used to measure the depth of evaporation. The diameter of the measuring cylinder will be 1/10 that of the pan and the area of the cylinder is kept 1/100 of that of the pan such that the depth of measurement by measuring cylinder is magnified 100 times. The cylinder, which is made of brass, is graduated from 0–200 mm. The 200 mm corresponds to 2 mm of depth of water in the pan. Evaporation pan readings are taken once (equidistant and accumulative) or twice a day (cyclic) at standard times at 08:30 and 17:30 hrs. Daily observations are taken in the morning hours.

Figure 2. U.S. Weather Bureau class A land pan

To prevent drinking of water by birds and animals, either some chemical repellants may be added to water, or the pan may be covered by a galvanised iron wire mesh (22 s.w.g., hexagonal mesh, 1.5 in between opposite sides). The screen tends to reduce pan evaporation by about 14% as compared to that from an un-screened pan. A correction factor of 1.144 is commonly applied to take care of the effect of nets. The pan is mounted on a wooden platform (100 mm high) to allow air circulation below the pan. The water level in the pan changes due to evaporation and rainfall and is used to estimate evaporation. A stilling well with a pointer gauge is installed in the pan. The amount of evaporation between two observations of water level in the pan is obtained by:

$$E = P \pm \Delta d \tag{2}$$

where P is the depth of precipitation during the intervening period and Δd is the depth of water added (+) to or removed (−) from the pan.

Since a pan is a small water body whose material is different from a natural body, its heat storage characteristics and air dynamics are different from a large water body. Therefore, evaporation from a pan is higher than a large open water body. Lake or reservoir evaporation is estimated by multiplying pan evaporation by a coefficient called pan coefficient:

$$E_R = K_P\, E_{pan} \tag{3}$$

where K_P is the pan coefficient, E_R is the depth of evaporation from a reservoir, and E_{pan} is the pan evaporation, both in mm/day. The value of pan coefficient depends on climate, location, season, size, and depth of the water body. This coefficient generally varies from 0.6 to 0.8. The lower values are typical of dry seasons and arid climates where the pan water temperature is less than the air temperature while higher values are typical in humid climates where the pan water temperature is higher than air temperature. In the absence of better estimates, an average value of 0.7 is generally used. Ramasastri (1987) computed open water evaporation using pan – lake coefficients for whole of India based on the evaporation data of 104 evaporimeters.

In 1970s, there were 72 evaporimeter stations in India. For 30 stations, data are available with IMD since 1959 and for 42 stations, data are available since 1961. IMD has installed a network of 232 evaporimeters (indagrimet.org). Besides, many evaporimeters are being maintained by project agencies, academic institutions, and research stations. But detailed information about the number and location of such evaporimeters is not available.

4.1.8. Pan Evaporation Data

Evaporation records from pans are frequently used to estimate evaporation from lakes and reservoirs and ET from an area.

Since information about monthly coefficients is not available in India, these coefficients are decided using the analogy of spatial variation from arid and humid regions to the conditions that prevail in India during winter and summer months. During the transition months, the ambient temperature and the pan temperature are considered to be the same. Also, as it is generally observed that a factor of 0.7 is accepted the world over as the coefficient on an annual basis, the monthly variation has to be plus or minus the value of the annual coefficient. According to these considerations, the monthly coefficients can be adopted as 0.6 in cold dry winter months, 0.8 in hot summer months and 0.7 in the transition months between the winter and the summer and the vice-versa. For this purpose, the parallel of 22° latitude has been taken as the demarcating line. The coefficients together with the months are given in Table 1.

Evaporation rates closely follow climatic seasons, and reach their peak in the summer months of April and May. The central areas of the country display the highest evaporation rates during this period. With the onset of monsoon, there is

Table 1. Variation of Pan Coefficient values with month

Location	Coefficient		
	0.6	0.7	0.8
North of 22° latitude	November–February	March–April and September–October	May–August
South of 22° latitude	December–January	February–March and September–November	May–August

a marked fall in the rate of evaporation. The annual potential evaporation ranges from 150 to 250 cm over most parts of the country. Monthly potential evaporation over the peninsula increases from 15 cm in December to 40 cm in May. In the North-East, it varies from 6 cm in December to 20 cm in May. It rises to 40 cm in June in West Rajasthan. After the onset of monsoon, evaporation decreases all over the country.

India Meteorological Department has published monthly and annual mean evaporation values of 30 stations in India. Rao et al. (1972) have presented monthly and annual evaporation maps for India based on evaporation observations recorded with wire mesh covered class A pans at about 80 stations. A map showing annual evaporation of India is presented in Figure 3. Typical features of evaporation for various months have been presented in Table 2.

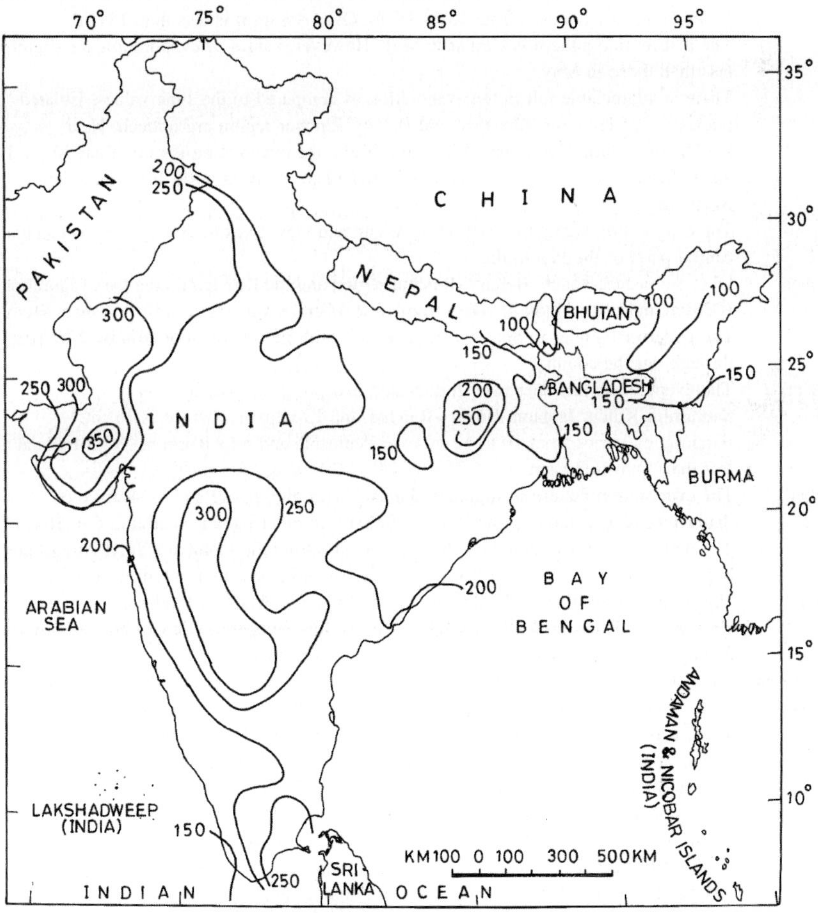

Figure 3. Map showing annual evaporation (cm) (From IMD Evaporation data of observatories of India)

Table 2. Typical features of evaporation for various months

Month/season	Evaporation variation over the country
January	Evaporation is highest (16 cm) over Saurashtra-Kutch. It is about 14 cm at Jalgaon in Maharashtra and Raichur–Bellary in North interior Karnataka. It is 8 to 10 cm over Rajasthan and Peninsular India, 6 cm over Uttar Pradesh and Bihar and less than 6 cm over Assam.
February	There is an increase in evaporation by 2 cm over previous month. The increase is 4 cm in Saurashtra-Kutch and Jalgaon and 6 cm over Raichur -Bellary.
March	The pattern is similar to January and February. Evaporation is 27 to 29 cm over Saurashtra-Kutch, Jalgaon and Raichur-Bellary. Over Rajasthan it varies from 15 cm to 20 cm and around 15 cm in Uttar Pradesh. Evaporation is less than 10 cm over Assam.
April	There is an overall increase in evaporation by 8 to 10 cm over the previous month.
May	Evaporation is the highest in this month. It varies from 12 cm over Assam to 50 cm over Jalgaon. Over Kota and Saurashtra-Kutch it is 40 cm. In north India, it varies from 20 to 35 cm. Gaya with its characteristic semi-arid climate is isolated with 35 cm. In the Peninsula it varies from 15 to 35 cm. Over Assam it is less than 15 cm.
June	The distribution pattern is similar to May. However, values of evaporation are slightly less than those in May.
July	There is appreciable fall in the evaporation as compared to the June values. Isolated pockets of 25 cm over Rajasthan and Bellary-Raichur region are noticed. Near Kovilpatti in Tirunalveli district in Tamil Nadu, the evaporation is more than 30 cm. It varies from 10 to 25 cm in the peninsula and 12 to 25 cm in north India including North-east.
August	The evaporation varies from 10 cm in Assam and west coast to 20 cm in Rajasthan and central parts of the Peninsula.
September	Over Saurashtra-Kutch, Bellary -Raichur region and the Kovilpatti area the evaporation is 20 cm and rest of India it varies from 10 to 16 cm. Over Assam, it is less than 10 cm.
October	The pattern is similar to September. However, with reduced evaporation by 2 to 5 cm through out the country.
November	The evaporation has decreased further and excepting for pockets over Saurashtra-Kutch, Jalgaon, Bellary-Raichur and Kovilpatti where it is 15 cm. Evaporation varies from 10 to 12 cm over Peninsula and 5 to 10 cm in North India and less than 5 cm in Assam.
December	The evaporation pattern is similar to January with high pockets (12–14 cm) over Saurashtra-Kutch, Jalgaon and Raichur-Bellary. In other parts it is around 6 to 10 cm.
Monsoon (June-October)	The pattern generally resembles those observed during the months of July, August and September. Evaporation over West-coast and Assam is less than 60 cm. It is more than 100 cm over Rajasthan and adjoining areas of Gujarat, Madhya Pradesh, and parts of Haryana and Uttar Pradesh. Isolated pockets of high evaporation could be seen around Kovilpatti in Tamil Nadu and Bellary -Raichur in Karnataka.
Non-monsoon (November-May)	The pattern very much resembles the monsoon. Evaporation is less than 100 cm over West-coast, East-coast and Assam and is around 125 cm over Rajasthan and adjoining areas. It is more than 100 cm around Kovilpatti in Tamil Nadu. Isolated pockets of high evaporation of 150 cm are seen around Jalgaon in Maharashtra and Bellary.
Annual	As is to be expected, the pattern is broadly similar to the monthly distribution, with centres of high evaporation over Saurashtra-Kutch, Jalgaon and Kovilpatti in Tamil Nadu. It is lowest, 100 cm over Assam. The 200 cm isoline covers a narrow strip of north-south tract from Punjab to Karnataka through Rajasthan and Maharashtra. Parts of Uttar Pradesh and Bihar States have evaporation ranging from 150 to 200 cm.

Source: Ramasastri (1987).

4.1.9. Evaporation Loss from Reservoirs

Evaporation loss from a water body depends upon climatic parameters of the area, i.e., temperature, sunshine duration, wind velocity, relative humidity, etc. Evaporation losses from reservoirs also depend upon the reservoir geometry and actual water surface area.

The National Commission on Agriculture (1976) estimated the annual evaporation losses from reservoir surfaces on the order of 50 km^2. The Water Management Forum, a national body of the Institution of Engineers (India), has indicated that in the Indian subcontinent the total evaporation loss of water from large, medium and small storage reservoirs would be around 60 km^3.

The National Water Development Agency (NWDA) has estimated evaporation losses from 29 reservoirs using live storage data. These data is given in Table 3.

Table 3. Observed Evaporation Losses from reservoirs

Project	Live storage capacity (Mm^3)	Evaporation loss (Mm^3)	Evaporation loss as % of live storage capacity
Koyna	1,947	186	10
Yeldar	809	123	15
Chitrakot	478	116	24
Bodhghat	3,715	196	5
Kutry I	1,360	211	16
Kutry II	1,391	370	27
Nugur I	1,391	370	27
Inchampalli	4,285	1,567	37
Andhra Valley	67	19	28
Pench	1,088	125	11
Lower Bhawani	777	73.88	10
Pykara	130.3	13.28	10
Parson's Valley	73.5	5.6	8
Upper Bhawani	85.3	3.14	4
Avalanche Emerald	153.9	3.28	2
Pillur dam	34.9	2.32	7
Nellithorai	14.05	2.83	20
Lower Sileru	380	18	5
Balimela	3,823	150	4
Machkund	192	76	40
Nagarjunsagar	6,923	536	8
Srisailam	7,075	722	10
Sriramsagar	2,322	268	12
Tungabhadra	3,701	396	11
Somasila	1,994	164.5	8
Lower Wardha	213	53	25
Jaykwadi	2,170	517	24
Vaigai	193	25	13
Polavaram	2,100	989	47
Total	48,885	7,302	15

Source: NWDA.

Table 4. Water losses due to evaporation (in km^3)

Particular	Year			
	1997	2010	2025	2050
Live capacity-major storage	173.73	211.44	249.15	381.47
Live capacity-minor storage	34.7	42.3	49.8	76.3
Evaporation from major reservoirs @15%	26.1	31.7	37.4	57.2
Evaporation from minor reservoirs @25%	8.7	10.6	12.5	19.1
Total evaporation loss	35.0	42.0	50.0	76.0

Source: NCIWRD (1999).

It was found that the average evaporation loss from hydropower reservoirs was 16% of the live storage and it was 14% of live storage for multipurpose projects. Using these data, a national average value has been estimated and found to be 15% of the live storage capacity. The evaporation losses from minor reservoirs have been estimated at 25% of their live storage capacity.

Using the norms of 15% of live storage capacity as evaporation losses by NWDA, the losses due to evaporation from major and minor storages have been computed. The details have been presented in Table 4.

Monthly distribution of evaporation loss reveals that potential evaporation is higher in the month of May over most parts of the country while the month of January witnesses the lowest value of evaporation. Further, estimates of evaporation loss in the country show that the annual evaporation from a water surface in the semi-arid and acid areas is as high as 2,000 mm. Annual average values of evaporation ranges from 1,400 mm to 1,800 mm over a large portion of the country. An examination of spatial variations of evaporation reveals that values exceeding 2,000 mm occur over west Rajasthan and parts of Saurashtra and Tamil Nadu, while less than 1,400 mm is reported at coastal Mysore, Bihar plateau and east Madhya Pradesh.

4.2. TEMPERATURE

Temperature directly affects two important hydrologic processes: evaporation and snowmelt. In hydrological studies, temperature of air, snow pack, soil, and water bodies is of interest. Temperature is a measure of the ability of the atmosphere and water to receive and transfer heat from other bodies. The temperature of a water body is also an indicator of its quality since it influences the amount of dissolved gases and the rate of chemical and biological activities.

Periodic observations of air temperature are made using thermometers, whereas continuous record is obtained using thermo graphs. Air temperature is normally measured 2 m above the earth surface. Four types of thermometers, dry bulb, wet bulb, maximum and minimum thermometers, are used. The dry bulb thermometer is used to measure temperature of the surrounding air. The wet-bulb thermometer is used to measure the temperature of the saturated air for determining the relative humidity and dew point of the surrounding air. With the help of dry bulb temperature and wet bulb depression, the relative humidity, vapor pressure or dew point are determined. In measuring air temperature, thermometers must be shaded from the sun without restricting natural ventilation. Measurements of air temperature should be accurate to within ±0.3 °C.

Maximum and minimum thermometers indicate the maximum and minimum temperatures attained by the surrounding air since the last measurement. Observations for these four temperatures are made once or twice a day at standard times at 08:30 and 17:30 hours (equidistant and instantaneous for dry and wet bulb and their derivatives; equidistant and statistic for maximum and minimum temperatures). Data from the thermograph is tabulated at hourly intervals (equidistant and instantaneous). In many hydrological studies, continuous temperature records are desirable and can be easily obtained using a thermograph in which changes in the temperature are recorded.

Temperature varies primarily with the magnitude of solar radiation and follows diurnal and seasonal cycles. It is influenced at particular times by the exchange of air masses and by cloudiness, which limits incoming radiation. Temperature varies with latitude which controls solar radiation (Figure 4), altitude, and proximity to oceans. Temperature of places near a large water body, such as sea, is moderated by its influence so that the annual and diurnal range is smaller. Generally, temperatures at nearby stations are strongly correlated. Normally, temperature decreases with altitude at a rate of approximately 0.6 °C per 100 meters for moist air and 0.9 °C for dry air. This rate is known as the lapse rate. The lapse rate for mean daily temperature for the month of January, April, July and October are shown in Figures 5 to 8 respectively.

4.2.1. Temperature Variation in India

Similar to precipitation, temperature also experiences marked variations over the Indian sub-continent. During the winter season from November to February, the temperature decreases from South to North due to the effect of continental winds over most of the country. The mean maximum temperature during the coldest months of December and January varies from 29 °C in some parts of the peninsula to about 18 °C in the North, whereas the mean minimum varies from about 24 °C in the extreme South to below 5 °C in the North. In higher Himalayan zones, temperature remains in sub-zero ranges for many months. The period from March to May is the time of continuous rise of temperature. The highest temperature in May and June in North India, particularly in the desert regions of the northwest may exceed 48 °C. Temperature of similar order can be expected in parts of Andhra Pradesh and Orissa. With the advent of southwest monsoon in June, there is a rapid fall in the maximum temperature all over the country, particularly in the central portions of the country. The temperature is almost uniform over the area covering two-thirds of the country, which gets good rains. From late September onwards, there is a fall

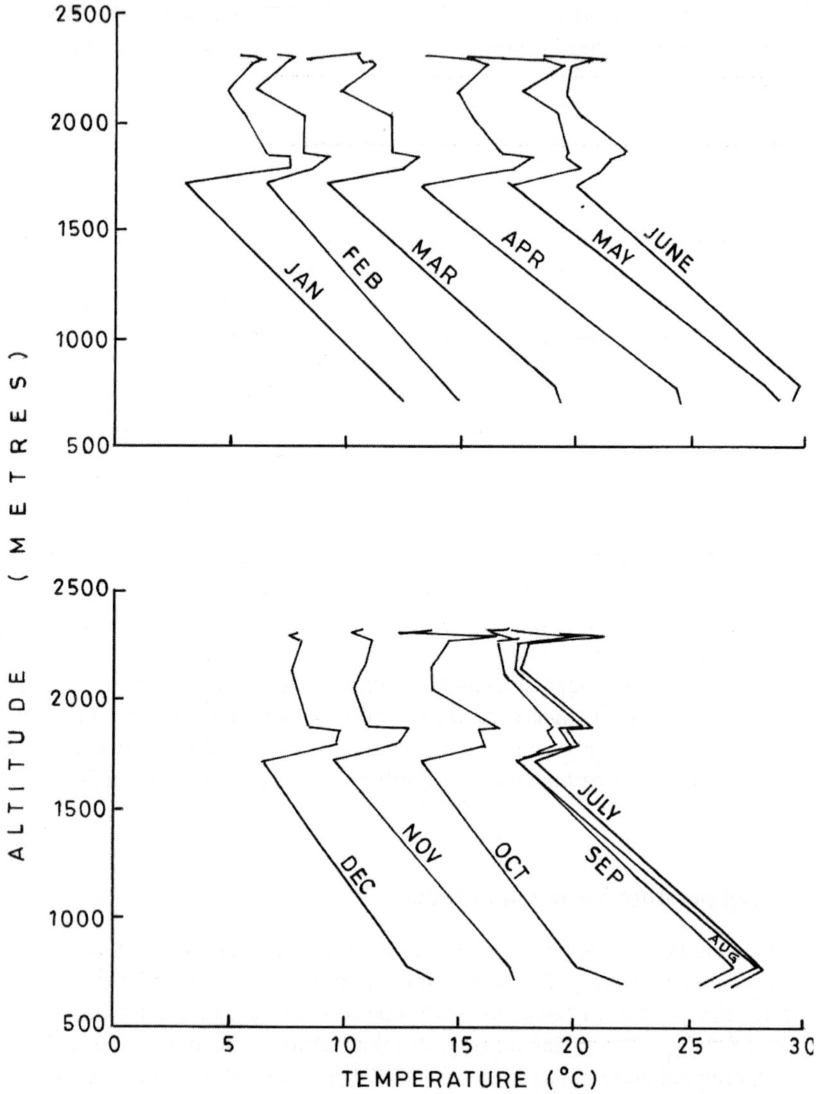

Figure 4. Temperature variation with latitude, altitude, and proximity to oceans

in temperature when the monsoon retreats from North India. In northwest India, the mean maximum temperature is below 38 °C and the mean minimum below 10 °C in the month of November. In the extreme North, temperature drops below freezing point in winter months. The highest maximum annual temperature contours of the country have been shown in Figure 9 and the lowest minimum annual temperature contours of the country have been shown in Figure 10.

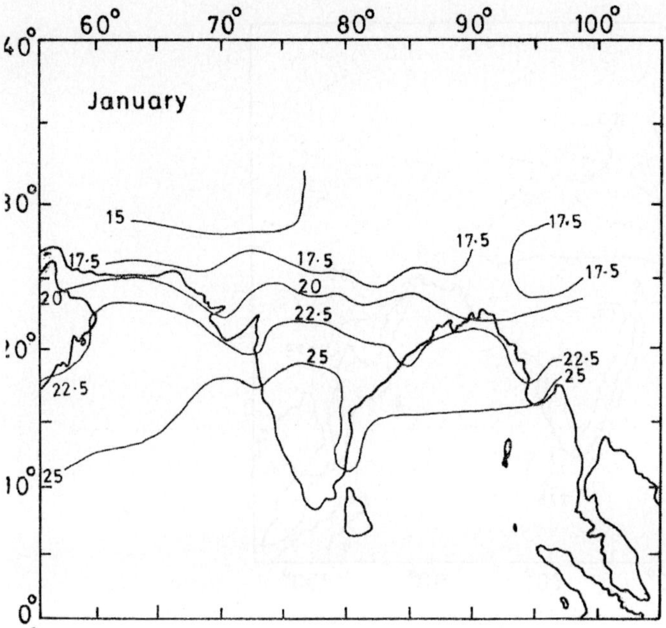

Figure 5. Mean daily temperature (°C) for January, reduced to sea level at ad-hoc lapse rate of 6 °C/km

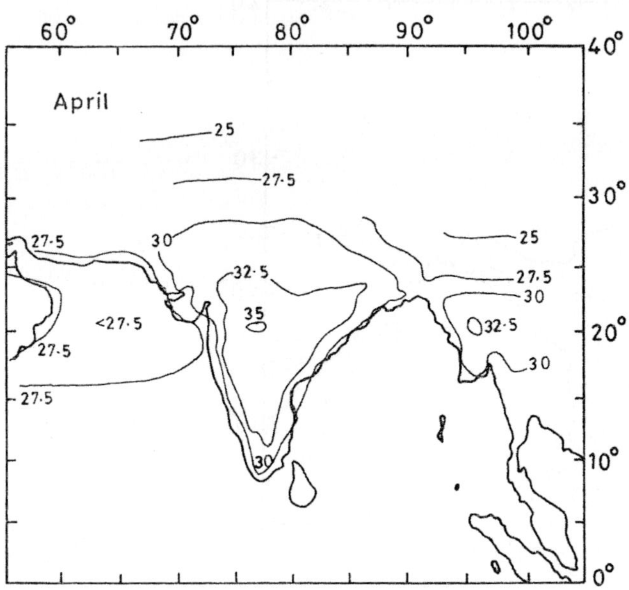

Figure 6. Mean daily temperature (°C) for April, reduced to sea level at ad-hoc lapse rate of 6 °C/km

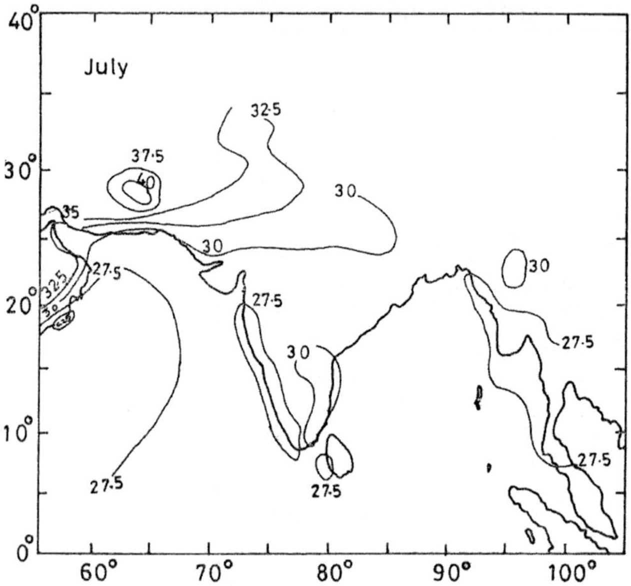

Figure 7. Mean daily temperature (°C) for July, reduced to sea level at ad-hoc lapse rate of 6 °C/km

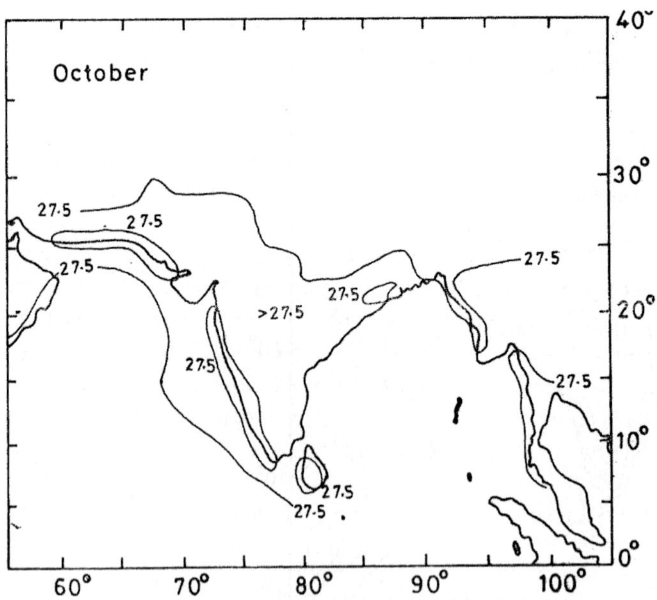

Figure 8. Mean daily temperature (°C) for October, reduced to sea level at ad-hoc lapse rate of 6 °C/km

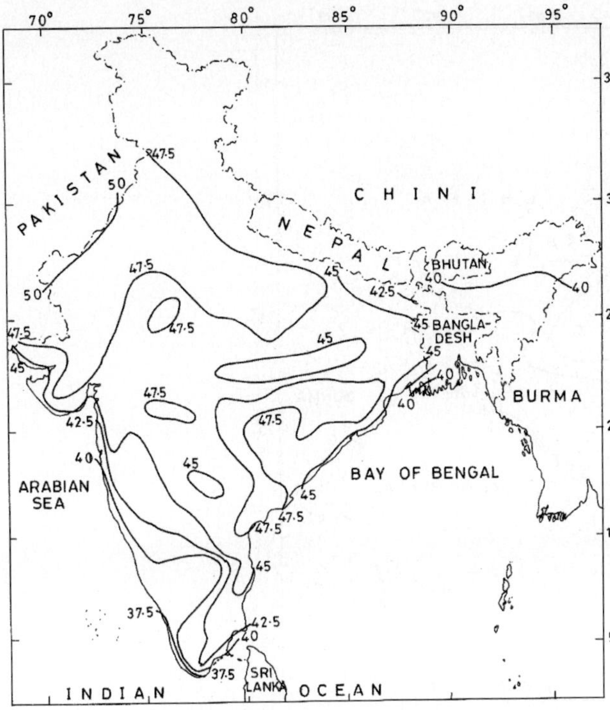

Figure 9. Map showing the highest maximum annual temperature contours of the country

The country has a reasonably good network of temperature recording stations which give a representative good picture of the variation of temperature in space and time. IMD has a network of 560 surface observatories where temperature is measured. The climatological tables of observatories in India (1931–60) issued by the India Meteorological Department contain normals of maximum and minimum annual temperatures (See Table 5). IMD has published the temperature maps for the months of January, April, July and October have also been published by IMD and are presented in Figure 11–14. As the maps of extreme temperatures are based on the data of 60 years or more, they could also be regarded as the highest, and lowest temperatures likely to be recorded. These maps and data should be adequate for getting a general picture of range and the distribution of temperatures in the country.

Figure 15 shows all-India mean series of seasonal and annual surface air temperatures, along with the trend line, based on data during 1901–2000 for a network of 34 stations. The time series indicates warming of 0.3 °C/100 years in the mean annual temperatures of the country as a whole. It can be seen that the past 1940, the cooling reported for the Northern Hemisphere is conspicuously absent in the Indian temperature data. It can also be seen that the warming in the mean annual

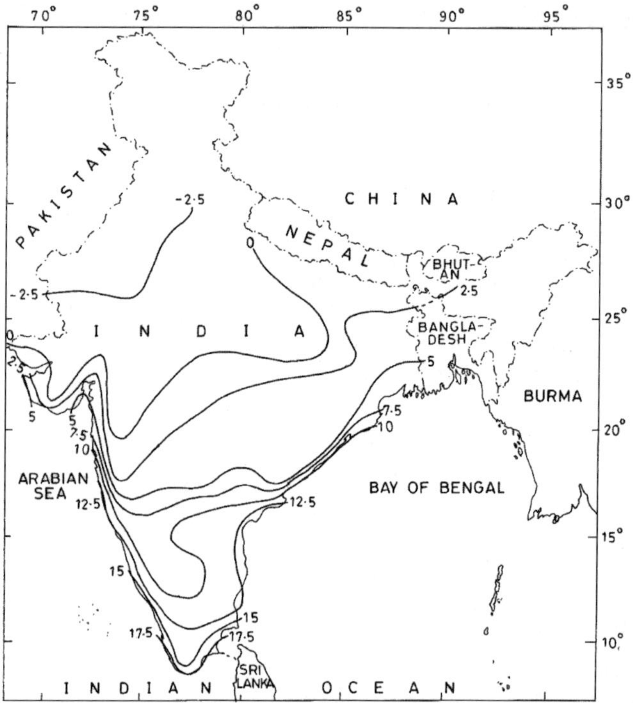

Figure 10. Map showing the lowest minimum annual temperature contours of the country

temperatures is mainly contributed by the post-monsoon and winter seasons. The monsoon temperatures do no show a significant trend on an all-India scale.

4.2.2. Variation of Temperature with Elevation

Very few studies have been reported on the distribution of temperature in the mountainous areas. A fixed value of temperature lapse rate (decrease of temperature with elevation) has been used by some authors for distribution of temperature in various elevation zones in a snow covered basin.

In a study dealing with the variation of temperature in the hills of Uttaranchal, it was observed that the lapse rate of temperature observed in both valley bottoms as well as along mountain slopes, was less than 6.5 °C/km, the mean environmental lapse rate. It was found that both the diurnal and annual range of temperature decreases with ascent from plains up to a height of about 2,400 m and thereafter it again increases attaining the highest value at higher elevations. Variation of temperature lapse rate ranging from 3.8 °C/km to 9.6 °C/km has been reported in the month of August on Chhota Shigri Glacier in Himachal Pradesh.

Singh and Singh (2001) studied the temperature lapse rate pattern in the Satluj catchment using data from three stations Rampur (1,066 m), Kalpa (2,439 m) and

Table 5. Table showing minimum/maximum temperatures of major cities in India

City	Elevation	Max/Min	Jan	Feb	Mar	Apr	May	Jun	Jul	Aug	Sep	Oct	Nov	Dec
Mount Abu	1,219 m	Max	19.3	21.2	25.3	29.4	31.5	29.1	24.3	22.5	24.1	26.6	24.1	21.2
		Min	9.3	11.5	15.9	20.0	22.3	20.5	19.3	18.3	18.4	17.4	13.5	11.2
Agra	169 m	Max	22.2	25.7	31.9	37.7	41.8	40.5	34.8	32.8	33.2	33.3	29.2	24.1
		Min	7.4	10.3	15.7	21.6	27.2	29.5	27.0	25.8	24.6	19.1	12.0	8.2
Ahmedabad	56 m	Max	28.7	31.0	35.7	39.7	40.7	38.0	33.2	31.8	33.1	35.6	33.0	29.6
		Min	11.9	14.5	18.6	23.0	26.3	27.4	25.7	24.6	24.2	21.2	16.1	12.6
Aurangabad	579 m	Max	29.3	31.8	35.6	38.3	39.8	34.7	29.5	29.2	29.8	31.6	30.3	28.9
		Min	14.3	16.1	20.1	23.8	25.0	23.2	21.8	21.2	20.9	19.6	16.1	13.8
Bangalore	921 m	Max	26.9	29.7	32.3	33.4	32.7	28.9	27.2	27.3	27.6	27.5	26.3	25.7
		Min	15.0	16.5	19.0	21.2	21.1	19.7	19.2	19.2	18.9	18.9	17.2	15.3
Bhopal	501 m	Max	25.7	28.5	33.6	37.8	40.7	36.9	29.9	28.6	30.1	31.3	28.5	26.1
		Min	10.4	12.5	17.1	21.2	26.4	25.4	23.2	22.5	21.9	18.0	13.3	10.6
Bodhgaya	113 m	Max	24.2	27.0	33.7	39.0	41.3	38.6	33.5	32.4	32.6	31.2	28.5	25.1
		Min	10.1	12.2	17.4	23.0	27.4	28.2	26.3	25.9	25.6	21.7	14.1	9.9
Bombay	11 m	Max	29.1	29.5	31.0	32.3	33.3	31.9	29.8	29.5	30.1	31.9	32.3	30.9
		Min	19.4	20.3	22.7	25.1	26.9	26.3	25.1	24.8	24.7	24.6	22.8	20.8
Calcutta	64 m	Max	26.8	29.5	34.3	36.3	35.8	34.1	32.0	32.0	32.3	31.8	29.5	27.0
		Min	13.6	16.5	21.5	25.0	26.5	26.7	26.3	26.3	26.1	23.9	18.4	14.2
Cherrapunji	131 m	Max	15.8	16.9	20.5	22.0	22.1	22.9	22.2	22.5	22.9	22.4	19.7	17.0
		Min	7.6	10.5	12.9	15.1	16.3	17.3	18.4	18.4	18.1	15.9	11.9	8.8
Cochin	Sea-level	Max	30.6	30.7	31.3	31.4	30.9	29.0	28.1	28.1	28.3	29.2	29.8	30.3
		Min	23.2	24.3	25.8	26.0	25.7	24.1	23.7	24.0	24.2	24.2	24.1	23.5
Darjeeling	2,134 m	Max	9.3	11.1	14.8	18.0	18.6	19.3	19.8	19.8	19.9	18.6	15.3	11.9
		Min	3.0	4.3	7.7	10.8	12.9	14.7	15.4	15.4	14.6	11.5	7.4	4.4
Dehradun	682 m	Max	19.1	21.4	26.4	32.1	36.2	35.3	30.4	29.5	29.6	28.2	24.7	20.9
		Min	6.1	8.2	12.4	17.0	21.5	23.6	23.1	22.7	21.3	16.1	10.3	7.0
Guwahati	55 m	Max	23.4	26.7	30.0	31.9	31.1	31.2	31.7	32.1	31.7	30.1	27.5	24.8

(Continued)

Table 5. (Continued)

City	Elevation	Max/Min	Jan	Feb	Mar	Apr	May	Jun	Jul	Aug	Sep	Oct	Nov	Dec
		Min	9.8	11.6	16.0	20.1	22.8	24.9	25.7	25.8	24.7	22.1	17.3	11.9
Goa	1.022 m	Max	31.5	31.8	31.8	32.8	32.8	30.5	28.9	29.1	29.4	31.1	32.8	32.7
		Min	19.4	16.7	23.1	25.2	26.8	24.5	23.9	24.0	23.6	23.4	22.3	21.0
Hyderabad	542 m	Max	28.6	31.2	34.8	36.9	38.7	34.1	29.8	29.5	29.7	30.3	28.7	27.8
		Min	14.6	16.7	20.0	23.7	26.2	24.1	22.3	22.1	21.6	19.8	16.0	13.4
Indore	564 m	Max	26.0	28.9	33.7	37.7	39.9	35.7	29.5	28.2	29.3	31.1	28.8	26.5
		Min	9.6	11.0	15.3	20.4	24.8	24.4	22.6	22.0	21.0	17.2	12.1	9.9
Jaipur	431 m	Max	22.0	25.4	31.0	36.5	40.6	39.2	34.1	31.9	33.2	33.2	29.4	24.4
		Min	8.3	10.7	15.5	21.0	25.8	27.3	25.6	24.3	23.0	18.3	12.0	9.0
Jaisalmer	225 m	Max	23.6	28.2	32.6	37.8	41.6	41.4	37.7	35.9	36.3	36.1	31.1	26.0
		Min	7.9	10.6	16.9	21.4	25.8	27.3	26.5	25.5	24.5	20.1	13.0	8.9
Jodhpur	221 m	Max	24.6	27.9	33.3	38.3	41.6	40.1	35.7	33.2	34.7	35.7	31.4	26.7
		Min	9.5	12.0	17.1	22.4	27.3	28.5	26.8	25.2	24.1	19.6	13.9	10.7
Kargil	3,048 m	Max	−4.2	−1.6	4.7	14.0	21.6	25.7	29.7	28.9	24.9	18.5	10.4	1.2
		Min	−13.3	−12.1	−5.3	3.4	9.4	13.4	17.7	17.2	12.5	15.4	−1.3	−7.9
Leh	3,170 m	Max	−2.8	0.8	6.4	12.4	17.1	21.1	24.7	24.4	20.9	14.2	7.8	1.6
		Min	−14.0	−11.8	−6.3	−1.2	2.8	6.7	10.2	9.6	5.4	−0.9	−6.6	−11.1
Lucknow	120 m	Max	23.3	26.4	32.9	38.3	41.2	39.3	33.6	32.5	33.0	32.8	29.3	24.8
		Min	8.9	11.5	16.3	21.8	26.5	28.0	26.6	26.0	25.1	19.8	12.7	9.1
Madras	26 m	Max	28.8	30.6	32.7	34.9	37.6	37.3	35.2	34.5	33.9	31.8	29.2	28.2
		Min	20.3	21.1	23.1	26.0	27.8	27.6	26.3	25.8	25.4	24.4	22.5	21.0
Madurai	107 m	Max	30.2	32.4	35.0	36.3	37.5	36.7	35.7	35.3	35.0	33.0	30.6	29.7
		Min	20.9	21.6	23.4	25.4	26.3	26.3	25.7	25.2	24.8	24.0	23.0	21.6
Mussorie	2,006 m	Max	10.1	11.9	16.2	21,0	24.8	24.0	20.8	20.2	20.0	18.7	15.8	12.7
		Min	2.5	3.7	7.2	11.8	15.0	16.4	16.0	15.5	14.3	11.1	7.4	4.3
Mysore	767 m	Max	28.3	31.2	33.5	34.0	32.6	29.0	27.3	27.9	28.7	28.4	27.4	27.0
		Min	16.4	18.2	20.2	21.4	21.2	20.2	19.7	19.6	19.3	19.6	18.3	16.5

Nagpur	311 m	Max	28.6	32.5	36.4	39.7	42.8	38.4	31.2	30.4	31.5	31.9	30.0	28.7
		Min	12.7	15.1	19.1	23.9	28.4	26.9	24.0	23.7	23.1	21.0	14.1	12.1
New Delhi	239 m	Max	21.3	23.6	30.2	36.2	40.5	39.9	35.3	33.7	34.0	33.1	28.7	23.4
		Min	7.3	10.0	15.1	21.0	26.6	28.7	27.2	26.0	24.6	18.7	11.8	8.0
Ootacamund	2,286 m	Max	19.9	20.6	21.9	22.0	21.8	18.0	16.4	17.3	18.2	18.7	18.9	19.7
		Min	5.0	6.3	8.4	10.2	11.2	11.0	10.9	10.9	10.4	10.0	8.3	6.3
Patna	53 m	Max	23.6	26.3	32.9	37.6	38.9	36.7	32.9	32.0	32.3	31.9	28.9	24.9
		Min	11.0	13.4	18.6	23.3	26.0	27.0	26.7	26.6	26.3	23.0	16.0	11.7
Pune	559 m	Max	30.7	32.9	36.0	37.9	37.9	32.0	27.8	27.7	29.2	31.8	30.8	30.1
		Min	12.0	13.3	16.8	20.6	22.6	23.0	22.0	21.5	20.8	19.3	15.0	12.0
Puri	6 m	Max	27.0	28.3	30.0	30.7	31.6	31.7	30.6	31.0	31.4	31.2	29.3	27.2
		Min	18.0	20.8	24.6	26.6	27.7	27.4	26.7	26.8	26.6	25.0	20.8	17.7
Shillong	1,496 m	Max	15.5	17.0	21.5	23.8	23.7	23.7	24.0	24.1	23.6	21.8	18.9	16.4
		Min	3.6	6.4	10.5	14.0	15.5	17.4	18.0	17.8	16.6	13.0	7.7	4.5
Shimla	2,205 m	Max	8.5	10.3	14.4	19.2	23.4	24.3	21.0	20.1	20.0	18.0	15.0	11.3
		Min	2.0	3.1	6.8	11.2	15.0	16.2	15.6	15.2	13.8	10.8	7.3	4.2
Srinagar	1,768 m	Max	4.4	7.9	13.4	19.3	24.6	29.0	30.8	30.0	28.3	22.6	15.5	8.8
		Min	−2.3	−0.8	3.5	7.4	11.2	14.4	18.4	17.9	12.7	5.7	−0.1	−1.8
Tiruchirapalli	78 m	Max	30.1	32.7	35.1	36.7	37.1	36.4	35.5	35.0	34.2	32.3	29.9	29.3
		Min	20.6	21.3	22.9	25.8	26.4	26.5	25.9	25.4	25.0	24.0	22.7	21.3
Trivandrum	61 m	Max	31.3	31.7	32.5	32.4	31.6	29.4	29.1	29.4	29.9	29.9	30.1	30.9
		Min	22.3	22.9	24.2	25.1	25.0	23.6	23.2	23.3	23.3	23.4	23.1	22.5
Udaipur	577 m	Max	24.2	27.6	32.3	36.0	38.6	35.9	30.7	29.3	30.9	32.0	29.1	26.3
		Min	7.8	9.7	15.1	20.2	24.9	25.3	23.9	22.9	22.1	18.9	11.0	8.3
Varanasi	81 m	Max	23.4	26.6	33.4	38.6	41.5	39.1	33.5	32.2	32.7	32.5	28.6	24.4
		Min	9.5	12.0	17.2	22.4	27.0	28.3	26.5	26.0	25.5	20.7	13.4	9.7

Source: IMD (1981).

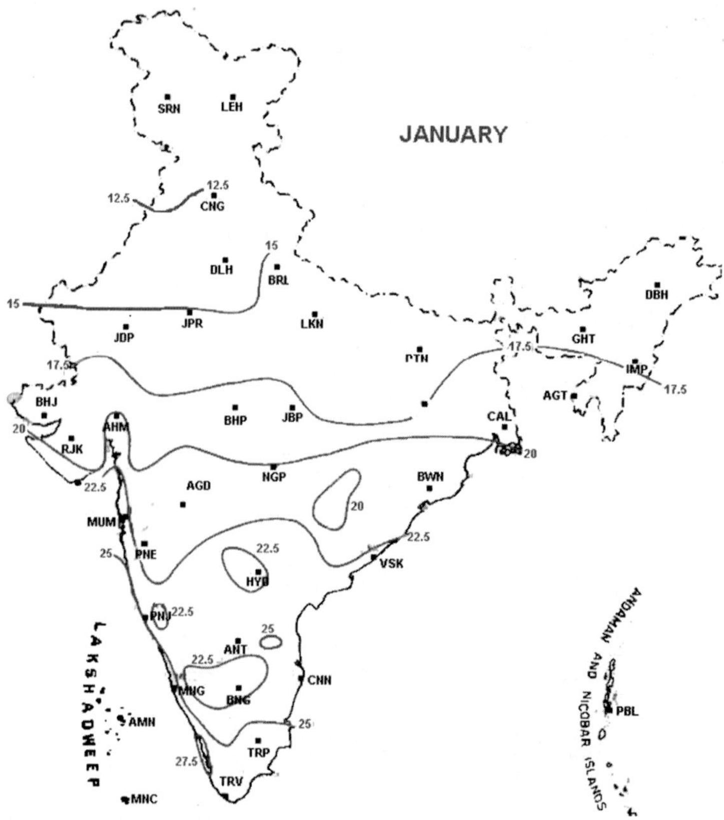

Figure 11. Mean temperature of India for January

Rakcham (3,130 m) located in a contiguous region. It was observed that mean monthly lapse rate varied between 6.7 °C/km to 8.9 °C/km for Rampur-Kalpa and between 3.0 °C/km to 5.3 °C/km for Kalpa to Rakcham. The average values for the snow melt season (March-June) for Rampur – Kalpa and Kalpa – Rakcham are 8.5 °C/km and 4.05 °C/km. respectively. The conclusions of the study should be viewed in light of the limited number of stations used in the study.

4.2.3. Water Temperature

Temperature of water is also input in many studies. Several types of thermometers are used for this purpose: the usual mercury thermometer or thermocouple thermometers, with voltmeter, with or without recorder. The former is used for direct temperature observations. Continuous records may be obtained with resistance or thermocouple elements. For most hydrological applications, precision of ±0.1 °C seems to be adequate.

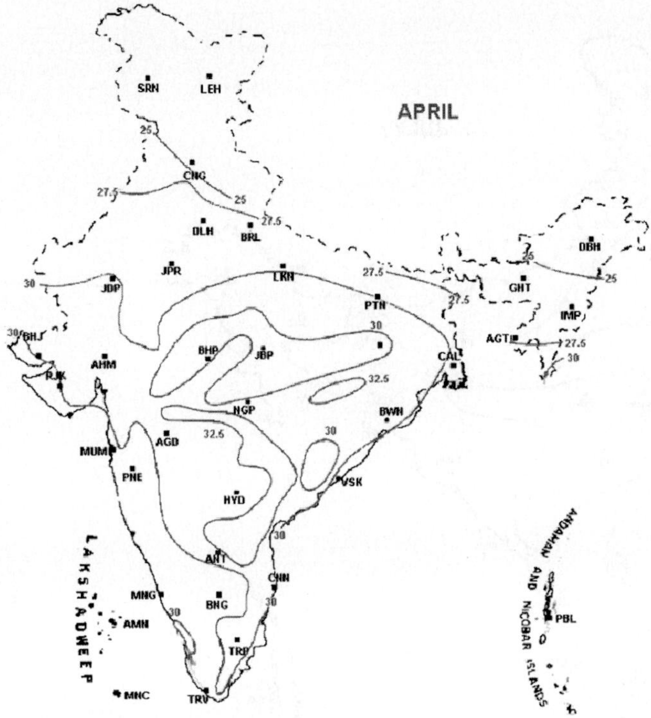

Figure 12. Mean temperature of India for April

4.2.4. Dew-Point Temperature

The temperature to which air must be cooled at constant pressure to reach saturation is called dew-point temperature or simply dew point, T_d. Further cooling will result in condensation of the excess moisture. Thus, dew point is a direct measure of the water vapor pressure. The difference between temperature and the dew point is a measure of the degree of saturation of air. The general application of dew point is made for estimation of precipitable water, height of cloud base, and forecasting of dew, fog, and frost. Following Linsley et al. (1989), the dew-point temperature can be estimated within 0.3 °C in the temperature range 40 to 50 °C as

$$T - T_d = (14.55 + 0.114T)(1 - RH) + [(2.5 + 0.007T)(1 - RH)]^3 + (15.9 + 0.117T)(1 - RH)^{14} \quad (4)$$

where T is in °C, and RH is the relative humidity expressed as a decimal fraction. The maximum persisting 24-hour dew point temperature (°C) in India is shown in Figure 16.

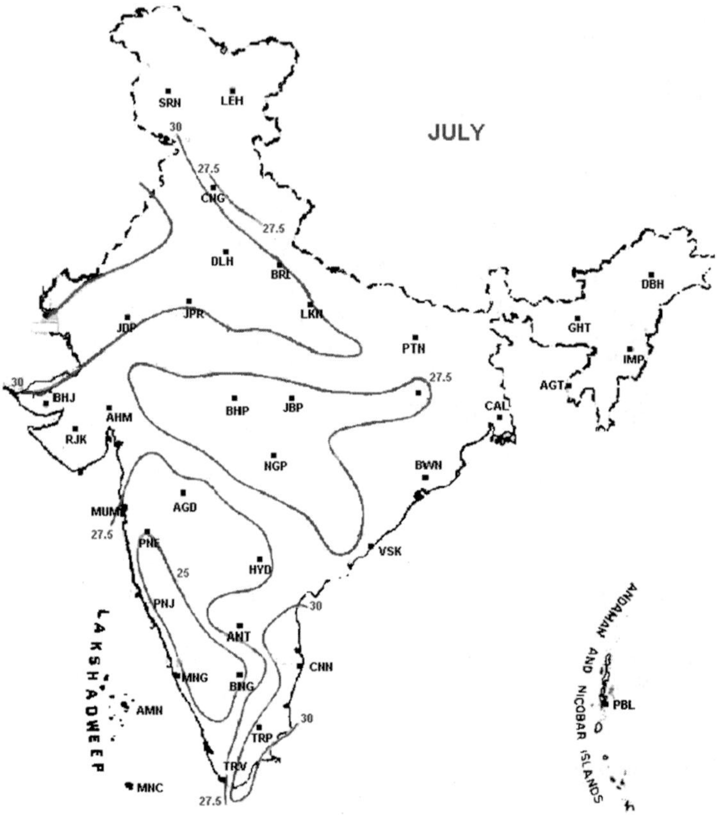

Figure 13. Mean temperature of India for July

4.3. HUMIDITY

Humidity is a general term used to indicate moisture in the atmosphere. Vapor pressure and humidity are intertwined. Absolute humidity (g/m^3) is the mass of water vapor contained in a unit volume of space, and can be determined from the equation of state for an ideal gas. The specific humidity (gm/gm) is the mass of water vapor contained within a unit mass of moist air.

4.3.1. Relative Humidity

Atmospheric humidity has a significant influence on evapotranspiration. The commonly used term *relative humidity* (RH) can be defined as the ratio (%) of the actual vapor pressure of the air to the saturation vapor pressure at the same pressure and temperature. Direct measurement of relative humidity is accomplished using a hygrograph. Indirect estimation of relative humidity is accomplished using

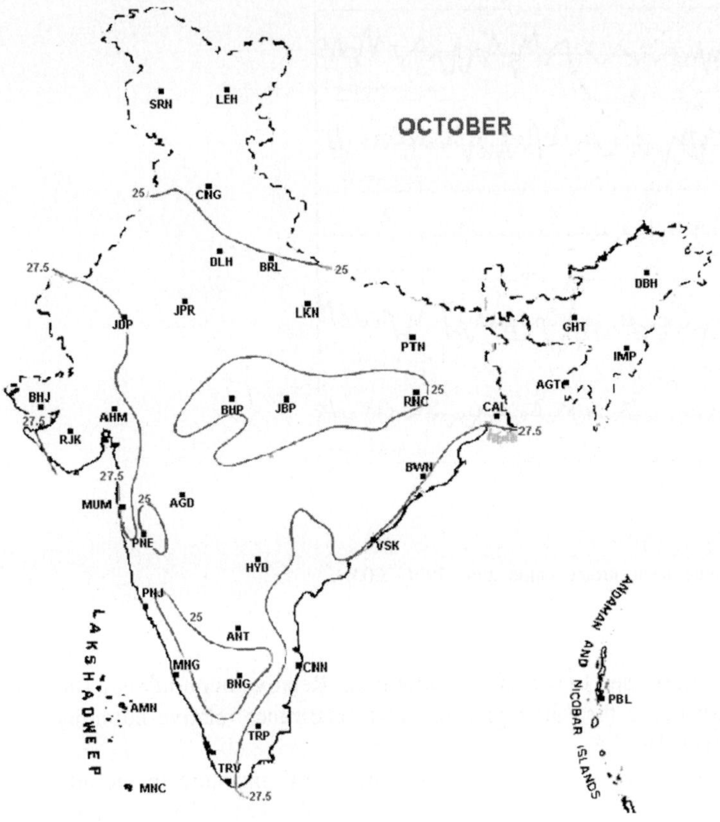

Figure 14. Mean temperature of India for October

dry and wet bulb temperature records. Wet-bulb thermometers must be carefully shielded from radiation. Further, there should be adequate ventilation to obtain a true wet-bulb temperature. The estimated humidity from the dry and wet bulb temperatures are obtained daily (equidistant and instantaneous) or twice daily at standard times at 08:30 and 17:30 hrs. The continuous record obtained from the hygrograph is tabulated at hourly intervals. Measurements of humidity are made at the same location where air temperature is measured.

In other words, RH is the ratio of the actual amount of water vapor present in the sample of air parcel to the maximum possible amount of water vapor in the sample at a given temperature. Generally, the percent value of relative humidity is used, i.e., the ratio is multiplied by 100. It may be expressed as

$$RH = e/e_s \tag{5}$$

where e_s is the saturated vapor pressure. Since most of the atmosphere is unsaturated, it is useful to estimate the degree of saturation. Saturation deficit is one of the

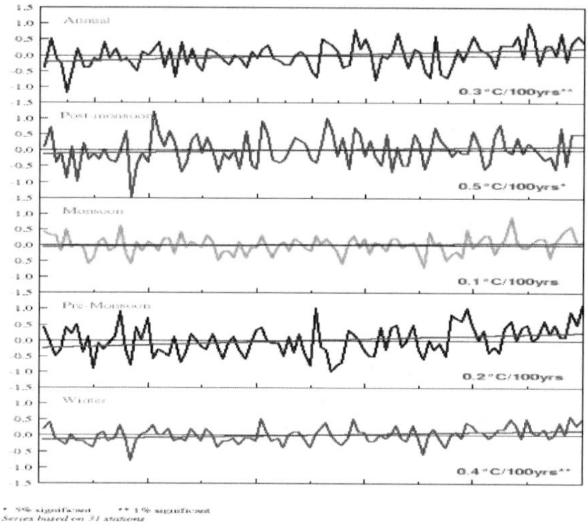

Source: IITM

Figure 15. All – India mean surface temperature (1901–2000)

important parameters used to compute evaporation. Relative humidity is a function of temperature because the vapor pressure that determines relative humidity is a function of temperature.

The relative humidity is not a direct measure of the moisture in the air. It is possible to have 100% relative humidity at 0 °C and 100% relative humidity at 30 °C. Even though the relative humidity is the same in both cases, the amount of water vapour in the atmosphere is far greater at 30 °C than at 0 °C., as seen from Figure 17.

In high latitudes, the relative humidity is found to be about 75–85% due to sustained low temperatures. In the equatorial belt, the relative humidity is also high (75–80%) because of high vapor pressure. In the subtropical belt which consists most of the deserts, a low humidity (20–40%) is observed.

Bosen (1958) has given a simple formula for computing RH in percent directly from air temperature T (°C) and dew-point temperature T^d(°C) as

$$RH = [(112 - 0.1T + T_d)/(112 + 0.9T)]^8 \quad (6)$$

In India, RH values have in general increased significantly over the past several decades. This has been attributed mainly to large increase in irrigated area over the country.

The relative humidity does not vary much over a short time. Places close to sea have higher RH and a smaller daily variation than inland locations. RH is assessed by joint measurements of dry bulb and wet bulb temperatures. The dry bulb refers

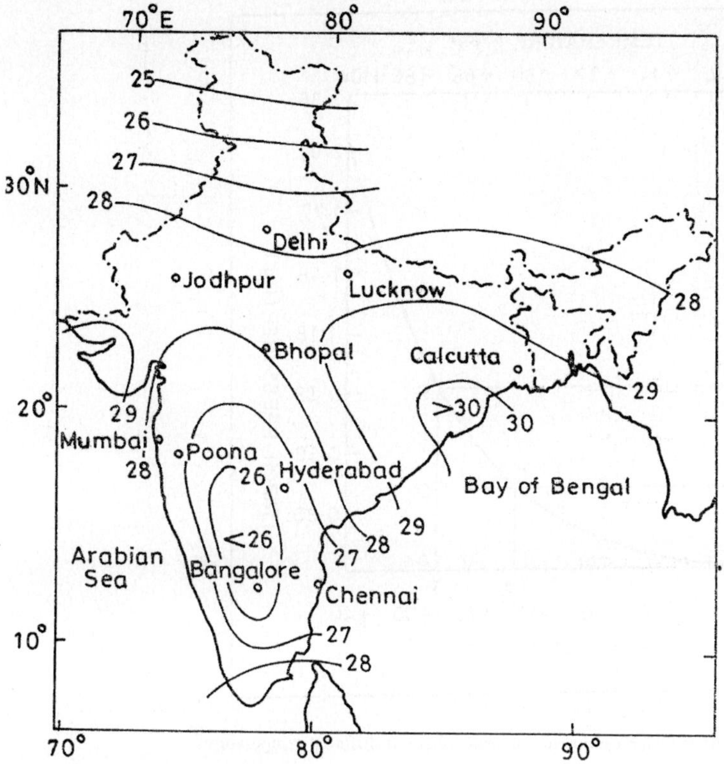

Figure 16. Maximum persisting 24-hr dew point temperature (°C) in India

to an ordinary thermometer. The wet bulb is covered with a clean muslin sleeve, tied around the bulb by a cotton wick which is dipped in a water container so that the wick and muslin are kept constantly moist. From these two measurements, the dew point temperature, and actual and saturated vapor pressures may also be calculated.

While the actual vapor pressure may vary little during the day, RH has a regular diurnal pattern with a minimum normally coinciding with the highest temperature. It also shows a regular seasonal variation. RH is calculated from the wet bulb depression (difference between dry and wet bulb readings). RH may also be measured continuously by means of a hygrograph in which humidity is registered on a chart.

Analysis of RH data over the country shows that RH has considerably increased over last 4–5 decades. One of the reasons behind this increase is tremendous increase in application of irrigation water over the agricultural area, a large percentage of which is evaporated. Figure 18 shows the relative humidity over India in the month of July.

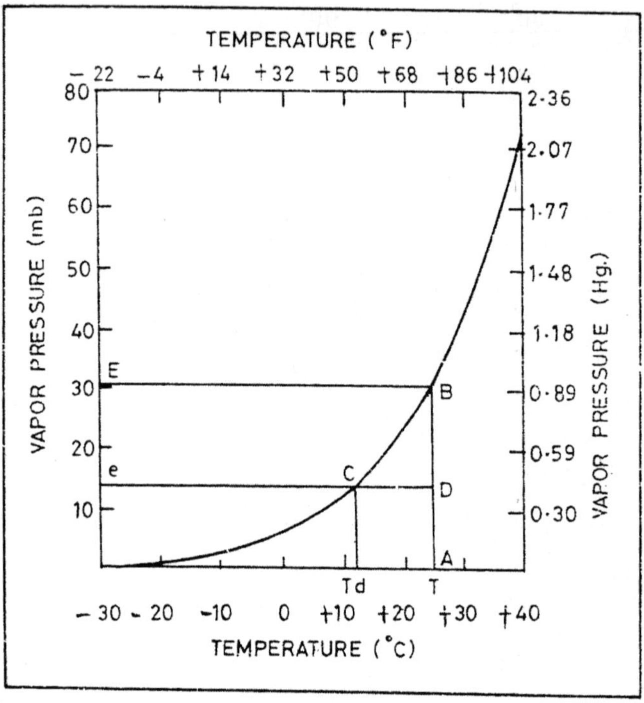

Figure 17. Amount of water vapour in the atmosphere at different temperatures

4.4. SUNSHINE

The sunshine duration is an input variable in the estimation of evapotranspiration in the absence of radiation measurements. The potential maximum sunshine duration depends on latitude and season; the actual sunshine hours vary due to clouds, fog, etc. In urban areas, the amount of bright sunshine may be reduced by atmospheric pollution and in coastal areas it may be reduced by sea mist.

Observation of sunshine duration is commonly made by Campbell Stokes sunshine recorder. It is a glass sphere mounted on a section of a spherical bowl. The sphere focuses Sun's rays on a card graduated in hours, held in the grooves of the bowl which burns the card when the Sun is shining. The recorder uses the movement of the Sun to form the time basis of the record. The card is changed daily after sunset. The lengths of burnt traces on the card indicate the sunshine duration and the data at the required resolution is tabulated.

The duration of bright sunshine is an important factor affecting crop growth. It would, therefore, be useful to have an idea of its distribution (Rao and Ganesan, 1972) which is as follows:

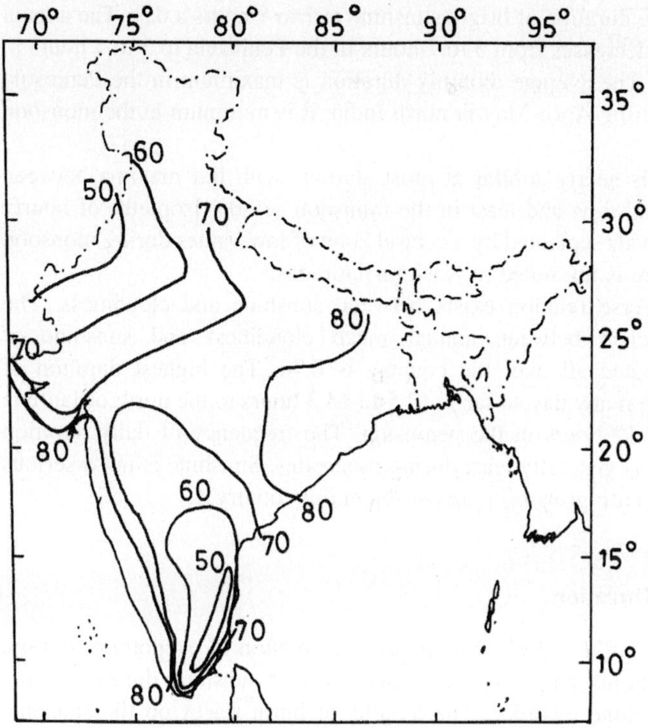

Figure 18. Relative humidity over India in the month of July

1. The daily average in January is more than 7 hours per day except in Kashmir. The western half of the peninsula and adjoining Gujarat areas have the highest average of more than 9 hours per day. Only in the extreme north and over Kashmir, the values are low and average in Srinagar (J & K) is only 2.5 hours per day. The low values in the extreme north are due to the passage of succession of western disturbances during the winter season when clouding is also high. With the advent of summer, the duration of bright sunshine shows a general increase and the average in April is 9 to 10 hours per day over most of the country. The highest averages are over Gujarat and the adjoining regions of Rajasthan and Maharashtra.
2. The pattern changes completely with the onset of the southwest monsoon in June and July. The West Coast has an average of less than 3 hours a day. Mumbai's average is only 2.5 hours. Assam and Bengal and places close to Chennai have averages of 4 to 5 hours a day.
3. By October, the southwest monsoon has withdrawn from most of the country. With a maximum of more than 10 hours over Western Rajasthan, sunshine decreases to less than 6 hours in the extreme northeast and southern parts of the country.

4. The annual average duration of bright sunshine is 7 to 9 hours a day. The annual range of sunshine decreases from 5 to 7 hours in the Peninsula to 3 to 4 hours in northern latitudes. The average monthly duration is maximum in the Peninsula in February and during April-May in north India; it is minimum in the monsoon season.
5. Diurnal variation is nearly similar at most stations with flat maxima between 10 to 15 hours local time and least in the monsoon season. Isopleths of hourly values show two ovals separated by a central zone of low values during monsoon months; this feature is not noted in Kashmir latitudes.
6. A pronounced inverse relation exists between sunshine and cloudiness. The correlation coefficient between annual mean cloudiness and sunshine of 57 stations distributed all over the country is 0.76. The highest duration of sunshine recorded on any day so far is 12.5 to 13.3 hours to the north of latitude 20 °N and 11.5 to 12 hours in the peninsula. The frequency of daily duration exceeding 9 hours is generally least during monsoons. Sunshine is not a serious limiting factor for crop growth in almost the entire country.

4.4.1. Sunshine Duration

The duration of sunshine is a useful variable for estimation of evapotranspiration. The instrument commonly employed for observation of sunshine duration is the Campbell Stokes sunshine recorder. The lengths of burnt traces on the sunshine card indicate the sunshine duration. Sunshine duration data is tabulated for each hourly duration in the day and is considered as of equidistant and accumulative nature.

4.5. WIND

4.5.1. Wind Speed and Direction

Wind speed and direction are inputs for calculation of evapotranspiration. Wind speeds are controlled by local pressure anomalies which in turn are influenced by temperature and local topographic features. Wind speed exhibits a wide variation not only from place to place but also during the day. The wind direction may influence evaporation if the surrounding environment has different humidity in different directions. Figure 19 shows the mean upper winds over India and neighbourhood at 200 mb during April and July.

Wind speed and wind direction are measured using anemometer and wind vane, respectively. Observations are made daily or twice daily at standard times at 08:30 and 17:30 hrs. Wind speed measurement may be instantaneous or, if wind run over a time interval is observed, then it is accumulative. Wind direction may influence measured evaporation totals if the surrounding environment in terms of wetness differs in different directions.

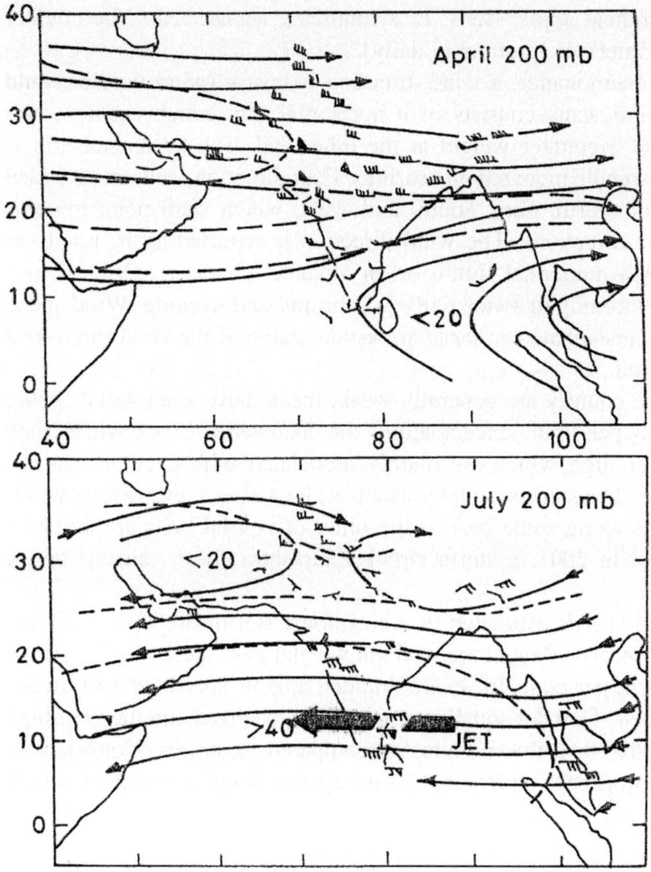

Figure 19. Mean upper winds over India and neighbourhood at 200 mb during April and July

4.5.2. Anemometer

Normally wind speed is measured at a height of 2 m above the ground surface. In India usually a cup type anemometer has been approved by the IMD and also specified by the Bureau of Indian Standrads. The number of rotations of the anemometer over a time interval is displayed by a counter or logged using a data logger and indicates the average speed over that interval. The data is useful in estimating the wave height in the reservoirs and for estimation of evaporation as well as estimation of the consumptive use of crops. Normally, the wind speed over a three-minute period is considered as instantaneous wind speed at that time. The daily wind run or the average wind speed is calculated from counter readings on successive days at the principal observation times. Accuracy of wind measurements by the three-cup anemometers is usually within ±0.5 m/s, which is acceptable for hydrological applications. In a totalizing anemometer, the counter must be read at fixed intervals (preferably daily).

By the side of the anemometer, a wind direction indicator (wind wane) should also be fixed. The wind wane consists of a horizontal arm which carries a flat tail fin at one end and a counter weight at the other end. The horizontal arm is mounted on a vertical spindle mounted on bearings. Four direction arms are provided which carry the letters, *N*orth, *E*ast, *S*outh, and *W*est, which shall point towards the corresponding true directions. The wind direction is reported as 16 points of the compass either as a numerical figure or an alphabet character. Observations are made daily in the morning or twice daily in morning and evening. Wind speed measurements may be instantaneous; these are accumulative if the wind run over a time interval is observed.

Surface winds in the country are generally weak, mean daily wind speeds being less than 10 to 15 km per hour, except along the Saurashtra Coast where they are stronger. Strong or high winds are mainly associated with cyclonic storms, depressions, thunder or dust storms, squally weather along coasts, etc. Strong winds due to cyclonic storms along some part or the other of coastal belts are almost a regular annual feature. In 2001, a storm struck Saurashtra coast, causing severe damages.

The maximum wind speeds estimated in the cyclonic storms are 150 to 250 km per hour. The northwesters of Bengal are well known and gust speeds in the squalls frequently exceed 60 km per hour. In severe thunder squalls, speeds of 160 km per hour have been recorded. Thunder squalls are generally localized and the very high winds in gusts last only for a few minutes as compared to storms. Considerable damage to crops and properties occurs when high speeds winds accompanied with thunder squalls blow over an area.

4.6. ATMOSPHERE PRESSURE

Atmospheric pressure influences the rate of ET and is a useful, though not a critical variable in its estimation. At any point it is the weight of the air column that lies vertically above the unit area. It is usually observed using a mercury barometer for instantaneous data and can also be continuously recorded using a barograph. Observations from barometer are made daily (equidistant and instantaneous) or twice daily (cyclic) at standard times at 08:30 and 17:30 hrs. The thermograph record is tabulated at hourly intervals corresponding to the standard timing of the daily observations (equidistant).

4.6.1. Atmospheric Water Vapor and Its Indices

Water vapor is ever present in the atmosphere although its quantum varies greatly in time and space. This fraction of water vapor is exceedingly important and is largely responsible for prevailing weather conditions. The amount of vapor in the air depends on the temperature. It is a maximum at high temperatures and a minimum at low temperatures. The amount of water in the atmosphere is related to temperature, as seen from Figure 17. The higher the temperature, the more the water vapor air can contain and transport. Because the specific gravity of water vapor is 0.622 of that of dry air, transfer of water vapor in the atmosphere is readily possible. When air is cooled to −30 °C, no water vapor can be carried by the air.

The minimum amount of water vapor that can exist in a space depends on the temperature and does not depend on the existence of other gases. This space becomes saturated when it contains the maximum amount of water vapor at a given temperature. The saturation vapor pressure is the pressure exerted by the vapor in a saturated space. To that end the vapor pressure over water is used most often. Corresponding to the given temperature, this is the maximum vapor pressure. The vapor pressure is expressed as

$$e = e_s - \frac{pc_p}{0.622H_v}(T_a - T_w) = e_s - \gamma(T_a - T_w) \tag{7}$$

4.7. RADIATION

Solar radiation is the primary source of energy. The solar constant is 1.94 cal/cm^2/ min at the edge of the earth's atmosphere with about 0.30 cal/cm^2/min reaching the earth's surface. The position of sun plays a primary role in determining the intensity of solar radiation to be received at a particular location at the earth's surface. The former is represented by the solar altitude or sun angle (Φ). The solar altitude is defined as the angle between sun rays and a tangent to the earth surface at the point of observation. The solar altitude is a complement to the zenith angle, θ, i.e., $\Phi = (90 - \theta)$. The zenith angle is an angle between the direct beam and the vertical from the ground. The important factors that determine the sun's altitude are the latitude of the place, day of the year and time of the day. Daily value of solar radiation incident on a unit horizontal surface at the top of atmosphere as a function of latitude are shown in Figure 20, while the monthly mean values are given in Table 6. Figure 21 shows the daily totals of Solar Radiation in langleys (cal/cm^2)/day at five different latitudes over the northern hemisphere for a transparent atmosphere

A radiometer is employed to observe total radiation received from sun and sky. Incident short-wave radiation on a horizontal surface can be measured with the help of a pyranometer. Typically, pyranometers are covered by glass domes which allow only the radiation in the range 0.3–3.0 μm to reach the pyranometer surface. To measure long-wave radiation, flat-plate radiometers are employed. These instruments measure radiation in all wavelengths. To determine long-wave radiation at a place, the total radiation received from sun and sky is subtracted from the short-wave radiation. Of course, measurements have to be carried at the same site. An instrument, known as net pyrradiometer, can measure the difference between total (short-wave and long-wave) incoming (downward) and outgoing (upward) radiation. A net pyrradiometer should be mounted at least 1 m above the representative vegetation cover (WMO 1994).

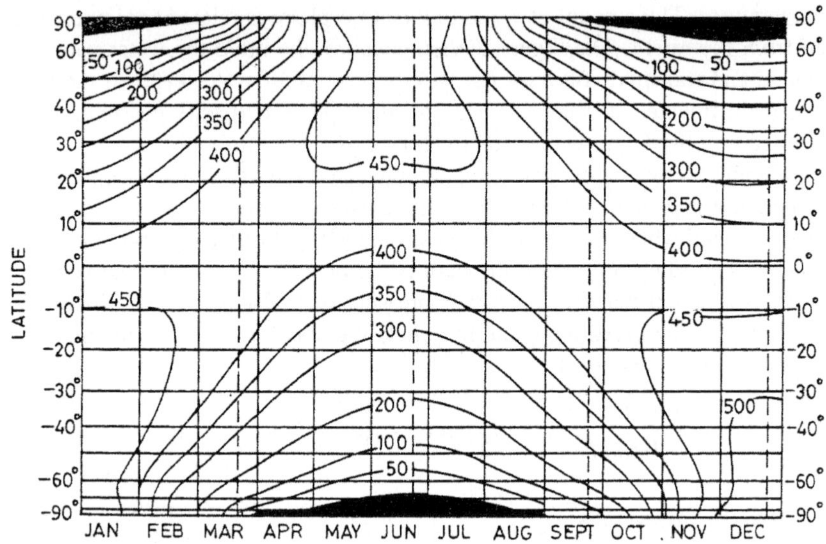

Figure 20. Daily Solar radiation in Wm^{-2} incident on a unit horizontal surface at the top of the atmosphere as a function of latitude and date

4.7.1. Effect of Cloud Cover on Short Wave Radiation

Clouds have the greatest attenuating effect on transmission of solar radiation through the atmosphere. They have an important influence on the both direct and diffuse radiation reaching the ground surface. The influence is highly variable and depends on the type and extent of cloud cover. However, measurements of characteristics

Table 6. Monthly Mean values of the extra-terrestrial radiation, $L_{p\ a}$t the top of the atmosphere in the unit W/m^2 (after Linacre and Hobbs, 1977)

Lat.	Jan	Feb	Mar	Apr	May	Jun	Jul	Aug	Sep	Oct	Nov	Dec
60 °N	40	101	194	319	14	68	450	362	248	133	60	27
50	105	171	259	364	436	475	463	397	304	201	126	89
40	175	239	316	396	453	476	469	423	350	264	195	160
30	245	301	363	423	456	470	466	437	388	321	265	229
20	310	354	400	436	450	453	452	438	413	366	323	296
10	367	398	424	433	431	424	426	429	425	405	376	360
0	416	430	434	421	400	387	390	407	425	429	419	410
10 °S	54	450	432	395	357	336	341	375	414	441	450	452
20	480	457	417	357	307	277	286	329	385	440	469	483
30	494	452	388	311	248	215	227	276	350	425	475	502
40	496	432	350	253	185	149	159	216	301	398	468	508
50	487	403	303	189	117	83	94	149	246	361	452	505
60	471	363	242	124	54	26	34	84	181	312	429	497

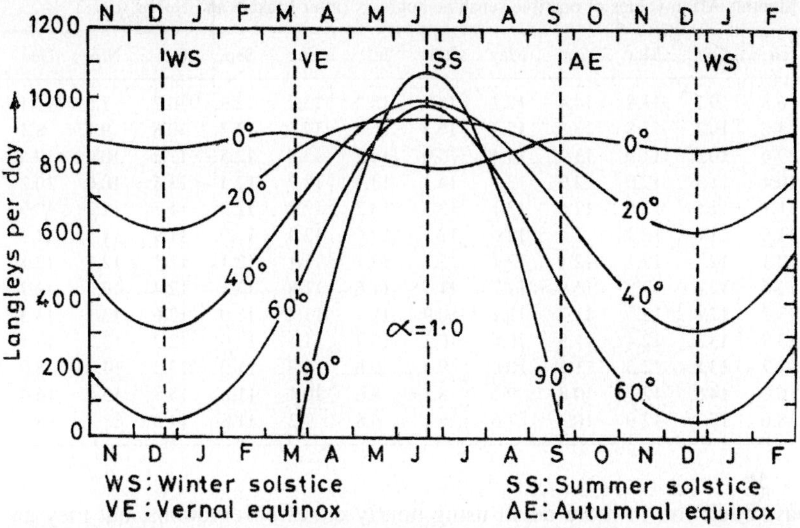

Figure 21. Daily totals of Solar Radiation in langleys (cal/cm^2)/day at five different latitudes over the northern hemisphere for a transparent atmosphere

required to determine the total reflectance and transmittance of cloud are scarce in literature. It is not yet possible to calculate these parameters directly for use in any predictive snowmelt model (Male and Granger, 1981). Several empirical regression relations between daily global radiation and observed fractional clouds or the hours of sunshine have been developed (Angstrom, 1924; Kimura and Stephenson, 1969; Paltridge, 1974; Thompson, 1976). Calculations based on these models show an error on higher side for daily radiation. However, the error is reduced as the number of days for calculations is increased. The following empirical expression based only on sunshine hours is widely used in long-term studies because of its simplicity:

$$G = I_t\left(a + b\frac{n}{N}\right) \tag{8}$$

where G is global radiation reaching the surface, I_t is the incident radiation at the top of the atmosphere, n is the duration of bright sunshine, N is the maximum possible sunshine duration, and a and b are the constants. In this equation, terms aI_t and I_t bn/N represent the diffuse and direct radiation, respectively. For an overcast sky, n/N tends to be zero and only diffuse radiation contributes to global radiation reaching the surface. Commonly used values for a and b are 0.25 and 0.50, respectively. Accordingly, Equation (8) may be written as

$$G = I_t\left(0.25 + 0.50\frac{n}{N}\right) \tag{9}$$

Under clear sky conditions, G equals 0.75 I_t, suggesting a daily mean transmission equal to 0.75. Several investigators have attempted to explain the attenuation of

Table 7. Monthly Mean values of possible sunshine hours, N (after Linacre and Hobbs, 1977)

Lat.	Jan	Feb	Mar	Apr	May	Jun	Jul	Aug	Sep	Oct	Nov	Dec
60°N	6.8	9.2	11.8	14.4	17.1	18.7	18.0	15.7	12.9	10.2	7.6	6.1
50	8.6	10.1	11.8	13.8	15.4	16.4	16.0	14.5	12.7	10.8	9.1	8.1
40	9.6	10.7	11.9	13.2	14.4	15.0	14.7	13.8	12.5	11.2	10.0	9.4
30	10.4	11.1	12.0	12.9	13.7	14.1	13.9	13.2	12.4	11.5	10.6	10.2
20	11.1	11.5	12.0	12.6	13.1	13.3	13.2	12.8	12.3	11.7	11.2	10.9
10	11.6	11.8	12.1	12.4	12.6	12.7	12.6	12.4	12.9	11.9	11.7	11.5
0	12.1	12.1	12.1	12.1	12.1	12.1	12.1	12.1	12.1	12.1	12.1	12.1
10°S	12.6	12.4	12.9	11.9	11.7	11.5	11.6	11.8	12.1	12.4	12.6	12.7
20	13.2	12.8	12.3	11.7	11.2	10.9	11.1	11.5	12.0	12.6	13.1	13.3
30	13.9	13.2	12.4	11.5	10.6	10.2	10.4	11.1	12.0	12.9	13.7	14.1
40	14.7	13.8	12.5	11.2	10.0	9.4	9.6	10.7	11.9	13.2	14.4	15.0
50	16.0	14.5	12.7	10.8	9.1	8.1	8.6	10.1	11.8	13.8	15.4	16.4
60	18.0	15.7	12.9	10.2	7.6	6.1	6.8	9.2	11.8	14.4	17.1	18.7

short-wave radiation in more detail using hourly cloud observations, but they are not of much relevance to hydrological studies in general and snowmelt studies in particular. The values of N are listed in Table 7.

4.8. AUTOMATED WEATHER STATION (AWS)

At inaccessible locations and where frequent visits could not be made, Automated Weather Stations (AWS) are frequently installed for data observation. Almost all the parameters needed for hydrological studies excepting pan evaporation can be monitored by AWS. These AWS monitor and record data at preset time intervals into a micro processor on board the AWS. Data from these AWS can be either transmitted in real-time through satellite or line of sight transmission. The data can also be periodically down loaded on to a computer from the micro processor on board the AWS. In India, numerous AWS have been installed at many places. No comprehensive database of these is available and most of the AWS have been installed for specific purposes. The availability of long-term AWS data, however, has been a problem since the infrastructure in terms of after-sales support is grossly inadequate. AWS are specialized devices and once the equipment develops a fault, it is quite difficult to rectify the fault.

CHAPTER 5
RUNOFF AND STREAMFLOW

Rivers have served as the lifeline for mankind and continue to do so. Streamflow data are the most important hydrological data for surface water analysis. Streamflow records are primarily continuous records of flow passing through a particular section of the stream. These data are analysed to determine the magnitude and variability of surface waters. They constitute input in planning, design, and operation of surface water projects and are also used in design of bridges and culverts, flood forecasting systems, and flood plain delineation. Before describing the use of streamflow data, it is helpful to know how these are observed.

5.1. ACQUISITION AND PROCESSING OF STREAMFLOW DATA

A network of stream gauging stations is established to collect data about surface water resources. The location of gauging sites depends on the purpose of data collection. If the site is needed for a specific project, the general location is in the vicinity of the project. However, if the objective is to study the general hydrology of a region, careful planning is required to identify locations so that optimum information is obtained for the resources deployed.

The number of gauging sites depends on the cost of installation and operation, the value of the data, watershed size, degree of development, objective of data collection, accuracy, hydrologic characteristics, etc. Some of these factors are interrelated. For example, large watersheds involve costlier projects and more data are needed.

River water level (gauge) and discharge are of immense use in water resources management. Gauge or river stage is the water level of a river at a given location and time measured with respect to some datum. A continuous observation of the river water level or stage can be made with comparative ease and economy. Discharge is the volume of flow passing through a section per unit time.

Measurement of discharge in a natural channel is comparatively difficult, time consuming, expensive and requires special skills. Therefore, the discharge at a site is measured less often. Usually, the stage is measured at short intervals while discharge is measured once each day. A relation between stage and discharge at a section, termed as rating curve, is used to transform the observed stages into discharges.

The Bureau of International Standards (BIS) has brought out standards dealing with measurement of flow in rivers. WMO and ISO have also brought out many publications related to streamflow measurement.

5.2. STREAM GAUGING NETWORKS

Every major stream should be gauged at or near its mouth and its major tributaries should also be gauged. According to WMO, the first gauging station is selected at the most upstream location where the drainage area is about 1,300 km^2. The second station is located at a point in the downstream direction where the drainage area is approximately doubled. The WMO recommendations for a minimum density of hydrometric stations are given in Table 1.

Stream gauging stations can be classified in the following categories:

Primary stations

Primary stations or key stations are the stations where measurements are continued for a long time to generate representative flow series of the river system and provide general coverage of a region.

Secondary stations

Secondary stations are essentially short duration stations that are set up only for such a length of period, which is sufficient to establish the flow characteristics of the river relative to those of a basin gauged by a primary station.

Special purpose stations

Special purpose stations are usually set up required for planning and design of projects or special investigations and are discontinued when their purpose is served.

5.2.1. Hydrological Observations

In India, Central Water Commission (CWC) is the central agency which operates a network of 877 hydrological observation stations all over the country. These stations are mostly located on major rivers of the country. The data measured at these stations are used for a variety of purposes such as estimation of yield of the river, water allocation, project planning and management, and flood forecasting. The data observed by field units is processed at various levels and then archived.

Table 1. WMO recommended minimum density of streamflow stations

Physiographic unit	Minimum densities per station (area in km^2 per station)
Coastal	2, 750
Mountainous	1, 000
Interior plains	1, 875
Hilly/undulating	1, 875
Small islands	300
Polar/arid	20, 000

For various national needs, several parameters are being monitored as described in Table 2.

Regarding the frequency of observations, in larger rivers, hourly observation of stage is usually sufficient. For small upland catchments whose data is needed used for special purposes or research, 15 or 30 minute observations may be used to adequately define the hydrograph shape. For the design of minor irrigation schemes and bridge /culvert on small catchments, 15 minute observations may be necessary. The frequency of observation also depends on the type of equipment – data at shorter time intervals may be obtained with automatic equipment while those at longer intervals are obtained at stations where gauging is manual.

5.2.2. Selection of Gauging Sites

After the general location of a gauging station has been determined, its precise location is selected to get the best conditions for stage and discharge measurement

Table 2. Basin-wise hydrological and sediment observation sites of CWC

States/Regions	Gauge	Gauge discharge (GD)	Gauge discharge and silt	GD and water quality	GD, silt, and water quality	Total
East-coast rivers of Andhra Pradesh	24	59	0	24	50	157
Brahmaputra basin	64	27	14	0	12	117
East-coast rivers of Tamil Nadu	0	3	0	13	14	30
East-coast -rivers of Orissa and West Bengal	27	15	0	1	24	67
Ganga basin, Damodar basin and Kangsabati	92	110	6	29	89	326
Indus basin	1	15	9	0	0	25
West-coast rivers of Kerala	0	0	0	3	16	19
Rivers of Meghalaya	0	4	0	0	0	4
West-coast rivers of Gujarat	18	25	0	9	32	84
Rivers of Mizoram and Manipur	5	5	1	0	0	11
Barak and other rivers of Tripura	4	11	11	0	0	26
West-coast rivers of Maharashtra, Goa and Karnataka	1	7	0	1	2	11
Total	236	281	41	80	239	877

Source: Central Water Commission.

and to develop a stable discharge rating. The ideal gauge site should satisfy the following criteria:

(a) The general course of the stream is straight for about 100 m upstream and downstream from the gauge site;
(b) The river should not be braided at the gauge site and all the flow must be confined to single stream at all stages;
(c) The stream-bed is not subject to scour and fill and is free of weeds;
(d) Banks are permanent, high enough to contain floods;
(e) The gauge site is far enough upstream from the confluence or from tidal effect to avoid any variable influence on the stage at the gauge site;
(f) A satisfactory reach for measuring discharge at all stages is available within reasonable proximity of the gauge site; and
(g) The site is readily accessible for ease in installation and operation of the gauging station.

An ideal site is rarely found for a gauging station and judgment has to be exercised in finalizing the site. The detailed guidelines for selection of sites are given in the BIS standard IS:2,752 and IS:5,119 (Part 1).

5.2.3. Measurement of Stage

Stages are measured with reference to a recognized datum, such as the mean sea level. The gauge height is usually expressed in thousandths of a meter. The water level is commonly measured using staff gauges; of late, autographic water level (chart) recorders, and digital type water level recorders have been installed at many sites. The non-recording gauges have low initial cost and are easy to install, but these require an observer and are less accurate. Sometimes, an automatic gauge and a non-recording gauge are maintained together.

According to the recommendations of Code IS:1,192, depth shall be measured at intervals close enough to define the cross-sectional profile accurately. Use of 15 verticals means risking introducing errors of importance and it would be safe to use 25 verticals to observe depth. Velocity observations, particularly with current meter, should be made simultaneously with the depth observations. IS:1,192 prescribes the detailed procedure to use current meters, their calibration, measurements in unsteady flow, etc., specifications and use of floats, computation of discharge, and determining uncertainties.

Staff Gauge

Staff gauges are either vertical or inclined. Vertical staff gauges are normally porcelain enameled iron sections, graduated every 10 mm. If river stage varies over a large range, the gauge consists of stepped sections (Figure 1) installed at different locations in a line normal to the flow. An inclined staff gauge is usually a graduated surface attached securely to a permanent foundation. The rock outcrops on the river bank also make good base for inclined staff gauge. The gauges should be located as close to the measuring section as possible, without affecting the flow conditions.

Figure 1. Sectional staff gauge

Staff gauges are manually read, generally each day in the morning in lean season and at (multi) hourly intervals during high flows.

Water Level Recorder

A water level recorder senses and records water level. The recorders can be classified as either analogue type or digital type, depending on the way the data are recorded. The analogue type recorders produce a graphic record of fluctuations of the water level with time.

The water level recorders are generally of shaft-angular-input type, and the angular rotation of the shaft is recorded. The depth of water surface is sensed for automatic recording by a float in a stilling well (Figure 2) which follows the rise and fall of the water level. A gas-purge system that transmits the pressure head of water in a stream to a manometer is known as a bubble gauge.

A water level recorder gives a continuous record of the water level on a chart from which values are manually extracted at desired intervals. The data from a digital water level recorder can either be at equal intervals of time, say at fraction of an hour, or only when there is a change in water level by more than a pre-set amount. The digital recorders store data in an electronic memory unit and these are downloaded to a computer. A large number of gauging stations have been provided with automatic water level recorders in the recent years.

5.2.4. Measurement of Discharge

Discharge is commonly expressed in cubic metres per second (m^3/s or cumec). The discharge at a site is a function of the cross-section area and flow velocity. The cross-section area is a function of the river stage. At most stations, discharge is measured once a day. Discharge measurement techniques can be

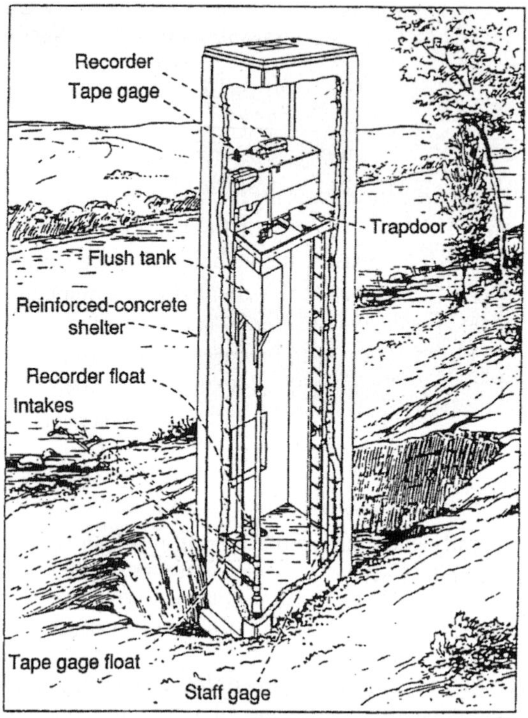

Figure 2. Stilling well for the float-type recorder

broadly classified as: (i) direct determination and (ii) indirect determination. There are many methods under each category.

i *Direct Determination of Discharge*
These are the methods, in which either discharge is directly measured or some variable on which discharge depends is measured. The commonly used methods are: velocity-area methods, dilution techniques, electromagnetic method, and ultrasonic method. The first two are described here.

Velocity-Area Methods: These methods involve measuring the flow area and velocity and these are multiplied to get discharge. Depending on the accuracy required, the width of the stream is divided into a number of vertical portions (Figure 3). In each of these, the velocity is measured at one or more points along the depth to get a representative velocity in that portion. The area of the individual portion can be easily calculated if the bed profile and stage are known. The velocity may be measured by a float, current meter, or by a moving boat.

A *float* is an article that floats on water, such as a wooden log, a bottle partly filled with water, or branch of a tree. For a float measurement, two cross-sections sufficiently far apart on a straight reach of channel are selected. A number of floats

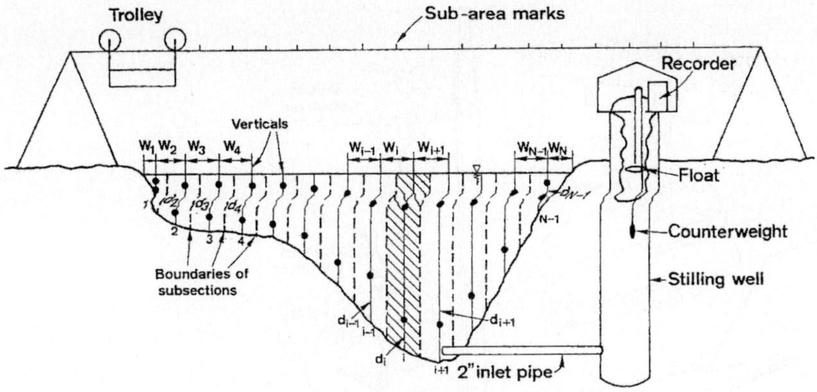

Figure 3. Schematic sketch for a velocity-area station

are introduced uniformly across the stream width a short distance before the actual upstream cross-section so that they loose inertia and move with the velocity of water when they reach the upstream cross-section. The position of each float with respect to distance from the bank is noted. A stopwatch is used to measure their travel time between the end cross-sections of the reach.

The velocity of the float is equal to the distance between the two cross-sections divided by the time taken by the float to cover this distance. The mean velocity in the vertical is equal to the float velocity multiplied by a coefficient whose value depends on the shape of the vertical-velocity profile of the stream and on the depth of immersion of the float with respect to depth. A coefficient of 0.85 to 0.90 is commonly used. The float method is not very reliable and its use is normally restricted when other methods can't be used.

Current meter is the most commonly used instrument to measure the velocity of flowing water. It consists of rotating element (rotor) whose movement is due to the reaction of the stream current. The angular velocity acquired by the rotor is proportional to the velocity of water. By placing a current meter at a point in a stream and counting the number of revolutions of the rotor during a time interval, the velocity of water at that point is determined. Current meters are of two types: those having a propeller rotating around a horizontal axis and those having a series of conical cups mounted around a vertical axis. Both types of current meters are used in India.

Horizontal-axis meters consist of a propeller mounted at the end of a horizontal shaft (Figure 4). The horizontal axis rotor with valves causes fewer disturbances to flow than vertical axis rotors. Furthermore, due to axial symmetry with the flow direction, the rotor is less likely to be entangled by debris than vertical axis rotors and the bearing friction is less compared to the vertical axis rotors. The vertical axis rotor with cups or valves can operate in lower velocities than the horizontal axis meters.

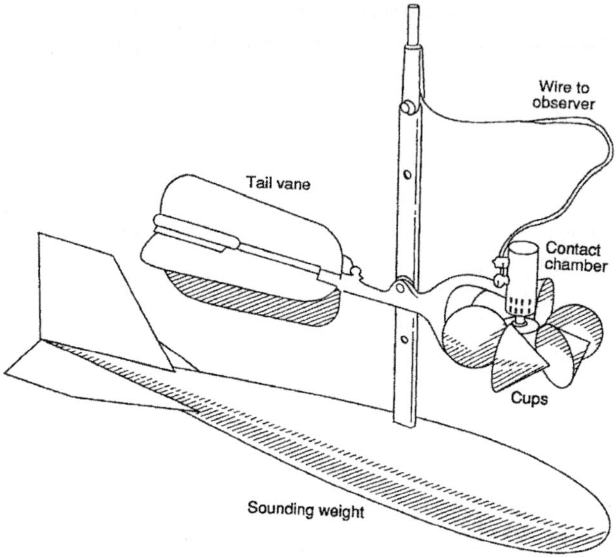

Figure 4. Price current meter

The current meter measurements are usually classified in terms of the means used to cross the stream during measurements, such as wading, cableway, bridge, or boat. Wading is possible in small streams of shallow depth only; the current meter is held at the requisite depth below the surface by an observer who stands in the water. In narrow well-defined channels, a cableway is stretched from bank to bank well above the flood level. A carriage moving over the cableway serves as the observation platform. Bridges are advantageous from the viewpoint of accessibility and transportation, although these are not the best locations hydraulically. The velocity measurement is performed on the downstream of the bridge to minimize the instrument damage due to drift and knock against bridge piers. Boats are most satisfactory for measurements in wide rivers.

In addition to the main current meter, a miniature (Pygmy) meter is used which is best suited for gauging when flow depth is less than 0.5 m and velocity is less than about 1 m/sec. For rivers greater than 10 m wide, at least 20 verticals be used for observation such that the discharge in any one segment does not exceed 10% of the total. Four to five verticals are preferred when channel width is about 1.0 m. Generally for most Indian conditions, an exposure time of 60 seconds can be adopted. If the velocities are very low, the exposure time should be increased to 100 seconds. Alternatively the time it takes to record 20 revolutions should be measured.

Moving Boat Method: In the moving boat technique, data are collected while the observer is aboard a boat traversing the stream along a pre-selected path, generally normal to the direction of flow. During the traverse, an echo sounder records the geometry of the cross-section and a continuously operating current meter senses the combined stream and boat velocities. The angle between the current meter, which aligns itself in a direction parallel to the movement of the water past and the pre-selected path, is also measured.

The velocity observed at each of the observation points in the cross-section (Figure 5), v_v, is the velocity of water past the current meter resulting from both stream flow and boat movement. It is the vector sum of the velocity of water with respect to the stream bed (v) and the velocity of the boat with respect to the stream bed (v_b). The velocity of streamflow can be obtained by measuring the angle α between the selected path of the boat and a vertical vane which aligns itself in a direction parallel to the movement of the water past it.

The flow velocity v, perpendicular to the boat path (true course) at each observation point 1, 2, 3, …, can be determined from the relationship

$$v = v_v \sin\alpha \tag{1}$$

Equation (1) yields that component of the stream velocity which is perpendicular to the true course even though the direction of flow may not be perpendicular.

Since the current meter is usually immersed at a depth of 0.5 m from the water surface, the velocity v corresponds to the surface velocity and not the average velocity in the vertical. This surface velocity is multiplied by a coefficient ranging from 0.85 to 0.95 to obtain the average velocity of flow at the section.

Computation of Discharge: After the cross-section has been selected, the width of the stream is divided into an adequate number of partial sections so as to have lesser variation between two adjacent verticals. If previous measurements have shown uniformity of cross-sections and the velocity distribution then fewer verticals may be taken. It is better if no partial section carries more than 5 to 10 percent of the total discharge. Figure 5 shows the cross section of a river in which verticals are drawn.

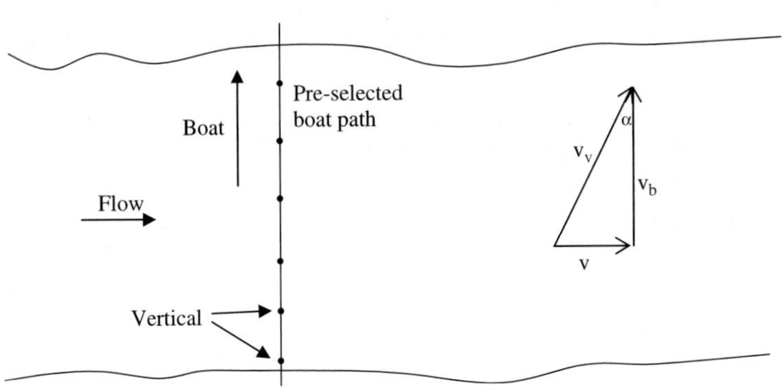

Figure 5. Moving boat method of discharge measurement

The velocity averaged over the vertical at each section is known. Considering the total area to be divided into (n–1) segments, the total discharge is calculated by the method of mid-section as:

$$Q = \sum_{i=1}^{n} (v_i a_i) \tag{2}$$

where Q is the total discharge, a_i is an individual partial cross-section area, and v_i is the mean velocity in that area. The area extends laterally from half the distance from the preceding observation vertical to half the distance to the next and vertically from the water surface to the sounded depth.

ii. *Indirect Determination of Discharge*

These methods make use of the relationship between the flow discharge and the depth at specified locations. The field measurement is restricted to the measurements of depths only. Two important indirect methods are flow measuring structures and slope-area method.

Flow measuring structures, such as notches, weirs, flumes, and sluice gates, are commonly used in laboratories or in field conditions. A typical setup consists of a reasonably straight approach channel, a downstream channel, and the structure itself. The structure having smooth surfaces should be rigid, water-tight, normal to the flow direction, and capable of withstanding peak flows without any damage to its body.

The basic principle behind the flow measuring structures is that these structures produce a unique control section in the flow. At these structures, the discharge Q is a function of the water-surface elevation measured at a nearby upstream location:

$$Q = f(H) \tag{3}$$

where Q is discharge (m^3/s), and H is the head of water (m) at the structure. The equation for weirs, for example, is

$$Q = K\,H^n \tag{4}$$

where K and n are constants. The flows unaffected by the downstream water are known as free flows. The flow that is affected by tail water conditions is known as drowned or submerged flow. Discharge under drowned conditions is obtained by applying a reduction factor to the free flow discharge. For a two-dimensional weir, the discharge is estimated as

$$Q = C_d \sqrt{g}\, b\, H^{1.5} \tag{5}$$

where C_d is the discharge coefficient, g is the acceleration due to gravity, and b is the crest width (m).

Slope-Area Method: In the slope-area method, discharge is estimated by observing the water surface slope and cross-section area. It is an indirect method of discharge estimation which is used when measurement by more accurate methods, such as the velocity-area method, is not possible. The accuracy of the slope-area method is less compared to the velocity-area methods.

A measurement reach is chosen for which three things are known: (i) The cross-sectional geometry and properties at its ends, (ii) the value of Manning's n, and (iii) water-surface elevations at the end sections. In the selected reach, a minimum of three cross-sections are generally desirable. As far as possible, the length of the reach should be such that the difference between water levels at the upstream and downstream gauges is not less than ten times the uncertainty in the difference. Slope is computed from the gauge observations at either end of the reach.

The mean velocity is established by using known empirical formulae which relate the velocity to the hydraulic mean depth, the surface slope corrected for the kinetic energy of the flowing water and the roughness characteristics. The discharge is computed as the product of the mean velocity and the mean cross-sectional area of the flow.

The resistance equation for uniform flow in an open channel, e.g., Manning's formula, can be used to relate the depths at either ends of a reach to the discharge. Figure 6 shows the longitudinal section of a river between two sections, 1 and 2. The head at a section consists of water surface elevation and the velocity head. The head loss is made up of two parts: (i) frictional loss and (ii) energy loss due to expansion or contraction. The friction slope can be written as

$$S_f = \frac{(h_1 - h_2) + \left[\frac{V_1^2}{2g} - \frac{V_2^2}{2g}\right](1-k)}{L} \tag{6}$$

where L is the reach length, k is the coefficient for energy loss; its value is 1 for contractions and 0.5 for expansions. According to Manning's formula, the mean velocity in reach 1–2 is calculated as

$$v_{1-2} = (1/n)R^{2/3}S^{1/2} \tag{7}$$

where R is the hydraulic mean depth, n is Manning's roughness coefficient, and S is the friction slope. If A is the cross-section area, then the discharge Q is

$$Q = (1/n)AR^{2/3}S^{1/2} = K\ S^{1/2} \tag{8}$$

The term $(1/n)AR^{2/3}$ is known as *conveyance* (K) of the channel and it depends on channel characteristics. As the flow in the reach may not be truly uniform, the average conveyance of the reach is expressed as the geometric mean of the conveyances of the two end sections (K_1 and K_2):

$$K = \sqrt{K_1 K_2} \tag{9}$$

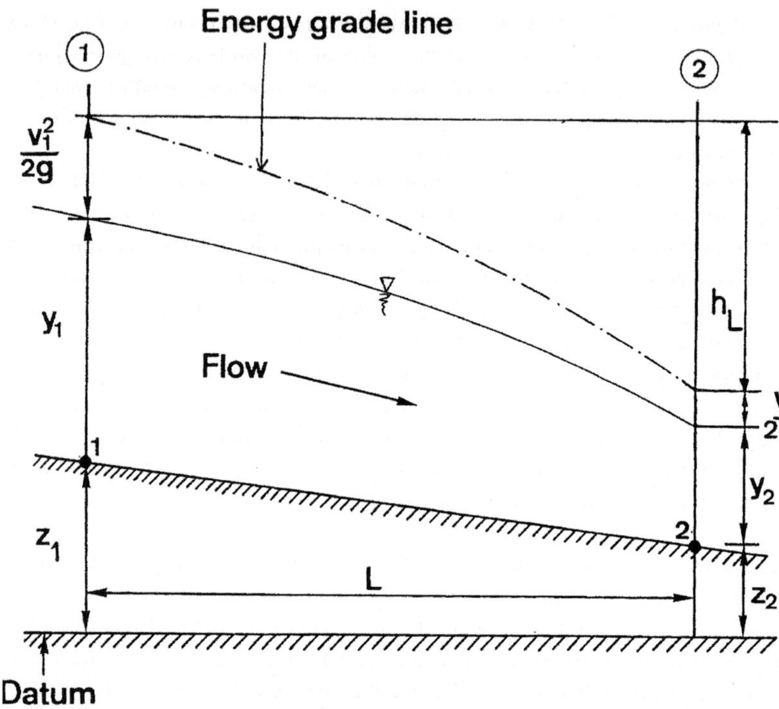

Figure 6. Channel reach for the slope-area method

The discharge can be calculated by

$$Q = K\sqrt{S} = \sqrt{K_1 K_2 S} \tag{10}$$

The slope-area method can be used with some degree of accuracy in open channels with stable boundaries, or in channels with relatively coarse bed material. This method may also be used in other cases, such as alluvial channels including channels with over-bank flow or non-uniform channel cross-sections, subject to the acceptance of large uncertainties involved in the selection of the value of the rugosity coefficient, such as Manning's roughness coefficient n.

The streamflow data are used for a variety of purposes. Some of these are computation of flow duration curves, unit hydrograph analysis, flood or low-flow frequency analysis, computation of the inflow to a reservoir, flow routing, and flood forecasting.

5.2.5. Stage-Discharge Relationship

Stream flow measurement normally involves: 1) obtaining a continuous record of river stage (water level) above a datum, 2) establishing the relationship between stage and discharge and 3) transforming the record of stage into a record of discharge. The measurement of discharge at a gauging station is costly and requires trained manpower, time, and special equipment, and therefore, usually discharge is not measured. The measurement of river stage is much easier and therefore, observations of river stage are commonly taken. Equipment are widely available to automatically measure and store river stage data at pre-determined intervals. The techniques for measurement of stage and discharge are discussed at length in books by Herschey (1978) and Rantz (1982), among others. Most countries including India have developed standards detailing the steps to be followed for such measurements. In addition, several international organizations such as ISO, WMO have also prepared standards and guidelines for the same.

Fortunately, there exists a relation between river stage and discharge at a cross section and this relation is known as rating curve or stage-discharge relation. A rating curve is developed by using the concurrent data of stage and discharge observed over a period of time. It is important that the data covers the range of stages that are likely to occur at the gauging station. At many stations, discharge is a function of river stage as well as other variables, including water surface slope, rate of change of stage, and bed features. If the stage-discharge relation is presented in tabular form, it is known as rating table.

The quality of computed stream flow data is determined by the quality of the stage-discharge relation. Since most hydrologic analyses, such as assessment of water yield and design of projects, are based on discharge data, the rating curve has important bearing on water resources planning and management. Therefore, the establishment of rating curve is important and requires familiarity with basic principles of stream flow hydraulics.

A channel whose flow characteristics do not change with time is termed as a stable channel. When these change with time, the channel is termed unstable. The stage-discharge relation changes in unstable channels. Therefore, sites without a stable control should be avoided, as far as possible. At a new site, initially discharge is measured over the expected range of stage to establish a rating curve. Later, discharge is measured at periodic intervals mainly to verify the established rating curve.

The factors that influence the rating curve can be broadly classified in two groups: natural and artificial. The natural factors include the geometry of the cross-section, the properties of bed and banks, the alignment of channel upstream of the gauging station, the properties of sediment being transported by the river, etc. The artificial factors include flow regulation structures, such as a weir, channel improvement works, a bridge, river training works, etc.

The combination of element(s) that control stage-discharge relation at a station is known as control. There are different types of controls: section and channel controls; natural and artificial controls; and complete, partial, and compound controls. When the geometry of a cross-section downstream of a gage constricts the flow or there is a break in bed slope (e.g., a fall), a section control is said to be effective. The constriction of flow may be due to a local rise of stream bed, due to a weir or dam, or due to reduced width (e.g., a bridge). If the relation is controlled by the geometry and roughness of a reach downstream, a channel control is said to exist. The length of the channel reach is directly proportional to discharge and inversely proportional to slope.

A complete control governs the stage-discharge relation over the entire range of stages at the gauging station. This is a rare occurrence. More common is a compound control in which a section control dominates lower discharges and another channel control dominates higher discharges. A weir or dam that does not get submerged at high discharges is a complete control. A partial control acts in concert with another control to influence the stage-discharge relation (Rantz, 1982). Artificial controls, as the name suggests, are man-made structures. These are section controls, expensive improvements in a long channel reach just to stabilize the rating curve are not justifiable. The stability of a rating curve depends on the attributes of controls – the relationship is stable if the control is stable. These attributes may change due to change in cross-section (e.g., erosion or deposition), and growth of vegetation. Most natural channels do not have a unique control for all stages. It is common to find section control for low-stages of flow and channel control for high stages.

When the stage is uniquely related to discharge, this is known as simple rating curve. In a compound rating curve, more than one curve is required. To establish a rating curve, the stage and discharge data are plotted on a graph paper, as shown in Figure 7, wherein stage is plotted on the Y-axis and discharge on the X-axis. Ideally, there should be a sufficient number of points, well distributed over the entire stage and discharge range. If the scatter of the plot is small, a smooth curve can be drawn through the points. The scatter in the data can be due to several reasons, including backwater effect, unsteady flow at the gauging site, scour of the bed and banks at the gauging site, or errors in observations.

A simple rating curve is commonly represented by a power equation that has the form

$$Q = a(H - c)^{b} \tag{11}$$

where Q is the discharge (m^3/s), H is the river stage (m), a and b are constants, and c is the stage (m) at which discharge is nil (known as the datum correction). This

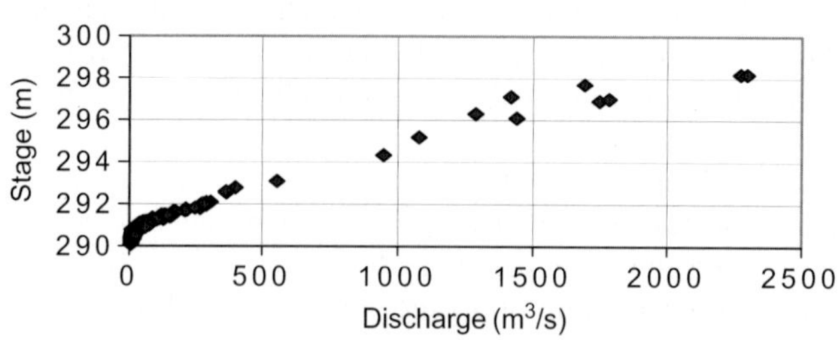

Figure 7. Plot of stage and discharge at a gauging station

equation plots as a straight line in the log domain. The method of least squares is commonly used to estimate the coefficients of the rating curve a and b in eq. (11). To estimate the datum correction c, iterations are performed by varying c and the value which yields best results is adopted. A rating curve should be tested for absence from bias and goodness of fit.

Compound Rating Curve

In many situations, a single curve may not be representative for the entire range of stages because the influence of roughness and boundary conditions is not the same at all stages. One may, therefore, require up to three curves – one for lower range of stages, one for middle range, and another for higher range. These segments may be combined by short transition curves. Besides, a compound rating curve is also necessary when the cross section consists of a narrow channel and flood plains on one or both sides. Such a cross-section is shown in Figure 8. At low water levels, flow is confined to the main channel but when river water level starts rising, at certain stage, the flow enters its flood plain. The conveyance characteristics of flood plains are usually quite different from the main channel and this causes a change in the slope in the rating curve. In such cases, two rating curves are established, the first corresponding to the flow in the main channel and the other for flow in the main channel plus flood plain.

The stage and discharge data are plotted on a log-log graph. The scatter is manually examined to identify break points. The stage-discharge relation can be developed following the procedure explained above, using the appropriate data set. A compound rating curve is shown in Figure 9.

In many situations, a single curve may not be representative for the entire range of stages – up to three curves may be required. One each for lower range of stages, for middle range, and for higher range.

Loop Rating Curve

In large rivers with flat bed slopes and significant changes in flow rates, the effects of unsteady flow on the rating curve may be significant. If a dam/weir or a confluence with a major river exists in the vicinity of the site downstream, the station may experience a variable backwater effect and in such conditions, discharge is not a function of river stage alone but it depends on the slope of the energy line as

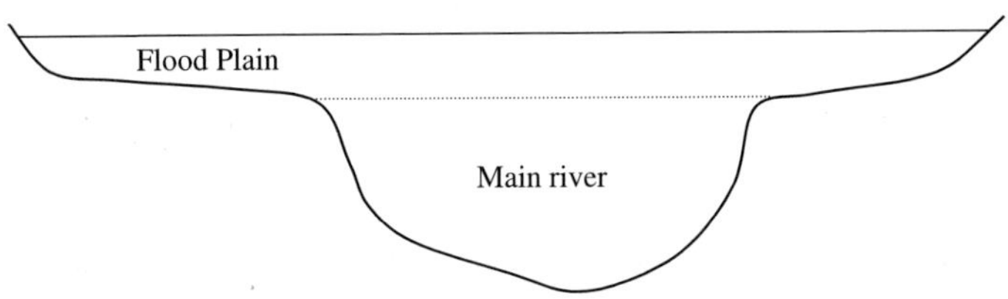

Figure 8. A river channel with flood plains on either side

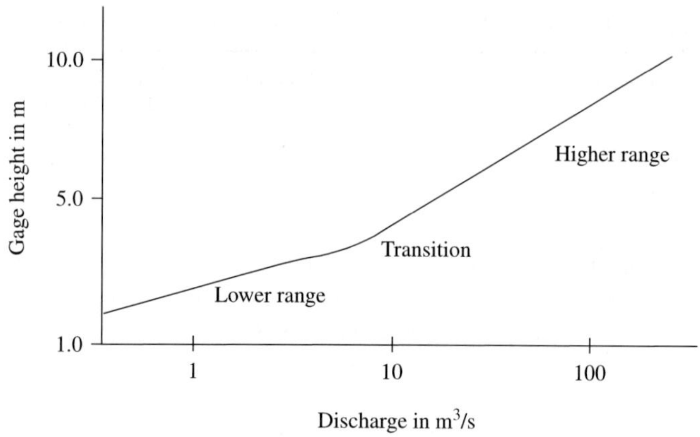

Figure 9. A compound rating curve

well. To measure the water surface slope, an auxiliary gage is installed. During the rising stage of a hydrograph, the velocity and discharge at a given stage are higher than the same at steady state flow. During hydrograph recession, the situation is reverse. This phenomenon is known as hysteresis and it gives a loop rating curve, i.e., discharge is not a unique function of stage. This happens because the slope of flood wave front during the rising stage is much steeper than during the steady-state condition and the situation is reverse during the falling stage. Such unsteady flow effects are not found in rivers where slopes are steep. Note that the shape of the loop varies from one station to another and the loop can be a complicated loop too.

Standards have been developed by many countries for establishing the stage-discharge relation. The international standard ISO:1,100/2 (see ISO, 1982) deals with determination of stage-discharge relations. Indian standard IS:15,119 (Part 2) describes the recommended practices for India.

5.2.6. Measurement of Sediments

Sediment is transported by moving water either as bed load or as suspended load. In India, bed load measurement is not being carried out as a matter of routine. Measurement of bed load in large stations is very difficult. In the absence of detailed measurements, Q_b is taken as a certain percentage of suspended sediments Q_s depending on concentration of suspended load and size of bed material and suspended sediment. To the suspended load must be added Q_b the bed load discharge at that station. Table 3 gives general guidance about this correction.

Since the suspended sediment load is estimated as a product of mean concentration of the sediment in motion and the corresponding discharge, silt measurement should normally be carried out at the same site where discharge is measured. It is to

Table 3. Bed Load Correction Table

Suspended load concentration	Stream bed material	Texture of suspended material	% of bed load in terms of measured suspended load
Less than 1,000	Sand	Similar to bed of stream	25 to 150
Less than 1,000	Compacted clay gravel cobbles, boulders	Small amount of sand	5 to 12
1,000–7,500	Sand	Similar to bed stream	10 to 35
1,000–7,500	Compacted clay, gravel cobbles, boulders	25% of sand or less	5 to 12
Greater than 7,500	Sand	Similar to bed stream	2 to 8
Greater than 7,500	Compacted clay, gravel cobbles, boulders	25% sand or less	5 to 15
Any concentration	Clay & silt unconsolidated	Silt & Clay	Less than 2

be noted that the concentration of suspended sediment changes from point to point in a cross-section and with time at the same point. According to the Indian Code IS:4,890, the concentration of suspended sediment is to be expressed in kg/m^3 of parts per million (by weight). Broadly, to obtain mean sediment concentration in motion at a vertical, the velocity distribution and sediment concentration curves are drawn. Product of concentration and velocity at the corresponding point gives the rate of sediment discharge curve. The area of the curve gives the sediment discharge. Division by the depth of flow gives the mean sediment discharge per unit area at the vertical. If the river is less than 30 m wide, three verticals will be enough. Fiver verticals should be chosen when river is 30–300 m wide and seven when the river is more than 300 m wide. Further, in a vertical, measurements can be taken at a single point or at multiple points. IS:4,890 also recommends use of depth integrating samplers.

The Indian Code recommends that bed load be estimated as a percentage, generally ranging from 5 to 20 percent, of the suspended load. However, measurements of bed load need to be undertaken particularly where high bed loads are anticipated. In India, CWC is carrying out sediment observations at 318 stations and State Governments are also operating some stations. This data is being published by CWC as annual sediment year books.

When upstream future development is expected, the projects under construction or which have the same priority of being taken up and completed as the project in question are considered for assessing the total sediment yield (IS-12,182-1987). The MOWR guidelines indicate that allowance shall be made for existing projects or projects under construction.

5.2.7. Sediment Rating Curve

The assessment of sediments being transported by a river is of vital importance in the design and management of water resources projects. Sediment rating curves are used to estimate the sediment load transported by a river. A sediment rating curve is a relation between the sediment discharge and the water discharge of a river. Stream gauging or a stage-discharge curve is used to obtain water discharge from measured stage. To determine sediment discharge of a stream, samples of water-sediment mixture are collected from the stream. A sampler can be chosen from a wide range of sediment samplers, depending upon the flow and site conditions. The samples are then analyzed in the laboratory to determine the sediment concentration.

Although the procedure for establishing a sediment rating curve is the same as that for a stage-discharge rating curve, there are additional complexities here. The sediment transported by a stream is influenced by many factors, such as velocity, depth, slope, and roughness of channel; characteristics of watershed, type; intensity and amount of precipitation; and properties of sediment. Besides, temperature has a significant influence on sediment transport – sediment discharge increases with decrease in temperature. Temperature affects the pick-up rate of sediment from the river bed. The effect could be so large that the bed material discharge may be twice as large at a temperature of 40 °F as at 80 °F.

When the methods and equipment for bed-material sampling were not developed, only suspended sediment was measured and used to develop the sediment rating curve. Usually, either the weight of the sediment or the sediment concentration is related to water discharge. Many times, these two forms are used interchangeably. McBean and Al-Nassri (1988) examined this issue and concluded that the practice of using sediment load versus discharge is misleading since the goodness of fit implied by this relation is spurious. They recommended that the relation be established between sediment concentration and discharge. Expectedly, the sediment concentration in a river is minimum near the water surface and it increases with depth. Silt and clay particles less than 0.005 mm in diameter are generally dispersed uniformly throughout the depth, but large grains are more concentrated near the bed.

An important feature of sediment transport is that sediment concentration increases fairly quickly with water discharge. As a result, the amount of sediment that is transported during a few days in a high flood may be equal to the amount that is moved during many months of fair weather. It is, therefore, of vital importance that representative data of high flood season are used to develop sediment rating curves.

A sediment rating curve is similar to a discharge rating curve, except that the relationship is established between water discharge and sediment concentration. The data of sediment discharge can be easily converted to sediment concentration if water discharge is known. Usually, the relation is expressed in the form of a power equation:

$$C = c\,Q^{d} \tag{12}$$

where C is the suspended sediment concentration (mg/l), Q is discharge (m^3/s) and c & d are constants. Conventionally, discharge and sediment concentration are plotted on a log-log graph paper and a straight line is drawn. A least squares method can be used to obtain the best fit line. Usually, the power equation is log transformed, and linear regression is applied to estimate the parameters. Typically, exponent d in eq. (12) lies in the range $2 < d < 3$.

In many situations, it may be difficult to find a simple relation between the sediment discharge and the water discharge. The reason it that the flow in a river at a given time may be due to very intense rainfall or due to moderate rainfall, the rain may be falling over an area of a catchment that might be susceptible to erosion or otherwise, the flow may be due to snowmelt, etc. and in each of these cases, the amount of sediment that is eroded and transported will differ. The sediment discharge and water discharge do not always increase or decrease simultaneously. For small basins, the peaks of the two may coincide but for large rivers, often the peak sediment discharge precedes the peak water discharge, depending on the hydrologic system of the watershed and the water velocity.

5.3. HYDROLOGIC DESIGN OF PROJECTS

Proper hydrologic design of the projects is necessary for better overall utilization of available resources in general and for better management and safety of structures. In India, water related projects are investigated, formulated and implemented by the concerned State Governments. CWC has been entrusted with the responsibility of examination of technical and economic feasibility of projects. Ministry of Water Resources, Govt. of India, has brought out "Guidelines for preparation of detailed project reports of irrigation multipurpose projects" in 1980s. This publication includes the guidelines/criteria which should be followed for preparation of the hydrology part of detailed project report (DPR). In addition, CWC has also brought out guidelines for preparation of River Basin Master Plan (1990). Apart from these guidelines, Bureau of Indian Standards (BIS) has also prepared a number of standards and codes of practices

For a storage based water development project, the main aspects of hydrology which are required to be studied are:

- Water availability study to understand the time series of inflows which can be expected in the reservoir,
- Flood studies to understand the floods which are likely to enter the reservoir. The reservoir should be safe while regulating the flood without endangering the dam,
- Reservoir routing studies to determine the spillway capacity and to fix the maximum water level of the reservoir,
- Sedimentation studies to determine the sediment inflow into the reservoir and to study their effect in reducing the reservoir capacity, and
- Simulation studies to study the performance of the reservoir under given inflow and demand pattern.

For planning any project, the first step is a correct assessment of water availability at the site of interest. This requires a sufficiently long sequence of data at that location; requisite length depending upon the type of storage, type of development and variability of inputs. In general, a longer period of data gives more confidence about the results. However, comparatively shorter length will suffice for within the year storage where the spill occurs almost every year and the critical period is of the duration of a few months. A longer data length is required for over-the-year storages. Flow duration curves are commonly used to determine water availability. Jain and Singh (2003) have described the methods to estimate water availability at a given location as well as the methods to determine required storage capacity of a reservoir.

Regional flow-duration models were developed by Singh et al. (2001) for nearly 1,200 potential micro-hydropower project sites located in 13 states of the Himalayan region covering almost the complete width of the country from Jammu and Kashmir in the west to north-eastern states in the east. The *Q*-quantile of dependability D (Q_D) for an ungauged watershed was derived assuming the following regional power relation:

$$Q_{mean} = CA^m \tag{13}$$

where A = watershed area; and C and m = coefficient and exponent, respectively, determined using regression. Using this relation, Q_D for an ungauged watershed of area A can be determined. Most of the historical flow data were available at 10-day intervals and were used for developing flow-duration models. At some sites, daily and monthly flows were available and these were transformed to 10-day flows. The selection of the 10-day interval in this study was primarily based on data availability.

Using the developed model, flows of the desired level of dependability can be estimated for an ungauged watershed located in the identified regions. The adequacy of the basic structure of the regional model was found to be fairly satisfactory for the watersheds of the states of Himachal Pradesh, West Bengal, Sikkim, South Assam, Meghalaya, Manipur, Nagaland, Mizoram, and Tripura. However, for the watersheds of Jammu and Kashmir and Uttar Pradesh, the model exhibited less than satisfactory performance and it performed poorly on the watersheds of Bihar and North Assam, requiring a cautious application. Note that the regional predictive relationship for mean flow used only the watershed area as its characteristic, ignoring many other characteristics because of their non-availability. Nevertheless, these types of studies are a way out when projects are to be planned in poorly gauged or ungauged regions.

After ascertaining water availability, the next important step in project planning is the estimation of design flood.

5.3.1. Design Flood Estimation

A design flood is a hypothetical flood (peak discharge or hydrograph) adopted as the basis in hydrologic design of project components. Economic, social, and other non-hydrologic considerations influence the philosophy of protection, and hence the design flood. When a sufficiently long flow series is available, frequency analysis is carried out to determine the floods of various frequencies. The other approach is to determine the design storm and thereby the design flood. For important projects, the results of various approaches are compared to finalize the design flood.

In principle, the selection of design flood is an assessment of risk involved against the cost of failure of a structure which has been designed to prevent any loss if a flood below the design magnitude arrives. The consequences of failure depend on the downstream conditions. In case of water resource projects where the consequences of failure could be catastrophic, design floods should also be estimated from probability considerations which enable evaluation of risk. Assessment of the probability of selected flood, however, involves establishing the relationship between the return period and the magnitude of rare floods. This relationship, if based on small sample of flood data, could be subject to serious errors. Application of frequency methods to abnormally high floods which have very low probability of occurrence should be made with utmost caution.

Hydrologic design of a project based on a flood of return period of 10,000 years may appear to be safe. Note that there are more than 4,000 dams in India with height more than 15 m or with storage more than 60 Mm^3. Therefore, a flood of 10,000 years return period could be expected to strike one of these dams every 3-year on the average. Further, the popular notion that failure cannot be tolerated is not achievable in real-life. The idea of calculated risk is in-built in structures to manage environmental systems including flood control projects. The objective should always be to avoid wasteful over-design and inefficient under-design. In the design of flood control projects, it is often assumed that death of humans is unacceptable implying that human life has an infinite value. Although this assumption is laudable, it is not practical as hundreds of lives are lost due to floods each year.

To find a practical solution to the problem, hydro-meteorologists have developed the concept of Probable Maximum Flood (PMF). However, this method suffers from a major disadvantage of giving an estimate of maximum flood based on data available up to the time of analysis and with no associated probability. Yevjevich (1972) has characterized the difference between the PMF method and frequency analysis method as being between "expediency" and "truth". Estimates through flood frequency methods are more relevant for lesser flood magnitudes where upper limits are based on adequate data/experience. Future flood events can exceed the specified PMF and the only way to better an estimate is to take into consideration further data available up to the time of review.

Since design floods often mark the difference between safety and disaster, utmost attention must be given to select and estimate the design flood that is most appropriate for a given case. Policy criteria have been laid down by most organizations for several applications and are followed unless there are compelling reasons for deviation in a particular case. Computation of design flood is an important part of hydrologic analysis and this is discussed here. But before that we define some important terms related to design flood estimation.

Probable Maximum Flood (PMF)

It is the flood discharge that may be expected from the most severe combination of critical meteorological and hydrological conditions that are reasonably possible in the region. The PMP is used in the design of projects, for example, dams where virtually complete safety from potential floods is sought.

Standard Project Flood (SPF)

It is the flood discharge that may be expected from the most severe combination of meteorological and hydrological conditions that are reasonable characteristic of the geographic area in which the study basin is located. Extremely rare combinations of those conditions are not considered. The peak discharge for a standard project flood is generally about 40 to 60 per cent of that for the probable maximum flood for the same drainage basin. The SPF is often used where failure of the structure would have some what less disastrous effect. For example, it is used in the design of flood control facilities whose failure might be disastrous.

The criteria for classification of dams are based on size of the dam and the hydraulic head. The classification for the dam is greater of the two indicated by the two parameters: gross storage and hydraulic head, as described in Table 4.

The criteria used for selecting the design flood for various hydraulic structures vary from one country to other. Table 5 gives a brief summary of the guidelines adopted by CWC (1969) to select design floods in India.

The design criteria for flood management schemes have evolved over the years, pooling the experiences and practices followed by various organizations and individuals. The prevalent hydrologic design criteria for determining the spillway capacity for India have been detailed in Indian Standard IS 11,223-1985, "Guidelines for fixing Spillway Capacity". According to these guidelines the inflow design floods that need to be considered for various functions of spillways are as discussed below.

(a) Inflow design flood for dam safety

It is the flood for which the dam should be safe against overtopping and structural failure. The criteria for classification of dams are based on the size and hydraulic head (see Table 4). A general practice is to adopt PMF as the design flood for large dams which have high hazard potential.

Spillway design flood is the maximum discharge that can be passed by a hydraulic structure without any damage or serious threat to its stability. Guidelines being followed in India in respect of spillway design flood are given in Table 5. For

Table 4. Classification of storage projects based on gross storage and hydraulic head

Classification	Gross storage	Hydraulic head
Small	Between 0.5 and 10 million m^3	Between 7.5 m & 12 m
Intermediate	Between 10 & 60 million m^3	Between 12 m & 30 m
Large	Greater than 60 million m^3	Greater than 30 m

Table 5. Guidelines for selecting design floods

S. N.	Structure	Recommended design flood
1.	Spillways for major and medium projects with storage more than 60 Mm^3	(a) PMF determined by unit hydrograph (UH) and probable maximum precipitation (PMP). (b) If (a) is not applicable or possible, flood frequency method with return period $T = 1,000$ years.
2.	Permanent barrage and minor dams with capacity less than 60 Mm^3	(a) SPF determined by UH and standard project storm (SPS) which is usually the largest recorded storm in the region. (b) Flood with a return period of 100 years. Choose (a) or (b), whichever gives higher value.
3.	Pickup weirs	Flood with a return period of 100 or 50 years depending upon the importance of the project.
4.	Aqueducts	(a) Waterway: Flood with $T = 50$ years. (b) Foundations and free board: Flood with $T = 100$ years.
5.	Project with very scanty or inadequate data.	Empirical formulae.

minor structures, a flood of 50 or 10-year frequency is adopted depending on the importance of the structure. Floods of larger or smaller magnitudes may be used if the hazard involved is high or low, respectively. In addition to the size, the relevant parameters to be considered in judging the hazard are:

- Distance and location of the downstream areas. Due consideration is given to the likely future developments, and
- The maximum carrying capacity of the downstream channel at a level at which catastrophic damage is not expected.

(b) Inflow design flood for efficient operation of energy dissipation works

The energy dissipation arrangements for the spillway may be designed for the best efficiency for a smaller inflow flood than the inflow design flood for the safety of the dam.

(c) Inflow design flood to check the extent of upstream submergence

The inflow design flood to check the extent of the upstream submergence depends on local conditions and the type of property and the effects of its submergence. Except for very important structures in the upstream like power houses, mines, etc. for which the levels corresponding to SPF or PMF may be used, smaller design floods and levels attained under these may suffice. In general, a 25-year return period flood for land acquisition and a 50-year return period flood for the built-up property acquisition may be adopted.

(d) Inflow design flood for the extent of downstream damage in the valley

The inflow design flood to check the extent of downstream damage depends on local conditions, the type of property and the effects of its submergence. For important facilities like power houses, the outflows under the inflow design flood for safety of dams and all gates operating conditions are relevant. Normally, the discharge relevant to check the acceptability of the downstream submergence may be smaller than that for power houses at or near the toe of the dam.

For important projects, dambreak studies are to be undertaken as an aid to determine the design flood. Where the professional judgment or studies indicate an imminent danger to present or future human settlements, the PMF should be used as the design flood. Any departure from the general criteria as above on account of larger or smaller hazard should be clearly brought out and recorded.

e) Design flood for fixing freeboard

The design of spillways and the size of flood control pool is determined using reservoir routing. Apart from safety, the economic considerations are also important. An optimum design criterion can be reached by compromising the cost and the risk factors. A flood of specific frequency is adopted depending on the functional importance, with judicious combination of safety and economy.

f) Design flood estimation for barrages

Weirs and barrages usually have small storage capacities and the risk of loss of life and property downstream would rarely be enhanced by failure of the structure. The loss of the structures by its failure would disrupt irrigation and communications that are dependent on the barrage. For barrages, the use of a 100-year return period flood or standard project flood whichever is higher, is the normal practice.

g) Design Flood Estimation for Road and Railway Bridges

For road bridges, the Indian Road Congress IRC: 5-1970, *Section-I General Features of Design* applies. According to this, the discharge for which the waterway of a bridge is to be designed shall be:

- the maximum flood observed for a period of not less than 50 years; or
- the discharge computed from an another recognized method applicable for that area; or
- the discharge found by the velocity area method; or
- the discharge found by UH method; or
- the maximum discharge fixed by the judgment of the engineers responsible for the design.

A comparison of the results of the above mentioned methods is to be made. A 50-year flood can be used for smaller bridges carrying railway lines of lesser importance. In the case of bridges carrying main and important rail lines, a 100-year return period flood is to be adopted as per the railway codes.

h) Design Flood Estimation for Cross-Drainage Structures

The cross drainage works can be classified under three broad categories: i) Structure for a carrier channel over a natural drainage, ii) Structure for a carrier channel underneath a natural drainage, and iii) Structure for a carrier channel crossing a natural drainage at the same level. The design flood to be adopted for minor cross-drainage works depends upon the size of drainage channels, the canals, the cost and importance of the structure.

Table 6. Recommended design criteria for flood control schemes

Situation	Guideline
Predominantly agricultural areas	25-year return period flood on small tributaries and 50-year on major rivers
Town protection works	100-year return period flood
Important industrial complexes, assets and lines of communications	100-year return period flood
Large cities	The maximum observed flood or sometimes the maximum probable flood

The BIS Code of practice for design of cross-drainage works [IS:7,784 (part-I), 1975] recommends that the design (of waterway) in such cases may be based on a 10- to 25-year frequency flood with increased afflux. However, the foundations and freeboard etc. should be checked for safely for increased afflux and velocities due to a 50-year or 100-year return period flood. For very large cross-drainage works, damage to which is likely to affect the canal supplies over a long period, the design should be based on the maximum probable flood.

As per the CWC criteria, waterways for canal aqueducts should be provided to pass a 50–100-year return period flood, but their foundations and freeboards should be safe for a flood of not less than 100-year return period.

Each site is unique in its local conditions, causes, and effects. Thus, the above mentioned norms may be taken as general guidelines. The designer may deviate from the norms and the criteria in special cases if justifiable on account of local conditions. In such cases, reasons be recorded and should have the acceptance of the approving authority.

Design Criteria for Flood Control Schemes: The design criteria for flood control schemes broadly recommend have given in Table 6.

The general practice is to adopt the flood corresponding to the applicable frequency or the observed maximum flood in the recent past, whichever is higher. In no case, the design High Flood Level (HFL) should be lower than the highest observed level on record. For small rivers carrying discharge up to 3,000 cumec, the design HFL should correspond to a 25-year return period flood. For a river carrying a peak flood above 3,000 cumec, the design HFL should correspond to 50-year return period. However, if the embankments are to protect big township, industrial areas or other places of strategic importance, the design HFL should generally correspond to a 100-year return period flood.

5.3.2. Methods of Design Flood Estimation

The following methods are generally used for design flood estimation:

a. Rational method,
b. Empirical method,
c. Flood frequency method,

d. UH technique, and
e. Watershed models.

The rational method, empirical method and flood frequency method are generally used for estimating the magnitude of flood peak. However, the UH technique and watershed models can give the design flood hydrograph, in addition to the magnitude of design flood peak. The use of a particular method depends upon: (i) the desired objective, (ii) the available data, and (iii) the importance of the project.

The rational formula is only applicable to small size (50 sq. km) catchments. The empirical formulae are essentially the regional formulae based on statistical correlation. The frequency analysis approach is the statistical methods to predict the flood peaks of a specified return period. The UH method is basically a rainfall runoff relationship normally applicable to moderate size catchments with area less than 5,000 sq. km.

5.3.3. Empirical Formulae for Design Flood Estimation

These are essentially regional formulae based on statistical correlation of the observed peak and important catchment characteristics. These empirical formulae are applicable only in the region for which they were developed in the range of flood peaks used; for other areas, they should be applied with caution. Table 7 gives some of the empirical flood formulae used in India along with the values of constants used and the limitations of the formulae. Although these were originally developed in FPS units, the equivalent formulas in metric units are given here.

The coefficient C in Dicken's formula differs from one region to another. The values of this co-efficient used for the computation of maximum flow for some of the catchments in UP are given in Table 8. Further, this coefficient for different return periods for some of these catchments is given in Table 9. Table 10 contains suggested values of C for different topographic classes.

The values of coefficient C of the Ryve's formula, as adopted in some studies for catchments in UP are given in Table 11.

Some other empirical formulae used in the past are the Craig formula, the Inglis formula, and the Rhind formula.

Whenever the hydrological records are inadequate for frequency analysis or UH analysis, the empirical formulae developed for that region are an alternative to estimate design flood at the project site. The empirical formulae are essentially the regional formulae based on statistical correlation of the observed peaks and important catchment properties. Most empirical formulae involve only one or two physical characteristics, of the catchment. Since flood peaks depend on many physical characteristics, these empirical formulae cannot be expected to give precise results. The use of empirical formulae is discouraged these days. Their use should be avoided for the estimation of design flood for the major structures.

Table 7. Commonly used empirical formulae

No.	Author	Formula in metric unit	Values of constants C in different regions		Limitations
1.	Dickens	$Q = C\,A^{3/4}$ Q in cumec, A in sq. km.	North-Indian plains	6.0	Generally applicable for moderate size basins in North and Central India
			North-Indian hilly regions	11–14	
			Central India	14–28	
			Coastal Andhra and Orissa	22–28	
2.	Ryves	$Q = C\,A^{2/3}$ Q in cumec, A in sq. km.	Area within 80 km from the east coast	6.8	Developed from a study of rivers in South India
			Area within 80–240 km from the coast	8.3	
			Limited area near the hills	10.2	
			Values estimated from observed data	Up to 40	
3.	Nawab Jung Bahadur	$Q = CA^{[0.92-(\log A)/14]}$ Q in cusec, A in sq. miles.	Yamuna River at Tajewala	2,500	More suitable for areas below 10000 sq. miles.
			Bhagirithi River at Pala	2,100	
4.	Creagar	$Q = 46CA^{[0.894A-0.048]}$ Q in cusec, A in sq. miles.	Yamuna River at Tajewala	100	
			Sarda River at Banbassa	100	

Table 8. Values of coefficient C in Dicken's formula used to compute the maximum flow for some catchments in UP

Location	Coefficient C in Dicken's formula in metric unit
Bhagirathi River at Pala	19.76
Yamuna River at Dakpathar	19.76
Sarda River at Banbassa	12.76
Ganga River at Hardwar	10.23
Ramganga River at Kalagarh	25.69
Yamuna River at Tajewala	14.40
Ken River at Gangao	11.26
Betwa River at Paricha	10.45
Karmnasa River at Silhat	24.65
Rihand River at Pipri	14.17

Table 9. Coefficient C in Dicken's formula corresponding to different return periods for some of the catchments in UP used for maximum flood flow computation

Return Period (years)	Values of Dicken's C in metric unit										
	Tons at Kishau		Yamuna at Tajewal		Bhagirathi at Tehri		Ganga at Haridwar		Sarda at Banbassa		Ramganga at Kalagarh
	A	*B*	*A*	*B*	*A*	*B*	*A*	*B*	*A*	*B*	
10	8.54	8.67	9.64	9.75	4.63	6.13	5.94	7.45	8.63	9.33	12.52
20	10.75	10.9	11.69	11.82	5.58	12.94	6.95	8.7	9.85	10.66	14.81
50	13.54	13.71	14.27	14.43	6.78	8.95	8.2	10.28	11.41	12.34	17.7
100	15.75	15.95	16.32	16.52	7.72	10.21	9.2	11.55	12.65	13.68	20.00
200	17.87	18.11	18.29	18.51	8.63	11.41	10.18	12.75	13.83	14.96	22.22
500	20.70	20.96	20.91	21.15	9.84	13.0	11.45	14.34	15.42	16.66	25.16
1,000	22.84	23.14	22.91	23.18	10.75	14.22	12.42	15.56	16.61	17.97	27.39

Note: A = When total catchment area is used including snow covered area B = When catchment area does not include perpetual snow area

Table 10. Coefficient C in Dicken's formula depending upon the catchments topography

Class of location	Conditions	Coefficient C in Dicken's formula in metric unit
Class I	Bare surface with precipitous hills	19.76–28.23
Class II	Catchment with hills on the skirts with undulating country up to the outfall	14.12–28.23
Class III	Undulating country with hard indurated clay soils	11.29–14.12
Class IV	Flat, sandy, absorbent or cultivated plains	2.82–7.06

Table 11. Coefficient C in Ryve's formula for some of the catchments in UP used for peak flow computation

Location	Coefficient C in Ryve's formula in metric unit
River Bhagirathi at Pala	41.22
River Yamuna at Dakpathar	41.22
River Sarda at Banbassa	30.54
River Yamuna at Tajewala	30.54

5.3.4. Rational Method

If rainfall of uniform intensity occurs over a basin beyond the time of concentration of the basin (which is the time taken for a drop of water from the farthest point of the catchment to reach the outlet) then the runoff will be constant at the peak value. The peak value of the runoff is given by the equation

$$Q_p = C\ i\ A \text{ for } t > t_c \tag{14}$$

where Q_p = peak discharge in m^3/s, C = co-efficient of runoff, A = Area of the catchment in sq. km, and i = intensity of rainfall. The above equation is the basic equation of the rational method. Using the commonly used units in metric system, the equation is written as:

$$Q_p = \frac{1}{3.6} C(i_{tc,p}) A \tag{15}$$

where $i_{tc,p}$ = the mean intensity of precipitation (mm/hr) for a duration equal to t_c (time of concentration), and an exceedence probability *p*. The application of the method for peak flood computation require three parameters, viz., t_c, $i_{tc,p}$ and C, which can be estimated as follows.

Time of Concentration

The time of concentration can be estimated using an empirical equation of the form:

$$t_c = C_{tL} \tag{16}$$

where C_{tL} and n are constant, L is the length of the main stream in km, L_{ca} is the length along the main stream from the outlet to a point opposite the C.G. of the catchment in km, S is the basin slope.

Another empirical equation, known as the Kirpich equation, is very much in use to estimate the value of t_c. This equation is given as:

$$t_c = 0.01947 L^{0.77} S^{-0.385} \tag{17}$$

where t_c = the time of concentration in minutes, L = the maximum length of travel of water in metres and S = Slope of the catchment = DH/L, DH = the difference in elevation between the most remote point on the catchment and the outlet in metres.

Rainfall Intensity (i)

The rainfall intensity – frequency – duration relationship for the given catchment can be used to obtain the rainfall intensity corresponding to a duration t and the desired probability of exceedence p (i.e., return period = 1/p).

Runoff Coefficient (C)

The coefficient C represents the integrated effect of the catchment losses and depends on the nature of the surface slope and rainfall intensity. Some typical values of C are given in Table 12 where the effect of rainfall intensity is not considered for the values of C. The rational method is commonly used for peak flow prediction in small catchments up to 50 km^2 in area. This method has found considerable application in design of urban drainage, small culverts, and bridges.

Table 12. Value of the coefficient C for different types of areas

Area			C value
Urban area	Lawns	Sandy – soil, flat, 2%	0.05–0.10
		Sandy – soil, steep, 7%	0.15–0.20
		Heavy – soil, average, 2–7%	0.18–0.22
	Residential Areas	Single family areas	0.30–0.50
		Multi units, attached	0.60–0.75
	Industrial	Light	0.50–0.80
		Heavy	0.60–0.90
	Streets		0.70–0.95
Agricultural Area	Flat: Tight clay	Cultivated	0.50
		Woodland	0.40
	Flat: Sandy loam,	Cultivated	0.20
		Woodland	0.10
	Hilly: Tight clay,	Cultivated	0.70
		Woodland	0.60
	Hilly: Sandy loam,	Cultivated	0.40
		Woodland	0.30

5.3.5. Design Flood Estimation Using UH

Taking into account the limitation in adopting the empirical formulae and the developments in hydrological science, Government of India had constituted a committee known as the Khosla committee to indicate a rational method to determine design discharge. The committee recommended systematic and sustained collection of hydro-meteorological data selected of catchments in different climatic zones of India to evolve approach for determination of design flood. The committee felt that design flood should be the maximum flood on record for a period of not less than 50 years. Where records over a period not much less than 50 are available, years the design flood should be 50-year flood determined from probability considerations on the basis of observed data. In case the data is inadequate, the design flood was recommended to be determined based on design storm. Since long term data on small and medium catchments is not available in the country, the approach based on design storm is frequently adopted.

The Khosla Committee recommended two schemes: Long Term Plan and Short Term Plan. Under the Short Term Plan, a method was devised to estimate the design flood peak based on unit hydrograph principle. Under the Long Term Plan, the country has been divided into zones which in turn are sub-divided in 26 hydro-meteorologically homogenous sub-zones of moderate sizes as shown in Figure 10 and listed in Table 13. The data collected from selected representative catchments in the sub-zone should be studied and analyzed in detail. Design of structures to control river flows must consider both extremes of runoff (that is, droughts and floods). Analyses are required to size the capacity of outlet works (spillways, bypasses, etc.)

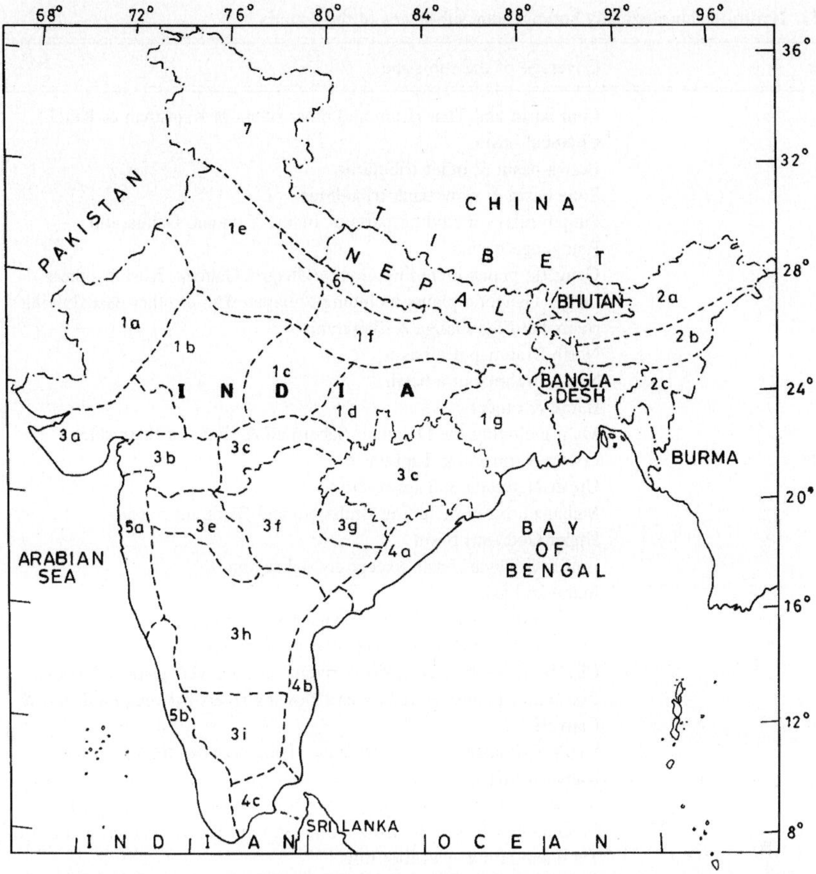

Figure 10. Hydro-meteorologically homogenous sub-zones of India

to cater for floods. It is often necessary in hydrologic design to have details of both the peak flows and the distribution of flow with time. In other words, the computed hydrograph may be needed, so that the runoff volume can be estimated.

Here the method for the estimation of design flood applicable to small catchments (less than 5,000 sq. km) based on UH approach is discussed. UH technique along with routing can be used to estimate the design flood for larger sized basins (greater than 5,000 sq. km) by dividing the basin into sub-basins. The UH method of determination of design flood hydrographs involves several steps. Derivation of design storm has been discussed in Chapter 3.

Basin Response

Under natural conditions of rainfall and drainage basins, the assumptions of linearity and of uniform distribution of rainfall excess in space are difficult to justify in large

Table 13. Hydro-meteorologically homogenous sub-zones of the country

Number	Coverage of the sub-zone
1(a)	Luni basin and Thar (Luni and other rivers of Rajasthan & Kutch)
1(b)	Chambal basin
1(c)	Betwa basin & other tributaries
1(d)	Sone basin & right bank tributaries
1(e)	Punjab plains including parts of Indus, Yamuna, Ganga and Ramganga basins
1(f)	Gangetic plains including Gomti, Ghagra, Gandak, Kosi & others
1(g)	Lower Gangetic plains including Subarnarekha & other east-flowing rivers between Ganga & Baitarani
2(a)	North Brahmaputra basin
2(b)	South Brahmaputra basin
2(c)	Barak & others
3(a)	Mahi including the Dhadhar, Sabarmati & rivers of Saurashtra
3(b)	Lower Narmada & Tapi basin
3(c)	Upper Narmada & Tapi basin
3(d)	Mahanadi basin including Brahmani and Baitarani rivers
3(e)	Upper Godavari basin
3(f)	Lower Godavari basin except coastal region
3(g)	Indravati basin
3(h)	Krishna sub-zone including Pennar basin except coastal region
3(i)	Cauveri & east flowing rivers except coastal region
4(a)	Circars including east flowing rivers between Mahanadi & Godavari
4(b)	Coromandal coast including east flowing rivers between Godavari & Cauveri
4(c)	Sandy Coroman belt (east flowing rivers between the Cauveri & Kanya Kumari)
5(a)	Konkan coast (west flowing rivers between the Tapi & Panaji)
5(b)	Malabar coast (west flowing rivers between Kanya Kumari & Panaji)
6.	Tarai sub-Himalayan foot hills
7.	J & K, Kumaon Hills (Indus basin)

catchments. Adaptation of UH Theory or models based on UH Theory to large catchments of over 5,000 sq. km in area requires special care to ensure that the assumptions are not seriously violated. Problems encountered in adaptation are:

- Uneven spatial distribution of rainfall is rule rather the than the exception. Storm events covering the entire catchment available for analyses are few and usually none the entire catchment whereas the design storm selected covers the whole catchment.
- Volume of direct surface runoff of observed flood events is comparatively very much lower than that of the design floods and the error involved in the use of UG derived from observed floods of smaller volume without adjustment for higher volume could be large.
- Gauge-Discharge observations of floods are available at only one or two sites. Consequently, most of the sub-basins into which the catchment may be divided will have to be considered as ungauged.

- Gauge-Discharge observations of floods at a site other than the dam site are available and method to transfer the information to dam site will be needed.

Through innovations and improvisations to suit each individual case keeping in view the data situation and objective, the above problems are to be effectively tackled.

For catchments with non-linear behaviour, Rao (1975) found that the practice of increasing the UG peak by 25% to 50% and modifying the UG when it is derived from smaller floods and the US Corps of Engineers practice of applying the normal UG to all unit periods of design storm except the largest to which UG with increased peak is applied are well known.

(*a*) *Rainfall Losses:* The usual variation in the spatial aerial distribution of storm rainfall in small basins may be taken into account when selecting infiltration indices and when deriving the hydrographs which reflect critical conditions. For larger drainage basins, however, the infiltration during time increments of the storm is proportional to the areas covered by rainfall intensities that exceed the infiltration capacity. Therefore, for larger drainage areas, assumptions concerning both the time and aerial distribution of rainfall intensities during the design storm must be made.

Now, increments of rainfall are first aligned to match the ordinates of the design UH so that the position of the maximum depth increment is matched with the maximum UH ordinate, the position of second largest depth increment is matched with the second largest UH ordinate and so on. The sequence of rainfall increments is then reversed to get a design sequence of precipitation increments.

Considering the approximations and assumptions involved in estimating the storm rainfall hyetograph, ϕ index approach is adequate for purposes of arriving at rainfall excess hyetograph in design flood studies. Intensity and temporal and spatial distribution of rainfall affects the ϕ index value. ϕ index is higher for high intensity rains not because of increased infiltration but due to the fact that area producing run off becomes greater as the rainfall intensity increases.

Detailed studies of storm rainfall flood runoff for a large number of drainage basins, most of them with areas between 50 to 500 sq. km located in various climatic sub-zones, have been made by Hydrology (Small Catchments) Directorate (CWC. 1973 to 1989). Modal values of ϕ-index ranging from 2.0 to 7.5 mm/hr depending on the climatic sub-zone have been recommended for assessing design floods of 50 year or 100 year return period required for design of railway and highway bridges across streams draining small to medium size catchments.

In estimating PMF for major structures, however, some conservatism is appropriate because of large investments and risk involved. Rao (1990) favoured the choice of lower ϕ-index values with 90% probability of exceedance instead of mode values and suggested ϕ-index values for application to PMP hyetographs of catchments in India as given in Table 14.

(*b*) *Calculation of effective rainfall*: Generally the infiltration index and the initial losses are derived from the available rainfall-runoff records for severe storms in the basin. Assuming the basin would be saturated at the time of design storm, the minimum infiltration rate and initial losses would be considered. The minimum

Table 14. Recommended ϕ-index values for application to PMP

Region		ϕ-index (mm/hr) for catchment areas (sq. km)		
		500–5,000	5,000–10,000	above 10,000
1.	Luni Basin (Rajasthan)	6.0	7.0	8.0
2.	Chambal Basin (Rajasthan)	4.0	5.0	6.0
3.	Ganga plains, Sone and Chambal (MP) basins	2.0	3.0	4.0
4.	Western Himalayas	1.0	2.0	3.0
5.	Brahmaputra Basin	1.0	2.0	3.0
6.	Mahi and Sabarmati Basins	2.5	3.5	4.5
7.	Upper Narmada and Tapi Basins	1.5	2.5	3.5
8.	Lower Narmada and Tapi Basins	2.0	3.0	4.0
9.	Mahanadi Basin	1.5	2.5	3.5
10.	Krishna Godavari & Pennar Basins	2.0	3.0	4.0
11.	Cauvery Basin	2.5	2.5	4.5
12.	East Coastal region	1.5	3.5	3.5
13.	West Coastal region	1.0	2.0	3.0

infiltration rate and minimum initial loss are used to compute the effective rainfall of a design storm. For this the initial losses must be subtracted first from the rainfall increments and thereafter a uniform loss rate equal to the minimum infiltration index is applied.

From a study conducted by CWC (1973) on rainfall-runoff correlation, the following relationships for the estimation of uniform loss rate (φ-index) are envisaged for flood producing storms and soil conditions prevalent in India:

$$R = a\, I^{1.2} \tag{18}$$

$$\varphi = (I - R)/24 \tag{19}$$

where R = runoff (cm) from a 24 hour rainfall of intensity I (cm/day), and a = coefficient which depends upon the soil type. Its typical values are given in Table 15.

In the absence of any other data, an assumption has to be made regarding the value of φ-index. The U.S. Army Crops of Engineers has recommended the use of infiltration index of 1.0 mm/hour for the first 12 hours, 0.8 mm per hour for the

Table 15. Typical values of Co-efficient *a*

Types of soil	Co-efficient *a*
Sandy soils and sandy loam	0.17–0.24
Coastal alluvium and silty loam	0.25–0.34
Red soils, clayey loam, grey and brown alluvium	0.42
Black cotton and clayey soils	0.41–0.46
Hilly soils	0.46–0.50

subsequent 12 hours, and 0.5 mm per hour thereafter. An alternative practice being used in India recommended by CWC is the use of constant loss rate of 1.0 mm per hour throughout the storm. The rates adopted for a particular project are primarily influenced by the soil type and land cover in the basin. Initial losses may be assumed to be zero during the period of design storm unless some sound evidence suggests another amount as initial losses.

(*c*) *Derivation of design UH*: Unit hydrographs are derived from the discharge records or by regional UH relationships. The UH technique based on Clark's approach may be preferred over the conventional techniques, such as Collin's method, etc. When a number of UHs have been developed for the project basin, a selection has to be made to determine the UH which will result in the best estimate of the design flood. Generally a normal UH is developed. Such a UH applies to normal or average storm patterns, and is considered suitable for use in computing design floods of all magnitudes but the probable maximum flood. The normal UH should be tested by reproducing observed hydrographs of major floods.

If the normal UH is computed from a recorded flood hydrograph that represents rainfall distributions and hydraulic conditions which are not likely to be greatly different from those during design storms, the derived normal UH can be used without modification. However, if the UH has been derived from a minor flood hydrograph, it may be necessary to modify it for use with the design storm due to differences in the aerial distribution of rainfall and hydrologic conditions between major and minor floods. Since the data for the determination of UH are usually limited to relatively minor floods, it is necessary to increase the peak ordinate of the normal UH to represent higher concentration of runoff. CWC recommended an increase of 25 to 50 percent in peak ordinate of the normal UH. This increase necessitates a general modification of the normal UH to preserve the unit volume. The modified UH is generally assumed to apply to all unit periods of the probable maximum storm. The U.S. Army Corps of Engineers practice, however, is to apply the normal UH to all unit periods of the probable maximum storm except the largest to which the modified UH is applied.

(d) The critical sequence of the effective rainfall are applied to the design UH to obtain the total design direct surface runoff hydrograph effective rainfall (in mm) in each period and are determined by multiplying by the time unit. The steps are:

- Multiply the first effective rainfall increment successively by each of the UH ordinates. The resulting quantities represent the ordinates of the direct surface runoff hydrograph produced by the first increment of effective rainfall.
- Compute the direct surface runoff hydrographs resulting due to second, third etc. increments of effective rainfall.
- Compute the total direct surface runoff from effective rainfall amounts by adding the direct surface runoff resulting due to different increments of rainfall in proper time relation being lagged successively by a time interval equal to the unit duration of the UH.

(e) To obtain the design flood hydrograph of total runoff, the baseflow expected during the design storm is estimated and added to the total direct surface runoff ordinates obtained in the previous step.

5.3.6. Regional UH Studies

Many studies have been undertaken to derive regional UH for various regions of India.

i. The small catchment directorate of CWC (1980) developed the following relationships between the one-hour UH parameters and physical characteristics of the catchments for Godavari basin subzone 3f:

$$t_p = 0.253(LL_c/\sqrt{S})^{0.45} \quad (20)$$

$$Q_p = 1.968(t_p)^{-0.842} \quad (21)$$

$$W_{50} = 2.30(Q_p)^{-1.108} \quad (22)$$

$$W_{75} = 1.356(Q_p)^{-1.007} \quad (23)$$

$$W_{R50} = 0.954(Q_p)^{-1.078} \quad (24)$$

$$W_{R75} = 0.581(Q_p)^{-1.035} \quad (25)$$

$$T_B = 4.572(t_p)^{0.90} \quad (26)$$

where Q_p is the peak discharge of UH in cumec; t_p is the time from the center of unit rainfall duration to the peak of the UH in cubic meter/second; W_{50} is the width of the UH measured at a discharge ordinates equal to 50% of Q_p in hours; W_{75} is the width of the UH measured at a discharge ordinate equal to 75% of Q_p in hours; W_{R50} is the rising side of the UH measured in hours at a discharge ordinate equal to 50% of Q_p in hours; W_{R75} is the rising side of UH measured in hours at discharge ordinate equal to 75% of Q_p in hours; T_B is the base width of the UH in hours; L is the length of the longest water course; L_c is the water course length from the outflow point to a point on the stream nearest to the centroid of the basin and S is the equivalent stream slope of the longest water-course. The one hour representative UH parameter and pertinent physiographic characteristics for 22 catchment of Godavari basin subzone 3F are given in Seth and Singh (1985–86).

ii. CWC (1982) also developed the regional UH relationships for Mahanadi basin subzone 3d analyzing the data of 16 catchments of the basin. One hour UH parameters were considered for developing the following relationships. The representative one hour UH parameters and pertinent physiographic characteristics for Mahanadi subzone 3d have been given by Seth & Singh (1985–86) as follows:

$$t_p = 1.97(LL_c/\sqrt{S})^{0.24} \quad (27)$$

$$Q_p = 1.12(t_p)^{-0.66} \quad (28)$$

$$W_{50} = 2.195(Q_p)^{-1.1008} \tag{29}$$

$$W_{75} = 1.221(Q_p)^{-0.95} \tag{30}$$

$$W_{R50} = 0.995(Q_p)^{-0.94} \tag{31}$$

$$W_{R75} = 0.532(Q_p)^{-0.93} \tag{32}$$

$$T_B = 5.72(t_p)^{0.77} \tag{33}$$

iii. Mathur & Kumar (1982) related the following physical parameters of 20 small and medium catchments with an objective to find out the most effective physical parameters representing the regional UH relationships.

$$S = \left(L / \sum_{i=1}^{n} L_i / S_i^{1/2})\right)^2 \tag{34}$$

where L_i is the length of the ith segment of the main stream (km); S_i is slope of ith segment of main stream (km/km); and S_i = land slope defined by

$$S_i = \frac{i\sum_{i=1}^{n} \frac{l_i + l_{i+1}}{2}(h_{i+1} - h_i)}{A} \tag{35}$$

where l_i is the length of ith contour (km); A is the catchment area (km^2), SL_C is the statistical stream slope from a point nearest to centroid, W_C is the Minimum width of the catchment from a point passing through centroid, and D is the drainage density (km^{-1}).

The above physical parameters for 20 catchments were given by Mathur and Kumar (1982), and Seth & Singh (1985–86). The multiple linear regression analysis of a total 31 combinations of physical parameters were considered, dropping one or several of them singularly and collectively. The relations between each parameter and basin lag have also been presented by Seth & Singh (1985–86).

iv. Huq et al. (1982) developed generalized synthetic UH relationships analyzing the data of 21 bridge catchments in Lower Gangetic Plains, Mahanadi Basin, Krishna Basin and Brahmaputra Basin. They have related the parameters of the representative UHs with a suitable combination of the following physical characteristics of the catchments using regression analysis:

$$Ft_p = 1.43\left(\frac{LL_C}{W_C}\sqrt{\frac{A}{S}}\right)^{0.38} \tag{36}$$

$$Q_p = 2.33A^{0.67}S^{0.38} \tag{37}$$

$$W_{50} = 2.33(Q_p)^{-1.10} \tag{38}$$

$$W_{75} = 1.321(Q_p)^{-1.22} \tag{39}$$

where, F = the form factor which is the ratio of the square of the length of the main stream to the total catchment area, i.e., L^2/A.

v. The small catchment directorate of CWC (1982) developed the following regional UH relationships for Krishna and Pennar Basins (subzone 3 h) relating the physical parameters of 21 catchments with their one-hour representative UH parameters:

$$t_p = 0.258\left(\frac{LL_C}{\sqrt{S}}\right)^{0.49} \tag{40}$$

$$Q_p = 1.017(t_p)^{-0.52} \tag{41}$$

$$W_{50} = 2.396(Q_p)^{-1.08} \tag{42}$$

$$W_{75} = 1.427(Q_p)^{-1.08} \tag{43}$$

$$W_{R50} = 0.750(Q_p)^{-1.25} \tag{44}$$

$$W_{R75} = 0.557(Q_p)^{-1.12} \tag{45}$$

$$T_B = 7.193(t_p)_{0.53} \tag{46}$$

The correlation coefficients obtained for the above equations are reasonable. The physical characteristics and one hour representative UH parameters for 21 catchments of subzone 3 h are given by Seth and Singh (1985–86).

vi. The Small Catchment Directorate of CWC (1984) derived the following relationships relating the physical parameters of the 23 catchments of upper Indo-Ganga Plains with representative 2-hour UH parameters:

$$Q_p = 2.030\left(\frac{L}{\sqrt{S}}\right)^{0.649} \tag{47}$$

$$T_p = 1.858(Q_p)^{-1.038} \tag{48}$$

$$W_{50} = 2.217(Q_p)^{-0.99} \tag{49}$$

$$W_{75} = 1.477(Q_p)^{-0.876} \tag{50}$$

$$W_{R50} = 0.812(Q_p)^{-0.907} \tag{51}$$

$$W_{R75} = 0.606(Q_p)^{-0.791} \tag{52}$$

$$T_b = 7.744(t_p)^{0.779} \tag{53}$$

In India, a number of studies have been carried out for the estimation of design floods for various structures by different organizations. Prominent among these include the studies carried out jointly by the CWC, Research Designs and Standards Organization (RDSO), and India Meteorological Department (IMD) using the method based on synthetic unit hydrograph and design rainfall considering physiographic and meteorological characteristics (CWC 1987).

Table 16. Values of Baseflow during flood season

Sl. No.	Region	Baseflow (cumec/sq. km)
1.	Luni, Chambal, Sone, Punjab plain, Gangetic plains, Upper Narmada & Tapi basins, Upper Godavari, Krishna, Cauvery, J&K, Kumaon Hills.	0.05
2.	Betwa, Mahi, Sabarmati, Lower-Narmada, and Tapi basins, Lower Godavari, Indravati Basin, East Coast, Terai (UP).	0.11
3.	Mahanadi basin, West coast.	0.22
4.	Brahmaputra basin	0.44

5.4. BASEFLOW

When computing design flood hydrographs for ungauged areas, the estimated baseflow rates of similar gauged basins expressed in cumec per sq. km of the catchment area are used to estimate the design baseflow. However, a study conducted for baseflow for small catchments revealed that baseflow during flood season varies from 0.05 cumec/sq. km to 0.44 cumec/sq. km depending upon the meteorological zones in which the basins are located. The values given in Table 16 were considered reasonable:

It is emphasized here that in terms of magnitude, the contribution of baseflow to the peak of a flood hydrograph is quite small.

5.5. FLOOD FREQUENCY ANALYSIS

Frequency analysis is performed to determine the chances of the likely occurrence of hydrologic events. This information is used to solve a variety of water-resource problems, for example, design of reservoirs, floodways, bridges, culverts, levees, urban drainage systems, irrigation systems, stream-control works, water-supply systems, and hydroelectric power plants, floodplain zoning, setting of flood-insurance premiums, etc. Although the frequency analysis of virtually every component of the hydrologic cycle is required, much attention is placed on frequency analyses of streamflow extremes and rainfall only.

The hydrologic data to be analyzed for frequency analysis must be treated in light of the objectives of the analysis, length and completeness of record, randomness of data, and homogeneity. The length of record should be more than 25 years for the derived distribution to be acceptable. The hydrologic data must have been controlled by a uniform set of hydrologic and operational factors. For example, the factors causing a winter rain flood are quite different from those during a spring snowmelt flood or a local cloudburst flood. These two types of floods should not be combined into a single record. Sometimes a hydrologic record may have gaps. Missing data may sometimes be estimated using regional analysis or by correlation with other hydrologic data in the region.

Hydrologic data are generally presented in chronological order constituting the complete duration series (CDS). For frequency analysis, CDS is seldom used because the hydrologic design of a project is normally dictated by only a few critical events. Therefore, hydrologic data can be selected in two ways: (1) partial duration series (PDS) and (2) annual duration series (ADS). PDS is comprised of the data exceeding a specified base level. In ADS, one value (usually the highest) is selected from each year. The two series are comparable if the record is longer than 10 years and either can be used.

Chow (1964) proposed the use of a frequency factor in hydrologic frequency analysis. If a hydrologic variable X is plotted chronologically in time, then a particular value x is found to be composed of two parts: namely, the mean, $\overline{x}$, and the departure from the mean Δx:

$$\mathrm{x} = \overline{x} + \Delta \mathrm{x} \tag{54}$$

The variable Δx can be expressed as the product of the standard deviation S and the frequency factor K. Therefore,

$$\mathrm{x} = \overline{x} + \mathrm{S\ K} \tag{55}$$

where K depends on the return period T and the PDF of X; K literally means the number of standard deviations above and below the mean to achieve the desired quantile. For a distribution, a relation between K and T can be derived. For two-parameter distributions, K varies with T. For skewed distributions, it varies with the coefficient of skewness (C_s) and is very sensitive to the length of record. The frequency factors for some commonly used distributions have been given by Jain and Singh (2003).

For many watersheds, streamflow data are either insufficient or non-existent at the sites of interest. The methods of frequency analysis using data from a single site will have then limited predictive value because of large sampling errors. To overcome the data deficiency, a regional frequency analysis is performed. By defining a region that is hydrologically similar in terms of the variable to be studied, data from several gauging sites within this homogeneous region are pooled together into a single regional frequency analysis. Examples of regional frequency analysis are estimation of design flood from rainfall-runoff relationship, prediction of flood peaks from the relation between observed values and drainage-basin characteristics, and estimation of rainfall depths and frequencies in ungauged areas from characteristics at well-gauged sites in the same area.

The first step in a regional analysis is to define the region itself. The definition of a region depends on the quantities to be estimated. Many methods are available to define a region that is homogeneous. For mean annual precipitation, large physiographic regions can be used, whereas for peak flow, the regions may be confined to drainage basins of certain sizes. Regional boundaries can be defined in terms of similarity of flood-frequency curves or flow curves. Homogeneity tests are used to check if flood-frequency curves in a region can be considered homogeneous.

The frequency distributions that are most commonly used in India include the normal distribution, log-normal distribution, extreme value distribution, generalized extreme value distribution, logistic distribution, log Pearson type-III distribution, exponential distribution, generalized pareto distribution, kappa distribution, and five parameter Wakeby distribution. RDSO carried out regional flood frequency studies using the USGS and pooled curve methods (RDSO 1991) for some of the hydro-meteorological subzones of India. Regional flood frequency studies have also been carried out at some of the academic and research Institutions. Kumar et al. (1999) developed regional flood frequency formula for seven hydro-meteorological subzones of Zone 3 of India: (1) Mahi and Sabarmati—Subzone 3(a); (2) Lower Narmada and Tapi—Subzone 3(b); (3) Upper Narmada and Tapi—Subzone 3(c); (4) Mahanadi—Subzone 3(d); (5) Upper Godavari—Subzone 3(e); (6) Lower Godavari—Subzone 3(f); and (7) Krishna and Penner—Subzone 3(h). For the combined zone 3, the relation proposed them is:

$$Q_T = \left\{-21.05 + 26.27\left[-\ln\left((1 - \frac{1}{T}\right)\right]^{-0.156}\right\} A^{0.64} \tag{56}$$

where A is the catchment area in sq. km.

Kumar and Chatterjee (2005) developed regional flood frequency relationships based on the L-moments approach for the catchments of the North Brahmaputra region of India. For this region, the GEV distribution was identified as the robust distribution. The following regional frequency relationship for GEV distribution was developed by them for the gauged catchments

$$Q_T = \left\{-11.67 + 12.48\left[-\ln\left((1 - \frac{1}{T}\right)\right]^{-0.025}\right\} \overline{Q} \tag{57}$$

where $Q_T = T$-year return period flood estimate (cumec), $\overline{Q} =$ mean annual flood peak of the (cumec) catchment. For the ungauged catchments of the regions, the following relation was proposed by Kumar and Chatterjee (2005)

$$Q_T = \left\{-51.05 + 54.6\left[-\ln\left((1 - \frac{1}{T}\right)\right]^{-0.025}\right\} A^{0.72} \tag{58}$$

5.6. FOOD FORECASTING

A nationwide flood forecasting and warning system has been established in India by CWC which has the responsibility of providing flood forecasting at the national level. CWC issues flood forecasts at 173 stations in the country of which 145 stations are for river stage forecast and 28 for reservoir inflow forecast. Currently, the system covers 70 rivers basins and 18 states/UTs. Figure 11 shows the number of CWC flood forecasting stations in different river basins in India.

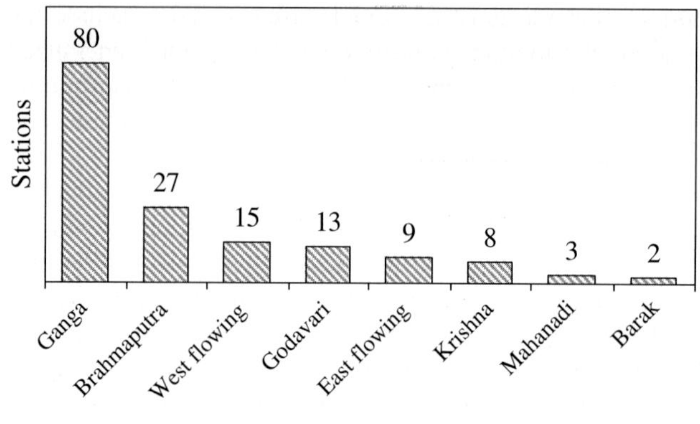

Figure 11. Number of CWC flood forecasting stations in different basins

Flood forecasting set-up in the country is proposed to be modernized by installing telemetry. During the 9th plan, two basins, namely, Chambal and Upper Mahanadi were covered. During the 10th plan, the basins proposed to be covered are Krishna, Godavari, Pennar, Lower Mahanadi, Brahmaputra (Siang and Jiadhal), Ghaghra, Rapti, Damodar, and Yamuna. Figure 12 shows statewise number of flood forecasting stations.

Regarding the techniques adopted for flood forecasting, gauge-to-gauge correlation is most widely used. Use of mathematical catchment models and statistical techniques with updating has been attempted in pilot studies for Yamuna basin, Damodar basin, etc. However, these techniques are still not widely used.

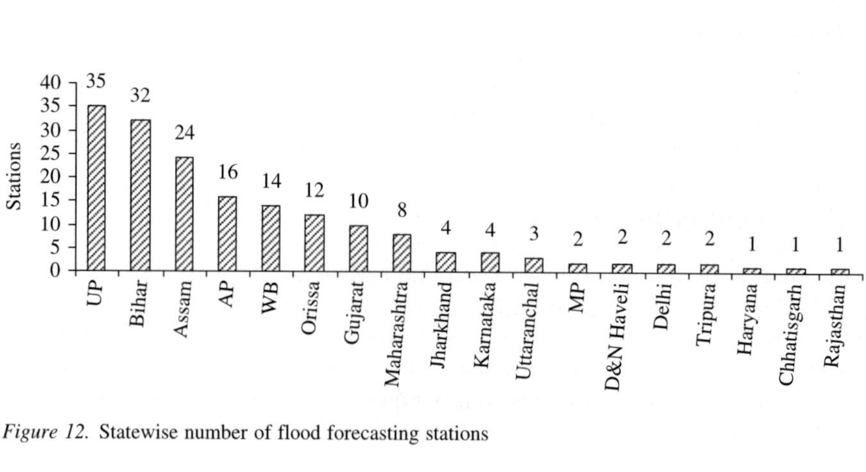

Figure 12. Statewise number of flood forecasting stations

CHAPTER 6
GROUNDWATER

6.1. INTRODUCTION

Ground water is often referred to as the "hidden" component of the hydrological cycle because it is not directly observable. Apart from the constraints which the uncertain knowledge of the ground water system may impose on its development, ground water has some obvious advantages over surface water: better protection against pollution; little treatment required before supplying to user; constant temperature; small distances between source and user; and a fairly steady supply due to large storage capacities. These advantages have stimulated the interest in exploiting the ground water and its conjunctive use with surface water. In many areas, the ground water resources are large and their occurrence has important bearings in water planning and management.

Besides climate and meteorological characteristics, the availability of ground water depends on the geology of a given area and its geomorphological features. Ground water is replenished by the precipitation, riverflow and return flow from irrigated fields. Ground water yield also depends on rock types. Granite and gneiss are better sources than charnockite. Deeply weathered and fractured zones along certain lineaments forms potential aquifers. These lineament zones are highly productive for construction of borewells. Hence, understanding of these factors is important to understand the occurrence and availability of ground water resources.

Since surface water resources in India are often contaminated, ground water plays an important role particularly as a source drinking water. Moreover, ground water has a number of unique features that render it particularly suitable as a water supply source:

- It is generally uncontaminated and can be applied directly without any treatment;
- It may be available close to place where it is required;
- It is dependable and relatively less affected by drought;
- Large storage, treatment and distribution can be avoided; and
- It is less expensive.

Currently, nearly 85% of India's population is dependent on ground water for their domestic demand, particularly as a source of drinking in rural areas. Ground water also plays an important role in agriculture and nearly 140 billion cubic meters

(BCM) of ground water is abstracted annually for use in irrigation. Many industries prefer to use ground water due to the features stated above.

Development of ground water systems implies pumping and an appropriate pumping scheme must be designed on the basis of hydrological conditions and the recharge capability of the system. The hydrogeological input is obtained from measurements on pumping and observation wells. Since the number of observation points is finite, the available information has to be interpolated in time and space in a consistent and reliable way. Hydrogeology of the area has important bearings on the occurrence, movement, and exploitation of ground water. Hence, the hydrogeology of India is discussed in the next section.

6.2. HYDROGEOLOGY OF INDIA

India is a vast country with varied hydro-geological situations resulting from diversified geological, climatological and topographic features. The geological history of India is very complex and diversified. The geological information available in different parts of the country is of varying reliability and detail (See Figure 1). Although the major part of the country has been surveyed and mapped, there are blanks in the geological map of India.

The rock formations which control the occurrence and movement of ground water range in age from Archaean to Recent and have varied composition and structure. The Archaeans including Dharwars and the associated gneisses and granites are the earliest rock systems found in the country. Nearly two-thirds of peninsular area is covered by these rocks. The Cuddapahs, Vindhyans, Gondwanas and the Deccan traps, which occupy the major parts of the remaining peninsular area were formed after the Archaeans. The physiography of India varies from rugged mountainous terrains of Himalayas, Vindhyachals, Eastern and Western Ghats, and Deccan plateau to the flat alluvial plains of the northern river valleys, coastal tracts, and the aeolian deserts in western India (See chapter 1). The evolution of Himalayas began during the mid-Miocene period and continued in phases up to the Pleistocene period. Most of the sediments in these mountains are older than this period.

The Peninsular India is occupied by the ancient shield of Archalan, Proterozoic and Vindhyan fold zones which extend into the ocean beneath the narrow strip of sedimentary platform cover. In the East Coast, the platform cover is shallow and with moderate gradient. The Western Coast of India is characterized by marginal deep followed by an upward rise. However, towards northeast, the peninsular platform continues and buckles down below the Bengal basin to great depths. The platform covers across Garo-Rajmahal stretch is moderately shallow with the basement of Archaean folded zone, again out cropping in Khasi-Garo and Mikir Hills. The major stratigraphic divisions of India are shown in the Table 1.

In India, early man is believed to have appeared about 5 lakh years ago, i.e., during the Middle Pleistocene period. The major rock systems are discussed in what follows.

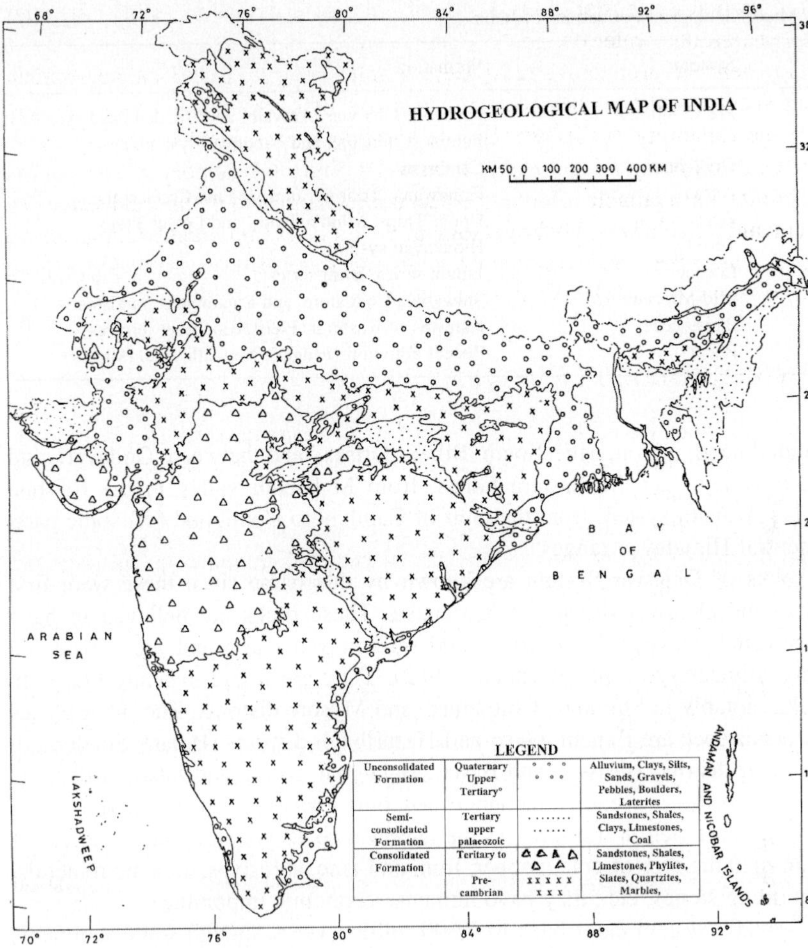

Figure 1. Hydrogeological map of India

6.2.1. Pre-Cambrian Era

The Pre-Cambrians of India comprise a great variety of metamorphosed and sedimentary rocks, some of which probably represent the first formed crust (Proterozic of the earth). In India, Pre-Cambrians are generally divided into *Archaean* (or Pre-Cambrian) and *Purana* (or Late-Precambrian). Pre-Cambrian rocks comprise of gneisses, schists, marbles, quartzites, granites, limestone, dolomites and phyllites.

(a) The Archaean Era

The Archaean rocks are found at many places: (i) the southern Peninsula in the form of the rocks of the Dharwar system in Karnataka and parts of Andhra Pradesh;

Table 1. Major Geological Formations of India

S. N.	System	Formation
1.	Pre-Cambrian	Archean (Dharwar, Aravalli, and Bundelkhand systems); Purana (Cuddapah and Vindhyan systems);
2.	Dravidian	Cambrian
3.	Aryan	Gondwana, Triassic, Jurassic, and Cretaceous
4.	Deccan Trap	Upper Traps, Middle Traps, and Lower Traps
5.	Tertiary	Himalayan systems
6.	Eocene	Jaintia series, Brail series
7.	Mid-Miocene	Shiwaliks, Cuddalore, and Rajamundri Sandstones
8.	Pleistocene	Karewas of Kashmir, Pleistocene river terraces
9.	Recent	Recent alluvium (bangar), New alluvium (khadar)

(ii) Bundelkhand region; (iii) South Bihar, Orissa, and between Godavari and Mahanadi valley; (iv) the Eastern Ghats, from Mahanadi valley to the Krishna valley; (v) Aravalli system from the Gulf of Cambay to Delhi; and (vi) some parts of the central Himalayan ranges.

The rocks of Dharwar System are known by this name since these were first studied in the Dharwar district of Karnataka. These rocks are believed to have been deposited between 2,100 and 3,500 million years ago and are considered to be of sedimentary origin (Krishnan, 1982). They are found in many places in Karnataka, notably in Shimoga, Chitaldurg, and Mysore districts. The other places of their occurrence are Ranchi, Gaya and Hazaribagh districts (Bihar), Sundergarh and Keonjhar districts (Orissa), and Bastar, Balalghat, Rewa, and Jabalpur districts (MP). These rocks are highly metamorphed into gneisses, schists, marbles, and quartzites and are rich in iron-ore. As the Dharwar rocks contain a large variety of important metallic ores of gold, copper, iron, and zinc, abrasives, ceramic minerals, gems, building stones, etc., they have immense economic importance.

The Aravalli System dates back to 3,500 million years, though some sediments were possibly incorporated later. The Satpura basin of metamorphosed sediments and igneous rocks is said to have developed between 860 and 1,000 million years. These rocks are found in Nagpur, Bhandara and Balaghat districts of Maharashtra and extend up to Chindwara and Raipur districts of M.P., and Sambalpur, Sundergarh and Singbhum districts of Orissa.

Bundelkhand gneiss is believed to have been formed more than 2,500 million years ago. Chiefly, it occurs in Bundelkhand area (Southern Uttar Pradesh), and Madhya Pradesh. These rocks are also found in Orissa, Bihar, Tamil Nadu, Karnataka, and Andhra Pradesh.

(b) The Purana Era

The rocks of this era can be sub-divided into the Cuddapahs and the Vindhyans. The Cuddapah System derives its name from the Cuddapah district of Andhra Pradesh where these types of rocks first developed 600 to 1,400 million years ago. Cuddapah rocks are found in the Cuddapah and Kurnool districts of Andhra Pradesh. They also occur around Belgaum district, in Chota Nagpur, Gaya, Monghyr and south Singhbhum districts of Bihar and Jharkhand; in Gwalior, Bastar, Bijawar, Jabalpur, and Rewa tracts of MP and Chhattisgarh. These rocks are also found in the Aravalli range. This rock system is rich in copper ore, nickel, jasper, asbestos, cobalt, and many types of marbles.

The Cuddapah and Kurnool systems of south India as well as Delhi and the Vindhyan system of North India are believed to belong to the Proterozoic period. The Cuddapah system consists of sedimentary rocks which have been subjected to varying grades of metamorphism. The stratigraphical succession of Cuddapah basin is: (i) Kistna Series, (ii) Nallamalai Series, (iii) Cheyair Series, and (iv) Papaghni Series.

The Vindhyan system derives its name from the Vindhyan mountains of central India. The rocks of this system extend from Dehri-on-Sone to Hoshangabad, and from Chittorgarh to Agra and Gwalior. They are found in Sone valley, Rewa district of MP, Sambalpur district of Orissa; in the Krishna valley and Palnad tracts in Andhra Pradesh; in Mahanadi valley in Chhatisgarh, Gulbarga and Bijapur districts in Maharashtra; Chittorgarh district in Rajasthan; and in the north-west of the Aravalli. Shales, slates, sandstones and limestones are the chief rocks. The Vindhyan system contains excellent building stones, limestone, and coal.

Chiefly it is composed of undisturbed sandstones, shales and limestones. The Vindhayan system stretches from Sasaram and Rohtas in the Western Bihar to Chittorgarh in Rajasthan. It occupies about 1 lakh km^2 area. The thickness of rocks of this system can reach up to 4,000 m.

6.2.2. The Dravindian (Cambrian to Middle Carboniferous) Era

The Cambrian system is found in the northern slopes of the Central Himalayan range between Spiti and Kulu, in Lahul, and Garhwal and Kumaon region of Uttaranchal.

6.2.3. The Aryan Era

The Aryan era, which comprises the rock formations from the Upper Carboniferous to Recent, is preserved fairly completely in the Peninsula, and in a perfect sequence in the Himalayan range along the entire northern border.

The *Gondwana system* derives its name from the Kingdom of the Gonds, an ancient tribe of Central India. The rocks forming Gondwanaland group were deposited in a series of large river or lake basins, which later sank among the ancient rocks trough faults. As the sediments accumulated, the loaded basins continued to subside resulting in thick deposits. In the north, the Gondwana rocks extend from the Damodar, the Sone, and the Narmada valley while the southern extent goes up to the Godavari valley. Gondwana group outcrops are found along the Himalayan foothills in Kashmir, Nepal, Bhutan and Assam. These are also seen in scattered outcrops in the Eastern Ghats, between Cuttack and Cape Comorin, in the Rajmahal hills, (Rewa) Madhya Pradesh, Saurashtra, and Kutch. The rocks of the Gondwanas are rich in coal, iron, copper, uranium, and antimony. Fireclay, sandstones, slates and conglomerates are used to construct buildings.

Triassic System is well developed in the northern Himalayan zone from Kashmir to Kumaon, particularly in Spiti-Kumaon belt, where it is termed as the Lilang system. It comprises of black limestones with shale intercalation and quartzites.

Jurassic System occurs in the Himalaya region (Spiti, Kumaon, Kashmir, Hazara), in Kutch (Gujarat), and Rajasthan. In the Zanskar range of Spiti, Garhwal and Kumaon, limestone occurs to a depth of 600 to 900 m. In Rajasthan, outcrops of Jurassic rocks can be seen in Bikaner and Jaisalmer.

The *Cretaceous System* was one of the most widely distributed systems in India and is represented by a variety of rocks, deposited in the land, sea and lakes. There were also igneous activities and lava flows. Upper Cretaceous system is also found in the Pondicherry -Tiruchirapalli belt. A number of small detached outcrops of marine Cretaceous, known as the Bagh beds (in Gwalior), occur along the Narmada valley.

6.2.4. The Deccan Traps

From the end of the Cretaceous till the beginning of the Eocene, there was intense volcanic activity in the Peninsula. There was an immense pouring of basaltic lava out of many fissures in the earth's crust. The first lava flows filled the topographical irregularities and the area was eventually converted into a volcanic plateau covering about a million sq. km in area. Lava traps attain their maximum thickness (up to 3,000 m) near the Bombay coast. It is estimated that the volume of molten rocks which have formed this trap exceed the bulk of the Himalayas.

The Deccan Traps are spread over Maharashtra, Saurashtra, Madhya Pradesh and parts of Deccan Plateau. They are found as far as Belgaum in the south, Rajahmundry in the south-east, Amarkantaka, Sarguja, and Jashpur in the east and Saurashtra, in the north-west. Hills of the Deccan Trap are noted in the Satpura area and in Rewa. The Deccan Traps can be further sub-divided into three groups: a) *the Upper Traps*, with an average thickness of 450 m, found in Maharashtra and Saurashtra, b) *the Middle Traps*, spread over Central India and Malwa, with a thickness of 1,200 m, and c) *the Lower Traps*, with a thickness of 150 m, found in Central India and Tamil Nadu.

The close of the Mesozoic era was marked by the outflowing lava, which had spread over vast areas of Western, Central and Southern India. The lava had spread as nearly horizontal sheets, the earliest flows filling up the irregularities of the pre-existing topography. The area occupied by the Deccan traps is about 520,000 km^2, including parts of Maharashtra, Saurashtra, Kutch, Madhya Pradesh, Central India and parts of Deccan. The Deccan Traps extend up to Belgaum in the south, Rajahmundry in the southeast, Amarkantak in the east, and Kutch in the northwest. Some geologists believe that the traps may have occupied some of the areas intervening between the main mass and the out lying patches, and that the original extent may well have over 1.5 million km^2. The Deccan Traps are the most extensive geological formations of Peninsular India.

The traps are a great treat treasure house of quartz, agate, amethyst; fluorspar, calcite and building stones useful for cement, concrete and road building material. These are also capped by ferruginous and aluminous laterite. The weathering of these volcanic rocks has given rise to the black cotton soil, which is very fertile and suitable for the production of cotton, groundnuts, and castorseeds.

6.2.5. The Tertiary System

This system is of immense significance for the geology of India. The events of the late Cretaceous period continued into the early Tertiary. The sea withdrew from various parts of the earth's surface so that the Tethys began to shrink gradually and eventually broke up. As a result of great volcanic activity, marine sediments in the Tethys up-heaved in stages and triggered the outstanding mountain building activity. It led to the beginning of formation of the Himalayas during the Cretaceous period.

In India, the Tertiary system is subdivided according to the phase of the Himalayan upheaval. Tertiary rocks are well developed in the Himalayan foothills from Kashmir through to the Brahmaputra valley. In Peninsular India, these are found in comparatively small areas in Saurashtra, Surat, Bharuch, Cambay region and Kerala on the west coast and in several places Godavari delta, Pondicherry; Cauvery basin-South Arcot, Thanjavur and Ramnathapuram districts, i.e. from southern Tamil Nadu to Orissa, and in Bikaner and Jaisalmer district of Rajasthan. Many great tectonic movements during this period led to the folding of the thick massive sediments and the rise of Himalayas took place in a series of five major movements.

1. The first rise (the Karakoram Stage) took place during the Upper Cretaceous when the Tethys was divided into a series of longitudinal ridges. This movement was also responsible for sub-division of the eastern gulf into the Assam and the Burma Gulf.
2. Karakoram Stage followed a period of comparative rest after which another upheaval took place during the Upper Eocene. This upheaval brought the deposition of the Murree, Nari, Gaj formations in the area which was marine in the south and brackish water in the north.
3. The third movement, which was probably the most powerful of all, occurred during the Middle Miocene times. This resulted in a considerable folding of the strata laid down in the Tethys into mountain ranges and by large granitic intrusions. Simultaneously, a long narrow trough (foredeep) was formed between the rising Himalaya and the Peninsular mass. Sediments from both sides were deposited in this trough which constitutes the Siwalik system. These sediments are largely of fresh water origin.
4. The fourth upheaval took place towards the close of the Pliocene. This heralded the beginning of the Pleistocene ice age. During this age, Pir Panjal rose up to its present height. Besides, some other ranges in the lesser Himalayas also attained greater heights although there was some subsidence here and there.

5. The final phase of the Himalayan upheaval took place in early Pleistocene when the Pir Panjal range rose. Some other ranges in the Lesser Himalaya were also elevated during this period.

Experts feel that the development of Himalayas is yet complete and it seems logical to infer that in time the alluvial deposits immediately in front of the Siwaliks will themselves be folded.

6.2.6. Eocene to Middle Miocene System

The upper part of the Eocene together with the Oligocene is present in Gujarat, western Rajasthan and Assam, where fossiliferous deposits are found. As the sea withdrew south in the late Oligocene time, marine conditions are replaced by deltaic conditions with river deposits.

In Assam, Eocene is represented by limestones and coal-bearing sandstones of]aintia series in the southern and eastern parts of the Shillong plateau. The *Brail series* in Assam have wide distribution in Surma valley and in the Naga hills. Such deposits also occur in the southern flank of the Pirpanjal, in western Rajasthan and in Gujarat. Coal and limestone are the important minerals which occur in this system.

Middle Miocene to Lower Pleistocene or the Siwalik System covers long stretches along the foot of the Himalaya from Haridwar to Brahmaputra valley. The rocks are made up of sandstones, grits, conglomerates, clays, and silts, which were deposited to a depth of 4,500 to 6,000 m in the shallow water basins. A major fault and three major thrusts separate the Shiwaliks from the older rocks. Shiwalik system is noted for its oil resources, lignite coal, clays, bauxite, and salt deposits.

6.2.7. The Pleistocene and Recent Formations

During the Pleistocene, there was an onset of ice-Age in India. In the Himalayas, there is an evidence of extensive glaciation up to 1,800 m, while glacial drift and terminal moraines cover hill-sides and valley floors down to 1,400 m. In Kashmir, four or five periods of glaciations with three inter-glacial periods, have been distinguished. First Glacial and Inter-glacial periods embrace the Lower Pleistocene in Kashmir. Middle Pleistocene included the Second Glacial and the Second Inter-glacial periods of erosion. This period was marked by heavy deposition of sediment in the Kashmir valley, consisting of boulder fans and thick fluvo-glacial deposits. Upper Karewas are thought to represent the Second Inter-glacial period. Of the same age are the lower Narmada beds. The upper Pleistocene includes the Third and Fourth glaciations and the intervening Third Inter-glacial period. Pir Panjal uplifted during the First and the Third Glacial periods, and the terrace deposits of Kashmir are marked by the end of the Fourth Glaciations.

The *Karewas* are considered to be the result of both the lacustrine and the fluvial deposits between 1,420 m thick in the Lower Karewas and about 600 m in the upper Karewas. Large stretches of alluvial terraces exist in the Sutlej valley composed of sands, gravels and clays. Pleistocene and recent deposits occur in the Narmada, the Tapi and Purna river valleys and along the upper Godavari and upper Krishna valley.

The most important Pleistocene geological formation is the Indo-Gangetic alluvium filling the great-depression between the foot of the Himalaya and the edge of the Vindhyan-Kaimur range. The deposits are up to 1,000 m thick. They are composed of gravels, sands and clays. The other alluvium *(bangar)* covers the higher ground, is dark coloured and contains carbonate of lime, usually occurring in nodules called *kankar*. The light-coloured new alluvium *(khadar)* is found near the present course of the rivers.

6.3. CHARACTERIZATION OF ROCK TYPES OF INDIA

From the view point of ground water characteristics, various rock types occurring in the country can be categorized in two types: Porous rock formations and Hard rock/consolidated formations. The description of these formations follows.

6.3.1. Porous Rock Formations

These formations can be further categorized in unconsolidated formations and semi-consolidated formations.

(a) Unconsolidated formations

The sediments comprising newer alluvium, older alluvium and coastal alluvium are by and large the important repositories of ground water. These are essentially composed of clays, sand, gravel and boulders, ferruginous nodules, kankar (calcareous concretions), etc. The beds of sand and gravel and their admixtures form potential aquifers. The aquifer materials vary in particle size, rounding and in their degree of assertion. Consequently, their water yielding capabilities vary considerably.

Extending from Jammu and Kashmir in the west to Tripura in the east, the piedmont zone of the Himalayas is skirted at some places by artesian aquifers under free flowing conditions. In the Indo-Ganga-Brahmaputra basin, the hydro-geological conditions indicate the presence of large quantities of fresh ground water at least down to 600 m or more below land surface which can be easily developed. High rainfall and good recharge conditions replenish ground water in this basin every year. The alluvial aquifers to the explored depth of 600 m have transmissivity values from 250 to 4,000 m^2/day and hydraulic conductivity from 10 to 800 m/day. Well yields of 40–100 litres per second (lps) are common but these can be more than 100 lps.

Area underlain by unconsolidated formations includes the Indus-Ganga-Brahmaputra Alluvial Tract, Bhabar and Tarai Belts, Central Tract, Fringe Belt Adjoining the Shield, Coastal Alluvial Zone, Gujarat alluvial plain, Ganga Delta, Intermontane Valleys, Intra-Cratonic Basins, and Aeolian Sand Tract. A brief description of these formations follows:

(i) Indus-Ganga-Brahmaputra Alluvial Tract: This tract occupies the depressed zone between the mountainous region of the north and the peninsular region of the south. Its western boundary is at Jammu and its eastern boundary in Assam. This alluvial is made up of the sediments brought by the Ganga and Brahmaputra River system. This zone is rich in ground water resources. This depression is considered to be an orogenic zone of downwarps which is in continuity with the earlier Siwalik orogeny trough. Recent studies have revealed that the thickness of the alluvial in the central part of Bihar is about 1,830 m and in the deltaic region near Calcutta, it is about 200 m.

(ii) Bhabar and Tarai Belts: The northern belt of the Indus-Ganga-Brahmaputra alluvial tract at the Himalayan foothills is characterized by coarse materials (principally boulder-gravel) in the nature of high level coalescing talus fans forming the piedmont terrain. It is referred to as *Bhabar in UP and Kandi* in J & K and Punjab. This belt is immediately to the south followed by Terai or the Sirowal belt of stratified alternating extensive bands of dominantly coarse sediments and clay. Between Bhabar and the confined aquifers of the Terai belt is the vast zone where prolific recharge takes place due to high rainfall and hilly streams. The presence of highly porous and permeable nature of fan formations ensures large supplies of ground water from the Terai belt for agricultural and industrial use. The Bhabar is characterized by deep water table (30 m or deeper), whereas the Terai has an upper unconfined aquifer and a lower interconnected system of confined aquifers. Large ground water reserves in the southern parts of Bhabar belt and the entire Tarai belt have been instrumental in the development of huge agricultural farms and associated industries in the Terai region of UP.

(iii) Central Tract: The central tract of the Indus-Ganga-Brahmaputra alluvium constitutes the major part of these three river basins. A series of (almost) parallel valleys and watersheds are present in Indus and Ganga basins. In the Indus basin, these valleys run approximately east northeast and west southwest. In the greater part of the Ganga basin, the valleys are oriented towards north-west south-east.

In the Brahmaputra basin, the thick alluvial deposits are mainly confined in the flood plains of the rivers and gravel beds in these deposits form the principal aquifers. In these aquifers, ground water often occurs under confined conditions and in favourable topographical locations. Th ground water potential of the aquifers in the Brahmaputra basin is of very high order.

(iv) Fringe Belt Adjoining the Shield: In the belt immediately fringing the peninsular shield, the thickness of the sediments is comparatively small: it generally does not exceed 150 m. The sediments are dominantly clayey, containing lenticular beds of granular material. *Kankarbeds* occurring at various depth ranges normally prove to be potential aquifers. In the western part of this belt, covering parts of Agra and Mathura districts in U.P., Delhi state, southern part of Rohtak, Gurgaon, Rewari and Mahenderagarh districts in Haryana, ground water is generally saline. Fresh water zones occasionally occur either as phreatic aquifers and generally do not exceed a depth of 20 m below ground level or as alternating beds with saline

horizons having limited areal extent. The deeper zones, as a rule, contain highly mineralized water.

Although the nature of sediments undergoes a change east of Bah, the quality does not pose a problem. The aquifers are of limited extent and the thickness of sediments usually does not exceed 100 m. Ground water development problems in this belt are of two types. First, the non-availability of extensive aquifer horizon and second, the presence of negative hydrological boundaries limit the yield potential of the aquifers.

(v) Coastal Alluvial zone: These include the east and west coast alluvial plains of Gujarat and the delta of Ganga and Brahmaputra. The zone also includes a belt of alluvial deposits that extends more or less continuously along the east coast from the Ganga delta to Cape Comorin. This belt is narrow and the maximum width does not exceed 80 km. At places, e.g., south of Chilka Lake in Orissa and near Pondicherry the width is considerably reduced. At the mouth of the Mahanadi, Godavari, Krishna and Cauveri Rivers, broad alluvial delta plains extend inland for a considerable distance. In this zone, the thickness of the alluvium varies and averages about 150 m. The sediments comprise sands of various grades which form promising aquifers. Often these occur under confined conditions.

Along the western coast, the alluvial deposits are discontinuous and are found in patches, mostly in the valleys of small streams draining Western Ghats. At the northern end, these plains gradually widen and merge with the alluvial plains of Gujarat. The thickness of the alluvium is over 100 m at a few places. At a few places, the alluvium contains pebbles and sand beds.

Gujarat has about 1,600 km long coast line from Lakhpat in Kuchchh district passing through Saurashtra and finally terminating at Umbergaon in Valsad district. The coastal tract of Kachchh and Saurashtra comprises a narrow stripe of 200–300 m thick Tertiary and Quaternary formations. This sequence rests over the Deccan Trap basalt. Here, limestone formation of 5 to 50 m thickness is the most prominent unconfined freshwater aquifer at the top. While the deeper aquifers in Gaj formation mostly hold saline water, the top freshwater aquifer in some parts of Saurashtra coast is affected by seasonal seawater intrusion at places where exploitation of ground water by industries and agriculture is high.

The coastal tract of North Goa district spread over about 15 km in length between Fort Aguada in South to Fort Chapora in the North. The Mandovi River Estuary in the south and the Chapora River in the North form the boundaries of watershed. The area is mainly covered by laterites and river alluvium and at some places the meta-sediments are exposed in the north. The ground water flow direction is in accordance with the general topography of the area. The aquifer is mainly shallow unconfined in nature occurring both in river alluvium as well as laterites exposed in the low lying areas.

Along Kerala coast the thick shallow aquifers in the alluvium are generally fresh with isolated pockets of saline water developed during summer near backwater channels and tidal rivers. The underlying aquifers in Tertiary formations, viz.,

Alleppey, Vaikom, Quilon and Warkali hold freshwater with limited brackish water zones. These aquifers extend westward up to the Arabian Sea (Thambi et al., 2005).

Along Tamil Nadu coast, the sedimentary formations of Mesozoic to recent ages overlie the crystalline basement. Deeper freshwater aquifers in the Tertiaries and Cretaceous, overlain by thick inherent saline water zone in South Arcot, Pudukkottai and Thanjavur districts have been found. In Pondicherry and Karaikal, deeper aquifers in Tertiaries hold freshwater. In the costal tracts of Pudukkottai district (Tamil Nadu), freshwater aquifers occur at depths around 250 to 300 m in the Cretaceous formations overlying the Archaean crystalline basement. The overlying Tertiary and Quaternary formations are mostly saturated with brackish to saline ground water.

In the coastal tracts of South Arcot district, Tamil Nadu, mainly occupied by thick Cuddalore sandstones of Tertiary age, the site at Porto Novo right on the coast line is located south of a fault zone. Here the crystalline basement is quite deep, around 2,000 m and the presence of fresh ground water beyond 160 m depth has been reported.

In the coastal tract of East Godavari district, Andhra Pradesh, towards east of Gautami Godavari river the subsurface geological sequence is of Tirupati sandstones, basaltic Traps and the Tertiaries – the Rajahmundry sandstones capped by alluvium. The depth to the top of Trap in the southern most coastal tract is about 800 to 900 m (Murthi and Ramakrishna, 1980). The maximum thickness of alluvium and underlying Rajahmundry sandstone and clays is about 300 m and 500 m, respectively. The Tertiaries and the Traps are exposed towards north of Bikkavolu. The sedimentary sequence in the coastal tract holds numerous potential aquifers but they are mostly saturated with brackish to saline (Raju et al., 1982; 1983).

In Orissa, the coastal tract occupied by sediments is about 330 km long. Resting on the Archaean gneisses, the Gondwanas and Tertiaries are overlain by recent to sub-recent sediments. The thickness of the alluvial capping on an average ranges from 50 to 150 m. The saline ground water zone has a minimum width of 15 km in the extreme northeast and maximum of 75 km in the central part of the Mahanadi delta (Das et al., 2004).

The coastal tract of Midnapur district, West Bengal encompasses a 90 km long strip between Haldia and Digha in the outfall area of the rivers Kasai and Subarnarekha. The area is occupied by more than 2,000 m thick column of Quaternary and Tertiary sediments. The Quaternaries are separated from the Territories by 'Grey Clay' marker horizon. Towards the coast there exists a near surface brackish/saline water saturated zone of thickness varying from 20 to 130 m in the Quaternary sediments. The aquifers in the underlying territories up to about 300 m hold fresh ground water. In northern parts of the Ganga delta, one can find productive aquifers at a depth ranging 75 to 120 m. In the central part, such aquifers occur at depths ranging from 90 to 200 m and near the coast, freshwater aquifers exist only below 250 m.

(vi) Intermontane Valleys: The Himalayan intermontane valleys are in their early stages of development. In contrast, the valleys of peninsular region are in a mature

stage of development. Except the valleys in the Narmada basin, peninsular valleys are broad and shallow, have a low gradient and have attained almost the base level of erosion.

Between the Pir-Panjal and Zanskar ranges lies the Kashmir valley. In this valley, the bed rock comprises rocks ranging in age from Cambro-Silurian to Jurassic. One can find thick deposits of Quaternary sediments whose thickness is more than 2,000 m. These sediments are dominated by fine-grained sands, loams and blue clays, river terraces locally known as *Karewas.* Also found here are river alluvium, pebble beds, and recent moraines.

The famous valleys of Paonta, Doon, Bahl and Nurpur are situated in the outer Himalayas of Uttaranchal and Himachal Pradesh. These constitute broad expanse of water-bearing formations, such as boulders, gravels and sand. The aquifers in central parts of the valley are confined to semi-confined. Good recharge potential and favourable hydro-geological conditions have led to large ground water storage in the area.

(vii) Intra-Cratonic Basins: The Narmada, Tapi and Purna basins of central India fall in this category. Note that there is a wide difference in the thickness of alluvial material in these basins. In geologic parlance, alluvial deposits comprising chiefly of sand and gravel have a thickness close to 100 m in the Narmada valley and there is a distinct possibility of large-scale development of ground water. In view of inadequate thickness of aquifers, the Purna valley offers limited ground water development possibilities. In the Tapi basin also, there is moderate ground water development in places where the aquifer thickness is sufficient.

(viii) Aeolian Sand Tract: Aeolian sands cover large areas in Western Rajasthan (Thar Desert) as well as some coastal tracts. In Rajasthan, the sand mantle is extremely thick, the depth is often of the order of 100 m. The height of dune may be up to about 60 m above the ground level. In coastal areas, the thickness of sand layer may be around 20–25 m.

(b) Semi-consolidated formations

The semi-consolidated formations are chiefly composed of shales, sandstones and limestone; sedimentary deposits belonging to Gondwana and tertiary formations are also included in this category. In Peninsular India, sandstones aquifers have high local potential. Elsewhere they have only moderate, potential and yield may be meager. In other places, they have moderate to very small potential. Though these formations have been identified to possess moderate potential, the physiography of the terrain normally restricts exploitation. Occasionally, the geologic conditions may give rise to artesian conditions, e.g., in parts of the Godavari valley, Cambay basin, arid parts of west coast, Pondicherry and Neyveli in Tamil Nadu. Potential semi-consolidated aquifers particularly those belonging to Gondwanas and Tertiaries have transmissivity values from 100 to $2,300\,m^2/day$ and the hydraulic conductivity from 0.5 to 70 m/day. Generally the well yields in productive areas range from 10 to 50 lps. Lathi and Nagaur sandstones in Rajasthan and Tipam sandstone in Tripura are moderately productive aquifers.

Area underlain by semi-consolidated sediments: Semi-consolidated sediments are found in Tertiaries and Mesozoics of Assam, Cambay basin, Eastern coastal areas, and Western Rajasthan. The mesozoic and tertiary formations of western Rajasthan lie in arid tracts. The Lathi group of sandstones is the most productive aquifers in western Rajasthan. Tertiaries and the Mesozoics of coastal areas include the Baripada bed, Cuddalore sandstones, Rajamundry sandstone, the Triuvakkari beds, the Bhuj beds, etc. These comprise a thick sequence of sea-ward dipping sediments which outcrop at surface at places and sometimes underlie the coastal alluvia. On the eastern coast, these constitute one of the most prolific aquifers occurring under artesian conditions. These often pose salinity problems in the Gujarat and Saurashtra coastal areas.

The rocks of the Gondwana system closely follow the course of some of the peninsular rivers. They also occur as outliers on northern side of the Indus-Ganga-Brahmaputra alluvium principally in Assam and Kashmir. They are chiefly made up of alternations of sandstones and shales with coal seams. The sandstones are generally argillaceous which restrict free movement of ground water. The presence of numerous planes of structural weaknesses in these formations normally favours ground water occurrence and movement. Poor permeability of the Gondwana precludes large-scale ground water development. The upper Gondwanas in certain areas yield good supplies of water. In such terrains ground water development is possible only where geomorphological conditions are favourable.

6.3.2. Hard Rock/Consolidated Formations

Hard rocks possess secondary porosity which is generally of two types: joints, fissures and fractures caused by tectonic features, and intergranular porosity due to action of natural weathering agents. Ground water regime in hard rocks is strikingly different from that of un-consolidated and semi-consolidated sediments. The ground water reservoir in hard rocks is generally not very deep; its depth and lateral extent depends upon the lithology and structure of the parent rock.

Storage of ground water in hard rock reservoirs is controlled by weathering and interconnected planes of structural weaknesses in the parent rock. Flow of ground water in the karstic limestones is often controlled by the nature of the solution channels than the topography. Although hard rocks have limited ground water potential, substantial yields may be obtained if wells are located at suitable sites on the basis of joints, cracks, lineaments, and other features of weakness.

Consolidated formations which occupy almost two-thirds of the country can be broadly classified into three types: i) igneous and metamorphic rocks excluding volcanic and carbonate rocks, ii) volcanic rocks, and iii) carbonate rocks. The nature, occurrence and movement of ground water in these formations are described below.

(i) Igneous and metamorphic rocks excluding volcanic and carbonate rocks

In this category, the most common rock types are granites, gneisses, charnockites, khondalites, quartzites, schist and associated phyllite, slate, etc. These rocks possess negligible primary porosities but are rendered porous and permeable due to presence of secondary porosity by fracturing and weathering.

In areas underlain by hard crystalline and meta-sedimentary rocks, viz. granite, gneiss, schist, phyllite, quartzite, charnockite, occurrence of ground water in the fracture system has been identified down to a depth of 100 m and even up to 200 m locally. The weathered material stores a considerable quantity of ground water in most of the granite/gneiss areas. Generally the fracture systems are hydraulically connected with the overlying saturated weathered matter. The transmissivity of the fractured rock aquifers varies from 10 to 500 m^2/day and the hydraulic conductivity from 0.1 to 10 m/day. Borewells tapping the fracture systems generally yield 1.0 to 10 lps.

Hydrogeology of Plutonic and Metamorphic Rocks: The occurrence and movement of ground water in plutonicigeous and metamorphic rocks is governed by secondary features, viz., joints, fissures, foliation planes and weathering. In these rocks, permeability decreases with depth, since the intensity of weathering and width of fractures also decrease with depth. The permeability of weathered material varies depending on the lithology of rock types and climatic conditions. The maximum depth of weathering could be about 35 m.

Importance of Fracture Studies: From hydrogeological point of view, two types of tectonic fractures, namely, tensile fractures and shear fractures, in crystalline rocks are of importance. Both are related to brittle deformation.

Because of their origin, tensile fractures which are related to brittle deformation have high storativity. They also form very good aquifers. Therefore this type of tectonic pattern should be considered in regional and local ground water development. Studies by Central Ground Water Board (CGWB) in parts of Tamil Nadu and Kerala States have shown that the yield of bore wells constructed in the tensile fracture zones varies from 3.6 to 221 lps while those located in the shear fractures ranges from 2.1 lps to 14.6 lps. Clearly, tensile fractures are more productive than shear fractures. Therefore, proper analysis of fracture types in hard rocks can help in successfully locating wells.

Geophysical investigations with contextual interpretations are effective in making assessments for ground water development. While surface geophysical techniques help define the negative and positive areas and zones before taking up the drilling, the post drilling borehole logging technique precisely identifies the zone to be tapped. The interpretation of geophysical data being contextual, a variety of problems related to exploration, development and conservation of ground water can be geophysically addressed and solutions achieved economically. The electrical resistivity and seismic refraction methods of geophysical exploration are widely used in ground water exploration in hard rock areas. Of these two, the electrical resistivity method is cheaper and is a direct method of ground water exploration.

(ii) Volcanic rocks

The basaltic lava flows are mostly horizontal to gently dipping and ground water occurrence in them is controlled by the contrasting water bearing properties of different lava flows. The topography, nature and extent of weathering, jointing, fracture pattern, and thickness and depth of occurrence of vesicular basalts are the important factors in the occurrence and movement of ground water in these rocks. Basalts or Deccan Traps usually have medium to low permeabilities depending on the presence of primary and secondary porosity. Under favourable conditions, bore wells in these rocks yield about 3 to 6 lps at moderate drawdowns. Transmissivity values of these aquifers is generally in the range of 25 to 100 m^2/day and the hydraulic conductivity varies from 0.05 to 15 m/day.

Hydrology of Basalts (Volcanic Rocks): Basalts are the most common volcanic rocks present in India and are of different geological ages. Deccan Trap, a suite of basaltic rocks occupies an area of about 5 lakh km^2. The occurrence of ground water in basalts depends upon primary and secondary features. The relevant primary features are: vesicles, lava tubes and lava tunnels, and inter-flow contacts. The important secondary features are fractures and joints, and weathering and laterisation.

The porosity of basalts may vary from less than 1% in massive basalts to as high as 85% in pumice. Porosity may be high on account of vesicles etc. but if these are not interconnected, permeability will be low. Permeability of basalt general decreases with increase in the age of rocks. The hydrological characters of basalts also depend upon the type of eruption and whether it was subaerial or submarine.

The Deccan trap basalts are less productive. Pumping tests in basalts of the Ahmednagar district of Maharashtra have given transmissivity values ranging from 15 to 150 m^2/day and of S from 0.01 to 0.13. Specific capacity of bore wells in Deccan Trap of Betul district (M.P) varies from 0.2 to 8.1 lpm/m/m of drawdown. Unit area specific capacity, obtained by dividing the specific capacity by the cross sectional area of dug well in basalts of Ahmednagar district varies from 1.0 lpm/m/m^2 in fractured basalts to 5 lpm/m/m^2 in vesicular basalts. The Deccan Traps are also widespread in Madhya Pradesh.

Almost the entire Peninsular India is occupied by a variety of hard and fissured formations, including crystalline, trappean basalt and consolidated sedimentary rocks (including carbonate rocks), with patches of semi-consolidated sediments in narrow intracratonic basins. Rugged topography combined with compact and fissured rock formations, to give rise to discontinuous aquifers, with limited to moderate yield potentials. The near-surface weathered mantle houses the important ground water reservoir and water circulates through the underlying fracture systems. In the hard rock terrain, deep weathered sediments, low-lying valleys and abandoned river channels generally contain an adequate thickness of porous material to sustain ground water development under favourable hydrometeorological conditions. Generally, the saturated fracture systems occur down to 100 m depth and occasionally yield even up to 30 liters per second (lps). The friable semi-consolidated sandstones also form moderate yielding aquifers, and auto flowing zones in these formations are not uncommon.

(iii) Carbonate Rocks

Carbonate rocks include limestone, marble and dolomite; limestones being more common occurrence extensively. The productivity of wells in carbonate rocks depends upon lithological, topographical and structural characteristics. The permeability of carbonate rocks varies considerably due to the presence of solution cavities and so does the yield of wells. The discharge from a single well from Vindhyan limestone in Jodhpur district (Rajasthan) was found to be 500,000 lph for practically no drawdown. The Pakhal limestones in Karimnagar district of A.P. have a transmissivity of 775 m^2/d and $S = 0.12$. Limestones in M.P. are also quite productive. However, carbonate rocks are not so productive everywhere. Productive limestone aquifers are found in Rajasthan and Peninsular India with yields ranging from 5 to 25 lps. Many springs exist in the Himalayan region in the limestone formations.

Specific capacity and well productivity studies in carbonate and other hard rocks help determine yield characteristics of wells tapping aquifers from varying depths. Wells tapping limestone show productivities between 0.02 and 5.93 lpm/m/m while for sandstone and shale the productivity varies from 0.03 to 0.25 and 0.003 to 0.14 lpm/m/m, respectively. In limestone, the highest productivity is shown by wells in the depth range of 30 to 50 m. The deeper wells show lower productivity.

6.4. INFILTRATION

Infiltration rate is defined as the volume of water entering into the soil at its surface per unit area per unit time and has the dimensions of velocity (L/T). This rate depends on a number of factors, viz., physico-chemical properties of the soil, vegetation and land use pattern, rainfall intensity and duration, and surface slope. Under special circumstances wherein the rainfall exceeds the ability of soil to absorb water, infiltration proceeds at a maximum rate, which is called soil's infiltration capacity.

Infiltration rate is of great interest to hydrologists, agriculturalists, irrigation engineers, etc. as it influences many hydrological processes, such as surface runoff, soil moisture, evapo-transpiration, ground water recharge and spring flow rates. The knowledge of infiltration properties can help agriculturists in adopting proper irrigation methods and irrigation schedule. Infiltration is one of the most important processes responsible for modifying precipitation and converting it to runoff and additions to soil moisture storage. The infiltration process and other hydrological processes are inter-related through a common dependence on soil moisture conditions.

The most important factors affecting infiltration are: soil properties (including structure, grain size distribution, porosity and compactness), soil moisture, land use characteristics, rainfall (amount and duration), soil surface slope, and climate. Singh (1992) has given a detailed description of these factors.

Infiltrometers are the simplest and the most convenient equipment to measure infiltration rates of soils. This involves artificial application of water to enclosed areas. In India, flooding type infiltrometers are frequently used. In these, water is applied on an enclosed area such that a constant head is obtained.

Infiltration studies have a variety of applications in water resources management. Designs of structures for flood mitigation and erosion control are based on estimates of peak discharge which requires prediction of infiltration rate. Water conservation procedures require computation of cumulative infiltration to produce estimates of runoff yield. Similarly, in exploiting water resources for plant growth from rainfall or irrigation, an assessment of cumulative infiltration becomes necessary for calculation of an optimal level of productivity. This assessment embodies efficient water use and maintains an acceptable level of erosion control.

6.4.1. Empirical Relations to Estimate Infiltration Rate

Many empirical and physical relationships have been developed to express infiltration as a function of clay content in the soil or amount of the total quantity of water infiltrated into the soil. Conceptual and empirical models have been developed by applying the principles governing soil water movement for simplified boundary or initial conditions. These models generally correlate infiltration to some measurable properties of the soil. This involves evaluation of parameters for a specific geographical location. The physically based models, on the other hand, are more complicated and use the theory of continuity of mass and soil water movement with certain simplified assumptions. The use of a particular equation or model depends on the intended purpose and the accuracy desired.

Empirical relationship for Ganga plains

In the Ganga plains, the following empirical relationship is used to compute infiltration capacity:

$$F = 0.40 \exp(-0.46A_{cp}) \tag{1}$$

where F = infiltration capacity, and A_{cp} = average clay content of soil in percent.

Amritsar formula

$$R_e = 3.9\,(p - 40.6)^{1/2} \tag{2}$$

where R_e = estimated ground water recharge in cm, and p = precipitation + supplement irrigation in cm.

Chaturvedi formula

$$R_e = 3.5\,(p - 38)^{2/3} \tag{3}$$

6.4.2. Infiltration Studies in Different States of India

India is vast country and expectedly the infiltration rates vary widely. The All India Soil and Land Use Survey Organization has carried out Infiltration studies at some places. The district wise/basin wise results of these studies are presented in the Table 2 and 3.

Table 2. District wise infiltration rates in India at the selected locations

State	District in which investigation was made	Soil type	Range of infiltration rate (cm/hour)
Andhra Pradesh	Adilabad	–	0.80–5.40
	Medak	–	3.00–8.20
	Guntur	–	1.70–10.90
Bihar	Santhal Pargana	–	0.80–12.00
Delhi	Mehrauli	–	0.80–5.80
Gujarat	Whole Gujarat	–	0.05–14.22
Jammu & Kashmir	Whole Jammu Region	Silty Loam, Clay Loam, Sandy Loam, Loam	0.30–5.00
Karnataka	Bidar	–	1.80–12.00
	Dharwar	Chlorite Schist, Shale	1.20–2.00
	Raichur	Granite Gneiss	0.30
	Gulbarga	Granite Gneiss, Limestone, Basalt	0.20–1.70
	Bijapur	Deccan Trap, Limestone, Mixed Sandy Soil	0.90–8.00
Madhya Pradesh	Bilaspur	–	0.8–11.8
	Raigarh	–	1.0–3.2
Maharashtra	Dhulia	–	1.40–2.00
	Shoalpur	–	0.80–1.90
Punjab	Ferozpur	Saline & Waterlogged soils	0.60–2.07
	Faridkot	Saline & Waterlogged soils	1.03–1.94
	Gurdaspur	Medium to Heavy Alkaline soil	0.16
	Amritsar	Medium to Heavy Alkaline soil	0.44–0.48
	Kapurthala	Medium to Heavy Alkaline soil	0.17–0.27
	Patiala	Medium to Heavy Alkaline soil	0.07–0.35
	Sangrur	Medium to Heavy Alkaline soil	0.21
Rajasthan	Western Rajasthan	Sandy Loam, Sandy clay Loam, Clay Loam, Loamy Sand	0.8–18.0
Uttaranchal	Almorah	–	5.4–10.9
	Pauri Garhwal	–	2.8–5.5
Uttar Pradesh	Etah	–	1.3–3.6
	Varanasi	–	1.41–6.15
	Bareilly	–	2.86–7.65
	Agra	–	0.86
	Farrukhabad	–	0.90
	Lucknow	–	1.16–2.80
	Aligarh	–	0.38–1.16
	Sultanpur	–	0.72–2.10
	Hardoi	–	0.54–3.24
West Bengal	Purulia	–	1.1–9.3

Source: NIH (1993–94).

Table 3. Range of infiltration rates in selected basins in India

Basin	Soil type	Range of infiltration rate (cm/hr)
Budigere Basin	Sand, Silt, Silty Clay, Red Loam	0.3–49.8
Pavanje Basin, Karnataka, Krishna	Laterite, Loamy sand, Sandy Soil, Sandy Clay, Clay	0.57–64.80
Malaprabha Sub Basin, Krishna	Laterite, Granitic Clay, Black Cotton, Red Sandy Soil, Silty Clay, Red Laterite, Clayey Loam, Black Clay, Loam, Clay	0.80–23.10
Ghataprabha Sub Basin	Red Sandy, Red Clayey, Loam, Mixed Sandy	0.90–16.50
Upper Luni Basin	Sandy clay, Clay loam, Silty clay loam, sandy loam, loam, Sand	3.50–22.99
Guhiya Watershed	Sandy clay loam, Silty Clay loam, sandy loam, loam, Sand	3.43–22.40
Hindon River Basin	Sand, Sandy loam, Silt Loam, Loamy Sand, Silt, Loam	0.30–51.90
Baithon watershed, Tehri Garhwal	–	0.35–7.05
Kanhar Catchment, Mirzapur (Ganga Basin)	–	0.48–16.65

Source: NIH (1993–94).

6.4.3. Seepage and Drainage

Canals are the major means to carry and deliver irrigation water in India. Seepage from an unlined canal starts as soon as water is filled in it. In unlined canals, a substantial percentage of transported water is lost to seepage. It has been estimated that seepage losses are about 45 percent of the water supplied at the head of the canal. According to the Indian Standard IS:9,452 (1980), the loss of water by seepage from unlined canals in India generally varies from 0.3 to $7.0\,m^3/per10^6\,m^2$. Obviously, losses are more in canals in alluvial areas. Estimates show that if the seepage loss is prevented, more than 6 M-ha of additional area could be irrigated.

Seepage loss from an unlined canal depends on the depth of water in the canal, depth to water table in the vicinity of the canal, width of the canal at the water surface, side slopes of the canal, distance of the governing drainage, and the coefficient of permeability of the porous medium. Lining of canals reduces the loss of water due to seepage; depending on the properties of the lining material, seepage occurs at a reduced rate. Good lining should be economical, impervious, hydraulically efficient, durable, and stable. Conventional lining using cement concrete, masonry, asphalt, pre-cast concrete and R.C.C. is expensive. Bentonite is available in abundance in Rajasthan, Bihar, and Kashmir and can be used for lining. Seepage losses from a canal may lead to water logging and soil salanisation.

6.4.4. Measurement of Seepage Losses from Canals

Methods for direct measurement of seepage losses from existing canals include the ponding, inflow-outflow, and seepage meter. Advanced methods, such as the tracer technique, electrical logging or resistivity measurement, and piezometric survey are also used in some cases.

The Indian standard code of practice IS:9,452 Part I, (1980) describes measurement of seepage losses from canals by the ponding method. In this method, a reach of the channel is isolated by temporary bunds or bulkheads and seepage from this section is measured. The method has been applied to canals with a discharge capacity of up to $150\,m^3/s$. Gravity flow or pumping may be used to fill the test pond, depending on the conditions that prevail at the site and the size of the canal. The rate of fall of the water surface within a few hours after the initial readings of the gauges shall provide an indication of loss rate. Seepage losses may be computed from observations after the steady state has been achieved.

To determine seepage losses by the inflow-outflow method following the Indian Standard IS:9,452 [Part II, 1980], the quantities of water that flow into and out of a canal reach are measured. The difference of these is attributed to seepage. Evaporation from the canal water surface and precipitation are considered in the computation. The length of the reach should be such that the loss from the reach is much more than the accuracy of measurements. When this method was used on canals in Maharashtra, the seepage losses were found to vary from 2.22 to 2.91 cumec per Mm^2 of the canal wetted surface.

Singh (1983) estimated seepage from the Janjokar minor and the Dabthua distributory by the inflow and outflow method and found it to be in the range of 1.17 to 2.14 cumec per Mm^2, respectively. Seepage loss from the Ferozepur Feeder canal was estimated by inflow-outflow method using current meters by Dhillon et al. (1986). It is a large capacity (design discharge = 316 cumec) double tile lined channel in Punjab. The average seepage flow from a reach 9.144 km long was found to be 2.70 ± 1.0 cumec per Mm^2 of wetted surface area (0.23 ± 0.09 m/day). This loss rate is inclusive of evaporation from the channel surface which was 0.01 m/day.

Plastics, manufactured as low-density polyethylene (LDPE) and high-density polyethylene (HDPE), has been used in India for controlling seepage through earthen structures since the 1950s. Punjab and Gujarat were the first states to use these for lining. UPIRI (1989) studied seepage losses from experimental irrigation channels and existing irrigation field channels under the following conditions: i) unlined channels, ii) channels with hard surface lining, iii) channels with LDPE film lining with varying thicknesses and earth cover over it, and iv) LDPE film lining covered with hard surface lining, such as brick lining, cement concrete tiles lining, and compressed clay tile lining. The seepage loss from an unlined reach of a distributory in Aligarh (UP) was found to be 1.139 cumec per Mm^2 surface area. The reach had average width of 1.37 m; side slope 1:1, average water depth 0.4 m and the discharge was 0.396 cumec. The seepage loss from a lined reach of this distributory was 0.295 cumec per Mm^2. Thus, the seepage reduction of about 74% was possible by LDPE lining.

6.5. GROUND WATER MONITORING NETWORK IN INDIA

Since ancient times, wells have been dug to tap water-bearing formations. A well may be used to observe the ground water table, if the well depth covers the range of water table level fluctuations and the litholog is known. If the wells in-use are used for observation, there should be sufficient gap between the last withdrawal and measurement to allow recovery of the water level in the well. Abstractions in the vicinity of an observation well should also be stopped sufficiently before the observation for the cone of depression to recover. A reference mark to measure water depth (levelled to a common datum) should be clearly marked near the top of the well.

When ground water is up to 15 m, observation wells can be constructed manually. In areas with unconsolidated sand, silt, or clay, small-diameter observation wells up to about 10 m depth can be constructed by the drive-point method. Deeper wells are drilled by the rotary or percussion methods. Because it is cheaper to drill small-diameter wells, observation wells with inner diameters ranging from 50 to 150 mm are common.

Ground water levels in observation wells are measured either manually or with automatic recording instruments. The most common manual method is by suspending a weighted tape, wire or rope from the surface at the well head and measuring the length of tape up to the water level. Several measurements may be made to get improved results. Water level depths can be measured to an accuracy of a few cm but this of course depends on the depth. As depth to water level increases, the length of tape/rope to be used increases, thereby increasing the weight and making the procedure cumbersome. Beyond depths of 50 m, the probability of errors begins to increase. Using the double-electrode system, water level up to about 150 m depth can be measured with ease.

Many types of automatic water-level recorders are in use these days. Such instruments should be portable, easily installable, capable both of recording under a variety of climatic conditions, should work unattended for long durations. They should also be able to measure ground water fluctuations at short intervals of time. Float actuated analogue recorder is often used in practice. The hydrograph is traced on a chart fixed to a drum and the recording drum or pen is driven by a spring or a clock mechanism. Automatic stage recorders used for stream gauging can be readily adapted to measure ground water levels. These instruments need some protection from extreme climatic conditions, accidents, and vandalism.

CGWB has a large network of stations for hydrological observations. It monitors ground water levels from a network of about 15,000 stations (mostly dug wells selected from existing wells, uniformly distributed throughout in the country). Table 4 gives the number of ground water hydrograph stations in different states and urban territories of India. Dug wells are being gradually replaced by piezometers for water level monitoring. Measurements of water levels are taken at these stations four times in a year in the months of January, April/May, August and November. Water samples are also collected every year during April/May for chemical analyses. The data so generated are used to prepare maps of ground water depths, water level

Table 4. Status of Ground Water Hydrograph Network Stations

Name of the States/Union Territories	Number of stations (in 1995)
States	
Andhra Pradesh	1,042
Arunachal Pradesh	17
Assam	371
Bihar	599
Goa	53
Gujarat	974
Haryana	521
Himachal Pradesh	78
Jammu & Kashmir	162
Karnataka	1,349
Kerala	651
Madhya Pradesh	1,350
Maharashtra	1,409
Manipur	25
Meghalaya	37
Mizoram	0
Nagaland	8
Orissa	1,122
Punjab	497
Rajasthan	1,414
Sikkim	0
Tamil Nadu	766
Tripura	37
Uttar Pradesh	1,514
West Bengal	836
Total States	14,832
Union Territories	
Andaman & Nicobar Islands	29
Chandigarh	14
Dadra & Nagar Haveli	7
Daman & Diu	61
Delhi	6
Lakshadweep	30
Pondicherry	16
Total U.T.s	163
Total for India	14,995

contours and changes in water levels during different periods and years. The data are also used to prepare long-term trends of rise and fall in water levels.

Deeper ground water level of over 50 m is observed in Piedmont aquifer in the Bhabar belt in foothills of Himalayas. In Western Rajasthan, ground water levels have depths ranging from 20 to 100 meters. Water levels range from 5 to 20 meters below land surface in peninsular region.

Considerable improvements in ground water observation networks have been made in nine peninsular states under the Hydrology Project Phase I. Details about the same are given in a later chapter.

6.5.1. Ground Water Quality Monitoring

Water in a dug well should be protected from rain, flood, or seepage of contaminated waters, which may pollute the well and then the aquifer. The quality of ground water is subject to change and deterioration as a result of the activities of man. Localized point sources of pollution include cesspools and septic tanks, leaks in municipal sewers and waste ponds, leaching from garbage dumps and sanitary landfills, runoff from animal feedlots, industrial waste discharges, cooling water returned to recharge wells, and leaks from tankers or pipelines. Larger geographical areas may suffer degradation of ground water quality because of irrigation water returns, recharge into aquifers of treated sewage or industrial effluents, and intrusion into fresh water aquifers from neighbouring sea water or other highly saline aquifers.

Water samples are frequently collected from wells. Bailer, a simple device to lift samples of water from a well, consists of a length of tube with a ball valve at the bottom. The bailer is tied to a cable and lowered into the well. The valve allows water to enter the sample but when the sampler is lifted, it prevents water from escaping at the bottom. For sampling at any depth below the water table, samplers with spring operated valves or grab samplers are employed. In some cases, selected zones in a well are temporarily isolated for sampling. Most Variables for ground water quality are more or less the same for surface water quality except for turbidity, which normally is not a problem in ground water.

The electrical conductivity of water increases with increasing salinity. This property is used to determine the concentration of minerals dissolved in water. The instrument used for measuring resistance may be a very simple transportable small resistance bridge that measures in situ the resistance of a water sample pumped or bailed from the well. In cases where a measurement of salinity at a certain point below the water table is required or if a chemical log of the whole column of water in the well is required, a salinity logger is used.

The issues related with ground water quality are discussed in detail in a later chapter.

6.6. GROUND WATER RESOURCES

When water is withdrawn from ground water, it results in drawdown of water table which is a temporary lowering of water table. In this situation, the water table recovers its position when the supply is replenished. Withdrawal differs from overdraft which may cause loss of storage capacity and may result in permanent lowering of water table.

Ground water resources can be classified as static and dynamic. The static resource is the amount of ground water available in the permeable portion of the aquifer below the zone of water level fluctuation. The dynamic resource is the amount of ground water available in the zone of water level fluctuation. The modes of ground water occurrence in the country can be broadly divided into two groups:

(a) Porous formation which comprise of unconsolidated and semi-consolidated sediments. Aquifers are generally extensive and interconnected resulting in moderate to very high potentials.
(b) Consolidated and fissured formations. Here, aquifers are discontinuous and as a result, they have limited yield potentials.

As described earlier, rock formations in India vary from Archean to Recent and the topography ranges from mountain terrains of Himalayas to the flat alluvial plains and coastal tracts. Due to diversified meteorological, topographical, and geological features, the ground water distribution over the country shows enormous variation. The plains in the Indus-Ganga-Brahmaputra basins extend over a distance of nearly 2,400 km from east to west. These plains constitute one of the largest ground water reservoirs in the world. Here, the aquifer systems are extensive, thick, hydraulically interconnected, and yields range from moderate to high. To the north of this tract all along the Himalayan foot hills is a narrow linear belt of Bhabhar piedmont deposits. Just south of this belt is the Tarai belt down slope with characteristic auto flowing conditions. Almost the entire peninsular part is covered by a variety of hard and fissured rock formations. The coastal and deltaic tracts form a narrow strip around the peninsula; some of these contain highly productive aquifers.

Three major groups of ground water structure are constructed in India. A dugwell has a diameter of about 1.0 m and is 10–15 m deep. It irrigates about 1 ha in the hard rock areas; this area can increase manifold depending upon aquifer properties and by installing a pump. Large diameter wells have a diameter of 2 to 3 m. Shallow tubewells have 10 to 15 cm diameter in alluvium and these are 15 to 20 m deep. Typical discharge is about 50 m^3/hr. The area irrigated is about 3 to 4 ha. In hard rock areas, these are called borewells and the area irrigated is about 2–3 ha. Deep public tubewells are about 100 to 150 m deep with about 20 cm diameter and are gravel packed. The discharge varies from 100 to 150 m^3/hr. The area irrigated is 40 to 80 ha and the cost can range between Rs. 5 to 9 lakh. These are constructed, operated, and maintained by government departments or corporation of the government.

6.6.1. Static Ground Water

The water that is present in the aquifers below the zone of water level fluctuation is known as static ground water. The amount of static ground water is estimated by:

$$\mathrm{SGW} = \Delta x^{*} A^{*} S_y \qquad (4)$$

where, Δx = thickness of aquifer between zone of water level fluctuation to limit of exploitation, A = areal extent of the aquifer, and S_y = specific yield of the aquifer. CGWB has estimated the static ground water resource of the country as 10,812 km^3. In addition about 10–15% of this quantity will be available as return flow from irrigation and domestic uses. As against this, the total withdrawal for all uses in the year 1990 was 552 km^3.

The quantum of ground water available for development is usually restricted to long term average recharge or dynamic resources. Presently there is no fine demarcation to distinguish the dynamic resources from the static resources. While water table hydrograph could be an indicator to distinguish dynamic resources, at times it is no so, particularly when water tables are deep. Sustainable ground water development requires that only the dynamic resources are tapped. Exploitation of static ground water resources could be considered during extreme scarcity conditions, that also only for essential purposes. The basin and state wise static fresh ground water resource are listed in Table 5 and 6 respectively.

6.6.2. Concepts Used to Quantify Ground Water Resource

To quantify the available ground water resource, two different concepts based on the hydrogeological situations are used: (a) Quantity concept – for unconfined (water table) aquifers (b) Rate concept – for confined aquifers.

(a) Unconfined (water table) aquifers
The useable ground water resource is essentially a dynamic resource which is (periodically) recharged by rainfall, irrigation return flows, canal seepage, etc. The most important component of aquifer recharge is the direct infiltration of rainwater, the quantity of which varies according to the climate, topography, soil and subsurface geological characteristics. A part of applied irrigation water reaches ground water depending on the efficiency of irrigation system and soil characteristics. Influent streams also recharge aquifers depending on the width of streams and the properties of river bed material. Other sources of recharge are percolation from canals, reservoirs, tanks, etc.

(b) Confined aquifers
Unconfined aquifers respond to ground water withdrawals differently than the confined aquifers. Confined aquifers yield water due to expansion of fluid volume and compaction of pore volume. By definition, a confined aquifer remains saturated. On the other hand, unconfined aquifers yield water by desaturation of the pore space as the water table declines. Water released from storage in an unconfined aquifer greatly exceeds that from a confined aquifer for equal of declines water level.

Ground water potential for the confined aquifers which are hydrogeologically separate from shallow water table aquifers is assessed by rate concept. The ground water available in a confined aquifer equals the rate of flow of ground water through it. The rate of ground water flow in an area can be estimated by using Darcy's law.

Most of the ground water development in India is up to a depth of about 50 m for ground water development. In the shallow region, ground water withdrawal is through dug wells, dug-cum-bore wells, and shallow tube wells. The exploitation in the deeper zone (50 m and more) is usually through public sector for community irrigation, municipal water supply, or industrial purposes. Deeper structures include heavy duty tube wells and bore wells. Ground water development has been intensive

Table 5. Static fresh ground water resource – Basinwise

S. N.	River basin	Static fresh ground water resource (km^3)		
		Alluvium/Uncon solidated rocks	Hard rocks	Total
1	Indus	1,334.9	3.3	1,338.2
2	Ganga-Brahmaputra-Meghna Basin			
2a	Ganga sub-basin	7,769.1	65	7,834.1
2b	Brahmaputra sub-basin	917.2	0	917.2
2c	Meghna (Barak) sub-basin	101.3	0	101.3
3	Subarnarekha	10.1	0.7	10.8
4	Brahmani-Baitarani	40.1	3.3	43.4
5	Mahanadi	108.4	11.3	119.7
6	Godavari	36	23.4	59.4
7	Krishna	13.6	22.4	36
8	Pennar	3.9	7.2	11.1
9	Cauvery	39.1	3.3	42.4
10	Tapi	4.3	3.2	7.5
11	Narmada	13.8	4.6	18.4
12	Mahi	9.7	2.9	12.6
13	Sabarmati	25.5	2.7	28.2
14	West Flowing Rivers of Kachchh and Saurashtra including Luni	103.1	10.1	113.2
15	West Flowing Rivers South of Tapi	5.4	5.8	11.2
16	East Flowing Rivers Between Mahanadi and Godavari	34.4	6.9	41.3
17	East Flowing Rivers Between Godavari and Krishna			
18	East Flowing Rivers Between Krishna and Pennar			
19	East Flowing Rivers Between Pennar and Cauvery	63.1	2.9	66
20	East Flowing Rivers South of Cauvery			
21	Area of North Ladakh not draining into Indus	Not Assessed		
22	Rivers draining into Bangladesh	Not Assessed		
23	Rivers draining into Myanmar	Not Assessed		
24	Drainage areas of Andman, Nicobar and Lakshadweep islands	Not Assessed		
	Total	10,633.0	179.0	1,0812.0

Source: CGWB (1999).

in the plains of the Indus-Ganga basin in Punjab, Haryana, and Uttar Pradesh. Similar situation is observed in hard rock regions in peninsular India.

Two types of situations of occurrence of confined aquifers may be encountered. In hard rock areas, the upper water table aquifer in the weathered zone is connected to the deeper fracture zone, which is semi-confined. In such situations, the assessment

Table 6. Static fresh ground water resources-Statewise

S. N.	States	Static Fresh ground water resource (km^3)		
		Alluvium/ unconsolidated rocks	Hard rocks	Total
1	Andhra Pradesh	76	26	102
2	Assam	920	0	920
3	Bihar	2,557	11	2,568
4	Gujarat	92	12	104
5	Haryana	420	1	421
6	Himachal Pradesh	13	0	13
7	Jammu & Kashmir	35	0	35
8	Karnataka	0	17	17
9	Kerala	5	6	11
10	Madhya Pradesh	14	27	41
11	Maharashtra	16	22	38
12	Orissa	162	13	175
13	Punjab	910	0	910
14	Rajasthan	115	13	128
15	Tamil Nadu	98	0	98
16	Tripura	101	0	101
17	Uttar Pradesh	3,470	30	3,500
18	West Bengal	1,625	1	1,626
19	Delhi	3	0	3
20	Chandigarh	1	0	1
	Total	10,633	179	10,812

Source: CGWB (2005).

procedure for unconfined aquifer accounts for full recharge and hence, no separate assessment is made for the confined aquifer.

In some alluvial areas, resource from a deep confined aquifer may be important. If the confined aquifer is hydraulically connected to the overlying shallow water table aquifer, it is a semi-confined aquifer. If there is no hydraulic connection to the overlying water table aquifer, the potential may have to be estimated by specific detailed investigations, taking care to avoid duplication of resource assessment from the upper unconfined aquifers.

Ground water flow available for development in a confined aquifer in the area can be estimated by using Darcy's law:

$$Q = T\ I\ L \tag{5}$$

where Q = Rate of flow through a cross-section of aquifers in m^3/day, T = Transmissivity in m^2/day, I = Hydraulic Gradient in m/km, and L = Average width of cross-section in km. The transmissivity may be computed from pumping test data of tubewells. Leakage from overlying or underlying aquifer may also be accounted for in the calculation of ground water available for development in a confined aquifer.

The tube well drafts tapping a deeper confined aquifer may be treated separately and may be accounted for at the time of quantitative assessment of deeper confined aquifer. The total draft of these tube wells may be taken as gross draft of which 30% may be taken recycled and may be added as recharge to water table aquifers. The utilizable recharge may be taken as 85 percent of the total ground water flow available for development.

Lateral flow in the confined aquifer may be computed by flow net analysis. However, for determine the optimum development of the confined aquifers, the recharge area of the confined aquifers may be demarcated. Now, the average annual recharge to the confined aquifer in this recharge area is estimated. The extent of development of this aquifer should be limited to this recharge.

Analysis of Pumping Test Data

Two important hydrological parameters, namely, transmissivity and storage coefficient are an essential input in assessment and management of ground water in an area. These parameters are commonly assessed through pump tests. Such a test is also called an 'aquifer test' because it is the aquifer, rather than the pump or well, which is tested. Pump test is also conducted to determine the performance characteristics of a well and pump. For a well-performance test, yield and drawdown are recorded. These data can be used to determining the specific capacity or the discharge drawdown ratio of the well, to select the type of pump, and to estimate the cost of pumping. The specific capacity gives a measure of the effectiveness or productive capacity of the well. Such a pumping test is sometimes termed a 'well test' because it tests the well rather than the aquifer.

The three common types of wells that are encountered in hard rock formations are large diameter dug wells, bore wells, and dug-cum-bore wells. As the hard rocks are anisotropic and heterogeneous, the ground water flow mechanism in such rocks is complicated. The type of ground water structure on which pumping test is carried out is also important in the choice of a particular method for the analysis of test data. The bore wells are usually of 6 cm to 9 cm in diameter and 30 to 60 m in depth. In fractured rocks, the following characteristics may be determined:

a. Horizontal hydraulic conductivity and confined storage coefficient of the fractured zone.
b. Effective vertical hydraulic conductivity and confined storage coefficient of weathered and fractured zone.
c. Horizontal hydraulic conductivity and specific yield of the weathered zone. These parameters can be determined by (i) Leaky aquifer theory (ii) Neuman's unconfined aquifer theory, and (iii) Numerical model. The conventional Jacob's straight-line method has been used by many workers to determine aquifer parameters in hard rocks.

Dug wells of large diameter are the primary source of ground water extraction not only in India but also in many Asian and African countries. There are over 9 million dug wells in India. The wide use of large diameter wells is mainly due to their low cost of construction and simplicity of operation and maintenance. Water from these wells is mostly manually drawn through buckets. These wells are also ideal in hard rock formations as they provide necessary storage for water during recuperation. Many studies have been conducted to better understand the hydraulics of large diameter wells to make the structures more efficient and to determine the aquifer parameters from pumping tests on such wells. The main problems in analyzing pumping test data from large diameter well are:

- Effect of well storage,
- Variation in discharge with time,
- Development of seepage face in unconfined aquifer,
- Partial penetration of well, and
- Anisotropic nature of the aquifer.

Due to these reasons, the conventional methods based on Theis' equation are not suitable for analyzing flow to large diameter wells. Analytical and numerical solutions of steady and unsteady flow to a well storage have been developed by several workers. For example, Chachadi et al. (1990) have used discrete kernel approach to analyze pumping test data from large diameter well.

Specific capacity determined from both recovery data on dug wells and drawdown data from bore wells can be used to determine the hydraulic characteristics of aquifers in hard rocks. Usually, specific capacity index in hard rocks decreases with increase in depth. This observation helps in determining the optimum depth of wells. The specific capacity of dug wells in different hard rocks formation is given in Table 7.

Well Spacing: Overexploitation of ground water can be avoided to some extent by specifying the density of wells, and the spacing between adjacent wells. Two approaches are used for this purpose. One uses the pumping test data to develop a relation between the duration of pumping and corresponding radii of influence in different hard rock terrains. The second approach is based on ground water recharge data to compute well density such that the total draft does not exceed the annual ground water recharge.

The first uses the formation characteristics, i.e., T and S, to compute the well spacing. Table 8 gives the results of above two approaches for determining well spacing and well density. This table indicates that the recharge method gives a

Table 7. Specific capacity of dug wells in different hard rocks formation

Formation	Sp. capacity (lpmin/m) (C)	Unit area sp. capacity = (C/Area) (lpmin/ m/m^2)	Sp. capacity index ($lpmin/m/m^2$) $= C/2\pi rh$)
Basalt	20.4–364	0.35–5.04	0.16–3.02
Limestone	30.4–348	0.53–19.83	0.16–14.14
Shale	28.4–39.3	0.53–1.22	0.26–0.37
Gneiss	15.5–228.6	0.23–10	–
Sandstone	54.3	0.97	0.60
Laterite	582.2	12.57	5.10

Source: Singhal and Singhal (1989).

Table 8. Well Spacing in Different Formations

Formation	Well spacing using radius of influence approach (m)	Recharge well spacing (m)	Well density approach (per km^2)
Greywackes	110–136	224–448	6 to 25
Basalt & related rocks	170–256	370–417	7 to 9
Limestones	126	415	7
Gneisses, Granites	134–382	191–492	5 to 35
Charnockites Literates (coastal areas)	130–275	190–225	25 to 35

Source: Singhal and Singhal (1989).

range of values between 190 to 492 m with the maximum well density in the range of 5 to 35 wells per sq. km.

6.6.3. Ground Water Flow

Movement of ground water from one point to another is caused by a difference in flow potential or 'head' between the points. Water moves from higher head to lower head. The area where ground water receives input is known as the recharge area. Infiltration of precipitation is the principal source of ground water recharge. Recharge varies from one place to another depending on rainfall properties, local topography, soil properties, geology, and hydraulic gradient. The speed of flow of ground water varies from one location to another depending on the geological and topographical factors. The normal range is between 1.5 m/year and 1.5 m/day. Weathering of rocks is the main factor that controls the movement and storage of ground water in the impervious basaltic rocks. Secondary porosity develops due fractures and joints; usually these are restricted only up to a depth of 40–50 meters.

In the plains of Punjab and Haryana, ground water flows from three sides: north, northeast and southwest to the central and eastern Haryana. This is because the topographical features have created a saucer like depression in this part. Geological Survey of India has made seismic refraction surveys which have revealed existence of a sub-surface ridge and strike in northwest-southeast direction which is the continuity of the Aravallis out-cropping in Mahendragarh and Gurgaon districts. This sub-surface ridge has important influence on ground water flow in this region.

In the Ganga basin, ground water movement follows the general drainage direction towards the Bay of Bengal. Consequently, the elevation of water table contours decrease as one moves from Uttaranchal towards the West Bengal. The *Bhabhar* zone contains piedmont of alluvial deposits along the Himalayan foothill regions. This belt is dominated by unconsolidated sand-boulder and clay boulder beds but at few places close to the foothills there exist Shiwalik formations and with limited thickness of the unconsolidated *Bhabhar*. From the south along the Son River, large quantity of ground water enters to join the flow in the main basin. Ground water flow in the Brahmaputra valley follows the same pattern. Brahmaputra River flows along a north-south fault bordering the western foothills of the Garo Hills and this continues into a fault present in the Himalayan foothills region near the Torsa Valley. These prevailing geological formations influence ground water occurrence and movement because the presence of faults determines the areas where ground water may enter and leave each stratum.

Another major aquifer system is present in the eastern parts of Chhattisgarh, surrounded by the Maikal range in the north and Nimgiri in the southeast. The Mahanadi and the Brahmani Rivers that flow towards the east and Wainganga River flowing towards the west provide good amount of water recharge to the aquifer.

Along the eastern coast of India, ground water movement is in the expected direction from northwest to the southeast. Extensive saline patches occur in Ramnad, Nellore and Krishna districts. In Ramnathpuram and Tirunelveli, the ground water occurs in unconfined aquifers and generally is of poor quality. In the west coast areas of Kerala and Karnataka the substratum is mostly lateritic and a good yield of ground water may be expected. Along this coast, ground water flow is from northeast to southwest coinciding with the water table contours of the area.

North of Malabar Coast in the adjoining areas of Mahableswar, ground water flows towards the northeast direction and moves to the Godavari valley. Another flow of ground water comes from northwestern and western sides coinciding with the Narmada and the Tapi River flows.

6.7. ESTIMATION OF GROUND WATER RESOURCE IN INDIA

Till the sixties, ground water potential in India was usually evaluated on sectoral, regional, or projects basis. Ministry of Agriculture prepared guidelines for estimation of ground water potential in early 1970s. In 1984, the *Ground Water Estimation Committee* (GEC) of CGWB recommended a methodology for estimation of ground water resources. A similar committee was constituted in 1990s to suggest improvements in GEC (1984) norms. This committee submitted its recommendation in 1997. Discussions in this section are mainly based on GEC (1997) recommendations, which are also under revision.

Two approaches were recommended by GEC (1984): ground water level fluctuation method and rainfall infiltration factor method. Retaining these approaches, GEC (1997) suggested several improvements. In the revised methodology, distinctions have been made in hard rock areas and alluvial areas, canal command areas, and recharge in monsoon season and non-monsoon season. GEC (1997) recommended that recharge due to rainfall in the monsoon season is to be estimated by ground water level fluctuation method. But if adequate data is not available, the rainfall infiltration factor method may be used. The usable ground water resource is the dynamic resource which is recharged annually by rainfall and other sources.

6.7.1. Ground Water Balance Equation

The water level fluctuation method employs the ground water balance equation. In gross terms, for any specified period, the equation for the unit (block, watershed, etc.) is:

$$\text{Input} - \text{Output} = \text{Storage change} \tag{6}$$

Here, input refers to recharge from rainfall and other sources and subsurface inflow into the unit; and output refers to ground water draft, evapotranspiration, baseflow to streams and subsurface outflow from the unit. Eq. (6) holds good for any period: a year, season, a month, etc. Typically, this equation is applied separately for different seasons (monsoon and non-monsoon) or crop seasons, such as Kharif, Rabi and summer seasons.

In Eq. (6), storage change (positive for storage increase, negative for storage decrease) is a function of the ground water level change and specific yield. Hence ground water level measurements at the beginning and end of the season/period are necessary to estimate storage change.

6.7.2. Unit for Ground Water Resource Assessment

Watershed with well-defined hydro-geological boundaries is the most appropriate hydrological unit for ground water resource estimation. It is advantageous to adopt this unit as the inflow/outflow across the boundaries is likely to be negligible in most cases. In hard rock areas, the hydro-geological and hydrological units normally coincide. This may not be the case in alluvial areas where the aquifers may traverse the basin boundaries. For classification of an area into alluvial or hard rock areas, the predominant hydrogeology of the unit is to be considered in Hard rock areas occupy about 2/3rd area of the country. In many states where the development unit is either a block or a taluka or a mandal, the final assessment of ground water potential may be on this basis to facilitate development programs. In the states where switch over to watershed is not immediately possible, the present practice of assessing the ground water potential on block/taluka/mandal basis may continue. GEC (1997) recommends that in each unit, ground water may be assessed once in three years. However, the ground water draft figures should preferably be updated every year.

GEC (1997) recommended a procedure to delineate subareas within the unit, which may be a watershed (hard rock areas) or a block/taluka/ mandal (alluvial areas). First, out of the total geographical area of the unit, hilly areas (slope greater than 20%) are separated as these are not likely to contribute to ground water recharge. However, the local topographical features such as valley, terrace, and plateau occurring within > 20% slope zone may be considered for recharge computations. Out of the remaining part, areas where the quality of ground water is beyond the usable limits should be identified and handled separately. The area with brackish/saline ground water be delineated and the resource of these areas be computed separately. Now, the remaining area is to be delineated as follows:

- Non-command areas which do not come under major/medium surface water irrigation schemes.
- Command areas under major/medium surface water irrigation schemes.

If necessary, further subdivision based on geomorphological and hydrogeological characteristics may be made.

6.7.3. Classification of Areas of Ground Water Resources

CGWB periodically estimates the ground water resources of India and publishes the results as *Ground Water Statistics*. This assessment gives an aggregate picture of ground water resources of the country. On the basis of annual availability of replenishable ground water, the entire country has been categorized in five classes: (i) areas of very low replenishable ground water resources (below 400 MCM/year), (ii) areas of low replenishable ground water resources (401-800 MCM/year), (iii) areas of moderate replenishable ground water resources (801-1,200 MCM/year), (iv) areas of high replenishable ground water resources 1,201-1,600 MCM/year), and (v) areas of very high replenishable resources (above 1,601 MCM/year). These are briefly described below.

(a) Areas of Very Low Replenishable Ground Water Resources
This category includes those areas which are either deficient in rainfall or hydrogeological conditions are not favourable for recharge. Hill areas of Himachal Pradesh and Meghalaya plateau are such examples of unfavourable conditions. This category includes mainly central and western Rajasthan, Himachal Pradesh, Meghalaya and southern Haryana besides pockets in Punjab, Bihar, Orissa, Tamil Nadu, and Kerala.

(b) Areas of Low Replenishable Ground water Resources
There are the areas which either do not have favourable infiltration due to geological constraints or lack in precipitation. Mainly these areas are:

(i) Northwestern India: comprising parts of Jammu and Kashmir, Western Punjab, Western Haryana, Eastern Rajasthan and Western Madhya Pradesh.
(ii) Western Coasts and Western Ghats: including Coasts of Maharashtra, Karnataka and Kerala along with the leeward side of the Western Ghats especially in Karnataka.
(iii) Eastern Parts: including most of the districts of Orissa, southern parts of Bihar, eastern parts of Assam and pockets of eastern Tamil Nadu.

(c) Areas of Moderate Replenishable Ground Water Resources
These areas are widely spread over the entire country without any definite pattern. Following are the prominent pockets: western Gujarat, south-western Madhya Pradesh, Uttaranchal, parts of north-western Uttar Pradesh, southern and north-western parts of Madhya Pradesh, and parts of Karnataka, Andhra Pradesh, and Tamil Nadu.

(d) Areas of High Replenishable Ground Water Resources
These areas are either the parts of the river basins or very close to them. Because of the close proximity to the major river systems these areas have large yield prospects. Following are the main areas: western parts of Jammu and Kashmir, south-western Punjab, western Haryana, north-eastern Uttar Pradesh, eastern parts of West Bengal, central parts of Maharashtra, and parts of Andhra Pradesh and Tamil Nadu.

(e) Areas of Very High Replenishable Ground Water Resources
These areas coincide with the major river basins which provide copious. These areas have both porous and fissured hydrogeological formations like, alluvium, basalt, sandstone, schist, marble etc. Such areas have large yield prospects, exceeding 150 m^3/hour in most of the parts. The major river basins of the country are much ahead in the agricultural production than the other parts because these are the promising areas of ground water occurrence.

6.7.4. Season-Wise Assessment of Ground Water Resources

Ground water recharge should be assessed separately for the non-command and command areas as well as for monsoon and non-monsoon season. A schematic view of ground water assessment philosophy is given in Figure 2. Total annual recharge is computed following eq. (7).

$$\begin{aligned}\text{Total annual recharge} = {} & \text{Recharge in monsoon season} \\ & \text{(from rainfall + other sources)} \\ & + \text{Recharge in non-monsoon season} \\ & \text{(from rainfall + other sources)} \qquad (7)\end{aligned}$$

Generally, a well hydrograph has a variation similar to a stream flow hydrograph with a peak followed by a recession limb. The recession limb in the post monsoon

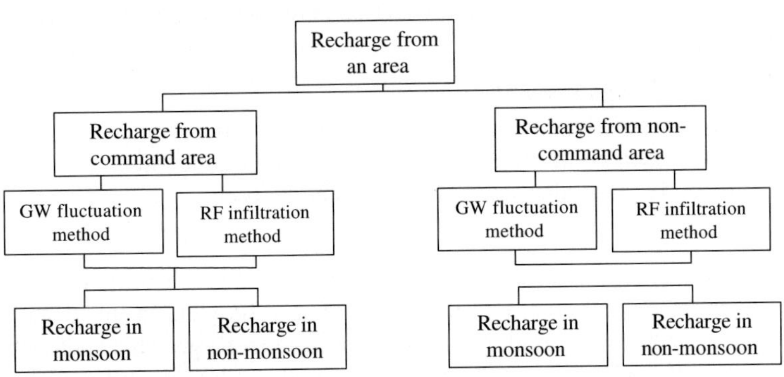

Figure 2. Schematic representation of components of ground water assessment

period, particularly in hard rock areas, has two distinct slopes: a steep slope limb from September-October to October-November and a gently sloping limb from October-November to May-June. The steeper limb indicates that of the total rise due to recharge during the monsoon period, a significant part is withdrawn after the end of monsoon. GEC (1997) recommended that the ground water recharge may be estimated on pre-monsoon (May-June) to post monsoon (October-November) water level fluctuations for the areas receiving rainfall from South-West monsoon. In areas such as Tamil Nadu where the predominant rainfall is due to North-East monsoon, recharge assessment may be based on pre-monsoon (October) to post-monsoon (February) water level fluctuations. hENCE, in these areas also an additional month is taken in the monsoon season, to account for the steep part of the recession limb.

With this background, we now discuss methodologies for ground water resource estimation in non-command and command areas.

6.7.5. Ground Water Assessment in Non-Command Areas

For this case, GEC (1997) recommended the use of ground water level fluctuation and specific yield approach since this method takes into account the response of ground water levels to inputs and outputs. However, this requires adequately spaced setting up of observation wells and water level records for a sufficiently long period. There should be at least three spatially well distributed observation wells in the unit, or one observation well per $100\,km^2$, whichever is more. If the unit comprises of both command and non-command areas, at least five observation wells must be present, such that at least two observation wells are available in each type of sub-area. Also, water level observations must be available for a minimum period of 5 years along with corresponding rainfall data in the unit. Regarding frequency of data, pre- and post-monsoon observations, preferably in successive years, are the minimum requirement. It would be ideal if monthly water level measurements are recorded so that the peak rise and maximum fall in the ground water levels can be determined. Efforts should be made to install continuous water level recorders at key representative locations inside the unit. If adequate data on ground water level fluctuations are not available, ground water recharge may be estimated using rainfall infiltration factor method.

a. Ground water level fluctuation method

This method is recommended for recharge assessment in the monsoon season. For non-command areas, recharge in the non-monsoon season is quite small and may be estimated empirically. Two alternate approaches can be used.

Case a:

Estimate specific yield from long duration pumping tests or based on norms for the particular hydrogeological area. Use this value of specific yield in the ground water balance equation for the monsoon season to estimate recharge. This approach is more suitable for alluvial areas or in hard rock areas where data about base flow in the dry season is not available.

Computation of recharge for the monsoon season: The ground water balance equation for the monsoon season in non-command areas is:

$$R_G - D_G - B + I_s + I = S \tag{8}$$

where R_G = gross recharge due to rainfall and other sources including recycled water, D_G = gross ground water draft, B = base flow into streams from the area, I_s = recharge from streams into ground water body, I = net ground water flow (inflow-outflow) into the area across the boundary, and S = ground water storage increase.

If the area under consideration is a watershed, the net ground water inflow may be taken as zero. If there is inflow and outflow across the boundary, the net inflow may be calculated using the Darcy law by delineating the inflow and outflow sections of the boundary. Besides such delineation, the calculation also requires estimate of transmissivity and hydraulic gradient across the inflow and outflow sections.

If the unit of assessment is a watershed in hard rock area, a single stream gauge station at the outlet of the watershed can provide the required data for the calculation of base flow. If adequate data is not available to estimate the baseflow and the baseflow is small, the baseflow term and recharge from stream in eq. (8) may be dropped.

After accounting for net ground water flow (I), recharge from streams (I_s) and base flow (B) in eq. (8), the resultant quantity refers to the possible recharge under the present status of ground water development in the area. Let $R = R_G - B + I + I_s$ is the possible recharge, which is gross recharge minus the natural discharges in the area in the monsoon season. Hence, eq. (8) can be rewritten as

$$R = S + D_G \tag{9}$$

Expressing storage increase S in terms of water level fluctuation and specific yield, eq. (9) becomes

$$R = h^* S_y{}^* A + D_G \tag{10}$$

where h = rise in ground water level in the monsoon season, A = area for computation of recharge, and S_y = specific yield. The recharge calculated from eq. (10) is the recharge from rainfall and other sources for the particular monsoon season. For non-command areas, the other recharge sources may be recycled water from ground water irrigation, tanks and ponds, and water conservation structures, if any (e.g., check dams, percolation tanks, nala bunds etc.). The recharge from rainfall is given by

$$\begin{aligned} R_{rf} &= R - R_{gw} - R_{wc} \\ &= h^* S_y{}^* A + D_G - R_{gw} - R_{wc} - R_t \end{aligned} \tag{11}$$

where, R_{rf} = recharge from rainfall, R_{gw} = recharge from ground water irrigation in the area, R_{wc} = recharge from water conservation structures, and R_t = Recharge from tanks and ponds.

Recharge from ground water irrigation (R_{gw}), from water conservation structures (R_{wc}) and from tanks and ponds (R_t) may be estimated based on the norms presented by GEC (1997). The recharge from rainfall for the particular monsoon season is estimated as per eq. (11).

Case b:

In this approach, the specific yield is estimated from ground water balance in the dry season, and then recharge is estimated from ground water balance in the monsoon season. The approach is suitable in hard rock areas where data about base flow in the dry season is available or this base flow is practically negligble.

The period January to May is recommended for dry season water balance, except in areas where predominant rainfall is in the North East monsoon. In these areas, the period March to May may be used for this purpose. Ignoring the net inflow term due to subsurface flow and assuming that the recharge from rainfall during the dry season is practically nil, the ground water balance in the dry season is

$$h^* S_y{}^* A = D_G - R_{gw} + B \tag{12}$$

where h = decrease in ground water level, D_G = gross ground water draft, R_{gw} = recharge recycled from ground water irrigation, and B = base flow from the area.

Recharge from ground water irrigation (R_{gw}) may be estimated based on GEC (1997) norms. Water conservation structures such as tanks, and ponds are expected to have negligible storage by the time the dry season commences. Hence recharge from these is ignored. The specific yield can now be calculated from equation (12) as

$$S_y = (D_G - R_{gw} + B)/(h^* A) \tag{13}$$

After determining specific yield from the water level fluctuation data in the dry season, the recharge in the monsoon season can be calculated from eq. (10), applying the water level fluctuation method for the monsoon season. The corresponding recharge from rainfall is given by eq. (11).

Estimation of normal recharge during monsoon season: The rainfall recharge obtained by eq. (11) is the recharge from rainfall in any particular monsoon season. This estimate is to be normalized for the normal monsoon season rainfall, which is the average of the monsoon season rainfall for the past 30 to 50 years. The normalization procedure requires that data on recharge and associated rainfall (r_i) are first obtained. To eliminate the effects of drought or surplus years, it is recommended that the rainfall recharge during monsoon season is estimated using eq. (11) not only for the year for which assessment is being made but also for at least four more preceding years. This will result in at least 5 pairs of data. If the assessment year and the four years preceding it were uniformly dry or wet, it is desirable to consider more than five years for normalization.

Let R_i be the rainfall recharge and r_i be the associated rainfall. The subscript i takes values 1 to N where N is usually 5. The rainfall recharge, R_i is obtained by eq. (14):

$$R_i = h_i{}^*S_y{}^*A + D_{G\,i} - R_{gw\,i} - R_{t\,i} \tag{14}$$

where R_i = rainfall recharge estimated for the ith year, h_i = rise in ground water level in the monsoon season for the ith year, S_y = specified yield, A = area for computation of recharge, D_G = gross ground water draft in monsoon season for the ith year, R_{gw} = recharge from ground water irrigation in the monsoon season for the ith year, R_{wc} = recharge from water conservation structures in the monsoon season for the ith year, R_t = recharge from tanks and ponds in the monsoon season for the ith year.

Among the pairs of R_i and r_i obtained above, those pairs that have R_i as negative or nearly zero should be omitted. Only those pairs in which $R_i > 0$ should be considered for further computations in the normalization procedure. It is also likely that all the computed R_i values are negative or nearly zero. In such a case, the water table fluctuation method should not be used, and the normal rainfall recharge during the monsoon season should be estimated by the rainfall infiltration factor method.

Using the pairs of data on R_i and r_i as described above, a normalization procedure was recommended by GEC (1997) to obtain the rainfall recharge corresponding to the normal monsoon season rainfall.

Estimation of normal recharge during non-monsoons season: The total recharge in the non-monsoon season is obtained as the sum of recharge from rainfall and from other sources in the non-monsoon season. The recharge from rainfall during the non-monsoon season may be estimated based on the rainfall infiltration factors given by GEC (1997), provided the normal rainfall in the non-monsoon season is greater than 10% of the normal annual rainfall. If the rainfall is less than this threshold value, the recharge due to rainfall in the non-monsoon season may be taken as zero.

Recharge during the non-monsoon season from other sources, namely, from ground water irrigation (R_{gw}) and from water conservation structures (R_{wc}) is to be estimated using the norms given by GEC (1997).

b. Recharge assessment based on rainfall infiltration factor

Clearly the ground water level fluctuation method cannot be used if adequate ground water level data are not available. In such a situation, recharge may be estimated based on the rainfall infiltration factor method. GEC (1997) norms to estimate recharge from rainfall and from other sources are presented in Section 6.8. Recharge from rainfall in monsoon season is given by

$$R_{rf} = f^*A^*\text{Normal rainfall in monsoon season} \tag{15}$$

where f = rainfall infiltration factor given in Section (6.8), and A = area of computation for recharge.

The same recharge factor may be used for both monsoon and non-monsoon rainfall with the condition that the recharge due to non-monsoon rainfall may be taken as zero, if the normal rainfall during the non-monsoon season is less than 10% of normal annual rainfall. In using the method based on the specified norms, recharge due to both monsoon and non-monsoon rainfall may be estimated for normal rainfall, based on recent 30 to 50 years of data. It is necessary to use data of adequately spaced rain gauge stations within and outside (nearby) the unit so that a correct value of rainfall is used.

For non-command areas, recharge from other sources corresponds to recharge from ground water irrigation and recharge from water conservation structures. These are to be estimated separately for monsoon and non-monsoon seasons based on the norms presented in Section 6.8. The total recharge is:

$$R(\text{normal}) = R_{rf}(\text{normal}) + R_{gw} + R_{wc} + R_t \tag{16}$$

where $R(\text{normal})$ = total recharge during monsoon season, $R_{rf}(\text{normal})$ = rainfall recharge during monsoon season for normal monsoon season rainfall, R_{gw} = recharge from ground water irrigation in the monsoon season, R_{wc} = recharge from water conservation structures in the monsoon season, and R_t = recharge from tanks and ponds in the monsoon season; all for the year of assessment.

c. Total annual recharge

The total annual recharge is the sum of recharge in the monsoon and non-monsoon seasons. In each season, the recharge comprises of recharge from rainfall and from other sources.

6.7.6. Ground Water Assessment in Command Areas

Recharge assessment in command areas may be done on the same lines as in non-command areas, except that two important additional components of recharge are to be considered, namely recharge due to seepage from canals and recharge due to return flow from surface water irrigation. In command areas, these two components may be significantly more than the recharge due to rainfall. Recharge from these sources may be significant both in monsoon and non-monsoon seasons. The specific yield may be estimated by applying water level fluctuation method in command areas in the dry season. If adequate data of water level fluctuations is not available, the method based on rainfall infiltration factor may be used.

Ground water level fluctuation method

As in the case of non-command area, the ground water level fluctuation method may be applied to estimate the recharge in the monsoon season for the command area also. In eq. (10) the recharge term includes the recharge in monsoon season from rainfall and other sources. For command areas, recharge from other sources includes contribution from canals, return flow from surface and ground water irrigation, storage tanks and ponds, and water conservation structures. The recharge from rainfall is given by

$$R_{rf} = h^{*}S_{y}{}^{*}A + D_{G} - R_{c} - R_{sw} - R_{t} - R_{gw} - R_{wc} \quad (17)$$

where, R_{rf} = recharge from rainfall, R_c = recharge due to seepage from canals, R_{sw} = recharge from surface water irrigation, R_t = recharge from storage tanks and ponds, R_{gw} = recharge from ground water irrigation, R_{wc} = recharge from water conservation structures, D_G = gross draft in the command area, h = rise in ground water level in the command area, and A = area of the command area for recharge assessment

In eq. (17), all quantities refer to the monsoon season only. For a particular command area, one or more of the recharge quantities from other sources may be zero. It may be noted that the net ground water inflow across the boundaries has been ignored. This may not be true especially in the case of alluvial areas where the choice of assessment unit (block/taluka) is based on administrative considerations. Hence, in such cases the component of net ground water flow across the boundaries should be included. This can be estimated as the product of gradient of ground water flow, transmissivity of the aquifer and the length across which flow takes place. The transmissivity value in these computations should be on the basis of long duration aquifer performance tests.

Eq. (17) gives the recharge in any particular monsoon season for the associated monsoon season rainfall. This estimate is to be normalized for the normal monsoon season rainfall which in turn is obtained as the average of the monsoon season rainfall for the recent 30 to 50 years. The normalization procedure to be followed has been described in GEC (1997) and involves the following:

a) Computation of a set of pairs of data on rainfall recharge R_i and associated rainfall, r_i for i = 1 to N in which N is at least 5.
b) Considering only those pairs of R_i and r_i in which R_i is greater than zero for further computations in the normalization procedure.
c) Dispensing with the water table fluctuation method if all R_i values are consistently negative or nearly zero, and instead adopting the rainfall infiltration factor method for computing the rainfall recharge in the monsoon season.
d) Using the pairs of data on R_i and r_i, and estimating the rainfall recharge for normal monsoon season rainfall condition by either of the two methods of normalization.
e) Comparing the rainfall recharge under normal monsoon season rainfall condition as obtained by the water table fluctuation method with that obtained by the rainfall infiltration factor method, and finally assigning the rainfall recharge value on the basis of a set of criteria so that unreasonably high *or* low estimates of rainfall recharge by the water table fluctuation method are avoided.

The computational procedure in this case is also similar to what has been described in Section 6.7.5, except that, in place of eq. (14), the expression to be used for the rainfall recharge term R_i is

$$R_i = h_i{}^{*}S_y{}^{*}A + D_{Gi} - R_{ci} - R_{swi} - R_{ti} - R_{gwi} - R_{wci} \quad (18)$$

where, R_i = rainfall recharge estimated for the ith year, h_i = rise in water level in the monsoon season for the ith year, S_y = specific yield, A = area for computation of recharge, D_{Gi} = gross ground water draft in the monsoon season for the ith year, R_{ci} = recharge due to seepage from canals for the ith year, R_{swi} = recharge from surface water irrigation for the ith year, R_{ti} = recharge from tanks and ponds for the ith year, R_{gwi} = recharge from ground water irrigation for the ith year, and R_{wci} = recharge from water conservation structures for the ith year.

Finally, the total recharge during the monsoon season for the conditions pertaining to normal monsoon season rainfall is

$$R(\text{normal}) = R_{rf}(\text{normal}) + R_c + R_{sw} + R_t + R_{gw} + R_{wc} \tag{19}$$

where R(normal) = total recharge during monsoon season and R_{rf}(normal) = rainfall recharge during monsoon season for normal monsoon season rainfall. Terms R_c, R_{sw}, R_t, R_{gw}, and R_{wc} represent recharge due to seepage from canal, surface water irrigation, tanks and ponds, ground water irrigation, and water conservation structures, respectively, all during the monsoon season for the year of assessment.

In the ground water level fluctuation method, a significant part of base flow is already accounted for by observing the post monsoon water level one month after the end of rainfall. The base flow in the remaining non-monsoon period is likely to be small, especially in hard rock areas. Further, detailed data for quantitative assessment of the natural discharge is not generally available. Considering these factors, GEC (1997) recommended that 5 to 10% of the total annual ground water potential may be set aside to account for natural discharges in the non-monsoon season. The balance will account for existing ground water withdrawal for various uses and potential for future development. This quantity is termed as the net annual ground water availability which should be calculated separately for non-command and command areas.

Normal recharge during non-monsoon season: The total recharge in non-monsoon season is the sum of recharge from rainfall and other sources in the non-monsoon season. The recharge from rainfall during the non-monsoon season may be estimated based on the rainfall infiltration factors given in section 6.8, provided the normal rainfall in the non-monsoon season is greater than 10% of the normal annual rainfall. If the rainfall is less than this threshold value, the recharge due to rainfall in the non-monsoon season may be taken as zero. Recharge during the non-monsoon season from other sources, namely from ground water irrigation (R_{gw}) and from water conservation structures (R_{wc}) are to be estimated from the norms given in Section 6.8.

6.7.7. Recharge Assessment using Rainfall Infiltration Factor

If adequate data of ground water levels are not available, the ground water level fluctuation method cannot be used. In such a situation, recharge may be estimated based on the rainfall infiltration factor method following GEC (1997) norms. The same recharge factor may be used for both monsoon and non-monsoon rainfall with the condition that the recharge due to non-monsoon rainfall may be taken as zero, if the normal rainfall during the non-monsoon season is less than 10% of normal annual rainfall. In using the method based on the specified norms, recharge due to both monsoon and non-monsoon rainfall may be estimated for normal rainfall, based on recent 30 to 50 years of data. It is necessary to have adequately spaced rain gauge stations within and outside the unit taken up for recharge computation. While adopting this method due weight should be given to the nearby raingauge stations.

For command areas, recharge from other sources corresponds to recharge due to seepage from canals, from surface and ground water irrigation, storage tanks and ponds, and from water conservation structures. These are to be estimated separately for monsoon and non-monsoon seasons. During each season, recharge takes place from rainfall and from other sources.

The recharge during the monsoon and the pre-monsoon seasons is summed up to obtain total annual recharge.

6.7.8. Ground Water Assessment in Saline Areas

In each unit, area with brackish/saline ground water is to be delineated and the ground water resource of these areas are be computed separately. However, in saline areas, due to non-availability of data, recharge assessment may be based on rainfall infiltration factor method. The areas underlain by saline ground water in different states of the country are shown in Table 9.

6.7.9. Ground Water Assessment in Water Level Depletion Zones

In some areas, ground water level shows a decline even in the monsoon season. The reasons for this may be: (a) There is a depletion in the ground water regime, with ground water draft and natural ground water discharge in the monsoon season

Table 9. Area Underlain by Saline Ground water in Different States

State	Total area of state (sq. km)	Area underlain by saline ground water EC > 4dslm	Annual replenishable recharge/yr
Haryana	44,212	11,438	2,452
Punjab	50,353	3,038	1,351
Delhi	1,485	140	32
Rajasthan	342,239	141,036	4,025
Gujarat	196,024	24,300	2,179
Uttar Pradesh	294,411	1,362	354
Karnataka	191,791	8,804	1,015
Total		190,118	11,408

Source: CGWB (2005).

(outflow from the region and base flow) exceeding the recharge, and (b) There may be an error in water level data due to inadequacy of observation wells.

If the water level data are found to be erroneous, recharge assessment may be made using rainfall infiltration factor method. If, on the other hand, water level data are assessed as reliable, the ground water level fluctuation method may be applied for recharge estimation. As h in eq. (10) will be negative, the estimated recharge will be less than the gross ground water draft in the monsoon season. It must be noted that this recharge is the gross recharge minus the natural discharges in the monsoon season. This implies that the area falls under the over-exploited category, with need for micro level study (see Section 6.7.11).

6.7.10. Apportioning Assessment from Watershed to Development unit

Where the assessment unit is a watershed, there is frequent need to convert the ground water assessment terms of an administrative unit such as block/taluka/mandal. To that end, recall that a block may comprise of one or more watersheds, in part or full. First, the ground water assessment in the subareas, non-command and command areas of the watershed may be converted into depth unit (mm) by dividing the annual recharge by the respective area. The contribution of this subarea of the watershed to the block is calculated by multiplying this depth with the area in the block occupied by this sub-area. This procedure must be followed to calculate the contribution from all the sub-areas of watersheds lying in the block to work out the total ground water resource of the block.

The total ground water resource of the block should be presented separately for each type of sub-area, namely for non-command areas, command areas and saline/brackish water areas, as in the case of the individual watersheds.

6.7.11. Studies for Critical and Over-Exploited Areas

In all areas which are categorized as semi-critical or worse, it is necessary to increase the density of observation wells. Micro-level studies are necessary in critical and over-exploited areas to reassess the ground water recharge and draft. Following approach was recommended by GEC (1997):

1. The micro-level studies in the critical and over-exploited areas in hard rock terrain should be based on watershed as a unit and not on administrative unit.
2. The area may be sub-divided into different hydro-geological recharge areas, discharge areas and transition zones and also on quality terms.
3. The number of observation wells should be increased to represent each such sub-area with at least one observation well with continuous monitoring of water levels.
4. Hydrological and hydrogeological parameters, particularly the specific yield, should be collected for different formations in each sub-area.

5. Details regarding other parameters like seepage from canals and other surface water projects should be collected after field studies instead of adopting recommended norms. Base flow should be estimated based on stream gaugings.
6. The data of number of existing structures and unit draft should be reassessed after fresh surveys and should match with the actual irrigation pattern in the sub-area.
7. All data available with CGWB, SGDs and other agencies including research institutions and universities etc. should be collected for the watershed/sub-areas and utilized for reassessment.
8. Ground water assessment for each sub-areas may be computed adopting the recommended methodology and freshly collected values of different parameters. The assessment may be made separately for monsoon and non-monsoon period as well as for command and non-command areas.
9. The ground water potential so worked out may be cross-checked with behaviour of ground water levels in the observation wells and both should match. If they do not, the factors that cause such an anomaly should be identified and the revised assessment should be re-examined.
10. Based on the micro-level studies, the sub-areas within the unit and the unit as a whole may be classified adopting the recommended norms for categorization.

6.8. NORMS FOR ESTIMATION OF RECHARGE

GEC (1997) has prescribed the following norms for estimating recharge.

Norms for specific yield

GEC (1997) accepted the norms presented in Table 10 for specific yield.

Recharge from rainfall

GEC (1997) recommended the norms to estimate recharge from rainfall and these are presented in Table 11.

Recharge due to seepage from canals

The norms for recharge due to seepage from canals are presented in Table 12.

6.8.1. Return Flow from Irrigation

The recharge due to return flow from irrigation may be estimated, based on the source of irrigation (ground water or surface water), the type of crop (paddy, non-paddy) and the depth of water table below ground level, using the norms provided in Table 13.

Recharge from storage tanks and ponds: Based on the average area of water spread, a depth of 1.4 mm/day may be used for the period in which the tank has water. If

Table 10. Norms for Specific Yield

Formation	Recommended value (%)	Maximum value (%)	Minimum value (%)
	Alluvial areas		
Sandy alluvium	16.0	20.0	12.0
Silty alluvium	10.0	12.0	8.0
Clayey alluvium	6.0	8.0	4.0
	Hard rock areas		
Weathered granite, gneiss, schist with low clay contents	3.0	4.0	2.0
Weathered granite, gneiss schist with significant clay contents	1.5	2.0	1.0
Weathered or vesicular, jointed basalt	2.0	3.0	1.0
Laterite	2.5	3.0	2.0
Sandstone	3.0	5.0	1.0
Quartzite	1.5	2.0	1.0
Limestone	2.0	3.0	1.0
Karstifield limestone	8.0	15.0	5.0
Phyllites, shales	1.5	2.0	1.0
Massive poorly fractured rock	0.3	0.5	0.2

Note: Usually the recommended values should be used, unless sufficient data based on field study is available to justify the minimum, maximum or other intermediate values.
Source: GEC (1997).

data on the average area of water spread is not available, 60% of the maximum water spread area may be used instead of the average area.
Recharge from percolation tanks: Half of gross storage, considering the number of fillings, with half of this recharge occurring in the monsoon season and the balance in the non-monsoon season.
Recharge due to check dams and nala bunds: Half of gross storage (assuming annual desilting maintenance exists) with half of this recharge occurring in the monsoon season, and the balance in the non-monsoon season.

6.8.2. Some Comments on GEC Methodology

A review of the GEC recommended methodology shows some drawbacks in it. The major drawback is that surface water and ground water are assessed separately and not conjunctively; this may result in double accounting and consequent over-estimation of total water availability. GEC methodology treats development units as independent and ignores interactions among them which is an important aspect. The methodology also ignores the water consumed by natural vegetation in the unit. Implicit objective of full development is another fundamental flaw. This assumption again neglects the dynamics of ground water between development units (generally blocks). Base flow depletion, saline water intrusion and declines in ground water level may all reach unacceptable levels much before full development is reached.

Table 11. Norms for Recharge from Rainfall

Formation	Recommended value (%)	Maximum value (%)	Minimum value (%)
	Alluvial areas		
Indo-Gangetic and inland areas	22.0	25.0	20.0
East Coast	16.0	18.0	14.0
West Coast	10.0	12.0	8.0
	Hard rock areas		
Weathered granite, gneiss, schist with low clay contents	11.0	12.0	10.0
Weathered granite, gneiss schist with significant clay contents	8.0	9.0	5.0
Granulite facies like Charnokite etc.	5.0	6.0	4.0
Vesicular, jointed basalt	13.0	14.0	12.0
Weathered Basalt	7.0	8.0	6.0
Laterite	7.0	8.0	6.0
Semi-consolidated sandstone	12.0	14.0	10.0
Consolidated sandstone, quartzite, Limestone (except cavernous limestone)	6.0	7.0	5.0
Phyllites, shales	4.0	5.0	3.0
Massive poorly fractured rock	1.0	3.0	1.0

Note:

1. Usually the recommended values should be used, unless sufficient information is available to justify the use of minimum, maximum or other intermediate values.
2. An additional 2% of rainfall recharge factor may be used in areas where watershed development with associated soil conservation measures is implemented. This additional factor is subjective and is separate from the contribution due to the water conservation structures such as check dams, nalla bunds, percolation tanks etc. The norms for the estimation of recharge due to these structures are provided separately. This additional factor of 2% is only provisional at this stage, and will need revision.

Table 12. Norms for Recharge due to seepage from canals

Unlined canals in normal soils with some clay content along with sand	1.8 to 2.5 cumec per million sq. m of wetted area or 15 to 20 ham/day/million sq. m of wetted area
Unlined canals in sandy soil with some silt content	3.0 to 3.5 cumec per million sq m of wetted area or 25 to 30 ham/day/million sq. m of wetted area
Lined canals and canals in hard rock area	20% of above values for unlined canals

Notes:

1. The above values are valid if the water table is relatively deep. In shallow water table and waterlogged areas, the recharge from canal seepage may be suitably reduced.
2. Where specific results are available from case studies, the above norms are to be replaced by norms evolved from these results.

Table 13. Recharge as percentage of application

Source of Irrigation	Type of Crop	Water table below ground level		
		< 10 m	10–25 m	> 25 m
Ground water	Non-Paddy	25	15	5
Surface water	Non-Paddy	30	20	10
Ground water	Paddy	45	35	20
Surface water	Paddy	50	40	25

Notes:
1. For surface water, the recharge is to be estimated based on water release at the outlet. For ground water, the recharge is to be estimated based on gross draft.
2. Where continuous supply is used instead of rotational supply, an additional recharge of 5% of application may be used.
3. Where specific results are available from case studies is some states, the above norms are to be replaced by norms evolved from these results.

Moreover, the amount of water required may be less than the amount of water available in many of the blocks.

Another feature of the assessment of ground water resources is that CGWB uses constant state-wide figures for depth of irrigation, irrespective of different crops, soil and climatic conditions. CGWB assumes that 30% of the gross draft returns to ground water zone. This, of course, is not correct in negative water balance areas of the country, which constitute a large part of the country. The cumulative impact of all the factors could introduce an error of more than 20% in the estimates of ground water resources.

6.9. GROUND WATER DRAFT

Ground water draft is defined as the "quantity of water withdrawn from the ground water reservoirs". The total quantity withdrawn is termed as gross draft. The annual ground water draft of a structure is computed by multiplying its average discharge and annual working hours. The number of working hours can be calculated by using the data of consumption of electrical or diesel energy. Ground water draft is also calculated by the irrigation requirement of crops in the command area of the structure. For working out ground water balance, 70% of gross extraction is taken which is known as net ground water draft. Balance 30% is presumed to return to ground water regime as seepage. Ground water draft of an area can be estimated by multiplying the number of wells of different types with the unit draft of each type of well.

Ground water draft can be estimated based on:

- Electric power consumption for agricultural pumpsets,
- Statistics of area irrigated by ground water and the associated crop water requirements, and
- Use of remote sensing data to obtain seasonal data on area of different irrigated crops in non-command areas where only ground water irrigation is used.

Gross ground water draft includes ground water extraction from all the existing structures during monsoon as well as non-monsoon period. This should preferably be based on the latest well census and updating the same based on the growth rate and/or ground water programs implemented by various agencies. To compute ground water draft, it is necessary to determine the average unit draft from different types of ground water structures based on the well yield capacity, its command area, and techno-economic viability.

Some of the dug wells and tube wells become non-operative either due to water table falling much below the depth of the well, due to silting, due to water quality problems, due to failure of the well, or due to expiry of useful life of the well. The ground water draft from these wells need not be considered. The number of these disused wells may be reflected by a depreciation factor which should be considered while estimating the ground water draft.

In view of the uncertainties in estimation of ground water draft by any method, it is clearly desirable to use more than one method to enable a cross check. The average draft for ground water structures in different states of India has been presented in Table 14. This table can be used as guideline for annual draft for different types of ground water structures in the states. In canal command areas, the present utilisation of wells is suboptimal. Note that the norms for ground water draft given in Table 14 are based on

Table 14. Average Annual Gross Draft for Ground Water Structures in Different States

S. N.	State	Type of ground water structure	Average gross unit draft (ham)
1.	Andhra Pradesh	Dugwell with Mhot	0.35
		Dugwell with Pumpset	0.65
		Borewell with Pumpset	1.3
		Shallow Tubewell	2.05
		Medium Tubewell	4.1
		Deep Tubewell	5.85
2.	Assam	Shallow Tubewell with Pumpset	3.0
3.	Bihar	Dugwell	0.6
		Private tube well with Pumpset	1.0
		Bamboo boring with Pumpset	0.75
		Deep tube well	30.0
4.	Gujarat	Dugwell with Pumpset	0.8
		Borewell with Pumpset	1.2
		Private shallow Tubewell	1.85
		Medium Deep Tubewell	6.0
		Deep Tubewell	30.0
5.	Haryana	Dugwell with Pumpset	1.5
		Private shallow Tubewell with Pumpset	1.81
		Deep Tubewell	15.0
6.	Himachal Pradesh	Medium Deep Tubewell with Pumpset	2.5
7.	Karnataka	Dugwell with Pumpset	0.9
		Borewell with Pumpset	1.7

the data for non-command areas, except in situation where wells in the command areas are optimally utilised.

6.9.1. Threats to Ground Water Quantity

An increased quantity of ground water is being withdrawn to meet the growing demands of population: for household uses, food, and other economic activities. The major threats to ground water are:

1. *Overdraft:* Any withdrawal in excess of safe yield (the amount that can be withdrawn without producing an undesirable result) is an overdraft. It occurs when ground water is withdrawn faster than recharge, resulting in falling water table. Consequently, wells go dry, pumping costs increase, and a part of the storage capacity may be permanently lost. Withdrawal of water from deeper zones can cause mixing of poor good quality water. In coastal zones, salt-water intrusion can occur.
2. *Subsidence:* is one of the outcomes of over-pumping. As the water table declines, water pressure is reduced. This causes the fine particles that held water to become compacted. In addition to permanently reducing storage capacity, the land above the aquifer can sink. This can damage property and fields.

In view of the above, guidelines are needed for extraction and use of ground water in a sustainable manner.

6.9.2. Guidelines for Ground Water Extraction and Use

Contribution of ground water for irrigation as well as industrial use and drinking has been on the increase during the last two decades. Indiscriminate extraction of ground water already poses the threat of aquifers going dry in some parts of the country. The Central and State Ground Water Boards have prepared ground water availability maps and have prescribed extraction rates to ensure that the extraction is balanced with recharge. The country has been zoned depending upon whether water is available in plenty, or it has already become scarce in the region. Extraction of ground water is prohibited in some regions where water depletion has already become critical.

The necessity and advantages of ground water development for domestic, irrigation and industrial uses has been widely appreciated in India. But, the indiscriminate use of ground water has led to serious situations of excessive drawdowns or mining of the aquifers. This has caused progressive and at places rapid lowering of the water table, consequent decline in the yield of wells, and intrusion of sea water in the coastal aquifers, etc. At the same time, in many canal command areas, because of increased recharge due to infiltration of excess irrigation water, water table is progressively rising, creating waterlogging and salinity problems. Hence, the ground water problems may be divided into two categories: quantity related and quality related. The coastal aquifers show wide variation in the water quality, both laterally and vertically.

6.10. ARTESIAN FLOW

Artesian water is that water which is confined beneath a relatively impermeable stratum such that if a well penetrates the confined zone, water will rise into the well to an elevation above that of the confined zone. The aquifer consists of a (thick) layer of sand or sandstone, which receives water at its exposed outcrop. Hydrostatic pressure in the confined aquifer is sufficient to raise water in wells to levels higher than the upper surface of the aquifer. The upward water movement is called artesian flow; its upper limit of rise is the hydrostatic pressure surface. Note that there is no specific pattern of artesian flow in the country. It is related with water table conditions and the conditions of the aquifers. In India, some areas where artesian ground water is observed are: a) Tarai Regions of Uttaranchal and Uttar Pradesh, b) Cambay Region (Kathiawar and Kutch Region) in Gujarat, c) Western Coastal Plain, d) Orissa, e) Parts of J & K, and f) Parts of Tamil Nadu.

6.10.1. Tarai Region of Uttar Pradesh

Tarai (or moist) region of Uttar Pradesh occurs south of the *Bhabhar*, adjacent to the northern limit of Indo-Gangetic plains. The Tarai belt is made up of clay, sandy clay mixed with Kankars, fine to coarse-grained sand, and gravels. Here ground level slopes southwards at a gentle slope ranging from 1 m to 3 m/km. The area is mostly drained by snow-fed perennial rivers which emerge from the Himalayas and is a prominent place for artesian flow. Artesian aquifers are found here at a depth of only a few meters below ground. Extensive availability of good quality ground water in this region has led to development of big farms where paddy is cultivated extensively.

6.10.2. Cambay Region

In the Pleistocene age, Kathiawar and Kutch experienced many earthquakes. As a consequence of these eruptions, the area experienced many uplifts and was substantially elevated above the sea level. The hydrostatic pressure in this confined aquifer is sufficient to raise water in wells to a level that is higher than the upper surface of the aquifer.

6.10.3. Western Coastal Plain

Another region where conditions are favourable to occurrence of artesian water is the coastal plains of the Western Ghats in South Maharashtra, Coastal Karnataka and Kerala. Here the seaward dipping strata contains many aquifers of sand and standstone. There are impermeable shells at the bottom. Very high rainfall in this area means there is enough water to recharge the aquifer.

6.10.4. Other Areas

Besides the above areas, other parts of the country like Jammu region, eastern plains of Tamil Nadu, and parts of Orissa also have favourable hydrogeological conditions for artesian flow.

6.11. CONJUNCTIVE USE OF WATER

The main objective of conjunctive use of surface water and ground water in a canal command is to optimally utilize water resources to maximize agricultural production per unit of water used without adversely affecting the environment. Various aims and objectives of the conjunctive use of water resources in an irrigation command have been enumerated by CWC (1995) and also by the National Commission for Integrated Water Resources Development Plan (NCIWRDP, 1999).

Different strategies of conjunctive use can be employed in an irrigation command (CWC 1995) depending on the spatial and/or temporal utilization of surface water and ground water. Strategies also differ for the use of good quality water or the saline water. Three strategies of conjunctive use in an irrigation command are as follows.

Strategy 1: Allocating parcels of land permanently to a particular use

Under this, separate areas of the command are permanently allocated for surface water or ground water use. The areas at higher elevation where it is difficult to take the canal water could be some such areas where ground water might be the sole source of irrigation. It is envisaged in this strategy that recharge from the surface water application will supplement the ground water and this will be utilized in an adjacent area allocated for ground water use.

This form of conjunctive use is effective where distance of the wells from the recharge area (surface irrigation) is so small that the ground water flow is sustained by the available gradient. This strategy is feasible in alluvial areas because of the appreciable movement of ground water. In hard rock areas and in clay soils, this strategy may not be feasible.

Strategy 2: Integrating surface water and ground water in time

In this strategy, surface and ground water resources are allocated in time such that in monsoon season, only surface water is used and in the non-monsoon season, only ground water is used. Since the same area is irrigated with surface and ground water at different times, ground water can use the same field channels that carry surface water. If private sources of ground water extraction are not available in the command, then augmentation tube wells are planned and operated in such a way that ground water carriage over long distances is avoided.

Combination of Strategy 1 and 2: Space & time integration

Under this strategy, some parcels of land are permanently allocated for surface water use, some parcels are permanently allocated for ground water use, and some parcels are supplied with surface water in one season and ground water in another. For parcels of land in which both ground and surface water are used, the intra-annual regime of uses can vary from year to year to take advantage of the stable regime of ground water.

6.11.1. Guidelines for Conjunctive use Planning

Recognizing the urgent need of conjunctive use, new projects which do not include conjunctive use of water are not recommended for Planning Commission's clearance. CWC (1995) gives general guidelines for planning conjunctive use. According to these guidelines, the quantification of water available for conjunctive use may have to be decided using appropriate methodologies. The steps include establishing a general ground water balance of the command area for "without conjunctive use project" conditions, delineate the area where ground water development is to be taken up based on the depth to water table and potential of aquifers, deciding the additional recharge that would become available in the command area in "with conjunctive use project condition", deciding the planned quantity of ground water use so as not to lead to progressive lowering or rising of water table and deciding the quantum of ground water use available for irrigation after considering the other (non-irrigation) uses of the planned ground water use, taking into account quality limitations such as the presence of brackish water. However, detailed action plans to implement the guidelines have to be drawn by respective states considering local conditions.

CWC (1995) suggests the use of NABARD guidelines for establishing preliminary ground water balance based on some rules of thumb for estimating recharge in command areas arising from seepage from canals, field channels and tanks and return flow from irrigation fields. In addition to estimating net annual ground water recharge, the report also stresses on taking into account: i) minimum necessary withdrawal in order to avoid large imbalance leading to large rise in ground water table, and ii) maximum permissible withdrawals with a view to maintain ecology and not allowing ground water to deplete, unless such depletion is likely to be beneficial due to the very high ground water table or rising tendency in the "without conjunctive use project" condition itself. The guidelines for extraction in command areas as percentage of additional recharge caused by the project are given in Table 15.

For the purpose of this table, a general long-term rise or fall of more than 0.2 m/year in case of alluvial condition and of more than 0.5 m/year in case of hard rock areas would qualify for classifying the trend as "rising" or "falling". In case an accurate ground water regime worked out by the specialists and tested and verified through modeling and field verification in both conditions is available, the maximum/minimum withdrawal can be worked out on the basis of the water balance studies instead of these percentages. Detailed studies are desirable in special areas having salinity problems. Further, economic considerations would finally decide the quantum of surface and

Table 15. CWC (1995) Guidelines for ground water extraction

Present ground water status		Minimum necessary additional withdrawal (%)	Maximum permissible withdrawal (%)
Depth of ground water	Trend		
Less than 2 m	Rising	70	100
Less than 2 m	Generally steady	50	80
Less than 2 m	Falling	30	60
2 m to 6 m	Rising	60	90
2 m to 6 m	Steady	40	70
2 m to 6 m	Falling	20	60
More than 6 m	Rising	50	80
More than 6 m	Steady	30	60
More than 6 m	Falling	0	40

ground water use in space and time. These guidelines may have to be modified in special situations such as coastal areas and areas with saline ground water, etc.

6.11.2. Status of Conjunctive use in India

With the launching of five-year plans in the year 1950–51, the objective was to achieve increased irrigation. While achieving the above goal, the integrated and conjunctive use of surface and ground water was not given the extent of attention and consideration it deserved. Most of the major irrigation projects were designed keeping in view the surface water inputs only. Hence, in course of time, the twin problem of water logging and soil salinity emerged, affecting agriculture production in the command areas. The Irrigation Commission (1972) noted the building up of water level in a number of irrigation projects in States of Bihar, Punjab, Haryana, and Uttar Pradesh.

The fourth five-year plan envisaged the coordinated use and management of both, surface water and ground water resources. The National Water Policy also stresses the integrated and coordinated development of surface water and ground water and their conjunctive use. It directs that conjunctive use planning should be envisaged right from the project planning stage and should form an essential part of the project. The Govt. of India had sanctioned six conjunctive use schemes in irrigation commands of the country [IGNP Stage-I command (Rajasthan), Sharda-Sahayak command (U.P.), Tungabhadra command (A.P. and Karnataka), Ghataprabha command (Karnataka), Hirakund command (Orissa), and Mahi-Kadana command (Gujarat)] so that the studies can yield some fruitful recommendations. The Central Ground Water Board (CGWB) was directed to carry studies and provide solution to reduce waterlogging in these six commands. The analysis was carried out during 1990–1996. Results indicated that economically feasible conjunctive use plans could be drawn and extended in these command areas. Further studies related to conjunctive use were completed by CGWB for the Nagarjuna Sagar project (A.P.); Indira Gandhi Nahar Pariyojana Stage – II (Rajasthan); and Kosi and Gandak canal command areas (Bihar).

6.11.3. Some Conjunctive use Studies Completed in India

Salient features of some studies carried out for conjunctive management in India are briefly described here.

a) Krishnagiri reservoir project

The Krishnagiri reservoir is located in Dharmapuri district of Tamil Nadu State of India. This reservoir has a catchment area of 5,397 km^2, water spread area of 12.85 km^2 and command area of 3,648 ha spread in 16 villages in Krishnagiri taluk. Normal rainfall of this area is 831 mm; available surface water resource is estimated at 6,046 ha-m and the ground water recharge is assessed at 5,635 ha-m. A conjunctive use study was been taken up to assess the ground water potential and utilize it along with surface water to raise maximum number of crops in the command and to assess the surplus water available for extending the command area. The study proposed to grow three seasonal crops and some annual crops in the command area. The requirement of water for all the crops put together was 7,214 ha-m. With the available surplus water and conjunctive use strategy, an additional 1,012 ha of command area can be brought under cultivation by extending the canals. The cropping pattern proposed in the additional 1,012 ha of command is one wet crop and two dry crops. The expected benefit from increased production will be 375 and 500 metric tons for wet crops and dry crop respectively. The benefit-cost ratio of the proposed strategy was 7.2:1.

b) Tungabhadra irrigation project

CGWB carried out conjunctive use studies have been carried out by the for the Tungabhadra right bank High Level Canal and Low Level Canal (HLC & LLC) command areas. The gross command area of these two canals is 6,354 sq. km which covers parts of Bellary district of Karnataka and Kurnool and Anantapur districts of Andhra Pradesh. The main source of irrigation is through Tungabhadra canal while irrigation through dugwells and tubewells is limited in the western and central part of the command. Indiscriminate use of surface/canal water has led to water logging and soil salinity.

Based on the canal water releases, annual surface water availability was assessed at 827 Mm^3 in the LLC and 784 Mm^3 in the HLC. Total water demands for all sectors worked out to be 1,073.23 Mm^3. The project area has few pockets of water logged area in the head reaches. There are a number of canal water shortage areas in the middle and tail ends of the canals. The immediate requirement in the area is to mitigate the problem of water deficit areas. The results indicate that by harnessing the ground water resource in pockets of canal water deficit areas, the water scarcity problem can be mitigated. This involves construction of

3,375 number of bore wells and 4,911 number of dug wells. For the surface water deficit areas which do not have sufficient ground water potential, it is imperative to develop ground water in head-reaches and supply canal water to deficit areas. The modeling studies have confirmed that ground water development plans can be implemented safely without any undesirable effect on the ground water regime.

c) Ghataprabha irrigation project

Ghataprabha River is one of the important tributaries of Krishna River. The study command area is served by Ghataprabha left and right bank canals. Ever since the introduction of canal irrigation in the area, there has been steady increase in the ground water potential. In the area, canal water is generally released in June/July. Waterlogging conditions set in by August and progressively extend over the area by November. One of the major objectives of the study was to control the problem of rising water table in the area by suitably adopting techniques of conjunctive use.

A mathematical model (MODFLOW) was calibrated and validated for the study area and five different scenarios of water use were analyzed. A conjunctive use plan was recommended that needs to be executed in stages. It has been estimated that an investment of Rs.40.5 crore is required for the left and right bank canal command areas for ground water development (construction of 20,619 bore wells). The study suggests setting up water use societies to maintain and operate bore wells, distribute water, collect water rates, settle disputes, etc.

d) Hirakud irrigation project

The Hirakud dam is a multipurpose project built across Mahanadi River in Orissa State, India. The Culturable Command Area (CCA) of the Hirakud project is 1,570.18 sq. km with an irrigation intensity of around 170%. Based on the monthly water table depth of nearly 130 monitoring wells, it was found that water logging conditions exist in an area of 174 sq. km during pre-monsoon and it practically covers the major part of the command during the post-monsoon period. Topographic set up, unlined canals, over-irrigation and prevailing paddy cultivation during the Kharif and Rabi seasons are mainly responsible for water logging conditions in the command. Utilizable ground water resources during Kharif and Rabi seasons have been estimated to be 508.04 and 764.44 Mm^3, respectively while the net draft from existing structures during Kharif and Rabi seasons is merely 2.35 and 4.35 Mm^3. Surface water availability during these seasons has been estimated at 1,361 and 1,495 Mm^3, respectively.

A linear programming model was formulated to optimally allocate the surface and ground water as per the cropping pattern. Attempts were also made to determine optimal cropping pattern for the maximum use of available surface and ground water to get maximum financial returns from irrigation. The study recommended use of 90% surface water and 10% ground water in the Hirakud command to rectify water logging conditions and mitigate water shortage in the tail reaches. Study also recommended that existing cropping pattern be modified with diversification of crops.

e) Sai Gomti interfluve tract – Part of Sarda-Sahayak irrigation project

The Sarda-Sahayak project in U.P. State, India was framed in 1967, commissioned in June 1974 and completed in the year 1985. The gross command area of project is 40.02 lakh ha and it was envisaged to irrigate 19.23 lakh ha achieving irrigation intensity of 96 %. With the commissioning of Sarda-Sahayak canal system, the ground water levels have risen in the command resulting in water logging conditions and spread of salt affected land. It has been noticed that existing availability and use of surface water in the upper and middle reaches of canals has caused significant seepage resulting in rise of ground water levels.

A study on conjunctive use of surface and ground water was taken in the Sai Gomati interfluve tract – part of Sarda-Sahayak irrigation project. The study area is bounded by Gomti River in North and by Sai River in South, covering an area of about 8,978 sq. km. Mathematical model studies were carried out using USGS-3D MODFLOW for devising the conjunctive use strategy and development of scenarios for various planning options. The study recommended installation of 21,375 shallow tubewells in the command to the existing battery of 56,625 tubewells at the total cost of Rs. 93.11 crore. Economic analysis was also attempted for the development of ground water strategy. The benefit-cost ratio "without project" and "with project" was calculated as 1.49 and 1.56 respectively.

f) Indira Gandhi Nahar Project Stage – I

Indira Gandhi Nahar Project (IGNP) was conceived in the year 1948 to use the water of Sutlej-Beas Rivers for irrigation in parts of Western Rajasthan. Total command area of project is 1.537 Mha. The construction works of the project were carried out in two stages. The IGNP Stage-I comprises CCA of 0.479 Mha. The first water in IGNP was released in October 1961.

After the introduction of irrigation in the command area, ground water levels were found to rise at an alarming rate of around one meter per year, resulting in water logging and soil salinity development. To overcome and mitigate these problems, the conjunctive use study was carried out in the Northwestern part of Thar Desert of Rajasthan which is underlain by wind blown sand dune deposits. Mathematical modeling studies were carried out for devising water use patterns under different development scenarios. Economic financial analysis for the development of ground water strategies were also carried out. The economically most viable strategy demonstrates that there is a need to install 10,023 number of shallow tube wells to lower the ground water table in the area with the total ground water development cost of Rs. 72.17 crore. An additional area of 0.1066 Mha can be brought under cultivation and the benefit-cost ratio for the scheme works out to be 1.03.

g) Optimal crop planning between Yamuna River and Eastern Yamuna Canal

The study area lies between the Yamuna River and Eastern Yamuna canal covering parts of Saharanpur, Muzaffarnagar and Meerut districts of UP. Total extent of the study area is 2,940 sq. km while cultivated area under irrigation is about 1,764 sq. km.

The main objective of this study was to determine the optimal cropping pattern and water allocation plan by conjunctive use of surface and ground water. A linear programming model was used to maximize the net annual benefits from the crops and determine 40 operating variables. Total net annual benefits from the developed optimal plan came out to be Rs. 178.89 crore with crop intensity of 152.63 %. In terms of water use and cost of cultivation, the optimal cropping pattern shows that crops such as potato, onion, tomato, groundnut, gram and bajara should be grown over larger areas, signifying greatest benefits, while rice, maize, cotton, jowar, urad and arhar should be grown over smaller areas, signifying small benefits. Crops such as wheat and sugarcane signify medium benefits.

h) Hindon – Kali Nadi Doab

The study area lies between Hindon and Kali Rivers in districts Saharanpur & Muzaffarnagar in UP. The study area covers about 950 sq. km and receives annual average rainfall of 704 mm during the monsoon period and 131 mm during the non-monsoon period. The annual ground water draft is 3,300 ha-m. A quantity of 56,470 ha-m of water is annually available for the existing crops. But the annual crop water requirement is 68,650 ha-m. Therefore, the present irrigation level in this study area works out to be 80 %.

The main objectives of the study were: a) to optimize allocation of surface and ground water and to maximize agricultural potential of the command area with the help of a linear programming model, and b) to develop aquifer response simulation model of the area to predict future behaviour of ground water table. The study concluded that available surface and ground water potential in the area can sustain only 80 % of the irrigation level. If 100 % irrigation is provided in the area, it is likely to result in decline of ground water table at the rate of 0.54 to 0.33 m per year. The study also recommended reduction of sugarcane area as this crop consumes more water in the non-monsoon season resulting in decline of ground water level.

i) Lakhaoti Branch of Madhya Ganga Project

A study for conjunctive management of surface and ground water in the Lakhaoti branch canal command of Madhya Ganga Project (UP) was carried out jointly by the Water Resources Development Training Centre, Roorkee, WALMI (UP) and Irrigation Department (UP). The study threw light on the future scenario of irrigation water requirement and ground water table behaviour in the area. The study developed a ground water model, a cost function for surface and ground water use, and optimization techniques for allocation of surface and ground water in space and time. The study developed the main canal operation strategy and evolved conjunctive use plan to minimize cost of providing irrigation water.

j) Mahi – Kadana canal command

The gross irrigated command area of 3,786.58 sq. km of Mahi – Kadana system was studied by CGWB for conjunctive use of surface and ground water. Surface water and ground water resources in the area have been estimated as 2,259 and 921 Mm^3 respectively. Irrigation consumes 95 % of the total water demands with a cropping intensity of 179 %. As per the study, full utilization of the available ground water and surface water is the most ideal situation of conjunctive use in this command. Surface water and ground water contributions come out to be 2,221 and 734 Mm^3 respectively. Ground water abstractions can be done through 8,355 dug wells and 855 shallow tubewells with annual drafts of 501 and 128 Mm^3, respectively.

Besides the above, studies related to conjunctive use have been completed by CGWB for the Nagarjuna Sagar project, A.P.; Indira Gandhi Nahar Pariyojana Stage – II, Rajasthan; and Kosi and Gandak canal command areas, Bihar.

6.11.4. Installation of augmentation tubewells (ATW)

Construction of augmentation tube wells is a form of conjunctive use. In the 1950s, it was observed that while the command of lower reaches of Western Yamuna Canal were facing scarcity of canal supply, waterlogging was developing in the upper reaches. In the year 1969, the Irrigation Department of Haryana chalked out a scheme to install deep tube-wells in the Western Yamuna canal command extending from Dadupur in Ambala District to Panipat in Karnal District. After creation of The Haryana State Minor irrigation (Tubewells) Corporation in the year 1970, the work related to the exploitation of underground water resources was transferred to it. Detailed investigations and ground water studies were undertaken and it was revealed that 3,300 cusecs of ground water could be safely exploited form Yamuna Basin without causing any appreciable fall in the water table in the region. On this basis, it was decided to construct a lined channel from Jagadhari to Munak to save losses and augment canal supplies from ground-water resources through deep tubewells. Thus, the concept of Augmentation Tube Wells came into being. Haryana was the pioneer State in India to make conjunctive use of ground water and surface water resources for irrigation. A total of 1,643 augmentation tube wells with capacity varying from 1 to 7 cusec were installed (Tanwar, 1997). Independent electric feeder lines were provided for continuous operation of these wells. This strategy resulted in the subsidence of water table in the region.

The work of the installation of deep tube-wells and Augmentation Canal Project was completed in December, 1972. An evaluation of this project (Saksena 2000) revealed that the objective of increasing canal water supplies at the Munak head of Western Yamuna Canal had almost been achieved. However, the lowering of water table has been observed within 1-2 km from the canal and lowering of pumps in case of private tube-wells had to be done. Subsequently, the Haryana government undertook construction of about 1,000 Augmentation tube-wells on Western Canal System to directly feed the canal.

In Gandak Canal Project of U.P., about 150 augmentation tubewells have been installed in 1990s to supplement canal water supplies. Similarly in Chambal irrigation system, 175 augmentation tubewells with an installed capacity of about 13 cumec have been constructed for conjunctive use of canal water and ground water. The wells discharge into various parts of the canal conveyance system including distributaries, minors and sub-minors.

The Chambal Irrigation System serves areas in the State of M.P. and Rajasthan. At Kota Barrage, water is diverted into two canals: Right Bank Canal serving M.P. and Left Bank Canal serving Rajasthan. Due to some deficiencies, hardly 50% proposed area could be irrigated. To augment supplies in the system, a project on Ambah Branch canal was taken. It includes construction of 175 augmentation tube-wells. The total installed pumping capacity would be about 47,000 m^3/hr (13 m^3/s). It is assumed that power supply limitations will restrict pumping in peak demand periods to 16 hour/day, giving a peak 15-day production of 11.2 MCM at the well heads, equivalent to a continuous discharge of about 8.7 m^3/s. The proposed cropping pattern would require approximately 2,300 hours of groundwater pumping during the period October through March.

SECTION 3

RIVER BASINS

CHAPTER 7

RIVER BASINS OF INDIA

Inland water resources of a country can be classified as rivers and canals; reservoirs; tanks & ponds; beels, oxbow lakes, derelict water; and brackish water. Other than rivers and canals, all the other water bodies cover an area of about 7 M-ha. Of the rivers and canals, Uttar Pradesh (including Uttaranchal) occupies the first place with the total length of rivers and canals as 31,200 km, which is about 17% of the total length of rivers and canals in the country. Other states following Uttar Pradesh are Jammu & Kashmir and Madhya Pradesh. Among the remaining forms of the inland water resources, tanks and ponds have maximum area (2.9 M-ha.) followed by reservoirs (2.1 M-ha).

Most of the area under tanks and ponds lies in the southern states of Andhra Pradesh, Karnataka and Tamil Nadu. These states along with West Bengal, Rajasthan and Uttar Pradesh, account for 62% of total area under tanks and ponds in the country. As far as reservoirs are concerned, major states like Andhra Pradesh, Gujarat, Karnataka, Madhya Pradesh, Maharashtra, Orissa, Rajasthan and Uttar Pradesh account for larger portion of area under them. More than 77% of area under beels, oxbow lakes and derelict water lies in the states of Orissa, Uttar Pradesh and Assam. Orissa ranks first as regards the total area under brackish water and is followed by Gujarat, Kerala and West Bengal. The total area of inland water resources is, thus, unevenly distributed over the country with five states namely Orissa, Andhra Pradesh, Gujarat, Karnataka and West Bengal accounting for more than half of the country's inland water bodies.

Before describing the river basins of India, it is helpful to note catchment areas, average annual water yields and lengths of selected major rivers of the world as given in Table 1. This data is insightful for comparison, to have a feel of the numbers, and to appreciate the relative magnitude of the key features. In terms of catchment area, Amazon has the first place followed Congo, and then Mississippi-Missouri. Amazon also occupies the first place as afar as mean annual discharge is concerned and this is again followed by Congo. Nile is the longest river of the world and Amazon comes at the second place.

Table 1. Catchment area, average annual water yield and length of selected major rivers of the World

Name	Catchment area, (km^2)	Rank by area	Mean annual discharge, (m^3/s)	Rank by discharge	Length (km)	Rank by length
Amazon, South America	6,915,000	1	174,900	1	6,400	2
Congo, Africa	3,820,000	2	39,000	2	4,700	8
Rio de la Plata-Parana, South America	3,100,000	4	22,900	5	4,876	7
Ob-Irtysh, Asia	2,990,000	5	12,500	16	5,410	6
Nile, Africa	2,870,000	6	3,100	34	6,671	1
Lena, Asia	2,490,000	8	16,100	11	4,400	11
Niger, Africa	2,090,000	9	6,100	27	4,200	14
Amur-Argun, Asia	2,051,000	10	11,000	17	4,444	10
Yangtze, Asia	1,826,000	11	21,800	6	6,300	3
Volga, Europe	1,360,000	12	7,700	22	3,531	20
Ganga, Asia	1,086,000		16,650	–	2,525	41
Brahmaputra, Asia	580,000		19,200	10	2,849	27
Mississippi-Missouri, North America	3,220,000	3	18,100	8	6,019	4
Yenisei-Angara, Asia	2,618,500	7	18,100	9	4,092	15
Mekong, Asia	1,525,000		14,100	13	4,200	13
Orinoco, South America	1,425,000		25,200	4	2,600	33
Mackenzie, North America	1,705,000		9,700	21	4,241	12
St. Lawrence, North America	1,302,800	14	13,000	14		–
Danube, Europe	787,500		6,400	26	2,858	26
Zambesi, Africa	1,330,000	13	16,000	12	2,700	30
Indus, Asia	930,000		6,600	25	2,900	25
Columbia, North America	645,000				2,000	55
Yukon, North America	825,000		6,800	24	2,849	28
Huang, Asia	1,258,700	15	3,300	33	5,464	5
Sao Francisco, South America	630,000		3,400	31	3,199	22
Euphrates, Asia	1,075,000		–	–	2,430	29
Murray-Darling, Australia	1,035,000		368	–	3,750	18

Source: Rand McNally (1998), Singh (1992), and others.

7.1. MAJOR RIVER GROUPS OF INDIA

Based on the topography, the river systems of India can be classified into four groups. These are: (i) Himalayan rivers, (ii) Deccan rivers, (iii) Coastal rivers, and (iv) Rivers of the inland drainage basin. The Deccan rivers are rainfed and therefore have very little flow during non-monsoon season; many of these are non-perennial.

The coastal streams, especially on the west coast, have small catchment areas and are short in length. Most of them are non-perennial. The streams of the inland drainage basin of Western Rajasthan are few and far in between. They flow for some time during monsoon only. Elaboration of these river categories is given in what follows.

7.1.1. The Himalayan Rivers

The Himalayan Rivers receive input from rain as well as snowmelt and glacier melt and, therefore, have continuous flow throughout the year. During the monsoon months of June to September, Himalayas receive very heavy rainfall and experience maximum snow melt and these are the periods when the rivers carry about 80% or more of the annual flows. This is also the time when these rivers are prone to flooding.

The main river systems in Himalayas are those of the Indus and the Ganga-Brahmaputra-Meghna. The Indus rises near Mansarovar in Tibet. Flowing through Kashmir, it enters Pakistan and finally falls in the Arabian Sea near Karachi. A number of important tributaries of Indus flow through India, namely, the Sutlej, the Beas, the Ravi, the Chenab and the Jhelum.

Bhagirathi and Alakhnanda are two important rivers that originate in Garhwal Himalayas. These join at Devprayag to form Ganga which is the most sacred river of India. This river traverses through Uttaranchal, Uttar Pradesh, Bihar, and West Bengal and thereafter enters Bangladesh. The important tributaries of Ganga are the Yamuna, the Ramganga, the Ghaghra, the Gandak, the Kosi, and the Sone. Many of these tributaries are mighty rivers themselves. Yamuna River is an important tributary of Ganga and its own important tributaries are Chambal and Betwa.

The Brahmaputra rises in Tibet where it is known by the name Tsangpo. It enters India in Arunachal Pradesh and after traversing through Assam, enters Bangladesh. Its important tributaries are the Dibang, Lohit, Subansiri, Manas and Teesta. Ganga and Brahmaputra rivers meet at Goalundo in Bangladesh. The Barak River, the head stream of Meghna rises in the hills in Manipur. The Meghna is the part of Ganga-Brahmaputra-Meghna-System. The combined Ganga-Brahmaputra River meets Meghna in Bangladesh and their huge volume of water flows into the Bay of Bengal.

Harnessing the waters of the major rivers that flow from the Himalayas is of paramount importance for India, Nepal, and Bangladesh. Through a close collaboration and pooling resources, huge benefits can be realized from flood control, assured irrigation, hydroelectric power generation, employment generation, and improvement of environmental quality. As these countries grapple with the political compulsions and realities, both domestic and international, of utilizing the flow of the Ganga-Brahmaputra system, precious water, largely unutilized, continues to flow to the sea. At times, it also inflicts large losses to life and property. A map of India showing important rivers is given in Figure 1. Table 2 shows the Principal Himalayan rivers of India.

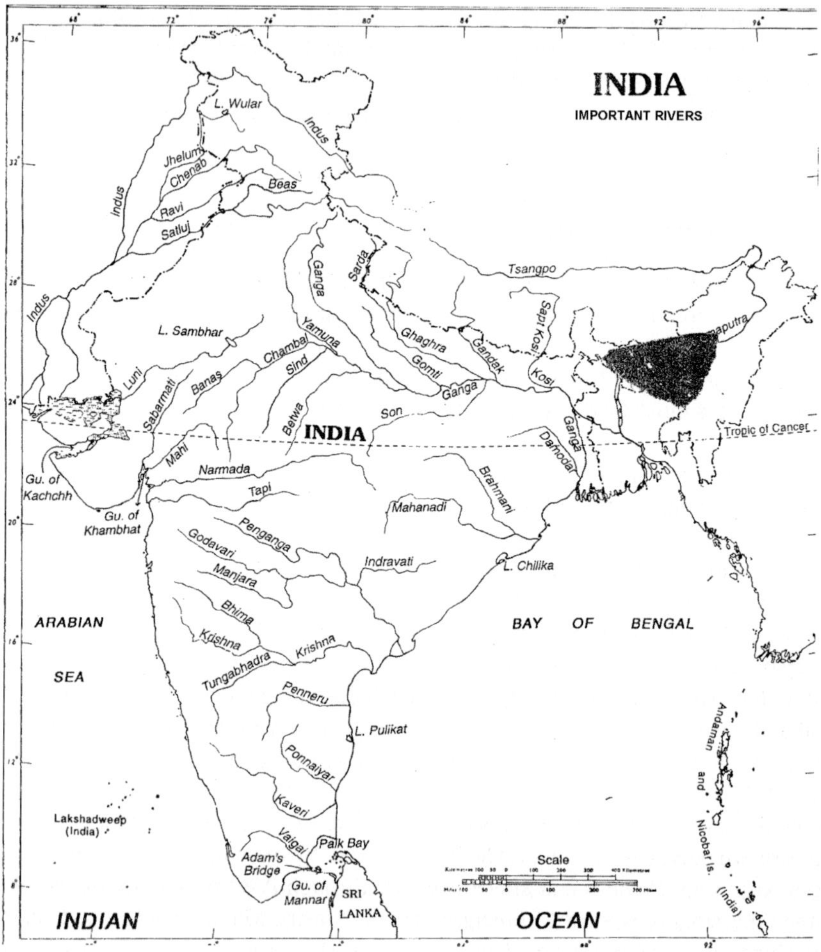

Figure 1. Important rivers of India

7.1.2. The Deccan Rivers

The rivers of Deccan can be further classified in two groups: west flowing rivers and east flowing rivers. The Narmada and the Tapi rivers flow westwards into Arabian Sea. The important east flowing rivers are the Brahmani, the Mahanadi, the Godavari, the Krishna, the Pennar, and the Cauvery. These rivers fall into the Bay of Bengal.

The Mahanadi, rising in the state of Madhya Pradesh, is an important river in the state of Orissa. In the upper drainage basin of Mahanadi, which is centered on the Chhattisgarh Plain, periodic droughts contrast with the situation in the delta region where floods damage the crops in the rice bowl of Orissa. Hirakud Dam, constructed

Table 2. Principal Himalayan rivers of India

Group No.	Order of magnitude (Area in km^2)	Name of river	Himalayan area included in catchment (km^2)
I	Above 256,000	Indus	265,728
II	Between 128,000 and 256,000	Brahmaputra	253,952
III	Between 38,400 and 128,000	Kosi	61,184
		Karnali	52,736
		Satluj	47,360
IV	Between 25,600 and 38,400	Gandak	37,376
		Jhelum	33,280
		Manas	30,720
		Chenab	26,880
		Raidak	26,112
V	Between 12,800 and 25,600	Ganga	22,784
		Luhit	20,480
		Subansiri	17,920
		Kali	16,128
		Beas	14,336
		Dibang	12,800
VI	Below 12,800	Tista	12,288
		Yamuna	11,520
		Ravi	7,936
		Rapti	7,680
		Ramganga	6,656
		Baghmati	3,840

Source: Chandra (1992).

in the middle reaches of Mahanadi, has helped in alleviating these adverse effects to a large extent by creating a reservoir.

The source of the Godavari is near Nasik, northeast of Mumbai (Bombay) in the state of Maharashtra, and the river follows a southeasterly course for 1,400 kilometers (km) to its mouth on the Andhra Pradesh coast. The Godavari basin is second in size only to Ganga; its delta on the east coast is also one of the country's main rice-growing areas. It is known as the *Ganga of the South* but despite the large catchment area, its discharge is moderate. The reason is medium depth of annual rainfall, for example, about 700 mm at Nasik and 1,000 mm at Nizamabad.

The Krishna rises in the Western Ghats and flows east into the Bay of Bengal. Its flow is not very large because of low rainfall in its catchment area: 660 mm annually at Pune. The Krishna is the third longest river in India.

The source of the Cauvery is in the state of Karnataka and the river flows southeastward. Its main tributaries are the Bhima, the Tungabhadra, the Ghatprabha and the Malaprabha. The waters of the river have been a source of irrigation since antiquity; in the early 1990s, an estimated 95% of the Cauvery flow was diverted for agricultural use.

The Narmada and the Tapi are the only major rivers that flow eastward into the Arabian Sea. The Narmada rises in Madhya Pradesh and crosses the state, passing swiftly through a narrow valley between the Vindhya Range and spurs of the Satpura Range. It flows into the Gulf of Khambhat (or Cambay). Tapi which is of shorter length follows a generally parallel course, between 80 km and 160 kilometers to the south of the Narmada, flowing through the states of Maharashtra and Gujarat on its way into the Gulf of Khambhat.

7.1.3. The Coastal Rivers

There are numerous coastal rivers which are comparatively small. While only handful of such rivers drain into the sea near the deltas of east coast, there are as many as 600 such rivers on the west coast. The West Coast rivers are important as they contain as much as 14% of the country's water resources while draining only 3% of the land. Figure 2 shows the Coastal and Inland rivers of India. A detailed description of these rivers is provided in Chapter 16.

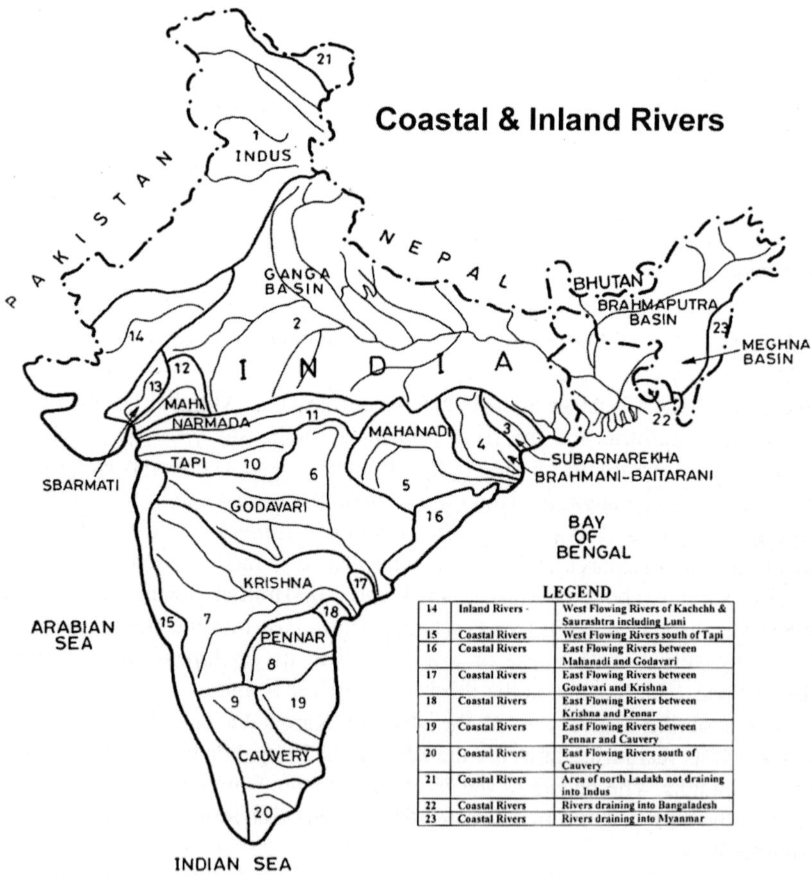

14	Inland Rivers -	West Flowing Rivers of Kachchh & Saurashtra including Luni
15	Coastal Rivers	West Flowing Rivers south of Tapi
16	Coastal Rivers	East Flowing Rivers between Mahanadi and Godavari
17	Coastal Rivers	East Flowing Rivers between Godavari and Krishna
18	Coastal Rivers	East Flowing Rivers between Krishna and Pennar
19	Coastal Rivers	East Flowing Rivers between Pennar and Cauvery
20	Coastal Rivers	East Flowing Rivers south of Cauvery
21	Coastal Rivers	Area of north Ladakh not draining into Indus
22	Coastal Rivers	Rivers draining into Bangaladesh
23	Coastal Rivers	Rivers draining into Myanmar

Figure 2. Coastal and inland rivers of India

7.1.4. Rivers of the Inland Drainage Basin

The rivers of the inland system, centered in western Rajasthan state, are few and frequently disappear in years of scant rainfall. A few rivers in Rajasthan do not drain into the sea. They drain into salt lakes or get lost in sands with no outlet to sea. These rivers are shown in Fig. 2 and are described in Chapter 16.

7.2. MYTHOLOGICAL NAMING OF RIVERS

Naming of rivers is as interesting a topic as the study of characteristics of the rivers. Water is essential for human survival and not a day passes in the life of an ordinary person when he does not consume it. Since rivers have been the most common source of water, life and culture of a place are inextricably woven with rivers. Our ancestors were keen observers of the behavior of rivers and named them in many ways. Color of water has been one motivation for naming them. Consequently, it is not surprising that there are many white rivers in the world. For example, *Rio Blanco* River in Argentina, and *Rio Branco* River in Brazil are *white rivers*. There is Blue Nile and there is White Nile in Africa and after their confluence the river becomes Nile. In USA also, one can spot a White River. Likewise, we have a *Red River* in Canada, *Rio Bermejo* River in Argentina, *Song Chay* River in Vietnam, *Son He* River in China and *Red River* in the USA. In a similar vein, there is a *Black River (Rio Negro)* in Brazil, Argentina and Chile. The *Hwang Ho* River in China is also called as Yellow River since its highly muddy water appears to be yellow. We may also add that this river is the most important among the colored rivers.

Social and cultural life in India is deeply influenced by religion and this influence can be vividly seen when one studies the names of Indian rivers and the mythological stories associated with their naming. In the following, mythological stories associated with some Indian rivers are narrated.

Bagmati River: It is also called *Bagvati* in *Vishnu Purana*. The *Svayambu* and the *Vardha Purana* call it the *Vagmati*. It is called *Bachamati* in Buddhist literature, because it was created by the word uttered by Buddha Krakuchhanda when the latter visited Nepal with his disciples from the Gaud-desa. Vdana mentions a river *Vaggumuda* which flowed to the east of the Vajji territory. This *Vaggamuda* seems to be the *Vagamati* of the present time. *Baghavati*, the name given to this river by *Vidyapati* seems to have some connection with the word Vyaghra (tiger). Tigers are found in abundance on its banks in the Nepalease Terai.

Bhima River: The Bhima River is also called *Punya Damini Bhima*. In South India, people give regards to the Bhima River the same way as they do to Ganga. The Bhima River originates from the Sahyadri hills. According to the legends, when Lord Shankar came near Bhima Shankar mountain after killing the demon Tripursur, he found that the Ayodhya's saintly king Bhimak was under penance at that place. King Bhimak begged for the blessings of Lord Shankar so that a pious river might originate from the sweat of Lord Shankar. Lord Shankar gave the desired blessings and accordingly, a river originated from his sweat. On the name of king Bhimak, this river was called as Bhima River. The river joins Krishna at Kurugadi in district Raipur, which is 25 km away from Gulbarg. A *jyotirling* (glorified symbol of Lord Shiva) namely, Bhima Shankar, and a religious place, namely Pandarpur, are located on the banks of this river.

Brahmaputra River: Brahmaputra River originates from the Mansarovar near Kailash range and is also known as Mahanad. Worth noting is the fact that the Brahmaputra River has a male name whereas all other majors rivers of India have female names. According to legends, Brahmaputra is the son of Lord Brahma. It is said that Shantanu, a famous ancient sage began a long meditation in an ashram in this area along with his beautiful wife Amodha. Amodha was so beautiful that Lord Brahma himself became enchanted by the beauty of Amodha and requested her to make love with him. But Amodha did not accept the Brahma's proposal. However, by that time Lord Brahma had become so excited that his semen discharged at that place. When Shantanu came to know about this, he inseminated the Brahma's semen in the womb of Amodha. Subsequently, Amodha gave birth to a son and he was called Brahmaputra. The tank near the ashram of sage Shantanu is known as the Brahm-kund. Another legend is that because Brahmaputra is the largest river in India, it carries a male name.

Brahmaputra is said to be the soul of Assam. It has seen the growth of the culture of Assam and is a witness of all the events big or small in the life of its inhabitants.

Chambal River: The ancient name of Chambal River was *Charmanvati*, meaning the river on whose banks leather is dried. In the ancient times, large-scale *Yagya* (prayer meetings in which also involve offerings to please God) used to be organized on the banks of this river and in these *Yagyas* animals were slaughtered and offered. According to Mahabharata, the color of river water would become red due to the flow of blood of the sacrificed animals and the skins of these animals were dried on the banks of the rivers. In due course of time, this river became famous as the river of 'chamda' (skin) and was named as *Charmanvati.*

Chandan River: It was also known by the name of 'Malini' and 'Chandana'. It is known as 'Chamba' in the Buddhist literature. According to the Kshetra Samasa, this river is called Sulakshini or Chandravati. According to the Jinavitasa, this river is named Aranyavaha or the torrent through the wilderness.

Gandak River: In the Muzaffarpur District of Bihar, this river is known as the *Narayani* and *Saligramini. Ramayana* mentions this river under the name of *Kalimahi.* The river is said to have been formed from the sweat of the cheeks of *Vishnu* when he performed austerities near its source. *Artemidoras* speaks of a certain affluent of the Ganga as breeding crocodiles and dolphins. He named it *Oidenes.* At the present time, Gandak is the only tributary of Ganga which breeds crocodiles. The Puranic tale of Gajehrajeha (fight between the elephant and crocodiles) is said to have taken place at its junction with the Ganga. Therefore, Oidenes of Artemedoros is no other than the Gandak.

Ganga River: The Ganga River has been considered as the most sacred river of India in *Puranas*. The word *Ganga* is considered as a synonym of pure and holy water. That is why the word is attached with the names of many other rivers in Central and South India. According to a mythological legend, Lord Brahma collected the sweat of Lord Vishnu's feet and created Ganga. Being touched by two members of the Trimurti (Brahma, Vishnu, and Mahesh), Ganga became very holy. The other synonyms of Ganga are *Vishnu Padee* (as the river is said to have originated from the feet of Lord Vishnu), *Mandakini, Devnadi, Sursari, Tripathga, Jahanvi, Bhagirathi*, etc. As mentioned in *Mahabharat*, when *Bhishm* was about to die, *Arjuna* was said to have extracted groundwater, namely, *Patalganga*, by shooting an arrow which made a hole in the ground and created a fountain.

The description of Ganga is available in *Rigveda* also. Several legends are famous about the origin of Ganga. It is said that the Ganga has originated from foots of *Lord Vishnu*. According to the second legend, Ganga was the daughter of mountain king *Himalaya*. As per *Devi Bhagwat*, Ganga has been called the wife of *Lord Vishnu*. According to *Mahabharat*, Ganga was the wife of *King Shantanu* as well as the mother of *Bhishm*.

According to a legend in the Ramayana, Sagara, the king of Ayodhya who had sixty thousand sons, once performed Ashvamegh yagya (a ritual for the good of his kingdom and to demonstrate his supremacy). During the rituals, the horse which is an integral parts of the ritual, was stolen by the jealous Indra and placed in the ashram (hermitage) of saint Kapila. Sagara sent all his sons all over the earth to search for the horse. They found it in the nether-world, in the ashram of saint Kapila. Assuming that the sage had stolen the horse, they hurled insults at him and interrupted his meditation. The saint became very angry and burnt all sixty thousand boys to death by the fire from his eyes. Since the final rites of these boys could not be performed, their souls continued to wander as ghosts. After many generations, Bhagiratha a descendant of Sagara, learnt about the fate of his ancestors and he vowed to bring Ganga to Earth so that her holy water could be used to liberate the souls of his ancestors and release them to heaven.

King Bhagirath left his kingdom to meditate and prey the Ganga River who was residing in the heavens to descends to the earth. Ganga could come down to the earth only after Lord Brahma (the Supreme God) permitted her to do so. Accordingly, Bhagirath prayed at a place in Uttarkashi where the Gangotri temple is situated these days. He prayed to Lord Brahma for a thousand years, requesting him to permit Goddess Ganga to come down to earth from heaven because only Ganga could release his ancestors' souls and allow them to go to heaven. On account of deep devotion of Bhagirath, ultimately Lord Brahma was pleased with him and granted his wish. But He told Bhagirath to pray to Lord Shiva, who alone could sustain the huge force of descent of Ganga. Accordingly, King Bhagirath prayed to Lord Shiva who, after some time, agreed to hold Ganga in his hair. Accordingly, Ganga descended from heavens on Lord Shiva's head and was soon trapped in his thick locks (Figure 3). In the process, the river water got further purified. One the request of King Bhagirath, Lord Shiva opened one of the locks

Figure 3. Ganga descending from the heaven in the locks of Lord Shiva

and the river reached the Earth. It is said that Ganga followed the chariot of Bhagirath to the place where the ashes of his ancestors were lying and released them from the curse. Alert reader will notice that this legend is a simple description of the hydrologic cycle. Perhaps King Bhagirath was an ancient civil engineer or a hydrologist!

Since Bhagirath brought Ganga to Earth, one headwater stream of Ganga is known as Bhagirathi. Further, the Hindi term *Bhagirath prayas* describes valiant efforts or difficult achievements of a person. A view of Bhagirathi River in the headwater regions is shown in Figure 4.

Ganga is also known by another name: Jahnavi. According to a different mythological story, when Ganga came down to earth long time ago, her fast moving waters created turbulence and destroyed crops in the fields. She also disturbed the meditation of a saint named Jahnu. Now, Jahnu was so much angry that he drank up all the water Ganga. This made the Devtas (semi-Gods) very sad and they prayed to Jahnu to release Ganga so that she could proceed on her mission. After their persistent prayers, Jahnu was pleased and he released Ganga water through his ears. On account of this, Ganga came to known by the name "Jahnavi" (daughter of Jahnu) also.

Figure 4. A view of Bhagirathi River

Kumbh Mela which is the largest religious gathering on Earth (attended by more than 80 lakh devotees) takes place after every 12 years at two places on the banks of Ganga River: Haridwar and Allahabad.

Ghaghara River: The name Ghaghara seems to have originated from the Sanskrit word *Ghaghara* a gurgling sound of water. Flow of this river used to produce this kind of sound.

Hooghly River: Shahjahan (the mughal king of Delhi) permitted the Portuguese to do trading in Bengal and they built a church in Bengal in 1590. The grass that grew around the Church was *Hugla*, and hence the name given to nearby river became *Ogolin*. With the passage of time, it became *Oglee* and eventually *Hooghly*.

Indus River: In Rig Veda, the deitified *Sindhu* is praised in many verses. Although some historians believe that the word Sindhu means a sea, the widely held view is that it refers to the Indus River. Therefore, Sindhu may be taken to mean the Indus River which is described as donor of gifts and owner of fertile fields. Our country came to be called Hindustan or India; these words are derived from the name Indus or Sindhu. The *Sindh* province of Pakistan derives its name from this river. Its name also figures in the national anthem of India. In the valley of this river rose the Indus Valley Civilization, which is the most ancient and highly developed civilization of this planet. A unique feature of this civilization is that it still survives despite numerous setbacks. Many web-sites (e.g., *www.harappa.com*, *www.archaeolink.com*) and books contain detailed information about the Indus Valley Civilization.

Kosi River: The Kosi River is a notorious river in the Ganga basin for changing its course. The river can be compared with the rivers of China which suddenly wash away large tracts of land. *Cunningham* identified this river with the *Tista*. According to Ramayana the river was named after *Kausiki* who was the sister of the sage Vishvamitra. Like her brother, she was a lady of short temper.

Krishna River: Krishna is a mighty east flowing river of peninsular India. It is the same river as Krsnavena in the Puranas or Krsnaveni in the Yoginitantra. It is also known as Kanhapenna in Jatakas and Kanhapena in the Hathigumpha inscription of Kharavela. The word *Krishna* also indicates dark color.

Lohit River: In the Assamese language, Lohit River is known as *Luit*. The word Luit comes from a Sanskrit word *Lohitya* meaning red river. When the red soil of this region would get eroded by the rain water and flow in the river, the river water would turn red and this gave the river its name. According to a legend, Ram who was the grandson of Maharshi Bhrangu and the son of Rishi Jagdagni, was a bright and obedient child. Once Renuka, the wife of Rishi Jagdagni, came to the banks of the river to fetch water. There she saw Chitrataka, a gandharva (semi-god) playing with his wives. Seeing this, Renuka's mind also got enamoured. When Rishi Jagdagni came to know about it, he was very angry and he ordered his son to kill his mother. The obedient son chopped his mother's head by a parshu (sharp axe) but the parshu got stuck to his hands. He tried many things, including long pilgrimages, but the parshu could not be removed. It finally came off when he took a bath in the Lohit River. In this process, Ram came to be known as Parshuram and the river which became red due to the blood of his mother was known as the Lohit River.

Mahanadi River: The literal meaning of Mahanadi River is large size river. The originating place of Mahanadi River, which is called as *holy Ganga* in Chhatisgarh and Orrisa is located near the Ashram of *Maharshi Shrangi*. It is said that once all the sages of this area came at this place for taking holy bath in *Mahakumbh*. The Maharshi was under meditation and penance at that time. The sages waited for several days to draw the attention of the Maharshi but the Maharshi's meditation was not disrupted. Thereafter, the sages went for the holy bath. While returning after the bath, all the sages brought some holy water with them. Finding that *Maharshi Shrangi* was still in the meditation, they filled the Maharshi's kamandal (vessel) with water, and returned to their native places. After some time, when the meditation of the Maharshi Shrangi was disrupted, the water of the kamandal fell down on the ground with the stroke of his hand. This water began to flow towards east and was converted into a stream. This stream was called as Mahanadi which is said to fulfill the desires of millions of people.

Mahi River: In Vayu Purana, this river is also known as Mahati. The name of the river seems to be derived from the lake from which it springs. This is often called the Mau or Mahu as well as the Menda. According to one legend, the Mahi is the daughter of the Earth and sweat of Indrdyumna, the King of Ujjain.

Narmada River: Narmada is also known as Narbada (Nerbudda). It is also called Reva. The name Reva has been mentioned in Puranas which has probably been derived from the Sanskrit root 'rev' to 'hop' owing to the leaping of the stream down its rocky bed. The Narmada River is also known by a few other names such as: Daksinaganga mentioned in Skanda Purana, Indija, Purvaganga, Mekaladrija Mekalasutra or Mekalakanyaka (Amarakosa) and Somabhava.

There are many legends regarding the origin of the Narmada. According to the one recorded by Beglar, Narmada was the name of the beautiful daughter of a shepherd living at the Amarkantaka. She used to carry her father's breakfast to him in the fields where he tended his cattle. On her way to and back from her father, the girl used to spend some time daily in a Yogin's (a person who practices Yoga) company whose ashram was by the road side. After sometime, the girl killed herself for some unexplained reasons. One day the Yogin, while in act of drinking bhang (a kind of drink), heard about the death of the poor girl. The cup of bhang stuck to his mouth and he died. A stream of water issued from his throat which is the Narmada River.

Another version is that the girl, finding herself pregnant with a child, committed suicide by throwing herself over the falls of Kapiladhara, and the river in which she died was named after her. According to another legend, the Narmada River sprang from the body of Lord Shiva.

Pun-Pun River: This river is mentioned in the Vayu and the Padma Puranas in connection with Gaya Mahatmya as the Punah-punah (again and again) of which Pun-Pun is the colloquial form. The river might have been called by this name because it was frequently in spate. The Puranas interpret the word Punahpuna in a spiritual sense, i.e., sins are removed again and again by offering oblations to the Pitras (forefathers) in the river.

Sarayu River: According to the Buddhist literature, the Sarayu or Sarju River is spelt as "*Sarabh*". Cunningham, in one of his maps, identified the Sarju with the Solomattis River mentioned by Megasthenes. Ptolemy names a river Sarobes which is identified by all scholars with the Sarayu.

Sipra River: The Sipra River is also called *Ksipra (Markandeya)*. It flows in the State of Madhya Pradesh. The river is famous for the sanctity associated with it. According to the legend, the river has originated from the blood of *Lord Vishnu*. In the time of Mughal King Akbar (15th century), it was believed that the river used to flow with milk. Probably this means that the region where it flowed was very prosperous.

Sindh River: In *Vishnu Purana*, the *Dasarna* River in Sindh has been identified as Sindhu River. The Sindh River is generally believed to be identical with the *Kali Sindh* River. *Mahabharat* refers to it as *Daksinasindh*. The *Meghdoot* refers to *Kalisindh* as Sindh. In *Varaha Purana* the *Kalisindh* River has been called as *Sindhupurana*.

Sone River: It has another name in *Amarkosha* where it is called *Hiranyavaha* because either its sand was of golden colour or because the river carried gold dust in its flow. In Hindi, gold is called *Swarna* or *Sona* which changes to Sone with the passage of time.

Subarnarekha or Suvarnarekha River: Subarnarekha was earlier known as Hiranyarekha. Both these words mean a golden streak. The name shows that the river brought gold in its flow and this belief still persists among the local people.

Tapi River: Also known by the name Tapti, Tapi was a daughter of Sun. Ptolemy named it *Nanagouna*. It is believed that Tapi rises from the sacred tank of Multai (Mulatapi, i.e., the source of Tapi). The Tapi has its name derived from *tapa*, 'heat' and according to local Brahmanas, it was created by the Sun to protect himself from his own warmth.

Teesta River: The literal meaning of the word Teesta is *Trishna* (desire) which never ends. In the Pali language, Teesta is called as Tanda. The legend of the Teesta River is mentioned in Kalika Purana. Among the other rivers of northern India, Teesta River is also called as the younger daughter of Himalaya. There is a legend about it in Kalika Purana. It is said that once Lord Shiva became pleased from the hard penance of the demons and gave them blessings. That particular demon was the devotee of Lord Shiva but he did not like Goddess Parvati, the wife of Lord Shiva. Parvati took this as an insult to her. As a result, a war started between the demon and Parvati. The demon was injured in the war and he was feeling thirsty. He requested Lord Shiva to save his life by quenching his thirst. Lord Shiva became pleased and from his inspiration, a stream of nectar-like milky water started flowing from the Parvati's breast. This stream was called as Teesta River and this river is fulfilling the desires of the people even today.

Tungabgadra River: Tungabhadra is a famous river of South India. It is also called Dhatri (midwife) of Vedas. While the Vedas were written in 'saptsindhav' or the country of seven rivers, their *bhashya* was written on the banks of Tungabhadra.

There is a legend about the origin of Tungabhadra River in Skand Purana. It is mentioned that once Lord Barah (an incarnation of Lord Vishnu who brought the Earth out of the pool of water after the great deluge) was resting at Barah Parvat. Suddenly, from his two front teeth, water started dropping. This water was converted into two streams. The stream originated from the left tooth was called as Tunga and the stream originated from the right tooth was called as Bhadra. The Tunga and Bhadra rivers are known as the sisters and they join together at Kundly and flow as the Tungabhadra River. Tungabhadra finally joins Krishna River, which is called as mother of these rivers. Its water is considered to be the best drinkable water in the world.

Yamuna River: The Yamuna River is regarded as sister of *Yama*, the God of death. *Bana* in his epic *Kadambari* calls the Yamuna River as *Kalindi*, because its water appears to be dark. People believe that those who take bath in the waters of this river become free from the fear of death.

7.2.1. The Mythological River – Saraswati

The ancient Indian holy scriptures of Vedas have frequent mention of a river named Saraswati. In the Sanskrit language, 'saras' means a lake or water body, and 'wati', means a female associated with it.

According to a belief, Saraswati River used to drain the north and northwest region of India. Saraswati is believed to have originated from the Har-ki-Dun glacier in west Garhwal in Uttaranchal. It flowed parallel to the river Yamuna for some distance and later joined it, proceeding south as the Vedic Saraswati. The seasonal rivers and rivulets, including Ghaggar, joined Saraswati as it followed the course of the present river through Punjab and Haryana. Sutlej River, the Vedic Shatadru, joined the river Saraswati as a tributary at Shatrana, approximately 25 km south of Patiala. Saraswati then followed the course of Ghaggar through Rajasthan and Hakra in Bhawalpur before emptying into the Rann of Kutch via Nara in Sindh province, running parallel to the Indus River.

In the verses of Rigveda, the most ancient of the four Vedas, Saraswati is described as a mighty river with many individually identified tributaries. Undoubtedly, the Saraswati River, carrying the flow of three perennial and numerous seasonal rivers, was a mighty river in the Vedic times. Saraswati was the seventh river of the Sindhu-Saraswati river system, hence the name 'Saptsindhu' for the region bounded by Saraswati in the east and Sindhu (Indus) in the west. Rigveda describes Saraswati as Ambitam, the best of the mothers; Naditam, the best of the rivers; and Devitam, the best of the goddesses. Recent findings do suggest that the Ghaggar-Hakra River once flowed with great fortitude, and was of major importance to the Indus Valley civilization. But this river subsequently dried up, probably due to diversion of its tributaries, perhaps before or around 1900 BC but may be much earlier.

In the enumeration of the rivers in Rigveda (one of the four vedas) 10.75, the order of rivers is Ganga, Yamuna, Saraswati, Shutudri. From the verses of vedas, it is quite clear that the river bearing the name 'Saraswati' flowed through Haryana and Rajasthan states. However, it is not settled whether this is the primal 'Saraswati' that the vedas frequently refer to.

The Saraswati River has been identified with various present-day or historical rivers, particularly the Ghaggar-Hakra River in India and Pakistan. There is a Saraswati (or Sarasvati), a small river in Haryana that flows into the Ghaggar River. There is also a Saraswati River in Gujarat. It flows from the south end of the Aravalli Hills southwestwards into the eastern end of the Rann of Kutch. Alternative suggestions include the Helmand River in Afghanistan, which historically bore the name 'Saraswati'. Further, there is also a river in Iran which has been given this name. Sometimes Saraswati also means the heavenly 'river' or the milky way and the river is personified as a goddess. Note that Saraswati is also the Goddess of learning in Hindu mythology. The goddess Saraswati developed independently from the river itself.

An organization known as the Saraswati Nadi Shodh Prakalp (SNSP) has started work to revitalize the Saraswati River. SNSP feels that the Saraswati River, adored in the Rigveda as a sacred river, is not a myth but truth. This has been established by researches undertaken by scientists, archaeologists and scholars over the last 15 years. SNSP has stated that over 80% of the 2,600 archaeological settlements of the Saraswati Sindhu Civilization dated to circa 3500 BC have been discovered on the banks of the Saraswati including some major sites.

The scientific investigations had established the causes and dates of the disappearance of this great river. Owing to tectonic disturbances around 2500 BC and 1900 BC, there was tilt in topography of Northwest India, resulting in migrations of rivers. The Sutlej migrated westwards to join the Indus and the Yamuna migrated eastwards to join the Ganga. Thus the major sources of glacier water were lost for the Saraswati, which was left dependent only on monsoon water from the Siwalik ranges.

As described elsewhere, Ganga and Yamuna rivers meet at a place known as 'sangam' in Allahabad also known as *Prayaga*. It is also believed that besides these two rivers, the third river joining at sangam is the Saraswati which flows underground. In this interpretation, Saraswati might have been an aquifer. An aerial view of Sangam at Allahabad is shown in Figure 5.

The identification of the Saraswati River of Vedas or the 'original' Saraswati has become embroiled in debates about the age of the Vedas and of the relation between Aryan culture and the Indus Valley civilization. Many researchers have attempted to chart the course of Saraswati River, of late using latest tools such as satellite data. However, despite large number of theories and hypotheses, there is no definite conclusion yet. Numerous web-sites, such as *http://www.haryana-online.com/saraswati.htm*, contain detailed description and folk lores associated with this enigmatic river.

7.3. RIVER BASINS OF INDIA

A river basin is the natural context in which water occurs and is perhaps the most appropriate unit for planning, development, and management of water resources. The drainage area of a system of rivers normally flowing into a common terminus

Figure 5. Sangam at Allahabad where Ganga and Yamuna meet. Photo courtesy http://en.wikipedia.org/wiki/Ganga

constitute a drainage basin. It is therefore convenient to assess water resources basin wise.

India is blessed with many rivers. On the basis of size, the river basins of India could be divided into three groups as mentioned in Table 3.

Besides these, some desert rivers flow for short distances and are lost in the desert.

According to the above classification, the number of major and medium river basins are 12 and 46, respectively, and these contribute nearly 92% of the total runoff in the country. Minor rivers account for about 8% of the total runoff. Of the major rivers, the Ganga – Brahmaputra Meghana system is the biggest with a catchment area of about 1.10 million km^2, which is more than 43% of the catchment area of all the major rivers in the country. The other major rivers with a catchment area more than 0.10 million km^2 are Indus, Godavari, Krishna, and Mahanadi. The catchment area of medium rivers is about 0.25 million km^2 and Subernarekha with 19,300 km^2 catchment area is the largest river among the medium rivers in the country.

Table 3. Classification of Indian river basins based on size

1	Major river basins	Basins whose catchment area is more than 20, 000 km^2
2	Medium river basins	Basins with a catchment area between 20,000 and 2, 000 km^2
3	Minor river basins	Basins with a catchment area below 2, 000 km^2

Table 4. Major river basins of the country

S. No.	Name of the River	Origin	Length (km)	Catchment Area (sq. km)	States falling in the catchment
1.	Indus	Mansarovar (Tibet)	1,114 +	321,289 +	J&K, Punjab, Himachal Pradesh, Rajasthan and Chandigarh UT
2.	a) Ganga	Gangotri (Uttar Kashi)	2,525 +	861,452 +	Uttaranchal, Uttar Pradesh, Himachal Pradesh, Haryana, Rajasthan, Madhya Pradesh, Bihar, West Bengal and Delhi UT.
	b) Brahmaputra	Kailash Range (Tibet)	916 +	194,413 +	Arunachal Pradesh, Assam, Meghalaya, Nagaland, Sikkim, West Bengal, Mizoram and Tripura
	c) Barak & other rivers flowing into Meghna, like Gomti, Muhari, Fenny etc.			41,723 +	Assam, Meghalaya, Nagaland, Manipur, Mizoram and Tripura
3.	Sabarmati	Aravalli Hills (Rajasthan)	371	21,674	Rajasthan and Gujarat
4.	Mahi	Dhar (Madhya Pradesh)	583	34,842	Rajasthan, MP and Gujarat
5.	Narmada	Amarkantak (Madhya Pradesh)	1,312	98,796	Madhya Pradesh, Maharashtra and Gujarat
6.	Tapi	Betul (Madhya Pradesh)	724	65,145	Madhya Pradesh, Maharashtra and Gujarat
7.	Brahmani	Ranchi (Bihar)	799	39,033	Madhya Pradesh, Bihar and Orissa
8.	Mahanadi	Nazri Town (Madhya Pradesh)	851	141,589	Madhya Pradesh, Maharashtra, Bihar, Chattisgarh and Orissa
9.	Godavari	Nasik (Maharashtra)	1,465	312,812	Maharashtra, Andhra Pradesh, Madhya Pradesh, Orissa and Pondicherry
10.	Krishna	Mahabaleshwar (Maharashtra)	1,401	258,948	Maharashtra, Andhra Pradesh and Karnataka
11.	Pennar	Kolar (Karnataka)	597	55,213	Andhra Pradesh and Karnataka
12.	Cauvery	Coorg (Karnataka)	800	81,155	Tamil Nadu, Karnataka, Kerala and Pondicherry
Total				2,528,084	

Table 5. Medium river basins of India

S. No.	Name of the River	Village/Distt. (Origin)	State	Length (km)	Catchment Area (sq. km)
West Flowing Rivers					
1.	Ozat	Kathiawar	Gujarat	128	3, 189
2.	Shetrunji	Dalkania	Gujarat	182	5, 514
3.	Bhadar	Rajkot	Gujarat	198	7, 094
4.	Aji	Rajkot	Gujarat	106	2, 139
5.	Dhadhar	Panchmahal	Gujarat	135	2, 770
6.	Purna	Dhosa	Maharashtra	142	2, 431
7.	Ambika	Dangs	Maharashtra	142	2, 715
8.	Vaitarna	Nasik	Maharashtra	171	3, 637
9.	Dammanganga	Nasik	Maharashtra	143	2, 357
10.	Ulhas	Raigarh	Maharashtra	145	3, 864
11.	Savitri	Pune	Maharashtra	99	2, 899
12.	Sastri	Ratnagiri	Maharashtra	64	2, 174
13.	Washishthi	Ratnagiri	Maharashtra	48	2, 239
14.	Mandvi	Belgaum	Karnataka	87	2, 032
15.	Kalinadi	Belgaum	Karnataka	153	5, 179
16.	Gangavati or Bedti (in upper reaches)	Dharwar	Karnataka	152	3, 902
17.	Sharavati	Shimoga	Karnataka	122	2, 209
18.	Netravati	Dakshina Kannada	Karnataka	103	3, 657
19.	Chaliar or Baypore	Elamtalvi Hills	Kerala	169	2, 788
20.	Bharathapuzha (known as Ponnani)	Annamalai Hills	Tamil Nadu	209	6, 186
21.	Periyar	Sivajini Hills	Kerala	244	5, 398
22.	Pamba	Devarmalai	Kerala	176	2, 235
East Flowing Rivers					
23.	Burhabalang	Mayurbahanj	Orissa	164	4, 837
24.	Baitarni	Keonjhar	Orissa	365	12, 789
25.	Rushikulya	Phulbani	Orissa	146	7, 753
26.	Bahuda	Ramgirivillage	Orissa	73	1, 248
27.	Vamsadhara	Kalahandi	Orissa	221	10, 830
28.	Nagavali	Kalahandi	Orissa	217	9, 410
29.	Sarda	Vishakhapatnam	Andhra Pradesh	104	2, 725
30.	Eleru	Vishakhapatnam	Andhra Pradesh	125	3, 809
31.	Vogarivagu	Guntur	Andhra Pradesh	102	1, 348
32.	Gundlakamma	Kurnool	Andhra Pradesh	220	8, 494
33.	Musi	Nellore	Andhra Pradesh	112	2, 219
34.	Paleru	Nellore	Andhra Pradesh	104	2, 483
35.	Muneru	Nellore	Andhra Pradesh	122	3, 734
36.	Swarnamukhi	Koraput	Orissa	130	3, 225
37.	Kandleru	Vinukonda	Andhra Pradesh	73	3, 534

(Continued)

Table 5. (Continued)

S. No.	Name of the River	Village/Distt. (Origin)	State	Length (km)	Catchment Area (sq. km)
38.	Kortalaiyar	Chinglepet	Tamil Nadu	131	3, 521
39.	Palar (including tributary Cheyyar)	Kolar	Karnataka	348	17, 871
40.	Varahandi	North Arcot	Tamil Nadu	94	3, 044
41.	Ponnaiyar	Kolar	Karnataka	396	14, 130
42.	Vellar	Chithri Hills	Tamil Nadu	193	8, 558
43.	Vaigai	Madurai	Tamil Nadu	258	7, 031
44.	Pambar	Madurai	Tamil Nadu	125	3, 104
45.	Gundar	Madurai	Tamil Nadu	146	5, 647
46.	Vaippar	Tirunolvolli	Tamil Nadu	130	5, 288
47.	Tambraparni	Tirunolvolli	Tamil Nadu	130	5, 969
48.	Subarnarekha	Nagri/Ranchi	Bihar	395	19, 296
				Total	248, 505

Source: CWC (1990) and MoWR (2004)

Lengthwise, Ganga is the longest river in India, followed by Godavari, and then Krishna. When the rivers are ranked by the catchment area, Ganga is the first, followed by Indus and Godavari. In terms of water yield, Brahmaputra is the first, Ganga is the second and Godavari is the third. The catchment areas of major and medium river basins, lengths of rivers and the states of origin are given in Tables 4 and 5 respectively. The basinwise information about the availability of surface water and ground water data can also be obtained from the website www.india-water.com. This site also contains some data that are available in public domain.

7.4. FRESHWATER RESOURCES

Although about three-fourths of earth is water, the estimated volume of fresh water in our rivers, groundwater, snow and ice, is about 2.5% only, the rest being sea/salt water. Most of the fresh waters are either in the form of ice and permanent snow covers in Antarctic/Arctic region (about 69%) or is stored underground in the form of deep underground basins/aquifers, soil moistures, etc. (30%).

The total usable fresh water supply to ecosystems and humans from river systems, lakes, wetlands, soil moisture and shallow groundwater is less than 1% of all fresh water (MoWR, 2003) and only 0.1% of all the water on earth. As per the World Health Organisation estimates, only 0.007% (this number is sometimes known as the James Bond number) of all water on the earth is readily available for human consumption globally.

7.4.1. Surface Water Resources of River Basins of India

The water resources of a basin are assessed based on stream flows measured at a terminal site (usually close to sea) on the river. Clearly, the accuracy of assessment depends upon the accuracy of discharge observations and the length of the data series. In many instances in India, this length is not adequate and the accuracy of observations is not known; the periods for which data are available are widely different in different basins. In India, rainfall data of longer length are usually available and are frequently employed to extend the runoff series. The average annual flow at the terminal point of a river is normally denoted as the water resources of the basin. Note that this refers to the availability of water with a probability of 50%. For the planning of water resources projects, the dependability at other levels, such as 75% and 90%, is needed. Such dependability is determined by using the time series of observations.

Most basins of the world are no longer virgin; the flow at the terminal site is significantly influenced by upstream utilization. The natural runoff of a basin could be computed by adding to the surface flow measured at the terminal site, the net export of surface water out of the basin, the net increase of the surface water storage, additional evapotranspiration caused by the use or storage of surface water, direct ground water flow from the river basin below or along the terminal site, the net export of ground water out of the basin, the net increase in ground water storage and soil moisture storage, and the additional evapotranspiration caused by use or storage of ground water. This is the general water balance approach, applicable to any basin for any period. However, if averages over a long time period are taken, the storage change would be zero or negligible. Also, assuming a case of no export or import, and neglecting the ground water flow below or along the terminal site, a simplification is possible. With this simplification, the average annual natural flow can be computed by adding to the average annual surface flow measured at the terminal site, the average annual extra evaporation/evapotranspiration due to use or storage of surface water and the average annual extra evaporation/evapotranspiration due to storage or use of ground water.

Most surface water flow of the Indian rivers occurs during the monsoon season of 4 to 5 months. Rao (1973) reported the country's annual surface runoff as 1,645 km^3. According to a recent estimate of the Central Water Commission, the surface water resources of the country are 1,869 km^3. These are statistical estimates based on available river flows. Recently, NCIWRD (1999) has also estimated water resources of different basins. According to them, the total surface water resources of India are 1,953 km^3. The average annual flow in Indian rivers of different basins is summarized in Table 6.

It is observed from Table 6 that some of the southern peninsular rivers like Cauvery, Pennar etc., are able to utilize almost all of their potential while some rivers in the northern region viz., Ganga and Brahmaputra etc., have significant amount of water which is not utilizable. Further, in Pennar and East flowing rivers between Pennar and Kanyakumari, the utilizable flow is shown to be more than the potential.

Table 6. Surface water resources potential of river basins (km^3) of India

S. No.	Name of the River Basin	Average annual potential in the river	As per NCIWRD (1999)	Estimated utilizable flow excluding ground water	Cultivable area, (thousands ha)
1.	Indus (Area in Indian Territory)	73.31	73.31	46.00	9, 638
2.	a) Ganga	525.02	525.02	250.00	60, 161
	b) Brahmaputra, Barak, and others	585.60	677.41	24.00	6, 145
3.	Godavari	110.54	110.54	76.30	18, 931
4.	Krishna	78.12	69.81	58.00	20, 299
5.	Cauvery	21.36	21.36	19.00	5, 523
6.	Pennar	6.32	6.32	6.86	3, 539
7.	East flowing and rivers from Mahanadi to Godavari and Krishna to Pennar	22.52	22.52	13.11	
8.	East flowing rivers between Pennar and Kanyakumari	16.46	16.6	16.73	
9.	Mahanadi	66.88	66.88	49.99	7, 994
10.	Brahmani & Baitarani	28.48	28.48	18.30	2, 360
11.	Subarnarekha	12.37	12.37	6.81	1, 194
12.	Sabarmati	3.81	3.81	1.93	1, 548
13.	Mahi	11.02	11.02	3.10	2, 210
14.	West flowing rivers of Kutch & Saurashtra including Luni	15.10	15.10	14.98	
15.	Narmada	45.64	45.64	34.50	5, 901
16.	Tapi	14.88	14.88	14.50	4, 292
17.	West flowing rivers from Tapi to Tadri	87.41	87.41	11.94	
18.	West flowing rivers from Tadri to Kanyakumari	113.53	113.53	24.27	
19.	Area of inland drainage in Rajasthan desert	Negligible	Negligible		
20.	Minor rivers draining to Myanmar (Burma) & Bangladesh	31.00	31.00		
Total		1, 869.00	1, 952.8	690.00	

Table 6 also reveals that the Ganga basin has the maximum utilizable surface water, i.e. approximately 50% out of the average annual runoff of 525 km^3 whereas the Brahmaputra with the largest average annual runoff of 585 km^3 contributes only 4% of utilizable flow of surface water. Also it shows that the river Sabarmati with average annual runoff of 3,812 km^3 has nearly half of its runoff as utilizable flow. Almost similar is the case with Subernarekha.

7.4.2. Utilizable Surface Water Resources

In majority of river basins, present utilisation is significantly high and is in the range of 50% to 95% of utilizable surface resources. But in the rivers such as Narmada and Mahanadi percentage utilisation is quite low; it is of the order of 23% and 34% respectively.

The distribution of water resources potential in the country shows that as against the national per capita annual availability of water as 2,208 m^3 the average availability in Brahmaputra and Barak is as high as 16,589 m^3 while it is as low as 360 m^3 in Sabarmati basin. Brahmaputra and Barak basin, with 7.3% of geographical area and 4.2% of population of the country, has 31% of the annual water resources. Per capita annual availability for rest of the country excluding Brahmaputra and Barak basin works out to about 1,583 m^3. A situation of availability of less than 1,000 m^3 per capita is generally considered as scarcity conditions. Cauvery, Pennar, Sabarmati, East flowing rivers and West flowing rivers are some of the basins which fall in this category.

There are two different ways in which utilization of water resources can be considered. It can be computed as the quantum of withdrawal of water from its place of natural occurrence such as river or ground water. Alternatively, it can be considered as the additional evaporation/evapotranspiration of the natural water that is caused by human activities such as diversion to command areas. The second approach is more appropriate for a scientific study of water balance of hydrological cycle. The former approach has been used more commonly in field studies in India. It is important to note at this stage that the estimates of water availability, potential, and utilization etc. should be treated as approximations. These numbers are likely to change as more data becomes available and new technologies are developed. This calls for re-review of the estimates at regular intervals, say 5–10 years.

Although the estimated surface water availability is of the order of 1,869 km^3, the amount of water that can be put to beneficial uses is much less. This is because India experiences monsoon climate implying that most (80–90%) of the annual runoff occurs during four monsoon months. As India has not been able to create very large storage capacity, most part of this flow goes as a waste, and in fact, some of it also causes flood damage. In view of immense variability of streamflows (see Table 7 which gives the monthly discharge of selected rivers as a percentage of their annual discharge), if more storages could be developed, it would be possible to considerably reduce the flood damage. Further these storages will permit carryover of water from wet years to dry years, providing additional benefits.

Within the premise of physiographic conditions, socio-political environment, legal and constitutional constraints, and technology development available at present, different authorities have estimated utilizable quantities of water from the surface flow differently. As per the recent estimate by the Central Water Commission, the utilizable annual surface runoff is about 690 cubic kilometers (= 36% of the total).

The storage space made available from the projects completed by 1995 is about 174 km^3 and the projects under construction would provide space to hold another

Table 7. Monthly discharge of selected rivers as percentage of annual discharge

Month	River system				
	Godavari	Krishna	Mahanadi	Narmada	Pennar
January	0.64	0.52	0.69	1.03	2.20
February	0.51	0.31	0.51	0.78	1.15
March	0.37	0.21	0.48	0.53	1.32
April	0.30	0.18	0.50	0.39	0.79
May	0.19	0.91	0.37	0.24	3.88
June	2.41	2.79	1.08	1.28	1.32
July	20.99	25.60	20.03	14.18	4.14
August	31.39	30.56	37.28	30.92	8.20
September	27.96	20.25	27.56	36.53	15.09
October	11.24	12.78	7.74	9.89	20.72
November	2.93	4.71	2.86	2.87	24.44
December	1.07	1.18	0.90	1.36	16.75

76 km^3 of water. This makes a total of about 250 km^3 of storage space, ignoring small structures. This is about 36% of possible storage space of 690 km^3. A part of the created storage (approximately 1 km^3 per year) is lost every year due to sedimentation. Clearly, the nation has a long way to go before the adequate capacity to regulate streamflows is built. To appreciate the growth in storage space, it is to be noted that this was 15.6 km^3 at the time of independence.

To harness the utilizable surface water, about 400 km^3 of live storage needs to be created. Thus, the development of surface water sector has another half way to go to achieve its full potential. As per the projections, 80% of the surface water and 85% of the ground water will ultimately be used for irrigation. The remaining 20% of surface and 15% of ground water for other uses will have to be appropriately allocated as per the demands for other sectors. An integrated planning for future demand and supply management needs to be taken up at national level.

7.4.3. Ground Water Resources of River Basins of India

The total annual ground water potential for a unit refers to the available annual recharge after allowing for natural discharge in the monsoon season through base flow and subsurface inflow/outflow. While deciding on the ground water available for future development, some provision needs to be kept for natural discharge in the non-monsoon season.

According to estimates, the basin wise total rechargeable ground water resources in India stands at 431.43 km^3/year. According to another recent state wise information, the ground water resource potential of the country is 433.882 km^3/year. The basin wise and state-wise ground water resource potential of the country is given in Table 8 and 9 respectively. The present level of development is about 42%.

Table 8. Ground Water Potential in River Basins of India (Pro Rata Basis) (Unit: km^3/year)

S. No.	Name of the Basin	Total replen-ishable ground water resources	Provision for domestic, industrial and other uses	Available ground water resources for irrigation	Net draft	Balance ground water potential available for use	Level of ground water devel-opment (%)
1.	Brahmani with Baitarni	4.05	0.61	3.44	0.29	3.16	8.45
2.	Brahmaputra	26.55	3.98	22.56	0.76	21.80	3.37
3.	Chambal Composite	7.19	1.08	6.11	2.45	3.66	40.09
4.	Cauvery	12.30	1.84	10.45	5.78	4.67	55.33
5.	Ganga	170.99	26.03	144.96	48.59	96.37	33.52
6.	Godavari	40.65	9.66	30.99	6.05	24.94	19.53
7.	Indus	26.49	3.05	23.43	18.21	5.22	77.71
8.	Krishna	26.41	5.58	20.83	6.33	14.50	30.39
9.	Kutch & Saurashtra Composite	11.23	1.74	9.49	4.85	4.64	51.14
10.	Madras and South Tamil Nadu	18.22	2.73	15.48	8.93	6.55	57.68
11.	Mahanadi	16.46	2.47	13.99	0.97	13.02	6.95
12.	Meghna	8.52	1.28	7.24	0.29	6.95	3.94
13.	Narmada	10.83	1.65	9.17	1.99	7.18	21.74
14.	Northeast Composite	18.84	2.83	16.02	2.76	13.26	17.20
15.	Pennar	4.93	0.74	4.19	1.53	2.66	36.60
16.	Subarnarekha	1.82	0.27	1.55	0.15	1.40	9.57
17.	Tapi	8.27	2.34	5.93	1.96	3.97	33.05
18.	Western Ghat	17.69	3.19	14.50	3.32	11.18	22.88
	Total	431.43	71.08	360.35	115.21	245.13	31.97

Source: IWRS (1998).

7.4.4. Allocation of Ground Water Resource for Utilization

The net annual ground water availability is to be apportioned between domestic, industrial and irrigation uses. As per the National Water Policy, requirement for domestic water supply is to be accorded the highest priority. This requirement can be computed based on population as projected to the year 2025, per capita requirement of water for domestic use, and relative load on ground water for urban and rural water supply.

The water available for irrigation use is obtained by deducting the allocation for domestic and industrial use from the net annual ground water availability.

Table 9. Ground Water Resource of India as on March 2003 (unit km^3/year)

S.No.	States	Total Replen-ishable Ground water Resource	Provision for domestic and industrial and other uses	Available ground water resources for irrigation	Net draft	Balance ground water Resources for future use	Level of ground water devel-opment (%)
States							
1	Andhra Pradesh	35.29	5.29	30.00	8.57	21.43	28.56
2	Arunachal Pradesh	1.44	0.22	1.22	–	1.22	Negligible
3	Assam	24.72	3.71	21.01	1.84	19.17	8.75
4	Bihar	26.99	4.05	22.94	10.63	12.31	46.33
5	Chhatisgarh	16.07	2.41	13.66	0.81	12.85	5.93
6	Delhi	0.29	0.18		0.12		
7	Goa	0.22	0.03	0.19	0.02	0.17	8.30
8	Gujarat	20.38	3.06	17.32	9.55	7.77	55.16
9	Haryana	8.53	1.28	7.25	8.13	0.00	112.18
10	Himachal Pradesh	0.37	0.07	0.29	0.03	0.26	10.72
11	Jammu & Kashmir	4.43	0.66	3.76	0.03	3.73	0.81
12	Jharkhand	6.53	0.98	5.55	1.84	3.71	33.13
13	Karnataka	16.19	2.43	13.76	4.76	9.00	34.60
14	Kerala	7.90	1.31	6.59	1.46	5.13	22.17
15	Madhya Pradesh	34.82	5.22	29.60	8.02	21.58	27.09
16	Maharashtra	37.87	12.40	25.47	9.44	16.04	37.04
17	Manipur	3.15	0.47	2.68	Negligible	2.68	Negligible
18	Meghalaya	0.54	0.08	0.46	0.02	0.44	3.97
19	Mizoram	1.40	0.21	1.19	Negligible	1.19	Negligible
20	Nagaland	0.72	0.11	0.62	Negligible	0.62	Negligible
21	Orissa	20.00	3.0	17.00	3.61	13.39	21.23
22	Punjab	18.66	0.87	16.79	16.40	0.00	97.66
23	Rajasthan	12.71	1.99	10.71	9.26	1.45	86.42
24	Sikkim	0.07	0.01	0.06	Negligible	0.06	Negligible
25	Tamil Nadu	26.39	3.96	22.43	14.45	7.98	64.43
26	Tripura	0.66	0.10	0.56	0.19	0.38	33.43
27	Uttar Pradesh	81.12	12.17	68.95	32.33	36.62	46.89
28	Uttaranchal	2.70	0.41	2.29	0.82	1.47	35.78
29	West Bengal	23.09	3.46	19.63	7.50	12.13	38.19
	Total States	433.24	71.14	361.98	149.82	212.78	41.53
UT's							
1	Andaman & Nicobar Islands	0.326	0.013	0.313	Negligible	0.313	Negligible
2	Chandigarh	0.030			0.025		

3 Dadra & Nagar Haveli	0.042	0.006	0.04	0.005	0.031	12.81
4 Daman & Diu	0.013	0.002	0.01	0.008	0.003	70.00
5 Lakshadweep	0.002			0.007		
6 Pondicherry	0.029	0.004	0.02	0.116	0.000	
Total UT's	0.442	0.025	0.384	0.160	0.348	
Grand Total	433.882	71.165	362.364	149.97	213.128	41.57

Source: CGWB (2005).

Potential for future irrigation development from ground water is determined by deducting the existing ground water draft for irrigation from the net annual ground water availability for irrigation. The resulting quantity is the net annual ground water availability for future irrigation development and this should be calculated separately for non-command areas and command areas. GEC (1997) recommended the following empirical relation if adequate data are not available:

$$\text{Allocation for domestic and industrial water requirement (mm/year)} = 22 * N * L_g \quad (1)$$

where N = population density in the unit in thousands per sq. km, L_g = fractional load on ground water for domestic and industrial water supply (≤ 1.0). In deriving eq. (1), it is assumed that water requirement for domestic and industrial use is 60 lpd per head.

Recently, CGWB has re-assessed the annual replenishable ground water resources to be about 434 km^3. The ground water available for irrigation is about 363 km^3; out of this the utilizable quantity (90% of available) is about 327 km^3. Thus the total utilizable ground water resource is 327 + 71 (domestic and other uses) = 398 km^3, as shown in Table 10.

The utilizable irrigation potential has been estimated as 64 million ha (M-ha) based on crop water requirement and availability of cultivable land. Out of this, the potential from natural rainfall recharge is 50.8 M-ha and augmentation from irrigation canal systems is 13.2 M-ha. The irrigation potential created from ground water in the country till 1993 was 35.4 M-ha.

Although the national scenario of ground water availability is favourable, there are certain pockets that face scarcity of water. This is because the ground water

Table 10. Total utilizable ground water resource of India

1.	Total Replenishable Ground Water Resource	434 km^3
2.	Provision for Domestic, Industrial & Other Uses	71 km^3
3.	Available Ground Water Resource for Irrigation	363 km^3
4.	Utilisable Ground Water Resource for Irrigation (90 % of S. N. 3)	327 km^3
5.	Total Utilisable Ground Water Resource (Total of S. N. 2 & 4)	398 km^3

Source: CGWB (1995).

development over different parts of the country is not uniform. It is quite intensive in some areas resulting in over-exploitation and leading to fall in water levels and salinity ingress in coastal areas. The declining water levels have resulted in failure of wells which require deepening of extraction structures.

7.4.5. Ground water development scenario

Ground Water Estimation Committee (GEC 1997) defined stage of ground water development (SGWD) as the ratio of net annual draft to utilizable resource:

$$\text{SGWD (\%)} = \frac{\text{Existing gross GW draft for all uses}}{\text{Net annual GW availability}} * 100 \tag{2}$$

The level of ground water development in an area is the ratio of net yearly draft to total utilizable ground water resources for irrigation:

$$\text{Level of ground water development} = \frac{\text{Net yearly draft}^{*}\ 100}{\text{Utilizable resource for irrigation}} \tag{3}$$

The SGWD is an index of the balance between ground water available and utilization. As SGWD approaches 100%, the potential for future development becomes meager. Over the country, SGWD varies from 1.33% in Himachal Pradesh to 93.85% in Punjab; it is 83.85% in Haryana. Ground water draft is very low in Himachal Pradesh since the land for agricultural development is limited. On the other hand, GW is under serious pressure in Punjab and Haryana where GW withdrawal exceeds annual recharge in most of the blocks. The large rate of withdrawal is largely attributed to the advent of paddy cultivation in these states. Water requirements for paddy are quite large and the expansion in area under paddy has led to tremendous growth in number of tubewells for irrigation. Likewise, the share of GW irrigation in the total irrigation in Tamil Nadu has risen steadily over time. During 1995, nearly 60 percent of utilizable ground water resources amounting to 2.2 Mham were tapped through minor irrigation structures. In Rajasthan, level of ground water development is nearly 51% and this is predominantly concentrated in the eastern parts of the State.

While interpreting the long-term trend of ground water levels, several points are need may be kept in view. If the pre- and post-monsoon water levels show a fairly stable trend, it does not necessarily mean that there is no scope for further ground water development. Such a trend indicates that there is a balance between recharge, draft and natural discharge in the unit. However, further ground water development may be possible, which may result in a new stable trend at a lower ground water level with associated reduced natural discharge. If the ground water resource assessment and the trend of long-term water levels contradict each other,

Table 11. Categorization of Areas of Ground Water Development

Category of areas	Stage of ground water development (%)
White	<65%
Grey	>65% but <85%
Dark	>85% but <100%

Source: GEC (1997).

this anomalous situation requires a review of the ground water resource computation, as well as the reliability of water level data.

GEC (1997) categorized areas on the basis of the level of ground water development taking into consideration the ratio of net yearly draft to utilizable ground water resources for irrigation. On the basis of this ratio, categorization of area is given in Table 11.

In dark areas, micro-level surveys are required to evaluate the ground water resources more precisely for taking up further ground water development.

Ground water assessment in India is carried out block wise, except Andhra Pradesh, Gujarat and Maharashtra where the assessment is carried out on the basis of mandals, talukas and watersheds. Out of 4,272 blocks in the country, 231 blocks have been categorized as "over-exploited" where the stage of ground water development exceeds the annual replenishable limit and 107 blocks are "dark" where the stage of ground water development is more than 85% (GGWB 2005). Besides, 6 mandals have been categorized as "over-exploited" and 24 as 'dark' out of 1,104 mandals in Andhra Pradesh. Similarly out of 184 talukas in Gujarat, 12 are "over-exploited" and 14 are 'dark' and out of 1,503 watersheds in Maharashtra, 34 are 'dark'. The state wise over-exploited and dark blocks/taluks/watersheds of India are listed in Table 12 below.

In India, the first large-scale ground water development for irrigation was taken up in 1934 with the initiation of a project for construction of about 1,500 public deep tubewells in the Ganga basin. Since the mid-sixties, there has been tremendous increase in exploitation of ground water as a source of irrigation. This increase came with the advent of high yielding varieties of crops. Government of India established the Agricultural Refinance and Development Corporation in 1963 with institutional investment for ground water development as one of the objective. This organization was later renamed as the National Bank for Agriculture and Rural Development (NABARD). Besides NABARD, Rural Electrification Corporation (REC) is also currently providing support for ground water development.

It is estimated that 70–80 percent of the value of irrigated production in India now depends on ground water irrigation. Thus, nearly two-fifths of India's agricultural output comes from the areas that are irrigated with ground water. Since agriculture and allied activities significantly contribute to India's GDP, the contribution of ground water to India's GDP is about substantial. As profits from ground water irrigation helped to spread new technology, some of these profits were invested

Table 12. Categorization of blocks/taluks/watersheds as over exploited and dark on all India basis

S. No.	States	No. of Assessment units (Blocks/ Taluks/ Watersheds)	Assessment units (Blocks/ Taluks/ Watersheds)			
			Over exploited		Dark/Critical	
			No.	%	No.	%
1	Andhra Pradesh	1,157	118	10.20	79	6.83
2	Arunachal Pradesh	59	0	0	0	0
3	Assam	219	0	0	0	0
4	Bihar	394	6	1.52	14	3.55
5	Chhatisgarh	145	0	0	0	0
6	Delhi	6	3	50	1	16.67
7	Goa	12	0	0	0	0
8	Gujarat	180	41	22.78	19	10.56
9	Haryana	111	30	27.03	13	11.71
10	Himachal Pradesh	69	0	0	0	0
11	Jammu & Kashmir	69	0	0	0	0
12	Jharkhand	193	0	0	0	0
13	Karnataka	175	7	4	9	5.14
14	Kerala	151	3	1.99	6	3.97
15	Madhya Pradesh	312	2	0.64	1	0.32
16	Maharashtra	2,316	154	6.65	72	3.11
17	Manipur	29	0	0	0	0
18	Meghalaya	39	0	0	0	0
19	Mizoram	12	0	0	0	0
20	Nagaland	52	0	0	0	0
21	Orissa	314	0	0	0	0
22	Punjab	138	81	58.70	12	8.70
23	Rajasthan	237	86	36.29	80	33.76
24	Sikkim	4	0	0	0	0
25	Tamil Nadu	385	138	35.84	37	9.61
26	Tripura	38	0	0	0	0
27	Uttar Pradesh	819	2	0.24	20	2.44
28	Uttaranchal					
29	West Bengal	275	0	0	61	22.18
	Total States	7,910	671	8.48	424	5.36
UT's						
1	Andaman & Nicobar Islands	1	0	0	0	0
2	Chandigarh	1	0	0	0	0
3	Dadra & Nagar Haveli	1	0	0	0	0
4	Daman & Diu	2	1	50	1	50
5	Lakshadweep	9	0	0	0	0
6	Pondicherry	4	1	25	0	0
	Total UT's	18	2	11.11	1	5.56
	Grand Total	7,928	673	8.49	425	5.36

Source: CGWB (2005).

back into ground water development, leading to "tubewell explosion" in many parts of India. Of course, this was facilitated by government's efforts to promote rural electrification, availability of cheap diesel, and institutional credit support from nationalized banks.

During the past four decades, there has been a phenomenal increase in the growth of ground water abstraction structures due to implementation of technically viable schemes for development of the resource, backed by liberal funding from institutional finance agencies, improvement in availability of electric power and diesel, good quality seeds, fertilizers, government subsidies etc. During the period 1951–92, the number of dugwells increased from 3.86 million to 10.12 million, that of shallow tubewells from 3,000 to 5.38 million and public bore/tubewells from negligible to 68000. The number of electric pumpsets has increased from negligible to 9.34 million and the diesel pump sets from 66,000 to about 4.59 million. There has been a steady increase in the area irrigated from ground water from 6.5 Mha in 1951 to 35.58 Mha in 1993. During VIII plan, it was anticipated that 1.71 million dugwells, 1.67 million shallow tubewells and 11,400 deep tubewells were expected to be added. Similarly number of electric pumpsets and diesel pumpsets is expected to rise by 2.02 million and 0.42 million respectively. Such a magnitude of ground water development requires realistic assessment of ground water resources to avoid any deleterious effects on ground water regime and to provide sustainability to the ground water development process. The state wise ground water structures and ground water draft is given in Table 13.

Table 14 gives the phenomenal growth of ground water extraction structures in India from 1950 onwards. During 1951 to 1997, the number of dug well increased approximately 2.5 times. While there were only 3,000 private tube wells in 1951, this number grew to more than 21 lakh by 1980. Between 1980 and 1997, the number again increased more than three times. Likewise, the number of public tube wells increased from 2,400 in 1951 to 90,000 by 1997. During this period, the cumulative irrigation potential due to ground water jumped from 6.5 M-ha to 46 M-ha and currently this potential exceeds 50 M-ha.

The significance of ground water in the economy emanates from the fact that agricultural yields are generally 30 to 50% higher in the areas irrigated with ground water compared to areas irrigated by water from other sources. This is primarily because a farmer using ground water has greater control over the supply of water than those who depend on other sources of irrigation. On a positive note, ground water irrigation stimulates investments in fertilizers, pesticides, and high-yielding varieties, leading to higher yields.

Development of ground water has led to increased "drought proofing" of India's agricultural. This phenomenon can be seen from the impact assessment of droughts. In the 1960s, ground water was a relatively insignificant source of irrigation, particularly in eastern India. In 1965–66, rainfall (June to September) was 20 percent below normal, leading to drought conditions. Food grain production declined by 19 percent at the national level over the previous year's output. In contrast, in 1987–88, rainfall dropped almost 18 percent below normal, but food grain production declined by only 2 percent over the previous year's level. Although the two droughts are not

Table 13. India: State wise Ground water Structures (in thousands) and Ground Water Draft (ham)

S. No.	State/ Union Territory	Dug wells	Shallow Tubewells	Public Tubewells	Ground water Draft (ham)
1	Andhra Pradesh	107,586.0	97,560.0	7,875.0	1,181,295.0
2	Arunachal Pradesh	–	–	–	–
3	Assam	–	45,140.0	2,007.0	47,147.0
4	Bihar	417,770.0	705,830.0	6,236.0	1,129,836.0
5	Goa	126.0	–	80.0	206.0
6	Gujarat	709,070.0	8,300.0	5,088.0	722,458.0
7	Haryana	42,420.0	434,980.0	1,799.0	479,199.0
8	Himachal Pradesh	3,570.0	374.0	199.0	4,143.0
9	J & K	1,700	2,020.0	172.0	3,892.0
10	Karnataka	512,650.0	35,020.0	–	547,670.0
11	Kerala	119,010.0	2,530.0	–	171,600.0
12	Madhya Pradesh	1,307,070.0	19,220.0	1,420.0	1,327,710.0
13	Maharashtra	1,116,020.0	100.0	–	111,612.0
14	Manipur	–	10.0	5.0	15 : 0
15	Meghalaya	–	790.0	3.0	793.0
16	Mizoram	–	–	–	–
17	Nagaland	450.0	–	4.0	454.0
18	Orissa	543,750.0	14,200.0	5,261.0	562,311.0
19	Punjab	93,470.0	622,600.0	1,861.0	7,117,931.0
20	Rajasthan	835,250.0	18,940.0	75.0	854,265.0
21	Sikkim	–	–	–	–
22	Tamil Nadu	1,446,630.0	134,970.0	–	1,581,600.0
23	Tripura	–	2,432.0	164.0	2,596.0
24	Uttar Pradesh	1,145,670.0	2,343,520.0	26,985.0	3516175.0
25	W. Bengal	49,180.0	262,970.0	3,726.0	315,876.0
26	Union Territories	16,190.0	24,700.0	288.0	41,178.0
27	Grand Total	9,485,856.0	4,776,246.0	63,248.0	14,325,349.0

Source: CGWB (1995).

Table 14. Number of groundwater wells and created irrigation potential from 1950 onwards

Time	Number of wells (in thousands)				Cumulative irrigation potential created by ground water (M-ha)
	Dug wells	Private tube wells	Public tube wells	Total	
March, 1951	3,860	3	2.4	3,865.4	6.50
March, 1980	7,786	2,132	33.3	9,951.3	22.00
March, 1985	8,742	3,359	46.2	12,147.2	27.82
March, 1990	9,407	4,754	63.6	14,224.6	35.62
March, 1992	10,120	5,379	67.6	15,566.6	38.89
March, 1997	10,501	6,743	90.0	17,334.0	45.73
March, 2002	–	–	–	19800 (projected)	50.00

Source: CGWB.

directly comparable, only marginal decline in production in 1987–88 compared to 1965–66 can be attributed to wider spread of irrigation in general and of ground water irrigation in particular.

7.4.6. Categorization of Areas for Ground Water Development

GEC (1997) suggested that the unit of ground water assessment can be categorized for development based on: (i) stage of ground water development, and (ii) long-term trend of pre-and post-monsoon ground water level. The following categorization was proposed based on these two factors.

Safe areas with potential for development: (a) areas where ground water resource assessment shows SGWD $\leq$ 70 percent and there is no significant long term decline of pre- or post-monsoon ground water levels; (b) areas where ground water resource assessment shows 70% $<$ SGWD $<$ 90% and both pre- and post-monsoon ground water levels do not show a significant long term decline. However, in these areas, caution may be exercised in planning future development with regard to quantum of additional ground water withdrawal.

Semi-critical areas for cautious ground water development: Areas where ground water resource assessment shows 70% $<$ SGWD $<$ 90%, and either pre- or post-monsoon ground water level shows significant long-term declines.

Critical areas: (a) areas where ground water resource assessment shows 70% $<$ SGWD $<$ 100%, and either pre-monsoon or post-monsoon ground water level shows a significant long-term decline; (b) areas where ground water resource assessment shows stage of ground water development resource assessment shows SGWD $<$ 100%; but both pre-monsoon and post-monsoon ground water levels shows a significant long-term decline; (c) areas where ground water resource assessment shows SGWD $>$ 100 per cent, but either pre-monsoon or post-monsoon ground water level does not show a significant long term decline.

Over-exploited areas: Where ground water resource assessment shows SGWD $>$ 100% and both pre-monsoon and post-monsoon ground water levels show a significant long-term decline.

GEC (1997) also recommended that future ground water development must be linked with water conservation measures especially in over-exploited and critical and semi-critical areas. In these areas, there must be some institutional, technological and financial restriction in the ground water extraction practices.

7.4.7. Available Ground Water Potential

Ground water being a dynamic and replenishable resource has to be estimated primarily based on the components of annual recharge which largely depends on hydrogeological and climatic conditions.

When we look at the availability at state level, the picture is satisfactory in terms of spatial distribution. Barring few pockets of northwest, northeast and Kerala in south, the entire country has more than 8,000.00 MCM/year (0.8 M ham) of ground water potentials for future use. But at the district level, the distribution of ground water balance narrates another story. In a large number of districts in different states especially, in Punjab, Haryana, Gujarat, Western Uttar Pradesh and Tamil Nadu, ground water is over-exploited.

Estimation of yield of aquifer requires subsurface investigations. These investigations are to be followed by detailed resource evaluation studies to determine the quantity of ground water and its quality. CGWB has conducted water balance studies in different parts of the country to quantitative determine the ground water potential.

7.4.8. Constraints in Future Ground Water Resources Development

Due to the precarious situation of ground water in India, the focus now needs to be on holistic management. There are numerous constraints in ground water resource development. In the following, the major constraints are enumerated.

1. By 2025, majority of the India's population will live in urban and peri-urban areas. With changes in life style, people and industries will require larger share of water. This implies that the share of agriculture in total water use has to be reduced through improved water management.
2. Urbanization is also creating an enormous pollution load on fresh water supplies and estuaries. The amount of pollutants thrown into the waterways is rapidly increasing and, at the same time, the flows of fresh water are decreasing as more water is diverted or evaporated through intensive use. Thus the concentration of pollutants is increasing.
3. Inefficient ground water management practices, coupled with unplanned and unregulated use is hampering sustainable ground water development.

Figure 6 represents comparison between total available replenishable ground water and ground water balance left in the basins for exploitation after being used for irrigation, industrial, domestic & other uses. Table 6 shows that Ganga basin possesses maximum amount of 170.99 cubic km/year of total available replenishable ground water & 96.37 cubic km/year is left for exploitation. Further, Subarnarekha basin possess minimum amount of 1.82 cubic km/year of total available replenishable ground water among all basins & only 1.4 cubic km/year is the balance of ground water available for exploitation/future use.

Detailed data about ground water resources of India has been given in Chapter 6.

7.5. DATA REQUIREMENT FOR WATER RESOURCES DEVELOPMENT (WRD) PROJECTS

One of the most important aspects in planning and operation of a water resources development project is to assess the availability of water and its time distribution, on long term as well as short-term basis. To ensure the success of a project, it is necessary to plan it such that desired quantity of water is available most of the time.

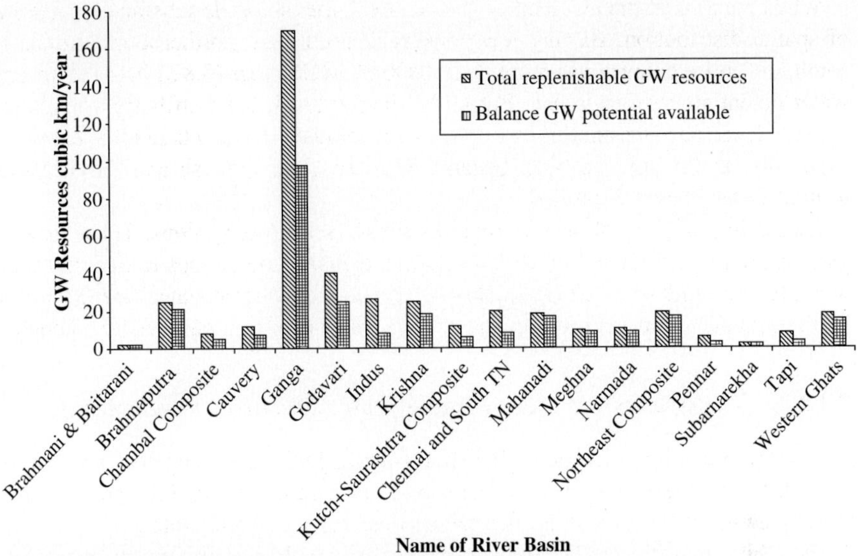

Figure 6. Total available replenishable ground water and ground water balance left in the basins for Exploitation. Source: Based on data from Ministry of Water Resources

Some shortage may be permitted to make the project cost effective and to have optimum utilization of the scarce water resources. In India, the normal practice is to plan an irrigation project for 75% dependable flows. The hydropower and drinking water supply schemes are planned for 90% and 100% dependable flows, respectively (Koche and Chawla, 2000).

The data requirements for typical water availability studies are summarized below.

(a) Runoff data of the desired specific duration (daily, 10-daily, or monthly, annual etc.) at the proposed site for at least 30 years; or
(b) Rainfall data of specific duration for at least 30 years for raingauge stations influencing the catchment of the proposed site as well as available runoff data of specific duration at the proposed site; or
(c) Rainfall data of specific duration of the catchment for the last 30 years for the proposed site and runoff and concurrent rainfall data of specific duration at a site in the same basin or a nearby basin provided orographic conditions and catchment properties are similar for 5 to 10 years or more.

In case of ungauged basins, catchment characteristics are utilized for estimating dependable flows. If runoff data do not pertain to virgin conditions because of construction of water resources structures upstream of the gauging site, the working tables of the upstream reservoirs are required. If the runoff data series consists of the records for the period prior as well as after the construction of the structure, the runoff series is considered to be non-homogeneous. Necessary corrections need to be made to the records belonging to the period prior to the construction of the structure so that all available runoff records become homogeneous.

CHAPTER 8

GANGA BASIN

Himalayas, the great mountain chains of Asia are the source of three major river systems of the world. These are the Indus, the Ganga (or the Ganges), and the Brahmaputra. The large and fertile plains of Indus and Ganga in Northern India of the Indian subcontinent have been the cradles of one of the greatest and still surviving ancient civilizations, the Indus Valley Civilization. The Ganga River, which occupies nearly one-third of the geographical area of India, is the most important and sacred river of India. So legendary has been the socio-economic, cultural and religious saga of this great river that the Indian mythology and history are full of stories and incidents woven around the river and numerous pilgrimages are dotted along its course.

8.1. GANGA-BRAHMAPUTRA-BARAK (MEGHNA) BASIN

The composite Ganga-Brahmaputra-Barak (GBB) basin covers nearly one-third of the land area of India. This region is composed of the Ganga, the Brahmaputra and the Barak basins. The Ganga and the Brahmaputra Rivers join in Bangladesh and thereafter flow under the name Padma. This river finally joins the Meghna River and the combined river outfalls into the Bay of Bengal. The combined basin of the Ganga, Brahmaputra, and Meghna rivers in India, Nepal, and Bangladesh is also known by the name *Greater Ganga Basin*.

Ganga and Brahmaputra together drain two basins of extraordinary variation in every sense: in terms of altitude, cropping pattern, climate, flora, fauna, density of human settlement, social and cultural life, etc. These two rivers meet in the world's largest and most active delta in Bangladesh. This highly fertile area is home to nearly 10% of the Earth's population or about 500 million people, the vast majority of them in the Ganges basin. This fact alone is enough to highlight the importance of the GBB basin. An index map of the basin is given in Figure 1.

A very large ecosystem on land having distinct types of vegetation and animal life is called a biome. Nilsson et al. (2005) have pointed out that the Ganga-Brahmaputra system, covering 10 biomes, encompasses the widest biodiversity on earth.

Planning and management of water resources in GBB Basin in Asia is a really challenging task in light of factors stated above and increasing consciousness for the

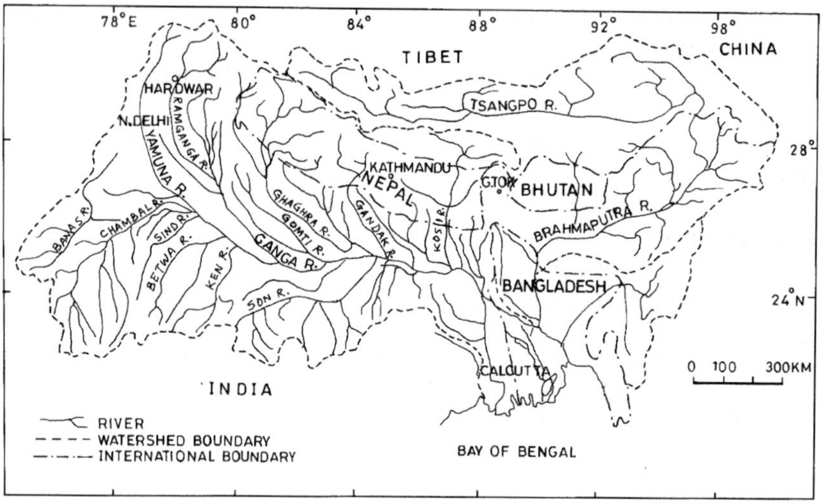

Figure 1. Index map of Greater Ganga basin

sustainable water governance. The GBB drains area of about 1,746,500 km^2 shared by five Asian countries: Bangladesh, Bhutan, China, India and Nepal (Table 1). Ganga and Brahmaputra, originate from the glacial sources in the Himalayan mountain range. Barak is a smaller rainfed tributary originating from the Naga hills of north-east India. The GBM is the 13th largest river basin in the world with runoff of about 1,400 BCM/year. At the global level, this outflow ranks third, after that of the Amazon and Congo. During the heavy rainfall period of the summer monsoon, this region drains to the Bay of Bengal, constituting most of the enormous volume of annual runoff, which in order of magnitude larger than the flow in the lean season. This results in regular annual inundation in the lower parts of the GBM, particularly in Bangladesh. The Himalayan crest line stands as a dividing line between the basins of these two long rivers in an approximately south-east to

Table 1. Area of the GBB basin by countries and geographical regions

Countries	Basin area in different regions (1,000 sq. km)				
	Mountain	Hills	Plains	Southern fringe	Total
Bangladesh	–	–	129.0	–	129.0
Bhutan	45.0	–	–	–	45.0
India	149.4	58.3	544.7	347.1	1,099.5
Nepal	118.9	21.1	–	–	140.0
Tibet (China)	333.0	–	–	–	333.0
Total	646.3	79.4	673.7	347.1	1,746.5

north-west direction. The highest peak on Earth, Mt. Everest (8,848 m) lies on this crestline in the GBB basin.

Table 2 gives areas of different countries in the basin of individual rivers as well as total areas of the rivers.

GBB is characterized by some unique ecological and socio-political diversities and complexities, which have resulted in unique difficulties in the utilization of its water resources. The problems of water management for sustainable development are not exactly similar in the individual river basins included in the GBB, in which almost 80% of the annual precipitation occurs during the three monsoon months of July to September. Another characteristic of the basin is a drastic decrease in the rainfall from east to west and south to north, across the Himalayan crestline, which creates a marked rain-shadow effect on the Tibetan plateau. The Meghalaya plateau, at places such as Cherrapunji, receives average annual rainfall exceeding 11,000 mm. Intense precipitation makes the GBB basin highly prone to floods, erosion, and river bed upgradation.

GBB has huge hydropower potential. Unfortunately, much of it is unexploited. The hydropower potential of the Ganga and its tributaries in India at 60% load factor is 10,715 MW; in Nepal, a potential of 85,000 MW is available. Further, Brahmaputra River has a potential of 34,920 MW and hardly 2% of it has been exploited so far.

A large temporal variation in precipitation in the course of the year generates large fluctuations in the flow characteristics of the rivers. Due to topography and spatial variations in climate and precipitation, there are water rich and water-scarce areas, generating conflicts over water sharing at various levels, among the countries, States and communities. This, in turn, has necessitated storage of the monsoon flow in the head water reaches of the mountainous landscape. Such storages besides providing for flood control will also help in irrigation, power generation, and urban water supplies. Notwithstanding the fact that such hydrological transformation can also transform the economy of the region, implementation of such interventions has been insignificant so far. Due to several reasons, storage dams in the Himalayas are being perceived by many people as a nuisance and hazard rather than an opportunity to utilize water resources.

The present chapter is devoted to the Ganga basin. Brahmaputra and Meghna basins form the subject matter of Chapter 9.

Table 2. Areas of different countries (thousand sq. km) in the basin

Country	Ganga	Brahmaputra	Barak/Meghna	Total
India	862.77	194.41	41.72	1,098.9
Bangladesh	46.60	47.00	36.00	129.60
Bhutan	–	45.00	–	45.00
Nepal	140.00			140.00
Tibet (China)	40.00	293.00	–	333.00
Total	1, 089.37	579.41	77.72	1,746.5

8.2. GANGA BASIN

Undoubtedly, the Ganga is the most sacred river of India – in fact natives typically call it *Ganga Maa* (or mother Ganga) or *Ganga ji* (or reverend Ganga). People of India believe that a bath in the holy waters of Ganga washes all the past sins of a person. If a few drops of the Ganga water are given to a person at the time of death, this is enough to elevate his soul to heaven. The first Prime Minister of India, Jawaharlal Nehru, described Ganga as, "She has been a symbol of India's age-long culture and civilization, ever changing, ever flowing, and yet ever the same Ganga."

Numerous pilgrimages are present all along the river. At the confluence of the Ganga and the Yamuna near Allahabad, bathing festivals attract hundreds of thousands of pilgrims. Other holy pilgrimage sites along the river include: Rishikesh and Haridwar, the places where the Ganga leaves the Himalayas; Allahabad, where the mythical Saraswati River is believed to meet the confluence of Ganga and Yamuna Rivers, forming 'Sangam'; the eternal city of Varanasi; and Patna. It is said that Lord Vishnu (Hari) himself used to take bath in Ganga (hence the name Haridwar – Hari + dwar or gate) to purify himself. Ganga is also regarded as the mother of *Bhishma* (the legendry character in Mahabharat).

Besides, Ganga River is the central theme of numerous films and popular songs of India. Its name prominently appears in the national anthem of India.

Water from the Ganga is used to cleanse any place or object for ritual purposes. Bathing in the river is believed to wash away one's all sins. To bathe in the Ganga is a lifelong ambition for natives who congregate in large numbers for river-centered festivals, such as the *Kumbh Mela* and numerous *Snan* (bath) festivals. It is believed that any water mixed with even the smallest amount of Ganga water at once becomes holy with healing powers. Hindus also cast the ashes of their dead in the river believing that this will take the souls of the deceased to heaven.

The catchment area of the Ganga falls in four countries, namely India, Nepal, Tibet (China), and Bangladesh. The major part of the geographical area of the Ganga basin lies in India. Many important tributaries of Ganga originate in the Himalayas in India and Nepal; Bangladesh lies in the deltaic region of the basin. The total length of the Ganga River is 2,525 km which makes it the 20th longest river in Asia and the 41st longest in the world (Philips World Atlas). The navigable length of Ganga River is 631 km which mostly lies in Bihar. An index map of the basin is given in Figure 2.

Although the headwaters region of Ganga in the Himalayas is dotted by a number of mighty tributaries, the Bhagirathi River that rises from the Gangotri glacier near Gomukh at an elevation of about 7,010 m above mean sea level (see Figure 3) is traditionally considered to be the source of Ganga River. The other main stream that originates in the Uttaranchal state of India is the Alakhnanda. Flowing downhill, Bhagirathi and Alakhnanda are joined by a number of streams, such as the Mandakini, the Dhuli Ganga, and the Pindar. These two rivers (Bhagirathi and Alakhnanda) meet at a place called Devprayag (see Figure 4) and thereafter the combined flow is known by the name Ganga.

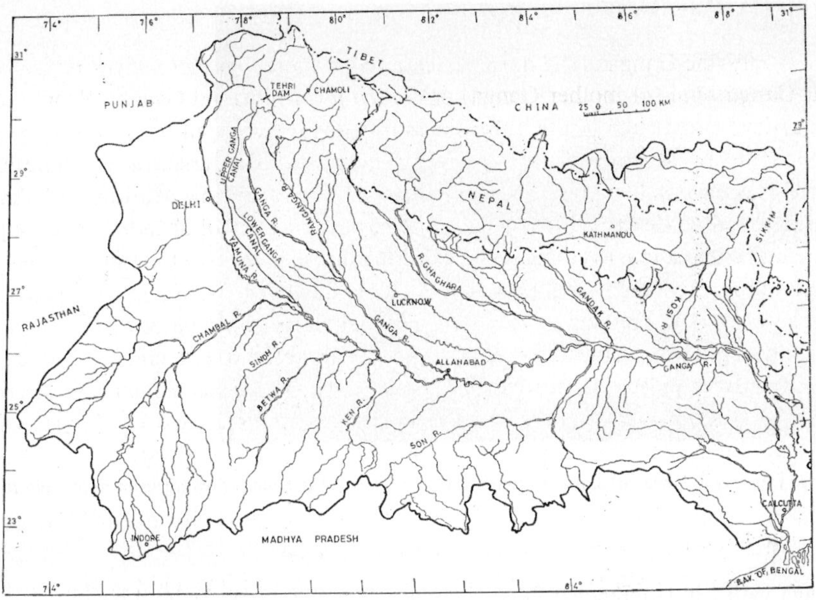

Figure 2. Index map of Ganga basin

Figure 3. Gomukh which is considered as the source of Ganga River

Figure 4. Devprayag where Bhagirathi (coming from left) and Alaknanda (flowing from right) join to form Ganga

Ganga valley in Uttaranchal State is a place of breath taking natural beauty. It is the home of the world renowned Valley of Flowers, a protected area. The famous high altitude Nanda Devi Biosphere Reserve is now a part of the Unesco's Global Network of World Heritage sites. This reserve will be soon included in the Man and Biosphere program to assess the global changes in biosphere reserves in mountains.

Ganga enters into plains near Haridwar and from here it flows in south/south-easterly direction. Yamuna is the most important tributary of the Ganga that joins it on the right bank at Allahabad. After confluence with Yamuna, the Ganga River flows in an eastward direction and is joined by a number of tributaries, such as the Ramganga, the Gomti, the Ghaghra, the Gandak, the Bagmati, the Kosi, the Sone and the Damodar.

The delta of Ganga is said to begin at a place known as Farakka where a barrage has been constructed to control river flow. At about 40 km downstream of Farakka, the river splits in two arms. The right arm, the Bhagirathi River, flows towards south and enters the Bay of Bengal about 150 km downstream of Calcutta. The left arm, known as Padma, turns towards east and enters Bangladesh. While flowing in Bangladesh, Padma meets the Brahmaputra River at a place known as Goalundo. The combined flow, still known as Padma, is joined by another mighty river, Meghna, at Chandpur, 105 km downstream of Goalundo. Further down, the river ultimately flows into the Bay of Bengal.

The Ganga basin extends over an area of 1,086,000 km^2. It lies between east longitudes 73° 30′ to 89° 0′ and north latitudes 22° 30′ to 31° 30′. The drainage area lying in India is 862,769 km^2 which is nearly 26.2% of the total geographical area of the country. Some tributaries, such as the Ghagra, the Gandak and the Kosi, drain areas in Nepal amounting to 190,000 km^2. The delta of the Greater Ganga basin covers an area of 56,700 km^2. The Ganga basin is bounded on the north by the

Himalayas, on the west by the Aravalis and the ridge separating it from the Indus basin, on the south by the Vindhyas and Chhotanagpur plateaus and on the east by the Brahmaputra ridge. The basin lies in the States of Uttaranchal, Uttar Pradesh, Madhya Pradesh, Bihar, Rajasthan, West Bengal, Haryana, Himachal Pradesh and the Union Territory of Delhi. The State-wise distribution of the drainage area is given in Table 3.

From a hydrological studies point of view, the entire run of Ganga River in India can be divided in three stretches or reaches. The upper reach extends from the origin to Narora, the middle reach from Narora to Ballia, and the lower reach from Ballia to its delta. The main physical sub-divisions of the Ganga basin are the Northern Mountains, the Gangetic Plains and the Central Highlands. Northern Mountains comprise the Himalayan ranges including their foothills. The Gangetic plains, situated between the Himalayas and the Deccan plateau, constitute the most fertile plains of the basin that are ideally suited for intensive cultivation. The central highlands lying to the south of the Great Plains consists of mountains, hills and plateaus intersected by valleys and river plains. They are largely covered by forests. Aravalli uplands, Bundelkhand upland, Malwa plateau, Vindhyan ranges and Narmada valley lie in this region.

Predominant soil types found in the sub-basin are sand, loam, clay and their combinations, such as sandy loam, loam, silty clay loam and loamy sand soils. The culturable area of the Ganga sub-basin is about 5.796×10^5 Mm^2, which is 29.5% of the total culturable area of the country.

Population

The Ganga basin is the largest basin in India, both catchment-wise, and population-wise. The total population in the basin as per 1991 census was estimated as 356.8 million, which is 42% of India as a whole. The average population density in the basin in 1991 was 414 persons per km^2 as against 267 for whole India. Besides, the density of cattle population is around 160 per km^2. As per 1991 census, the

Table 3. State-wise distribution of the drainage area of Ganga River in India

State	Drainage area (km^2)
Uttaranchal & Uttar Pradesh	294,410
Madhya Pradesh	199,385
Bihar	143,803
Rajasthan	112,490
West Bengal	72,618
Haryana	34,271
Himachal Pradesh	4,312
U.T. of Delhi	1,480
Total	862,769

Source: WG (1999).

basin had 111 urban centers with a population of more than 0.1 million and out of these, 7 cities had a population exceeding a million.

Two of the world's largest industrial cities with population exceeding 10 million, Kolkata (population 13.2 million, 2001 census) and Delhi (population 12.8 million), lie in Ganga basin. Other big cities in the basin are: Kanpur (population 2.69 million), Lucknow (population 2.26 million), Patna (population 1.71 million), Agra (population 1.32 million), Meerut (population 1.17 million), Varanasi (population 1.21 million), and Allahabad (population 1.05 million). A view of Ganga at Varanasi (or Banaras, a pre-historic city, well-known for its religious significance and a place noted for education) is shown in Figure 5. Besides these, an array of medium sized towns are scattered throughout the main stream of the Ganga. Not only do they put their immense pressure on the basin's freshwater supplies, but also have resulted in large-scale downstream pollution from the discharge of untreated urban wastewater and from agricultural chemicals. In addition, there are 68 gross polluting industrial units along the course of the river and its tributaries which discharge untreated industrial effluents.

8.2.1. Hydrology of Ganga Basin

The headwater reaches of the river receive a considerable part of precipitation as snow and some mountain peaks in the region are permanently snow covered. The average annual rainfall in the basin varies from 35 cm at the western end to nearly 200 cm near the delta.

The average annual discharge of the Ganga, the Brahmaputra and the Meghna rivers is 16,650, 19,820, and 5,100 m^3/s, respectively. The average annual flow of Ganga at Farraka is about 5.25×10^5 Mm^3. There are large variations in the flow of Ganga with time. Snow and glacier melt during the hot months (March to June) and provide large summer flows to Ganga and its tributaries. The maximum discharge in these rivers is observed during monsoon months (June to September). At Goalundo, the average annual flow of the Ganga River is 11,470 m^3/s. The

Figure 5. Ganga River at Varanasi

maximum and minimum flow at this site is 70,934 and 1,161 m^3/s (Chaturvedi and Rogers, 1985). The peak flow at Farakka in 1971 was estimated at 70,500 m^3/s. A line diagram of Ganga and its main tributaries is given in Figure 6 which also gives the average annual flows at various gauging points/ tributaries in million cubic metres. There are several features which immediately attract attention. The south bank tributaries (left side of figure) contribute less than the northern tributaries; the discharge at Delhi, far upstream of Yamuna, is relatively quite small; the total flow of the Brahmaputra is greater than that of the Ganga; and significantly, no figure is given for discharge down the River Bhagirathi. What the simple figure does not show is: the huge fluctuation between monsoon flows and the low season, when many of the south bank rivers simply dry up, and the north bank rivers have a vastly reduced flow; the fluctuation from year to year; and current and projected storage dams and their capabilities, which could smooth out the irregularities on an annual or inter-annual basis.

The surface water resource potential of the Ganga and its tributaries in India has been assessed at 525 billion m^3 out of which 250 billion m^3 is considered to be utilizable. Based on the 1991 census, the per capita water availability in the basin was nearly 1,471 m^3 per year. Although the Ganga basin is bestowed with abundant water resource, its occurrence/availability both in quantity and quality

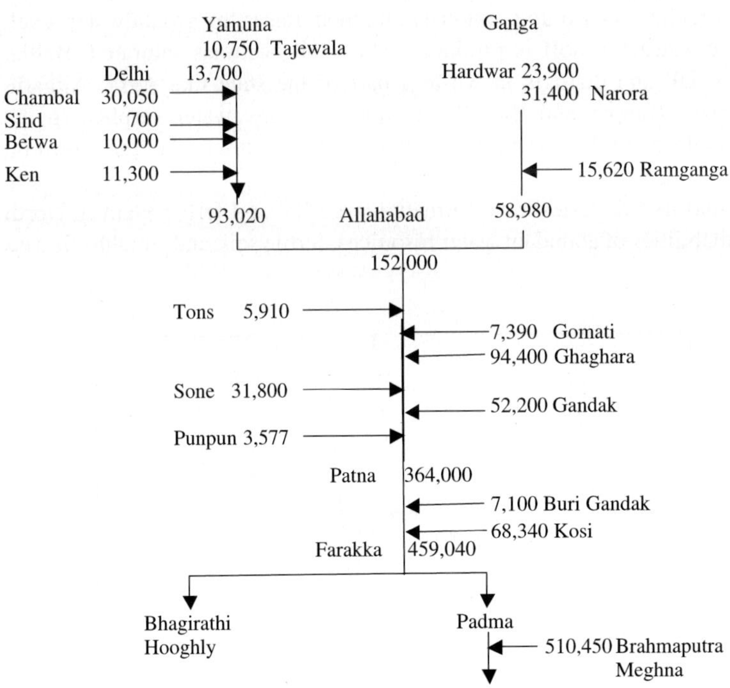

Figure 6. Line diagram of Ganga and its major tributaries. Numbers are average annual flows (MCM)

is not uniformly distributed either spatially or temporally. More than 75% of the annual rainfall occurs in monsoon months of June to September. As a result, large areas are subjected to floods on one hand and droughts on the other.

The storage potential of the Ganga Basin in India has been identified at $8.446 \times 10^4\,Mm^3$. However, till 1995 a total of $3.68 \times 10^4\,Mm^3$ of the storage space could be created. Water resources development schemes to create storage of $1.706 \times 10^4\,Mm^3$ are under construction and projects to provide another $2.956 \times 10^4\,Mm^3$ of storage are in the pipeline. The total replenishable ground water resource of the Ganga basin is estimated at $1.71 \times 10^5\,Mm^3$ out of which, about $4.86 \times 10^4\,Mm^3$ was being utilized by 1999. The Ganga River carries one of the world's highest sediment loads, equal to nearly 1,451 million metric tons per annum. During the lean season, the discharge in Ganga at Narora could be as low as 321 cumec. According to Chapman and Thompson (1995) Further down at Kannauj, it is 1,542 cumec and at Kanpur 1,679 cumec. At Allahabad, where Yamuna meets Ganga, lean season flow has been recorded as 1,870 cumec; at Varanasi it is 4,120 cumec, at Patna it is 5,693 cumec, and at Mungher the lean season flow is about 7,250 cumec. Figure 7 shows the annual discharge of Ganga at Farakka for the period 1950 to 1985.

Except in the Uttaranchal and Uttar Pradesh Himalayas, PE exceeds precipitation and water deficit conditions prevail in the entire Ganga basin. Only during the monsoon season, especially during the early rainy season, rainfall first recharges soil moisture, which was used up previously and then, raises the groundwater level. Surface and sub-surface runoff is produced when soil becomes saturated. Ballia, Pipri in eastern UP and districts in western part of the state like Agra, Aligarh, Meerut, Mainpuri, Kanpur and Bareilly do not show any water surplus. In the foothills of Himalayas in the Tarai areas, water surplus occurs during the monsoon period.

The Ganga and its tributaries have formed a large flat and fertile plain in North India. The availabilities of abundant water resources, fertile soil, and suitable climate

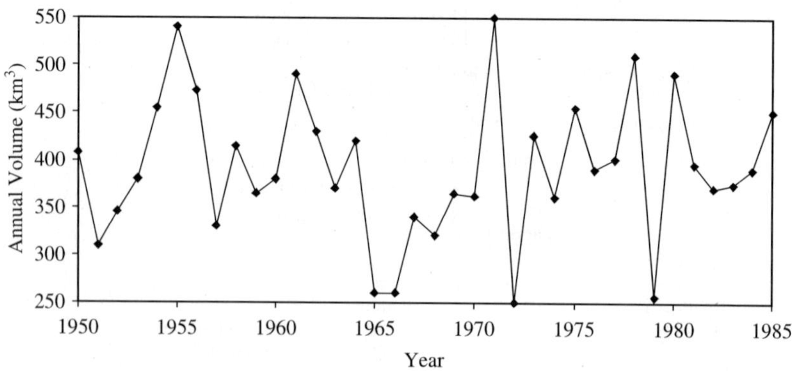

Figure 7. Annual discharge of Ganga at Farakka

have given rise to a highly developed agriculture based civilization and one of the most densely populated regions of the world. The net sown area in the Ganga basin in India is around 44 million hectares (M-ha) and the net irrigated area is 23.41 M-ha. Migration of the tributaries draining the eastern part of the basin has resulted in conspicuous back-swamp and meander bolt deposits. These sedimentological features play a dominant role in the hydrodynamics of the region.

The hydroelectric potential of the Ganga basin has been assessed as 10,715 MW at 60% load factor. Out of the 142 identified schemes in the basin, 22 schemes with a total installed capacity of 2,437 MW are in operation and 12 schemes with an installed capacity of about 2,716 MW are in various stages of construction.

8.3. MAJOR TRIBUTARIES OF GANGA

The principal tributaries joining the Ganga River through its 2,525 km course are Tons, Yamuna, Ramganga, Ghaghra, Gandak, Kosi, Mahananda, Pun-Pun, Kiul, Burhi Gandak and Sone. Chambal and Betwa are the two important sub-tributaries of Yamuna. The drainage area of different river systems within Ganga Basin are given in Table 4.

The major tributaries of Ganga River are described next.

Table 4. Drainage area of different tributaries of Ganga

S.N.	Name of the river system	Drainage area sq. km
1	Main Ganga including Karmnasa Baya, Bagmari-Pagla	113,163
2	Yamuna including Chambal, Betwa & Ken	363,082
3	Sone	71,259
4	Ghaghra	57,647
5	Ramganga	32,493
6	Damodar including Khari-Gangur-Ghia	31,220
7	Gomti	30,435
8	Rupnarayan including Haldi, Rasulpur & Kangsabati	23,760
9	Mahananda	17,440
10	Tons	16,860
11	Kiul-Harohar	16,661
12	Kosi	11,070
13	Burhi-Gandak	10,150
14	Punpun	8,530
15	Mayurakshi-Babla	8,530
16	Gandak	7,620
17	Ajay	6,050
18	Jalangi	5,640
19	Badua-Chandan	4,840
20	Bagmati	3,720
21	Adhwara	2,600
22	Kamla-Balan	2,980
23	Tidal rivers	15,650

8.3.1. Yamuna River

The Yamuna River is the biggest tributary of the Ganga River. It is also considered as a sacred river in India. According to the mythological legends, Yamuna was the daughter of Surya (the Sun God), and sister of Yama (the God of Death). A popular belief is that those who take a dip in its holy waters are not tormented by the fears of death. The Yamuna River is intimately connected to Lord Krishna's life. Lord Krishna sanctified the Yamuna River from the beginning of His *avtar* (incarnation) in the world. While his father Vasudeva was crossing the Yamuna with baby Lord Krishna for a safe place at the other bank of the river, the river was in spate. Legend says that the moment the rising water touched the feet of Lord Krishna, the river went into a recession.

The Yamuna River originates from the Yamunotri Glacier near Banderpoonch peaks (38° 59′ N 78° 27′ E) in the Mussourie range of the lower Himalayas at an elevation of about 6,387 meters above the mean sea level in district Uttarkashi (Uttaranchal). A view of the Yamunotri temple is given in Figure 8. It is said that the temple of Yamunotri was built in the last decade of the 19th century. A hot water pool is present at Yamunotri and the water is so hot that people cook rice and potato by putting them in cloth bags and dipping the bag in the hot water.

Arising from the source, the Yamuna River flows through a series of valleys for about 200 km in lower Himalayas and then emerges into Indo-Gangetic plains. In the upper reaches, the main valley is overlooked by numerous hanging valleys, carved by glaciers. The gradient of the river is steep here and the entire geomorphology of the valley has been carved by the erosive action of the river water. In the headwater reach of 200 km, Yamuna draws water from several major streams. The

Figure 8. A view of the Yamunotri temple (Source: www.yap.nic.in)

combined stream flows through the Shivalik range of hills of Himachal Pradesh and Uttaranchal states of India and enters into plains at Dak Pathar in Uttaranchal where the river water is regulated by a weir and is diverted into a canal for power generation. On the right side of the Yamuna basin is the Mussourie spur-along which lies the sprawled hill station of Mussourie (also known as the Queen of Himalayas).

From Dak Pathar, the Yamuna flows through the famous Sikh religious shrine of Poanta Sahib. Figure 9 shows the bed of Yamuna River near Dak Pathar. Flowing through Poanta Sahib, it emerges from the foothills of Kalesan, north of Tajewala. It reaches Hathnikund/Tajewala in the Yamuna Nagar district of Haryana state, where the river water is diverted into Western Yamuna canal and Eastern Yamuna canal for irrigation. During dry season, practically no water flows in the river downstream of Tajewala barrage and the river remains dry in several stretches between Tajewala and Delhi. Ground water accrual and contributions from seasonal streams again regenerate the river. Yamuna River enters Delhi near Palla village after traversing for about 224 km.

A canal known as the Satluj Yamuna link (SYL) canal joining Satluj with Yamuna is under construction here. This canal was to transfer Haryana's share of 3.5 MAF of water from the Indus basin. The state of Haryana has completed its portion of the canal but Punjab is yet to complete its portion. Punjab Government is not in favour of construction of this canal. Recently, the Punjab legislature passed an act known as the Punjab Termination of Agreement Act 2004 whereby the earlier agreements have been declared as null and void.

Further downstream, there is a barrage at Wazirabad which supplies drinking water to the city of Delhi. Generally, the flow in the river downstream of the Wazirabad barrage is almost nil in dry season because the available water is not adequate to meet the demand of Delhi. Yamuna River flow downstream of the Wazirabad barrage largely consists of untreated or partially treated domestic and industrial wastewater contributed by numerous drains along with the water

Figure 9. The bed of Yamuna River near Dak Pathar

transported by Haryana Irrigation Department from the Western Yamuna Canal (WYC) to the Agra Canal via the Nazafgarh Drain and the Yamuna. About 22 km downstream of the Wazirabad barrage, the Yamuna water is diverted into the Agra Canal for irrigation through the Okhla barrage. Generally, water flow through the barrage during the dry season is nil. Whatever water flows in the river beyond the Okhla barrage is contributed through domestic and industrial wastewaters generated by East Delhi, Noida and Sahibabad and joins the river through the Shahdara drain.

At Tajewala, the ratio Q_{max}/Q_{min} (where Q_{max} is flood peak discharge and Q_{min} is the lowest flow rate in the year) comes to about 40, in comparison to 33 of Indus, 34 of Ganga and 4 of Amazon. This large ratio is indicative of a wide temporal variation in the flow. During non-monsoon period the entire inflow at Tajewala is diverted to canal systems leaving the river dry. However downstream of this point the river starts picking effluent seepage from ground water reservoir.

Further downstream, Yamuna flows through the Agra city which is famous for Tajmahal (see Figure 10), a white marble wonder. Shortly afterwards, it passes through another historical city, Mathura. The total length of the Yamuna from its origin to Allahabad (confluence with Ganga) is 1,376 km and the drainage area is 366,223 sq. km. The Yamuna is a mighty river in itself and has a number of tributaries. In its first 170 km stretch, the tributaries Rishi Ganga Kunta, Hanuman Ganga, Tons and Giri join the main river. Later, big rivers, such as the Chambal, the Sind, the Betwa, and the Ken, join it.

The catchment of the Yamuna River system covers parts of Uttar Pradesh, Uttaranchal, Himachal Pradesh, Haryana, Rajasthan, Madhya Pradesh & Delhi states. The area of the Yamuna catchment lying in different states is shown in Table 5.

For the purpose of flood forecasting, tracking of Yamuna River begins at Poanta in Garhwal area on the confluence of rivers Tons, Pawar and Giri. The next important site is at Tajewala followed by Kalanaur and Mawi before Delhi. It roughly takes about 60 hours for the water to reach Delhi from Tajewala, which enables issue

Figure 10. The Tajmahal, Agra

Table 5. The catchment of Yamuna River in various states

Name of state	Yamuna catchment area in the state (sq. km)	% of catchment area
Uttar Pradesh (including Uttaranchal)	74,208	21.5
Himachal Pradesh	5,799	1.6
Haryana	21,265	6.5
Rajasthan	102,883	29.8
Madhya Pradesh	14,028	40.6
Delhi	1,485	0.4

of warnings at least two days in advance. Table 6 shows statewise allocation of utilizable flows of Yamuna River.

The Interstate Agreement also envisages that a minimum flow of 10 cumec shall be maintained in Yamuna downstream of Tajewala and Okhla head works throughout the year for ecological considerations. It is also assessed that a quantum of 680 MCM is not utilizable due to flood spills. The allocation of available flows amongst the beneficiary states is regulated by the Upper Yamuna River Board. In a year when the surface-water availability is more than the assessed quantity, the surplus availability will be distributed amongst the states in proportion to their allocation. However, in a year when availability is less than the assessed quantity, first the drinking water allocation of Delhi will be met and balance will be distributed amongst Haryana, UP and Rajasthan in proportion to their allocation.

The Yamuna River can be segmented in five distinguished independent segments due to characteristic hydrological and ecological conditions as shown in Table 7.

i. Physiographic and Geomorphology of Himalayan Segment (Upper Yamuna catchment)

In the Himalayan segment (Upper Yamuna catchment), the drainage system and the characteristics of landforms are closely interdependent and inter-related. The Upper Yamuna catchment falls into 3 well defined physiographic belts: the Lesser Himalaya, The Siwalik, and the Doon Valley.

Lesser Himalayas: Elevation in this region ranges from 4,000 m to 1,000 m. This region has a mild topography with gentle slopes and deeply dissected valleys which

Table 6. Statewise allocation of utilizable flows of Yamuna River

State	Allocation in MCM
Haryana	5,730
Uttar Pradesh	4,032
Rajasthan	1,119
Himachal Pradesh	378
Delhi	724
Total	11,983

Table 7. Different Segments of Yamuna River

Name of segment	Location	Length
Himalayan Segment	From origin to Tajewala Barrage	172 km
Upper Segment	Tajewala Barrage to Wazirabad Barrage	224 km
Delhi Segment	Wazirabad Barrage to Okhla Barrage	22 km
Eutriphicated Segment	Okhla Barrage to Chambal Confluence	490 km
Diluted Segment	Chambal Confluence to Ganga Confluence	468 km

suggest that rivers and streams are still furiously at work. The upper part of this area has high mountains, most of which have seasonal snow capped peaks and glaciated ranges. Prominent glaciers are Bandarpunch, Jamadar Bamak and Deokhera Bamak. The retreating movement of these glaciers and their tributary glaciers may still be observed in the form of 'V' shaped valleys with moraines and smooth and aggradational slopes.

Low to moderately high mountains occur between the altitudes of 1,000 m and 3,000 m. Actually, the Lesser Himalayas region is a massive mountainous tract with a series of ridges and spurs divided by river valleys. The slope varies from 25 to 50% but may rise even up to 80%. The rivers and their tributaries have carved out entrenched valleys with steep slopes in higher reaches, while flatter valleys are found in lower reaches. At a number of places, rivers have formed depositional terraces. A large part of this region is made up of sedimentary rock formations intruded by granite and basic volcanic rocks. The northern part has been subjected to low-grade metamorphism.

Shiwalik Hills: The Shiwalik Hills are formed as a result of intense dissection by fine textured pattern of drainage lines. They are a long prominent ridge trending NW-SE with altitudes ranging from 750 to 1,500 m. The main ridge is composed of a gentle northern slope and a steeper southern slope. A water divide is located more or less half way through this ridge. It is drained with numerous parallel to sub-parallel streams flowing towards north or south in consequent entrenched channels. The Shiwalik or the outer Himalayan range is a youthful range separated from the less Himalaya by the main boundary fault.

Doon Valley: The Doon Valley is a long tectonic synclinal structure of the outer Himalaya. It lies within the ranges of lesser Himalaya to the north and the Shiwalik range of the Outer Himalaya towards north. It is a low-lying region between the two ranges with altitudes not more than 500 m to 750 m. The Doon gravels composed of boulder and gravel beds with this clay bands constitute the piedmont slopes. Stream frequency and drainage density is low here because of the poor development of drainage, possibly due to porous and permeable characteristics of the bed rocks.

ii. Climate and precipitation

The Himalayas exercises a dominating influence on climate in the northern region of the Upper Yamuna catchment. In this region, winters are very cold, while summers are moderate. The average annual rainfall varies between 1,500 mm to 400 mm. The entire catchment comes under the influence of south-west monsoon and a major part of rainfall is received between June and September. Winter rainfall is scanty and occurs between December and February. In the lower part of Yamuna basin, temperatures are relatively moderate. In summers, temperature frequently exceeds 40 ° C.

iii. Tributaries of Yamuna River

The tributaries of Yamuna account for 70.9% of the catchment area; the balance of 29.1% area directly drains into the Yamuna River or is occupied by smaller streams. Further, the catchment area of Yamuna amounts to 40.2% of the area of Ganga Basin and 10.7% of the land area of India.

The important tributaries of the Yamuna River are the Tons, the Chambal, the Hindon, the Sarda, the Betwa and the Ken. Other small tributaries of the Yamuna River include Rishiganga, the Uma, the Hanuman Ganga, the Giri, the Karan, the Sagar and the Rind. The main Yamuna and Tons are fed by glaciers, viz., the Bandar Punch Glacier and its branches and originate from the Great Himalayan range. Many smaller streams in the Yamuna basin, for example, the Chautang, the Sahibi, the Dohan, the Kantili, the Bapah and the Banganga end up in the sandy tracts. A brief description of the important tributaries of the Yamuna River is given in the following sections.

Tons River: The Tons River is the longest tributary of the Yamuna River. It flows through Garhwal, the western part of the Himalayan state of Uttaranchal. The river originates at an elevation of 3,900 m and joins the Yamuna below Kalsi. With its source in the 6,315 m high Bandarpunchh Mountain, it is the biggest tributary of the Yamuna. In fact, Tons carries more water than the Yamuna itself.

Kali River: The Kali River originates from the Doon Valley in the western part of Uttaranchal. This river is named as Kali, possibly because the colour of the river water is *kala* (black). The basin covers a drainage area of approximately 750 sq. km within the latitude of 29° 13′ 30″ N and longitude of 77° 32″ 45E. From its origin up to the confluence with Hindon River, a tributary of Yamuna, the river travels a distance of approximately 150 km through Saharanpur and Muzaffar Nagar, Meerut, and Ghaziabad Districts. Despite a significant drainage area, mostly lying in plains, the river does not carry any significant flow. The city of Muzaffar Nagar is situated on the left bank of the Kali River.

Kali is a highly polluted river. People living in about 1,700 villages and a dozen towns along the river from Saharanpur to Kannauj are exposed to polluted water. Levels of pollutants including heavy metals and industrial sludge have reached alarming proportions in the river and even handpumps drawing water from a depth of 35 meters are found to contain pollutants. The presence of a salt of cyanide has resulted in over 100 species of fish getting extinct from the river over the past 40 years. Similarly, the release of heavy metals and toxic material from various industries and slaughter houses into the river has led to chronic ailments among people living along it.

Hindon River: Hindon is an important tributary of Yamuna River. In fact, this river is sandwitched between two major rivers: Ganga on the left and Yamuna on the right. Hindon originates from upper Shiwalik (Lower Himalayas). It lies between the latitude 28° 04′ to 35° 05′N and longitudes 77° 08′ to 77° 04′E. It is a purely rainfed river with catchment area of about 7,083 sq. km. This river has a total run of about 400 km. The width of Hindon River ranges from 20 m to 160 m.

Banganga Basin: The Banganga River originates in the Aravali hills, near Arnasar and Bairath in Jaipur District. It flows towards the south up to the village of Ghat, then east through partly hilly and partly plain terrain. The length of the river is 240 km.

The main tributaries of the Banganga River are Gumti Nalla, Suri River, Sanwan and Palasan Rivers. The Gumti Nalla basin is located between latitudes 25° 00′ 30″ and 25° 53′ and longitudes 75° 52′ and 76° 01′. The Gumti Nalla originates in the hills near Bhaimpura village in Jaipur District, flows for 24 km in a north-easterly direction and joins the main river near the Talo village. The catchment area of the basin is 102 km^2.

- *Suri River*: The Suri basin is located between latitudes 26° 44′ and 26° 53′ and longitudes 76° 25′ and 76° 30′. The Suri River originates in the hills near Kanst village in Dausa District, flows north for 18 km and then northeast for 10 km before joining the main river near the village of Kailai. The catchment area of the basin is 76 km^2.
- *Sanwan River*: The Sanwan basin is located between latitudes 26° 59′ and 27° 22′ and longitudes 76° 16′ and 76° 46′. The Sanwan River originates in the hills near Angri village in Alwar District, flows 29 km southwards, up to Sirsa Devi Bund, then 66 km eastwards before joining the main river near the village of Juthiara. The catchment area of the basin is 660 km^2.
- *Palasan River*: The Palasan River originates in the hills near Rajpura village in Alwar District and joins the main river near the village of Indiana, after flowing in a south-easterly direction for 24 km and in an easterly direction for 24 km. The catchment area of the basin is 539 km^2 and it is located between latitudes 27° 02′ and 27° 18′ and longitudes 76° 25′ and 76° 49′.

Chambal River: The Chambal River, called Charmanvati in ancient times, is the largest of the rivers flowing through and Rajasthan State. This tributary of Yamuna is 960 km long. The total area drained by the Chambal up to its confluence with the Yamuna is 143,219 sq. km out of which 76,854 sq. km lies in Madhya Pradesh state, 65,264 sq. km in Rajasthan state and 1,101 sq. km in Uttar Pradesh. The Chambal basin lies between the longitudes 73° 20′ E and 79° 15′ E and latitudes 22° 27′ N and 27° 20′ N.

The Chambal River rises in northern slopes of Vindhyan Mountains, about 15 km West-South-West of Mhow in Indore District in Madhya Pradesh state at an elevation of about 853 m. The river flows first in a northerly direction in Madhya Pradesh for a length of about 346 km and after passing the historic fort of Chaurasigarh, it flows in a generally north-easterly direction for a length of around 225 km through Rajasthan. The Chambal flows for another 217 km between Madhya Pradesh and Rajasthan and further 145 km between Madhya Pradesh and Uttar Pradesh. The Chambal enters U.P. near the Charak Nagar village and flows for about 32 km before joining the Yamuna River, south east of the Sahon village in Etawah district at an elevation of about 122 m. Three dams, namely, Gandhisagar, Rana Pratap Sagar and Jawahar Sagar, have been constructed in its basin.

From the source down to its junction with the Yamuna, the Chambal has a fall of about 732 m. Out of this around 305 m is within the first 16 km reach from its source. It falls for another 195 m in the next 338 km, where it enters the gorge past the Chaurasigarh Fort. In the next 97 km of its run from the Chaurasigarh Fort to Kota city, the bed falls by another 91 m. In the rest of its 523 km run, the river passes through the flat terrain of the Malwa Plateau and later in the Gangetic Plain with an average gradient of 0.21 m/km.

In the reach of 96 km from 344 km to 440 km from its source, the Chambal River flows through a deep gorge, while lower down, there are wide plains. The Gandhisagar Dam is located near the center of this reach. As there is a deep gorge immediately upstream of the dam, the reservoir has a large storage capacity despite its comparatively low height. For next 48 km the river flows through Kundal Plateau, the Rana Pratap Sagar dam is constructed at the lower end of this reach, about 1.6 km upstream of Chulia falls. Again, the topography permits fairly good storage upstream of the dam. Further down, The Jawahar Sagar dam is located in the middle of the Kota gorge. The Kota Barrage is located near the Kota town, where the river emerges from the gorge section into the Plateau. The tributaries that join Chambal upstream of the gorge include Sipra, Choti, Kalisindh, Sivana, Retam, and Ansar. The total area draining at the Kota Barrage is 27,319 sq. km. Lower down in the reach between the Kota Barrage and its junction with the Yamuna, the tributaries joining Chambal are the Kalisindh, the Banas, Parvathi, Seep, Kund and Kuwari. Three reservoirs, namely, Patanpur on the Parbati River, Mohanpura on the Newaz River, and Kundaliya on Kalisindh River have been proposed to increase power generation at the Gandhi Sagar and Ranapratap Sagar reservoirs.

The Chambal drainage area resembles a rectangle up to the junction of the Parvathi and Banas Rivers with the Chambal flowing along its major axis. On its south, east and west, the basin is bounded by the Vindhyan mountain ranges and on the North West by the Aravallis. Below the confluence of the Parvathi and Banas, the catchment becomes narrower and elongated. In this reach, it is bounded by the Aravalli mountain ranges on the North and the Vindhyan hill range on the south. Until the 1970s, the ravines in the lower portion of Chambal were infested with dacoits.

The Chambal is a rainfed catchment. The annual average rainfall up to the Gandhisagar dam site is 860 mm. The area is located in the central part of India and it experiences extremes of climate. The temperature in the hottest months of April to June occasionally goes up to 40 °C in the shade and falls to a minimum of 2 ° C during the coldest months. The relative humidity ranges between 30 to 90 percent during the year.

Tributaries of Chambal River

- *Alnia:* The Alnia River originates in the north-western slopes of Mukindwara hills, and flows for about 58 km before joining the Chambal near Notana village in Kota District. The catchment area of the Alania River is 792 km^2.
- *Kalisindh:* It originates in the northern slopes of the Vindhya hills. Flowing in M.P., it enters Rajasthan near Binda village in Jhalawar District. After flowing for about 145 km in Rajasthan, it joins Chambal near Nonera village in Kota District. The catchment area of the Kalisindh River is 7,944 km^2 and in Rajasthan State, it extends over parts of Jhalawar and Kota Districts. The Parwan River is an important tributary of Kalisindh River. The Parwan originates in the Malwa Plateau and after flowing for about 186 km in M.P., it enters Rajasthan near Kharibor village in Jhalawar District. It joins Kalisindh near Ramgarh village in Kota District. The catchment area of the Parwan River is 2,892 km^2.
- *Mej:* The Mej River originates in Mandalgarh Tehsil in Bhilwara District and joins the Chambal River near Bhaius Khana village in Kota District. The catchment extends over Bhilwara, Bundi, and Tonk Districts, with the total area being 5,860 km^2.
- *Chakan:* The Chakan River is formed by the confluence of many small rivulets. The river flows in a south-easterly direction and joins the Chambal near Karanpura village in Sawai Madhopur District. The catchment is situated in Sawai Madhopur, Tonk, Bundi and Kota Districts 789 km^2 area.
- *Parwati River:* The Parwati River originates in the northern slopes of the Vindhyan hills in M.P. where it forms a boundary between MP and Rajasthan for about 18 km, and then enters Rajasthan near Chatarpura village in Baran District. Thereafter, it flows for about 83 km in Rajasthan before again forming the boundary between MP and Rajasthan for a distance of about 58 km up to Pali village in Kota District, where it joins the Chambal. The river catchment in Rajasthan is situated in Kota and Jhalawar Districts. Major tributaries of the Parwati River are Lhasi, Berni, Bethli, Andheri, Retri, Dubraj, Bilas and Kunu.
- *Kunu River:* The Kunu River originates north of Guna town in MP. It flows for 48 km in M.P. before entering Rajasthan near Museri village in Baran District. The river re-enters M.P. again and flows for about 24 km. It then again enters Rajasthan near Gordhanpura village in Kota District and joins Chambal River. The catchment in Rajasthan lies in Baran District. The catchment area of the river is 726 km^2. The tributaries of the Kunu River are Karal and Rempi.
- *Banas River:* The Banas River originates in the Khamnor hills of the Aravali range (about 5 km from Kumbhalgarh) and flows along its entire length through Rajasthan. Banas is a major tributary of the Chambal River, the two rivers meeting near village Rameshwar in Khandar Block in Sawai Madhopur District. The total length of the river is about 512 km and the catchment area is 45,833 km^2. The main tributaries of the Banas River are Berach and Menali on the right bank, and Kothari, Khari, Dai, Dheel, Sohadara, Morel and Kalisil on the left bank.

The Banas River itself has many big tributaries. The Berach River originates in the hills northeast of Udaipur city. It flows northeast for about 157 km in Udaipur, Chittorgarh and Bhilwara Districts before joining Banas near Bigod village in Mandalgarh Tehsil of Bhilwara District. The catchment area of the river is $7,502\,km^2$, which lies between 73° 25′ and 75° 02′ east longitudes and 24° 29′ and 25° 14′ north latitudes. The Berch flows in a hilly region up to Badgaon reservoir and then through plains. This river receives flow from Ayar, Wagli Wagon, Gambhiri and Orai Rivers.

The Mashi River, a tributary of the Banas originates near Kishangarh in Ajmer District. The catchment area of the river is $6,335\,km^2$. It flows east and then south for about 96 km in partly hilly and partly plain areas of Ajmer and Tonk Districts before joining Banas River near Tonk.

Khari is another tributary of the Banas which originates in the hills near Deogarh in Rajsamand District. Its catchment area is $6,268\,km^2$. It flows northeast for about 192 km through Udaipur, Bhilwara and Ajmer Districts before joining the Banas River near Chosala village in Ajmer District. Another tributary of the Banas River, the Dai River originates in the southeastern slopes of the Aravali range, near Nasirabad Tehsil of Ajmer District. Its catchment area is $3,015\,km^2$. It flows southeast for about 40 km and east for about 56 km in Ajmer District and for a short reach through Tonk District, before joining the the Banas River near Bisalpur village in Tonk District.

The Morel River is a tributary of the Banas River which originates in the hills near Dharla and Chainpura villages in Bassi Tehsil of Jaipur District. The catchment area of the river is $5,491\,km^2$, lies between 75° 42′ and 76° 56′ east longitudes and 26° 14′ and 27° 9′ north latitudes. It flows southeast for 29 km, then southwest for 35 km, up to the confluence with the Dhund River, and then southeast for 83 km in Jaipur and Sawai Madhopur Districts, before joining the Banas River near Hadoli village in Karauli Tehsil of Sawai Madhopur District. The Dhund, Kankrauli and Kalisil are the major tributaries of the Morel River.

The Kothari River, another tributary of Banas, originates in the eastern slopes of the Aravali range near Horera village in Bhilwara District. The catchment area of the river is $2,341\,km^2$, lying between 73° 47′ 30″ and 75° 3′ 30″ East longitudes and 73° 47′ 30″ and 75° 3′ 30 North latitudes. The river flows through Rajsamand and Bhilwara Districts for about 51 km in a hilly region, and 100 km through plains, before joining the Banas near Nandrai village in Bhilwara District.

Betwa River: The Betwa River is a tributary of Yamuna River. Its basin extends from longitude 77° to 81 °E and latitude 23° 8′ to 26° 0′ N. The Betwa River originates at an elevation of 470 m in the Bhopal District in Madhya Pradesh. After traversing a distance of 590 km, the river joins the Yamuna River near Hamirpur at an elevation of 106.68 m. The total catchment area of the Betwa River is 46,580 sq. km of which 31,971 sq. km (68.64%) lies in Madhya Pradesh and 14,609 sq. km (31.36%) lies in Uttar Pradesh. The basin is saucer shaped with sandstone hills around the perimeter. The river has 14 principal tributaries out of which 11 are completely in Madhya Pradesh and 3 lie partly in Madhya Pradesh and Partly in Uttar Pradesh. The Halali and Dhasan Rivers are the important tributaries of the Betwa River. The Halali is the largest tributary having a length of 180.32 km. In the entire basin the rainfall varies from 100 cm to 140 cm in upper reaches and from 80 cm to 100 cm in lower reaches. The average annual rainfall in the Betwa basin is 110 cm.

Ken River: Ken is an inter-state river, flowing through the States of Madhya Pradesh and Uttar Pradesh. Its basin lies between north latitudes 23° 20′ and 25° 20′ and east longitudes of 78° 30′ and 80° 32′. The river originates near the Village Ahirgawan in Jabalpur District of M.P. at an altitude of 550 m above mean sea level and joins the Yamuna River, near Chilla Village in U.P., at an elevation of about 95 m. It forms the common boundary between Panna and Chhattarpur Districts of M.P. and state boundary between Chhattarpur District (M.P.) and Banda District (U.P.). The river has a total length of 427 km, out of which 292 km lies in M.P., 84 km in U.P, and 51 km forms the common boundary. The total catchment area of the Ken River basin is 28,058 sq. km, out of which 24,472 sq. km lies in Madhya Pradesh and the balance 3,586 sq. km in Uttar Pradesh.

Tributaries of Ken River: The important tributaries of the Ken River are Sonar, Bearma, Kopra, Bewas, Urmil, Mirhasan, Kutni, Kail, Gurne, Patan, Siameri, Chandrawal, Banne, etc., among others. The longest tributary is Sonar which is 227 km in length and lies wholly in M.P. In terms of catchment area also, Sonar is the largest tributary with a catchment area of 12,620 sq. km.

The Sonar sub-basin is located fully in Madhya Pradesh between north latitudes of 23° 20′ and 23° 50′ and east longitudes of 78° 30′ and 79° 15′. It is a leaf shaped elongated catchment, with an average width of about 40 km. The Sonar basin is bounded by Bearma basin (a sub-basin of the Ken River) on the east side, by Dhasan basin (a sub-basin of Betwa River) on the west side and the Vindhyan ranges on the south. Sagar and Damoh are the major districts falling in this sub-basin and parts of Panna, Chattarpur and Raisen Districts also fall in the basin. The total catchment area of the Sonar basin is 6,550 sq. km. The major tributaries of Sonar are Bewas, Dehar, Kaith and Baink on the left bank and Kopra and Bearma on the right bank.

8.3.2. Ramganga River

Ramganga is the first major tributary joining Ganga. It rises at an altitude of about 3,110 m in the lower Himalayas near the Lohba village in the Garhwal district of Uttaranchal. The length of the Ramganga River from the source to the confluence with the Ganga is 596 km. During its course, the river flows through a mountainous terrain and has a number of falls and rapids. The river enters the plains at Kalagarh near the border of the Garhwal district, where the famous Ramganga dam has been constructed. Beyond Kalagarh, the river flows in a southeasterly direction and finally joins the Ganga on its left bank near Kanauj in the Fategarh district. The river flows entirely in the states of Uttaranchal and Uttar Pradesh. The catchment area of the basin is about 32,493 sq. km. The important tributaries that join the Ramganga River are the Kho, the Gangan, the Aril, the Kosi, and the Deoha (Gorra).

8.3.3. Gomti River

The Gomti River originates near Mainkot, about 3 km east of the Pilibhit town in Uttar Pradesh, at an elevation of 200 m. The river drains the area between Ramganga and Ghaghra systems. The total length of the river is about 940 km and it flows entirely in the State of Uttar Pradesh. The total drainage area of the river is 30,437 sq. km. The river flows through Sahajahanpur, Kheri, Lucknow, Barabanki, Sultanpur, Faizabad, Jaunpur, Varanasi and Ghazipur districts before merging into the Ganga in Audihar in Jaunpur. Lucknow, the capital city of Uttar Pradesh, is situated on the banks of the Gomti River. The main tributaries of the Gomti River are the Gachai, the Sai, the Jomkai, the Barna, the Chuha and the Sarayu.

The Gomti-Kalyani doab is a fertile area, irrigated by canals, tube wells and open wells. This doab lies in the Barabanki district of U.P. and is bounded by Kalyani River on the North and the Gomti River and its tributary on the South covering an area of about 146,526 ha. On the west, the area extends up to the Sarda Sahayak feeder channel and on the east up to the confluence of the Gomti and Kalyani Rivers. The area lies between Longitude 81° 0′ E to 80° 37′ E and Latitude 26° 38′ 15″ N to 27° 17′ 00″ N. The area is drained by the Gomti and Kalyani Rivers and their tributaries. Doab area has a subtropical and monsoonic climate. Most of the precipitation occurs during the months of July, August and September. The normal rainfall of the area is 1,004 mm.

The area is part of the Indo-Gangetic plain and its formation consists mainly of sand, silt, clay and occasional kankar bands. These materials are found intermixed and beds are variable both in lateral extent and thickness. At places an admixture of these formations is also found. The structure and thickness of sand beds are such that these form very potential aquifers.

Major cities situated on its banks are Lucknow, Sitapur, Hardoi, Bara Banki, Rae Bareli, Pratapgarh, Sultanpur and Jaunpur. The water quality of the Gomati River at Lucknow and Jaunpur has been found quite unsafe. According to the World Health Organization (WHO) standards, there should not be more than 5,000 bacteria in 100 cc of water. But in the Gomti River between Dalliganj bridge and Hanuman Setu (bridge), the bacteria count has soared to 1.75 lakh per 100 cc of water. Rough estimates show that 25 nullahs, including Sarkata, Patanullah, and Wazirganj, pour around 6.5 million litres of effluents daily into the Gomti. Besides, three cremation grounds, Guala, Murdahia and Bhainsa Kund, on the banks of the river also add to the pollution. Estimates show that an average flow of the Gomti River is 1,500 million litres per day (MLD). During rains, it reaches 45,000 MLD while in summer it falls to 500 MLD. Everyday, the Lucknow city draws around 200 MLD of water from Gomti. During summer 210 MLD of effluents are released daily in the remaining 300 MLD of water, thus making it unworthy of even bathing, leave alone drinking.

8.3.4. Ghaghra River

The Ghaghra River originates at an elevation of 4,800 m near Mansarover Lake. The river is also known as Manchu and Karnali in Nepal. After flowing for about 72 km in a south-easterly direction, the river enters Nepal. Ghaghra enters into India at Kotia Ghat near Royal Bardia National Park, Nepal Ganj, where it is known as the river Girwa for about 25 km. A barrage called Girijapuri Barrage has been constructed at the end of the Girwa. Below the barrage the river attains the name of the Ghaghra. The total catchment area of the Ghaghra River is 127,950 sq. km, out of which 45% falls in India. The Sarda River is the important tributary of Ghaghra River, which forms the boundary between India and Nepal for some distance. The Sarju, Rapti and Little Gandak are the other important tributaries of the Ghaghra River. The Ayodhya city is located on the banks of the Sarju River, near Lucknow. This city is famous as the capital of king Dashrath who was the father of Lord Rama. The total length of Ghaghra River before its confluence with Ganga River (at Doriganj downstream of Chhapra town in Bihar) is 1,080 km. The Ghaghra River carries more water than Ganga before its confluence.

The width of the river can be appreciated by Figure 11. This river has changed its course many times; at a few places it has shifted by as much as 7.0 km. The critical locations along Ghaghra in India where major shifting has occurred are Tanda, Ayodhya, Golabazar, Barhaj and Bansdih. Heavy mining of sand from its course and variation in flow is the possible reasons for shifting in river Ghaghra.

8.3.5. Karamnasa River

The Karmnasa River, a tributary of Ganga River, originates at an elevation of 350 m near Sarodag on the northern face of the Kaimoor range in the Mirzapur district of Uttar Pradesh. It flows in a north-westerly direction through the plains of Mirzapur

Figure 11. Ghaghra River at Tanda (Photo courtesy Dr Sanjay Jain)

and joins the Ganga River near Chanusa. Its tributaries are the Durgavati and the Chandraprabha, the Karunuti, the Nadi, and the Khajuri. The length of the river is 192 km, out of which 116 km lies in Uttar Pradesh. The balance of 76 km forms the common boundary between Uttar Pradesh and Bihar. The total drainage area of the Karmnasa River with its tributaries is 11,709 sq. km.

8.3.6. Sone River

The Sone River is an important right bank tributary of the Ganga River. The river originates at an elevation of 600 m at Sonbhadra in the Maikala range of hills in Madhya Pradesh. The total catchment area of the basin is 71,259 sq. km. The important tributaries of the Sone River are Rihand, Kanhar, Ghaghar, and Koel. The Rihand dam has been constructed on the Rihand River. The total length of the river is 784 km, out of which about 500 km lies in Madhya Pradesh, 82 km in Uttar Pradesh and the remaining 202 km in Bihar. The river meets the Ganga River about 16 km upstream of Dinapur in the Patna district of Bihar.

8.3.7. Punpun River

The Punpun River is an important right bank tributary of the Ganga River in lower reaches. The Punpun River originates from Chottanagpur hills of the Palamau district at an elevation of 300 m in Bihar and lies between east longitudes of 84° 10′ to 85° 20′ E and north latitudes of 24° 11′ to 25° 25′ N. The river mostly flows in a northeast direction and finally joins the Ganga River at Fatawh, about 25 km downstream of Patna. This 200 km long river is mostly rainfed and carries little discharge during non-monsoon period. It has a number of tributaries; namely, the Butane, the Madar, and the Morhar draining the Chotanagpur plateau. Punpun often causes heavy flood damages on the eastern side of Patna city.

The shape of the Punpun River basin is roughly trapezoidal. The catchment area of the Punpun catchment is about 8,530 sq. km, which is about 1% of the total area of the Ganga basin in the country. The agricultural area under the Punpun River basin is about 5,000 sq. km. Broadly, the geology of the area varies from granite, gneiss, charnokites in the hills to the recent alluvium in the plains. The average annual rainfall for the basin is 1,181 mm. The total gross recharge in the Punpun River system is 1.6 lakh ha-m. However, only 70% of the gross recharge can be utilized.

8.3.8. The Kiul

The Kiul River rises in the Chotanagpur plateau at an elevation of 605 m and flows first in an easterly direction close to the southern base of the Girdeshwari hills and later in a northerly direction. After that it flows towards north-easterly direction up to Lakhisarai and joins Ganga near Surajgarha in the Monghyr district. In the course of its run, the river traverses a total length of 111 km and drains an area of

16,580 sq. km. The Harohar, the Barnar, the Azan and the Ulan are the tributaries of Kiul River, Harohar being the most important.

8.3.9. Gandak River

The Gandak River originates near the Nepal-Tibet border at an altitude of 7,620 m to the north-east of Dhaulagiri and flows about 100 km in a south-easterly direction in Nepal. It is known as Kali in Nepal. After that it debauches into the plains of the Champaran district of Bihar at Trivani. A number of tributaries, such as the Mayangadi, the Bari, the Trisuli, the Panchnad and the Sonhad join the river. The total length of the river from its source to outfall into the Ganga is 630 km of which 380 km lie in Nepal and Tibet. The total drainage area of the river is 46,300 sq. km of which 7,620 sq. km is in India. A barrage has been constructed on this river at Tribeni in Bihar. A canal takes off from this barrage and irrigates an area of 1.5 million ha.

8.3.10. Burhi Gandak

The Burhi Gandak River, known as the Sikrahana in its upper reaches, originates in the Champaran district of Bihar from the springs of the Someshwar hills at an elevation of 300 m. After flowing for a distance of about 56 km, the river takes a southerly turn where it is joined by the Dubhara River and the Tour River. From this point forward, the river takes a south-easterly direction and flows through the Muzaffarpur district for about 32 km. In this reach, the river spills over its banks and a number of spill channels take off and rejoin it later. The river joins Ganga close to the town of Monghyr. The total length of the river is 320 km. The drainage area of the river is 10,150 sq. km.

8.3.11. Kosi River

Kosi is a major tributary of the Ganga River which originates at an altitude of 7,000 m in the Himalayas. The Kosi, known as Kaushiki in the Sanskrit literature, is a frequently referred river in India. It is a perennial stream whose three main tributaries, the Sun Kosi from the west, the Arun from the north, and the Tamur from the east meet at Tribeni to form the Sapt Kosi. The Arun Kosi is the biggest of the three streams. Mt. Everest and Mt. Kanchenjunga, the two highest mountain peaks in the world, lie in the catchment of the Arun Kosi River. The total drainage area of the Kosi River is 74,500 sq. km out of which 11,000 sq. km lie in India.

The Kosi basin lies between 85° and 89° east longitudes and 25° 20′ and 29° north latitudes. In the north, it is bounded by the ridges separating it from the Tsangpo (Brahmaputra) River while the Ganga River forms its southern boundary. The eastern boundary is the ridge line separating it from the Mahananda catchment. The western boundary is the ridge line separating it from the Gandak/Burhi Gandak catchment.

The Kosi catchment can be divided into two distinct parts, one lying in Tibet across the great Himalayan range and other to the south of it. The trans-Himalayan portion is a high plateau which is not as jumbled a mass of hills as the great Himalayan mountain ranges running generally in an eastward direction separated by cross ribs. All the peaks and valleys above 4,900 meters are covered with perennial snow, except steep slopes where its retention is not physically possible. The snow line in winter is roughly at 3,000 meters while during summer it is at about 4,500 meters. The great snowfields that contribute to the river discharge in spring generally lie within 3,000 to 4,500 meters. The area within India has flat topography which is confined in Bihar. Nearly 80% of the Kosi catchment is in Nepal and Tibet. About 77% of the area is under cultivation.

In the plains where Kosi forms the delta, it has a general slope from north to south and west to east, being steeper in the north and flatter in the south. In the north, the slope is 56–75 cm per km but decreases to 6 cm per km beyond Bhaluhi. The entire lower catchment is nearly a plain country only broken by numerous old beds of Kosi. Innumerable depressions are found in plains where water accumulates for the most part of year. The Kosi delta is of conical shape with contours running almost circumferentially with center located in the vicinity of Belka hill. The total catchment area of Kosi basin is 95,156 sq. km of which 20,376 sq. km lies in India. Kamla Balan and Bagamati are two major tributaries on the right side of the river in Bihar.

Tributaries of Kosi River

Bagmati River: Originating in the Shivpuri range of hills in Nepal at an elevation of 1,500 m, the Bagmati River flows in a westward direction draining the Kathmandu valley. The river enters India at Rasulpur in the Muzaffarpur district. The river finally joins the Kosi near the Barna village on the border of the Darbhanga and Sahasra districts in Bihar. The total drainage area of the river is 13,400 sq. km of which 6,320 sq. km lies in India. Kathmandu, the capital city of Nepal, is located on the banks of this river and the famous Pashupatinath temple in Nepal is also situated on the banks of the Bagmati River. Due to the availability of nutritious silt, the fertility of the Bagmati River basin is quite high.

Kamala River: The Kamala River originates in the Mahabharat range of hills in Nepal at an elevation of 1,200 m and flows in a southerly direction. A number of tributaries join the Kamala River during its run. It passes through a gorge above Chauphat and debouches into the Terai area of Nepal at Chisapani. The Kamala enters the Indian territory near Jayanagar in the Darbhanga district of Bihar. In the lower reaches, it follows the course of the Balan and is, therefore, also known as the Kamala-Balan. Flowing in a south-easterly direction, it joins the Kosi on the border of the Darbhanga and Sahasra district.

Problems in Kosi Sub-basin

- *Shifting*: The Kosi River was earlier known as "the river of sorrow" in ancient times because of its frequent shifting. The residents in the Kosi catchment in

Bihar had to suffer a lot of destruction due to the unpredictable lateral movement of the river. During the last 200 years, the river has laterally moved by 112 km from Purnea to its present position. The Kosi project was undertaken in 1954 to prevent the lateral shift. This has been done by confining the river to a defined course with the help of embankments and river training works.

The catchment of the Kosi River and its tributaries are affected badly by severe floods almost every year. As the river with tributaries, Kamla Balan, Bagmati and Adhwara Group emerge from the hills and enter into the Bihar plains, they deposit huge amounts of silt, thereby reducing the capacity of channels and increasing the severity of flooding in the plains of the Kosi basin. After long studies, the Kosi project was taken up to reduce the losses due to floods as well as to increase irrigation facilities in the area. The Kosi project has greatly mitigated the severity of flooding in the Kosi River system.

8.3.12. The Mahananda River

The Mahananda River originates in the hills of the Himalayas in the Darjeeling district of West Bengal at an elevation of 2,100 m. The first 20 km of its course lies in the hills of Darjeeling. Thereafter, it flows in a south-westerly direction forming the boundary between India and Bangladesh. The Balsan, Mechi, Ratwa and the Kankai are the main tributaries of the Mahananda River. The total drainage area of the Mahananda River is 20,600 sq. km out of which 11,530 sq. km lies in India. The river finally enters Bangladesh and joins the Ganga at Godagiri.

8.3.13. The Mayurakshi River

This river has its origin on the slopes of Trikut hills about 43 km upstream from Dumka in Bihar. At its origin it is known as Matihara. In its course towards the south-east, it is joined by a number of rivulets, streams and tributaries. The prominent among its tributaries are Bhurburi, Dhobai, Pusaro, Tepra, Bhamri, Dauna and Sidheswari. The catchment area of Mayurakshi River is 8,530 sq. km.

8.3.14. The Damodar River

The Damodar River rises in the Palamau hills of Choota Nagpur at an elevation of about 609.75 m. It flows in a south-easterly direction entering the deltaic plains below Raniganj. Near Burdwan, the river abruptly changes its course to a southerly direction and joins Hooghly about 48.27 km below Calcutta. Its slope during the first 241.35 km is about 1.89 m/km, during the next 160.9 km about 0.568 m/km and during the last 144.8 km about 0.189 m/km. The total length of the river is nearly 547 km and a total drainage area is 22,005 sq. km. The principal tributary Barkar joins the Damodar after traveling for about 241 km. The catchment area of the river above the confluence is fan shaped and as such is susceptible to the concentration of flood flows but the catchment below the confluence is narrow and has an average width of 16.09 km. The upper catchment is rough hilly areas denuded of forest and vegetal cover and is subject to erosion while the lower catchment is silt covered and fertile. There are no irrigation facilities in the upper catchment and the cultivation is solely dependent on monsoon rains. However the lower portion had irrigation from the Anderson Weir situated at Rhondia on the Damodar nearly 19 km below the Durgapur Barrage.

Damodar basin lies in the states of Jharkhand and West Bengal. Based on the 2001 census, the population of the basin is close to 14.25 million. For this population, domestic demand has been estimated at 338.46 MCM/year. This demand is likely to increase to 507 MCM/year by the year 2021 (Roy et al. 2004). For the industrial sector, the present water demand is 663 MCM/year which is about 15.36% of the utilizable water potential of the basin. This demand is likely to increase to about 884 MCM/year by the year 2021 due to new thermal power plants, brick fields, rice mills, and cold storages etc.

Two irrigation canals, the Left Bank Main Canal and the Right Bank Main Canal take off from the Durgapur barrage. The cultivable command area in the basin is 3,940 sq. km. Water demand for agricultural uses in the Damodar basin for the year 2001 was nearly 652.41 MCM/year. This demand is projected to sharply rise to 1,948 MCM/year by the year 2021. At present, the infrastructure for irrigation is highly inadequate in the basin.

The climate of the area consists of very mild winters and hot-wet summers. During the months of December to March the general flow of surface air is north-easterly. It is of continental origin and is less humid. During June to September, the flow of wind is southwesterly, from sea to land, and this period has high humidity. Transition seasons of the hot weather months of April and May and the retreating monsoon months of October and November lie between the above two principal seasons. In general the months of June to October are known as monsoon season and nearly 90% of the mean annual rainfall of about 1,250 mm occurs in this period. The months of November to May are known as non-monsoon or dry period when rainfall is small. For the five sub-catchments, namely, Tilaiya, Konar, Maithon, Panchet, and Durgapur, the average annual precipitation is 1,117.7 mm, 1,320.8 mm, 1,141.7 mm, 1,142.0 mm, and 1,320.8 mm respectively.

Heavy rainfall occurs generally in the Damodar Valley due to:

1. The monsoon depression from head bay moving in a north west direction over the catchment or west-north-west direction passing near the catchment area., and
2. The land low developing over Gangetic West Bengal or Bihar Plateau and the remaining stationary or moving slowly across or near the Damodar Valley.

In the past, the Damodar River used to cause frequent and immense flood damages in Bihar and West Bengal, so much so that the river came to be associated with sorrow and sufferings. To overcome these, the then government decided to build a series of dams on the pattern of developments in the Tennessee Valley in the USA. Accordingly, five dams were constructed in the Damodar Valley, namely, Maithon, Panchet, Konar, Tilaiya and Tenughat and a barrage was constructed at Durgapur. Further, the Damodar Valley Corporation (DVC) was established in 1948 for management of the Damodar Valley water resources. The dams in the Damodar system serve for different purposes, viz., domestic and industrial water supply, flood control, irrigation, and hydropower generation.

8.3.15. The Ajay River

The catchment of the Ajay River spreads between latitudes 23° 25′ N and 24° 35′ N and longitudes 86° 15′ E and 8815′ E. The Ajay River system originates in the low hills near Deoghar in the Santhal Pargana District of Jhakhand and flows in a south-easterly direction passing through the Monghyr District and Birbhum and Burdwan Districts of West Bengal. Ultimately the river falls into the Bhagirathi River at Katwa about 216 km above Calcutta. The Ajay River system lies between the Mayurakshi on the north and Damodar and the Banka/Khari River system on the south. The habitation in the catchment on the whole is dense and requires a reliable flood warning system which can provide timely warning to the people with sufficient lead time to take precautionary measures which can save lives and their valuables.

The Ajay River traverses a total length of 299 km, 24 km being in Monghyr, 102 km in District Santhal Pargana, 22 km along the boundary of Santhal Pargana and Burdwan, 115 km along the boundary between Singhbhum and Burdwan and the rest of the total length falls in the Burdwan district of West Bengal. It meets the Bhagirathi near Katwa. The river has a catchment area of 6,050 sq. km. The various tributaries of the Ajay River are Darua, Pathro, Jainti, Hinglo, Tumuni, Kane, Kanur and Kundur.

The catchment area of the Ajay River is long and narrow. The river course has remained more or less firm and there is no evidence of marked changes in the course. The bed slope of the river in different reaches varies considerably. The slope of the main river is flatter than that of the tributaries and slope in the plains of West Bengal is much flatter than that in its upper reaches.

The Ajay catchment lies in the path of tropical depressions or cyclonic storms which form in the Bay of Bengal in the monsoon period and move generally in the north-west direction. Abnormally heavy rain spells are generally associated with these storms during monsoon. The average annual rainfall in the river catchment varies from 1,280 mm to 1,380 mm; higher values being associated with hills. Nearly 75% to 80% of the rainfall is concentrated in the four monsoon months of June to September.

The river system is divided into three reaches from the point of view of floods. The upper reach is almost hilly, having comparatively steep slope. The middle reach is subjected to occasional flooding caused due to breaches and overtopping of the embankment during high floods in the river. Whenever the river is in high stage, flood locking of its tributaries takes place, which causes heavy spill on both sides of the rivers. Even during moderately high discharge the lower reach is the worst flood prone reach of the river and it suffers from very frequent inundation.

8.4. THE GANGA DELTA

The delta of the Greater Ganga basin is one of the largest in the world and is known by the name Sunderbans after the Sundari trees which are in abundance in this area. The silt deposits of the delta cover an area of 60,000 sq. km. The river courses in the delta are broad, carrying a vast amount of water. The intense rains in the monsoon season cause most of the delta region in Bangladesh to flood, leaving the villages that are built on artificially raised land isolated. On the seaward side of the delta are swamplands and the Sunderban forests which are protected areas. The peat found in the delta is used for fertilizer and fuel. The water supply to the river depends on the rains brought by the monsoon winds from July to October and the melting snow from the Himalayas during the period from April to June. The delta also experiences strong cyclonic storms before and after the monsoon season which can be devastating.

Long ago, the delta used to be densely forested and inhabited by many wild animals. These days, the delta area is intensely cultivated to meet the needs of the growing population. Consequently, many wild animals have disappeared, although the Royal Bengal Tiger still lives in the Sunderbans. The fish population in the rivers remains high and provides an important part of the inhabitants' diet. The bird life in the Ganga basin is also prolific.

8.5. MAJOR WATER RESOURCES DEVELOPMENT PROJECTS IN THE GANGA BASIN

The development of water resources projects in the Ganga basin has a long history. There are a large number of water resources development projects in the basin. A few projects are also being planned in collaboration with the Government of Nepal. In order to boost the utilization of waterways, some segments of the Ganga River have been declared as national waterways. A description of the major works is given in what follows.

8.5.1. Ganga Canal Systems

i. Upper Ganga Canal (UGC)

A remarkable work that includes many great civil engineering structures is the Upper Ganga Canal system whose construction was initiated by the then British government in the 1840s to ward off drought and famine in the western part of the current the Uttar Pradesh state. Northern India experienced a famine in 1837 and 1840 when, according to estimates, the population of the area fell by 20%. Completed in 1854, UGC was to irrigate an area of 0.7 million ha. This canal was the reason why the first engineering college in India, the Thomason College of Civil Engineering was set up at Roorkee, close to Haridwar. In due course of time, this college was upgraded to an engineering university and then converted into an IIT. Today, Roorkee is the home to a number of engineering organizations dealing with water.

At the time of commencement of UGC Project, there was no science of soil mechanics and hydraulic engineering was also in its primitive stages. There was no existing large extensive canal system in the country and the world by that time. In formulation and execution of the project and in projecting the layout and design of main canal and its major structures, Sir Proby Cautley adopted the simple scheme of taking the help of nature and maintaining a compromising attitude, specially in taking the canal across the powerful streams and rivers coming in the alignment and also used his engineering skill in producing simple and bold designs with locally available material (brick, lime and surkhi). Although in the first project report, UGC was proposed as a navigation canal, finally it was constructed as an irrigation canal.

UGC draws its supply from the Ganga River at Haridwar. Prior to the construction of the permanent head works, water was being forced into the canal by means of temporary weirs constructed of wooden crates filled with boulders. Originally, the width at the head was 61 m and the depth at full supply level nearly 3.3 m. Sir Cautley placed the head of the canal at Mayapur, a place well settled and safe from the attack of Ganga River. Currently, the head works are situated at Bhimgoda. Figure 12 shows a pictorial view of Haridwar.

The reach of the canal from the head works to 32nd km may well be classified among the greatest feats of irrigation engineering in India. The alignment traverses numerous drainages track and the arrangements for negotiating these drainages are complicated by the fact that the canal bed is at places far below and at other far above the general level of surrounding country. Four large torrents are crossed in this reach while several smaller ones are admitted into the canal. The steep slope of the country moreover necessitates the negotiation of over 18 m of bed fall in those first 32 km. UGC flows in deep cutting up to the 19th km. In the 10th km, it encounters the Ranipur torrent which is carried over it on a masonry super-passage. The Pathri torrent is passed over the canal on second super passage.

At the 21st km, the Ratmau torrent is passed across the canal, being admitted directly into the channel on one side and escaped again through a weir. At the 30th km comes the Solani aqueduct which is the finest work on the canal (see Figure 13).

Figure 12. A pictorial view of Haridwar. Ganga is flowing in the upper part and the UGC at the middle

Figure 13. Old Solani aqueduct of the Upper Ganga Canal near Roorkee

Solani aqueduct was constructed with the special inspiration of the great work of Alcantara aqueduct in Portugal, constructed during the years 1,713 to 1,732 and universally accepted to be a stupendous monument of modern art and engineering in Europe. At the time of commissioning Ganga canal in 1854, Solani Aqueduct was ranked as one of the most remarkable massive structure of brick masonry in the whole world.

It cannot, however, be pretended that the design of UGC was perfect by any means. Indeed no sooner was the full supply of the channel first admitted, several serious faults became apparent. In many places, the bed slope was too steep, the masonry falls generated excessive velocities below them, causing heavy scour in bed and banks, and the general layout of the distributary system had some drawbacks. Considerable amount of remodelling was necessary before these defects were removed, but in other respects, the work even now remains substantially as Cautley built it. Noteworthy is the fact that at the time it was constructed, experience of artificial canals was limited to works of only about a third the size of the Ganga Canal. But the mistakes appear to be insignificant when compared with the enormous advances in hydraulic engineering, which the scheme represented. The engineering judgment of planners and designers of UGC was truly marvelous.

The Upper Ganga Canal system is a leading irrigation system in India. It extends over an area of 24,000 sq. km bounded by natural or man made water courses. The Ganga River is on the eastern side, the Hindon River and the Yamuna River on the western side, and the lower Ganga canal on the southern side. The Upper Ganga Canal takes off from the Ganga at the Bhimgoda weir near Haridwar. The main canal is 290 km long and is one of the most exquisitely constructed civil engineering structures. In view of expansion of the command area, the discharge capacity of the canal was recently augmented to 297 m^3/s. The command area of the canal is now about 0.924 million ha.

The command area of the system is located between 27° N and 30° N latitudes and 77° 15′ E and 78° 40′ E longitudes covering the districts of Saharanpur, Muzaffarnagar, Meerut, Ghaziabad, Bulandshahar, Aligarh, Mathura, Agra, Etah, and Mainpuri. A major project to revitalize the head reaches of the canal was completed in 2003. A parallel canal has been constructed upstream of Roorkee and a new aqueduct has been constructed over the Solani River.

A number of small rivers, such as Kali, Karwan, Solani and West Kali, flow and interact with the ground water system in the canal command. The Upper Ganga Canal system has a large expanse and the main canal as well as its branches are an important source of recharge as they all are unlined. The extraction from ground water is through pumping by public and private tubewells.

UGC system is fed through a headwork complex with a regulation at Mayapur and a diversion weir at Bhimgoda across the Ganga River. The important branches of the system are the Deoband branch (taking off on the left bank at 35 km), the Anupshahr branch (taking off on the left bank at 80 km), the Mat branch (taking off on the right bank at 177 km), and the Hathras branch (taking off from the Mat branch on the left bank at 80 km). The system is unlined and has a network of 115 distributaries. The canal system is connected with natural drains/rivers to discharge the surplus water of the canal.

Climatically the area belongs to a dry sub-humid to moist-humid category. The normal annual rainfall varies from 1,050 mm in the north to 650 mm in the south. Around 90% of annual rainfall occurs in the monsoon season (June to October). The annual pan evaporation for the area is about 150 cm. The temperature varies from 3° C to 4° C in January to 43° C to 45° C in May or June.

The scope of the system has been considerably altered since it was first constructed. The Lower Ganga Canal (LGC) which was opened in 1878, intersected by the main branches of the UGC, the Etawah and Kanpur branches and the tail portions of these are now officially included in the LGC system. In their place, however, three new important branches, the Deoband, Mat branches have been added to UGC. In the length of its channels, the UGC is still largest in India. The system comprises 910 km of main canal and branches and 5,280 km of distributaries or 6,190 km of channels in all.

There are many small hydropower plants on UGC which utilize its falls to generate power. A hydroelectric power station was built on the Ganga canal at Bahadrabad, 16 km from the site of the head works, where a fall of 5.8 m was available. Incidentally, a famous research station is situated at Bahadrabad which is involved in model studies of WRD projects. Another station is situated at Mohammadpur near Roorkee.

ii. Madhya Ganga Canal

The Madhya Ganga Canal takes off from the Ganga at the Raoli Barrage, about 11 km east of Bijnor in the U.P. state. Raoli Barrage, completed in 1994, is 583 m long; the normal water level of the pond is 221.5 m. The capacity at the head of this 115 km long canal is 234 m^3/s and its distributaries are 1466 km long. The usage of water is: Anupshar branch 25.5 cumec, Lakhaoti Branch 63 cumec, Upper Ganga

Canal 58 cumec, Parallel Mat Feeder 74 cumec, and losses 13.5 cumec. Madhya Ganga Canal provides irrigation to paddy crop in 114,000 ha as well as augments supply to the UGC system. The Lakhoti branch canal takes off at chainage 82 km from the Madhya Ganga Canal. Its discharge at the head is 63 cume and it is 74 km long. Minors from this canal provide irrigation over about 192,000 ha land in Neem-Kali Doab in Aligarh and Bulandshahar Districts.

Rivers in many parts of India carry enormous water during rainy season which can be used for irrigation for Paddy and to recharge ground water aquifers. This gave rise to the concept of Monsoon canals. Several such canals have been planned and constructed, viz., (i) the Eastern Ganga Canal (ii) the Madhya Ganga Canal, and (iii) the Parallel Lower Ganga Canal. These canals are examples of conjunctive use of water.

iii. Lower Ganga Canal (LGC)

To irrigate the lower portion of Ganga – Yamuna doab, a project was sanctioned in 1872. Work commenced in that year for constructing a new canal from the Ganga River with head further down the river. LGC system comprises a weir across Ganga at Narora (near Aligarh), some 6 km below Rajghat, and the canal takes off from the right bank of the river. The weir, which is 1,158 m long, is fitted with falling shutters and enables the level of the normal cold weather supply of the Ganga to be raised by 3.05 m to feed the canal. Under sluices, consisting of 42 vents each of 2.14 m span are provided on the right flank of the weir. The canal head which is set at right angle to the sluices has 30 bays of equal width.

The main canal is 100 km long and irrigates 0.5 million ha. The Lower Ganga Canal has a discharge capacity of 156 m^3/s. It was completed in 1879. It serves the districts of Mainpuri, Etah, Farrukhabad, Etawah, Kanpur, Fatehpur and Allahabad in central U.P. These canal systems are irrigating a large area of the Ganga-Yamuna Doab. But, of course, there are still large tracts of culturable area which do not have irrigation facilities.

The LGC system has 1,060 km of main canal and branches and 5,015 km of distributaries. In fact UGC and LGC form a single system. A considerable proportion of channel comprised in the lower system belonged originally to the upper and a supply of water is regularly passed from the later to the former. Viewed in this light, the Ganga canals form the largest irrigation system in the world. The length of the channel, contained in it, is no less than 12,240 km and it irrigates an area of the size of one million ha.

A map showing network of these canals is given in Figure 14.

iv. Agra Canal

The committee which proposed the construction of LGC, also suggested that a canal should be taken off from the Yamuna at or near its junction with the Hindan river below Delhi, to irrigate some more area in the Ganga-Yamuna watershed. Investigation showed, however, that the district lying on the left bank of river Yamuna were already or could be fairly commanded by the Ganga canal through Mat branch, whereas on the right bank, there existed a very precarious tract which

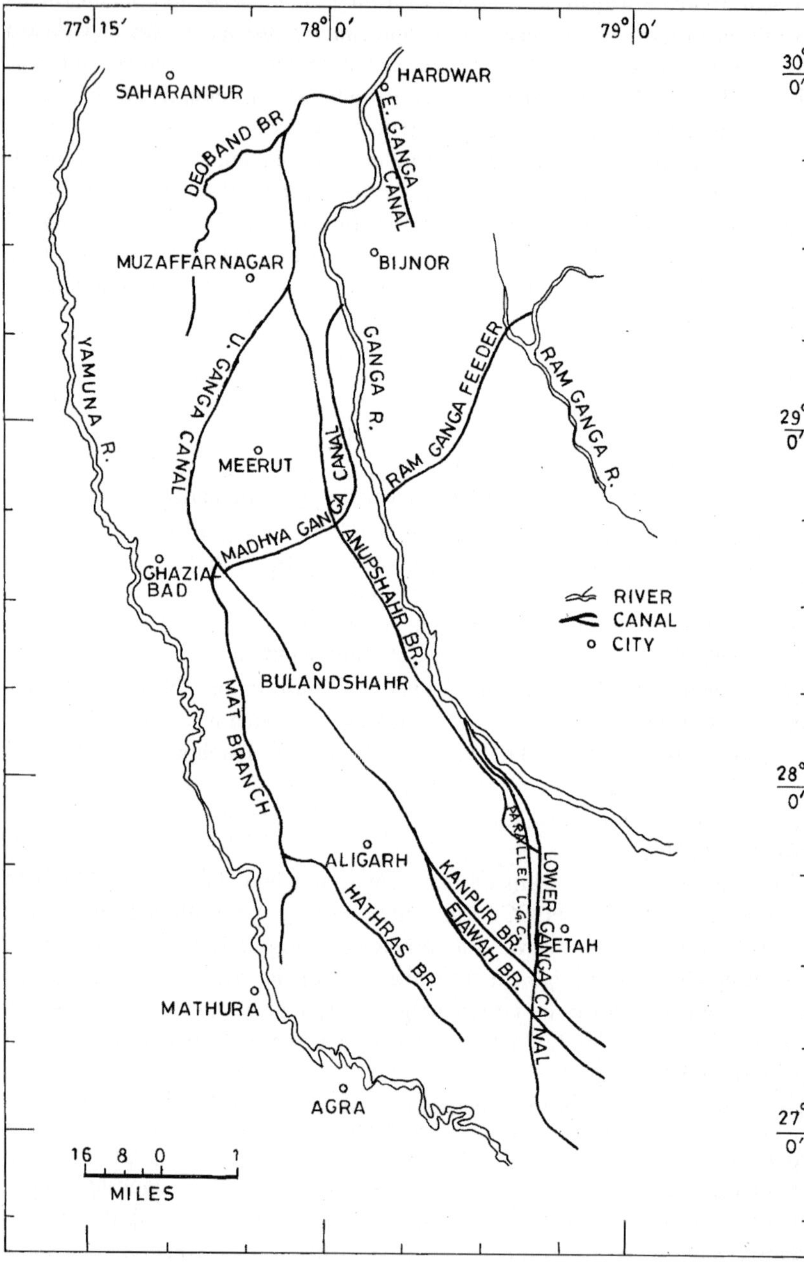

Figure 14. Network of canals in upper Ganga basin

was greatly in need of irrigation. The site chosen for the head works was at Okhla, 11 km below Delhi and upstream of the junction of the Yamuna and Hindan. To make the water of the Hindon available for the canal, a regulating weir, consisting of 39 vents each of 3.2 m span, was constructed across that river, by means of which its discharge can be diverted through an artificial channel, known as Hinan cut into the Yamuna immediately above the Okhla weir. A further supplementary volume can be obtained from the Ganga canal via the Jani escape, which connects the Ganga Canal with Hindan at point 48 km above the site of the diversion works.

The canal was constructed for a cold weather full supply discharge of 30.8 m^3/s with a depth of 2.13 m and for a discharge of 56.0 m^3/s during the rainy season with a depth of 3.05 m, the width at the head being 21.4 m. It terminates at its 116th km at which point the Agra Navigation channel, 25 km long, connected it with Yamuna at Agra. The distribution system comprises of 1,440 km of distributaries.

In the first 13 km of its course, the canal crosses three important torrents, which bring down to the Yamuna the drainage of the rocky hills on the right bank of the canal. The floods from these hills rise rapidly but are of short duration. To obviate the necessity of large drainage crossing, embankments are provided for two of them on the right of the canal. These reservoirs have storage of 4.5 MCM. The floods are received by them and the water stored until it can be passed out gradually. The third torrent is carried under the canal through a large siphon.

v. East Ganga Canal Project

The East Ganga Canal Project envisages the utilization of surplus water of the Ganga River during the Monsoon season from the existing Barrage at Bhimgoda Haridwar for providing irrigation to 105,000 ha of paddy crop, mainly in the Bijnor district (99.64 thousand ha), the Haridwar district (360 ha), and the Moradabad district (5,000 ha). The gross command area of the project is 3.01 lakh ha, out of which 2.33 lakh ha is cultivated. The proposed intensity of irrigation in Kharif is 45% which will produce an additional amount of 36.0 lakh quintals of paddy crop.

Before introduction of this project, the area was being irrigated by perennial supplies and by the Upper Ganga and Lower Ganga Canals. The original project was planned in the year 1976 for an estimated cost of Rs. 48.46 crore and the construction work was started in the year 1980–81.

The soil in the command of the East Ganga canal is generally light loam, except in northern areas of the Malin-Chhoya, Chhoya-Ban, Ban-Ganga and Ganga-Kho doabs falling in the Najibabad, Kiratpur and Kotwali blocks. The loamy soil in most of the blocks is highly suitable for rice cultivation and similar soils in the Mirzapur and Saharanpur districts produce bumper harvests of paddy. Irrigation can develop easily as soon as water is made available in abundance. At present the existing sources of irrigation in the Bijnor district are mainly the state tube wells, private tube wells and a small canal system called the Bijnor canal group.

The East Ganga Canal command has monsoonal climate. Monsoon generally starts in the last week of June and lasts up to September. The mean annual precipitation at Bijnor is 1,073 mm. About 90% of the total annual rainfall is received during June to September and the rest 10% in the remaining months. The winter rains are insufficient to meet the water requirement of crops in the area and there is always a need for irrigation water to grow Rabi crops. Uncertain irregular rain causes floods in the area.

8.5.2. Yamuna Canal Systems

The Yamuna River emerges from the hills near Tajewala where water is diverted for irrigation by the Western and Eastern Yamuna Canals. The Yamuna flows further 280 km down to Okhla near Delhi from where the Agra Canal takes off.

Eastern Yamuna Canal

The earliest canals in Northern India of which any record exists are the old canals that take off from both banks of the Yamuna. The Eastern Yamuna canal was originally constructed during the Mughal dynasty, probably in the reign of Muhammad Shah (1719–1748 A.D.), and a Royal preserve at Ranup, on the left bank of the Yamuna, is said to have received water from the canal. The works were, however, soon abandoned. They were partially restored by one of the Rohilla Chiefs in about 1780 A.D. and water was brought to Saharanpur. Probably the extended canal ran during one season at most. It was not provided with masonry works and its excessive bed slope would have led to total retrogression. The canal was later realigned by the British early in the 19th century. As was the usual practice in the case of old works of this nature, the courses of rivers and drainage lines had been utilized as much as possible and all high ground carefully avoided. The alignment of the canal, as remodeled by the British, followed this old practice to a considerable extent but a large number of masonry works were introduced.

The canal has its head on the Yamuna on its eastern bank, a point not far from the head of the Western Yamuna canal. Originally, the supply required was directed by means of temporary spurs into the eastern channel of the river and delivered at Naushera, where the excavated canal commenced and where arrangements were made for surplusing any water not required back in the river. When the canal was opened in 1830, it was realized that the gradient was too much steep, rapids forming in the bed at many points seriously threatening the masonry works. To reduce the slope, masonry falls were constructed at suitable intervals and channel was in part realigned. The distribution system had also a faulty design and was operated and had gradually to be remodelled. Many of the evils experienced in the Western Yamuna Canal repeated themselves in the Eastern Yamuna Canal system such as:

i) The bad alignment of the main canal and of its distributaries

ii) The over irrigation of certain areas led to water logging here also.

The construction of the weir for the Western Yamuna canal at Tajewala in 1875–79 resulted in a modification of the methods adopted for obtaining the supply for the Eastern Yamuna canal. A new head and regulator for the latter was built in connection with the weir and supply channel excavated from Tajewala to Naushera, the volume in the river was divided between the two canals in fixed proportions, the Government of the United Provinces making an up-keep of the weir and its connected works. The salient features of Eastern Yamuna canal are given in Table 8.

Table 8. Salient features of Eastern Yamuna Canal

Length of main canal	207.561 km
Length of distributaries	1,287.20 km
Canal capacity	84.95 cumec
Command area	1, 618.8 Mm^2

Western Yamuna Canal

Emperor Feroz Shah Tughlaq constructed the Western Yamuna Canal (WYC) to divert water to the hunting grounds in Hansi-Safidon area in Haryana in 1355 A.D. After the death of Emperor Tughlaq in 1388, the canal fell in disuse. Mughal Emperor Akbar got the canal renovated in 1568. It was further improved by Emperor Shahjahan in 1628. Note that the Hasli canal was also constructed at the same time to bring Ravi water to Lahore; it was later extended by the Sikhs to bring water to the sacred tank at Amritsar (CBIP 1965). During the British period, the canal was realigned and repaired from 1821 onwards. After a long lull in fresh water development, GW development in the area began in 1960s. It was noticed that there is decline in GW level as well as decline in river discharge. Conjunctive use of SW & GW is being practiced in the area by completion of augmentation canal project.

Upper Yamuna Basin covers an area of 25,000 sq. km. In the canal area, tropical climate dominates. In summers, temperature can rise as high as 47° C; during winters, they can go as low as −1° C. The average rainfall in the area is 746 mm/year; 84% of it falls during summer monsoon. WYC command area is located between the north latitudes 28° 20′ & 30° 28′ and east longitudes 75° 48′ & 77° 35′.

The total length of the WYC with all its branches is 325 km. In addition, there are about 32 distributaries and 95 minors, the combined length of which is 1,220 km. To augment canal supplies and prevent water logging in adjacent tract, a large number of augmentation wells were constructed along WYC. Further, to prevent seepage losses along WYC and to further augment its supply, a lined augmentation canal, 69 km in length was constructed recently, taking off from Yamuna Nagar and out falling in WYC at Munak. Heavy-duty wells constructed along this canal direct about 14–15 cumec of ground water to surface water canal system. In addition, there are similar augmentation wells constructed along Delhi parallel Branch and the Narwana Branch Karnal link.

In the entire reach between Tajewala to Wazirabad the river behaves as a mature river and meanders within the flood plains, the width of which increases as the slope decreases. The maximum width of the flood plain is around 3 km. In order to protect the habitations and cultivated area from flood protection, embankments have been constructed along the river, more of them along the right bank.

Ground Water Situation: The area of mountainous sub-basin is underlain by Siwalik and older formations which are semi-consolidated and consolidated in nature and are composed of sand rocks, shale and boulder conglomerates. These are poor repositories of ground water and effectively make no ground water contribution to sub-basin. The plain tract lying south of the Siwalik zone forms a part of

Indo-Gangetic alluvial plains of recent origin. The thickness of the alluvial deposit (as deduced from geophysical evidence) is small along the fringe of peninsular mass but progressively increases towards northwards and is maximum in the foredeep area lying immediately south of the Himalayan zone. The alluvial plains are underlain by loose unconsolidated river borne sediments and form very good repository of ground water. Below a particular level, not far below the land surface, the alluvium is saturated with ground water. Water table occurs between 10 to 60 m BGL in the sub-mountainous tract but generally lies between 3 m to 10 m BGL in rest of the area. The aquifer system lying closest to the land surface is in unconfined condition. At deeper levels, particularly below regionally or sub-regionally extensive poorly permeable layers, the ground water occurs in semi-confined to confined conditions. It is expected that with increasing depth, the alluvium could get more and more consolidated because of the increasing over burden and hence have reduced porosity and permeability. The unconfined aquifer, which is quite potential, generally bears an effluent relation with the surface drainage.

Major Aquifer systems: Aquifers have been demarcated in this area based on exploratory drilling and bore-hole logging. Except in areas close to Delhi where the bedrock occurs at a shallow depth, four distinct groups of permeable horizons were identified.

The Aquifer Group 1 (AG1) (composed of different sand and clay layers) extends from water table to 167 m below ground level (BGL). This is composed of relatively coarser sediment and contains fresh formation water. In general, this aquifer behaves as unconfined aquifer displaying delayed yield phenomenon. The effects of leakage are negligible. AG1 is underlain by a clayey horizon 10 to 15 m which is regionally extensive except around sites Nagal, Tikaula, Labhkari, Newal and Ambheta. At these sites, AG1 seems to be directly in contact with Aquifer Group 2 (AG2). AG2, composed of different sand and clay lenses occurring at variable depths (ranging from 65 to 283 m BGL) has distinguishing characteristics of its own at places and at others is separated because of distinguishing features of the overlying and underlying groups. The sediment of this group is less coarse and kanker is encountered at sites located west of river Yamuna. The quality of water is reasonably fresh except in south-western parts. Seven successful long duration pumping tests with observation wells were conducted. The aquifer behaved as confined at 3 sites and as semi-confined at rest 4 sites.

AG2 is underlain by another clayey horizon which is considerably thick at places and appears to be regionally extensive except around sites Sikka and Kheri. A local aquifer group of limited thickness is found enclosed within thin clay around sites Nitli and Baleri. Aquifer Group 3 (AG3) comprises of thin sand layers alternating with thicker clay layers occurring at variable depths varying between 197 m to 346 m BGL, underlies the above groups. The granular material of this group is generally finer in texture and more so in southerly direction. Kankar occurs in southern parts. The parameters of aquifer groups of WYC area are given in Table 9.

AG3 is underlain, in turn, by a thick clayey horizon which in turn is underlain by another permeable granular horizon. Aquifer Group 4 was not fully penetrated.

Table 9. Parameters of Aquifer Groups of WYC area

Parameter	Gen. range	Average values
	Aquifer Group – 1	
Transmissivity (m^2/day)	800–5,210	2,200
Lateral Hydraulic Conductivity 'K' (m/day)	14–47	24
Specific yield (S_y)%	6–24	12
	Aquifer Group – 2	
Transmissivity (m^2/day)	750–1,050	700
Lateral Hydraulic Conductivity 'K' (m/day)	4–11	7.2
Storativity	5.6×10^{-4} to 1.7×10^{-3}	1.0×10^{-3}
Specific yield (S_y)%	3.35×10^{-4} to 2.7×10^{-3}	1.9×10^{-3}
	Aquifer Group – 3	
Transmissivity (m^2/day)	345–830	525
Lateral Hydraulic Conductivity 'K' (m/day)	3.5–10.7	7.1
Storativity	6.6×10^{-4} to 2.4×10^{-4}	4.5×10^{-4}

Behaviour of Ground water: The master slope of water table is from north to south with lateral slopes away from the ground water ridges. There is a prominent ground water trough roughly along the river Yamuna, but this departs Westwards in areas between Panipat to Delhi. A general down-valley shift of contours in post-monsoon period in response to recharge can be observed.

The area between the Eastern Yamuna and Western Yamuna canals south of latitude 30° behaves as a hydro-geological unit with no flow across boundaries but some inflow from north. The piezometric levels of the lower portions of the AG1 show similar behavior to that of the phreatic surface with minor departures. It appears that there is flow in the vertical direction within AG1; and in parts between Yamuna and WYC this is found to be upwards. No GW flow occurs in AG1 from across the eastern boundary but generally outflow takes place from Western boundary. Substantial vertical leakage is expected to be occurring between AG1 and AG2 and much of the flow converging towards the trough may be leaking upwards to AG1.

In general in the submontaneous tract and the areas along the major canals and part of the central area in the south, AG1 has potentiometric head of 0 to 4 m above the AG2. All the observations fit well with the fact that both AG1 and AG2 receive recharge in Bhabhar tract and areas along the major canals and discharge the same to Yamuna drainage.

In major part of the area between the canals, AG3 and AG4 have a higher head over AG2. However, only in southern parts of AG3 and AG4 are likely to provide leakage to AG2 above. The study of daily hydrographs of river levels along with those of ground water levels in the river tract for one year indicate a general sympathetic behavior and a good hydraulic connectivity of the river with ground water storage along the banks. It was concluded from the study that influent–effluent relation of river with the areas in immediate vicinity of the two banks changes with places and times and may not always coincide with the regional relation of ground water body with the river, on either bank.

The soils in sub-mountainous region are classified as reddish soils. South of this zone roughly up to Panipat the soils are classified as 'Tropical Arid Brown'. In area south of Panipat the soils are arid brown soils. The soils in the basin, west of Yamuna River are mostly sandy loam to loam with medium permeability. In Karnal district considerable sedimentation and alkalinization of the soil has occurred due to rise in water table by canal seepage and from excessive use of irrigation water.

The area is predominantly an agricultural tract, with about 76% area under cultivation. The major crops grown in the area are wheat (58% area), rice (20% area), Gram (10% area), Maize (13% area), Jowar (4% area), Barely (3% area), Bajra (7% area), and sugar cane (19% area).

Betwa Canal

The sanction of the estimate of the Betwa canal in 1881 marked the opening of yet another and a very important era in the history of irrigation works in India, namely the era of protective works or the works designed primarily for the protection, of precarious areas against famine, the direct return obtainable from them being the secondary consideration.

The Betwa canal was the first protective work in India which was constructed in the United Provinces. A report was prepared in 1868 which established the practicability of such a canal for irrigation of the triangular area in the Jalaun district formed by three rivers: Yamuna, Pahuj and Betwa. The subsoil water throughout this tract was at immense depth. The project was completed in 1893.

The head works of the canal are situated on the Betwa River near Parichcha, 27 km from Jhansi. The river at this point has a discharge of 23,000 m^3/s. Canal regulator has five bays. This weir forms a reservoir in the river channel and impounds 48 MCM of water at crest level. In 1899 automatic shutters were erected which increased the capacity of the reservoir to 61 MCM.

The main canal is 30 km long. At its termination it bifurcates into Hamirpur and Kathound Branches. In 1904–05, the Kathound branch was remodelled to carry discharge of 17.0 m^3/s. But the greatest weakness of the canal lay in the general inefficiency of its cold weather supply and in 1905 works was commenced on the construction of a supplementary reservoir at Dhukwan 40 km above the Parichha. The Dhukwan weir is 1,196 km long and has maximum height of 17.4 m. The work was completed in the year 1910.

Dhasan Canal

The Dhasan and Bearma are tributaries of Betwa, flowing from the east. The canal lies in Hamirpur district between a triangular area made by above rivers. The project was sanctioned in 1905. Two dams have been constructed upon Dhasan River, the upper one at Pahari, the lower at Lachaura some 11 km further down. Both dams have same maximum height of 16 m, the Pahari dam is 580 m, the Lachaura dam 542 m long. Gates, 2.5 m high, are erected on the crests and effective storage of

78 MCM at Pahari and 15 MCM at Lachaura are thus obtained. Both dams are of concrete, with masonry facing up and downstream.

The Dhasan canal has a head discharge of 20 cumec and a bed width of 13.7 m. With its 3 branches, it has a total length of 170 km and feeds 300 km of distributaries. There is only one masonry work of importance on the main line the Kohina Nala aqueduct. The canal was opened for irrigation in 1910.

Ken Canal

To irrigate the watershed between the Ken and the Bhagin, the Ken canal has been constructed. It was sanctioned in 1903 and came into operation in 1908. It consists primarily of a weir across the Ken at Bariarpur, some 100 km south of Banda, a main canal 59 km long, and two branches with a connected distributory system.

The Bariarpur weir has a crest length of 512 m and a maximum height of 8.0 m above the solid rock on which it is founded. The weir is capable of impounding 14 MCM of water. The canal takes off direct from this reservoir and is designed to carry a normal supply 22.5 cumec, which can be increased to 28 cumec at time of intense demand. The steep slope of the country necessitated a large number of masonry falls – there are 22 such falls in the first 13 km of the main canal. Two principal works for cross drainage being Majhgawan and Mawapura aqueduct.

To supplement the supply in canals, the Gangao dam has been constructed. This dam is situated on the Ken River some 50 km above the head works of the canal. The work was taken up in 1911 and completed in 1917. The dam is of masonry, of the same design as those in the Dhasan and Bariarpur. It is 740 m long with a maximum height of 16 m and is capable of 76 MCM of water at crest level. Ken is very formidable river, carrying in flood a discharge of 17,000 cumec and consequently both at Bariarpur and Gangao the works are on a scale of considerable magnitude. The canal system has 138 km of main canal and branches and 413 km of distributaries. In addition to large works smaller storage schemes were also provided in the Bundelkhand area.

Ghagar Canal

Construction of the Ghagar canal was commenced near the end of 1912 and completed in 1918. The main feature of the scheme is the masonry dam at Dhandraul, which has been constructed across a gorge where Ghagar River pierces a low line of hills on its way to join the Sone. This dam, which is 305 m long and 20.72 m high forms in conjunction with an earthen embankment, 5.23 km long of a mean height of 7.6 m, a reservoir capable of storing more than 140 MCM of water.

Two low saddles in the hills provide means of escape with their crests at the full supply level of the lake while the third, which forms the main escape is divided into 12 bays each of 6.1 m span. A supplementary dam, 152 m long and 13.7 m high, on the neighbouring Karamnasa River diverts the water of that river also into the reservoir through the so called Karmnasa cut, thereby supplementing the supplies in the Ghagar. From this reservoir, the canal system consisting of 100 km of main canals and branches and 120 km of distributaries is fed. The canal is crossed by 48 drainages. The steep slope of the country necessitated the provision of numerous masonry falls in the beds of various channels.

Sarda Canal

The scheme consists of two parts: the Sarda canal proper and the Sarda Kichcha Feeder which leaves it at about 11 km. The former comprises a comprehensive project for irrigation of the north western district of Oudh, while the latter assures a supply to the extension of the existing Rohilkhand canals. The head works and the first 11 km of the canal are common to both. Thereafter, the Sarda canals runs in a southerly direction, while the feeder flows through the Tarai, the low-lying land at the foot of Himalayas.

The head works of the combined project are situated on the Sarda River a few km below the point where it debauches from the hills. At this place, the river forms the boundary between (British) India and Nepal; the government of Nepal had courteously consented to a small exchange of territory so as to permit the British Government jurisdiction over the land upon which the left abutment of weir and left bank training works are situated. A channel is built on the left flank to irrigate certain area of Nepal. The weir and sluices have been designed to pass a maximum flood of 11,300 cumec. The head regulator and first 11.26 km of combined canal were constructed to carry 233 cumec.

The Sarda canal proper, below bifurcation consists of main canal with a length of 28.15 km, after which it bifurcates into three branches. The project comprises 769.10 km of main canal and branches, 5,422.33 km of distributaries and 160.90 km of escapes or 6,352.33 km of channel in all.

The Upper Sarda Barrage is located in Banbassa of Nainital district for purpose of directing water in Sarda main canal for irrigation and power generation. Design flood discharge was 16,900 m^3/s. Length of barrage is 598 m and it has 4 under sluice bays. The barrage was completed in 1928.

The 1920 Sarda river agreement between the British Indian Government and Nepal guaranteed 11.3 m^3/s for the summer *(Kharif)* irrigation and 9.25 m^3/s for the winter *(Rabi)* irrigation. Trishuli and Devighat hydroelectric projects on the river Trishuli, Chatra canal and renovation and extension of the Chatra canal were completed by the Government of India, which bore the complete costs of these projects. At least on the Indian side it is felt that all these projects, fully funded by India, have provided large benefits to the two countries.

8.5.3. Tehri Dam Project

The Ganga River system has substantial hydropower potential, besides being a vast source for irrigation in the Gangetic Plains. The hydropower potential of the Ganga and its tributaries has been assessed as 10,715 MW. So far, very small amount of this potential has been exploited. There are no storages on the main river and its two arms Bhagirathi and Alaknanda. To harness the hydropower potential of the river, a major storage dam is being constructed on Bhagirathi at a point 1.5 km downstream of Tehri town in the State of Uttaranchal. To meet the peaking power needs, another dam is planned downstream of Tehri at Koteshwar. It will cater for pumped storage operation also. The entire Tehri Hydro Power Complex, as envisaged, would have an installed capacity of 2,400 MW and would provide substantial benefits by way of irrigation and domestic water supply.

The Tehri dam project was first conceived in 1949 and was sanctioned by the Planning Commission of India in 1972 when approval of the proposal for a 260.5 m rock-filled dam at Tehri was given in. Tehri dam is located on the outer Himalaya in the Tehri-Garhwal district of Uttaranchal in the Ganga Basin. The dam will be the fifth highest dam in the world with a height of 260.5 meters and spread over an area of 45 sq. km in the Bhagirathi and Bhilangana valleys near Tehri town. The dam, when finished, will completely submerge the Tehri town and nearly 100 villages will be totally or partially submerged. Nearly one lakh persons have been displaced and resettled because of the dam and a whole new city, known as New Tehri has been developed. The Government of India sanctioned the project in 1972. It is a multipurpose project, the main purposes of the dam being hydropower, irrigation, flood control, fisheries & tourism. The designed annual irrigation from the reservoir is 2,700 million sq. meter. The Tehri dam on the Bhagirathi (under construction) will provide a live storage capacity of 2,613 million cubic m (gross storage capacity of 3,540 million cubic m) to be used for power generation and irrigation. The project is being executed by The Tehri Hydro Development Corporation with its headquarters at Rishikesh. A view of Tehri dam is given in Figure 15.

The average river flows in the Ganga near Haridwar in monsoon season (July to Sept.) range between 2,000 to 3,000 cumec and the lean flows are about 100 cumec. The average river slope up to Haridwar is 20 m/km. The salient features of the project are given in Table 10. The main features of the project are:

- A 260.5 m (above the deepest foundation) high earth and rockfill dam creating a live storage of 2,615 Mm^3.
- An underground powerhouse of 1,000 MW(4×250 MW) with conventional turbine generating units.
- Another underground powerhouse of 1,000 MW(4×250 MW) with reversible pump turbine set (pumped storage plant).
- A 103.5 m high concrete dam (which will function as a balancing reservoir) with a surface power house of 400 MW(4×100 MW) at Koteshwar about 20 km downstream of Tehri dam.
- A transmission system for evacuation of power generated at Tehri and Koteshwar through 765 kV lines to Meerut.

The project would generate about 4,300 million units of energy on the 90% dependable water availability and about 5,000 trillion units on the average water availability. The peak load power generation is 2,400 MW. Besides, the project will provide irrigation facilities to 2.7 lakh ha of area in the command of existing canal systems off taking from the Ganga downstream of Haridwar. The project will also provide about 10 cumec of water to Delhi to meet its future domestic requirements.

Figure 15. A view of Tehri dam (under construction)

The total cost of project including the cost of transmission system but excluding the cost of modernizing and extending the canal distribution system is about Rs.5,060 crore.

The Tehri dam project area is seismically active and falls in Zone-IV of the seismic zoning map of India, which corresponds to intensity VIII on MM Scale. Most of the past earthquakes have magnitudes of 5–7 on the Richter scale. The dam is designed for a probable earthquake of magnitude 8.0. A site specific assessment of seismicity has been made for detailed designs.

Catchment Area Characteristics

The catchment area at the dam site is 7,511 sq. km. Out of this 2,323 sq. km is snow bound. This catchment varies in elevation from 9,600 m to 600 m in a length of 187 km. The valley is narrow and moderately forested. The catchment area below the perpetual snow line (4,880 m) is divided into following the three types:

i) Agriculture land (1,240 sq. km).
ii) Forest land (2,400 sq. km).
iii) Soyam land lying between the above two types being mainly used by villagers for cattle grazing, etc. (1,500 sq. km).

The spread of Tehri dam reservoir will be about 42 sq. km at the full reservoir level of 830 m. The length of reservoir along the Bhagirathi River is 44 km and along the Bhilangana River, it is 25 km. The Tehri dam will affect 22 villages fully and 74

Table 10. Salient features of the Tehri Dam Project

1. Reservoir	
M.W.L.	835 m
F.R.L.	830 m
Dead Storage Level	740 m
Gross Storage	3,540 MCM
Live Storage	2,615 MCM
2. Main Dam	
Type	Earth & Rockfill
Top Level	839.50 m
Height	260.50 m above deepest foundation level
Width at river bed	1,141 m
Length and width at top	575 m, 20 m, Flared 25 m on abutments
3. Diversion Tunnel	
Type	Horse Shoe
On Left Bank	2 number, 11.30 m dia, 1,774 & 1,778 m long
On Right Bank	2 number, 11.30 m dia, 1,298 & 1,429 m long
Diversion Flood	8,120 Cumec
4. Spillway	
(A) Chute Spillway	
Crest Elevation	815.00 m
Design Discharge	15,480 Cumec
No. & Size of Bays	3,10.50 m each
(B) Right Bank Shaft Spillway	
Crest Elevation	830.20 m
Design Discharge	3,879 Cumec
Intake Type	2 Number, Ungated Funnel Shaped
(C) Left Bank Shaft Spillway	
Crest Elevation	815.00 m
Design Discharge	3,750 Cumec
Intake Type	2 Nos. Gate Weir type Intake
5. Power House (Underground 2 Number)	
Installed Capacity	2,000 MW
Conventional Units	4 × 250 MW
Reversible Units	4 × 250 MW
Design Head	231.5 m
Gross Head	188 m
Head Race Tunnel	4 Number, 8.50 m dia

villages partially besides submerging the Tehri Town. Out of the total submerged area, about 1,600 ha is the cultivated land, about 1,600 ha is the forest land and the remaining is uncultivated. In addition, some area will be acquired for the Tehri Town, project colony, workshop, stores and roads, etc. which affect 13 additional villages. The project will displace about 13,840 families.

Hydrology

The snow-bound catchment has little rainfall but it contributes runoff in non-monsoon period due to snow melt. The rest of the catchment has the annual precipitation varying from 101.6 to 263.0 cm. More than eighty percent of the annual precipitation occurs during the monsoon period causing occasional floods. These floods often cause soil erosion bringing heavy sediment loads in the river. The river discharge at the dam site generally varies from 30 to 2,000 cumec, the minimum being in January and the maximum in August.

The probable maximum flood (PMF) has been worked out at 15,540 cumec which corresponds to a frequency of 1 in 10,000 years and has been adopted as the design flood. The routed discharge for which spillway structures have been designed has been worked out to be about 13,100 cumec.

The Debate over the Tehri Dam

Opposition to the Tehri dam began in 1976 when the Anti-Tehri Dam Committee was formed by local opponents mainly in respect of the question of displacement, compensation and rehabilitation. The opposition on environmental grounds came in prominence in 1978 when a massive landslide dam-burst occurred in the upper catchment of the river producing devastating floods to some distance downstream of the dam site. A social activist, Mr. Sunderlal Bahuguna was the leader of this movement which led to the establishment of a governmental working group for the assessment of the environmental impact of the proposed dam.

The report of the governmental working group further stressed the issue of seismicity in the Himalayan region as a major factor to be considered for dam safety. In 1990 a more detailed investigation was undertaken by the Environmental Appraisal Committee (EAC) of the Ministry of Environment. Three important factors that were examined by the committee were: compensation and resettlement for the involuntarily displaced population, siltation and economic life of the dam, and the seismic risk associated with large dams in the tectonically active Himalayas. These factors are not specific to the Tehri dam alone and will apply equally in the case of any other large dam in the Himalayas. It may be added that the construction of large dams to store water to augment lean-period flow is a solution to many water related problems in the lower parts of Ganga Basin.

Intense public debate on the Tehri dam has centered around seismicity and dam safety in Himalayas. As expected, there was divergence in expert opinions on the seismicity issue which caused considerable confusion. While this debate over the desirability or otherwise of large dams in Himalayas has generated much controversy, it has also generated a wealth of literature. Anti-dam campaigners also tried to the sensationalize campaign by highlighting the scenario of a possible dambreak-generated flood that might wipe many districts of Western UP.

The opposition to the Tehri dam began on the question of displacement and compensation for the affected people and constructing a dam in earthquake prone Himalayan region (Valdiya 1991). But it grew beyond the resettlement and safety aspects and later ecological and environmental issues dominated the debate. A wide range of literature on the broad policy issues has been generated by the Tehri Dam debate (Bandyopadhyay, 1990) and some of this could be a rich information base for future projects. In this way, it can serve as a vital information base and important background for the evolution of overall policy guidelines for decision making on other big dams proposed in the Himalaya. Admittedly, some of the questions that have been raised in the Tehri dam debate are of vital significance in finding appropriate solutions to the question of optimal utilization of the Himalayan water resources on a sustainable basis and good governance.

The possibility of occurrence of an earthquake of magnitude 7.0 or higher on Richter Scale in the Himalayan region has been widely accepted. On the basis of the risk associated with possible damage from earthquakes, the wisdom behind the decision in favour of the construction of large dams in the Himalayas has been questioned by many. Debate has been going on for years about the adequacy of the design of the Tehri dam to sustain the dam against the maximum credible earthquake expected in the area. The occurrence of a major earthquake in 1991 in the upper catchment areas of the dam was a shot in the arms of the opponents. It may be added here that Tehri dam is an earth and rock fill dam and that such dams can withstand earthquakes of quite high magnitude. One may recall that the construction of the Sardar Sarovar dam was also disrupted by activists on many counts.

With the closure of tunnels in Oct. 2005, first filling of the reservoir commenced. Trial run of the turbines of the power house was conducted in March 2006.

8.5.4. Lakhwar Dam

The Lakhwar dam site is proposed on the Yamuna River at latitude 80° 31′ 3″ N and longitude, 77° 56′ 58″ E in the Dehradun district of Uttaranchal, about 2 km south-west of the Lakhwar village, which is about 28 km upstream of the Dak Pathar Barrage.

Initially in 1967 it was envisaged to construct a 176 m high concrete gravity dam at Lakhwar and an underground power house on the right bank of the river with an installed capacity of 158 MW. Along with this, another auxiliary project involving construction of 60 m high concrete gravity dam at Vyasi was also envisaged. But in December 1972 the proposal was modified and now the plan is to construct:

1. A 192 m high concrete cored gravity dam across the Yamuna River near the Lakhwar village by which a head of 169 m would be created. This head will be utilized by a powerhouse at the foot of the dam with an installed capacity of 300 MW. For the main dam, the deepest anticipated foundation level is at an elevation of 608 m, with top level at EL 800.00 m. The length of the dam at the top level of EL 800.00 m will be 450.0 m. The capacity of reservoir, behind Lakhwar dam, at full reservoir level of EL 796.00 m will to be 580×10^6 cum.
2. An auxiliary cored concrete gravity dam, 5 km downstream of Lakhwar, at the Vyasi with a height of 58 m from its deepest foundation level. From this dam water will be conducted through twin power tunnels to a surface powerhouse at Hathiari with an installed capacity of 240 MW.
3. The discharge released from the Hathiari powerhouse will be balanced by a proposed barrage at Katapathar 2.25 km downstream of the Hathiari powerhouse.

The entire Lakhwar Vyasi Scheme is divided into two stages. The first stage involves the Lakhwar dam and the Lakhwar underground powerhouse, while the 2nd stage involves the Vyasi dam, Hathiari powerhouse, and Katapathar barrage. The National Hydroelectric Power Corporation will be constructing the project. However, preparation of DPR for the project is stalled for want of consent of the Govt. of UP & NCT of Delhi to fund the irrigation & drinking water component of project, respectively.

8.5.5. Tapovan Vishnugarh Project

The Tapovan Vishnugarh Project is one of the Mega projects being constructed on the Dhauliganga River in the Chamoli district of Uttaranchal state. The construction work on this 520 MW project began in February 2005. The approximate cost of the project is 2,158 crore. After completion the project will generate 2,558.37 million units per year on 90% dependability. This project is being constructed by the National Thermal Power Corporation (NTPC).

Vishnuprayag hydroelectric project on Alaknanda River in Chamoli District of Uttranchal is a run-of-the-river consisting of a 63 m barrage and an underground powerhouse. This project is likely to commence power by the year 2007 and will generate about 200 crore units each year.

8.5.6. Ramganga Multipurpose Project

The Ramganga dam on the tributary by the same name has created a reservoir at Kalagarh with a live storage capacity of 2,190 million m^3. The Ramganga dam is situated about 3 km upstream of the Kalagarh village, 45 km from Dhampur, Bijnor District, Uttar Pradesh. It is about 110 km to the North East of the Moradabad city. The exact location of the dam site is latitude 29° 31′ 13″ North and Longitude 78° 45′ 35″ East. Its construction began in 1961. The Ramganga project comprises a 127.5 m high earth dam, a power house, irrigation outlet system and a vast canal system. It is one of the first few high earth dams in India constructed close to areas prone to seismic activity. The dam has been constructed over weak rocks, having relatively complex geological features, in thickly wooded forests infested by wild animals. The valley where the dam exists is known by the name of Patli Dun. The catchment area at the dam is 3,134 km^2. The reservoir has a live storage capacity of 2,450 MCM at FRL 365.3 m and the MDDL is at 317 m.

Two earthen dams, one 125.6 m high on the main Ramganga River and the other 72.2 m high Saddle dam on one of its tributaries Chui Sot, comprise the Multipurpose Ramganga River Project. The Saddle dam is about 3 km away from the main dam on the North Eastern rim of the reservoir. The area is seismically active and falls in a seismic belt. A feeder channel takes off from the dam to provide extra supplies of water to the upper and lower Ganga canal and the Agra canal, besides direct irrigation. A total of 666,000 ha cropped area will be irrigated by this project.

The Ramganga Reservoir creates a lake submerging 55 sq. km in the famous Corbett National Park. The reservoir does not only add to the beauty of this famous wild life sanctuary but also helps in developing pisciculture and a bird sanctuary. Apart from these, the development of boating facilities, crocodile pockets and the project area provides an attraction to the tourists. The project is located in industrially undeveloped area of Pauri Garhwal and has provided a great impetus to its economic and social development. The life of the reservoir has been estimated as 100 years and the design sedimentation rate is 4.25 ha-m/100 sq. km/year.

The multipurpose project has brought in additional benefits of flood control and power generation. The existing resources of power in western U.P. are the hydroelectric power stations located on falls on the Upper Ganga and Sarda Canals and a few thermal stations. The major power projects included in the third five-year plan were the 90 M.W. thermal station at Harduaganj and Yamuna Hydro Electric scheme stage I and II. The existing Ganga, Sarda and the Yamuna power stations have to depend on river supplies, there being no storage reservoirs.

The water available for utilization from the Ramganga reservoir is estimated to be 0.215 M ha-m annually. With this water, 5.75 lakh hectares of additional land can be brought under irrigation by means of a network of canals. This added irrigation has improved agriculture in 14 districts of U.P. and three lakh tons of additional food grains are being produced annually. It has provided substantial flood protection to the districts of Moradabad, Rampur, Bareilly, Shahjahanpur, Bijnor and Farrukhabad.

The Ramganga Project has provided irrigation to new areas and has increased the firm power available in the combined Ganga-Sarda-Yamuna Ramganga grid. A powerhouse with an installed capacity of 198 MW comprising 3 units of 66 MW each is located at the toe of the dam. With mean annual inflow of 2,683 MCM, the project has a firm power of 38 MW. Ramganga power house, in conjunction with the existing Ganga-Sarda grid, the Yamuna stage I and II and existing thermal power station, is expected to increase the firm power of the grid by 82.25 MW. The Ramganga powerhouse output is nearly 450 million units of power per year.

Irrigation outlets have also been provided for emergency when the powerhouse is not functioning and irrigation demands have to be met. The water released from the powerhouse flows down the Ramganga for 25.6 km. It is diverted into a feeder channel at Harevelli where a 416.3 m long barrage has been constructed across the Ramganga River. The river and the barrage allow for the balancing of daily fluctuation of load as the Ramganga power House is mainly as a peaking station. The Feeder Channel, with a full supply discharge of 135 cumec, is 74 km long and discharges into the Ganga River opposite Garhmukteshwar. The augmented supplies of the Ganga River are picked up at Narora in the Bulandshahar district and utilized for the Lower Ganga Canal.

The Ramganga River used to cause flood havoc in the past in the districts of Moradabad, Bijnor, Rampur, Bareilly, Shahjahanpur and Farrukhabad, during monsoons. Due to the storage of river water in the Ramganga Reservoir, the flood havoc has been eliminated in most of the years. In addition to direct advantages 5.5 cumec water would also be supplied to Delhi to overcome the shortage of drinking water supply.

8.5.7. Projects in the Vicinity of Dehradun

A number of medium and small projects have been constructed in the region around Dehradun.

Dhalipur hydropower plant: Dhalipur hydropower plant is located on Yamuna River, 5 km from Herbertpur, in Dehradun District, Uttaranchal. The project utilizes tail water of upstream Dhakrani power house which uses Yamuna waters diverted by Dak Patthar barrage into Dakrani power channel. The rate of inflow is 198.24 m^3/s. The power house has 3 units of 17 MW each. It has a firm power of 39 MW. This project was commissioned in 1965–70.

Giri diversion project: Giri is a diversion project on Giri River, tributary of Yamuna River, located 25 km from Paonta Sahib in Sirmaur district in Himachal Pradesh. The barrage is 163 m long. Commissioned in 1978, Giri power house has 2 units of 30 MW each.

Chibro hydropower project: Chibro hydropower project is located near Ichari diversion dam on Tons River, a tributary of Yamuna, 67 km from Dehradun, in Dehradun District, Uttaranchal. The catchment area at the dam is 4,890 km^2. The height and length of the dam is 59.25 m and 155 m respectively. It has a small pond of live storage capacity of 5.11 MCM. The power house has 4 units of 60 MW each and produces a firm power of 98 MW.

Khodri power house: Downstream to Chibro is the Khodri power house, 52 km from Dehradun, in Dehradun District, Uttaranchal. The project uses the tail water of Chibro powerhouse (5.6 km upstream) for power generation. Khodri powerhouse has 4 units of 30 MW each and is able to generate a firm power of 51.5 MW. This project was commissioned in 1984.

Chilla hydropower project: Chilla hydropower project is located downstream of Pashulok diversion barrage on Ganga River, 4 km from Haridwar, in Dehradun District. The catchment area at the Pashulok barrage is 21,400 km^2. The maximum and minimum pond level is 336.5 m and 333.15 m respectively. The barrage is designed for a designed flood of 13,200 m^3/s. The power house has 4 units of 36 MW each and a firm power of 143 MW. This project was commissioned in 1980–81.

Khara hydropower house: Khara hydropower house is located on Ahsan River, tributary of Yamuna River, 50 km from Saharanpur, in Saharanpur District, Uttar Pradesh. The hydropower project utilizes tail water of upstream Kulhal on Asan River. The mean rate of inflow here is 186 m^3/s. The power house with 3 units of 24 MW each and has a firm power of 62 MW was commissioned in 1992.

Maneri Bhali I hydropower project: Maneri Bhali I hydropower project comprises of 39 m high and 127 m long Maneri diversion dam on Bhagirathi River, 150 km from Rishikesh, in Uttarkashi District, Uttaranchal. The catchment area at the dam is 4,024 km^2. The project was completed in 1984. A small reservoir with live storage capacity of 0.60 MCM has been created behind the dam. The power house has 3 units of 30 MW each which produce a firm power of 42 MW.

Tanakpur Barrage: Tanakpur Barrage was constructed in 1992 on Sarda River, a tributary of Ghaghara which in turn is a tributary of Ganga. The powerhouse is 12 km away from Tanakpur in Udhamsingh Nagar District, Uttaranchal. Tanakpur barrage is 475.30 m long and its design flood is 19,824 m^3/s. The pond level of the barrage is 246.7 m. Tanakpur power house has 3 units of 40 MW each. It has a firm power of 53 MW.

Kishau dam: Kishau dam is proposed on the Tons River, a tributary of the Yamuna River, at about 95 km from Dehradun. A concrete gravity dam about 236 meters high will be constructed at this place. The total extra electric power generated will be to the tune of 1,688.19 × 10^6. On completion of the project, 1.015 maf of irrigation water will be available, out of this only 0.515 maf will be utilized by the Eastern Yamuna Canal to irrigate 97,076 ha of extra land and the rest of 0.500 maf will be made available for drinking water to Delhi State.

The cost has been estimated to be about Rs. 35,662 million (December 1998 price level). A provision of 372 MGD (about 1.7 million cubic m/day) has been earmarked for Delhi's use from the storage of this dam.

8.5.8. Jamrani Multipurpose Dam Project

This dam is situated about 10 km from Kathgodam. The proposed dam is 130.6 m high roller compact concrete structure. In the first phase a barrage on the Gola River and a Feeder Channel and rejuvenation of the existing canal system is proposed. After completion, 89,702 hectares land will be irrigated, 52.93 million m^3 drinking water and in future 15 MW hydro-electric power house is proposed.

8.5.9. Rihand Dam

The Rihand Dam was constructed on the Rihand River in the Sonbhadra district of Uttar Pradesh in the year 1962. The objective of the project was to provide water mainly for generation of hydropower necessary for speedy development of agriculture and industry in the backward areas of Eastern and South-Eastern parts of the State of Uttar Pradesh.

The river above the dam site drains an area of 13,333 sq. km and has a length of 257.50 km. The stream slopes down from an elevation of about 915.50 m in the upper valley to 190.5 m at the dam site. The region is hilly and is covered with vegetation. At Chopan in the Sonbhadra district (U.P.), the Rihand River joins the Sone River.

The Rihand dam is located at a latitude of 24° 12′ 30″ N and a longitude of 83° 03′ E. It comprises a 90 m high and 934.2 m long concrete gravity dam near the Pipri village. The dam impounds 10,608.32 million cubic meters (Mm^3) of water at the full reservoir level (FRL) of 268.22 m. A powerhouse with 6 units of 50 MW each is also constructed at the toe of the dam on the right bank of the river to provide a firm power of 105 MW with a total annual power output of 912 million units.

The reservoir can attain an MWL of 271.52 m during the passage of PMF of 13,339 cumec. The reservoir has a dead storage of 1,628.38 Mm^3 below R.L. 236.22 m and a live storage of 8,979.94 Mm^3 between R.L. 236.22 m and the FRL. Its water spread area is 469.40 sq. km at FRL which lies in both Uttar Pradesh (347 sq. km) and Madhya Pradesh (122 sq. km).

The average annual rainfall in the Rihand basin is 1,422 mm while the average annual runoff is 475 mm. The Rihand River experiences heavy floods during the monsoon season and has little discharge during the remaining part of the year. The maximum and minimum runoffs have been estimated to be 8,993 Mm^3 and 3,503 Mm^3, respectively, the average annual runoff being 6,328.47 Mm^3.

Although the original project envisaged generation of only hydroelectric power, these days the reservoir water is used mainly for the generation of thermal power as the coal is available in adjacent areas. The following thermal power stations have been installed at the periphery of the reservoir:

a. Anpara Super Thermal Power House of 2,000 MW,
b. Renu Sagar Power Station of 600 MW,
c. Shakti Nagar Super Thermal Power Station of 2,000 MW,
d. Vindhya Nagar Super Thermal Power Station of 2,000 MW, and
e. Rihand Nagar Super Thermal Power Station of 1,000 MW.

The reservoir is encroached upon by construction of several ash dykes and other structures near its periphery. Consequently, the capacity of the reservoir has reduced by 44.7 Mm^3. Thus, the net original capacity of 10,563.62 Mm^3 and the net live storage capacity of 8,979.94 Mm^3 is available for regulation of water. The reservoir could be filled up to FRL only in the years 1964, 1971 and 1995 and it was well below FRL during the rest of the years. Consequent upon installation of several thermal power stations around the periphery, the reservoir level is now not allowed to fall below 252.98 m. Therefore, the new dead storage level is 252.98 m.

The provision of dead storage in the reservoir is made considering the sedimentation in 140 years of its operation. During this period, the live storage was expected to reduce from 8,979.94 Mm^3 to 8,185.49 Mm^3. However, the capacity survey of the reservoir carried out before the monsoon of 1995 indicates that the sediment deposition has reduced the live storage to 8,009.94 Mm^3 in only 33 years of operation. This gives a sedimentation rate of 2,918 m^3/sq. km/year against the assumed value of 904 m^3/sq. km/year. The large difference in the assumed and the estimated rates is partly attributed to inaccuracies in the original capacity surveys.

8.5.10. Rajghat Dam Project

Rajghat dam in Bundelkhand region provides irrigation facilities for area in Uttar Pradesh and Madhya Pradesh States. The financial burden will be borne by both states under the central command of the Betwa River Board. This project has been cleared by the Central Government. The canal network is proposed to be borne by individual state governments. The pump canal system of the Betwa, Gurshahai, and Bargaon is being considered.

8.5.11. The Halali Dam

The Halali dam, also known as Samrat Ashok Sagar project, was constructed across the Halali River which is a tributary of the Betwa River. The dam site is located both in the Raisen and Vidisha districts of Madhya Pradesh, 40 km away from Bhopal and 16 km away from the Salamatpur railway station. The Halali River rises around Bhopal at an altitude of about 487.68 m above msl and joins the Betwa near Vidisha. After flowing for a length of about 38 km NE, it enters a narrow gorge with high hills on both sides. The dam has been constructed at this gorge at 23° 30′ N latitude and 77° 30′ E longitude. The rock at the dam site belongs to Upper Vindhyans and consists of massive to current bedded, course grained sand stone, grits, conglomerates together with minor bands of shales and pockets of soft friable sand rock.

Halali is a rolled-filled earthen dam, 945m long with a maximum height of 29.57m above the foundation level. The catchment area of the project is 699 sq. km with a maximum rainfall in the area as 1680 mm and the average rainfall of 1,108 mm. About 25% of the catchment area is hilly and the rest is in plains. The project envisages a gross command area of 374.19 sq km with CCA as 279.24 sq km. The intensity of irrigation in the command is 135%. The design rate of sedimentation is 0.476 mm/year for gross storage. The top bund level of the reservoir is fixed at 466.32 m and MWL is fixed at 464.19 m. The water spread at FRL is 52.59 sq. km. The width of the earthen dam at its top is 4.57 m. Additional spillway with a 41.15m length is provided at RL 459.61 m.

8.5.12. Gandhisagar Dam

Three important storage reservoirs constructed in the basin include Gandhisagar, Jawaharsagar and Rana Pratapsagar cascade which provide a live storage of 8,500 million cubic m. The barrage at Kota diverts water into canals on either side, irrigating a total of 0.57 million ha agriculture area.

Gandhisagar is the main storage dam constructed across the Chambal River, intercepting a catchment area of about 23,025 sq. km. The dam serves as a backup storage for power generation in the Gandhisagar, Rana Pratapsagar and Jawaharsagar dams and irrigation through canal systems taking off from the Kota Barrage. The dam is 64.63 m high, straight gravity masonry type with an installed capacity of 115 MW and an irrigation potential of 7.57 lakh ha.

The Gandhisagar is the upper most dam in the series of three dams and a barrage of the Chambal Valley Project. The catchment area intercepted up to various structures is given in Table 11:

Table 11. Catchment area intercepted up to various structures

Name of structures	Catchment area (sq. km)
Gandhisagar Dam	23, 025
Between Gandhisagar & Ranapratap Sagar Dam	2, 103
Between Ranapratap Sagar Dam & Jawahar Sagar Dam	2, 054
Between Jawahar Sagar Dam & Kota Barrage	137
Total	27, 319

The live and dead reservoir storages provided for at planning stage were 6,910 Mm^3 and 836 Mm^3, respectively with a gross storage of 7, 746 Mm^3. The original reservoir submergence area at FRL was 680 sq. km.

The Gandhisagar dam was constructed in 1960 as a 64.63 m high straight gravity masonry dam 514 m long with a 182.93 m central spillway and five power blocks on its right along with non-overflow blocks at both flanks. The installed hydropower capacity is 115 MW and the irrigation potential created is 7.57 lakh ha. The dam is located at a latitude of 24° 44′ N, longitude of 75° 33′ E, 8 km north-east of Bhanpura. The spillway section consists of 10 spans each of 18.3 m length and 9 sluice piers 7.927 m wide, accommodating 9 sluices of 3.05m × 7.62m each with their sill level at 363.872 m and steel crest gates 18.3m × 8.54m with crest at 391.46m. The discharging capacity of the spillway is 13,705 cumec at MWL. At the foot of the power intake is the power house which has five generating units. The gross storage capacity of the reservoir was assessed as 7,746 MCM at the time of first impounding.

The Full Reservoir Level (FRL) and the Maximum Water Level (MWL) of the dam is 400 m, and the Dead Storage Level (DSL) is 381.0 m. Sedimentation surveys were conducted in the reservoir area in 1975 and 1989. Surveys conducted in 1975 indicated a reduction of the gross capacity by 333 MCM over a period of 15 years. Those conducted in 1989 indicated a further reduction of 419 MCM in the gross storage.

A hydrographic survey using modern equipment was conducted during March-October, 2001 up to the maximum water level. The rate of sedimentation adopted at the project planning stage was 3.6308 ham/100 sq. km/year. However, the average rate of sedimentation during the first 41 years, on the basis of 2001 survey, works out to 5,508 ham/100 sq. km/year. The anticipated feasible life of the reservoir works out to be 125 years, as against the planning stage stipulation of 100 years.

The gross storage capacity of Gandhisagar dam was assessed from toposheets at the planning stage as 8450 M cu m, with full reservoir at 400 m. Subsequently, based on aerial photographs and contour surveys, the gross storage was refixed at 7,746 M cu m. In all subsequent publications, this value is considered as original gross capacity. The surveys conducted from time to time have indicated a progressive reduction of the storage capacity as given in Table 12.

Table 12. Storage in the Gandhisagar dam on various dates

	Project planning stage	1960–61 Reassessment	1975 Resurvey	1989 survey
Gross Storage at FRL	8,450 MCM	7,746 MCM	7,413 MCM	7,323 MCM
Live Storage	7,620 MCM	6,910 MCM	6,827 MCM	6,798 MCM
Dead Storage	830 MCM	836 MCM	586 MCM	525 MCM

8.5.13. Rana Pratap Sagar Dam

Rana Pratap Sagar is a masonry gravity dam on Chambal River, 54 km from Kota, in Chittaurgarh District, Rajasthan. The catchment area at the dam is 24,576 km^2. The height and length of the dam is 38.3 m and 1,143 m respectively. The reservoir has a live storage capacity of 2,899.5 MCM at FRL 352.9 m and the MDDL is at 343 m. The power house has 4 units of 43 MW each, producing a firm power of 54 MW with mean annual inflow of 49,538 MCM. RRVPN Ltd. commissioned the project in 1968–69.

8.5.14. Jawahar Sagar Dam

Jawahar Sagar is a straight gravity concrete dam on Chambal downstream of Rana Pratap Sagar, 36 km from Kota, in Kota District, Rajasthan. The catchment area at the dam is 26,880 km^2. The height and length of the dam is 37 m and 336 m respectively. This reservoir has a live storage capacity of 7.40 MCM. Its FRL is at 298.78 m and MDDL is at 295.78 m. With 3 units of 33 MW each, the power house produces a firm power of 32 MW. This project was completed in 1972–73.

8.5.15. Chambal Valley Project

The Chambal Valley Project would provide water for both irrigation and industrial purposes in the south eastern part of the State, primarily in the districts of Kota, Baran, Bundi and Sawaimadhopur. Three dams and a barrage on the Chambal River have been created with a canal network in Rajasthan & Madhya Pradesh. The project will irrigate approximately 5 lakh ha. The salient features of the Chambal Canal System have been given in Table 13.

8.5.16. Obra Dam

Obra is an earth and rockfill dam constructed across Rihand River, 33 km from Robertsganj, Uttar Pradesh. The catchment area at the dam is 546.5 km^2. The height and length of the dam is 29 m and 2000 m, respectively. Commissioned in 1970–71, the Obra powerhouse has 3 units of 33 MW each and has a firm power of 21 MW.

Table 13. Salient features of Chambal Canal System

Name of canal	Length (km)	Design/head discharge	CCA (ha)	Design irrigated area (ha)	Tehsil
Right main canal	179	110.40	68,400	47,800	Shopepur & Vijapur
Lower main canal	53	53.80	29,200	20,300	Sabalgarh & Joura
Ambah branch canal	171	35.40	132,100	92,500	Sabalgarh, Joura, Ambah & Bhind
Morena branch canal	38	14.20	48,600	34,100	Morena
Asan outfall canal	3	41.00	–	–	–
Bhind main canal	82	35.70	96,200	60,700	Gohad, Bhind & Mehgaon
Mau branch canal	48	9.90	40,100	28,100	Gohad

Source: WG (1999).

8.5.17. Bansagar Tons

Bansagar Tons hydropower project is located on Beehar barrage on Tons River and Beehar River, tributaries of Yamuna, 50 km from Rewa, in Rewa District, Madhya Pradesh. The catchment area at the Bansagar dam is 8,648 km^2, at Tons bridge is 4,457 km^2, and at Beehar Bridge is 1,637 km^2 respectively. The height and length of the Beehar barrage is 9.1 m and 138 m respectively. The maximum and minimum pond levels of the barrage are 180 m and 277 m respectively. The power house has 3 units of 105 MW each. It has a firm power of 35 MW with mean annual inflow of 2,100 MCM and 700 MCM at tons bridge and Beehar Bridge respectively. MPEB commissioned the project in 1991–92.

Bansagar Tons II and III

Bansagar Masonary gravity dam on Sone River in Rewa District, Madhya Pradesh. At the dam site, the catchment area is 18,648 km^2. The height of the dam is 63 m and the live storage capacity of the reservoir is 5,410 MCM at FRL 341.65. Corresponding to a 90% dependable year, the annual inflow of the dam is 4,840 MCM. Bansagar Tons II and III powerhouses are located at a distance of 8 km and 70 km from Rewa. The Bansagar Tons II power house has 2 units of 15 MW and Bansagar Tons III has 3 units of 20 MW each. Bansagar Tons II and Bansagar Tons III projects have initial phase firm power of 12.9 and 16.4 MW and final phase firm power of 3.3 and 13.3 MW respectively. MPEB commissioned the project in 2001–02.

8.5.18. Parbati Dam

The Parbati dam is situated across Parbati River in the Chambal basin in the Baseri tehsil in the Dhaulpur district of Rajasthan. It is located nearly 50 km away from the district headquarters and almost 15 km from Bari town. The geographic coordinates of dam are 26° 37′ 36″ N latitude and 77° 26′ 52″ E longitude. The area experiences extreme temperature variation with minimum and maximum temperatures as 1 °C to 49 °C, respectively. The average annual rainfall in the area is nearly 67 cm. Farmers in the area grow wheat, gram, barley, pulses, *bajra*, etc. The forest in the area falls under the dry deciduous variety. Physiographically, the area is characterized by a dissected plateau and alluvial plain region. The geological sequence of the area forms part of vindhyan supergroup, which is essentially made up of sedimentary rocks including sandstones, shales and limestones.

The catchment and gross command area (GCA) of the Parbati project are approximately 780 sq. km and 325 sq. km, respectively. The project has a live storage capacity of 102.893 MCM as per the hydrographic survey carried out during the impounding year 1963. The length of the main canal is 58 km. The project command is restricted to main course on its southern side. The design rate of sedimentation is 0.157 mm/year.

8.5.19. Matatila Reservoir

Matatila dam was constructed in the year 1956 across Betwa River, a tributary of the Yamuna River. The dam lies at 25° 6′ 15″ N latitude and 78° 23′ 00″ E longitude. It is located in Lalitpur tehsil of Matatila district, Uttar Pradesh, at about 56 km from Jhansi.

Matatila is an earthern dam 6.6 km long with masonary spillway of ogee shape. The height of the dam is 24.40 m. It has 23 vertical lifting gates and 4 sluices. The dam has a live storage capacity of 1,019.40 MCM and dead storage capacity of 113.30 MCM. The total capacity of the reservoir at FRL 308.46 m is 1,132.70 MCM with a water spread area of 142.43 m^2. It is a multipurpose dam which provides facilities for irrigation, water supply and fish cultivation.

8.5.20. Ramsagar Dam

The Ramsagar dam is situated across the Bamani River, a tributary of the Ramsagar River (which is a tributary of the Chambal River) in Bari tehsil, the Dhaulpur district of Rajasthan. It is located nearly 40 km away from the district headquarters and almost 8 km from the Bari town. The geographic coordinates of the dam are 26° 35′ N latitude and 77° 35′ E longitude. The area experiences extreme temperature variations with minimum and maximum temperatures as 1 °C to 49 °C, respectively. The average annual rainfall in the area is nearly 67 cm. Major crops in the area are wheat, gram, barley, pulses, *bajra*, etc. The forest in the area falls under the dry deciduous variety. Physiographically, the area is characterized by a dissected plateau and alluvial plain region. The geological sequences of the area are part of Vindhyan supergroup, which is essentially made up of sedimentary rocks, including sandstones, shales and limestones.

The catchment and gross command area (GCA) of the Ramsagar project are approximately 176 sq. km and 62 sq. km, respectively. The project has a gross and live storage capacity of 30.83 and 29.39 MCM, respectively as per the hydrographic survey carried out during year 1905, which is the impounding year. The length of the main canal is 11.27 km. The design rate of sedimentation is 0.81 mm/year.

8.5.21. Massanjore Reservoir

The Massanjore reservoir is located on the Mayurakshi River upstream of the confluence of Sidheswari with Mayurakshi. The total length of the Mayurakshi River up to the dam site is 70 km from its origin, out of which 43% lies in the reservoir waterspread area. The reservoir was created in 1954 by the construction of a stone masonry dam at Massanjore in the Santhal Parganas in the district of Bihar about 24 km from the West Bengal border to irrigate the lands in West Bengal and Bihar and to generate 4,000 kW of hydro power. The lake extends over 30 km upstream in the Bihar State. The 1,859.62 sq. km catchment of the river above the dam is leaf shaped with no appreciable vegetal cover. Rolling and undulating in nature with scattered hillocks the catchment comprises various types of lands.

Forests constitute only 6% of the catchment area. The vegetation is generally limited to hill tops and slopes. About 44% of the catchment area comprises lands under paddy cultivation.

8.5.22. Dhauliganga Power Project

The Dhauliganga project has been constructed in the Pithoragarh district of the Uttaranchal state. Under this project a 56 m high concrete dam on the Dhauliganga River, a 750 m long diversion tunnel, and a 5.3 km long head race tunnel have been constructed. The total installed power capacity of the underground power house is 280 MW (4*70 MW). Constructed by the National Hydropower Corporation (NHPC), this project has started generation of electricity in 2006. This hydropower project will generate 1,134 million units of electricity each year. The beneficiary states of this project are Delhi, Punjab, Himachal Pradesh, Rajasthan, Chandigarh, Uttaranchal, Uttar Pradesh, and Jammu and Kashmir. The estimated cost of the project is Rs. 1,578 crore.

8.5.23. The DVC System

To control the Damodar River, the Damodar Valley Corporation (DVC) was set up in 1948. Construction work on four dams at Tilaiya, Konar, Maithon and Panchet was completed in 1953, 1955, 1957 and 1959, respectively. Along with these, a barrage at Durgapur was completed in 1958. One more dam was constructed at Tenughat at a later date. First four reservoirs, viz., Maithon, Panchet, Konar and Tilaiya, come under the purview of the Damodar Valley Corporation (DVC), whereas the Tenughat reservoir is being operated by the Bihar Government. The operation of the Durgapur barrage is being controlled by the Government of West Bengal. Further, the Central Water Commission has the responsibility to operate the Maithon and Panchet reservoirs. These dams are being operated for flood control, municipal and industrial water supply, irrigation, and power generation. A line diagram of the system is shown in Figure 16.

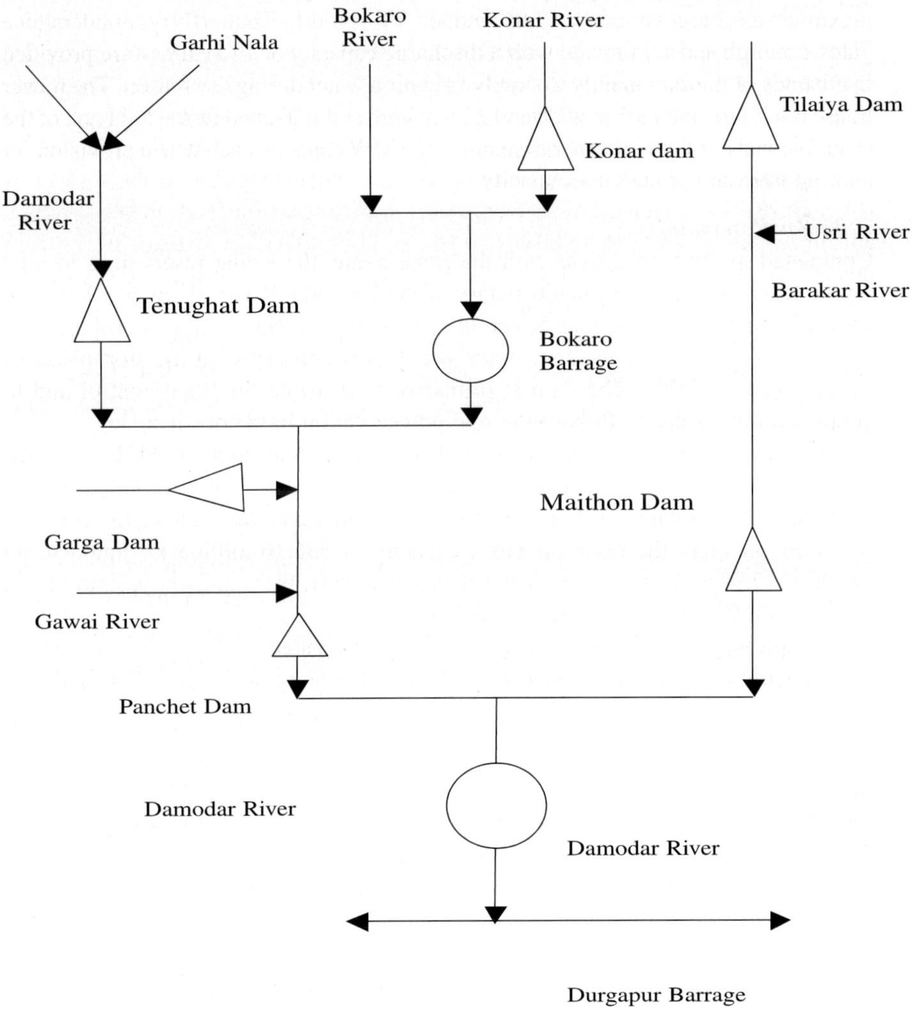

Figure 16. Line diagram of the DVC system

i. Tilaiya Reservoir

The Tilaiya dam is located in the upper reaches of the Barakar River about 64.4 km downstream of its source. The river at this site passes through a narrow gorge approximately 91.4 m wide with banks steeply rising about 45.7 m on either side. The dam was completed in 1952. Barakar River rising from the hilly forests of the Hazaribagh district at an elevation of 610 m has a catchment area of 984 km^2, comprising mainly forests, pastures, cultivated lands and wastelands. The annual rainfall in the area is 127 cm. The reservoir has a water spread area of 5,921 ha at FRL with a volume and mean depth of 394 MCM and 6.65 m respectively.

The Talaiya dam is a concrete gravity dam with a maximum height of 30.2 m above the river bed. The spillway has 14 tainter type crest gates of 3.05 m*9.1 m, with a maximum discharge capacity of 3,852 cumec. Two modified butterfly type undersluice gates 1 m high and 1.7 m wide with a discharge capacity of 14.2 cumec are provided in the body of the dam mainly to supply irrigation water during dry season. The power house is 41.5 m long 19.5 m wide and 22.6 m high and is located on the left bank of the river. It consists of two of generating units of 2 MW capacity each with a provision for a third future unit of the same capacity.

ii. Konar Reservoir

Completed in 1955 on a river with the same name, the Konar reservoir is located on the Konar River, a seasonal tributary of the Damodar River, about 30.6 km from its confluence with Damodar River in the district of Hazaribagh in Bihar. This dam was the second of the four dams which was completed in the first phase of development of DVC. The dam is primarily responsible for flood control and to supply cooling water to Bokaro thermal power station in the downstream.

The Konar earth and concrete dam has a catchment area of 997 km^2 which comprises thick jungles, wastelands and cultivated areas. The total length of the dam was 4.00 km with earthern embankment flanked on either side of the concrete structure. At FRL, the reservoir has a capacity of 336.16 million m^3 and a mean depth of 12.97 m. In the basin of the Konar reservoir, the main soil is sandy loam to clay with 0.3 to 0.9% of organic carbon.

The spillway was provided with 9 tainter crest gated each of 10.4 m × 9.9 m. Two 2.3 m dia. undersluices were embedded in the body of the dam. The spillway has a maximum capacity of 6,792 cumec. Two power stations with a total capacity of 35 MW have been proposed.

iii. Maithon Reservoir

The Maithon dam is a concrete cum earthen dam located on Barakar River at a distance of 26 km from Asansol in Bardhaman District, West Bengal. The dam was constructed for flood conrol, irrigation and power generation. This reservoir came into being in 1957. At the dam site, the catchment area is 6,249 sq. km. The average annual basin precipitation is 114 cm and the average annual run off is 2,616 million cubic m. In the basin, and the maximum observed flood (June 1949) was 7,085 cumec. For the project, the spillway design flood is 14,736 cumec. The live storage capacity at the dam is 566 MCM at FRL 152.4 m and the MDDL is at 132.5 m.

Table 14. Salient features of the Maithon Reservoir

	Elevation (meters)	Storage (Million cu m)
Minimum draw down level	132.62	165.30
Spillway crest	140.24	397.29
Maximum conservation pool	146.35	738.59
Maximum flood control pool	150.91	1,120.30
Full and maximum pool	152.44	1,275.77
Top of dam	156.09	

For generation of hydropower, 3 units of 20 MW each have been installed. It has a firm power of 16 MW with mean annual inflow of 2,615 MCM. Salient features of the reservoir are stated in the Table 14.

It can be seen that a separate storage is earmarked at the Maithon dam for flood control storage amounting to 539.39 million cu m (between EL 146.34 m to 152.44 m. The land has been acquired up to RL 150.91 m.

iv. Panchet Dam

The Panchet dam, an earthen dam with concrete spillway, was commissioned in 1958 on the Damodar River located at a distance of 20 km from Asansol in Jharkhand. The height and length of the dam are 47.85 m and 2,345 m respectively. The reservoir traps a catchment area of 10,961 sq. km. The average annual basin precipitation is 114 cm and the average annual run off is 4,540 million cubic m. At the dam site, the maximum observed flood (June 1949) was 8,558 cumec. For the project, the spillway design flood that was adopted was 17,853 cumec. Two units of 40 MW have been installed at the powerhouse for generation of hydropower. Some important features of the reservoir are given in the Table 15.

At the Panchet dam also, flood control storage amounting to 1,083 million cubic m (between EL 125.0 m to 135.67 m) has been earmarked. The land acquisition has been made up to RL 129.57 m. Houses have been acquired up to RL 132.62 m at the Panchet.

v. Operation guidelines of the Maithon and Panchet Dams

The conservation storage level for the Maithon and the Panchet is 146.34 m and 125 m, respectively. The operation of the reservoirs during monsoon season below the conservation storage (i.e., 146.34 m and 125 m) consists in releases to meet the

Table 15. Salient features of Panchet Reservoir

	Elevation (m)	Storage (Million cu m)
Minimum draw down level	119.50	170.26
Spillway crest	123.47	312.15
Maximum conservation pool	125.00	392.36
Maximum flood control pool	132.62	1, 058.62
Full and maximum pool	135.67	1, 475.65
Top of dam	139.33	

downstream requirements. The operation curves guide the flows such that on the first of October, the reservoir will be at the monsoon storage level of 146.34 m for the Maithon and 125 m for the Panchet. When the reservoir level rises or shows a tendency to rise above 146.34 m and 125 m for the Maithon and the Panchet reservoirs, flood control operation commences and they shall cease as soon as the level comes down to RL 146.34 m and 125 m.

As regards the operation of the Panchet reservoir, the reservoir may be drawn down to EL 119.51 m to meet the power requirements in June, when the monsoon flows arrive in June the reservoir will be, if at a lower level, allowed to build up to RL 121.95 m as quickly as possible after allowing releases for the essential downstream and power requirements.

On the 1st of July, the Panchet reservoir is to be kept as near to RL 121.5 m as possible for ensuring generation of hydro-electric power. The reservoir level shall be kept between curves A & B during July, August and closer to curve A during September and as far as possible at the monsoon storage level RL 125 m during October.

Guidelines for releases from the Maithon and Panchet dams taken together for flood control operations during June, July, August and September are:

- Limit the combined outflow to the safe limit indicated by the West Bengal, except that the outflow shall not be less than the actual inflow or 1,983.7 cumec (70,000 cusecs) whichever is less, till 20% of the combined flood reserve is used up.
- Limit the combined outflow to 3,400 cumec (120,000 cusec) while using the combined flood reserve between 20% to 50%.
- Limit the combined outflow to 4,534 cumec (160,000 cusec) while using the combined flood reserve between 50% to 70%.
- Limit the combined outflow to 5,668 cumec (200,000 cusec) while using the combined flood reserve between 70% to 100%.
- Balance outflow with inflow when all available combined flood reserve is used up, when the Maithon and Panchet reaches 150.91 m and 132.92 m, respectively, and the combined inflow is more than 5,667 cumec (200,000 cusec).

Emergency operation

- Flood starts when 50% of the flood reserve is occupied.
- Between 50% to 70% occupation, the inflow and outflow is balanced up to a maximum of 4,534 cumec (160,000 cusec).
- Between 70% to 100% occupation, half inflow is released subject to a maximum of 7,085 cumec (250,000 cusec) and a minimum of 5,668 cumec (200,000 cusec).
- At 100% occupation balance the outflow with inflow.

Dry weather operation: The water available from monsoon storage and dry season flows are utilized for meeting the committed requirements of Kharif, Rabi, Industrial and drinking water and hydro-electric power requirements.

Allocation of surplus, if any, after meeting the committed requirements will be made in November, the surplus water will be allocated for irrigating Bore paddy (Jan 21 to April 30), hydro-electric power and other requirements and a working table is prepared on the basis of allocation of waters for different uses. The reservoir is operated as per the working table.

8.5.24. Farakka Barrage

Farakka is a gauging site very close to the India-Bangladesh border. At Farakka the river has a stable cross-section with a long history of monitoring. In Figure 6 information on the annual volume of water flow of the Ganga is shown. The hydrograph does not show any trend in the hydrological characteristics of the Ganga from 1950 to 1985. The annual fluctuation is extremely high, but there is no obvious change over the recorded time period.

Near the Indo-Bangladesh border, a barrage has been constructed at Farakka to divert water to the river Hooghly to increase and maintain adequate depths of flow for navigation and operation of Calcutta port. Farakka barrage was completed in 1975.

8.5.25. Other Projects

In addition to the above there are several other reservoirs in the basin. Some other important projects on the tributaries of the Ganga include Sarda sagar on the Sarda, and Mayurakshi on the Mayurakshi River. A brief description of a few follows.

Baigul reservoir

Baigul (Sukhi) is a small tributary of the Ganga originating from the foothills of Kumaon Himalayas. In the year 1967, the river was harnessed for irrigation purposes to form the Baigul reservoir. The dam is located at 28° 56′ N and 79° 40′ E, and provided with four gates and two sluices for outflow. The catchment area of dam is 305 km^2, comprising of wooded forests and fields. The reservoir has an area of 2,995 ha at FRL of 211 m above MSL. The maximum spillway discharge amounts 566 m^3/s.

Baghla reservoir

Baghla reservoir is a small irrigation impoundment, created on the rivulet Barica in 1952. Situated at a distance of 55 km southwest of Allahabad, the reservoir covers 250 ha at the FRL of 119 m above MSL. The water retention capacity is 9.58 million m^3 at FRL and 0.141 million m^3 at DSL. Catchment comprises 32 km^2 of hilly land, receiving a rainfall of 100 cm.

Renuka Dam

This is a proposed project in Himachal Pradesh. The expected total cost of the project is Rs.1,224.64 crore. A provision of 275 MGD (about 1.25 million cubic m/day) has been earmarked for Delhi's use in the proposed Renuka Dam project.

Kosi Project

Kosi is one of the highly silt-laden rivers in the world. The resultant lateral migration of Kosi has caused considerable misery in Bihar and because of this, it came to be known as Bihar's 'river of sorrow'. There was considerable pressure in India to control the Kosi River, especially after a major flood in 1954. As a response, a Kosi river barrage was built below Chatra in Nepal, along with a pair of embankments to confine the Kosi to its existing course. The Kosi embankments were completed in 1959 and the barrage in 1963. The cost of the barrage and the associated works were borne by India. Nepal received about 10 MW of power, some flood protection and a bridge over the barrage which facilitated east-west communications in that part of Nepal.

The Kosi, the Karnali and the Pancheshwar projects are three mega projects which have received a lot of attention and discussion over the past three decades. Of these, the Kosi high dam is the oldest project, which was proposed by the Government of India in 1950s. The project was originally proposed to have a 239 m high dam at Barakshetra in Nepal, and a downstream barrage at Chatra, also in Nepal. The 850,000 ha-m storage has considerable flood attenuation capacity, reducing a peak flow of 24,050 m^3/s to 5,660 m^3/s, thereby providing substantial flood control benefit. The project would generate 1,800 MW of power and irrigate large tracts of land in Nepal and Bihar.

Karnali dam

The original Karnali dam proposal suggested a height of 240 m for the dam and a power potential of 4,500 MW. The Himalayan Hydro Development Company (HHDC) in 1988 proposed raising the height of the dam to 262 m so that the power potential might be increased from 4,500 MW to 10,000 MW. Another suggestion by HHDC is to phase the Karnali Project with an upstream run-of-the-river hydroelectic project and storage development on the Bheri and Seti rivers, which are tributaries of the Karnali. These supplements to the Karnali Project would generate an additional 5,400 MW of electricity. The 1988 cost of the Karnali Project was estimated at about US$ 4.4 billion; 95% of the power generated by the Karnali Project is to be exported to India.

Pancheshwar Project

The Pancheshwar project on the Mahakali River is of interest to both India and Nepal. The Mahakali River is a western boundary river between India and Nepal. The project can generate 2,000 MW and provide Irrigation benefits. India has completed the Investigation of the Pancheshwar site on its side of the border and wants the project to move ahead fast. However, Nepal is still investigating the conditions on its side of the border. Between the two, Karnali will be a completely Nepalese project and the Pancheshwar a joint project between India and Nepal.

Gandak Project

The Gandak Project was designed to irrigate 0.96 million ha in nine districts of Bihar, 56,650 ha in Nepal, some area in Uttar Pradesh and supply 15 MW of hydroelectric power to Nepal. This project was commissioned in 1971 and was declared complete in 1985. This project was also fully financed by the Government of India.

The Sarda canal provides protective irrigation to nearly 0.6 million ha area. Some other major projects under construction are Rajghat on the Betwa, Bansagar on the Sone and Lakhwar-Vyasi on the Yamuna. The National Hydroelectric Power Corporation is running a 120 MW Tanakpur Project and also signed an agreement with the Uttaranchal state government for constructing an 850 MW BHEL project on the Alakhnanda River in February 2005. Storage capacities of the other large existing and under construction projects having a storage capacity of 10 Million cubic meter or more is summarized below as Tables 16 and 17, respectively.

Table 16. Salient features of some existing projects in Ganga Basin

Name of the project	State	Completed in	Gross storage capacity (million cubic meter)	Live storage capacity (million cubic meter)	Designed annual irrigation (Million sq. m)	Installed capacity (MW)
Ajan	Bihar	–	–	24.71	32.40	–
Anrej	Bihar	–	30.40	26.89	19.00	–
Badua	Bihar	1966	128.28	109.78	425.10	–
Belharan Reservoir	Bihar	–	15.55	14.56	23.20	–
Botane	Bihar	1961	–	48.33	67.10	–
Bilasi Reservoir	Bihar	–	28.74	24.67	40.00	–
Chandan	Bihar	1976	136.92	124.58	627.50	–
Durgawati	Bihar	1957	–	257.70	–	–
Kohira	Bihar	1961	36.85	27.71	103.20	–
Kharagpur lake	Bihar	1956	16.50	16.28	50.60	–
Latratu	Bihar	–	46.86	41.35	33.00	–
Lilajan	Bihar	1961	218.68	188.71	226.80	–
Lower Kiul	Bihar	1961	–	251.05	259.20	–
Malay	Bihar	–	32.32	28.52	–	–
Murway	Bihar	1971	18.05	15.05	49.00	–
Nalkari Dam	Bihar	1968	105.00	85.00	–	–
Nakti	Bihar	–	14.70	13.64	28.30	–
Oriya Dam	Bihar	–	–	17.20	36.40	–
Palna	Bihar	–	12.33	11.34	23.00	–
Phulwaria Reservoir	Bihar	–	58.29	50.66	–	–
Sunder Reservoir	Bihar	1977	25.66	19.24	96.00	–
Tarlow	Bihar	–	15.33	12.84	95.00	–

(Continued)

Table 16. (*Continued*)

Name of the project	State	Completed in	Gross storage capacity (million cubic meter)	Live storage capacity (million cubic meter)	Designed annual irrigation (Million sq. m)	Installed capacity (MW)
Tahle	Bihar	–	191.16	154.16	25.50	–
Tenugha	Bihar	1973	280.57	216.44	–	–
Upper Sakari	Bihar	–	330.86	251.23	–	–
Aoda	M.P.	1934	55.70	45.00	48.60	–
Arania Bhadurpur	M.P.	1986	25.24	22.36	27.50	–
Bairarpur	M.P.	–	79.00	62.00	435.10	–
Bainswar	M.P.	1971	–	90.00	25.70	–
Barodia	M.P.	1961	–	14.58	9.10	–
Beniganj	M.P.	1970	28.00	26.00	41.70	–
Bila Nadi Tank	M.P.	1973	56.68	51.95	122.70	–
Bilas	M.P.	–	56.00	53.00	–	–
Bohita	M.P.	1987	17.96	17.46	4.40	–
Chappi	M.P.	1972	15.74	13.54	20.20	–
Chillar	M.P.	1970	34.80	31.00	34.40	–
Devendra Nagar	M.P.	1967	10.87	10.09	24.00	–
Dholawad	M.P.	–	54.30	50.00	64.40	–
Doraha	M.P.	1986	18.06	15.59	28.30	–
Girla	M.P.	1987	14.50	13.38	3.90	–
Gomukh	M.P.	–	37.51	35.16	61.00	–
Gurma Tank	M.P.	1980	39.16	36.05	72.40	–
Harri Tola	M.P.	1987	14.22	12.22	47.10	–
Harsi	M.P.	1935	230.50	224.50	404.90	–
Hathikhenda	M.P.	1960	16.28	15.22	2.40	–
Jarmora Tank	M.P.	1980	18.54	17.34	30.40	–
Jassiya (Koncha)	M.P.	1972	21.55	18.72	38.00	–
Jhumka	M.P.	1985	25.80	23.40	29.60	–
Kaketo	M.P.	1935	80.48	79.95	56.10	–
Kaliasote	M.P.	–	35.38	34.41	75.00	–
Kanchan	M.P.	–	14.00	13.00	38.50	–
Kerwan	M.P.	1975	25.03	22.57	40.50	–
Kethan	M.P.	1976	19.15	17.57	25.30	–
Kotwal Bhind	M.P.	1927	–	37.74	–	–
Kunwarpur	M.P.	1977	16.75	15.62	42.50	–
Madosagar	M.P.	1929	–	124.20	–	–
Mansurwari Tank	M.P.	1982	12.34	11.22	38.40	–
Marhi	M.P.	1981	14.24	12.74	20.50	–
Mala	M.P.	1929	19.96	16.87	26.30	–
Mola Tank	M.P.	1939	–	17.68	36.90	–
Morwan	M.P.	1965	16.50	15.76	26.70	–
Nandanwara	M.P.	–	35.50	22.65	18.20	–
Pagara Tank	M.P.	1927	113.39	66.73	22.27	–
Paronch	M.P.	–	20.60	16.69	22.40	–
Piplia Kumar	M.P.	1978	–	15.15	14.70	–
Rampur	M.P.	1917	24.14	18.08	28.30	–
Rangwan	M.P.	–	–	157.50	170.90	–

Ratapani Tank	M.P.	1964	15.14	14.50	17.80	–
Shahib Khedi	M.P.	1977	12.46	10.30	14.40	–
Tigra	M.P.	1917	420.00	400.00	32.20	–
Tillar	M.P.	–	53.15	47.30	72.00	–
Umrar Tank	M.P.	1981	18.90	16.70	28.30	–
Ajan	Rajasthan	–	–	16.45	84.20	–
Alnia	Rajasthan	1961	44.24	36.78	56.40	–
Arwar	Rajasthan	1959	47.91	47.77	49.00	–
Badagaon	Rajasthan	1969	–	31.52	34.40	–
Bade Talab	Rajasthan	–	–	14.41	12.80	–
Bagoloia	Rajasthan	1955	19.43	18.81	19.40	–
Bardha	Rajasthan	–	–	47.04	33.00	–
Baretha	Rajasthan	–	–	52.68	38.30	–
Baroch Dam	Rajasthan	1970	30.46	28.76	32.00	–
Badagaon	Rajasthan	1967	31.51	30.17	34.00	–
Bhimlat	Rajasthan	1958	10.82	10.76	8.80	–
Bhupal Sagar	Rajasthan	–	–	18.55	21.10	–
Buchchora	Rajasthan	–	–	15.09	12.30	–
Bundika Gothra	Rajasthan	1957	28.46	19.82	24.00	–
Chaparwar	Rajasthan	–	–	34.18	50.10	–
Chandsen	Rajasthan	–	–	14.70	21.00	–
Chittali	Rajasthan	–	–	12.52	31.70	–
Chariya	Rajasthan	–	–	23.68	24.00	–
Dheel Bundh	Rajasthan	–	–	25.20	35.70	–
Dhill Sagar	Rajasthan	1939	–	25.24	33.50	–
Dugari	Rajasthan	–	–	19.23	15.60	–
Galwa	Rajasthan	1960	48.74	47.26	53.00	–
Gambhiri	Rajasthan	1956	76.46	65.13	62.00	–
Gopalpura	Rajasthan	1978	32.65	31.98	60.80	–
Gudha	Rajasthan	1958	95.66	93.67	80.00	–
Jetpura	Rajasthan	1977	18.54	16.38	23.40	–
Juggar	Rajasthan	1956	28.07	24.61	36.80	–
Kalakh Sagar	Rajasthan	–	–	16.45	23.40	–
Kalisil	Rajasthan	1953	41.68	37.24	35.20	–
Khari	Rajasthan	1957	58.94	33.28	28.10	–
Madho Sagar	Rajasthan	–	–	22.62	25.90	–
Madri	Rajasthan	1964	18.41	16.57	32.00	–
Meja	Rajasthan	1956	96.22	91.69	94.20	–
Mandal	Rajasthan	–	13.88	11.88	13.60	–
Mangalasar	Rajasthan	–	19.02	15.97	21.40	–
Mansarovar	Rajasthan	1950	18.52	13.05	19.70	–
Mashi	Rajasthan	1960	48.14	35.11	30.00	–
Morah Sagar	Rajasthan	–	–	12.94	44.00	–
Morel Bundh	Rajasthan	1956	67.68	70.68	119.60	–
Moti Sagar	Rajasthan	–	12.54	12.26	14.20	–
Nahar Sagar Dam	Rajasthan	1905	23.14	20.10	22.70	–
Namana	Rajasthan	1956	21.24	20.64	45.00	–
Nand Samand	Rajasthan	–	–	121.24	13.00	–
Narainsagar	Rajasthan	1968	19.94	19.82	28.00	115
Nardi Budiya	Rajasthan	–	–	14.16	20.30	–
Nehar Sagar	Rajasthan	1958	23.17	20.10	33.10	–
Orai Dam	Rajasthan	1967	35.28	32.79	46.00	–

(*Continued*)

Table 16. (Continued)

Name of the project	State	Completed in	Gross storage capacity (million cubic meter)	Live storage capacity (million cubic meter)	Designed annual irrigation (Million sq. m)	Installed capacity (MW)
Paibalapur	Rajasthan	1956	11.66	10.08	10.00	–
Parwan	Rajasthan	1960	–	36.00	–	–
Raisagar Bundh	Rajasthan	–	12.54	10.95	–	–
Rajsamand	Rajasthan	1971	107.20	98.71	28.60	–
Ramgarh	Rajasthan	1901	–	58.98	48.60	–
Sainthal	Rajasthan	–	12.86	13.70	19.60	–
Sardar Samand	Rajasthan	1906	–	88.23	85.60	–
Sareri Dam	Rajasthan	1957	55.75	55.07	53.00	–
Silised	Rajasthan	–	13.93	12.12	17.40	–
Surwal	Rajasthan	1956	–	13.50	40.80	–
Tordi Sagar	Rajasthan	1887	–	47.11	83.70	–
Udai Sagar	Rajasthan	1700	31.13	27.60	19.40	–
Umed Sagar	Rajasthan	1903	–	18.60	25.40	–
Urmila Sagar	Rajasthan	–	573.00	534.80	16.80	–
Adwa	Uttar Pradesh	1984	88.10	83.70	169.80	–
Adhrawara	Uttar Pradesh	1955	60.60	58.24	88.80	–
Arjun	Uttar Pradesh	1957	69.20	63.80	107.70	–
Balan & Tons	Uttar Pradesh	1961	261.05	195.39	–	–
Balimiki	Uttar Pradesh	1958	38.35	37.78	62.70	–
Barwa	Uttar Pradesh	1923	214.92	23.28	54.60	–
Barwar	Uttar Pradesh	1923	33.39	31.72	34.50	–
Barwa Sagar	Uttar Pradesh	1968	–	10.20	–	–
Baur	Uttar Pradesh	1967	103.35	98.38	189.00	–
Chandra Prabha	Uttar Pradesh	1960	93.39	90.00	146.10	–
Chandrawal	Uttar Pradesh	1972	34.71	30.86	43.10	–
Dhandhraul	Uttar Pradesh	1917	130.15	129.33	–	–
Dhukwan	Uttar Pradesh	1909	–	58.70	–	20
Dungia	Uttar Pradesh	1918	28.32	20.25	–	–
Gangao Dam	Uttar Pradesh	1915	56.49	56.49	376.40	–
Garai Upper Storage	Uttar Pradesh	1915	27.04	14.80	–	–
Ghagar Main	Uttar Pradesh	1917	152.50	129.46	271.10	–
Gularia Dam	Uttar Pradesh	1968	83.00	80.00	12.80	–
Haripura	Uttar Pradesh	1975	72.75	54.56	199.90	–
Jamni	Uttar Pradesh	1973	92.90	84.10	124.80	5
Jirgo Reservoir	Uttar Pradesh	1958	150.00	131.70	259.00	–
Kabrai	Uttar Pradesh	1955	14.61	13.30	19.40	–
Kohargaodi	Uttar Pradesh	1930	–	12.52	32.30	–
Lachura	Uttar Pradesh	1910	15.31	11.00	970.00	–
Lalitpur	Uttar Pradesh	1953	96.78	80.00	75.80	–
Majhgawan	Uttar Pradesh	1917	26.80	25.05	37.70	–
Meja Reservoir	Uttar Pradesh	1965	303.22	283.43	212.00	–
Musakhand	Uttar Pradesh	1967	113.26	110.37	222.90	–
Nagwa	Uttar Pradesh	1952	45.50	45.39	272.40	–

Nanak Sagar	Uttar Pradesh	1962	09.70	162.76	537.10	–
Nangadh	Uttar Pradesh	1960	101.90	76.42	348.00	–
Naugarh Dam	Uttar Pradesh	1968	101.88	87.81	–	–
Ohan	Uttar Pradesh	–	–	38.00	–	–
Pahari	Uttar Pradesh	1913	79.34	46.00	–	–
Pahuj Reservoir	Uttar Pradesh	–	–	15.00	–	–
Paricha	Uttar Pradesh	1885	–	78.60	4, 525.50	–
Pilli Dam	Uttar Pradesh	1968	55.20	47.73	114.90	–
Rangwan	Uttar Pradesh	1957	163.67	154.00	376.40	–
Saprar	Uttar Pradesh	1956	75.25	65.65	170.00	–
Sarda Sagar-I	Uttar Pradesh	1962	493.00	364.00	677.50	–
Sarda Sagar-II	Uttar Pradesh	1962	567.00	417.00	739.30	–
Shazad Dam	Uttar Pradesh	–	127.10	118.98	200.00	–
Sirsi	Uttar Pradesh	1958	206.44	197.67	60.00	–
Tumaria Extension	Uttar Pradesh	1964	168.10	152.86	161.90	–
Upper Khajuri	Uttar Pradesh	1962	48.17	42.43	72.80	–
Hinglow Dam	West Bengal	1976	17.07	14.15	123.80	–
Kangsabati	West Bengal	1965	1, 036.00	900.40	590.00	–

Table 17. Salient features of some projects under construction in Ganga Basin

Name of the project	State	Gross storage capacity (Million m^3)	Live storage capacity (Million m^3)	Annual irrigation (Million m^2)	Installed capacity (MW)
Anjan	Bihar	23.26	21.29	–	–
Auranga	Jharkhand	502.00	442.65	754.00	–
Barnar	Bihar	79.95	70.54	220.00	–
Batan	Bihar	64.30	59.18	60.00	–
Dhansing Totli	Bihar	12.10	10.39	29.50	–
Durgabati	Bihar	287.70	257.70	360.00	–
Jalkund	Bihar	308.00	231.00	–	–
North Koel	Jharkhand	1, 171.63	968.15	130.00	–
Oriya Dangro	Bihar	–	14.08	36.40	–
Orni Reservoir	Bihar	51.56	44.60	96.00	–
Punasi	Bihar	141.35	113.16	240.00	–
Sugthana Reservoir	Bihar	23.44	18.26	36.40	–
Torai Reservoir	Bihar	36.32	32.61	80.00	–
Upper Kiul	Bihar	97.98	84.25	150.00	–
Upper Shank	Bihar	278.72	239.87	71.10	–
Bah	Madhya Pradesh	86.40	76.52	145.00	–
Banki	Madhya Pradesh	18.42	17.07	34.40	–
Barcharnalla Tank	Madhya Pradesh	18.00	17.00	33.00	–
Barnai	Madhya Pradesh	12.00	11.00	20.80	–
Bunda Nalla Project	Madhya Pradesh	15.00	14.00	40.00	–
Chaldu Tank	Madhya Pradesh	49.19	38.66	100.00	–
Dudhi	Madhya Pradesh	26.66	22.40	40.00	–

(Continued)

Table 17. (*Continued*)

Name of the project	State	Gross storage capacity (Million m^3)	Live storage capacity (Million m^3)	Annual irrigation (Million m^2)	Installed capacity (MW)
Ghorapachar	Madhya Pradesh	12.41	11.57	17.80	–
Ghunghotta	Madhya Pradesh	110.50	83.60	130.50	–
Kaohan Dam	Madhya Pradesh	28.02	25.42	–	–
Lakhundar	Madhya Pradesh	–	33.90	60.00	–
Makroda	Madhya Pradesh	46.64	41.00	57.80	–
Mahan-T	Madhya Pradesh	104.00	100.00	197.40	–
Naren Irr. Project	Madhya Pradesh	20.54	18.32	28.30	–
Rampura Khurd	Madhya Pradesh	13.81	11.83	33.00	–
Sind	Madhya Pradesh	–	13.44	370.00	–
Sindh River Project	Madhya Pradesh	96.60	82.80	376.50	–
Bassi	Rajasthan	23.20	20.24	32.50	45
Bhimsagar	Rajasthan	–	75.20	99.90	–
Bilas	Rajasthan	–	22.95	27.00	–
Bilaspur Project	Rajasthan	–	36.00	–	–
Galwa-II	Rajasthan	67.12	65.68	66.80	–
Gausunda	Rajasthan	75.47	66.98	70.00	–
Harishchandra Sagar	Rajasthan	–	72.00	161.80	–
Hindlot	Rajasthan	22.60	20.38	–	–
Kothari Stage-II	Rajasthan	26.00	21.50	34.00	–
Panchana	Rajasthan	59.45	52.61	99.80	–
Piplot	Rajasthan	–	18.00	–	–
Wagan	Rajasthan	–	37.56	53.70	–
Birpur	Uttar Pradesh	–	39.40	–	–
Gaunta Dam	Uttar Pradesh	28.81	24.28	49.70	–
Kanhar	Uttar Pradesh	323.00	127.54	320.00	–
Kanota	Uttar Pradesh	14.15	12.45	–	–
Mejha	Uttar Pradesh	303.22	227.25	–	–
Pathkavli	Uttar Pradesh	163.00	122.25	80.00	–
Saznam	Uttar Pradesh	81.54	71.60	72.70	–
Urmil Tank	Uttar Pradesh	116.60	111.50	80.00	–

8.5.26. Proposed Hydroelectric Projects in Uttaranchal

An agreement has been made between Govt. of Uttaranchal & some Central Public Sector Power Undertakings [National Hydroelectric Power Corporation (NHPC) Ltd., Tehri Hydro Development (THDC) Ltd., Satluj Jal Vidut Nigam (SJVN) Ltd., and National Thermal Power Corporation (NTPC) Ltd.] for construction of several hydroelectric power projects. A list of these projects is given in Table 18.

Koti Bhel Hydropower Stage–I A: This project is proposed at Muneth Village near Dev Prayag in Tehri Garhwal District of Uttaranchal. A 82.5 m high concrete gravity dam is proposed to store and divert water for an underground power house

Table 18. Hydroelectric Projects Proposed or being executed in Uttaranchal

Name of project	Capacity in MW	Location	Constructing agency
Bonkang Bailing	330	Pithoragarh	THDC Ltd
Chungar Chal	240	Pithoragarh	NHPC Ltd
Devra Mori	33	Uttarkashi	SJVN Ltd
Devsari Dam	300	Chamoli	SJVN Ltd
Garba Tawaghat	630	Pithoragarh	NHPC Ltd
Gohana Tal	60	Chamoli	THDC Ltd
Jadhganga	50	Uttarkashi	THDC Ltd
Jakhol Sankari	35	Uttarkashi	SJVN Ltd
Jhelam Tamak	60	Chamoli	THDC Ltd
Karmoli	140	Uttarkashi	THDC Ltd
Karmoli Lumti Tuli	55	Pithoragarh	NHPC Ltd
Lata Tapovan	162	Joshimath	NTPC Ltd
Maleri Jhelam	55	Chamoli	THDC Ltd
Rupsiabagar Khasiyabar	260	Pithoragarh	NTPC Ltd

that will have 3 units of 80 MW each (total 240 MW). The estimated cost of the project is Rs. 1,264. Preparation of DPR is in progress. *Koti Bhel Hydropower Stage – I B*: This project is proposed at Village Pali near Dev Prayag in Tehri Garhwal Districts of Uttaranchal. Here, a 83 m high concrete gravity dam with an underground power house having 4 units of 70 MW each will be constructed to generate 280 MW (4*70 MW) of power. Preparation of DPR of this project is in progress.

Koti Bhel Hydropower, Stage – II: It is proposed at Village Kaudiyala near Rishikesh in Pauri & Tehri Garhwal Districts of Uttaranchal. The proposal is to construct a 85.0 m high concrete gravity dam and an underground power house with 8 units of 55 MW (total capacity = 440 MW). Preparation of DPR is in progress and the project is likely to cost Rs. 2,577 crore.

8.6. GANGA-BRAHMPUTRA-BARAK BASIN STUDY

Chaturvedi and Rogers (1985) have presented the results of extensive studies on the Ganga–Brahmaputra–Barak basin in the Indian sub-continent. The following discussion is based on their book detailing the study. The *Greater Ganga system* is the second largest international river basin in terms of runoff, second only to the Amazon basin in South America. The peak outflow from the system at its estuary is $141,000\,m^3/s$. It carries about 127.61 m-ha-m of water in the Bay of Bengal each year of which about 80% is in the monsoon season. The monthly mean flow of Ganga River at its tail end reaches up to 57,000 cumec. The north-eastern boundary of this system is formed by Himalayas which are geologically young mountains and it is bounded by Vindhyachal mountains in the south. A low watershed separates the basin from the Indus basin in the west. The extensive alluvial plains of Ganga basin are part of this system. The delta of the Greater Ganga system covering an area of 56,700 sq. km is one of the biggest in the world.

The Greater Ganga system has wide diversities in physiographic- geographical characteristics, topography, soil and land use as well as socio-economic aspects. The basin is like an elongated bowl with very high steep mountains in the north, comparatively low mountains in the south and east and a very flat fertile alluvial plain in between. The region has also witnessed rapid growth in population over the last few decades; there has been tremendous urbanization and demands for water have risen rapidly. The development in the basin has been largely on an ad-hoc basis. Although the region has huge water potential, due to various reasons including its international character, most of this potential remains unutilized. More than 80% of annual precipitation takes place in four months of monsoon; the area receives solid as well as liquid precipitation. The physiographic and meteorological characteristics of the system coupled with monsoon concentrated precipitation lead to heavy floods. Since Himalayas are geologically young and erodible mountains and have very steep slopes, the high flows also carry high sediment loads. Most of the storage sites and hydroelectric potential lie in Nepal and the North-East part of India. The emphasis on WR development in the past was on flow diversions limited to low flows.

The schematic diagram of the system used in the coordinating model is given in Figure 17 There are 49 river–schematic nodes of interest at which junctions, diversions, ground water pumping, and return flows occur. In the first stage of preliminary screening, a single linear model was constructed to explore and coordinate all the demands placed on the system. This model was used as a tool to explore the various goals to be obtained and the constraints on development. The idea was that once the system has been fully explored by this inexpensive model it can be broken up into smaller, more manageable pieces for a more complete analysis. The model considered five reservoirs and 75 irrigation works as already developed. In the second stage, the entire basin was decomposed into smaller systems. Two types of decomposition were planned. In the first case of hydrologic decomposition, the system was divided into nine sub-basins. The operation of each sub-basin was optimized by varying the irrigation level under the given water releases and energy target levels. Each sub-basin then reports to the central organisation, its optimal irrigation level and search for energy production, shadow prices and their effective ranges on the water and energy target levels. The master problem was then solved by maximising incremental irrigation areas and the total irrigation level and energy production for the whole basin is computed. This process is done in an iterative fashion. In the second type of decomposition, the basin was decomposed by political units. The problem was analysed under two schemes: flow quota scheme and resource allocation scheme. These schemes were worked out by three algorithms. The first algorithm requires that the minimum flow at any point in the Ganga should be greater than the sum of the flow quota fixed by the central government for all the states above that point. Under the second algorithm, water that leaves an upstream state may be used by a downstream neighbor, provided that

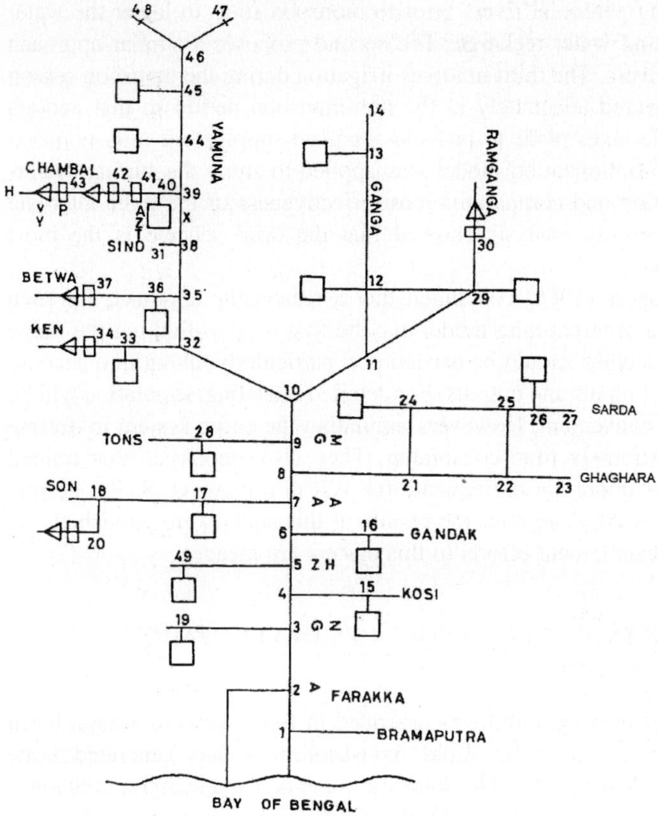

Figure 17. Schematic diagram of Greater Ganga System

this neighbor allows a flow greater than or equal to the sum of all the upstream states quota into the next state. In the third algorithm, the restriction of algorithm two was lifted.

The authors emphasised the conjunctive use of surface and ground water for this system pointing out that ground water is a major user of energy and surface water is a rich source of energy through hydroelectric generation. In the conjunctive use study, it was proposed that infiltration may be increased during the monsoon season by heavy pumping during non-monsoon season and thus creating ground water storage. The extent and date of pumping was estimated such that it is replenished in 120 days of the monsoon season and equilibrium is achieved. The areas suitable for ground water storage were identified and it was concluded that in this scheme of underground storage of flood waters, the total potential irrigation in the Ganga basin may be limited by the area of irrigable land rather than water supply. Various alternative schemes of ground water recharge were proposed. The first involves pumping heavily along perennial rivers prior to monsoon so as to lower the water table and induce ground water recharge. The second proposes a similar approach along non-perennial rivers. The third involves irrigation during the monsoon season with groundwater lowered adequately in the non-monsoon period so that enough ground water recharge takes place to provide adequate supplies for non-monsoon months. A simulation-optimization model was applied to study the surface water-ground water interaction and comparative cost effectiveness of the three alternate approaches. The sensitivity analysis showed that the third scheme is the most attractive.

Chaturvedi and Rogers (1985) concluded that a reasonable approach for such large systems is that a programming model may be first used to find out the range for which simulation studies should be carried out, particularly taking into account the stochastic nature of inputs and outputs. For detailed modeling, simulation will be most convincing and convenient. However, simulating the entire system in diverse conditions will be extremely time consuming. They also emphasize that trained manpower is the most important prerequisite for WR development. Sadly, despite a passage of more than 20 years after the results of this study were published, the additional resource development efforts in this system are meager.

8.7. PROBLEMS IN WATER RESOURCES DEVELOPMENT OF GANGA BASIN

Highly productive irrigated agriculture is practiced in fertile soils of Ganga basin since time immemorial and recent rapid industrialization has generated large demands for water and hydropower. The domestic water demand in high-population-density urban areas scattered throughout the Ganga basin has outstripped the supplies. Bangladesh which lies at the downstream end, requires a mechanism that can augment the dry season flow as well as control floods. The irrigation infrastructure in Ganga basin, as elsewhere in the country, is getting old. A number of large irrigation projects are over 100 years old. During the past 50 years, many of these have been modernized by replacing the weirs and anicuts by gated barrages and repairing the canal systems.

Large dams on the perennial rivers are an inevitable solution to the above problems. A few smaller projects have been already completed and several larger ones are planned in the Himalayan catchments of the northern tributaries to the Ganga and Brahmaputra. There is no doubt that storage of even a fraction of the huge flow and generation of hydroelectricity would go a long way in advancing economic transformation and development within the GBM. In short, these dams have been described as panacea for the poverty in the Himalayan region. However, decisions regarding construction of these dams is no longer guided by hydrology and engineering; other factors have started influencing the decisions.

The basic problem in utilizing water resources in the Ganga basin is that in relation to the relatively large annual flow in the basin, the storage capacity of existing and foreseeable reservoirs in India is not large enough to permit conservation of flows during high flow season. The live storage capacity of all reservoirs in the Ganga basin is less than one-sixth of the annual flow, which does not permit a significant degree of flow regulation. Lean season flows in the basin without an adequate storage backup are not sufficient to meet the requirements for various demands while monsoon flows are so high that the Ganga and its tributaries remain in spate almost every year.

Many of the diversion works are not backed by any large upstream storage. Therefore the supply of water for irrigation is limited by the flow of the rivers. Only a few tributaries, namely Chambal, Betwa and Sone have large reservoirs and are relatively better developed.

A list of the important projects in Ganga Basin with storage back up (if any) is given in Table 19.

Table 19 shows that although there is a vast canal networks in Upper Ganga basin, there are no large existing reservoirs except on Ramganga. After the Tehri dam is completed in near future, Ganga River will be partly regulated. Tehri will also help in irrigating new areas of about 3 lakh ha besides stabilizing the existing system. Tehri will also supplement water supply to Delhi through the Upper Ganga Canal.

In the Yamuna system, the possibility of any upstream reservoir in near future is bleak. Although the Lakhwar Vyasi Dam is under construction for over 3 decades, no significant progress has been made. Similarly, two other reservoirs, namely Kishau and Renuka are in pipeline for more than 10 years. Gandhisagar dam on Chambal River with a storage capacity provides regulated supplies to Rana Pratap Sagar, Jawahar Sagar and Kota Barrage.

Rihand dam on a tributary of Sone River is the other large reservoir in the system. Rihand is primarily a hydropower station and is operated to meet the peaking demands. The power releases from the project supplement Sone Barrage particularly

Table 19. Important Irrigation Schemes in the Ganga Basin

Name of the project	Area benefited	Major upstream storage dam		
		Existing	Under construction	Proposed
Upper Ganga Canal	North West U.P.	Nil	Tehri	–
Madhya Ganga Canal	Western U.P.	Nil	Tehri	–
Lower Ganga Canal	Western U.P.	Ramganga	–	–
Western Yamuna Canal	West U.P./Delhi	Nil	Lakhwar	Kishau
Eastern Yamuna Canal	West U.P./Delhi	Nil	Lakhwar	Kishau
Sharda Sahayak Canal	North East U.P.	Nil	–	Pancheshwar
Saryu Canal	North East U.P.	Nil	–	Karnali (in Nepal)
Sone Canal	Bihar	Rihand	Bansagar	–
Gandak Canal	Bihar	Nil	–	–
Kosi Canal	Bihar	Nil	–	–
Kota Barrage	Rajasthan/M.P.	Gandhisagar	–	–

during dry season. Bansagar an interstate project of UP, Bihar and Madhya Pradesh is nearing completion. It will further improve the water availability at Sone barrage.

Sharda Sahayak and Saryu Nahar in U.P. and Kosi and Gandak Projects in Bihar also do not have any back up storage. Despite being very attractive hydropower projects, Karnali and Pancheshwar projects, little headway is made in construction. Kosi and Gandak dam projects are still on drawing boards. There appears to be poor coordination between India and Nepal on joint execution of projects. Delay in implementing the projects will result in escalation of costs. Irrigation potential developed from surface water is around 45% and that from ground water is around 80% in Bihar while the same in U.P. is around 60% and 90% respectively.

After the formation of Jharkhand and Uttaranchal States, the hydropower potential in U.P. and Bihar is only 403 and 60 MW respectively. This has been developed to the extent of 70 to 80 percent. The hydropower potential in Uttaranchal is about 9,341 MW of which only 9% is developed. Similarly in Jharkhand, only 15.73% of the total 478 MW has been developed. This underlines the need for more large dams to be taken up in Uttaranchal for hydropower generation and these will also provide regulated releases to stabilize existing irrigation systems in Ganga and Yamuna basins.

No large reservoir has been constructed on Ken River which is a major tributary of Yamuna. Ken multipurpose project has been under consideration for several decades but no progress was made in absence of agreement on sharing of its waters between UP and MP. This project will also transfer surplus water to Betwa basin where construction of Rajghat and Matatila have affected the development in the upstream reaches. More about this issue can be found in a later Chapter.

Development of large reservoirs in southern tributaries of Ganga has not been satisfactory (Hasan, 2005). The North Koel dam project remains to be completed for several decades for some environmental hitches. Similarly Auranga dam is yet to be taken up for implementation. Bue even pending projects are completed, after the total live storage capacity created in Ganga basin will be about 52 BCM. This will be less than 10% of the average annual runoff and less than 20% of the utilizable runoff. Therefore, sincere efforts are needed to complete the major storage dams which have already been identified in Uttaranchal, U.P., Bihar and M.P. Of late, the problems encountered in constructing dams in India are increasing. A majority of good storage sites of the basin are in Nepal and this is a cause of delay in the construction of dams on the northern tributaries of Ganga. Agreements with Nepal are to be made for implementation of identified projects.

In the Ganga basin, the flooding problem is mainly confined to the middle and terminal reaches. In general, the severity of the problem increases from west to east and from south to north. The worst flood affected states in the Ganga basin are Uttar Pradesh, Bihar, and West Bengal. In Uttar Pradesh, flooding is largely confined to the eastern districts where the rivers that cause flooding include the Sarada, the Ghagra, the Rapti, and the Gandak. The major causes of flooding here are drainage congestion and bank erosion. North Bihar is in the grip of floods almost every year due to the spillage of rivers. In West Bengal, floods are caused due to the drainage problems as well as tidal effects. The Ganga Flood Control Commission was set up by the Government of India for flood management in the Ganga Basin.

The Ganga basin has an extremely high density of population. This dense population, coupled with high growth rate is expected to generate huge demand for additional water in Ganga basin. Further, industries are rapidly growing in the region. This will create substantial additional water demand as well as problems of water quality. This scenario will be similar in all the countries in the Ganga basin: Nepal, India, and Bangladesh. In the absence of a well-coordinated water resources developing and sharing agreement, the local, regional, and international conflicts may become critical. Nepal and India should come to agreements on joint developments of major water projects. Between Bangladesh and India, the sharing of the flow in the Ganga below Farakka has evaded a long term solution. In the overall interest of sustainable development in the region, early resolution of these problems is very much needed.

8.8. WATER QUALITY ASPECTS

In addition to the quantity related problems, an increasing level of pollution from urban and industrial areas has also created a sharp decline in the quality of Ganga water. The problem has arisen largely due to the discharge of untreated urban wastes and industrial effluents from the cascade of large and medium cities located along the course of Ganga and its tributaries. Although Ganga is considered as a holy river in mythology, people do not hesitate while dumping domestic and industrial waste in the river.

Numerous cities located in the Ganga basin generate and discharge huge quantities of wastewater, a large portion of which eventually reaches the river through natural drainage system. The industrial complexes which have sprung up along the rivers also dump considerable pollution loads into the river. Over the years, the Ganga and its tributaries have become the channels of transport of industrial effluents and the drains for the wastewater of the cities. It is estimated that some 900 million litres of sewage is dumped into the Ganga every day; three-fourths of the pollution in the Ganga is from untreated municipal sewage. In particular the middle reach of the basin between Kanpur and Buxar is the most urbanized and industrialized, as also the most polluted segment of the basin. Municipal and industrial wastes with dangerous concentration find entry into the watercourse in this segment and pose a grave threat to society.

In the hilly reaches up to Rishikesh, Ganga water is quite clean except for sediments. From Rishikesh onwards, disposal of sewage into Ganga begins. Besides the municipal waste of Rishikesh and Haridwar, industrial units discharge partly treated effluents into the river. Haridwar City has a population of 1.5 lakh and nearly 60,000 people visit the city every day on an average. This number rises to a few lakh on important religious days and may go up to 15 lakh on the auspicious days during Kumbha Mela (fair). Considerable lengths of sewer lines are clogged by silts that flow in from the adjoining hills.

Further downstream from Haridwar, Ganga flows through Bijnor, Garhmukteshwar, Narora and Kannauj. Here, water is not much polluted as these two towns do not have any large industry. A note worthy feature of this area is considerable quantity of baseflow that joins the river in this reach during the post-monsoon season. Moving downstream, the situation changes for the worse at Kanpur from the quality point of view. Sewage from the city (population 2.7 million) coupled with untreated toxic waste discharge from about 150 industrial units results in severe damage to water quality. The mean value of DO at 3 mg/l at Jajmau, near Kanpur, reflects the levels of pollution caused by discharge from 80 tanneries and other industries.

At Allahabad with population of more than a million, municipal wastes are the major contributor to river pollution. Yamuna whose water is highly polluted joins Ganga at Sangam. Large volume of municipal and industrial waste is dumped in the river at Varanasi, a city with approximately 1.2 million population, The Varuna River, which joins the Ganga in the vicinity of Varanasi, receives waste from many drains. Besides, due to the religious belief that those who die in Varanasi are sure to go to heaven, on average, more than 40,000 dead bodies are cremated on the river bank and the ashes and remains are dumped in the river.

Entering in Bihar, a number of industries (including fertilizer and oil refining) have come up along Ganga River. Patna is the most populous city whose wastes are dumped in the river. At Kolkota in West Bengal, the Hooghly (Ganga) river basin is highly populated as the waster from numerous industries as well as municipal sewage is dumped in the river.

In view of the magnitude of water quality problems in the Ganga basin, two actions plans were launched by the government of India: the Ganga action plan and the Yamuna action plan. These are discussed next.

8.8.1. Ganga Action Plan

The pollution load dumped in the river by human interference is a serious health hazard to the dense population of the basin. In recognition of the magnitude of this problem and realizing the importance of water quality as a cardinal element of river management, the Government of India started the planning and execution of several programs to check the pollution of the river Ganga from Rishikesh to Diamond Harbour and its tributaries. An ambitious programme, known as the Ganga Action Plan (GAP), was initiated in 1985. The main objectives of the plan are to improve the quality of the river water and to establish self-sustaining city authorities with the capacity to install and maintain treatment plants. Under the GAP, the government attempted to build a number of waste treatment facilities and to collaborate with a number of voluntary organizations. Pollution abatement works for the Ganga River had been taken up in 25 class I towns (population exceeding one lakh) along the main Ganga River under the three basin States of U.P., Bihar, and West Bengal. The main objectives of the GAP were:

- Reduction of pollution load on the river and improving the water quality as a result thereof, and
- Establishment of domestic/municipal wastewater treatment systems with emphasis on resource recovery to make such systems self-sustainable as far as possible.

In addition, GAP was to serve as a model to demonstrate the methodology of improving the water quality of the other polluted rivers and water bodies of the country to their designated best use class. A multi-pronged approach was adopted to achieve the objectives of the GAP. Similar other plans are in various stages of implementation in some tributaries of the Ganga.

The major polluting industries on the Ganga are the leather industries, especially near Kanpur, which use large amounts of Chromium and other chemicals and much of it finds its way into the meager flow of the Ganga. With the lax monitoring and enforcement by the government, the possibility of immediate control of pollution is limited. However, industry is not the only source of pollution. Sheer volume of waste – estimated at nearly 1 billion litres per day – of mostly untreated raw sewage – is a significant factor. Also, inadequate cremation procedures contribute to a large number of partially burnt or unburnt corpses floating down the Ganga. In addition, animal corpses can also be seen floating in the river.

The first phase of the plan envisaged major works such as renovation of sewage pumping and treatment plants, setting up of new treatment plants to produce energy, manure and biogas and laying down sewage disposal systems where they do not exist at present. According to the reports, all the sewage pumping stations in Rishikesh and Haridwar are in order and pollution of the river has been largely checked. At Kanpur, work has been initiated for cleaning the main stream and renovation and restoration of the existing sewage pumping station and treatment plants has been taken up. Similar works have been taken up at Varanasi, Allahabad, and Patna. Table 20 gives desired and the real water quality levels for Ganga River at various locations. It can be noted that there is no significant improvement in the river water quality with time, in fact the quality has deteriorated at many places. Further, total coliform remains a serious problem at majority of the locations.

The Ganga River dolphin (*platanista gangetia*) is found in India, Nepal, Bhutan and Bangladesh, in Ganga, Brahmaputra, Meghna, Karnaphuli and Hoogli river systems. A unique feature is that these dolphins are blind and their eyes have no lenses. They use a sophisticated echolocation system to navigate and find food. They eat shrimp and fish from the mud in river bottoms. They are solitary creatures and are only found in fresh water. The Ganga River dolphin is an endangered species which are hunted by humans for meat and oil.

8.8.2. Yamuna Action Plan

In recent years, the Yamuna has become highly polluted due to various reasons and this has adversely affected human health and bio-diversity of the eco-system. Like other rivers, the main cause of pollution of the Yamuna River is discharge

Table 20. Desired and Existing (at various times) Water Quality Levels for Ganga

Location	Desired class	Observed class & critical parameters				
		1997	1998	1999	2000	2001
Ganga at Rishikesh U/S, U.P	A	B, Totcoli	B, Totcoli	C, Totcoli	NA	C, Totcoli
Ganga at Haridwar D/S, U.P	B	C, Totcoli	C, Totcoli	C, Totcoli	NA	C, Totcoli
Ganga at Garhmukteshwar, U.P	B	E, BOD, Free Ammonia	NA	D, BOD	NA	D, BOD, Totcoli
Ganga at Narora (Bulandsahar), U.P.	B	NA	NA	D, BOD, Totcoli	NA	D, BOD, Totcoli
Ganga at Kannauj U/S (Rajghat), U.P	B	D, BOD, Totcoli	D, BOD	D, Totcoli	D, BOD, Totcoli	D, BOD, Totcoli
Ganga at Kannauj D/S, U.P	B	D, BOD, Totcoli	D, BOD	D, Totcoli	D, BOD, Totcoli	D, BOD, Totcoli
Ganga at Bithoor (Kanpur), U.P.	B	D, Totcoli	NA	D, Totcoli	D, Totcoli	D, Totcoli
Ganga at Kanpur U/S (Ranighat), U.P	B	D, BOD, Totcoli	NA	D, Totcoli	D, Totcoli	D, Totcoli
Ganga at Kanpur D/S (Jajmau Pumping Station), U.P	B	D, BOD, Totcoli	D, BOD	D, Totcoli	D, BOD, Totcoli	D, BOD, Totcoli
Ganga at Dalmau (Rai Bareilly), U.P.	B	D, Totcoli	D, Totcoli	NA	NA	D, Totcoli
Ganga at Allahabad (Rasoolabad), U.P.	B	D, BOD, Totcoli	E, Totcoli, Free Ammonia	C, Totcoli	NA	C, Totcoli
Ganga at Allahabad D/S (Sangam), U.P.	B	D, BOD, pH, Totcoli	E, pH, Totcoli, Free Ammonia	D, Totcoli	NA	D, BOD, Totcoli
Ganga at Varanasi U/S (Assighat), U.P	B	D, BOD, Totcoli	D, BOD	D, Totcoli	NA	D, Totcoli
Ganga at Varanasi D/S (Malviya Bridge), U.P	B	E, DO, BOD, Totcoli	E, DO, BOD	D, BOD	NA	D, BOD, Totcoli
Ganga at Trighat (Ghazipur), U.P	B	D, BOD, Totcoli	D, BOD	D, BOD	NA	D, BOD, Totcoli
Ganga at Buxar, Bihar	B	D, BOD	D, Totcoli	D, Totcoli	D, Totcoli	C, Totcoli

Ganga at Khurji, Patna U/S, Bihar	B	D, Totcoli	D, Totcoli	D, Totcoli	D, Totcoli	C, Totcoli
Ganga at Patna D/S (Ganga Bridge),Bihar	B	D, Totcoli	D, Totcoli	D, Totcoli	D, Totcoli	D, Totcoli
Ganga at Rajmahal, Bihar	B	D, Totcoli	D, Totcoli	D, Totcoli	D, Totcoli	D, Totcoli
Ganga at Diamond Harbour, West Bengal	C	B	B	NA	D, Totcoli	D, Totcoli
Ganga at Garden Reach, West Bengal	C	B	B	NA	D, Totcoli	D, Totcoli
Ganga at Howrah-Shivpur, West Bengal	C	D, Totcoli	B	NA	D, Totcoli	D, Totcoli
Ganga at Serampore, West Bengal	C	B	B	NA	D, BOD, Totcoli	D, BOD, Totcoli
Ganga at Baharampore, West Bengal	B	B	B	NA	D, Totcoli	D, Totcoli
Ganga at Palta, West Bengal	B	D, BOD	B	NA	D, BOD, Totcoli	D, Totcoli
Ganga at Dakshineshwar, West Bengal	B	D, BOD	B	NA	D, BOD, Totcoli	D, BOD, Totcoli
Ganga at Uluberia, West Bengal	D	B	B	NA	D	D, DO, BOD, Totcoli

Note: NA- Not Available
Source: Central Pollution Control Board.

of untreated wastewater into the river from the towns located along its banks. To arrest river pollution, certain measures have been taken by the Government of India (GOI) in 12 towns of Haryana, 8 towns of Uttar Pradesh, and Delhi under an action plan. This plan, known as the Yamuna Action Plan (YAP), is being implemented since 1993 by the National River Conservation Directorate (NRCD) of the Ministry of Environment and Forests. YAP has been framed to prevent pollution of the Yamuna River. The towns in Haryana are: Yamuna Nagar, Karnal, Panipat, Sonepat, Gurgaon, Faridabad, Chhachhrauli, Indri, Radaur, Gharaunda, Gohana, and Palwal. In Uttar Pradesh, the focus towns are: Saharanpur, Muzaffar Nagar, Ghaziabad, Noida, Vrindavan, Mathura, Agra, and Etawah.

Delhi alone contributes around 3,296 MLD (million litres per day) of sewage by virtue of drains outfalling in Yamuna. Due to the low perennial flow and the huge quantity of waste it receives, Yamuna is one of the most polluted rivers of the country. Ten to fifteen years ago, a large quantity of Delhi's sewage was used to irrigate nearby agricultural lands. However, today agricultural lands have been converted into residential colonies. Most of these colonies are unplanned and provision of drainage of waste water is usually not made. More than 3.5 lakh people live in the 62,000 *jhuggis* (hutments) that have come up on the flood plains of the Yamuna River and its embankments. The Government of India has prepared plans to rebuild and repair the sewerage systems and the drains that empty into the river over the next five years. However, there are no plans at present to shift the jhuggis.

Rapid urbanization in the Delhi area has further compounded the pressure on the sewerage system. With the population of Delhi increasing from 0.4 million in 1911 to 13.9 million in 2001, there is an ever-increasing pressure on the water resources. Trunk sewers are in poor condition, there is a shortage of sewage treatment capacity and lack of sanitation facilities in unsewered areas of Delhi which account for nearly 50% of the population. These are the factors that are responsible for the continued pollution of the Yamuna in Delhi. Although the water treatment capacities have increased from 159 MLD in 1951 to nearly 2,951 MLD by 2002, the average shortfall in 2002 was about 68 MLD. Further information about the Yamuna Action Plan is available at the web-site *www.yap.nic.in.*

Table 21 gives the desired and the actual water quality class and critical parameters for Yamuna River at various locations. Similar to Ganga, there is no significant improvement in the river water quality of Yamuna with time and the quality has deteriorated at many places. Further, total coliform remains a serious problem at majority of the locations including the locations in Uttaranchal. Planning Commission (2006) reported the following values for Yamuna at Okhla (Delhi): total dissolved solids 569 mg/l, BOD 52 mg/l (norm 3 mg/l), coliform count 85,000 maximum probable number (MPN) per 100 ml (norm 5,000 MPN/100 ml).

Despite huge investments made in the two projects, these have not been very effective in improving the river water quality.

8.9. GROUND WATER ASPECTS IN TERAI AND GANGA PLAIN ZONES

In the Terai region, groundwater occurs both under confined and unconfined conditions. The clayey sand and sandy formation yield groundwater under an unconfined condition down to an average depth of 50 m. Confined groundwater occurs in sand-gravel beds at depths exceeding 50 m. The wells tapping confined aquifer show both flowing and non-flowing conditions. The Terai aquifers show a strong confining condition. The piezometric head in the flowing wells ranges between 6.60 and 8.90 m above ground level and in the non-flowing wells from 1.65 to 11.20 m below ground level. The tubewells yield 25 to 55 1/ s for a pressure head varying between 1.5 m and 8.7 m in the case of flowing wells while the non-flowing wells discharge 8.9 to 38 l/s at drawdowns from 4 to 9 m.

Table 21. Desired and Existing (at various times) Water Quality Levels for Yamuna

Location	Desired class	Observed class & critical parameters				
		1997	1998	1999	2000	2001
Yamuna at U/S Dak Patthar, Uttar Pradesh		NA	B, Totcoli	D, Totcoli	D, Totcoli	C, Totcoli
Yamuna at U/S of Lakhwar Dam, U.P.	A	NA	D, Totcoli	D, Totcoli	NA	C, Totcoli
Yamuna at Mathura U/S, U.P.	B	D, BOD, Totcoli	D, DO, Totcoli	E, BOD, Totcoli, Free ammonia	D, BOD, Totcoli	D, BOD, Totcoli
Yamuna at Mathura D/S, U.P.	B	D, BOD, Totcoli	D, BOD, Totcoli	E, DO, BOD	E, DO, BOD, Totcoli	D, BOD, Totcoli
Yamuna at Agra U/S, U.P.	C	D, BOD, Totcoli	D, Totcoli	D, BOD, Totcoli	D, BOD, Totcoli	D, BOD, Totcoli
Yamuna at Agra D/S, U.P.	C	NA	D, Totcoli	D, Totcoli	NA	NA
Yamuna at Etawah, U.P.	C	D, BOD, Totcoli	E, DO, BOD	D, BOD, Totcoli	D, BOD, Totcoli	D, BOD, Totcoli
Yamuna at Allahabad D/S (Balua Ghat), U.P	B	D, T.Coliform	E, Totcoli, Free Ammonia	C, Totcoli	NA	C, Totcoli
Yamuna at Allahabad. U.P.	B	C, Totcoli	D, Totcoli	D, Totcoli	D, Totcoli	D, Totcoli
Yamuna at Mazawali, U.P.	C	E, DO, BOD, Totcoli	D, BOD, Totcoli	D, BOD, Totcoli	E, DO, BOD, Totcoli, Free ammonia	E, DO, BOD, Totcoli
Yamuna at Bateshwar, U.P.	C	D, BOD, Totcoli	D, BOD, Totcoli	D, BOD, Totcoli	D, BOD, Totcoli	D, BOD, Totcoli
Yamuna at Juhikha (After Confl. With Chambal), U.P.	C	C	C	D, Totcoli	D, BOD, Totcoli	D, BOD, Totcoli
Yamuna at Mathura U/S		D, BOD, Totcoli	D, DO, Totcoli	E, BOD, Totcoli, Free ammonia	D, BOD, Totcoli	D, BOD, Totcoli

Note: NA- Not Available.
Source: Central Pollution Control Board.

Table 22. Groundwater potential (Mm^3) in Indo-Nepal basins

Name of basin	Annual total utilizable potential	Annual resource for irrigation (@ 85% of total)
Gandak & Burhi Gandak	5,057	4,298
Kosi & Mahananda	9,281	7,889
Ghaghra including Gomti	35,645	30,298

8.10. GROUND WATER IN GANGA PLAINS

The Ganga plain in North Bihar is rich in groundwater resources. In all, three major saturated groundwater zones have been identified down to the drilled depth of 690 m. The lower granular zone forms the deep aquifer in the depth range of 300 to 440 m. The extensive granular zones, four in number, occurring within the middle clays in a depth range of 50 to 200 mbgl together constitute the intermediate aquifer and the upper granular zone forms the shallow aquifer extending from 20 to 70 mbgl. In addition to these three aquifers, a water table aquifer also exists.

The groundwater potential of various Indo-Nepal river basins is shown in Table 22. Obviously, considerable irrigation potential from groundwater sources exists in these plains in North Bihar and eastern Uttar Pradesh.

8.11. EPILOGUE

By no means, the task of restoration of the Ganga to its old pristine glory will be easy. Nevertheless, this will be worth the trouble because the sustainable development of a large geographical area and population crucially depends on it.

CHAPTER 9

BRAHMAPUTRA AND BARAK BASIN

The North-East Region (NER) of India covers an area of 27.23 M-ha (8.11% of the country's area) consisting of seven states of Arunachal Pradesh, Assam, Meghalaya, Manipur, Mizoram, Nagaland, and Tripura. Popularly, these states are also known as the *Seven Sisters*. The area is characterized by hills with valleys in between; in all about 65% of the land mass comprises of hills. The terrain coupled with heavy rainfall of the tune of 2,000–4,000 mm causes situations of floods in the valleys and water scarcity in the hills.

Brahmaputra and Barak are the two major river basins in NER. The ground water resource in this region consists of total of 35 BCM. The total water resource including rivers from Tripura, Manipur & Mizoram is 647.83 km^3. Table 1 shows the details of water resources potential in the region.

Clearly, such a huge water resources potential provides lot of scope for developmental activities for hydropower, navigation, pisiculture, irrigation, recreation, and so on. However, due to many reasons, the development of water resources in this region is in nascent stage as compared to other parts of the country. This chapter provides a detailed description of water resources and their development in Brahmaputra basin.

9.1. THE BRAHMAPUTRA BASIN

As per the Hindu belief, Brahmaputra means 'son of the creator, Lord Brahma'. According to Tibetans, the source of this river lies in the Kanglung Kang Glacier near the Kailash range of Himalayas located in the south-western part of the Tibetan plateau at an elevation of 5,300 m (82°10′E, 30°30′N) near Konggyu Tso Lake. Here the river is called Tamchok Khambab Kangri. Many snowfed streams join the river from the passes at about 60 km southeast of Mansarovar Lake. The Brahmaputra River traverses a distance of 2,880 km through three countries, namely, China, India, and Bangladesh, before joining the Bay of Bengal. It has a catchment area of 580,000 sq. km, an average annual discharge of 19,820 cumec, an average annual sediment load of 735 million metric tonnes, and a specific flood discharge of 0.149 cumec/ sq. km. The Brahmaputra River basin with the river system from its source to outfall is shown in the Figure 1.

Table 1. Water Resources Potential in Northeastern Region

Basin	Average annual SW potential (km^3)	GW potential (km^3)	Total potential (km^3)
Brahmaputra	537.2	27.9	565.1
Barak	48.4	1.8	50.2
Potential in Tripura, Manipur, and Mizoram	31.0	1.5	32.5
Total	616.6	31.2	647.8

Among the top ranking large rivers of the world, the Brahmaputra River is fifth with respect to discharge, fourteenth with respect to the drainage area, twenty-fifth with respect to length, second with respect to sediment load, and first with respect to specific discharge.

The catchment area of the river falls in four countries. Although the main river does not flow through the Kingdom of Bhutan, 96% of Bhutan's area falls under this basin. The basin is of irregular shape: the maximum east-west length is 1,540 km and the maximum north-south width is 682 km. The basin lies between 23 °N to 32 °N latitude and 82 °E to 97°50′ E longitude. The part of the Tibetan plateau falling under the basin has an elevation varying from 3,000 to 5,000 m and is dotted with numerous glaciers.

The Brahmaputra valley is long and narrow; it is 640 km long and the width varies from 64 km to 90 km. The valley is bounded in the north by high Himalayan mountain ranges, in the east by the Patkai hill ranges, in the south by the lower (Assam) hill ranges and in the west, it is contiguous with the plains of Bangladesh. The southern (Assam) mountainous region under the basin is comprised of parts

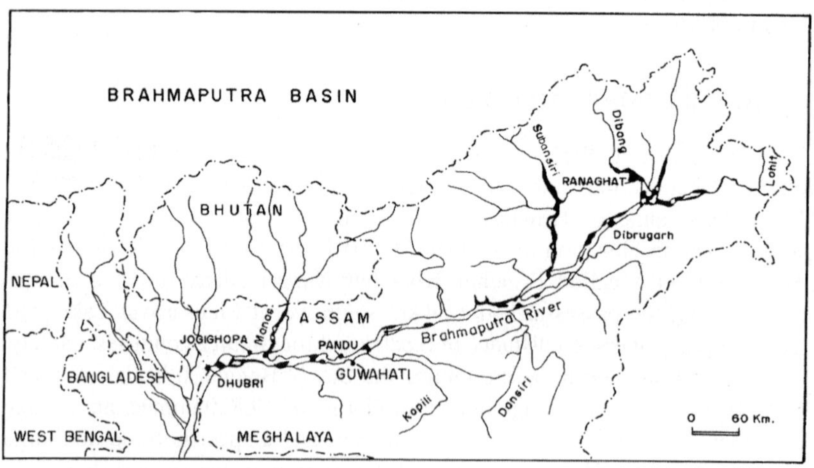

Figure 1. The Brahmaputra basin

of Naga hills, Mikir hills, North Cachar hills, Khasi hills and Garo hills lying in continuity in the east-west direction from Patkai hill ranges up to the Bangladesh border.

The Brahmaputra River characteristics, including the local name, length, catchment area and gradient of the river from source to its outfall and nature of topography at different parts through which the river flows, are furnished in Table 2. In India, the Brahmaputra basin covers parts of six states: Arunachal Pradesh, Assam, Nagaland, Meghalaya and West Bengal and the whole of Sikkim. The statewise distribution of the Indian part of the catchment area is given in Table 3. The culturable area of the basin is 12.15 M-ha which is 6.2% of the culturable area of the country.

The Brahmaputra basin covers 6 topographic regions falling in 4 countries. These regions are described in Table 4.

Table 2. Reachwise distribution of length, catchment area, gradient of the Brahmaputra River and nature of topography through which the river flows

Location (Country the river flows through)	Name river and length (km) in that reach	Catchment area in that part (sq. km) and % of total area	Gradient		Topography of the area
			Reach with Elevation (m)	Average Gradient	
Tibet (China)-(Upper Reach)	Tsangpo (1,625)	293,000 (50.52%)	Source EL = 5, 300 m) to Indo-China Border (EL 660 m)	1:385	High Tibetan Plateau
India (Middle reach)		195,000 (33.62%)			
i) Arunachal Pradesh	Siang in 1st part & Dihang in 2nd part (278)		i) Indo-China border to Kobo (EL 120 m)	1:515	Himalayan mountain region
ii) Assam	Brahmaputra (640)		ii) Kobo to Indo-Bangladesh border (EL 28m)	1:690	Brahmaputra Valley
Bhutan	Main river does not flow here	45,000 (7.76%)	–	–	Himalayan mountain
Bangladesh	Brahmaputra, Jamuna, Padma and then Meghna (337 km up to mouth)	47,000 (up to confluence with Ganga (8.10%)	i) First 60 km. from India Border	1:1,140	Plains including coastal belt
			ii) Next 100 km	1:1,260	
			iii) Next 90 km	1:2,700	
			iv) Rest up to sea	1:3,700	

Table 3. Statewise Distribution of the Indian part of the Brahmaputra catchment area

S. N.	Name of the State	Area of the state (sq. km)	Catchment area in the state (sq. km)	Percentage of state area in the basin	Percentage of total catchment area in India
1.	Arunachal Pradesh	83, 578	81, 600	97.6	41.85
2.	Assam	78, 523	70, 700	90.0	36.26
3.	Meghalaya	22, 489	11, 800	52.5	6.05
4.	Nagaland	16, 523	10, 900	65.0	5.59
5.	Sikkim	7, 300	7, 300	100	3.74
6.	West Bengal	87, 853	12, 700	14.5	6.51
	Total		195, 000		100

Table 4. Topographic distribution of Brahmaputra basin area

S. N.	Nature of Topography	Basin Area under the Topography (sq. km)	Percentage of Total basin area	Geographical Location
1.	High Tibetan Plateau	293,000	50.5	Southern Part of the Tibet province of China.
2.	High Himalayan mountains	137,050	23.6	Part of Himalayan kingdom of Bhutan and of 3 states of India: Arunachal Pradesh, West Bengal and Sikkim.
3.	Brahmaputra Valley	56,200	9.7	Part of Assam State of India.
4.	Lower (Assam) Mountainous Region	37,200	6.4	Part of 3 states of India: Nagaland, Assam and Meghalaya.
5.	Plains	56,550	9.8	Part of the 2 plains districts of West Bengal (India) and part of Bangladesh.
6.	Coastal Region	Negligible		Coastal region of Bangladesh.

9.1.1. The Brahmaputra River System

The course of the Brahmaputra River can be divided into three reaches: upper, middle, and lower. A Flow diagram of the Brahmaputra River and its tributaries has been shown in Figure 2.

Upper Reach: In its upper reach, the river flows 1,625 km from the source to the Indo-China border through Tibetan plateau with an elevation from 3,000 m to 5,000 m, mainly in the east direction almost parallel to the Himalayan mountains and north thereof. Here, the river is known as Tsangpo, which means 'the purifier'. After flowing for 80 km in an easterly direction, Tsangpo meets two big rivers,

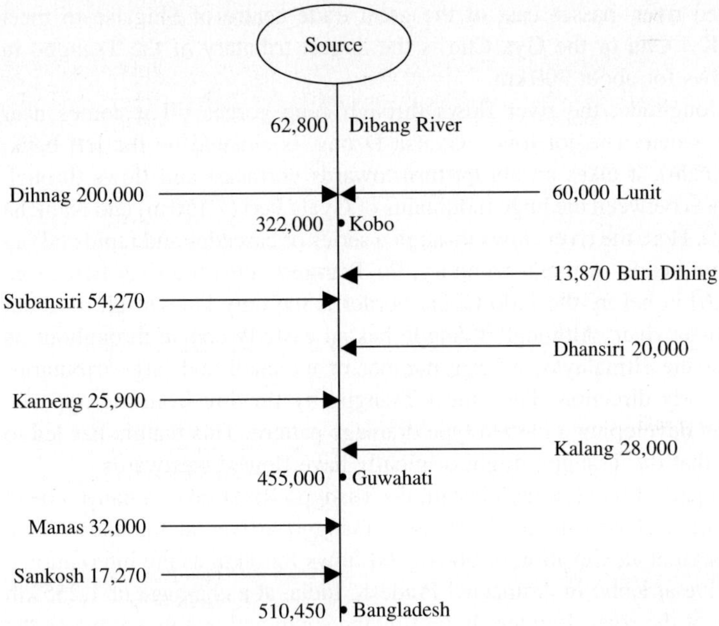

Figure 2. Flow Diagram of the Brahmaputra. The numbers represents average flow in cumec

viz., Mayum Chu and Chema Yung Dung. The main river is about 160 km north of the Himalayas. The drainage area spreads to a maximum of 80 km to the south and 135 km to the north from the Tsangpo.

Important glaciers that form the northern boundary of the basin from west to east are: Kailash 6,100 m (Longitude 82 °E to 84 °E), Lunkar 6,000 m (Longitude 84° to 84°30′ E), Kangchung Kangri and Lapchung 6,000 m (Longitude 84°30′ E to 86°25′ E) and Nyenchen Tanglha 7,088 m (Longitude 86°25′ E to 95 °E). In addition to rainfall, the river in this reach is also fed by snow and glacier melt.

The catchment area of the Brahmaputra River up to the Indo-China border is 293,000 sq. km; this part of the catchment falling under high Tibetan plateau is 50.5% of the total catchment area. The basin here is long and narrow; the maximum length in the east-west direction is 1,540 km and the maximum width in the north-south direction is 310 km. This part of the basin is bounded in the south by Himalayan ranges and in the north-west side initially by the Kailash ranges of Himalayas and then by the Nyenchentanglha mountain ranges.

The 650 km reach from Pindzoling to Gyatsa Dzong is one of the most remarkable inland navigable systems in the world where boats ply at an altitude of 3,659 m and more. In this part, the average width of the river is more than 2 km. Along this reach, the major tributaries joining the Tsangpo are the Shap Chu, Nayang Chu, Rang Chu, Yarling Chu on the right bank and the Tong Chu, Shang Chu, Kyi Chu on the left bank. Of these, the Nayang Chu and Kyi Chu are much bigger. Nayang

Chu, a snow fed river, passes east of the great trade centre of Shigatse to meet Tsangpo. The Kyi Chu or the Gya Chu is the largest tributary of the Tsangpo in Tibet which flows for about 300 km.

After 92 °E longitude, the river flows through deep gorges till it comes near 93 °E longitude where a major town, Gyatsa Dzong, is situated on the left bank. Near Gyala (Linzhi), it takes an abrupt turn towards northeast and flows through stupendous gorges between the huge mountains of Gyala Peri (7,150 m) and Namcha Barwa (7,755 m). Here the river flows through a series of cascades and rapids taking numerous hairpin bends. A major tributary, Po-Tsangpo, joins the river here from north. About 1.6 km before the Indo-China border is the only known fall of about 24.4 m on the main river. Although Tsangpo has an easterly course throughout its run the north of the Himalayas, a large number of its small and large tributaries flow in the westerly direction. They meet Tsangpo by flowing from the opposite direction thereby developing a barbed type drainage pattern. This feature has led to the speculation that the Tsangpo might originally have flowed westwards.

In the eastern part of the Tibetan plateau, the Tsangpo River takes a hairpin bend around the Namcha Barwa mountain ranges. Tsangpo arrives at the Indo-China border near Monku at an elevation of 660 m and flows for 5 km as the international boundary to arrive at Kobo in Arunachal Pradesh, India, at a chainage of 1,255 km from the mouth of the river. It enters India flowing south and is now known as the Siang River.

Many important places are situated on the bank of Tsangpo River and its tributaries. Lhasa, the capital city of Tibet, lies on the bank of Kyi-chhu River, a tributary of the Tsangpo River. Shigatse, the great trade centre and second largest city of Tibet with the great Monastery of Tashi Lhumpo, lies at the confluence of the tributary Nyang-chhu and Tsangpo.

Middle Reach: In the middle reach, from the Indo-China border to the Indo-Bangladesh border, Brahmaputra flows 918 km through India. Of this, 278 km is through the mountainous state of Arunachal Pradesh and the next 640 km is through the valley in the state of Assam. In Arunachal Pradesh, the river is known as Siang in the upper reach and Dihang in the lower reach. The river crosses Himalayas through deep gorges traversing 226 km from the Indo-China border up to the Pasighat town flowing mainly in the southern direction.

From Pasighat up to the Indo-Bangladesh border for a length of 692 km, the river passes through alluvial plains. All along its course in this region, the river flows in a braided form and the main course oscillates from one bank to the other, forming many islands and sand chars. Most of these sand chars are not habitable as they get submerged during high stage. During monsoon months, almost all the braided channels join to form one vast sheet of moving water. The average gradient of the river from the Indo-China border to Kobo within the state of Arunachal Pradesh is 1:515.

Dihang River enters into the plains of Assam at Kobo where it meets two major trans-Himalayan tributaries, Dibong and Lohit, coming from northeast and east, respectively. From here onwards, the combined river is known as Brahmaputra. At the tri-junction of the Dihang, Lohit and Debang, the contribution from the Dihang is about 31.63% of the total discharge. From Kobo, the river flows first in the southwest direction and then in the west direction. The lengths of the river reaches from Kobo to its mouth and to Pasighat are 918 km and 226 km, respectively. The average gradient of the river in this reach is 1:515. River terraces are noticed along the river stretch between Yinkiong and Pasighat. Some of the terraces are 250 m high above the river bed.

Further down, the river passes through towns of Tinsukia, Dibrugarh, North Lakhimpur, Sibsagar, Jorhat, and Golaghat. In this reach, the river is a highly braided channel, except at Pandu where it is constricted into a single channel of 1.2 km wide. This is the narrowest point in the entire reach. In this reach, the river is navigable throughout the year and has the important towns of Tezpur, Naogaon, and Guwahati. Many tributaries meet the river here; the biggest amongst them is the Jia Bhareli which contributes about 4.9% of the total discharge of the Brahmaputra. Downstream of Pandu, many tributaries, notably, the Pagladia, Manas, Champamati, Saralbhang (Gaurang), and Sankosh join the Brahmaputra.

Majuli Island (Latitude 26°53′ N and Longitude 94 °E), the biggest river island of the world, is the most remarkable feature of the valley reach of the Brahmaputra. It is formed by bifurcation of the Brahmaputra into two branches: the Kherkatiya Suti on the north and the Brahmaputra (Dihing) on the south. Majuli is not a normal alluvial island – it was a piece of land known as Majali, on the south bank of the Brahmaputra till the year 1750. An extreme flood caused the Brahmaputra (Luit) to divert its course southwards on the east of Majali, thereby joining it with the Dihing River. Subsequently the Brahmaputra diverted its main flow through the lower reach of the Dihing, and the land in between the two rivers became the Majali Island which is now called Majuli.

Majuli Island is inhabited by 150,000 people and is under constant threat of bank erosion especially during the monsoon season. Owing to active bank erosion, the area of the island has reduced from 1,300 sq. km as in 1950 to 900 sq. km at present. Figure 3 gives a view of Brahmaputra River near Guwahati, Assam.

Figure 3. Brahmaputra River near Guwahati

Tributaries in Middle Reach: A number of tributaries join the main river course in this reach. The two north bank tributaries, Subansiri and Manas are trans-Himalayan rivers. Besides these, the major part of the catchment area of the other north bank tributaries, such as Ranganadi, Borgong, Jia-Bharali and Sankosh, lie deep inside the Himalayan mountainous terrain of Bhutan and Arunachal Pradesh. Details about the river and the basins of the north bank tributaries are furnished in Table 5. Details about the river and the basins of the south bank tributaries are furnished in Table 6.

Subansiri is among the largest tributaries of the Brahmaputra. Gold was commercially mined from its basin till the recent past. Hence it derived this name (Subarna = Gold in Sanskrit). It is a snowfed perennial trans-Himalayan river. Draining an area of $37,700\,km^2$, Subansiri has a maximum discharge of 18,799 cumec and minimum of 131 cumec. It contributes 7.92% of the total yield of the Brahmaputra.

Manas is the biggest tributary of this reach which has a maximum discharge of 7,641 cumec. It contributes 5.48% of the total discharge of the Brahmaputra. Another big river of this reach is Sankosh which contributes 2.81% of the total discharge of the Brahmaputra.

Table 5. Some parameters of major north bank tributaries of the Brahmaputra River in the middle reach in Assam (India)

S. N.	Name of Tributary	Distance of confluence (km) from mouth of Brahmaputra	Altitude at source (m)	Length (km)			Catchment Area		
				In Hills	In Plains	Total	In sq. km	% in hills	% in plains
1.	Simen	917			–				
2.	Jiadhol	877	1,247	–	130	99	1,346	22.7	77.3
3.	Subansiri	820	5,389	312	–	442	37,000	95.7	4.3
4.	Ranganadi	811	3,440	–	21	–	2,940	76.2	23.8
5.	Borgong	719	700	21	66	42	550	63.6	36.4
6.	Jia-Bharali	675	–	198	–	264	11,843	71.9	28.1
7.	Gabharu	635	–	–	–	–	295	19.3	80.7
8.	Belsiri	617	–	–	–	–	751	24.6	75.4
9.	Dhansiri (N)	607	–	–	75	–	956	34.8	65.2
10.	Noa-nadi	567	–	–	–	–	366	18.6	81.4
11.	Nonoi	552	–	–	–	120	860	23.8	76.2
12.	Barnadi	542	–	10	102	112	739	17.2	82.8
13.	Puthimari	509	3,750	74	116	190	1,787	33.4	66.6
14.	Pagladiya	507	1,300	19	180	197	1,820	24.2	75.8
15.	Manas	422	4,900	270	105	375	41,350	85.9	14.1
16.	Champamati	400	–	–	60	–	1,038	13.2	86.8
17.	Gaurang	380	–	–	57	–	1,023	18.5	81.5
18.	Tipkai	377	–	–	–	–	1,744	9.8	90.2
19.	Sankosh	337	7,300	214	107	321	10,345	92.4	7.6

Table 6. Some parameters of the south bank tributaries of Brahmaputra confluencing in the middle reach in the Assam State

S. N.	Name of tributary	Chainage of confluence with Brahmaputra (km from mouth)	Altitude at source (m)	Length (km)			Catchment Area		
				In hills	In plains	Total	Total (km^2)	Percent in hills	Percent in plains
1.	Dibong	982	4,267	136	57	193	12, 270	96.5	3.5
2.	Lohit	977	4,876	310	84	394	23, 400	79.6	20.4
3.	Dibru	929	–	0	149	149	1, 852	0	100.0
4.	Buri-Dihing	877	2,300	–	–	363	8, 730	56.8	43.2
5.	Desang	852	2,594	80	150	230	3, 950	45.7	54.3
6.	Dikhow	842	1,823	104	96	200	4, 370	78.4	21.6
7.	Jhanji	832	1,416	53	55	108	1, 349	64.7	35.3
8.	Dhansiri (S)	757	–	252	102	354	12, 580	51.3	48.7
9.	Kopili	557	1,630	127	129	256	20, 068	79.1	20.0
10.	Kulsi	477	–	99	48	147	4, 005	77.0	23.0
11.	Krishnai	445	–	39	26	65	1, 615	80.0	20.0
12.	Jinari	437	–	32	28	60	594	69.0	31.0
13.	Jinjiram	337	–	0	160	160	3, 467	70.4	29.6

The characteristics of the north bank tributaries are quite different than those of the south bank tributaries because:

(i) North bank tributaries originate from higher Himalayan ranges and the river gradients are much steeper than those of the south bank tributaries,
(ii) Rainfall in the north bank tributary basins is much higher,
(iii) Catchment areas of north bank tributaries are much bigger,
(iv) North bank tributaries travel through geologically younger Himalayan ranges which contain unconsolidated sedimentary rocks and hence carry higher sediment load, and
(v) Most of the north bank tributaries are flashy.

Lower Reach: It covers the lower portion of the river for a length of 337 km from the Indo-Bangladesh border to its outfall into the Bay of Bengal. The whole portion of the lower reach of the river falls within Bangladesh. Near the Indo-Bangladesh border, the Brahmaputra River takes a turn towards south and flows mostly in the southern direction with a little eastward trend. The river flows 225 km from the Indo-Bangladesh border up to Goalundo.

In the reach from the confluence of the Tista River near Bahadurabad to Goalundo, the Brahmaputra River is known as Jamuna. At Goalundo, the Brahmaputra (Jamuna) joins another major river, Ganges-Padma, coming from west and the combined river flows as Ganga-Padma for 80 km. Near Rajabari, a little south of the tropic of cancer, another major tributary Meghna coming from north-east joins it and the combined river flows for 32 km as Meghna River. A little downstream, the Meghna River trifurcates in three channels forming a delta. The three channels are:

Sandwip (the east channel), Shahbaz (the central), and Tutulia (the western channel). These channels ultimately outfall into the Bay of Bengal forming broad estuaries.

Practically, there is no notable tributary on its east (left) bank in this reach, except the Meghna River. Important tributaries on its west (right) bank are Raidak, Dharla, Tista, and Atrai. The river gradient in this reach varies from 1:11,340 near the Indo-Bangladesh border to 1:37,700 near its mouth.

Tributaries in lower reach

Some important tributaries of the lower reach are discussed below.

Raidak River: This is one of the main right bank tributaries which is known as Wong in its upper reach in Bhutan. It debouches into the plains in the Jalpaiguri district of the West Bengal state. The Raidak River confluences with the Brahmaputra River at a chainage 327 km in Bangladesh.

Dharla River: An important right bank tributary which originates in the Sikkim state and is known as Jaldhaka River in the upper reach. The river traverses 186 km through Sikkim, Bhutan, then districts of Darjeeling, Jalpaiguri and Cooch Behar of West Bengal and finally enters into Bangladesh. Dharla confluences with the Brahmaputra River at a chainage of 307 km in the Rangpur district of Bangladesh.

Teesta River: Teesta is the largest of the rivers of North Bengal which rises in the Himalayas in North Sikkim. Running through narrow gorges for nearly 138 km, it debouches into the plains of the Jalpaiguri district at Sivok. It has a number of tributaries, many of them are mountain torrents. Draining an area of 12,540 sq. km of which 3,017 sq. km lies in North Bengal, it joins the Brahmaputra near the Rangpur town in Bangladesh after traversing a distance of 309 km.

9.1.2. Barak River

Barak (Meghna) River originates from Japvo mountain of Manipur hills at an altitude of 3,015 m and flows south through mountainous terrain up to Tipaimukh near the tri-junction of the three states: Assam, Manipur and Mizoram. Here, the river takes a hairpin bend and debouches into the plains of Cacher district of Assam and forms the border of Assam and Manipur states up to Jirimat, a little upstream of Lakhimpur. The river then flows through the Barak valley of Assam. From the source to the Indo-Bangladesh border, the Barak River flows for 564 km.

The drainage area of the sub-basin lying in India is 41,157 sq. km which is nearly 1.38% of the total geographical area of the country. It lies between east longitudes 90°10′ to 95°7′ and north latitudes 21°58′ to 26°24′. It is bounded on the north by the Barail range, on the east by the Naga and Lushai hills and on the south and west by Bangladesh. The bed gradient of the Barak River is very flat and varies from 1:10,000 in the upper reach to 1:20,000 in the lower reach. There are two major physiographic regions in the sub-basin, namely, the hilly region and the plains. The plains are thickly populated and extensively cultivated. Predominate soil types found in the sub-basin are laterite and red and yellow soils. The culturable area in the sub-basin 0.893 M-ha which is only about 0.5% of the culturable area of the country. The state wise distribution of drainage area is given in Table 7.

Table 7. Statewise distribution of drainage area of Barak River

State	Drainage Area (km^2)
Meghalaya	10,650
Manipur	9,550
Mizoram	8,280
Assam	7,224
Tripura	4,725
Nagaland	728
Total	41,157

9.2. HYDROMETEOROLOGY OF THE BASIN

Hydrometeorology of the Brahmaputra basin varies greatly owing to the following reasons:

(i) The basin has wide variations in latitude, longitude and altitude. About half of the basin area lies in magnificent high Tibetan plateau just north of high Himalayan ranges.
(ii) The Himalayas pass through the basin, dividing it into two distinct climatic regions.
(iii) Orography plays a vital role in the hydrometeorology of the basin. The basin covers six distinct topographical regions, including two distinct orographic regions. Small and medium mountain ranges also influence hydrometeorology of the basin.
(iv) The presence of the Bay of Bengal on the south greatly influences the basin hydrometeorology.
(v) Different parts of the basin come under the influence of different winds in different seasons. Examples are continental air moving under Siberian anti-cyclones, trade winds, monsoons, westerly disturbances, hot wind from south west Asia, cyclones and depressions that develop over the Bay of Bengal and many a times pass over the basins.
(vi) The presence of numerous glaciers and permanent snow-covered regions.
(vii) The influence of strong winds in high altitude regions.
(viii) Wide variations in temperature owing to non-uniform surface heating.
(ix) Wide variations in vegetative cover and retention of soil moisture.
(x) Wide variations in transparency and insolation, depending on the altitude of the place.

9.2.1. Basin Climate

The great Himalayan mountain ranges with a width varying from 200 km to 300 km lie in the east-west direction just south of the Tibetan plateau and divide the basin into two distinct climatic zones (Figure 4). The northern part of the basin under the high Tibetan plateau, falls under climatic Zone III and is classified as ‘Mountain

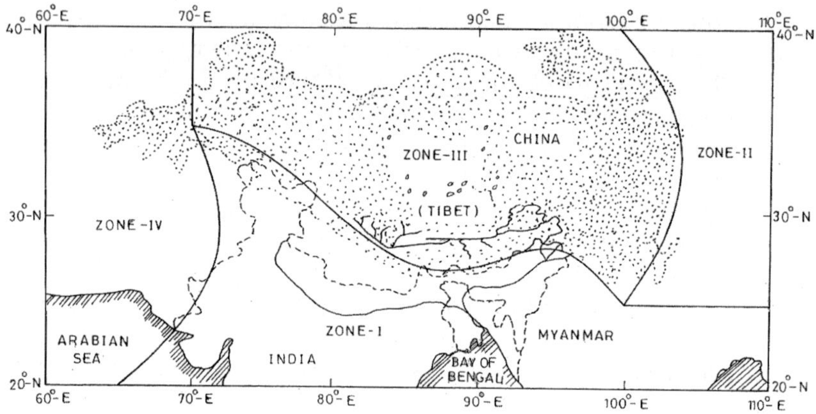

Figure 4. Climatic zones

Climate', which is cold, dry and arid. The southern part of the basin falls under climatic Zone I and is classified as 'Tropical Monsoon Climate'; the climate of this zone is relatively warm and humid, and experiences high rainfall, especially during June to September under the influence of the southwest monsoon.

The basin experiences four distinct seasons in a year: winter (December to February), summer or pre-monsoon (March to May), rainy season or monsoon (June to September), and autumn or post-monsoon (October and November). There is a wide seasonal variation of temperature under the influence of different prevailing winds. The minimum temperature in the plains and valley areas of the basin varies from 9 °C in the western part to 4 °C in the north-eastern part.

The western part of the Brahmaputra valley and the northern part of the plains of Bangladesh experience hot summer in April and May. During summers, the average maximum temperature in this part of the basin remains above 35 °C, whereas the maximum temperature is 40 °C.

The basin area under Zone I experiences 20% to 35% of the average annual rainfall during this summer period owing to northwesterly storms that generally occur in a few days mostly in the afternoons. Sometimes this part of the basin also experiences widespread heavy rainfall for two to three days resulting from deep depressions that develop over the Bay of Bengal and pass over the basin. In general, the humidity and rainfall in the Brahmaputra basin area under Zone I are relatively high.

The months of October and November, after the retreat of the southwest monsoon, constitute the autumn or post-monsoon period. The weather during autumn remains very comfortable with relatively clear sky and moderate temperatures. On average these two months experience 3% to 6% of the average annual rainfall.

The annual number of cloudy days in the Tibetan part of the basin is comparatively less; it is 98 days in Lhasa. The northeastern part of the Brahmaputra valley and lower mountainous regions of the eastern part of Arunachal Pradesh and Assam remain under cloud cover for more than 60% of the days in a year. The average number of cloudy days in the eastern part of the valley at Dibrugarh is 241 days, whereas it is 191 days at Guwahati located at the west central part of the valley.

The number of rainy days in the Tibetan plateau is less, although it experiences light drizzle accompanied by snowfall in some of the days during winters. The number of rainy days annually at Lhasa is 48. In the Brahmaputra valley, the number of rainy days increases from west to east. The number of rainy days at Guwahati, the capital city of the Assam state, is 115 days, whereas it is 172 days at the Dibrugarh city located on the eastern side of the valley.

The low altitude rain bearing clouds brought in by the southwest monsoon from the Indian Ocean and the Bay of Bengal during late May move in the northeast direction and get intercepted by the southern (Assam) hill ranges and causes extremely heavy rainfall along the Cherrapunji (see Figure 5) and Mawphalang Pynurala belt of the Meghalaya state. The wettest town of Cherrapunji is located in the Khasi hills, just south of the basin boundary towards the windward side of the southwest monsoon winds. The clouds that pass over the 1,800 m high mountain ridges of the Garo and Khasi hills of Meghalaya (which means the abode of clouds) enter into the Brahmaputra basin and widespread rainfall takes place in the valley and the mountain reaches of Bhutan and Arunachal Pradesh. The intensity and duration of rainfall increases towards the foothills of the Himalayas and more towards northeast. In the Tibetan plateau, rainfall occurs in the months of July and August; rainfall decreases as the monsoon winds advance towards the northwest. The central and western parts of the plateau are cold, dry and arid.

9.2.2. Climatic Variation

Climatically the basin can be divided mainly in two zones: a) area north to the Himalayan mountain ranges (Zone III) classified as 'Mountain Climate'; and b) the southern part of the basin (Zone I) classified as 'Tropical monsoon climate'. Characteristic features of these zones are described below.

Figure 5. Cherrapunjee (locally names as 'Shora'), the wettest place on Earth

Climate of the basin area under zone III: Mountain climate

This zone encompasses an area of 293,000 sq. km which is about 50.5% of the total area. About 1,625 km long portion of the Brahmaputra River from the source to the Indo-China border flows through this zone. The great plateau of Tibet, lying at 3,000 m to 5,000 m above the mean sea level is known as the roof of the world. The plateau is wide and undulating with snow covered mountain ranges and dotted with numerous glaciers and lakes. Some of the glaciers extend into deep valleys.

About 53,000 sq. km area of the basin remains under permanent snow cover; the snow cover area increases as winter advances and in December and January, snow covers almost whole of this zone. In winters this zone comes under the influence of continental winds under Siberian anticyclone and experiences severe cold. The great Himalayan mountain ranges shield the southern part of the basin from this Siberian cold air. January is the peak winter month and minimum recorded temperature at Lhasa is −16 °C.

In months from June to September, the southwest monsoon moves northeast and is obstructed by the great Himalayan mountain ranges. The monsoon winds penetrate into southeastern part of this climatic zone and precipitate. The amount and rate of precipitation decreases rapidly as the moisture-laden air advances in the northwest direction. The central part of the basin under this zone is extremely dry and arid.

Climate of the basin area under zone I: Tropical monsoon climate

The southern part of the basin covering an area of 287,000 sq. km falls under this climatic zone. This part of the basin area is comprised of an area of 195,000 sq. km in India, 45,000 sq. km in Bhutan, and 47,000 sq. km in Bangladesh. The river reach for a length of 1,255 km from the Indo-China border to its outfall into the Bay of Bengal falls under this zone.

The general climate of this zone is tropical, humid and experiences a high amount of rainfall under the influence of southwest monsoon (June to September). The winter months from December to February are relatively dry and experience about 2% to 5% of the average annual rainfall. The three months, March to May, are generally known as pre-monsoon months. October and November constitute the post-monsoon period and experience 3% to 6% of the average annual rainfall.

9.2.3. Seasonal Weather Variation

The weather of the basin varies greatly from place to place, depending on its latitude, longitude, altitude, location, orography and distance from the sea. Monsoon is the most important factor responsible for seasonal weather variation. On the basis of monsoon variations, the year is divided into four seasons.

Winter Season: Three months from December to February are winter months. In the Tibetan plateau and high altitude places of Arunachal Pradesh and Bhutan, winter lasts longer - from November to March. The intensity of winter is much more severe. In this season, the days are bright and sunny but nights are cold.

Fine weather of this season gets occasionally disturbed by the western disturbances which bring light rainfall and severe cold in the southern part of the basin.

Summer Season: The three months of summer from March to May constitute the period of rising temperature and decreasing air pressure. Dry hot wind blows from west to east over the northern and western parts of the basin under Zone I; sometimes these are associated with dust storms. At times, this zone experiences heavy rainfall for two to three days, resulting from depressions that develop in the Bay of Bengal. This period is also called as pre-monsoon and experiences 20% to 35% of the average rainfall. Summer in the Zone III is moderate and experiences very small amounts of rainfall.

Rainy Season (Southwest Monsoon Season): The period of four months from June to September is the rainy season when the basin experiences 65% to 80% of the average annual rainfall under the influence of southwest monsoon. These monsoon winds arising from the Indian Ocean and the Bay of Bengal move from the southwest to the northeast direction and enter into the Brahmaputra valley by the first week of June. These monsoon winds get obstructed by the Patakii and Arakan mountain ranges in the east, Himalayas in the north and southern (Assam) mountain ranges in the south. As a result the monsoon winds gradually move towards the west. The Brahmaputra valley, the Lower (Assam) mountain ranges and the foothills of Himalayan ranges experience heavy rainfall. The rainfall gradually decreases towards higher mountain ranges. Orography plays an important role. Sometimes, cyclones develop over the Bay of Bengal and move over the Brahmaputra basin. Under their influence part of the basin under Zone I experiences heavy rainfall.

Autumn (Post-Monsoon): The period of two months, October and November, just after the monsoon and before the onset of winter, is the autumn season. In this period, the monsoon winds start retreating. From the middle of October the temperature begins to fall. The weather during this season is characterized by high day temperatures, clear sky and pleasant nights. This period experiences 3% to 6% of the annual average rainfall.

9.2.4. Typical Annual Weather Phenomenon

The northeastern part of the Brahmaputra valley and the lower reaches of Arunachal Pradesh experience more rainfall and the area remains under clouds about 67% of the days in a year. The number of cloudy days gradually reduces towards the western side of the basin. The Tibetan plateau generally remains clear and bright; only 10% to 25% of the days are cloudy days.

The western side of the Brahmaputra valley experiences more thunderstorms than the eastern side. The number of thunderstorm days at Dibrugarh, located at the eastern side of the valley, is only 31 days annually, whereas it is 62 days at Guwahati (located towards the western side). The Tibetan plateau also experiences thunderstorms annually in about 15% of the days.

Table 8. Annual weather phenomenon at some stations within the Brahmaputra basin

Station	Average annual number of days			
	Cloudy	Thunder storm	Dust storm	Foggy
Lhasa	98	53	0	41
Dibrugarh	241	31	1	52
Tezpur	147	16	0	7
Guwahati	191	62	5	45

The Brahmaputra basin experiences foggy nights in winters. Some areas near the main river experience dust storms in February and March. The average annual number of cloudy days, thunderstorm days, dust storm days and foggy days of four stations Lhasa, Dibrugarh, Tezpur and Guwahati are furnished in Table 8.

9.2.5. Temperature Variation

The basin experiences wide seasonal variations of temperature. The recorded mean and recorded daily maximum and minimum temperatures of four stations are furnished in Table 9. The monthly means of daily maximum and minimum temperatures of the above four stations are furnished in Table 10. The basin can be divided into five regions on the basis of temperature and its variation as discussed below.

Region I: Tibetan Plateau: Here the average minimum temperature during December and January comes down to −12 °C to −13 °C at Lhasa and recorded minimum temperature is −16 °C in January. The average maximum temperature varies from 25 °C to 28 °C during May to August. The high altitude areas always remain under snow cover. The insolation is also large in this plateau.

Region II: Himalayan Mountainous Region: The 200 to 300 km wide belt of the Himalayan mountain ranges passes through the basin. The states of Arunachal Pradesh and Sikkim of India and Kingdom of Bhutan fall in this region. As the

Table 9. Temperature and mean relative humidity at some stations in the Brahmaputra basin

Station	Temperature (in °C)				Mean relative humidity (%)	
	Mean Daily		Recorded		At 0800 Hrs.	At 1700 Hrs.
	Max.	Min	Max.	Min.		
Lhasa	16	1	32 (June)	−16 (Jan)	69	–
Dibrugarh	27	18	39 (June)	4 (Feb)	88	76
Tezpur	29	19	38 (May)	6 (Feb)	85	70
Guwahati	29	19	40 (April)	5 (April)	84	69

Table 10. Monthly mean of daily maximum and minimum temperature (in °C) of some stations located in the Brahmaputra basin

Month	Lhasa		Dibrugarh		Tezpur		Guwahati	
	Max.	Min.	Max.	Min.	Max.	Min.	Max.	Min.
January	14	−12	25	7	25	9	26	7
February	16	−12	27	9	28	10	29	8
March	18	−8	31	12	33	13	34	11
April	22	−4	32	15	34	17	36	17
May	25	1	35	18	35	19	35	19
June	28	4	35	21	35	22	35	22
July	27	6	35	22	35	23	35	24
August	26	4	35	23	35	24	35	24
September	24	4	34	21	35	23	34	23
October	22	−5	33	17	33	18	33	18
November	17	−9	29	12	30	13	30	13
December	14	−13	26	8	26	9	27	8

altitude varies from 1,000 m to 5,000 m, the temperature variation is also wide. The windward side of the mountains experiences heavy rainfall which creates temperature differences between windward and leeward sides. Again, insolation results in wide differences of temperature between sunny and shadow areas in high altitude regions above 3 km, where the atmosphere is free from dust, and air pollution and hence insolation plays an important role.

Region III: Valley and Northern Plains: The Brahmaputra valley in Assam and the northern part of Bangladesh fall under this region. The western side of this temperate region (area west of 93 °E longitude) experiences heat wave during April and May. The average maximum temperature remains above 35 °C and the recorded maximum temperature is 40 °C. The temperature in the eastern side of the valley is slightly moderate. Dibrugarh, Tezpur, and Guwahati are situated in the Brahmaputra valley, and the average and maximum observed temperatures of these stations are furnished in Tables 9 and 10.

Region IV: Southern (Assam) Mountainous Region: Parts of the southern (Assam) mountainous region south of the Brahmaputra valley from where the southern tributaries of the Brahmaputra originate and traverse fall under this temperate region. The altitude of this region varies from 500 m to 3,000 m; the areas are moderately cool in summer and very cold in winter. In some days of December and January in high altitude places, temperature drops below the freezing point. Most of the places above 1,500 m experience ground frost in peak winter months. The region does not experience any snowfall.

Region V: Coastal Belt: This area is influenced by the ocean effect. The area also experiences a few cyclones that develop from deep depressions over the Bay of Bengal. Generally, cyclones hit the coast during pre- and post-monsoon periods. The area does not experience extreme weather.

9.2.6. Humidity

Depending on latitude, altitude, sunshine hour, the influence of different prevailing winds, precipitation, and ground moisture, humidity show a large variation. The average humidity of places under climatic Zone-I is relatively high but it is low under Zone-III. The central part of the Tibetan portion of the basin under Zone III is relatively dry owing to its high altitude, coolness and location. Low pressure and strong wind increase dryness.

The mean relative humidity of four stations are furnished in Table 9. It shows that the mean relative humidity at 8 hrs of the three stations Dibrugarh, Tezpur and Guwahati, located in Zone-I is above 80% whereas it is 69% at Lhasa in Zone-III.

9.3. SOILS OF THE BRAHMAPUTRA BASIN

The major groups of alluvium-derived soils of the Brahmaputra valley are Entisols, Inceptisols and Alfisols. Climate, particularly rainfall and temperature, have the dominant role in formation of soils of the basin. Topography modifies the influence of climate on these soils. The duration of soil forming processes also leads to the variations in the profile characteristics. The upland soils are more developed due to the origin of the alluvium in pleistocene age whereas other soils are less developed due to their origin on younger alluvium.

Generally, soils of uplands are dark brown to yellowish brown. These colours of the soils indicate their better drainage conditions. The fluctuations in the depth of water tables in the profile causes the variation in the depth of mottles as the soils remain temporally saturated during rainy season. The recent and old flood plain soils, the channel soils, and low lying soils of upland have characteristic gleyed colours associated with wetness. The poorly drained soils have dark gray to pale brown colour. Some of these soils are flooded every year. The paddy growing soils of the basin also develop characteristics of grayish matrix colour.

The soils of the Brahmaputra valley have wide textural variations ranging from sand to clay. The flood plain soils are predominantly silty loam to sand. The upland soils are loam to clay loam at the surface horizons and clay loam to clay at the lower horizons. The alluvium near the Brahmaputra River is newer and stratified. In these flood plain soils, coarser materials are deposited first, followed by finer materials.

9.3.1. Physico-Chemical Characteristics

Soils of the Brahmaputra basin have a great diversity in their physical and chemical properties. Flood plain soils contain a higher percentage of sand at lower depths. The sand content of upland soils is less than 28% at the surface and it decreases at lower depths. The fine and very fine fractions constitute the major portion of the total sand. Soils near the main river are coarser in texture while those away from the river have finer texture. The silt content of the soils ranges from 30–75%. The depth-wise distribution of silt fraction in the profiles of flood plain soils is irregular but a gradual decrease of silt in upland soils is observed. In most of the soils, the content of medium silt exceeds that of coarse and fine silt.

The soils developed on uplands, old flood plain, lower horizons of some recent flood plains and upper horizon of old channels contain a higher amount of clay (21–58%). Clay increases with depth in upland and old flood plain soils. A significant negative correlation between silt and clay contents on the upland profiles and old flood plain profiles indicates that clay is formed by the weathering of silt.

9.3.2. Bulk Density, Saturation Percentage and Moisture Holding Capacity

The bulk density of the soils ranges from 1.36 to 2.10 gm/cm^3. Relatively more developed upland soils of the basin have a higher bulk density than the other soils. The saturation percentage of the soils varies from 25.7 to 54.7%; the field capacity varies from 3.1 to 29.2%.

The organic matter content in the surface horizon of the soils of the Brahmaputra basin varies from 0.60 to 2.9% and its value decreases with depth. The organic matter content has been found to increase with the increase in rainfall and decrease with the increase in the temperature.

Soils of the middle region of the Brahmaputra basin contain free $CaCO_3$ and carbonate nodules at lower depths. The flood plain soils contain carbonate throughout the profile, indicating a lack of carbonate leaching. The calcium carbonate content of these soils is higher at the surface horizons which indicates recent deposition of alluvium derived from calcium rich rocks. Low pH, high rainfall, and moderate to well-drained conditions of these soils enhance dissolution and leaching of carbonates. Therefore, the electrical conductivity (EC) of the Brahmaputra basin soils is low. The free iron oxide content of the soils ranges from 0.57 to 5.28% and that of the aluminum oxides ranges from 0.31 to 3.95%.

The pH of the recent flood plain soils of the Brahmaputra basin ranges from 7.6 to 8.0 and that of the old channel soils, old flood plain soils and upland soils ranges from 5.5 to 8.4, 4.4 to 7.5 and 4.8 to 6.4, respectively. The pH of these soils decreases with increase in the intensity of rainfall and with the increase in age.

Higher rainfall and temperature increase the base status and the Fe and Al oxides and decreases the silica ratio of the soils. In general, the silica ratio decreases with depth in most of the upland soils. As the soils of eastern region of the basin experience a higher rate of weathering, they contain a higher amount of iron and aluminum. The recent flood plain soils of the valley also contain a higher amount of Fe and Al.

9.3.3. Mineralogy of Soils

The heavy minerals (sp. gr. > 2.5) in sand fractions of the soils ranges from 0.1 to 6.1%. The chlorite, biotite, amphibole and pyroxene together constitute the major proportion of heavy minerals of sand fractions of soils of the basin. Clay mica is the dominant mineral in clay fractions of flood plain soils (53–61%). Chlorite constitutes 8 to 10% of the clay fractions of the flood plain soils of the basin. Kaolinite is another dominant mineral constituting 36–40% of the clays.

The major groups of the alluvium-derived soils of the Brahmaputra basin are entisol, inceptisol and alfisol. Inceptisols comprise 45% of the soils followed by entisols and alfisols which represent 36.6 and 12.3 per cent of the soils of Assam, respectively. The dominant soils of Assam are grouped under 5 classes and 7 subclasses of land capacity. Nearly 69% of the land comes under the arable class suitable for agriculture with proper management.

9.3.4. Organic Matter and Nitrogen Content

The organic matter reserve of the Brahmaputra valley soils ranges from low to high. It is high in bheel and virgin soils. In continuously cultivated soils, the organic content has decreased to an alarming level. The average organic matter content of new alluvium, old alluvium and upland soils are 0.97%, 1.02% and 2.70%, respectively. The organic matter in the profiles shows a gradual decrease with depth.

The total Nitrogen content in the top soils of old alluvium vary from higher to low, while the new alluvial soils contain a medium level of total N. The average value of the total N content in the valley has been reported to be 0.1%. The amount of the total N in new alluvial soils varies between 0.013% and 0.135%. On the other hand, the N content of the old alluvial soil is in the range 0.034 to 0.165%.

Top soils of the old alluvium have low to deficient levels of phosphate. In general, the available P_2O_5 content in old alluvial soils is less than 0.01% (1% citric acid soluble). The top soils of new alluvium are relatively rich in phosphate. The soil acidity is the main reason of the low level of phosphorus available for plants in these soils. The top soils of the new alluvium are richer in potash, compared with the top soils of the old alluvium. The average values of the available K_2O content of new alluvial soils of Lakhimpur, Nagoan and Kamrup districts are 0.006%, 0.014% and 0.002%, respectively.

In the Brahmaputra valley soils, the percentage of the total (acid soluble) MgO is relatively greater than that of CaO, but the available (replaceable) Mg is lower than Ca. In recent alluvium and low-lying soils exchangeable Ca^{++} and Mg^{++} are dominant. Cations of clay surface and their content increase down the profile of upland soils. In old alluvium derived soils, only exchangeable Mg^{++} increases with the depth of profile. The status of exchangeable Ca and Mg in the surface soils of the Brahmaputra valley soils of Nagoan, Tezpur, Guwahati and Jorhat is low with values ranging between 73 to 148 and 3.6 to 39, respectively. The Ca:Mg ratio of these soils varies between 1.41 and 12.17.

Crop yields in most of the arable soils of the Brahmaputra Valley can be considerably increased by adequate manuring combined with proper cultural practices. Most of the old alluvial soils need to be adequately manured, particularly with nitrogenous and phosphatic fertilizers to increase crop yield. Organic manures, such as cowdung, compost, oil cake and green manures, have been found effective for most crops in these soils.

9.3.5. Fertility of Tea Soils

In the Brahmaputra valley tea is grown on land with an elevation between 50 to 120 m above the mean sea level. The tea soils of the Brahmaputra valley, which occupy 196,173 ha, are mostly alluvial and vary widely in texture. Most tea soils of the Brahmaputra valley have low to medium status of organic carbon. About 54% of the tea sections have medium to high level of nutrient index for soil organic matter; the rest are with low nutrient index. A typical tea garden in the valley can be seen in Figure 6.

9.4. HYDROGEOLOGY

Hydrogeologically, the Brahmaputra basin can be divided into two distinct categories: (a) dissected alluvial plain and (b) the inselberg zone. The first category is represented in the flood plain extending from south of the sub-Himalayan piedmont fan zone in the north to right up to the main rock promontory of the Garo hills and the Shillong plateau. There are a large number of buried channels consisting of gravel and sands, natural levees of sands and silts, back swamps/bils of silts and clays. The inselberg zone is characterized by fractured, jointed, and weathered ancient crystalline rocks with inter-hill narrow valley plains, consisting of thin to occasionally thick piles of assorted sediments.

9.4.1. Aquifers

Subsurface lithological data indicate two broad groups of aquifers: (i) shallow water table and (ii) deeper water table or confined ones, separated by a system

Figure 6. A tea garden in Brahmaputra valley in Assam

of aquicludes. The occurrence of groundwater in the hard rock inselberg areas is controlled by foliations, fractures/joints, and weathered zones. The physiographic features, types of drainage, depth and duration of precipitation also play an important role. Many of the shallow open wells go dry by the month of March.

In the tertiary sediments, ground water occurs in the sandstones and along the fractures and solution cavities in the limestones. Ground water occurs both under water table and confined conditions. In the piedmont and alluvial fan deposits ground water occurs generally under water table conditions. The water table has, in general, a southerly slope corresponding roughly to the surface topography. The ground water yield from the older alluvium varies from 500–700 lpm for a drawdown of about 3 m. The estimated transmissibility and coefficient of storage are of the order of 793 lpm/m and 8.2×10^{-3}, respectively.

On the north bank, in the dissected flood plains between Abhayapuri and Beki-Chaulkhowa section, ground water generally occurs under unconfined or water table condition, at places it is perched where a thin veneer of sand or silt rests on the pockets of clay. In the younger or newer alluvium, ground water occurs both under water table and confined conditions. Depths to water levels vary from ground level to 10m below the land surface. The depth to the tube wells ranges from 30 to 200 m and the depth to water ranges from 6m below the land surface to 2 m above the land surface. The yield of the deep tubewells ranges from 2 to 4 kl/minute for a drawdown of 3m to 6m. The transmissivity of the aquifers varies from 69 to 1,600 lps/m and the storage coefficient is of the order of 3.52×10^{-2}.

Within the inselberg zone, at places shallow unconfined aquifers occur right from the ground level down to a depth of about 30 m. The sediments are mainly fine to coarse sand and silt. In these units, the first saturated granular zone of medium to coarse sand and gravel with occasional cobbles, pebbles occur between 10 and 20 m below the ground. In the Chapar geomorphic units, a top clay layer of 15 to 18 m thick occurs. The deeper aquifers with intervening thin clay lenses occur in a composite zone ranging in thickness from 5 to 80 m below 25–30 m. Throughout the area, the shallow and deeper aquifers are more or less interconnected. Water levels are comparatively shallower in the flood plain and the inselberg zone, the depth to water level varying mostly from less than 2 to 6 m below ground level but in the older Chapar Formation, deeper water level exists.

9.4.2. Aquifer Properties

The sediments in the lower Assam plain of the Brahmaputra can be grouped into belts from north to south as described below:

1) *Piedmont zone (Bhabar belt-Pleistocene):* The piedmont terraces consist of talus fans having all the general characteristics of foothill terraces all along the sub-Himalayan belt. The terraces comprise rock fragments, boulders, pebbles, and ill-sorted sand and minor clay. These fans hardly retain any percolating water and the ground intake from precipitation and runoff migrates rapidly down to the terraces below. Adequate ground water exists in the area under water table

as well as confined conditions. The average depth to water is 3–3.5 m during March-April. Flowing conditions in the case of a few tube wells indicate water at depths from 26–38 m with thickness of water column ranging from 3 to 14 m. The southern extremity of the Bhabar belt is characterized by water table at shallow depths of 1 to 2.5 m.

2) *The Tarai zone (Pleistocene):* The linear Tarai tract fringes the piedmont zone throughout, commencing from the spring line. It occurs below the piedmont zone comprising terraces of brown clay, coarse to medium sand with occasional pebbles and gravel beds. The water table persists within depths of 1–2 m for about 4–5 km down south of the spring line. This belt forms the main recharge zone for aquifers in shallow depths in the sub-Tarai zone. Near surface, aquifers are under the water table condition.
3) *The present day flood plain deposits and their abandoned channels*: They occur between the Tarai belt and the Brahmaputra River. The sediments are mainly gravel, sand, silt and clay. The near-surface sand and gravel beds constitute the aquifers which contain ground water under water table conditions. Deeper aquifers are often under confined condition.

On the southern periphery of the alluvial plain near the northern flank of Brahmaputra River, a progressive increase in the depth to water from 2 to 5 m is noticed. In the central part of the area where alternating sand and clay beds occur below the topmost clayey horizon up to a thickness of 18 m, at least 3–4 significant confined aquifers have been tapped from depths of 20–53 m. In the southern part of the alluvial plain, where sandy sediments predominate, small diameter tube wells tap water from the aquifers between the 10–26 m depth. The aquifers are under semi-confined to free condition. Deeper aquifers down to a 60 m depth are also under semi–confined condition, as no impervious horizon of any importance exists within this depth. The piezometric level varies in the alluvial plain from 1.4 to 5.9 m below the land surface. Many tube wells in the sedimentary plain have shown artesian flow.

9.4.3. Chemical Quality of Ground Water

The ground water is generally mildly alkaline with a pH value ranging from 6.5 to 8.5. The total dissolved solids are low. The chloride (10–40 ppm) and bi-carbonate (50–350 ppm) values are considerably low. The iron content ranges from a fraction of a ppm to as high as 50 ppm, having higher values in the north eastern sector, and varies both laterally and depth wise. The total hardness as $CaCO_3$ generally varies from 50 to 300 ppm, and the specific conductance at 25 °C generally varies from 150 to 650 mhos/cm.

Although ground water is available in plenty, because of high iron content, its quality affects its potable nature. At greater depths, say 30 m, ground water is free from much of iron. In some places, good quality water is available at 14–30 m depth. The temperature of the ground water is, in general, in the range of 23–25 °C during winter, 24–26 °C during pre monsoon period, and 27–28 °C during peak monsoon when the water table rises within 2 to 2.5 m of surface. Owing to stagnation, the condition of ground water of the back swamp lakes shows a higher temperature range of up to 31–34 °C.

9.5. GROUND WATER RESOURCE POTENTIAL

The total annual replenishable ground water potential of Brahmaputra basin is assessed to be 26.55 BCM. Basin wise and state wise details of ground water potential are given in Table 11.

Percentage of ground water development is very low; only about 3.5%. Clearly, there is large scope of development of ground water resources in the NER but of course, the demand are also very low.

9.6. CROPS AND CROPPING PATTERN

The major crops raised in Assam include rice, wheat, maize & other cereals, pulses, oilseeds, jute, sugar cane, and potato. The horticultural crops grown in the state are fruit crops, plantation crops, tuber crops, spices, and vegetables. Commercial cultivation of ornamentals has been initiated recently. Rice is by far the most important crop of Assam, occupying an area of 25.80 lakh hectares or about 80 % of the annual cropped area. Of this area, winter or Kharif rice (known as 'Sali') occupies 70% of the total area under rice. Sali is cultivated from June-July to November-December. Next in importance is the 'ahu' or autumn rice which occupies 25% of the total area under rice. The 'ahu' crop season ranges from March-April to July. The third crop of rice occupies only 5% of the rice area in the

Table 11. Ground water potential (BCM) NE Region

Basin / state	Total Replenishable GW resource	Available for domestic use	Available for irrigation	Net draft	Balance available
Brahmaputra Basin	26.55	3.99	22.56	0.76	21.80
		State wise potential			
Arunachal Pradesh	1.44	0.22	1.22	0.00	1.22
Assam	24.72	3071	21.01	0.94	20.07
Manipur	3.15	0.47	2.68	Negligible	2.68
Meghalaya	0.54	0.08	0.46	0.02	0.44
Mizoram		Data Not Available			
Nagaland	0.72	0.11	0.62	Negligible	0.62
Tripura	0.66	0.10	0.56	0.19	0.37

Percentage of ground water development is very low; only about 3.5%. Clearly, there is large scope of development of ground water resources in the NER but of course, the demand are also very low.

state. 'Boro' and early ahu crop are raised in this season, lasting from December-January to April-May. The area covered under high yielding varieties is a little over 40% of the total cropped area.

The area under wheat in the state increased from 5,000 ha in 1965–66 to 93,000 ha in 1985–86. But this has shown a declining trend since 1990–91. The area under maize (corn) and other cereals is rather low, being only 21,000 and 15,000 hectares, respectively.

Pulses are cultivated primarily during the Rabi season, the important pulses being black gram, green gram, pea, lentil, and rajmah. About 10–15 thousand ha, out of the total area of 1.2 lakh ha under pulses, are cultivated during the Kharif season, the crops raised include arhar (pigeon pea), summer green gram and summer black gram. The productivity of pulses is rather low, being around 450–550 kg/ha. Pulses are grown in the rain-fed areas only. Edible oilseed crops grown in the state include rape and mustard, niger, soyabean, sesamum and ground nut, while linseed and castor are the non-edible crops. However, rape and mustard alone occupy 92% of the total area under oilseeds which is around 3.5 lakh ha. The productivity of oilseeds is very low, since the crops are rainfed, grown during the Rabi season. Jute is the primary fiber crop cultivated in around one lakh ha. The production of jute ranges from 8.5 to 10.5 lakh bales annually. Sugarcane is another important crop, grown in about 40,000 ha, annual production being around 14.5 to 15.5 lakh tons. The area under potato is around 635,000 ha, the average yield being about 6 tons/ha.

A large number of horticultural crops are raised in the valley. Important horticultural and plantation crops include orange, banana, pineapple, papaya, limes and lemons, ginger, turmeric, chilies, tapioca, sweet potatoes, arecanut, coconut, betel vine, black pepper, an array of summer and winter vegetables, and onion, etc. The area under fruits is around 0.78 lakh hectares; the total production being about 9.2 lakh tons. Similarly, the total production of vegetables is around 19 lakh tons, the crops occupying an area of about 1.55 lakh hectares.

Mixed farming is usually practiced by farmers – rearing livestock and raising of crops are common. Growing of horticultural crops is primarily confined to homesteads, except pineapple and oranges which are raised commercially. Over 90% of the total cropped area is rainfed. The Brahmaputra River has made the valley fertile and has contributed to the favourable moisture regime in the area. The high rainfall also helps rainfed farming.

The cropping intensity in the state of Assam is around 125%. The lack of irrigation is the primary cause for such a situation. Production and productivity of the major crops during 1991 to 1993 are shown in Table 12.

As for the cropping pattern, Ahu rice (summer paddy) followed by Kharif rice (Sali or winter paddy) and jute followed by Kharif rice are the two primary cropping patterns in low lying areas. In the rainfed uplands, Ahu/olitorins jute followed by mustard/wheat is the common pattern, where double cropping is practiced. Mainly due to sandy loam soil of Goalpara, Kamrup, Nowgong and Barpeta districts, more areas are under double cropping.

Table 12. Production and productivity of four major crops of Assam

Crops	1990–91		1991–92		1992–93	
	Production	Productivity	Production	Productivity	Production	Productivity
Rice	32.70	1,313	31.97	1,261	33.0	1,308
Rape & Mustard	2.95	535	3.03	586	2.90	476
Pulses	1.21	433	1.26	472	1.19	475
Jute	866	1,632	8.67	1,537	10.34	2,011

Production is in lakh tonnes (Jute in lakh bales), productivity is in kg/ha.

9.6.1. Land Tenure and Farm Size

The land holdings in the Brahmaputra valley are small (average size: 1.47 ha). Many socio-economic problems of the farmers are due to the large number of fragmented, small and marginal holdings. The share tenancy is quite widespread. As per the world agricultural census, 16% of the holdings are wholly rented, while another 9% are partially rented. The 59.66% of the land holdings are of size 1 hectare or less, 22.62% of size 1–2 ha and 9.04% of 2–3 ha.

9.6.2. Demography

Historically, the Brahmaputra Valley has been highly plural in demographic composition. It has been a shared homeland of various racial, religious, ethnic, linguistic and cultural groups. As per the 1991 Census, the population of the Brahmaputra Valley in Assam was 19,109,300 persons. As per the provisional Census Report of 2001, the population of Assam has reached 26,638,407 persons. During the post-colonial period, the Brahmaputra Valley had shown a rapid growth of population up to 1991 Census. The population of the north-eastern states has gone up from 10.36 million in 1951 to 31.55 million in 1991. The staple diet for people in the region is rice. However, the population growth rate declined substantially from 1991 onwards. The literacy rate, too, has increased from 56.08%in 1991 to 66.08% in 2001.

The density of population in Assam has increased from 286 per sq. km in 1991 to 340 per sq. km in 2001. However, the density of population in the Brahmaputra Valley is much higher than the average of the state.

9.7. PRECIPITATION

Depending on the amount of precipitation, the basin may be broadly divided into two regions:

(i) Tibetan plateau that falls under climatic zone-III.

(ii) Southern part of the basin that falls under climatic zone-I.

The southwest monsoon causes widespread rainfall in the basin south of Himalayas under climatic zone I during June to September. About 65% to 80% of the average annual rainfall takes place in this period. Rainfall during July and August is the highest and amounts to more than 33% of the annual rainfall. Orography plays an important role in areal distribution of rainfall. Rainfall is very high on the southern slopes and the foothill region of Himalayas in Bhutan and Arunachal Pradesh.

The monsoon rainfall is generally widespread and at places it is heavy. In most places, monsoon rainfall is received in showers of one to three hours duration. At some places, heavy showers of six hours or more duration are also observed. Such situations generally persist for 2 to 3 days but may continue for 4 to 5 days.

The Tibetan plateau and higher reaches of mountain ranges above 3,000 m receive snowfall during December to February. The south-eastern part of the Tibetan plateau receives some monsoon rainfall in July and August. The amount of rainfall decreases gradually towards western and central part of the plateau.

The average annual rainfall in the basin varies from less than 400 mm to more than 6,000 mm. The mean annual rainfall over the whole catchment excluding the Tibetan part is around 2,300 mm. Within the valley reach in Assam, rainfall is more towards the northeast and gradually decreases towards the west. The mean monthly and average annual rainfall, and the average number of rainy days in a year at four stations are furnished in Table 14.

Depending on the prevailing climate and influence of different winds mainly monsoon, the year may be divided into the following four periods of rainfall:

I: Winter rainfall from December of February,
II: Pre-monsoon rainfall from March to May,
III: Monsoon rainfall from June to September, and
IV: Post-monsoon rainfall from October to November.

Under the influence of western disturbances, the basin gets small amounts of rainfall in winter months. The area under Zone I experiences occasional winter rainfall amounting to 2% to 5% of the average annual rainfall. Higher reaches of Arunachal Pradesh and Bhutan receive greater amounts of rainfall during these winter months, amounting to about 10% of the average annual rainfall. The Tibetan part of the basin under Zone III experiences snowfall under the influence of continental cold air that blows owing to Siberian anticyclones. The Tibetan plateau also receives light winter rainfall, although the total amount is not substantial.

The pre-monsoon months from March to May experience rainfall mainly from isolated thunderstorms occurring over small areas for short duration. Scattered thunderstorms are also observed on some days over large areas. Generally such activities do not persist for more than a day or two. Widespread heavy rainfalls are sometimes observed during late May due to cyclones over Zone I. In general, this pre-monsoon period, experiences 20% to 35% of the average annual rainfall. Higher reaches of Arunachal Pradesh and Bhutan experience about 35% of the average annual rainfall, the southern portion of the catchment experiences 20% to 30%. Such rainfalls give rise to flooding in some tributaries or the main river itself.

Some portions of the Brahmaputra valley in Assam and the adjoining mountainous regions receive heavy to very heavy rainfall. Sometimes rainfall continues for 7 to 8 days during monsoon months under synoptic meteorological conditions. Such spells generally take place under four meteorological situations, namely: (i) break monsoon situation; (ii) northwesterly to northern movement of monsoon depressions or storms from the Bay of Bengal; (iii) the formation and the movement of land depressions over northeast; and (iv) the upper air cyclonic circulation over the basin. Under such situations, one-hour rainfall of 70 mm is common in some areas. Such a spell of heavy rainfall gives rise to flooding in the concerned tributaries or the main river.

Monthwise numbers of rainy days of four stations are furnished in Table 13. Note that Lhasa experiences only 48 rainy days in a year compared to 172 days at Dibrugarh. The rain shadow areas of the Mikir hills in the south-central part of the basin experiences a low of 95 days. Almost 65% of the rainy days occur in the monsoon season and 20% in the pre-monsoon season. The number of rainy days in the southern periphery of Arunachal Pradesh is around 120 and it increases to 170 at the northern end. In the high altitude northern part of Arunachal Pradesh, the number of rainy days in winter and pre-monsoon period is considerably high, although the total amount of rainfall is not much.

The October and November months are the post-monsoon period and experience very low rainfall, of the order of 3% to 6% of the annual value. Rainfall in this period occurs generally from isolated thunderstorm or from the cyclones that develop. The average rainfalls of the northern and southern tributaries of the Brahmaputra River in the middle reach for the four seasons are furnished in Table 14 and Table 15, respectively.

Table 13. Mean monthly rainfall and rainy days of four stations in the Brahmaputra Basin

Month	Lhasa		Dibrugarh		Tezpur		Guwahati	
	Rainfall (mm)	Rainy Days	Rainfall (mm)	Rainy Days	Rainfall (mm)	Rainy Days	Rainfall (mm)	Rainy Days
January	3	1	38	7	15	3	10	2
February	13	1	62	10	27	6	4	5
March	8	1	103	14	48	8	25	6
April	5	1	241	18	153	12	145	10
May	25	3	307	19	271	19	236	15
June	53	8	500	22	308	18	312	18
July	122	13	536	24	348	19	312	20
August	89	10	451	22	331	18	261	16
September	66	7	352	21	210	16	167	15
October	13	2	152	8	104	6	71	5
November	3	1	53	4	23	3	14	2
December	0	0	16	3	6	2	4	1
Annual	400	48	2,811	172	1,844	130	1,561	115

A day having rainfall more than 0.3 mm is termed as a rainy day.

Table 14. Average rainfall of some north bank tributary basins of the Brahmaputra in the middle reach

S.N.	Name of River	Average Rainfall (in mm)				Annual
		Dec.-Feb. (Winter)	Mar.-May (Pre-monsoon)	June-Sept. (Monsoon)	Oct.-Nov. (Post-monsoon)	
1.	Simen	–	–	–	–	3,680
2.	Jiadhol	–	–	–	–	3,500
3.	Subansiri	108	461	1,450	131	2,150
4.	Ranganadi	88	604	1,601	174	2,467
5.	Borgong	–	–	1,455	–	2,216
6.	Jia-Bharali	60	471	1,529	157	2,217
7.	Gabharu	–	–	–	–	1,582
8.	Belsiri	38	384	1,147	99	1,668
9.	Dhansiri (N)	56	463	1,345	133	1,997
10.	Noa-Nadi	–	–	–	–	1,722
11.	Nanoi	50	539	1,359	126	2,074
12.	Barnadi	–	–	–	–	2,037
13.	Puthimari	46	608	1,407	93	2,154
14.	Pagladiya	39	689	1,820	111	2,659
15.	Manas	52	418	1,286	113	1,869
16.	Champamati	–	–	–	–	2,667
17.	Gaurang	–	–	–	–	3,123
18.	Tipkai	–	–	–	–	3,264
19.	Sankosh	–	–	–	–	2,863

Table 15. Average rainfall of some south bank tributary basins of the Brahmaputra in the middle reach

S.N.	Name of River	Average Rainfall (in mm)				Annual
		Dec.-Feb. (Winter)	Mar.-May. (Pre-monsoon)	June-Sept. (Monsoon)	Oct.-Nov. (Post-monsoon)	
1.	Dibong	394	1, 415	1,701	350	3,860
2.	Lohit	216	840	1,776	159	2,991
3.	Dibru	96	669	1,466	140	2,371
4.	Buri-Dihing	113	631	1,567	175	2,486
5.	Desang	110	613	1,535	174	2,432
6.	Dikhow	113	594	1,553	178	2,438
7.	Jhanji	80	654	1,277	186	2,197
8.	Dhansiri (S)	58	449	1,116	144	1,767
9.	Kopili	42	376	1,194	123	1,735
10.	Kulsi	25	378	1,152	117	1,672
11.	Krishnai	14	404	1,138	157	1,713
12.	Jinari	28	835	1,842	345	3,050
13.	Jinjiram	31	653	1,937	226	2,847

Several closed cells of maxima of more than 4,000 mm exist. One such closed cell of maxima (Cell-I) is noticed in the area around Pasighat between latitude 27 °N to 28°45′ N and longitude 93 °E to 95°30′ E. The second such cell (Cell-II) is in the western part of Assam in the foothill region of Bhutan hills between latitude 26°15′ N and 27°15′ N and longitude 88°30′ E to 90 °E at and around the Sankosh basin. The third closed cell of maxima (Cell-III) exists in the Khasi hill region along the Cherrapunji and Mawphalang Pynurala belt of the Meghalaya state between latitude 24°30′ and 25°30′ N and longitude 89°30′ E and 91°30′ E. Extremely heavy rainfall takes place in this cell. The town of Cherrapunji located in this cell just south of the basin boundary and towards the windward side of the southwest monsoon experiences the average annual rainfall of 10,800 mm and is the wettest place in the world. Rainfall between cells I and II falls sharply to less than 2,000 mm in the Mikir and adjoining hills of the southern Assam ranges.

There are a number of rain shadow areas in the basin. Rainfall is less than 1,500 mm in major parts of high altitude regions of Bhutan and Arunachal Pradesh. The eastern part of the Lohit district of Arunachal Pradesh (east of 95°30′ E longitude and north of 28° N latitude) is a rain shadow area where the average annual rainfall is less than 1,000 mm. Another rain shadow area is in the north western part of Arunachal Pradesh adjoining Bhutan, where the average annual rainfall is less than 1,000 mm. Rainfall in the Assam valley varies from 1,500 mm in the western side to more than 4,000 mm in the northeast corner. The variability of annual rainfall over the basin is 15 to 20%.

The variability of winter rainfall is less than 40 percent in the north-eastern part of Arunachal Pradesh; it increases to about 65 percent in the south-eastern part of the basin and 85 percent in the south-western part. The variability of pre-monsoon rainfall is about 30% over the basin, except the Tibetan part. The variability of monsoon rainfall over the basin, except Tibetan part, is about 15 to 20 percent, whereas the variation in individual months is 30 to 40 percent. For the post monsoon season, the variability is 40 to 50 percent in the eastern part of the catchment and it increases to 80 percent in the western part.

Exceptionally heavy rainfall occurs over Assam and neighbouring hill states when westerly waves moving across Nepal-Assam Himalayas happen to synchronize with any of the synoptic situations. Under such situations, heavy to very heavy rainfall of isolated nature takes place. The maximum rainfalls of 1-hour, 3-hour, 6-hour, 12-hour, 24-hour, etc. are recorded at different places and are mentioned below.

a) 1-hour maximum recorded rainfall: Rainfall of nearly 70 mm/hour is common in areas favourable for heavy rainfall. The highest recorded 1-hour rainfall is 97.5 mm at Saralpara under Tikpai sub-basin located in the Cell-II of maxima where the average annual rainfall is above 4,000 mm. One-hour rainfall of above 60 mm is also seen in the Cell-I near Pasighat.

b) 3-hour maximum recorded rainfall: Rainfall of 60 mm and above in a 3-hour duration is quite frequent in areas favourable for heavy rainfall. The three-hour rainfall of 200 mm is recorded in Cell-II of maxima at and around Saralpara. The maximum recorded 3-hour rainfall in Cell-I of maxima at and around Pasighat is 120 mm.

Table 16. Maximum recorded 24-hour rainfall in different months at four stations

Month	Maximum 24 hours rainfall (in mm) at station			
	Lhasa	Dibrugarh	Tezpur	Guwahati
January	3	35	29	29
February	38	46	39	53
March	13	56	62	56
April	8	197	84	71
May	20	113	148	94
June	28	202	137	145
July	28	165	140	233
August	28	223	143	188
September	23	122	103	115
October	15	105	141	84
November	23	53	60	36
December	0	56	19	17
Recorded max. 24 hour rainfall	38	223	148	233

c) 6-hour maximum recorded rainfall: Six-hour maximum rainfall of 300 mm has been recorded in Cell-II whereas it is 150 mm in Cell-I. The recorded 6-hour rainfall in areas of minima is around 60 mm.

d) 24-hour maximum recorded rainfall: Rainfall of 200 mm and above in 24 hours is frequent in areas favourable for rainfall, such as the region around Pasighat and around Sankosh basin. Rainfall of more than 300 mm occurs rarely. A cell of maxima of 500 mm in 24-hours is observed in Cell-II and a cell of 300 mm is observed in Cell-I near Pasighat. Adjacent to this, there is another cell of 250 mm rainfall near Likabali and Gerukamukh in the Himalayan foothills region. Very high rainfall in some places has occurred under synoptic meteorological conditions in rare cases. A few such recorded 24-hour rainfall amounts are:

(i) At Jamduar 805.5 mm on 7.9.76
(ii) At Jamduar 603.2 mm on 5.10.77
(iii) At Roing in A.P. 745.0 mm on 16.8.77

The 24-hour recorded rainfall for different months of the year for four stations are furnished in Table 16.

9.8. DISCHARGE

The average annual discharge of the river at its confluence with the Ganga River at Goalundo in Bangladesh is 19,820 cumec. The recorded maximum and minimum discharges at Guwahati (catchment area 424,100 sq. km, or 73.12% of the catchment area) are 72,794 cumec and 1,757 cumec, respectively. The return period of the maximum discharge is 10 years. Three discharge measuring sites of the river in Tibetan plateau are Shigatse (the new name is Rikaze), Chushuldzong (Qushui Gogge) and Tsela Dzong (Ziang Zhong) located at an altitude of 3,816 m, 3,565 m and 2,957 m, respectively. The maximum recorded discharges at these three sites are 3,380 cumec, 6,230 cumec and 10,200 cumec, respectively. The maximum recorded discharge at Tsela Dzong is 14% of the maximum discharge at Guwahati, whereas the catchment area at Tesladzong is 44% of Guwahati. The Pasighat gauging site (chainage 1,029 km) is located in Arunachal Pradesh, just at the foothills, where the catchment area is 249,000 sq. km (42.93% of the total) and the observed maximum discharge is 29,640 cumec (40.72% of Guwahati). There are three discharge measuring sites in the valley reach of Assam: Bessamara, Pandu (Guwahati city) and Jogighopa.

Typical discharge hydrographs of the Brahmaputra River for five stations, Chushuldzong, Tseladzong, Pasighat and Guwahati are shown in Figure 7. The hydrographs of Pasighat and Guwahati show that the river experiences a number of floods from June to mid October. The peak annual discharge was observed at Guwahati on 18th September; this wave had peak discharge at Pasighat on 15th September, i.e., the flood wave took three days to travel from Pasighat to Guwahati. The peak discharge at Guwahati is nearly three times the peak discharge at Pasighat. The percentage contributions to the annual yield of the river at Guwahati by the main river at Pasighat and other main tributaries are given in Table 17.

The Brahmaputra is a perennial river. Snow melt and base flow contributions sustain its lean period flow from December to March. The main contribution to runoff of the river is from surface runoff. The maximum and minimum discharges of the Brahmaputra River at different sites are furnished in Table 18.

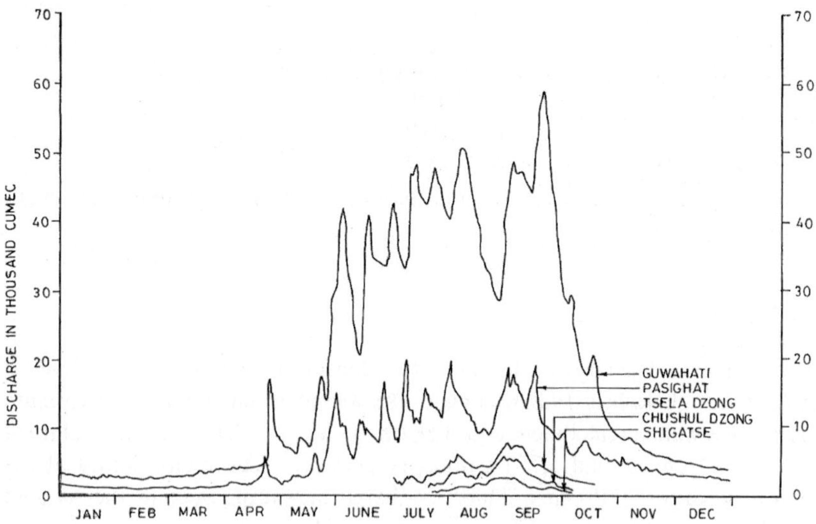

Figure 7. Discharge hydrographs of the Brahmaputra River at different sites

Table 17. Contributions to the annual yield at Guwahati by the main river at Pasighat and main tributaries

By the main Dehang River at Pasighat	34.00%
By the Dibong River	7.65%
By the Lohit River	9.50%
By the Subansiri River	10.30%
By tributaries joining between Pasighat and Guwahati	38.55%
Total	100.00%

Table 18. Recorded maximum and minimum discharge of the Brahmaputra River at different sites

Name of site (State, Country)	Chainage in km from the mouth	Catchment area in sq. km	Recorded discharge (cumec)	
			Max.	Min
Shigatse (Tibet, China)	2,105	88,620	3,380	–
Chushuldzong (Tibet, China)	1,933	113,500	6,230	–
Tseladzong (Tibet, China)	1,515	186,730	10,200	–
Pasighat (Arunachal Pradesh, India)	1,029	249,000	29,640	1,076
Guwahati (Assam, India)	542	424,100	72,794	1,757

In the middle reach, the major contribution to the flow of the main river comes from the right bank tributaries, such as Subansiri, Jia-Bharali, Dhansiri (North), Manas and Sankosh and from the left bank tributaries, such as Dibong, Lohit, Burhi-Dihing, Desang, Dikhow, Dhansiri (South), Kopili, etc. The recorded maximum and minimum discharge of major and left bank tributaries are furnished in Table 19.

There are four discharge observation sites in the lower reach in Bangladesh, namely, Chilmari, Bahadurabad, Sirajgonj, and Nagarbari. At Bahadurabad, the average discharge during the monsoon period is 31,850 cumec and the average dry season (December to March) discharge is 6,510 cumec. The average discharge during July and August is 39,900 cumec. The minimum observed discharge is 3,120 cumec.

9.8.1. Flooding

The problem of flooding is acute in the Brahmaputra valley and the possible solutions are complex. On average, flooding affects an area of 1 million hectares in Assam. The main reasons for frequent floods in the Brahmaputra valley are the narrowness of the valley, high rainfall, and heavy encroachment of flood plains. Besides, seismic activity in this region is a continuous process. Earthquakes of magnitudes of about 5 on the Richter scale are quite frequent. The earthquakes

Table 19. Observed maximum and minimum discharge of some major tributaries of the Brahmaputra River in its middle reach (in India)

Name of tributary	Gauging site	Observed discharge (cumec)	
		Maximum	Minimum
A. Right Bank Tributaries			
Subansiri	Chawidhowghat	18,799	138.00
Ranganadi	Pahumaraghat	1,911	13.50
Jia-Bharali	N.T. Rd. Cr.	9,939	53.00
Dhansiri (N)	N.T. Rd. Cr.	2,092	10.50
Pagladiya	N.T. Rd. Cr.	1,737	1.70
Manas	Mathanguri	10,842	142.00
Sankosh	Srirampur Rly. Bridge	6,391	87.00
B. Left Bank Tributaries			
Dibong	Jiagaon	11,205	176.00
Lohit	Digarughat	12,350	209.00
Burhi-Dehing	Khowang	2,217	15.30
Dikhow	A.T. Rd. Cr.	1,379	5.40
Dhansiri (S)	Numuligarh	2,296	9.30
Kopili	Dharamtul	1,942	17.80

of 1897 (8.7 Richter scales) and 1950 (8.6 Richter scale) had seriously disturbed the drainage of the Brahmaputra valley. Highly braided nature of the Brahmaputra, coupled with soil strata of its banks, steeper slopes and high sediment load are the main cause of its excessive erosive activity. These coupled with the practice of Jhumming cultivation in some areas is responsible for the high sediment load in the rivers.

During the last few years, floods have created havoc in the region and have put the people to untold miseries. The floods of 1987, 1988, 1992 and 1995 have been particularly severe in the Brahmaputra valley. The entire embankment system was put to maximum strain resulting in 102 breaches and cuts in 1987 and 220 in 1988. A tentative assessment of flood damage in Assam during the decades 1953 to 1995 is summarized in Table 20.

Since completely staying away from the flood plains in the NE region is neither possible nor desirable, mankind should be willing to regulate developmental activities in the flood plains. With the launch of the National Policy on Flood Control in 1954, the tempo of construction of embankments had increased rapidly. Due to this, the length of embankments increased from 6,000 km in 1954 to 15,675 km in 1990 besides improvement of 30,857 km of drainage channels. The years corresponding to the major flood events along with corresponding discharges at Pandu (near Guwahati, catchment area 424,100 sq. km, 73.12% of the whole) are given in Table 21.

The extent of flood depends on the width and depth of flow. The average lift at a braided reach near the Dibrugarh city (where the river width is 8.03 km) is 3.65 m. It is 9.91 m at a nodal point near Guwahati where the river width is 1.20 km (which is the minimum within the valley in Assam). After the cessation of monsoon by mid

Table 20. Tentative assessment of flood damage of Assam during 1953 to 1995

Theme	Total	Maximum
Area affected in million hectares	41.66	3.82 (1988)
Population affected in million	98.10	10.47 (1987)
Damage to crops in million hectares	5.08	1.13 (1988)
Loss in term of crore rupees	3,288.31	334.10 (1988)
Damage to houses – number	3,327,189	4,998 (1988)
Damage to houses – value in terms of rupees	296.80	103.92 (1988)
Cattle lost	431,537	108,913 (1987)
Human lives lost	1,724	226 (1988)
Damage to public utilities in rupees (crores)	832.42	225.82 (1988)
Total damage to crops, houses and public utilities (million rupees)	44,175.30	6,638.40

Table 21. Maximum observed floods at Pandu (Guwahati)

Year	Discharge (cumec)	Year	Discharge (cumec)
1957	57,733	1958	61,320
1960	57,985	**1962**	**72,794**
1966	57,570	1980	55,092
1988	61,015		

October, the water level and discharge of the river start receding and attain the lowest annual water level (LWL), sometime in the late February or early March. The recorded minimum discharge of the river at Guwahati during this period is 1,757 cumec.

Considering the seriousness and complexity of the flood problem, the Brahmaputra Flood Control Commission was set up in 1971 to prepare a comprehensive plan of flood control and implement it. The flood management works so far adopted have been mainly embankments, drainage channels, town protection works, and erosion control works. The works completed as part of flood management during the last three decades have been described in Table 22.

Table 22. Works completed as part of flood management during the last three decades

Brahmaputra Valley		Barak Valley	
i) Embankments		i) Embankments	
a) Brahmaputra	984 km	a) Barak	214 km
b) Tributaries	2,568 km	b) Tributaries	524 km
ii) Anti-erosion Schemes	421	ii) Anti-erosion Schemes	81
iii) Drainage Channels	600 km	iii) Drainage Channels	247 km
iv) Sluices	56	iv) Sluices	29

The difficult terrain of the Brahmaputra basin and insufficient hydrological and geological data have made investigations of multi-purpose reservoirs both difficult and time consuming. Nevertheless, the flood control measures implemented so far have afforded a reasonable protection to an area of 16.35 lakh ha out of the total flood-prone area of 31.50 lakh ha. But for the protection works, Dibrugarh town would have ceased to exist.

9.8.2. Observed Floods and their Causes

On average, the Brahmaputra River experiences 4 to 5 major floods annually during the monsoon months. The two rainiest months, July and August, experience most of the major floods. On rare occasions, the river also experiences floods in the later part of May. On an average, the flood level remains above the danger level for 39 days in a year at Guwahati; the maximum recorded period is 70 days. In some years, severe flooding results in huge losses of property, buildings, crops, live stocks, and crops besides dislocation of normal life and communication networks. Such floods occur as a result of heavy to very heavy rainfall in the catchment area. A monthwise distribution in the percentage of occurrences of historic floods within the valley are furnished in Table 23.

The percentage occurrence of major floods owing to such four typical meteorological situations in the month of June, July and August are 25%, 35% and 37%, respectively. The Brahmaputra River is turbulent and aggrading in nature; it is highly braided in the valley reach and in Bangladesh. The slope of the river in the state of Assam is much steeper than required for a river of its size. During flood, the river is burdened with a heavy sediment load, and a huge quantity of water flows down the river with a high velocity attacking its banks.

Generally, water level remains above the danger level for long periods in the valley reach, thereby creating acute flood congestion. Consequently, tributaries within the valley experience back flow from the main river when it is in high stage; the back flow may extend up to 25 km from the confluence with the Brahmaputra River. Under such circumstances, tributaries cannot discharge their own flow.

Table 23. Occurrences of floods in the Brahmaputra River in the valley reach during monsoon months

Month	Percentage Occurrence of Historic Floods owing to				
	Break monsoon situation	Movement of depression / storms from Bay of Bengal	Movement of land depression	Upper air cyclonic circulation	Total occurrences
June	2	10	5	8	25
July	10	10	7	8	35
August	18	5	7	7	37
Sept.	–	–	3	–	3
Total	30	25	22	23	100%

Table 24. Number and duration of floods of some major tributaries of the Brahmaputra at its middle reach

S.N.	Tributary	Annual Number of Floods		Annual Flood Duration (Days)	
		Max.	Min.	Normal	Max.
1.	Burhi – Dihing	5	2	19	41
2.	Desang	9	2	12	36
3.	Dikhow	5	Nil	3	13
4.	Jhanji	3	1	2	4
5.	Dhansiri (S)	10	1	–	55
6.	Kopili	45	3	32	68
7.	Puthimari	7	1	9	16

When tributaries together with the main river experience flooding, it becomes devastating. The normal duration for which the water level of the Brahmaputra River remains above the danger level annually is from 40 to 50 days but in some years it extends up to 70 days. The characteristics, such as annual maximum and minimum number of floods and normal and maximum flood duration of some important tributaries, are furnished in Table 24.

The main factors causing extensive floods in the valley are:

(i) adverse physiography of the valley;
(ii) heavy rainfall in the catchment of the tributaries joining in the valley reach;
(iii) excessive sediment load owing to frequent earthquakes and landslides and erosion of the top soil by heavy rainfall; and
(iv) inadequate waterway owing to encroachment of riverine areas.

9.8.3. Gauge Hydrograph

The gauge hydrographs of four stations, Dibrugarh, Guwahati, Goalpara and Dhubri, are shown in Figure 8. Note that the water level starts rising gradually from the month of March but rapidly from May. The danger level of the four stations is also shown in the figure. Note that the duration for which the water level remains above the danger level (DL) gradually increases towards the downstream side. The specific flood discharge of the Brahmaputra River at Bahadurabad in Bangladesh is 0.149 cumec/sq. km.

In the Brahmaputra reach in Assam, the observed HFL of all the stations are much higher than the DL. The observed HFL against the DL of six stations are shown in Table 25. The return period of the observed HFL is gradually reduced from 108 years for Dibrugarh to 21 years for Dhubri. The observed HFL and the DL of major tributaries of the Brahmaputra River within valley reach (middle reach) are shown in the Table 26.

The record shows that the observed HFL of the tributaries are all much above the DL of the site. It clearly indicates that the tributary rivers also experience high and devastating floods.

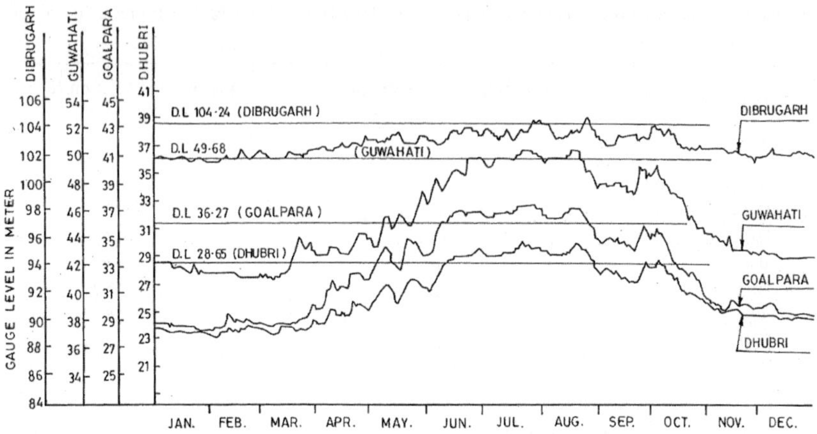

Figure 8. Gauge hydrographs of the Brahmaputra River at four sites within the middle reach

9.8.4. Travel Time

On the average, a flood wave takes 5 days 3 hours to travel through the valley portion from Pasighat (foothill) to Dhubri (near the Indo-Bangladesh border), a distance of 672 km. In this reach the average river gradient varies from 1:3700 near Pasighat to more than 1:8,875 near Dhubri. The average time taken by the flood wave to travel (time lag) reachwise the valley portion is furnished in Table 27.

9.9. SEDIMENT TRANSPORT

The sediment yield of the Brahmaputra River is very high and its position is second amongst the large rivers of the world. The sediment yield of the river at Guwahati is 0.0755 ha-m/sq. km of the catchment area; the maximum and minimum sediment concentration values are 679 and 36 ppm, respectively. The average annual sediment load of the river at Bahadurabad in Bangladesh is 735 million metric tonnes; the

Table 25. Observed highest flood level of the Brahmaputra River in the valley reach

Station	Chainage from the Mouth (km)	Danger Level (m)	HFL (m)	Return period of HFL (Years)
Dibrugarh	917	104.24	105.95	108
Neamati	809	85.04	86.84	32
Tezpur	672	65.38	65.66	30
Guwahati	542	49.68	51.04	25
Goalpara	432	36.27	36.88	23
Dhubri	357	28.65	29.97	21

Table 26. Observed highest flood level and danger level of some tributaries of the Brahmaputra River within the valley reach (middle reach)

Sl. No.	Tributary	Observation Site	Danger Level (m)	Observed HFL (m)	Year of Occurrence
1.	Subansiri	Chowaldhowaghat	100.43	101.31	1972
2.	Ranganadi	Pahumaraghat	–	95.92	1979
3.	Jia-Bharali	N.T. Road Cr.	77.00	80.89	1970
4.	Dhansiri (N)	N.T. Road Cr.	–	84.14	1979
5.	Puthimari	N.T. Road Cr.	51.82	54.20	1973
6.	Pagladiya	N.T. Road Cr.	52.75	55.51	1970
7.	Manas	N.T. Road Cr.	47.55	49.88	1974
8.	Sankosh	Srirampur	–	46.71	1968
9.	Lohit	Dhola	128.27	129.49	–
10.	Burhi-Dihing	Khowang	102.11	103.51	1980
11.	Desang	N.H. Cr.	92.00	93.88	1980
12.	Dikhow	A.T. Rd. Cr.	93.30	95.60	1979
13.	Dhansiri (S)	Golaghat	89.50	90.68	1971
14.	Kopili	Dharamtul	56.31	57.85	1966

Table 27. Reachwise travel time (time lag) of flood wave of the Brahmaputra River from Pasighat to Dhubri in the valley (middle reach) portion

From Station (chainage in km)	To station (chainage in km)	Distance along the river (km)	Travel time of flood wave (Hrs)
Pasighat (1029)	Dibrugarh (917)	112	12
Dibrugarh (917)	Neamati (809)	108	24
Neamati (809)	Tezpur (672)	137	24
Tezpur (672)	Guwahati (542)	130	24
Guwahati (542)	Goalpara (432)	110	24
Goapara (432)	Dhubri (357)	75	15
Total		672	123

average maximum concentration of the river is 1,180 ppm. The percentage break down of the surface sediment load at Bahadurabad is 12.5% coarse, 14.2% medium and 73.3% fine.

The composition of the surface sediment load is 12.5% coarse, 14.2% medium, and 73.3% fine. The tributaries that originate from the high mountains carry high sediment loads as these flow through geologically young Himalayan mountain ranges of unconsolidated sedimentary rocks with steep slopes and the area falls in one of the most earthquake prone belts. In comparison, the tributaries originating from the southern (Assam) hill ranges carry less sediment load as these hills are geologically much older, stable, of lesser height, and have flatter bed slopes. Heavy bank erosion by the Brahmaputra River takes place at different reaches.

In the valley reach in Assam, the sediment yield of the north bank tributaries is approximately 5 to 6 times higher than that of the south bank tributaries. Over

70% of the annual average sediment load of the Brahmaputra River at Guwahati is contributed by the main river and two tributaries, Dibong and Lohit.

Among the north bank tributaries, Subansiri, Jia-Bharali, Manas and Sankosh carry high sediment loads, while Burhi-Dihing, a south bank tributary, carries the highest sediment load. Other tributaries, such as Kopili, Dhansiri (S) and Disang, carry significant amounts of sediment load. The average annual suspended sediment load of some important tributaries of the Brahmaputra River within its middle reach in the valley is furnished in Table 28. The bed load of the Brahmaputra River is generally considered as 15% of the suspended sediment load.

Most of the tributaries of the Brahmaputra River form alluvial deltas just at the foothill region. Most of the suspended sediment carried by the rivers from the mountainous region gets deposited in the river bed. Sometimes rivers change their course in this region when old channels get filled up with sediment. The north bank tributaries are more prone to such problems. The main reasons of high sediment load in the Brahmaputra River are:

(1) Geologically Brahmaputra is a young river whose present configuration took shape only during Pleistocene and recent times.
(2) The catchment area of the river falls in a seismic belt. The basin experiences a number of earthquakes of moderate intensity every year and landslides and slips occur owing to these yields high sediment loads in rivers. The basin experienced severe earthquakes in the past; the earthquakes of 1762, 1822, 1865, 1869, 1897 and 1950 were notable. The earthquakes of 1897 and 1950 were amongst the severest. These two earthquakes disturbed the topography and the drainage system to a great extent, resulting in heavy sediment load and bank erosion. The river is still in the process of attaining stability.
(3) The river and its tributaries traverse through unconsolidated sedimentary rocks of the Himalayas and high amounts of sediment are washed down by the monsoon rains.
(4) High intensity rainfall erodes the top soil of the catchment area; landslides and slips also occur in many places which add high sediment load in the river.

Table 28. Average annual suspended sediment load of some important tributaries of the Brahmaputra River

Name of tributary	Observation site	Average annual suspended sediment load (ha-m)
Dihang	Ranaghat	7,694
Dibong	Jiagaon	2,620
Luhit	Digarughat	4,650
Subansiri	Bhimpuraghat	2,497
Jia-Bharali	N.T. Road Cr.	2,143
Manas	Mathanguri	2,038
Sankosh	Srirampur	594
Burhi-dining	Khowang	594
Dhansiri (S)	Numuligarh	422

(5) Steep slopes of river and its tributaries in the mountainous reaches bring down sediment.
(6) Forest fires also expose the soil to erosion.
(7) Gradual encroachment of forest areas.

Besides the above natural causes, some human activities also contribute to high rates of sediment load; one such human activity is 'Jhumming' of shifting cultivation which adds sediment load to a great extent. Jhumming is extensively practiced by tribal communities living in the mountainous regions of the catchment area especially in the Arunachal Pradesh and Southern Assam hill ranges and is done by burning the forest and exposing the soil for cultivation. About 1.35 million hectares of forest land remains under Jhum cultivation at one time or the other.

9.9.1. Braiding and Bank Erosion

The Brahmaputra River does not display a stable morphology. The alluvial plains of the river are nearly 96 to 112 km wide and 644 to 805 km long. The bed is about 10 km wide. The bed slope is very gentle, varying from 1 in 6,600 to 1 in 9,900, and at some places, the river bed lies below the mean sea level.

The channel is highly braided and short-term channel migration is quite drastic; rates of movements as high as 870 m a year being common. The most significant bank line modifications take place during the falling river stage, when sediment is deposited as in the channel, causing migration of the thalweg.

The Brahmaputra River is highly braided in the reach downstream from Pasighat. The width of the Brahmaputra River from Kobo to the Indo-Bangladesh border varies from 6 km to 18 km except in nine reaches (nodal points) where it traverses through deep and narrow throats. The width of the river at the nodal points varies from 1.2 km to 5.1 km. The reachwise average width, maximum width and width at nodal points of the Brahmaputra River are furnished in Table 29.

Heavy bank erosion by the Brahmaputra River at different reaches takes place owing to excessive sediment load, erodible nature of bank material, and formation of char islands. Active bank erosion is generally observed to take place both upstream and downstream of the nodal points and also in the downstream reaches of the confluence of major tributaries. The instability of the Brahmaputra River coupled with silt and sand strata of its banks is also responsible for considerable bank erosion in its valley reach. There is a tendency of the Brahmaputra River to shift southward within the valley reach; this tendency has become more prominent after the great earthquake of 1950 which raised the whole land mass of the northeastern part of the valley by 3 to 4 m. This southward thrust has initiated widespread erosion in the south bank near the Dibrugarh town which is continuing even after construction of different anti erosion schemes. A few kilometers downstream of the nodal point near Guwahati, the river is observed to have a northern migration since 1920 and active bank erosion has taken place in the Nalbari and Barpeta districts of Assam. The situation is different further downstream of the next nodal point near Jogighopa

Table 29. Width of the Brahmaputra River in different reaches from Kobo to the Indo-Bangladesh border

S.N.	Reach (Chainage in km from the mouth) and reach length	Width of the river (km)			
		Mean	Maximum (chainage in km)	At nodal point (chainage in km)	Place
1.	Kobo (977) to Dihingmukh (877), 100 km	8.19	14(967)	4.8(977)	Kobo
2.	Dihingmukh to Dhansirimukh (747), 130 km	8.18	13(870)	5.1(864) 3.7(832)	Near Desangmukh Bechamara
3.	Dhansirimukh to Pandu (535), 212 km	8.12	13.1(730)	4.4(727) 3.6(674) 1.2(535)	Gamiri Near Tezpur Paudu
4.	Pandu (535) to Dhubri (357), 178 km	7.93	18.5(503)	2.4(517) 2.4(422)	– Jogighopa

where the river shows a tendency of migration towards the south. The south bank in this reach is facing active bank erosion.

Erosion is more prominent during recession stages of the flood waters. Owing to the variation of the river width along its course, the velocity of flow also changes and as a result scouring and siltation also take place. In the last few years, owing to active bank erosion at some affected reaches, the river width has increased from 8 km to 10 km and in some other reaches from 15 km to more than 18 km.

9.10. IRRIGATION

Water resources in the Brahmaputra region are largely under-utilized. Large extent of availability of these resources on temporal and spatial scale provides an opportunity for improving life style of the population residing in the region. The basic demands of the region are drinking water, irrigation, hydropower and navigation. Water resources of this part of the country are almost free from pollution except for small biological contamination. Even with rigid norms of drinking water, the water required for drinking purpose is negligible.

The geographical area of Assam, Meghalaya, Nagaland, and Arunachal Pradesh, the four states adjoining the Brahmaputra Valley, is about 20 million hectares. The area under all crops in these four states has been reported as 4.5 million hectares. Most of this area (3.6 million hectares) is in Assam which has 237,000 hectares under tea. Out of the 4.5 million hectares of the cropped area, only 0.7 million hectares are irrigated. The main crops in the region are rice, wheat, maize, pulses, oilseeds, sugarcane, certain vegetables, jute, and mesta. The total food grain production for all crops, excluding tea but including sugarcane, is around 6.2 million tons. The production of rice in Assam in 1994–95 is about 3.4 million tons. The average yield of about 1,300 kg/ha is much less than that in other parts of the country.

Unlike other states of India, irrigation was hardly practiced for agricultural production in the north-eastern region. In the mid sixties, a few flow irrigation projects were developed in Assam. The cultivatable area in the region has been estimated to be about 7 million hectares. The irrigation potential has been estimated to be 3.6 million ha.

There is no shortage of water for irrigation. But due to limitations of topography and frequency of floods in lower reaches of the basin large scale irrigation is not feasible in Arunachal Pradesh and Assam. However, due to hilly terrain, the drainage of water is very fast thus causing water stress during dry spell. Apart from the huge surface water resources in the Brahmaputra valley, the utilizable ground water has been estimated to be over 2 million ha-m. Even with the maximum provision for irrigation, the annual water requirement is not likely to exceed 6×10^6 ha-m. The total irrigation potential of the region is 4.26 M ha of which only about 20% has been utilized. Surface water resources can be tapped either by diversion structures or by pumping for irrigation. The planning for irrigation has to be sub-basin wise to utilize the full irrigation potential.

9.11. HYDROPOWER

The hydropower potential of the Brahmaputra and Barak basin in NE states has been estimated at 34,920 MW at 60% load factor. This amounts to 41.5% of the potential of all the rivers of India. The river along with its tributaries has great potential for hydropower development. It is estimated that Arunachal Pradesh alone has a hydropower potential of 26,756 MW @ 60% load factor and only 125 MW (0.47%) has been developed so far. Overall for the region, despite the availability of large hydropower potential, current development is very small. Out of the available potential, only less than 2% has been developed so far. A few public sector corporations have now been involved in development of hydropower potential. These projects will also moderate floods during monsoon and augment lean season flow. Table 30 details sub-basin wise hydropower potential for the north-eastern region.

The large chunk of hydropower in the NE region cannot be developed without a large export outlet from this region. CEA has identified three primary reasons for the extremely poor development of the hydroelectric resources of the NE region: (i) difficult terrain (ii) lack of demand, and (iii) non-availability of bulk transmission corridors. It is noted that land communication between the Indian mainland and the NE region is through a narrow chicken neck of about 10 km width. Detailed investigations have been carried out to identify project sites for hydropower development.

Manas and Sankosh are two major tributaries of Brahmaputra having their catchments in Bhutan and with suitable storage sites for hydropower development. An ambitious project, Sankosh dam (4,000 MW) is under consideration of Bhutan

Table 30. Sub-basin wise estimated hydro potential of Brahmaputra basin

Basin/Rivers	No. of schemes identified	Potential at 60% load factor (MW)
Dihang-Dibang	28	13,615
Lohit	11	4,152
Subansiri	25	6,893
Upper Brahmaputra	19	789
Kameng	34	1,982
Kalang (Kopili)	16	510
Teesta	30	3,021
Lower Brahmaputra	03	50
Barak and neighboring rivers	60	3,908
Total	226	34,920

Source: WG (1999).

Government. At present India has bilateral arrangement only in respect of 336 MW Chukha hydroelectric project with Bhutan. For bilateral projects, the present Indian initiative is focused on Karnali, Pancheswar, and Saptakoshi projects in Nepal and Sankosh project in Bhutan.

Installed capacity of hydropower projects and gross energy generation in 2000–01 in NE Region is shown in Table 31. Note that of the total installed capacity, 37.42% comes from hydropower projects.

A number of small hydropower schemes have also been identified in the north-east region. Table 32 gives the details of such schemes for various states.

Table 31. Status of major hydropower projects in North Eastern Region of India

Installed Capacity (MW) as on 31/03/2001			Gross Energy Generation (MU)
State Agency	Hydropower	Total power	
Arunachal Pradesh	23.60	39.40	–
Assam	2	622.70	1,059.208
Manipur	2.80	12.10	–
Mizoram	5.40	24.50	–
Nagaland	3.30	6.56	–
Meghalaya	185.20	185.20	657.519
Tripura	16	86.6	328.748
Total in State Sector	238.30	977.06	
A. NEEPCO	325.00	703.66	2,559.063
B. NHPC	105.00	105.00	551.06
Total in NE Region 668.30 (37.42%) 1,785.72 (100%)			

Source: NEEPCO

Table 32. Identified small hydropower schemes in NE Region

State	Capacity < 3 MW		Capacity 3–15 MW		Total	
	Number	Capacity, MW	Number	Capacity, MW	Number	Capacity, MW
Arunachal Pradesh	433	382.31	49	460.72	482	843.03
Assam	38	50.00	8	68.00	46	118.00
Manipur	91	59.75	4	29.88	95	89.63
Meghalaya	83	41.00	13	97.50	96	138.50
Mizoram	73	42.32	13	101.00	86	143.32
Nagaland	67	26.89	17	117.50	84	144.39
Tripura	8	9.85	–	–	8	9.85
Total	793	612.12	104	874.60	897	1, 486.72

Source: NEEPCO

9.11.1. Completed Projects

In the following, we describe selected important completed projects of this region.

Doyang: Doyang is a rockfill dam on the Doyang River, a tributary of Brahmaputra, 30 km from Wokha in Nagaland. The catchment area at the dam is 2,606 km^2. The height of the dam is 92 m and the reservoir has a live storage capacity of 535 MCM at FRL 333.00 m. With mean annual inflow of 1,355 MCM, Doyang power house has 3 units of 25 MW each and has a firm power of 21 MW. NEEPCO commissioned the project in 2001.

Kopili: This project consists of Umrong dam on Umrong River, a tributary of Brahmaputra River. The power house is located at a distance of 70 km from Lanka in Assam. The catchment area at the dam is 1,318 km^2. The height of the dam is 30 m. The reservoir has a live storage capacity of 55.5 MCM at FRL 609 m and the MDDL is at 592.83 m. Kopili power house has 4 units of 50 MW each, with mean annual inflow of 2,860 MCM. NEEPCO commissioned the project during 1988–97 which has a firm power of 55 MW.

Khandong: This dam on Khandong River, a tributary of Brahmaputra, is located at a distance of 70 km from Lanka in Assam. The height of the dam is 66 m and its catchment area is 1,256 km^2. The reservoir has a live storage capacity of 129.5 MCM at FRL 719.3 m; the MDDL is at 704 m. Khandong power house has 2 units of 25 MW each, with mean annual inflow of 2,860 MCM and provides firm power of 13 MW. NEEPCO commissioned the project in 1984.

Loktak: Loktak Hydroelectric station involves a barrage, namely Ithai barrage, near Loktak Lake, 39 km from Imphal in Manipur. The catchment area at the barrage is 1,000 km^2. The height and length of the barrage are 11 m and 59 m respectively. The pond has a live storage capacity of 397 MCM at maximum water level 768.5 m and the minimum water level is at 766.2 m. Loktak power house has 3 units of 35 MW each, with mean annual inflow of 679 MCM. It has a firm power of 42 MW. NHPC commissioned the project in 1983.

Rammam II: This hydroelectric station uses water from Rammam and Lodhama weirs constructed on Rammam River and its tributary Lodhama Khola; these are the tributaries of Teesta River. The power house is located at a distance of 77 km from Darjeeling, West Bengal. The intercepted catchment area is 209 km^2. Four units of 12.75 MW each have been installed with a firm power of 14 MW. West Bengal State electricity Board commissioned the project in 1995–96.

Ranganadi: A concrete gravity dam has been constructed on Ranganadi River, a tributary of Brahmaputra River, 50 km from Zir City in Arunachal Pradesh. The catchment area at the dam is 1,894 km^2. The height of the dam is 68 m and the reservoir has a live storage capacity of 21.28 MCM at FRL 567 m; its MDDL is at 560 m. Doyang power house has 3 units of 135 MW each, with maximum and minimum annual inflow of 5,080 and 2,048 MCM respectively. It has a firm power of 166 MW. NEEPCO commissioned the project in 2002.

Rangit III: Rangit III hydroelectric station was constructed in the year 2000 by NHPC on Greater Rangit River, a tributary of Teesta River. The dam is located at a distance of 130 km from Siliguri in South Sikkim District. The catchment area at the dam is 979 km^2 and the dam is 47 m high. The reservoir has a live storage capacity of 1.175 MCM at FRL 639 m and the MDDL is at 627 m. The power house has 3 units of 20 MW each, with mean annual inflow of 0.696 MCM. It has a firm power of 39 MW.

Teesta Canal Power House I, II, III: Three power houses have been constructed on Teesta canal, namely Teesta PH I, II, and III. Teesta canal emerges from Mahananda barrage, which has been constructed on Teesta River, a tributary of Brahmaputra River in Siliguri, West Bengal. The Power houses have been constructed on the canal at 5 km, 21 km and 31 km from barrage, near Siliguri town in Jalpaiguri District, West Bengal. The catchment area at the Mahananda barrage is 675 km^2. The design flood of Mahanada barrage is $2,265\,m^3/s$. All the three power houses have 3 units of 7.5 MW each. WBSEB commissioned the project in 1997–99.

Teesta Low Dam Stage – III Hydroelectric Project: It is being constructed on Teesta River in Darjeeling distt. (W.B). It is located near New Jalpaiguri. The project will have 32.5 m high barrage and the power plant will have installed capacity of 4×33 MW or 132 MW to annually generate 594 MU in a 90% dependable year. This project is estimated to cost Rs.770 crore. Work is in progress and is anticipated to be completed by March 2008.

Teesta Low Dam Stage – IV Hydroelectric Project: This project is under construction near Siliguri town in Dist. Darjeeling, West Bengal. Here, a 45 m high concrete gravity dam will be constructed. The catchment area for the project is 8,021 sq. km. With FRL of 182.25 m and MDDL of 179.00 m, the reservoir has a gross storage of 36.63 MCM and live storage of 7.91 MCM. For power generation, the underground power house will have Kalpan Turbines, 4×40 MW each which will work at gross head of 25.8 m. Its annual generation is likely to be 720 MU (90% dependable year). Work is expected to be complete by September 2009 at a cost of Rs.1,061 crore.

Umiam III: This is the third in the series of projects with Umiam I of 36 MW and Umiam II of 18 MW capacity. Umiam III hydroelectric project consists of Kyrdemkulai concrete gravity dam on Umiam & Umtru Rivers, tributaries of Brahmaputra. The power house is located at a distance of 52 km from Shillong in District Ri Bhoi of Meghalaya. The catchment area at the dam is 520 km^2 and its height is 27.5 m. The reservoir has a live storage capacity of 2.78 MCM at FRL 681.23 m and the MDDL is at 675.58 m. Umiam III power house has 2 units of 30 MW each. It has a firm power of 16 MW. Meghalaya State Electricity Board commissioned the project in 1979.

Umiam IV (Nongkhylle): Umiam IV hydroelectric project consists of Nongkhylle dam across Umtru River. The power house is located at a distance of 55 km from Shillong in District Ri Bhoi of Meghalaya. The catchment area at the dam is 150 km^2. The height of the dam is 43 m. The reservoir has a live storage capacity of 4.94 MCM at FRL 503 m and the MDDL is at 496 m. Umiam IV power house has 2 units of 30 MW each, with mean annual inflow of 52 MCM. It has a firm power of 15 MW. Meghalaya State Electricity Board commissioned the project in 1992.

9.11.2. Projects Under Construction/Investigation

Many hydropower schemes are under investigation in the north-east region of India. There are immense possibilities of cooperation with neighbouring countries for hydropower generation. India and Bhutan have agreed to build a 1,020 MW hydroelectric project on the Wangchu River in Bhutan. Surplus power from the Rs. 14bn (US$ 400 M) project will be sold to India, which will finance 60% of the cost through a grant and the balance through a loan. The dam will be handed over to the Bhutan Government two years after completion.

A list of projects under construction in the region is given in Table 33.

A discussion of some important projects that are under planning/investigation follows.

Pagladiya Dam: This is a multipurpose project whose construction work is likely to be initiated soon and is scheduled for completion in 2008. The project which has been cleared by the Technical Advisory Committee of the Ministry of Water Resources, Government of India, involves construction of an earthen dam of 26 m height over Pagladiya River, a tributary of Brahmaputra, in Nalbari District of Assam. The dam will intercept an area of 570 sq. km. The Pagladiya River has been responsible for recurrent floods in the north bank of the Brahmaputra. Benefits from Pagladiya dam include flood moderation in 40,000 ha area, irrigation benefits to 54,000 ha, and incidental power generation by a plant of 3 MW capacity. The project cost has been estimated at Rs. 541 crore.

Siang (Dihing) Multipurpose Dam Project: Originally, the Siang Dam project proposal envisaged a multipurpose dam with 20,000 MW installed capacity to

Table 33. Projects under construction in NE region

Project	Capacity (MW)	State
Dhansiri + LBP	120	Assam
Karbi Langpi (Lower Borpani)	100	Assam
Kopili Stage II	25	Assam
Likim- Ro	24	Nagaland
Loktak Downstream	90	Manipur
Ranganadi	405	A. Pradesh
Teesta Stage V	510	Sikkim
Tuirial	60	Mizoram
Tuirini	35	Mizoram
Tuivall	35	Mizoram
Umiam- Umtru St-V	30	Meghalaya

Source: CEA, NEEPCO

generate hydropower. Now, the proposal has been split in three dams: (i) a rockfill dam of 257 m height near Pugging village on the Siang River with installed capacity of 11,000 MW, (ii) a 154 m high rockfill dam near Raying village, 15 km upstream of Kaying on Siyom River with installed capacity of 750 MW, and (iii) another 45 m high rockfill dam upstream of Pasighat near Rotung village with installed capacity of 1,500 MW. These three smaller dams have been planned to avoid submergence of important towns. Investigations for the dams are in progress.

Siyom Project: This proposed project will be located near Village Raying (near Along) in West Siang District of Arunachal Pradesh. A 188m high, concrete face rockfill dam is proposed. Underground power house with 1,000 MW capacity (4×250 MW) will be constructed and annual generation will be 3,640.95 MU in a 90% dependable year. The DPR for the project has been completed.

Subansiri Multipurpose Dam Project: Similar to Siang, the original Subansiri project proposal envisaging a large dam in Arunachal Pradesh with 4,800 MW installed capacity has been recently modified. Now the proposal is to construct three smaller rockfill dams: (i) a 265 m high dam on Subansiri river near Menga village, upstream of Daporijo with installed Capacity of 2,500 MW, (ii) a dam of same height (265 m) on Kamla River near Tamen with installed capacity of 2,000 MW, and (iii) a dam of 110 m height on Subansiri River near Gerukamukh having installed capacity of 1,500 MW. In addition to hydropower, these dams are expected to give irrigation benefit to an area of 25,000 ha. Besides, these will also help in flood moderation, navigation, pisciculture, and recreation. Detailed surveys and investigations for the projects are under way.

The Lower Subansiri Hydropower Project is under construction in lower Subansiri/ Dhemaji Distt. (Arunachal Pradesh & Assam) near Nagaon at an estimated cost of Rs.6,285 crore. A concrete gravity dam, 116 m in height is envisaged. The proposed generation capacity is 2,000 MW (8×250 MW) and expected annual generation is 7,422 MU in 90% dependable year. The project is expected to be completed by 2010.

Teesta Stage – V Hydroelectric Project (under construction) in East Sikkim District in Sikkim. A 96.45 m high concrete gravity dam with underground power house containing 3 units of 170 MW each (total 510 MW) is being constructed. It annual generation will be 2,573 MU (90% dependable year). Work is scheduled for completion by 2007 at an estimated cost of Rs. 2,198 crore.

A private company has won the contract to construct the 360 MW Teesta VI Hydropower Project. This project is the last scheme of six stages of cascade development on the Teesta River. The investment by the Government of Sikkim in the project will be in the form of equity to the extent of 26 %.

Tipaimukh Dam Project: It envisages construction of a 162.8 m high rockfill dam across the Barak River near Tipaimukh village on Manipur-Mizoram border, intercepting a catchment area of 12,758 sq. km. The project will generate electricity by 6 units of 250 MW each. It is expected to yield irrigation benefits to 98,844 ha and flood control benefits to 1.50 lakh ha area in Cachar, Karimganj & Hailakandi districts of Assam. Besides, indirect benefits like navigation, pisciculture, recreation etc. are also expected.

Table 34 lists selected projects that are under survey and investigation in the north-east region

9.12. ENVIRONMENTAL AND OTHER ASPECTS

In this section, we discuss several topics which are of interest from environmental point of view.

9.12.1. Forest

The forests of Assam contain a great diversity of flora and fauna and other natural ecosystems. There are about 51 different forest types and subtypes occurring in the region, forming an amalgam of Indian, Malayan and Tibetan elements. Five major groups of forests in Assam have been identified: 1) Tropical Wet Evergreen Forest; 2) Tropical Semi-evergreen Forest; 3) Tropical Moist Deciduous Forest; 4) Littoral and Swamp Forest; and 5) Tropical Dry Deciduous Forest.

Forestry has been playing a very intimate role in the social lives of the people living in the region. During the Ahom dynasty, a specific administrative branch looked after the management of timber. As of 1992, the Reserved Forests in Assam was 17,581 sq. km. By the year 2000, the area under wildlife sanctuary and national parks was around 3,000 sq. km. In the late 1980s, the state of Assam had a forest cover of 22 percent. The percentage of forest cover in the plains was around 17.2 percent, while in the hills it was 42 percent. The reserved forests account for 19.14 percent of the state's total geographical area.

Table 34. Some project proposals that are under investigation in north-east region

Name of the Project	State	Dam height (m)	Location	River	Installed capacity (MW)	Other benefits
Lohit Multipurpose Project	Arunachal Pradesh	272	Mompani	Lohit	2,000	Flood control, pisciculture, and recreation
Kameng Multipurpose Project	Arunachal Pradesh	200	Near Bhalukpung	Kameng	1,100	–
Jadukata Multipurpose Project	Meghalaya	120	9 km south-west of Nongatain town	Jadukata	450	Flood control
Someswari Multipurpose Project	Meghalay	63	6 km upstream of Nanhgalbibra town	Someswari	130	Flood control and irrigation
Bairabi Multipurpose Dam Project	Mizoram	60	Near Bairabi	Dhaleswari tributary of Barak	75	
Noa-Dehing Multipurpose Dam Project	Arunachal Pradesh	57	Near Miao	Noa-Dihing, a tributary of Brahmaputra	75	Flood control, irrigation to an area of about 8,000 ha
Debang Multipurpose Dam Project	Arunahcal Pradesh	165	Near Munli (Ichi)	Debang river	12*250 MW = 3,000	Flood control, recreation, navigation, and pisciculture
Kulsi Multipurpose Dam Project	Meghalaya – Assam	60	1.3 km. upstream of Ukium village	Kulsi River, a tributary of Brahmaputra	36	Irrigation to area of 39,000 ha in Assam.

9.12.2. Wildlife

The Brahmaputra valley has legendary sites for unique wildlife. Assam, harbouring the big five mammals-Rhino, Tiger, Wild Buffalo, Gangetic Dolphin and Elephant, has been in limelight for its role in the conservation. The spotted deer, which is found in India, has its last distribution range in western Assam. The Malayan Sun Bear is found in the southern part of the Brahmaputra valley as the Hoolock Gibbon. Similarly, the Golden Langur is found in a few pockets in northern Assam.

At the beginning of the year 2000, Assam had 5 national parks and over 10 wildlife sanctuaries. The National Parks are: Kaziranga (430 sq. km), Manas (500 sq. km), Nameri (200 sq. km), Dibru-saikhowa (340 sq. km) and Orang (79 sq. km). The salient wildlife sanctuaries are: Pabitora, Barnadi, Burachapori, Panidihing, Hollangapar Gibbon Sanctuary, Chakrasila, Bordoibam-beelmukh, Garampani, Padumoni-Bherjan-Borajan, and Laokhowa. The Manas National park, declared as a World Heritage Site, is one of the important Tiger Reserves in India. Recently, Nameri National Park has been declared as a Tiger Reserve. Dibru-saikhowa National Park (740 sq. km) has been declared as a Biosphere Reserve.

The Kaziranga National Park (KNP) is a place of international importance because of its mega-diversity in flora, fauna and ecosystem. The KNP is a healthy wetland system. Being situated in the flood plain of the Brahmaputra River, the soil of the park is rich in alluvial deposits. Flood is an annual phenomenon submerging 50–75 percent of the park area. The flood waters generally start receding after 8–12 days. In the KNP, 478 species of birds reside out of which 25 species are globally threatened. The Great Indian One Horn Rhino (*Rhinoceros unicornis*) is one of the three Asian Rhino species found in Assam. The KNP holds almost two-thirds of the Indian Rhino (Figure 9) Population. The state of Assam harbours more than 70% of the Great Indian Rhino Horn population of the world and provides shelter to over 400 tigers and 5,600 wild elephants.

Figure 9. The Great Indian One Horn Rhino in its natural habitat at KNP

9.12.3. Wetlands

Wetlands are regarded as one of the most productive ecosystems in the world. Wetlands include marshes, ponds, lakes, rivers, streams, reservoirs, and marine waters of the shoreline. Although highly efficient, wetland ecosystem is very fragile and easily disturbed by human actions.

The wetlands in Assam are primarily fresh water and riverine in nature. Assam has 3,519 wetlands as summarized in Table 35. Out of these, 2,278 wetlands are with low turbidity, 346 with moderate and 178 wetlands with high turbidity. The total fish production from Assam's wetlands is 1.55 lakh tons per year.

The total area under water bodies in the north eastern region is around 3.32 lakh ha. Yet the production of the fish is very small and most of the fishes are procured from other parts of the country. Planned development of water bodies provides tremendous scope for development of fresh water fish potential.

9.12.4. Water Quality

The major ion chemistry of the Brahmaputra is characterized by high bi-carbonate content and source rock influence. While higher values of TSM than TDS during monsoon indicate predominance of physical weathering over chemical weathering, chemical weathering is relatively more pronounced during the dry season. On average, 60% of the bicarbonates in the Brahmaputra water come from silicate weathering and the rest from the carbonates.

Surface suspended sediments range from fine sand to clay, the size fraction greater than 12 μm constituting an important size population. Surface suspended sediments are moderately to poorly sorted with greater amounts of finer material in the distribution, particularly during the rainy season. The detrital contribution in

Table 35. Types of Wetland in Assam

Wetland Type	Number	Area (ha)	Percent
Natural			
Lake/Pond (locally called 'Beel'	690	15, 494.0	15.30
Ox-bow Lake (locally called 'Era suti')	861	15, 460.6	15.27
Waterlogged (seasonal)	1,125	23, 431.5	23.15
Swamp/Marsh (locally called 'Pitoni/Jolah')	712	43, 433.5	42.91
Total	3,388	97, 819.6	96.66
Man-made			
Reservoirs	10	2, 662.5	2.60
Tanks (locally called 'Pukhuri')	115	749.5	0.74
Total	125	3, 412.0	3.34
Grand Total	3,513	101, 231.6	100.00

Source: Assam Remote Sensing Application Centre.

the form of Quartz, Feldspar and Mica make up more than 80% of the mineralogy. Chlorite, Illite, and Kaolinite constitute about 95% of the clay minerals.

Sediment chemistry does not reveal any marked spatial or temporal variation. However, spatial variation in the metal/aluminum ratio in some cases has been observed. This may be attributed to continuous addition of freshly eroded materials by a large number of tributaries at different points of the main stem. Since there is hardly any major industry in the catchment area and the amounts of sediment carried by the tributaries are enormous, there is not enough ground to link any occasional rise in metal concentration to any point source.

The major sources of carbon in the Brahmaputra basin are deforestation and fossil fuel consumption. In the case of nitrogen, precipitation, soil erosion, drainage and fertilizer application constitute the main sources. Phosphorous carried by the suspended load makes up a substantial part of the P loading in the Brahmaputra. A comparison of different forms of P between suspended and bed sediments indicates a decrease in organic-P, Fe-P and total-P after deposition. Since much of the suspended sediment-transported P is not available biologically, land management practices to minimize sediment input to the river should produce no significant reduction in the biological availability of P. The total flux of P from the Brahmaputra ($150–270 \times 10^{10}$ gm P/year) constitutes nearly 5% of the global flux and is about three times greater than that carried in dissolved form. Considering the mobilization of P by fertilizer use, it is reasonable to assume that the P concentrations will further increase.

Suspended sediments play a significant role on the overall solute and sediment biogeochemistry of the river. The biogeochemistry of Ganga and Brahmaputra is markedly different with respect to HCO_3, SO_4, Cl, SiO_2, Ca and a few heavy metals, such as Cu, Fe, and Zn. It may be possible that apart from the natural factors, such as geology, topography, vegetation, and precipitation, this is also owing to the difference in the intensity of utilization of the two rivers and anthropogenic impacts on their watersheds which may be different, having a direct reflection on their geochemical behaviour.

9.12.5. Navigation

The total length of rivers for navigation in this basin is around 3,880 km. Brahmaputra from Sadiya in Assam to Bangladesh border is known as National Waterway Number 2. In the past, the Brahmaputra River served as the principal route for transportation of goods between Calcutta and places in the Assam valley. Regular steamer services were operated by the River Steam Navigation (RSN) Company from 1863. This company was merged with the India General Steam Navigation Company in 1899 to form the Joint Steamer Companies. They provided freight service between Calcutta and 16 riverside towns from Dhubri to Dibrugarh. Outgoing traffic from the valley to Calcutta was also heavy.

Following the 1950 earthquake, steamer services beyond Neamatighat became difficult. The government formed a public sector company called the Central Inland

Water Transport Corporation in 1957 (CIWTC). The CIWTC is continuing to operate the service now. However, with the opening of the national highway and conversion of the meter gauge rail link to broad gauge, the traffic on the river route got diverted to road and rail. The traffic potential on the river route is estimated to be about 1.1 million tons by the year 2004–2005. Till 1996–97, the capacity utilized was only 2,117 (20%of the available). Thus there exist prospects for development of navigation facilities in the region. Efficient and quicker turnaround is likely to improve the traffic potential between Calcutta and the Assam valley. The planned release of water from major storages will improve the draught in the river during the lean season by 1.5 to 2 meters. This will help provide a more efficient and economical freight traffic on the river route.

9.13. CONCLUDING REMARKS

The hydrology of Brahmaputra River is unique in that its basin enjoys a variety of seasons and climate ranging to very cold to tropical. It can serve as one of the most productive natural laboratories for hydrologic investigations.

CHAPTER 10

INDUS BASIN

10.1. THE INDUS BASIN

According to a popular theory, the majority of inhabitants of North India are the descendents of the Aryans who came to this part of the world thousands of years BC. The river which is known as the Indus these days was called *Sindhu* by the Aryans. The people who lived across this river were called Hindus (a word derived from Sindhu) and the land was called Hindustan, the ancient name of India. The fertile planes of Indus were the seats where the Indus valley civilization arose about 2,500 BC. In ancient times, Indus was second only to Ganga River in terms of cultural and commercial importance for the region. Its importance for India was considerably reduced after the partition of the country in the year 1947.

The Indus River, one of the mightiest and the longest rivers in the world, originates from Mount Kailash in Tibet on the north side of the Himalayas at an altitude of 5,486 m. The total length of the river is 3,199 km (Basit Masud 2003). It bisects the physical territory of Pakistan longitudinally. From its origin to the Guddu Barrage in Pakistan, it is called the Upper Indus, while downstream from the barrage it is known as the Lower Indus. The Karakoram and the Haramosh ranges bound the basin on the north, the Himalayas on the east, the Sulaiman and the Kirthar ranges on the west and the Arabian Sea on the south.

In Upper Indus Basin, the principal tributaries are the Kabul, the Swat and the Kurram on the right bank and five major tributaries on the left bank in Punjab, namely, Jhelum, Chenab, Ravi, Beas, and Satluj. These five tributaries have given the present name to the land: Punjab (*punj* = five & *ab* = river), the Land of Five Rivers. The total length of the tributaries of Indus is 5,600 km.

The basin extends over an area of 1,165,500 km^2 and lies in Tibet (China), India, Pakistan, and Afghanistan. The drainage area lying in Pakistan is 692,700 km^2. The area lying in Afghanistan and China is 15,100 km^2. The drainage area lying in India is 321,289 km^2 which is nearly 9.8% of the total geographical area of the country. The river travels about 2,880 km up to its outfall into the Arabian Sea. The length of the river in India is 1,114 km. The Index map of the Indus basin is presented in Figure 1.

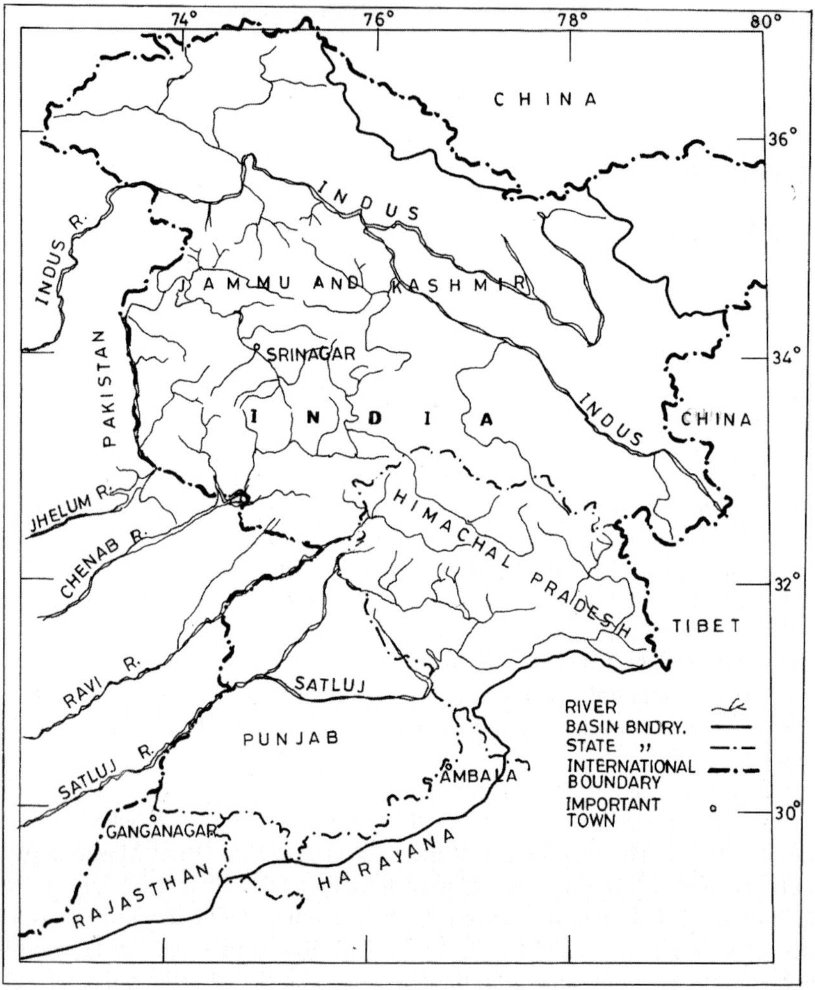

Figure 1. Index map of Indus Basin

In India, the Indus basin lies in the States of Jammu and Kashmir, Himachal Pradesh, Punjab, Rajasthan, Haryana and the Union Territory of Chandigarh. It is situated between longitudes 72°35′ E to 79°50′ E and latitudes 28°52′ N to 37°20′ N. The state wise distribution of the basin area is given in Table 1.

The upper part of the basin, which lies in Jammu and Kashmir and Himachal Pradesh is dominated by mountain ranges and narrow valleys. In Punjab, Haryana and Rajasthan the basin consists of vast plains, which are the fertile granary of this country. The submontane, brown hill and alluvial soils are the principal soil types found in the basin. The culturable area of the basin is 9.6 Mha, which is about 4.9% of the total culturable area of the country. The mean annual flow of the Indus basin

Table 1. Statewise drainage area of Indus basin (in Indian Territory)

State	Drainage Area (km^2)
Jammu and Kashmir	193,762
Himachal Pradesh	51,358
Punjab	50,306
Rajasthan	15,814
Haryana	9,939
Chandigarh	110
Total	321,289

rivers amounts to about 187 cubic km. There is a significant contribution from snowmelt. For the Satluj basin, the contribution of snow and glacier melt to annual runoff is about 60% (Singh and Jain 2002).

The only significant tributary inflow to the lower Indus River is from the Khirther and Sulaimanki range hill torrents which find their way to the river partly through Baran Nai and partly Manchhar Lake, one of the largest fresh water lakes in Asia. Sometimes, flood protection measures in the basin suffer heavy damage from the unmanaged flood flows from these hill torrents.

10.2. TOPOGRAPHY OF INDUS BASIN

The upper part of the basin lying in Jammu and Kashmir and Himachal Pradesh mostly consists of mountain ranges and narrow valleys. In Punjab, Haryana and Rajasthan the basin consists of vast plains, which are the fertile granary of the country. The principal soil types found in the basin are submontane, brown hill, and alluvial soils.

The net sown area in the J & K state is 0.748 Mha, roughly equally divided between the two divisions: Jammu and Kashmir. However, due to harsh climate, the area sown more than once is very low in Kashmir region. Due to the same reason, the cropping intensity in much high in Jammu region as compared to Kashmir. Only about one-fourth area is Jammu region is irrigated but it produces 70% of the total food grain production of the state. The details of crop area and production in J & K have been given in Table 2.

Table 2. Crop Area and Production in J & K

Particulars	Jammu	Kashmir	Total
Net sown area(m ha)	0.39	0.34	0.748
Area sown more than once (m ha)	0.29	0.07	0.367
Cropping intensity (%)	174.4	120.6	149.1
Percent irrigated area (%)	25.6	55.8	41.5
Food grain production (mt)	0.77	0.34	1.12
Estimated requirement (mt)	0.88	1.09	2.02

The Kashmir valley in Indus basin is gifted with exotic natural scenic beauty and water bodies. On account of these, poets call it a *paradise on the earth*. These rivers and lakes water bodies are of great ecological and socio-economic significance. The Dal, Wullar, Nagin, Mansar, and Khajjiar lakes are famous for their scenic beauty. A detailed discussion on lakes is given in a later chapter.

10.3. KANDI BELT

Kandi belt is the name given to the sub-montane region of the Himalayas. In the J & K State, Kandi Belt is 10 to 30 km wide, stretching from Akhnoor in the west to Kathua in the east. Its area is 811 km^2, covered in two districts – Jammu and Kathua. About 57% of total area of these districts is under Kandi belt. In Punjab state, Kandi Belt covers northern districts of Gurdaspur, Hoshiarpur and Ropar, situated at foot of Lower Shivaliks and extending into adjoining states of Haryana and HP. Most villages of this region face water scarcity during summer. Upper portion of Kandi belt consists of low hills covered by shrubs and forest and lower foot hill terrain has cultivated lands and gully beds. Kandi Belt has undulating topography, steep and irregular slopes, erodible and low water retentive soils and terrain badly dissected by numerous gullies. The annual mean rainfall in the region is approximately 1,000 mm. Figure 2 shows the Kandi belt in Jammu and Kashmir.

The major water related problems of the Kandi Belt are:

- Although the area has a number of seasonal streams, water availability is a problem,
- Ground water is inadequate and is not economically accessible because aquifers are generally deep with low recharge,
- The area has severe problem of soil erosion and nutrient losses; high intensity storms aggravate the problems,

Figure 2. Kandi belt in Jammu and Kashmir

Figure 3. Contour cultivation in Indus basin

- Other irrigation systems, based on transported water, are not feasible due to undulating topography and light textured soils, and
- Productivity of crops is low and unstable.

Ponds have played a crucial role in water management in the Kandi belt, and were the main source of drinking water till 1960s. Almost all Kandi villages have a big pond to meet the domestic needs through out the year. These big ponds were constructed with masonry work on three sides; the fourth side is left open for the water to flow in. These ponds were generally dug adjacent to a seasonal rivulet to divert part of flood water into it. For example, the Jammu District has 249 ponds while the Kathua District has 116 ponds.

The solution to these problems require conservation of basic natural resources, i.e., soil and water so that productivity can be improved and demands of the society can be appropriately satisfied. Rain water harvesting and management allows more time for rainfall or runoff to infiltrate into the soil and recharge ground water. Runoff velocity is restricted in the permissible limits by breaking the land slope into several short ones. The methods for the same include bunding, land leveling, terracing, vegetative bunding, contour cultivation, cover cropping, ridging and furrowing. Terrace farming on hill slopes (see Figure 3) is a popular method to conserve soil and water. Evaporation losses can be reduced by following practices like mulching.

10.4. CLIMATE

Large variations in temperature are noticed in the Indus basin. The upper part of the basin experiences harsh winters with a lot of snowfall while the lower part has comparatively mild winter but hot summer. At Srinagar which is located in the upper part of the basin in the Kashmir valley, the minimum temperature in the month of Jan. is −6.7 °C while the maximum temperature in July climbs up to 35 °C. Table 3 gives the temperature ranges and rainfall at selected location in the basin in Jammu region.

Table 3. Salient climatic features of Jammu Division in Indus basin

S. N.	Location	Temperature (°C)		Rainfall (mm)
		Maximum	Minimum	
1.	Jammu	18.7–32.7	7.0–26.2	1,300
2.	Kathua	18.5–39.4	8.0–26.0	1,450
3.	Udhampur	18.0–38.0	6.5–25.4	1,070
4.	Rajouri	16.5–36.0	5.3–23.0	1,065
5.	Doda	16.5–37.5	6.0–23.2	460
6.	Poonch	15.3–32.0	4.3–21.4	951

Table 4. Monthly rainfall and pan evaporation at Jammu

Month	Rainfall (mm)	Pan evaporation (E) (mm)	R – E (mm)
Jan	35.0	15.4	+19.6
Feb	50.7	28.3	+22.4
March	57.8	116.1	–58.3
April	34.9	202.8	–167.9
May	12.5	186.0	–173.5
June	49.0	180.0	–131.0
July	318.3	135.6	+182.7
Aug	497.8	113.1	+384.7
Sep	135.8	102.6	+33.2
Oct	78.1	75.8	+2.3
Nov	25.2	44.6	–19.4
Dec	5.2	25.0	–19.8
Total	1, 300.3	1, 225.3	

In Table 4, monthly rainfall and pan evaporation at Jammu are listed. Note that August is the wettest month and April is the driest. During the period from March to June, rainfall is much smaller than pan evaporation.

10.5. TRIBUTARIES OF THE INDUS RIVER

In this section, the important tributaries of Indus are described. A network diagram of Indus Basin is shown in Figure 4.

10.5.1. Jhelum River

The Jhelum River is known in Kashmir as *Veth*. It originates from Verinag, a spring at the bottom of a high scarp of a mountain spur, at the upper end of Kashmir valley. The river has several tributaries notably the Liddar, the Sind and the Pohru

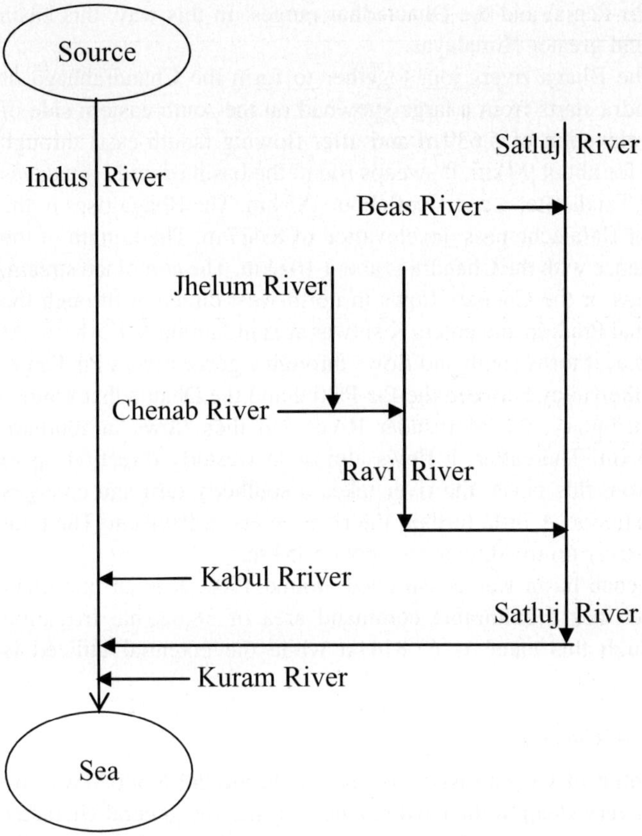

Figure 4. Flow Diagram of Indus Basin

rivers, which rise in Kashmir. The river joins the Chenab at Trimmu after a flow of 322 km. Up to the Indo-Pakistan border, the river flows through the Kashmir valley for a length of 402 km and crosses the Pirpanjal range through a deep gorge. The catchment area up to the Indo-Pakistan border is 34,775 sq. km.

10.5.2. Chenab River

The Chenab River is one of the five main rivers of the great Indus System. The major part of the Chenab catchment lies in India; its lower reach including the confluence with the Indus River is in Pakistan. In India, the Chenab basin is located in the Western Himalayas between latitudes 30° to 34° N and longitudes 74° to 78° E. It spreads over the two states of Himachal Pradesh and Jammu & Kashmir which comprise the extreme western sector of Himalayas. Upper half of this basin is located between the Zanskar and the Pir-Panjal ranges whereas the lower half is located between the Pir-Panjal and the Dhauladhar ranges. In this way, this basin covers outer, middle and greater Himalayas.

The Chandra and the Bhaga rivers join together to form the Chandrabhaga or the Chenab. The Chandra starts from a large snowbed on the south-eastern side of Baralacha Pass at an elevation of 5,639 m and after flowing (south-east) through snow clad barren area for about 90 km, it sweeps round the basin of mid Himalayas and joins the Bhaga at Tandi after a course of about 185 km. The Bhaga rises in the north-western slopes of Baralacha pass the elevation of 8,477 m. The Length of the Bhaga up to the confluence with the Chandra is about 105 km. The combined stream, known as Chandrabhaga or the Chenab, flows in north-west direction through the Pangi valley of Himachal Pradesh and enters Kishtwar area in Jammu & Kashmir. At Bengawar near Kishtwar, it turns south and flows through a gorge across Pir-Panjal range and then enters the valley between the Pir-Panjal and the Dhauladhar ranges. It receives its major tributary, the Marusudar River and then flows in southern direction for about 25 km. Thereafter, it flows almost in westerly direction up to the Salal Dam site. From this place, the river takes a southerly turn and emerges out into plains near Akhnoor. A little further, the river enters in Pakistan. The total length of the Chenab River up to Akhnoor is about 535 km.

Ranbir Canal in Chenab basin was constructed around 1905. It is an important irrigation project which has a culturable command area of 38,623 ha. Irrigation potential created through this canal is 67,814 ha while the potential utilized is 65,458 ha.

Hydropower Potential of Chenab River

The hydropower potential of Chenab River is very high, around 8,300 MW. The gradient of the river is very steep in the head reaches – it has the general character of torrent with a gradient of 10 m/km. In the lower reaches, the gradient gradually reduces as one moves downstream, typical values being 3–4 m/km.

Tributaries of Chenab River

The Chenab basin consists of many valleys or sub-basins which contribute considerable amount of runoff to the main river. The main tributaries in its passage up to Kishtwar are Thirot, Shedi, Sohal, Lidder and Marusudar. Among these, Marusudar is the biggest tributary which meets Chenab at Bandalkot. Between Kishtwar and Akhnoor, Chenab receives the water of Neeru, Pugal, Bagi, Bachleri, and Ana tributaries.

Tawi Sub Basin

The Tawi sub-basin is a part of Chenab basin in Western Himalayas. The basin is contained in between 32°35′ to 33°05′ North Latitude and 74°35′ to 75°45′ East Longitude. The river flows for about 141 km up to the point where it enters Pakistan. Its origin lies in the Kalikundi Glacier. The catchment area of Tawi is 2,168 sq. km which falls mostly within the Districts of Jammu and Udhampur with a small portion in Doda.

The upper part of the Tawi sub-basin is covered by hard granite intrusive rocks and the lower part by loose and soft Shiwalik rocks. The average height of the basin is 2,200 m – it varies from 400 m to 4,000 m above mean sea level. The slope of the basin is from East to West in upper part, while North-East to South-east in the lower part. Out of total area, forest comprises of about 31% and the rest is barren, fallow grazing, or cultivable waste.

In the Tawi basin, July and Aug. are generally the wettest months with about 55% rainfall and November is the least rainy month with about 2–3% of total rainfall. Tawi River experiences heavy flood in July and Aug. Monsoon starts from 1st July with heavy thunder showers and continues up to mid September. Normal annual rainfall in the basin varies from 111 cm to 150 cm.

The Tawi River is endowed with vast water resources and has considerable hydropower potential. Upper part of its catchment is snow fed while the lower part is predominantly rainfed. It has nine major tributaries which carry discharge mostly in monsoon period. The maximum discharge of Tawi was 12,200 cumec in September 1988 at Jammu; the minimum discharge is about 8–11 cumec. Low discharge is experienced during the month of October, November and December. There is rise in the discharge of the river in the month of March and early summer due to melt of snow in Kalikundi glacier valley.

10.5.3. Beas River

In the Greek literature, Beas was known once known as the Hyphasis. This river marks the most easterly extent of the conquests of Sikandar (Alexander the Great, king of Macedonia) in 326 BC. This river rises at an elevation of 3,960 m in the Rohtang Pass in the Punjab Himalayas in central Him*a*chal Pradesh. Subsequently, it flows south through the Kulu valley, receiving tributaries from the flanking mountains, and then enters the Punjab plains to meet the Satluj at Harike. The total length of the Beas River is 460 km and catchment area is $20{,}303\,km^2$. The estimated hydropower potential of Beas River in the Himachal Pradesh state has been given in Table 5. Pong Dam is a major existing project in the valley.

Near Mandi town, the catchment area of the Beas is approximately $5{,}000\,km^2$. The annual fluctuation of water flow is very high. The variability of high flow in the summer months is remarkable. The river flow in summer mainly consists of monsoonal run-off combined with snow-melt discharge. The low flow in winter is more or less constant. Neither high flow nor low flow nor annual average of discharge follows any significant trends whatsoever.

10.5.4. Ravi Basin

The Ravi River rises near the Rohtang Pass in the Kangra district in Kulu (HP). It flows westward through a triangle formed by the junction of the Pirpanjal and Dhaola Dhar ranges. After crossing the Shiwaliks, it enters the Punjab plains at Madhopur and later enters Pakistan 26 km below Amritsar. From its source to the

Table 5. The hydropower potential of Beas River in the Himachal Pradesh state

S. N.	Name	River/Khad	MW	Status
1.	Allian – Dhungan	Allaain Nallah	192	Under Construction
2.	Malana	Malana Khad	86	Under Operation
3.	Parbati Stage – I	Parbati Khad	750	Under Construction
4.	Parbati Stage – II	Parbati Khad	800	Under Construction
5.	Parbati Stage – III	Parbati Khad	501	Under Construction
6.	Larji	Beas River	126	Under Construction
7.	Beas Satluj Link	Beas River	990	Under Operation
8.	Uhl Stage – I (Shanan)	Uhl Khad	110	Under Operation
9.	Bassi	Uhl Khad	60	Under Operation
10.	Uhl Stage – III	Uhl Khad	100	Under Construction
11.	Neogal	Uhl Khad	15	Under Construction
12.	Binwa	Binwa Khad	6	Under Operation
13.	Baner	Baner Khad	12	Under Operation
14.	Khauli	Khauli Khad	12	Under Construction
15.	Gaj	Gaj Khad	10.5	Under Operation
16.	Pong Dam	Beas River	360	Under Operation
17.	Patikari	Bakhli Khad	16	Under Construction
18.	Sainj	Sainj Nallah	100	Under Investigation
19.	Gada Gosain	Tirthan Nallah	25	Under Investigation
20.	Baragaon	Baragaon Nallah	10.5	Under Construction
21.	Dhaulasidh	Beas River	80.0	Under Construction
22.	Lambadug	Beas River	15.0	Under Construction
23.	Trithan	Beas River	18.0	Under Investigation
	Total		4, 495	

Source: Web site of HP State Electricity Board.

Indo-Pakistan border, the river has a length of about 370 km and the total catchment area is 14,442 sq. km. The estimated hydropower potential of Ravi River in the Himachal Pradesh state has been given in Table 6.

i. Ujh sub basin

The Ujh River is a tributary of Ravi. The head waters of Ujh lie in the Kailash mountains at an altitude of 4,300 m near the Bhaderwah hills of Jammu province. Ujh flows for a distance of nearly 100 km before it joins Ravi below Nainkot in West Pakistan. Just upstream of the dam site, four streams, Bhini, Sutar, Dunarki and Talan join Ujh at a place named Panjtirthi. Bhini and Ujh are perennial rivers. The remaining three streams flow only during the rainy season. The catchment area of Ujh River at the dam site is 990 sq. km. The catchment is hilly and rugged, varying in altitude from 510 m to 4,300 m. Areas having an altitude of 2,000 m and above which constitute about 20% of the catchment area are generally snow bound for most of the winter.

The mean annual temperature of the southern part of the catchment is 23 °C and of the eastern portion of the catchment, it is 16 °C. The temperature at higher altitudes in the northern part is expected to be low. The climatic conditions vary from semi-arid to humid from south to northern parts of the catchment.

Table 6. Hydropower potential of Ravi Basin

S. N.	Name	River	Capacity in MW	Status
1.	Bara Bhangal	Ravi River	200	Under Investigation
2.	Bajoli Holi	Ravi River	200	Under Investigation
3.	Holi	Holi Nallah	3	Under Construction
4.	Kutehar	Ravi River	260	DPR Ready
5.	Gharola	Gharola Nallah	0.05	Under Operation
6.	Kugti	Budhil Nallah	45	Under Construction
7.	Harsar	Budhil Nallah	60	Under Construction
8.	Bharmour	Budhil Nallah	45	Under Construction
9.	Budhil	Budhil Nallah	70	Under Construction
10.	Tundah Stage – II	Tundah Nallah	30	Under Investigation
11.	Tundah Stage – I	Tundah Nallah	15	Under Investigation
12.	Hibra	Ravi River	231	Under Construction
13.	Chamera Stage – II	Ravi River	300	Under Construction
14.	Sal Stage – I	Sal Nallah	6.5	Under Investigation
15.	Sal Stage – II	Sal Nallah	2	Under Operation
16.	Bhuri Singh Power house		0.45	Under Operation
18.	Chamera Stage – I	Ravi River	540	Under Operation
19.	Siul	Siul Nallah	13.0	DPR Ready
20.	Saikhoti	Baira Nallah	17.0	Under Construction
21.	Chamba	Ravi River	125	Under Investigation
22.	Chamera Stage – III	Ravi River	231	Under Construction
	Total		2, 361	

Source: Web site of HP State Electricity Board.

There are two rainy seasons one from December to March associated with the passage of western disturbances and the other mid-June to mid-September due to south-west monsoon. The rainfall in October and November is generally small in amount. The cold season precipitation from December to March is chiefly due to western disturbances which advance from Persia and Baluchistan across Northern India. These disturbances occasionally give very stormy weather with stormy winds on the higher elevations giving much snow. In April and May, thunder storms are occasionally observed giving light to moderate showers of rain. The south-west monsoon is a predominant feature in this region with 50 years normal of annual rainfall being 1,400 to 1,600 mm around the Ujh catchment.

ii. Baira Nalla sub basin

The Baira Nalla sub-basin is a part of Ravi basin in Western Himalayas. This sub-basin is located above Tissa in Chamba district of Himachal Pradesh. The basin is bounded between latitude 32° 47′ to 33° 02 and longitude 75° 57′ to 76° 23′. The catchment is on the southern slopes of Pir-Panjal range in the Western Himalayas. The area of catchment is 585 sq. km; its elevation ranges from 5,321 m to 2,693 m. The catchment is densely forested – the main tree specie being Deodar. Agriculture is the main land use. However, there are a few pastures and uncultivated areas and barren hill slopes. The mountain slopes are steep and are susceptible to land slips due to rains. The soils are degraded and loose and are easily prone to erosion. Consequently, the sediment flow in the river is high.

The catchment is partially snow covered during winter. The catchment is affected due to western disturbances during winter when precipitation is mainly in the form of snow. The average annual rainfall of the area is about 1,122 mm. Winter temperatures are low and are generally below zero during the months of December to February.

10.5.5. The Satluj River

The Vedic river Shutudri has rejuvenated as Satluj in modern times. The Satluj River rises as Langchen Khabab in the lakes of Mansarover and Rakastal in the Tibetan Plateau from the southern slopes of Mount Kailash at an elevation of about 4,570 m above mean sea level (msl) and forms one of the main tributaries of Indus River. It travels about 322 km in the Tibetan province of Nari-Khorsam forming a plateau by successive deposits of boulders, gravel, clay and mud. Satluj enters into the Indian territory near Khab, a narrow cut into massive rock structure. The flow of Satluj, arising mainly from snow and glaciers, has cut a valley about 914 m deep through these deposits. Hereafter, the river flows through Himachal Pradesh and Punjab states of India. After flowing in north-westerly direction, it changes direction towards south-west and covers another 322 km up to Bhakra gorge, where a 225 m high straight gravity dam (Bhakra/Govind Sagar) has been constructed. The lower catchment largely drains directly into the reservoir and the higher slopes drain through tributaries. The important tributaries of the Satluj River are the Soel khad, Alseed khad, Ali khad, Gamrola khad, Ghambhar khad, Seer khad, Sukhar khad, Sarhali khad, and Lunkar khad.

Near the Nangal Town, Satluj enters the Anandpur Dun, a valley/plain area between the Shiwalik and the outer range of the Himalayas. This valley runs from Nangal in the North to Kakrala village in the South over a distance of about 50 km and has an average width of 10 km. With elevations ranging between 366 m and 278 m above MSL, it has a North-South gradient of 2 m/km. The river flows along the valley's longer axis finally to leave it near Ropar. The Soan Nadi joins the Satluj in the upper sections of this valley from the North-West and the Sirsa Nadi merges with it in the Southern part of the valley left bank (Eastern bank). Due to its general gradient, Satluj along with its tributaries runs through a braided course. Elongated strips of land between the river and the peripheral hills have a general slope towards the Satluj. These parts of Dun are traversed by a large number of seasonal torrents, locally called Khads, which descend quickly from the neighboring hills. Some of the important streams which contribute their flow to the Satluj on its way to Roper are Donala Khad, Dabawali Khad, Charan Ganga Khad, Lohand Khad and Kundlu Ki Khad. Some small flashy streams also outfall in the Satluj from the right-bank above Roper. After flowing sluggishly through Anandpur Dun, the Satluj debauches from the Shiwaliks just above Roper, and emerges on to the plains of Punjab. There used to be a weir at Roper with falling shutters and under-sluices for the diversion of water into Sirhind canal. This was later replaced with a barrage as a component of the Bhakra-Nangal Project in fifties. Another canal, named as Bist Doab canal, takes off from the right-bank of the river. Several natural streams and man-made drains join the Satluj between Roper and Ferozepur.

There is a group of streams below Roper, which flow in a NE–SW direction. Siswan Nadi is another important seasonal stream, which initially flows NE to SW, but gradually turns NW to merge with the Satluj near Khizarpur village after traversing a distance of over 40 km over the plains. Immediately under the high bank along the old course of the Satluj runs a perennial stream called Budha Nala, which rises at Chamkaur in Roper district. It enters Ludhiana District near Bahlopur. Flowing close to the town of Ludhiana, it flows into the Satluj in Jagraon Tehsil, a few km east of the Ludhiana-Firozpur district boundary. East Beas and West Beas merge with the Satluj from the right bank, upstream of its confluence with the Beas, which joins the Satluj at Harike (see Figure 5). A number of surface drains have been constructed to facilitate drainage of the catchments in the plains. These outfall in the Satluj and contribute to its discharge during the rainy season of July to September.

Like all Punjab rivers, the Satluj constantly shifts its course. During the twenty years (1882 to 1903) it has moved by about 1.6 km at several points in the Ludhiana and Samrala Tehsils and about a mile towards the North in Jagraon Tehsil. Some shifting of river course (up to 3 km) has been observed in Satluj in Punjab. The places where major shifting has occurred are Phillaur, Noormahal and Nakodar. In 1988, a heavy rainfall had resulted in acute flooding downstream of Bhakra and this flooding also caused shifting of Satluj river. At some places, mining of sand is also the reason behind shifting.

i. Topography of the Satluj basin

The total geographical area of Satluj River up to Bhakra dam is about 56,980 km^2, of which about 37,153 km^2 lies in Tibet. The remaining about 19,827 km^2 area lies

Figure 5. Confluence of Satluj and Beas Rivers near Harike (photo courtesy Dr. Sanjay Jain)

in the Indian territory. Indian part of the Satluj basin is elongated in shape. The shape and location of this basin is such that a major part of the basin area lies in the greater Himalayas where heavy snowfall is experienced during winters. The catchment area lies between longitudes 76° 22′ E to 78° 42′ E and latitudes 31° 13′ N to 32° 23′ N.

A major portion of the Satluj basin lies in the greater Himalayan range. The elevation of the catchment varies widely from about 500 m to 7,000 m above msl, although only a very small area exists above 6,000 m. The mean elevation of the basin is about 3,600 m. The gradient is very steep near its source and gradually reduces downstream. Owing to large differences in seasonal temperatures and great range of elevation in the catchment, the snowline is highly variable, descending to an elevation of about 2,000 m during winter.

Over the past several years a number of small hydroelectric projects had come up in the stretch between Rampur and Wangtu. This nearly 80-km stretch of valley is under intense development. Schemes at Ghanvi (22 MW) and Bhaba (120 MW) are complete. Nathpa Jhakri Hydroelectric Project (a 1,500 MW scheme) is one of the largest hydroelectric projects of the country. Work is in progress on three more hydroelectric projects, Baspa Stage-I (150 MW), Baspa Stage-II (300 MW) and Karchham-Wangtu (600 MW).

The salient features of the Satluj catchment are summarized in Table 7.

ii. Climatic Conditions in Satluj basin

The great contrast in the topographical relief results in a variety of climate in the Himalayas. Principal controls producing such differences are those of altitude, local relief and mountain barrier effect. The most important factors controlling the weather and climate in the Himalayas are the altitude and aspect. Largely due to variations in altitude, the climate varies from hot and moist tropical climate in lower valleys to cool temperate climate at about 2,000 m and tends towards polar climate as the altitude increases beyond 2,000 m. Depending upon broad climatic conditions, the following four seasons prevail over the basin.

The Winter season (December to March): This season extends from December to March. During this season, precipitation is caused by extra-tropical weather system of mid-latitude region (also called western disturbances) originating from Caspian

Table 7. Salient features of Satluj catchment

Reach	Catchment area (km^2)	Elevation range (m)	Average annual rainfall (mm)	Major contribution to streamflow
Tibetan plateau	37, 050	4,000–6,000	Nil	Snow and glacier
Spiti Valley	7, 084	3,300–5,300	Scarce	Snow and glacier
Namgia to Rampur	6, 490	3,000–4,800	Little	Snow and rainfall
Rampur to Suni	2, 068	1,200–3,000	1,000–1,500	Rainfall
Suni to Kasol	700	900–2,000	910–1,630	Rainfall
Kasol to Bhakhra	3, 108	600–2,000	1,520	Rainfall

Sea and moving eastward through Iran, Afghanistan and Pakistan. The precipitation during this season is generally in the form of snow in the greater Himalayas, snow and rain in the middle Himalayas, and light to moderate rain over outer Himalayas and north Indian plains.

The Pre-monsoon season: This season lasts for about 3 months from April to June and is considered as transit period between winter and south-west monsoon. Light to moderate rains are essentially caused by convective storms. Convection increases because of increasing trend of temperature in the Himalayan region in this season.

The Monsoon season (July –September): In this season moist air currents from the Bay of Bengal cause precipitation over the Himalayas. This is the season of abundant rain and rivers are generally flooded. Snow and glaciers at very high altitudes continue melting during this season. The monsoon normally starts withdrawing from this region towards the end of September.

The Post-monsoon season (October–November): In this season, clear autumn weather prevails and there is generally little rainfall. This is the driest season in the entire Himalayas as well as in the plain areas.

iii. Streamflow Characteristics of Satluj Basin

The streamflow of the Satluj River consists of the contribution from rain, snow and glaciers and respective contribution of each component varies with time of the year. Generally, snowmelt contribution starts from March and lasts until June/July depending upon the snowpack water equivalent accumulated in the preceding winter season and prevailing temperatures in the summer season. As the summer season progresses, the snowmelt contribution increases continuously and after some time, it exceeds the rainfall component. Thus, in the pre-monsoon season (April-June), a major part of streamflow is generated from seasonal snow. During monsoon season (July-September), monsoon rains producing higher discharges in the river augment flow. Generally, high discharges/floods are observed in the months of July and August and these are essentially due to heavy rain in the lower part of the basin. Usually, the end of May/June ablates seasonal snow accumulated on glaciers during winter season and glaciers start contributing to streamflow thereafter. Glaciers contribute to their maximum in the months of July and August. As such, glacier-melt runoff contribution lasts till September/October. In the post-monsoon season, streamflow is believed to be partly from glaciers and some occasional rain events. The minimum streamflow is observed during winters, because no melting takes place due to lower temperature regime.

Singh and Bengtsson (2004) studied sensitivity of Satluj basin to climate change. Changes in distribution of melt runoff were found to be more pronounced in summer when the decrease is expected to be about 10% for a temperature rise of 2 °C. Annual decrease was little less, about 5%. However, if only the lower and middle part of the basin is considered where snow disappears in summer, the reduction could be about 27%.

The estimated hydropower potential of Satluj River in the Himachal Pradesh has given in Table 8.

Table 8. Hydropower Potential in Satluj Basin

S. N.	Name	River	Capacity in MW	Status
1.	Rongtong	Rongtong Khad	2.5	Under Operation
2.	Kealing Lara	Spiti	40	Under Investigation
3.	Lara Project	Spiti	60	Under Investigation
4.	Mane Nadang	Spiti	76	Under Investigation
5.	Nadang	Spiti	80	Under Investigation
6.	Lari Sumate	Spiti	104	Under Investigation
7.	Sumte Kothans	Spiti	130	Under Investigation
8.	Chango Yangthang	Spiti	140	Under Investigation
9.	Yangthang Khab	Spiti	400	Under Investigation
10.	Khab Pooh	Satluj	340	Under Investigation
11.	Pool Spillo	Satluj	300	Under Investigation
12.	Jhangi Thopan	Satluj	300	Under Investigation
13.	Thopan Powari	Satluj	400	Under Investigation
14.	Shongtong Karcham	Satluj	225	Under Investigation
15.	Baspa Stage – I	Baspa	210	Under Investigation
16.	Rukti	Rukti Khad	1.5	Under Operation
17.	Baspa Stage – II	Baspa	300	Under Construction
18.	Karcham Wangtoo	Satluj	1, 000	Under Construction
19.	Bhaba Aug. P/H	Shango Khad	3	Under Construction
20.	SVP Bhaba	Bhaba Khad	120	Under Operation
21.	Nathpa Jhakri	Satluj	1, 500	Under Operation
22.	Ghanvi	Ghanvi Khad	22.5	Under Operation
23.	Rampur	Satluj	403	DPR Ready
24.	Nogli Stage – I	Nogli	2.5	Under Operation
25.	Chaba	Nauti Khad	1.75	Under Operation
26.	Kol Dam	Satluj	800	Under Construction
27.	Bhakra Dam	Satluj	1, 200	Under Operation
28.	Keshang	Keshang Khad	160	Under Construction
29.	Sorang	Sorang	100	Under Investigation
30.	Tidong	Tidong Khud	100	Under Investigation
31.	Karang	Karang Khud	16	Under Investigation
32.	Ropa	Ropa Khud	80	Under Investigation
33.	Ghanvi-II	Satluj	7	Under Construction
34.	Barhal	Satluj	9	Under Investigation
	Total		8, 633.75	

Source: Web site of HP State Electricity Board.

10.6. SURFACE WATER POTENTIAL OF INDUS RIVER

An average annual surface water potential of 73.3 km^3 has been assessed in this basin. Out of this, 46.0 km^3 is utilizable water. Culturable area of the basin is about 9.6 M ha, which is 4.9% of the total culturable area of the country. The seasonal flow in each sub-basin of Indus river is given in Table 9.

The present use of surface water in the basin is 40.0 km^3. Live storage capacity in the basin has increased significantly since independence. From just about 0.01 km^3

Table 9. Seasonal flow of Indus River and its tributaries

Name of the river	Percentage of the annual mean flow during			
	April–June	July–September	October–December	January–March
Indus	31	54	8	7
Jhelum	44	36	8	12
Chenab	28	56	7	9
Ravi	30	51	8	11
Beas	15	67	10	8
Satluj	23	62	9	6

in the 1st Plan period, the total live storage capacity of the completed projects has increased to 13.8 km^3. In addition, a storage quantity of over 2.4 km^3 would be created on completion of projects under construction. An additional storage to the tune of over 0.3 km^3 would become available on execution of projects under consideration. The hydropower potential of the basin has been assessed as 19,998 MW at 60% load factor.

10.7. WATER RESOURCES DEVELOPMENT IN INDUS BASIN IN INDIA

For utilization of water resources of the Satluj, the Beas and the Ravi, multipurpose projects such as Bhakra, Nangal, Pong, Pandoh and Ranjit Sagar have been constructed. The installed hydroelectric power generation capacity of these dams taken together is 3,420 MW. Irrigation benefits have been provided to 54 M ha; most of it in Punjab, some extended to Haryana and further to desert districts of Bikaner and Jaisalmer in Rajasthan. Several towns and villages in Punjab, Haryana, Western Rajasthan, Union Territory of Chandigarh and also the national capital Delhi in the Yamuna basin are supplied water for drinking purposes from the Bhakra-Beas system.

Except Nangal having a negligible storage in the balancing reservoir, the other dams have live storage capacity of 16,843 MCM to cater to various demands and control floods in Punjab. The area protected from flood is yet to be properly quantified. The submergence in most cases is in the Himachal Pradesh, affecting about 500 sq. km area.

The Indus basin has two classic cases of interlinking its rivers. The oldest is a link canal of 283 cusec capacity constructed in 1955 for diverting surplus water of the Ravi at Madhopur headworks to the Beas near Pathankot. Transfer of 4,716 MCM Beas waters into the Satluj has been made at Pandoh in Himachal, about 21 km upstream of Mandi town through a 13.1 km long tunnel of 255 cusec capacity. The link tunnel outfalls into the Satluj at Slapper. These have proved useful in augmenting water for generation of power and other uses.

The interlinking of the basin's three rivers has also facilitated better regulation of river flows through integrated operation of reservoir at Bhakra, Pandoh, Pong, and Ranjit Sagar. The system feeds a large irrigation network of 900 km long canal from Harike to Gadra, known as the Indira Gandhi Canal, its 8,800 km distributaries and 6,500 km field channels. Another canal, the Satluj-Yamuna Link (SYL) to transfer Indus water to Haryana is under construction.

A number of dams, weirs, barrages, river-interlinks, large network of canals and distributaries have been constructed on Indus basin rivers with massive investment after Independence. Punjab has been the main beneficiary and became the granary of India. Although Haryana is not inside the Indus basin, it has been allotted some water of Indus in accordance with old agreements, awards and the Punjab Reorganization Act, 1966. Himachal Pradesh is an upper riparian basin state of these three rivers contributing major flow from perennial snow cover, higher rainfall and large drainage area of 51,358 sq. km.

Harnessing the water resources of the Indus, the Jhelum and the Chenab within the Indian catchment of 178,406 sq. km is negligible, though the Indo-Pak Treaty allows the use of some flow for existing and specified future purposes. Only a couple of hydroelectric projects in the Chenab basin at Salal in Jammu and Kashmir and Hurot in Himachal have been completed. Work on Dulhasti power project in Jammu and Kashmir is in progress. The Jhelum has no major project although a flood storage dam is needed to control inundation in the rainy season and to regulate inflow to the Wular Lake for inland navigation. Uri is an important hydropower project in this basin, location being the Baramula district in J & K. The project envisages a concrete gravity dam and a power house with installed capacity of 240 MW.

Flowing through the Ladakh region, the main Indus River enters virgin into Pakistan except for small abstractions to meet day-to-day water requirements of thinly populated local habitations. Chenab is the largest river in terms of flow, originating in Himachal Pradesh. Its upstream tributary, the Chandra, could be linked with the Beas through a tunnel near Manali to supplement the SYL and the Indira Gandhi Canal supplies via the Beas-Satluj link for the benefit of Haryana and Rajasthan. A similar possibility exists in the downstream for diversion of still larger quantity of Chandra-Bhaga waters to Budhil nala near Kugti and its further transfer to the Indira Gandhi Canal through the Ravi-Beas Link.

10.8. WATER RESOURCES PROJECTS IN INDUS BASIN

10.8.1. Bhakra Nangal System

Bhakra Beas system is one the most prestigious and showcase project of India. The composite Bhakra-Nangal project consists of the Bhakra dam constructed on Satluj River in the State of Himachal Pradesh and Nangal barrage constructed on the same river downstream of Bhakra Dam in the State of Punjab. A distance of around 13 km separates these two projects. A schematic diagram of Bhakra Beas System is given shown in Figure 6.

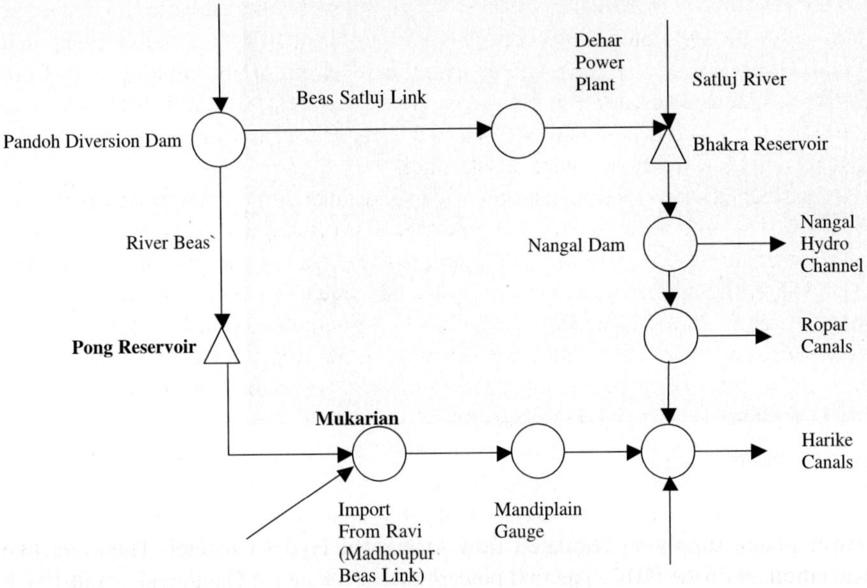

Figure 6. Schematic Diagram of Bhakra Beas System

i. Bhakra Dam

The Bhakra dam (bhakra.nic.in) is a concrete gravity dam with a total height of 225.55 m above the deepest foundation level, completed in 1963. It creates a lake called Gobind Sagar with total storage capacity of 0.9621 M ha-m at an elevation 515.11 m and having a surface spread of 16,868 hectare (41,680 acre). An ogee type of spillway is provided at the center of the dam with its crest at an elevation 501.4 m. The spillway discharge is controlled by four 15.24 m long radial gates. Facilities for release of water for irrigation and power production consist of sixteen numbers of 2.64 m horse shoe type river outlets and 4.572 m diameter steel penstocks, respectively. The river outlets are arranged in two tiers of eight each in the central spillway section. Generally water for irrigation is provided through discharges meant for power generation. However, when the irrigation demand or when the demand at Nangal pond is more than the power releases the river outlets are operated. The overflow spillway and river outlets together can take care of around 11,327 cumec of water. A view of Bhakra dam has been shown in Figure 7.

The flow of Satluj River at Bhakra is supplemented by diversion of Beas water through Beas-Satluj link, which takes off from Pondoh dam across Beas River. The Bhakra power plants are operated in conjunction with Gunguwal and Kotla Power Houses on the Nangal Hydel channel.

ii. Nangal Barrage

Nangal Barrage diverts the water of Satluj River into the Nangal Hydel Channel (NHC) for power generation and irrigation. It acts as a balancing reservoir with storage about 19.74 MCM to smoothen the variations in releases from the Bhakra

Figure 7. The Bhakra Dam

power plants supplying regulated flow to Nangal Hydel Channel. There are two powerhouses on the NHC. The first powerhouse is located at Ganguwal, about 19 km from Nangal and second is situated at Kotla at a distance of 9.6 km downstream of Gunguwal. A head of about 28.4 m is available at each of these power houses. A view of Nangal Barrage is shown in Figure 8.

iii. Bhakra canal systems

Bhakra canals: The Bhakra Main Link (BML) takes off from the tail end of the Nangal Hydel Channel at Ropar and is aligned towards Tohana in the Hissar District. It is a lined channel of 172 km long with a full supply capacity of 354 cumec. The Narwana, Fatehabad and the Bhakra main branches take off from the Bhakra Main Line and through distributaries, irrigate areas of Punjab, Haryana and Rajasthan.

Figure 8. Nangal Barrage on Satluj River

Bist Doab Canal: It takes off from the right bank of River Satluj at Ropar just upstream of Ropar Head Works with full supply discharge of 45.3 cumec. This canal serves a gross area of 0.25577 million ha.

iv. Beas-Satluj Link

A link between Beas and Satluj rivers could be beneficial in many ways: a) for the production of hydropower at Dehar Power Plant, b) augmenting storage in Bhakra reservoir for generation of power, c) to meet irrigation demands for part of Haryana, which otherwise could not be commanded from the Beas at Pong. Also this additional diversion into the Satluj augments the firm power at Bhakra where two sets of power plants of 1,050 MW installed capacity are in operation. Beas-Satluj Link (BSL) Project diverts about 4,700 MCM of water annually from Beas River into Satluj River.

Beas river water is diverted from Pandoh dam into the Satluj River near village Dehar, upstream of Bhakra reservoir through a 38 km long water conductor system. The head due to an elevation difference of more than 335 m is utilized to generate 990 MW power. Water of Beas river at Pandoh dam is carried through a tunnel up to Baggi, through an open hydel channel up to Sundarnagar, and through a tunnel from Sundarnagar to Slapper. In between the hydel channel and Sundarnagar-Slapper tunnel, a Balancing Reservoir (BR) has been constructed at Sundarnagar with limited storage capacity to ensure regulated water supply to meet the fluctuating and peaking load requirement for power generation at Dehar Power House.

The Pandoh-Baggi Tunnel (PBT), 7.62 m diameter and 13.1 km length carries 255 cumec water and outfalls into concrete lined hydel channel. Water flows through this 1.8 km long open channel into Sundarnagar Balancing Reservoirs (BR). The Sundarnagar BR has 3.7 MCM live storage capacity and takes care of fluctuating water demand at the Dehar power House. Sundarnagar Slapper tunnel (8.53 m dia, 12.38 km long) terminates into surge shaft. The tunnel carries 404 cumec of water for running six generating units of Dehar Power Plant (DPP) and flows into Satluj River. The DPP located on the right bank of river Satluj produces 990 MW power at its full capacity.

Beas, Satluj and other Himalayan rivers are known for their high silt content. The tributaries of river Beas, viz. Tirthan, Parboti and Sarvori meet the Beas upstream of Pandoh dam. Silt laden Beas river water from Pandoh dam flows through tunnel and hydel channel into the BR. The large quantity of silt carried by the Beas settles in the BR at Sundarnagar reducing its holding capacity. Hence, silt disposal from BR is carried out through dredging. Silt dredger with a capacity of 500-800 m^3 of solids/hr with 15% to 20% of silt by volume is installed in the BR. The dredged slurry is discharged into the Sukheti khad flowing adjacent to the BR. The Sukheti Khad confluences Beas River near Mandi town after flowing a distance of 21 km through a relatively flat terrain.

Silt load in the river system: Out of total catchment of 5,278 sq. km of the Beas, 784 sq. km has been identified as a high silt yielding area, specifically in the sub-watershed of Parvati, Bhunter and Larji. It was anticipated that a total silt load of 407 ha-m would annually reach Pandoh dam and would fill the dead storage of the dam in 27 years. However, silt in Pandoh dam was filled up to the spillway crest in a period of nine years after commissioning of the project. Operational guidelines for post delta stage were prepared by BBMB for bed load passes. According to these guidelines, flushing and dredging operations were started in 1986 to restrict the entry of sediment into PBT. It was also anticipated that on an average 215 ha-m of suspended load would enter the water conductor system of BSL and settle in BR, which has a live storage of 369 ha-m. Out of 162 ha-m of silt entering in the BR, about 80 ha-m of silt settles in it annually.

Beas Unit –Pandoh Dam: The Beas-Satluj link scheme consists of a 76.2 m high rockfill diversion dam at Pondoh on Beas River in Mandi district (HP). The Pandoh dam was constructed in 1977 at Pandoh, 21 km upstream of Mandi town on Mandi-Kulu road in Himachal Pradesh. The reservoir has a live storage of 18.56 MCM. The conveyance system comprises 13.2 km long, 7.62 m diameter tunnel from Pondoh to Baggi and 11.4 km long hydel channel from Baggi to Sundernagar with a maximum capacity of 212.5 cumec. From the tail of the canal, water is led to the Satluj power plant near Dehar. There are six turbines to generate power, each with an installed capacity of 165 MW.

Beas Project-II Pong Dam: Pong is a multipurpose earth & rockfill dam on Beas River, 40 km from Mukerian, Mukerian District, Himachal Pradesh. It is located in the Himalayan foot-hills downstream of Pandoh dam. It has a central impervious core with sand and gravel shell zones on either side. The dam is 132.6 meter high from the deepest foundation level and about 100.6 meter high above the river bed. The catchment area at the dam is $12,560\,\text{km}^2$. The height and length of the dam is 133 m and 1,951 m respectively. The reservoir has a gross storage capacity of 8,570 MCM and live storage of 7290 MCM at FRL 426.72 m and the MDDL is at 384 m. At FRL, the water spread area covers about 260 sq. km. Pong power house has 6 units of 60 MW each, with mean annual inflow of 15,338 MCM. It has a firm power of 156 MW. BBMB commissioned the project in 1978–83.

When the water level reaches 426.7 m the spillway radial gates would be raised. From irrigation and power generation considerations, the minimum level of permissible water level has been fixed at an elevation of 384 m. Between the FRL at 426.7 m and the dead storage level of 384 m a capacity of 0.729 M ha-m for controlled irrigation and power generation is available. An overflow spillway with six bays of 14.4 m (47.5 ft each has been provided on the left abutment of the dam with the crest at an elevation of 416 m. The discharge over the spillway is controlled by six 14.5 m wide and 12.34 m high radial gates with a discharging facility of 12,375 cumec at the highest flood level of 433.12 m.

v. Hydropower plants in Bhakra system

There are several hydropower plants under Bhakra System in Indus basin. The installed capacity of each hydropower plant has been presented in Table 10.

Table 10. The installed capacity of hydropower plants under Bhakra system

Name of the hydropower plant	Installed Capacity	(MW)
Bhakra (Right Bank)	$3 \times 132 + 2 \times 157$	710
Bhakra (Left Bank)	5×108	540
Ganguwal	$2 \times 24.2 + 1 \times 29.25$	77.65
Kotla	$2 \times 24.2 + 1 \times 29.25$	77.65
Dehar	6×165	990
Pong	6×60	360
	Total	2, 755.30

vi. Sharing of Benefits

The sharing is as follows:

a) Distribution of water

i) Sutlej water: As per the Bhakra Nangal Agreement of 1959, the share of Punjab, Haryana and Rajasthan is 57.88%, 32.31% and 9.81% respectively.

ii) Ravi Beas water: According to the agreement of 1981 and the distribution approved by BBMB in 1982, the surplus Ravi Beas water (after taking out pre-partition utilization) is distributed among Punjab, Haryana and Rajasthan in the ratio of 30%, 21% and 49%, respectively. Delhi and Jammu & Kashmir have been given fixed shares of 0.2 MAF and 0.65 MAF as per 1981 agreement. No water has been allocated either from Bhakra Nangal or from Beas Project to HP. The Board has been approving the proposals from HP regarding supply of water for irrigation and drinking water purposes out of the Bhakra Nangal and Beas projects.

b) Distribution of Power

Bhakra Complex: 15.22% of available energy less the requirement of Common Pool Consumers and transmission losses is distributed to Rajasthan. The remaining 84.78% to Punjab, Haryana, New H.P. and UT of Chandigarh in the proportion 54.5%, 39.5%, 2.5% and 3.5% respectively.

Dehar Power Plant: Available power less 15 MW to HP allocated by Government of India, 4% transmission losses and project supply at Sundernagar, Slapper and Pandoh is distributed to Punjab, Haryana and Rajasthan in the ratio 48%, 32% and 20%, respectively.

Pong Power Plant: Available power less project supplies at Talwara and 4% transmission losses is shared by Punjab, Haryana and Rajasthan in the same ratio as the sharing of cost, viz., 24.9%, 16.6% and 58.5%.

vii. Operation of Bhakra and Pong Reservoirs

The BBMB is operating the reservoirs at Bhakra and Pong in an integrated manner in the best interest of the Partner States. BBMB has constituted a technical committee for this purpose, comprising Chief Engineers from Irrigation Departments of the partner States, Technical Members from the State Electricity Boards and Chief Engineer (Indus Basin), Chandigarh, Central Water Commission, with Director (Agriculture), Punjab as a Special Invitee, to decide the operation of reservoirs under its control. The committee after taking a stock of reservoir levels, inflows, irrigation & power requirements, decides releases from the reservoirs for the following month on a 10-daily basis at the monthly meetings. The same are reviewed by the BBMB at its quarterly meetings.

10.8.2. Indira Gandhi Nahar Project (IGNP)

Indira Gandhi Nahar Project is a standing example of how large inter-basin transfers have brought about all round socio-economic growth with overall improvement in the ecology and environment of the region. Under the Indus Water Treaty, the waters of three eastern rivers viz. Satluj, Beas and Ravi were allocated to India. As the land to be benefited in India lies mostly to the east and south of these rivers, the three rivers had to be interlinked and the water conveyed to canal systems for serving vast agricultural tracts in Western India. The main storage on Satluj is at Bhakra while that on Beas is at Pong. Bhakra system provides irrigation to 26.3 lakh ha of new area besides stabilization of existing irrigation on 9 lakh ha. The Beas-Satluj link is 37.25 km long, of which 25.45 km passes through a tunnel under difficult rock formations. The capacity of the tunnel is 254.70 cumec. Another dam on Ravi namely, Ranjit Sagar dam will provide additional water to Beas and also generate a large amount of power. Subsequently, it was decided to provide 9.36 BCM of water to Rajasthan Canal (Indira Gandhi Nahar) for irrigating the areas of Thar Desert.

Transfer of surplus waters of Ravi, Beas and Satluj to Rajasthan right up to Jaisalmer and Barmer through Indira Gandhi Nahar Pariyojana has eliminated drought conditions, provided power benefits and transformed desert waste land into an agriculturally productive area by bringing irrigation and vegetation to about 2 Mha area. Contribution in agricultural production due to implementation of the project is worth Rs. 1,750 crore annually. Canal water is also being used to meet domestic needs. The project has dramatically changed the living standard and socio-economic conditions of the people in the region.

The post independence era of the country has witnessed rapid strides in creation of irrigation potential, resulting in substantial increase of agricultural production. The IGNP (formerly known as The Rajasthan Canal Project) with a command area of 15.43 lakh hectare is the largest irrigation and drinking water project to cater five districts in north-western Rajasthan. The main canal gets water from the Satluj River in Punjab through a feeder canal which takes off from Harike Barrage, constructed at a point down stream of its influence of Beas and Satluj Rivers.

The entire project comprises of the main Indira Gandhi Canal, nine branches, three lifts, and 21 district distributaries with a length of 7,150 km. The project has been divided in two stages. Under the first phase, Rajasthan feeder, the main canal up to 195 km, the Suratgarh low level, and Namshera branches were completed. Under the second stage the construction of remaining portion of the main canal from Chhatargarh to Mohangarh has been completed. The irrigation potential from the IGNP project is assessed as 13.87 lakh ha. A view of Indira Gandhi Canal is shown in Figure 9.

Figure 9. A view of the main Indira Gandhi Canal

The IGNP is a gigantic canal project to carry 524 cumec water from the Harike Barrage in a 204 km long feeder canal in Punjab, to the vast Great Indian Desert, known as the Thar Desert, in Western Rajasthan. The canal network is spread in an area of about 60 km wide and 1,000 km long belt. It consists of 204 km of feeder, 450 km main canal, 8000 km of distribution networks and several thousand km of lined water courses, to spread over a gross command area of 2.5 Mha and provide irrigation to a culturable command of 1.55 Mha.

The salient features of the IGNP are:

- It will provide additional irrigation in 964,000 hectares,
- Will deliver drinking water for 14 million humans, besides a large cattle population,
- Will help afforesation in an area of 362,000 hectares,
- Would provide fodder for 5.2 million units of cows or equivalent animals,
- Is expected to provide direct employment to 500,000 persons on regular basis, and
- Will enables exploitation of mineral resources and industrialization.

The project was conceived by the great Indian civil engineer, Kanwar Sain, around the year 1940 and construction was started in the year 1958. Since then, the project has gone under considerable modifications and revisions. It is still (year 2006) under construction near tail areas. Planning, design and construction of canal system is managed by a high powered Canal Board with many advisory and technical committees.

The main canal is a contour canal with distribution network and irrigation on the right side only. Few lift schemes are provided on the left side. Although the main canal was initially conceived as unlined, design was subsequently modified and it was constructed as a lined canal as it passes through sandy desert soils. High cuttings of about 20 m above bed level as well as heavy bed filling of more than 4 m are encountered in the course of the canal.

No construction materials except the desert sand are available along the canal or even within 100 km. There are no rivers or stone hillocks nearby. However, there are some stone hillocks near Ratangarh at a distance of 200 to 300 km from main canal and in tail areas after 450 km of canal (near Mahangarh) at a distance of about 50 km from the tail. Even the coarse sand (locally known as Bajri) required for cement mortar is available at a distance of 200 to 300 km away from canal in deep quarries of Shivbari (Bikaner) and Bap (Phalodi). However, clay soil for manufacture of tiles/bricks is available in small pockets in depression in between sand dunes, at distances varying from 5 to 100 km.

The geology of the area is completely concealed under the thick blanket of dunal sand and alluvium; no rock exposures are seen on the surface. The lithology of deep bore holes, dug cum bore hole and piezometers in the area reveal that stratigraphical unit in the area ranges in age from alluvium of quaternary group consolidated sedimentaries of paleozoic group.

In the head reaches, the depth of the canal is limited to 6.5 m for stability of sandy soils, operational problems and easiness in construction. It gradually decreases in tail. Internal side slopes of 1:2 (V:H) were considered safe for sandy soils and provided for depth from 6.5 m to 5 m, throughout the entire 450 km length of the main canal. Bed slope has also been restricted to 1 in 12,000, because of long length of canal and to have sufficient command. Even with this flat slope, the drop in water level is 54 m from head to tail. It is uniform from head to tail in 450 km length. Thus velocities are also very much limited in the entire length from 1.5 m/s to 1.2 m/s. Bed width varies practically from 11 times the depth at head to 2 times the depth at tail. Single tile lining in bed and double tile lining on sides has been adopted. Burnt clay tile lining is provided up to 365 km and thereafter P.C.C. block lining is adopted till the end.

10.8.3. Nathpa Jhakri Hydroelectric Project

The 1500 MW Nathpa Jhakri Hydropower Project (NJHP) is a prestigious project commissioned by Satluj Jal Vidyut Nigam Limited (SJVN). SJVN (Formerly Nathpa Jhakri Power Corporation Limited) was incorporated in 1988 as a joint venture of the Govt. of India and the Govt. of Himachal Pradesh. Work on NJHP began in 1993. The project includes a 60.5 m high dam and underground desilting complex. While the dam is located at Nathpa village in Kinnaur District, the power house is located on the left bank of Satluj River at Jhakri in Shimla District. NJHP has many unique features. It has one of the largest underground desilting complexes of the world, one of the deepest surge shafts of the world, and an underground power house with large a cavern of 222 m X 20 m X 49 m, housing 6 Francis turbines of 250 MW each. NJHP will generate 6,700 Million Units of electrical energy in a 90% dependable year. The first unit of the project was commissioned in October 2003. SJVN is also associated with many other projects besides NJHP. Details about Satluj Jal Vidyut Nigam Limited are available at sjvn.nic.in.

Flash floods in August 2000 led to extensive loss of time and money to the project, and caused severe destruction to the dam and power house.

10.8.4. Other Projects

Anandpur Sahib: Anandpur Sahib hydropower project is located on Anandpur Sahib Canal that takes off from Nangal barrage on Satluj River, 8 km from Anandpur Sahib in Ropar District, Punjab. Two hydropower houses namely Anandpur Sahib I and Anandpur Sahib II have been constructed under this scheme. Both the powerhouses have 2 units of 33.5 MW each with a total installed capacity of 134 MW. It has a firm power of 105 MW. Punjab State Electricity Board (PSEB) commissioned the project in 1985.

Baira Siul Project: Baira hydropower project is an earth core and rockfill dam on Baira River supplemented by Bhaledh and Siul Rivers, tributaries of Ravi River in Indus basin. The dam is located in Chamba District, Himachal Pradesh near Pathankot. The catchment area at the dam is 1,038 sq. km out of which 660 sq. km lies in Baira river basin and the rest lies in Bhaledh River basin. The height and the length of the dam are 53 m and 160 m respectively. The reservoir has a live storage capacity of 1,270 MCM at FRL 1,122.15 m and the MDDL of the reservoir is at 1,113 m. Baira surface power house has 3 units of 60 MW each, with mean annual inflow of 1,060 MCM. It has a firm power of 39 MW and the annual generation is 750 million units in 90% dependable year. NHPC commissioned the project in 1981 at a cost of Rs. 142.5 crore.

Baspa II: Baspa II hydropower project has been constructed on a gated barrage on Baspa River, a tributary of Satluj River in Indus basin. The project is located 210 km from Shimla in Kinnaur District, Himachal Pradesh. The catchment area at the dam is 969 sq. km. The barrage has a live storage capacity of 75 ha-m at FRL 2,531.50 m and the minimum pond level is at 2,527.50 m. The power house has 3 units of 100 MW each, with annual inflow of 99,175 ha-m and 146,737 ha-m on 90% and 50% dependable year respectively. The project was commissioned in 2003.

Bassi Project: Bassi power project is located in Jogindernagar district in Himachal Pradesh. The project utilizes tail waters of Shanan reservoir located in the upstream of Bassi on Uhi River, tributary of Beas River. Bassi power house has 4 units of 15 MW each. It has a firm power of 16 MW. HPSEB commissioned the project in 1970–81.

Chamera I: Chamera I is a major project consisting of a concrete arch gravity dam on Ravi River, 25 km from Dalhousie and 80 km from Pathankot in Chamba District, Himachal Pradesh. The catchment area at the dam is 472.5 sq. km. The height and the length of the dam are 140 m and 295 m respectively. The reservoir has a live storage capacity of 110 MCM and mean annual inflow of 1,273 BCM. Its FRL and MDDL are 760 m and 747 m. Chamera I underground power house has 3 units of 180 MW each. It has a firm power of 160 MW and in a 90% dependable year, the annual generation is 1,664 million units. NHPC commissioned the project in 1994 at a cost of Rs. 2,114 crore.

Chamera power station – II: It is located near Pathankot, in Distt. Chamba, Himachal Pradesh. It has a 39 m high, 118.50 m long concrete gravity dam and an underground power house containing 3 units of 100 MW each. In a 90% dependable year, the annual generation is 1,500 MU. This project was completed by NHPC in 2004 at an estimated cost of Rs.1,930 crore.

Lower Jhelum: Lower Jhelum dam has been constructed on Jhelum River near Warikhah in Baramulla District, Jammu & Kashmir. Lower Jhelum power house has 3 units of 35 MW each. It has a firm power of 62 MW. J&KPDC commissioned the project in 1978–79.

Malana: Malana is a concrete gravity dam completed in 2001 on Malana River, a tributary of Parbati River of Beas basin. The dam is located 20 km from Bhuntar in Kulu District, Himachal Pradesh. The catchment area at the dam is 4,725 sq. km. The height and the length of the dam are 18 m and 305 m respectively. The reservoir has a live storage capacity of 0.249 MCM at FRL 1,893 m; the MDDL has been fixed at 1,879 m. Malana power house has 2 units of 43 MW each, producing a firm power of 12 MW. It has a mean annual inflow of 403 MCM.

Mukerian: Mukerian is a hydropower project, located on Shah Nahar Canal diverted from Beas River, in Hoshiarpur District, Punjab. A barrage namely Shah Nahar barrage has been constructed 5 km down stream of Pong dam in Hoshiarpur District, Punjab. Four hydropower projects namely Mukerian I, Mukerian II, Mukerian III, and Mukerian IV have been constructed under this scheme. The Shah Nahar barrage has a pond level of 330.7 m. The length of the barrage is 562 m. The barrage has been constructed for a design flood of $11,073\,m^3/s$. Two powerhouses have 3 units of 15 MW and other two powerhouses have 3 units of 19.5 MW each. Thus the Mukerian power house has a firm power of 137 MW. Punjab State Electricity Board (PSEB) commissioned the project in 1983–89.

Sanjay Bhaba: Sanjay Bhabha hydropower project consists of a gated weir, namely Bhabha weir, located on Bhabha Khad, tributary of Satluj River, 190 km from Shimla in Kinnaur District, Himachal Pradesh. The catchment area at the weir is 280 sq. km. The pond has a storage capacity of 0.304 MCM at maximum water level 2425.20 m. The power house has 3 units of 40 MW each, with mean annual inflow of 480 MCM. It has a firm power of 33 MW. HPSEB commissioned the project in 1989.

Salal Project: Salal I is a 113 m high, 450 m long rockfill concrete dam on Chenab River, 90 km from Jammu in Udhampur District, Jammu & Kashmir. The catchment area at the dam is 21,500 sq. km. The height and the length of the dam are 113 m and 630 m respectively. The FRL of the dam is at 487.68 m. Salal sub-surface power house has 3 units of 115 MW each. With mean annual inflow of 21,000 MCM, its annual generation is 2,038 MU. It has a firm power of 227 MW. NTPC commissioned the project in 1987. At 1987 price level, the project cost was Rs. 6,212.1 million.

Salal Power Station – II: also has a capacity of 345 MW (3 × 115 MW) and annual generation of 1,063 million units. Its 1st unit was commissioned in 1993, 2nd unit in 1994, and the 3rd unit in 1995.

Shanan Project: Shanan dam has been constructed on Uhi River and its tributary Lambadug River, tributaries of Beas River in Indus basin. It is located 4 km from Jogindernagar in Kangra District, Himachal Pradesh. The catchment area at the dam is 381 sq. km. Shanan power house has 4 units of 15 MW each including one extra unit of 50 MW.

Thein Dam (Ranjit Sagar): Ranjit Sagar Hydropower house is located at Thein earth and rockfill dam on Ravi River, 24 km from Madhopur barrage, in Gurdaspur District, Punjab. The reservoir behind the Thein dam is known as Ranjit Sagar reservoir. The catchment area at the dam is 6,086 km^2. The height and length of the dam is 160 m and 617 m respectively. The reservoir has a live storage capacity of 2,344 MCM. The power house has 4 units of 150 MW each. It has a firm power of 129 MW. Punjab State Electricity Board (PSEB) commissioned the project in 2000.

UBDC Project: Upper Bari Doab Canal (UBDC) is a hydropower project, located on Madhopur barrage, on Upper Bari Doab Canal diverted from Ravi River, 12 km from Pathankot in Gurdaspur District, Punjab. The hydropower complex is a mixture of three power projects, namely UBDC I, UBDC II, and UBDC III. The Madhopur barrage is 774 m long and has a normal pond level of 348.5 m. The catchment area at the project is 6,086 km^2. The barrage has been constructed for a design flood of 17,750 m^3/s. All the three powerhouses have 1 unit of 15 MW and 1 unit of 15.45 MW each with a total installed capacity of 91.35 MW. It has a firm power of 55 MW with mean annual inflow of 8,609 MCM. Punjab State Electricity Board (PSEB) commissioned the project in 1971–91.

Upper Sindh II: Upper Sindh II dam has been constructed on Sindh Nallah and Wangath Nallah tributaries of Jhelum River. It is located 40 km from Srinagar in Srinagar Kangan District, Jammu & Kashmir. The catchment area at the dam is 927 sq. km out of which 697 sq. km lies in Sindh basin and the rest 230 sq. km in Wangath basin. The tail waters of Upper Sindh I are diverted into Upper Sindh II for power generation in the project. Upper Sindh II power house has 3 units of 35 MW each. JKPDC commissioned the project in 2000–02.

Uri Project: Uri is a 20 m high 93.5 m long barrage, located on Jhelum River, 8 km from Baramulla in Kashmir North District, Jammu & Kashmir. The catchment area at the dam is 12,570 sq. km. The reservoir has a live storage capacity of 0.36 MCM and mean annual inflow of 8,400 MCM. The FRL of the reservoir is 1,491 m. Uri underground power house has 4 units of 120 MW each, It has a firm power of 213 MW and annual generation of 2,663 million units in a 90% dependable year. NTPC commissioned the project in 1997 at a cost of Rs. 3,300 crore.

Small Hydro Projects: Licences have been given to set up 4 × 5 MW small hydroelectric power projects in Kangra District of Himachal Pradesh. The projects are: 5 MW Baner III, 5 MW IKU II, 5 MW Drinidhar, and 5 MW Upper Khaul. Another private company has won the bid for the 70 MW Budhil Project in Chamba district. The power plant will have two units of 35 MW each and will generate 313.33 million units of power per annum.

10.8.5. Projects Under Construction

In addition to the above several other projects under construction or consideration.
Baglihar Hydro-Electric Project: The hydropower potential from the four major river basins in the state of Jammu and Kashmir, namely, Chenab, Jhelum, Indus, and Ravi has been estimated to be in the range of 11,000 MW. The Chenab basin by itself has the largest potential, with an estimated capacity of 8,000 MW (at 60% load factor). Eight hydropower projects with a total capacity of 5,320 MW have already been identified along its track in Jammu and Kashmir State. However, only about 10% of this huge renewable source of energy on Chenab has been exploited up to now. Baglihar is one of such projects with ultimate installed capacity of 900 MW to be developed in two stages of 450 MW each. It is located across Chenab River in Doda District. Pakistan has raised objections against this project under the Indus Water Treaty.
Bursar Project: Investigations are in progress for the 1,020 MW (4×255 MW) project consisting of a 252 m high rockfill dam and an underground power house near Hanzal Village (near Kishtwar) in Doda District of Jammu & Kashmir. Annual generation is expected to be 2,018 MU (in a 90% dependable year) at an estimated cost of Rs. 4,378 crore.
Chamera Stage – III Project: It is being constructed in Chamba Distt. in Himachal Pradesh. A 68 m high concrete gravity dam on Ravi River is being constructed along with an underground power house consisting of 3 units of 77 MW each (total 231 MW). Completion is scheduled by August 2010 at estimated cost of Rs. 1,406 crore.
Chutak Project (Proposed): This project is located near Minji Village in Kargil District of Jammu & Kashmir. A 47.5 m long diversion barrage with an underground power house having 4 units of 11 MW each are proposed.
Dulhasti Project: It involves a 65 m high, 186 m long concrete gravity dam, being constructed in Doda District in J & K. An underground power house containing 3 units of 130 MW each (= 390 MW) will generation 1,928 MU annually at an estimated cost of Rs. 3,560 crore when it is complete (expected date is end of 2006).
Kishanganga Project: This is a proposed run-of-the-river hydropower project on Jhelum River. It will be located near Kralpore Village (near Bandipore) in Baramulla District of Jammu & Kashmir. Its underground power house will have an installed power production capacity of 330 MW yielding an annual generation of 1,350 MU in a 90% dependable year. Like the Baglihar Project, construction of this project has also been opposed by Pakistan claiming

that the diversion tunnel of this project will cause reduction in flow of Neelam River which will be detrimental to Pakistan's interests.

Kol Dam: This project is being constructed by the National Thermal Power Corporation (NTPC) across Satluj River, 4 km upstream of Dehar Power Plant in Bilaspur District. Kol Dam will be a 163 m high rock and gravel fill dam with clay core. Its crest level will be at 648 m. Kol reservoir will have MDDL at 636 m, FRL at 642 m, and MWL at 646 m. Four units, each of 200 MW capacity will yield total installed capacity of 800 MW. The estimated annual generation in a 90% dependable year will be 3,054 GWHr and the estimated cost of the project is Rs. 6,300 crore. Work on the project began in the year 2000.

Nimoo Bazgo Project: It is proposed near Alchi village in Leh District of Jammu & Kashmir. A 57 m high 247.9 m long concrete gravity dam is proposed to be constructed with a surface power house. The project will have installed capacity of 45 MW (3 X 15 MW) yielding an annual generation of 239.33 MU in a 90% dependable year.

Pakal Dul Project: It is located near Village Pakal (near Kishtwar) in Doda District of Jammu & Kashmir. It will have a 167 m high concrete face rockfill dam and an underground power house with capacity of 1,000 MW (4X 250 MW), giving an annual generation of 3,387 MU (in a 90% dependable year). The project is likely to cost Rs. 5,577 crore. The DPR for the project has been completed.

Parbati Stage – II Project: It is being constructed near Kiratpur in Distt. Kullu, Himachal Pradesh. The project will have a 85 m high concrete gravity dam and a surface power house containing 4 Pelton Turbine Generating units of 200 MW each. These turbines will generate 3,108.66 MU of energy annually (90% dependable year). The estimated cost of the project is Rs. 3,920 crore and it is likely to be commissioned by September 2009.

Parbati Stage – III Project: This project being constructed on Sainj River in Kullu district, Himachal Pradesh consists of a 43 m high rockfill dam with a 34.5 m orifice type spillway. The underground powerhouse will have 4 units of 130 MW each. It will produce 1977 million units of power in a 90% dependable year. The latest estimated cost of the project is Rs. 2,305 crore. Work has commenced and is expected to be completed by November, 2010.

Sewa Project: The Sewa project is being constructed on Sewa River, a tributary of Ravi River, in Basholi tehsil of Kathua River in Jammu and Kashmir. Here, a hydropower project, known as Sewa stage-II is under construction. The project envisages 53 m high concrete gravity dam, a tunnel and three units of 40 MW each in the under-ground powerhouse. The estimated cost of the project is Rs. 665.46 crores. The project is likely to be completed by August 2007 (The Times of India newspaper, August 11, 2004). The installed capacity of power plant will be 120 MW and it will generate 533.52 MU of energy annually.

Sewa Stage – II Project: Is located near Pathankot on Sewa River (a tributary of Ravi) in Kathua District, J&K State. The project consists of a 53 m high concrete gravity dam. Is power house will be equipped with 3×40 MW vertical Pelton turbine units with rated net head of 560 m. Annual generation is expected to be

534 MU (90% dependable year). The estimated cost of the project is Rs. 665 crore and it is likely to be completed by the end of 2007.
Uri Stage – II Project: This project is located in Uri Tehsil of Baramulla district in Jammu & Kashmir. The project consists of a concrete gravity dam, 52 m high, 173.2 m long with spillway consisting of 4 bays of 9.0 m each. An underground power house of 132 m length, 15 m width, and 41 m height is being constructed to accommodate 4 units of 60 MW capacity each. Annual energy generation from the plant to the tune of 1,124 MU (90% dependable year) is expected. This project is scheduled for completion by Nov. 2009 at an estimated cost of Rs. 1,725 crore.

10.8.6. Water Bodies In Indus Basin

In addition to the reservoir projects, there are many lakes in Indus basin. The important lakes in have been described in Chapter 19. However, brief information about these lakes is summarized in Table 11.

10.9. DAMS AND WATER TRANSFER IN INDUS BASIN

To utilize India's share of water, it was necessary to build storages on three rivers to conserve flood waters and divert it to semi arid lands in Punjab and Haryana and desert areas of Rajasthan. The Ranjit Sagar reservoir on Ravi was completed in year 2000. With the completion of this dam, over 90% of the water is being utilized in the region for various purposes. Table 12 briefly summarizes the main projects.

It is readily seen from the table that annual value of energy produced alone is nearly seven times the capital cost of Bhakra dam project. This highlights the high benefits of early implementation of such WRD projects. In addition to irrigation, power, water supply other benefits of the project include development of industries, and communications. Since the value of benefits as well as cost goes on increasing with time, it is never too late to take up construction of such projects that continue to give service for hundreds of years.

In Himachal Pradesh and Jammu & Kashmir in Indus basin, irrigation potential is limited but there is high potential for hydropower. In Jammu and Kashmir against a hydropower potential of 7,487 MW at 60% load factor only 515 MW was developed till November 2002 (Hasan, 2005) which is only 6.88%. Similarly in Himachal Pradesh against a potential of 11,647 MW only, 2.036 MW was developed which is about 17.5%. Work is in progress on many large hydropower projects in Himachal Pradesh, which will help in harnessing the remaining potential in reasonable time. The live storage permitted under Indus Water Treaty on Chenab and its tributaries has not been completely utilized so far. Storage projects would not only generate hydropower but the regulated flow would greatly enhance power generation from the run of the river schemes in the downstream.

Table 11. Details of Water bodies in Jammu & Kashmir (Area wise)

SN	Name of Water Body	Type	Nearest Place	District	Elevation (m)	Water Spread Area (sq. km)		Perimeter (km)	
						Pre-Monsoon	Post-Monsoon	Pre Monsoon	Post Monsoon
1.	Pangong Tso	Lake	-na-	Leh	4,250	273.454*	296.564	185.251	184.554
2.	Amto Gor	Lake	-na-	Leh	-na-	174.080	174.385	65.032	65.846
3.	Tso Morari	Lake	Karzok	Leh	4,527	115.845*	141.054	61.823	74.705
4.	Wular	Lake	Bandipur	Baramulla	1,580	50.15	55.05	66.07	49.06
5.	Hygam Jhil	Swamp	Tarazu	Baramulla	1,569	15.127	18.745	27.517	30.303
6.	Dal	Lake	Dalgate	Srinagar	1,586	13.186	13.629	21.978	21.047
7.	Kar Tso	Lake	-na-	Leh	4,537	Snow	10.536	Snow	20.018
8.	Nambi Narku	Marshy	Narkur	Srinagar	1,583	4.570	5.951	9.459	10.300
9.	Kyun Tso	Lake	Chila	Leh	5,000	Snow	5.420	Snow	10.823
10.	Tazangkuru Tso	Lake	Tsakshang	Leh	4,800	Snow	5.063	Snow	9.449
11.	Startsapu Tso	Lake	Kurjok	Leh	4,337	Snow	4.866	Snow	11.753
12.	Manasbal	Lake	Gratabal	Baramulla	1,600	2.591	3.486	9.346	8.981
13.	Un-named	Marshy	Nusu Ghat Kalhum	Baramulla	1,581	2.138	2.390	5.644	6.024
14.	Rakh Malang	Marshy	Safapora	Baramulla	1,538	1.845	2.035	5.332	5.441
15.	Kyule Tso	Lake	Kyule	Leh	5,400	Snow	1.864	Snow	5.605
16.	Un-named	Marshy	Kulul	Pulwama	1,591	-na-	1.639	-na-	6.075
17.	Yaye Tso	Lake	Yayala	Leh	4,800	Snow	1.637	Snow	4.869
18.	Gangabal Lake	Lake	Naranag	Srinagar	3,569	Snow	1.634	Snow	6.420
19.	Yusup Tso	Lake	Zulungpha	Leh	5,400	Snow	1.542	Snow	6.205

(Continued)

Table 11. (Continued)

SN	Name of Water Body	Type	Nearest Place	District	Elevation (m)	Water Spread Area (sq. km)		Perimeter (km)	
						Pre-Monsoon	Post-Monsoon	Pre Monsoon	Post Monsoon
20.	Nagin lake	Lake	Sadrabal	Srinagar	1,604	0.776	0.907	4.385	4.815
21.	Rakhi Gandka Shah	Marshy	Barthana	Srinagar	1,583	0.701	0.820	3.853	4.119
22.	Mansar	Lake	Mansar	Udhampur	700	0.629	0.747	3.198	3.413
23.	Hokar sar	Lake	Zainakot	Baramulla	1,584	0.475	0.732	3.095	3.414
24.	Vishan Sar	Lake	Vishansar	Baramulla	3,677	Snow	0.635	Snow	4.077
25.	Godsar	Lake	Godsar Nala	Baramulla	3,810	Snow	0.607	Snow	3.755
26.	Tso Kur	Lake	–	Leh	4,225	0.357	0.549	2.43 1	3.156
27.	Gundi-I-Khalil	Lake	TaraZu	Baramulla	1,569	0.388	0.535	2.442	2.840
28.	Krishan Sar	Lake	Vishansar	Baramulla	3,819	Snow	0.448	Snow	2.900
29.	Tar sar	Lake	Pahilmazar	Anantnag	3,785	-na-	0.430	-na-	3.398
30.	Salnai sar	Lake	Lolgul Gali	Baramulla	4,055	Snow	0.425	Snow	2.892
31.	Nund kol	Lake	Dandlod	Srinagar	3,607	Snow	0.351	Snow	2.854
32.	Madmatti sar	Lake	Waril Gali	Baramulla	4,449	Snow	0.348	Snow	2.316
33.	Konsar Nag	Lake	Phutin Pansaisali	Pulwama	3,585	-na-	0.329	-na-	2.565
34.	Surinsar	Lake	Surinsar	Udhampur	736	0.302	0.324	2.283	2.123
35.	Bod Sar	arshy	Nambalbal	Srinagar	1,593	0.228	0.318	2.024	2.342
36.	Marsar	Lake	Nagaberan	Srinagar	3,849	-na-	0.271	-na-	2.926
37.	Bodhsar	Lake	Bodhsar	Budgam	3,956	Snow	0.262	Snow	2.201

*partly covered with snow, Snow: fully covered with snow. NA: data not available.

Table 12. Brief summary of the main projects

Particulars	Name of Dams				
	Bhakra	Nangal	Pong	Pandoh	Ranjit Sagar
Year of Completion	1963	1954	1974	1977	2000
Cost (Rs. Crore)	218	27	326	449	3,200
		Benefits			
Annual Generation in Million units	7,000	1,000	1,800	3,600	1,500
Value of energy Rs. 2 /unit	1,400	200	360	600	300
New Area irrigated (lakh Acre)	65	40	8	–	0.32
Cost of agriculture product in crore rupees	465	–	350	100	140

Source: Hasan (2005).

10.10. WATER RESOURCES DEVELOPMENT IN THE INDUS BASIN IN PAKISTAN

Pakistan and India share the river waters of the Indus Basin under the Indus Basin Treaty signed in 1960. The treaty allocates the three western rivers, Jhelum, Chenab, and Indus to Pakistan, apart from some limited use in the state of Jammu and Kashmir in India. Unrestricted use of waters of three eastern rivers, namely Ravi, Satluj, and Beas can be made by India while only run of the river schemes for hydropower development and limited storage and riparian irrigation is permitted on Chenab and Ravi.

As part of Indus treaty works, Pakistan developed a gigantic irrigation system, known as the Indus Basin Replacement Project, to cater to the needs of the areas which got deprived of irrigation when three eastern rivers of the Indus System were allocated to India. The project included construction of a huge network of link canals to divert water from the western rivers to the eastern rivers, two major storage reservoirs, and a number of barrages. Pakistan built ten links, six barrages and two dams namely Mangla dam and Tarbela dam, during the post-treaty period of 1960–1970. Its Indus Plains host most of its population and the world's largest irrigation network commanding an area of 345 lakh acres which contribute to 90% of its national agricultural production. A brief description of Tarbela and Mangla dams follows.

10.10.1. Tarbela Dam

The major storage reservoir on the Indus is located at Tarbela where an earthfill dam was completed in 1976. The dam has a live storage capacity of 11.5 cubic km

Table 13. Salient features of the Mangala dam

Gross storage capacity	$7,253.13\,\text{Mm}^3$
Live storage capacity	$6,587.027\,\text{Mm}^3$ (at FRL 366.37 m)
Area of the Reservoir at FRL	$257.025\,\text{Mm}^2$
Catchment area	$33,655.45\,\text{km}^2$

for irrigation and hydropower production. It is 147.8 m high and 2,743 m long with a live storage capacity of 9.74 MAF and installed capacity of 3,478 MW of hydropower generation.

10.10.2. Mangla Dam

The first large earthfill dam, namely, Mangla Dam on River Jhelum in Pakistan was completed and commissioned in 1967. The dam is 115 m high with installed capacity of 1,000 MW to generate power. This multipurpose reservoir also stores water for irrigation and controls floods. The Salient features of the Mangala dam are given in Table 13.

10.11. FLASH FLOODS IN THE INDUS RIVER BASIN

Heavy concentrated rainfall in the catchment during the monsoon season, which is sometimes augmented by snowmelt flow and rainfall – primarily outside the Indus Plain, generally cause floods. Monsoon currents, originating in the Bay of Bengal, and consequent depressions often result in heavy downpour in the Himalayan foothills and Koh Hindu Kush, occasionally producing destructive flash floods along one or more of the main rivers of the Indus System. However, in some cases, exceptionally high floods have been caused by the formation of temporary natural dams by landslide or glacier movement and their subsequent collapse.

In the upper part of the Indus Basin system, floodwater spilling over the river banks generally returns to the river. However, in the lower part of the Indus River, which flows at a higher elevation than the adjoining lands for the most part, spills do not return to the river. This increases the extent and period of inundation, resulting in more damage. Although flood protection is provided by the embankments at many locations in the upper areas, bunds have been breached at times. Such breaches often cause greater damage than would have occurred without the bunds because of their unexpected nature and intensification of land use following the provision of flood protection.

Floods in Satluj in Year 2000

On the early morning of August 1, 2000, very high flood was observed in Satluj River which caused enormous damage in Rampur (It was once the capital of the princely state of Bashahr, perhaps the largest in area among the Shimla Hill States. Rampur is famous for Lavi Fair) and to the Nathpa Jhakri Hydro-electric Project. The other major loss was the delay in completion of the project.

River reach from Rampur to Wangtu is very fragile – the valley is narrow, both sides steeper in gradient. Extensive felling of trees started as early as 1850 in the Satluj valley by British contractors who took benefit of the poor condition of the hill chiefs.

During July-August 1993, a landslide from the right bank of Satluj obstructed the flow of Satluj near Nathpa, 185 km from Shimla, creating an artificial lake. Water of the lake entered the 120 MW Bhabha Powerhouse causing it heavy losses. After great effort, the rocks blocking Satluj were blown into pieces by dynamites, releasing water from the lake.

On the evening of August 11, 1997, the Panvi Khud, a small rivulet originating in the Shatul Ghati brought tons of debris with it and blocked Satluj for several days. Six major bridges left no trace cutting the Kinnaur from rest of the country and eroding about 6 km of the NH 22. A big lake was formed at Wangtu due to blockage of the Satluj. On the same night, the Andhra Khud flowing in the same mountain range washed away a roadside township of Chirgaon. On the same night, considerable damage was inflicted by the Nogli Khud. Andhra Power House (17 MW) and Nogli Power House (2.5 MW) suffered heavy losses and did not generate electricity for about a year.

Land slides

Heavy rainfall in the mountain areas is also responsible of large-scale landslides. There are instances when large volumes of earth moved to river beds, creating an earthen dam which blocked the flow of river. This leads to the formation of a lake behind this temporary dam. Breakage of this dam gives rise to flooding in the downstream areas. Such a landslide, for example, took place in year 2004 and blocked the flow of Peerchu River in Tibet, a tributary of Satluj River. Similar instances of landslide and dam formation have been observed in the Ganga valley also.

10.12. WATER QUALITY OF THE TRIBUTARIES OF INDUS BASIN IN PUNJAB

Water pollution in the Indus occurs through three sources: municipal wastewater discharges, industrial wastewater discharges, and agriculture return flows through drainage structures. Treatment plants (oxidations ponds), provided in many cases, are not maintained properly. As a result, the wastewater does not receive the desired degree of treatment. The parameter of major concerns is the discharge of organic matter. This causes depletion of dissolved oxygen of the river water. In extreme cases, when the assimilative capacity of the river is exceeded, anaerobic (septic) conditions result. This could be a problem in the river, during the months of low dilution (December and January). At present, water shortages are feared during the incoming crop season. Under anaerobic conditions, iron and manganese become more soluble and become a potential source of ground water pollution. In addition, due to high coliform content, use of water for drinking purposes, without appropriate treatment, would result in water-borne diseases, such as malaria, typhoid, cholera & dysentery. Industrial waste water discharges, depending upon the nature of industry, comprise wide-ranging variables. This includes organic matter; ions like sodium, potassium, calcium, magnesium, carbonates and bicarbonates, chloride; other inorganic variables, such as fluoride, silica, cyanide; metals like cadmium, chromium, copper, mercury, lead, zinc, nickel, etc. The return-agriculture flow characteristics include salinity, total dissolved solids, sodium adsorption ratio (SAR), nitrates, phosphates and pesticides. The status of water quality of rivers of Punjab is given in Table 14.

Sampling Stations

1. Satluj : Harike Reservoir,
2. Ghaggar: Confluence with Saraswati River,
3. Beas : Goindwal Sahib,
4. Ravi : Headworks at Madhopur.

Indiscriminate and unscientific irrigation has disturbed the equilibrium between inflow and outflow of water in command areas of main beneficiary states leading to excess salt concentration. The area affected by salinity in Punjab is about 490,000 ha and in Haryana 197,200 ha. The over-irrigated command of the Indira Gandhi Canal is also suffering from waterlogging and salinity. Misuse and quality degradation is bound to affect the productivity of land and cause shortage of water.

Table 14. Status of Water Quality of rivers of Punjab

S. No.	Parameter	Satluj	Ghaggar	Beas	Ravi
1	Temperature °C	16	16	16	14
2	PH	7.7	7.6	7.8	7.8
3	Conductivity (μmho)	378	424	342	202
4	Nitrogen (NO_2+NO_3)	1.0	2.34	1.4	.04
5	DO (mg/l)	7.7	5.8	7.8	9.0
6	BOD (mg/l)	1.8	28.0	4.2	0.4
7	COD (mg/l)	6.4	57.6	14.4	1.6
8	Cl^- (mg/l)	20	54.0	23.0	10
9	So_4	14	30	16	8.0
10	Na	4.2	21.2	14.6	1.8
11	Fecal Coliform	170	500	500	0.0
12	Turbidity (NTU)	22	62	24	7.0
13	Total Coliform	500	9,000	5,000	7.0
14	TDS	340	396	302	194

Source: Punjab Pollution Control Board Report.

Table 15. Details of Hydroelectric projects being given to private sector by Govt. of HP

Name of the Project	District	River Basin	Tentative installed capacity
Part I: Projects for which P.F.R. is ready			
Category-I			
Chanju-I	Chamba	Chanju/Ravi	25 MW
Chanju-II	Chamba	Chanju/Ravi	25 MW
Category-II			
Youngthang Khab	Kinnaur	Spiti Satluj	261 MW
Bara Bangahal	Chamba	Ravi	200 MW
Bajoli Holi	Chamba	Ravi	180 MW
Gondhala	Lahaul & Spiti	Chandra/Chenab	144 MW
Bardang	Lahaul & Spiti	Chenab	114 MW
Chhatru	Lahaul & Spiti	Chandra/Chenab	108 MW
Part II: Projects Identified			
Category-I			
Mane Nadang	Lahaul & Spiti	Spiti/Satluj	70 MW
Lara	Lahaul & Spiti	Spiti/Satluj	60 MW
Ropa	Kinnaur	Roopa/Satluj	60 MW
Kuling Lara	Lahaul & Spiti	Spiti/Satluj	40 MW
Bharai	Bilaspur	Bharai/Satluj	5.5 MW
Miyar	Lahaul & Spiti	Miyar/ Chenab	90 MW
Tinget	Lahaul & Spiti	Miyar/ Chenab	81 MW
Teling	Lahaul & Spiti	Chandra/Chenab	61 MW
Patam	Lahaul & Spiti	Miyar/ Chenab	60 MW
Rupin	Shimla	Rupin/Yamuna	39 MW
Category-II			
Chango Youngthang	Kinnaur	Spiti/Satluj	140 MW
Symte Kothang	Kinnaur	Spiti/Satluj	130 MW
Lara Sumta	Kinnaur	Spiti/Satluj	104 MW
Reoli/Dugli	Lahaul & Spiti	Chenab	715 MW
Dugar	Chamba	Chenab	360 MW
Gyspa	Lahaul & Spiti	Bhaga/Chenab	240 MW
Sach-Khas	Lahaul & Spiti	Chenab	210 MW
Seli	Lahaul & Spiti	Chenab	150 MW
Tandi	Lahaul & Spiti	Chenab	150 MW
Rashil	Lahaul & Spiti	Chenab	150 MW

Efficient water management practices by reducing the water allowance and irrigation frequency are necessary to check these problems. The water thus saved can provide critical irrigation to more desert areas in Rajasthan. Despite optimum development, the principles of easy access, fair use, justice and social equity have not been adequately addressed in all adjoining states.

The govt. of Himachal Pradesh has decided to give the following Hydroelectric Projects to the private sector for operation and has invited tenders for the same. The details of these projects have been given in Table 15.

CHAPTER 11

NARMADA BASIN

Narmada, the largest west-flowing river of the Indian peninsula, is an important river of India. This chapter describes the basin of this river.

11.1. THE NARMADA BASIN

The Narmada River, rises in the Amarkantak Plateau of Maikala range in the Shahdol district of Madhya Pradesh at an elevation of 1,057 meters above mean sea level at a latitude 22°40′ north and a longitude of 81°45′ east. The river travels a distance of 1,312 km before it falls into Gulf of Cambay in the Arabian Sea near Bharuch in Gujarat. The first 1,079 km of its run are in Madhya Pradesh. In the next length of 35 km, the river forms the boundary between the States of Madhya Pradesh and Maharashtra. Again, in the next length of 39 km, it forms the boundary between Maharashtra and Gujarat. The last length of 159 km lies in Gujarat. The index map of the basin is shown in Figure 1.

The Narmada basin extends over an area of 98,796 sq. km and lies between longitudes 72°32′ E to 81°45′ E and latitudes 21°20′ N to 23°45′ N. The statewise distribution of the drainage area is given in Table 1. The catchment area up to Sardar Sarovar dam is 88,000 sq. km.

The basin is bounded on the north by the Vindhyas, on the east by the Maikala range on the south by the Satpuras and on the west by the Arabian Sea. Most of the basin is at an elevation of less than 500 meters above mean sea level. A small area around Panchmarhi is at a height of more than 1,000 meters above mean sea level.

The Narmada River has a number of falls in its head reaches. At 8 km from its source, the river drops 21 to 24 m at Kapildhara falls (see Figure 2). At 0.4 km further downstream, it drops by about 4.6 m at the Dhuandhara falls (see Figure 3). Its first major tributary, the Burhner joins the Narmada from the left at the 248th km of its run. Flowing in a generally south-westerly direction in a narrow and deep valley, the river takes several pin-head turns. At the 286th km from the source, it turns northwards and hardly a km further downstream it receives the Banjar, another major tributary from the left. Narmada flows past Mandla town in a number of channels called *Sahasradhara*. Close to Jabalpur city, 404 km from the source, the

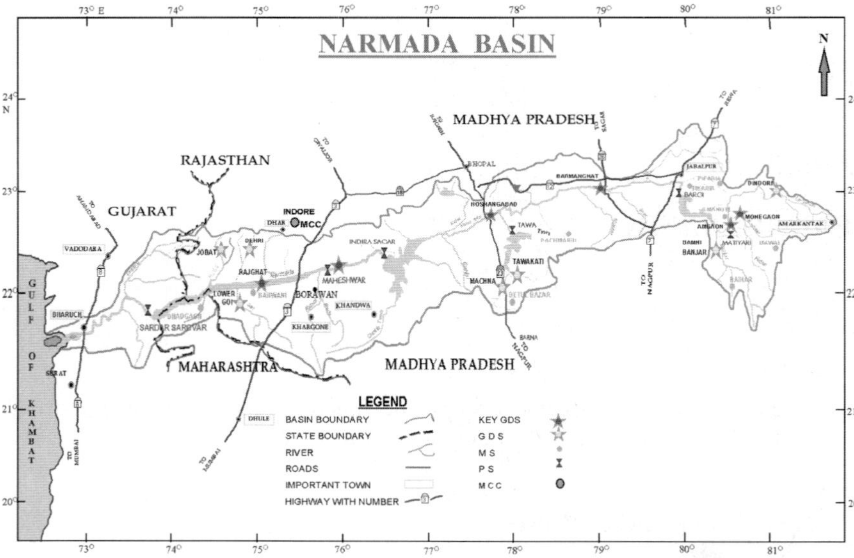

Figure 1. Index map of the Narmada basin [Source: NCA (2000)]

river drops nearly 15 m at the picturesque *Dhuandhara* falls, after which it flows through a narrow channel carved through the famous marble rocks of Bhedaghat.

Emerging from the marble rocks, the Narmada enters the upper fertile plains and at the 464th km of its run receives the Hiran, a major right bank tributary. Continuing to flow in a westerly direction through the upper plains the river receives several tributaries, such as the Sher, Shakkar, Dudhi, Tawa, Ganjal from the left and the Tendoni, Barna, Kolar from the right. Flowing further west, the river enters the middle plains near Panghat in East Nimar district. At Mandhar, 806 km from the source and at Dhardi, 47 km further downstream, the river drops over falls of 12 m at each place. At the 966th km from the source, nearly 6.4 km downstream of Maheshwar, the Narmada again drops by about 6.7 m at the Sahastradhara falls. During its journey through the middle plains, it receives the Chhota Tawa, the Kundi from the left and the Man from the right.

Table 1. Statewise drainage area of Narmada basin

State	Drainage Area (sq. km)
Madhya Pradesh	85,859
Maharashtra	1,538
Gujarat	11,399
Total	98,796

Figure 2. Kapildhara water fall on Narmada, about 8 km from its origin

Further west, after traversing through the middle plains, the river has two falls of 12 m each: at Nandhar (806 km from the source) and at Dhardi 47 km downstream from Nandhar. At 966 km from source, nearly 6.4 km downstream of Maheshwar, the Narmada again drops by about 6.7 m at the Sahasradhara falls. Flowing further west, the river enters the lower hilly regions and flows through a gorge, receiving the Goi

Figure 3. Dhuandhara falls on Narmada River near Jabalpur city

from the left and the Uri, the Hatni from the right. The 113 km long gorge is formed by the converging of the Vindhyas from the north and the Satpuras from the south towards the river. Emerging from the gorge, the river enters the lower plains and meanders in broad curves till it reaches Broach. The Karjan from the left and the Orsang from the right are the important tributaries joining the river in this reach. Beyond Broach, the valley widens into an estuary. Finally, the river enters the Gulf of Cambay.

The Narmada is navigable in its lower reaches. The tidal influence is felt up to 48 km upstream from the mouth of the river in the Gulf of Cambay. Broach is an important river port. Between the mouth of the Narmada and the city of Broach, fairly large sized barges ply for the transport of goods. For another 32 km above Broach, navigation is possible in the sandy reach. Beyond this point, on account of rocky outcrops and rapids, navigation is not possible. The tributaries of the Narmada are not navigable.

11.1.1. Population

On the basis of the 1991 census and the percentage of the area of each district lying in the basin to the district as a whole, the total population in the basin was about 16.733 million (Gupta 2001). The State-wise distribution is: Madhya Pradesh 12.820 million, Maharashtra 0.313 million, and Gujarat 3.600 million.

The important urban centers in the Narmada basin are Mandla, Jabalpur, Hoshangabad, Khandwa and Khargone in Madhya Pradesh, and Bharauch in Gujarat. There are 40 large and 70 medium scale industrial units operating besides large number of small scale units in the basin. (NIH, 1999). Jabalpur is the most populous city in the basin with a population exceeding two lakh. The average density of population in the basin is 107 persons per sq. km against the figure of 182 for India as a whole. The districts of Raisen and Mandla are thinly populated and have 66 persons per sq. km. Of the total population in the basin, nearly 81% live in the rural areas and 19% live in urban areas.

11.2. MAJOR TRIBUTARIES AND SUB-BASINS

Narmada River has 41 tributaries. Of these, 22 are on the left bank and 19 on the right. The important tributaries/sub-basins of the Narmada are described briefly in what follows.

11.2.1. Barna River

The Barna rises in the Vindhya range in the Raisen district of Madhya Pradesh, east of Barkhera village, at an elevation of 450 m. Barna rises at a north latitude of 22°55′ and an east longitude of 77°44′. It flows for a total length of 105 km in a south-easterly direction to join the Narmada near the Dimaria village. Barna is a right bank tributary of Narmada River. It drains a total area of 1,987 sq. km. The distance of confluence with Narmada from the source is 605 km.

Ground elevation in the Barna sub-basin ranges from 300 m near the Bareli gauging site to 630 m in the uppermost part of the basin. The hilly portion of the basin is predominately covered by dense deciduous forest. The portion of the catchment downstream to the Barna Dam site comes under its command.

The soils in the basin are medium to deep black cotton soils except in the hilly region where skeleton soils have been reported. In the Barna sub-basin, four distinct seasons occur during each year. They are: i) cold weather, ii) hot weather, iii) south-west monsoon, and iv) post-monsoon. The cold weather season commences in December and continues till the end of February. It is characterized by bright cloudless days and clean nights and piercing winds. Frost is known to occur occasionally. Hail too is not uncommon. There is a slight precipitation in the basin during this season. The mean annual temperature in the cold weather varies from 17.5 °C to 20 °C. The hot weather starts in March and continues up to the middle of June. May is usually the hottest month. This season is generally dry except for occasional thunderstorms. The mean annual temperature during the hot weather varies from 30 °C to 32.5 °C, with maximum temperature some times touching 48 °C. The south-west monsoon sets in by the middle of June and withdraws by the first week of October. June to September are the rainiest months. Nearly 90% of the annual rainfall is received during the five monsoon months from June to October. In the south-west monsoon, the temperature ranges 27.5 °C to 30 °C. In the post-monsoon season temperature between 25 °C to 27.5 °C are experienced. Average annual rainfall in the sub-basin is about 1,130 mm.

The Bareli gauging site is located at about 25 km downstream of the Barna dam and has the catchment area of about 1,590 sq. km. Another gauging site, namely, Sultanpur is located upstream of the dam and the catchment area up to the gauging site is about 414 sq. km. At this site, discharge is measured only during monsoon months as the river is usually dry at other times. The flow at the Bareli gauging site mostly depends on the discharge from the Barna dam during non-monsoon season. A small river, Ghoghar, joins the Barna River at 8 km upstream from the gauging site.

Barna Dam was constructed across Barna River in Raisen District. A small tank, namely Palakmati tank is situated in the catchment which intercepts runoff of an area of 85.47 sq km.

11.2.2. Ganjal River

The Ganjal rises in the Satpura range in the Betul district of Madhya Pradesh, north of Bhimpur village at an elevation of 800 m at a north latitude of 22°0′ and an east longitude of 77°30′. The Ganjal River flows for 89 km in a north-westerly direction to join the Narmada near Chhipaner village. Ganjal sub-basin lies between the latitudes 21°58′ N to 22°25′ N and longitudes 77°17′ E to 77°45′ E and covers an area of $1,930\,km^2$. The entire sub-basin forms part of two districts, Hoshangabad and Betul. The Morand River, a tributary of the Ganjal River joins it just upstream of Chhidgaon and has a length of about 121 km.

Topographically, the Ganjal sub-basin can be divided into three distinct zones: (i) low land, (ii) slopes, and (iii) upland. The part of the sub-basin having elevation less than 400 m above mean sea level can be considered as low land. Hill slopes are characterized by elevation ranging from 400 m to 550 m and the part of sub-basin having elevation more than 550 m can be regarded as upland. About 63% area of the sub-basin is covered by dense forest, 12% by agriculture, 19% by waste land and 6% by open forest. In the down stream areas, agricultural activity is carried out. Upstream of this agricultural zone, vast open forest exists. Apart from this, most of the basin is covered by dense forest and waste land.

There are three main types of soils in the sub-basin, viz. medium black, laterite and shallow black. Shallow black soil lies in the lower reach and the upstream area of the basin. Most of the sub-basin is covered by medium black soil.

11.2.3. Chhota Tawa River

The Chhota Tawa River, a left bank tributary of Narmada River rises in the Satpura range in the west Nimar District of Madhya Pradesh near Kakora village at north latitude 21°31′ and east longitude 75°50′. It flows for a total length of 169 km in a north eastern direction to join the Narmada River at its 829 km run from source, north of Purni village. Basin of Chhota Tawa lies between east longitudes 85°50′ to 77°11′E and north latitude 21°27′ to 22°11′N. The Chhota Tawa is next in size to the Tawa among the left bank tributaries and drains a total area of 5,051 sq. km. The southern border of the basin is the water divide line of Satpura mountain range. The basin gradually slopes down in the north-north-east direction. A river gauging site has been set-up at Ginnore where the catchment area is 4,816 sq km and he elevation of the gauging station is 218 m. Average slope of the river is 1:400.

The basin has two types of soil. In the main part of the basin covering central and lower reaches, soil is of medium black variety. Texture of the soil is silty clay loam and soil depth varies from 50 to 100 cm. Soil of the upper reach of the catchment is shallow black variety. Texture of the soil is clay loam and soil depth varies from 25 to 50 cm.

Upper and some part of the central region are under forest cover amounting to about 32% of the basin area. About 60% area is culturable and remaining part is under shrub, suitable for grazing. Good quality teak wood grows in upper reaches of the basin. Main crops under agriculture in the basin are wheat, pulses (mainly Arhar), linseed, sesame and rice. Wheat is the most important crop in the basin. Large quantity of cotton is also grown in the basin. Groundnut is another cash crop that grows in the basin.

Climate of the basin is humid and tropical. Average annual rainfall of Jamtara and Chhota Tawa sub-basins are 1,480 mm and 855 mm respectively. South-west monsoon from June to September accounts for 86% and 88% of the average annual rainfall respectively for Jamtara and Chhota Tawa sub-basins. July and August are

Table 2. Maximum and minimum temperatures at Khandwa and Punasa towns

Station	Jan.–Mar		Apr.–June		July.–Sept.		Oct.–Dec.	
	Max.	Min.	Max.	Min.	Max.	Min.	Max.	Min.
Khandwa	38.0	11.8	41.5	24.3	30.5	22.6	31.3	11.1
Punasa	38.8	11.9	42.9	24.2	31.1	22.9	33.1	12.0

the two rainiest months. The maximum and the minimum temperatures at Khandwa and Punasa towns in the four quarters of the year are given in Table 2.

11.2.4. Hiran River

Hiran River rises in the Bhanrer range in the Jabalpur district of Madhya Pradesh near the Kundam village at an elevation of 600 m. Geographical coordinates of its origination are at a north latitude of 23° 12′ and an east longitude of 80° 27′. The river flows in a generally south-westerly direction for a total length of 188 km to join the Narmada from the right near Sankal village. Hiran has the distinction of being the biggest right bank tributary of the Narmada and drains a total area of 4,792 sq. km. The sub-basin covers part of areas of Jabalpur and Damoh Districts of M.P.

Hiran sub-basin has a flat topography and higher drainage density. The topographic elevations vary from about 600 m above sea level to about 320 m above mean sea level. About 57% of the area is under agriculture and the dense forests occupy nearly 36% of the total area of the basin. About 80% of the agriculture is practiced on low land area, the elevation of which is less than 400 m above mean sea level. However, on the hill slope, some agriculture is also practiced. Nearly 5% of the total basin area is wasteland and eroded forests which exist at higher elevations. Among the major crops sown are wheat and paddy. Paddy is grown during monsoon season and wheat during winter.

Low hills with medium forest cover are found near the source area at Kundam; areas in the valley have a mixed cover of fields and open forests. Central region near Sihora is dominated by flat land where agriculture is carried out. The lower part of the basin is generally flat, fertile farm land. Black cotton soils are extensively found in the basin and they occur in different depths in different region. The low land area (elevation 400 m) is mainly deep black cotton soil on which agriculture is practiced. Most of the forests lie on red sandy soil which exists at higher elevations. However, a part of the forests exist on shallow black cotton soils and red loamy soils. The depth of the soil varies from 0.5 m to as much as 12.0 m. The black cotton soil in low land is very deep, but on hill slope the depth of black cotton soil is much less. Typically, the water table varies between 4 m to 10 m. The water table depth varies on an average from 5.5 m in May/June to 3.5 m in Nov./Dec. Draw down of 0.26 m to 8.2 m have also been observed seasonally.

At the outlet, the river is gauged at Patan road bridge. A small tank, known as the Bahoribund tank was built in 1923 in this sub-basin. Its storage capacity has been reduced to a large extent by siltation. The catchment of the tank is about 109 sq. km.

11.2.5. Jamtara Sub-Basin

Jamtara sub basin is a part of Narmada River containing 399 km of the main Narmada River from its source up to the discharge measuring site Jamtara. In this part, elevation ranges from 1,100 m in the upper part of the basin to 450 m near the Manot gauging site and 360 m at Jamtara. The sub-basin falls in the upper hilly zones of Narmada basin. The catchment area of the basin up to Jamtara discharge site is 16,575 sq. km. Jamtara sub-basin basin lies between east longitudes 79°45′ to 81°45′ E and north latitude 21°20′ to 23°45′ N.

From the source of the river to a distance of 269 km lies the Manot gauging site where catchment area is 4, 980 km^2. Here, Narmada flows in a generally northwesterly direction but just upstream of Manot, it turns in a loop to the south. Narmada has a number of falls in its head reaches. Two major tributaries, Burhner and Banjar, join Narmada from left at 248 km and 287 km near Manot and Mandla respectively.

Topographically, the basin can be divided into three distinct levels: low land areas, hill slopes or semi-hilly areas, and upland or hilly areas. The upper reaches of Jamtara sub-basin are well-forested and are covered with good quality Sal and Teak woods; North, east, central part of the basin is mainly covered with Sal wood forest whereas Southern and Western part is covered with Teak woods. Occasionally, dense forest cover in seen in upper reaches and flat farmland is more evident in the lower reaches. Flat agricultural areas containing bunded fields are interrupted by low hills with a medium to dense forest cover.

The soil of Jamtara basin is loamy clay, in general with about 60% area having red and yellow variety, about 25% area having deep black variety and rest portion having medium black variety. At the end of the dry season, deep cracks are seen in the soil which permit rapid infiltration of the early monsoon rains, thereby inhibiting runoff. In the upland forest areas, surface runoff may be generated at earlier stage in monsoon season because of shallower soils. The runoff may also be more concentrated in small channels, whereas in the flat low land areas, sheet flow may be more prevalent.

The climate of the basin is humid and tropical, although extremes of heat and cold are often encountered at places. The maximum and the minimum temperatures of Mandla and Jabalpur Town in four quarters are given in Table 3.

Table 3. Maximum and Minimum temperature in Jamtara Sub-basin

Station	Jan.–Mar.		Apr.–June		July–Sept.		Oct.–Dec.	
	Max.	Min.	Max.	Min.	Max.	Min.	Max.	Min.
Mandla	34.9	9.0	40.2	19.6	29.1	21.7	29.4	6.8
Jabalpur	36.2	10.1	42.1	21.0	30.6	23.1	30.5	7.7

11.2.6. Kolar River

The Kolar rises in the Vindhyachal range in the Sehore district of Madhya Pradesh near Bilqisganj village at an elevation of 550 m at a north latitude of 23°7′ and an east longitude of 77°17′. It flows for a total length of 101 km in a south-westerly direction to join the Narmada from the right, south of Nasrullahganj. During its course, Kolar drains a total area of 1,347 sq. km. The Kolar sub-basin is located in the latitude range of 22°40′ to 23°08′ and longitude 77°01′ to 77°29′ in two districts, Sehore and Raisen.

Topographically, the Kolar basin can be divided into two distinct zones. The upper four-fifth part having elevations ranging from 600 m to 350 m is predominantly covered by deciduous forest (dense and open). The boundaries of catchment are mild sloped at the northern and of the basin. The river debauches to plains from this area upstream of Jholiapur through ramp shaped southward sloping topography. The soils are skeleton to shallow in depth except near channels where they are relatively deep. The rock outcrops are easily visible at many places. In this area, the rocks are weathered and deep fissures can be seen. The channel beds are rocky or graveled. The thin soils get saturated even during low intensity rains and water moves through the fissures rapidly. Agricultural activity is carried out in relatively large areas in the north western part (adjacent to Ichhawar) and in small pockets elsewhere in which the main crops are wheat and grams.

The lower part of the basin consisting of flat-bottomed valley narrowing towards the outlet and having elevations ranging from about 350 m to 300 m is predominantly cultivable area. The soils are deep in the area and have flat slopes. The places where agricultural activity is carried out have bunded fields in which water is impounded during the monsoon period. Part of this area comes under the command of Kolar Dam. This dam has been constructed near Lawakheri to provide drinking water to Bhopal and for irrigation. Downstream of the dam, a barrage has been built at Jholiapur.

11.2.7. Orsang River

The Orsang rises in the Vindhya range of the Jhabua district of Madhya Pradesh, near the Bhabra village at an elevation of 300 m, at a north latitude of 22°30′ and an east longitude of 74°18′ and flows for a total length of 101 km in a south-westerly direction to join the Narmada from the right, near Chandod. It drains a total area of 4,079 sq km and is next in size to the Hiran amongst the right bank tributaries.

11.2.8. Sher River

The Sher rises in the Satpura range near Patan in the Seoni district of Madhya Pradesh at an elevation of 600 m at a north latitude of 22°31′ and an east longitude of 79°25′. Sher flows in a north-westerly direction for a total length of 129 km. It meets Narmada from the left near Brahmand. The Sher drains a total area of 2,901 sq. km.

Sher sub-basin lies in the districts of Narsimhpur, Chindwara and Seoni in Madhya Pradesh. It is identified with hilly terrain, heavily intersected by streams and rivers. The third and higher order streams in the catchment are ephemeral. The vegetation of the basin consists of forest of medium density, scrubland, pockets of cultivation on undulating land and some denuded land. The average annual rainfall at two rainfall stations of the sub-basin, namely, Narsimhpur, and Lakhandan is reported as 1,241 and 1,269 mm, respectively.

11.2.9. Tawa River

The Tawa, the biggest left bank tributary of Narmada, rises in the Mahadeo hills of the Satpura range in the Chhindwara district of Madhya Pradesh near the Cherkathari village at an elevation of 900 m. Its origin is at a north latitude of 22°13′ and an east longitude of 78°23′. Taw River flows in a generally north-westerly direction for a total length of 172 km to join the Narmada from the left, north-east of Hoshangabad. The Denwa is its important tributary. The Tawa drains a total area of 6,333 sq. km. A dam, namely Tawa dam, has been constructed in this catchment.

11.2.10. Other Tributaries

Salient features of the other tributaries of Narmada are given in Table 4.

11.3. CLIMATE IN NARMADA BASIN

The climate of the basin is humid tropical ranging from sub-humid in the east to semi-arid in the west with pockets of humid or per humid climates around higher hill reaches.

IMD is maintaining class 1 observatories at 18 locations in and around Narmada basin where the observations of dew point, temperature are made twice a day at 08:30 IST and 17:30 IST. The list of observatories along with their date of starting is shown in Table 5.

The Tropic of Cancer crosses the Narmada basin in the upper plains area and a major part of the basin lies just below this line. The climate of the basin is humid and tropical, although at places extremes of heat and cold are often encountered. In a year, four distinct seasons occur in the basin. They are: (i) cold weather, (ii) hot weather, (iii) south-west monsoon and (iv) post-monsoon.

Cold weather Season: The cold weather season, which commences in December and continues till the end of February, is characterized by bright cloudless days and clean nights and piercing winds. Frost is known to occur occasionally; hail too is not uncommon. There is slight precipitation in the basin during the season.

Hot weather Season: The hot weather starts in March and continues up to the middle of June. May is usually the hottest month. This season is generally dry except for occasional thunder-storms.

Table 4. Details of selected tributaries of Narmada

Name	Bank on which joins Narmada	Length (km)	Catchment area (km^2)	Details of origin	Place of joining Narmada and distance of this place from source of Narmada (km)
Banjar	Left	184	3, 626	Rises in the Satpura range, near Rampur village, Durg district, Chhattisgarh, elevation 600 m, latitude 21°42′ N, longitude 80°50′ E.	Near Mandla, 287
Burhner	Left	177	4, 118	Rises in the Maikala range, south-east of the Gwara village, Mandla district, MP, elevation 900 m, latitude 22°32′ N, longitude 81°22′ E	Near Manot, 248
Dudhi	Left	129	1, 541	Rises in the Mahadeo hills of the Satpura range, west of the Chhindi village, Chhindwara district, MP, elevation 900 m, latitude 22°23′ N, longitude 78°45′ E	North-west of Nibhora, 575
Goi	Left	129	1, 891	Rises in the Satpura range, near village Dhavdi, West Nimar district, MP, elevation 600 m, latitude 21°40′ N, longitude 75°23′ E	West of Barwani village, 1,038
Hatni	Right	80	1, 942	Rises in the Vindhya range in the Jhabua district of Madhya Pradesh, east of Kanas at an elevation of 450 m at a north latitude of 22°32′ and an east longitude of 74°40′ and flows for a total length of 81 km in a southerly direction to join the Narmada from the right	Near Kakrana, 1,075
Karjan	Left	93	1, 489	Rises in the Satpura range, Surat dist., south of Nana village, elev. 300 m, latitude 21°23′ N, longitude of 73°35′ E	East of Sinor village, 1,199
Kundi	Left	121	3, 820	Rises in the Satpura range in West Nimar dist., MP, near Tinshemali village at an elev. 600 m, latitude 21°25′ N, longitude of 75°45′ E	Near Mandleshwar 943
Man	Right	89	1, 528	Rises in the Vindhya range in the Dhar dist. MP, near Dhar town at elev. 500 m, latitude 22°33′ N, longitude of 75°18′ E	North of the Talwara Deb village, 992

(Continued)

Table 4. (Continued)

Name	Bank on which joins Narmada	Length (km)	Catchment area (km^2)	Details of origin	Place of joining Narmada and distance of this place from source of Narmada (km)
Shakkar	Left	161	2,292	Rises in the Satpura range, east of the Chhindi village, Chhindwara district, MP, elevation 600 m, latitude 22°23′ N longitude 78°52′ E	north-west of Paloha, 546
Tendoni	Right	118	1,632	Rises in Vindhya range, Raisen dist. MP, east of Sodarpur village, elev. 600 m, latitude 23°22′ N, longitude 78°33′ E	Near Bhatgaon village, 602
Uri	Right	74	1,813	Rises in Vindhya range, Jhabua dist. MP, near Kalmore, elev. 450 m, latitude 22°36′ N, longitude of 74°47′ E	Near Nisarpur, 1,029

Source: NIH (1999).

South-West Monsoon: The south-west monsoon sets in by the middle of June and withdraws by the first week of October. Months from June to September are the rainiest months. During this season, the weather is somewhat sultry and oppressive, especially in areas adjoining the Narmada River.
Post-Monsoon Season: In the post-monsoon season, a few thunderstorms occur, especially in October. Thereafter, the weather clears up and dry pleasant weather prevails throughout the valley.

11.3.1. Rainfall in Narmada Basin

According to the records maintained by the India Meteorological Department, there were ten rain-gauges in 1867 in the entire Narmada basin. The number rose to 21 rain-gauges in the year 1891, the year from which published rainfall data are available. Thereafter, there has been a steady growth of the rain-gauge network in the basin. The number of raingauge stations in the basin was 205 in 1980. Of these, nearly 120 raingauge stations have data for more than 40 years. About 50 self-recording raingauge stations (SRRG) are maintained by IMD or other agencies like the flood forecasting division of CWC, state irrigation departments, etc.

The normal annual rainfall for the basin works out to 1,178 mm. South-west monsoon (June to October) is the principal rainy season accounting for nearly 94% of the annual rainfall. About 60% of the annual rainfall is received during July and August months. Table 6 shows the monthly distribution of normal annual rainfall in the basin. The rainfall is heavy in the upper hilly and upper plains areas of the

Table 5. List of Observatories in and around Narmada Basin

District	Observatory	Lat.	Long.	Elevation (m)	Date of Starting
Bilaspur	Pendra Road	22°46′	81°54′	625	23.6.1903
Balaghat	Malajkharid	22°00′	80°42′	581	01.7.1978
Jabalpur	Jabalpur	23°12′	79°57′	393	01.1.1869
Shahdol	Umaria	23°32′	80°53′	459	24.4.1931
Mandla	Mandla	22°35′	80°22′	443	15.6.1950
Seoni	Seoni	22°05′	79°33′	619	01.1.1870
Narsimhapur	Narsimhapur	22°57′	79°11′	356	10.2.1962
Chindwara	Chindwara	22°06	79°00′	685	01.6.1908
Hoshangabad	Hoshangabad	22°46′	77°46′	302	01.1.1870
Hoshangabad	Pachmarhi	22°25′	78°26′	1,075	01.5.1870
Sehore	Bairagarh (Bhopal)	23°17′	77°21′	523	09.7.1927
Raisen	Raisen	23°19′	77°50′	440	13.6.1970
Betul	Betul	21°52′	77°56′	653	15.9.1948
Devas	Kannod	22°40′	76°44′	353	17.3.1969
Nimar	Khandwa	21°50′	76°22′	318	01.1.1877
Khargone	Khargone	21°49′	75°37′	251	27.9.1969
Indore	Indore	22°43′	75°48′	567	25.2.1877
Dhar	Dhar	22°36′	75°18′	583	18.2.1973
Jhabua	Alirajpur	22°17′	74°24′	293	01.7.1955
Broach	Broach	21°42′	73°00′	17	25.2.1969
Surat	Surat	21°12′	72°50′	12	29.1.1947
Dhulia	Nandurbar	21°20′	74°15′	206	11.7.1949
Jalgaon	Jalgaon	21°03′	75°34′	201	01.9.1936

Source: NIH (1999).

basin. It gradually decreases towards the lower plains and the lower hilly areas and again increases towards the coast and southwestern portions of the basin. The annual rainfall in the upper part of the catchment is more than 1,400 mm and in some pockets it exceeds 1,650 mm. From the source to Sardar Sarovar dam, the coefficient of variation varies from 19% to 37%.

Table 6. Monthly distribution of normal annual rainfall

Month	Rainfall (mm)	Percent of annual rainfall
June	152.4	13
July	392.4	33
August	314.8	27
September	199.7	17
October	40.6	3.4
Non-monsoon months	78.1	6.6
Total	1,178.0	100.0

Table 7. Seasonal normals of rainfall (cm) for Narmada basin

Jan–Feb		Mar–May		Jun–Sep		Oct–Dec	
Rainfall (cm)	No. of rainy days	Rainfall (cm)	No. of rainy days	Rainfall (cm)	No. of rainy days	Rainfall (cm)	No. of rainy days
2.9	1	2.7	5	110.8	36	6.7	8

Source: Rakhecha & Soman (1992).

The seasonal normal rainfall for Narmada basin is given in Table 7. The monthly normal of rainfall in two adjoining/in-basin cities is shown in Table 8.

11.3.2. Temperature

In cold weather, the mean annual temperature varies from 17.5 °C to 20 °C and in hot weather from 30 °C to 32.5 °C. In the south-west monsoon, the temperature ranges from 27.5 °C to 30 °C. In the post-monsoon season, temperatures range from 25 °C to 27.5 °C. The maximum and minimum temperatures for a few representative towns in the Narmada basin are given in Table 9, which clearly indicate the extent of variations:

11.3.3. Evaporation

Little data on evaporation is available in the basin. There are a few agro-meteorological observatories located in the basin. The potential evaporation depths for Kolar sub-basin in the basin are given in Table 10.

11.3.4. Real-Time Data Acquisition System

A Real-time Data Acquisition System (RTDAS) has been planned by the Narmada Control Authority (NCA) for the Narmada basin. This system consists of 96 remote stations which will be linked with a master control centre for acquisitions of hydro-meteorological data in real-time. The data obtained through this system will be used, among other things, for reservoir regulation. A master control centre will be set-up at Indore at the headquarters of NCA. Data communication will be based on Indian communication satellite from INSAT series (METSAT). Category wise, these will be five project stations, six key gauge and discharge stations, seven gauge and

Table 8. Average monthly rainfall for selected stations in Narmada basin

City	Elevation	Jan	Feb	Mar	Apr	May	Jun	Jul	Aug	Sep	Oct	Nov	Dec
Bhopal	501 m	6.0	4.0	9.0	6.0	14.5	173.5	501.0	278.0	265.0	42.0	26.0	5.0
Indore	564 m	6.0	4.0	2.0	3.0	13.0	147.0	282.0	207.0	164.0	31.0	15.5	7.0

Table 9. Maximum and Minimum Temperature for a few cities in/near Narmada Basin

City	Elevation	Max/ Min	Jan	Feb	Mar	Apr	May	Jun	Jul	Aug	Sep	Oct	Nov	Dec
Ahmedabad	56 m	Max	28.7	31.0	35.7	39.7	40.7	38.0	33.2	31.8	33.1	35.6	33.0	29.6
		Min	11.9	14.5	18.6	23.0	26.3	27.4	25.7	24.6	24.2	21.2	16.1	12.6
Bhopal	501 m	Max	25.7	28.5	33.6	37.8	40.7	36.9	29.9	28.6	30.1	31.3	28.5	26.1
		Min	10.4	12.5	17.1	21.2	26.4	25.4	23.2	22.5	21.9	18.0	13.3	10.6

Source: IMD (1972).

Table 10. Monthly potential evaporation for Kolar sub-basin in Narmada Basin

Month	Evaporation depth (m)	Month	Evaporation depth (m)
January	0.125	July	0.256
February	0.14	August	0.137
March	0.175	September	0.167
April	0.278	October	0.146
May	0.45	November	0.128
June	0.306	December	0.093

Source: NIH (96-97a).

discharge stations, and eight meteorological stations. The parameters that will be measured include rainfall, wind speed and direction, ambient temperature, relative humidity, water level, solar radiations, and evaporation.

The initial phase comprises of 26 remote stations (8 meteorological stations, 13 gauge and discharge sites, and 5 project stations). The RTDAS network is an interactive data communication system under star configuration (Srivastava and Sinha, 2003).

11.4. SOILS AND LAND USE

Reconnaissance soil surveys made in connection with the Bargi, Sardar Sarovar, Barna and Tawa projects indicated that the Narmada basin consists mainly of black soils. The different varieties are deep black soil, medium black soil and shallow black soil. In addition mixed red and black soil, red and yellow soil and skeletal soil are also observed in pockets. Of these deep black soil covers the major portion of the basin.

11.4.1. Forests and Agriculture

In the basin, forests occupy nearly 31,670 sq. km which is 32.1% of the total area and the cultivable area is about 59,000 sq. km which is about 60% of the total area. Out of the total culturable area, nearly 4.49 M-ha are annually cultivated. About 4.5% of the cultivated area is irrigated annually. Wheat is the most important irrigated crop in the basin covering nearly 28.1% of the total irrigated area. Wherever irrigation facilities exist, perennial and eight-monthly crops are cultivated. Cultivation is by a system of rotation of crops and the major crop seasons are the Kharif and the Rabi.

11.4.2. Land Use and Agricultural Practices

The culturable area in the basin is about 3.02% of the total culturable area of India. The total cropped area in the basin forms 2.92% of the total cropped area in the country. The area under irrigated crops is about 4.47% of the cropped area in the basin. The state-wise general pattern is as under.

Land Use in Madhya Pradesh: Of the gross irrigated area of nearly 132,400 hectares, 31.4% is under rice, 36.5% under wheat, 5.2% under sugarcane, 4.4% under gram, 0.8% under cotton and the rest under other crops. The other irrigated crops are jowar, bajra, maize, barley, pulses, fruits, vegetables, linseed, rape, mustard, tobacco and fodder crops. Food and non-food crops cover about 98.3% and 1.7% of the irrigated cropped area, respectively.

Land-use in Maharashtra: Of the gross irrigated area of 7,900 hectares, 51.9% is under wheat, 6.3% under rice, 2.5% under sugarcane, 5.1% under cotton, 1.3% under grain and the rest under other crops. The other irrigated crops are jowar, bajra, maize, pulses, condiments, spices, groundnut, sesamum, tobacco and fodder crops. Food and non-food crops cover about 88.6% and 11.4% of the irrigated cropped area, respectively.

Land-use in Gujarat: Of the gross irrigated area of 73,200 hectares, 49.8% is under cotton, 11.3% under rice, 10.7% under wheat, 0.4% under sugarcane and the rest under other crops. The other irrigated crops are jowar, bajra, maize, barley, condiments, spices, rape, mustard, fruits, vegetables, tobacco, and fodder crops. Food and non-food crops cover about 37.8% and 62.2% of the irrigated area respectively.

To sum up, of the total irrigated area in the basin, nearly 23.6% is under rice, 28.1% under wheat, 17.7% under cotton, 3.5% under sugarcane, 2.8% under gram and the rest under other miscellaneous crops. Food and non-food crops cover about 77.2% and 22.8% of the irrigated area respectively. Table 11 summarizes land use details in Narmada basin.

11.5. WATER RESOURCES OF NARMADA BASIN

We first discuss the stream gauging network in Narmada basin.

11.5.1. Stream Gauging Network

Systematic observations of gauge and discharge were started in Narmada basin only in 1947 by the then Central Waterways, Irrigation and Navigation Commission. The main river Narmada is now gauged at nine sites at Manot, Mandla, Jamtara,

Table 11. Land-use details in the Narmada Basin (area in km^2)

Item	State			
	Madhya Pradesh	Maharashtra	Gujarat	Total
Gross area	85,859	1, 538	11, 399	98,796
Area under forests	29,370	690	1, 610	31,670
Cultivable area	49,840	800	8, 370	59,010
Net sown area	36,810	800	7, 380	44,990
Area sown more than once	2,410	70	150	2,630
Total cropped area	39,220	870	80	40,170

Bermanghat, Hoshangabad, Mortakka, Mandleshwar, Barwani, and Garudeshwar, where daily gauge and discharge observations and hourly gauge observations during monsoon season (June-Oct.) are made. Only a few principal tributaries in the Narmada are gauged. In addition to the CWC sites, state governments have also established sites to measure streamflows.

Like the peninsular rivers of India, the Narmada rises in the latter half of June and the flow reaches its maximum in the months of August and September. Thereafter, it begins to fall in October and reaches its lowest level just before the monsoon. A quick glance of observed data shows that a maximum flow of 39,644 cumec was recorded at Garudeshwar on the 17th of September, 1950 and a minimum of 10.3 cumec on the 10th June 1952.

Some details of the cross-section of Narmada River at selected gauging sites are given in Table 12.

Rating curves were developed at several gauging sites of Narmada using the procedure described in Chapter 5. A power equation was fit as flows:

$$Q = a(H - c)^b \tag{1}$$

Where, Q is discharge in m^3/s and H is river stage in m.

Table 12. Some details of the cross section of Narmada River at selected gauging sites

Name of the gauging site	Zero of gauge (m)	Width (m)	Channel slope	Area (m^2) at 15 m depth
Manot	442	170	0.00125	2, 592
Jamtara	360	280	0.00033	3, 715
Bermanghat	306	320	0.00038	3, 140
Hoshangabad	282	700	0.00023	8, 831
Mortakka	150	670	0.00053	6, 291
Mandleshwar	138	600	0.00046	10, 064
Garudeshwar	12	500	0.0001	6, 682

Source: NIH (1999).

Table 13. The parameters defining rating curves for various sites

Name of the gauging site	Parameters			Datum (m)
	a	b	c	
Manot	99.467	1.769	0.50	86
Jamtara	85.046	1.795	2.30	360
Bermanghat	98.428	1.73	4.00	306
Hoshangabad	173.183	1.858	1.90	282
Mortakka	605.09	1.54	2.60	150
Mandleshwar	331.209	1.715	0.90	138
Garudeshwar	250.00	1.66	1.70	12

Source: NIH (1999).

The parameters for the rating curve at some gauging sites are given in Table 13. A stream Network model for Narmada basin is presented in Figure 4.

11.5.2. Surface Water

The surface water potential of the Narmada river system has been assessed at different times by different authorities. In 1949 when the basinwise water resources of the country were assessed on the basis of Khosla's formula, the annual runoff of the Narmada river system was estimated to be 49,241 MCM.

In 1960, the total annual runoff of the Narmada river system was assessed as 40,088 MCM. In 1965, the Narmada Water Resources Development Committee, set up by the Government of India, reported the total annual runoff of the Narmada River system at Garudeshwar site as 44,331 MCM. Table 14 summarizes the annual observed runoff in selected tributaries of Narmada.

The 75% dependable flow in the basin has been assessed at 34,537 MCM or 28 MAF by the Narmada Water Disputes Tribunal. On this basis, the shares of the four party States in this quantity were fixed as: M.P. 18.25 MAF (22,511.01 MCM), Gujarat 9 MAF (11,101.32 MCM), Rajasthan 0.5 MAF (616.74 MCM) and Maharashtra 0.25 MAF (308.37 MCM). The runoff factor for the 75% dependable flow worked out to be 0.29.

11.5.3. Ground Water

Six major aquifers are found in the basin. Among these, the Archaeans are found in many districts: Balaghat, Seoni, Chhindwara, Betul, Dewas, and Jhabua. The Chhattisgarh aquifer system is found in some parts of the Durg district. The Vindhyans systems are extensively found in the basin and are found in Khargone, Khandwa, Dewas, Raisen, Sagar and Damoh districts. Another major geologic feature, the Gondwanas partly cover Shahdol, Jabalpur, Chhindwara, Betul and Hoshangabad districts. Deccan traps cover a very extensive area in the form of

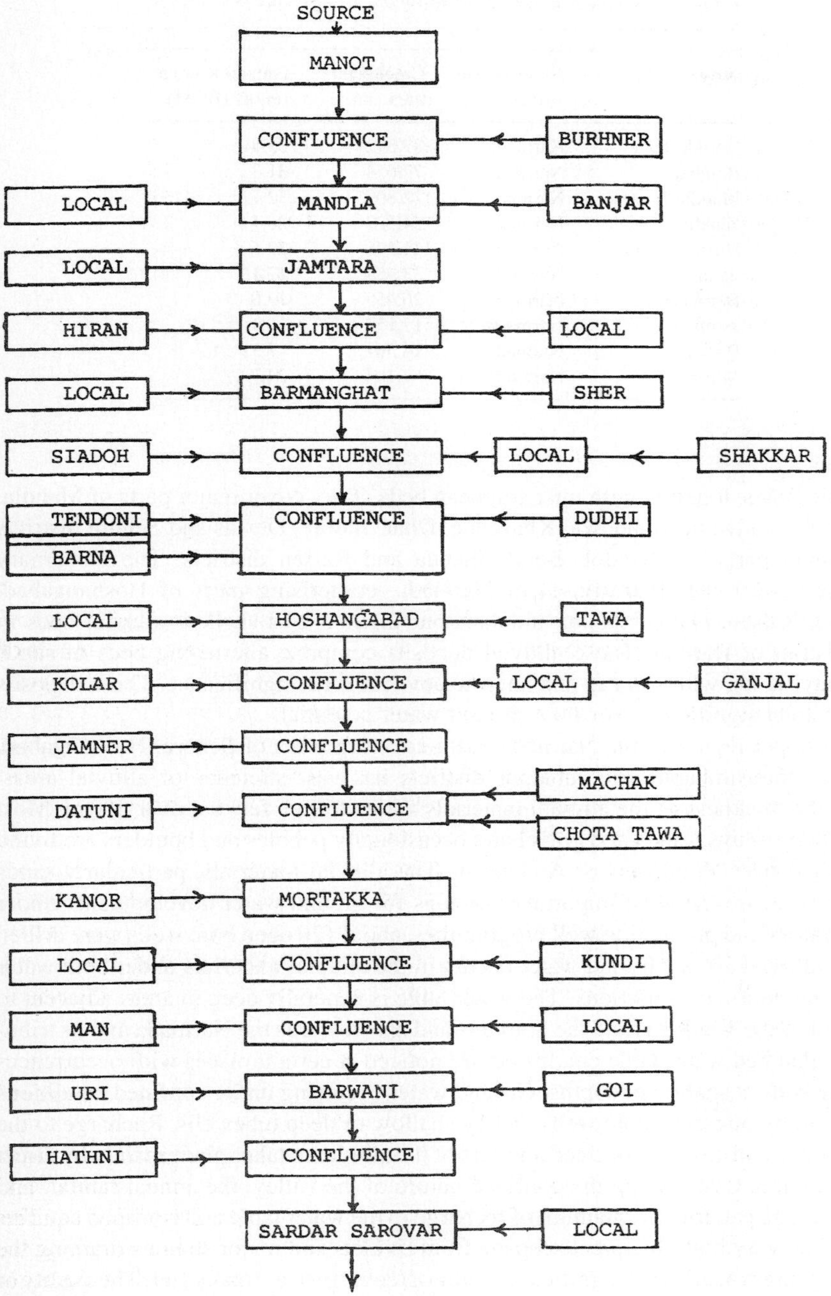

Figure 4. Stream Network of Narmada basin (Source: NIH 1999)

Table 14. Annual average observed runoff at selected CWC sites in Narmada basin

Name of the site	Name of the stream	Catchment area (km^2)	Annual average runoff (BCM)
Garudeshwar	Narmada	87,892	35.4
Rajghat	Narmada	77,674	31.3
Mandleshwar	Narmada	72,809	32.7
Handia	Narmada	54,027	26.1
Hoshangabad	Narmada	44,548	22.6
Sandia	Narmada	33,954	17.1
Barmanghat	Narmada	26,453	13.0
Jamtara	Narmada	17,157	10.1
Bijora	Narmada	14,561	7.5
Mortakka	Narmada	67,184	24.7

basaltic flows, together with inter-trappean beds. They cover major parts of Mandla, Seoni, Chhindwara, Khandwa, Khargone, Dhar, Indore, Dewas and Sehore districts and small parts of Shahdol, Betul, Jhabua and Raisen districts. The quaternary deposits cover the river alluvia of Narmada, comprising parts of Hoshangabad, Sehore, Raisen, Narasimhapur and Jabalpur districts and shallow alluvial beds in the district of Damoh. These alluvial deposits comprise alternating beds of sands and clays along with soil cap and lateritic cover of local significance. These deposits are of great significance for their ground water potential.

The upper region of the Narmada basin covering parts of Sehore, Hoshangabad, Raisen, Narasimhapur and Jabalpur districts has vast stretches of alluvial areas. Here, the thickness of the alluvial materials varies from a few cm over 550 m. Most commonly, clays, sands and gravel and occasionally pebbles and boulders are found deposited over Vindhyans or Archaeans. The alluvial materials, particularly sands and gravels are the most important horizons for ground water development. Under exploratory and production well programmes, about 120 deep bore wells were drilled in the alluvial areas. Ground water occurs in the alluvial materials under both water table and confined conditions. The water table is generally deep in areas adjacent to the foot of the Vindhyas and the Satpuras and also towards the Narmada and its tributaries. Perched water table conditions are noticed at certain places with occurrences of clay beds at shallower depths. Ground water occurring under confined conditions is tapped by dug-cum-bore wells and by shallow or deep tubewells. Recharge to the water table and also the confined aquifers of the alluvium, takes place mainly by direct precipitation. Considering the confined nature of the valley, the annual rainfall and the drainage pattern, the quantum of recharge to the water table and confined aquifers should be very high. Incipient seepage from rivulets and major streams draining the recharge areas also form a significant source of recharge to ground water. The quality of ground water both is generally suitable for irrigation. In the following areas, promising aquifers have been proven after investigations by various agencies:

(i) The Padghal area located to the south of the Narmada, in Hoshangabad district, covering 100 sq. km.
(ii) The Powerkheda-Ari-Babai area located to the south of the Narmada, in Hoshangabad district, covering 650 sq. km.
(iii) The Khapuria Kalan-Tonga area in Raisen district, lying north of the Narmada, covering about 600 sq. km.
(iv) The Mahuakhera-Godarwara area in Hoshangabad and Narasimhapur districts, south of the Narmada, covering about 2,000 sq. km.
(v) The Dhobi-Kasikheri area in Narasimhapur district, north of the Narmada, covering 120 sq. km.
(vi) The Shahpura-Bhedaghat area in Jabalpur district, north of the Narmada, covering 400 sq. km.

Systematic ground water studies in the basin have been intensified recently for obtaining precise data on the seasonal fluctuation of the water table, estimate of aquifers and their parameters, the chemical quality of ground water and recharge-discharge relationship, etc. Under this programme, the drilling of observation wells and setting up permanent hydrographic stations are being undertaken. The trends of groundwater table in the Dhar and Jhabua districts in Narmada basin are presented in Figure 5.

The fertile Narmada valley has tremendous potential in ground water of excellent quality and there is great scope for the development and utilization of this water for a variety of purposes.

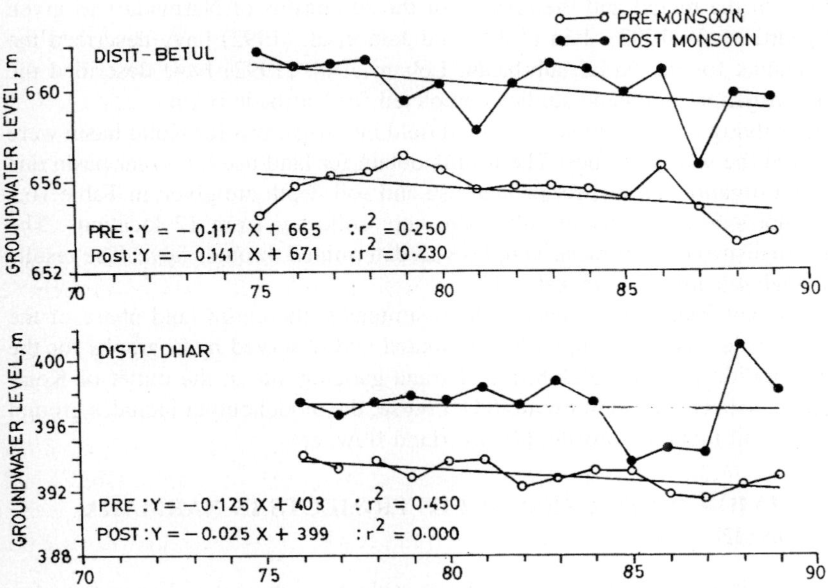

Figure 5. Ground water level fluctuation for pre- and post-monsoon seasons (NIH, 1999)

11.5.4. Hydroelectric Power

The hydroelectric power potential of the basin has been assessed as 2,027 MW at 60% load factor. The details about major hydropower and other projects in the basin are given in a later section. There are thermal stations at Jabalpur and Satpura.

11.6. WATER QUALITY

Table 15 gives existing and desired water quality class at many places in the basin at different times. It can be seen that as compared to the other rivers, the quality of Narmada water is quite good. Of concern is the fact that even near the point of origin, the quality of river water was in class 'C' in the year 2001 while it was in class 'B' in earlier years. As was observed for most other rivers, in case of Narmada also, BOD and Total Coliform are critical parameters.

11.7. MODELING STUDIES

Six sub basins of Narmada River were chosen for simulation of hydrologic response using the SHE model. These sub basins include Kolar, Barna, Sher, Ganjal, Hiran and Narmada up to Manot. The Systeme Hydrologique Euoropeen (European Hydrological System) or SHE is an advanced, physically based, distributed catchment modelling system. The system has developed by Danish Hydraulic Institute, the British Institute of Hydrology, UK and SOGREAH (France) with the financial support of the commission of the European Communities. A detailed description of the model and the results for the sub-basins of Narmada was given by Refsgaard et al. (1992). Jain (1990) and Jain et al. (1992) have described the detailed results for the Kolar sub-basin. Lohani et al. (1992) have described the results of irrigation command studies completed for Narmada basin.

While applying the SHE model, detailed field investigations for Kolar basin were made to find the soil properties. The results obtain for land use for Kolar basin due to field investigations in respect of land use and soil depth are given in Table 16.

The samples from the Kolar sub-basin were collected from 12 locations. The particle size distribution of these samples was determined through sieve. The results of these analyses are given in Table 17.

SHE is a physically-based model which simulates, the entire land phase of the hydrologic cycle. As an example, the simulated and observed hydrographs for the model calibration period (1983–85) at Satrana gauging site at the outlet of Kolar sub-basin have been shown in Figure 6 Of course, the model output includes ground water table, soil moisture profile, ET, overland flow, etc.

11.8. MAJOR WATER RESOURCES PROJECTS IN NARMADA BASIN

Prior to the independence, there was no large-scale development of Narmada water, partly because the region was not subject to any serious famines. Moreover because the soil in the basin is generally retentive and capable of producinga fairly good

Table 15. Desired and existing water quality levels for Narmada

Location	Desired class	Observed class & critical parameters				
		1997	1998	1999	2000	2001
Narmada near source at Amarkantak, M.P.	A	B DO, BOD	B DO	NA	B	C DO, Totcoli
Narmada at Mandla near road bdg., MP	C	B	B	B	B	B
Narmada at Sethanighat, M.P.	B	D BOD, Totcoli	C Ph, Totcoli	D BOD, Totcoli	D BOD, Totcoli	D BOD, Totcoli
Narmada at Narsinghpur, M.P.	C	B	B	B	D BOD	B
Narmada at Hoshangabad U/S, M.P.	B	C Totcoli	C pH, Totcoli	C Totcoli	D BOD, Totcoli	C Totcoli
Narmada at Hoshangabad D/S, M.P.	B	D BOD, Totcoli	D pH, BOD, Totcoli	C Totcoli	E pH, Totcoli	D BOD, Totcoli
Narmada at D/S of Omkareshwar, M.P.	B	B	B	B	B	B
Narmada at Mandleshwar, M.P.	C	B	B	B	B	B
Narmada at Maheshwar, M. P.	B	B	B	B	B	B
Narmada at Badwani, M.P.	C	B	B	B	B	B
Narmada at Garudeshwar, Gujarat	A	D BOD, Totcoli	A	NA	NA	C Totcoli
Narmada at Chandod, Gujarat	C	C	A	NA	NA	C
Narmada at Panetha, Gujarat	C	C	A	NA	NA	B
Narmada at Bharuch, Zadeshvar, Gujarat	C	D BOD	B	NA	NA	C

* NA- Not Available. Source: www.cpcb.nic.in.

harvest even in years of scanty rainfall any large-scale irrigation development in the basin had not been taken up seriously in the past. Only a few medium projects, with total irrigation potential of about 40,500 ha, existed in the basin.

The Master Plan on the utilization of the Narmada waters was submitted to Khosla Committee by the Madhya Pradesh Government. It proposed that by constructing 19 dams on Narmada and its tributaries, irrigation can be provided to 3.1 million ha area by utilizing 29,295 MCM of water. The Rajasthan State, whose desert areas in Barmer and Jalore districts are close to the tail reaches of the (then proposed)

Table 16. Soil depth in Kolar basin

Land use class	Percent of basin	Soil depth
Agriculture on deep soil	2.9	8.0
Agriculture on moderate deep soil	11.9	1.7
Agriculture on medium deep soil	8.2	1.0
Agriculture on shallow soil	19.1	0.5
Forest upland	3.9	0.3
Forest lowland	46.9	0.3
Wasteland	8.1	

Source: Jain et al. (1990).

Table 17. Particle size distribution

Sample no.	Particle size distribution (percent by weight)			
	Clay	Silt	Sand	Gravel
1	40.79	32.32	26.51	0.48
2	17.20	14.85	54.73	13.22
3	25.84	27.04	35.24	11.88
4	21.00	45.00	19.50	14.50
5	45.00	39.50	12.90	2.60
6	56.40	40.10	2.82	0.68
7	55.60	33.90	8.03	2.47
8	39.10	33.60	21.32	5.98
9	36.37	27.13	22.21	14.29
10	29.00	27.50	8.29	35.21
11	12.50	40.00	20.00	26.75
12	69.00	27.75	2.86	0.39

Source: Jain et al. (1990).

high level Narmada Canal, requested the Committee to allocate the Narmada waters for irrigating more than 0.4 million ha falling in their State. Subsequently, a list of 31 major projects was finalized for the Narmada basin. These are listed in Table 18.

According to the latest planned development program, in all 29 major, 135 medium and 3,000 minor projects will be constructed to irrigate 46 lakh ha of land and generate power with an installed capacity of about 3,590 MW (Srivastava and Sinha, 2003). Out of the 29 major projects, Matiyari (Dhoba Toria), Tawa, Sukta, Kolar & Barna Projects have been completed. Bargi Dam is operational; the left bank canals system of Bargi is being constructed. The remaining projects are planned to be completed in two phases. In the first phase, Indira Sagar (Narmada Sagar), Omkareshwar, Maheshwar, Bargi Diversion (R.B.C.), Man and Jobat projects are proposed to be completed. The remaining 18 projects will be constructed in second phase. The Narmada Water Dispute Tribunal has allocated

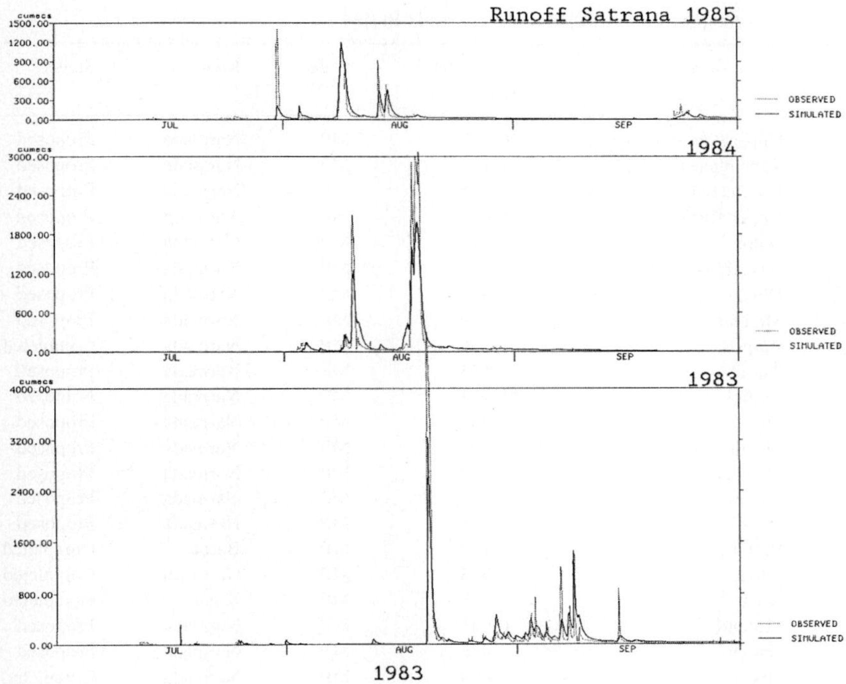

Figure 6. Simulated and observed hydrographs for the model calibration period (1983–85) at Satrana gauging site, Kolar sub-basin

18.25 M.A.F. (22,511 million cubic meter) of water to Madhya Pradesh (Table 19) subject to the condition that the allocation would be reviewed after 45 years, i.e., by 2024–25.

The water use by the projects would be as shown in Table 19.

In the following, the important WRD projects in the Narmada basin are described.

11.8.1. Sardar Sarovar Project

Narmada basin is the subject of one of the largest basin development schemes in the world. The Sardar Sarovar Project in Narmada basin, India, is one the most ambitious but controversial projects of recent times. In view of the immense importance, the salient features of the project, its benefits, environmental impacts and opposition are described in detail in the following.

Tapping the resources of the Narmada has been the dream of political leaders and development planners for decades. Large parts of Gujarat and Rajasthan face recurrent droughts and there have been instances when water had to be transported by trains to save the people from famine. The idea of constructing dams on the Narmada River was first suggested in 1946. For quite some time, this idea could not

Table 18. Thirty-one projects envisaged in the Narmada Basin

S. N.	Name of project	Catchment area (km^2)	State	River	Status
1	Upper Narmada	1, 243	MP	Narmada	Proposed
2	Raghavpar (Hydel)	6, 160	MP	Narmada	Proposed
3	Rosa (Hydel)	4, 312	MP	Narmada	Proposed
4	Upper Burhner	1, 606	MP	Narmada	Proposed
5	Halon	715	MP	Narmada	Proposed
6	Basania (Hydel)	9, 583	MP	Narmada	Proposed
7	Dhobatoria	1, 854	MP	Narmada	Proposed
8	Matiyari	159	MP	Narmada	Proposed
9	Bargi	14, 556	MP	Narmada	Completed
10	Ataria	554	MP	Narmada	Proposed
11	Chinki	22, 414	MP	Narmada	Proposed
12	Sher	881	MP	Narmada	Proposed
13	Machrewa	470	MP	Narmada	Proposed
14	Shakkar	1, 479	MP	Narmada	Proposed
15	Sitarewa	202	MP	Narmada	Proposed
16	Dudhi	808	MP	Narmada	Proposed
17	Barna	1, 176	MP	Barna	Completed
18	Tawa	5, 983	MP	Narmada	Completed
19	Kolar	508	MP	Kolar	Completed
20	Morand	1, 041	MP	Narmada	Proposed
21	Ganjal	436	MP	Narmada	Proposed
22	Sukta	469	MP	Narmada	Completed
23	Chhota Tawa	969	MP	Narmada	Proposed
24	Indira Sagar	61, 642	MP	Narmada	Proposed
25	Omkareshwar (UC)	64, 880	MP	Narmada	Proposed
26	Maheshwar (UC)	69, 184	MP	Narmada	Proposed
27	Upper Beda	544	MP	Narmada	Proposed
28	Man	690	MP	Narmada	Proposed
29	Lower Goi	1, 119	MP	Narmada	Proposed
30	Jobat	792	MP	Narmada	Proposed
31	Sardar Sarovar (UC)	88, 000	MP	Narmada	Proposed

UC: Under construction.

materialize because the states did not agree on the distribution of the river water. The then Prime Minister of India, Mr. Jawaharlal Nehru, laid the foundation stone for the Sardar Sarovar dam, a multipurpose project which is the terminal dam of the basin-wide scheme in 1961. This project was delayed and in 1965, a committee was appointed by the Government of India to prepare a detailed plan for the development of the Narmada basin. The committee recommended the construction of a dam and a canal in Gujarat and twelve major projects in Madhya Pradesh. The two principal dams proposed were the Indira Sagar Dam and the Sardar Sarovar. The recommendations of the committee were endorsed by the Government of Gujarat but rejected by the Governments of Madhya Pradesh and Maharashtra. Subsequently in 1969, the issue was referred to the Narmada Water Disputes Tribunal which

Table 19. Utilisation by Major Water Resources Projects in Narmada Basin, MP

S.N.	Category	Area to be irrigated (Lakh ha)	Water used	
			(M.A.F)	Million cubic m
1	29 Major Projects	14.15	11.35	14, 000
2	135 Medium Projects	6.7	2.89	3, 562
3	3,000 Minor Projects	6.7	2.51	3, 100
4	Total	27.55	16.75	20, 660
5	Domestic & Industrial use	–	1.5	1, 850
6	Grand Total	27.55	18.25	22, 511

Source: NVDA (2005).

was established under India's Interstate Water Disputes Act of 1956. The Tribunal considered the issues for a decade and made its final award in 1979.

This award, which provides for diversion of 11,718.25 million m^3 (9.5 million acre-feet, MAF) of water from the reservoir into a canal and irrigation system, has formed the basis for construction of the current Sardar Sarovar Project. Finances for this ambitious project were secured in 1985 when the World Bank entered into credit and loan agreements with the Governments of Gujarat, Madhya Pradesh, and Maharashtra. It provided U.S. $ 450 million for the construction of the dam and the canal. The construction of the dam began in earnest in 1987. Another major project under construction upstream of Sardar Sarovar is the Indira Sagar project.

Sardar Sarovar Project is a multipurpose interstate project of 4 States (Madhya Pradesh, Gujarat, Maharashtra and Rajasthan) being implemented by Govt. of Gujarat. It is an ambitious and technologically complex irrigation scheme which is to draw upon the flow of the Narmada River to alleviate the water needs of large areas of the state of Gujarat. The project, which is one of the largest water resource projects ever undertaken in India, includes a dam, a riverbed powerhouse, a main canal, a canal powerhouse, and an irrigation network. Its projected impact extends over a large area, and it will potentially affect 25–40 million people. The components of the project are designed to irrigate a vast area of Gujarat and Rajasthan (although not a basin state, was also later allocated a share of its waters), and to provide drinking water to areas of central and northern Gujarat. The water is to be delivered by creating a storage reservoir on the Narmada River with a full reservoir level of 138.684 m (455 feet), along with an extensive canal and irrigation system.

The project comprises construction of a 163 m high and 1,200 m long concrete gravity dam across Narmada River near village Kevadia of Distt. Baroda. The top of the dam will be at 146.5 m and the spillway crest level at 121.92 m. The M.W.L. of the Dam is EL 140.21 m. The live storage capacity of the reservoir will be 5,800 MCM. The 458 km long lined canal will irrigate 17.92 lakh hectares of land in Gujarat and will also provide 616 MCM share of water to Rajasthan. The installed capacity of the Bed River Power House is 120 MW and that of Canal Head Power

House is 250 MW. The project will also cater domestic water supply needs of 135 towns and 8,215 villages of Gujarat.

At the SSP dam, the maximum spillways discharge capacity is 84,950 cumec. There are 23 bays in the main spillway whose crest gates are 18.30 m × 16.76 m. In the auxiliary spillway, there are 7 bays with radial crest gates of size 18.30 m × 18.30 m.

Based on the analysis and after considering the possibilities of transposition and the different alternate synthetic combination of storms for design purposes, the storms given in Table 20 were recommended for Sardar Sarovar Project.

The dam is being constructed in a hilly region, and the reservoir created behind the dam will resemble a narrow lake extending from the dam over 200 km upstream, submerging approximately 37,533 ha of land in three states: Gujarat, Maharashtra, and Madhya Pradesh. Out of this, 11,300 ha is agricultural, 13,385 ha forest land, and the rest consists of river bed and waste land. While the full impact of the project remains in dispute, it is generally acknowledged that 248 villages will be submerged (33 villages of Maharashtra, 19 of Gujarat, and rest in MP) mostly partially, affecting about 100,000 people. Many of these people, especially in Gujarat and Maharashtra, are considered to be 'tribals' and have no formal title to their land. The number of families likely to be effected due to submergence, based upon 1991 census, are estimated as 40,727; out of these 33,014 are in Madhya Pradesh. Gujarat will be required to resettle 14,124 families of Madhya Pradesh in the command area of the project in Gujarat. The remaining 18,890 families will be resettled in Madhya Pradesh.

A large number of farmers, about 140,000 according to an estimate, will lose land to the canal and irrigation systems. In addition, thousands of people living downstream will find their lives affected by the project. Weigh this up against the benefits: irrigation of 1.8 million ha, 1,450 MW of hydroelectric power, drinking water to 135 towns and 8,215 villages (some of these suffer frequent droughts), flood protection for 210 villages with an aggregate population of 750,000 and other less important benefits. The area that will be submerged is about 1.65% of the area that will get benefits. The ratio of population displaced to the population benefited is 1:37. Generation of wealth in an area also contributes to general economic development of the area.

Table 20. Design storms recommended for Sardar Sarovar

No. of day/Date	Depth in mm
1st day (28.08.1973)	28.0
2nd day (29.08.1973)	63.0
3rd day (30.08.1973)	80.0
4th day (05.09.1970)	61.0
5th day (06.09.1970)	72.0

Source: NIH (1999).

The Narmada main canal will be the largest of its kind in the world, extending 450 km to the Rajasthan border and crossing 19 major rivers and 244 railway lines or roads. With 31 branch canals, the aggregate length of the distribution system will be 75,000 km which will require approximately 80,000 hectares of land. The main canal will be 250 meters wide at the head and 100 m wide at the border: the capacity of this canal system is such that it will be able to empty the proposed reservoir storage in less than two months. The canal will also transport Narmada water to Saurashtra and Kutch region of Gujarat which are drought prone areas. Many wild life sanctuaries and parks will get water from the project. Figure 7 shows a view of the main Sardar Sarovar dam (under construction).

Only one-sixth of the project cost is for construction of the dam. An additional equal amount is required for hydro-power installation at the dam and canal bed powerhouse. The other one-third of the cost is for the main canal and the rest is for development of the irrigation network in the command area. For different levels of irrigation efficiency, the internal rate of return was between 16.77 and 21.88 on economic prices of the inputs and outputs and the corresponding benefit-cost ratios between 1.59 and 3.29 (Chitale, 1997). The acceptable levels are 9 and 1, respectively.

The Controversy

The complexity of this project allows for considerable dispute about how to best calculate and compare the projected costs and anticipated benefits. The project's

Figure 7. Main Sardar Sarovar dam (Under construction)

proponents emphasize the enormous benefits it is expected to bring to millions of people at the cost of displacing comparatively few. They ask that these projected benefits – the provision of drinking water to as many as 40 million people and the irrigation of 1.8 million hectares of land – be weighed against the relatively small number of people who will be displaced from land. From their perspective this balance tilts heavily in favour of the project.

The opponents of the project question both the projected benefits and the cost of the project. They argued that the irrigation benefits have been vastly overestimated and that adequate drinking water may never reach the most needy drought-prone areas of Gujarat, such as Kutch. (It may be noted that water has already reached drought prone areas of Gujarat and has considerably eased the situation). According to them, the economic costs are based on unrealistic figures and have been grossly underestimated. Also, the human and environmental costs have been vastly underestimated or ignored by mis-reporting the number and the extent to which people will be affected by the project, and disregarding the costs of cultural disruption that will occur when tribal people are moved from their traditional lands. In their opinion, the number of people to be affected must include those living in the submergence area, those displaced by construction of infrastructure, those affected by the canal, and those living downstream whose lives and livelihoods will be affected. The reservoir submergence has ignored the adverse effect on the forest cover.

In the Sardar Sarovar Project, the major concern about compensation has focussed on the category of people identified as "oustee". An oustee is an individual "whether landed or landless, who since at least one year prior to the date of publication of the notification under the relevant Indian Land Acquisition Act, has been ordinarily residing, or cultivating land, or carrying on any trade, occupation, or calling or working for gain in Gujarat, Madhya Pradesh, and Maharashtra, who would be displaced from his usual habitat due to the carrying out of the Project." Two factors affect the compensation to which the Sardar Sarovar Project oustee is entitled. One is the assessment of an oustee's right to compensation, which is complicated by disparity between the way the government administers, registers, and taxes land and the way people conceive of and use resources. Secondly, the R&R policies of the three states of Gujarat, Madhya Pradesh, and Maharashtra vary in the compensation they give to oustees. The policy of Gujarat is considered to be the most 'lucrative'. Consequently, the oustees want the states of Madhya Pradesh and Maharashtra to match their policies with those of Gujarat.

The three states have different norms for treatment of "major" sons (sons over the age of 18), encroachers (those residing on and cultivating land to which they do not have legal title), and the landless. The Gujarat policy makes no distinction between landed and landless oustees and offers full benefits to major sons. According to the award of the Narmada Water Disputes Tribunal, all people displaced by the Sardar Sarovar reservoir have the right to settle in Gujarat, if they so desire. Naturally, so long as Gujarat's resettlement and rehabilitation benefits are significantly better than those offered by the other two states, most displaced people are apt to take this option. Hence, Madhya Pradesh resettlement plans anticipate that only 10 percent of the displaced people from that state will remain in the state. An unequal compensation is seen to violate the spirit of the Tribunal Award which provides that all oustees may remain in their home states. To do so under current policies would entail a financial sacrifice for some, while relocation to Gujarat would mean for many "a long cultural journey." Therefore, while the right of choice still exists in principle, the disparity in benefits means a choice between migration to another state or a lower standard of living.

The concern of compensation dominates the discussions of R&R; there are disagreements about what constitutes full, fair, and appropriate compensation, and further disagreements about whether all the oustees will or can be fully, fairly, and appropriately compensated under prevailing circumstances. The proponents argue that displacement should be treated as a development opportunity and that project-affected people should not only regain their standard of living but also be treated as the first beneficiaries of the project. Clearly, high costs attached to R&R reduce the cost-benefit ratio, making it more difficult to raise political and financial support. Even in cases where project benefits make it possible to offer attractive R&R packages, high compensation is opposed by the decision-makers fearing that this will set a precedent of high awards which could not be met by all future projects. Note that the oustees mostly belong to marginal and disempowered communities and R&R requires that all people affected by the project must 'improve or at least regain the standard of living they were enjoying prior to their displacement'.

The Protests by NGO's

Meanwhile, a number of Indian Non-Governmental Organizations (NGOs) began opposing the project, mainly on the grounds of environmental, human (tribal) rights, the skewed economics of the project, etc. The *Narmada Bachao Andolan* (NBA) is the main opponent. They were later joined by several foreign NGOs. Some NGOs, such as *Action Research in Community Health Association* (ARCH), who were originally opposed to the project because of insufficient compensation for the affected population, later accepted the improved measures and supported the continuation of construction.

Originally, the campaign against the Sardar Sarovar Project revolved around resettlement issues. Earlier disagreement between the states, mainly about water allocation and the height of the proposed dams, had been resolved amicably by the middle of 1974. After certain clarifications, this was made an award by a Tribunal under the Inter-State Water Disputes Act at the end of 1979. This Tribunal also set out a resettlement and rehabilitation scheme which, at that time, was considered very liberal though the World Bank insisted on even higher standards. Landless farm laborers and so called "encroachers" were to receive 2 ha compensation, often more than the area possessed by the communities among whom they were to be resettled. It had been planned to resettle some of the displaced people on degraded forest land. In the meantime a new "Forest Conservation Act" had been passed in 1980, forbidding any forest land to be diverted to other purposes, unless specifically sanctioned by the Central Government. This led to a shortage of land needed for resettlement.

Gradually, NGOs started to employ radical protest techniques, such as marches, hunger strikes, traffic disruptions, and intimidation of those wanting to be resettled. They even instituted a "mass drowning rather than relocate" campaign at the first village threatened by the rising waters. Much was made of the fact that many of the so called "oustees" were still tribal. A lot of incitation of these relatively primitive people took place. The proponents, however, claim that the tribal families have shown a desire to avail of this development opportunity. The NGOs also started canvassing for foreign support through newspaper articles, talks, and petitions to heads of the governments, the UN, and the donor agencies. All this led to Japan withdrawing its financial support to the project. The World Bank appointed an Independent Review Mission, the first of its kind in the history of the Bank. The report of this mission was interpreted variously by various people; it was criticized by many, e.g., Alagh and Buch (1997), and apparently further complicated the matter. After the World Bank withdrew from the project, the Government of India decided to proceed without external help.

The opponents managed to mobilize a substantial number of people in India and elsewhere against the project, but a large scale popular support for it was shown by about a million people turning up for a pro-project demonstration. Public issue of bonds was floated by the Government of Gujarat to mobilize funds for the project. All the issues were over-subscribed, showing public support to the project.

Project Status

More than 98% of the excavation and 90% of concreting for the main dam were over by 2000. The matter regarding the final height of the dam was considered by the Supreme Court of India. On October 18, 2000, the Supreme Court of India delivered its judgment on the Sardar Sarovar Project. In a 2 to 1 majority judgment, it allowed immediate construction on the dam up to a height of 90 m. Further, the judgment authorized construction up to the originally planned height of 138 m in 5-meter increments subject to receiving approval from the Relief and Rehabilitation Subgroup of the Narmada Control Authority. In 2002 summer, the dam height was allowed to be raised to 95 m. By June 2004, construction of the dam up to EL 110.64 m was completed. Canal head powerhouse has been completed and all five units of 50 MW are ready. With the lowest block of the dam attaining an elevation of 110m, 450,000 ha area are receiving irrigation water and power generation will also commence. As per the schedule, the project in all respects is expected to be completed by 2010.

First unit of riverbed powerhouse is likely to begin work in 2006. Construction of Narmada Main Canal up to 357 km is over. As per the plans, the dam is expected to be completed to the EL 138.68 m (top of the dam) by 2006.

Water Supply through Narmada Canal Major urban centers in Saurashtra region have been badly hit with underground water level in most parts going down to 800 to 1,000 feet. Even at that level, it is mostly brackish water. In summers, Rajkot city, the main nerve center of the Saurashtra region, gets just about a couple of hours supply twice a week. Jamnagar city, the second largest in the region, gets once in three days, and Junagarh every fourth day.

A huge pipeline network based on the SSP's main canal has been envisaged to supply drinking water. The configuration of the system is shown in Figure 8 Only 790 of the total 8,215 villages to be given drinking water from the SSP are in Ahmedabad and the Panchmahals districts. The rest are in Saurashtra, Kutch and North Gujarat. Similarly, 120 towns from the total 135 to be supplied drinking water are in the three regions. Twelve are in Ahmedabad and three in the Panchmahals. Water from the Mahi and the Narmada rivers is to be taken through a pipeline from the Pariyej lake in Kaira district where the water will flow through the Mahi canal and will be distributed at different points in the Saurashtra region to supply 275 million litres of water per day to 1,625 villages and some towns in Ahmedabad, Amreli, Bhavnagar and parts of Rajkot districts.

An escape for Mahi Right Bank Canal (MRBC) had been created from Narmada main canal at chainage of 150 km. The branch canals named, Shedhi and Limbhasi take off from MRBC and end up at Pariej and Kaneval tanks, respectively. A trunk line takes off from Pariej/Kaneval reservoirs and reaches Pipli pumping station. The 103 km long Saurashtra Branch Canal (SBC) takes off from Narmada Main canal at chainage 256.88 km. A pumping station at Dhanki lifts water from SBC at chainage 67 km and supplies it to Maliya Branch Canal (MBC). The total length of all trunk pipelines is 2,722 km. The main pumping stations to lift water for

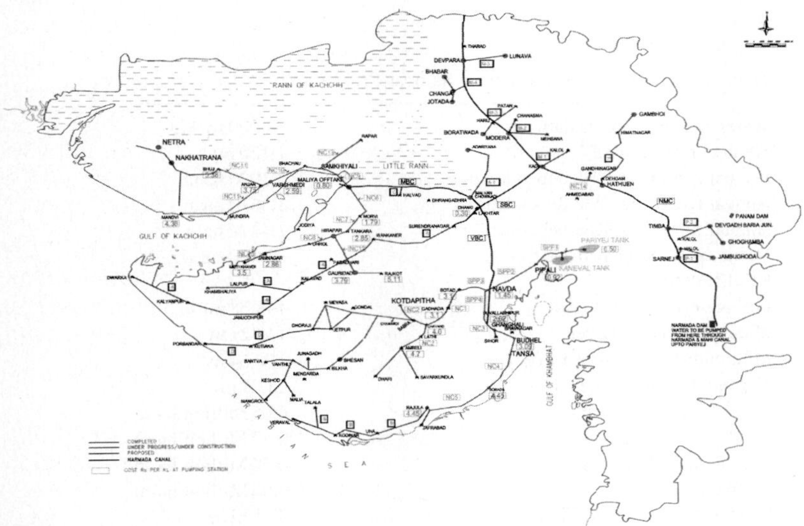

Figure 8. Configuration of Narmada Canal based water supply system

the Narmada-Mahi based Saurashtra pipeline projects are: two pumping stations at Pariej and one at Kaneval. A well organized administrative set up has been created for the drinking water supply projects. *Pani Panchayats* are being created at village level and they are responsible for O&M of village level facilities with technical assistance from Gujarat Water Supply and Sewerage Board (GWSSB). A SCADA (Supervisory Control And Data Acquisition) System for better control on O & M and running the system with optimum efficiency is being planned. Further details of the system are available in Talati and Kumar (2005).

Significantly, the cumulative benefits of the project are not reflected in the dam's command area. At present 40 percent water from whatever little is received by all major dams in Saurashtra – has to be reserved for drinking water. Once that problem is taken care of, more water will be available from these dams for irrigation. This way, actually irrigation spread will improve, while the SSP is already going to irrigate nearly 18 lakh ha of land.

Many websites contain information about this project, for example, Sardar Sarovar Narmada Nigam Ltd. (*www.sardarsarovardam.org)*, Narmada Control Authority (*www.nca.nic.in*), Narmada Valley Development Authority *http://www.nvda.nic.in/sardarsarovar.htm* and an NGO's site *www.narmada.org/sardarsarovar.html*.

Some relevant features of the Sardar Sarovar Project are given in Table 21.

11.8.2. Indira Sagar Project (ISP)

The Indira Sagar (also known as Narmada Sagar) Project (Punasa, Khandwa District) is a multipurpose project of M.P. on the Narmada River upstream of

Table 21. Salient features of Sardar Sarovar Project

Watershed area above dam site	88,000 sq. km
Mean annual rainfall	1,120 mm
Annual run-off at 50% dependability	4.10 M ha m
Annual run-off at 75% dependability	3.36 M ha m
Annual run-off at 90% dependability	2.44 M ha m
Designed flood (1 in 1000 years)	87,000 cumec
RESERVOIR	
Full Reservoir Level (FRL)	138.68 m
Maximum Water Level	140.21 m
Minimum Draw Down Level	110.64 m
Normal Tail Water Level	25.91 m
Gross Storage Capacity	0.95 Million ha m
Dead Storage Capacity	0.37 Million ha m
Live Storage Capacity	0.58 Million ha m
Annual evaporation	0.06 Million ha m
Land Submergence at FRL	37,533 ha
Number of villages affected at FRL	244
Number of families affected at FRL	40,963

Sardar Sarovar Project. Its importance arises from the fact that the projects proposed downstream of it, i.e., Omkareshwar, Maheshwar and Sardar Sarovar would be able to attain their full potential of irrigation and power generation only after getting regulated releases from ISP.

The Indira Sagar Project envisages construction of a 92 m high and 653 m long concrete gravity dam with a surface power house of 1,000 MW installed capacity (8 × 125) and a 249 km long canal to provide irrigation in 1.23 lakh ha of C.C.A. in the districts of Khandwa and Khargone. On completion, a reservoir of 9,744 MCM live storage capacity will be created. Submergence is likely to affect 249 villages, 30,739 families, 40,332 ha of forest and 510,016 ha of other lands. Compensatory afforestation has to be done in 80,945 ha area and catchment area treatment in 30 sub-catchments totaling 62,975 ha in the directly draining area, chargeable to the project. The salient features of Narmada Sagar Project are given in Table 22. A view of the Indira Sagar Project is shown in Figure 9.

Table 22. Salient features of Narmadasagar Project

Particulars	Details
Catchment area (km^2)	61,880
Inflow from immediate upstream (MCM)	740.09
Gross storage (MCM)	12,211.45
Live storage (MCM)	9,745.00
Dead storage (MCM)	2,467.00
Design flood (cumec)	65,147
Maximum water level (m)	263.35
Full reservoir level (m)	262.13
Minimum drawdown level (m)	245.06

Figure 9. Indira Sagar Project (Under Construction)

Based on the analysis made by NIH and after considering the possibilities of transposition and the different alternate synthetic combination of storms for design purposes, the following storms were recommended for Narmada Sagar which was agreed upon by IMD and CWC. The design storms recommended for Narmada Sagar are given in Table 23.

11.8.3. Omkareshwar Project

The Omkareshwar multipurpose project is situated downstream of ISP on the main Narmada River, near village Mandhata, District Khandwa. The project envisages construction of a 53 m high and 949 m long concrete dam with gated spillways to irrigate 1.468 lakh ha of C.C.A. through a 142 km long L.B.C. and a 64 km long R.B.C. and a 83 km lift canal. Riverbed powerhouse of 520 MW installed capacity (8×65 MW) is proposed on the right bank and the annual generation is likely to be 1166 MU in a 90% dependable year. The reservoir will have approximately 300 MCM live storage capacity, the submergence will affect 30 villages, 5,829 ha forestland, 4,059 ha of private and revenue lands and 3,024 families. Compensatory afforestation in 11,660 ha and catchment area treatment in 79,886 ha has to be done. The cost of the project at 2002 price level is estimated at Rs. 2,225 crore. The construction work is in progress and is expected to be over by Feb. 2008. The power component of project is being taken up by a joint venture with NHPC. The salient features of Omkareshwar Project are given in Table 24.

11.8.4. Maheshwar Hydel Project (M.P.)

The Maheshwar hydel project, located about 40 km downstream of Omkareshwar multi-purpose project on the main Narmada near Mandleshwar town of Madhya Pradesh, envisages construction of a 35 m high concrete dam with a 670 m long spillway. The dam will have earthen flanks on the left and right banks of lengths 1,573 m and 464 m, respectively. There will be a surface powerhouse of 400 MW (10×40 MW) on the right bank.

This project has been proposed in the private sector and the work has been awarded to Shri Maheshwar Hydel Power Corporation Limited (SMHPCL).

Table 23. Design storms recommended for Narmada Sagar

No. of day/Date	Depth in mm
1st day (28.08.1973)	33.8
2nd day (29.08.1973)	100.6
3rd day (30.08.1973)	99.8
4th day (15.07.1944)	90.2

Source: NIH (1999).

Table 24. Salient features of Omkareshwar Project

Particulars	Details
Catchment area (km^2)	64,880
Inflow from immediate upstream (MCM)	1,973.57
Gross storage (MCM)	1,500.00
Live storage (MCM)	811.63
Dead storage (MCM)	688.28
Design flood (cumec)	43,639
Maximum water level (m)	202.7
Full reservoir level (m)	201.2
Minimum drawdown level (m)	193.5

A power purchase agreement was signed between SMHPCL and MPEB in 1994. The revised estimate of the project has pegged to cost to Rs. 1,569 crores. The project work is under progress. The salient features of Maheshwar Hydel Project are given in Table 25.

11.8.5. Rani Avanti Bai Sagar (Bargi) Project

Bargi (later renamed as Rani Avanti Bai Sagar Project) is a major masonry earth scheme in the head reaches of Narmada River. The project consists of the Bargi dam on the Narmada River near village Bargi in the Jabalpur district 43 km away from Jabalpur city. The latitude and longitude of the dam are 22°56′30″ N and 79°55′30″ E, respectively. The catchment area up to dam site is 14,556 sq. km and the average annual rainfall in the catchment is 1,414 mm. It is a 69 m high and 5,337 m long composite gravity dam completed in 1980's. The maximum height of the masonry dam is 69.80 m while that of earth dam is 29 m. The catchment area at the dam site is 14,556 sq. km. The gross, live and dead storage capacity of the reservoir is 3.92 billion cubic meter (B Cum), 3.18 B Cum and 0.740 B Cum

Table 25. Salient features of Maheshwar Hydel Project

Particulars	Details
Catchment area (km^2)	69, 184
Inflow from immediate upstream (MCM)	2, 797.53
Gross storage (MCM)	488.46
Live storage (MCM)	28.37
Dead storage (MCM)	462.55
Design flood (cumec)	49, 695
Maximum water level (m)	164.9
Full reservoir level (m)	162.8
Minimum drawdown level (m)	162.15

Source: Gupta (2001).

respectively. The maximum water level, full reservoir level and the dead storage level of the reservoir are at 425.70 m, 422.76 m and 403.55 m, respectively. The estimated life of the reservoir is 100 years.

The project has been envisaged as a multipurpose scheme meant to serve for water supply for domestic and industrial purposes, irrigation and hydropower generation. One canal system (left bank canal system) is nearing completion and one power plant (River bed power plant) of 2 units of 45 MW each has already been completed for utilizing the stored water of the Bargi reservoir. It has a firm power of 45 MW. The left bank canal takes off from the left flank of the Bargi dam and covers a distance of 137.2 km. The design capacity of the canal is 124.65 cumec. The command area under Bargi LBC lies in Jabalpur and Narsinghpur districts of Madhya Pradesh. There are 4 branch canals, 11 distributaries, 16 minors and 4 sub-minors serving the complete command.

The gross and culturable command area of the left bank canal is 2.574 and 1.57 lakh ha, respectively. Other canal system (right bank canal system) is under construction. The purpose of the right bank canal is to transport 116 M Cum of water for domestic use and around 2,300 M Cum of water for irrigation and interbasin transfer annually. One power house (Canal power house) with a capacity of 15 MW is also proposed to be constructed in the left bank canal system. Annual water requirement from the reservoir through the left bank canal for domestic water supply and irrigation is 54 M Cum and 2,160.1 M Cum, respectively. In addition, annual firm energy requirement from the reservoir is 363 Mkwhr. It has a firm power of 45 MW.

The peak of the design flood hydrograph is 51,510 cumec. According to the reservoir operation manual for the Bargi dam, no provision is made for flood moderation by operation of this dam. The length of river channel between the Bargi dam and the Hoshangabad city is approximately 265 km. For discharging excess water during the flood season, a 385.72 m long spillway has been provided in the centre of the masonry section. 21 nos. of radial gates of size 13.71 m length and 15.25 m height have been provided on this spillway. The shape of overflow section of the spillway has been designed to conform to the shape of lower nappe of water flowing over a sharp crested vertical edge. The spillway has been designed to pass a flood hydrograph having the base period of 7 days and peak inflow of 45,296 cumec. The design flood hydrograph is based on the unit hydrograph concept. The maximum water level at the Bargi dam site has been limited to 424.28 m because of the Mandla township upstream of the Bargi dam.

The water availability at the dam site was estimated based on the discharge data at the Jamtara gauge site. The Jamtara gauge site is located 16 km downstream of the Bargi dam site and has a catchment area of 16,576 sq. km. Systematic gauging has been done at this site since 1949. The average annual inflow at the dam site is 7,197 M Cum. There are 13 raingauge stations in the catchment area. In some of the raingauge stations, rainfall data for nearly 89 years has been recorded. The average annual rainfall in the catchment up to Jamtara is 1,414 mm. Rainfall in the basin mostly occurs during the monsoon months (July to October). During monsoon, 94.09% of the total rainfall of the calendar year occurs.

The reservoir has been classified as hilly. The shape of the reservoir is almost longitudinal. Its longest periphery from the axis is about 80 km. The width at 16 km and 25 km from the axis is 16 km and 3.2 km, respectively.

11.8.6. Tawa Dam

The first major work to be undertaken in the Narmada basin was the Tawa Project. The Tawa project was constructed in 1974 on the Tawa River near village Ranipur in Hoshangabad district. Tawa is a left bank tributary of Narmada River and the catchment area up to dam site is 5,983 sq. km. The dam is operated to supply water for irrigation and municipal uses. The average annual rainfall for the catchment is 1,564 mm. and the 75% dependable flow is 3,075 MCM. The gross storage capacity at FRL (355.397 m) is 2,310 MCM while the live storage capacity is 2,050 MCM. The maximum water level for this dam is 356.66 m. The peak of design flood hydrograph for the dam is 36,800 cumec. The actual observed maximum flood was 24,300 cumec in 1961.

It is a composite dam of earth and masonry 1,630 m long with two dykes of 185 m length each and of a maximum height of 57.95 m. The gross storage in the reservoir is 2,311 m cu. m of which the live storage is 2,087 m cu. m. The dam is operated to supply water for irrigation and municipal uses. There are canals, one on each bank. The left bank canal is 120 km and the right bank canal 76.85 km long. The total irrigation is expected to be of the order of 331,854 ha. The project will provide irrigation to about 121,406 ha in the first phase. There is a facility to generate hydropower through a power house with installed capacity of 13 MW.

Waterlogging study in Tawa basin was made by NIH in 1988. The results show that an area of 80 sq. km was affected by waterlogging and about 140 sq. km area where water table lies in between depth of 1 to 3 m was prone to waterlogging. It is suggested that periodic assessment of waterlogging using remotely sensed data should be carried out on a regular basis.

11.8.7. Barna Dam

The dam is located near village Bari of Tehsil Bareli (near National Highway 12, connecting Bhopal to Jabalpur) in Raisen district and was completed in 1978. It is located on Barna River which is a right bank tributary of Narmada River. The total catchment area of the dam is 1,176 sq. km. The gross storage capacity at FRL (348.55 m) is 539.00 MCM and the live storage capacity is 455.80 MCM. Barna dam is 432 m long and 47.7 m high at the deepest section. The main canal will be 38 km long and it will irrigate 60,290 ha. The design flood of the dam is 13,557 cumec. The maximum discharge observed at the dam site in 1965 was 11,480 cumec.

Barna project is an irrigation project of Madhya Pradesh. On the right of the main dam, there is a saddle dam from which a 0.67 m long joint water carrying canal emerges. From this canal two branch canals take off from the left and right banks for irrigating the command area falling on its left and right banks, respectively. The irrigation infrastructure has greatly improved the irrigation potential in the command and benefited the farmers.

The project does not provide flood protection. However, through judicious and cautious reservoir operation, peaks can be moderated to some extent to subside the damage and submergence of life and properties in the downstream of the dam site, especially near Barlei township.

11.8.8. Kolar Dam

Kolar dam has been constructed near the Lawakheri Village. It has a gross storage capacity of 270 MCM and live storage of 260 MCM. Water from the reservoir is being used to provide drinking water to the city of Bhopal which lies at a distance of 30 km towards north. Water from the dam is envisaged to provide irrigation to an area of 610 million sq. m. For this purpose, a barrage has been constructed in the basin near Jholiapur from where two canals take off.

In addition to the above projects, there are some other small projects and projects are under construction. Salient features of existing and under construction water resources projects with a live storage capacity equaling and exceeding 10 MCM have been presented in Table 26 and Table 27, respectively.

Table 26. Salient features of selected existing projects in Narmada Basin

Name of the Project	State	Year of completion	Gross storage capacity (million m^3)	Live storage capacity (million m^3)	Designed Annual Irrigation (million m^2)	Installed capacity (MW)
Sukhi	Gujarat	-	176.85	166.60	250.00	–
Wadhwan	Gujarat	1910	12.84	12.00	142.00	–
Bahori Bund	Madhya Pradesh	1929	36.84	34.51	64.80	–
Chandrakeshar	Madhya Pradesh	1976	30.08	28.47	70.40	–
Depalpur	Madhya Pradesh	1952	39.93	37.94	9.30	–
Dukri Kheda	Madhya Pradesh	1956	12.26	11.51	26.70	–
Pariat	Madhya Pradesh	1927	20.06	18.50	–	–
Sakalda	Madhya Pradesh	1987	15.65	12.67	–	–
Sampna Tank	Madhya Pradesh	1956	16.92	14.30	385.00	–
Satik	Madhya Pradesh	1964	19.80	18.36	14.60	–
Simrar Tank	Madhya Pradesh	1967	12.94	12.29	25.00	–
Satpura Dam	Madhya Pradesh	1967	95.60	71.70	–	–
Sukata	Madhya Pradesh	1983	89.40	78.14	176.40	–

Table 27. Salient features of selected Under Construction Projects in Narmada Basin

Name of the Project	State	Gross storage capacity (million m^3)	Live storage capacity (million m^3)	Designed Annual Irrigation (million m^2)
Ani	Gujarat	37.20	34.10	23.00
Dholi	Gujarat	13.12	11.47	–
Karjan	Gujarat	657.00	583.30	780.00
Men	Gujarat	37.90	34.34	40.50
Saktha	Gujarat	89.37	75.30	–
Shankra	Gujarat	11.68	10.33	16.40
Choral	Madhya Pradesh	23.92	19.23	39.00
Dajla-Dewada Project	Madhya Pradesh	56.35	50.29	121.00
Matiyari	Madhya Pradesh	56.80	51.12	101.00
Sakalda	Madhya Pradesh	15.26	12.36	–

11.9. FLOODS, WATERLOGGING AND DRAINAGE

In Madhya Pradesh, Narmada flows in a deep channel with high banks, which are not frequently over-topped. Below the Hiranphal-Navagam gorge, Narmada spreads out over the flat Gujarat plains and flows as a wide stream with low banks for the rest of its course down to the Gulf of Cambay. Floods do occur frequently in this reach and cause damage to life and property.

Regional flood frequency analysis of two hydro-meteorological sub-zones of zone 3, namely (i) Lower Narmada and Tapi sub-zone 3(b); and (ii) Upper Narmada and Tapi sub-zone 3(c), was carried out by NIH (NIH 1999). Long term annual maximum peak flood data records and areas of various bridge sites under these sub-zones formed inputs to these studies. Summary statistics of catchments of the sub-zones 3(b) and 3(c) are given in Table 28.

Based on analysis, the following regional flood formulae for the two sub-zones were determined:

Table 28. Salient features of various catchments of the sub-zones 3(b) and 3(c)

Sub-zone	Area of sub-zone (km^2)	Number of bridge sites in the sub-zone	Range of catchment area of bridge sites (km^2)	Range of mean annual peak flood (m^3/s)	Range of record length (Years)
3 (b)	77,000	19	17.22–1017.00	34.95–558.29	12–28
3 (c)	86,353	15	41.80–2110.85	111.95–1730.53	14–30

Sub-zone 3(b)

$$Q_T = \left[61.3 \left(-\ln \left(1 - \frac{1}{T} \right) \right)^{-0.20} - 46.9 \right] A^{0.46} \tag{2}$$

Sub-zone 3(c)

$$Q_T = \left[52.2 \left(-\ln \left(1 - \frac{1}{T} \right) \right)^{-0.11} - 44.3 \right] A^{0.67} \tag{3}$$

11.9.1. Flood Control by Reservoirs in Upper Narmada Basin

The runoff factor for the 75% dependable flow of Narmada works out to be 0.29. The annual rainfall in the upper part of the Narmada catchment is more than 1,400 mm and in some pockets it exceeds 1,650 mm. The Narmada catchment up to Hoshangabad received unprecedented rainfall in 1999. In the catchment up to Hoshangabad, three major dams have been constructed: Bargi, Barna and Tawa. Consequently, a large part of the Hoshangabad city remained flooded for considerable time, causing severe damages.

The reservoir regulation manuals for Bargi dam recommends filling of the reservoir in stages and rule levels which are to be maintained at fortnightly intervals during the monsoon season have been specified. According to this manual, the reservoir level has to be filled up to 410.0 m by 30th July and up to 422.76 m (FRL) by 15th September.

According to the reservoir operation and maintenance manual for Bargi, the reservoir does not provide for flood protection. However, through judicious and cautious reservoir operation, peak flood can be moderated to some extent. During very high floods when reservoir water level is at FRL, the incoming floods are to be passed over spillway taking care that the rate of outflow does not exceed the rate of inflow to the reservoir. The absorption between FRL (422.76 m) and MWL (425.7 m) is 884 MCM. According to this manual if the rate of rise of reservoir water level is restricted to 10 cm per hour, the incoming flood can be safely moderated and regulated for a prolonged period of about 30 hours without crossing MWL.

In the second half of September 1999, a severe storm occurred in the upper parts of Narmada Basin. The normal monthly rainfall for the Bargi catchment for the month of September is 208 mm. In 1999, the average rainfall over the Bargi catchment from 13.09.1999 to 19.09.1999 was 337 mm. According to the operation manual, the Bargi dam is to be kept at 422.76 m on September 15 and in 1999, the actual reservoir level on September 15 was 422.85 m. For the Tawa dam, the normal annual rainfall is 1,564 mm while the average rainfall in Tawa catchment from 12.09.1999 to 22.09.1999 was 610 mm.

The observed discharges at Hoshangabad during September 18–23 were in the vicinity of 25,000 cumec and this high flow has caused flooding in the city and severe damages. The discharge at Sandia was probably higher than this; it was so high between 16 to 22 Sept. that measurements could not be taken. The one-day PMP for the upper Narmada catchment is about 60 cm as given in the meteorological atlas by Indian Institute of Tropical Meteorology, Pune. Daily rainfall depths exceeding 180 mm have taken place in this storm.

As noted above, the live storage capacity of the Barna dam is very small and no significant control of flood is possible by regulation of this dam. The catchment area at Bargi dam is 14,556 sq. km, at Tawa 5,983 sq. km and the free catchment up to Hoshangabad is 23,337 sq. km.

In the worst case scenario when one-day PMP takes place over the entire catchment, the flow at Hoshangabad would be 222,000 cumec. In case the rain is over the catchment of Tawa and Bargi the flow at Hoshangabad will be 102,946 cumec. Similar calculations were carried out for the rainfall at the rate of 90% of PMP etc. It was found that when the one-day rainfall is about 25% of PMP and it takes place only over the Tawa and Bargi catchments, the flow at Hoshangabad would be close to 25,700 cumec which is of the same order as the discharge experienced in the September 1999 floods. It is also seen that if this rainfall (15 cm) in one day takes place over the intermediate catchment, it will lead to peak discharge over 42,500 cumec which, when routed, would give a peak discharge of 29,779 cumec at Hoshangabad.

In the storm of September 1999, daily rainfalls of magnitude up to 180 mm were observed at many stations. For example, the rainfall at Sandiya on 16 September was 185 mm, at Mandla on 17 September it was 182 mm, at Hirdayanagar it was 170 mm on 17 September and at Gadarwara on 15 September, rainfall was 160 mm. Thus, rainfall at the rate of over 150 mm per day can be expected in this catchment though it will not be uniformly spread over an area of about 15,000 sq. km. Nevertheless the possibility of generation of flood of such a high order cannot be ignored.

The volume under the discharge hydrograph at Hoshangabad during September 11–23, 1999 was 18,455.6 MCM. Compared to this, the combined live storage capacity of Bargi (3,180 MCM) and Tawa (2,050 MCM) is 5,230 MCM. Therefore, even if the entire live storage of both these dams is available for moderation of floods, the floods that were experienced in September 1999 cannot be significantly moderated and very few options are available in such an eventuality. It is also important to note that the free catchment area at Hoshangabad is 23,337 sq. km. Approximate calculations show that the flows generated by this catchment alone can cause havoc in the Hoshangabad city in the event of a major storm taking place over this free catchment.

For flood management in this catchment, it will be necessary to adopt a combination of structural and non-structural measures. The following measures can be considered to manage the flooding problem at Hoshangabad city:

1. It would be necessary to install and operate a flood forecasting system for the basin. This will help in efficient utilization of the available storage space in the dams for flood moderation. Based on the forecasts, reservoir levels can be pre-depleted to make available additional space to regulate the floods.

2. A warning system to issue flood warnings sufficiently in advance needs to be put in place so that the damages to life and movable property can be minimized.
3. Options to increase the carrying capacity of river channel at critical locations should be explored.
4. The construction of embankments at selected locations will help in protecting the vulnerable areas.
5. It would be necessary to create additional storage space in the basin in the light of the award of NWDT.

11.10. THE NARMADA WATER DISPUTES TRIBUNAL (NWDT)

The Govt. of India constituted the Narmada Water Dispute Tribunal on 6th Oct., 1969 to adjudicate the water dispute regarding the sharing of water resources of Narmada River. The Tribunal deliberated over the issues and gave the final decision in December, 1979. The highlights of the decisions of the Tribunal are:

(i) The utilizable quantum of waters of the Narmada River at Sardar Sarovar dam-site on the basis of 75% dependability was assessed as 28 MAF (34,537 MCM).

(ii) The shares of the four party States in this quantity was fixed as: M.P. 18.25 MAF (22,511.01 MCM), Gujarat 9 MAF (11,101.32 MCM), Rajasthan 0.5 MAF (616.74 MCM) and Maharashtra 0.25 MAF (308.37 MCM).

(iii) The Full Reservoir Level of Sardar Sarovar Dam was fixed at 138.68 m.

(iv) The power benefits from the project are to be shared as: Madhya Pradesh 57%, Maharashtra 27%, and Gujarat 16%. The cost of power component of the project is to be shared by Gujarat, M.P. and Maharashtra in the ratio of their benefits in power.

(v) M.P. has to make a uniform release of 8.12 MAF (10,015.86 MCM) ex-Maheshwar to meet the requirements of Gujarat and Rajasthan from Sardar Sarovar Dam in a normal year having 28 MAF (34,537 MCM) flow.

(vi) The Indira Sagar project is to be taken up by MP and completed with a FRL of 262.13 m (860 ft) either concurrently or earlier than the construction of Sardar Sarovar.

(vii) Gujarat is required to credit to MP each year 17.63% of the expenditure on Indira Sagar dam.

(viii) The allocation of Unit I – Dam and Appurtenant works cost between irrigation & power was done as 43.9% and 56.1%. The irrigation component of the project is to be shared by Gujarat and Rajasthan in the ratio of water allocation for the dam and canal.

(ix) A machinery, viz., Narmada Control Authority (NCA) was set up for implementation of the decision of the Tribunal. A Review Committee was also set up under Union Ministry of Water Resources to review decisions of the Authority.

(x) For efficient, economical and early execution of Unit I & III of Sardar Sarovar Project, a construction advisory committee was set up. The Review Committee was empowered to decide on issues on which there is disagreement in the construction committee.

(xi) The decisions of the Tribunal are subject to review after a period of 45 years from the date of publication of the final award in the gazette of Govt. of India. This means

(xii) Although Rajasthan is not a riparian state as far as Narmada is concerned, it was allocated a part of Narmada water with the concurrence of MP, Gujarat and Maharashtra.

11.10.1. Rules of Regulation

As per the NWDT award, NCA shall frame the rules of regulation and water accounting within the ambit of given guidelines. The procedure for preparing this account and rules of regulation were spelt out by the Tribunal. The Authority shall also ensure implementation of the orders of the Tribunal in respect of quantum and pattern of regulated release by M.P.

The water year is to be reckoned from 1st July to 30th June of the next calendar year. The Authority has to determine the volume of water flowing in the Narmada River and its tributaries in a water year. The utilizable flow in excess or falling short of 34,537 MCM (28 MAF) is to be shared by the four party States in the same ratio as allocation of utilizable quantum of Narmada waters at 75% dependability, i.e., 34,537 MCM.

The requirements at Sardar Sarovar have to be met by releases by MP ex-Maheshwar and by inflows from the intermediate catchment between Narmadasagar and SSP surplus to the requirements of MP and Maharashtra. The releases ex-Maheshwar works out to approximately 10,000 MCM and uniform monthly releases of nearly 83,500 MCM would make up this amount. Although it would be fully known only in October whether the year is normal, surplus or deficit, the releases by MP in the filling period would have to be more or less on the basis of a normal year. As the months of July and early part of August are crucial for Kharif sowing, it is important that regulatory arrangements are made to ensure due share to various parties.

The Authority was directed to review the ten-day releases made by M.P. at least once a month and more often if considered necessary for directing any change in the releases. NCA would also ensure by directing the releases by MP that there is sufficient utilizable water in Sardar Sarovar at all times to meet the requirements of the next ten-days subject to water being available in the storage of M.P. after taking into account the proportionate requirements of M.P. For this purpose, Gujarat and Rajasthan are required to intimate their requirements of the 10-day period well in advance.

The Narmada Control Authority was directed by the Tribunal to determine, from time to time, the volume of water stored by each state in reservoirs and other storages and it may, for that purpose, adopt any device or method. Further, the water available in the live storage of the various reservoirs on 30th June shall be reckoned as an inflow to be shared in the next water year.

Note that the apportionment relates to actual withdrawals and not consumptive use. The available utilizable waters on any date will be inclusive of return flows and exclusive of evaporation losses in various reservoirs. The Tribunal has assessed the utilizable quantity of water at 75% dependability on the Narmada at Sardar Sarovar dam site as 28 MAF (34,537 MCM) assuming the inflow at 75% dependability as 27.01 MAF (33,316.29 MCM) and bringing this up to 28 MAF (34,537 MCM) considering regeneration, carryover and evaporation loss. As per the NWDT report, evaporation loss is 4 MAF (4,934 MCM), regeneration or return flow (+)2 MAF (2,467 MCM) and carry over (+)3 MAF (3,700 MCM). The Tribunal did not specify how regeneration is to be worked out but as per the NWDT report, regeneration including return flow should be taken as 10% of irrigation use in upstream major, medium and minor projects in any month with a lag of one month. Further, 60% of the water used for domestic and industrial purposes within Narmada basin may be taken as return flow uniformly available throughout the year.

The surplus water shall first be utilized for filling up the reservoirs to their capacity and further extra surplus water should be utilized for irrigation and other purposes only after this has been ensured.

After meeting the storage requirements and withdrawals, the surplus waters in the filling period which would go waste to sea even without generating power can be allowed to be utilized by party States to the extent they can. Gujarat is required to inform NCA and the designated representative of all the concerned states whenever water starts going waste to sea as also when flows cease. During the period of such flows, the party States, whose reservoirs are spilling and the spill water cannot be stored elsewhere, may utilize such flows from the said reservoirs as they like and such utilization will not count either towards allotment of supplies to them nor will it establish prescriptive rights.

11.10.2. Resettlement and Rehabilitation Activities

The Resettlement and Rehabilitation Policy for the affected persons of Sardar Sarovar Project (SSP) is based on the decisions and final orders of the Narmada Water Disputes Tribunal (NWDT) Award. Considering the socio-economic and cultural background of the population being displaced and with a view to improving the living conditions of these people, all the three participating States have formulated their own policies which contain more liberal provisions than those envisaged in the Narmada Water Disputes Tribunal (NWDT) Award.

In pursuance of the decision taken by the NCA, a frequency of 1 in 100 year flood has been adopted for working out the backwater effect and preparing the submergence schedule. The original submergence schedule which was based on the Revised Implementation Schedule (RIS) of 1989 is being reviewed from time to time based on the actual approved yearly construction programme. A mechanism has been evolved to link the progress of R&R works and dam construction keeping in view the interim orders/ directions of the Supreme Court of India. Accordingly, the programmes of dam construction are being approved on year to year basis after reviewing the progress of R&R works.

Out of total of 40,827 project affected families (PAFs), house plots have been allotted to 10,471 families and agricultural land has been allotted to 9,998 families up to 31st January, 2000.

In Sardar Sarovar Project, out of the total targeted area of 179,180 ha for catchment area treatment of directly draining sub-watersheds an area of 128,230 ha has been treated up to January, 2000. Compensatory afforestation work has been carried out in an area of 45,525 ha against a total target of 46,358 ha. This work is being done in lieu of 17,943 ha taken by the Sardar Sarovar Project for submergence and resettlement and rehabilitation in Maharashtra. Baseline data have been collected, action plan has been prepared and preventive and curative measures are being implemented in project and command areas and at rehabilitation sites as far as the health related issues are concerned. Epidemiological surveillance studies are also making progress.

In Indira Sagar Project, out of a total targeted area of 62,975 ha for catchment area treatment of directly draining sub-watersheds, an area of 44,276 ha has been treated up to September, 1999. Compensatory afforestation work has been carried out in an area of 67,188 ha against a total target of 80,945 ha. This work is being done in lieu of 41,420 ha used by the Indira Sagar Project for submergence, residential colony, power house complex, dam, saddle dam, and approach roads.

CHAPTER 12

TAPI, SABARMATI AND MAHI BASINS

Three important rivers of central India are described in this chapter: Tapi, Sabarmati, and Mahi. These rivers outfall in the Arabian Sea, close to each other and share some common characteristics. These were the reasons of describing these in a shared chapter.

12.1. THE TAPI BASIN

The Tapi basin is situated in the northern part of the Deccan Plateau and extends over an area of 65,145 km which is nearly 2% of the total geographical area of the country. The catchment area up to Ukai dam is 62,225 sq. km. Nearly 80% of the basin lies in State of Maharashtra. The basin lies between east longitudes of 72°38′ to 78°17′ and north latitudes of 20°5′ to 22°3′. It is bounded on the north by the Satpura range, on the east by the Mahadeo hills, on the south by the Ajanta range and the Satmala hills and on the west by the Arabian Sea. Bounded on the three sides by the hill ranges, the Tapi River along with its tributaries flows over the plains of Vidarbha, Khandesh, and Gujarat and covers large areas in the State of Maharashtra and small areas in Madhya Pradesh and Gujarat. The basin has an elongated shape with a maximum length of 587 km from east to west and the maximum width of 210 km from north to south. Perimeter of the basin is about 1,840 km. The State-wise distribution of the drainage area is given in Table 1. An Index map of the Tapi basin is presented in Figure 1.

There are two well-defined physical regions, in the basin, namely, hilly region and the plains. The hilly regions comprising Satpura, Satmalas, Mahadeo, Ajanta and Gawilgarh hills are well forested. The plain covers the Khandesh areas, which are broad and fertile suitable for cultivation primarily, the basin consists of black soils. The coastal plains of Gujarat are composed of alluvial clays with a layer of black soil above. The culturable area of the basin is about 4.29 Mha which is 2.2% of the total culturable area of the country. The forest cover is about 25% of the area in the basin.

Physiographically, the area is a basaltic landscape with major physiographic units of plateau lands, escarpments, hills, piedmont plains, colluvio-alluvial plains and valley plains.

Table 1. State-wise distribution of the Tapi basin

State	Basin area falling in state (sq. km)	Percent of total basin area
Madhya Pradesh	9, 804	15.00
Maharashtra	51, 504	79.10
Gujarat	3, 837	05.90
Total	65, 145	100.00

The entire Tapi basin sea can be divided in three sub-basins: Upper Tapi Basin up to Hathnur [confluence of Purna with the main Tapi (29,430 sq. km)], Middle Tapi Basin from Hathnur up to the Gidhade gauging site (25,320 sq. km), and Lower Tapi Basin from the Gidhade gauging site up to the sea (10,395 sq. km). The annual rainfall for the upper, middle, and lower Tapi basins for an average year is 935.55 mm, 631.5 mm, and 1,042.33 mm respectively.

In Table 2, the areas of various districts that fall in the upper, middle, and lower Tapi Basins are given. Note that the Jalgaon district lies almost completely in the basin.

12.1.1. Tapi River

The Tapi River is the second largest west flowing river of Indian Peninsula. The river rises near Multai in the Betul district of Madhya Pradesh at an elevation of about 752 m and before joining the Gulf of Cambay, it traverses a total length of about 724 km. Of the total length, 228 km lies in Madhya Pradesh, 228 km in Maharashtra, 214 km in Gujarat and the remaining 54 km forms the common boundary between Madhya Pradesh and Maharashtra. The river flows generally east to west through Madhya Pradesh, Maharashtra & Gujarat States. The drainage

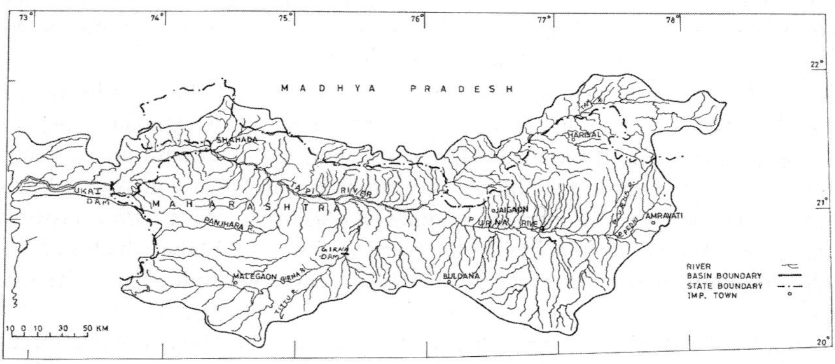

Figure 1. Index map of Tapi basin

Table 2. Area (%) of different districts within three sub-basins

District	State	Percent area
	Upper Tapi Basin	
Akola	Maharashtra	63.57
Buldhana	Maharashtra	62.39
Amravati	Maharashtra	65.03
Jalgaon	Maharashtra	5.32
Betul	M.P.	40.91
Khandwa	M.P.	35.74
	Middle Tapi Basin	
Jalgaon	Maharashtra	91.45
Aurangabad	Maharashtra	9.73
Nasik	Maharashtra	39.56
Dhule	Maharashtra	44.58
Khargaon	M.P.	8.16
	Lower Tapi Basin	
Dhule	Maharashtra	34.97
Khargaon	M.P.	5.46
Surat	Gujarat	47.35
Bharuch	Gujarat	12.24

density of the river is moderate to high. Broadly speaking, the pattern is dendritic with a sub-parallel system existing adjacent to stream courses, particularly in the alluvial belt. In the lower reaches, river widens considerably. Low tides come right up to Kakrapar weir.

There are 5 major, 27 medium and 364 minor existing projects; 3 major, 24 medium and 123 minor ongoing projects; and 3 major, 4 medium and 197 minor projects are proposed in the Tapi basin up to Ukai dam.

The utilizable water from Tapi River at Ukai dam has been estimated by Central Water Commission (CWC) to be 14,500 MCM. According to the agreements, the upstream utilization by riparian States of Maharashtra and Madhya Pradesh will be 5,420 MCM and 1,980 MCM, respectively. The remaining quantity on the order of 7,100 MCM can be utilized by Gujarat. The water requirements for irrigation, municipal and industrial uses from the Ukai reservoir are of the order of 4,546 MCM and 1,000 MCM respectively.

12.1.2. Major Tributaries

The Tapi River has 14 major tributaries having a length more than 50 km. Four right bank tributaries, namely Vaki, Aner, Arunawati, and Gomai have their origin in Satpura ranges and flow generally in South-West direction. They are comparatively of shorter length and individually drain small areas as they descend down the steep slopes. Ten important left bank tributaries, namely Nesu, Amravati, Buray, Panjhara, Bori, Girna, Waghur, Purna, Mona, and Sipna drain into the main Tapi River. The left bank tributaries rise in Gawaligarh hills, Ajanta hills, the Western Ghats, and the Satmalas. These rivers are of comparatively long with large individual drainage areas. The Purna and the Girna rivers together account for nearly 45% of the Tapi basin area. A brief description of major tributaries is given next. Figure 2 Shows the stream network of Tapi basin.

The Purna

The Purna River is the principal tributary of Tapi River. Purna is the only river in the upper basin, which has a perennial flow. The river rises in the Betul district in Gawaligarh hills of the Satpura range at an elevation of 900 m at north latitude of 21°38′ and an east longitude of 77°36′. The Purna River traverses first in a southwesterly direction for about 60 km through hills and forests, and then it enters in the Purna plains. After traversing a length of 274 km in a generally westerly direction, the Purna joins the Tapi north-west of Edalabad. The Pedhi, the Katpurna, the Murna, the Mun and the Nalganga are the main left bank tributaries and the Arna, the Chandrabhaga and the Wan are the principal right bank tributaries of Purna River. The total drainage area of Purna River is 18,929 sq. km and it joins Tapi at 282 km of Tapi's run.

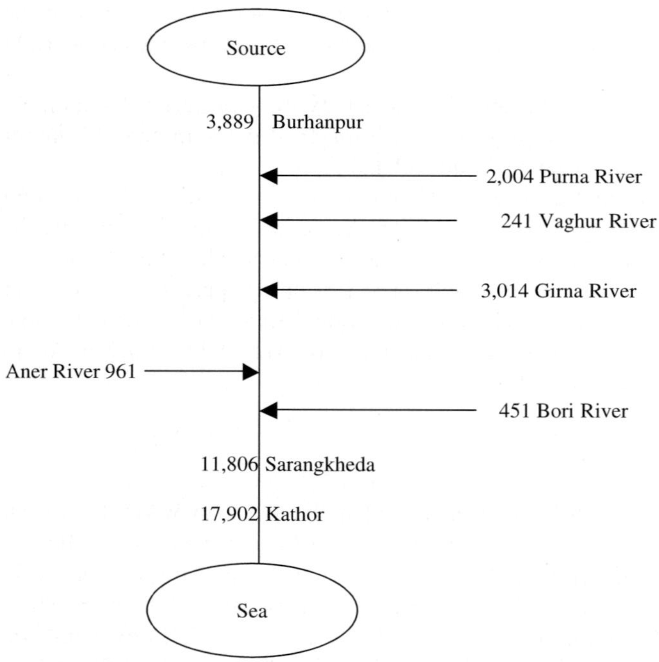

Figure 2. Stream network of Tapi basin

The Vaghur River

The Vaghur River rises in the Ajanta hills at an elevation of about 751 m at north latitude of 20°29′ and an east longitude of 75°42′ and flows in a generally northerly direction for a total distance of 96 km before joining the Tapi River on the left, north-west of Bhusawal. The total drainage area of Vaghur River is 2,592 sq. km and the point of confluence is at 312 km from the source of Tapi.

The Girna River

The Girna River rises in Western Ghats at an elevation of 900 m at north latitude of 20°44′ and an east longitude of 73°51′. Girna has its upper catchment in the eastern slopes of Sahyadri mountains. The river flows first in an easterly direction up to Jaamda and then turns north. It takes a westerly turn at Nandra to join the Tapi from the left near Nanded. At this point, Tapi has already run for 340 km. Before Joining the Tapi River, the Girna River traverses a distance of 260 km. The drainage area of the river is 10,061 sq. km, which is nearly one-sixth of the total catchment of the Tapi. Girna is the second biggest tributary of Tapi in terms of catchment area. The waters of the Girna are used for irrigation in Nasik and Jalgaon districts. Due to heavy rainfall in the catchment, Girna has important influence on the floods in the Tapi basin.

The Bori River

The Bori River rises in the Malegaon sub-division of Nasik district at north latitude of 20°48′ and an east longitude of 74°22′ at an elevation of about 600 m. The river flows first in an easterly direction and then in a northerly direction to join the Tapi from the left, east of Betavad. The river traverses a distance of 130 km and drains an area of 2,580 sq. km. The waters of the Bori are used for irrigation in Dhulia and Jalgaon districts. The confluence of Bori with Tapi falls at 386 km of the run of Tapi.

The Panjhra River

The Panjhra River rises near Pimpalner from the crest of the Sahyadri hills at an elevation of 600 m at north latitude of 20°52′ and an east longitude of 73°55′. After flowing east for about 99 km, it takes a turn to the north and joins the Tapi from the left, south of Thalner. The river traverses a distance of 138 km before joining the Tapi. The drainage area of the Panjhra River is 3,257 sq. km.

The Buray River

The Buray River rises in the Satmala hills at an elevation of 600 m at north latitude of 21°10′ and an east longitude of 74°4′ and traverses a distance of 87 km before joining the Tapi River from the left, north-east of Sindkheda. The drainage area of the river is 1,419 sq. km.

The Aner River

The Aner River is the longest right bank tributary of the Tapi. It rises at an elevation of 600 m at north latitude of 21°23′ and an east longitude of 75°45′ from the southern slopes of the Satpura hills and traverses in a generally south-westerly direction for a distance of 94 km to join the Tapi from the right, south of Hol. The drainage area of the Aner River is 1,702 sq. km.

The Arunavati River

The Arunavati River rises at an elevation of 450 m at north latitude of 21°33′ and an east longitude of 75°11′ from the southern slopes of the Satpura hills. It traverses in a generally south westerly direction. The river flows for a distance of 53 km to join the Tapi from the left, east of the Virdel village. The drainage area of the Arunavati River is 935 sq. km.

The Gomai River

The Gomai River also rises from the southern slopes of the Satpura hills at an elevation of 600 m at north latitude of 21°47′ and an east longitude of 74°46′. It traverses in a generally south-westerly direction for a distance of 58 km to join the Tapi from the right near village Parkasha. The drainage area of Gomai River is 1,148 sq. km. Gomai is the last major tributary which joins Tapi when it has flown for 481 km.

12.1.3. Topography, Physiography and Geology of Tapi Basin

The Tapi basin has two well-defined physical regions, viz., the hilly regions and the plains. The hilly regions cover the Satpura, the Satmala, the Mahadeo, the Ajanta and the Gawilgarh hills and are well forested. The plains cover the Khandesh plains which are broad and fertile areas suitable for cultivation. The culturable area of the basin is considered as the total of the land under miscellaneous crops and trees, current fallows, other fallows, culturable wasteland and net area sown.

The basin in Madhya Pradesh is mostly covered with Deccan trap lava flows. The other formations found in the basin are alluvium, lower Gondwana, Cuddapah system, Bijawar series, and granites gneiss. Most of the area of Tapi basin falling within Maharashtra state is full of cuts & valleys. Lands on the right side of the river lying on southern slopes of Satpura hills consist of black soils. The soil cover is deep and rock is found at greater depths. Lands on the left side of the river on northern slopes of Sahyadri consist mainly of dykes & red murrum soil and are rocky in most parts.

12.1.4. Soils of Tapi Basin

The soils in the Tapi basin up to Ukai dam can be broadly classified into 3 groups: 1) coarse shallow soils, 2) medium black soils, and 3) deep black soils. The area covered by these three groups of soils in the basin is given in Table 3.

Table 3. Area covered by various soils in Tapi basin

S. N.	Type of soil	Districts covered
1.	Coarse shallow soils	Betul, Khandwa, Khargon, Amrawati, Akola, Buldhana, Jalgaon, Dhule, Aurangabad and Nasik.
2.	Medium black soils	Khandwa, Amrawati, Akola, Buldhana, Jalgaon, Dhule and Nasik.
3.	Deep black soils	Amravati, Akola, Buldhana, Jalgaon, Dhule, Nasik, Surat and Bharuch.

Coarse Shallow Soils

These soils have developed primarily from the basaltic Deccan traps and have been considerably affected by natural processes of weathering and erosion. Their depth is generally between 25 cm to 50 cm and seldom more. Their texture from surface to sub-surface varies from silty loam to clay. Their organic matter content is usually poor and they are moderately drained.

Medium black soil

These soils have developed from Deccan traps and cover the largest area of the basin. Their depth is generally between 50 cm to 1 m. These soils contain higher lime reserve and are alkaline in reaction. These soils are fair in their contents of phosphates and potash but low in organic matter and nitrogen.

Deep black soils

These soils are found along the Purna River and in the middle & lower reaches of Tapi River. These soils have originated primarily from decomposition of trap rocks of hilly ranges. The depth of this soil varies from 1 m to 6 m. The soils have very high clay content, montmorillonite predominating and not easily workable during monsoon. The soil reaction varies from neutral to alkaline.

12.1.5. Land Use in Tapi Basin

The major land use of the basin in the year 1995–96 is presented in Table 4. Cultivated area (around 60%) and forests (>20%) cover the major part of the catchment. The important crops grown in the basin are cotton, jowar, bajra, oilseeds, wheat, paddy, tuar, black gram, fodder crops, vegetable, fruit, and sugarcane. The Tapi basin, as observed from the NOAA satellite, is presented in Figure 3.

12.1.6. Rainfall and Climate of Tapi Basin

The annual average rainfall in the Tapi basin is 830 mm; the maximum being 2,030 mm. Normally, the south-west monsoon sets in by the middle of June and withdraws by mid-October. About 90% of the total rainfall is received during the

Table 4. Major land uses in Tapi basin (for the year 1995–96)

Land use category	Area (sq. km)
Forest area	14, 788.72
Barren/Uncultivable area	2, 737.97
Non-agricultural area	2, 002.84
Cultivable waste land	719.21
Permanent pasture	2, 312.33
Miscellaneous crops/trees	118.43
Fallow land	1, 781.46
Gross sown area	46, 069.57
Gross irrigated area	5, 741.84

monsoon months, of which 50% is received during July and August. There are 70 rain gauge stations in and around the basin up to Ukai dam.

The climate of the basin is characterized by hot dry summer and winter. Owing to topographical characteristics, the climate is variable. In winter, the minimum temperature varies from 10 °C to 14.5 °C. May is the hottest month with temperature varying from 38 °C to 48 °C. In fact, the Purna sub-catchment is one of the hottest regions of India. Eight IMD observatories at Betul, Amrawati, Akola, Khandwa, Buldhana, Jalgaon, Malegaon and Surat are located in and around the basin.

Table 5 gives some important meteorological data and the evapo-transpiration estimates for the three sub-basins of Tapi.

12.1.7. Population and Agriculture

As per the 1991 census, the population of Tapi basin was 12.576 million of which, the rural population was 9.132 million and the urban population was 3.444 million. The density of population in the year 1991 in Madhya Pradesh, Maharashtra and

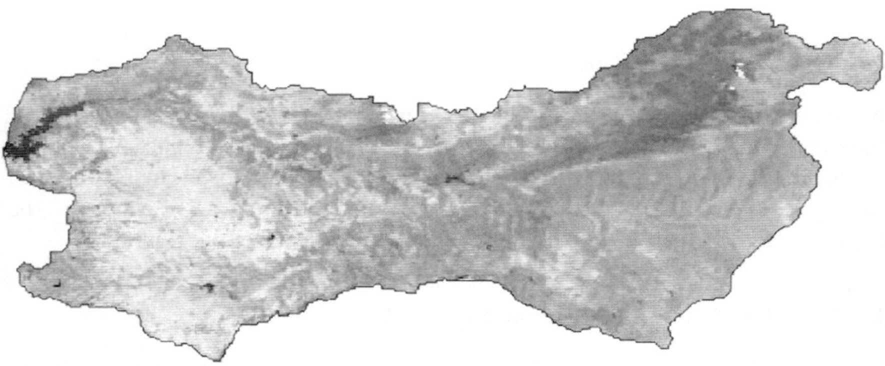

Figure 3. A view of Tapi basin up to Ukai dam from NOAA Satellite. Ukai reservoir can be clearly seen near the top left edge

Table 5. Meteorological data and evapo-transpiration estimation for sub-basins of Tapi

Month	Max_temp °C	Min_temp °C	Max._Hum %	Min_Hum %	Wind speed km/day	Cloud cover Oktas	PET (mm/day)
January	33	8.5	55	28	110.4	1.7	3.93
February	36.9	10.4	42	21	124.8	1.3	5.12
March	40.6	13.8	35	18	146.4	1.4	6.55
April	43.7	20.1	32	17	172.8	1.8	7.94
May	45.1	23	39	18	295.2	2.2	10.58
June	42.4	22.3	67	43	333.6	4.9	8.71
July	36.5	22	81	66	278.4	6.4	5.42
August	34.3	21.8	84	68	266.4	6.6	4.65
September	35.7	20.9	80	60	208.8	5.0	5.18
October	36.2	15.2	65	40	112.8	2.7	4.74
November	33.8	11.6	57	34	91.2	2.1	3.76
December	32.4	9	58	33	88.8	1.8	3.33
January	33.5	6.7	58	28	110.4	1.7	3.95
February	37.6	8.2	43	19	124.8	1.2	5.18
March	41.7	12	36	15	146.4	1.3	6.67
April	45	18.7	36	14	172.8	1.6	8.12
May	45.8	23.5	54	19	295.2	1.5	10.64
June	42.9	22.5	73	44	333.6	4.6	8.73
July	37.7	22.4	85	65	278.4	6.6	5.43
August	34.8	21.7	88	72	266.4	6.8	4.41
September	36	20.2	85	60	208.8	5.4	5.05
October	36.9	13.4	68	37	112.8	2.5	4.82
November	35	10.5	59	32	91.2	2.1	3.82
December	32.9	7.6	62	34	88.8	1.9	3.36
January	35.5	10.3	65	39	165.6	1.0	5.02
February	38.1	11.5	62	33	170.4	0.8	6.1
March	41	15.7	64	32	180	1.2	7.09
April	42.4	20.3	66	38	199.2	1.4	8.16
May	41.9	23.5	68	55	285.6	2.7	8.35
June	37.4	23.4	79	70	324	5.6	6.12
July	34.4	23.3	88	79	302.4	6.9	4.27
August	33.2	23.2	89	79	264	7.1	3.75
September	36.4	22.5	86	70	189.6	5.1	4.79
October	38.7	19.4	72	49	148.8	2.0	5.51
November	37.7	15.5	61	43	158.4	1.4	5.16
December	35.9	12.1	65	43	172.8	1.3	4.87

Gujarat was 128 per sq. km, 250 per sq. km, and 204 per sq. km, respectively. The livestock population of the basin was estimated at 7.0 million.

Agriculture is the main occupation of the people of the basin. The upper reaches of the basin are covered by dense forests. The livestock population of the basin is 70 lakh as per 1991 census. Mainly animals found in the basin are buffaloes, cows, goat and poultry are found in the basin.

The net sown area for the year 1994–95 was 3,751,418 ha and gross sown area was 4,523,699 ha which is 60.29% and 72.70%, respectively of the total geographical area of the basin. Cotton, Jowar, Bajra, Oilseeds, Wheat, Paddy, Tuar, Black gram, Sugarcane, Vegetables and Fruits are the main crops grown in the basin.

12.1.8. Water Resources Development in Tapi Basin

Twelve G&D sites are maintained by CWC in Tapi basin, viz. Dedtalai, Burhanpur, Lakhpuri, Gopalkheda, Yerli, Dapuri, Savkheda, Malkheda, Morane, Gidhade & Sarangkheda located upstream of the Ukai dam and Ghala G&D site located downstream of the Ukai dam. The Kathor G & D site is located downstream of the Ghala G & D site and is maintained by the Government of Gujarat. Table 6 gives average flow at some gauging sites.

The utilizable water from Tapi River at Ukai dam has been estimated by Central Water Commission (CWC) to be 14,500 MCM. According to the agreements among the riparian states, the upstream utilization by States of Maharashtra and Madhya Pradesh will be 5,420 MCM and 1,980 MCM respectively. The balance quantity of 7,100 MCM can be utilized by Gujarat. Preliminary investigations by the Government of Gujarat show the water requirements for irrigation and D&I (domestic and industrial) uses from the Ukai reservoir are about 4,546 MCM and 1,000 MCM respectively. Therefore, the surplus quantity at Ukai reservoir has been estimated to be 1,554 MCM. Water is exported from this basin from Ukai dam and Kakrapar weir. The proposed Par-Tapi-Narmada interlinking scheme plans to transfer 1,554 MCM from the Ukai dam to meet the demands in water deficit areas in North Gujarat.

There are 5 major, 27 medium and 364 minor existing irrigation projects in the basin with annual irrigation of 357,959 ha utilizing 2,717 MCM of water. Most of these projects are located in Maharashtra portion. Construction of 3 major, 24 medium and 123 minor projects is going on while 3 major, 4 medium and 197 minor projects are proposed to be constructed in future. Hathnur dam, Kakrapar weir, Ukai dam, Girna dam, and Dahigaon weir are some of the important projects in the basin. Important existing major and medium hydraulic structures along with their capacities are presented in Table 7.

Table 6. Average flow (in $10^9\,m^3$) at different CWC gauging stations (Catchment area >5,000 km^2)

Name of the site	Name of the stream	Catchment area (km^2)	Annual average runoff (BCM)
Burhanpur	Tapi	8,487	5.4
Gopalkheda	Purna	9,500	1.0
Yerli	Purna	16,517	1.8
Dapuri	Girna	8,901	0.5
Savkheda	Tapi	48,136	7.1
Gidhada	Tapi	54,750	6.2
Sarangkheda	Tapi	58,400	7.7
Ghala	Tapi	63,325	5.7

Table 7. Major/Medium existing Projects in Tapi basin

Name of Project	Gross Storage (10^6 m^3)	Live Storage (10^6 m^3)
	Upper Tapi Basin	
Sonkhedi Tank	5.45	4.59
Sukta	1,347	1,311
Chandora	18.2	16.48
Katepurna	97.67	86.35
Nalganga	76.2	69.32
Uma	14	11.68
Nirguna	349	28.85
Morna	44.74	41.46
Gyanganga	36.26	33.93
Mos	17.5	15.14
Paltag	9.09	7.51
Man	39.76	36.83
Thoran	8.48	7.9
Hathnur	388	255
Total	2,134.64	1,926.04
	Middle Tapi Basin	
Girna	608.45	523.55
Manyad	53.95	40.27
Bori	40.3	25.15
Suki	50.16	39.85
Abhora	7.44	6.02
Boker Bari	7.09	6.54
Agnawati	3.74	2.76
Tondapur	4.63	4.64
Aner	103.23	56.38
Karwand	33.84	31.15
Panjhra	43.41	35.63
Malangaon	13.02	11.35
Kanholi	11.79	8.45
Burai	21.33	14.21
Arunawati	27.78	14.97
Rangawali	15.02	12.89
Nagasakya	15.62	11.24
Haran bari	34.78	27
Total	1,095.58	872.05
	Lower Tapi Basin	
Ukai	8,510	7,092
Kakrapar	51.51	36.57
Lakhigav	38.8	37.41
Ver	4.9	4.61
Total	8,605.21	7,170.59

From the on-going major, medium, and minor projects, anticipated utilization is expected to be about 1,495.75 million m^3.

In the next section, we describe the Ukai Project which is a major WRD project in Tapi basin.

12.1.9. Ukai Project

The Ukai is the largest multipurpose project so far completed in Gujarat state. The Ukai dam is located across Tapi River near Ukai village of Fort-Songadh taluka in Surat district. Its catchment is located between longitudes 73°32′25″ to 78°36′30″ E and latitudes 20°5′0″ to 22°52′30″ N. The dam is located at about 29 km upstream of the Kakrapar weir. The total catchment area of the Ukai reservoir is 62,225 sq. km, which lies in the Deccan plateau. The catchment of the dam covers large areas of 12 districts of Maharashtra, Madhya Pradesh, and Gujarat. The districts that lie in the catchment include Betul, Hoshangabad, Khandwa and Khargaon of Madhya Pradesh; Akola, Amravati, Buldhana, Dhule, Jalgaon and Nasik of Maharashtra and Bharuch and Surat of Gujarat state.

The dam comprises a 4,927 m long earth-cum-masonry dam. The reservoir has a live storage capacity of 0.71 Million ha-m at FRL 105.156 m. The earth dam is 80.77 meter high above the lowest foundation level, while the masonry dam is 68.68 meter high above the lowest foundation. The storage extends irrigation facilities to an area of 1,522 sq. km under the Ukai left bank canal and Ukai right bank canal system besides firming up irrigation in an area of 2,275 sq. km under the command of pick up weir at Kakrapar. Thus, the total irrigation facilities provided are for 3,797 sq. km under Ukai and Kakrapar.

The Ukai head works were completed in the year 1972 after a construction period of seven years. There are 22 radial gates of 15.545 × 14.783 m size each. The length of the spillway is 425.195 m. The area under submergence due to the construction of reservoir (at FRL) is 601.30 sq. km, out of which 303 sq. km is in agricultural land, 223 sq. km is forest land and the remaining 75.30 sq. km is wasteland. In Ukai dam, 5.5 km^3 of water is stored against the gates. This amounts to nearly 78% of the live storage. Thus, a large quantity of water is not stored against a static wall (the dam) but against a 'live' and moving device (gates). In such circumstances, operation of gates is very critical not only for the structural safety but also for flood control. This arrangement of storing large quantity of water behind gates has saved large chunks of land from permanent submergence thereby reducing the cost of the dam. Srisailam dam (Chapter 14) is another large dam where enormous volume of water is held behind gates.

The 75% dependable yield of Tapi River at Ukai dam site is 11,350 MCM. Out of this, the share of Gujarat is 3,947 MCM or 38.4%. This water is utilized to:

1. Provide irrigation in 152,000 ha in the command of Ukai Left Bank and Right Bank canals,
2. Firm up irrigation in 227,500 ha in the command of Kakrapar canals, and
3. Generate hydropower up to 300 MW for 8–10 hours a day to meet the peak demands.

Initially, the Ukai dam toe hydro-power station with an installed capacity of 300 MW was planned as a peaking station and it was to be operated only for 8 to 10 hours in a day. In the initial years, with no upstream utilization in Maharashtra, the project was expected to generate about 990 MU annually. This was expected to reduce to 745 MU after 30 years and further to 587 MU after 60 years of operation, due to upstream development leading to reduced inflows to the reservoir. The actual performance of power house indicates that the stipulated generation of 900 MU was exceeded in 12 out of 25 years, the maximum annual generation reaching 1,233 MU in the year 1977–78. Table 8 shows the quantum of water utilized for power generation at Ukai dam from 1990 to 1999. The reservoir also provides partial flood control in the downstream areas; Surat is the major city being protected.

12.1.10. Floods in the Tapi Basin

During monsoons, the Tapi River is frequently in spate and floods occasionally cause havoc in the plains of lower reaches. In the post-monsoon period, the discharge in the river is quite low, say of the order of 300 cumec. In the basin, rain storms typically move from east to west; this also is the general flow direction in the basin. Most floods occur during the period July to September, though occasionally they may occur in the second week of October. The maximum number of floods has been found to take place in the month of August.

Floods were a frequent phenomenon in Tapi River at Surat before the Ukai Dam was constructed. During 1876 to 1970, danger level was crossed at the Hope Bridge, in Surat on 19 instances. This implies a frequency of once in every five years. High floods were experienced during three consecutive years from 1882 to 1884. There was a very high flood in August 1944 and actually two almost similar flood peaks occurred in the same week. A heavy flood, only slightly lower than that in 1944, was experienced again in 1945. Consecutive large floods were also observed in 1958 and 1959. Heavy floods were witnessed in 1968, 1969, and 1970. The floods of September 1959 and August 1968 were catastrophic. The floods of 1994 and 1998 also caused considerable damage to the Surat city and other low-lying areas on the downstream.

Table 8. Water utilized for power generation at Ukai Dam

Year	Electricity generated (million kwhr)	Water utilized for electricity generation (million m^3)
1990–91	676.00	5, 575.33
1991–92	338.00	2, 701.32
1992–93	873.00	6, 697.80
1993–94	485.00	3, 823.79
1994–95	790.00	5, 920.70
1995–96	387.00	3, 725.11
1996–97	546.00	4, 588.55
1997–98	861.00	6, 648.46
1998–99	913.00	8, 288.99
Average	652.11	5, 330.00

Operation of Ukai Reservoir for Flood Control

CWC had prepared detailed guidelines for flood control operation of the Ukai reservoir in the year 1975. The discussion in this section is drawn from (Patel and Patel 2003). These guidelines had suggested restricting the outflow from the dam to 24,070 cumec (8.5 lakh cusec) for floods of peak up to 36,800 cumec (13 lakh cusec). Higher floods could also be negotiated, provided the initial reservoir level at the time of impinging of flood is at EL 101.19 m. But since the floods can occur even by the beginning of October, maintaining EL 101.19 might result in the reservoir remaining vacant if higher flows do not take place and in that eventuality future demands will remain unsatisfied. To resolve the problems, reservoir is filled to the maximum level of 103.33 m by the end of August and is gradually raised up to FRL 105.15 m thereafter. But in the process, there is a risk of not completely filling the reservoir. It was also suggested that the reservoir level should not be allowed to rise above the FRL of 105.15 m for passing floods up to 36,800 cumec. Further, it was presumed that weather and inflow forecasts would be available during floods to make the best use of available storage space. CWC guidelines suggested that the flood forecasting and dissemination facilities need to be strengthened to improve the quality of forecasts and to increase the warning time to the maximum possible extent. Availability of inflow forecasts permits the reservoir operator to make pre-releases and create additional flood control capacity without the risk of losing the conservation storage.

The guidelines stipulated that when floods higher than 36,800 cumec occur at or near FRL, the outflows will exceed 28,080 cumec and the reservoir level may rise up to MWL, depending upon the size of the flood. It was also brought out that it would not be possible to contain the outflow within the permissible value of 28,317 cumec (10 lakh cusec) when the floods between SPF and PMF occur or when the reservoir is operated under emergency situation, without the aid of flood forecasting. The maximum flow may reach the designed maximum outflow of nearly 45,300 cumec (16 1akh cusec).

Regulation of 1994 and 1998 Floods

Abnormally big floods occurred in Tapi basin in the year 1994 and 1998. On 26.8.1994, the reservoir level at the dam was at 103.54 m (339.70 ft) and the inflow was 2,315 cumec (82,000 cusec). Gradually, the inflow increased and reached a maximum of 24,267 cumec (8.57 lakh cusec) on 8th Sept. at 18:00 hr. The flood was regulated by allowing the reservoir to rise from RL 104.98 m (344.42 ft) to RL 105.39 m (345.77 ft) and peak release was 14,385 cumec (5.08 lakh cusec). The reservoir level depleted and reached FRL 105.15 m (345 ft) on 9.09.1994 at 08:00 hr. Thus the water level was kept above FRL for 22 hours so as to release less water and provide flood relief to the downstream areas. During this flood event, the total volume of inflow was 2,570 million m^3 and the volume released was 1,744 million m^3. The maximum water level reached at Nehru Bridge, Surat, was 10.0 m.

Till 1994, the warning level and danger level of Tapi River at Surat City were 10.18 m and 11.18 m respectively. But in the 1994 flood, the peak release from Ukai dam was close to 14,900 cumec (5.25 lakh cusec). Consequently, river water level rose to 10.2 m and this caused extensive submergence and heavy damages in Surat City. After the monsoon of 1994, the matter was reviewed and the new warning and danger levels were fixed at 8.5 m and 9.5 m, respectively. These levels correspond to discharge of 11,328 cumec (4.0 lakh cusec) and 13,027 cumec (4.6 lakh cusec). Note that the discharge corresponding to the warning level is considered to be the safe carrying capacity of the river at that location.

In terms of volume and peak, the flood of 1998 was much bigger than the flood of 1994. The first advance advisory flood warning for the heavy incoming flood was received from CWC, on 15th September 1998 at 07:05 hr, informing that in the next 42 to 48 hours, approximately $2,500\,\mathrm{Mm}^3$ of water was likely to enter Ukai reservoir due to heavy rainfall in Maharashtra and Madhya Pradesh. It was also informed that the total volume of this flood may be of the order of 4,000 million m^3. At the time of this warning, the water level-in the reservoir was 104.1 m and the storage capacity available up to FRL (105.15 m) in the reservoir was only $456\,\mathrm{Mm}^3$.

To accommodate this flood by pre-depleting the Ukai reservoir, release at the rate of 1,453 cumec was started from the dam at 10:00 hr on 15-9-98. The release was increased subsequently with receipt of each new forecasts indicating that flood was building up. During this event, the maximum observed inflow was of the order of 29,817 cumec at 19:00 hr on 16-9-98. Against this, the maximum release from the dam was at 19,765 cumec at 22:00 hr on 16-9-98. While routing the flood, the maximum reservoir level attained was 105.47 m, which is 0.32 m above FRL. During this event, the reservoir level was above FRL and in the vicinity of 105.46 m for over 72 hours.

During flood season, the Ukai Reservoir is being operated in consultation with CWC. The last gauging station being monitored by CWC on Tapi River is at Gidhade, 175 km upstream from Ukai dam. The catchment area between Gidhade and Ukai Dam which drains into Ukai dam Reservoir is 7,475 sq. km. From this area, no rainfall or run off data are available to the Ukai dam operating authorities till the water reaches the dam. Further, the water spread area of Ukai reservoir at FRL is 600 sq. km and any rainfall over this area directly contributes to the inflow at dam site without any loss. During the flood in September 1998, it was observed that many times, the flood received at Ukai was much higher than the forecasts. Partly, this can be attributed to heavy rainfall in the catchment area between Gidhade and Ukai dam.

An analysis of data (Patel and Patel, 2003) shows that on 16th Sept. 1998, CWC had forecasted an inflow volume of $1,460\,\mathrm{Mm}^3$ while the volume received at the dam site was 1,748 million m^3 which is 20% higher than forecast given by CWC. Further, total 4,321 million m^3 water was received at the dam site till 06:00 hrs of 19th Sept. 1998. Out of this, only 727 million m^3 was stored in the reservoir and remaining 3,686 million m^3 was released. In the initial response on 15th Sept. 1998, 1,159 million m^3 water was received at dam site out of which, $403\,\mathrm{Mm}^3$ was stored in the reservoir which is 35% of the inflow and remaining $756\,\mathrm{Mm}^3$ was released in the downstream.

As against the peak inflow of 29,817 cumec, peak outflow was 19,765 cumec due to which the maximum water level of Tapi River at Nehru Bridge at Surat on 17.9.1998 was 11.4 m. Had the unregulated flood of 29,817 cumec reached Surat, the river water level at Nehru Bridge might have touched 11.90 m. Thus, due to regulation of Ukai project, the peak flow at Surat was significantly reduced. However, even the peak release of 19,765 cumec inundated low lying areas in and around Surat City and caused extensive damages. In view of this, the Government of Gujarat decided to review the flood control operation policy of Ukai dam.

Ever increasing need for water for irrigation and power for industries in Ukai command and looking to the performance of the reservoir regarding flood control as described above, prompted the government to review the operation policy of Ukai dam in the changed circumstances so as to optimize the irrigation, power and flood control benefits. New rule curve levels have been determined after a detailed study and these recommendations are given in the second column of Table 9. Further, CWC has tried to optimize the rule levels for the reservoir by minimizing the total penalties due to violation of storage and channel flow targets and the optimized rule levels are given in the third column of Table 9.

Patel and Patel, (2003) report that the safe carrying capacity of the Tapi River at Surat has alarmingly reduced from the design value of 24,070 cumec (8.5 lakh cusec) to only 11,320 cumec (4 lakh cusec) due to the following reasons:

a) Non-completion of flood embankments on both sides of Tapi River upstream of Surat City,
b) Extensive encroachment on river flood plains in and around Surat City,
c) Extensive siltation in river channel in and around Surat City, and
d) Afflux created due to the construction of Siganpore weir just upstream of Surat City.

12.1.11. Other Projects in Tapi Basin

In addition to Ukai, there are some other existing projects in the Tapi basin. Salient features of the existing projects whose live storage capacity is more than 10 million cubic meter have been presented in Table 10.

Table 11 gives salient features of those projects which are under construction and whose live storage capacity is more than 10 million cubic meter.

Table 9. Rule levels recommended by Task group for Ukai Reservoir

Date	Recommended rule level (m)	Optimized rule level (m)
July 1st	97.85	97.50
August 1st	101.50	101.00
September 1st	103.63	103.00
September 15th	103.63	–
Oct. 1st	105.15	105.15

Table 10. Salient features of selected existing projects in Tapi basin

Name of the Project	State	Year of completion	Gross storage capacity (million m^3)	Live storage capacity (million m^3)	Designed annual Irrigation (million m^2)
Kakrapar	Gujarat	1953	52.30	44.00	199.81
Ver-II	Gujarat	–	38.94	37.33	61.20
Sampur	Madhya Pradesh	1956	14.94	11.21	38.50
Sawpna	Madhya Pradesh	1957	16.94	14.31	39.00
Aner	Maharashtra	1978	103.27	58.91	71.80
Bari	Maharashtra	1977	39.45	25.02	59.60
Burai	Maharashtra	1983	21.33	14.21	28.30
Chankapur	Maharashtra	1911	80.00	74.59	64.20
Girna	Maharashtra	1969	608.45	523.55	572.10
Gyan Ganga	Maharashtra	1970	42.80	38.90	42.50
Haranbasi	Maharashtra	1980	34.77	33.02	67.10
Karvand	Maharashtra	1970	33.84	31.15	55.70
Kelzan	Maharashtra	1981	17.10	16.22	35.40
Morna	Maharashtra	1971	44.71	41.46	22.00
Malangaon	Maharashtra	1971	13.02	11.33	68.80
Manyad	Maharashtra	1973	53.93	40.26	59.00
Mas	Maharashtra	1982	17.50	15.04	87.40
Nalganga	Maharashtra	1963	76.20	70.54	58.90
Nirguna	Maharashtra	1976	32.29	28.83	33.40
Purme Pada	Maharashtra	1955	13.53	12.96	113.90
Panzara	Maharashtra	1973	43.50	35.80	31.30
Rangavali	Maharashtra	1982	15.02	12.89	56.90
Sudki Kalepurna	Maharashtra	1974	97.63	92.88	25.60
Suki	Maharashtra	1977	50.17	39.86	550.00
Uma	Maharashtra	1982	14.01	11.68	
Upper Tapi Stage-I	Maharashtra	–	388.00	255.00	

Table 11. Salient features of selected projects under construction in Tapi basin in Maharashtra

Name of the Project	Gross storage capacity (million m^3)	Live storage capacity (million m^3)	Designed annual irrigation (million m^2)
Bahula	20.03	16.33	46.50
Mun	39.76	35.86	–
Punad	39.75	36.99	170.00
Shahanoor	47.85	45.85	95.60
Waghur	258.04	181.30	240.00
Wan	84.91	79.28	180.00

12.1.12. Groundwater Potential

Table 12 contains groundwater related information for the three sub-basins of Tapi. As can be seen from this table, the normal natural ground water recharge is the

Table 12. Ground water recharge and potential for three sub-basins of Tapi

Sub-basin	Normal natural recharge to GW (10^6 m^3)	Canal irrigation recharge to GW (10^6 m^3)	Provision for D&I & other uses (10^6 m^3)	Utilizable GW resources for irrigation (10^6 m^3)	GW resource for future use (10^6 m^3)
Upper Tapi	3, 425.16	194.82	1, 057.95	2, 305.79	1, 901.39
Middle Tapi	2, 335.24	473.92	901.86	1, 716.61	952.39
Lower Tapi	1, 068.53	518.60	324.83	1, 136.06	974.68

highest in the Upper Tapi sub-basin while the recharge due to canal irrigation is the maximum in the Lower Tapi sub-basin.

12.1.13. Water Quality

Table 13 reports the status of water quality indicators at selected locations in the Tapi basin at five different time periods. At almost all places, the observed quality was worse than the desired quality. BOD and total coliform were the parameters that frequently aggravated the situation.

12.2. THE SABARMATI BASIN

The Sabarmati basin extends over an area of 21, 674 km^2 which is nearly 0.66 percent of the total geographical area of India. It lies between east longitudes of 72°15′ to 73°49′ and north latitudes of 22°15′ to 24°53′. It is bounded on the north and north-east by the Aravalli hills, on the east by the ridge separating it from the Mahi basin, on the south by the Gulf of Cambay and on the west by the ridge, separating it from the basins of minor streams draining into Rann of Kutch and the Gulf of Cambay. The basin lies in the States of Gujarat and Rajasthan. The State-wise distribution of drainage area is given in Table 14.

The important soil types found in the basin are black, alluvial and sandy soils. The culturable area of the basin is about 1.55 M-ha which is 0.8 percent of the total culturable area of the country. An index map of Sabarmati basin is given in Figure 4.

12.2.1. The Sabarmati River

The Sabarmati River is one of the four main rivers which traverse the alluvial plains of Gujarat. It rises in the Aravalli hills at a north latitude of 24°40′ and an east longitude of 73°20′ in the Rajasthan state at an elevation of 762 meters near the popular shrine of Amba Bhavani. After traversing a course of about 48 km in Rajasthan, the river enters the Gujarat State. At the 51st, km of its run, the Wakal River joins it from the left, near village Ghonpankhari. Flowing in a generally south-west and winding among jungle covered hills over a bed strewn with shingles

Table 13. Desired and existing water quality levels for Tapi

Location	Desired class	Observed class & critical parameters				
		1997	1998	1999	2000	2001
Tapi at Nepanagar, M.P.	A	B BOD, Totcoli	B Totcoli	B Totcoli	B Totcoli	B Totcoli
Tapi at Burhanpur, M.P.	A	B Totcoli	D BOD, Totcoli	B BOD, Totcoli	B BOD, Totcoli	B Totcoli
Tapi at Hathnur, M.P.	A	A	B Totcoli	B Totcoli	B BOD, Totcoli	B Totcoli
Tapi at Ajnand Village, Maharashtra		D BOD, Totcoli	D BOD, Totcoli BOD, Totcoli	D BOD, Totcoli	D BOD, Totcoli	D BOD, Totcoli
Tapi at Bhusawal U/S, Maharashtra	C	D BOD	D BOD	D BOD	D BOD	D BOD
Tapi at Uphad Village, Maharashtra		D BOD	D BOD	D BOD	D BOD	D BOD
Tapi at Ukai, Sherula Bridge, Gujarat	C	D Totcoli	D BOD	NA	NA	NA
Tapi at Mandavi, Gujarat	C	D BOD	D BOD	NA	NA	NA
Tapi at Kathore, (NH-8 Bridge), Gujarat	C	D BOD, Totcoli	D BOD	NA	NA	NA

NA- Not Available. Source: Central Pollution Control Board.

and boulders, at the 67th, km of its run, it receives the Sei River from the right near Mhauri and then the Harnav River from the left at about 103rd, km from the source, before it enters Dharoi reservoir. Emerging from the dam it passes through

Table 14. Statewise drainage areas of Sabarmati basin

State	Drainage area (km^2)
Gujarat	17,550
Rajasthan	4,124
Total	21,674

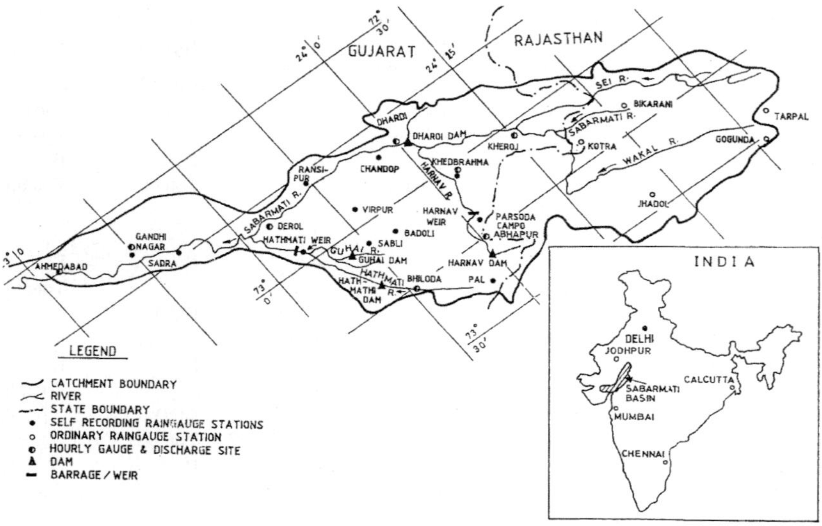

Figure 4. Index map of Sabarmati basin

the plains and is joined on its left at about 170 km from its source by the Hathmati River. Continuing to flow south-westwards, the river passes through Ahmedabad at about 165 km downstream of Dharoi dam. Further 65 km downstream, another tributary, the Watrak River joins it from the left. Flowing for a further distance of 68 km, the river outfalls into the Gulf of Cambay in the Arabian Sea.

The total length of the river from the head to its outfall into sea is 419 km of which about 48 km are in Rajasthan and the remaining 371 km in Gujarat. The river meets the following tributaries during its course of flow:

- Sei (catchment area 883 sq. km),
- Wakal (catchment area 1,893 sq. km),
- Harnav (catchment area 865 sq. km),
- Hathmati (catchment area 1,574 sq. km), and
- Watrak (catchment area 8,638 sq. km).

The Sabarmati River runs in a valley with the ground rising on both sides. The drainage area assumes a shape of a fan covering the part of Rajasthan state and parts of Sabarkantha, Ahmedabad, Banaskantha, Mehsana, Surendranagar and Kaira districts of Gujarat state. Sabarmati River has been subjected to severe pressure due to the fast pace of urban and industrial growth of Ahmedabad's urban agglomeration, especially after the 1960s. The topography of the Sabarmati basin is hilly in the early reaches up to Dharoi after which the river flows mostly in plains. On an average, the river empties 44,775 MCM of water in the Arabian Sea annually. The studies carried out by the Govt. of Gujarat reveal the possibilities of utilizing 9,692.2 MCM of water for irrigating 132,420 ha of land and for power generation to the extent of 2,899 KW at 60 % load factor. At times, the Sabarmati River sends down very heavy floods and some of these have caused devastation in Ahmedabad and villages lower down, destroyed crops, carried away cattle, changed the course of the delta channels and filled up harbour with silt. The highest known floods have occurred in 1875, 1941, 1950 and 1973.

12.2.2. Tributaries of the Sabarmati River

A brief description of various tributaries of the Sabarmati basin is given in what follows. Figure 5 Shows the stream network of Sabarmati basin.

Sei River

The Sei River rises from the south western spurs of the Aravalli hills and mostly flows in Rajasthan state. The river is formed by the confluence of several rivulets originating from the western slopes of Aravalli hills. The Sei River meets Sabarmati after traversing a course of 102 km. The catchment area of the Sei is 883 sq. km which is hilly with steep slopes. The average annual rainfall in the catchment is 675 mm. The total catchment area of Sei dam is 331.66 sq. km and normal expected yield at dam site is 52.032 MCM. Most of the runoff occurs in the monsoon season. The winter rainfall is very small. However, in good rainfall years, some flow remains in the river up to March-April. A dam, namely Sei dam, has been constructed on this river.

Wakal River

The Wakal River rises from the south western spurs of the Aravalli hills at a north latitude of 24°46′ and an east longitude of 73°23′. After traversing a course of 158 km, it meets the Sabarmati River. The catchment area of the Wakal is 1,893 sq. km. The basin is hilly, covered with forests. The basin covers the Sabarkantha district of Gujarat state and part of Udaipur district of Rajasthan state.

Harnav River

The Harnav River rises from Aravalli hills near village Ghadvas at north latitude of 24°12′ and an east longitude of 73°16′. After traversing 61 km, the Harnav River meets the Sabarmati River. The catchment area of Harnav River is 865 sq. km. The catchment area in the first 32 km is hilly and thereafter the basin is plain. The average rainfall in the catchment area is 80 cm. The main tributaries of Harnav River are Kaluri and Kusumba rivers draining an area of 311 and 124 sq. km, respectively. The catchment area up to the dam site is hilly and the remaining portion is generally in plains.

For gauge & discharge measurements, there are three gauging sites in the basin at Abhapur, Harnav-1 and Khedbrahma. One storage reservoir with dual purpose of flood control irrigation and three pickup weirs to cater for irrigation have been constructed across the river. All the three weirs are located downstream of Harnav dam and upstream of Khedbrahma.

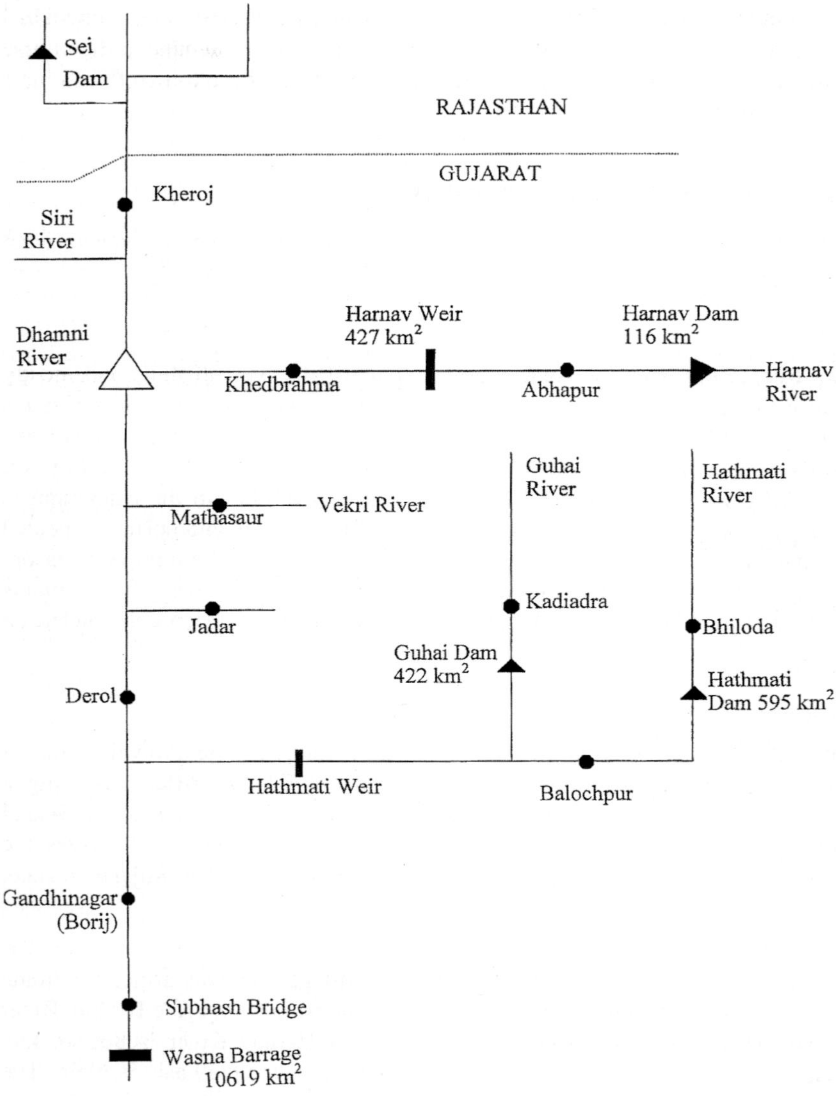

Figure 5. Stream network of Sabarmati basin

Hathmati River

The Hathmati River rises from the Gujarat Malwa hills south western foothills of the Rajasthan range near Godad at a north latitude of 23°55′ and an east longitude of 73°29′ in Sabarkantha district. After traversing a course of 98 km, it meets the Sabarmati River near Ged, 20 km south west of Himatnagar in Sabarkantha distt.

The catchment area of Hathmati River including its tributaries is 1,574 sq. km. The elongated catchment consists of 70% hilly and thinly wooded jungles and about 30% cultivated and inhabited lands. The two main tributaries of Hathmati are Bodoli and Guhai having catchment areas of 119 and 505 sq. km, respectively. The catchment area up to the confluence of Guhai and Hathmati is a hilly region. The average annual rainfall in the catchment is 86 cm.

On the Hathmati River, a dam, namely Hathmati dam, and a pick up weir, namely Hathmati weir or Himatnagar weir, have been constructed for getting the irrigation facilities from the tributary. Hathmati dam is located in district Sabarkantha and has the dual purpose of irrigation and flood control. The weir is located on the downstream of dam but upstream of confluence of Hathmati with Sabarmati. The purpose of the weir is irrigation. On the Guhai River, a storage dam has been constructed for irrigation and flood control. Gauge and discharge measurements are taken in the basin at Bhiloda, Balochpur, Kadiadra and Himatnagar weir.

Watrak River

It originates in Aravalli hills in south Rajasthan an. After flowing for about 178 km in south-west direction, it joins Sabarmati near Pala village, about 34 km from Ahmedabad. Eru, Mazam, and Shedhi are important tributaries of Watrak. The catchment area of Watrak is about 1,114 sq. km out of which, 336 sq. km lies in Rajasthan and the rest in Gujarat. In the headwater region, the river runs in hilly and forested tracts.

12.2.3. Major Projects in Sabarmati Basin

The major water resources development projects in the Sabarmati basin are described next.

12.2.4. The Sei Dam

Sei dam has been constructed on Sei River in the Rajasthan state near village Teja Ka Bas, Kotra tehsil of district Udaipur. This project is also known as Jawai reservoir project. The latitude and longitude of the dam site are 24°23′ N and 73°11′8″ E. The main objective of this project is to supply water for irrigation in Jawai command area. The water from the catchment up to this dam site is not available for the Dharoi reservoir command.

12.2.5. The Harnav Dam

In the head reaches of Harnav River, a gated dam, known as Harnav dam, has been constructed. Also known as Harnav-II dam or Vanaj dam, the dam is located near Vanaj village, in Vijaynagar taluka, Sabarkantha district and has dual purpose of irrigation and flood control. The catchment area at the dam site is 116 sq. km. The dam was completed in the year 1990. The FRL and HFL of the reservoir are at level 332.00 m and 336.85 m, respectively. The dead storage and live storage capacity of the Harnav dam is 1.70 and 19.97 million cubic meter respectively. The crest level of the head regulator for release of water for irrigation from the dam is at 317.50 m.

Three weirs have been constructed between the dam site and Khedbrahma: (i) Mamrechi weir situated 3 km downstream of Harnav dam near Abhapur (ii) Kenyatta weir situated 12 km downstream of Harnav dam at Attarsumba Ashram, and (iii) Chhapra weir or Harnav weir situated 25 km downstream of Harnav dam. The Mamrechi and Kenyatta weirs were constructed by the former princely states. The third weir, namely Chhapra weir, was constructed in 1958 as Harnav-I. There are three gauging sites on Harnav River at Khedbrahma (downstream of Harnav weir) at Abhapur (downstream of Harnav dam) and at Harnav weir. The water requirements at the three weir sites are met from the Harnav dam and the flow from the intermediate catchment. Harnav dam is operated to meet the irrigation demands from these weirs. The Harnav River spills its water directly in the Dharoi reservoir.

Since there is no rain gauge station in the catchment area of the Harnav dam-and no gauging site upstream of the Harnav dam, estimation of inflow in the reservoir is made on the basis of rate of rise or fall of water level in the reservoir.

12.2.6. The Harnav Weir

Also known as Harnav-1 or Chhapra weir, the weir is located in district Sabarkantha downstream of Harnav dam and upstream of Khedbrahma. The weir was completed in the year 1959. The catchment area at the weir site is 427 sq. km. This single purpose project has been constructed to cater for irrigation. The spillway of the weir is 191 m long. To supply water in the command area of the Harnav weir, the water at the weir site is diverted in the Damavas and Patera tanks, which have combined, live storage capacity of 2.067 MCM. Thus, water is spilled to the Dharoi reservoir from the Harnav weir after filling the Damavas and Patera tanks to their capacity.

12.2.7. The Dharoi Dam

The Dharoi dam is the most important structure of the Sabarmati basin. It is located on Sabarmati River near village Dharoi in Kheralu taluka of district Mehsana, 103 km from the source of the river. The latitude and longitude of the dam are 24°00′ N and 72°52′ E, respectively. Its purposes are water supply to the cities of Ahmedabad and Gandhinagar, irrigation, flood control and power generation. The catchment area of the river at the dam site is 5,540 sq. km. The dam was completed in the year 1976. The FRL and HFL of the dam are at a level of 189.59 m and 175.87 m respectively. The dead storage and live storage capacity of the reservoir (as per revised capacity plan after 50 years) are 131.99 and 775.89 MCM, respectively. A view of Dharoi dam has been shown in Figure 6.

Figure 6. A view of Dharoi dam

12.2.8. Hathmati Dam

The Hathmati dam is located near Fatehpur village in Bhiloda taluka, Sabarkantha district, and serves dual purpose of irrigation and flood control. The catchment area at the dam site is 595 sq. km. The dam was completed in the year 1972. Along with the Hathmati dam, the Indrasi dam for providing necessary storage and the Navalpur waste weir for disposing off the flood water have also been constructed. The FRL and HFL of the reservoir are at a level of 180.74 m and 183.18 m, respectively. The dead storage and live storage capacity of the Hathmati dam are 3.90 and 148.93 million cubic meters, respectively. The crest level of the head regulator taking off water for irrigation from the dam is at 170.69 m.

The Indrasi dam has been built on the Indrasi River for providing enhanced storage for the Hathmati dam. The dead storage and live storage capacity of this reservoir at a level of 178.76 m (as per revised fresh capacity plan, 1988) are 0.392 and 18.80 million cubic meters, respectively. Above the level 178.76 m, the Hathmati reservoir and the Indrasi reservoir merge together. One head regulator has also been provided in this dam for releasing water for irrigation. The crest level of this head regulator is at a level of 172.21 m.

To dispose off the flood water from the Hathmati reservoir, the Navalpur waste weir has been constructed. The crest level of this ogee shaped weir is at a level of 180.74 m and it is ungated. The length of the spillway is 241 m and its discharging capacity at HFL is 2,943.2 cumec.

12.2.9. Guhai Dam

The Guhai dam is located on Guhai River near village Khandial in Himatnagar taluka of district Sabarkantha, 39 km from the source of the river. The latitude and longitude of the dam are 23°42′00″ N and 73°3′24″ E, respectively. Its purposes are irrigation and flood control. The catchment area of the Guhai River at the dam site is 422 sq. km. The dam was completed in the year 1990. The FRL and HFL of the dam are at a level of 173.00 m and 173.77 m respectively. The dead storage and live storage capacity of the reservoir are 5.30 and 57.04 MCM, respectively.

The culturable command area of this project is 11,465 ha and the annual water requirement from this reservoir for irrigation is 44.702 MCM.

12.2.10. Hathmati Weir

The Hathmati weir is located on the downstream of Hathmati dam and Guhai dam but upstream of confluence of Hathmati River, with Sabarmati River. The ogee shaped weir was remodelled in 1972–73. It has the sill level at 134.078 m and an overflow section 306.367 m long. The total catchment area of the weir is 1,357 sq. km. The design discharge of canal head regulator is 19.4 cumec. This weir is used to divert the flow of Hathmati River into the B, C and D zones (See below) of the command area of the Hathmati system for irrigation purposes. During floods, the excess flow of Hathmati River passes over the weir to join the Sabarmati River.

Hathmati canal system is one of the oldest irrigation systems of Gujarat state. The command area of this project comprises of four zones:

(i) Zone A: New areas in Bhiloda and Himatnagar Taluka of Sabarkantha District under direct command of the reservoir.
(ii) Zone B: Area of existing Hathmati canal system getting water through Himatnagar weir.
(iii) Zone C: New area beyond Bokh between Sabarmati and Khari rivers in Sabarkantha, Gandhinagar and Ahmedabad districts.
(iv) Zone D: Command area of existing Khari cut canal system. The main canal of zone A takes off from the Hathmati dam.

The capacity of the main canal in initial reach is 15.581 cumec. An escape has been provided at chainage 1,425 of main canal to release water in the Hathmati River for B, C and D zones. Upstream of the Hathmati weir, two ancient storage structures, known as Limla dam and Karol dam have been constructed for storing excess water at the Hathmati weir. This water is to be released in the command areas of B, C and D zones. The storage capacity of the Limla dam is 10.28 million cubic meter and that of Karol dam is 7.504 million cubic meters. Thus, water in excess of the irrigation demands of the B, C and D zones at the Hathmati weir can be diverted to fill these two storage structures for use in case of scarcity before spilling the water in Sabarmati River.

A head regulator of 4.816 cumec capacity has been provided in Indrasi dam for releasing water for B, C and D zones. The water released from Hathmati reservoir in the Hathmati River for B, C and D zones either through Indrasi head regulator or through escape of main canal of zone A is picked up at Himatnagar weir and is diverted in the main canal of B-zone.

During floods, water of Hathmati River is diverted in the main canal of zone B at Himatnagar weir and water is stored in Limla dam which is filled through Bokh feeder, taking off from Hathmati canal (zone B) and Karol dam which is being fed through Hathmati main canal and Karol feeder. This stored water is released through head regulator in river Khari for Khari cut canal (zone D) and the same is picked up at Raipur weir.

The main canal of zone C takes off from Bokh feeder of Hathmati main canal of zone B. As such, water diverted in main canal of zone B at Himatnagar weir is utilized in zone B, zone C, and zone D.

12.2.11. Watrak Project

Watrak dam is an earthen dam constructed near Pahadia village, about 19 km upstream of Dhupal Weir. This dam was completed in 1984 with gross and live storage capacities of 177 and 154 million m^3. The height of the dam above the deepest foundation to the spillway crest is 36 m. Spillway of the dam is ogee shaped with six gates and has been designed for a flood of 12,798 cumec. FRL of the dam is at 136.25 m and MDDL is at 126.39 m. Two canals take off from the dam to supply water for irrigation. The right bank canal runs for about 23.5 km and irrigates an area of 3,258 ha while the left bank canal has two branches and irrigates an area of 15,103 ha. At FRL, the submergence area is 32 sq. km.

12.2.12. Kalpsaar Project

The Kalpsaar project envisages construction of a dyke in the Gulf of Khambat, to isolate a portion of the gulf, and convert it into a large sweet water lake. As such this project is not in the Sabarmati basin. The project will provide water, generate electricity, and the rail/road on the dyke will also reduce the distance between Mumbai and Saurashtra by 225 km. A noteworthy feature of Kalpsaar is that the dam and the lake are both out in the ocean, and not on the land, so there is no submergence of any land area and consequently displacement etc. In addition, it is also important being tidal power project, one of the alternative and renewable sources of energy.

Besides the above, there are some more existing water resources projects in the basin. Salient features of existing water resources projects with a live storage capacity of 10 MCM and above have been presented in Table 15.

12.2.13. Water Quality

Table 16 reports the status of water quality indicators at selected locations in the Sabarmati basin at five different time periods. Note that at almost all the places, the

Table 15. Salient features of selected existing projects in Sabarmati basin

Name of the Project	Year of completion	Gross storage capacity (million cubic meter)	Live storage capacity (million cubic meter)	Designed annual irrigation (Million Sq. meter)
Bhogava – I	1959	18.51	16.78	32.40
Bhogava – II	1959	23.32	16.79	7.20
Mazam	–	43.86	77.00	62.90
Meswa Main Dam	1964	82.00	28.03	172.00
Thoriali	1958	30.16	154.33	42.10

Table 16. Desired and existing water quality levels for Sabarmati

Location	Desired class	Observed class & critical parameters				
		1997	1998	1999	2000	2001
Sabarmati at Kheroj Bridge, Gujarat	A	D BOD, Totcoli	D pH, BOD, Totcoli	C pH, BOD, Totcoli	NA	C Totcoli
Sabarmati at Gandhi Nagar Chiloda Bridge, Lekawada, Gujarat	C	D BOD, Totcoli	D Totcoli	D BOD, Totcoli	NA	NA
Sabarmati at Dharoidam, Gujarat	A	D BOD, Totcoli	D pH, BOD, Totcoli	C BOD, Totcoli	NA	D BOD
Sabarmati at Ahmedabad at V.N. Bridge, Gujarat	C	Below-E Conductivity	Below-E Conductivity	Below-E Conductivity	Below-E Cond, Fammonia	Below-E pH, Conductivity
Sabarmati at Railway Bridge Ahmedabad, Gujarat	C	D BOD, Totcoli	E BOD, Totcoli, Free Ammonia	E DO, Totcoli	NA	E DO, BOD, Totcoli
Sabarmati after conf. with Meshwa at Vautha (near Dhokla),Gujarat	D	E	Below-E Conductivity	Below-E Conductivity	Below-E DO, Cond., Fammonia	Below-E Conductivity
Sabarmati at Vill. Miroli Taluka Dascroi, Ahmedabad, Gujarat	D	E DO	Below-E DO, Conductivity	Below-E DO, Fammonia, Conductivity	Below-E Cond., Fammonia	NA

* NA- Not Available. Source: Central Pollution Control Board.

observed quality was much below the desired quality. In particular, at Ahmedabad, the observed quality was in the 'E' class and the river is more or less like a drain. The flow is very small during the summer months. Sabarmati River is notorious for being almost depleted of dissolved oxygen during summer or pre-monsoon months. However, the situation has improved after the Narmada canal has started supplying water to some places in the basin due to dilution effect. It is also a matter of concern that at some places, the quality of river water is below 'E' class.

12.3. THE MAHI BASIN

The Mahi basin extends over an area of 34,842 km which is nearly 1.1 percent of the total geographical area of the country. It lies between east longitudes 72°15′ to 78°15′ and north latitudes 22°0′ to 22°40′ N respectively. The Mahi River originates in the Mahi Kanta hills in the Vindhyachal range, in the western part of Madhya Pradesh. It is bounded on the north and north-west by the Aravalli hills, on the east by the ridge separating it from the Chambal basin, on the south by the Vindhayas and on the west by the Gulf of Cambay. The basin lies in the States of Rajasthan, Gujarat and Madhya Pradesh. The statewise distribution of drainage area is given in Table 17. An index map of the Mahi basin is presented in Figure 7.

The upper part of the basin in Rajasthan and Madhya Pradesh and comprises mostly of hills and forests except the lower half in M.P., which is fairly plain. The central part lying in Gujarat consists of developed lands. The lower part of the basin lying in Gujarat is flat and fertile and well developed alluvial tract. Important soil types in the basin are red and black soils. The culturable area of the basin is about 2.21 Mha which is 1.1 percent of the total culturable area of the country.

The Mahi River originates in the northern slope of the Vindhyachal ranges in MP at an elevation of nearly 500 m and flows for about 583 km before outfalling into the Arabian Sea. It flows southwards for about 120 km in MP before entering Banswara District of Rajasthan. The river makes a 'U' shaped loop in Rajasthan before entering Gujarat and finally discharges into the Gulf of Cambay. Being a monsoonal river it is found that the inflow from seven dry months is less than 5% of the total inflow. The principal tributaries of the Mahi River are Som, Jakham, Moran, Anas, and the Bhadar.

Table 17. Statewise drainage area of Mahi River

State	Drainage area (km^2)
Rajasthan	16, 453
Gujarat	11, 694
Madhya Pradesh	6, 695
Total	34, 842

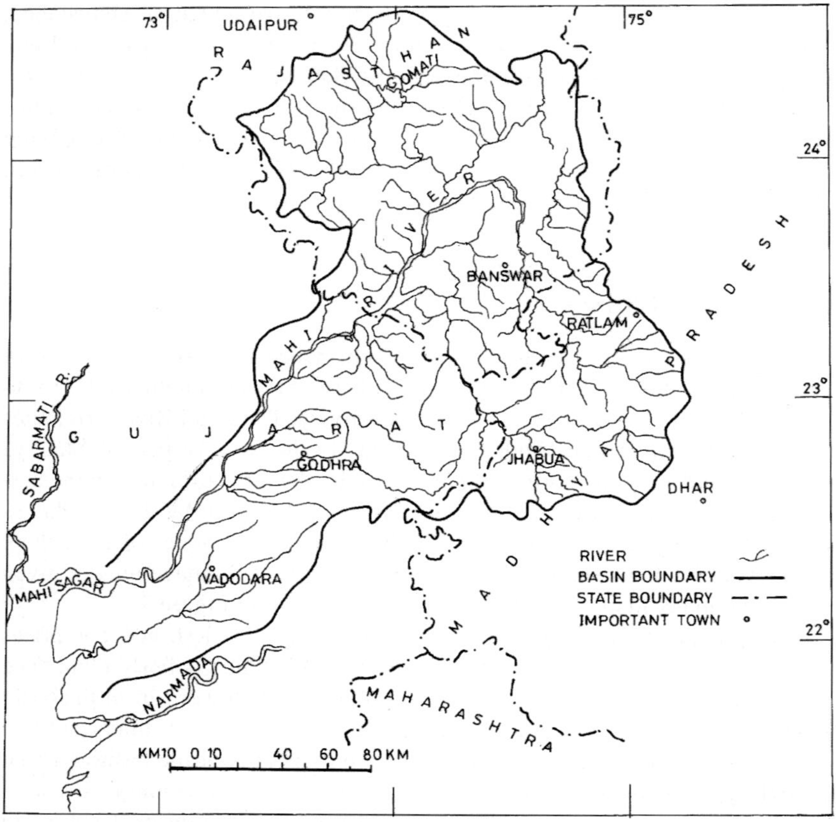

Figure 7. Index map of Mahi basin

The mean annual rainfall over the Mahi basin is around 700 mm, of which 94% falls during four monsoon months.

12.3.1. Major Tributaries of the Mahi River

In the following, we describe the major tributaries of Mahi River. Figure 8 Shows the stream network of Mahi basin.

The Som

The River Som originates in the hills near the village Som in Kherwara Tehsil of Udaipur District at an elevation of 600 m. It flows southeast through a hilly region and joins river Mahi near the village Baneshwar. The entire catchment lies in Udaipur and Dungarpur Districts. The tributaries of the Som River are Tidi, Gomti and Jakham. The catchment area of the Som sub basin is $6,443\,km^2$.

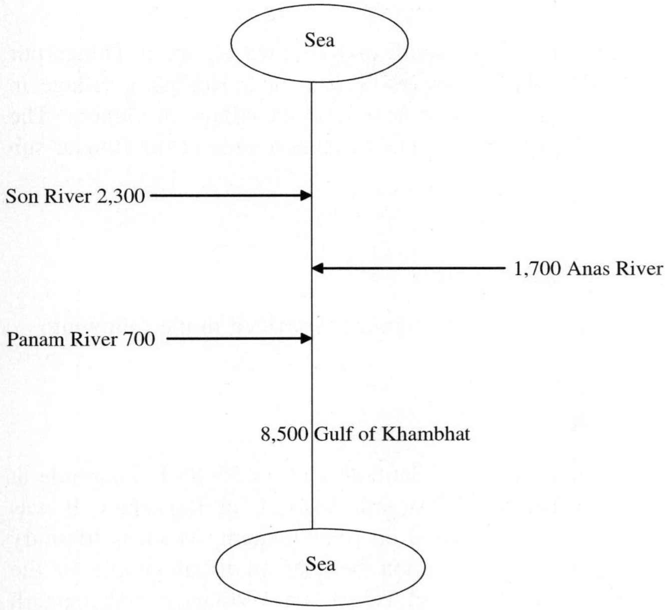

Figure 8. Stream network of Mahi basin

The Jakham

The River Jakham originates southwest of the hills near Chotti Sadari in Chittorgarh District. It flows through the hilly region of Udaipur District in a south-western direction and joins river Som near Bilara village. The Sub-Basin is situated in Chittor and Udaipur Districts. Karmai and Sukli are the major tributaries of the Jakham River. The catchment area of the Jakham sub basin is $2,318\,\text{km}^2$.

The Moran

River Moran originates in the southern hills of Dungarpur town. It flows through Dungarpur District and joins Mahi River near Galiakot village. The total catchment of river lies in Dungarpur District. Karmai and Sukli are the tributaries of The Moran River. The catchment area of the Moran sub basin is $735\,\text{km}^2$.

The Anas

The River Anas originates in the northern slopes of the Vindhyan ranges, near Amber village in MP. It flows in a north-western direction, enters Rajasthan near Meledikhera village and joins Mahi River about 15 km downstream of Galiakot village. The Sub-Basin in Rajasthan is situated in Banswara District. The Hiran is the tributary of the Anas River. The catchment area of the Anas sub basin is $1,441\,\text{km}^2$.

The Bhadar

The river Bhadar, originating in the hills south of Kanguwa village in Dungarpur District, flows from north to south and enters Gujarat near Kokhakra village in Dungarpur District, where it joins the Mahi near Karanta village in Gujarat. The catchment is situated in Dungarpur District. The catchment area of the Bhadar sub basin is $6,047\,\text{km}^2$.

12.3.2. WRD Projects in Mahi Basin

Selected important WRD projects in Mahi basin are described in the following.

12.3.3. Jakham Reservoir

Jakham reservoir is located at 24°10′30″ N latitude and 74°35′30″ E longitude at village Anuppura, Pratapgarh Tehsil, Chittorgarh District, in Rajasthan. It was completed in the year 1986. It is constructed on river Jakham, which is tributary of river Mahi. The project provides irrigation benefits to tribal people of the area consisting of 104 villages of Dhariawad Tehsil and 3 villages of Pratapgarh Tehsil. The area near the dam is hilly and consists of waste land, hence a pickup weir at Nagalia, which is 13 km away from dam is constructed from which main canal emerges.

The project has a catchment area of 1,010 sq. km with average yield as 212.376 MCM. It lies in Pratapgarh and Chotisadri Tehsils of Chittorgarh in agroclimatic zone IVB. The irrigated area in catchment covers 95.85 sq. km (9.49%) and the unirrigated cultivable area is 366.32 sq km (36.27%). Around 45% of the catchment area has sufficient soil cover for raising crops and this area may contribute to soil erosion to Jakham reservoir. The Forest area covers 23.52% of catchment area. Soils in the area vary considerably. The soils are silty loam to clay loam, occasionally clay, and grayish brown to dark grayish brown. Soils are deep and occur as buried pediments. These appear to be derived from alluvium of phyllites and limestone.

The catchment area falls in typical sub-tropical sub-humid to humid climate conditions, characterized by mild winter and moderate summer with high relative humidity during the months of July-September. Mean monthly minimum temperature ranges from 11 °C (Jan.) to 26 °C (June) and mean monthly maximum temperature ranges from 21.8 °C (Jan) to 43.8 °C (May). The average annual rainfall varies between 800–900 mm. Mean annual PET varies from 1,301 to 1,400 mm.

The Top of dam is at RL 373.0m with HFL at RL 371.65m. The dead storage level of reservoir is at RL 332.0m. The length of dam is 253.0 m with maximum height as 81.0m. Two main canals, Right Main Canal (RMC) with length as 23.76 km and Left main canal (LMC) of 39.90 km length emerge from the pickup weir. The canals are lined with design discharge of 3.533 cumec (RMC) and 7.92 cumec (LMC).

12.3.4. Panam Dam

Panam dam in Santrampur Taluka of Panchmahals district in Gujarat was first impounded in the year 1977. It is a multipurpose project constructed on Panam River, a tributary of Mahi. Panam River originates in Devgadh Baria Taluka of Panchamahals district. After flowing mostly through hills in North – westerly direction, it joins Mahi River about 25 km downstream of Panam dam. The dam site is located at 23°3′00″ North latitude and 73°42′00″ East longitude in village Kel-Dezar. Nearest railway station is Godhra, which is 45 km from dam.

Total catchment area at the dam site is 2,312 sq. km. The catchment is mostly hilly and covered with forests except near the dam site where it is relatively flatter. The reservoir has a designed gross reservoir capacity of 737.987 MCM with live capacity being 689.567 MCM. Top of dam is at RL 131.41 m, with HFL at 128m. This masonry dam is 56.36 m high and 269.45 m long, with 182 m long spillway. There are 10 radial gates of size 14.93 m × 11.28 m. Irrigation potential of Panam dam is 493.70 sq km. Soils of the catchment are derived from rocks like quartzites, schists and phyllites. Deep soils cover about 79% of the culturable.

Project area experiences tropical climate with minimum temperature of 12 °C in January and maximum temperature of 39 °C in May. Average annual rainfall in the area is 940 mm. About 80% of the rainfall occurs during July and August. Panam dam has total GCA of 582.73 sq. km with CCA as 411.16 sq. km. The command is entirely located on the left bank of river.

12.3.5. Mahi Bajaj Sagar Project

Mahi Dam is a straight masonry dam on Mahi River, located at a distance of 56 km from Banswara in Banswara District, Rajasthan. The length of the masonry and earthen parts of the dam is 435 m and 2,627 m respectively. The reservoir behind the 43 m high dam is known as Mahi Bajaj Sagar reservoir. The catchment area of the river up to the dam site is 6,149 sq. km. Mahi Bajaj Sagar reservoir has a live storage capacity of 2,070 MCM at FRL of 280.75 m and the MDDL of the project is at 259 m.

Two power stations namely Mahi I and Mahi II have been constructed under the development plan. Mahi I power house, located at Mahi dam, has 2 units of 25 MW each. Mahi II power house is located at a distance of 45 km from the dam site. Mahi II power house uses the tail waters of upstream Mahi I power house. Mahi II power house has 2 units of 45 MW each. Mahi I and Mahi II power projects have the firm power of 15 MW and 23 MW respectively. The minimum and maximum annual inflow of the Mahi I project is 435 MCM and 6,618 MCM respectively. RRVPN Ltd. commissioned the projects in 1986 and 1989 respectively.

12.3.6. Kadana Project

Kadana is masonry cum earthen dam on Mahi River, located at a distance of 80 km from Godhra in Panchmahal District, Gujarat. The catchment area at the dam is

$25,520\,km^2$ and the mean annual inflow of the reservoir is 596.432 MCM. The height and length of the dam is 66 m and 575 m respectively. The reservoir behind the dam has a live storage capacity of 1,203 MCM at FRL at 127.7 m and the MDDL is at 114.3 m. Kadana power house has 4 units of 60 MW each. It has been commissioned by the Gujarat Electricity Board in 1990–98. On an average, about 2,900 MCM water is used per year to produce 316 million units at the power house.

There are many other existing water resources projects in the basin. Salient features of the projects with a live storage capacity of 10 MCM and more are presented in Table 18.

Table 19 gives the salient features of selected projects that are under construction in the basin.

Table 18. Salient features of selected existing projects in Mahi basin

Name of the Project	State	Year of completion	Gross storage capacity (million m^3)	Live storage capacity (million m^3)	Designed annual irrigation (million m^2)	Installed capacity (MW)
Badar	Gujarat	–	46.72	38.00	89.00	–
Hadaf	Gujarat	–	32.26	25.02	58.20	–
Karad	Gujarat	1964	35.38	33.00	45.40	–
Machhen Nala	Gujarat	–	37.91	29.17	30.30	–
Patadungri	Gujarat	1950	41.06	39.64	45.60	–
Umaria	Gujarat	–	13.54	11.67	21.90	–
Wankhori	Gujarat	1958	41.88	36.22	2, 631.50	–
Wankleshwar Bhay	Gujarat	–	14.30	12.67	–	–
Haro	Rajasthan	1958	11.91	11.06	–	–
Jaismand	Rajasthan	1700	414.89	172.27	118.90	–
Lodhisarkanaka	Rajasthan	1958	11.94	10.05	11.00	–
Surwania	Rajasthan	1963	13.20	11.66	12.00	–

Table 19. List of the Projects under construction in Mahi Basin

Name of the Project	State	Gross storage capacity (million cubic meter)	Live storage capacity (million m^3)	Designed annual irrigation (Million m^2)
Edalwada	Gujarat	11.33	10.53	13.80
Dholwad	Madhya Pradesh	54.29	49.94	64.00
Daya Irrigation	Rajasthan	12.40	11.21	12.80
Somkagdar	Rajasthan	35.20	32.70	49.50
Somkamla Amba	Rajasthan	118.65	106.15	136.10

12.3.7. Water Quality

Table 20 gives the status of water quality indicators at selected locations in the Mahi basin at five different time periods. There does not seem to be much problem from water quality point of view in the basin.

Table 20. Desired and existing water quality levels for Mahi

Location	Desired class	Observed class & critical parameters				
		1997	1998	1999	2000	2001
Mahi at Badnawar, M.P.	A	NA	B Totcoli	B Totcoli	B Totcoli	B Totcoli
Mahi (D/S) conf with R. Chap (Under Sagwara-Sarhi Rd. Bdg.), Rajasthan	C	C	B	C	C	B
Mahi after conf. with Anas at Pardi (Banaswada), Gujarat.	A	C BOD, Totcoli	A	NA	NA	NA
Mahi near Rajasthan Border at Kadana Dam, Gujarat	C	C	A	NA	NA	B
Mahi at Virpur, Gujarat	C	D BOD	A	NA	NA	C
Mahi at Sevalia, Gujarat	C	C	A	NA	NA	C
Mahi at Vasad, Gujarat	C	C	B	NA	NA	C

* NA- Not Available. Source: Central Pollution Control Board.

CHAPTER 13

MAHANADI, SUBERNAREKHA AND BRAHMANI BASINS

In this chapter, Mahanadi, Subernarekha, Brahmani, and Baitarani basins are described. Among these, Mahanadi is the biggest and the most important.

13.1. THE MAHANADI BASIN

The Mahanadi basin extends over an area of 141,589 km^2 which is nearly 4.3% of the total geographical area of the country. It lies between east longitudes 80°30′ to 86°50′ and north latitudes 19°21′ to 23°35′. It is bounded on the north by the Central India hills, on the south and east by the Eastern Ghats and on the west by the Maikala range. The upper basin is a saucer shaped and mostly lies in Chhattisgarh state. The basin lies in the States of Orissa, Bihar, Chhattisgarh, and Maharashtra. The basin is circular in shape with a diameter of about 400 km and an exit passage of about 160 km length and 60 km breadth. The statewise distribution of the drainage area is given in Table 1. A map of the Mahanadi basin is given in Figure 1.

Physiographically, the basin can be divided into four regions, namely, the northern plateau, the Eastern Ghats, the coastal plain and the erosional plains of central table land. The first two are hilly regions. The coastal plain is the delta area which is highly fertile. The central table land is the central interior region of the basin, traversed by the river and its tributaries. The basin has a culturable area of about 79,900 km^2 which is about 57% of the basin area and four percent of the total culturable area of the country.

13.1.1. The Mahanadi River

The Mahanadi River is one of the major rivers which flow from west to east and finally drain into the Bay of Bengal. The water potential of the Mahanadi River is next to Godavari in the peninsular rivers. The Mahanadi River commences in southeastern Chhattisgarh as a small stream draining the eastern part of the Chhattisgarh Plain.

Table 1. Statewise distribution of drainage area of Mahanadi River

State	Drainage area (km^2)
Chhattisgarh	75, 136
Orissa	65, 580
Bihar	635
Maharashtra	238
Total	141, 589

Source: Orissa Government.

The Mahanadi River rises in a pool, 6 km from Pharsiya village near Nagri town in Raipur District of Chhattisgarh at a height of 442 m. Initially for about 56 km, the river flows towards west in a shallow valley between low, scattered hills. Four small streams join the river near Kanker where it takes a sharp turn to northwest. While flowing about 113 km in the same direction, at Rajim, the Pairi River joins from the right. About 13 km above Sheorinarayan, the first major tributary Sheonath joins the Mahanadi near Khargahni in Bilaspur District. Beyond this confluence, the Mahanadi takes an easterly course for a length of about 138 km. At Sherorinarayan the tributary Jonk joins from the right. About 17 km further down near Mahuadih,

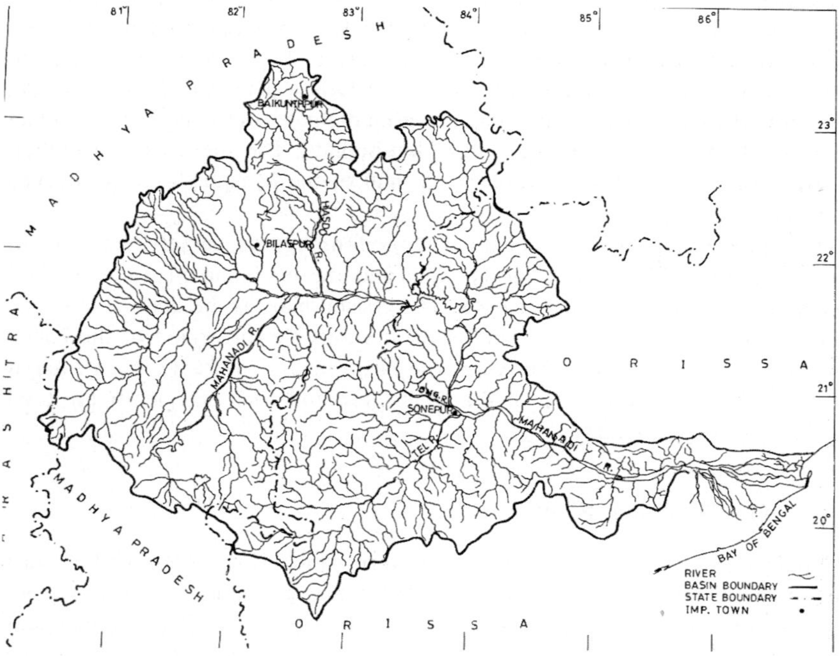

Figure 1. Index map of Mahanadi basin

the Hasdo joins from the left. It forms a braided course of about 21 km long while receiving the Barai River on its left. Further down, it is joined by the Mand River from the left at Chandarpur. After traversing a further length of 28 km, Mahanadi leaves Chhattisgarh and enters Orissa.

The Ib River joins from the left near Bagra, now flowing into the reservoir created by the Hirakud dam which is located across Mahanadi 10 km from Sambalpur city. Below Sambalpur, the river turns south, then splits near Charpali into two channels which unite again near Dhama. Between Hirakud and Chiplima, a distance of about 27 km, the bed is rocky with a number of rapids. About 11 km below Dhama, the Mahanadi enters the district of Bolangir and flowing in a southerly and south-easterly direction for a length of about 45 km it reaches Sonepur. About 11 km upstream of Sonepur, the Ong falls into the Mahanadi from the right. Near Sonepur, the river takes a gradual turn to the south-east, and is joined on its right by the second biggest tributary, the Tel. The Eastern Ghats mountain chain starts from this place. At Haudh, the river again splits into two arms. Beyond Athmalik, the valley narrows down sharply and between Jamudeli and Baramul, for a distance of about 23 km, the river flows through the extremely narrow Satkosis Gorge. Tikarpara village is about 6 km below the start of this gorge. This gorge ends at Baramul and the river crosses the Eastern Ghats. The valley of the Mahanadi between Baramul and Baideshwar is flat with scattered hillocks. Below Baramul the river widens again, attaining a width of about 1.6 km. At Kantilo, it turns east-north east. Near Pathpur, it flows through the Kaimundi gorge. Below this gorge, the river widens again and takes a sharp turn to the left, finally to emerge into the delta at Naraj, 11 km west of Cuttack. Below Naraj the river splits into two channels, the Katjuri and the Birupa. Splitting into numerous branches, the Mahanadi River falls into the Bay of Bengal, near a place known as False Point.

Mahanadi flows for a total length of about 851 km of which, 357 km is in Chhattisgarh and the balance of 494 km is in Orissa. The river enters Orissa State below Baloda Bazaar and crosses the Eastern Ghats to enter the Plains of Orissa near Cuttack. It finally debouches into the Bay of Bengal through a series of branches. The famous pilgrimage city of Puri is at the mouth of one of these branches.

13.1.2. Tributaries of Mahanadi River

The Seonath, the Jonk, the Hasdo, the Mand, the Ib, the Ong, and the Tel are the principal tributaries of Mahanadi. Important tributary/sub-basins and their catchment areas are shown in Table 2. A description of major tributaries is given in what follows. Flow Diagram of Mahanadi Basin has been presented in Figure 2.

i. The Seonath

The Seonath River is the longest tributary of the Mahanadi River. It rises in an undulating region with numerous small groups of hills at an elevation of about 533 m near Kotgal and flows for about 383 km to join the Mahanadi on its left near Kharghand. The main sub tributaries of the Seonath are the Kharhara, the Tandula

Table 2. Important sub-basins of Mahanadi and their catchment areas

Name of the tributary/sub-basin	Area (km^2)	Percentage of area of Mahanadi basin
Seonath	30,761	21.72
Jonk	3,484	2.47
Hasdeo	9,856	6.96
Mand	5,200	3.67
Ib	12,447	8.79
Upper Mahanadi	21,652	15.29
Ong	5,128	3.62
Tel	22,818	16.12
Middle Mahanadi	12,654	8.93
Lower Mahanadi	17,589	12.43
Total	141,589	100.00

Source: WG (1990).

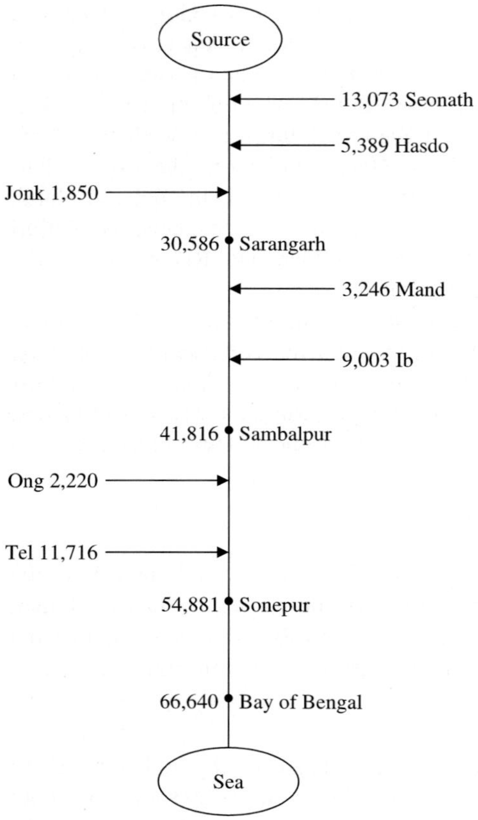

Figure 2. Flow Diagram of Mahanadi Basin. The numbers represent average annual flow in cumec

and the Kharun on the right bank and the Surhi, the Hanp, the Agar and the Arpa on the left bank. The total drainage area of the Seonath River is about 30,761 km^2 which is nearly 22% of the total drainage area of the Mahanadi basin.

ii. The Jonk

The Jonk River originates from the Khariar hills in the Kalahandi district of Orissa State at an elevation of about 762 m. The Jonk River flows for a length of about 196 km to join the Mahanadi on its right at Sheorinarayan. The total drainage area of the river is about 3,484 km^2.

iii. The Hasdo River

The Hasdo River rises at an elevation of about 915 m at a place nearly 10 km north of Sonhat in the Sarguja district of Chhattisgarh. The river traverses a distance of about 333 km to join the Mahanadi on its left near village Mahuadih. The total drainage area of the river is about 9,856 km^2. Gej River is the principal sub tributary of the Hasdo River.

iv. The Mand River

The river originates at an elevation of about 686 m in the Sarguja district of Chhattisgarh and flows for about 241 km to its confluence with the Mahanadi on its left near Chandarpur. The total catchment area of the river is about 5,200 km^2.

v. The Ib River

The Ib River originates in the hills near Pandrapat, in the Raigarh district of Chhattisgarh at an elevation of about 762 m. The total length of the river is about 251 km and it falls now into the Hirakud Reservoir on its left. The total drainage area of the river is 12,447 km^2. It is rain-fed river and hence nearly 80% of runoff occurs during the monsoon months of June to October. Geologically, granite and gneisses underlie the catchment's soil mantle. A mixed red and black soil dominates the region surrounding the reservoir. For about 40 km in the lower reaches, a part of the catchment is submerged under the Hirakud Reservoir.

vi. The Ong River

Ong is another important tributary of the Mahanadi River. It rises at an elevation of 457 m on a hill in the northern outskirts of the south-north running range of mountains situated to the right of the Jonk River. The Ong River traverses a total length of about 204 km to join the Mahanadi on its right about 11 km above Sonepur. The drainage area of the Ong River is 5,128 km^2 lying in the interior of the main basin. The climate of the Ong basin is mainly dry sub-humid. There are four distinct seasons in the year: (i) the cold weather, (ii) the hot weather, (iii) the south-west monsoon, and (iv) the post-monsoon. The mean minimum temperature ranges between 1 °C and 13.7 °C. The maximum temperature ranges from 38 °C over the hills to 43 °C. In the plains, south-west monsoon is the principal rainy season. During this period the basin receives over 90% of its total annual rainfall. A variety of soils are found in the basin: red, yellow and black.

vii. The Tel River

The Tel River originates in plain and open country in the Koraput district of Orissa, about 32 km to the west of Jorigam. The river traverses a total length of 296 km to join the Mahanadi River on the right bank, 1.6 km below Sonepur. The total drainage area of the Tel River is about 22,818 km^2.

13.1.3. Climate of Mahanadi Basin

To the north, in the region of the Mahanadi River the climate is predominantly sub-tropical with summer temperatures of around 29 °C and winter temperatures of 21 °C. The normal time of onset of monsoon over the basin is the first week of June. The bulk of the precipitation (800 to over 1,600 mm) over the basin falls in the period from July to September while during January to February, precipitation received is less than 50 mm.

The meteorology and climatology of the catchment are significantly influenced by the geographical location of the catchment with respect to the Bay of Bengal, where from most of the weather systems originate. Also, the orography of the Eastern Ghats, influence the rainfall pattern over the catchment to a great extent. The south-west monsoon normally sets over this area in the first week of June and withdraws in the first week of October. The south-west monsoon (June-October) accounts for nearly 91% of the annual rainfall. December is the driest month contributing less than 10% of annual rainfall.

i. Rainfall

The normal (1901-60) annual rainfall of the Mahanadi catchment is 141.7 cm. The average normal annual rainfall of the catchment above Hirakud is 139.6 cm and the corresponding value for catchment below Hirakud is 145.8 cm. Historical records show that the highest monthly and annual rainfalls were 1,405.9 mm in June 1936 and 3,669.8 mm in 1944 respectively at Bulandarpara.

Besides the observatories maintained by IMD, about 200 raingauge stations are maintained by the concerned States. Although distribution of the stations is fairly even, their number is not sufficient considering the hilly terrain which occupies the major part of the catchment. Rainfall data at some of the observatories and raingauge stations are available from past 80 years.

ii. Temperature

In the Mahanadi catchment, May is the hottest month and December the coldest. The diurnal range of temperature is the maximum during February and March; it is less during July and August. At Raipur and Sambalpur, the temperature varies from 12 °C to 40 °C while at Cuttack, it varies from 14 °C to 40 °C. Temperature variation at Puri which is closer to sea is from 16 °C to 32 °C.

iii. Evaporation

Pan evaporation is being observed at four stations in the Mahanadi basin, namely, Labhandi, Hirakud, Bolangir and Cuttack. Out of these, long-term pan evaporation data is available for Labhandi and Cuttack stations. The average monthly pan evaporation of these stations is given in Table 3.

iv. Soils and Land Use

The main soil types found in the basin are red and yellow soils. Mixed red and black soils occur in parts of the Bolangir, Sambalpur, and Sundargarh districts of Orissa. Laterite soil is found in the lower parts of Orissa. The deltaic soil is found in the coastal plains of the Mahanadi. Black soil and sandy soil with "Kankar" are the main soils found in the part of basin lying in Chhattisgarh.

Except in the Chhattisgarh and coastal plains, the basin has an extensive area under forests. Forest and agriculture are the main stay of the people in the interior parts of the basin. The Chhattisgarh and coastal plains, with a high incidence of rainfall, are predominantly rice growing areas.

v. Stream-Gauge Network

The gauge and discharge measurements are being made at several locations on Mahanadi River as well as its tributaries. The list of such gauging stations is given in Table 4.

13.1.4. Water Availability in Mahanadi Basin

An average annual surface water potential of 66,900 Mm^3 has been assessed in this basin. Out of this, 50,000 Mm^3 is utilizable water. The present use of surface

Table 3. Pan evaporation data (cm) of two stations in Mahanadi Basin

Month	Station	
	Labhandi	Cuttack
January	3.0	3.3
February	4.9	4.4
March	6.8	5.3
April	10.9	6.6
May	14.6	7.5
June	11.8	5.6
July	5.7	3.9
August	4.9	3.4
September	4.1	3.9
October	3.7	3.6
November	3.0	3.3
December	2.4	3.0

Table 4. Gauge-Discharge sites on Mahanadi River and its tributaries

S.N.	Name of Gauge-Discharge Site	River	Catchment Area at the site (km^2)	Annual average runoff (BCM)	Year of Starting
1	Rajim	Mahanadi	8,760	3.09	1971
2	Ghorari	Mahanadi	13,209	–	1963
3	Shrinarayaneo	Mahanadi	46,620	–	1961
4	Saradih	Mahanadi	57,029	–	1957
5	Basantpur	Mahanadi	57,780	24.43	1971
6	Sambalpur	Mahanadi	83,698	–	1926
7	Khairmal	Mahanadi	115,514	–	1957
8	Tikarpara	Mahanadi	124,000	52.95	1947
9	Naraj	Mahanadi	131,720	–	1926
10	Baronda	Pairi	3,225	–	1977
11	Kotni	Seonath	6,990	2.47	1977
12	Simga	Seonath	16,060	6.02	1971
13	Jondhra	Seonath	29,645	10.62	1979
14	Andhiyarkoro	Hamp	2,210	0.42	1977
15	Ghatora	Arpa	3,035	1.51	1977
16	Bamnidhi	Hasdeo	9,730	4.80	1971
17	Rampur	Jonk	2,920	1.62	1971
18	Kurubhata	Mand	4,625	3.02	1977
19	Sundergarh	Ib	5,870	3.85	1977
20	Deogaon	Ib	8,676	–	1957
21	Guchapalli	Ong	3,444	–	1976
22	Salebhata	Ong	4,650	2.04	1971
23	Kesinga	Tel	11,960	5.70	1977
24	Kantamal/Deoli	Tel	19,600	10.40	1971

water in the basin is 17,000 Mm^3. Live storage capacity in the basin has increased significantly since independence. From just about 800 Mm^3 in the pre-plan period, the total live storage capacity of the completed projects has increased to 8,500 Mm^3. In addition, a substantial storage quantity of over 5,400 Mm^3 would be created on completion of projects under construction. Additional storage to the tune of over 11,000 Mm^3 would become available on execution of projects under consideration. The hydropower potential of the basin has been assessed as 627 MW at 60% load factor.

NIH (1986) estimated water availability at five sites in the Mahanadi Basin. Table 5 summarizes the results of this study.

The Small Catchment Directorate of CWC has derived synthetic unit hydrographs for many sub-zones of India as described in Chapter 5. For the sub-zone 3-d that falls in the Mahanadi basin, the equations established by regression analysis are given in Chapter 5. In addition to UH ordinates, the concerned report also gives 50-year 1-hour, 3-hour, 6-hour, 9-hour, 12-hour, 15-hour, 18-hour, and 24-hour rainfall maps. These maps are used to determine design storm for the catchment in question.

Table 5. Estimates of water availability for various sites in Mahanadi Basin

Name of site (River)	Catchment area (km^2)	Water availability in Mm^3 for percentage of time the flow is equal or exceeded		
		50%	75%	90%
Salebhata (Ong)	4,650	2,246.90	1,723.20	1,406.10
Kantamal (Tel)	19,600	11,147.50	9,121.30	7,632.80
Hirakud (Mahanadi)	83,400	41,345.50	34,106.40	24,390.30
Tikarpara (Mahanadi)	124,000	65,636.00	54,838.20	43,248.00
Naraj (Mahanadi)	131,720	69,340.90	57,667.90	44,948.80

13.1.5. Major and Medium Projects in Mahanadi Basin

Given the large size of the Mahanadi basin and plenty of available water, it is natural that a number of projects have been constructed for utilization of these resources. Brief description of the major projects follows.

i. Hirakud Dam

Hirakud is one of the earliest and prestigious major multi-purpose river valley projects in India after independence. Commissioned in 1957, the reservoir is situated a little downstream of the confluence of Mahanadi with its tributary Ib, 15 km upstream of Sambalpur town. Situated within the geographical ordinates of 21°30′ and 21°50′ N, and 83°30′ and 84°05′ E, the reservoir has a water spread area of 719.63 km^2 at FRL. The 1,248 m long masonry dam is 61 m high and this, along with the earthen dams, has a combined length of 25.8 km. At the dam site, the maximum annual runoff was 91,900 Mm^3 while the minimum annual runoff was 12,400 Mm^3. The total catchment area up to the dam is 83,400 km^2. With gross storage capacity of 5,818 Mm^3, this is one of the biggest reservoirs. The spillway capacity at FRL is 41,428 cumec.

The average annual rainfall in the region is 152 cm. More than 65% of the vast catchment area stretching over the central Indian plateau is fertile land area. The salient features of Hirakud Reservoir are given in Table 6.

The project has been designed to serve three purposes, namely flood control, irrigation and power. In addition it was planned to supplement supplies to the old irrigation system in the Mahanadi delta. Now, the reservoir serves the irrigation needs of 2,640.38 km^2 of land. The water released through power house irrigates further 4,360 km^2 of CCA in Mahanadi delta. A hydropower plant at the dam has 307.5 MW of installed capacity. Besides, the reservoir produces a fish crop of 350 ton every year. The reservoir space is also used to provide flood protection to 9,500 km^2 of delta area in district of Cuttack and Puri. A view of Hirakud dam has been shown in Figure 3.

ii. Ravishankar Sagar

The Ravishankar Sagar project (RSP) dam was constructed in 1978 on the Mahanadi River, at 20°38′ N, 81°34′ E. RSP is about 92 km south of the city of Raipur. It is

Table 6. Salient features of Hirakud reservoir

Elevation at FRL	192.024 m
Elevation at DSL	179.830 m
Gross storage capacity	8,136 Mm^3
Live storage capacity	5,818 Mm^3
Dead storage capacity	2,318 Mm^3
Water spread area at FRL	719.63 km^2
Length of the masonry dam	1,248 m
Number of sluices	64
Number of crest gates	34
Crest level of spillway dam	185.928 m
Maximum spillway capacity	41,428 cumec

a multi-purpose reservoir which serves irrigation, hydro-electric power-generation and the industrial requirements of the Bhilai Steel Plant. At the FRL of 348.7 m, the reservoir surface area is 95.40 km^2. The total catchment area is estimated at 3,620 km^2, of which 625 km^2 is intercepted by the upstream dam, Dudhawa and

Figure 3. A view of Hirakud dam

486 km^2 by Murumsilli reservoir. At full level, the reservoir storage capacity is 909 Mm^3. The maximum depth of the reservoir is about 32 m.

The off-taking channels carry water in the order of 11,000 to 30,000 cumec in the peak season. However, the outlets are rarely completely closed. Water level fluctuates by 3 to 5 m in a year. The sediments of Ravishankar Sagar are poor in nutrients and organic matter.

iii. Dudhawa Reservoir

The Dudhawa reservoir is situated at 81°45′21″ E longitude and 20°18′1″ N latitude across Mahanadi River near Dudhawa village about 21 km west of Sihawa near the origin of Mahanadi river and 29 km east of Kanker. The reservoir is in Dhamtari district of Chhattisgarh state. The construction of the project started in 1953–54 and it was commissioned in 1963–64. This reservoir is designed to supply water to Ravishankar Sagar Project complex thereby increasing its irrigation potential. Water will also be provided to additional culturable areas under the command of the existing Mahanadi Tandula canal system. The maximum height of this earthen dam is 24.53 m and length is 2,906.43 m. Two subsidiary bunds of the dam have heights of 6.61 m and 2.83 m and lengths 568.42 and 426.70 m, respectively. The catchment area of the reservoir is 625.27 km^2 and gross command area is 566.80 km^2. At the Full Reservoir Level (FRL), the submergence area of the reservoir is 44.80 km^2.

iv. Sondur Reservoir

The Sondur reservoir is constructed at 82°6′ E longitude and 20°14′ N latitude across Sondur River a tributary of Mahanadi. Located near Gram Machka, Nagri block, Dhamtari district of Chhattisgarh state, the dam was constructed in the year 1988. The catchment area of Sondur River up to the dam site is 518 km^2. Major portion of the catchment lies in Dhamtari district of Chhattisgarh and Koraput district of Orissa state. Sondur project comprises of a 3.33 km long composite dam. This consists of a 191.25 m long masonry dam at the center which includes overflow and non overflow portions; the rest is earthen dam at both the flanks. The spillway has 5 radial gates of size 15m × 10m each. Irrigation sluice is provided at the left flank.

The project is also designed to supply water to RSP complex through Dudhawa reservoir thereby augmenting the irrigation potential of RSP complex for irrigation. It would also provide irrigation to about 122.60 km^2 of Kharif and Rabi crops in Sihawa Nagri block. The designed rate of sedimentation in gross storage is 0.357 mm /year.

Anicuts built on Mahanadi and Baitarani Systems are functional as part of Orissa Canal system, which currently irrigates 3,500 km^2 in Mahanadi and Baitarani basins mostly and also in Brahmani basin by exporting water from Mahanadi. Two of the canals of "Orissa Canal System" starting from Birupa weir (on Birupa River, a branch of Mahanadi River) namely Kendrapara and Pattamundai canals start from the right of the river and proceed to irrigate to the right of Birupa. The third, High

Level Canal, Range I originating from the Left of the weir crosses Birupa-Brahmani watershed to enter the Brahmani basin.

Two tributaries, namely Ong and Tel, join Mahanadi downstream of Hirakud dam. Both these carry large volumes of flow during monsoons. The average monsoon runoff at Tikarpara site is 65,636.00 MCM. In absence of any large storage dam lot of flow of Mahanadi goes to the sea and a small portion is utilized in Mahanadi delta. There is an imperative need for construction of a large terminal storage to conserve the precious resource and utilize it to meet the reasonable needs of the basin and transfer the surplus to water short areas. A dam site at Manibhadra dam has been identified and investigated but no progress seems to have been made. Opposition to the proposed dam is on account of large population displacement and other reasons. Live storage capacity of the proposed dam is 6,000 Mm^3. Studies have shown that Mahanadi has surplus water (Govt. of Orissa has not agreed with this conclusion) that is proposed to be transferred from Manibhadra reservoir to the Dowlaiswaram barrage on the Godavari. In fact, the proposed Manibhadra dam is the starting point of the peninsular component of NWDA's ILR proposal.

v. Hasdeo Bango

Minimata Hasdeo Bango is a multipurpose storage reservoir on Hasdeo River, tributary of Mahanadi River, 70 km from Korba, in Korba District, Chhattisgarh. The catchment area at the dam is 6,730 km^2. The masonry gravity dam is 87 m high. The FRL and the MDDL of the reservoir are 359.66 m and 329.79 m and it has a live storage capacity of 3,040 MCM. Mean annual inflow to the reservoir is 3,540 MCM. The power house has 3 units of 40 MW each and a firm power of 20 MW.

vi. Tandula

This is an important project of Chattisgarh state. The dam is located in Balod tehsil of Durg district at about 5 km from the Balod city. A dam was completed on the confluence of Sukha Nala and Tandula River in 1921, with a catchment area of 827.2 sq. km. The gross, live, and dead storage capacities of the reservoir are 312.25 MCM, 302.28 MCM and 99.67 MCM respectively. For the reservoir, the highest flood level, the FRL, and MDDL are 333.415 m, 332.19 m, and 320.445 m. A canal takes of from the dam to provide irrigation to 68,219 ha of Kharif crop. Main canal and distributaries run for about 110 km and the length of minors is 880 km. The monsoon rainfall in the command is about 1,293 mm.

In the 1950s, it was realized that the Tandula reservoir is unable to meet the demands of the command and hence a reservoir, named Gondali reservoir, was created on Jujhara Nala in 1957 and a supplementary canal of 9 km length was constructed to supply water from Gondali reservoir to Tandula reservoir. After construction of Bhilai Steel Plant in 1956, water is being supplied to this plant from the Gondali reservoir and supply for irrigation has been stopped.

There are many other existing projects in the basin and some other are under construction. Salient features of existing and under construction water

Table 7. Salient features of selected existing projects in Mahanadi basin

Name of the Project	State	Year of completion	Gross storage capacity (Mm^3)	Live storage capacity (Mm^3)	Designed annual irrigation (km^2)
Gondli	Chhattisgarh	1964	101.48	96.75	30.80
Kedar Nalla Project	Chhattisgarh	1969	16.79	15.71	46.60
Keshwa Nalla	Chhattisgarh	1963	18.03	17.86	38.50
Khapri	Chhattisgarh	1911	13.02	11.57	42.50
Khakhara	Chhattisgarh	1967	171.52	141.50	–
Khurung Tank	Chhattisgarh	1958	195.31	192.38	404.90
Kinkari	Chhattisgarh	1980	16.22	15.15	56.70
Koncha	Chhattisgarh	1970	21.56	18.71	37.60
Kumhari	Chhattisgarh	1927	11.61	11.34	26.30
Kunaria	Chhattisgarh	–	22.00	17.50	55.10
Madiyan	Chhattisgarh	1963	12.02	11.57	25.10
Maniyari Tank	Chhattisgarh	1931	151.29	147.61	307.70
Murum Silli	Chhattisgarh	1923	165.00	162.00	–
Pairy	Chhattisgarh	–	216.79	199.08	728.50
Paralkot	Chhattisgarh	1923	66.36	63.63	145.80
Pilosaiki Project	Chhattisgarh	–	19.20	14.40	35.90
Pindrawan	Chhattisgarh	1909	10.74	10.61	25.90
Saroda	Chhattisgarh	1964	31.14	30.14	73.60
Sikasar	Chhattisgarh	1978	217.00	198.80	728.50
Surhi Tank	Chhattisgarh	1963	22.14	21.78	63.90
Budha Budhani	Orissa	1976	14.98	12.50	43.80
Damar Bahal	Orissa	1983	22.30	19.32	37.80
Lurada	Orissa	1990	31.98	28.29	660.00
Russel konda	Orissa	1901	51.66	38.75	–
Saipala	Orissa	1982	21.28	18.35	31.60
Surada	Orissa	1902	35.16	28.29	–
Sundar Dam	Orissa	–	26.35	23.55	74.10
Upper Dahuka	Orissa	–	29.60	23.44	27.40

resources projects with a live storage capacity of $10\,Mm^3$ and above are presented in Table 7 and Table 8, respectively.

13.1.6. Water Quality Aspects

Based on investigations, CPCB has determined water quality classes of various reaches of Mahanadi River and its tributaries at different times. This classification along with critical parameters for each reach is given in Table 9. In general, the quality of river water in upper reaches is quite good and often the observed class is better than the desired class. However, in downstream reaches, the quality is degraded and BOD and total coliform are identified as the critical parameters at most of the places.

Table 8. Salient features of selected projects under construction in Mahanadi basin

Name of the Project	State	Gross storage capacity (Mm^3)	Live storage capacity (Mm^3)	Design annual irrigation (km^2)	Installed capacity (MW)
Arpa Project	Chhattisgarh	389.34	338.89	–	–
Balar Reservoir	Chhattisgarh	39.62	36.62	55.70	55
Chirrpani Project	Chhattisgarh	51.02	50.25	91.10	–
Gongha Reservoir	Chhattisgarh	34.01	30.07	76.90	–
Glaj Tank	Chhattisgarh	30.31	23.17	–	–
Kasarteda	Chhattisgarh	72.76	63.70	111.20	–
Kelo	Chhattisgarh	175.40	144.00	256.70	–
Khanher Phakut	Chhattisgarh	21.57	19.38	34.40	–
Kharang Tank	Chhattisgarh	195.21	192.38	404.80	–
Matia Moti	Chhattisgarh	29.43	26.48	65.00	–
Piparia	Chhattisgarh	45.70	40.60	60.70	–
Chiroli Dam	Orissa	300.00	189.00	345.00	–
Hariharjhar	Orissa	79.88	58.68	137.00	–
Lower Indra Project	Orissa	256.81	250.00	–	–
Upper Jonk Project	Orissa	110.00	104.00	165.90	–
Upper Suktel	Orissa	12.30	1,069	15.20	–

Table 9. Desired and existing water quality levels for Mahanadi

Location	Desired class	Observed class & critical parameters				
		1997	1998	1999	2000	2001
Mahanadi at Rudri U/S at Dhamtori Reservoir, M.P.	E	A	A	B	B	B
Mahanadi at U/S of Rajim, M.P.	D	B	B	B	A	B
Mahanadi at Sheorinarayan Village, M.P.	C	B	B	D BOD	D BOD	B
Mahanadi at Kharad, M.P.	C	B	B	B	B	B
Mahanadi after confl. With River Mand, MP		B	B	D BOD	B	B
Mahanadi at Interstate Boundry, M.P.	C	B	B	D BOD	B	B
Mahanadi at Hirakud Reservoir, Orissa	C	D BOD	D BOD	B	C	B
Mahanadi at Braj Raj Nagar U/S, Orissa		D BOD	D BOD	D BOD	C	C
Mahanadi at Sambalpur U/S, Orissa	C	D BOD	D BOD	D BOD	D BOD, Totcoli	C
Mahanadi at Sambalpur D/S, Orissa	C	D BOD	D BOD	D BOD	D BOD, Totcoli	D BOD, Totcoli
Mahanadi D/S (after confl. with R. Ong Sonepur U/S), Orissa	C	B	B	D BOD	C	B

Mahanadi after confl. With Tel River (Sonepur D/S),Orissa	C	D BOD	D BOD	D BOD	D BOD	D BOD, Totcoli
Mahanadi at Tikarpara, Orissa	C	D BOD	D BOD	D BOD	D BOD	C
Mahanadi at Narsinghpur, Orissa	C	D BOD	D BOD	D BOD	B	B
Mahanadi at Cuttack U/S, Orissa	C	D BOD	D BOD	D BOD	D BOD	C
Mahanadi at Cuttack D/S, Orissa	C	D BOD	D BOD	D BOD	D BOD, Totcoli	D BOD, Totcoli

* NA- Not Available. Source: www.cpcb.nic.in

13.2. SUBERNAREKHA SYSTEM

Subernarekha, Burahbalang and small east flowing rivers between the Ganga and the Baitarni, are covered under this group. The spatial coverage of this basin extends over an area of 29,196 km^2 and lies between east longitudes 85°10′ to 87°25′ and north latitudes 20°45′ to 23°30′. Lying in the northeast corner of the peninsular India, the region covers relatively large areas in the states of Bihar and Orissa and a somewhat smaller area in West Bengal. The State-wise distribution of the drainage area is given in Table 10. An Index map of Subernarekha basin is given in Figure 4.

The basin is bounded on the north and the west by the Chhotanagpur plateau, on the south by the ridges separating it from the Baitarani basin, and on east by the ridge separating it from the Kasai valley. The basin has a rectangular shape with a maximum length of 305 km in a north-west to south-east direction and a maximum width of 123 km in a north-east to south-west direction. There are two major topographical divisions in the basin, namely (i) the northern plateau and (ii) the coastal plains. The plateau region covers District Purulia of West Bengal and Ranchi and Singhbhum districts of Bihar. Part of the Mayurbhanj district of Orissa lying in the basin is hilly and well forested. The coastal plains lying in the basin cover parts of Balasore district in Orissa and of the Midnapur district of West Bengal. The general slope of the basin is from north-west to south-east. There are

Table 10. Statewise distribution of drainage areas of Subernarekha system

State	Drainage Area (km^2)
Bihar	13,685
Orissa	11,964
West Bengal	3,547
Total	29,196

Source: WG (1999).

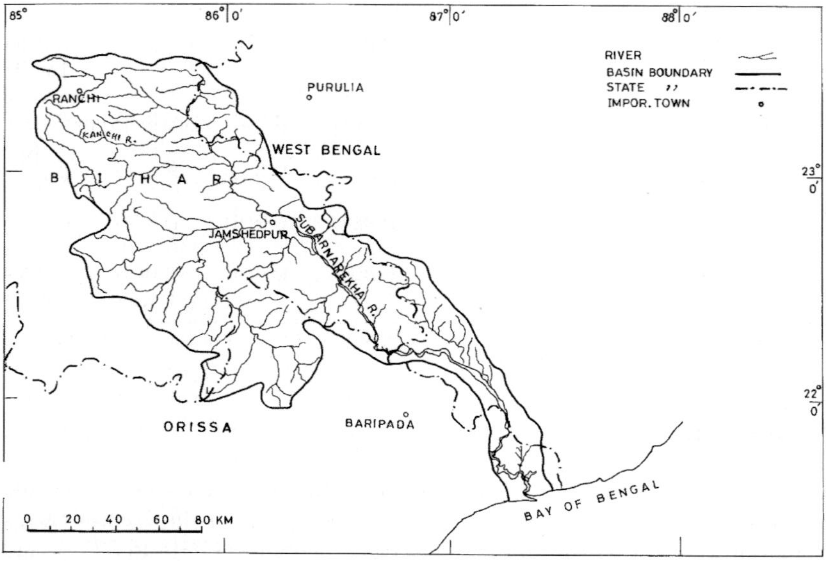

Figure 4. Index map of Subernarekha basin

four sub-basins of which those of the Subernarekha basin has an elongated shape, while the Burhabalang basin is triangular in shape with the main river as the base and the apex near Kalamgadia village in the Mayurbhanj district.

The various river systems in the Subernarekha basin and their catchment areas are given in Table 11.

The Subernarekha River basin presents a classic example of conflict among competing uses of water both sectorally and across regions. Fresh water from the river is used by industry as a direct process input and as a disposal agent for the dilution of effluents; by agriculturists for irrigation; and by the household sector for drinking. As long as the resource was abundantly available, it could be treated as a free good. Recently, increasing water scarcity and alarmingly high pollution levels in the Subernarekha River have caused widespread concern in the region major uses of water now find themselves competing for this scarce resource, leading to conflicts amongst various stakeholder groups.

Table 11. Rivers in Subernarekha system and their catchment areas

Name of the river	Catchment area in km^2
Subernarekha River including Rasulpur and Piehabani streams	19,296
Two small streams between Subernarekha and Burhabalang	2,418
Burhabalang River	4,837
Two small streams between Burhabalang and Baitarani Rivers	2,645

13.2.1. Subernarekha River

The Subernarekha River originates near Nagri village (Ranchi district) in the Chhotanagpur plateau of Jharkhand State at an elevation of about 610 m at latitude 23° 18′ N and longitude 85° 11′ E. Subernarekha literally means *golden line*. According to legends, in ancient times, gold was being extracted in the river's bed close to its origination point near Piska village, close to Ranchi. Before falling in the Bay of Bengal, the river flows through Ranchi and Singhbhum districts of Jharkhand, Midnapore district of West Bengal and Balasore district of Orissa. Subernarekha River flows for a total length of 395 km. Out of this, 269 km lies in Bihar, 64 km in West Bengal, and 62 km in Orissa. The Subernarekha basin covers an area of 19,300 km^2. This area is nearly 0.6% of the total national river basin area and yields 0.4% of the country's total surface water resources. With regards to statewise distribution, Subernarekha drains an area of 14,600 km^2 in Jharkhand (75.6% of total catchment area), 2,500 km^2 (12.9%) in Orissa and 2,200 km^2 (11.5%) in West Bengal. Its important tributaries are the Kanchi, Karkari and Kharkai. Flow diagram of Subernarekha Basin has been presented in Figure 5.

The topography of the Subernarekha basin varies from steep hills to flat coastal plains, plateaus, uplands and central plains. The basin has three distinct seasons: winter, summer and rainy. Rainfall varies greatly – both annually and seasonally and the average annual rainfall is 1,500 mm. During monsoons, the basin may receive a spell of copious rainfall followed by a long dry period. Of the 28,609 million m^3(Mm^3) of rainfall, about 57% is lost by evapotranspiration and 15% infiltrates

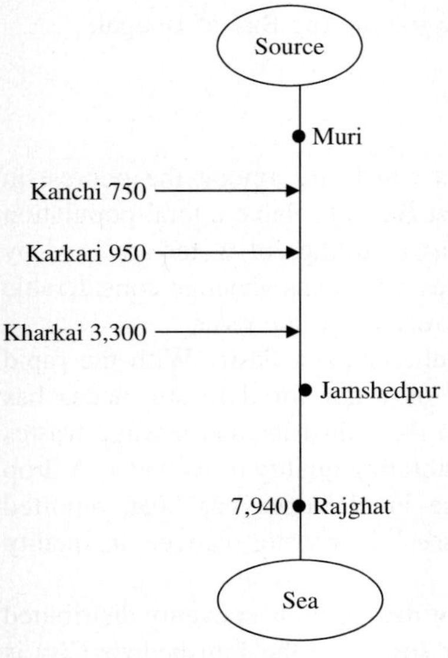

Figure 5. Flow Diagram of Subernarekha Basin

into ground. Balance 7,940 Mm^3 of water flows past Rajghat station into the Bay of Bengal annually. This yield corresponds to approximately 0.4 MCM/km^2 of the basin area per year. This low yield is caused by the fact that the Subernarekha is a rain-fed river of peninsular origin, and the average annual rainfall of 1,500 mm varies on a monthly and geographical basis.

In the major part of the basin, ancient Precambrian, igneous and metamorphic rocks are exposed. Younger geological formations, namely tertiary gravel, Pleistocene alluvium and recent alluvium, are found in the lower reaches of the basin, south-east of Ghatsila. Red soil is the dominant soil type of the basin – it is derived from old bedrocks and is mostly of residual and alluvial origin. In the lower valleys and coastal plains, river-borne alluvial deposits of unconsolidated materials are found.

13.2.2. Burhabalang River

The Burhabalang River rises south of Simphalgarh village in the Mayurbhanj district at an elevation of about 800 m and flows for a length of 164 km to join the Bay of Bengal. Two small streams flow between Burhabalang and Baitarani Rivers. The Northern stream rises near Nilgiri village in the Balasore district at an elevation of 150 m and flows for a total length of 22 km to join the Bay of Bengal. The second stream rises north of Dantur village in the Balasore district at an elevation of 300 m and flows for a length of 54 km to join the Bay of Bengal.

Two small streams flow between Subernarekha and Burhabalang. The northern stream rises west of Lakshman Nath village in the Balasore district at an elevation of 45 m and its total length is 38 km, and it joins Bay of Bengal. The second stream rises south-west of Gopiballabhpur village in the Midnapur district at an elevation of 100 m and flows for a length of 77 km before joining the Bay of Bengal.

13.2.3. Demography and Economy

Subernarekha River passes through three states which are among the poorest in India. These states – Jharkhand, Orissa and West Bengal – have a total population of about 150 million. Apart from irrigation, vast quantities of water are used by industries and towns in the basin. These industries and towns generate considerable quantities of wastewater, which is discharged directly into the river.

Ranchi is the most important town in the Subernarekha basin. With the rapid increase in population of the city, withdrawal of water for domestic needs has risen sharply over the years. The effluents from the industries and sewage wastes emanating from the industrial township are deteriorating quality of its water. A drop in dissolved oxygen, accompanied by decrease in plankton, has been reported below the discharge point. Due to natural processes, river water recovers its quality further downstream.

As expected, water supply in the urban areas of the region is unevenly distributed and leaves much to be desired. Domestic water supply in the Jamshedpur City is intermittent. Residents in TISCO township have a 100% piped water supply while the urban centres of Jugsalai, Mango and Adityapur face water crises. In most cases, water is supplied for 6–8 hours/day. In the other areas, supplies are for even shorter periods and water may even is supplied only twice or thrice a day. In Mango NAC, piped water supply is provided only 4 days a week. The demand for water is bound to increase because of population growth. The decadal growth rates of population in the Subernarekha basin states (Jharkhand, Orissa and West Bengal) range from 15% in Orissa to 23% in Jharkhand. Demand for irrigation water could also – significantly increase in future as currently only 8% of total cultivable area is irrigated.

About 75 million people in these three states (mainly in Jharkhand) live below the poverty line. An emerging shortage of water, serious water-quality problems, and a lack of any policy and institutional framework characterize this river basin. An increase in the quality and quantity of water would contribute significantly to improving the quality of life in this region.

13.2.4. Surface Water Resources of Subernarekha

The upper portion of the Subernarekha and its tributaries run through the fertile lands of Bihar, with farming in this region mainly dependent upon insufficient and seasonally variable rainfall. The water resources of the Subernarekha river system remain largely undeveloped. Besides containing fertile lands, the upper basin also includes extensive mineral resources and a number of industries have therefore been established along the banks of the river. Table 12 represents the annual average observed runoff for Gauging sites of CWC in Subernarekha basin.

Average annual rainfall in the Subernarekha basin is on the order of 1,250 mm with the maximum and minimum on record being 1,420 and 1,150 mm, respectively. About 90% of this rainfall is received during the South–West monsoon season, i.e., June to October. The Subernarekha river basin includes of two reservoirs [Chandil and Icha], and two barrages [Kharkai and Galudih]. Each reservoir and barrage has two canals; namely, left bank canal (LBC) and right bank canal (RBC). On the right bank canal of Galudih barrage (GRBC), there are three small command area reservoirs [Haldia, Jambhira, and Baura]. This canal has two branch canals; namely, Betnoti branch canal (BBC), and Subernarekha branch canal (SBC). The salient features of the reservoirs are given in Table 13. There is also a small hydroelectric power plant on the Chandil left bank canal (CLBC). The irrigation and municipal

Table 12. Annual average observed runoff at selected CWC sites in Subernarekha basin

Name of the site	Name of the stream	Catchment area (km^2)	Annual average runoff (BCM)
Adityapur	Kharkai	6,309	3.18
Ghatsila	Subernarekha	14,176	7.39
NH5 Govindpur	Burhabalang	4,495	3.45

Source: CWC (2002).

Table 13. Salient Features of Reservoirs in Subernarekha River Basin

Description	Catchment area (km^2)	Storage		
		Live (Mm^3)	Dead (Mm^3)	Gross (Mm^3)
Chandil dam	5,646	1, 264	204	1,468
Icha dam	2,849	925	117	1,042
Kharkai barrage	5,815	–	–	–
Galudih barrage	13,629	–	–	–
Haldia dam	55	45.88	2.12	48
Jambhira dam	76.50	214.10	1.18	216
Baura dam	47	259.71	1.90	262

and industrial water supply requirements during the off-peak period of the power plant are routed through a proposed small bypass canal. The CLBC also acts as a water carrier to deliver 12.25 MCM (million cubic meters) to West Bengal. Live storage in Chandil reservoir mainly belongs to Bihar, a small amount of 30.75 MCM, belongs to West Bengal.

13.2.5. Water Resources Development Projects in the System

There are a few water resources projects in the system and their description follows.

i. Subernarekha multipurpose system

The Subernarekha multipurpose system comprises of:

- Subernarekha multipurpose Project (SMP), Phases I and II (Jharkhand).
- Subernarekha Irrigation Project (SIP), Orissa.
- Subernarekha Barrage-cum-Dolong Dam project (SBDP), West Bengal.

The SMP comprises of two dams (Chandil and Icha), two barrages (Kharkai and Galudih) and a network of canals from these, and a canal from the Galudih barrage to carry water to Orissa; all these lie in Jharkhand state. Three small storage reservoirs and a network of canals from these reservoirs are in Orissa. The main objectives of the SMP are: i) to provide reliable water supply to agricultural lands of 1,600,900 and 50 km^2 in Jharkhand, Orissa and West Bengal, respectively, ii) to provide 740 MCM/year of water for M&I uses in Jharkhand; and iii) to reduce flood damage in Orissa and West Bengal by providing 463 MCM of flood-storage capacity for the Chandil dam. Orissa and West Bengal will construct embankments in their respective territories along the flooding reaches of the river and generate 30 MW of hydroelectric power through medium, mini and micro-hydroelectric projects located at various points of the canal system.

Phase I of the SMP envisaged the partial construction of the Chandil dam (50%) with its main canal (30%) partial construction of the Icha dam (20%) and its right and left bank canals (20%); and partial construction of the Galudih Barrage (36%) and its right bank canal (22%). It also envisaged the construction of three distributaries under Kharkai Barrage (not included in Phase I) and partial completion of the micro-distribution system (210 km^2) besides establishing a Water and Land Management Institute (WALMI), irrigation research stations, and pilot water management and pollution control programs. None of the components was planned for full completion in Phase I of the project. This was due to several reasons including: (1) insistence by the co-basin states to take up all components simultaneously; (2) the inability of the World Bank to fund an implementation plan for more than 4 years; and (3) the physical impossibility of completing all these components in 4 years.

Started in 1982–83, SMP multipurpose project was planned for irrigation, hydropower generation and water supply at 75, 95 and 100% reliability, respectively. However, attempts to implement all project components simultaneously and consequent delays has rendered the project infeasible and has reduced its economic viability, the economic rate of return in Bihar had decreased from 17.8% in 1981–82 to 8.6% in 1989. A lack of funds also has been a major problem though a significant amount has been spent (Rs. 9,528 million by June 2001). However, no benefits for the irrigation or household sectors have been generated. The lack of irrigation benefits can be attributed to the failure to deliver irrigation water because of the incomplete canal system. In addition, the increasing demand for irrigation, drinking water, industrial uses and other uses has further complicated the situation.

Although the estimates of project benefits in all categories are very large (See Table 14), significant benefits have accrued only in terms of the mitigation of flood damage and the generation of non-farm employment. In terms of physical achievements, the Chandil dam and Galudih barrage are almost complete, but other components are still incomplete. Partial storage is available in the Chandil reservoir and that is providing M&I water supply to the industrial city of Jamshedpur.

Phase I of SMP was implemented with financial assistance from the World Bank. It was to be followed immediately by Phase II. However, the implementation of Phase II got delayed and it is now in progress. The components are in different stages of completion.

ii. Getalsud reservoir

Getalsud reservoir is located at 23°27′N and 85°33′E, across the Subernarekha River, 40 km east of Ranchi city, now the capital of Jharkhand State. Completed in 1971, this multi-purpose reservoir was envisaged mainly to meet municipal water demands of Ranchi town, industrial needs of the Heavy Engineering Corporation and other factories. The reservoir has a maximum surface area of 34.59 km^2 and a capacity of 288.5 Mm^3 at the FRL of 595.70 m. At the lowest level, the area is only 9.87 km^2. Getalsud dam has a catchment area of 717 km^2 and is located at about 50 km from the origin of the Subernarekha River. The height of the dam is 35.5 m. Two power houses, namely Subernarekha I & II, have been commissioned. Both the power houses have one unit of 65 MW each and have a firm power of 9.5 MW and 5.0 MW respectively.

Table 14. Subernarekha Multipurpose project (Phase I): direct benefits pertaining to the entire scheme when complete

SN	Indicator	Appraisal estimate	Closing date estimate	Estimate at full development
1	Directly benefiting farm families (number)	130,000	0	232,000
2	Increased irrigated area (ha)	255,000	0	255,055
3	Increased food grain production (1,000 tonnes / year)	700	0	700
4	Increased municipal and industrial water supply (MCM/year)	700	0	607
5	Moderation to flood damage (Rs. million/ year)	18.5	21.2	21.2
6	Generated farm employment (number of jobs)	138,000	0	190,000
7	Generated non-farm employment (number of jobs)	24,800	27,715[a]	39,300
8	Value added in agriculture (Rs million/year)	1,485	0	2,254
9	Reduction in farm families below poverty line (percent)	53	42[b]	58

[a] pertains only to project construction employment;
[b] due only to project construction works

In addition to the Getalsud project, some more projects exist or are under construction in the basin. Salient features of existing and under construction water resources projects in the basin with a live storage capacity of 10 Mm^3 and above are summarized in Table 15 and Table 16, respectively.

13.2.6. Tri-Partite Agreement For Sharing of Subernarekha Waters

Notwithstanding rich water resources, Subernarekha basin is subject to frequent droughts. Being an interstate river, any development in the Subernarekha basin requires

Table 15. Salient features of selected existing projects in the Subernarekha system

Name of the Project	State	Year of completion	Gross storage capacity (Mm^3)	Live storage capacity (Mm^3)	Design annual irrigation (km^2)
Hatia	Jharkhand	1962	31.21	29.65	–
Largara	Jharkhand	–	22.20	21.36	6.40
Palna Reservoir	Jharkhand	–	12.41	11.35	–
Baladia	Orissa	–	66.61	49.34	–
Kalo dam	Orissa	–	29.61	24.67	–
Kharkhai Dam	Orissa	1970	69.88	61.68	–
Sunei	Orissa	–	55.51	46.88	152.00

Table 16. Salient features of selected projects under construction in Subernarekha basin

Name of the Project	State	Gross storage capacity (Mm^3)	Live storage capacity (Mm^3)	Designed Annual Irrigation (km^2)	Installed capacity (MW)
Jharijara Reservoir	Jharkhand	22.45	19.80	40.50	–
Subernarekha	Jharkhand	1,963.00	1,619.00	2,090.00	30
Bhankbal	Orissa	27.14	20.96	98.00	–
Palan Reservoir	Orissa	12.41	11.39	–	–

joint agreement among the co-basin states: Jharkhand, Orissa and West Bengal. The signing of a Tri-Partite Agreement (TPA) in 1978 by three co-basin states (Jharkhand was a part of Bihar at that time) marked the beginning of the preparation of a comprehensive plan for the development of the basin through the Subernarekha Multipurpose System. Studies of socioeconomic benefits, environmental impact, dam safety, planning for operation and maintenance, management information system, and hydrometeorology have been undertaken in the context of these agreements.

TPA is mainly based on the availability of water up to Kokpara (UK) and beyond Kokpara (BK) (Table 17). The 75% water year dependable flow to Kokpara (catchment area 15,369 km^2) has been assessed at 5,550 MCM. This quantity is allocated to Bihar, Orissa, and West Bengal states in the ratios of 32:12:1. The shortages, if any, during a year, are to be shared in the same ratio, but surpluses, if any, will be shared in the ratios of 8:1:11, respectively. The 75% water year dependable flow below Kokpara (catchment area 4,631 km^2) is assessed at 1,418 MCM, and is allocated to the respective states in the ratios of 2.7:2.9:5.9. Similar ratios are applicable in shortage or surplus water years.

Table 17. Highlights of Tripartite Agreement up to Kokpara (UK) and Below Kokpara (BK)

Important features	Water share ratio		
	Bihar	Orissa	West Bengal
Up to Kokpara (UK)			
(i) Assessed 75% water year dependable flow as 5,550 MCM	32	12	123*
(ii) Surplus over the assessed 75%	8	1	11
(iii) Shortage below the assessed 75%	32	12	1
Beyond Kokpara (BK)			
(i)Assessed 75% water year dependable flow as 1,418 MCM	2.7	2.9	5.9
(ii) Surplus and shortage with respect to the assessed 75%	2.7	2.9	5.9

* 61.50 MCM upstream of Chandil dam+12.25 MCM through CLBC for Kharif and Rabi+30.75 MCM below Kokpara through storage space in Chandil dam + 18.50 MCM below Kokpara from run-of-river=123 MCM.

Table 18. Tripartite Agreement for water at Icha Dam

Important features	Water share ratio	
	Bihar	Orissa
838 MCM storage capacity of dam	5.0	1.8 (through Galudih Right Bank Canal)
310 MCM utilization upstream of dam	1.0	4.0
Surplus and shortage with respect to the assessed 75%water year dependable flow	5.0	1.8

Note: Cost of dam to be shared in the ratio of 5.0:1.8.

At the Icha reservoir (Table 18), the live storage capacity is to be shared between Bihar and Orissa in the ratio of 5.0:1.8, respectively. The same ratio holds for sharing of water in surplus or shortage years and for sharing of the reservoir cost. The sharing of water at Icha begins on 1st July every year, with the Kharif irrigation (June to October) having the highest priority. At Galudih barrage (Table 19), during non-monsoon periods, a total of 197 MCM is to be shared in the ratio of 3:5 between Bihar and Orissa. The water share ratio during shortage years is also 3:5, respectively.

To provide a quick overview, Table 20 lists the Culturable Commanded Areas (CCA) and capacities of important canals of the basin.

Table 21 provides details of the upstream and downstream water diversion requirements at individual dams and barrages.

13.2.7. Water Pollution Problems

A Central Pollution Control Board study estimates that 0.4 MCM of untreated or semi-treated water containing 61 tons of biological oxygen demand is discharged

Table 19. Tripartite Agreement Features at Galudih Barrage

Important features	Water share ratio	
	Bihar	Orissa
After Icha dam construction, use of run-of-river from barrage:		
(i)197 MCM during non-monsoon	3	5[a]
(ii) Shortages	3	5[a]
Before Icha dam construction, use of run-of-river from barrage:		
(i)197 MCM during non-monsoon	3	5[a]
(ii) All excess over 197 MCM during non-monsoon	0	1[a]

[a]Through Galudih Right Bank Canal.

Table 20. Culturable commanded areas and capacities of important canals in Subernarekha system

Site name	CCA (km^2)		Capacity (MCM/year)	
	LBC	RBC	LBC	RBC
Chandil	663.54	115.18	2,652.00	505.00
Icha	148.94	507.94	629.16	1,575.57
Kharkai	4.98	138.77	69.71	568.38
Galudih	69.81	45.68	306.53	4,012.65
Haldia	15.70	39.50	47.30	117.63
Baura	323.55	–	867.24	–
Jambhira	311.09	35.30	851.47	119.99
SMC	–	–	3,737.01	–
SBC	73.39	–	346.89	–
BBC	228.36	–	1,892.16	–

Note: LBC = left bank canal; RBC = right bank canal; SMC = Subernarekha main canal; SBC = Subernarekha branch canal; and BBC = Betnoti branch canal.

daily into the river from all urban sources. The contribution of rural areas is approximately 50,000 kg of biological oxygen demand load. This and the uncontrolled discharge of metallic, non-metallic and toxic substances and effluents from mines have resulted in degraded river water quality. Toxic and radioactive substances are found mainly in the lower segments of the river.

Table 22 shows the desired and observed water quality classes and critical parameters for different locations in Subernarekha basin at different times. It can be seen that at three locations, the observed quality is below the desired class and total coliform is a critical parameter for this basin.

13.3. THE BRAHMANI BASIN

The Brahmani Basin lies between latitude 20°28′ to 23°35′ N and longitude 83°52′ to 87°30′ E. It lies in the districts of Raigarh and Sarguja in Chhattisgarh, Ranchi and Singhbhumi in Jharkhand and Sundergarh, Deogarh, Sambalpur, Angul, Dhenkanal, Keonjhar, Jajpur and Kendrapara in Orissa. Brahmani basin is situated between Mahanadi Basin (on the right) and Baitarani Basin (on the left). Chhotanagpur Plateau in the East and South bound the basin, in the north a ridge separates it from Mahanadi basin, and to the east of the basin lie the Bay of Bengal and the Baitarani basin. Out of the total basin area of 39,269 km^2, the major parts of the basin covering 22,516 km^2 (57.34% of the basin) falls in Orissa state. Further, about 15,406 km^2 (39.23% of the basin area) falls in Jharkhand state and about 1,347 km^2 which is only 3.43% of the basin area falls in the Chhattisgarh state.

District wise geographical area and rural and urban population of the basin is given in Table 23. The index map of Brahmani and Baitarni basin is given in Figure 6.

Table 21. Upstream and Downstream Annual Irrigation and Municipal and Industrial Water Supply Requirements (in MCM)

Site	Bihar		Orissa		West Bengal	
	Irrigation	Water Supply	Irrigation	Water Supply	Irrigation	Water Supply
			Upstream			
Chandil dam	655.50	100.00	–	–	52.70	8.80
Icha dam	62.00	9.00	216.60	31.40	–	–
Kharkai Barrage	413.40	60.00	–	–	–	–
Galudih Barrage	24.50	4.00	–	–	–	–
			Downstream			
Chandil LBC	606.61	521.63	–	–	12.25	–
Chandil RBC	104.71	51.08	–	–	–	–
Icha LBC	194	36	–	–	–	–
Icha RBC	518	36	–	–	–	–
Kharkai LBC	11	47	–	–	–	–
Kharkai RBC	106	35	–	–	–	–
Galudih LBC	79	63	–	–	–	–
Galudih RBC	35	25	1,044	–	–	–
			Direct Command Area			
BBC	–	–	218.57	–	–	–
SBC	–	–	70.28	–	–	–
DC	–	–	66.44	12.91	–	–
			Command area reservoir			
Haldia dam	–	–	49.02	–	–	–
Jambhira dam	–	–	304.78	–	–	–
Baura dam	–	–	322.00	–	–	–
Haldia LBC	–	–	15.30	0.54	–	–
Haldia RBC	–	–	37.83	1.37	–	–
Jambhira LBC	–	–	33.80	10.82	–	–
Jambhira RBC	–	–	297.89	1.23	–	–
Baura LBC	–	–	309.82	11.25	–	–

Note: BBC = Betnoti branch canal; SBC = Subernarekha branch canal; DC = direct command; Water supply pertains to municipal and industrial water supply.

13.3.1. Brahmani River System

Brahmani River is the second largest river in the State of Orissa. In fact, two headwater streams, namely Sankh River and South Koel River originate in Chhattisgarh and Jharkhand states, respectively. After the confluence of Sankh River and South Koel River at Vedvyas (in Orissa at latitude 22°48′ N and longitude 84°14′ E, at an elevation of 200 m), the combined river is known by the name Brahmani. The Brahmani River flows through the heart of Orissa till it joins the Bay of Bengal at Dhamara mouth. The Flow diagram of the Brahmani basin is given in Figure 7.

The South Koel River originates near village Nagri in Ranchi district of Jharkhand at an elevation of about 700 m at latitude 23°20′ N and longitude 85°12′ E. Karo River a major left bank tributary joins the South Koel at a distance of 221.25 km

Table 22. Desired and existing water quality levels for Subernarekha

Location	Desired class	Observed class & critical parameters				
		1997	1998	1999	2000	2001
Subernarekha at Ranchi,(Tatisilwai) Bihar	C	D Totcoli	D Totcoli	D Totcoli	D Totcoli	D Totcoli
Subernarekha at Jamshedpur, Bihar	C	C	A	C	C	D Totcoli
Subernarekha at D/S Jamshedpur,(Tata Nagar),Bihar	C	E DO, BOD, Totcoli	D Totcoli	D Totcoli	D Totcoli	D Totcoli
Subernarekha at Chandil Bridge, Bihar	C	C	C	C	C	C

*NA – Not available. Source: Central Pollution Control Board.

from origin, just south of Gudri in Singhbhum district. At the confluence, Koel intercepts a catchment of 13,378 km^2 out of which 1,438 km^2 lie in Orissa and 11,940 km^2 in Jharkhand. The Koel River enters Orissa at 262 km of its run. Another right bank tributary named Deo joins South Koel at 288 of its run km in Orissa and the Sankh River joins after another 16 km.

Table 23. Geographical Area and Population within Basin

State	District	Area (km^2)		Population 2001 (in thousands)		
		Geographical	Within basin	Rural	Urban	Total
Orissa	Sundargarh	9,712	5,717.77	680.98	588.06	1,269.042
	Keonjhar	8,303	1,743.56	166.32	56.83	223.145
	Sambalpur	6,657	1,371.05	81.36	0.00	81.36
	Deogarh	2,940	2,529.11	238.57	20.09	258.655
	Angul	6,375	4,235.38	760.29	147.00	907.294
	Dhenkanal	4,452	3,968.66	819.18	92.79	911.967
	Jajpur	2,899	1,836.14	932.90	37.61	970.509
	Kendrapada	2,644	1,114.41	439.26	74.13	513.388
	Orissa Total		22,516.08	4,118.86	1,016.50	5,135.36
Jharkhand	Lohardega	1,490	625.60	169.75	46.20	215.954
	Simdega	8,820	3,927.34	569.32	33.96	603.282
	Gumla		4,420.23	773.60	39.79	813.39
	Ranchi	7,573	2,470.17	690.94	32.35	723.287
	W. Singhbhum	5,290	3,962.56	536.79	63.58	600.37
	Jharkhand Total		15,405.90	2,740.40	215.88	2,956.283
Chhattisgarh	Sarguja	22,337	196.00	70.91	0.00	70.91
	Jashpur	6,154	1,150.80	275.41	20.19	295.6
	Chhattisgarh Total		1,346.80	346.31	20.19	366.5
	Basin Total		39,268.78	7,205.57	1,252.57	8,458.14

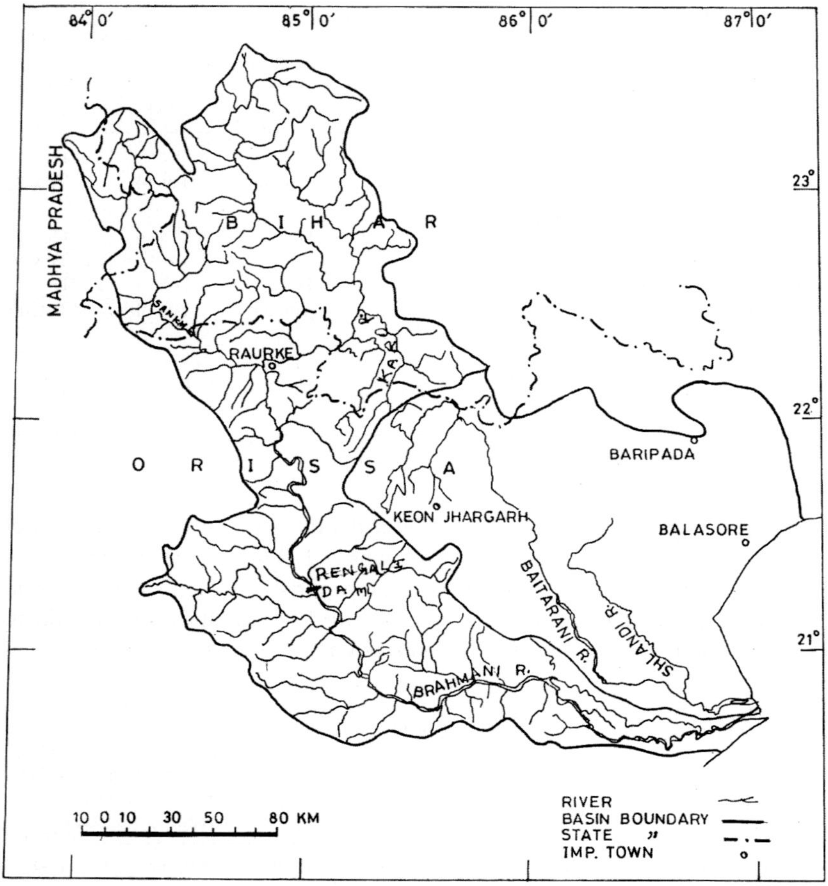

Figure 6. Index map of Brahmani and Baitarni basin

The Karo River originates in Chhotanagpur Plateau of Jharkhand near Nagri in Ranchi district at an elevation of 700 m at Latitude 23°17′ N and Longitude 85°08′ E. It drains an area of $2,784\,\text{km}^2$ and flows for a length of 125 km before it joins South Koel.

The Sankh River rises near village Lupungpat in Ranchi district at an elevation of 1,000 m, the latitude and longitude of origin being 23°14′ N and 84°16′ E. The river flows for a distance of 67.5 km in Jharkhand before entering Chhattisgarh. After flowing for about 50 km in Chhattisgarh, it again enters Jharkhand and flows for 78 km before finally leaving Jharkhand. The Sankh River flows for a distance of 45 km in Orissa before it meets Koel. Thus, the total length of Sankh River is 240.5 km. Its catchment area is $7,353\,\text{km}^2$ out of which $1,422\,\text{km}^2$ lie in Chhattisgarh, $4,472\,\text{km}^2$ in Jharkhand, and $1,456\,\text{km}^2$ in Orissa.

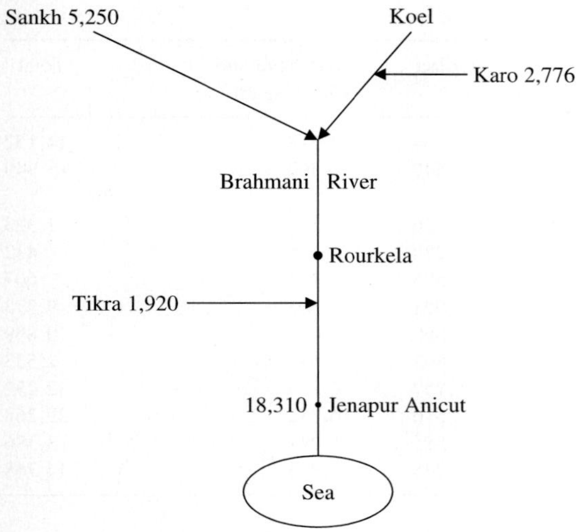

Figure 7. Flow diagram of Brahmani basin

After the confluence at Vedvyas, the Brahmani River heads towards the southeast direction and traverses a total length of 461 km before it joins the Bay of Bengal. It drains a total catchment area of 39, 269 km^2. Deltaic region of Brahmani begins near village Gadamadhupur Jenapur of Jajpur district, where the Kalamitra Island divides the river into two branches. The south branch is known as the Brahmani main and the north branch is called Kharsuan. These two branches again join. Further down, Birupa River, a distributary of Mahanadi River joins Brahmani River and still further down, Baitarani River joins it. About flowing for about 15 km, the river falls in the Bay of Bengal. Table 24 shows land use data of the Brahmani Basin.

13.3.2. Climate of The Basin

i. Rainfall

Brahmani Basin has tropical climate with a fairly hot summer and moderately cold winter. Three well-defined seasons are observed: summer, winter and rainy. In the small region close to the coast, climate is influenced by sea which has a moderating effect while the hilly areas are somewhat cooler due to their altitude.

Rain is the only mode of precipitation in the basin. Average annual rainfall of the basin is 1,305 mm. The range of minimum and maximum rainfall is 969 mm to 1,574 mm. The highest annual rainfall observed was at Sukinda station, district Jajpur, in the year 1986 and was 2,489 mm. District Keonjhar received the lowest rainfall. The basin receives 90% of its total annual rainfall from the south west monsoon and the bulk of it (more than 75%) occurs during the monsoon period from June-October. In the remaining months, only occasional showers are received.

Table 24. Land Use Details of Brahmani Basin (area in km^2)

Land Use		Orissa	Jharkhand and Chhattisgarh	Total
Forests		9,534	4,618	14,152
Misc. tree crop & grooves not included in agriculture land		442	507	949
Permanent Pastures etc.		926	397	1,323
Land not available for cultivation		286	146	432
Land under Reservoir		503	104	607
Culturable Waste and Fallows		5,068	4,305	9,373
Net area	Rainfed	3,860	6,039	9,899
under	Irrigated	1,897	636	2,533
cultivation	Total	5,757	6,675	12,432
Total Basin Area		22,516	16,752	39,268
Gross Irrigated Area		2,585	1,001	3,586
Gross cropped area		6,945	7,823	14,768

Source: Mohile and Gupta (2003).

Thunderstorms are quite frequent during monsoon, the intensity being more in the plains. A typical value of potential evaporation for the basin is 150 cm per annum.

ii. Temperature

In the basin, the maximum temperature rises to 47 °C during summer while the minimum during winter may be as low as 4 °C. Temperatures in the coastal region are moderate but humidity is higher. As per the observations at the IMD stations, the cloud cover is maximum during July and August and minimum during Dec. to Feb.

The coastal part of the Brahmani basin lies in the track of cyclonic storms that originate in the Bay of Bengal during the months of April to November and move westwards. The storms developing in April-May and in October-November are somewhat stronger and some of these were devastating to life, property, crops and vegetation. The wind velocity is higher in the months of April, May and June and lower in the months of Dec. and Jan.

13.3.3. Water Resources of the Basin

i. Surface water yield

The river Brahmani with its large catchment of varying characteristics (Agricultural land 32%, forest 36%, and culturable waste 23%) yields a runoff of 400 to 550 mm from the average monsoon rainfall of 1,200 mm.

The major tributaries Sankh, Koel, and the main river are gauged by CWC and State government at five locations, whereas the other tributaries are also gauged, mainly at project locations. From the gauged data, (CWC 2002), the average yield at the head of the delta is 18, 318 Mm^3. The catchment area of the basin is 39, 269 km^2 against the Jenapur area of 33, 955 km^2 and the total yield would be 22, 000 Mm^3. The Orissa Water Planning Organization has estimated the average annual yield of Orissa portion as 10, 661 Mm^3 from a basin area of 22, 516 km^2. The 75% dependable yield for the entire basin is assessed as 17, 204 Mm^3 (NWDA, 1988) and 8, 277 Mm^3 is for the Orissa portion (DOWR, 1999).

ii. Ground water availability

Based on soil characteristics and rainfall pattern, which consistently increases from 1,200 mm in the coast to 1,400 mm in the western region, the ground water availability is assessed as 5, 170.66 Mm^3 (Table 25). Essentially about 23, 000 km^2 of agricultural land in the basin states are potential infiltration zone.

Since ground water utilization is low; much of this recharge reappears in the river system.

iii. Water budget

Monthly average runoff at three gauge stations, namely, Panposh, Gomlai, and Jenapur, in the basin are given in Table 26.

Water withdrawals, uses and returns for various uses in Brahmani Basin have been assessed by Mohile and Gupta (2003). The summary information is given in Table 27.

Table 25. Annual Ground Water Availability in Brahmani Basin

State	District(s)	Basin area (km^2)	Gross recharge (Mm^3)	Recoverable recharge (85%) (Mm^3)
Jharkhand	Ranchi, Gumla, Simdega, Lohardega	11, 443.34	1, 651.98	1, 404.18
	Singhbhum	3, 962.56	657.09	558.53
Chatishgarh	Sarguga	196.00	20.65	17.55
	Raigarh	1, 150.80	212.06	180.25
Orissa	Sundergarh	5, 717.77	554.79	471.57
	Sambalpur	1, 371.05	150.38	127.82
	Deogarh	2, 529.11	226.42	192.46
	Keonjhar	1, 743.56	154.35	131.20
	Anugul	4, 235.38	547.18	465.10
	Dhenkanal	3, 968.66	481.53	409.30
	Jajpur	1, 836.14	465.95	396.06
	Kendrapada	1, 114.41	48.28	41.04
	Total	39, 268.79	5, 170.66	4, 395.06

Source: Orissa Government.

Table 26. Monthly average runoff at three gauge stations in the basin

Month	At Panposh		At Gomlai		At Jenapur	
	Confluence of South Koel & Sankh. Catchment area = 19,448 km^2		Upstream of Rengali reservoir. Catchment area = 21,950 km^2		Head of Delta. Catchment area = 33,955 km^2	
	In %	In Mm^3	In %	In Mm^3	In %	In Mm^3
June	7.14	778	7.55	793	4.33	793
July	20.94	2,282	24.37	2,559	17.52	3,209
August	29.60	3,226	31.27	3,284	29.12	5,334
September	23.70	2,583	22.99	2,414	21.74	3,982
October	7.01	764	6.86	720	9.09	1,665
November	3.73	407	2.48	260	4.26	780
December	1.98	216	1.19	125	3.16	579
January	2.66	290	1.08	113	2.82	517
February	1.36	148	0.72	76	2.28	418
March	0.48	52	0.54	57	2.15	394
April	0.73	80	0.41	43	1.63	299
May	0.67	73	0.54	57	1.90	348
Total	100.00	10,900	100.00	10,502	100.00	18,316

Source: CWC (2002).

The 75% dependable water resources at Rengali dam site have been assessed as 9,140 m^3; this corresponds to flow of the year 1965–66. Further, it is assessed that water utilization upstream of the project is about 3,191 m^3.

13.3.4. Irrigation and Agriculture

Agriculture is the primary occupation and the most important economic activity of the inhabitants. Most people in rural areas, who constitute 80% of the basin population are engaged in agriculture. The rest 20% of the population earn their

Table 27. Water withdrawals and uses (Mm^3/Year) for various uses in Brahmani Basin

Use	Withdrawals (from)			Consumptive use
	Total	SW	GW	
Irrigation	4,049	3,623	426	560
Domestic – Rural	260.74	0.00	260.74	130.37
Domestic – Urban	63.98	51.18	12.80	12.80
Industrial	322.44	257.95	64.49	64.49
Total	4,696	3,932	764	768

livelihood working in industries and other business. Double cropping pattern is prominent in Kendrapara district followed by Sambalpur and Keonjhar

Rice is the most common crop grown, accounting for over 86% of the gross sown area of the basin. Pulses account for 7% of the gross sown area and oil seeds occupy 5.45%. The basin has abundant mineral resources but few industries have been set-up.

About 300 Mm^2 in Brahmani basin was getting irrigated from Mahanadi Canal system in 1960–61 and another 50 to 100 Mm^2 from minor irrigation in the whole basin. Overall rainfed agriculture was practiced over $8,600\,Mm^2$.

Anicuts were built at Jenapur on the Brahmani River and at Jokadia on the Kharsuan River in 1870–80 for irrigation and navigation along the coast. The Jenapur Anicut has gone into distress and has become defunct. The high level canal originates in Mahanadi basin traverses across the neck of the Brahmani-Baitarani plains and irrigates large tracts of land in the Brahmani Baitarani basin. Although the Birupa weir storage provides some confidence and supply in emergencies, basically the system is a Kharif irrigation system, depending on normal stream flow. The total area irrigated in the Brahmani Basin through HLC Range I and Pattamundai Canal is about 300 Mm^2.

The Brahmani Basin with a drainage area $39,269\,km^2$ has a few major & medium irrigation projects. Irrigation development has been accelerated as a plan program. As only 10% to 15% of the area is receiving dependable irrigation, the introduction of high yield, hybrid, and technologically superior varieties crop is not done. The Rengali Dam (in Orissa) built on Brahmani, intercepting $25,250\,km^2$ is a major project, providing flood protection to the lower basin and generating hydropower. The irrigation potential of the project is yet to be utilized as the distribution system off-taking from the Samal Barrage (35 km downstream of the dam) is still in progress. Table 28 depicts the arable land & irrigated land in Brahmani basin. Medium Projects numbering 7 in Orissa and 7 in Jharkhand have resulted in almost 10% of the basin area getting irrigation.

The agricultural land use pattern in the Brahmani basin is shown in Table 29.

13.3.5. Flood & Drainage

During floods, the Brahmani River turns into a large turbulent channel posing threat to life and property. The maximum flood in the river has been recorded as 24,246 cumec on 20.8.75 at Pankapal site. The highest gauge level at the site in that event was recorded to be 24.78 m against the danger level of 23.00 m. Since the Rengali Multipurpose Project has come up, it has been moderating floods in the lower reach covering an area of $14,019\,km^2$. Out of this, deltaic stretch of $4,000\,km^2$ is mostly vulnerable. At some locations, raising and strengthening of flood embankments has also been taken up to control flooding.

Below Jenapur, Brahmani River bifurcates to Brahmani (Kimiria) on its right and Kharsuan on its left. These two rivers again join 100 km downstream. An anicut was built on the Brahmani arm and another at Jokadia was built on Kharsuan in 1890,

Table 28. Arable land & Irrigated land in Brahmani basin

District	Geographical area (km^2)	Area within basin (km^2)	Culturable land (km^2)	Irrigated land (Kharif) in km^2
Sundargarh	9, 712	5, 786.08	2, 589	197.32
Deogarh	2, 940	2, 512.37	682	118.66
Sambalpur	6, 657	1, 371.05	281	81.52
Angul	6, 375	4, 225.94	1, 566	257.38
Dhenkanal	4, 452	3, 956.91	1, 815	370.76
Keonjhar	8, 303	1, 723.48	401	72.11
Jajpur	2, 899	1, 824.75	1, 047	458.44
Kendrapara	2, 644	1, 115.5	646	338.82
Sub Total		22, 516.08	9, 027	1, 895.01
Ranchi	7, 573	2, 470.17		
Lohardega	1, 490	628.14	7,489	
Gumla & Simdega	8, 820	8, 343.57		
West Singhbhum	5, 290	3, 960.03	2, 671	
Sub-total		15, 405.91	10, 160	600.36
Sarguja	22, 337	423.70	46	0.7
Raigarh	6, 154	923.10	747	16.6
Sub-total		1, 346.80	793	17.3
Total		39, 268.79	19, 980	2, 512.67

from the left of which the High Level Canal takes off for irrigation and navigation. This canal finally discharges into Baitarani. This canal has since become defunct.

The entire flood spill of the Brahmani – Kharsuan rivers flows to the sea over a 10 to 20 km wide and 70 km long flat flood plain. The entire delta of Brahmani – Kharsua covering 3, 500 km^2 area is flood prone. To protect the densely populated Aul area, a 70 km long ring bund was constructed, protecting 250 Mm^2 of agriculture land, and nearly 15 lakh population.

Table 29. Gross Area (in km^2) under Different Crops in Brahmani Basin

Crop	Irrigated condition	Un-irrigated condition	Total area
Rice	2, 858.57	7, 948.73	10, 807.30
Wheat	85.56	21.95	107.51
Other Cereals (millets/ ragi)	0	742.15	742.15
Pulses (BGH grams)	136.78	1, 667.27	1, 804.05
Sugarcane	16.39	0	16.39
Oil crops (groundnut, mustard, til)	215.66	497.34	713.00
Summer crop/ vegetables	37.25	211.71	248.96
Maize	21.66	111.29	132.95
Roots & tubers (potato)	8.79	216.04	224.83
Fruit trees	–	949.80	949.80
Total	3, 380.66	12, 366.28	15, 746.94

13.3.6. Domestic & Industrial Water Use

Nearly 1.252 million people in Brahmani Basin live in 41 urban centers. Total water consumption in major towns in the basin is about 97,000 kL/day and waster generation is about 74,500 kL/day. Current withdrawals of water for domestic use are given in Table 30.

Brahmani basin with its rich mining and agricultural resources and cheap labour offers ideal ground for establishment and operation of industrial units. The establishment of an integrated steel plant at Rourkela in Sundergarh district during late 1950s was the harbinger for large-scale industrialization of the area. Many industries in the basin are small-scale units that have no significant water requirement, wastewater generation and environmental implications.

Parts of the Brahmani basin have witnessed rapid urbanization in the last four decades. Urban conglomerates such as the Rourkela city and its suburbs, Talcher and its satellite towns, Rengali township, Dera Colliery township, Angul, Dhenkanal and Keonjhar districts have rapidly grown only during the last three decades due to industrial development or construction of dams.

Brahmani Basin has vast potential for development of inland fisheries in reservoirs, ponds, tanks, and canals. The predominant species found in Brahmani River are Carps, Catfishes, feather-backs, forage fishes, prawns & Hilsa. The entire stretch of Brahmani, Kimiria and Kharsuan have flood plains with over bank zones which are sheltered locations for breeding. Total annual fish catch is estimated at 24,000 M.T.

Table 31 enumerates the water demand & waste water generation from major industries in Brahmani Basin. The Current withdrawal of water for industrial use is given in table 32.

Table 30. Current withdrawals of water for domestic use in Brahmani basin

SN	Water use type	Per capita annual demand (m^3)	Total Population (in Million)	Withdrawals (in Mm^3)
		Domestic (Urban)		
1	Municipal use (Urban area) @ 140 LPCD	51.10	1.25	63.98
	Total withdrawal for Urban Use			63.98
		Domestic (Rural)		
1	Drinking water & sanitation needs (rural) @ 70 LPCD	25.55	7.21	184.09
2	Live stocks (Cattle & Buffalos) @ 50 LPCD	18.25	3.82	69.68
3	Other live stocks (pigs, sheeps, goats, etc.) @ 10% of SN 2	–	–	6.97
	Total withdrawal for rural use			260.74

Table 31. Water demand & waste water generation from major industries in Brahmani Basin

SN	Name of the Industry	Products	Water consumption KLD	Wastewater generation KLD
1	Rourkela Steel Plant	Iron & Steel	265,580	120,000
2	Rourkela Steel Plant	Fertilizer CAN	28,807	7,920
3	Fertilizer Corporation	Fertilizer (Urea)	45,883	16,608
4	National Aluminum Company – Smelter Unit	Aluminum	5,066	4,900
5	National Aluminum Company – Captive Power Plant	Electric Power	135,000	90,000
6	ORICHEM Ltd.	Sodium dichromate, Basic Chromate Sulphate, Yellow Sodium Sulphate	170	10
7	Talcher Thermal Power	Electric Power	13,227	6,483
8	Talcher Super Thermal Power Plant NTPC Kaniha	Electric Power	137,099	52,080
		TOTAL	630,832	298,001

Source: OPCB, Orissa

13.3.7. Water Quality Aspects

Brahmani, a mighty river during the monsoon, turns in the summer into more or less a stagnant pool of water held in deep gorges and pot holes in the river bed. The river becomes incapable of washing down the pollutants, which are discharged into it from the nearby industries, towns & villages. Greater part of the Brahmani River (below Panposh) is highly polluted.

The Brahmani Baitarni River basins hold extremely rich mineral resources. Consequently, many industrial units including the famous Rourkela Steel Plant as well as a number of fast growing townships are located here. This basin was essentially a virgin natural environment till India's independence with marginal human interference. Industrialization which started in late fifties was slow in the beginning. From the water quality point of view, the basin area is still far cleaner than other urban industrial complexes in the country. Table 33 shows the desired and observed water quality classes and critical parameters for different locations in Brahmani basin at different

Table 32. Current withdrawal of water for industrial use

SN	Industrial water Use	Withdrawal (Mm^3)
1	Consumption of water by major industries	230.32
2	40% of above (lump sum) for other industries in the basin	92.13
	Total Withdrawal	322.44

Table 33. Desired and existing water quality levels for Brahmani

Location	Desired class	Observed class & critical parameters				
		1997	1998	1999	2000	2001
Brahmani at U/S Panposh, Orissa	C	D BOD	D BOD	C	C	C
Brahmani at D/S Panposh, Orissa	C	D BOD	D BOD	D BOD	D BOD, Totcoli	D BOD, Totcoli
Brahmani at Rourkela D/S, Orissa	C	NA	D BOD	D BOD	C	D Totcoli
Brahmani at Bonaigarh, Orissa	C	D BOD	D BOD	B	D BOD	C
Brahmani at Rengali, Orissa	C	D BOD	B	D BOD	B	B
Brahmani at Samal, Orissa	C	D BOD	D BOD	D BOD	B	C
Brahmani at Talcher U/S, Orissa	C	D BOD	D BOD	D BOD	D BOD	C
Brahmani at Kamalanga, Orissa	C	D BOD	D BOD	D BOD	D BOD	D BOD
Brahmani at Bhuban, Orissa	C	D BOD	D BOD	D BOD	D BOD	C
Brahmani at Dharmashala, Orissa	B	D BOD	D BOD	D BOD	D BOD, Totcoli	C Totcoli
Brahmani at Pattamundai, Orissa	B	D BOD	D BOD	D BOD	D BOD, Totcoli	C Totcoli

* NA- Not Available. Source: Central Pollution Control Board

times. While the observed quality is below the desired class in general, it is better at a few places. Here too, BOD and total coliform are critical parameters.

13.3.8. Master Plan

Govt. of Orissa, Dept. of Irrigation formulated a master plan for the Brahmani basin with the objective of utilizing the large water potential. Incidentally Orissa with 4% of the country's area is endowed with 11% of the country's water wealth. The Brahmani basin with 18,318 MCM yield is estimated to have a total irrigation potential of $4,493\,\text{km}^2$ from major and medium irrigation schemes. The minor irrigation potential is $1,217\,\text{km}^2$. One major project and 21 medium projects are envisaged in the basin.

The state of Jharkhand has a total agricultural area of about $29,740\,\text{km}^2$. As per the assessment of 2nd Bihar Irrigation Commission, $12,765\,\text{km}^2$ irrigation potential may be created in Jharkhand through major and medium Irrigation Schemes. By

Table 34. Irrigation & Water Need by 2025 in Brahmani Basin

Project types	Jharkhand		Chhattisgarh	
	Area (km^2)	Water need (Mm^3)	Area (km^2)	Water need (Mm^3)
Major/ Medium	1,884.11	1,884.11		
Minor	1,004.86	653.16	20.09	13.06

the end of March 2001, 2,005 km^2 of potential had been created. The ultimate irrigation water requirements of Jharkhand and Chhattisgarh from future projects have been assessed as given in Table 34.

13.4. THE BAITARANI RIVER BASIN

The Baitarani River, one of the major rivers of Orissa, drains an area of 14,218 km^2. The basin lies mostly in the State of Orissa excluding 736 km^2 in Singhbhum District of Jharkhand State. In Orissa part, the Districts Keonjhar, Mayurbhanj, Balasore, Cuttack, Sundergarh and Dhenkanal fall in the Baitarani basin of which Keonjhar District covers the major portion of the basin area. The Baitarani River originates from Guptaganga hills in Keonjhar District of Orissa, about 2 km from Gonasika village, at an elevation of 900 m at latitude 21°31′ N and longitude 85°33′ E. Initially the river flows in northern direction for about 80 km and then takes a sudden right turn. In this reach, the river serves as a boundary between Jharkhand and Orissa states up to the confluence of Kangira River.

While flowing towards south, the River enters the plains at Anandapur and further downstream meets the deltaic zone at Akhudapada, where it gets bifurcated. Further, it meets the Brahmani and is renamed as Dhamara and joins the Bay of Bengal after traveling a distance of 360 km. There are 64 tributaries of Baitarani River out of which 35 join on the left side and 29 on right side. The prominent tributaries are Kangira, Khairi, Bhandarn, Deo, Kanjihar, Sita, Kusai and Salanadi. Annual average observed runoff at some of the CWC gauging sites of Brahmani and Baitarni basins are given in Table 35.

Table 35. Annual average observed runoff at CWC sites in Brahmani and Baitarni basin (Catchment area > 5,000 km^2)

Name of the site	Name of the stream	Catchment area (km^2)	Annual average runoff (BCM)
Jaraikela	Koel	9,160	5.09
Panposh	Brahmani	19,448	14.07
Gomlai	Brahmani	19,820	11.73
Jenapur	Brahmani	33,955	16.98
Anadpur	Baitarani	8,570	4.69

During flood, the Baitarani River turns into a large turbulent stream posing potential threat to life and property. The maximum observed flood has been recorded as 4.36 lakh cusec in the year 1960 at Birdi G & D site. At present there is no flood moderating project completed in the Baitarani main stream. Only a few major and medium projects have been completed. These are Salanadi, Kanjhari, Remal and Akhuapada projects having total command area of $1,323\,\text{km}^2$. In addition to the above, the ongoing projects are Kanupur and Deo with command area of $395\,\text{km}^2$. There are some proposed projects and Bhimkund project at Udaipur is an important major project that is in the stage of investigation.

Based on topography and hydrometeorology, the Baitarani River basin may be classified in three groups: Upper, Middle, and Lower Baitarani basins.

13.4.1. Upper Baitarani Basin

The upper Baitarani basin covers an area of $5,792\,\text{km}^2$ of which $736\,\text{km}^2$ lies in Jharkhand. This is mostly mountainous and rocky and lies in higher altitude. The upper portion of this basin is in the northern plateau and Eastern Ghats. The rock of the basin belongs to iron ore series of the upper Dharwar system of the eastern archean group. The major tributaries are Khairi-Bhandan, Deo that join Baitarani in the sub-basin.

13.4.2. Middle Baitarani Basin

The middle Baitarani basin with an area of $4,333\,\text{km}^2$ is entirely in the State of Orissa. This portion is partly hilly and partly plain. The major tributaries are Kanijhari, Kushei, Kantamuli and Sim.

13.4.3. Lower Baitarani Basin

The lower Baitarani basin covering an area of $4,093\,\text{km}^2$ mostly consists of Salandi and Matai River basins. Physiographically this sub-basin is in the deltaic region and situated in fertile plain of Cuttack and Balasore Districts. Major portion of the sub-basin is made up of alluvial soil.

13.4.4. Geology

The geological formations in and around the Upper Baitarani are of two main series, the iron ore series and younger Kolhan series. The iron ore series are represented by mica, hornblende, schist, gneiss, phyllite, chert and jasper which along with Singhbhum granite constitute the surrounding country rock. The Kolhan series comprise mainly flat bedded Kolhan, sand stone and conglomerate. The sand stone usually form the flat tapped hills over the granite terrain in this area. The generalized geological break up of the south Singhbhum and Keonjhar Districts is New Dolerite, Kolhan series, Singhbhum series, and Iron ore series.

13.4.5. Physiography and Climate

The Baitarani basin is an oval shaped basin having drainage area of $14,218\,\text{km}^2$. The basin lies in the Singhbhum District of Jharkhand and Keonjhar, Dhenkanal, Mayurbhanj, Sundergarh, Cuttack and Balasore Districts of Orissa. The upper Baitarani basin is about 700 m above mean sea level and therefore the climate is extreme in nature. The middle Baitarani basin is partly hilly and partly plain, and the lower Baitarani basin is in coastal area. The effect of the sea is very much felt in the lower basin in coastal plain.

The rainfall in the basin is mostly from south-west monsoon and lasts from June to October. About 80% of the annual precipitation occurs during these months. The rainfall is caused mostly by depressions in Bay of Bengal. The annual rainfall varies from a maximum of 1,595 mm to a minimum of 745 mm and average rainfall is 1,187 mm. The maximum recorded temperature of Keonjhar District in summer is 48.5 °C and minimum in winter is 6 °C.

The relative humidity is minimum in the months of April and May and maximum in the months of August and September. The maximum and minimum humidity are of the order of 83.08% and 39.63% respectively.

As per the data of the IMD station at Keonjhar, the minimum and maximum wind speeds are 5 kmph and 75 kmph respectively. The maximum cloud cover is observed in the months of June and July whereas minimum is in the months of December and January.

In the Baitarni River basin in central part of India that covers a large area in the state of Orissa and a smaller part of Jharkhand, water surplus occurs in the southwestern part of the basin. Water deficit is very low and is concentrated only in western and central parts of the basin, which vary from 70 mm to 140 mm. Moreover, due to the low water deficit and high amount of water surplus, the entire river basin provides a maximum of 400 mm of water of the river as runoff as a good amount for groundwater recharging.

13.4.6. Existing and Ongoing Projects in Brahmani and Baitarani Basins

After independence, many water resources development projects have come up in these basins. Brief details of important projects are given below.

i. Hirakud canal system

Large tracts in Sambalpur District are irrigated by this major irrigation system. These works were built after independence. The flood waters of Mahanadi River are stored at Hirakud Dam and are distributed for irrigation in later months. A part of the stored water gets transported to the Brahmani basin.

ii. Rengali dam

Rengali is an earthen and rockfill dam constructed on Brahmani River, located at a distance of 80 km from Angul in Angul District, Orissa. The height and length of the dam is 70.5 and 1,040 m respectively. Rengali reservoir is the second largest man-made lake in Orissa, covering 37,840 ha at full reservoir level and 28,000 ha at the mean level. Its FRL is at 123.5 m and the MDDL is at 109.72 m. Regali reservoir stores 4,400 to 5, 150 Mm^3 of water with mean annual inflow of 14,900 MCM. The reservoir has a catchment of 25, 250 km^2, comprising mostly forests and wasteland. Rengali power house has 5 units of 50 MW each and a firm power of 53 MW. The water temperature (20.25 to 29.0 °C) and pH (8.2) are conducive to reasonably high organic productivity.

Rengali dam has the capacity to moderate the high flood of 5 to 6 lakh cusec to 3 to 4 lakh cusec and to 2 lakh cusec or less in approximately 30% of time. Even with this moderation, growing a Kharif crop over 500 Mm^2 of good land (low land of elevation up to 10 m) has become unviable.

Orissa has planned to utilize some more water resources in the second phase of Rengali Irrigation Project. Till date only Rengali dam across Brahmani and Mandira dam across Sankh have been constructed. Water from Mandira reservoir is being used for the Rourkela Steel Plant.

Rengali Irrigation Project: This is a barrage scheme constructed across Brahmani River at Samal in Angul district, 35 km downstream of Rengali dam project. The project is designed to pick up flood releases and tail race water from Rangali dam and divert the same as per requirements through two canal systems. The CCA includes both flow and lift irrigation. The salient features of Rengali Irrigation Project are given below:

Free catchment at barrage site:	4, 780 km^2 (between Dam & barrage)
Total catchment:	30, 030 km^2
Mean Annual Rainfall:	1, 570 mm

For this project, the Gross Command Area is 3, 364 km^2 while the culturable command area (CCA) is 2, 591 km^2. The intensity of Kharif Irrigation is 100% and that of Rabi is 90%. The Kharif cropped area is 2, 591 km^2 and Rabi cropped area is 2, 332 km^2.

iii. Kansabahati irrigation project

This is a reservoir project constructed across Badajore Nalla, a tributary of Brahmani River in Sundergarh district near village Kadambahal. The reservoir intercepts a catchment area of 179 km^2 and is designed to provide irrigation to a CCA of 5,050 ha. Kharif irrigation is for 4,615 ha and Rabi irrigation for 2,430 ha. The anicut lies within Rajgangpur Block area.

iv. Pitamahal irrigation project

A reservoir project with earthen and masonry dam across the Pitamahal River in Sundergarh district intercepts a catchment area of 102 km^2. The project is designed to irrigate 4,261 ha of CCA which includes 2,630 ha in Kharif season and 1,631 ha in Rabi season. The gross and live storage capacity of the project are 27.68 and 24.00 Mm^3 respectively.

Table 36. Salient particulars of the existing Projects in Brahmani and Baitarani basin

Name of the Project	State	Gross storage capacity (Mm^3)	Live storage capacity (Mm^3)	Designed annual irrigation (km^2)	Installed capacity (MW)
Nandini	Jharkhand	19.28	16.54	40.80	–
Paras	Jharkhand	18.49	16.90	34.70	–
Kanjhari Dam	Orissa	40.52	34.52	133.90	–
Mardira	Orissa	542.78	370.08	–	–
Roomal	Orissa	19.28	15.70	64.00	–
Salnadi	Orissa	565.00	556.35	601.40	3

v. Gohira irrigation project

This is a reservoir project with 3,550 m long earth dam built near village Ambaghat in Deogarh district. It intercepts a catchment area of 236 km^2 and is designed to irrigate 8,100 ha in Kharif season and 5,265 ha in Rabi season in Reamal block of Deogarh district.

vi. Aunali irrigation project

This project is constructed across Aunali Nalla in Chhendipada block of Angul district near village Patrapada. It intercepts a catchment area of 150.22 km^2 and is designed to irrigate 1,188 ha in Kharif season.

vii. Dadaraghati irrigation project

It is a reservoir project with a rock-fill dam constructed near village Rangathali in Angul district, across Gambharia Nalla. This reservoir intercepts a catchment area of 102 km^2 and is designed to irrigate 4,514 ha CCA with an annual intensity of irrigation of 140%. The gross and live storage capacity of the project are 23.44 and 19.74 Mm^3 respectively.

Table 37. Salient particulars of the Projects Under construction in Brahmani and Baitarni basins

Name of the Project	State	Gross storage capacity (Mm^3)	Live storage capacity (Mm^3)	Designed annual irrigation (km^2)
Chirgaon Reservoir	Jharkhand	14.57	13.17	29.00
Kans Reservoir	Jharkhand	23.24	21.46	33.00
Kansabahal	Orissa	40.42	28.72	81.10
Kusei Dam	Orissa	71.96	56.74	104.80

Table 38. Desired and observed water quality classes for Baitarani

Location	Desired class	Observed class & critical parameters				
		1997	1998	1999	2000	2001
Baitarani at Joda, Orissa	C	D BOD	D BOD	D BOD	B	C
Baitarani at Anandpur, Orissa	C	D BOD	D BOD	D BOD	C	C
Baitarani at Jajpur, Orissa	C	D BOD	D BOD	D BOD	D BOD	D Totcoli
Baitarani at Chandbali, Orissa	C	D BOD	D BOD	D BOD	D BOD	D BOD, Totcoli
Baitarani at Dhamra, Orissa	C	D BOD	D BOD	D BOD	D BOD	D BOD, Totcoli

* NA- Not Available. Source: Central Pollution Control Board.

viii. Derjang irrigation project

It is a reservoir project with zoned earth-fill dam constructed across Lingra Nalla in Angul district. It intercepts a catchment area of 394 km^2 and irrigates 6,478 ha of CCA. The gross and live storage capacity of the project are 45.28 and 41.55 Mm^3 respectively.

ix. Ramiala irrigation project

It is a reservoir project with rolled filled homogeneous earth dam constructed across Ramiala River in Dhenkanal district near village Budhibil. Ramiala dam intercepts a catchment area of 328 km^2 and is designed to irrigate 9,600 ha CCA with annual intensity of irrigation of 162.5%. The gross and live storage capacity of the project are 86.00 and 69.40 Mm^3 respectively.

In addition to these, salient features of some other existing projects with a live storage capacity of 10 Mm^3 and above are presented in Table 36. Table 37 gives salient features of some projects that are under construction.

13.4.7. Water Quality Aspects

According to the studies made by CPCB, the observed water quality status of Baitarani River and its tributaries and the desired class are given in Table 38. It is observed from this tale that the observed class in mostly ‘D’ while the desired class is ‘C’.

CHAPTER 14

KRISHNA AND GODAVARI BASINS

14.1. THE KRISHNA BASIN

The Krishna is the second largest eastward draining interstate river in Peninsular India. It rises in the Mahadev range of the Western Ghats at an altitude of 1,337 m near Mahabaleshwar in Maharashtra State, about 64 km from the Arabian Sea. It flows for a distance of 305 km in Maharashtra, 483 km in Karnataka and 612 km in Andhra Pradesh before finally out falling into the Bay of Bengal. Thus the length of the river is about 1,400 km and it flows across the whole width of the peninsula, from west to east, through Maharashtra, Karnataka and Andhra Pradesh states. The basin lies between latitudes 13°07′N and 19°20′N and longitudes 73°22′E and 81°10′E. On the north, the basin is bounded by the range separating it from the Godavari basin, on the south and east by the Eastern Ghats and on the west by the Western Ghats. The basin extends over an area of 258,948 km^2, which is nearly 8% of the total geographical area of the country. An index map of the Krishna basin is given in Figure 1.

The basin is roughly triangular in shape with its base along the Western Ghats, the apex at Vijayawada and the Krishna itself forming the median. The western border of Krishna basin is formed by an unbroken line of ranges of the Western Ghats. Most other parts of this basin are comprised of rolling and undulating terrain. The state-wise distribution of drainage area is given in Table 1.

The Tungabhadra, Narayanpur, Srisailam, Nagarjunasagar and Prakasam barrage, are the existing major projects in the basin. Jurala and Almatti are the ongoing major projects while Pulichintala is the proposed major project.

Water balance studies carried by NWDA (by sub-dividing the whole basin into 12 sub-basins) have shown that the basin will be water deficit to the tune of 3,235 Mm^3 in the ultimate development scenario.

14.1.1. Tributaries of Krishna River

Thirteen major tributaries join the Krishna River along its course, out of which six are right bank tributaries and seven are left bank tributaries. All the major tributaries draining the base of the triangle fall into the Krishna River in the upper two-thirds of its length. Among the major tributaries, the Ghatprabha, the Malprabha and

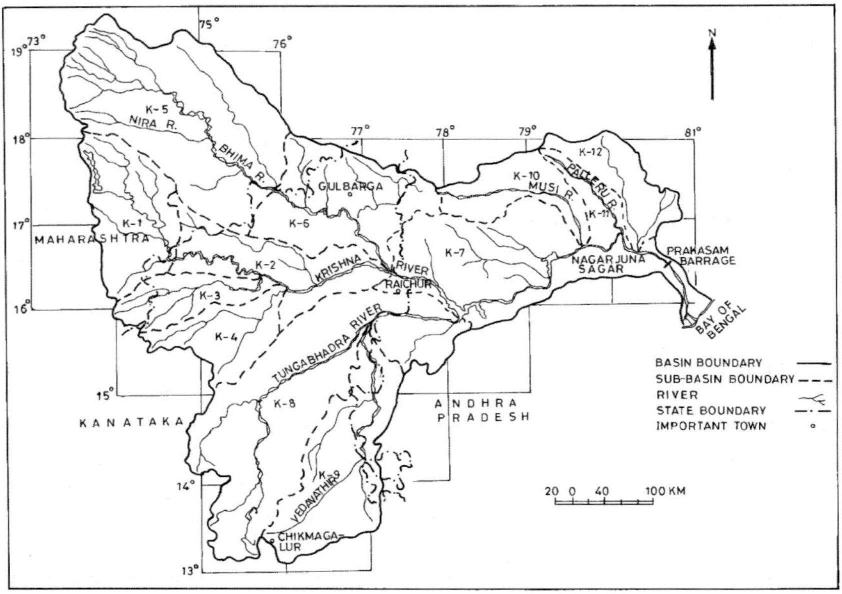

Figure 1. Index map of Krishna Basin

the Tungabhadra are the principal right bank tributaries which together account for 35.45% of the total catchment area, whereas the Bhima and the Musi are the principal left bank tributaries which together account for 35.62% of the total catchment area. The catchment area, length and elevation of source of the tributaries are given in Table 2. Flow Diagram of Krishna basin is given in Figure 2.

In the following, a detailed description of the principal tributaries is given.

The Tungabhadra river

Tungabhadra River is formed from the union of the two rivers, namely, Tunga and Bhadra, which together rise in Varahagiri in the Western Ghats of Karnataka State at an altitude of about 1,196 m. The two rivers confluence at a village called Kudali

Table 1. Statewise area of Krishna Basin

State	Length (km)	Drainage area (sq. km)	Percentage of total area
Maharashtra	305	69,425	26.8
Karnataka	483	113,271	43.7
Andhra Pradesh	612	76,252	29.5
Total	1,400	258,948	100.0

Table 2. Details of major tributaries of Krishna River

SN	Name of tributary	Bank	Elevation of source (m)	Length (km)	Catchment area (sq. km)
1.	Koyna	Right	4,719	118	4,890
2.	Panchganga	Right	1,020	74	2,575
3.	Dudhganga	Right	870	103	2,350
4.	Ghataprabha	Right	884	283	8,829
5.	Malaprabha	Right	793	304	11,549
6.	Bhima	Left	945	861	70,614
7.	Tungabhadra	Right	610	531	71,417
8.	Dindl	Left	718	178	3,490
9.	Peddavagu	Left	707	109	2,343
10.	Halia	Left	708	112	3,780
11.	Musi	Left	661	265	11,212
12.	Paleru	Left	515	152	3,263
13.	Munneru	Left	238	195	10,409

near Shimoga. The united Tungabhadra River flows for about 531 km in a generally northeasterly direction, through Mysore and Andhra Pradesh and joins the Krishna at an elevation of about 264 m beyond Karnool. The length of the river is 786 km. The important tributaries of the Tungabhadra River are the Varada, the Hagari, the Vedavati, and the Kumudvati. The total drainage area of the Tungabhadra is 71,417 km^2. The mean annual rainfall in the Tungabhadra basin is 884 mm (NIH, 1992).

The catchment area of Tungabhadra sub-basin can be demarcated into three zones depending on the vegetative growth viz., (i) The Western Ghat belt from Agumbe to Honnali with thick forest and heavy rainfall, (ii) thin vegetative cover from Honnali up to Harihar with moderate rainfall, and (iii) very thin vegetative growth with bare topped hills beyond Harihar and up to Mallapuram with scanty rainfall. The land use in the catchment consists of forest (14.5%), cultivation (59%), pastures (9%), wasteland (12%); the rest (5.5%) is fallow land (KERS 1985).

Drainage density of the catchment is $0.44\,km/km^3$ and the average slope is 6%. On the basis of topography, vegetation and the type of soil the value of runoff coefficient is 0.21 The river derives major portion of its flow in its initial course of 206 km, draining from the Western Ghats.

The annual river flow varies from 8,412 Mm^3 to 17,148 Mm^3, the average being 11,427 Mm^3. Although the volume of rainfall is quite high, the rate of soil erosion is low because of good forest cover. In addition, the soils are mainly lateritic, which are not easily susceptible to erosion when there is vegetation cover.

The storage projects existing in the area are: (1) Bhadra Reservoir across Bhadra near Lakkavalli, (2) Tunga anicut across Tunga near Gajanur, (3) Hagari-Bommanhalli reservoir across the river Hagari, (4) Bhadra anicut near Bhadravathi, (5) Dharma reservoir across the river Varada, and (6) Anjanapura reservoir across

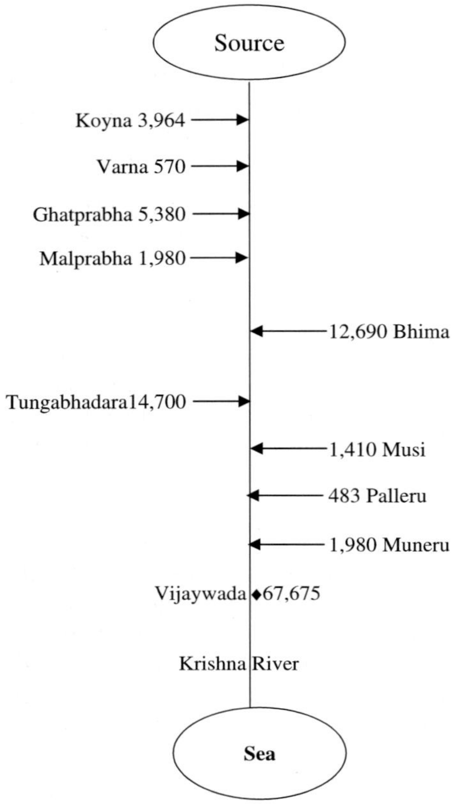

Figure 2. Flow Diagram of Krishna basin

Kumudvathi. In addition, the catchment area is intercepted with a large number of minor tanks. Counting the major and medium reservoirs in the catchment, the intercepted catchment is approximately 6,740 km^2 and the independent catchment is approximately 21,440 km^2.

The Malprabha river

Malprabha is a right bank tributary of Krishna River. The Malprabha catchment lies between North latitudes 15°00′ and 16°12′ and east longitudes 74°14′ and 76°05′. The Malprabha River originates from the Chorla Ghats, a section of the Western Ghats, at an elevation of about 792 m about 35 m south-west of Belgaum District of Karnataka. The river flows east and north-west and joins Krishna at Kapila Sangam in the Bijapur District at an elevation of about 488 m. Malprabha traverses a length of 306 km before meeting the Krishna River. The Bennihala and the Hirehalla are the principal tributaries of the river Malprabha. The total catchment area of the Malprabha including its tributaries is 11,549 km^2, which lies wholly in the State of Karnataka.

To harness the waters of the Malprabha River a dam has been constructed at Niviluteerth, Belgaum District to impound 1,377 Mm^3 water. The reservoir catchment covers an area of 3,300 km^2.

Physiography: The Malprabha catchment is approximately triangular in shape. The terrain is flat to gently undulating except for a few hillocks and valleys. The northern boundary is the common ridge between the Malprabha and the Ghatprabha rivers and the eastern boundary is the common bridge between the Malprabha, the Krishna and the Tungabhadra rivers. The southern and western boundaries are the common ridge between the Malprabha and the west flowing rivers. The important rock formations in the sub-basin are: (i) sedimentary rock formation (Kaladgi group) comprising limestone, shale and quartzites, (ii) Schistose rock formations (Dharwad super group) comprising granite, gneiss and crystalline rocks. The important soil types found in the basin are black soils, red soils, laterite and lateritic soils, alluvium, mixed soil, red and black soil and saline and alkaline soils.

Climate: Three main seasons prevailing in the catchment are: the summer from March to April, the monsoon from May to November, and the winter from December to February.

Rainfall is mainly received by the south-west monsoon. The rainfall in the non-monsoon period is insignificant. The average annual rainfall of the catchment for the periods from 1901–02 to 1930–31, 1931–32 to 1948–49 and 1949–50 to 1984–85 were 718 mm, 775 mm and 815 mm respectively. Except the monsoon months, the climate of the catchment is generally dry. The mean relative humidity is high during the south-west monsoon season and comparatively low during the non-monsoon period. In summer, the weather is dry and the humidity is low.

During the monsoon season, winds flow from the south-west or west. In the non-monsoon season, winds from north-east and south-east are common. During the south-west monsoon, sky is heavily clouded. During the remaining part of the year, clear or lightly clouded sky prevails.

The Bhima river

The Bhima River is a major left bank tributary of the Krishna River, which also rises in the Western Ghats and flows south-eastwards through Maharashtra and Mysore. It falls into the Krishna River at about 26 km north of Raichur at an altitude of 343 m. The total length of the river before joining Krishna River is 861 km. The main tributaries of the Bhima are the Mula, the Mutha, the Nira, the Ghod, the Man and the Sina. The total catchment area of the Bhima sub-basin is 76,614 km^2.

Twenty medium and major reservoir projects have been constructed on the Bhima River and its tributaries in Maharashtra. Most of the headwater reservoirs are ungated. The dams in Bhima system are operated to meet various conservation requirements and for controlling the flood in the Bhima valley. The Bhima project, also known as Ujjani dam, is the biggest dam in the system that has been constructed on the Bhima River in the Krishna basin in Solapur district, Maharashtra state. The main purposes of this dam are irrigation and flood control. Pandarpur is an important religious place which is to be protected against floods. The 15 major projects are:

1. Pimpalgaon joge
2. Yedgaon
3. Manikdoha
4. Wadaj
5. Dimbhe
6. Chaskaman
7. Pawana
8. Varasgaon
9. Panshet
10. Khadakwasla
11. Bhatghar
12. Bhama Askhed
13. Veer
14. Ghod
15. Ujjani

Following five lakes are operated by the Tata group and are known as Tata lakes:
1. Andhra 2. Mulshi 3. Shirota 4. Lonawala 5. Valvan

The current practice of operating the dams in this system is based on thumb rule as scientific analysis of the operation procedure for the reservoirs has not been carried out. This implies sub-optimal management of reservoirs in the basin. Water scarcity and flood situations occur quite frequently. Figure 3 shows the line diagram of the Bhima system.

The Solapur district critically depends on the Ujjani reservoir for drinking and irrigation water supply, with almost 80% of its water coming from the reservoir. Ujjani irrigates some 350 Mm^2 of land in Solapur through various lift irrigation schemes and another 1,500 Mm^2 benefit through water discharged into the Bhima

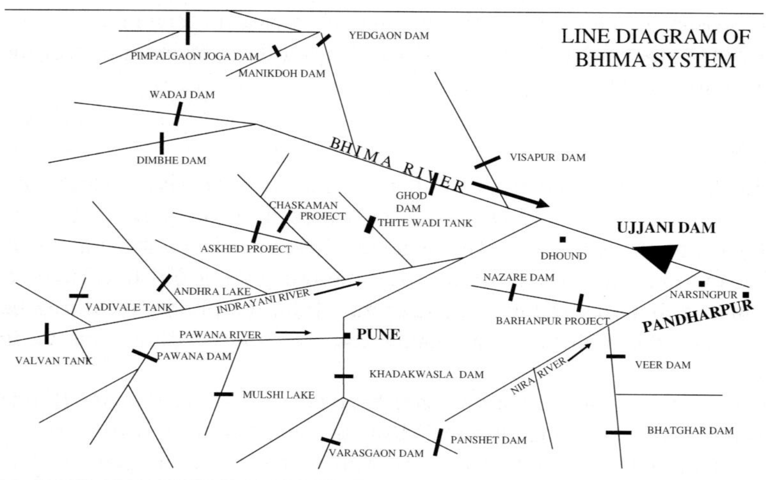

Figure 3. Line diagram of Bhima system

River. The Ujjani catchment is spread over 14,856 km^2. However, in some recent dry years, the ready-to-use water stock in Ujjani, with a capacity of 1,517 Mm^3 was nearly over.

The Ghatprabha river

Rising from the Western Ghats at an altitude of 884 m, the Ghatprabha River is one of the southern tributaries of the Krishna River. The catchment area of Ghatprabha lies between latitude 15°45′ and 16°25′ N and longitude 74°00′ and 75°55′ E. The Ghatprabha River flows eastwards for a distance of 283 km before joining Krishna at Kudalisangam, about 35 km north-east of Kaladgi at an elevation of 500 m. The river flows for about 60 km in the Ratnagiri and Kolhapur Districts of Maharashtra before entering the Belgaum District of Karnataka. In Karnataka, it flows through Belgaum and Bijapur Districts and joins the Krishna about 16 km from Almatti. The catchment area of Ghatprabha including its tributaries is 8,829 km^2.

Tributaries of Ghatprabha River: The principal tributaries of Ghatprabha River are the Tampraparni, and Hiranyakeshi and the Markandeya. The Tamraparni, rising in Maharashtra flows through the Sindudurg district. It forms the boundary between Maharashtra and Karnataka for 6 km and after a run of 19 km in Karnataka, joins Ghatprabha on the left bank. The Markandeys, rising in Maharashtra flows for 8 km and after a run of 66 km in Karnataka, joins the Ghatprabha on the right bank.

A dam has been constructed at Hidkal in Hukkeri taluk to impound 2,202 Mm^3 water running two canals on either bank and coupled with weirs and lift irrigation schemes on the foreshores.

Physiography and Geology: The sub-basin is approximately triangular in shape. Its northern boundary is the common ridge between the Krishna and the Ghatprabha rivers; the southern boundary is the common bridge between the Ghatprabha and the Malprabha rivers. Except for isolated hillocks and valleys, most of the catchment area is flat to gently undulating.

The geological formations found within the sub-basin are (i) Deccan trap of tertiary age, (ii) Sedimentary formations known as 'Kaladagi group' comprising limestone, shale and quartzities, (iii) Schists, Gneiss and other crystallizing rocks, and (iv) Laterite rocks. The soils generated from the formations are mostly permeable. As the surface is covered with moorum, runoff is moderate in the catchment.

Hydrogeological studies carried out in the sub-basin reveal that ground water occurs in all the geological formations, viz., the Dharwar Schists, Peninsular Gneisses, quartzites and alluvial deposits. The occurrence and movement of groundwater in these rocks is controlled by the nature and extent of weathering and the presence of joints and fractures in them. The sub-soil water table generally does not exceed 9 m below ground water level under normal conditions. The ground water development in the sub-basin is from open wells and dug-cum-borewells.

Climate: The climate of the catchment is marked by a hot summer and a mild winter. In the western parts of catchment, the summer season is mild but it is severe in the other parts. On certain days during summers, the day temperature rises up to

41 °C. The monsoon sets early in June and withdraws by the end of October. With the onset of monsoon in early June, there is an appreciable drop in the day temperature. The winter season lasts from November to December. Generally December is the coldest month with the mean daily maximum and minimum temperature being 29.3 °C and 13.9 °C respectively. April is generally the hottest month with mean daily maximum and minimum temperatures of 35.7 °C and 19.5 °C. Expectedly, the relative humidity is high during the south-west monsoon and low during the non-monsoon period. In summer the weather is dry and humidity is low.

Wind velocities are generally low with some increase during the late summer and in the monsoon season. From April to Sept., wind blows mainly from the south-west and west. In October, it is from the directions between north and east but on some days from south-west or west. During November and December, wind is mostly north-easterly or easterly. South-westerly and westerly appear in January and from February onwards the easterlies decrease in frequency and by April the afternoon wind blows predominantly from the west and south-west.

During the south-west monsoon, the sky is heavily clouded. Cloud cover decreases in the post-monsoon period. From December to February, the sky is generally clear with occasional light clouds. Cloud cover increases from April onwards. The sunshine percentage in the catchment varies from 21 to 96.

The Musi river

The Musi River is a major left bank tributary of Krishna, having its origin in the hills of Anathagiri near Vikarabad, Rangareddy District, A.P. It flows through Hyderabad city and runs mostly west to east until the Aleru River joins it. Flowing southwards, it meets the Krishna River near Wazirabad at an elevation of about 61 m. When it confluences with Krishna river, Musi River has already flown for 267 km.

The river has a rocky and very steep fall. It brings very heavy and sudden floods during the monsoon. During the year 1908, Musi swelled up in high floods and submerged a major portion of Hyderabad city and many villages on its banks, and caused severer damages to the property and life. Immediately after this disaster, the problem of controlling flood received the attention of Government of Hyderabad. Consequently two major reservoirs Himayatsagar and Osmansagar were constructed on the two branches of river Musi called Easa and Musa (Musi). The storages of these two reservoirs are exclusively being utilized for water supply to the twin cities of Hyderabad and Secunderabad. In recent years also, the city of Hyderabad has seen flooding and unplanned urban growth is one of the reasons for this.

14.1.2. Climate of Krishna Basin

The climate of the Krishna basin is dominated by the southwest monsoon, which provides most of the precipitation for the basin. High flow in the river occurs during the months of August–November and the lean flow season is from April to

May (at Vijayawada). Climatic types range from per-humid through dry sub-humid in the west through semi-arid in the central and eastern parts of the basin. The south-central part of the basin is truly arid.

On an average, annual rainfall in the Krishna basin is 784 mm. The southwest monsoon sets in by middle of June and withdraws by the middle of October. About 90 percent of the annual rainfall is received during the monsoon months; more than 70% occurs during July, August and September.

Except the monsoon months, the climate of the catchment remains dry. From the climatological observations, it is seen that the mean daily maximum temperature in the basin varies from 27.7 °C to 40.4 °C and the mean daily minimum temperature varies from 20.6° to 27.2 °C.

The relative humidity in the basin ranges from 17 to 92 percent. Mean relative humidity is high during the monsoon period and comparatively low during the post monsoon period. In summer, the weather is dry and the humidity is low.

Winds in the basin are generally light with some increase in force during the later half of summers. The catchment is influenced by south-west winds during the monsoon season. In the post monsoon season, they blow from north-west to north. In the winter season the winds blow from the north-west and south-west directions. In the Krishna basin, wind speed varies from 4.0 to 21.7 km/hr.

Sky is generally heavily clouded during the monsoon season. Cloudiness decreases sharply in the post monsoon months. During rest of the year, sky is clear or lightly clouded. The cloud cover in the basin varies from 0.8 oktas to 8.00 oktas.

14.1.3. Geology of Krishna Basin

The geology of the Krishna basin is dominated in the northwest by the Deccan Traps, in the central part by unclassified crystallines, and in the east by the Cuddapah Group. The Dharwars (southwest central) and the Vindhian (east central) form a significant part of the outcrops within the unclassified crystallines. Krishna delta is predominantly formed by Pleistocene to recent material. Figure 4 shows the valley of Krishna River in Eastern Ghats.

14.1.4. The Deltaic Plain

The Krishna delta is situated between latitudes 15°42′ to 16°30′ N and longitudes of 80°30′ to 81°15′ E with its head at Vijayawada. After cutting the Eastern Ghats, the river forms a deltaic plain which is nearly 95 km wide and covers an area of about 4,736 km^2. The main river splits in four distributaries which debauch into the Bay of Bengal. The first channel begins near Avanigodda and the three main distributaries are the Golumuttapaya, the Nadimieru and the Main channels. A weir at the head controls the flow within the deltaic plain. Vast amounts of sediment material has been added at the mouths of the distributaries during the past 50 years leading to the formation of river mouth bars and barrier islands with associated

Figure 4. Valley of Krishna River in Eastern Ghats

back island lagoons. As the delta progressed, these lagoons were filled in with finer grained sediments. From Vijayawada to the Bay, the average slope is 20 cm/km. The Krishna delta has large tracts of mangrove swamps along the coast with maximum concentration surrounding the three main distributaries.

Tidal flats occupy a considerable area of the lower deltaic plain especially between the Golumuttapaya and Avanigodda distributaries (Div Island). In fact, the tidal flats may be the product of a degraded inter-distributary bay between two (now abandoned) former channels.

14.1.5. Surface Water Potential of Krishna Basin

Average annual surface water potential of this basin has been assessed at 78.1 km^3. Out of this, 58.0 km^3 is utilizable water. Culturable area in the basin is about 0.203 Mm^2, which is 10.4% of the total culturable area of India. The hydropower potential of the basin has been assessed as 2,997 MW at 60% load factor.

At present, the utilization of surface water in the basin is 50.0 km^3. Past few decades have seen significant increase in storage capacity in the basin. From just about 3.2 km^3 in the pre-plan period, the total storage capacity of the completed projects has increased to 34.5 km^3. In addition, the projects under construction would add storage quantity of over 4.9 km^3 on completion. Figure 5 shows the gross flows of Krishna River at Vijayawada. This figure does not show any major trend in the data. The annual average observed runoff at some CWC sites in Krishna basin is given in Table 3.

14.1.6. Water Quality Aspects

Based on the systematic sampling of river water at many locations in the basin, its suitability for various purposes is determined by CPCB and the results are listed in

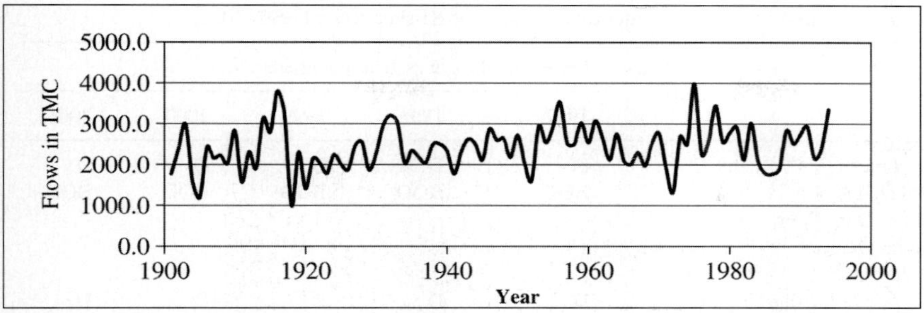

Figure 5. Gross flow of Krishna River at Vijayawada

Table 4. In general, the quality is not as per the desired class and BOD remains the most critical parameter. At some places, DO and total coliform are also causing problem.

14.1.7. Major Existing Water Resources Projects in Krishna Basin

A number of dams and barrages have been constructed and are under construction in the Krishna basin to utilize water resources. Nagarjunasagar and Srisailam are

Table 3. Annual average observed runoff at some CWC sites in Krishna basin

Name of the site	Name of the stream	Catchment area (km^2)	Annual average runoff (BCM)
Arjunwad	Krishna	12, 660	46.6
Kurundwad	Krishna	15, 190	65.2
Galagali	Krishna	22, 560	99.0
Huvenhedgi	Krishna	55, 150	100.1
Krishna Agraharam	Krishna	132, 920	203.2
Pondugala	Krishna	221, 220	184.0
Wadenapalli	Krishna	235, 544	130.2
Vijayawada	Krishna	251, 360	102.1
Dhond	Bhima	11, 660	32.5
Narsingpur	Bhima	22, 856	31.5
Takali	Bhima	33, 916	32.6
Yadgir	Bhima	69, 863	57.0
Wadakbal	Sina	12, 092	56.8
Hariahalli	Tungabhadra	14, 582	97.5
Ollenur	Tungabhadra	33, 018	80.4
Mantralayam	Tungabhadra	60, 630	84.6
Bawapuram	Tungabhadra	67, 180	61.5
T. Ramapuram	Hagari	23, 500	5.7
Damercherla	Musi	11, 501	8.9
Bhupasamundram	Vadavathi	15, 026	4.6

Table 4. Desired and existing water quality status of Krishna River (1997–2001)

Location	Desired class	Observed class & critical parameters				
		1997	1998	1999	2000	2001
Krishna at Mahabaleshwar Dhom Dam Near Koina Dam, Maharashtra	C	D BOD	D BOD	D BOD	D BOD	D BOD
Krishna at Krishna Bridge, Karad, Maharashtra	C	D BOD, Totcoli	D BOD	D BOD	D BOD	D BOD
Krishna at Kurundwad in Kolhapur, Maharashtra	B	D BOD	D BOD	D BOD	D BOD	D BOD, Totcoli
Krishna at Sangli, Maharashtra	C	D BOD	D BOD	D BOD	D BOD	D BOD
Krishna at U/S of Ugarkhurd Barrage, Karnataka	C	NA	C	C	C	C
Krishna at D/S of Narayanpura Dam, Karnataka	C	NA	C	NA	NA	C
Krishna at Tintini Bridge, Karnataka	C	C	C	C	C	C
Krishna at Rajapur Weir, Maharashtra	B	D DO, BOD	D BOD	D BOD	D BOD	D BOD
Krishna at D/S Of Devasagar Bdg., Karnataka	C	C	NA	C	C	C
Krishna after Confl. with Tungabhadra, Sangameshwaram A.P.	C	D BOD	NA	NA	D BOD	NA
Krishna at Vijaywada, A.P.	C	C	C	C	NA	NA

* NA - Not Available.
Source: Central Pollution Control Board.

two large terminal reservoirs on Krishna. These two have enough storage to regulate the inflows received in Andhra Pradesh for irrigation and hydropower. The left and right bank canal systems of Nagarjunasagar extend up to Godavari and Pennar delta and are notable examples of intra-basin water transfer. Telugu Ganga canal taking off from the Srisailam reservoir and carrying water up to Tamil Nadu for Chennai water supply is another praiseworthy example of inter-basin transfer with cooperation amongst the states.

A brief description of the important projects in Krishna basin is given here.

Upper Krishna project stage – 1

The first stage of the project comprises of two components:

- A dam across the Krishna River near Almatti village in Bagewadi taluk of Bijapur district to storing water up to RL 512.20 m for providing irrigation to an area of 160 Mm^2.
- Another dam across the Krishna River, at Narayanapur (downstream of Almatti Dam) near Bachihal and Siddapur village in Muddebihal taluk of Bijapur district. A Left Bank Canal from the dam will provide irrigation under to 4,250 Mm^2 area and the water utilization will be 3,368 Mm^3. The salient features of the Narayanpur dam are given in Table 5.

Narayanpur or Upper Krishna Stage I dam is located downstream of Almatti dam after confluence of the Malaprabha and Krishna rivers. A major part of Upper Krishna command is covered under the Narayanpur canals for which, the main supplementing storage would be at Almatti. Narayanpur dam is a composite dam, located at north latitude of 16°10′00″ and an east longitude of 76°21′00″ near village Siddapur in Muddebihal Taluka of Bijapur district. Narayanpur dam is 10,637.52 m long and 29.72 m high. The catchment area of the reservoir is 47,850 km^2. The Project provides water for irrigation to the drought prone areas of Bijapur, Bagalkote, Gulbarga, Raichur and Koppal Districts. This project has been taken up in two stages. A view of Narayanpur dam can be seen in Figure 6.

Upper Krishna project stage – 2

The second stage of the project envisages raising FRL of Almatti Dam to 524.26 m to utilize further quantum of 1,907 MCM for providing irrigation to an additional area

Table 5. Salient features of Narayanpur dam

Particulars	Details
75 % dependable yield	22,914.40 Mm^3
Gross Storage capacity	1,072.08 Mm^3
Live Storage capacity (above MDDL)	863.04 Mm^3
Dead Storage capacity (below sill)	203.03 Mm^3
Utilization of water	3,369.70 Mm^3
MWL of dam	492.25 m
FRL of dam	492.25 m
MDDL of dam	481.58 m
Sill level of dam	462.90 m
Location of spillway	Central
Length of spillway	459.00 m
Discharge from spillways	37,922 cumec
Type and number of crest gates	Radial, 30
Size of crest gates	15 × 12 m
Submergence area	132.06 Mm^2
Villages affected	41
Population affected	48,125
Irrigable area	4,087.47 Mm^2

Figure 6. A view of Narayanpur dam

of 1,972 Mm^2. However, the state of Maharashtra claims that in case the reservoir water level rises beyond 519.6 m, there will be submergence in Maharashtra which a violation of the stipulation that the reservoir should not cause any submergence in its territory.

Almatti dam of Upper Krishna project is located on the Krishna River about 10 km downstream of the confluence of its tributary Ghatprabha. The Almatti dam is located at a north latitude of 16°19′48″ N and east longitude of 75°53′15″ E near village Almatti in Bagewadi Taluka of Bijapur district. This is a 49.29 m high composite dam of length 1,565 m. The catchment area of the dam is 35,925 km^2. Irrigation and hydropower generation are planned at Almatti dam apart from ensuring releases for the Narayanpur dam which is downstream to it. The salient features of the dam are given in Table 6. A view of Almatti dam can be seen in Figure 7.

The power house at Almatti has 5 units of 55 MW each and one unit of 15 MW, totaling 290 MW. The MDDL varies from 505.97 to 511.16 m and the tail race level is 489.0 m. For the 55 MW units, the rated head is 26.6 m and for the 15 MW unit, it is 24.0 m.

Srisailam dam

The Srisailam project, renamed as 'Neelam Sanjiva Reddy Sagar' in the honour of the former president of India, was originally planned as hydroelectric project by the Govt. of Andhra Pradesh. Subsequently, the domestic water supplies to Chennai and irrigation benefits to upland areas have been included. This is a part of the scheme for integrated development of the water resources of river Krishna in the state of Andhra Pradesh. The dam is located near the famous shrine 'Srisailam' known as "South Benaras" after the confluence of Tungabhadra and Bhima rivers

Table 6. Salient features of Almatti dam

Particulars	Details
75 % dependable yield	21,057.54 Mm^3
Gross Storage capacity	3,485 Mm^3
Live Storage capacity (above MDDL)	2,986 Mm^3
Dead Storage capacity (below sill)	353.11 Mm^3
Top of dam	528.756 m
FRL of dam	519.6 m
MDDL of dam	506.87 m
Location of spillway	Central
Length of spillway	486.50 m
Flood lift of spillway	15.24 m
Discharge from spillway	31,000 cumec
Number of crest gates of spillway (radial)	26
Size of crest gates of spillway	15 × 15.24 m
Sill level of river sluice	495.30 m
Submergence area	242.30 Mm^2
Villages affected	136
Population affected	180,000
Length of Almatti left bank canal	103 km
Irrigation from Almatti left bank canal	162.00 Mm^2

with the Krishna. Srisailam village is about 200 km south of Hyderabad and the Srisailam dam is situated at about 869 km downstream of the origin of Krishna River where the catchment area is 211,700 km^2. A view of Srisailam dam can be seen in Figure 8.

Figure 7. A view of Almatti dam

Figure 8. A view of the Srisailam dam and reservoir

The project involves the construction of a masonry dam of straight gravity type with an overall length of 512 m and a maximum height of about 144 m from the deepest foundation level. The reservoir formed behind the dam has a storage capacity of 8,720 Mm^3. The spillway portion is 266.4 m long having 12 bays of 18.3 m clear span each and controlled by 18.3 m × 16.7 m radial gates. It has a discharging capacity of 37,380 m^3/s. The non-overflow of the dam is on either side of spillway portion. At FRL of 269.75 m, the storage capacity is 8,723 Mm^3. The dead storage level of the dam is 260.3 m and live storage is 7,080 Mm^3. The catchment area at the dam is 203,600 km^2.

Two powerhouses are located near the dam. At the right bank powerhouse, 7 units of 110 MW each have been installed while the left bank powerhouse has 6 units of 150 MW each. The powerhouses were submerged in a flood in 1998 which caused extensive damage to them. When the Srisailam reservoir touches FRL, the backwater extends up to the Kurnool town. Fishing is also carried out in the reservoir and some recreation facilities are being setup. From the Srisailam reservoir, planned annual irrigation is 3,100 Mm^2, utilizing about 2,209 Mm^3 of water. A noteworthy feature of this dam is that nearly 5.9 km^3 of water is stored against the gates. Thus, a large quantity of water is not stored against a passive wall but against a 'live' and moving device. Recall that Ukai dam (Chapter 12) is another large dam where enormous volume of water is held behind gates.

In summer months, the river flows dwindle down to less than 30 m^3/s. The maximum flood observed in Krishna River at the dam site is of the order of 18,500 m^3/s during monsoon period (June-October). During the construction, a problem of deciding river diversion arrangements arose. The maximum river discharge data for fifteen consecutive years were studied to select the optimum discharge which would give the largest working period. Finally, diversion arrangements were designed for about 850 m^3/s capacity which ensured about 195 working days during the period from Nov–June.

The diversion arrangements finally adopted comprised of: A 9.14 m diameter circular tunnel of 567 m^3/s capacity through the left abutment; a diversion channel of 15.24 m bed width to carry the balance of 283 m^3/s; and two semi-permanent concrete upstream and downstream cofferdams to divert the river flows for isolation of the construction area.

Papanasi temple: When the Srisailam project was planned, the Papanasi and Sangameswara temples happened to fall in its submergence area. To save these ancient monuments, these were cut and moved block-by-block. At Alampur in Mehbubnagar District, the temples were reconstructed exactly the same way as they were at the original site. Shifting of Papanasi temple is a fine example of how water resources development can be carried out while protecting the cultural heritage of the country. Recall that the famous Abu Simbel temple in Egypt was also shifted in the same manner when the Aswan Dam was constructed on Nile River in Egypt.

Pulichintala project

The Pulichintala project was originally investigated as an irrigation project. But due to construction of Nagarjunasagar dam, this project was not taken up as the entire ayacut that was originally to be irrigated under Pulichintala project was covered by Nagarjunasagar project. The present proposal is to construct a dam to stabilize irrigation in the existing ayacut in the Krishna delta for the paddy. Hydropower generation by utilizing the irrigation releases for the delta is also planned by using installed capacity of 60 MW.

At the proposed site, the catchment area is 240,733 km^2. The FRL of the reservoir will be at 53.34 m and MDDL at 42.67 m. Pulichintala reservoir will have gross storage capacity of 1,296 Mm^3 and a live storage of 1,026 Mm^3.

Nagarjunasagar project (NSP)

The Nagarjunasagar project is the largest and highest masonry dam (125 m) in the world. It is situated downstream of Srisailam reservoir on the main Krishna river in Andhra Pradesh. It is a multipurpose project with irrigation, hydropower and flood control components.

The Nagarjunasagar dam is Located near Nandikonda village, Pedavoora Mandal, Nalgonda district, latitude is 16°34′24″ N and longitude 79°18′47″ E. It is one of the dams whose height is more than 100 m; it is 124.663 m above the deepest foundation level. The catchment area at the dam at 215,193 km^2; the annual rainfall in the catchment is 889 mm, the maximum observed flood is 30,050 cumec, and the design flood (return period 1,000 year) is 58,340 cumec. For the reservoir, the FRL, DSL, and MDDL of the reservoir are at 179.83 m, 121.92 m, and 156.36 m, respectively. For this reservoir, the maximum storage is 11,555 Mm^3 and the live storage is 6,940 Mm^3. Spillway at the masonry dam is 470.916 m long and has its crest at 166.421 m. It is equipped with 26 gates of size 13.71 m × 13.14 m. When the Nagarjunasagar reservoir is full, its backwater extends up to the Srisailam dam and covers an area of 285 sq. km. A view of Nagarjunasagar dam can be seen in Figure 9.

Figure 9. A view of Nagarjunasagar dam

Nagarjunasagar project complex has substantial capacity for hydropower generation. It has one conventional and seven reversible units, each with 110 MW capacity. The right bank canal power house has 3 units of 30 MW each and left canal power house has 2 units of 20 MW each.

On an average, the Nagarjunasagar project will annually provide irrigation to a command of 12,480 Mm^2 ha by utilizing 7,465 Mm^3 water. The work on the project was completed in 1974. The project comprises a dam with two canals taking off on either side. The Nagarjunasagar Right Main (Jawahar) Canal is 203 km long and envisages creating irrigation potential in an area of 4,750 Mm^2 in Guntur and Prakasam districts. The Nagarjunasagar Left Main (Lal Bahadur) Canal is 179 m long and envisages creating irrigation potential in an area of 4,200 Mm^2 in Nalgonda, Khammam and Krishna districts. Each of these canals can carry maximum discharge of 311.5 cumec. The irrigation potential created by June 1999 was 8,100 Mm^2.

Ghatprabha dam

The Ghatprabha project comprises three stages. The first stage consists of a 71 km long left bank canal from the Dupdhal weir constructed across Ghataprabha River in 1897 near Dupdhal for providing irrigation to an extent of 425 Mm^2. The second stage comprises of extension of left bank canal from km 72 to its full length of 109 km and a dam across Ghatprabha River near Hidkal, up to a partial height of RL 650.14 m. The Ghatprabha dam is located near Hidkal town in Hukkeri taluk of Belgaum district at latitude 16°9′0″ N and longitude 74°38′0″ E. At the dam site, the catchment area is 1,412 km^2 and the water yield is 1,970.38 Mm^3.

Ghatprabha dam has a live storage capacity of 1,387 Mm^3 to provide irrigation to a total area of 1,396 Mm^2, inclusive of the area under stage I. The third stage involves raising the FRL of Hidkal dam to its final level of RL 662.94 m (RL 2,175.00 feet) thereby creating gross storage of 1,448 MCM and constructing a 202 km long Right Bank Canal and 86 km long Chickkodi Branch Canal to irrigate 1,913.86 Mm^2. Thus, the total command area under the project comes to 3,310 Mm^2. Figure 10 shows a view of the dam. The salient features of the reservoir are given in Table 7

Tungabhadra project

Tungabhadra is the largest tributary of the Krishna River which contributes an annual discharge of 14,700 Mm^3 at its confluence with the main river. The Tungabhadra dam is located at Mallapuram, 5 km away from Hospet in the Bellary district. The latitude and longitude of the dam are 15°15′0″ N and 76°21′0″ E, respectively. With full reservoir level of 497.74 m above MSL, the reservoir extends over 378.14 Mm^2. The catchment up to the dam site is 28,168 km^2. The annual rainfall in the upper catchment of the river is 104 cm. Minor rivers that feed the Tungabhadra River are dammed at many places, creating small to medium sized reservoirs, such as Vanivilas Sagar and Anjanapur, and several large tanks, such as Shantisagar and Madag. The river carries large amount of silt and therefore, silt deposition in the reservoir is high. This has reduced the capacity of the reservoir by 13.5% in its first decade of existence. The salient features of the reservoir are given in Table 8. A view of Tungabhadra dam can be seen in Figure 11.

The climate at the reservoir site is mainly dry (humidity 80.7% to 93.7%); the average monthly maximum and minimum air temperatures ranging from 31.0 to 39.5 °C and 13.8 °C to 22.3 °C, respectively. The water remains warm (23.1 to 29.5 °C) throughout the year.

Figure 10. A view of Ghatprabha dam (Hidkal dam)

Table 7. Salient features of the Ghatprabha (Hidkal) dam

Particulars	Details
Gross Storage capacity	1,448.69 MCM
Live Storage capacity (above MDDL)	1,387 MCM
Dead Storage capacity (below sill)	60.31 MCM
Height of Dam	60 m
Type of Dam	Composite
FRL of dam	662.94 m
MDDL of dam	633.83 m
Length of spillway	149.35 m
Discharge from spillway	4,613 cumec
Crest gates of spillway (radial)	10
Size of crest gates of spillway	12.19×7.62 m
Submergence area	63.38 Mm^2
Villages affected	22
Population affected	15,660
Power Generation	32 MW
Irrigable area	3, 310.00 Mm^2
Length of Left bank canal	109 km
Capacity of Left bank canal	80.7 cumec
Area of Left bank canal	1, 618.80 Mm^2
Length of Right bank canal	202 km
Capacity of Right bank canal	66.56 cumec
Area of Right bank canal	1, 691.29 Mm^2
Withdrawals by canals	2,110.45 MCM
Reservoir losses	92.60 MCM
Gross utilization	2,203.05 MCM

Vanivilas sagar project

Vanivilas Sagar dam is situated on the Vedavati River in Chitradurga district, about 104 km northeast of the Babudan Hills (the source of the Vedavati River). Created in 1901, it is one of the oldest reservoirs in the State. Total Catchment area at the dam site is 5,374.00 km^2 and the independent catchment area is 1,554 km^2. Mean annual precipitation for the catchment is 600.00 mm and mean annual run-off at dam site is 294.49 Mm^3. Maximum flood discharge at the dam site has been estimated to be 1,000 cumec. The salient features of the reservoir have been given in Table 9. At full reservoir level of 652.28 m, the impoundment has a water spread area of 87.63 km^2. The latitude and longitude of the dam are 13°16′0″ N and 75°16′20″ E, respectively. In the headwaters area, the river receives high rainfall to the tune of 375 cm per annum. However, the precipitation in the local catchment is just 60 cm per year, as the reservoir is situated in the semi-arid plains.

The reservoir water is uniformly warm throughout the year (22.3 to 26.3 °C) and no thermal stratification develops. The pH of the water usually lies in the range 8.4–8.5.

Table 8. Salient features of the Tungabhadra project

Particulars	Details
Yield	11,978.026 Mm^3
Gross Storage capacity	3,737.82 Mm^3
Live Storage capacity	3,308.54 Mm^3
Dead Storage capacity (below sill)	65.128 Mm^3
Height of Dam	35.36 m
Length of Dam	2,449 m
Type of Dam	Composite
FRL of dam	497.74 m
MDDL of dam	477.01 m
Type of Spillway	Central
Length of spillway	701 m
Peak Discharge from spillway	18,408 cumec
Crest gates of spillway (radial)	33
Size of crest gates of spillway	18.29 m × 6.10 m
Power Generation	99 MW
Irrigable area	3,627.95 Mm^2
Length of Left bank canal	227 km
Capacity of Left bank canal	198 cumec
Irrigable Area of Left bank canal	2,439 Mm^2
Length of Right bank canal	251 km
Capacity of Right bank canal	71 cumec
Irrigable Area of Right bank canal	375.04 Mm^2
Withdrawals by canals	3,383.86 Mm^3
Reservoir losses	353.96 Mm^3
Gross utilization	3,737.82 Mm^3

Figure 11. A view of Tungabhadra dam

Table 9. Salient features of Vanivilas Sagar project

Particulars	Details
Full reservoir level	652.28 m
Storage capacity at FRL	850.30 Mm^3
Water spread area FRL	87.63 Mm^2
Dead storage level	630.950 m
Dead storage capacity	47.80 Mm^3
Live storage capacity	802.50 Mm^3
Length of dam	405.4 m
Villages affected	32
Length of High Level Canal	9.60 km
Design discharge of High Level Canal	8.825 cumec
Irrigable area of High Level Canal	4.47 Mm^3
Length of Left Bank Canal	48.00 km
Design discharge of Left Bank Canal	8.85 cumec
Irrigable area of Left Bank Canal	60.59 Mm^3
Length of Right Bank Canal	46.40 km
Design discharge of Right Bank Canal	8.85 cumec
Irrigable area of Right Bank Canal	56.29 Mm^3

Bennihora project

Bennihora project is a major river valley projects across Bhima River, a tributary of Krishna River. The dam is located in Chithapur Taluk, Gulbarga District. Its latitude is 17°27′00″ N and longitude 77°01′00″ E. The catchment area of the reservoir is 2,204.09 km^2. The salient features of the dam are given in Table 10.

Bhadha reservoir project

This is a major multipurpose river valley project across the Tungabhadra River. The dam is located near Lakkavalli village in Tarikere Taluk, of Chickamagalur District, and its latitude is 13°42′00″ N and longitude 75°38′20″ E. The catchment area of the reservoir is 1,968 km^2. The salient features of the dam are given in Table 11.

Bhima irrigation project

The Bhima project is a major river valley projects across Bhima River, a tributary of Krishna River near, near Ujjani Village in Solapur District. The dam is located at latitude 18°4′26″ N and longitude 75°07′15″ E. The gross catchment area of the reservoir is 14,856 km^2 and the free catchment area is 9,766 sq. km. salient features of the dam are given in Table 12.

Hipparagi barrage

The Hipparagi Barrage project is located near village Hipparagi, in Jamakhandi Taluka of Bijapur district. The latitude and longitude of the barrage are

Table 10. Salient features of Bennihora Project

Particulars	Details
Utilization	162.82 MCM
Catchment Area	2,204 sq. km
Gross Storage Capacity	149.97 MCM
Live Storage Capacity	140.68 MCM
Dead Storage Capacity	9.29 MCM
Type of Dam	Composite
Length of Dam	2,270 m
Tank Bund Level	441.71 m
Full Reservoir Level	438.89 m
Spillway Discharge	8,800.80 cumec
Spillway Gates	7 Nos. (15 m × 11.5 m)
Submergence Area	24.73 Mm^2
Villages Affected	9
Population Affected	9,338
Length of Left Bank Canal	66 km
Length of Right Bank Canal	82 km
Gross Command Area	248.59 Mm^2
Net Command Area	202.34 Mm^2

Table 11. Salient features of Bhadha Reservoir Project

Particulars	Details
75% dependable yield	2,396.45 Mm^3
Gross Storage Capacity	2,024.65 Mm^3
Dead storage Capacity	240.69 Mm^3
Live storage Capacity	1,783.96 Mm^3
Total Submergence area	112.51 Mm^2
Forest area	7.17 Mm^2
Cultivable area	32.75 Mm^2
Villages affected	28
Allocation of Water as per Krishna Water Disputes Tribunal Award	1,747.15 Mm^3
Gross command area	1,628.18 Mm^2
Cultivable command area	1,215.00 Mm^2
Irrigable area	1,055.70 Mm^2
Type of Dam	Composite
Length of Dam	1,708 m
Height of Dam above river bed level	59.13 m
MWL	657.75 m
Full Reservoir level	657.75 m
Minimum draw down level	636.40 m
Dead storage level	631.50 m
Type and length of Spillway	Ogee, 82.30 m
Discharging capacity of Spillway	3,020 cumec
Type of Crest Gates	Vertical
Size of Crest Gates	7.62 m × 18.28 m
Installed capacity for hydropower	160 MW

Table 12. Salient features of Bhima Project

Particulars	Details
Gross Storage Capacity	3,114 Mm^3
Dead Storage Capacity	1,700 Mm^3
Live Storage Capacity	1,414 Mm^3
Total submerged area	29,000 ha
Villages submerged	82
Annual Utilization	2,410 Mm^3
Length of Dam	2,467 m
MWL of Dam	497.58 m
FRL of Dam	496.83 m
Type and length of Spillway	Ogee, 608 m
Discharging capacity of Spillway at FRL	15,717 cumec
Number and Size of Crest gates	41,12 m × 6.5 m
Length of Left bank canal	126 km
Irrigable area of Left bank canal	68,840 ha
Length of Right Bank Canal	112 km
Irrigable area of Right Bank Canal	44,100 ha

16°33′00″ N75°10′00″ E, respectively. The catchment area at the barrage is 22,699 km^2. The salient features of the project are given in Table 13.

Malprabha project

This project is located on the Malprabha tributary of Krishna River, near village Navilutheertha, in Saundatti Taluka of Belgaum district. The latitude and longitude of the barrage are 15°49′00″ N and 75°6′0″ E, respectively. The catchment area of the reservoir is 2,564 km^2. Salient features of the dam are given in Table 14. A view of Malprabha dam can be seen in Figure 12.

Upper tunga project

This project is located near Shimoga Taluka of Shimoga district, 100 m downstream of existing Anicut. Salient features of the dam are given in Table 15.

Koyna dam

Koyna is multipurpose masonry gravity dam on Koyna River, located at a distance of 20 km from Chiplun, in Ratangiri District, Maharashtra. The catchment area at the dam is 891.78 km^2. The height of the dam is 85.35 m. The reservoir has a live storage capacity of 2,662 MCM at FRL which is at 650.85 m and the MDDL is at 609.50 m. The mean annual inflow at Koyna dam is 4,745 MCM. Four power houses have been constructed under this scheme. Koyna I & II power houses have 4 units of 65 MW and 4 units of 75 MW each with a total installed capacity of 560 MW. Koyna III power house has 4 unit of 80 MW each and Koyna IV power house has 4 units of 250 MW each.

Table 13. Salient features of Hipparagi Barrage

Particulars	Details
75 % dependable Yield	17,393.91 Mm^3
Gross Storage Capacity	169.90 Mm^3
Live Storage Capacity	138.75 Mm^3
Dead Storage Capacity	31.156 Mm^3
Withdrawals by canals	308.65 Mm^3
Reservoir losses	25.76 Mm^3
Water supply	28.31 Mm^3
Total Utilization	362.74 Mm^3
Type of Dam	Composite
Length of Dam	5,463.00 m
Height of Dam	26.00 m
MWL of Dam	531.40 m
FRL of Dam	531.40 m
MDDL of Dam	517.47 m
Length of Spillway	368.78 m
Discharge capacity of Spillway	19,810 cumec
Crest gates of Spillway	23 of size 8.22 m × 13.72 m
Submersion Area	29.98 Mm^2
Villages affected	28
Population affected	49,364
Irrigable area	596.90 Mm^2
Foreshore lift canal	
a) Ainapur	
i) lift	33.14 m
ii) length in km	76
iii) command area	267.41 Mm^2
b) Haliyal	
i) lift	37.63
ii) length in km	94
iii) command area	329.49 Mm^2

Markendaya project

The Markendaya project is located on Ghatprabha tributary of Krishna River, near village Shirur in Hukkeri Taluka of Belgaum district. The latitude and longitude of the barrage are 16°02′00″ N and 74°38′33″ E, respectively. The catchment area at the project side is 432 km^2. The salient features of the dam are given in Table 16.

Singatalur lift irrigation

The Singatalur Lift Irrigation project is located on Tungabhadra River near village Hammige, in Mundaragi Taluka of Gadag district. The latitude and longitude of the barrage are 15°02′00″ N and 75°50′0″ E respectively. The catchment area at the project is 19,850 km^2. The salient features of the dam are given in Table 17.

Table 14. Salient features of Malprabha Project

Particulars	Details
Yield	1,205.44 Mm^3
Gross Storage Capacity	1,068.39 Mm^3
Live Storage Capacity	866.25 Mm^3
Dead Storage Capacity	95.99 Mm^3
Irrigable Area	2,181.91 Mm^2
Submerged Area	135.78 Mm^2
Village affected	43
Population affected	41,000
Type of Dam	Masonry
Height of Dam	56 m
Length of Dam	69 m
MWL of Dam	633.83 m
FRL of Dam	633.83 m
MDDL of Dam	623.93 m
Location of Spillway	Central
Length of Spillway	85.34 m
Discharging capacity of Spillway	5,236 cumec
Type of Gates of Spillway	radial
Gates of Spillway	4 of size 15.24 m × 12.19 m
Length of Right Bank Canal	138.00 km
Capacity of Right Bank Canal	58.10 cumec
Command Area of Right Bank Canal	1,399.21 Mm^2
Length of Left Bank Canal	168.00 km
Capacity of Left Bank Canal	38.91 cumec
Command Area of Left Bank Canal	531.34 Mm^2

Krishna irrigation project

The Krishna Irrigation Project covers Satara and Sangli Districts in Southern Maharashtra. The project consists of two storage dams across the Krishna River (Dhom) and its tributary Venna River (Kanher) for irrigation in downstream semi-arid and arid zones as well as industrial water supply to Wai and Karad towns.

The Krishna Irrigation Project (K.I.P.) was cleared for implementation in the year 1966. The proposed water utilization of the Krishna Irrigation Project with three storage reservoirs was 1,051.54 Mm^3. However, in view of the reduced quantum of water from the Krishna basin allotted to Maharashtra State as a result of Krishna tribunal award (year 1974–75), it was decided to reduce the utilization of Krishna irrigation project to 849.62 Mm^3.

The Krishna irrigation project's catchment area of the storages and command area lies in between latitude 17°0′ N and 18°0′ N; and longitude 73°45′ E to 74°45′ E. This command area can be classified in the three groups: (1) Having normal annual rainfall 700 to 800 mm; (2) Having normal annual rainfall 400 to 500 mm (semi arid zone), and (3) Having normal annual rainfall below 400 mm (arid zone). Kanher and Dhom command area falls under the 1st category. Arphal command area up to 40 km falls under the 1st category, from 40 km to 100 km comes under the 2nd

Figure 12. A view of Malprabha dam

category, and beyond 175 km of the Arphal left bank canal, the area is under the 3rd category.

Annual rainfall in the command area varies from 700 to 800 mm; it is less than 400 mm in the tail reach. Nearly 75% of annual rain falls during the months of June to September. Average annual runoff at Dhom reservoir site is 423.30 Mm^3. Catchment area at Dhom dam site is 217.56 km^2. Average annual runoff at Kanher reservoir (Venna River) site is 483.60 MCM and the catchment area at this dam is 204.69 km^2.

Part of the command area of this project is under semi-arid and arid zone. Demand for irrigation water in arid zone from Krishna irrigation project has increased due to increase in irrigation area. Reservation of water from Krishna irrigation project storages for the industrial area (Wai and Karad towns) has also increased. But the reduced water allocation for Krishna irrigation project necessitated the revision of water planning for the project whole. The 75% dependable yield from the two storages viz. Dhom and Kanher is approximately 822.15 MCM. The quantum of water available is not adequate to cater to all demands of extending irrigation facilities (Dhom command, Kanher command, Arphal command) and reserving water for industrial area (Wai and Karad). Hence it is necessary to use the available water optimally.

Table 15. Salient features of Upper Tunga project

Particulars	Details
Type of dam	Composite
Length of dam	770 m
FRL of dam	588.24 m
Top level of dam	592.20 m
Canal off take	584.45 m
Length of canal on left bank	339.50 m
Type of Spillway	Ogee with roller bucket
Length of Spillway	311 m
Crest level of Spillway	583.50 m
Type of Crest gates	Radial
Number of Crest gates	22
Size of Crest gates	11.75 m × 4.74 m
Length of non-over flow section	335.00
Sill level of river sluice	573.00
Number of gates of river sluice	3
Size of gates of river sluice	2.5 m × 4.0 m
Length of embankment	192 m
Top width of embankment	7.5 m
Top level of embankment	593.2 m
Number of Gates in Left side irrigation sluice	5
Size of Gates in Left side irrigation sluice	2.5 m × 1.829 m
Sill level of Left side irrigation sluice	582.168 m
Number of Gates in Right side irrigation sluice	2
Size of Gates in Right side irrigation sluice	2.5 m × 1.219 m
Sill level of Right side irrigation sluice	582.778 m
Irrigable area	947.00 Mm^2
Water utilization	349.35 Mm^3

Accordingly, the project planning was changed by deleting storage scheme at Borkhal; reducing the water losses in conveyance by lining the canals; deleting some of the command area under the Arphal left bank canal; diluting the crop pattern and extending the irrigation facilities to the more command area; and reserving the water for industrial area at Wai and Karad. Jowar, groundnut and pulses are major Kharif crops, and the major Rabi crop is Jowar (local) or hybrid variety; and well irrigation for perennial crop is only practiced in the Dhom and Kanher command area.

Osmansagar reservoir

The Osmansagar reservoir is situated near Gandipet village in Rangareddy district, AP, at latitude and longitude 17°22′30″ and 78°22′0″ respectively. It was constructed across Musi River in the year 1920 with live capacity of 156.81 MCM and gross storage capacity of 180.54 MCM. The designed maximum flood discharge is 2,971 cumec. Though the main purpose of the reservoir at the time of construction was to absorb flood water but it is now serving as a water supply source to the twin

Table 16. Salient features of Markendaya Project

Particulars	Details
Average yield	124.31 Mm^3
Gross Storage Capacity	104.66 Mm^3
Dead Storage Capacity	14.47 Mm^3
Live Storage Capacity	70.20 Mm^3
Total submerged area	8.98 Mm^2
Villages submerged	9
Population affected	2,258
Gross Utilization	113.27 Mm^3
Irrigable area	191.05 Mm^2
Type of Dam	Gravity dam in concrete
Length of Dam	1,356 m
Height of Dam	39.10 m
MWL of Dam	704.00 m
FRL of Dam	704.00 m
MDDL of Dam	690.55 m
Dead Storage Level of Dam	688.85 m
Type of Spillway	High ogee
Length of Spillway	112.50 m
Discharging capacity of Spillway	3,728.00 cumec
Type and number of Crest gates	Radial, 7 gates
Size of Crest gates	8.00 m × 12.7 m
Length of Left bank canal	12 km
Irrigable area of Left bank canal	8.90 Mm^2
Length of Right Bank Canal	80 km
Irrigable area of Right Bank Canal	182.15 Mm^2

Table 17. Salient features of Singatalur Lift Irrigation Project

Particulars	Details
75 % dependable yield	7,204.94 Mm^3
Type of weir	Ungated concrete barrage
Length of weir	402.00 m
Height of weir	3.00 m
Irrigable area	194.25 Mm^2
MWL of Dam	508.75 m
FRL of Dam	507.00 m
Minimum water level of Dam	501.00 m
Type of Spillway	Low ogee
Discharging capacity of Spillway	14,724.62 cumec
Type of Crest gates	Vertical lift type
Number of gates	27
Size of gates	12.0 m × 3.0 m

cities of Hyderabad and Secunderabad. The catchment area up to Osmansagar is 738.15 sq. km comprising 637.14 sq. km free and rest of 101.01 sq. km intercepted. The salient features of the reservoir are given in Table 18.

According to the sedimentation study made by Vishwantaham and Eashwaraiah (2004) through remote sensing, the current live storage capacity of Osmansagar reservoir is 100.19 MCM and the loss of live capacity works out to 56.63 MCM over a period of 82 years. The average annual loss of live capacity is approximately at the rate of 0.7 MCM/year.

Prakasam barrage

This barrage is the terminal structure on the Krishna River to meet the delta requirements in the Krishna basin. It has been named after a former chief minister of the Andhra Pradesh. The barrage is located near Vijayawada in Andhra Pradesh. At the barrage, the catchment area is 257,078 km^2 and the pond level is 17.4 m. Prakasam barrage provides irrigation to a command of 4,450 Mm^2 utilizing 5,132 Mm^3 of water.

Other existing projects

In addition to the above, there are several existing water resources projects in the basin. Salient features of existing water resources projects with a live storage capacity of 10 Mm^3 and above have been presented in Table 19.

14.1.8. Major Ongoing Water Resources Projects in Krishna Basin

Telugu ganga project (TGP)

Telugu Ganga Project (TGP) is a prestigious on-going project in this basin. Telugu Ganga Project is an interstate project formulated to utilize flood water of Krishna and Pennar rivers to irrigate 2,327.025 Mm^2 in drought prone areas of Kurnool, Cuddapah and Chittoor districts of Rayalaseema and uplands of Nellore districts in A.P., besides conveying 15 TMC (424.80 MCM) of Krishna water of Chennai City. The scheme consists of 408 km long canal from Srisailam dam up to the Andhra Pradesh and Tamil Nadu border. The project commenced in 1983 and the target date of project completion was 2005. The project also envisages power generation at three locations: Velugodu (9 MW), Chennamukkapalli (15 MW), and Kandaleru (99 MW).

Table 18. Salient features of Osmansagar Reservoir

Full reservoir level	545.59 m
Maximum water level for absorbing flood	550.16 m
Water spread area	27.22 sq. km
Capacity at FRL	180.54 MCM
Capacity at MWL	328.45 MCM
Live storage capacity at FRL	156.82 MCM
Flood storage	147.92 MCM
Free catchments area	637 sq. km

Table 19. Salient features of selected existing projects in Krishna basin

Name of the Project	State	Year of completion	Gross storage capacity (million cubic meter)	Live storage capacity (million cubic meter)	Designed annual irrigation (million Sq. Meter)	Installed capacity (MW)
Bhariravuni Tippa	Andhra Pradesh	1961	74.24	65.29	68.00	–
Dindi Project	Andhra Pradesh	1943	5, 901.00	58.14	127.50	–
Gajuladinne	Andhra Pradesh	1979	48.14	42.47	129.50	–
Himayat Sagar	Andhra Pradesh	1926	216.60	104.80	–	–
Jutpally	Andhra Pradesh	1965	94.12	85.48	9.80	–
Koilsagar Project	Andhra Pradesh	1955	69.83	64.46	43.00	–
Kotepallyvagu	Andhra Pradesh	1969	44.52	36.85	3.30	–
Lakhnapur	Andhra Pradesh	1968	92.85	85.48	10.40	–
Lanka Sagar	Andhra Pradesh	1968	18.81	17.26	12.30	–
Musi Dam	Andhra Pradesh	1961	136.94	130.26	167.30	–
Pakhal	Andhra Pradesh	1919	–	82.50	38.40	–
Paliar	Andhra Pradesh	1928	72.45	66.50	79.70	–
Pendlipakla	Andhra Pradesh	1938	15.32	14.53	13.70	–
Sarala Sagar Project	Andhra Pradesh	1959	15.05	13.91	16.80	–
Siddapur	Andhra Pradesh	1919	24.74	18.56	2.50	–
Wyra Project	Andhra Pradesh	1930	70.13	59.39	70.40	775
Ambligola	Karnataka	–	12.39	11.70	29.50	–
Bhadra	Karnataka	1963	2, 023.00	1, 635.00	1, 060.00	58
Chamdrampalli	Karnataka	1971	34.20	31.40	52.20	–
Dharma	Karnataka	1964	23.00	21.10	27.90	–
Gayatri Reservoir	Karnataka	1963	27.53	18.10	9.50	–
Jamradhalla	Karnataka	1968	37.00	28.00	15.40	–
Ramanahalli	Karnataka	1960	12.86	11.68	19.40	–
Ashti	Maharashtra	1976	–	22.99	47.70	–
Bhatghar	Maharashtra	1892	672.67	665.59	–	18
Budhiyal	Maharashtra	1966	–	19.02	42.50	–
Chandhani	Maharashtra	1966	20.70	15.20	20.20	–
Ekruk	Maharashtra	1891	–	61.17	26.10	–
Ghod	Maharashtra	1957	216.31	170.74	246.00	–
Hirani	Maharashtra	1966	12.57	11.17	16.60	–
Hingani	Maharashtra	1978	45.51	31.97	–	–
Kanhar	Maharashtra	1988	286.00	271.68	91.80	–
Khadakwasla	Maharashtra	1979	85.92	62.59	620.00	–
Khairi	Maharashtra	–	15.11	13.74	28.00	–
Khasapur	Maharashtra	1956	19.81	15.84	21.40	–
Kolkewadi Dam	Maharashtra	1975	36.22	11.22	–	40
Kurnur	Maharashtra	1970	35.24	32.26	36.40	–
Lonawala	Maharashtra	1916	–	11.52	13.10	72
Mangi	Maharashtra	1966	32.69	31.70	31.20	–
Manikdoh	Maharashtra	1986	308.06	288.07	–	–
Mehekari	Maharashtra	1966	16.13	12.98	40.50	–
Mhaswad	Maharashtra	1988	47.91	46.21	40.50	–
Morna	Maharashtra	1985	21.18	15.16	51.70	–

(*Continued*)

Table 19. (*Continued*)

Name of the Project	State	Year of completion	Gross storage capacity (million cubic meter)	Live storage capacity (million cubic meter)	Designed annual irrigation (million Sq. Meter)	Installed capacity (MW)
Mulshi	Maharashtra	1929	–	522.00	81.60	150
Nazare	Maharashtra	1974	16.17	10.50	32.00	–
Panshet	Maharashtra	1971	303.96	256.00	–	–
Pathari	Maharashtra	1905	11.87	11.61	63.00	–
Pawana	Maharashtra	1975	305.00	274.00	–	10
Radhanagari	Maharashtra	1954	236.79	219.00	265.60	–
Shefal	Maharashtra	1901	17.55	16.93	–	–
Shirwata	Maharashtra	1920	185.98	130.41	–	–
Sina	Maharashtra	1990	67.95	52.30	73.60	–
Thokarwadi	Maharashtra	1922	363.70	360.91	29.00	–
Tisangi	Maharashtra	1966	26.16	24.46	40.90	–
Tulashi	Maharashtra	1978	98.29	91.92	57.10	–
Vir	Maharashtra	1961	278.53	265.78	161.10	–
Visapur	Maharashtra	1936	33.22	25.72	53.20	–
Wadaj	Maharashtra	1981	36.00	33.20	–	–
Walwan	Maharashtra	1916	72.50	54.37	–	–
Yadegaon	Maharashtra	1978	93.42	79.37	–	–

The project benefits include irrigation for $1,112.925\,\text{Mm}^2$ in Kurnool and Cuddapah districts with Krishna flood waters amounting to $820.99\,\text{Mm}^3$ and irrigation for $1,214.10\,\text{Mm}^2$ in Nellore and Chittoor districts with Pennar flood waters: $849.30\,\text{Mm}^3$. For the Chennai city, water requirement from Krishna basin will be $424.65\,\text{Mm}^3$. Out of this, the share of Andhra Pradesh, Karnataka, and Maharashtra each will be $141.55\,\text{Mm}^3$. The components include four balancing reservoirs: Velugodu Balancing Reservoir, S.P.V. Balancing Reservoir, Somasila Reservoir, and Kandaleru Reservoir.

Other under construction projects

In addition to the above, there are several under construction water resources projects in the basin. Salient features of under construction water resources projects with a live storage capacity of $10\,\text{Mm}^3$ and above have been presented in Table 20.

14.1.9. Krishna Water Dispute Tribunal

In 1969, the government of India constituted the Krishna Water Dispute Tribunal to adjudicate upon the water dispute regarding the Krishna River. The KWDT gave its award in 1973 and it was published in 1976. Details about the award are given in Chapter 21.

Table 20. Salient features of selected Under construction projects in Krishna basin

Name of the project	State	Gross storage capacity (MCM)	Live storage capacity (MCM)	Designed annual irrigation (million sq. m)	Installed capacity (MW)
Gajuladinne	Andhra Pradesh	127.43	121.20	128.00	–
Varadarajaswamy Gudi	Andhra Pradesh	11.02	10.20	34.80	–
Amarja	Karnataka	44.01	40.07	89.00	–
Bennithora Project	Karnataka	155.06	145.78	200.00	–
Hagaribommanhalli	Karnataka	57.00	50.00	29.80	–
Hirehalla	Karnataka	47.23	41.56	80.20	–
Lower Mullamari	Karnataka	49.13	39.90	97.10	–
Maskinalla	Karnataka	13.11	10.70	28.30	–
Narihalla	Karnataka	22.92	20.86	14.10	–
Rangayyanadurga	Karnataka	14.20	13.10	–	–
Upper Mullamari	Karnataka	21.23	17.55	32.80	–
Bori	Maharashtra	37.46	30.72	104.50	–
Chaskaman	Maharashtra	238.17	210.99	390.00	–
Chikotra	Maharashtra	42.80	39.99	46.90	–
Dimbhe	Maharashtra	382.22	353.91	–	–
Dudhganga	Maharashtra	701.78	663.83	650.00	15
Jaggamhatti	Maharashtra	27.79	26.29	31.40	–
Jawalgaon	Maharashtra	34.75	25.19	53.40	–
Kadvi	Maharashtra	71.24	70.56	92.20	215
Kasari	Maharashtra	78.56	77.96	94.60	215
Kasarsai	Maharashtra	13.75	12.62	–	–
Kumbhi	Maharashtra	76.88	76.49	88.90	27
Patgaon	Maharashtra	76.26	75.79	83.60	–
Pauna	Maharashtra	–	244.00	–	5
Sankh	Maharashtra	19.93	14.87	–	–
Urmodi	Maharashtra	82.94	76.73	90.40	–
Veer Baji Pasalkar	Maharashtra	374.00	275.00	–	11
Varna	Maharashtra	963.97	799.73	1, 140.00	27
Wadivala	Maharashtra	32.48	22.00	36.30	–
Yerlawadi	Maharashtra	32.82	19.61	–	–

14.2. THE GODAVARI BASIN

Godavari is the largest river in Peninsular India and third largest in India. Godavari is held in reverence as "Vridha Ganga" or "Dakshin Ganga". Holy places are located on the banks of the river at Nasik and Bhadrachalam. Godavari rises in the Sahyadris near Triambakeswar, about 80 km from the shore of Arabian Sea, at an elevation of 1,067 m in the Nasik district of Maharashtra. Kumbh Mela which attracts millions of devotees is organized at Nasik after every 12 years. After flowing for about 1,465 km in a general south-easterly direction through Maharashtra and Andhra Pradesh, Godavari falls into the Bay of Bengal north of Rajahmundry. The basin lies between latitudes 16°16′0″ N and 23°43′ N longitudes 73°26′ E and 83°07′ E. The basin extends over an area of 312, 813 km^2, which is nearly 10% of the total geographical area of the country. It is bounded on the north by the Satmala Hills, the Ajanta Range and the Mahadeo Hills, on the east and south by the Eastern Ghats and on the west by the Western Ghats. The state-wise distribution of the catchment area is shown in Table 21. Important tributaries of the Godavari are the Pravara, the Purna, the Manjra, the Maner, the Penganga, the Wardha, the Pranhita, the Indravati and the Sabari. The Jayakwadi project, Sriram Sagar project and Cotton barrage (Dowleswaram) are the important projects existing in the basin. The proposed major projects are Bhopalpatnam, Inchampalli and Polavaram. An index map of Godavari basin is given in Figure 13.

About 64 km from its source, Godavari receives the waters from Dharna, on its right bank and a short distance lower down the Kadwa joins it from the left. The combined waters of the Pravara and Mula which rise in the hills of Akola join the river about 217 km from its source. About 338 km lower down, while still in Maharashtra, the river receives the combined waters from the Purna and Dudhna rivers and after a further 138 km at the border of Maharashtra and Andhra Pradesh, the waters of the Manjira River join it from the south. At this point, Godavari flows at an elevation of about 329 m.

The Pranhita River, conveying the combined waters of Penganga, Wardha and Wainganga, which drain Nagpur and southern slopes of the Satpura ranges, falls into the Godavari about 306 km below its confluence with the Manjira. Forty-eight km lower, the waters of the Indravathi join the river. Both the Pranhita and the Indravati are major rivers in their own right. The last major tributary is the Sabari from Orissa, which falls into the Godavari, 100 km above Rajahmundry.

Further below Rajahmundry, the river branches off into two main streams – the Gautami Godavari on the east and Vasishta Godavari on the west. Further down, a branch Vainateyam splits from the Vasishta Godavari at Gannavaram 22 km from the coastline. All these streams run down to sea through arid alluvial delta formed over the ages by the mass of silt that has been deposited. The basin is roughly triangular in shape and the river itself runs practically along the base of the triangular.

Table 21. Distribution of catchment area in Godavari basin

S. N.	State	Drainage area (sq. km)	% of the total basin area
1.	Maharashtra	152, 199	48.6
2.	Andhra Pradesh	73, 201	23.4
3.	Madhya Pradesh	31, 821	10.0
4.	Chhattisgarh	33, 434	10.9
5.	Orissa	17, 752	5.7
6.	Karnataka	4, 406	1.4
	Total	312, 813	100.0

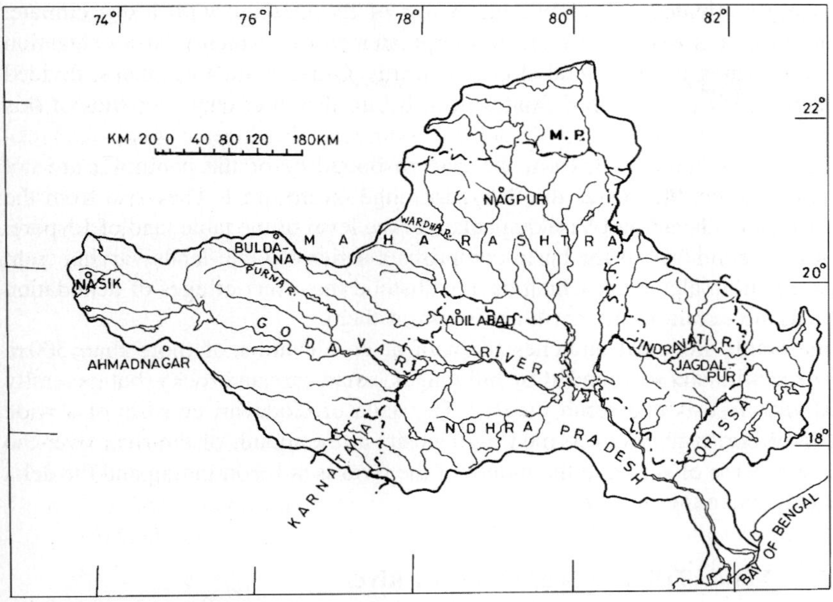

Figure 13. Index map of Godavari basin

14.2.1. Topography

The Godavari basin is bounded on the north by the Satmala hills, on the south by the Ajanta range and the Mahadeo hills, on the east by the Eastern Ghats and on the west by the Western Ghats. Except for the hills forming the watershed around the basin, the entire drainage basin of the river comprises rolling and undulating country – a series of ridges and valleys interspersed with low hill ranges.

The upper reaches of the Godavari drainage basin are occupied by the Deccan Traps containing minerals, hypersthene, augite, diopside, enstatite, magnetite, epidote, biotite, zircon, rutile, apatite and chlorite. The middle part of the basin is principally Archean granites and Dharwars composed of phyllites, quartzites, amphiboles and granites. The downstream part of the middle basin is occupied mainly by the Cuddapah and Vindhyan metasediments and rocks of the Gondwana group. The Cuddapahs and Vindhyan are quartzites, sandstones, shales, limestones and conglomerates. The Gondwanas are principally detritals with some thick coal seams. The Eastern Ghats dominate the lower part of the drainage basin and are formed mainly from the Khondalites which include quartz- feldspar- garnet- sillli-manite gneisses, quartzite, calc-granulites and charnockites. In the coastal region the tertiary Rajahmundry sandstones crop out.

The western edge of the basin is an almost unbroken line formed by the Sahyadri range of the Western Ghats from 600 to 2,100 m height. It has the heaviest rainfall and the dampest climate in the basin. Hardly 50 to 60 km east of the Ghats lie the sparsely cultivated and undulating plains of the Deccan, with a dry climate. The interior of the basin is a plateau, the greater part of which is at an elevation of 300 to 600 m with its general slope eastwards. Great undulating plains, divided from each other by flat topped ranges of hills, are the chief characteristics of this plateau.

The Eastern Ghats which form the eastern boundary of the peninsula are not well-defined or continuous as the Sahyadri range on the west. They rise from the plains of East Godavari and Visakhapatnam to the level of the table land of Jeypore. The northern boundary of the basin comprises a series of table-lands varying from 600 to 1,200 m in elevation, which have withstood the effect of ages of denudation better than the terrain to the north and south of them.

To the south, lie great stretches of plain at an elevation of more than 300 m interspersed with and surrounded by hill ranges, some bare and rocky, but generally covered with forests and scrub jungles. The delta of Godavari consists of a wide belt of river borne alluvium formed by deposits at the mouth of the river over the ages. The process of silting at the mouth of the river is still continuing and the delta is gradually extending into the sea.

14.2.2. Major Tributaries of Godavari River

The Godavari basin was divided into 12 sub-basins by the Godavari Water Disputes Tribunal. These are: (i) the Upper Godavari (from the source to its confluence with Manjira) (G-1), (ii) the Pravara (G-2), (iii) the Purna (G-3), (iv) the Manjira (G-4), (v) the Middle Godavari (from its confluence with Manjira to its confluence with the Pranhita) (G-5), (vi) the Maner (G-6), (vii) the Penganga (G-7), (viii) the Wardha (G-8), (ix) the Pranhita (G-9), (x) the Lower Godavari (from its confluence with the Pranhita up to the Sea) (G-10), (xi) the Indravati (G-11) and (xii) the Sabari (G-12). A flow diagram of Godavari River is shown as Figure 14.

The largest tributary of the Godavari is the Pranhita (inclusive of Penganga and Wardha) with about 34% coverage of drainage area. The Pravara, Manjira and Maner are right bank tributaries covering about 16.1%, the Purna, Pranhita, Indravathi and Sabari are important left bank tributaries, covering nearly 59.7% of the total catchment area of the basin. The Godavari in the upper, middle, and lower reaches make up for the balance 24.2%. The particulars of the catchment area, length, elevation of the source points of the river and its tributaries in the order of their occurrence along the length of the main river are shown in Table 22.

Major tributaries of Godavari flowing through well-established drainage networks are the Pravara, the Purna, the Manjra, the Penganga, the Wainganaga, the Wardha, the Pranahita, the Indravati, and the Sabari. A brief description of the principal tributaries is given in the following.

The Pravara river

The Pravara River originates in the Western Ghats at an altitude of about 1,067 m. After traversing a distance of about 200 km in the easterly direction it falls into

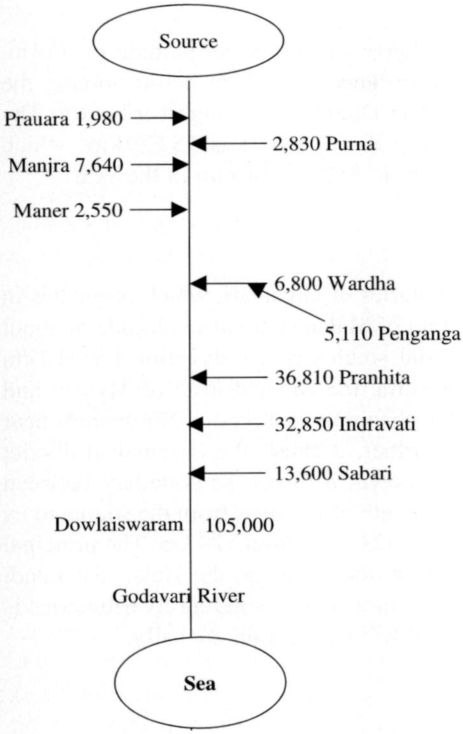

Figure 14. Flow diagram of Godavari River

the Godavari River near Newasa at an altitude of about 457 m. The Mula River is the right bank tributary of Pravara River. The drainage area of the Pravara River is 6, 537 km^2 lying entirely in Maharashtra, but its principal source of supply is only about 32 km length of the Western Ghats.

Table 22. Important Tributaries of Godavari

SN	Tributary	Elevation of Source	Length (km)	Catchment area (sq. km)
1	Pravara	1, 050	208	6, 537
2	Purna	838	374	15, 579
3	Manjira	823	686	30, 844
4	Maner	533	225	13, 106
5	Pranhita	640	721	109, 079
6	Indravathi	914	536	41, 665
7	Sabari	1, 372	418	20, 427
8	Main Godavari including minor tributaries	1, 067	1, 465	75, 576

The Purna river

The Purna River originates in the Ajanta Range of hills at an altitude of 838 m. It flows in a south-easterly direction for a distance of 373 km before joining the Godavari at an elevation of about 351 m. The Dudna is its longest tributary. The catchment area of the Purna River, including its tributaries is 15,579 km^2 which lies in a low rainfall zone which gets 711 mm to 889 mm of rain in the year.

The Manjara river

The Manjara River is one of the major tributaries of Godavari, which originates in the Balaghat Range of hills in the Bhir district of Maharashtra at an altitude of about 823 m. The river flows in a general east and south easterly direction for 512 km through the Osmanabad district of Maharashtra, the Bidar district of Mysore and the Medak district of Andhra Pradesh before changing its direction Northwards near Sanga Reddipet. After flowing for 75 km further, it enters the Nizamabad district of Andhra Pradesh and from 102 km lower down, it forms the boundary between Maharashtra and Andhra Pradesh. The total length of the river from the source to its confluence with the Godavari at an altitude of 323 m is about 724 km. The principal tributaries of the Manjara River are the Tirna, the Karanga, the Halai, the Lendi and the Maner. The catchment area of the Manjra River including its tributaries is $30,844\,km^2$ lying in a zone which gets about 635 mm of rain annually.

The Pranhita river

The Pranhita is the most important tributary of the Godavari River which gets water from two important tributaries, namely the Wardha and the Wainganga. The combined waters of the Wainganga and the Wardha are called the Pranhita, which traverses for 113 km along the border between Maharashtra and Andhra Pradesh before falling into the Godavari at an elevation of 107 m. The catchment area of the Pranhita and of all its tributaries is $109,077\,km^2$ and lies in a medium rainfall zone of 889 mm to 1,600 mm (annual).

The Wardha river

Wardha is one of the right bank tributaries of Pranhita River. The Wardha sub basin lies between latitude 19°18′ N and 21°58′ N and longitudes 77°20′ E and 79°45′ E. Wardha originates at an altitude of 777 m in the Betul district of Madhya Pradesh and enters Maharashtra about 32 km from its source. After traversing a distance of 528 km, it joins the Wainganga at an elevation of 146 m. The major left bank tributaries of the Wardha are the Kar, the Wena, The Jam and the Erai and the right bank tributaries are The Madu, the Bembla and the Penganga. The drainage area of the Wardha River is $24,087\,km^2$ and throughout its course, the river flows through dense forests. The average annual rainfall for the entire sub-basin is 1,000 mm approximately.

The Penganga river

The Penganga River originates at an altitude of 686 m in the Buldana Range in Maharashtra and after traversing a distance of 676 km in a generally east-south-east direction, it joins the Wardha River at an elevation of 174 m. Except in its uppermost reaches in which the terrain is barren and hilly, the river passes through dense forests of Yeotmal and Nanded districts. The important tributaries of the Penganga River are the Pus, the Arna, the Aran, and the Waghari. The drainage area of the Penganga River and its tributaries is 23,898 km^2.

The Wainganga river

The Wainganga River originates at an altitude of 640 m in the Seoni district of Madhya Pradesh. The major tributaries of the Wainganga are the Pench, the Kanhar, the Bagh and the Andhari. The total drainage area of Wainganga and its tributaries is 61,093 km^2.

The Indravati river

The Indravati originates at an altitude of 915 m in the Thuamul-Rampur plateau in Kalahandi district of Orissa on the western slopes of the Eastern Ghats. It flows westward through the Koraput district of Orissa and the Bastar district of Madhya Pradesh. The important tributaries of the Indravati River are the Podagada, the Kapur, the Muran, the Narangi, the Baordbig, the Kotri, the Bandia and the Nandira (Berudi). After flowing through a number of rapids, the river emerges into plains near Khatiguda village of Nowrangpur district. The total drainage area of Indravati River including its tributaries is 41,665 km^2 lying in a relatively high rainfall zone with about 1,524 mm of annual rainfall. It joins the Godavari River at an elevation of about 82 m at about 531 km from its source.

The Sabari river

The Sabari or the Kalab is the last major tributary of the Godavari. The river originates at an altitude of 1,372 m in the Sinkaram hill range of the Eastern Ghats. It flows for short distances in a north, north-westerly and westerly direction and joins the Godavari, about 418 km from its source, at an altitude of 25 m. The Sileru or Machkund is the most important tributary of the Sabri River originates at an elevation of 1,219 m and flows for a length of about 306 km before joining the Sabari River. The total catchment area of the Sabari including its tributaries is 20,427 km^2.

14.2.3. Soils

The principal soil types in the Godavari basin and adjoining areas are (i) black soils (regur), (ii) red soils, (iii) laterites and lateritic soils, (iv) alluvium, (v) mixed red and black soils and red and yellow soils and (vi) saline and alkaline soils. The soils in the basin are generally fertile.

14.2.4. Climate of Godavari

The climate of the Godavari drainage basin has high humidity throughout the year effected by the northeast and southwest monsoons. The delta region is semi-arid with an average annual rainfall of 1,042 mm and a maximum temperature in May of 37.3 °C. The coldest month is January with a mean daily maximum temperature of 26.9 °C and a mean daily minimum temperature of 19.2 °C.

Three distinct seasons occur in the Godavari basin, viz. (i) the hot weather, (ii) monsoon and (iii) the winter. The summer season is from mid-February to the end of May. In the hot weather, the heat is unbearable in the central, northern and eastern regions. The weather is comparatively less hot in the westernmost parts of the basin. The south west monsoon sets in by mid June and ends by mid October. During this period the basin receives about 84% of its total annual rainfall. The cold weather season in the entire basin, from mid October to mid February is generally pleasant, the western and the north eastern regions being colder than the rest of the basin.

Temperature

The Godavari basin has a tropical climate. The mean annual surface temperature in the Western Ghats area is about 24 °C. It increases gradually towards the east and attains a maximum of 29.4 °C on the east coast.

During January, a typical winter month, the mean daily minimum temperature going from west to east increases from 15 °C on the Western Ghats to about 18 °C on the east coast; The mean daily maximum temperature generally exceeds 30 °C in the western part of the Godavari basin and is slightly less than 30 °C in the eastern part;

Maximum and minimum temperatures in the basin gradually increase as one moves from west to east. Temperatures are the highest in April-May, which are typical summer months. The maximum temperature increases from 35 °C in the west to 40 °C in the middle of the plateau, though it goes down again to 35 °C on the east coast. The minimum temperature increases from 22 °C in the west to 26 °C in the east coast.

During July, a typical monsoon month, the minimum temperature increases from 20 °C in the Western Ghats to 26 °C near the east coast and the maximum temperature increases from 27 °C in the Western Ghats to 33 °C near the east coast. During October, a typical post-monsoon month, the minimum temperature is 23 °C near the coast. The mean daily maximum temperature is a little above 39 °C over the entire basin.

Rainfall pattern in the basin

The Godavari basin receives its maximum rainfall during the Southwest monsoon. The monsoon currents strike the west coast of the peninsula from west and south-west; meets the Western Ghats or Sahyadri range which presents almost an uninterrupted barrier ranging from 610 m to 2,134 m in height. Before surmounting this barrier the currents deposit most of their moisture on its windward side, and then sweep across the interior of the peninsula on the easterly course. Rainfall is governed largely by the orography of the area, which leads to variation in the amount of precipitation. The monsoon currents follow the eastward slope of the country from the crest of the Ghats, which form the watershed. Conditions in the interior are, therefore, somewhat unfavorable for heavy precipitation except in association with the depression from the Bay of Bengal. The Northeast part of the Godavari basin also receives some rain in association with monsoon depressions, which move west-northwest across the Orissa coast.

The Godavari receives flow from a length of about 129 km of the high rainfall zone in the Western Ghats where the annual rainfall varies from 1,000 to 3,000 mm. East of the Western Ghats, the rainfall decreases rapidly to less than 600 mm along a line running approximately from Chitradurga through Sangli and Pune to a point Northeast of the line connecting Kurnool, Raichur, Bijapur and Ahmadnagar. East of this line the rainfall again gradually increases to about 900 mm towards the East coast.

The Godavari basin as a whole receives 84% of the annual rainfall on an average, during the Southwest monsoon, which sets in mid June and ends by mid October. January and February are almost entirely dry in the Godavari basin; the rainfall during these two months being less than 15 mm. During the next three months, up to end of May, rainfall varies from 20 mm to about 50 mm in most parts of the basin. The Indravati and Pranhita sub-basins receive up to 86% and 88% of the annual rainfall during the same period due to influence of the cyclonic storms which predominantly pass through these sub-basins. Particulars of average rainfall in the different sub-basins are given in Table 23.

14.2.5. Water Quality Aspects

Table 24 gives the results of systematic sampling of river water quality carried out by CPCB at different times. It can be seen that at almost all places, the quality

Table 23. Rainfall Pattern in the Sub-basins of Godavari

S.N.	Name of Sub-basin/tributary	Rainfall during		Annual rainfall (mm)	Monsoon rainfall percent of annual rainfall
		June-sept. (mm)	Oct-may (mm)		
1	Upper Godavari	645	125	770	84
2	Pravara	476	130	609	79
3	Purna	660	137	797	83
4	Manjira	696	150	846	82
5	Middle Godavari	809	146	955	85
6	Maner	762	170	932	82
7	Pranhita	1, 196	167	1, 363	88
8	Lower Godavari	929	280	1, 209	77
9	Indravati	1, 366	222	1, 588	86
10	Sabari	1, 137	295	1, 432	79
11	Godavari Basin	953	79	1, 132	84

Table 24. Desired and existing water quality levels of Godavari River (1997–2001)

Location	Desired class	Existing class & critical parameters				
		1997	1998	1999	2000	2001
Godavari at U/S of Gangapur Dam, Nasik, Maharashtra	• B	D DO, BOD	D BOD	D BOD	D BOD	D BOD
Godavari at Panchavati at Ramkund, Maharashtra	B	D DO, BOD	D BOD	D BOD	D BOD	D BOD
Godavari at Nasik D/S, Maharashtra	B	D DO, BOD	D BOD	D BOD	D DO,BOD	D BOD, Totcoli
Godavari at Dhalegaon, Maharashtra	B	D DO, BOD, Totcoli	D BOD, Totcoli	D BOD	D BOD	D BOD
Godavari at Nanded, Maharashtra	B	D DO, BOD	D BOD, Totcoli	D BOD	D BOD	D BOD, Totcoli
Godavari at Raher, Maharashtra	C	D BOD	D BOD	D BOD	D BOD	D BOD
Godavari at Mancherial, A.P.	C	D DO, BOD	NA	NA	NA	NA
Godavari at Polavaram, A.P.	C	B	NA	NA	D BOD	D BOD
Godavari at Rajahmundry U/S, A.P.	C	NA	NA	NA	D BOD	D BOD
Godavari at Rajahmundry D/S, A.P.	C	Below-E BOD, pH	NA	NA	D BOD	D BOD

• NA- Not Available.
Source: Central Pollution Control Board

of Godavari water is below the desired class. For example, at many places, the desired class was B or C while the existing class was D.

14.2.6. Major Water Resources Development Projects in Godavari Basin

The water resources potential in Godavari basin has been assessed by Central Water Commission to be 110.54 km^3. The utilizable surface water is about 76.3 km^3; the replenishable ground water is about 45 km^3. There is a vast potential for irrigation development and hydropower generation in the basin. The present utilization is of the order of only 40 km^3 in the case of surface water and 6 km^3 in the case of ground water.

Water resources of Upper Godavari catchment are fully utilized up to Sriram Sagar Dam. Downstream of this dam, Godavari is joined by many major tributaries, namely, Pranhita, Indravati and Sabari which carry large volumes of flood waters during monsoon. Table 25 gives monsoon yield of some of Godavari's tributaries at selected locations.

Table 25 shows that Pranhita, Indravati and Sabari contribute a huge amount of flow to Godavari in its lower reaches. Since there is no large storage reservoir in this area, most of this water (except a small quantity diverted at Godavari barrage to meet the requirements of delta irrigation) goes to the sea unutilized during the monsoon period. Of course, the scope of additional new irrigation is limited due to typical topography of the lower Godavari basin. Therefore, it is important to construct storages at Inchampalli and Polavaram to utilize huge quantity of water going waste to the sea every year. Potential storage sites for hydropower generation have also been identified on Indravati at Bhopalpatnam and Bodhghat. These storages will also regulate flows for utilization in the downstream areas. The surplus waters of lower Godavari could be transferred to the water deficit areas after meeting all the requirements of the basin in Andhra Pradesh.

Table 25. Annual average observed runoff at Important CWC sites in Godavari basin

Name of the site	Name of the stream	Catchment area (km^2)	Annual average runoff (BCM)
Polavaram	Godavari	307, 800	81.50
Koida	Godavari	305, 460	81.58
Konta	Sabari	19, 550	14.84
Injarum	Sabari	12, 925	11.42
Perur	Godavari	260, 200	54.82
Pathagudem	Indravati	40, 000	21.17
Medadapalli	Indravati	24, 212	40.75
Chindnar	Indravati	17, 270	8.36
Tekra	Pranhita	108, 780	33.85
Sirpur	Pranhita	47, 500	9.89
Bamni	Pranhita	46, 020	1039
Penganga Bridge	Pranhita	18, 441	3.04
Ghugus	Pranhita	21, 429	4.05
Hivra	Wardha	10, 240	0.86
Ashti	Pranhita	50, 990	25.27
Pauni	Pranhita	35, 520	13.50
Satrapur	Kanhan	11, 100	2.19
Somanpally	Maneru	12, 991	1.17
Mancherial	Godavari	102, 900	5.03
Yelli	Godavari	53, 630	1.79
Purna	Purna	15, 000	0.45
G R Bridge	Godavari	33, 934	0.75
Dhalegaon	Godavari	30, 840	0.71

Source: CWC (2002)

Prior to independence, only a few irrigation projects were constructed in Godavari basin. Important among these are Godavari delta system (with Dowlaiswaram weir as head works). Nizamsagar reservoir, Kadana dam and Pravara dam. After independence, under various five-year plans a large number of multi-purpose and irrigation projects were taken up. The most important among them are the Jayakwadi, Sri Ram Sagar and Godavari Barrage (by remodeling the existing Dowlaiswaram) weir. Prominent among the proposed major projects in the basin are Bhopalapatnam on Indravati, Inchampalli and Polavaram on Godavari.

Inchampalli

The Inchampalli project is proposed on the Godavari River about 12 km downstream of the confluence of Indravati with the Godavari River in Andhra Pradesh. It is a joint project among the States of Maharashtra, Madhya Pradesh and Andhra Pradesh. It is a multi purpose project envisaging irrigation benefits for upland areas, generation of hydropower, navigation facilities in the river, development of pisciculture and providing recreation benefits, besides mitigating flood hazards. Flows in abundance are available at Inchampalli, as it is just downstream of the place where two major tributaries, Pranhita and Indravati join the Godavari River. The catchment area of the dam is 269,000 Mm^2. The FRL and MDDL of the reservoir will be 112.77 m and 106.98 m respectively. The gross storage capacity and live storage capacity of the reservoir will be 8,959 and 4,098 MCM respectively. The annual irrigation from the dam is 950 Mm^2 and annual utilization is 620, 000 Mm^2. For hydropower generation install capacity at the dam is 875 MW. The Salient features of Inchampally Project are given in Table 26.

Polavaram

The Polavaram project is planned downstream of Inchampalli after the confluence of another major tributary the Sabari with the Godavari River. It is also a multi-purpose project for irrigation, hydropower, and water supply to Vizag city. The catchment area of the dam is 307, 000 Mm^2. The FRL and MDDL of the reservoir are at 45.72 m and 41.15 m, respectively. The gross storage capacity and live storage capacity of the reservoir is 4,945 and 2, 043 Mm^3 respectively. The project has been planned to utilize the significant quantum of flows that would be received from Sabari and power releases and spills from Inchampalli for its own uses and also for regulating releases for the Godavari delta. The annual irrigation from the dam is 4,720 Mm^2 and annual utilization is 3, 823, 000 Mm^2. For hydropower generation install capacity at the dam is 720 MW. The salient features of Polavaram Project are given in Table 27.

Dowleswaram barrage (cotton barrage)

The Dowleswaram Barrage is the terminal project on Godavari, located downstream of Polavaram, catering to the needs of Godavari delta. This 3,500 m long barrage built in the mid 19th century is supposed to be Asia's largest barrage. It was named as Cotton Barrage after Sir Arthur Cotton, who built the barrage and who is fondly

Table 26. Salient features of Inchampally Project

Particulars	Details
Proposed Utilization from Ayacut in Kharif	635.53 Mm^2
Proposed Utilization from Ayacut in Rabi	700.11 Mm^2
Hydropower production	975 MW
Water Availability	15,885 Mm^3
Water Requirement for Irrigation	2,265.6 Mm^3
Water Requirement for Hydropower	9,912 Mm^3
75% Dependable yield	10,025 Mm^3
Area for submergence at FRL 112.770	970 Mm^2
FRL	112.770 m
M.D.D.L.	106.980 m
Gross storage	10,375.88 Mm^3
Live Storage	9,950.23 Mm^3
Type of Dam	Masonry
Type of Spillway	Ogee
Length of Spillway	1,278 m
No & Type of gates	57, Radial Type
Size of gates	18 m × 12.37 m
Length of R/B Canal	293 km
Irrigation command area	513.95 Mm^2
Number of villages affected	229
Population affected	1 Lakh

Table 27. Salient features of Polavaram Project

Particulars	Details
Catchment area	306, 643 km^2
Proposed Utilisation from Existing Ayacut	NIL
Proposed Utilisation from New Ayacut	2, 910 Mm^2
Hydropower production	720 MW
Water Required for project	8, 535 Mm^3
Water supply to Vishakhapatnam city	663.82 Mm^3
Diversion to Krishna River	2, 265.6 Mm^3
FRL	45.72 m
M.D.D.L.	41.15 m
Gross storage at FRL	3, 388.20 Mm^3
Live Storage at MDDL	3, 381.408 Mm^3
Type of Dam	Earth cum Rock fill
Length of Dam	2,310 m
Type of Spillway	Ogee
Length of Spillway	897.50 m
No & Type of gates	44, Radial Type
Size of gates	16 m × 20 m
Length of L/B Canal	181.50 km
Length of R/B Canal	174 km
Number of villages affected	276
Population affected	117,034

remembered and revered in regard to his yeoman services to the upliftment of the people in the area. This barrage has completely transformed the famine and poverty wracked areas in the Godavari Delta into a prosperous place. Sir Cotton is credited with preparation of a grand plan for development of water resources of Peninsular India but this could not materialize due to some reasons. The catchment area at the barrage is 312,800 million sq. m. The gross storage capacity of the pond at FRL 13.81 m is 10 Mm^3. The annual irrigation from the barrage is for 9,800 Mm^2 area for which, on an average, annually 777,400 Mm^3 of water is utilized.

Karanja project

The Karanja Reservoir Project is a major project constructed on Manjra River, a tributary of Godavari River. The dam is located near Byalhalli in Bhalki Taluk in Bidar district. The catchment area of the dam is 2,025 km^2. The length of the dam is 3,480 m with FRL as 584.15 m. Salient features of the Karanja project are given in Table 28.

Sriram sagar project (SRSP)

Although the Telangana area of Andhra Pradesh State is endowed with great natural water resources, such as the Godavari and Krishna Rivers, their tributaries and

Table 28. Salient features of Karanja Project

Particulars	Details
Yield	271.59 Mm^3
Gross Storage capacity	217.78 Mm^3
Live Storage capacity	207.17 Mm^3
Dead Storage capacity	10.591 Mm^3
Level at top of dam	589.15 m
Maximum Water Level	587.00 m
Crest Level	574.15 m
Sill Level	575.15 m
Peak Spillway Discharge	13,282.36 cumec
Spillway Gates	6 of size 15 m × 10 m
Submergence Area	56.73 Mm^2
Villages affected	7 Full and 2 Partial
Population affected	9,080
Length of Left Bank Canal	31 km
Head Discharge of Left Bank Canal	1.982 Cumec
Ayacut of Left Bank Canal	32.38 Mm^2
Length of Right Bank Canal	131 km
Head Discharge of Right Bank Canal	16.935 Cumec
Ayacut of Right Bank Canal	283.29 Mm^2
Length of Fore Shore Lift Canal	24 km
Head Discharge of Fore Shore Lift Canal	1,756 Cumec
Ayacut of Fore Shore Lift Canal	40.47 Mm^2
Gross Command Area	489.68 Mm^2
Net Command Area	356.14 Mm^2

extensive fertile lands, it largely remains undeveloped partly due to lack of assured irrigation facilities. The extent of Irrigation is small and even this is by means of numerous small tanks, which are dependent on uncertain rainfall. The completion of Sriram Sagar Project Stage-I is expected to make up and set right the shortfalls and imbalances in the economy of the region and would greatly contribute to the well being and prosperity of the people of the region. Thus, this project has great importance for the development of this region.

The Sriram Sagar Project is a multipurpose project, located across the Godavari River near Pochampad of Nizamabad District in Andhra Pradesh at a distance of 200 km from Hyderabad. The dam is located at a latitude of 18°58′ N and a Longitude of 78°20′0″E. The catchment area at the dam site is $91,760\,\text{km}^2$. The water spread of the reservoir is $453\,\text{km}^2$, with a capacity of 3.17 BCM (Subramanyam, 1979). The reservoir utilizes 1,869 BCM of water to irrigate 0.23 M-ha of land in the Districts of Karimnagar and Nizamabad, of which one-third would be under wet cultivation and the rest under dry crops such as maize, jowar, chillies and pulses.

The masonry spillway is designed for a maximum discharging capacity of 45,307 cumec with MWL at 333.146 m. The arrangement for energy dissipation of the spillway consists of slotted roller bucket for the first 20 spans from left end of the spillway and ski jump bucket in the remaining 22 spans. The left and the right earthen dams are of rolled type. Three $2.438 * 3.657$ m river sluices have been provided to serve for diversion purposes during dam construction and to serve as permanent low level river outlets. Four sluices of $2.438 * 3.657$ m size have been provided for south canal with sill at 307.850 m.

The command area of the SRSP consists of undulating terrain with extensive granite rocks. The area comprises of prominent ridges and valleys formed by gully erosion; it slopes towards the Godavari River and is drained by many small streams that empty into Godavari and its distributaries. The majority of the soils are sandy loams. Though soil conditions support easy drainage, there are a number of low-lying areas which require systematic drainage facilities.

The main rock types occurring in command area are pink and gray granites, with fine to coarse-grained texture. Most of the ridge and relief areas are developed into poorly to moderately weathered formations. Low lying and plain areas are developed into moderately to highly weathered formation. The thickness of weathering extends up to 11 m. It is observed that weathering intensity decreases with depth and generally basement is encountered without fracturing. Hence dug wells are feasible and bore wells are not feasible in general.

The area is drained towards Godavari River. The drainage pattern is dendritic. Undulating topography is characteristic of the area and hence distributaries are aligned on ridges. The maximum and minimum elevations of the area are 300 to 200 m above mean sea level respectively. The normal monsoon rainfall in the area is 986 mm.

The predominant soil type of the area is red sandy loams of good permeability. Black sandy, silty loams are found in low-lying areas stretching along the streams and also in the commands of irrigation tank. Paddy is the principal crop which is grown extensively in the areas receiving irrigation from canal and tank water.

Sugarcane is grown but quite sparsely. In areas that do not receive canal water, paddy is grown under well irrigation.

Water table depth varies from shallow ground level in canal fed areas to 10 m in tail end areas. In the areas receiving canal supplies, the wells existing prior to the project are not being put to use now. These areas are under sugarcane cultivation. Wells are found to be excavated to maximum depths of 2.5–4.0 m to meet the water demand during canal closures. In the tail-end areas, wells are used in both the seasons. The salient features of Sriramsagar project is given in Table 29.

Nizamsagar project

The Nizamsagar project is a multipurpose project, constructed in Nizamabad district in Telengana area of Andhra Pradesh. The project was completed on Manjira River in 1931, a tributary of Godavari, which was a single state river. The canals and distribution system was completed in late 1935. It is a masonry dam, 3.2 km long and 48.15 m high above deep foundation. At the time of its completion, this project was one of the largest in the state but also one of the biggest schemes in India.

The live storage capacity of the reservoir is 724.736 Mm^3, irrigating an ayacut of 967.233 Mm^2. The filling period is generally from July to August and depletion period is from September to June. The installed capacity of the power house at the dam is 15 MW.

Lower manair reservoir

Manair River is a tributary of Godavari River and the reservoir is located at latitude 18°24′ N and longitude 79°8′ E in Karimnagar district of Andhra Pradesh state. The total catchment area of the river up to confluence with Godavari River is 13,106.25 km^2. The lower Manair dam was constructed in 1985 across Manair River at tits 110th km and at its confluence with Mohedamada River. The dam is located at 18°8′24″N and 79°8′6″ E. The total catchment area up to the Lower Manair dam site is 6,464 sq. km; the free catchment area at the Lower Manair Dam is

Table 29. Salient features of Sriramsagar Project

Particulars	Details
Length of Masonry Dam	14.6 km
Bed level	325.240 m
Bed width	51.0 m
Proposed utilization from existing ayacut	890.31 Mm^2
FRL	332.537 m
Gross storage	3,171.84 Mm^3
Live Storage at MDDL	2,322.24 Mm^3
Number of gates	42,
Size of gates	15.24 m × 10 m
Number of villages affected	29
Population affected	38.529

$1,797.46\,km^2$ and the rest is intercepted. The original capacity of the reservoir at F.R.L. 280.46 m is $680.648\,Mm^3$. The spillway has 20 gates of $15.24\,m \times 7.31\,m$ and has been designed for maximum flood of 14,158 cumec.

There are no sediment observation stations on Manair River. Based on Khosla's formula, sediment inflow is assessed at the rate of 4.97 ha.m/year/100 sq. km.

Kaddam reservoir

The Kaddam reservoir was constructed across Kaddam River at its 80th kilometer of run in the Adilabad district of Andhra Pradesh. After traversing a distance of 6.5 km from the dam, Kaddam joins the Godavari River. The latitude and longitude of the dam are 19°07′ N and 78°47′ E. The dam was constructed during the year 1958 and was remodified in the year 1965. The total drainage area of the Kaddam River up to the dam site is 2,631 sq. km. The original capacity of the reservoir at full reservoir level 213.300 m is $215.80\,Mm^3$.

Upper indravati project

This project, constructed on the Indravati River in Orissa is one of the major multipurpose transbasin diversion projects in India. It involves diversion of the waters of the Indravati River, a tributary of the Godavari River into the Mahanadi basin for power generation and irrigation. The whole scheme envisages construction of 4 dams (Indravati masonry dam; Podagada earth dam; Kapur earth dam; and Muran concrete, masonry and earth dam) and 8 earthen dykes (4 on left and 4 on right). The main dam is on the Indravati River in Nowrangpur and Kalahandi districts of Orissa; the other three being on its three tributaries: the Podagad, the Kapur, and the Muran. All these form a single reservoir, connected together through two link channels, to generate, in all, 1,990 million kWhr of electricity per year and simultaneously to annually irrigate $2,185\,Mm^2$ from the releases through the power house. The reservoir has gross storage capacity of 2,300 MCM and live storage of 1,486 MCM. At FRL, the reservoir covers and area of 112 sq. km. At the dam site, the catchment area of Upper Indravati basin is 1,153 sq. km.

The distinct feature of this project is the trans-basin diversion of water from, Godavari Basin up to river Hati (Mahanadi basin) for power generation and irrigation. Tailrace release from powerhouse is picked up at Hati Barrage constructed across river Hati at Mangalpur in Kalahandi district with two canals taking off on either bank. Left main canal is 52 km long and commands an ayacut 49,078 ha. 83 km. long right main canal provides irrigation to 27,191 hectares. Another lift canal is under consideration for irrigating CCA of 33,027 ha.

Water from the reservoir is to be conveyed through a water conductor system for power generation. Thereafter, the release through 7.8 km long tail race channel discharges into the Hati stress, a tributary of the Tel River in Mahanadi basin and is picked up by a weir across the Hati for utilization in irrigation. While designing this project, the aim was to provide as much irrigation as the available resources can provide. With the calculated runoff, the project could provide irrigation of about $1,214.10\,Mm^2$.

For power generation, the power house is connected to reservoir by a water conductor system consisting of a pressure tunnel, a surge tank and five numbers of penstocks. The tunnel has a length of 4,215 m and is of 7 m diameter. The tunnel discharges into a differential surge tank of 18 m diameter and 96.75 m deep. There are 5 penstocks of 3.05 m diameter each. The annual power potential of this project is 1,990 million KWH at 100 percent load factor. The installed capacity of 600 MW consists of 5 units of 120 MW, each operating under an average load of 374 m. It is a surface power house.

The tail race discharge is utilized for irrigating $1,093\,\mathrm{Mm}^2$ of C.C.A. in Sadon and Dharamgarh subdivisions of Kalahandi District. The intensity of irrigation is 200 percent. The project is located in Kalahandi and Koraput districts which are among the most undeveloped districts of Orissa and are predominantly inhabited by adivasis and tribals. Although there is immense scope of industrial growth and agricultural development, the progress is slow as the rich mineral resources of the districts are not yet properly tapped. When the full potential of this project is utilized, it will undoubtedly accelerate growth of economy of these two districts.

Although droughts are frequent in these districts, even when rainfall in good, rainfall distribution is in variance with the crop water requirements and artificial irrigation is the only means to increase crop production. There is no other project in these districts which has such a high irrigation potential. Further, this region suffers from shortage of electricity. Even after the full utilization of the additional power from the Balimela Power Station (360 MW, 1,980 MKWhr), Talchur Thermal Expansion (220 MW, 1,012 MKWhr) and Rengali Power Station (100 MW, 523 MKWhr) that are now under construction and the Upper Kolab Project, shortage of power with both in energy and capacity will exist. A view of Indravati project has been shown in Figure 15.

Bhopalpatnam Reservoir: This is a proposed reservoir on Indravati River, a tributary of Godavari River. The gross storage, dead storage and live storage capacity of the

Figure 15. A view of Indravati project

reservoir will be 8,368.00, 549.00 and 7, 819.00 Mm^3 respectively. The FRL and MDDL of the reservoir will be 200.254 and 176.48 m respectively. The maximum and minimum dependability has been assessed at 21,969.00 and 3, 290.00 Mm^3 respectively.

Balimela dam

The Balimela Project on the Sileru River in the state of Orissa in India consists of a 70 m high earthfill dam and 3 earthen dykes on the saddles in the left abutment hill. The project is located at a distance of 35 km from Malkangiri in Godhra in Malkangiri District, Orissa. The catchment area at the dam is 4,910 km^2. At FRL 462.7 m, the reservoir has a live storage capacity of 3,610 MCM. The spillway is located on a saddle to the right. Half of the stored water is diverted to another valley and a head of 275 m is created for power generation. The other half quantity of water is used for power generation at successive power stations on the Sileru River itself. Balimela power house has 6 units of 60 MW each, with mean annual inflow of 5,190 MCM. It has a firm power of 161 MW. It was commissioned during 1973–77. A view of Balimela dam is shown in Figure 16.

Mula irrigation project

Mula is a major irrigation project on the Mula River, a tributary of Pravara, which in turn is a tributary of Godavari. The Mula project is located at longitude

Figure 16. A view of Balimela dam

74°34′30″E and latitudes 19°1′30″N. This multipurpose project caters for irrigation and municipal (Ahemdnagar city) water supply. It provides industrial water supply to defense units, sugar factories, and other industries. The project infrastructure consists of an earthen dam with gated spillway at Baregaon-Nandur in Rahuri Taluka of Ahmednagar district. The FRL, MDDL and MWL of the reservoir are 552.30 m, 534.21 m and 553.21 m respectively. The Gross storage, live storage, and dead storage of the reservoir are 736.32 Mm^3, 608.88 Mm^3 and 127.44 Mm^3 respectively. The water spread area of the reservoir at FRL is 56.397 km^2. Project's right bank and left bank canals with an extensive distribution system provides irrigation to an area of 80,810 ha. Major portion of command has black cotton soil. The rainfall is very scanty in this area; average rainfall is 50 cm in lower catchment and evaporation is moderate.

Pench projects

The Pench River is a tributary of Kanhan River which lies in the Godavari River Basin. There are two major projects on the Pench River. Pench hydropower project comprises of Totaladoh masonry cum concrete dam on Pench River, 3 km from Totaladoh, in Nagpur District, Maharashtra. The catchment area at the dam is 4,275 km^2. The height and length of the dam is 75 m and 680 m respectively. The reservoir has a live storage capacity of 1,249 MCM at FRL 490 m and the MDDL is at 464 m. The power house has 2 units of 80 MW each. It has a firm power of 34 MW with annual inflow of 1,857 MCM in a 75% dependable year. MSEB commissioned the project in 1986–87.

Pench Irrigation Project comprises a storage-cum-diversion dam, 23 km downstream of Totladoh Dam on Pench River to impound releases through the tail race discharge after power generation at Totladoh reservoir. It has lined canals on both the banks, envisaging irrigation of 104,476 ha area in Nagpur and Bhandara districts. Besides irrigation, this project also provides 145 MCM of water for Nagpur water supply and 87 MCM for Koradi Thermal Power Station through its right bank canal. The temperature in the area rises above 36 °C in summer and goes below 20 °C in winter. The average annual rainfall is 1,051 mm. Free catchment area between Totladoh and Kamthi Khedi is 388.0 sq. km with an inflow from free catchment of 112.6 MCM. The Kamthi Khedi diversion dam near Parsheoni Village has a gross storage capacity of 230.00 MCM resulting in live storage of 180.00 MCM. It is 2,248 m long, with height 44.50 m, above cut-off trench in case of earth dam and 45.5 m in case of masonry dam above the lowest foundation level. The dam comprises of a central spillway with earth dam on right flank and earth dam in left saddle. The spillway has 16 gates of size 12 m × 8 m.

Upper kolab

Upper Kolab is a straight masonry gravity dam completed in 1990 on Kolab River, 5 km from Jeyapore in Korapt District, Orissa. The catchment area at the dam is 1,630 km^2. The height and length of the dam is 55 m and 631 m respectively. The reservoir has a gross storage capacity of 1,215 MCM and live storage capacity of 935 MCM at FRL 858 m, and mean annual inflow of 1,803 MCM; its MDDL is at

844 m. At FRL, the reservoir water spread covers 114 sq. km. Peak of design flood hydrograph is 10,020 cumec. The spillway has 11 radial gates of 12.2 m × 12.2 m with crest level at 845.8 m. Two canals take off from the reservoir: right bank canal is known as the Jeyapore main canal which is 58.83 km long with discharge at head 98.1 cumec, and the left bank canal is known as the Padampur canal which is 12.47 km long with 2.79 cumec discharge at the head. Upper Kolab power house has 4 units of 80 MW each. With a design head of 261 m, it has a firm power of 111 MW. OHPC Ltd commissioned the project in 1988–93.

Projects in sileru (machkund) basin

Three important projects of this basin are described here. Machkund project comprises of Kalaput dam, constructed on Machkund River (Sileru River is known by this name in upper reaches). The project is located at a distance of 65 km from Jeypore, in Onukudelli District, Orissa. The catchment area at the power house is 1,932.85 km^2. This 60.7 m high dam has created a reservoir with live storage capacity of 893 MCM at FRL 839 m and the MDDL is at 819 m. At the dam site, mean annual inflow has been estimated at 1,023.7 MCM. For power generation, it has 3 units of 17 MW each and 3 units of 23 MW each respectively. This project, commissioned in 1955–59, has a firm power of 81 MW.

Upper Sileru hydropower project is located on Guntawada masonry gravity dam on Sileru River, 230 km from Visakhapatnam in Andhra Pradesh. The height and length of the dam are 31 m and 625 m respectively. The reservoir has a live storage capacity of 109.5 MCM at FRL 414.5 m and the MDDL is at 406.3 m. The power house has 4 units of 60 MW each. It has a firm power of 58 MW with mean annual inflow of 2,000 MCM.

Lower Sileru hydropower project is located on Donkarayi masonry gravity dam on Sileru River also known as Machkund in upper reaches, 94 km from Bhadrachalam, in Khammam District, Andhra Pradesh. The catchment area at the dam is 2,254 km^2. The height and length of the dam are 71.5 m and 1,399 m respectively. The reservoir has a live storage capacity of 380 MCM at FRL 316 m; its MDDL is at 291.2 m. The power house has 4 units of 115 MW each. It has a firm power of 154 MW with mean annual inflow of 90.6 MCM. This project was commissioned in 1976–78.

Upper wainganga

This project, later renamed as Sanjay Sarovar Project, is a major irrigation scheme in Seoni and Balaghat Districts of Madhya Pradesh. It envisages the construction of a composite earth and masonry dam with maximum height of 42.67 m across Wainganga River near Bhimgarh village at latitude 22°22′51″N and 70°30′20″E. The construction of the dam was completed in 1995. The dam intercepts a catchment area of 2,008 sq. km. At FRL of 519.38 m, the reservoir has gross and live storage capacities of 507 MCM and 410 MCM. In the catchment of the dam, the maximum and minimum annual rainfalls are 1,748 mm and 647 mm, the average being 1,225 mm. The 75% dependable yield at the dam site is 703.1 Mm^3. Ten radial type gates of size 15.24 m × 10.67 m have been provided at the spillway whose crest lies at 508.71 m.

Other projects

In addition to the above, there are several existing and under construction water resources projects in the basin. Salient features of existing and under construction water resources projects with a live storage capacity of 10 MCM and above have been presented in Table 30 and Table 31, respectively.

14.2.7. Flood of 1986 in Godavari Basin

In the Godavari basin, the normal rainfall over the individual sub-basins varies widely. Godavari basin upstream of Nanded receives about 860 mm in an average year while Indravati sub-basin receives 1,580 mm. During monsoon period, the eastern half of the basin receives about 750 to 1,500 mm of rain every year while in the rest of the area, except a narrow Ghats strip, annual rainfall is between 450 to 750 mm. Consequently eastern half of the basin is more prone to floods as compared to the west.

Floods in the Godavari River are the result of heavy rains over the basin. Severe storms are mostly associated with the depressions/cyclonic storms of Bay of Bengal (BoB) origin. Unprecedented floods occurred in the Godavari basin during 13–20 August 1986 which were caused by a depression originated in the BoB.

The famous flood of 1986 was caused by two storms that occurred in quick succession. A low pressure area was formed over East Madhya Pradesh on 5th August 1986. Moving in a westerly direction, it turned into deep depression on 8th morning and centered close to Surat in Gujarat (Northwest of the Godavari basin). On 10th morning it was centered near latitude 23 °N and longitude 66°30′ E (Pandharinath, 1987). It weakened considerably on 11th August. Although this storm did not cause flooding, it sufficiently saturated the basin. The flood causing storm originated as a low pressure area over North–West BoB on 9th August 1986. This system became well-marked on 10th morning and concentrated into a depression on the same evening, with central region near latitude 18 °N and longitude 88 °E. It moved in a westerly direction and intensified into deep depression and centered near latitude 18 °N and longitude 86°30′ E on 11th morning. Moving in a west-north-westerly direction it crossed north Andhra coast near Kalingapatnam on 12th night. It centered over eastern part of the basin near Karaput on 13th morning. Moving in a north-westerly direction it moved over the extreme eastern part of the basin weakened, into a depression and centered near Raipur on 14th morning. It further weakened and formed as a well marked low over Northwest Madhya Pradesh on 15th morning. To appreciate the relative amounts of rain that fell over the basin in the flood event, Table 33 contains the normal rainfall of monsoon months as well as for the storm events at a few stations over the Godavari basin. Table 32 shows the Flood stages at various gauging sites of Godavari River during August 1986.

Table 30. Salient features of selected existing projects in Godavari basin

Name of the Project	State	Year of completion	Gross storage capacity (MCM)	Live storage capacity (MCM)	Designed annual irrigation (Million sq. m)	Installed capacity (MW)
Boggulavagu	Andhra Pradesh	–	11.52	10.34	20.60	–
Buggavanka	Andhra Pradesh	–	14.32	12.04	34.40	–
Donkaravi Dam	Andhra Pradesh	1974	1,670.00	1,252.50	–	25
Jalaput	Andhra Pradesh	1959	970.00	892.00	–	25
Kinnerasani	Andhra Pradesh	–	237.86	207.85	–	–
Lakhnawaram Lake	Andhra Pradesh	1909	60.46	60.46	28.70	–
Nallavagu Project	Andhra Pradesh	1969	21.13	18.49	19.60	–
Peddavagu	Andhra Pradesh	1987	13.85	11.70	64.00	–
Pocharam	Andhra Pradesh	1922	51.55	42.52	40.50	–
Ramadhugu	Andhra Pradesh	1964	18.00	14.84	11.80	–
Ramappa Lake	Andhra Pradesh	1919	–	82.50	21.00	–
Salivagu	Andhra Pradesh	1964	15.87	14.64	12.80	–
Singur	Andhra Pradesh	–	850.00	566.25	160.00	–
Swarna	Andhra Pradesh	1978	42.00	35.83	26.90	–
Vottigedda	Andhra Pradesh	–	89.09	79.93	98.00	–
Ari Tank	Madhya Pradesh	1952	15.30	12.89	42.50	–
Gangulpara Tank	Madhya Pradesh	1958	1,111.32	10.93	34.50	–
Karad	Madhya Pradesh	1963	15.40	13.69	–	–
Nahaleswara Tank	Madhya Pradesh	1967	16.14	14.55	45.10	–
Oon	Madhya Pradesh	1966	19.80	18.36	1.60	–
Paralkot	Madhya Pradesh	1973	66.30	63.60	145.80	–
Sarathi	Madhya Pradesh	1923	17.06	16.16	34.80	–
Thanwar Project	Madhya Pradesh	1987	139.00	129.00	182.10	–
Adhala	Maharashtra	1974	30.00	27.60	50.90	–
Alandi	Maharashtra	1983	29.52	27.46	44.90	–
Asolamedha	Maharashtra	1918	92.70	74.62	99.20	–
Bagh River Project	Maharashtra	1977	65.08	48.67	336.70	–
Sirpur	Maharashtra	1976	203.77	192.45	–	–
Bhandardara	Maharashtra	1923	312.40	307.31	–	17
Badalkasa	Maharashtra	1923	17.39	16.45	40.50	–
Bor	Maharashtra	1967	138.67	127.40	61.90	–
Chandai	Maharashtra	1978	11.48	10.49	36.10	–
Chandpur	Maharashtra	1915	31.12	23.24	48.60	–
Chargaon	Maharashtra	1976	20.16	17.51	21.20	–
Chorkhamra	Maharashtra	1923	21.05	20.80	40.50	–
Chulband	Maharashtra	1976	19.10	16.54	40.40	–
Darna	Maharashtra	1912	226.87	219.82	258.70	–
Dheku	Maharashtra	1962	13.53	12.16	27.10	–
Dina Nadi Project	Maharashtra	1977	61.17	55.95	113.60	–
Ekburji	Maharashtra	1964	14.12	11.97	24.30	–
Gangapur	Maharashtra	1954	215.76	203.76	232.60	–
Galhati	Maharashtra	1965	16.37	13.83	23.00	–
Ghirani	Maharashtra	1968	25.08	22.46	28.30	–
Ghorazari	Maharashtra	1987	45.19	38.00	38.50	–

(Continued)

Table 30. Continued

Name of the Project	State	Year of completion	Gross storage capacity (MCM)	Live storage capacity (MCM)	Designed annual irrigation (Million sq. m)	Installed capacity (MW)
Girija	Maharashtra	1986	24.50	21.53	38.00	–
Goki	Maharashtra	1980	50.22	42.70	78.00	–
Itiadoh	Maharashtra	1970	288.83	225.12	400.80	–
Jayakwadi	Maharashtra	1976	2, 909.00	2, 171.00	2, 780.00	12
Karadkhed	Maharashtra	1973	12.39	11.01	26.90	–
Karanjwan	Maharashtra	1974	175.56	166.22	448.80	–
Kardwhed	Maharashtra	1977	12.30	11.01	–	–
Kalyan	Maharashtra	1986	15.36	12.22	19.00	–
Karpara	Maharashtra	1976	26.82	24.90	31.10	–
Kanhdi	Maharashtra	1978	22.21	20.48	15.40	–
Khair Bandha	Maharashtra	1915	16.79	15.95		–
Khelna	Maharashtra	1966	12.60	11.08	24.30	–
Koradi	Maharashtra	1980	22.58	15.12	48.00	–
Kundalika	Maharashtra	1986	46.30	34.00	44.00	–
Kundrala	Maharashtra	1974	11.76	10.41	15.60	–
Manar	Maharashtra	1964	138.32	128.68	244.70	–
Mangarh	Maharashtra	1972	210.00	171.50	17.00	–
Makardhokada	Maharashtra	1977	21.35	19.93	33.70	–
Manjra	Maharashtra	1984	250.70	173.32	236.90	–
Masoli	Maharashtra	1982	34.08	26.94	27.50	–
Naleshwar	Maharashtra	1918	12.35	11.18	16.90	–
Navegaon	Maharashtra	1987	45.94	29.59	–	–
Ozarkhed	Maharashtra	1982	67.96	60.32	58.90	–
Palkhed	Maharashtra	1975	–	21.22	549.30	–
Pandharabodi	Maharashtra	1967	13.80	13.04	24.60	–
Pus	Maharashtra	1972	113.92	91.25	93.60	–
Ramtek	Maharashtra	1913	105.15	104.26	109.20	–
Saikheda	Maharashtra	1969	38.51	27.18	31.20	–
Sidheswar	Maharashtra	1962	250.00	81.00	615.00	–
Sindhaphana	Maharashtra	1965	12.59	10.80	17.80	–
Sonal	Maharashtra	1981	20.27	16.93	31.60	–
Sukhana	Maharashtra	1968	21.34	18.49	25.10	–
Tawaraja	Maharashtra	1984	20.50	16.96	40.40	–
Tirana	Maharashtra	1969	22.00	18.62	21.80	–
Tiru	Maharashtra	1974	23.31	15.95	32.90	–
Upper Dudhana	Maharashtra	1965	15.37	13.00	34.90	–
Umari	Maharashtra	1972	231.00	172.10	11.80	–
Waghadi	Maharashtra	1979	41.11	35.37	58.40	–
Wan	Maharashtra	1969	25.18	21.91	52.60	–
Wunna	Maharashtra	1968	23.56	21.64	12.10	18
Yeldari	Maharashtra	1967	966.00	810.00	–	65
Nalkangeri Dam	Orissa	–	39.48	30.84	–	–
Umerkota Project	Orissa	1965	29.61	27.14	50.70	–

Table 31. Salient features of selected under construction projects in Godavari basin

Name of the Project	State	Gross storage capacity (MCM)	Live storage capacity (MCM)	Designed annual irrigation (Million m^2)	Installed capacity (MW)
Sathunal Project	Andhra Pradesh	35.14	28.61	76.80	–
Talliperu	Andhra Pradesh	20.67	14.46	98.00	–
Velleru	Andhra Pradesh	682.40	406.00	580.00	–
Vottivagu	Andhra Pradesh	82.06	74.96	98.00	–
Chulkinala Project	Karnataka	26.57	15.06	40.50	–
Kalisarar	Madhya Pradesh	32.47	27.75	10.00	–
Kanhargaon	Madhya Pradesh	25.50	21.60	389.00	–
Adan	Maharashtra	78.32	67.25	107.20	–
Amal Nalla	Maharashtra	22.70	21.20	39.00	–
Anjana Palsi	Maharashtra	15.55	13.71	25.50	–
Arunavati	Maharashtra	211.85	175.40	310.00	–
Bawanthadi	Maharashtra	298.75	204.79	250.00	–
Bham	Maharashtra	75.42	69.76	–	–
Bhavali	Maharashtra	44.75	40.79	450.00	–
Channa	Maharashtra	14.79	13.99	21.20	–
Dham	Maharashtra	72.46	62.51	102.10	–
Dongargaon	Maharashtra	14.17	12.44	12.50	–
Erdha	Maharashtra	15.27	11.96	25.10	–
Gosikhurd	Maharashtra	732.41	552.44	1,900.00	–
Human	Maharashtra	304.76	273.26	360.00	–
Isapur	Maharashtra	1,250.00	950.00	–	–
Karwa	Maharashtra	59.59	52.91	100.00	–
Karwappa	Maharashtra	32.57	31.58	52.50	–
Khekaranala	Maharashtra	26.13	23.71	33.10	–
Lendi	Maharashtra	207.95	159.66	200.00	–
Lower Dudhana	Maharashtra	344.80	242.00	300.00	–
Lower Pus	Maharashtra	81.10	59.67	96.80	–
Lower Terna	Maharashtra	133.56	122.46	200.00	–
Lower Wardha	Maharashtra	253.34	216.87	530.00	–
Majalgaon	Maharashtra	453.64	373.64	938.90	–
Masalga	Maharashtra	14.68	11.95	24.30	–
Mukane	Maharashtra	139.76	122.48	–	–
Nand Storage	Maharashtra	66.00	55.60	–	–
Pakadigudam	Maharashtra	13.31	11.80	37.10	–
Pothra	Maharashtra	38.43	34.73	63.10	–
Punegaon	Maharashtra	20.39	17.58	670.00	–
Shivana Takali	Maharashtra	113.61	42.42	74.90	–
Tambapuri	Maharashtra	21.27	19.63	47.80	–
Tultuli	Maharashtra	186.76	160.45	300.00	–
Upper Pravara	Maharashtra	252.54	244.62	670.00	–
Upper Wardha	Maharashtra	786.40	614.69	800.00	–
Vishnupuri	Maharashtra	–	83.55	–	–
Wedgaon Storage	Maharashtra	160.42	138.72	–	–
Waki	Maharashtra	58.45	55.96	–	–

Table 32. Flood stages at various gauging sites of Godavari River during August 1986

SN	Name of gauging site	Initially level attained	Date & time	Peak level attained	Date and time	Date and time when water level receded below danger level
1	Tekra	DL 13.0 m	12.08.1986 (21.00 hrs)	19.6 m historical	15.08.1986 (03.00 hrs)	18.08.1986 (15.00 hrs)
2	Perur	DL 13.0 m	12.08.1986 (24.00 hrs)	19.3 m	15.08.1986 (03.00 hrs)	18.08.1986 (24.00 hrs)
3	Bhadrachalam	DL 16.2 m	13.08.1986 (03.00 hrs)	23.0 m all time high	16.08.1986	19.08.1986 (12.00 hrs)
4	Kunavaram	DL 13.4 m	12.08.1986 (15.00 hrs)	24.4 m	16.08.1986 (21.00 hrs)	After 20.08.1986
5	Dowlaiswaram	DL 4.3 m	13.08.1986 (09.00 hrs)	6.55 m all time high	16.08.1986	20.08.1986 (03.00 hrs)

Source: Pandharinath (1987).
DL: Danger Level

Table 33. Normal average depth of rainfall and actual flood period rainfall depths in various sub-basins of the Godavari River basin

SN	Name of sub-basin	Normal average depth of rainfall (cm)				Flood period average depth of rainfall (cm)	
		June	July	August	September	9–11 August 1981	8–14 August 1986
1	Godavari basin upstream of Nanded	14.3	19.3	13.9	17.1	2.3	5.47
2	Manjira	12.5	20.0	17.9	22.1	0.6	10.07
3	Penganga	17.5	28.2	20.5	18.1	9.0	16.65
4	Wardha	16.9	33.3	24.2	18.9	10.0	22.77
5	Wainganga	18.9	44.6	35.8	21.1	13.3	23.71
6	Indravati	22.1	45.7	40.6	26.8	11.6	26.35
7	Sabari	20.3	38.9	37.8	25.7	3.5	31.74
8	Godavari basin Downstream of Nanded	16.3	29.5	23.9	20.3	2.8	35.45
9	Whole Godavari basin	17.25	32.67	26.74	21.0	6.77	22.00

Source: Pandharinath (1987).

In addition to the above system, an east-west oriented trough in the lower and mid-troposphere was observed running across the basin on 8th, 9th and 10th. Subsequently this trough was observed only in the mid-troposphere till 12th and became unimportant on 13th. Under the influence of these two systems, heavy to very heavy rain falls were reported at many places over the eastern and central part of the basin.

According to available records since 1881, the highest gauge level recorded at Dowlaiswaram was 5.95 m recorded on 15th August 1953. But this record was surpassed on 16th August 1986 when a gauge level of 6.55 m was recorded. The flood level at Bhadrachalam touched an all time high record of 24.4 m on 16th August 1986.

Available records show that the floods during August 1986 affected 18 districts of Andhra Pradesh. Besides taking a toll of 161 human lives and over 4,000 heads of cattle, it damaged about 12.5 lakh acres of cropped area, affected 2,321 villages and inflicted damage to about 1.05 lakh houses. A 70 m breach took place in the flood embankment of Vasistha River, a tributary of Godavari, at Gopalapuram in East Godavari district. On the night of 18th August 1986 Dowlaiswaram anicut was breached, inundating 12 villages. At Palocole, the railway track was under 2 m of water while Dowlaiswaram approach road was under 3.3 m of water. Polavaram and Kunavaram towns were completely submerged under floodwaters. About 5.5 lakh people had to be evacuated to safer places.

CHAPTER 15

CAUVERY AND PENNAR BASINS

15.1. THE CAUVERY BASIN

The Cauvery River, also known as Dakshin Ganga or 'Ganga of South', is one of the major interstate peninsular rivers of South India. It is the fourth largest river in the Indian peninsula next only to Godavari, Mahanadi, and Krishna. The Cauvery River rises in the Western Ghats and flows in eastwardly direction passing through the states of Karnataka, Tamil Nadu, Kerala and Pondicherry before it drains into Bay of Bengal. The basin lies between latitudes 10°05′ N and 13°30′ N and longitudes 75°30′ E and 79°45′ E. It is bounded on the west by the Western Ghats, on the east and south by the Eastern Ghats and on the north by the ridges separating it from the Tungabhadra (Krishna) and Pennar basins. The total length of the river from source to its outfall into Bay of Bengal is about 800 km. Of this, 320 km is in Karnataka, 416 km is in Tamil Nadu and 64 km forms the common boundary between Karnataka and Tamil Nadu States. The principal tributaries of the river are the Kabini, the Suvarnavathi, the Shimsha, the Arkavathi, the Chinnar, the Palar, the Bhavani, the Noyil, the Tirumanimuttar, the Amaravathi and the Ponnanai Ar. Krishnarajasagar, Mettur and Grand Anicut are the existing major projects in the basin.

The Cauvery basin extends over an area of $81,155\,\text{km}^2$, which is nearly 24.7% of the total geographical area of the country. An index map of the basin is given in Figure 1. The basin lies in the States of Tamil Nadu, Karnataka, Kerala and Pondicherry. The shape of the basin is somewhat rectangular with a maximum length and breadth of 360 km and 200 km, respectively. The statewise distribution of drainage areas is given in Table 1.

Physiographically, the basin can be divided into three parts: the Western Ghats area, the Plateau of Mysore and the Delta. The delta area is the most fertile tract in the basin. The principal soil types found in the basin are black soils, red soils, laterites, alluvial soils, forest soils, and mixed soils. Red soils occupy large areas in the basin. Alluvial soils are found in the delta areas. The culturable area of the basin is about $58,000\,\text{km}^2$ which is about three percent of the culturable area of the country.

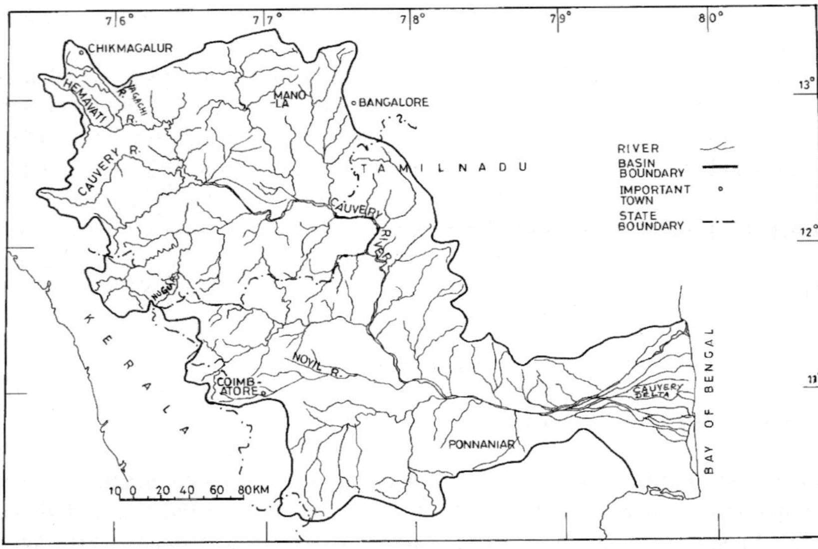

Figure 1. Index map of Cauvery basin

15.1.1. The Cauvery River

The Cauvery (Kaveri) River is one of the sacred interstate rivers of India which originates at Talakaveri on the Brahmigiri Hills (12°75′N latitude and 74°34′E longitude) of the Coorg district in Karnataka State at an elevation of 1,341 m. The river flows down from the hills in a series of rapids and cascades and is joined at the foot of the hills by the Kannike stream at Bhagamandala. The course of the river through Coorg district is very tumultuous.

The Krishnarajasagar dam has been constructed across the Cauvery about 19 km north-west of Mysore City. The Shivasamudram Falls, which are the second largest waterfalls in India, are on the course of the Cauvery River. At Sivasamudram, the river divides into two branches and falls through a height of more than 91 m in a series of major falls, notably, the Gagana Chukki and the Bhara Chukki. Hydroelectric energy is produced at these falls. Further, the picturesque Hogenakal falls lie on the border

Table 1. Statewise distribution of drainage area of Cauvery basin

State	Drainage area (km^2)	Percentage of area
Tamil Nadu	43, 867	54.1 %
Karnataka	34, 273	42.2 %
Kerala	2, 866	3.5 %
Pondicherry	149	0.2 %
Total	81, 155	100%

between Karnataka and Tamil Nadu. Below the falls, near the town of Salem, the 154 meter high Mettur Dam impounds water in the Stanley Reservoir.

The important tributaries joining Cauvery in the Coorg district are the Kakkabe, the Kadanur and the Kummahole. Cauvery then enters the Mysore district where important tributaries joining the river from the left are the Harangi, the Hemavathi, the Shimsha and the Arkavathi. The tributaries joining it from the right are the Lakshmanathirtha, the Kabbani, and the Survanavathi. Further down, the river enters Tamil Nadu state where the many tributaries, namely, the Bhavani, the Noyil, and the Amravathi join it. The salient features of the major tributaries in the Cauvery basin are given in Table 2. The flow Diagram of Cauvery is given in Figure 2.

15.1.2. Sub Basins Under Cauvery Basin

The entire Cauvery catchment can be divided into 16 sub-basins, viz., (1) the Upper Cauvery [from the source to the Krishnarajsagar (KRS) dam], (2) the Kabini, (3) the Shimsha, (4) the Arkavathi, (5) the Middle Cauvery (the catchment of main Cauvery from KRS dam at the upstream end to just below the confluence of Arkavathi River with Cauvery), (6) the Suvarnavathi, (7) the Palar, (8) the Chinnar, (9) the Bhavani, (10) the Noyil, (12) the Tirumanimuttar, (11) the Amaravathi, (13) the Ponnanai

Table 2. Salient features of tributaries of Cauvery River

SN	Name of the tributary	Catchment area (km^2)	Length (km)	Origin and altitude	Sub-tributaries	Basin state(s)
1	Harangi	717	50	Pushpagiri Hills of Western Ghats 1,067 m		Karnataka
2	Hemavathy	5,410	245	Ballarayana Durga in western Ghats, 1,219 m		Karnataka
3	Kabini	7,040	230	Westren Ghats in Kerala, 2,140 m	Taraka, Hebballa, Nugu, Gundal	Karnataka, Kerala & Tamil Nadu
4	Suvarnavathy	1,787	88	Nasrur Ghat range		Karnataka & Tamil Nadu
5	Lakshmanathirtha	–	131	Westren Ghats, 1,950 m	Ramathirtha	Karnataka
6	Shimsha	8,469	221	Tumkur district, 914 m	Veeravaishnavi, Kanihalla, Chickkhole, Hebbahalla, Mullahalla, & Kanva	Karnataka
7	Arkavathy	4,351	161	Nandidurga 1,480 m	Kumaudavathy, Manihalla & Kuttehole, Vrishabhavathy	Karnataka & Tamil Nadu

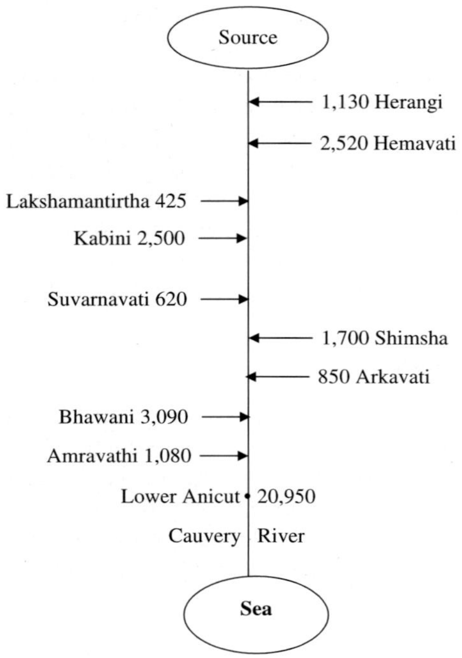

Figure 2. Flow Diagram of Cauvery River

Ar., (14) the Upper Coleroon, (15) the Lower Coleroon, and (16) the Cauvery Delta. The sub-basin wise drainage area, rainfall, runoff and groundwater potential are given in Table 3. Table 4 represents annual average observed runoff at important CWC gauging sites in Cauvery basin.

Upper cauvery sub-basin

The Upper Cauvery Sub-basin lies between latitudes 10°54′ and 13°21′N and longitudes 75°30′ and 76°36′E comprising the catchment of the Hemavathi, the Laxmanthirtha and the main Cauvery River up to KRS dam. The total length of the Cauvery River in this sub-basin is about 224 km. The lengths of the Harangi, Hemavathi and Laxmanthirtha, all tributaries of Cauvery River in this sub-basin from their origin to their confluence with the Cauvery are 48 km, 245 km, and 131 km, respectively.

The Upper Cauvery sub-basin drains an area of $10,619\,\text{km}^2$ which is about 13% of the total catchment area of the Cauvery basin. The catchment of the Upper Cauvery sub-basin lies wholly in Karnataka covering parts of the Chickmaglur, Kodgu, Hassan, Mandya, and Mysore districts.

The Harangi River: The Harangi rises in the Pushpagiri Hills in Coorg district and after draining a catchment area of $717\,\text{km}^2$, it joins Cauvery near Kudige in the Somawarpet taluk. Heavy rainfall from the south-west monsoon is received in the catchment area of Harangi.

Table 3. Sub-basin wise Area, average annual Rainfall, Runoff at 75% Water Year Dependable Flow and Ground Water Potential

S. N.	Name of Sub-basin	Area (km^2)	Rainfall (mm)	Runoff (75% water year dependable flows, Mm^3)	Ground water potential (Mm^3)
1	Upper Cauvery	10, 619	1, 025	5, 394	578.5
2	Kabini	6, 810	1, 097	3, 641	386.4
3	Shimsha	8, 469	656	619	506.0
4	Arkavathi	4, 351	451	287	103.5
5	Middle Cauvery	2, 676	424	330	205.6
6	Suvarnavathi	1, 787	438	238	63.9
7	Palar	3, 214	469	105	139.7
8	Chinnar	4, 061	653	312	177.6
9	Bhavani	6, 154	908	1, 917	187.7
10	Noyil	2, 999	504	225	54.2
11	Amaravathi	8, 280	572	898	308.0
12	Tirumanimuttar	8, 429	536	649	350.1
13	Ponnanai Ar	2, 050	542	191	207.4
14	Upper Coleroon	3, 082	656	589	252.6
15	Lower Coleroon	1, 378	815	224	120.6
16	Cauvery Delta	6, 566	810	1, 051	224.6

The Hemavathi River: Hemavathi is one of the principal tributaries to join the Cauvery on its northern bank. It rises in Ballalarayanadurga in the Western Ghats in the Mudigere Taluk of Chickmagalur district of Karnataka. The Western Ghats region is mountainous area covered with thick vegetation. The Hemavathi joins the Cauvery on its left bank after traversing a length of 193 km in Hassan and Mudigere Districts of Karnataka in the water spread of Krishnarajasagar reservoir near Akkihebbal. The important tributaries of the Hemavathi are the Yagachi and the Algur. The drainage area of the Hemavathi River is 5, 410 km^2.

The Hemavathi River, in its early reaches, passes through a very heavy rainfall region in the vicinity of the Kotigere and Mudigere. The Yagachi River flows in a meandering course along NNW-SSE to SSE-NNE directions, and joins the Hemavati

Table 4. Annual average observed runoff at CWC sites in Cauvery basin (Catchment area >5, 000 km^2)

Name of the site	Name of the stream	Catchment area (km^2)	Annual average runoff (BCM)
Musiri	Cauvery	66, 243	14.7
Nallamaranpatti	Amaravathi	9, 080	2.3
Kodumudu	Cauvery	53, 233	15.0
Savandapur	Bhawani	5, 776	1.2
Urachikottai	Cauvery	45, 235	13.4
Biligundulu	Cauvery	36, 682	13.0
T. K. Halli	Shimsha	7, 890	1.3
T. Narsipur	Kalini	7, 000	5.1

River at Gorur. The Algur River joins Hemavathi from south near Algur. Numerous small streams also join the Hemavati River all along its course. The annual rainfall of the area varies from 762 to 5,080 mm with an average annual of 2,972 mm.

The Kabini sub basin

The Kabini River is the most important south-bank tributary of the Cauvery in its upper reach. The Kabini sub-basin lies between latitudes 11°29′ N and 12°20′ N and longitudes 75°48′ E and 75°54′ E. The Kabini River rises in the Western Ghats in the Wyand district of Kerala State at an elevation of about 2,140 m above mean sea level and after draining an area of $6,810\,km^2$, it joins the Cauvery at Tirumakudal Narasipur in Mysore district. The length of the river is 210 km. Kabini River drains a hilly and forested catchment in the upper reaches where rainfall is heavy.

Shimsha sub-basin

The Shimsha sub-basin lies between the latitudes 12°18′ N and 13°30′ N and the longitudes 76°15′ E and 77°19′ E. The Shimsha River, one of the important tributaries of Cauvery, rises in the south of Devarayanadurga hill in Tumkur district. After flowing southwest in the initial reach, it turns to southwards and then to east. Thereafter pursuing a southerly course, the river enters the Mandya district. Moving further, it finally takes southeasterly course and joins the Cauvery, a few km below the Shivasmudram falls. The total length of the Shimsha River from its origin to confluence with Cauvery is about 200 km. The Shimsha sub-basin has a catchment area of $8,469\,km^2$, which constitutes 10.4% of the Cauvery basin area. The entire catchment area of the sub-basin lies in the Karnataka State.

Arkavathi sub-basin

The Arkavathi sub-basin lies between the latitudes 12°16′ N to 13°23′ N and longitudes 77°11′ E to 77°42° E. The Arkavathi River is one of the important tributaries of the Cauvery River. It rises at Nandidurga hills in Chikballpur Taluk of Kolar district. After flowing in the southwest direction in the initial stage, it receives flow of Kumudvathi River. From this point, the Arkavathi River flows in a southerly direction up to Ramanagaram and turns towards southeast. It then flows in the same direction up to the confluence of its tributary Suvarnamukhi on the left bank. Thereafter, it flows in southerly direction and receives the water of Kuttlehole from the left near Kanakapura town. Further, it flows down and receives the waters of Doddahalla from the left and then finally joins Cauvery at Kungedoddi. The total length of the Arkavathi River is about 150 km from its origin to its confluence with Cauvery. The Arkavathi sub-basin has a catchment area of $4,351\,km^2$ that constitutes 5.4% of the Cauvery basin area. The major portion of the sub-basin amounting to $4,184\,km^2$ lies in Karnataka state and the rest $167\,km^2$ falls in Tamil Nadu.

Middle Cauvery sub-basin

The Middle Cauvery sub-basin lies between the latitude 11°52′ N to 12°48′ N and longitude 76°30′ E to 77°29′ E comprising the catchment of main Cauvery from Krishnarajasagar dam to the state boundary just below Mekedatu gorge on the main River.

Below the Krishnarajasagar dam, the Cauvery River continues to flow eastwards for 15 km up to Srirangapatnam and then changes its course south-eastwards. It receives an important tributary, viz., the Kabini on its right bank at Triumakudal Narasipur and another tributary, namely, the Suvarnavathi joins it from the right at Talakad, about 25 km downstream. The river then takes a north-east direction and receives the Shimsha from the left below Sivasmudram. After this point the river starts cutting through the Eastern Ghats and from a width of nearly one kilometer, the cross-section narrows considerably. It then flows in cascades through a gorge. At Sivasmudram, the river gets divided into two branches and falls through a height of more than 91 m in a series of falls and rapids. The two major falls are the Ganga Chukki and Bhara Chukki. The fall of river at this point is being utilized for the generation of hydro electricity power. The Sivasmudram power station built in 1902 is one of the earliest hydroelectric power stations in Asia.

The two branches of the river join after the falls and flow through a gorge, known as Mekadatu gorge. This gorge is quite narrow. After flowing through this gorge, Cauvery continues its eastward journey and forms the boundary between Karnataka and Tamil Nadu states for a distance of about 64 km. Another left bank tributary, viz., the Arkavathi joins the river just before it enters into Tamil Nadu state. The total length of the Cauvery River is about 130 km from Krishnarajasagar dam to the state boundary below Mekedatu gorge. Middle Cauvery sub-basin has catchment area of $2,676\,\text{km}^2$, which constitutes 3.03% of the Cauvery basin area. The entire catchment area of the sub-basin lies in the Karnataka State.

The Suvarnavathi sub basin

The Suvarnavathi River is one of the southern tributaries of the Cauvery in its upper reaches. It is the second tributary joining on right bank. The Suvarnavathi sub-basin lies between latitudes 11°35′ N to 12°10′ N and longitudes 76°46′ E to 77°12′ E. Suvarnavathi originates in the Nasurghat range of hills situated in the south eastern portion of Mysore district near Gajjala hatti valley and flows northwards through Chamrajasagar and Yelandur taluks. Two streams, viz., Niredurgihalla and Araikaduhalla join together near Badibadga to form the river Suvarnavathi or Honhole. This river, after flowing for a further distance of 11 km, is joined by a tributary known as the Chikkahole. After flowing further for a distance of about 15 km, it is joined by another tributary, the Yenehole from left side. The Suvarnavathi River finally joins Cauvery on its right side at Talakad in the Kollegal taluk. From its origin to its confluence with Cauvery, the total length of Suvarnavathi River is about 88 km. The Suvarnavathi sub-basin covers an area of $1,787\,\text{km}^2$ in the states of Karnataka and Tamil Nadu. It forms 2.2% of the area of the Cauvery basin.

Palar sub-basin

The Palar River is one of the southern tributaries of the Cauvery. The Palar sub-basin lies between the latitudes 11°35′ N and 12°14′ N and longitude 77°10′ E and

77°50′ E. The Palar River rises in the hill ranges of Satyamangalam Taluk of Periyar district and flows northwards till it receives a small tributary, namely Moranur Halla from west where river turns perpendicularly to the east and finally joins the Cauvery on right side near the upstream of Mettur reservoir. The Palar River forms the common boundary between Karnataka and Tamil Nadu in Mysore and Periyar districts, respectively, for about 45 km. The drainage area of Palar sub-basin is 3,214 km^2, lying in the states of Karnataka and Tamil Nadu.

Chinnar sub-basin

The Chinnar sub-basin lies between the latitudes 11°45′ N to 12°45′ N and longitudes 77°25′ E to 78°20′ E. The sub-basin comprises the catchments of four independent streams, namely Chinnar, Doddahalla, Nagavathi and Thoppaiar. The Chinnar River is the main among them. The catchment area of the Chinnar sub-basin as a whole is 4,061 km^2, which constitutes 5% of the total catchment of the Cauvery basin. Most of the sub-basin area lies in Dharmapuri district in Tamil Nadu at an elevation ranging from 300 to 900 m above mean sea level.

Bhavani sub-basin

The Bhavani River is one of the tributaries of Cauvery River in its mid-reach whose sub-basin lies between latitudes 10°56′3″ N and 11°46′14″ N and longitudes 76°24′41″ E and 77°41′11″ E. The Bhavani River rises at an altitude of about 2,634 m in the Billimala range of Nilgiri hills in Tamil Nadu. The total drainage area of the river is 6,154 km^2 and it flows for a distance of 216 km before joining the Cauvery from the right at Bhavani in Coimbatore district. The important tributaries of the Bhavani are the Siruvani, the Kundah, the Coonoor and the Moyar.

Noyil sub-basin

This sub-basin comprises the catchment of Noyil River including the catchment of its small tributaries, viz., Sanganurpallan, Vannattangarai, Nallar and Chinnakarai. The Noyil sub-basin lies between latitudes 10°54′ N and 11°19′ N and longitudes 76°39′ E and 77°56′ E. The sub-basin area includes a part of the command area of Lower Bhavani project canal, Kalingarayan channel and Perimbikulam main canal. The Noyil sub-basin is bounded on the north by Bhavani sub-basin, on the south by Amaravathi sub-basin, on the east by Cauvery River and on the west by the Western Ghats. The river flows entirely in Tamil Nadu and the basin is spread over the districts of Coimbatore, Periyar and Tiruchirapalli. The area of the basin is 2,999 km^2, which constitutes 3.7% of the area of the Cauvery basin.

Amaravathi sub-basin

The Amaravathi River is among the main tributaries of the Cauvery River in its mid reach. It is a right bank tributary next to Noyil, downstream of Mettur dam in Tamil Nadu. The Amaravathi River originates from Naimakad at an elevation of 2,300 m in the Southern Ghat in Devikulam taluk of Iddukki district of Kerala State. A number of streams join the river in Kerala before its entry into Tamil Nadu. The

Amaravathi sub-basin lies between latitudes 10°6′ N and 11°2′ N and longitudes 77°3′ E and 78°6′ E. It is bounded by Noyil sub-basin in the north, Vaigai basin in south, the southern part of the Western Ghats in the west, and the Cauvery River in the east. Except that part of the upper hilly catchment of the sub-basin which lies in Kerala, the rest of the sub-basin is spread over Tamil Nadu. The area of the basin is 8, 280 km^2 which constitutes 10.2% of the area of the Cauvery basin.

Tirumanimuttar sub-basin

The Tirumanimuttar sub-basin lies between latitudes 10°36′ N and 11°55′ N and 77°27″ E and 78°41′ E. The basin comprises the catchment of Cauvery River below Mettur dam on both sides up to Upper Anicut including the sub-catchment of Sarabhanga Nadi, Tirmnanimuttar, Pungar and Ayyar, but excluding the catchments of Bhavani, Noyil and Amaravathi rivers. The area of the Tirumanimuttar sub-basin is 8, 429 km^2, which is 10.39% of the total area of Cauvery basin. The sub-basin lies mostly in Salem and Tiruchirapalli districts of Tamil Nadu.

Ponnanai Ar sub-basin

The Ponnanai Ar sub-basin comprises the catchment of the Ponnanai Ar and its tributaries on the right side of the Cauvery River between Upper Anicut and Grand Anicut. The Ponnanai Ar sub-basin is bounded on the north by Upper Coleroon subbasin, on the west by Tirumanimuttar sub-basin, on east by Cauvery Delta and on the south by the basin covering the area between the Cauvery and Vaigai river basins. The sub-basin lies between North latitudes 10°25′50″ and 10°53′25″ and East longitudes 78°08′00″ and 78°50′00″. The Ponnanai Ar River originates in the scattered hills near Kadavur in Kulittalai taluk of Tiruchirapalli district and flows in a northeast direction through Kultittala, Manapparai and Tiruchirappali taluks. In the middle reach, the river Ponnanai Ar is also known as Ariyar and it is known as Kodamurutti Ar in the lower reach. The Ponnanai Ar sub-basin covers an area of 2, 050 km^2 which is 2.53% of the area of the Cauvery basin.

Upper coleroon sub-basin

The Upper Coleroon Sub-basin covers the direct catchment of the Coleroon River around the Upper Anicut and lies between the latitudes 10°50′ N and 11°15′ N and longitudes 78°35′ E and 79°27′ E. The sub-basin also comprises of the catchment of a few independent streams such as Upper Marudaiyar, Nandiyar, Nari Odai, Andi Odai etc. The area of the sub-basin as a whole is 3, 082 km^2, which constitutes 3.8% of the total catchment area of the Cauvery basin. Most of the sub-basin area lies in Tiruchirapalli district but a sma1l part lies in Thanjavur district of Tamil Nadu.

Lower coleroon sub-basin

The Lower Coleroon sub-basin with a catchment area of 1, 378 km^2 covers the catchment of the Coleroon River below the lower Coleroon Anicut and it extends up to its confluence with the sea. This sub-basin area lies entirely in Tamil Nadu state between latitudes 11°08′ N and 11°25′ N and longitudes 79°13′ E and 79°48′ E. The area of this sub-basin is $1,378\,\text{km}^2$ which is about 1.7% of the total catchment of the Cauvery basin. Most of the sub-basin area lies in South Arcot and Tiruchchirappalli districts and a small part lies in Thanjavur district.

Cauvery delta

The Cauvery delta comprises the command area of the Vennar branch, Cauvery branch and part of the Grand Anicut canal irrigation system. The sub basin also covers a part of the command area of Kattalai canal scheme and the New Kattalai high-level canal scheme. The Cauvery Delta sub-basin is bounded on the north by the Upper Coleroon and Lower Coleroon sub basins, on the west by the Tirumanimuttar sub-basin and Ponnanai Ar sub-basin, on the south by Palk Strait and the area covered by the streams between Cauvery and Vaigai and on the east by the Bay of Bengal. The sub-basin lies between latitudes 10°17′ N and 11°22′ N and longitudes 78°48′ E and 79°53′ E.

Downstream of the Grand Anicut (Upper end of Cauvery Delta sub-basin), the Cauvery River subdivides itself into two main branches, namely, Cauvery and Vennar System, which get further sub-divided into 36 rivers to feed the delta through a network of channels and branches, distributaries and sub-distributaries. The Cauvery Delta is spread over $6,566\,\text{km}^2$ which is 8.09% of the total area of the Cauvery basin.

15.1.3. Climate and Geology of the Cauvery Basin

Cauvery basin experiences tropical climate. Here, the main climatic feature is the monsoon rain. The north-east monsoon provides the greater portion of the annual precipitation. The far north-western part of the drainage basin has a per-humid climate which passes eastwards into humid, moist sub-humid, dry sub-humid and semi-arid zones. The recorded maximum and minimum temperatures are 44 °C and 18 °C respectively. The basin experiences four distinct seasons, namely: south west monsoon; north east monsoon; cool, dry winter; and hot, dry summer. These four seasons fall from June to September, October to December, January to February, and March to May respectively.

Rainfall

Over the basin, the highest rainfall is received along the western border of the basin during the southwest monsoon. The eastern side of the basin gets most of the rain during the northeast monsoon. Depressions in the Bay of Bengal affect the basin in the monsoon, causing cyclones and widespread heavy rains. To observe climatic variables, IMD has established 11 observatories in the basin and there are 352 raingauging stations in an d around the basin.

Geology

The geology of the drainage basin is predominantly formed from Precambrian rocks, principally the Dharwars, Peninsular granitic Gneiss, Charnockites and the Closepet

Granite. The Dharwar metamorphics mainly comprise of phyllites, slates, schists with chlorite, biotite, garnet, and hornblende. Accompanying these are greenstones and quartzite. The Closepet Granite of the upper reaches of the Cauvery basin is a pink granite consisting mainly of quartz, plagioclase, microcline, perthite, and subordinate hornblende. Over the main basin, the peninsular granites and gneisses comprising of biotite granitic gneiss, hornblende granitic gneiss are widely found. The Charnockites are confined to the Nilgiri Range in the central part of the drainage basin. These are represented by gabbros, olivine norites, and pyroxene. Cretaceous sediments crop out in the coastal region and consist of conglomeratic sandstone, coralline limestone, and shale.

15.1.4. Surface Water Resources

Water balance studies for 75%, 50%, 90% and 100% dependable flows

Patil (2004) has carried out sub-basin wise monthly water balance studies for Cauvery basin, with and without considerations of ground water for a 75% dependable year. The sub-basin wise water balance studies for 90%, 100% for water deficit year and 50% water year dependable flows for water surplus year were also carried out.

The sub-basin wise results of water balance are as follows.

i. *Upper Cauvery Sub-basin*: By comparing monthly water balances in the Upper Cauvery sub-basin, it is found that the months of July and May are surplus and months of September, December and February are water deficit in terms of the water needs in the sub-basin, with the consideration of ground water availability. The maximum surplus of water is 704 Mm^3 which occurs in the month of July for the 75% dependable flow and the maximum water deficit of 291 Mm^3 occurs in the month of August for the 100% water year dependable flow considering ground water availability. The amount of annual surplus and annual deficit water has decreased from 3,940 Mm^3 and 286 Mm^3 to -262 Mm^3 and $-1,935$ Mm^3 in the sub-basin with the ground water availability for 50%, 75% 90% and 100% water year dependable flows, respectively.

ii. *Kabini Sub-basin*: In the Kabini sub-basin, months of June, October and May have surplus water in comparison to the water needs in the sub-basin, with the consideration of ground water and the months of December, January, February and March are water deficit. The maximum water surplus is 446 Mm^3 in the month of May for 100% dependable flow and the maximum water deficit is 203 Mm^3 in the month of January for 75% dependable flow. The sub-basin faces water shortage in the month of August.

iii. *Shimsha Sub-basin*: In the Shimsha sub-basin, in most months, water availability is in excess than water requirements; very few months are deficit in water if the ground water availability is considered. The maximum water surplus of 372 Mm^3 is in the month of October for 50% water year dependable flow and maximum water deficit of 4 Mm^3 is in the month of April for 100% water year dependable flow with the ground water availability considerations.

iv. *Arkavathi Sub-basin*: In the Arkavathi sub-basin, in all the months there is deficit in the water availability in comparison to the water needs, with the consideration of ground water availability. For 90% water year dependable flow, the maximum water deficit of $226\,Mm^3$ occurs in the month of September.

The amount of annual water deficit in the sub-basin is $594\,Mm^3$, $713\,Mm^3$, and $796\,Mm^3$ with the ground water availability for 50%, 75%, and 90% water year dependable flows, respectively.

v. *Middle Cauvery Sub-basin*: In the Middle Cauvery sub-basin, the months of June and May are deficit in the water availability in comparison to the water needs, and other months have surplus in the water availability. The maximum water surplus of $58\,Mm^3$ occurs in the month of September for 50% water year dependable flow.

The amount of annual surplus water is $331\,Mm^3$, $269\,Mm^3$, $215\,Mm^3$ and $128\,Mm^3$ in the sub-basin with the ground water availability for 50%, 75%, 90% and 100% water year dependable flows, respectively.

vi. *Suvarnavathi Sub-basin*: In the Suvarnavathi sub-basin, the months of September, October and November are surplus for 75% and 50% water year dependable flows. In the other months, there is a deficit in the water availability in comparison to the water needs. The maximum water surplus of $19\,Mm^3$ is in the month of October in 50% water year dependable flow and for 90% and 100% water year dependable flows, sub basin has water deficit throughout the year.

vii. *Palar Sub Basin*: The months of November to May have deficit in the water availability in the Palar sub-basin, with the consideration of ground water availability. The maximum water surplus of $29\,Mm^3$ is found in the month of September in 50% water year dependable flow and maximum water deficit of $18\,Mm^3$ is in the month of August in 100% water year dependable flow. In the month of August, the sub-basin becomes water deficit in 90% and 100% water year dependable flow from water surplus in 75% and 50% water year dependable flows.

viii. *Chinnar Sub-basin*: In the Chinnar sub-basin, all the months are deficit in the water availability, with the consideration of ground water availability. The maximum water deficit of $5,294\,Mm^3$ is in the month October in 50% water year dependable flow. The amount of annual water deficit in the sub-basin is $-15,787\,Mm^3$, $-12,932\,Mm^3$, $-13,012\,Mm^3$ and $-13,155\,Mm^3$ with the ground water availability for 50%, 75%, 90% and 100% water year dependable flows, respectively.

ix. *Bhavani Sub-basin*: In the Bhavani sub-basin, the months of October, November and December are surplus and the months of June, August, February, March and April are deficit in the water availability, with the consideration of ground water availability. The maximum water surplus, $157\,Mm^3$ is in the month of October in 50% water year dependable flow and maximum water deficit of $144\,Mm^3$ occurs in the month of August in 100% water year dependable flow with the ground water availability considerations.

The annual surplus has decreased and turned to annual deficit from $239\,Mm^3$ to $-247\,Mm^3$, $-652\,Mm^3$ and $-947\,Mm^3$ in the sub-basin with the ground water availability for 50%, 75%, 90% and 100% water year dependable flows, respectively.

x. *Noyil sub-basin*: In the Noyil sub-basin, the months of October, November and December have surplus water while the months of June, July, August, September, January, February, March, April and May are deficit in the water availability when considering ground water availability. The maximum water surplus of 108 Mm^3 is in the month of October in 50% water year dependable flow and maximum water deficit of 45 Mm^3 is in the month of June in 50% water year dependable flow with the ground water availability considerations:

The amount of annual deficit water is −40 Mm^3, −52 Mm^3 and −64 Mm^3 in the sub-basin with the ground water availability for 75%, 90%, and 100% water year dependable flows, respectively.

xi. *Amaravathi sub-basin*: In the Amaravathi sub-basin, the months of October and November are surplus and the months of June, July, August, January, February, March and April are deficit in the water availability, with the ground water availability consideration. The maximum water surplus of 195 Mm^3 is noted in the month of November in 50% water year dependable flow and maximum water deficit of 172 Mm^3 is in the month of March in 50% water year dependable flow. In the month of December, the sub-basin becomes water deficit in the event of 90% and 100% water year dependable flows from water surplus in 50% and 75% water year dependable flows.

The amount of annual deficit water is −304 Mm^3, −454 Mm^3, −547 Mm^3 and −711 Mm^3 in the sub-basin with the ground water availability for 50%, 75%, 90% and 100% water year dependable flows, respectively.

xii. *Tirumanimuttar Sub-basin*: In the Tirumanimuttar sub-basin, the months of October, November and December are surplus and other months are deficit in the water availability, with the ground water availability consideration. The maximum water surplus of 531 Mm^3 is in the month of October for 50% water year dependable flow and in the month of March, deficit is 82 Mm^3 for 90% water year dependable flow.

The amount of annual surplus water has decreased from 583 Mm^3 to 219 Mm^3, 76 Mm^3 and 17 Mm^3 in the sub-basin with the ground water availability for 50%, 75% 90% and 100% water year dependable flows, respectively.

xiii. *Ponnanai Ar sub-basin*: In the Ponnanai Ar sub-basin, all the months are surplus in the water availability, with the consideration of ground water availability except the months of April and May in 100% water year dependable flow. The maximum water surplus 85 Mm^3 is in the month of November in 50% water year dependable flow, and maximum water deficit of 2 Mm^3 is in the month of April in 100% dependable flow.

The amount of annual surplus water has decreased from 244 Mm^3 to 217 Mm^3, 157 Mm^3 and 61 Mm^3 in the sub-basin with the ground water availability for 50%, 75%, 90% and 100% water year dependable flows, respectively.

xiv. *Upper Coleroon Sub-basin*: In the Upper Coleroon sub-basin, the months of October, November and December are surplus and the month of June, is deficit in the water availability, with the ground water availability consideration. The maximum water surplus of 192 Mm^3 is observed in the month of November in 50%

water year dependable flow and maximum water deficit of 37 Mm^3 is in the month of June in 100% water year dependable flow. In the month of August the sub-basin becomes water deficit in 90% and 100% water year dependable flows from water surplus in 50% and 75% year dependable flows.

xv. *Lower Coleroon Sub-basin*: In the Lower Coleroon sub-basin, there is no shortage of water in comparison to the water needs in any month. The maximum water surplus is 102 Mm^3 in month of September in 50% water year dependable flow.

xvi. *Cauvery Delta Sub-basin*: In the Cauvery Delta sub-basin also there is no deficit in the water availability in comparison to the needs, with the ground water availability consideration, except in the month of May in 90% and 100% dependable water years. The maximum water surplus 192 Mm^3 is in the month of September in 50% water year dependable flow.

15.1.5. Ground Water Potential

The Sub basin wise ground water potential of the Cauvery Basin is presented in Table 5.

It can be noticed from the table that total replenishable ground water resources of the basin are 10,357 MCM. With a net draft on 4,937 MCM, about 3,866 MCM of the resource is available for future use. Among the sub-basins, Noyil and Suvarnavathi have the smallest quantity of ground water left for future use.

Table 5. Sub basin wise ground water potential at Cauvery River Basin in India

Name of the Sub basin	Ground Water (Mm^3)				
	Total replenishable resource	Provision for domestic & industrial use @15%	Available resource for irrigation	Net draft	Balance for future use
Upper Cauvery	777.5	116.6	660.8	82.3	578.5
Kabini	715.3	107.3	608.0	221.6	386.4
Shimsha	929.7	139.5	790.3	284.3	606.0
Arkavathi	423.5	63.5	360	256.5	103.5
Middle Cauvery	286.9	43.0	243.8	38.2	205.6
Suvarnavathi	169.2	25.4	143.8	79.9	63.9
Palar	341.5	51.2	290.3	150.6	139.7
Chinnar	511.6	76.7	434.9	257.3	177.6
Bhavani	667.3	100.1	567.2	379.5	187.7
Noyil	406.4	61.0	345.4	291.2	54.2
Amaravathi	1, 048.7	157.3	891.4	583.3	308.0
Tirumanimuttar	1635.2	245.3	1, 389.9	1, 039.9	350.1
Ponnanai Ar	434.9	65.2	369.7	162.2	207.4
Upper Coleroon	618.8	92.8	526.0	273.4	252.6
Lower Coleroon	408.9	61.3	347.5	226.9	120.6
Cauvery Delta	981.0	147.2	833.9	609.3	224.6
Total Cauvery	10, 356.5	1, 553.5	8, 803	4, 936.5	3, 866.5

15.1.6. Major Water Resources Development Projects in the Basin

The utilization, area irrigated and districts benefited by the completed major & medium projects are given in Table 6. Description of selected major projects follows.

Krishnarajasagar dam

This is an existing major multipurpose project on Cauvery in Karnataka State. The latitude and longitude of the dam are 12°25′30″ N and 76°34′30″ E. It is located about 19.3 km from Mysore City near Kannambadi village, Srirangapatna Taluk, and district Mandya. The dam is situated below the confluence of the Cauvery River with two of its main tributaries, the Hemavathy and the Lakshmanathirtha. The catchment area of the dam is 10,619 km^2. The flow of the river at the site of the dam fluctuates from a normal high flood of 2,832 cumec during the monsoon season, to less than 3 cumec during the summer. The highest flood in the river which occurred in the year 1924 was 10,787 cumec.

KRS dam is a 2,620 m long gravity dam of stone masonry in *surki* mortar. The length of the dam at the top is 2621 m. There is no overflow spillway. The floods are disposed off through 152 sluice gates situated at different elevation in the body of the dam. Height of the dam above the bed level is 39.8 M. The reservoir formed by the dam has a gross and live storage capacity of 1,368.847 Mm^3 and 1,244.21 Mm^3 respectively. The water spread area of the reservoir at full reservoir level is 129 km^2. The reservoir fills during June to September period and supplies water during October to May period. The length of left bank canal is 45.92 km and the length of right bank canal is 32 km. The irrigation demand from the dam is 1, 325 Mm^3. Some other salient features of the project are given in Table 7. A View of KRS dam is shown in Figure 3.

Table 6. Utilization, area irrigated by completed major and medium projects in Cauvery basin

S. N.	Name of the project	Year of completion	Utilization in Mm^3	Irrigation in Mm^2	Districts benefited
	Major Projects				
1	Krishnaraja Sagar	1944	1, 732.57	793.12	Mandya, Mysore
2.	Nugu	1959	217.99	105.26	Mysore
	Medium Projects				
1	Byramangala	1945	28.31	16.19	Bangalore
2.	Chikkahole	1969	19.82	16.50	Mysore
3.	Gundal	1980	39.63	40.48	Mysore
4.	Hebbala (H.D. KOTE)	1972	11.32	12.14	Mysore
5.	Kanwa	1946	33.97	20.76	Bangalore
6.	Mangala	1970	16.99	8.50	Tumkur
7.	Marconahalli	1941	113.24	45.60	Tumkur
8.	Nallur Amanikere	1987	8.49	13.00	Mysore
9.	Suvarnavathy	1984	101.92	33.00	Tumkur

Table 7. Salient features of KRS Project

Design yield	5, 351 Mm^3 at 50% dependability
Full reservoir level (FRL)	38.04 m
Minimum drawdown level (MDDL)	22.56 m
Dead storage level	18.29 m
Sill level of canal sluices	
a) Right Bank low level	Canal Sluice 18.29 m
b) Left Bank low level	Canal Sluice 18.29 m
c) Sill level of river sluice	3.657 m
Length of Dam	2,620.365 m
Height of Dam	39.867 m above river bed level
Maximum release capacity	9,794 cumec through sluice gates

The credit to construct this project goes to the eminent engineer of India, Dr. M. Visweswaraiah. In recognition to his services to the nation, Dr. Visweswaraiah's birthday, September 15, is celebrated as engineer's day in India. The project was commenced during the year 1911 and commissioned during 1932. The expenditure incurred on the project was Rs. 91 billion. An area of 10,785 ha of culturable land and 25 villages were submerged due to this project. The reservoir is named after the late Sri Krishna Raja Wadiyar in whose illustrious reign its construction was undertaken. KRS was constructed with the twin objectives:

1. To ensure a steady supply of water for generating Hydro-Electric Power at Sivasamudram.
2. To supply water for irrigation of about 48,563 ha of land situated in the arid tracts of Mandya district.

Three canals take off from the dam. The main canal is the 44.8 km long left bank high level canal, named as Visweswaraiah canal after the well known engineer

Figure 3. A view of KRS dam

Dr. M. Visvesvaraiah. This canal has a capacity of 67.5 cumec and provides irrigation to 77,806 ha. The two low level canals have much smaller capacities. The 32 km long low level right bank canal, also known as Varuna canal, has a capacity of 7 cumec and provides irrigation to 1,534 ha. The low level left bank canal is 21 km long, has a capacity of 1.42 cumec and irrigation potential of 151 ha. The dam also provides flood control.

In addition to the main benefit of providing reservoir backed assured irrigation to a large area, a notable feature associated with KRS is the Vrindavan gardens. This is an exquisite garden located just below the KRS dam, and has a beautiful biological park, and a number of fountains run from the hydraulic head generated by the dam. The KRS dam supplies water for the upkeep of the garden which is a major tourist attraction of South India. A view of Vrindavan garden of Mysore is shown in Figure 4.

Mettur dam (Stanley reservoir)

The flow in Cauvery is mainly dependent on the south-west monsoon which influences large parts of its upstream catchment in Karnataka and Kerala, in the months of July and August. It was seen that the flow in the river after the south-west monsoon receded, was not sufficient to support the agriculture in the Cauvery delta. The north east monsoon was less dependable and failed frequently, with the result that the delta irrigation suffered badly, often just at the time when the crops were in the maturing stage.

Mettur dam is a remarkable project. Situated at 11°49′ N latitude, it is located at 52 km from Salem, in Salem District, Tamil Nadu, 270 km downstream of the KRS and just downstream of the Karnataka-Tamil Nadu border where the river emerges from the Eastern Ghats. Situated at an elevation of 243 m, it is a 85 m long masonry dam and a 1,615 m long earth dam section, the maximum height of the dam being

Figure 4. Vrindavan garden of Mysore. This place is a major tourist attraction of South India

70.4 m. Rising above the Cauvery River bed, the dam is constructed across two hills of the Eastern Ghats. Mettur reservoir has catchment area of 42,217 km^2, a live storage capacity of 2.65 BCM. At FRL 241 m, the water spread area is 153.5 sq. km and its MDDL is at 219 m.

In the present form, Mettur Reservoir was formulated by Col. Ellis in 1910. Construction began in July 1925 and was completed by 1934. The Mettur Reservoir is also called Stanley reservoir; it got this name after Sir George Frederick Stanley, the then Governor of Madras, who inaugurated the project. At the time of its construction, Mettur was the highest masonry dam in Asia and the largest in the world. It was planned to store the high flows during the south-west monsoon and distribute them evenly throughout the irrigation period, thus firming up the irrigation provided by Grand anicut canals. It is a landmark project that has helped stabilize irrigation in the Cauvery delta area and has yielded consistent benefits for the past 75 years. This project has helped stabilize existing irrigation in 1.665 M-ha of area and has extended irrigation to 121,810 ha of new area. Constructed primarily to stabilize irrigation in Thanjavur delta, the dam caters to about one third of the irrigated area of Tamil Nadu besides generating hydroelectric power of 200 MW (4 units of 50 MW each). Stanley is the largest reservoir in Tamil Nadu.

In the reservoir, the highest and lowest water temperatures recorded are 32 °C (in May) and 24.2 °C in (January), respectively. Dissolved oxygen in the surface layer in most of the months is above 4 mg/l with the maximum supersaturated conditions at the surface are accompanied by a depletion of oxygen in the hypolimnion. The water is alkaline, the pH values range between 7.5 to 8.8. Total hardness as $CaCO_3$ varies in a range of 86–128 mg/l suggesting the lake as a hard water one.

Even during the winter season, air temperature in this part of the country does not drop below 15 °C, keeping the water relatively warm. With the onset of summer, when the top layer warms up, there is no cool water below to offer any thermal resistance. Moreover, the release of cooler water from the bottom layer through the outlets of the dam removes any disparity in temperature between the top and bottom layers.

Lower Mettur Power Houses I/II/III/IV: In this system, there are 4 power houses, namely Lower Mettur Power House I, II, III, and Lower Mettur Power House IV. The power house I is located at Chekkanur Reservoir, 10 km downstream from Mettur dam. Power house II is located at Nerunjipettai Reservoir, 20 km downstream from Mettur dam. Power house III is located at Koneripatti Reservoir, 16 km down stream from Bhavani dam, and power house IV is located at Urachikottar Reservoir, 5 km downstream from Bhavani dam. These power houses are located on Cauvery River in Erode District, Tamil Nadu. The lengths of the dams are 948 m, 1,478 m, 710 m and 1,085 m respectively. All the four power houses have 2 units of 15 MW each, producing a firm power of 46 MW.

Grand anicut

An anicut (a kind of small barrage) known as the Grand Anicut was constructed across Cauvery River by the kings of Chola dynasty some 2,000 years ago, mainly to provide irrigation to large areas in Thanjavur district. With some minor changes, the Grand anicut is still in use and is one of the oldest in-use irrigation structures in the world that have been successfully serving their purposes; this one meeting the requirements of Cauvery delta area in Tamil Nadu. The Grand Anicut is located downstream of Mettur reservoir. The irrigation demand from the anicut is $9,670\,\text{Mm}^3$.

Nugu reservoir

This reservoir is located near Beerwal village, H.D. Kote Taluk, Mysore District in Kabini sub-basin. The longitude of the dam is 76°27′00″ and the latitude is 11°58′15″. At the dam site, the catchment area is 984 sq. km. Nugu is an earthen dam on right side with central masonry spillway. Table 8 gives the salient features of the Nugu Reservoir Project. A view of Nugu dam is shown in Figure 5.

Kundah I

Kundah I hydropower project is located at Avalanchi and Emerald masonry gravity dam, on Avalanchi and Emerald streams, 32 km from Udhagamandalam, in Nilgiris District, Tamil Nadu. The catchment area at the dam is $58.5\,\text{km}^2$. The height and

Table 8. Salient features of Nugu Reservoir Project

Design yield	$283.28\,\text{Mm}^3$
Storage	
(i) Gross	$154.10\,\text{Mm}^3$
(ii) Dead	$15.50\,\text{Mm}^3$
(iii) Live	$138.6\,\text{Mm}^3$
Full Reservoir level (FRL)	R.L. 725.42 m
Minimum draw down level (MDDL)	RL 709.57 m
Dead storage level, sill level of canal sluice	707.14 m
Height of dam above river bed level.	35.8 m
Capacity of spillway	566 cumec (2 vertical gates of 9.144 m × 6.1 m)
Halsoor Anicut	
Location	Near Halsoor village, 2 km downstream of the dam
Sub-basin/Taluk/District	Kabini/H.D.Kote Taluk/Mysore.
Length of weir	63.5 m
Discharging capacity of weir	849 cumec
Capacity of Head Regulator	
1) Nugu Right Bank Canal	4.25 Cumec
2) Nugu Left Bank Canal	1.98 Cumec
Minimum pond level	RL 693.27 m (Crest level of Anicut)
Description of Canals: (Main Canal)	
Length of RBC (in km)	Nugu High Level Canal 101.5 km (87 km up to Narasambudi tank), 14.5 km beyond.
Authorized Head Capacity	18.85 Cumec

Figure 5. Nugu dam in Cauvery basin

length of the dam is 57 m and 366 m respectively. At FRL 1,985.7 m, the reservoir has a live storage capacity of 154 MCM; the MDDL of the dam is at 1,943.1 m. Mean annual inflow to the reservoir is 108.2 MCM. The powerhouse has 3 units of 20 MW each to produce a firm power of 27.5 MW.

Kundah II

Kundah II hydropower project is located at Kundah Palam gravity dam constructed across Kundah River, 52 km from Udhagamandalam, in Nilgiris District, Tamil Nadu. The catchment area at this 32 m high dam is 113.96 km^2. The reservoir has a live storage capacity of 0.849 MCM; its FRL is at 1,624.6 m and the MDDL is at 1,609.34 m. Five units of 35 MW each have been installed at the powerhouse to produce a firm power of 63.5 MW.

Kundah III

Kundah III hydropower project is located at Pegumbahallah Forebay dam and Nirallapallam diversion weir, on Pegumbahallah Nirallapallam River, 77 km from Coimbatore, in Nilgiris District, Tamil Nadu. The catchment area at the Pegumbahallah dam is 41.44 km^2. The height of the Pegumbahallah dam is 55.5 m. The Pegumbahallah reservoir has a live storage capacity of 0.679 MCM at FRL 869 m; the MDDL is at 835 m. Power is generated by 3 units of 60 MW each yielding a firm power of 37 MW.

Kundah IV

Kundah IV hydropower project is located on Pillur dam on Bhavani River, 85 km from Coimbatore, in Nilgiris District, Tamil Nadu. The 88 m high dam has a catchment area of 1, 191 km^2; the dam is 357 m long and the mean annual flow to the dam is 685 MCM. The reservoir has a live storage capacity of 28 MCM at FRL 427 m and the MDDL is at 396 m. The power house has 2 units of 50 MW each. It has a firm power of 10 MW.

Harangi dam

The Harangi dam is located in the Upper Cauvery sub-basin near Hudgur village in Somwarpet Taluk of District Coorg. The latitude of the dam is 12°29′30″ N and longitude is 75°54′20″ E. At the dam site, the catchment area is 419.58 sq. km. Harangi is a masonry dam with central spillway and non-overflow section and there are earthen dams on either flank. Length of the dam is 845.8 m. Table 9 gives the salient features of the Harangi dam. A view of Harangi dam is shown in Figure 6.

Hemavathi reservoir

The Hemavathi reservoir is located on Hemavathi River near Gorur village in Taluk and district Hassan in the Upper Cauvery sub-basin. The latitude of the dam is 12°45′0″ N and the longitude is 76°03′0″ E. At the dam site, the catchment area is 2810 sq. km. It is a masonry dam with central spillway and earthen flanks on either bank. Table 10 gives the salient features of the Hemavathi project. A view of Hemavathi dam is shown in Figure 7.

Kabini reservoir project

The Kabini dam was constructed on the river with the same name in the Kabini sub-basin near village Beechanahally in Mysore district. The latitude and longitude

Table 9. Salient features of Harangi dam

Design yield at 50% dependability (in TMC)	1, 113.07 Mm^3
Storage (in TMC)	
i) Gross	240.74 Mm^3
ii) Dead	12.1 Mm^3
iii) Live	228.65 Mm^3
Levels of Storage	
i) Full Reservoir Level (FRL)	RL 871.42 m
ii) Minimum drawn down level (MDDL)	RL 850 m
iii) Dead Storage Level	RL 849.78 m
Height of Dam above River bed level	50 m
Length of Spillway	67 m
Capacity of spillway	3,400 cumec
Description of Canals	
Length of Left Bank Canal (in km)	153
Length of Right Bank Canal (in km)	137.50
Length of Lift Canal (in km)	50
Length of Branch canals under Right Bank canal (km)	103.5
Authorized Head Capacity of Left Bank Canal (at head up to 11.28 km)	55.35 cumec
Right Bank Canal (off take 11.28 km of Left Bank Canal)	42.45 cumec

Figure 6. A view of Harangi dam

Table 10. Salient features of the Hemavathi project

Designed yield	$2,228.57\,Mm^3$ (at 50% dependability)
Storage	
i) Gross	$1,050.83\,Mm^3$
ii) Dead	$38.04\,Mm^3$
iii) Live	$1,012.8\,Mm^3$
Reservoir evaporation losses	$99.13\,Mm^3$
Particulars of Dam	
Length	4,692 m
Height above river bed level	44.5 m
Length of spillway	91.44 m with 6 radial gates of sizes 10.67 m × 9.144 m each.
Capacity of spillway	3,622.4 cumec
Description of Canals	
i) Length of Hemavathi Left Bank Canal	214 km (it is two seasonal canal)
ii) Hemavathi Right Bank Canal	91 km (One Seasonal)
iii) Hemavathi Right Bank High Level Canal	106 km (One Seasonal)
iv) Length of Branch Canal under Hemavathi Left Bank Canal	
a) Tumkur Branch Canal	240 km
b) Nagamangala Branch Canal	78.5 km
Authorized Head Capacity	
i) Hemavathi Left Bank Canal	113.2 cumec
a) Tumkur Branch Canal	40.44 cumec
b) Nagamangala Branch Canal	24.34 cumec
ii) Hemavathi Right Bank Canal	9.34 cumec
iii) Hemavathi Right Bank High Level Canal	25.67cumec

Figure 7. A view of Hemavathi dam

of the dam are 11°50′30″ N and 76°20′17″ E. At the dam site, catchment area is 2,141.90 sq. km. Salient features of the project are given in Table 11. A view of Kabini dam is shown in Figure 8.

15.1.7. Ongoing and Other Projects

The utilization, area irrigated and expenditure under ongoing major & medium projects are given in Table 12.

In addition to the above there are several other reservoirs in the basin. Storage capacities of the other large Completed and under construction projects having a storage capacities of 10 Million cubic meter or more is summarized below as Table no. 13 and 14 respectively.

Table 11. Salient features of the Kabini Reservoir Project

Design yield	2, 767.06 Mm^3 (at 50% dependability)
Storage	
i) Gross	552.85 Mm^3
ii) Dead	99.7 Mm^3
iii) Live	453.15 Mm^3
iv) Sill level of Right and Left Bank canal sluice	685.50 m
v) Sill level of river sluice	676.35 m
Designed Silt charge per year	2.83 m^3/sq. km of catchment area
Ayacut (in acres)	
i) Left Bank flow canal	12.14 Mm^2
ii) Left Bank lift canal	–
iii) Right Bank flow Canal	445.16 Mm^2
iv) Right Bank lift Canal	421.7 Mm^2

Figure 8. A view of Kabini dam

15.1.8. Water Quality Aspects

Several studies were made by the Central Pollution Control Board to ascertain the status of water quality in the Cauvery River. The results show that at many places, the quality of water was quite poor compared to what was the desired class. The summary results of the analysis at various locations in Cauvery River from 1997–2001 is given in Table 15.

Table 12. Utilization, area irrigated and benefits under completed medium projects in the Cauvery basin

S. N.	Project	Year of commencement	Benefitted area (in ha)	Utilization (In TMC)	Benefitted Taluk/district
			Medium Projects		
1	Arkavathy	1975	6,232	3.14	Kanakapura
2	Chicklihole	1978	865	1.11	Somwarpet
3	Huchannakoplu L.I.S	1991	3,360	0.56	Holenarsipura/Hassan
4	Iggalur	1979	4,047	5.10	Channapatna, Maddur, Malavalli
5	Kamasamudra L.I.S	1985	3,916	0.80	Holenarsipura/Hassan
6	Manchanabele	1970	2,433		Magadi & Ramanagar
7	Taraka	1970	7,040	3.84	H.D. Kote
8	Uduthorehalla	1978	6,597	1.23	Kollegal
9	Votehole	1977	7,487	2.40	Alur, Belur/Hassan

Table 13. List of the Completed projects in Cauvery Basin

Name of the Project	State	Year of completion	Gross storage capacity (MCM)	Live storage capacity (MCM)	Designed Annual Irrigation (Million m^2)	Installed capacity (MW)
Anjanapura Dam	Karnataka	1938	39.76	20.00	67.40	–
Byramamgala	Karnataka	1945	21.15	19.04	16.20	–
Chamaraja Sagar	Karnataka	1933	89.65	86.05	–	–
Chilkole	Karnataka	1967	–	10.53	16.10	–
Gundal	Karnataka	1974	27.42	23.00	42.50	–
Hebbahalla	Karnataka	1963	11.95	10.80	12.10	–
Kanva	Karnataka	1946	22.73	21.45	20.20	–
Macromahalli	Karnataka	1941	68.00	60.00	60.70	–
Moti Talab	Karnataka	–	21.90	16.42	–	–
Suvarnavathy	Karnataka	1975	–	35.60	28.30	–
Amaravathy	Tamil Nadu	1958	114.61	112.37	87.00	20
Avalanche & Emerald Dam	Tamil Nadu	1960	156.20	152.80	–	70
Chinnar	Tamil Nadu	1977	14.15	13.98	7.60	–
Lower Bhawani	Tamil Nadu	1955	928.80	907.80	837.70	34
Manalar	Tamil Nadu	1978	13.58	12.55	–	–
Makurthy Dam	Tamil Nadu	1938	50.98	50.71	–	–
Palar Porandalar	Tamil Nadu	1978	–	43.19	40.80	–
Parsons Valley	Tamil Nadu	1966	19.25	17.19	–	12
Pillur Dam	Tamil Nadu	1966	44.40	34.97	–	25
Porthinund Dam	Tamil Nadu	1966	60.00	45.73	–	–
Pykara	Tamil Nadu	1935	56.80	55.22	–	17
Sandyanalla Dam	Tamil Nadu	1963	26.62	23.11	–	–
Upper Bhavani	Tamil Nadu	1965	101.20	85.20	–	16
Upper Dam	Tamil Nadu	1968	16.31	14.92	24.50	–

Table 14. List of the Under Construction projects in Cauvery Basin

Name of the Project	State	Storage capacity (MCM)		Designed annual irrigation (million sq. m)	Installed capacity (MW)
		Gross	Live		
Arkavathy	Karnataka	44.90	40.18	62.30	–
Manchanable	Karnataka	35.66	28.87	38.50	–
Taraka	Karnataka	112.00	82.00	89.00	–
Uduthorehalla	Karnataka	21.20	16.24	66.00	–
Votehole Reservoir	Karnataka	42.76	36.68	74.90	–
Karapuza	Kerala	76.50	72.00	93.00	–
Kuttiyadi	Kerala	120.70	113.00	360.00	75
Kodaganar	Tamil Nadu	–	12.30	41.20	–
Noyyal Orathupalayam	Tamil Nadu	–	17.45	42.00	–

Table 15. Desired and existing water quality levels for Cauvery

Location	Desired class	Observed class & critical parameters				
		1997	1998	1999	2000	2001
Cauvery at Napokulu Bdg (D/S), Karnataka	A	C Totcoli	C pH, Totcoli	C Totcoli	C Totcoli	C Totcoli
Cauvery at Kushal Nagar U/S (Near Baichanahalli), Karnataka	A	C Totcoli	C Totcoli	C Totcoli	C Totcoli	C Totcoli
Cauvery at KRS Dam, Balamurikshetra, Karnataka	C	A	B	D BOD	C	B
Cauvery at Sri Rangapattanna, D/S of Road Bdg., Karnataka	B	NA	D BOD	D BOD	C Totcoli	C Totcoli
Cauvery at D/S of Karekuara Village, Karnataka	C	C	C	C	C	C
Cauvery at Sathyagalam Bridge, Karnataka	C	A	D BOD	D BOD	C	B
Cauvery at Mettur, Tamil Nadu	C	D DO, Totcoli	C	B	C	B
Cauvery at 1km. D/S of Bhavani River confluence Tamil Nadu		D Totcoli	D Totcoli	D Totcoli	C	C
Cauvery at Pallippalayam, Tamil Nadu	C	D Totcoli	D Totcoli	C	C	C
Cauvery At Erode near Chirapalayam, Tamil Nadu		D Totcoli	D Totcoli	D Totcoli	C	C
Cauvery at Velore Near Kattipalayam, Tamil Nadu		D Totcoli	C	B	C	B
Cauvery at Mohanur near Pattaipalayam, Tamil Nadu		C	C	B	B	B
Cauvery at Thirumukkudal-Confluence Point of R. Amravati, Tamil Nadu		C	C	C	C	B
Cauvery at Musiri, Tamil Nadu	C	C	C	C	C	C
Cauvery at Tiruchirappalli U/S, Tamil Nadu	B	D DO, BOD, Totcoli	D BOD, Totcoli	D Totcoli	C Totcoli	D Totcoli

Cauvery at Tiruchirappalli D/S, Tamil Nadu	B	D Totcoli	C Totcoli	C Totcoli	C Totcoli	C Totcoli
Cauvery at Trichy, Grand Anicut, Tamil Nadu	B	C Totcoli	C Totcoli	C Totcoli	NA	C Totcoli
Cauvery at Thanjavur, Tamil Nadu	C	C	NA	NA	C	B
Cauvery at Coleroon, Tamil Nadu		C	C	D Totcoli	C	C
Cauvery at Pitchavaram, Tamil Nadu		C	C	D Totcoli	D Totcoli	E DO

*NA- Not Available. Source: Central Pollution control Board.

15.2. THE PENNAR BASIN

The Pennar basin extends over an area of 55,213 sq. km which is nearly 1.7% of the total geographical area of the country. Out of the total area of the basin, 6, 937 km^2 lies in Karnataka and 48, 276 km^2 in Andhra Pradesh. The basin lies between east longitudes of 77°04′ and 80°10′ and north latitudes of 13°16′ and 15°52′. It is bounded on the north by the Erramala range, on the east by the Nallamala and Velikonda ranges of the Eastern Ghats, on the south by the Nandidurg hills and on the west by the narrow ridge separating it from the Vedavati valley of the Krishna basin. The basin lies in the States of Andhra Pradesh and Karnataka. An index map of the entire basin showing various sub-basins is given in Figure 9. The statewise distribution of drainage area is given in Table 16.

15.2.1. The Pennar River

The Pennar or the Uttara Pinakini is one of the major rivers of Indian peninsula flowing east and draining into Bay of Bengal. It is the next largest river to Cauvery in the Peninsula. Pennar River is locally known as Penneru; it is also called 'Henne' which means Penna in Telugu language. The name Pinakini is associated with '*Pinaka*' the bow of Siva or Nandeeswara, the presiding deity of Nandi hills, the place of origin of the river.

Pennar rises at east longitudes of 77°36′ and north latitudes of 13°33′ in the Chenna Kesava mountain range in the state of Karnataka. After flowing in north western direction for a distance of 48 km through the Kolar and Tumkur districts of Karnataka, the river enters Andhra Pradesh in the Anantpur district. At 115th km, the river re-enters Pavagada taluk of Karnataka and at 128th km again, it enters Kalynadurg taluk of Anantapur district of Andhra Pradesh. At its 336th km, the river forces its way through Gandikota gorge, one of the magnificent gorges in the world. The river emerges from Velikonda range of Eastern ghats at its 467th km and enters the plains below Somasila. After traversing for a distance of 597th km, the river joins the Bay of Bengal. The basin

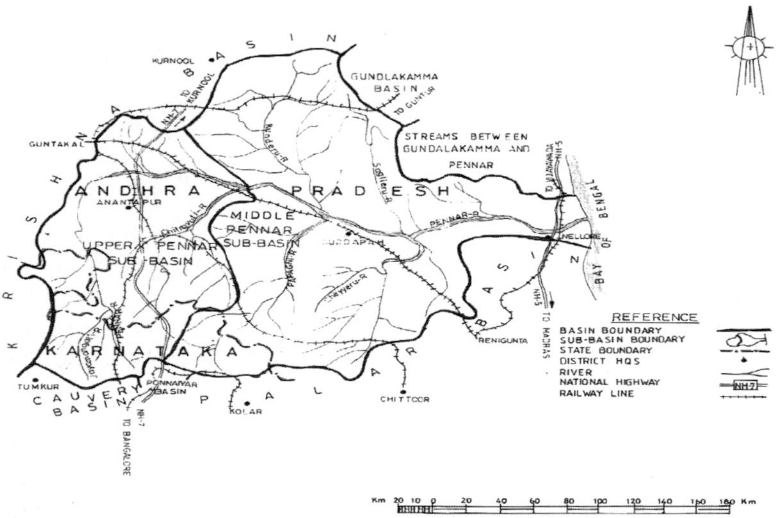

Figure 9. Index map of Pennar Basin

of Pennar River lies between latitudes 13°16′ and 15°52′ and longitudes 77°4′E and 80°10′E. The major tributaries of the river are the Jayamangali, the Kunderu, the Sagileru, the Chitravathi, the Papangi, and the Cheyyeru.

Two of its tributaries, Kumudavali and Jayamangali, join it in its 69th and 82nd kilometer from its source, respectively. After traversing 67 km through Anantpur district in Andhra Pradesh, Pennar re-enters Karnataka and flows across the Tumkur district for about 13 km to emerge in the Anantpur district again. Beyond the confluence of the Jayamangali, the river runs almost northwards for a distance of 146 km passing through the plains. Further, the river diverts in the east near Ponnahobalam and flows through the Marutia and Katrimala reserve forest ranges. The river Chitravadi, a major tributary, falls into the Pennar on its right at the 336th km from the source. Further, two major tributaries, the Kunderu and the Papagni, meet with Pennar near Kamalpuram. The river continues to flow in a South-easterly direction, cuts across the Nallamala hill range and passes through the town of Siddhavattam. After receiving the Sagileru it turns eastwards and near Boyanapalli, it is joined by the Cheyyaru flowing from the right. The river finally

Table 16. Statewise drainage areas of Pennar basin

State	Drainage area (km^2)	Percentage of basin area
Andhra Pradesh	48, 276	12.60
Karnataka	6, 937	87.40
Total	55, 213	100.00

falls into Bay of Bengal. The length of the river is 597 km, of which about 61 km is in Karnataka and remaining 536 km in Andhra Pradesh. It is estimated that the total sediment load transported by Pennar into the Bay of Bengal is 6.9×10^6 ton. Flow Diagram of Pennar Basin is given in Figure 10.

15.2.2. Climate of Pennar Basin

The catchment receives rainfall both during the south-west and north-east monsoons. The rainfall during the non-monsoon period is not significant. The north-east monsoon [October through January] provides a little precipitation but the predominant rain falls when the southwest monsoon [June through September] occurs. Post monsoon cyclonic activity in the Bay of Bengal during September and October produces an increased rainfall in the coastal region. The mean annual rainfall within the drainage basin varies from about 550 mm around Anantapur area to 900 mm around Nellore.

From the temperature records, it is seen that the mean maximum daily temperature varies from 40.3 °C observed at Cuddapah to 34.7 °C observed at Arogyavaram and

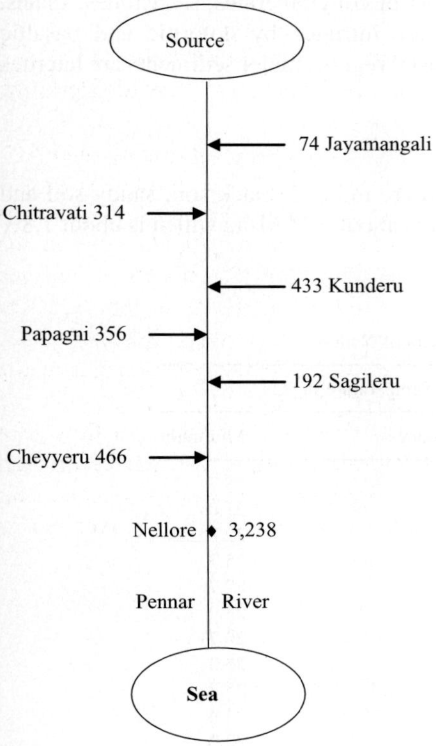

Figure 10. Flow Diagram of Pennar Basin

the mean minimum daily temperature varies from 20 °C observed at Nellore to 15.3 °C observed at Arogyavaram. In general, humidity is high during the monsoon period and moderate during non-monsoon period. The relative humidity in the catchment of Pennar ranges from 21 to 84 percent. The mean monthly maximum and minimum temperatures at Nellore hydro-meteorological station are given in Table 17.

Winds are generally light to moderate with some strengthening in the monsoon season. The catchment is influenced by winds from south-west and north-west during the period from May to September and from north-east and south-east during the period from October to April. The wind speed in the catchment varies from 4.3 to 3 km/hr.

Geology

The geology of the drainage basin is predominantly formed from Archean rocks, principally granitic intrustives into metamorphic schists. The Archeans in this region comprise biotite and hornblende granite-gneisses, granodiorite, diorite, and pegmatite. Of secondary importance are the Dhawar metamorphics comprising of phyllites, slates, schists with chlorite, biotite, garnet, staurolite, kyanite, silllimanite and hornblende. In the central part of the basin, the dominant rocks belong to the Cuddapah and Kurnool groups consisting of conglomerates, sandstones, shales, dolomites, limestones and cherts. These are intruded by doloritic and basaltic igneous materials in many places. In the coastal regions major sediments are laterites and recent alluvium.

Soils

The important soil types found in the basin are red soil, black soil, sandy soil and mixed soil. The culturable area of the basin is about 3.54 M-ha which is about 1.8% of the culturable area of the country.

Table 17. Monthly Temperature Data of Nellore

Month	Mean Temperature in °C	
	Maximum	Minimum
January	29.6	18.7
February	31.9	21.0
March	34.6	22.8
April	37.4	25.3
May	40.1	27.7
June	38.6	28.0
July	36.1	26.7
August	35.6	26.3
September	34.8	25.7
October	32.6	24.4
November	29.6	22.2
December	28.7	20.2

15.2.3. Sub-Basins of Pennar Basin

The Pennar basin can be divided in three sub-basins, namely the Upper Pennar, the Middle Pennar, and the Lower Pennar. The areas of various sub-basins are given in Table 18.

15.2.4. Upper Pennar Sub-Basin

The upper Pennar sub-basin includes the area of the basin from the source to the confluence with Chitravati River and it also includes the catchment of Jayamangli River. The length of Pennar River in this sub-basin is 336 km and the catchment area is 19, 700 km^2. This sub-basin occupies 35.7% of the total basin area. Out of this, the area in Andhra Pradesh is 14,666 and it is 5, 034 km^2 in Karnataka.

Topography, physiography, and other features

More than 10% of the area of this sub-basin lies in hilly and dense forested area. The principal mountain chains are mainly Durga ranges, the Nagasamundra hills, and the Mutsukota hills. The highest peak with an elevation of 1,148 m is located in Pavagada taluk of Karnataka. There are many long bridges and a number of small streams. The sub-basin is nearly rectangular in shape; the western and northern boundary is the common ridge of Krishna and Pennar basins.

Most of the sub-basin is underlain by crystalline Archean rocks, Dharwar super group, Cuddapah group of rocks belonging to Proterozic rocks age, and recent to sub-recent soils and alluvium. Groundwater is present in almost all the geological formations, the occurrence being mainly dependent upon the development of secondary porosity through weathering and fracturing. The alluvium occurs in a narrow patch along either side of the Pennar River and it comprises of sands of various grades and clay mixed with Kankar.

Climate

The sub-basin experiences hot weather from mid February to the end of May. The monsoon season begins in early June and lasts till the end of November and is followed by winter season from December to mid February. This sub-basin receives rainfall from south-west and north-east monsoons. The average annual rainfall varies from 811 mm at Chikballapur to 448 mm at Chinnakottapalli. The rainfall during the non-monsoon season is insignificant.

Table 18. Areas of various sub-basins under Pennar basin

Sub-basin	Area in km^2
Upper Pennar	19,700
Middle Pennar	16,953
Lower Pennar	18,583

April and May are the hottest months in this sub-basin with the maximum temperature nearing 38.4 °C. The minimum temperature is observed in December and January and it is 17.2 °C. Maximum relative humidity observed during October month is 76% and maximum wind velocity of 18.9 km/hour has been reported in the month of July. Due to arid climate, evapotranspiration takes place at high rate: it is nearly 200 mm at Anantpur during May.

Soils

The dominant soil groups in this sub-basin are red earth which are found in Anantpur and Kurnool district and occupy nearly 53.5% of the sub-basin area. The other important soils are black soils (16.5% of the area), red loamy soil (15.9% of the area), and red sandy soil (9% of the area). Red earths are dark reddish brown to brown in colour, deep to very deep and loamy sand to clay loam in texture. These moderately well drained soils possess satisfactory water holding capacity and crops like ragi, jowar, pulses and groundnut can be grown on these. The black soils are dark brown to grayish brown in colour, deep to very deep and finer in texture. These are overly-drained and have satisfactory water holding capacity. Red loamy soils are dark brown to reddish brown in colour, deep to very deep and coarse to medium in texture. These are well-drained and have poor water holding capacity. The major crops are groundnut, pulses, ragi, rice and fruits and vegetables. Not much year-to-year variation is observed in the culturable area in the sub-basin.

15.2.5. Middle Pennar Sub-Basin

This sub-basin lies between latitudes 13°24′ to 15°52′ N and longitudes 77°42′ to 78°50′ E. It covers an area of Pennar Basin from its confluence with the Chitravathi River to its confluence with the Papagni River. The area of this part of catchment is 16,953 sq. km which is 30.5% of total basin area. Out of this area, 1,877 sq. km lies within the Kolar district of Karnataka and the remaining 15,076 sq. km lies in the Anantpur, Chittoor, Cuddapah, Kurnool and Prakasam districts of Andhra Pradesh. The major tributaries of this sub-basin are Kundaru, Rageiu and Papagni rivers.

Topography

The shape of Middle Pennar sub-basin nearly resembles a parallelogram whose length in North-South direction is about 260 km and width in the East-West direction is about 65 km. About 41% of the area of this sub-basin lies in the Kurnool district and about 31% in the Cuddapah district. These two districts covering nearly 72% of the area are hilly districts of Andhra Pradesh with Kolar plateau, Chitravathi, Muttasukota and Erramala hilly range. Besides, there are a number of isolated hills. The highest hill is the Horselay hill located in the south-east part of the sub-basin with an elevation of 1,314 m. Hills and forests cover nearly 20% of the sub-basin.

Rocks of Pre-Cambrian age like archeans, composite gneisses, shales, limestones and quartzites of Kurnool and Cuddapah system of rocks are the major geological formation of this sub-basin. Ground water is found to occur in all geological formations: the occurrence and movement are controlled by the major extent of weathering and the presence of joints and fractures.

Climate

This sub-basin experiences the severe hot weather from the middle of February to the middle of June. This is followed by monsoon season lasting till October during which southwest monsoon rains are received. The retreating monsoon also causes some rains during November. The period from December to mid February is relatively cool and dry. The annual rainfall in the sub-basin varies from 497 mm to 788 mm.

As noted above, the basin experiences high temperatures during summer, the maximum temperature reaches about 40 °C in the month of May while the minimum is around 16.6 °C in the month of December. Monthly evaporation of the order of 222 mm has been reported for the month of May at Kurnool observatory of IMD. The basin is fairly dry, typical relative humidity values are about 75% at 8:30 hours during the monsoon season at the Kurnool observatory, while the values at 17:30 hours nearly of the order of 24% during April and May. The sky is moderately to heavily clouded and overcast on some days during the monsoon period and during the rest of the year, sky is generally clear or lightly clouded.

The average annual reference evapo-transpiration for the Pennar basin is 1,840 mm.

Stream flow data

There are two gauge and discharge sites in the sub-basin at Mylavaram dam site and at Sitarampuram (Alladipalli). Considering the yield of the sub-basin and imports to it, the overall water availability at 75% dependability was assessed at 2,763 MCM; it is 3,180 MCM at 50% dependability.

15.2.6. Lower Pennar Sub-Basin

The area below the middle Pennar basin and the outfall to the sea falls under this sub-basin. The total area under lower Pennar Sub-basin is 18, 583 km^2. Rainfall in this area is much high compared to the other parts of the basin.

15.2.7. Major Tributaries of Pennar

The principal tributaries of the Pennar River are the Jayamangali, the Kunderu, the Sagileru, from the left and the Chitravati, the Papagni and the Cheyyeru from the right.

The Sagileru river

The Sagileru River is a major tributary of Pennar that joins it on the left side. The Sagileru River rises in the Nallamalla range at an altitude of about 700 m near Cumbum in Ongole District. Sagileru flows towards south in the Districts of Ongole and Cuddapah of Andhra Pradesh; it flows in general in southward direction for a total distance of 141 km. It joins Pennar River in Sidhout Taluk of Cuddapah District from the left bank near Venkatesampalli. The Sagileru basin is located between latitude 14°28′ to 15°34′ N and longitude 78°46′ to 79°10′ E. The total geographical area of the basin is about 3,203 sq. km and falls in Ongole and Cuddapah districts 960 sq. km and 2,243 sq. km respectively.

The Sagileru basin is characterized by sub-dendritic type of drainage pattern. The total length of Sagileru River along its course is about 162 km and the river runs almost in central part of the basin from North to South. There are only small tributaries to this River, worth mentioning streams being Enamaleru, Tadukuvagui and Maderu. The Enamaleru stream drains into Sagileru River at 49th km of run. The Tadukuvagu and the Maderu streams join the Sagileru at 68th km and 111th km respectively. Sagileru is a non-perennial river – most of discharge in the river is observed in the months from June to November. After January, the river becomes almost dry.

The basin comprises of rather undulated country with deep valleys between two big hill ranges on either side throughout the length. Most of the area in the basin lies between contours 500 ft (152.4 m) to 2,500 feet (762 m).

Climate: The climate of Sagileru basin is generally regarded as unpleasantly hot which is probably due to the early setting in of high temperatures. The average maximum temperatures in April and May are 42 °C and 43 °C respectively.

The Sagileru basin lies in medium rainfall zone as the annual average rainfall in the basin is 767 mm. The basin is under the influence of both the southwest and northwest monsoons. From the rainfall pattern it is observed that the southwest monsoon sets in during middle of June. It brings fair quantity of rains to the basin up to the end of September. There are heavy rains in the upper portion of the basin during this period. The northeast monsoon breaks in October and the rains continue till December. These rains are heavier in lower portion of the basin than the earlier rains. On the whole, the incidence of rainfall during southwest monsoon is greater than northeast monsoon in the basin.

Physiography: The Sagileru basin comprises of shales and phyllites of Pullampet formation, occupying a major portion. Quartizites and limestones of the same stage occur in patches, occupying very little part of the basin. Bairenkonda quartzites occupy the northwestern, southwestern and eastern parts of the basin to a limited extent. The general trend of the formation is north-south direction with little variation of 10 to 15, dipping 60 to 80 due east. The shales and phyllites are well bedded and jointed in east-west, northwest-southeast, northeast-southwest directions. Fractures extending to depths of 30 to 70 meters below ground level are common in this formation except near mounds and foot hills. Bairenkonda quartzites are hard and massive and rarely comprises of fractured zones.

Cropping pattern: The principal crops grown in the basin in descending order of the areas covered by them are Paddy, Groundnut, Bajra, Jowar, Ragi, vegetables and sugarcane. Paddy is the wet irrigated crop and is grown mainly in the Kharif season, i.e. from July to November. The second crop of paddy during Rabi season is taken in a very small area in the ayacut of only a few tanks. The sugarcane is perennial and is another wet irrigated crop which covers a very negligible area in the basin. The remaining crops, viz. groundnut, Bajra, Jowar, Ragi and vegetables come under dry irrigated crops, i.e. only supplementary irrigation is provided to these. The groundnut and vegetables are grown in both the seasons of Kharif and Rabi while Bajra and Jowar are grown mainly in Kharif season only; Ragi is grown mostly in Rabi season.

Soils: The principal soil types in the Sagileru basin are: (1) Red loam, (2) Red sand, (3) Black loam, and (4) Black clay. Red soils cover a major portion of the basin. Texturally, red soils comprise of course sandy loam, fine sandy loam and loams. These are sufficiently permeable to be well drained. Black soils are alluvial soils and occur in minute extent in the basin.

Projects: Two medium projects have been constructed across the Sagileru River: (1) Upper Sagileru project (USP), and (2) Lower Sagileru Project (LSP). The Upper Sagileru Project, constructed in 1896 near Diguva Thamballapalli village in Badvel Taluk is an anicut across the river. The project does not have any direct command area. However, through its left bank canal, it feeds 10 tanks which in turn irrigate an area of about 2,210 ha. The lower Sagileru project constructed in 1960 in Badvel Taluk of Cuddapah District is a storage reservoir. This project also does not have any direct command area but feeds 34 tanks through its left bank canal. The right bank canal of this project is not yet operational.

The Jayamangali river

The Jayamangali River rises in the Tumkur district of Karnataka and joins Pennar on the left bank after traversing a distance of 77 km in northeasterly direction. The total drainage area of the river is 1,282 sq. km. The Survarnamukhi River is an important tributary of the Jayamangali River.

The Kunderu river

The Kunderu River rises in the Kurnool district of Andhra Pradesh and after traversing a distance of 205 km, it meets Pennar from the left bank near Kamalapuram. The total drainage area of the river is 8,057 sq. km which is about 15.55% of the total catchment area of Pennar River.

The Chitravati river

The River Chitravati rises near Chikballapur town in the Kolar District of Karnataka and before joining the Pennar River it traverses a distance of 218 km in northeast direction. The river falls into the Pennar from the right near Gandaluru. The drainage area of the river is 5,908 sq. km which is about 12.4% of the basin area.

The Papagani river

The Papagani River rises near Sidlaghatta town in the Kolar district of Karnataka State and joins Pennar from the right bank. The length of the river is 205 km and the drainage area of this major tributary is 7,423 sq. km which is about 14.14% of the total basin area.

Table 19. Annual average observed runoff at some CWC sites in Pennar basin

Name of the site	Name of the stream	Catchment area (km^2)	Annual average runoff (BCM)
Nagalamadite	Pennar	5, 050	0.2
Siddavattam	Pennar	38, 663	0.0
Tadapatri	Pennar	12, 482	0.3

The runoff in the river is being measured by CWC at several places. The annual average observed runoff at some CWC sites in Pennar basin is given in Table 19.

15.2.8. Land Use in Pennar Basin

Table 20 gives area under major land uses in the Pennar basin. Forests account for 21% of the area whereas nearly 12% area falls under barren land. Net sown area is 36% of the total basin area while total culturable area is about 55%. In the basin, double crops are taken over very small, about 1.66% area.

The population of the basin was 9.7 million in 1991 which grew to 11.65 in year 2001. According to the projections, the population is likely to be 18.7 million by the year 2050. In the basin, the rural population is 60% of the total.

15.2.9. Water Resources of Pennar Basin

The Pennar basin receives rainfall from the south-west as well as the north-east monsoons. Many depressions in the Bay of Bengal or Arabian Sea result in moderate

Table 20. Land use in Pennar basin

S N	Land use	Area (ha)	Percent of area
1	Forest	1, 159, 161	21.00
2	Barren land	637, 221	11.54
3	Land put to non-agricultural use	481, 342	8.72
4	Permanent pastures and other grazing lands	208, 299	3.77
Sub-total		2, 486, 023	45.03
5	Land under miscellaneous crops and trees	69, 541	1.26
6	Culturable waste	255, 735	4.63
7	Other fallows	365, 109	6.61
8	Current fallows	378, 500	6.86
9	New sown area	1, 966, 392	35.61
Sub-total (Culturable area)		3, 035, 277	54.97
10	Area sown more then once	91, 814	1.66
11	Gross sown area	3, 058, 206	37.27
12	Total area of Pennar basin	5, 521, 300	100.00
	Irrigation from existing, on-going and proposed projects	1,073,195 ha	

rainfall in the Pennar basin. While the upper part of the basin is served mostly by the south-west monsoon, the lower catchment is served by the north-west monsoon. Pennar basin receives less rain fall in upper zones while the rainfall is more between Nallmalais and the sea. The average monsoon rainfall is 540 mm in the head reaches which increases to 1,040 mm in the tail end. As a result of this distribution of rainfall, much of the rainwater remains un-utilized and flows to the sea while the upper reaches of the basin face severe water shortage. In the Andhra Pradesh area of the basin, the average rainfall is 600 mm. Andhra Pradesh state utilizes about 1,220 MCM from the Pennar River for irrigation above Somasila and about 708 MCM below Somasila. Another 1,415 MCM will be put to beneficial use on completion of the projects in progress. Of this, 850 MCM is for utilization outside the Pennar basin under the Telugu Ganga project. Irrigation development in the upper reaches of Pennar basin will serve the drought prone areas much better.

The upper Pennar sub-basin has negligible surface storage, while the middle and lower Pennar have Mylavaram and Somasila dams respectively. Upper Pennar sub-basin has an existing import of 387 Mm^3 of water from the Krishna basin through Tungabhadra High Level Canal. Similarly, in the middle Pennar basin there is an existing import of 919 Mm^3 of water through Kurnool-Cuddapah canal from Tungabhadra sub-basin of the Krishna basin. There is an existing export of 73 Mm^3 of water from the lower Pennar sub-basin.

Though floods are not very frequent in the Pennar basin, soil erosion is on the high side due to fairly steep slopes, arid climate, and lack of vegetation. Already some areas have become so eroded now that the soil cover is not adequate for agriculture.

Much of the flow below Somasila reservoir may continue to remain un-utilized for beneficial use because large extent of the catchment east of the Eastern Ghats is endowed with higher rainfall and the topography offers very little scope to impound the flood flows. There is limited use in small irrigation tanks and very small abstractions at Sangam and Nellore anicuts. Large quantum of water is consumed by evaporation and evapo-transpiration by vegetation. There are many tanks in the area for irrigation and domestic use in villages. Table 21 gives water withdrawals for domestic and industrial uses in three different sub-basins for a recent year. The total withdrawal stands at 1,029.86 MCM. Industrial sector accounts for 684.22 MCM or about 66% of these withdrawals.

Table 21. Withdrawals for Domestic and Industrial needs (MCM) in Different Sub-basins (Year 1999–2000)

Item	Upper pennar sub-basin	Middle pennar sub-basin	Lower pennar sub-basin
Domestic, Rural	101.51	56.36	61.70
Domestic, Urban	48.54	37.01	40.52
Industrial	292.06	187.20	204.96
Total	442.11	280.57	307.18

The gross area under irrigation in the Upper Pennar sub-basin was 189,732 ha in the year 1999–2000 which is 18.66% of the gross cropped area. Irrigation is mainly from minor schemes like tanks and wells. The canal irrigation is mainly through imported waters. In the Middle Pennar sub-basin, the gross area under irrigation during 1999–2000 was 191,054 ha which is 26.5% of the gross cropped area. Existing irrigation in the sub-basin is mainly from minor irrigation schemes.

Pennar is a water deficit basin. To fulfill the growing water needs of the agriculture and industrial sectors, there is an existing import of water in the basin. Furthermore, under inter basin water transfer schemes, transfer of water is recommended to Pennar basin from the adjoining basins.

15.2.10. Water Availability and Demand

Table 22 shows an assessment of water availability and demand in Pennar Basin. As per this assessment, if water is used in a sustainable manner, the total availability for irrigation is 10,270 MCM while the demand for food sector is about 15–20% more than this quantity.

The base flow in the Pennar River is showing a decreasing trend with at time and it is much less than what used to be in the past. Similarly the total outflow into the sea is also showing a decreasing trend. In the basin, around 30–35% irrigation is by surface water and rest is by ground water. There is a very heavy and increasing withdrawal from surface and ground water to fulfill the demands. In future, the demands for domestic, industrial, and agriculture sector are likely to increase significantly and this is likely to worsen the water availability situation.

Table 22. Water Availability and Demand in Pennar Basin

Water availability			Surface water (Mm^3)	Ground water (Mm^3)
1	Utilizable water		6,900	4,930
2	Import		3,388	–
3	Export		(−)495	–
4	Net available		9,793	4,930
		Water Demand		
i.	Domestic requirement	Urban population @ 200 lpcd	546	
		Rural population @ 120 lpcd	246	246
		Total	792	246
		Consumptive use @ 30%	238	74
ii.	Industrial demand	Average	8,261	
		Consumptive use @ 30%	2,478	
iii.	Environmental flow releases @ 1 % of annual flows		44	
	Total demand (other than agriculture)		2,760	74
	Water available for irrigation (2/3 of ground water)		7,033	3,237

15.2.11. Water Quality

Rivers in semi-arid areas are characterized by wide variations in the annual flows and poor quality of water. Pennar River also has a semi-arid catchment. Water of the Pennar River up to Anantpur and Cuddapah is not of good quality for irrigation and drinking due to large quantities of carbonates and bicarbonates. The fluoride concentration is also high due to the presence of soluble salts and fluorides from the rocks and soils in the catchment. Due to poor quality of water, yield from irrigated crops is very poor. In case of paddy, only special saline resistant varieties could be grown with low yields. While the soils nearer the ridges are generally good, nearer the valleys the soils are saline or alkaline due to water logging and deposition of salts. However, the quality of water of the tributaries of Pennar is good and hence there is good irrigation development in the lower reaches. Kundu River, a major tributary of the Pennar carries water of K.C. Canal which draws good quality water from Tungabhadra. Even though the catchments of Tungabadra and Pennar are close by (the basin of Vedavati lies in between), there is large difference in the quality of water of these rivers.

Table 23 gives the desired and observed class of water of Pennar River at different times. Not much data are available for this river but whatever data are available give a grim picture as the observed quality was found to be much worse than the desired.

The Pennar water possess high silt load during monsoon period resulting acute drinking water problems for people in rural areas who directly depend on it. In Andhra Pradesh, ground water occurs under unconfined and semi confined conditions. Rainfall is the principal source of recharge; the others being percolation of river water during high flow periods and seepage of irrigation water. During summer (low flow) period, ground water contributes to baseflow. Among the cations and anions present in the ground water sodium and chloride are predominant in Andhra Pradesh region. Concentration of cations and anions are in the order: Na > K > Ca > Mg; Cl > SO_4 > F. Non-carbonate hardness is present in the region.

The indices of salt-water contamination like Mg/Ca, Na/Ca indicate that the ground water in the entire area is slightly contaminated with sea water. Based on this fact, some people have hypothesized that this area was probably inundated by seawater in the past. Seawater from the Bay of Bengal is the main contributor to

Table 23. Desired and existing water quality classes for Pennar

Location	Desired class	Observed class & critical parameters		
		1997	2000	2001
Pennar before confluence with Chitravathi, Tadpatri, Unganoor, A.P.	C	D BOD	NA	D BOD
Pennar after conf. with Cheyyuru, Somasile, A.P.	C	D BOD	NA	C
Pennar at Siddvata, Nellore, A.P.	C	D BOD	D BOD	NA

* NA – Not Available. Source: Central Pollution Control Board.

salinity in the coastal areas and this is caused by the reduction in the Pennar water flow. The problem of saline water intrusion in the fresh water zone gets severe during the dry period when the Pennar and its tributaries face a drastic fall in river flow. This situation is worsening and might lead to a terrible environmental hazard in the future unless a suitable remedial action is initiated.

15.2.12. Major Water Resources Development Projects

In this section, the major water resources projects of the Pennar basin are described.

Table 24. Some features of Somasila Project

General data			
Ayacut	167,670 ha (414,000 acres)		
Kharif	111,375 ha (275,000 acres)		
Rabi	56,295 ha (139,000 acres)		
Culturable Command Area (CCA)	1.64 lakh ha (4.05 lakh acres)		
Proposed utilization under the project			
Kharif (Stabilization of Existing command)	996 Mm^3 (35.15 TMC)		
Rabi a) Dry	292 Mm^3 (10.308 TMC)		
b) Wet	87 Mm^3 (3.09 TMC)		
Total	1, 375 Mm^3 (48.548 TMC)		
Main canal	North feeder channel	South feeder channel	Kavali canal
Length	72.92 km	76.2 km	67.5 km
FSL at Head	80.92 m	85.38 m	33.84 m
Maximum discharging capacity	10.76 cumec	10.48 cumec	28.32 cumec

Source: GoAP (1984).

Table 25. List of the Completed projects in Pennar Basin in Andhra Pradesh

Name of the project	Year of completion	Storage capacity (MCM)		Designed annual irrigation (Million sq. m)
		Gross	Live	
Arnair	1958	52.44	48.80	–
Bahuda Reservoir	1972	11.26	10.70	11.50
Gandipalem	1987	53.00	49.42	64.00
Mid Pennar dam	1968	141.58	134.50	312.00
Pedderu Stage-I	–	14.97	14.12	25.70
Rallapadu	1966	31.29	30.80	44.00
Swarnamukhi Project	1959	47.67	42.43	99.40
Upper Pennar Dam	1959	51.28	49.57	38.80

Table 26. List of the Projects under construction in Pennar Basin in Andhra Pradesh

Project	Storage capacity (MCM)		Designed annual irrigation (Million sq. m)
	Gross	Live	
Buggavanka Project	14.32	10.35	–
Cheyyeru Project	63.47	44.89	52.00
Meddileru	19.38	15.03	32.40
Valligaleru Reservoir	131.47	69.42	–

Somasila reservoir

This is an operational major project on Pennar River in Andhra Pradesh for stabilizing irrigation in Pennar delta. The dam is located near Somasila Village, Atmakur Taluk of Nellore District at Latitude 14°29′15″ N and Longitude 79°18′25″ E. Catchment area at dam site is 50, 492.5 km^2. It is an earth and rockfill + masonry and concrete dam whose maximum height above deepest foundation level is 38 m. Somasila has also been integrated as a component of the Telugu Ganga canal project which is proposed to carry water from Srisailam to Chennai city and provide irrigation benefits to the en-route areas. The gross storage capacity at FRL 100.58 m and live storage capacity of the reservoir are 2,210.00 MCM and 1,994.00 MCM respectively. At the maximum water level of 101.80 m the storage capacity is 2,483 MCM and at dead storage level of 82.30 m, it is 214 MCM. At FRL, the reservoir water spread area is 212.28 km^2. The designed annual irrigation from the dam is 44.00 Million sq. meter. The irrigation and water supply demand from the dam is 1,453 and 409 Mm^3 respectively. The annual volume reliability for irrigation and water supply is 0.533 and 0.917 respectively.

An ogee type spillway, 236.21 m with crest level at 86.868 m, has been provided to dispose flood waters. The spillway has 12 radial gates, 15.24 m × 13.72 m and the maximum discharging capacity at FRL and MWL is 19,680 cumec and 22,375 cumec respectively. Table 24 gives some other details about the project.

Sathanur reservoir

Sathanur reservoir was created in 1957 on the Pennar River at Sathanur village in Tiruvannamalai Sambuvarayar district. The reservoir has a capacity of 228.91 million m^3 at FRL of 222.2 m and a mean depth of 11.4 m. It covers 2,010 ha area at the FRL.

Other reservoirs

Salient features of the other important completed and under construction projects having storage capacities of 10 MCM or more is summarized in Table 25 and 26 respectively.

CHAPTER 16

OTHER BASINS AND ISLANDS

In this chapter, we describe several small rivers that flow in various parts of India. Most of these rivers have small catchment areas and flow essentially during the monsoon season. However, occasionally the catchments of these rivers receive intense rainfall and cause flooding. Of particular interest are the rivers in the Western Ghats area, which carry substantial quantity of flow as their catchments receive very high amount of rainfall.

16.1. WEST FLOWING RIVER SYSTEM

The Western Ghats are a principal geographical barrier in the path of the Arabian Sea branch of the Southwest monsoon, and this interruption in flow of mountain is largely responsible for the heavy rainfall over the western coastal belt. The Southwest monsoon season (June to September) is the principal rainy season in which over 90% of annual rainfall takes place. Heavy rainfall in this belt has given rise to many rivers of short length but high discharge.

In this region, most rivers originate at an elevation ranging from 400 m to 1,600 m, close to the Western Ghats ridge. The rivers generally flow westward and meet the Arabian Sea after a short run varying from 50 km to 300 km. The rivers are very steep in the upper reaches and fairly steep in the middle reaches. It is only when the rivers are near the sea that they have relatively flat gradients and form moderately wide flood plains.

The list of west flowing rivers, their important tributaries and the states through which these flow is given in Table 1. Sharavathi is an important river of this part which originates in Western Ghats near Ambutirtha in Tirthahlli Taluk of Shimoga District. It flows for only 128 km covering an area of 2,771 sq. km but has immense hydropower potential. The river has a vertical drop of about 253 m near Jog which is the highest water fall in India. Sharavathi joins the Arabian Sea near Honavar in Uttar Kannada District. It covers a varied terrain and climate and supports a rich diversity of flora and fauna.

In addition to the above main river basins, there are many free catchments between the identified river basins which have small streams that directly drain into the Arabian Sea. These free catchments are close to the sea and are at lower elevation.

Table 1. West flowing rivers and their tributaries

Name of the river/ tributary	Catchment area (km^2)	Origin	Altitude (m)	Length of the river (km)	Sub-tributaries	State through which the river flows
Mahadayi/ Mandavi	2,032	Western Ghats, Belgaum district	600	87	Maderi	Karnataka, Goa
Kalinadi	4,188	Western Ghats, Bidi village	600	153	Pandhari, Tatti-halla and Nagi	Karnataka
Gangavalli (Bedthi)	3,574	Western Ghats south of Dharwad	700	152	–	Karnataka
Aghanashini (Tadri)	1,330	Western Ghats Near Sirsi	500	84	–	Karnataka
Sharavathi	3,592	Western Ghats Humacha in Shimoga district	700	122	–	Karnataka
Chakra	336	East of Kodachadri in Shimoga district	600	52	Kollur	Karnataka
Varahi (Haladi)	759	Kavaledurga in the Shimoga district	600	66	–	Karnataka
Netravathy	3,222	Bellarayan in the Dakshina Kannada district	1,000	103	Gundiahole, Kumaradara and Shisiahole	Karnataka
Barapole (Valapat-tanam)	1,867	Brahamagiri Ghat Reserve Forest in Coorg district	900	110	–	Karnataka & Kerala

Some important independent catchments in the West Flowing River system are described in the following sections.

16.1.1. Independent Catchment between Sharavathi and Chakra River

Many small streams flow in this area and join the Arabian Sea. Kollur River, Ghantihole, Venkatapur, Baindurhole, Shankargundi, Kumbarhole and Yedamavinahole are the important streams in this region. The entire catchments of the streams flowing in this region lie in the state of Karnataka.

16.1.2. Independent Catchment between Varahi and Netravathi River

The independent catchment comprises of streams namely Swarna, Seethanadhi, Mulki river, Pavanje, Nadisalu, Gurpur,Yennehole and Madisalhole. Catchment areas of the streams falling in this region entirely lie in the state of Karnataka.

16.1.3. Independent Catchment between Netravathy and Chandragiri

The main streams draining this region are Chandragiri (Payaswani) and Shiriya Rivers. The Chandragiri River rises west of Mercara in Coorg District of Karnataka State at an elevation of about 600 m. Pyayaswani River originates from Patti Ghats reserve forest in Coorg District of Karnataka at an elevation of 1,350 m. The two river joins together at Machipana about 15 km upstream of their out fall point into Arabian Sea near Kasaragud. It drains catchment area of 1,406 km^2 out of which 836 km^2 lies in Karnataka state and the balance portion in Kerala state.

16.2. WEST FLOWING RIVERS OF KACHCHH AND SAURASHTRA INCLUDING LUNI AND INLAND DRAINAGE

This is a fairly large basin whose total area is 334, 390 km^2. It lies between east longitudes 68°7′ and 75°50′ and north latitudes 20°40′ and 29°25′. Located in Western India, this area covers large parts of Rajasthan, Gujarat and Diu (UT) as well as (very small) inland drainage area of Punjab. The state-wise distribution of the drainage area is given in Table 2.

The basin is bounded on the north and east by the ridge separating it from the Indus, the Ganga and the Sabarmati basins, on the south by the Arabian Sea and on the west by Pakistan. The shape of the basin is irregular; it has a maximum length of 1,045 km^2 in a NE-SW direction and a maximum width of 470 km in a NW-SE direction.

Although many rivers flow in this region, the more important are the Luni, the Shetrunji, the Bhadar, the Machhu, the Rupen, the Saraswati, the Banas, the Mitti and the Damanganga. The chief characteristics of the rivers flowing in Rajasthan

Table 2. Statewise catchment areas of west flowing rivers of Kachchh and Saurashtra

State	Drainage area (km^2)
Rajasthan	193,359
Gujarat	140,932
Diu	45
Punjab (inland drainage)	54
Total	334,390

are that after flowing for some distance, they disappear in the desert. A description of the major rivers follows.

16.2.1. Luni River

Luni river basin is located in south-western Rajasthan, between latitudes 23°41′ and 27°05′ and longitudes 71°04′ and 74°42′. It is bounded by the arid western districts in the west, by Banas basin in the east, Shekhawati basin in the north, and Sukli and West Banas basins in the south. The basin extends over parts of Ajmer, Barmer, Jalore, Jodhpur, Nagaur, Pali, Rajsamand, Sirohi and Udaipur Districts. The total catchment area of the basin is 37,363 km^2. Luni River originates in the western slopes of the Aravalli range at an elevation of 550 m near Ajmer. After flowing for about 495 km in a south-westerly direction in Rajasthan, the river disappears in the marshy land of Rann of Kutch. In the basin, elevations range from 890 m to 10 m (near the outlet). Nearly half of the basin is occupied by rugged mountains where soils are shallow.

Annual rainfall over the Luni basin varies between 600 mm to 300 mm; the mean annual rainfall is computed as 320 mm, of which about 97% falls during the four monsoon months (June-September). The number of rainy days in a year is quite low, around 14. Being located in arid region, annual average pan evaporation is 2,640 mm which is eight times the mean rainfall. Flow in the river is observed only during a few days in a year.

The main tributaries of Luni on the left are Sukri, Mithri, Bandi, Khari, Jawai, Guhiya and Sagi whereas only Jojari River joins it on the right side.

i. Sukri river

The Sukri basin is located between latitudes 25°00′30″ and 25°53′ and longitudes 72°36′ and 73°42′, covering an area of 3,036 km^2. The Sukri River is formed by the confluence of several small nallahs (streams) – Ghanerav Nadi, Muthana ka Bala, Magai Nadi etc., originating from the Aravalis in Pali and Udaipur Districts. Sukri flows for about 110 km in a south-east to north-west direction and feeds Bankli Dam on the way. It joins Luni River near the village Samdari in Barmer District. This sub-basin covers parts of Jalore, Pali, and Barmer Districts.

ii. Khari river

The Khari basin, with catchment area of 1,232 km^2, is located between latitudes 25°18′ and 25°46′ and longitudes 73°20′ and 73°50′. The Khari River is formed by the confluence of small streams, namely, Somesar and Khari Kherwa. Somesar River (originally known as Sumer Nadi) originates in the western slopes of the Aravali range near the village Somesar in Pali District. The stream Umrawas Ka Nalla, originating in the western slopes of the Aravali near village Kanklawas, joins Sumer Nadi. Kotki Nadi, originating from Dewair Reserved Forest Bhakar also joins Sumer Nadi after flowing for about 30 km. After all these small streams join, the river is called Khari. It joins Bandi River downstream of Hemawas reservoir after flowing for about 25 km.

iii. Mithari river

The Mithari basin is located between latitudes 25°00′ and 25°30′ and longitudes 72°52′ and 73°29′. The Mithari River is formed in the south-western slopes of the Aravali range in Pali District by confluence of many local nallahs. The river flows through Jawai, Bali and Falna Districts for about 80 km in a north-west direction before vanishing in sandy plains near village Sankhwal in Jalore district. The catchment extends over parts of Pali and Jalore Districts and covers a total of 1,644 km^2.

iv. Bandi (hemawas) river

The Khari and Mithari Rivers join near Bombadra pickup weir to form the Bandi River. It joins the Luni near the village Lakhar in Pali District after flowing for about 45 km. The Bandi basin is located between latitudes 25°15′ and 25°55′ and longitudes 72°56′ and 73°57′. The catchment, situated in Pali District, covers an area of 1,685 km^2.

The Guhiya River originates in the hillocks near the villages Khariyaniv and Tharasani in Pali District. It joins Bandi River near village Phekariya in Pali District. The basin is located between latitudes 24°45′ and 26°14′ and longitudes 72°58′ and 74°14′. The catchment area of the basin is 3,835 km^2 and it falls in Pali district. The main tributaries of Guhiya River are Raipur Luni, Radia Nadi, Guria Nadi, Lilri Nadi, Sukri and Phunpharia Bala.

v. Jawai river

The Jawai basin is located between latitudes 24°43′ and 25°34′ and longitudes 72°31′ and 73°24′. The Jawai River, and its main tributary Sukari, originate in the western slopes of the Aravalis, in Udaipur District. The river flows generally in a north-west direction for about 96 km, before joining River Khari near the village Sayala in Jalore District. The catchment area is 2,976 km^2.

Sukri (Sayala to Luni): The Jawai River, after its confluence with the Khari River is called Sukri. The basin is located between latitudes 25°05′ and 25°26′ and longitudes 71°44′ and 72°22′. It flows in a southwest direction for about 80 km before joining the Luni River near Golia village. The catchment covering 1,615 km^2 extends over parts of Jalore and Barmer Districts. The main tributaries of Sukri River are Krishnawati and Kameri.

vi. Jojari river

The Jojari basin is located between latitudes 26°07′ and 26°43′ and longitudes 73°08′ and 74°00′. The Jojari River originates in the hillocks near village Pondlu in Nagaur District. It flows for about 83 km from north-east to south-west before joining Luni River near the village Khejadli Khurd in Jodhpur District. The catchment extends over Jodhpur and Nagaur Districts and covers an area of 2,571 km^2. A number of small streams join this river in the upper reaches.

vii. Sagi river

This basin is located between latitudes 24°46′ and 25°09′ and longitudes 71°40′ and 72°26′. The Sagi River originates in the south-western slopes of Jaswantpura hills in Jalore District. It flows initially for about 72 km northwest and then southwest before joining the Luni near Gandhav village in Barmer District. Kari Nadi is the only tributary of River Sagi. The catchment extends over parts of Jalore and Barmer Districts and covers an area of 1,495 km^2.

16.2.2. Machhu River

Machhu is one of the major rivers of Saurashtra region of Gujarat. It rises near Bhadla in the Rajkot district of Gujarat at an elevation of 275 m at North latitude 22°11′ and East longitude 71°6′ flowing in a generally northern, north-western course and disappears in the little Runn of Kutch downstream of Malia. On the way, a few rivers and tributaries join it. Important among them are Jumbudi, Banaiya, Patalia Vonkala, Asoi, Maha and Matelio. The total length of the Machhu River is 161 km and its total catchment area is 2,331 km^2.

Machhu irrigation scheme was first contemplated as early as 1924 by the authorities of former Wankaner state. The scheme was envisaged construction of two reservoirs on the main Machhu River. This scheme intended to provide irrigation facilities to the villages of Wankaner and Morbi Talukas which are often hard hit by severe drought.

The average rainfall for the Machhu catchment is 52.20 cm. The maximum rainfall during any one year in the catchment has been at Sardhana during the year 1917, being 139.88 cm. The minimum rainfall observed at Kuwadva during 1928 was 7.08 cm. The runoff factor for the catchment is 20%. The daily mean minimum and maximum temperature as observed at the Rajkot IMD station, are 10.7 °C (Jan) and 45.5 °C (May) respectively. The wind velocity is highest during the month of May to August.

The soils in the command area are mostly of residual origin derived from basaltic Deccan trap and have montmorillonite as the clay material.

16.2.3. Damanganga River

The Damanganga River originates in Sahyadri hill ranges in Nasik district of Maharashtra at elevation 930.5 m above mean sea level. Two major distributaries, Daman and Vag, after traveling through a course of 79 km and 61 km respectively, meet near village Matunji to form Damanganga River. For its major course, it runs through Maharashtra, then enters Gujarat to finally join Arabian Sea. The gross catchment area of the river Damanganga at the dam site is 1,813 km^2. The average rainfall in the area is 220.2 cm with maximum being 378.0 cm.

16.2.4. Water Resources Projects in the Basin

i. Machhu-I dam

This dam is located on river Machhu near Jalsika village in Wankaner taluka of Rajkot district. The location of the dam is 57 km from the source of the river. The town of Wankaner, one of the seats of former princely states, lies 22 km downstream of Machhu-I dam. The dam was completed in the year 1958. The catchment area up to the dam site is 735 km^2. This project has been designed for gross and live storage capacities of 72.7 Mm^3 and 70.8 Mm^3 respectively with FRL at 135.35 m.

The Machhu-I dam has been conceived as a reservoir impounding water for the purpose of irrigation. The command area of the project lies on the left bank of Machhu River in the Wankaner and Morbi talukas of Rajkot district. The gross command area of this project is 182.18 Mm^2 and the culturable command area is 104.09 Mm^2.

The whole of the catchment area gradually rises towards the source in the north-eastern direction, i.e., Mandva hills from where the river takes its origin. Out of the 735 km^2 of the catchment area at the dam site, 36.3 km^2 have already been intercepted by a tank at Aria, 35.4 km upstream of the dam site.

The reservoir is intended to provide irrigation water all the year round. Hence a provision for the evaporation and absorption losses that occur in the reservoir throughout the year has been made. Taking evaporation and absorption losses as 1.83 m for the entire year, an allowance of 8.67 Mm^3 of water has been made in the reservoir. The spillway of this project was designed to pass a flood of 2,595 cumec with HFL at 137.46 m. Clear overfall weir of 487.68 m in the river portion with water cushion has been provided.

ii. Machhu-II dam

This dam is located on river Machhu near village Jodhpur in Morbi taluka of Rajkot district at a distance of 103 km from the source of the river. The total catchment area up to the dam site is 1,928 km^2. The gross and live storage capacities of this dam are 100.41 and 90.73 Mm^3 respectively with FRL at 57.32 m. The Machhu-II dam has been conceived as a reservoir impounding water for conservation purposes like irrigation and municipal water supply and for flood control. The maximum observed flood for this project is 13,026 cumec while the peak of design flood hydrograph is 18,548 cumec. However for the purpose of spillway design, the peak flood of 26,420 cumec has been adopted. This dam was breached in August, 1979 floods. The dam has been rebuilt now.

The towns of Morbi and Malia, lie 9 km and 46 km downstream of the Machhu-II dam respectively. The Morbi town is situated on the left bank of the river whereas the Malia town is situated about 1.5 km away from the left bank. The area under the command lies on the left bank of river Machhu. Geographically, the command lies between latitudes of 22°46′ N and 22°57′ N and longitudes of 70°52′ E and 70°40′ E. The area has more or less flat topography which is characteristic of the coastal low-lands. The water stored in the reservoir is utilized mainly during

Kharif & Rabi seasons. The area where this scheme has been taken up is much affected by famine and hence the Govt. of Gujarat has taken up this scheme as a protective scheme.

The catchment area up to Machhu-II dam site is partly hilly and partly cultivated and fan shaped. Most of the annual rainfall in the catchment area falls in the monsoon months from June to September. The average annual rainfall in the Machhu – II catchment is 60.2 cm. The maximum temperature reaches around 43 °C in the hottest month of May whereas the lowest temperature falls to 6 °C in the coldest month of January.

It may be pointed out that in this particular case, the capacity of the reservoir is governed by the submergence of lands and the probable submergence of a part of the railway line from Rafaleshwar to Dhuva on the Novlakhi Wankaner line. The lowest level of the top of railway embankment at certain places is at R.L. 59.44 m.

iii. Damanganga Dam

Damanganga dam was constructed in the year 1983 across Damanganga River near village Madhuban of Valsad district. The dam site is 30 km away from Vapi and 60 km from Valsad. The dam is designed to impound gross storage of 567 Mm^3 at RL 79.584 m. The live storage capacity of the dam is 502.00 Mm^3. The length of masonry dam is 352 m and length of earthen dam is 2,376 m. The maximum height of dam above river bed on ground is 57.50 m. The spillway length is 191.11 m. Ten radial gates of size 15.55 m × 14.02 m are provided. The designed annual irrigation from the dam is 570 Mm^2.

iv. Mitti dam

The Mitti dam is basically an irrigation scheme located on River Mitti near Village Trambau in Abdasa Taluka of Kachchh District in Gujarat. The latitude and longitude of the site are 23°20′0″ amd 68°49′30″ respectively. The scheme drains a catchment area of 468.79 km^2 and serves a gross command area of 55.67 Mm^2. The gross storage, dead storage and live storage capcities of the dam is 17.40 Mm^3, 2.68 Mm^3, and 14.72 Mm^3 respectively.

v. Palitana dam

The Palitana dam, which is also known as Shetrunji irrigation scheme was constructed in the year 1959 across river Shetrunji near village Nani – Rajasthali of Palitana taluka of Bhavnagar district. The dam site is 10 km away from Palitana city on Palitan Talaja road.

The dam is designed to impound gross storage of 415.414 Mm^3 at RL 55.53 m. The length of masonry dam is 769.79 m and total length of earthen dam is 3,126.86 m. The maximum height of dam above river bed on ground is 25.31 m. The gross catchment-area of the Shetrunji River at the dam site is 4,317 km^2 and intercepted catchment is 389 km^2. The catchment area of the river is partly hilly and partly plain. The drainage area up to village Padargadh is hilly and full of vegetation.

Shape of catchment is fan type with maximum length being 105 km and width about 72 km. Designed rate of sedimentation is 0.476 mm/year for gross storage.

The mean annual rainfall in the watershed is 55.20 cm with the maximum being 96.10 cm and the minimum being 17.20 cm. In the command, the maximum rainfall is 50.80 cm. The mean annual flood at the dam site is 4,183 cumec. A 645.56 m long spillway is designed to pass discharge of 7,080 cumec. Four gates of size 1.51 m × 1.83 m on right flank and 2 gates on left bank are provided.

vi. Other projects

There are some more existing and under construction water resources projects in the regioin. Salient features of existing and under construction water resources projects with a live storage capacity of 10 MCM and above have been presented in Table 3 and Table 4 respectively.

16.3. WEST FLOWING RIVERS SOUTH OF TAPI

All west flowing rivers from Tapi to Kanyakumari can be placed in this group. The total area of this region is $113,057\,km^2$ which lies between longitudes 72°43′ E and 77°35′ E and latitudes 8°4′ N and 21°10′ N. The basin is located in southern and western India and covers practically the whole of Kerala, Goa, Daman, Dadra & Nagar Haveli and parts of Tamil Nadu, Karnataka, Maharashtra and Gujarat. The state wise distribution of the drainage basin is given in Table 5.

The region is bounded on the north by the ridge separating it from the Tapi basin, on the east by the Western Ghats, on the south by the Indian Ocean and on the west by the Arabian Sea. The basin is highly irregular in shape and has a maximum length of 1,555 km in NW-SE direction and a maximum width of 125 km in NE-SW direction.

The basin is traversed by as many as 115 small and fairly big rivers, of which one in Tamil Nadu, 32 in Kerala, 10 in Karnataka, 3 in Goa, 11 in Maharashtra and 5 in Gujarat are important rivers. The list of 98 rivers, (45 rivers systems between Tapi to Tadri and 43 rivers systems from Tadri to Kanyakumari) along with the catchment area, rainfall and average annual flows are given in Table 6.

The Achencoil River basin in Kerala lies between latitude 9° to 10 ° N and longitude 76° to 77 °E, covering an area of $1,484\,km^2$. Achencoil flows for 128 km. It discharges a part of its flows into the Arabian Sea through a flood regulator and the major part flows to the Vembanad estuary. Achencoil basin receives an annual average rainfall of 2,600 mm, mainly from the two monsoons: the Southwest monsoon (June to September) contributes about 60% of the rainfall and the Northeast monsoon (October to November) about 30%. Spatially, the average annual rainfall increases from about 2,400 mm per year in the coastal region to 3,200 mm in the highland region. The average annual yield of the basin is 2,793 MCM. This basin has highly undulating topography; the land use varies from evergreen forests and plantation crops (mainly rubber and tea) in the highland regions to rice and coconut in the midland and coastal regions.

Table 3. Salient features of selected existing projects in West Flowing Rivers of Kutch And Saurashtra Including Luni basin

Name of the project	State	Year of completion	Gross storage capacity (Mm^3)	Live storage capacity (Mm^3)	Designed annual irrigation (Mm^2)
Aji Dam	Gujarat	1957	32.28	30.00	11.30
Aji-II	Gujarat	–	18.87	18.48	23.80
Aji-III	Gujarat	–	65.17	50.96	66.20
Bawal	Gujarat	1978	26.73	24.00	–
Bhadar	Gujarat	1969	237.80	223.70	171.50
Bhimdad	Gujarat	1951	31.18	11.07	11.50
Bhuki	Gujarat	–	15.88	14.58	13.50
Brahmani	Gujarat	1954	75.01	71.39	38.50
Chaparwadi	Gujarat	1946	18.52	14.71	29.40
Chimanbhai Lake	Gujarat	1906	17.88	13.40	–
Dantiwada	Gujarat	1969	464.40	445.00	445.20
Dhatarwadi	Gujarat	1976	32.71	26.56	24.80
Demi-I	Gujarat	1964	21.52	19.52	21.90
Demi-II	Gujarat	–	18.91	18.31	24.20
Deo	Gujarat	–	84.09	76.33	96.10
Fodarness Dam	Gujarat	1970	23.64	22.26	–
Fulzar	Gujarat	1961	14.89	13.76	12.20
Gandali	Gujarat	1956	11.35	10.87	9.40
Ghee	Gujarat	1953	–	13.82	8.30
Ghelo	Gujarat	1968	13.35	12.00	31.50
Godhatad	Gujarat	–	14.70	13.12	6.60
Goma	Gujarat	1976	18.52	16.46	25.30
Hiran	Gujarat	1966	38.58	35.00	26.30
Jhanjeshwari	Gujarat	1977	10.64	10.00	40.50
Jhuj	Gujarat	–	28.65	27.58	58.90
Kaila	Gujarat	1955	13.98	13.20	8.80
Kalubhar	Gujarat	–	26.94	21.64	510.70
Kanakvati	Gujarat	1955	14.61	13.20	15.60
Karmal	Gujarat	–	12.70	10.03	40.40
Kashwati	Gujarat	1978	13.98	13.08	6.10
Kelia	Gujarat	–	18.10	17.35	8.80
Kharo	Gujarat	–	15.25	12.50	–
Khodiyar	Gujarat	1965	–	40.35	–
Machmundri	Gujarat	1963	31.84	27.50	4.50
Maduvanti	Gujarat	1976	12.14	11.73	21.90
Mahagarh	Gujarat	1970	60.00	45.00	–
Malan	Gujarat	1957	14.47	13.87	23.30
Moj	Gujarat	1966	53.06	50.11	48.20
Munjiasar	Gujarat	1957	14.73	12.00	13.40
Nara	Gujarat	1975	41.03	38.83	29.80
Niruna	Gujarat	1963	32.20	11.02	19.40
Nyari-II	Gujarat	–	13.00	11.70	4.70
Pauna	Gujarat	1953	13.70	13.14	10.10

Phoophal	Gujarat	1974	59.61	51.68	79.10
Rajawadla tank	Gujarat	1902	22.87	11.20	2.40
Rajawal	Gujarat	–	33.01	27.40	103.00
Ranghola	Gujarat	1952	44.51	42.93	35.00
Rawal irri. Scheme	Gujarat	1960	23.02	19.62	58.30
Rawal-II	Gujarat	–	27.00	24.00	41.80
Rojki	Gujarat	1962	12.00	11.02	15.40
Rudramata	Gujarat	1962	64.74	56.00	30.00
Sananadaro	Gujarat	1955	12.27	11.75	6.80
Sankroli	Gujarat	1963	10.99	10.30	12.20
Sasoi	Gujarat	1954	51.03	46.78	30.60
Singoda	Gujarat	–	36.44	35.17	60.70
Sukhbhadar	Gujarat	–	44.13	37.22	73.90
Suvi	Gujarat	1966	14.28	11.90	13.40
Tapar Dam	Gujarat	1975	14.71	14.50	–
Und	Gujarat	–	72.50	65.48	73.00
Varjani	Gujarat	–	10.35	10.05	–
Vartudam	Gujarat	1969	13.31	11.64	26.90
Venu-II	Gujarat	–	22.59	16.15	61.20
Vijaya Sagar	Gujarat	1948	20.82	19.18	18.40
Waidy	Gujarat	1984	13.60	12.60	20.10
Bankli	Rajasthan	1955	–	34.55	35.10
Hemarvas	Rajasthan	–	–	62.56	20.20
Jawai	Rajasthan	1954	207.37	193.37	176.10
Jaswant Sagar	Rajasthan	1973	–	32.83	38.20
Kharda	Rajasthan	–	–	15.15	19.40
Ora	Rajasthan	–	22.66	22.29	31.90
Raipurluni	Rajasthan	1964	12.54	10.90	14.20
Sardar Samant	Rajasthan	–	–	88.22	85.60
West Banas Dam	Rajasthan	1962	39.08	36.22	40.80

Table 4. Salient features of selected under construction projects in West Flowing Rivers of Kutch and Saurashtra Including Luni basin

Name of the Project	State	Gross storage capacity (Mm^3)	Live storage capacity (Mm^3)	Designed annual irrigation (Mm^2)
Aji-IV	Gujarat	22.94	15.14	40.00
Amipur	Gujarat	–	29.45	67.70
Datardi	Gujarat	12.20	10.60	–
Goma	Gujarat	32.00	30.70	25.20
Hiran-II	Gujarat	38.58	34.29	95.10
Machhundri-II	Gujarat	31.50	27.50	64.80
Mazarri	Gujarat	18.10	13.75	–
Medhacreck	Gujarat	–	37.95	–
Mukteshwar	Gujarat	39.90	31.71	58.30
Sani	Gujarat	55.08	46.58	41.20
Sipu	Gujarat	177.80	156.00	320.00

Table 5. Statewise catchment areas of west flowing rivers South of Tapi

State	Drainage area (km^2)
Tamil Nadu	4, 694
Kerala	35, 930
Karnataka	24, 764
Goa	3, 700
Maharashtra	32, 806
Dadra & Nagar Haveli	490
Daman	65
Gujarat	10, 608
Total	113, 057

Table 6. Salient features of west flowing rivers south of Tapi

S. N.	Basin	Average annual rainfall (mm)	Catchment area (km^2)	Average-annual flow (Mm^3)
	Rivers System from Tapi to Tadri			
1	Purna	1, 521	2, 322	1, 587
2	Ambica	1, 999	2, 715	3, 528
3	Damanganga	1, 750	1, 382	1, 572
4	Auranga	2, 200	172	246
5	Par	2, 076	688	942
6	Varoli (Jogni)	2, 200	238	340
7	Free catchment between Varoli and Vaitarna	2, 000	805	1, 047
8	Vaitarna	2, 794	3, 647	6, 700
9	Ulhas	3, 556	3, 804	8, 793
10	Free catchment between Ulhas & Parval	2, 200	266	380
11	Parvel	2, 962	803	1, 546
12	Patalganga	3, 366	940	2, 056
13	Amba	3, 050	740	1, 467
14	Kundalika	3, 143	825	1, 685
15	MandaI Jansi	2, 655	1, 505	2, 597
16	Savitri	3, 493	2, 889	6, 558
17	Bharaja	3, 473	383	639
18	Ilne	3, 200	74	154
19	Jog	3, 511	385	379
20	Vashishti	3, 391	2, 238	4, 933
21	Free catchment between Vashiti & Shastri	3, 000	225	439
22	Shastri	3, 260	2, 174	4, 607
23	Free catchment between Shastri & Kajvi	2, 900	279	526
24	Kajvi	2, 550	762	1, 263
25	Free catchment between Kajvi & Machkundi	3, 000	94	183

26	Machkundi	2, 150	833	1, 164
27	Kodavali (Rajapur)	3, 216	665	1, 390
28	Vaghotan,	2, 500	903	1, 467
29	Kharda (Deogarh)	2, 500	455	739
30	Piyali	3, 000	86	168
31	Achra	3, 207	297	619
32	Gad	2, 600	891	1, 506
33	Karli	2, 750	753	1, 346
34	Free catchment between Karli & Talwada	3, 000	246	480
35	Talvada	3, 225	137	287
36	Free catchment between Talwada & Terekhol	3, 125	96	195
37	Terekhol	3, 578	530	1, 233
38	Chapora (Tillari)	3, 578	530	1, 233
39	Mandvi (Mahadayi)	3, 134	2, 032	4, 139
40	Rachol	2, 959	772	1, 485
41	Sal	2, 800	344	626
42	Free catchment between Sal & Kalinadi	2, 900	446	841
43	Kalindi	2, 436	4, 188	6, 631
44	Gangavali (Bedthi)	2, 039	3, 574	4, 737
45	Agnashini (Tadri)	2, 956	1, 330	2, 556
			Sub-total	**87, 009**
	River Systems from Tadri to Kanyakumari			
46	Sharavathy	3, 169	3, 592	7, 399
47	Free catchment between Banduruholi & Sharavathy			
48	Banduruholi	4, 000	1, 041	2, 707
49	Free catchment between Banduruholi & Chakra			
50	Chakra	4, 082	336	892
51	Haladi (Varahi)	4, 587	759	2, 263
52	Sita			
53	Free catchment between Swarna & Mulki			
54	Mulki	3, 800	3, 067	7, 576
55	Free catchment between Mulki & Gurpur			
56	Gurpur			
57	Netravathy	5, 363	3, 222	12, 813
58	Manjeshwar	3, 600	90	210
59	Uppala	3, 600	250	585
60	Shiruja	3, 800	587	1, 450
61	Morgal	3, 800	132	326
62	Chandragiri	3, 809	1, 406	3, 481
63	Chittari	3, 600	145	339
64	Nileswar	3, 700	190	457
65	Kariangode	4, 000	561	1, 459
66	Kaviyi	3, 600	143	335
67	Peruvamba	3, 700	300	722

(Continued)

Table 6. (Continued)

S. N.	Basin	Average annual rainfall (mm)	Catchment area (km^2)	Average-annual flow (Mm^3)
68	Ramapuram	3, 600	52	122
69	Kuppam	3, 900	539	1, 366
70	Vallapattanam	3, 735	1, 867	4, 533
71	Anjarakadi	4, 000	412	1, 071
72	Tellicherry	3, 900	132	335
73	Maha	4, 000	394	1, 024
74	Kuttiadi	3, 900	588	1, 478
75	Korapuzha	3, 800	624	1, 541
76	Kallai	3, 200	96	200
77	Chaliyar	3, 201	2, 923	6, 082
78	Kadalundi	3, 201	1, 122	2, 335
79	Terurpuzha	2, 800	117	213
80	Bharatapuzha	2, 276	6, 186	7, 054
81	Keechari	2, 800	401	730
82	Puzhalkal	2, 800	234	426
83	Karuvannur	2, 800	1, 054	1, 900
84	Chalakudy	4, 290	1, 704	3, 277
85	Periyar	2, 919	5, 398	12, 210
86	Moovattupuzha	3, 385	2, 004	4, 409
87	Meenachil	3, 400	1, 272	2, 811
88	Manimala	3, 800	847	2, 092
89	Pamba	3, 600	2, 235	5, 230
90	Achenkoil	2, 600	1, 484	2, 793
91	Pallickal	2, 800	220	400
92	Kallada	2, 800	1, 699	3, 092
93	Ithikkara	2, 600	642	1, 085
94	Ayroor	2, 300	66	99
95	Vamanapuram	2, 400	687	1, 072
96	Mamom	2, 000	114	148
97	Karamana	2, 000	703	914
98	Neyyar	2, 000	497	646
		Sub-total		113, 702
	Total			200, 711

Source: WG (1999).

16.3.1. Water Resources Projects in the Area

Some important projects located in this region are described next.

Tillari

Tillari is a masonry dam on Tillari River, 19 km from Chandgad in Kolhapur District, Maharashtra. The catchment area at the dam is 47.59 km^2. The height and length of the dam are 38 m and 485 m respectively. The reservoir has a live storage capacity of 113.24 MCM at FRL 750.2 m and the MDDL is at 737 m. With one unit of 60 MW, Tillari power house has a firm power of 16 MW. MSEB commissioned the project in 1986.

Bhivpuri

Bhivpuri hydropower project is located on Thokarwadi Rubble masonry gravity dam on Andhra River, a tributary of Indarayni River, 18 km from Karjat, in Raigad District, Maharashtra. The corresponding reservoir is known as Andhra Lake. The catchment area at the dam is 124 km^2. The height and length of the dam is 58 m and 571 m respectively. The reservoir has a live storage capacity of 325.64 MCM at FRL 668 m. Initially, the power house had the installed capacity of 48 MW, which was uprated to 6 units of 12 MW each in 1961. In 1998–99 the old plant was replaced with 3 units of 24 MW each. The project has a firm power of 32.5 MW and mean annual inflow of 247.32 MCM.

Bhira and Bhira pumped storage projects

These projects are located on Mulshi rubble masonry dam on Mula and Neela Rivers. These projects are located at a distance of 30 km and 27 km respectively from Kolad, in Raigad District, Maharashtra. The catchment area at the dam is 247.6 km^2. The height of the Mulshi dam is 47 m. The reservoir has a live storage capacity of 522.76 MCM at FRL 607 m and the MDDL is at 590 m. Former power project has 6 units of 25 MW each whereas the later has a single unit of 150 MW. The Bhira hydropower project has a firm power of 86 MW with mean annual inflow of 685 MCM.

Bhira tail race project

One more power house has been constructed to utilize the tail water of Bhira power house and Kundlika River, known as Bhira tail race. This power house is located 32 km from Mangaon, in Raigad District, Maharashtra. The catchment area at the dam is 245 km^2. The power house has 2 units of 40 MW each and has a firm power of 21 MW with mean annual inflow of 608 MCM.

Vaitarna

Vaitarna hydropower project is located near Vaitarna and Alwandi masonry and earthen dam on Vaitarna and Alwandi Rivers, 30 km from Ghoti, in Nashik District, Maharashtra. The catchment area at the dam is 160.8 km^2. The height and length of the dam is 47 m and 555 m respectively. The reservoir has a live storage capacity of 35 MCM at FRL 603.5 m and the MDDL is at 580 m. The power house has a unit of 60 MW. It has a firm power of 11 MW with mean annual inflow of 635 MCM. MSEB commissioned the project in 1976.

Kalinadi

Kalinadi Hydropower station utilizes water from the Bommanhalli masonry gravity dam constructed on Kalinadi River. The power house is located at a distance of

4 km from Dandeli in Uttara Kannada District, Karnataka. The catchment area at the power house is 636 km^2. The height and length of the dam are 28.96 m and 1,025 m respectively. At FRL 438.38 m, the reservoir has a live storage capacity of 83.90 MCM. Kalinadi power house has 6 units of 135 MW each. It has a firm power of 341 MW with a mean annual inflow of 735 MCM. KPCL commissioned the project in 1979–84.

Kadra

Kadra is a masonry gravity and earth dam on Kalinadi River located a distance of 35 km from Karwar in Uttara Kannada District, Karnataka. The catchment area at the dam is 432 km^2. The height of the dam is 40.50 m. The reservoir has a live storage capacity of 209 MCM at FRL 34.5 m and the MDDL is at 27 m. Kadra power house has 3 units of 50 MW each, producing a firm power of 40 MW. KPCL commissioned the project in 1997–99.

Kodasalli

Kodasalli is a 49 m high concrete gravity earth dam on Kalinadi River located a distance of 70 km from Karwar in Uttara Kannada District, Karnataka. The catchment area at the dam is 1,049 km^2. The reservoir has a live storage capacity of 178.82 MCM at FRL 75.5 m and the MDDL is at 62.5 m. Kodasalli power house has 3 units of 40 MW each and a firm power of 38 MW. It was commissioned in 1998–99.

Pykara Singara

Pykara Singara hydropower project on Glenmorgan masonry & gravity dam on Pykara and Mukurthy Rivers, located at a distance of 36 km from Udhagamandalam in Nilgiris District, Tamil Nadu. The height and length of the dam is 29 m and 250.6 m respectively. At FRL 1,972.56 m, the reservoir has a live storage capacity of 6.113 MCM. Pykara Singara power house has 3 units of 6.65 MW each, 2 units of 11 MW each, and 2 units of 14 MW each with a total installed capacity of 70 MW. It has a firm power of 46 MW. TNEB commissioned the project in 1932–54.

Kadamparai

Kadamparai is a masonry gravity and earth dam on Kadamparai River, tributary of Aliyar River, located near Pollachi in Coimbatore District, Tamil Nadu. The catchment area at the dam is 83 km^2. The height of the dam is 63.5 m. The reservoir has a live storage capacity of 27 MCM at FRL 1,149 m and the MDDL is at 1,112 m. Kadamparai power house has 4 units of 100 MW each. It has a firm power of 10 MW. TNEB commissioned the project in 1987–88.

Sabarigiri

Sabarigiri hydroelectric station is located on the Kakki concrete gravity dam constructed on Pambas and Kakki Rivers. The power house is located 72 km from

Pathanamthitta in Quilon District, Kerala. The catchment area at the power house is 316 km^2. The height and length of the dam is 110 m and 336 m respectively. The reservoir has a live storage capacity of 446.5 MCM at FRL 981 m and the MDDL is at 908 m. It has 6 units of 50 MW each with a mean annual inflow of 824 MCM. It has a firm power of 139 MW. Kerala State electricity Board commissioned the project in 1966–67.

Kakkad

Kakkad hydroelectric station has two concrete gravity dams, namely Moozhiyar and Veluthodu, constructed on the rivers with the same names. Moozhiyar and Veluthodu streams are the tributaries of Pamba River. The power house is located in Pathanamthitta District, Kerala. The catchment area at the power house is 358.5 km^2 and the mean annual inflow is 77.59 MCM. The length of both dams is 192 m. It has 2 units of 25 MW each with a firm power of 29 MW. Kerala State electricity Board commissioned the project in 1999.

Supa dam

Supa is a concrete gravity dam on Kalinadi River located 22 km from Karwar in Uttara Kannada District, Karnataka. Supa dam is the main storage for the Kalinadi Hydroelectric project in Karnataka. It is a 101 m high concrete gravity dam. This dam creates main storage reservoir for the entire project with a total capacity of 4,300 Mm^3. The catchment area at the dam is 1,057 km^2 and its height is 101 m. The length of the dam of 322 m which is divided in 20 wedge-shaped blocks to suit the geology of the terrain and to facilitate speedier construction. The water impounded in Supa reservoir is being utilized for power generation at Nagajhari Power house which has 2 units of 50 MW each. It has a firm power of 54 MW with mean annual inflow of 3,393 MCM.

Varahi Project

Varahi project consists of a hydropower station near the Hulical earth dam constructed on Varahi River. The power house is located 35 km from Kundapur in Udupi District, Karnataka. The catchment area at the dam is 214 km^2. The height and length of the dam are 35.9 m and 660 m respectively. The reservoir has a live storage capacity of 15 MCM at FRL 564 m. It has 2 units of 115 MW each with a mean annual inflow of 198 MCM. It has a firm power of 121 MW. KPCL commissioned the project in 1989–90.

Idamalayar

Idamalayar is a concrete gravity dam on Idamalayar River, tributary of Periyar River, 81 km from Ernakulam in Ernakulam District, Kerala. The catchment area at the dam is 381 km^2. The height and length of the dam is 91 m and 373 m respectively. The reservoir has a live storage capacity of 1,018 MCM at FRL 169 m and the MDDL is at 115 m. Idamalayar power house has 2 units of 37.5 MW each, with mean annual inflow of 1,370 MCM. It has a firm power of 37 MW. KSEB commissioned the project in 1987.

Linganamakki

Linganamakki is a masonry gravity earth dam on Sharavathy River, 35 km from Sagar, in Shimoga District, Karnataka, completed in 1964. The dam is located at 14°41′24″ N latitude and 74°50′54″ E longitude and has a catchment area of 1,992 km^2. The height and length of the dam is 61.28 m and 2,749 m respectively. Very high rainfall is received in the catchment of this dam; the recorded maximum rainfall is 8,150 mm. In addition, inflow is also received from Chakra and Savahaklu reservoirs which are linked with Linganamakki through a canal. Linganamakki reservoir has a live storage capacity of 4,276 MCM; its maximum water level is 555.04 m, FRL is at 554.43 m, and the MDDL is at 522.73 m. It has caused submergence of an area of 326 sq. km. Eleven radial gates of 15.24 m × 7.31 m have been provided at the spillway, the design flood peak being 8,070 cumec. From Linganamakki, water flows through a channel to the Talakalale balancing reservoir.

The power house at the site is known as Mahatma Gandhi (Jog Falls) power house. It has 4 units of 12 MW each and 4 units of 18 MW each with a total installed capacity of 120 MW. A view of Jog Fall is given in Figure 1.

Sharavathy

Sharvathy hydroelectric station has been constructed near the Talakalake masonry gravity dam constructed on Sharvathy River. The power house uses the tail water

Figure 1. A panoramic view of Jog Falls

of Linganamakki Dam constructed on Sharvathy River. The project is located at a distance of 30 km from Sagar, in Shimoga District, Karnataka. The catchment area at the dam is 46 km^2. The height and length of the dam is 62.5 m and 353.6 m respectively. The reservoir has a live storage capacity of 54 MCM at FRL 516.64 m and the MDDL is at 515.11 m. It has 10 units of 89.1 MW each. It has a firm power of 545 MW.

Aliyar

Upper Aliyar is a masonry gravity dam on Aliyar River, 35 km from Pollachi, in Coimbatore District, Tamil Nadu. The catchment area at the dam is 121 km^2. The height and length of the dam is 81 m and 316 m respectively. The reservoir has a live storage capacity of 25.91 MCM at FRL 770 m and the MDDL is at 730 m. The power house has a single unit of 60 MW each. It has a firm power of 12 MW with mean annual flow of 55 MCM. TNEB commissioned the project in 1970.

Idukki

Idukki is a concrete arch gravity dam on Periyar Cheruthoni and Killivally rivers, located at a distance of 25 km from Thodupuzha in Idukki District, Kerala. The catchment area at the dam is 649 km^2. The height and length of the dam is 169 m and 366 m respectively. The reservoir has a live storage capacity of 1,459.7 MCM at FRL 732.62 m and the MDDL is at 694.94 m. Idukki power house has 6 units of 130 MW each. It has a firm power of 230 MW with mean annual inflow of 1,360 MCM. A view of Idukki dam is shown in Figure 2.

The Cherthono tributary is dammed by a concrete gravity dam, about 135.7 m high and 651 m long. This dam was completed in 1976.

Kodayar

Kodayar is masonry gravity dam on Kodayar River, located at a distance of 25 km from Kulasekharan in Coimbatore District, Tamil Nadu. The catchment area at the dam is 29 km^2. The corresponding reservoir is known as Upper Kodayar. The height and length of the dam is 88 m and 181 m respectively. The reservoir has a live storage capacity of 73 MCM at FRL 1,326 m and the MDDL is at 1,296 m. Kodayar power house has a single units of 60 MW each. It has a firm power of 26 MW with mean annual inflow of 75 MCM.

Sholayar

Sholayar is a masonry gravity dam on Sholayar River, 65 km from Chalakudy in Trichur District, Kerala. The catchment area at the dam is 186.5 km^2. The height and length of the dam is 58 m and 431 m respectively. The reservoir has a live storage capacity of 149.2 MCM at FRL 812.68 m and the MDDL is at 779 m. Sholayar power house has 3 units of 18 MW each, with mean annual inflow of 1,710 MCM. It has a firm power of 27 MW. KSEB commissioned the project in 1966–68.

Figure 2. The majestic Idukki dam

Sholayar I

Sholayar I Hydro electric project is located near Sholayar masonry & earthen dam on Sholayar River, 25 km from Valparai and 90 km from Pollachi in Coimbatore District, Tamil Nadu. The catchment area at the dam is 120.32 km^2. The height and length of the dam is 105 m and 3,622 m respectively. The reservoir has a live storage capacity of 138 MCM at FRL 1,003 m and the MDDL is at 960 m. Sholayar I power house has 2 units of 35 MW each, with mean annual inflow of 1,034 MCM. It has a firm power of 28 MW. TNEB commissioned the project in 1971.

Kuttiyadi

Kuttiyadi is a masonry gravity dam on Kuttiyadi River, 55 km from Kozhikode in Kozhikode District, Kerala. The catchment area at the dam is 39 km^2. The height and length of the dam is 40 m and 229 m respectively. The reservoir has a live storage capacity of 29.6 MCM at FRL 758.04 m and the MDDL is at 737.62 m. Kuttiyadi power house has 3 units of 25 MW each, with mean annual inflow of 205.30 MCM. It has a firm power of 28 MW. KSEB commissioned the project in 1972.

Salient features of some more existing and under construction water resources projects with a live storage capacity of 10 MCM and above are presented in Table 7 and Table 8, respectively.

Table 7. Salient features of selected existing projects across West flowing Rivers South of Tapi

Name of the project	State	Year of completion	Gross storage capacity (Mm^3)	Live storage capacity (Mm^3)	Designed annual irrigation (Mm^2)	Installed capacity (MW)
Tapi To Tadri						
Anjunam	Goa	–	44.82	44.20	16.30	–
Bommanhalli Dam	Karnataka	1977	96.90	83.90	19.40	–
Thattihalla Dam	Karnataka	1980	264.00	249.00	–	810
Barvi	Maharashtra	1984	460.40	430.40	–	–
Khopoli	Maharashtra	–	–	186.90	–	70
Natuwadi	Maharashtra	1983	28.08	27.23	45.00	–
Surya	Maharashtra	–	285.31	276.35	270.00	4
Tansa	Maharashtra	1892	183.60	145.09	–	–
Tulshi	Maharashtra	1879	13.00	10.40	58.00	–
Vehar	Maharashtra	1860	41.50	31.72	–	–
Tadri To Kanyakumari						
Chakra Dam	Karnataka	1985	222.00	133.00	–	–
Savahaklu Dam	Karnataka	1980	94.36	65.80	–	–
Talakalale	Karnataka	1966	130.00	54.00	–	891
Anayirankal	Kerala	1965	49.84	48.99	–	11
Chalakudy	Kerala	1965	153.47	150.80	393.80	–
Mmeenkara	Kerala	1962	–	11.33	109.30	–
Challiar	Kerala	1965	–	13.70	–	–
Kakki Dam	Kerala	1966	455.02	447.41	416.80	410
Kallada	Kerala	–	504.90	487.92	920.00	–
Kannira Puzha	Kerala	1977	61.10	59.50	220.00	–
Madupatty	Kerala	1956	55.22	51.22	–	–
Malampuzha	Kerala	1955	236.69	226.96	420.09	–
Mangalam	Kerala	1962	25.47	24.67	68.80	–
Neyyar	Kerala	1959	106.08	101.15	179.50	–
Pamba Dam	Kerala	1967	39.22	31.15	490.00	–
Panniar Dam	Kerala	–	–	52.00	–	30
Peechi Dam	Kerala	1957	110.44	108.17	231.70	–
Ponmudi	Kerala	1963	51.59	47.40	93.70	–
Poringal Kuthu	Kerala	1957	32.00	30.30	393.80	60
Pothundi	Kerala	1968	50.91	43.89	109.30	–
Vazhani	Kerala	1958	18.12	16.65	71.30	–
Wadakkanchery	Kerala	1977	–	18.12	35.60	–
Walayar	Kerala	1956	18.90	18.40	64.70	–

(Continued)

Table 7. (Continued)

Name of the project	State	Year of completion	Gross storage capacity (Mm^3)	Live storage capacity (Mm^3)	Designed annual irrigation (Mm^2)	Installed capacity (MW)
Chittaratadnakma	Tamil Nadu	1974	27.80	20.85	14.93	–
Parambikulam	Tamil Nadu	1967	504.66	379.71	101.00	–
Periyar	Tamil Nadu	1897	443.55	299.35	57.87	–
Thirumurthy	Tamil Nadu	1967	54.80	49.39	80.83	–

16.4. EAST FLOWING RIVERS BETWEEN MAHANADI AND GODAVARI

Turning attention to the other side of the peninsular triangle, this group covers the east-flowing rivers between Mahanadi and Godavari and extends over an area of 49,570 km^2 which lies between longitudes 81°15′E to 85°30′E and latitudes 16°55′N to 20°18′N. The basin lies at the east coast of the peninsular India and covers large areas in the States of Orissa and Andhra Pradesh. The state-wise distribution of the drainage areas is shown in Table 9.

The basin of rivers in this zone bounded on the north, the west and the south by the various ranges of the Eastern Ghats and on the east by the Bay of the

Table 8. Salient features of selected under construction projects in West flowing Rivers South of Tapi

Name of the Project	State	Gross storage capacity (Mm^3)	Live storage capacity (Mm^3)	Designed annual irrigation (Mm^2)	Installed capacity (MW)
Tapi To Tadri					
Salauli	Goa	246.34	239.13	229.80	–
Dandeli Dam	Karnataka	78.23	24.42	–	–
Bhatsa	Maharashtra	957.10	923.12	430.00	10
Deogad	Maharashtra	100.43	98.02	83.50	–
Gad	Maharashtra	66.62	64.73	–	21
Hetavana	Maharashtra	127.51	127.00	128.30	–
Talamba	Maharashtra	297.15	285.83	160.00	–
Wandri	Maharashtra	36.51	34.71	30.60	–
Tadri To Kanyakumari					
Bijjur	Karnataka	65.00	16.42	18.20	–
Mani Dam	Karnataka	1,018.47	928.68	–	14
Attapady	Kerala	65.00	60.70	83.80	–
Chimmony	Kerala	179.24	172.49	260.00	–
Muvathupuzha	Kerala	42.00	31.00	520.00	–
Puyaukutty-I	Kerala	1,226.00	1,021.50	–	–
Vamanapuram	Kerala	81.60	80.00	180.10	–

Table 9. Statewise catchment areas of east flowing rivers between Mahanadi and Godavari

State	Drainage area (km^2)
Orissa	25,665
Andhra Pradesh	23,905
Total	49,570

Bengal. The basin, which is irregular in shape has a maximum length of about 182 km in the northwest-southeast direction and a maximum width of 476 km in the northeast-southwest direction.

The basin can be divided into two major topographical divisions: the hill ranges of the Eastern Ghats and the coastal plains. The hill ranges are well-forested. The plains extending from the eastern slopes of the Ghats slope gently towards the Bay of Bengal. The entire zone can also be divided into 13 sub-basins: 1) Small streams between the Mahanadi and the Rushikulya draining into the Chilka lake; 2) The Rushikulya River; 3) Small stream - between the Rushikulya and the Bahuda; 4) The Bahuda; 5) Small streams between the Bahuda and the Vamsadhara; 6) The Vamsadhara; 7) The Nagavali; 8) Small streams between the Nagavali and the Sarada; 9) The Sarada; 10) The Varaha; 11) The Tanadava; 12) The Eluru; and 13) Small stream between the Eluru and the Godavari.

The Vamsadhara, the Rushikulya, the Nagavali and the Sarada Rivers in the basin are the more important.

16.4.1. Vamsadhara Basin

The Vamsadhara River originates from the hills near village Lanjigarh of Kalahandi district of Orissa. After flowing for a length of 160 km in Orissa, it forms the boundary between Orissa and Andhra Pradesh and then traversing through Andhra Pradesh, falls into the Bay of Bengal. Out of the total catchment area of 11,500 sq. km, 9,400 sq. km lies in Orissa. This river ranks 5th to Mahanadi in terms of its water potential and flood producing capacity. The total length of river from head to ocean is 380 km out of which 276 km falls in Orissa and 104 km in Andhra Pradesh. The major tributaries are Harabhangi, Badanalla, Mandratanaya and Sanonae.

Harbhangi river

The Harabhangi River is a tributary of Vamsadhara River which originates from Ramagiri of Ganjam district in Orissa state at an altitude of 1,100 m. The catchment area intercepted at dam site is 572 sq. km. The catchment is almost fan shaped and more than 50% of its area is covered by forest and the area is thinly populated. Harbhangi traverses 36 km in A.P. before falling into Bay of Bengal. Nearly 78% of its catchment area lies in Orissa.

16.4.2. Rushikulya River Basin

The Rushikulya river

The Rushikulya River is one of the medium seized east flowing rivers in Peninsular India. The larger portion of the river falls in Ganjam and Gajapati districts, and few portion in Pun and Phulbani districts of Orissa state. It is an important river of Orissa having a drainage area of about 82,100 Mm^2 which is approximately 5.71% of the total area of the State. The drainage basin spread over the parts of the Districts of Ganjam, Phulbani and Puri lies within the geographical coordinates of 19°07′ N to 20°19′ N latitudes, and 84°01′ E to 85°06′ E longitudes.

The Rushikulya River originates from Ararha bity and Kutrabor hills of Eastern Ghats at an elevation of 500 m in Ganjam and Phulbani Districts. It flows in the south eastern direction through Purushottampur and falls to the Bay of Bengal near Chhatrapur in Ganjam District. The length of the main stream from its origin to the outfall is about 175 km. The principal tributaries of the Rusikulya River are Padma and Ghodahado on the right and Badanadi, Baghua, Dhanei and Kharakhari on the left side.

The climate of the basin is moderate having four distinct seasons namely: Winter (December to February), summer (March to May), south-west monsoon (June to September) and post monsoon period (October and November). The maximum and minimum recorded temperatures of the basin are 45 °C and 12 °C respectively. Relative humidity during the summer monsoon season varies from 88% to 93%. The average annual rainfall in the basin is about 1,235 mm, of which about 80% occurs during south-west monsoon season. Besides the monsoon rains, cyclonic rains which have caused severe floods have also been observed in recent years. In certain years, serious drought has also occurred due to inadequate rainfall.

16.4.3. Nagavali Basin

The Nagavali basin is surrounded by the Mahanadi on north, Godavari basin on west, Champavathi, Peddagedda, Kandivalasa River basins and Bay of Bengal in south, and Vamsadhara River basin in the east. The total catchment area is about 8,845 km^2. Out of this, the catchment area in Orissa state is 4,040 km^2 and in Andhra Pradesh State it is 4,805 km^2. The maximum length of the basin from north to south is about 210 km lying between latitudes 18°11′ N to 19°45′ N. The maximum width of the basin is about 110 km in between longitudes 82°53′ E to 84°2′ E.

The Nagavali River rises in the Eastern Ghats in the Orissa State. After flowing in south eastern direction for about 40 km, it takes a turn towards southerly direction and flows in this direction for 25 km. There after, where the river crosses the Visakhapatnam-Raipur railway line and continues to flow in the same direction and parallel to the railway line for another 22 km where it passes one and half km east of Rayagada. The river continues to flow for another 20 km in the same direction and almost parallel to the railway line and Parvathipuram-Rayagada road.

It enters Andhra Pradesh State limits at a point 6 km south of Jimidepeta Railway Station. A number of minor hill streams fall into the main river in the Orissa State limits.

After entering Andhra Pradesh State, the Nagavali river takes a south easterly direction and continues to flow in this direction till it falls in the Bay of Bengal about 10 km south east of Srikakulam town near Dibbalapalem village. Description of its main tributaries is given next.

Janjavathi river

The Janjavathi River takes its origin in Koraput District of Orissa State and flows for about 60 km in easterly direction. It joins the Nagavali River on the right side near Devara Gumpa in Komarada Mandal of Vizinagaram District.

Vottigedda river

The Vottigedda River takes its origin near the Andhra Pradesh-Orissa State border near Addapi at an altitude of 1,060 m above sea level and flows in southerly direction for about 50 km. It joins the Nagavali River on the left side near Chidimi in Veeraghattam Mandal, Srikakulam District.

Vegavathi river

The Vegavathi River is the major tributary of the Nagavali River which takes its origin in Arakuvalley Mandalam in Visakhapatnam District at an altitude of 1,600 m above sea level. The river flows almost in easterly direction for about 80 km and is joined by the Suvarnamukhi River on the left side at a place called Kottisa. The Suvarnamujkhi River is the tributary of the Vegavathi River which takes its origin in Koraput District of Orissa State.

16.4.4. Sarada River Basin

This is a small river that drains the area between Eastern Ghats and the Bay of Bengal in the north-eastern coastal area of Andhra Pradesh. The drainage area of the Sarada River basin is 2,590 km^2, which lies between longitude 82°30′ to 83°07′ E and latitude 17°25′ to 18°17′ N. The river is bounded by Anakapalli minor drainages and Naravagedda minor basin on the eastern side, Sileru River (a tributary to Godavari) on the Northern side and Vardha river basin on the Western side. The entire basin is situated in the 19 administrative mandals of Visakhapatnam District and one mandal of Vijayanagaram District.

The maximum length of the basin from east to west is 63 km and from north to south, it is about 95 km. The Sarada River originates in the Anantgiri reserved forest area of Eastern Ghats near Lakshmipuram village of Chodavaram mandal. The river flows in the southern direction until it reaches Anakapalli, it then takes a southwestern turn from Anakapalli to Medupaka and thereafter flows in southerly direction until it joins the sea. The river is not perennial; it being a hill stream is characterized by occasional floods. The drainage pattern of the basin was observed as dendrite type. The elevation of the river at its origin is about 1,447.80 m.

The basin consists of Pleistocene unconsolidated formations near sea coast and archean complex of unclassified crystalline khondalities and granites. The major soil types in this basin are: red loamy soil, red sandy soil, and coastal sands and alluvial soils. Red sandy soils cover the largest area in the basin. The acidic granites, gneisses, quartzites and felpathic are subordinate rock types, rich in iron and magnesium bearing minerals. These are responsible particularly for the formation of red soils, and at certain places for formation of yellow, grey or even black colored soils. Coastal sands and alluvial soils occur in the coastal belt of the basin.

Expectedly, the coastal part of the basin is highly humid. The basin experiences a marked fall in the temperature in the month of November. The highest temperature of the basin observed at the Anakapalli was close to 44 °C and lowest temperature is about 12 °C. During the winter months, the mean temp ranges from 29 °C to 17 °C. During the summer months, the mean temperature ranges from 38 °C to 25 °C.

The basin lies in medium rainfall zone and the rainfall varies from 700 to 1,000 mm. Most of the rainfall in this region is received during south-west monsoon from June to September.

16.4.5. Suddegedda Basin

This basin lies between Mahanadi and Godavari, between latitudes 17°09′10″ N and 17°30′45″ N and longitudes 82°08′30″ E and 82°19′15″ E. It occupies an area of 526 km^2. The basin covers Prathipadu Mandal and parts of Rajavommangi, Pithapuram and Gollaprolu Mandals. Nearly 57 villages fall in the basin area.

A number of ephemeral hill streams working their way through the undulatory hilly tracts join together to form the stream. Gokavaram "Yeti Calva" or 'Kuntidevi vagu: Further downstream, this river is known as Suddagedda. The drainage pattern of the basin is dendritic to subdendritic. The Suddegedda basin has a drainage density of 1.13 km/km^2. Subbareddysagar Project, a major tank across the above stream, has a storage capacity of 5.1 Mm^3.

In the basin, elevation ranges from 700 m to 20 m. Although this is not a large variation in absolute terms, looking at the elevation range in Eastern Ghats and size of the catchments, this variation is quite large. The basin area has a series of hill in the northern side which are a part of Eastern Ghats. The general slope of the basin is towards south-southeast.

The basin area enjoys tropical climate with hot summers and cold winters. Although the area receives rainfall from both the monsoons, more than half of the rainfall is contributed by southwest monsoon (June and September) while the remaining rainfall is by north-west monsoons during the months of October and November. The average annual rainfall is 1,059 mm.

The predominant soils in the basin are black clay and red and light brown sandy soils. Towards the northern part of the basin, red soils are predominant in the hilly tracts and valley portions while in the middle part of the basin, light brown soils and towards the southern part black soils are predominant.

The main crops grown in the area are paddy, banana, sugarcane and commercial crops like chillies and cotton. The northern part of the basin is mostly occupied by cashew and mango orchards. Paddy and banana are grown utilizing irrigation water under Subbardeddysagar project and ground water. Rainfed crops like pulses and gingelli are also grown in the area. The total irrigated area from surface water sources is 69.81 Mm^2, out of which an area of 17.58 Mm^2 is irrigated from the Subbareddysagar reservoir and 52.23 Mm^2 gets water from minor irrigation tanks.

16.4.6. Water Resources Projects

A number of minor irrigation structures exist on some of these tributaries which include: Bhanjanagar reservoir; Sorada reservoir; Madhoborida diversion weir; Janivilli diversion weir; Sarisomuli diversion weirs; Ghodahado Daha Project; Ramanadi Project; Baghua Stage I Project; Dhanei Project; Jayamangal Project; and Hirdharbati projects.

Harabhanghi irrigation project

The Harabhanghi irrigation project is a reservoir scheme across Harabhangi River located at 19°29′58″ N latitude and 84°8′12″ E longitude in Ganjam district, 250 km away from Orissa State capital Bhubaneswar. The aim of the project is to supply the irrigation water through canal system to the command area lying in Ganjam district, Orissa. Ganjam district has got immense water resource so that such type of major project has been planned to cater the water requirement for irrigation.

Being an interstate river between Orissa and Andhra Pradesh, an agreement was made in 1962 between the Government of Orissa and Andhra Pradesh regarding the utilization of water of Vamsadhara basin. According to this agreement reached in the year 1962, "the yield of Vamsadhara basin is just sufficient to meet the requirement of both the states. The water of Vamsadhara basin may consequently be utilized by A.P. and Orissa on a fifty:fifty basis." The 75% of dependable water available at Gotta barrage (A.P.) is estimated to be about 3,100 MCM. Thus Orissa's share comes to 1,550 MCM. Harabhangi project is one such project in the state that utilizes its share of water in Vamsadhara basin.

The basin receives about 85% of the annual rainfall from SE monsoon during June to October. The maximum, minimum and average rainfall in this area are about 1,917 mm, 799 mm, and 1,132 mm respectively. The Live storage capacity of reservoir is 17,200 ha-m and the reservoir envisages irrigating about 9,650 ha of C.C.A. in Ganjam district.

16.5. EAST FLOWING RIVERS BETWEEN GODAVARI AND KRISHNA

These rivers flow towards east and join the Bay of Bengal between Godavari and Krishna, in total covering an area of 12,289 km^2. The basin lies between longitudes 80°30′ E to 81°45′ E and latitudes 16°10′ N to 17°30′ N. This whole area falls in the State of Andhra Pradesh.

The Godavari River bounds the area on the north and the east, the Krishna River basin lies to the west and the Eastern Ghats and the Bay of Bengal on the south. The area has an irregular shape with the maximum length of 155 km in north-south direction and 125 km in east and west direction. Topographically the area can be divided in two distinct classes: the hilly area of the Eastern Ghats and the plains. There are five small basins in the area, namely, 1) a small stream between Gundalkamma and Musi; 2) Musi; 3) Paleru; 4) Manneru; and 5) a small stream between the Manneru and the Pennar. The Gundlakamma and the Manneru are the important basins.

16.5.1. Gundlakamma River Basin

The Gundlakamma River is a small river of about 264 km length which originates in the Nallamali hills of Mallamalai forest near Gundla Brahmeswaram Village, in Nandyal taluka of Kurnool District at an altitude of 680 m. Gundlakamma flows through deep ravines and thickly grown natural forests and hilly tracts up to Cumbum tank situated in Cumbum village in Prakasam District. The river flows in north-east direction up to the confluence of Konduleru River, then takes a turn towards east up to the Pittambanda village where it joins of Konkeru River. From here onwards, the river flows in the south-east direction and finally joins Bay of Bengal near Pallipalem village. The river's total catchment area is 8,195 km^2, including the area drained by its tributaries Jamaleru, Venumuleru, Mekaleru, Teegaleru, Duvvaleru, Rallavagu, Konduleru, Pasupaleru, Konkeru, Chilakaleru, Voleru, etc. The Gundlakamma basin is bounded by Vogeru vagu, Romperu on east side, by Nallamalai hill range on the western side, Krishna basin on the northern side and Musi river basin on the southern side and flowing eastwards into Bay of Bengal.

The climate of coastal part of the study area may be broadly classified under tropical coastal type and rest is of steppe type. According to Koppan, the climate is tropical Savannah in upper part and dry season in high sun period in the rest of the area. The daily mean temperature is about 27.5 °C. Mean maximum temperature is around 32.5 °C. Mean minimum temperature is about 22.5 °C. Highest maximum temperature is about 47 °C and lowest minimum temperature is about 14 °C. The mean diurnal range of temperature is about 10 °C.

For the Gundlakamma basin, the annual rainfall is 800 mm. Of this, about 25 mm falls during January and February, about 75 mm during March to May, about 400 mm during the monsoon months of June to September and about 300 mm during

October to December. The annual runoff is about 200 mm. At times, the basin gets heavy rainfall due to cyclones forming in Bay of Bengal.

16.6. EAST FLOWING RIVERS BETWEEN KRISHNA AND PENNAR

The area occupied by the east flowing rivers between Krishna and Pennar extends over 24,649 km^2 and lies between longitudes 78°45′ E to 80°52′ E and latitudes 14°38′ N to 16°22′ N. The entire area lies in state of Andhra Pradesh.

The Nallamala hills and the Velikonda range of the Eastern Ghats bound the area on the west, a ridge separates it from the Krishna basin on the north, another ridge separates it from the Pennar basin on the south, and the Bay of Bengal extends on the east. The shape of the basin is irregular and it has a maximum length of 213 km in a west-east direction and a maximum width of 186 km in the north-south direction. There are two major topographical divisions in the basin: the hilly area of the Eastern Ghats, and the plains. The plains extending from the eastern slopes of the Ghats slope gently towards the Bay of Bengal. The basin can be divided into nine sub-basins in the basin, of which the Gundlakamma and the Manneru are the more important.

Various river systems in the basin from the north to the south are: Three small streams up to the Vogarivagu; The Vogarivagu; a small stream between the Vogarivagu and the Gundlakamma; the Gundlakamma; a Small stream between the Gundlakamma and the Musi; the Musi; the Paleru; the Manneru; and a small stream between the Manneru and the Pennar.

16.6.1. Water Resources Projects

There are several existing and under construction water resources projects in the small basins from Mahanadi to Godavari and Krishna to Pennar. The salient features of these projects with a live storage capacity of 10 MCM and more have been presented in Table 10 and Table 11, respectively.

16.7. EAST FLOWING RIVERS BETWEEN PENNAR AND CAUVERY

The region covers the east flowing rivers between the Pennar and the Cauvery extending over a large area of 64,751 km^2. The basin lies between east longitudes 77°35′ to 80°21′ and north latitudes 11°15′ to 14°30′. Lying in peninsular India, the area covers large areas in the State of Andhra Pradesh, Karnataka, Pondicherry and Tamil Nadu. The State-wise distribution of the area is shown in Table 12.

On the eastern side of the area lies the Bay of Bengal. On the remaining other sides, it is bounded by the various ranges of the Eastern Ghats. These are the Velikonda Range, the Nagari Hills, the Javadi Hills, the Shevaroy Hills, the Chitteri Hills, the Kalrayan Hills, the Kollaimalai Hills, the Pachai Malai Hills etc. The area

Table 10. Salient features of selected existing projects in existing east flowing Rivers from Mahanadi to Godavari and Krishna to Pennar basin

Name of the project	State	Year of completion	Gross storage capacity (Mm^3)	Live storage capacity (Mm^3)	Designed annual irrigation (Mm^2)
Cumbum Tank	Andhra Pradesh	1896	95.65	64.00	32.00
Gesthani Dam	Andhra Pradesh	1969	94.10	89.85	56.10
Kanigiri	Andhra Pradesh	1906	134.99	114.07	372.40
Konam Reservoir	Andhra Pradesh	1981	24.03	23.60	28.90
Maddigedda	Andhra Pradesh	1980	13.73	12.27	16.00
Mopadu	Andhra Pradesh	1923	–	59.33	44.00
Paleru Reservoir	Andhra Pradesh	–	73.00	66.50	26.60
Raiwada	Andhra Pradesh	–	101.98	92.66	24.00
Tammileru	Andhra Pradesh	1978	84.95	76.45	32.00
Vamsadhara Stage-I	Andhra Pradesh	–	539.44	465.25	200.00
Varaha Dam	Andhra Pradesh	1975	13.17	12.21	17.60
Baghua-I	Orissa	1977	49.34	32.18	–
Daha	Orissa	1979	28.00	21.95	68.70
Godahade	Orissa	1976	22.20	18.50	62.00
Salia	Orissa	1977	59.75	48.00	110.40

has an irregular shape; it has a maximum length of about 290 km and a maximum width of about 360 km. There are three major topographical divisions in the basin, namely: the hill ranges of the Eastern Ghats; the table land or the plateau region; and the coastal plains. The basin can be divided into 12 sub-basins, of which the

Table 11. Salient features of selected under construction projects in existing east flowing Rivers from Mahanadi to Godavari and Krishna to Pennar basin

Name of the project	State	Gross storage capacity (Mm^3)	Live storage capacity (Mm^3)	Designed annual irrigation (Mm^2)
Andra	Andhra Pradesh	27.76	26.40	37.70
Gundlakamma	Andhra Pradesh	98.60	52.36	–
Janjhavathi Reservoir	Andhra Pradesh	96.30	78.56	40.20
Madduvalasa	Andhra Pradesh	95.56	93.62	60.00
Thandwa	Andhra Pradesh	140.46	121.60	131.30
Tharakaramathirtha Sagaram Project	Andhra Pradesh	82.94	68.15	–
Vengalaraja Sagaram	Andhra Pradesh	47.59	42.35	67.20
Vathinagu	Andhra Pradesh	82.06	74.96	–
Yarralakalwa Project	Andhra Pradesh	–	114.67	95.00
Yeleru Reservoir Scheme	Andhra Pradesh	679.60	538.00	404.70
Badnella Irrigation Project	Orissa	82.50	67.00	137.40
Baghua Phase-Ii	Orissa	37.50	31.00	43.80

Table 12. Statewise catchment areas of east flowing rivers, between Pennar and Cauvery

State	Drainage area (km^2)
Andhra Pradesh	16,478
Karnataka	6,173
Tamil Nadu	41,759
Pondicherry	341
Total	64,751

Palar, the Ponnaiyar and the Vellar are the more important. The shape of the Palar basin is like a rhombus. The Ponnaiyar basin is elongated in shape, whereas the Vellar basin is fan shaped.

The major rivers that flow in the area (listed from the north to the south) include: the Kunleru; the Swarnamukhi; Small streams draining into the Pulicat Lake; the Araniar; the Kortalaiyar; the Cooum; the Adyar; the Palar; Minor streams between the Palar and the Gingee; the Gingee; the Ponnaiyar and the Vellar.

A few important rivers are described in the following.

16.7.1. Palar Basin

The Palar (also known as Palaru) basin is located at a latitude of 12°15′N to 13°40′N and at a longitude of 77°50′E to 80°15′E. The basin is bounded by the Papagni, Bahuda and Punchu Basins of Andhra Pradesh and the Madras basin of Tamil Nadu on the north, by the Ponnaiyar basin on the west, by the Tonidaru basin on the south and by the Bay of Bengal on the east. The basin is quite wide at its head reaches and tapers down to a narrow width of about 50 km on the coastline. Out of the total catchment area of 18,409 km^2, about 2,616 km^2 lie in Karnataka, 4,387 km^2 in Andhra Pradesh and 11,406 km^2 in Tamil Nadu.

The Palar (Palaru) River originates in the Kolar district of Karnataka and after traversing a distance of 30 km through Andhra Pradesh, it enters Tamil Nadu on its 130th km near Vaniyambadi. The Kavundi aru River joins the Palar River near Pallikonda. From this point, the river widens tremendously and maintains nearly the same width till it meets the Sea. Palar River is an ephemeral channel that has a flow for only a few days in a year and during the major part of the year, it remains totally dry.

The Cheyyaru, Kamandalaru, and Kavundiarum are the important tributaries of the Palar River. The Kiliyaru River across which Madurantakam tank has been formed joins Palar near the Sea mouth. The Cheyyaru River originates from the southern slopes of the Javandu hills near Komattiyur and after traversing through Chengam, Polur and Tiruvettipuram, it finally joins Palar River near Walajabad. The Cheyyaru anicut has been built across the river, about 10 km south-east of Arni. Another anicut namely, the Uttiramerur anicut has also built on the Cheyyaru River about 10 km east of Tiruvettipuram. The Kamandalaru River is a small stream which originates from the Java hills and after traversing through Arni town, it joins the Cheyyaru River, just downstream of the Cheyyaru anicut.

The Kiliyaru River is a small tributary of the Palar River. It originates between Chetput and Vandavasi. The Madurantakam tank, one of the largest tanks of Tamil Nadu has been built on this river. The Kiliyaru River joins the Palaru River about 20 km upstream of the mouth of Palaru River.

16.7.2. Ponnaiyar (Pennayaru) Basin

The Ponnaiyar River, an interstate river, is one of the largest rivers of the state of Tamil Nadu (TN), often reverently called 'little Ganga of the South'. The river has supported many a civilizations of peninsular India across the history and continues to playa vital role in supplying precious water for drinking, irrigation and industry to the people of the states of Karnataka, Tamil Nadu and Pondicherry.

The Ponnaiyar basin is located in between the north latitudes of 11°30′ to 13°40′ N and east longitudes of 77°30′ to 79°50′ E. The basin is bounded by the Cauvery basin from the west and south, and the Palaru and Tonidaru basins from the east and north, respectively. The catchment area of the basin is 15,865 km^2 out of which 3,597 km^2 lies in Karnataka.

The Ponnaiyar River rises from Chik-Ballapur in Karnataka at the south-eastern slopes of Chennakesari hills at an altitude of about 1,000 m above mean sea level. The river enters Tamil Nadu after traversing 85 km from its origin. A number of large and small tributaries such as Chinna Aru, Markanda Aru, Mattur Aru, Vaniaru, Pambaru, Kallaruc join the river during its way, before it falls in the Bay of Bengal. A number of large irrigation tanks have been constructed across the Pennaiyaru River at Chik-Ballapur, Sidlaghat, Vijayapur, Hoskoto and Whitefield in Karnataka.

The river flows for a major portion through wooded country, deep ravines and narrow gorges, with the country at the sides rising steeply to a height of about 90 m above the riverbed. Ponniyar River has practically no flow in the river except during Northeast monsoon months (October–December) or the less precipitous southwest monsoon (June–August). During the rainy spells, there are flash floods in Ponniyar. This has aptly given rise to a proverb in the Tamil language "the Ponnaiyar will rise and fall even before the butter melts".

The Ponnaiyar basin is predominantly built up with granite and gneisses rocks of archean period. The granite is of very good quality and extensive outcrops and masses of it are commonly found. The chief components of rocks are hornblende and feldspar. Foliation is seldom seen. In the plains of reserve forest, quartz is found commonly. The diamond granite is also found in scattered pockets in the areas of Chitteri hills in Dharmapuri and Krishnagiri sub-divisions. Charnokite rocks of archean period are also seen in some areas. At the tail end of the basin, pockets of sandstone, clays pebble of tertiary period, and limestones of cretaceous period are found. Alluvium and sand-dunes of quaternary period are also seen at a few places.

The catchment falls under the tropical belt. The climate in general is hot; April and May being the hottest months of the year when the temperature rises to 34 °C.

November and December are the coldest months for this basin when the ambient temperature falls to 22 °C.

The average annual rainfall over the catchment varies from 800 mm to 1,000 mm. The catchment gets rainfall both from southwest and northeast monsoons. Most of the rainfall received in this tract is from June to December but October and November are the rainiest months. A considerable amount of dew falls during December to February.

Krishnagiri dam

The Krishnagiri dam is located on the Pennaiaru river at a north latitude of 12°28′ N and east longitude of 78°11′ E. The dam was completed in the year 1957. The river bed level at the dam site is 464.634 m above mean sea level. The catchment area and length of the main stream is 5,429 km^2 and 150 km respectively. It is an earthen dam with gravity stone masonry. The height of the dam is 22.866 m and the maximum design flood is 4,250.8 cumec. The full reservoir level is 483.232 m with a capacity of 66.83 MCM.

Sathanur dam

The Sathanur dam is located on the Pennaiaru river at a north latitude of 12°11′ N and east longitude of 78°50′ E. The dam was completed in the year 1958. The river bed level at the dam site is 183.490 m above mean sea level. The catchment area and length of the main stream is 10,826 km^2 and 245 km respectively. It is an earthen and gravity dam with masonry. The height of the dam is 41.15 m and the maximum design flood is 5,663 cumec. The full reservoir level is 483.232 m with a capacity of 66.83 Mm^3. The total length of the dam is 779.8 m of which the masonry section is 418.5 m and rest is the earthen dam. The masonry dam is designed as a gravity structure, 40.9 m above the riverbed and 44.7 m above the deepest foundation level.

Memathur regulator

The Memathur regulator exists across Manimukta River in Vellar basin at about 5 km below the confluence of Gomukhi and about 14 km north-west of Vridhachalam town. It was constructed during the year 1873. The total catchment area of the regulator is 1,683.50 km^2. It irrigates an area of 24.20 Mm^3.

16.7.3. Vellar Basin

The Vellar basin is an important river basin in Tamil Nadu situated between latitude 11°13′ N to 12°00′ N and longitude 78°13′ E to 79°47′ E. The basin lies in between Ponnaiyar River in the north and Cauvery River in the south with the total catchment area of 7,659 sq. km. The basin can be divided in two distinct sub-basins. There is no exchange of surface water between the two sub-basins.

The Vellar River flows through the Dharamapuri (a small portion), Salem, Trichirapalli, and South Arcot districts of Tamil Nadu. The river originates at an elevation of 900 m near the Tumbal village in the Chittori Hills of the Eastern Ghats in the Salem district of Tamil Nadu. In the initial reaches, the river is known as Ammapalayam and it flows in a south-west direction for about 21 km until another tributary Kariyakoil joins it. Thereafter, the river attains the name Thumbal and continues to flow in the south-west direction till the tributary Vasistanadi joins it on the right bank. The Vasistanadi in its headwater reaches is known as Anaimadavu. It generally flows towards south direction and meets Thumbal on its right bank. After this confluence, the river flows in a south-east direction and attains the name Vasistanadi. It crosses the Salem-Kallakurchi road near Talaivasal at about 69 km from its origin. The river continues to flow in the south easterly direction for a further distance of about 28 km and meets the Swetanadi. The river then takes an easterly direction and crosses the Trichi-Chennai Trunk road over Toludur regulator located at 4 km downstream of the confluence with Swetanadi.

After the confluence of Swetanadi, the river attains its name Vellar. In its further course, Vellar receives the tributaries Chinar and Anaivari Odai on its right bank. From the point of infall of Anaivari Odai, the river takes a north direction, flows for about 13 km, then takes south-east direction and flows for another 15 km up to the infall of the Manimukta nadi on its left bank. The river then flows in an easterly direction, crosses the Kumbakonam Villipuram road through Sethiathope regulator and finally empties into the Bay of Bengal near Portonovo in the Chidambaram Taluk of South Arcot district.

The tributaries of the Vellar River are Vasistanadi, Swetanadi, Gomukhi River, Manimukta River, Chinar River and Anairvari River.

The Vellar basin experiences tropical monsoon climate, with much variation in temperature, humidity and evaporation throughout the year. During summer months the maximum temperature is reported to 40 °C, whereas in winter months the minimum temperature goes down to about 20 °C. The monsoon season in the basin is from June to September in which significant rainfall is observed in the basin. The mean annual rainfall in the basin varies from 825–1,390 mm. Most of the rainfall in the basin occurs in the monsoon months and less rainfall experiences in non monsoon period.

Manimukta reservoir

The Manimukta reservoir is formed across Manimukta tributary of Vellar River at about 5 km downstream of the confluence of Mani and Mukta sub-tributaries and about 6 km from Kallakurichi Town. The dam was completed in the year 1970. It has a catchment area of 484 km^2. The live storage capacity of the reservoir is 19.89 Mm^3 and it irrigates an area of 17.20 Mm^2.

Gomukhi reservoir

The Gomukhi reservoir is located on the tributary of Vellar River of the same name, at about 16 km to the north-west of Kallakurichi Town. The reservoir was completed in the year 1965. It has a catchment area of 293 km^2. The live storage capacity of the reservoir is 15.86 Mm^3 and it irrigates an area of 2,023.43 ha.

16.7.4. Tondiaru Basin

The Tondiaru basin is located at a latitude of 11°55′ N to 12°30′ N and at a longitude of 79°05′ E to 80°05′ E. The basin is bounded by the Palaru Anicut on the north, by the Ponnaiyar basin on the west and south and by the Bay of Bengal on the east. The catchment area of the basin is 3,637 km^2 which lies entirely in Tamil Nadu state. Two small rivers, namely, Onguraru and Tondiaru drain in this basin separately.

The average annual rainfall in the basin is 1,078 mm. The annual infiltration and annual runoff in the basin has been estimated at 102 mm and 112 mm respectively.

Onguraru river

The Ongur aru River originates from the south of Vandavasi and after traversing through a distance of about 50 km, it falls into the Bay of Bengal in the backwaters north of Marakkanam.

Tondiaru river

The Tondiaru River originates near Chetput. After 50 km of its origin, the Varadaru River joins it from the right. The Vidur reservoir has been constructed at the downstream of the confluence. The command area of the Vidur Reservoir is 1,298 km^2. Below the Vidur reservoir, the Tondiaru River is also known as Chenjiaru River, which falls into the Bay of Bengal about 6 km south of Pondicherry. The Vidur dam is located at a north latitude of 12°04′ N and east longitude of 79°35′ E. The dam was completed in the year 1960. Vidur is an earthen dam with stone masonry. The height of the dam is 13.42 m and the maximum design flood is 1,912.9 cumec. The FRL of reservoir is 33.232 m with a capacity of 17.145 Mm^3.

The only station in the basin where the flows in the Tondiaru River is gauged is at the Vidur dam. The river bed level at the dam site is 26.83 m above mean sea level. The catchment area and length of the main stream is 1,298 km^2 and 50 km respectively.

Palaru anicut

The Palaru anicut, located on Palaru River was constructed in the years 1855–58. The Anicut is located at a north latitude of 12°53′ N and at an east longitude of 79°24′ E. The catchment area and length of the main stream is 10,293 km^2 and 152 m respectively. The effective length of Palaru anicut is 778.96 m. The river bed level & crest level of anicut is 148.78 m & 151.335 km respectively. The maximum observed flood discharge at the site was 4,829 m^3/s in 1903.

Ponnai anicut

The Ponnai anicut, located on Pennai Aru River was completed in 1897. The Anicut is located at a north latitude of 13°07′ N and at an east longitude of 79°16′ E. The catchment area and length of the main stream at the anicut is 1,725 km^2 and 90 m respectively. The effective length of Ponnai anicut is 216.46 m. The river bed level and crest level of anicut is 218.293 m & 220.918 km respectively. The maximum observed flood was 1,685 m^3/s in 1930.

Aliyabad anicut

The Aliyabad anicut on Kamandalaru River was completed in 1866. The Anicut is located at a north latitude of 12°38′ N and at an east longitude of 79°08′ E. The catchment area and length of the main stream is 223 km^2 and 25 m respectively. The effective length of Aliyabad anicut is 61.59 m. The river bed level & crest level of anicut is 197.256 m & 198.759 km respectively. The maximum flood observed was 1,025 m^3/s in 1936.

Cheyyaru anicut

The Cheyyaru anicut on Cheyyaru River was completed in 1852. The Anicut is located at a north latitude of 12°35′ N and at an east longitude of 79°22′ E. The catchment area and length of the main stream is 1,841 km^2 and 110 km respectively. The effective length of Cheyyaru anicut is 70.63 m. The river bed level & crest level of anicut (without falling shutters) is 115.854 m and 117.689 m respectively. The maximum flood observed was 2,587 m^3/s.

Uttiramerur anicut

The Uttiramerur anicut on Cheyyaru River was completed in 1956. The Anicut is located at a latitude of 12°40′ N and at a longitude of 79°41′ E. The catchment area and length of the main stream is 3,885 km^2 and 155 km respectively. The river bed level & crest level of anicut is 69.512 m & 70.774 m respectively.

Nedungal anicut

The Nedungal Anicut is located on the Pennai aru river at a north latitude of 12°23′ N and east longitude of 78°15′ E. The Anicut was completed in the year 1877. The river bed level and crest level of the anicut is 461.280 m and 464.076 m respectively. The catchment area and length of the main stream is 5,670 km^2 and 165 km respectively. The length of the anicut is 278.05 m and the maximum design flood is 2,000 cumec. The maximum flood observed was 3,043 cumec in 1903.

Tirukkoilur anicut

The Tirukkoilur Anicut is located on the Pennai aru river at a north latitude of 11°57′ N and east longitude of 79°16′ E. The Anicut was completed in the year 1879. The river bed level and crest level of anicut is 81.250 m and 82.927 m respectively. The catchment area and length of the main stream is 12,317 km^2 and 300 km respectively. The length of the anicut is 456.41 m. The maximum flood observed was 7,651 cumec in 1903.

Ellis choultry anicut

The Ellis Choultry Anicut is located on the Pennai aru river at a latitude of 11°54′ N and at a longitude of 79°25′ E. The Anicut was completed in the year 1950. The river bed level of anicut is 54.255 m. The catchment area and length of the main stream is 12,940 km^2 and 320 km respectively. The maximum flood observed was 6,569 cumec in 1909.

16.8. EAST FLOWING RIVERS SOUTH OF CAUVERY

The area where east-flowing rivers between Cauvery and Kanyakumari are located lies between longitudes 77°09′ E to 79°17′ E and latitudes 8°5′ N to 10°30′ N. The total catchment area of these rivers is 35,026 km^2, which lies completely in the state of Tamil Nadu. The area is bounded by the Varushanad hills, the Andippatti hills, the Cardamom hills and Palani hills on the west, by the Indian Ocean on the south, by the Palk-Strait, Palk Bay and the Gulf of Mannar on the east and the ridge, which separates it from the Cauvery basin on the north. Shape of the area is irregular; it has a maximum length of 236 km in the northwest-southeast direction and a maximum width of 275 km in the northeast-southwest direction. There are two major topographical divisions in the basin: the hilly area; and the plains. The area can be divided into 11 sub-basins, of which the Vaigai and the Tambraparni are the most important. The Vaigai basin is elongated in shape, whereas the Tambraparni basin is fan-shaped.

The various river systems in the area north to south are: the Vellar; a small stream between the Vellar and the Varshalei; Varshalei; a small stream between the Varshalei and the Vaigai; the Vaigai; the Gundar; a small stream between the Gundar and the Vaippar; the Vaippar; small streams between the Vaippar and the Tambraparni; the Tambraparni; and small Streams south of the Tambraparni up to Kanyakurnari.

16.8.1. Agniaru Basin

The Agniaru basin is located between latitudes of 09°55′ N to 10°48′ N and longitudes of 78°14′ E to 79°30′ E. The basin is bounded by the Cauvery basin from the north and by the Pambaru basin from the south. The catchment area of the basin is 4,463 km^2, which lies entirely in Tamil Nadu State.

The basin is drained by three rivers Agniaru, Ambuliyaru and Vellaru. The Agniaru drains Pudukottai–Vallam-Orathanadu highlands. The length of the river is 80 km. The Ambuliyaru River is a small stream draining the Alangudi Plateau. The Vellaru River originates just north of Tuvarankurichi and traverses a total distance of 135 km through Pudukkottai, Arimalam, and Arantangi. All the three rivers finally fall into Palk Strait.

The mean annual rainfall for the Agniaru basin is 1,040 mm. The maximum rain falls during the month of October. The annual evaporation for the Agniaru Basin is 928 mm, which is about 89% of the total rainfall. The annual infiltration for the Agniaru basin is 51 mm, which is about 5% of the total rainfall. The mean surface water potential has been estimated to be 316.87 Mm^3. Assuming 0.5 as the extraction factor, the ground water potential works out 91.47 Mm^3 for the entire basin. The surface runoff in the basin is 61 mm with base flow of 10 mm.

16.8.2. Pambaru Basin

The Pambaru basin is located between latitudes of 09°45′ N to 10°25′ N and longitudes of 78°10′ E to 79°10′ E. The basin is bounded by the Cauvery basin to its west, by the Agniaru basin to its north and by the Kottakkarairu basin to its south. The area of the basin is 3,488 km^2, which lies entirely in Tamil Nadu.

The Pambaru River originates in the northern slopes of Alagar Malai at an altitude of 830 m. In the head reaches, the river is known as Tirumanimutharu. The river traverses through Kallai and Devakottai and falls into the sea just north of Sundarapandiyan Pattinam. The Pallaru River, a tributary of the Pambaru River drains Kondangikuttu, Karandamalai and Ayilur reserve forests and joins Pambaru River just south of the village Kandramanikkam.

The mean annual rainfall for the Pambaru basin is 941 mm. Maximum rainfall takes place during the north-east monsoon month of October. The annual evapotranspiration for the Pambaru Basin is 878 mm, which is about 93% of the total rainfall. The annual infiltration for the Pambaru basin is 40 mm, which is about 4% of the total rainfall. In the basin, surface water potential has been estimated to be 104.5 Mm^3. Assuming 0.5 as the extraction factor, the ground water potential works out 57.5 Mm^3 for the basin. The surface runoff in the basin is 23 mm with base flow of 7 mm.

16.8.3. Periyar-Vaigai System

The Periyar-Vaigai system is one of the oldest irrigation systems in India. It is a trans-basin scheme, which came into existence towards the end of the nineteenth century. The system consists of two reservoirs namely, the Periyar reservoir on the Periyar river in the Kerala state, the Vaigai reservoir on the Vaigai river in the state of Tamil Nadu, and the irrigation command areas in the Vaigai basin. The Periyar River, which originates on the western slope of the Western Ghats, flows westwards and discharges into the Arabian Sea. The Vaigai River, which originates on the eastern slope of the Western Ghats, flows east and discharges into the Bay of Bengal. The index map of the system is shown in Figure 3.

The system is benefited from both the South-West monsoon and the North-East monsoon. The Periyar catchment receives rainfall during the South-West monsoon. The Vaigai catchment and the command area receive rainfall during the North-East monsoon. The natural flows in the Vaigai basin (east of the Western Ghats) were fully utilized by the end of the nineteenth century and water shortage was experienced in the basin. This led to the construction of Periyar reservoir and the Periyar-Vaigai trans-basin scheme which made it possible to divert the waters from

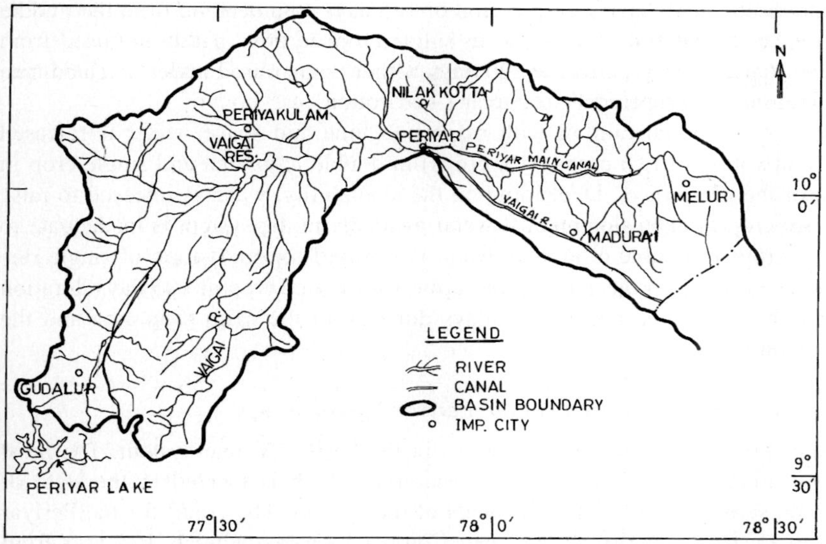

Figure 3. Index map showing Periyar-Vaigai System

the Periyar basin in the state of Kerala to the Vaigai basin in the state of Tamil Nadu for supplementing the irrigation in the command area of the system.

The irrigated command area of the Periyar-Vaigai system covers parts of the Madurai and Ramanathapuram districts of Tamil Nadu and is located on the plains between the Western Ghats and the Bay of Bengal. It covers a gross area of 1,300 Mm^2 of which 734 Mm^2 is cultivable. Madurai, the second largest city in Tamil Nadu is located on the fringe of the project area.

The area has a tropical monsoon climate; the normal annual rainfall of the Periyar being about 2,000 mm. Major portion of the rainfall in this area falls during the South-West monsoon from June to September. In clear contrast to this, the normal annual rainfall for the Vaigai reservoir catchment is about 750 mm. This area receives a major portion of the annual rainfall during the North-East monsoon from October to December.

January and February are the months of cold and dry weather while April and May months have hot and dry weather when the maximum daily temperature sometimes exceeds 40 °C. Although the project area experiences two monsoons in a year, the distribution pattern of rainfall is not suitable to grow crops without irrigation. The actual pan evaporation is high, nearly 2,400 mm per year. Evaporation rates in the Vaigai catchment in the months from March through August are of the order of 7 to 8 mm per day.

Paddy is the principal crop raised in the project area. The farmers use short duration varieties of paddy for the first crop, which has a crop period of 105 days commencing from 1st June. For the second crop, medium duration varieties of paddy are used which have a crop period of 135 days commencing from the middle of September to the end of January. In single cropped area, paddy is raised from mid August with a crop period of 120 days. Other crops raised under irrigated area are groundnut (2 to 3 percent), sugarcane and cotton.

In a year of normal rainfall with respect to time and space, water is released for the cultivation of two crops (rice crops) in double crop area and single crop in the rest of the pre-project. Depending on the availability, water is released to raise single rice crop in extension area. If water available in the system is inadequate to raise two crops in double crop area, water is released for cultivation of single rice crop in the entire pre-project area. The farmers grow rice crop of 105 days duration in the double crop area and of 120 days duration in the main crop area and the extension area.

Major water resources projects in the Periyar-Vaigai system

There are three main hydraulic structures in the Periyar-Vaigai system. The most important among them is the Periyar reservoir, which is located in the state of Kerala and satisfies most of the demands of the system. The credit for the Periyar Dam goes to the noted British engineer, Colonel John Pennyquick. Releases from the Periyar reservoir are picked up at the Vaigai reservoir, which is a balancing reservoir. Upstream of Vaigai reservoir, the releases from the Periyar reservoir are required to satisfy the irrigation demands of the Cumbum valley. Peranai regulator is located downstream of the Vaigai dam. Vaigai reservoir was constructed across Vaigai River during 1954–59. Its catchment area is 2,253 sq. km and capacity at FRL is 194.78 MCM.

Periyar Irrigation System: After coming out of the Periyar power house, the releases of the Periyar reservoir and the natural flows of the Vaigai basin flow through the Cumbum valley. The Cumbum valley agriculture areas are irrigated through the channels taking off from 15 anicuts constructed across the rivers Vairavanar and Suruliyar. This command area enjoys the perennial supply of water from the Periyar reservoir in addition to the natural inflows. It consists of two distinctive areas: a) The old command of 6,017 ha, served by 17 channels and b) New command of 2,082 ha, served by recently constructed P. T. Rajan channel.

The old command is designed for two paddy crops (June–February): the first coinciding with the South-West monsoon; and the second with North-East monsoon. The new command is designed for single crop during September–March.

Vaigai Irrigation System: The command area of the Vaigai irrigation system lies downstream of the Vaigai reservoir and is in the rain shadow region of the Western Ghats. The total command area of this system is 76,736 ha. Paddy is the principal crop grown in this area. The command area consists of pre-project area of 58,827 ha and extension area of 17,909 ha. The pre-project area and the extension areas are distributed in PMC and TMC. Of the pre-project area, 53,312 ha lies in PMC and 5,515 ha in TMC. In the pre-project command area of PMC, double crops are cultivated in 18,227 ha as per the present practice. In other area, single crop is cultivated. Of the total extension area, 15,567 ha lies under PMC command and 2,342 ha lies under TMC. In the extension area, single rice crop is cultivated. Pre-project area and extension area are together known as the post-project area. The double crop areas lie between the Peranai regulator and the Kallandiri regulator. The single (main) crop areas get distributed in the entire system. As far as normal irrigation supply is concerned, the pre-project areas have priority for irrigation water over the extension areas.

In addition, the command area has a number of system tanks which store the runoff from their own catchment areas and the irrigation return flows. There are 251 tanks up to the Kallandiri regulator which is located downstream of the Peranai regulator. The water available in the tanks is used for irrigation. This irrigation system is mainly operated for cultivation of rice crop in the command area. The command area consists of double crop area and the single crop area on the downstream of the Vaigai reservoir.

Various demands of the system

The various demands of the system include Periyar Irrigation System Demands (Cumbum valley), Vaigai Irrigation system demands (PMC and TMC) and the Madurai water supply demands. While releasing water from the Periyar reservoir, normally highest priority has been given to the Cumbum valley irrigation demands and the Madurai city water supply demands. Next priority has been given for meeting single crop area demands under the Vaigai irrigation. If water is still available, then the double crop requirements and the extension area demands are satisfied.

Lower Periyar: Lower Periyar is a concrete gravity dam on Periyar River, located at a distance of 32 km from Kothamangalam in Idukki District, Kerala. The catchment

Table 13. Salient features of selected existing projects in east flowing Rivers Between Pennar and Kanyakumari

Name of the project	State	Year of completion	Gross storage capacity (Mm^3)	Live storage capacity (Mm^3)	Designed annual irrigation (Mm^2)	Installed capacity (MW)
Chittar-I	Tamil Nadu	1970	17.28	11.13	129.50	–
Chittar-Ii	Tamil Nadu	1970	28.55	16.99	–	–
Chalavasam Tank	Tamil Nadu	1986	35.00	31.26	–	–
Pachiparai	Tamil Nadu	1906	152.36	126.02	259.00	–
Poondi Reservoir	Tamil Nadu	1944	–	77.87	–	–
Peruvaripallan	Tamil Nadu	1971	17.50	13.12	–	–
Perunchani	Tamil Nadu	1952	–	81.84	–	–
Red Hills	Tamil Nadu	–	81.00	71.00	–	–
Servalar	Tamil Nadu	1986	35.00	31.26	–	–
Thanbraparani	Tamil Nadu	1943	155.80	147.34	–	–
Vidur	Tamil Nadu	1959	–	17.13	13.00	–
Wellingdom	Tamil Nadu	1923	73.40	60.01	112.00	–

Table 14. Salient features of selected under construction projects in east flowing Rivers Between Pennar and Kanyakumari

Name of the Project	State	Gross storage capacity (million cubic meter)	Live storage capacity (million cubic meter)	Designed annual irrigation (Million Sq. Meter)	Installed capacity (MW)
Anaikuttam Reservoir	Tamil Nadu	11.16	11.08	18.20	–
Kelevarapalli	Tamil Nadu	13.61	13.17	32.40	–

area at the dam is 181.3 km^2. The height and length of the dam is 32 m and 284 m respectively. The reservoir has a live storage capacity of 4.55 MCM at FRL 253.00 m and the MDDL is at 237.74. Lower Periyar power house has 3 units of 60 MW each. It has a firm power of 57 MW. The project was commissioned by the KSEB in 1997.

In addition to the above, there are several existing and under construction water resources projects in the basin. Salient features of existing and under construction water resources projects with a live storage capacity of 10 MCM and above have been presented in Table 13 and Table 14, respectively.

16.9. RIVERS DRAINING INTO BANGLADESH

There are a number of minor rivers originally flowing through Tripura and Mizoram that enter into Bangladesh before finally draining into Bay of Bengal. The important rivers from Tripura are Manu, Khowai, Gomati and Muhuri. Karnafulli River flows through Mizoram and enters Bangladesh. The total catchment area of the basin lies in Mizoram and Tripura is 10,031 km^2. The state-wise distribution of the area is given in Table 15.

Runoff in the rivers have been measured by CWC. Table 16 represents the annual average observed runoff at selected CWC sites in East & West flowing Rivers (Catchment area $> 5,000\,\text{km}^2$).

Table 15. State-wise distribution of drainage area of minor rivers of Tripura and Mizoram

State	Drainage area (km^2)
Mizoram	4,266
Tripura	5,765
Total	10,031

Table 16. Annual average observed runoff at selected CWC sites in East & West flowing Rivers (Catchment area > $5,000\,\text{km}^2$)

Name of the site	Name of the stream	Catchment area (km^2)	Annual average runoff (BCM)
Purushottampur	Rushikulya	7,112	1.34
Kashinagar	Vamsadhara	7,820	1.67
Srikakulam	Nagavali	9,500	1.86
Thammavaram	Gundlakamma	7,864	6.3
Chengalpattu	Palar	16,230	1.6
Arcot	Cheyyar	10,174	1.7
Vazhavachanur	Ponniyar	10,780	2.0
Villupuram	Ponniyar	12,900	1.8
Paramakudi	Vaigai	6,796	0.9
Kumbidi	Bharathapuzha	5,755	23.7

16.10. HYDROLOGY AND WATER RESOURCES OF ISLANDS

There are two main groups of Islands in India: Andaman and Nicobar group and Lakshadweep group.

16.10.1. Andaman and Nicobar Islands

This group of islands is located in the Bay of Bengal, between 92° to 94 °E longitudes and 6° to 14 °N Latitudes. The area of Andamans district is 6,408 sq. km and that of the Nicobars district 1,841 sq. km. The total land area of the group is 8,249 sq. km. Out of this, the Andaman District occupies 6,408 sq. km. and the Nicobar District 1,841 sq. km. In this predominantly rural setting, the rural area is 8,232.36 sq. km while the urban area is only 16.64 sq. km. In these islands, altitude varies from sea level to 732 meters. The highest point in Andaman islands is Saddle Peak (North Andaman Island) at 732 m and in Nicobar islands, the highest point is Mount Thullier (Great Nicobar Island) at 642 m. Port Blair is the administrative capital of the islands.

Andaman and Nicobar Islands appear in history as old as the period of Lord Rama. Mythologically, the name Andaman was presumed to be derived from Hanuman, who was known to the Malays as Handuman. Hanuman, the noted friend of Lord Rama is believed to have landed here on the way to Sri Lanka, in search of Sita. Since prehistoric times, these islands were the home of aboriginal tribes. The tribes of, the Andaman group of islands are the Great Andamanese, Onges, Jarawas, and Sentinatese, all of Negrito origin, while the tribes of Nicobars are the Nicobarese and Shompens, both of Mongoloid stock.

In recent times, the existence of these islands was first reported in the 9th century by Arab merchants, who sailed past them on their way to the straits of Sumatra. The first western visitor was Marco Polo who called it the land of the head hunters. Marathas had captured these islands and remained in control till the death of Maratha

Admiral Kanhoji Angre in 1729. The first settlement by the British took place in 1789, which was later abandoned in 1796. The second settlement was basically a penal settlement, taken up in 1858, after the First War of Independence, followed by the settlement of convicts, Moplas, some criminal tribes from Central and United Provinces, refugees from erstwhile East Pakistan, Burma and Sri Lanka as well as ex-servicemen. In a way the islands were inhabited until mid 19th century mainly by aboriginal tribes whose main occupation was hunting using bows and arrows. Current population of these islands is about 278,000 people which amounts to 34 persons per sq. km.

Andaman & Nicobar Islands is a Union Territory, stretched over an area of more than 700 km from north to south with 36 inhabited islands. Once a hill range extending from Burma (Myanmar) to Indonesia, these undulating islands are covered with dense forests and endless variety of exotic flowers and birds. These islands receive annual rainfall of about 3,000 mm, mainly in two periods: May to Mid-September and November to January. There is not much variation in temperature. Mean minimum temperature is 23 °C and mean maximum temperature is 30 °C. Relative humidity varies between 70% and 90%.

The Andamans are a group of more than 500 islands many of which are still uninhabited. The islands stretch into the territory of some 750 km from north to south in the Bay of Bengal. These islands are now being developed. Figure 4 show the index map of Andaman & Nicobar Islands.

Topography

The topography of the islands is hilly and abounds in tropical flora. These are endowed with evergreen thick forests and tropical trees with mangrove swamps on the water's edge; the forest cover is 86%. Many parts of the islands are hilly. The beaches have white sand, coral reef, and unpolluted sparkling clear water. The sandy beaches on the edge of meandering coastline are fringed with coconut-palms that sway to the rhythm of the sea. The sea around the islands offers excellent scope for water sports. The rare flora and fauna, underwater marine life and corals, with crystal clear water and mangrove-lined creeks, offer a dream-view of the rare gifts of nature. Figure 5 gives a view of the main island.

About 50% of the forests have been set aside as Tribal Reserves, National Parks and Wildlife Sanctuaries, which are inviolate. Luxuriant mangroves, perhaps the richest in the world, occupy nearly 11.5% of the territory. More than 150 plant and animals species are endemic in nature. Mahatma Gandhi Marine National Park is rich in corals, varieties of colored fishes, sea turtles etc., besides other marine life. It is a bird's paradise – more than 271 varieties of birds inhabit the idyllic landscape, out of which 39 are endemic. Megapode, Swiftlet, Hornbill and Nicobar Pigeon are some of the specialties of the Andaman & Nicobar Islands.

Total length of Andaman Islands is 467 km. Their maximum width is 52 km while the average width is 24 km. Nicobar Islands are smaller is size with total length of 259 km and the maximum width of 58 km. In the Andaman group,

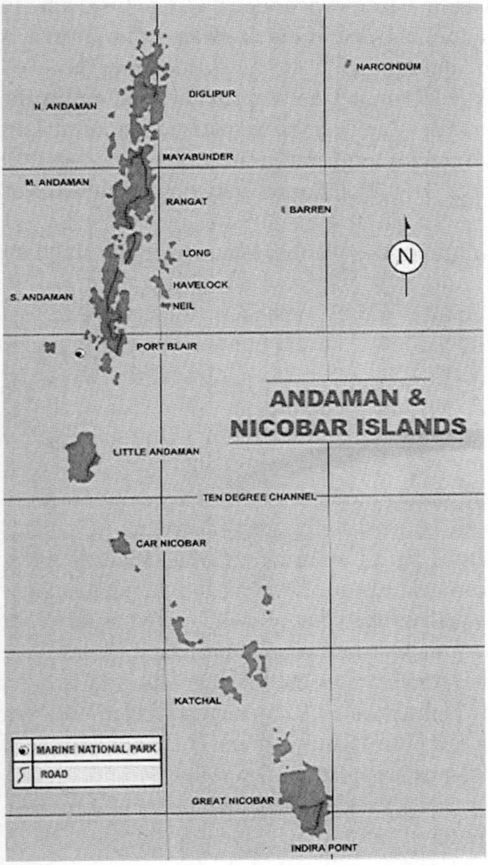

Figure 4. Index map of Andaman & Nicobar islands

Figure 5. A view of the Andaman Island

the biggest island is the Middle Andaman Island whose area is 1,536 sq. km. In the Nicobar group, the biggest island is the Great Nicobar Island whose area is 1,045 sq. km.

The normal rainfall at Port Blair is 3,180 mm. Due to proximity of sea at the islands, temperature variation is very small – mean minimum temperature at Port Blair is about 22.5 °C and the mean maximum temperature is about 30 °C. At Port Blair, the mean relative humidity is about 81%. Population statistics for the islands is given in Table 17.

There are number of small streams in Andaman, Nicobar Islands. Obviously, their length is very small.

Further detailed information about these islands is available at *www.Andaman.nic.in.*

16.10.2. Lakshadweep Islands

The Lakshadweep Islands are irregularly scattered in the Arabian Sea between 8° to 12°30′ north latitude and between 71° to 74° east longitudes. There are 36 islands including 12 tolls, 3 reefs and 5 submerged banks covering an area of 32 km^2 land, 4,200 km^2 of lagoon and 40,000 km^2 of oceanic zone. Among the 36 islands, only 10 are inhabited and the rest are seasonally inhabited or uninhabited islands. At present, except Pitti Island, no other area of Lakshadweep is protected by law.

These palm-trees covered, coral islands have beautiful lagoons and are an attractive and peaceful place. Lakshadweep consists of 36 islands located between 200 km and 300 km off the Kerala coast. These islands are coral islands, northern extension of the Maldives. Ten of the islands are inhabited: Andrott, Amini, Agatti, Bitra, Chetlat, Kadmat, Kalpeni, Kavaratti (headquarters), Kiltan, and Minicoy. Lakshadweep is a Union Territory of India.

Lying 400 km west off the Malabar Coast – called Lakshadweep, comprising of approximately 36 coral islands and 12 atolls. Out of these only 11 islands are inhabited. These are some of the finest, unexplored coral islands in the world, although facilities and infrastructure are not well developed. The Lakshadweep islands are amongst the finest coral islands in the world. The clear aquamarine lagoons protected by a reef, make one feel cozy and safe from the ocean waves. The islands are comparatively tiny and flat with only coconut palms.

Table 17. Population statistics for the Andaman and Nicobar islands

	1971	1981	1991	2001
Total Population	115,133	188,741	280,661	356,265
Rural Population	88,915	139,107	205,706	
Urban Population	26,218	49,634	74,955	

16.10.3. Daman and Diu

Daman earlier called 'Damao', is situated on the west coast of India, bordered by the Arabian Sea. It is surrounded by the Kolak River on the north, Gujarat on the east, Kalai River on the south and Gulf of Cambay on the west. This place has very low altitude, about 12 m. Being close to sea, it has mild climate. In summer, the maximum temperature goes up to 36 °C and the minimum annual temperature is 20 °C.

The island of Diu, an erstwhile Portuguese colony, is situated off the Saurashtra coast of Gujarat bordering Junagadh district. The word 'Diu' is derived from the Sanskrit word 'Dweep'. Pandavas during their fourteen years of exile are believed to have passed a few days at a place known as Mani Nagar. According to mythology, Diu was ruled by the great king Jallandhar who was a daitya (Demon) and was killed by Lord Vishnu with his Sudarshan Chakra. During the period from the 14th to 16th century, Diu was a major sea port and a naval base.

This island occupies an area of 38.5 sq. km and its altitude is 29 m. Temperature at this place varies between 38 °C (maximum in summers) to 20 °C (minimum in winters). Mean annual rainfall is close to 70 cm. Diu is famous for its beautiful beaches such as Ahmedpur, Mandvi, Chakratirath, Jallandhar, and Gomtimata.

EMINENT WATER RESOURCES PROFESSIONALS WHO HAVE MADE OUTSTANDING CONTRIBUTIONS IN INDIA

Bharat Ratna M.Visvesvaraya (1861–1962): Engineer, Statesman, Educationist, Administrator and a Great Humanist. Responsible for pioneering work in the field of water resources in the Southern States of Karnataka, Maharashtra and Andhra Pradesh. His birthday which falls on 15th September is celebrated as Engineers' Day in India.

Dr. A. N. Khosla (1892–1984): Outstanding engineer and administrator. He rose to become the Chairman (Central Water Commission), the Vice Chancellor of the University of Roorkee, and the Governor of Orissa State. He is credited for construction of several major projects including the Bhakra Dam. He developed the theory for *Design of Weirs on Permeable Foundations*.

Dr. K.L. Rao: An excellent engineer who went on to become the Minister of Irrigation and Power, Govt. of India. He was responsible for design of many dams, notably Hirakud and Nagarjunasagar. He conceived inter basin transfer of water on a grand scale through the Ganga–Cauvery link.

Sir Arthur Cotton: Famous for the construction of Dowaleshwaram anicut (barrage) across Godavari River in mid 19th century. Prior to the construction of barrage, Godavari delta was frequently visited by famines. In his honour, this barrage is also known as Cotton barrage and Cotton is revered as a saint.

Dr. Kanwar Sain: Former Chairman Central Water Commission. He is credited for construction of the Indira Gandhi Canal, Kosi Project, Flood Protection of Dibrugarh, and conceived the Farakka Barrage. He was also closely associated with the Mekong Project.

K.K. Framji: Another famous hydraulics engineer of India. He was the Secretary General of ICID for nearly 25 years (1963–1988) and was the joint winner of 'Olympia Prize' 1985. He was the Chairman of ISO Technical Committee TC 113 for a very long time.

Sir Proby Thomas Cautley (1802–1871): The man who constructed the Upper Ganga Canal in 1850s, a marvelous irrigation project which was instrumental in socio-economic transformation of Western Uttar Pradesh. A boys' hostel at IIT Roorkee is named after him.

Gerald Lacey (1887–1983): Developed widely used equations for design of stable channels which helped in canal design.

Dr. Bharat Singh: He was a Professor in Civil Engineering at the University of Roorkee who rose to become the Vice-Chancellor of the same university. He has written three text books on irrigation and water resources and numerous papers. Prof. Singh was a member of National Commission for Integrated Water Resources Development, a member of the panel constituted by the then Prime Minister for Cauvery Water Dispute, and a member of expert panels of several major projects in India. An outstanding water resource engineer, he is a Fellow of the National Academy of Engineering and has won several awards.

SECTION 4

WATER USES, PROJECTS, PROBLEMS, AND GOVERNANCE

CHAPTER 17

MAJOR USES OF WATER IN INDIA

Since independence various measures have been taken for development of water resources in India under the Five Year Plans. A noteworthy achievement has been self-sufficiency in food grain production. From 50 million tons in pre-plan era, food grain production reached 211 million tons in 2001–02. This praiseworthy achievement has been possible mainly due to development of irrigation by major, medium, and minor project and advancements in agriculture technology. Area covered by irrigation till end of IX Plan stood at 90.93 million hectares (M-ha) as against 9.70 M-ha in the pre plan era.

17.1. THE DEMAND FOR WATER

The demand for water for a use is the quantity of water required to be supplied for that specific use. For example, for agricultural use, demand includes consumptive requirement of crops, percolation losses, canal seepage, and evaporation. In case of hydropower generation, although no water is consumed, the practice in India is to consider the evaporation loss from the reservoirs as the use of water for this purpose.

While estimating present demand by different sectors, management policies, availability of technology and other resources are required to be considered along with the prevalent practices. For future demand projections, emerging practices and technologies which are likely to be in practice in the future socio-economic environment have to be considered. The effective water supply is the amount of water actually given and received by the user sectors. Only a part of the quantity of water given to a particular use is consumed and the rest of the water (return flow) drains away and can be reused by the downstream users, (usually after some treatment).

17.2. WATER RESOURCES REQUIREMENT

The availability of water in India shows a great deal of spatial and temporal variability. The increase in population and expansion of economic activities inevitably leads to increasing demands for water use for diverse purposes. Therefore, overall national planning and resource management in respect of water with emphasis on allocation of priorities among the diverse uses is necessary.

Water use efficiency is presently estimated to be around 40% for canal irrigation and about 60% for ground water irrigation schemes. India's per capita water availability per year (1991 census) was estimated at 2,209m^3 against the global average of 9,231m^3.

Surface water is either used in-stream for hydropower, recreation, navigation etc. or is diverted for off-stream use. Ground water is mostly used for irrigation or for domestic requirements. The consumptive uses of water are: (a) Irrigation for agriculture, (b) rural and municipal water supply, and (c) industrial water supply. The Indian economy has traditionally been agriculture based and the principal consumptive use of water is for irrigation. On the total water use in 1990, the share of agriculture was 83%, followed by domestic use (4.5%), energy (3.5%) and industrial use (2.7%). The remaining 6% were for other uses including environmental requirements. At the time of independence, it was of crucial importance to develop irrigation to increase agricultural production for making the country self-sustained and for poverty alleviation. Accordingly, irrigation sector was assigned a very high priority in the 5-year plans. Giant schemes like the Bhakra Nangal, Hirakud, Damodar Valley, Nagajunasagar, Rajasthan Canal project etc. were taken up to increase irrigation potential and maximize agricultural production.

The long term planning has to take into account for the growth of population. A number of individuals and agencies have estimated the likely population of India by the year 2025 and 2050. In Chapter 2, the topic of population was discussed in detail. According to the estimates adopted by NCIWRD (1999), by the year 2025, the population is expected to be 1,333 million in high growth scenario and 1,286 million in low growth scenario. For the year 2050, high rate of population growth is likely to result in about 1,581 million people while the low growth projections place the number at nearly 1,346 million.

In India, the food grain availability in the year 2004 was around 525 gm per capita per day whereas the corresponding figures in China and USA were 980 gm and 2,850 gm respectively. Assuming that the level of consumption will rise with time due to socio-economic factors like better standard of livings, NCIWRD (1999) estimated that the food grain requirements per capita per year would be 218 kg for the year 2025 and 284 kg for the year 2050. After considering factors such as feed requirement, losses in storage and transport, seed requirement, and buffer stock, the projected foodgrain and feed demand for 2025 would be 320 million tons (high demand scenario) and 308 million tons (low demand scenario). The requirement for the year 2050 would be 494 million tons (high demand scenario) and 420 million tons (low demand scenario).

The required rate of growth in foodgrain production can be achieved through extension of irrigated areas and by increasing the grain yield per unit area assuming that there may not be any significant increase in net sown area. It has been established that productivity of irrigated areas is at least double, if not more, compared to un-irrigated areas in respect of wheat and rice crops. The present average productivity of irrigated land is about 2.5 t/ha and less than 1.0 t/ha for rain-fed lands. Therefore, irrigation will be the prime input for increasing the foodgrain output. Based on the trends in this sector, these levels are likely to go up to at least 3.5 and 1.5 t/ha respectively by 2050 AD. In fact, given sufficient attention, yield of the order of 5.5–6.0 also seen feasible.

The sector-wise requirement of water is being discussed in the following sections.

17.3. DOMESTIC WATER NEEDS

Community water supply is the most important among the requirements of water and it is about 5% of the total water use. The volume of water being utilized for domestic needs is far less than that used for irrigated agriculture. Estimates show that about 7 km^3 surface water and 18 km^3 ground water is being used for water supply in urban and rural areas. Organized water supply and sanitation programs have not yet covered the entire country. Under the International Drinking Water Supply and Sanitation Decade Program launched in 1981, the aim was to provide adequate drinking water facilities to 90% of the urban population and 85% of rural population, and sanitation facilities to 50% of urban population, and 5% of rural population.

While about 82% of the population has access to safe drinking water supply in rural areas, the accessibility in urban areas is around 85%. However, in most of the cities and towns, the supply is grossly inadequate, particularly in slums inhabited by the poorer sections of the society. Use of booster pumps to draw water from municipal networks is common. Besides, the quality of water supplied through these networks is very poor and people are increasingly installing small devices in their homes to clean water before consuming.

Along with the increase in population, another important change from the point of view of water supply is the higher rate of urbanization. As per the projections, the higher is the expected growth in population, the higher would be urbanization. It is expected that nearly 61% of the population will reside in urban areas by the year 2050 in high growth scenario and about 48% in low growth scenario. So the expansion of water supply network for drinking, domestic use and various civic amenities has to be accordingly planned. Currently, most houses that are being constructed in the outskirts of cities make their own arrangement for use of ground water for domestic needs.

Different organizations and individuals have given different norms for water supply in cities and rural areas. The figure adopted by the NCIWRD (1999) were 220 liter per capita per day (lpcd) for classes I cities. For the cities other than class I, the norms are 165 lpcd for year 2025 and 220 lpcd for the year 2050. For rural areas, 70 lpcd and 150 lpcd have been recommended for the year 2025 and 2050. Based on these norms and projection of population, it is estimated that by the year 2050, the water requirements per year for domestic use will be 90 km^3 for low demand scenario and 111 km^3 for high demand scenario. It is expected that surface

Table 1. Basic water requirements for human needs

Purpose	Rural Areas (Litres/ capita /day or lpcd)	Urban Areas (lpcd)
Drinking water	5	5
Cooking	3	5
Ablution	6	–
Bathing	15	55
Washing of utensils, clothes & household	11	45
Flushing of toilets/sewer	–	30
Total basic water requirement (BWR)	40	140

Source: WG (1999).

and ground water sources will meet about 70% of urban water requirement and 30% of rural water requirement.

The core needs of domestic use of water are for drinking, cooking, washing and bathing. Non core needs are for toilet flushing, sewer flushing, washing clothes, water for lawns, etc. The union ministry of works and housing has fixed minimum norms for various basic human needs as given in Table 1.

The National Commission on Urbanization, Govt. of India, has suggested minimum norms for use of water. According to the commission, even in the worst drought conditions and even in the poorest colonies, at least 70 liters of water must be delivered per day to sustain the human life at a minimum standard of hygiene.

In India, recommended norms for water supply are based on the population size. These are given in Table 2.

Shri Rajiv Gandhi, the Prime Minister of India during 1984 to 1989, had constituted several technology missions with well defined objectives. The National Drinking Water Mission was one of them. Recently, all drinking water schemes

Table 2. Recommended norms for water supply

S. N.	Population	Recommended water supply norm (lpcd)
	Less than 20,000	
1	a. Population served by stand posts	40
	b. Population provided with pipe connections	70
2	20,000 to Less than 100,000	100
3	100,000 to Less than 1,000,000	100 (with no sewerage system) 135 (with sewerage system)
4	1,000,000 and above	150
5	Rural and hills (per elevation difference of 100 m)	40 or one hand-pump for 250 persons within a walking distance of 1.6 km
6	Rural – additional water for cattle in Desert Development Programme (DDP) areas.	30

Source: WG (1999) and Planning Commission (2006).

have been brought under the Rajiv Gandhi National Drinking Water Mission. Besides providing drinking water to total population in the country, emphasis will on tackling water quality. At present, only 30% of rural households have access to safe sanitation facilities. The total sanitation campaign now operates in 452 districts.

The requirements of drinking and municipal water supply to metropolitan and other important towns in the country have already become critical. Among the metropolitan cities, total water requirement of Delhi is 800 million gallon per day (MGD). Delhi, which is situated on the banks of the Yamuna, gets scarcely 25% of its needs from the river. The balance needs are met by releases from Bhakra Dam to the west, and Ramganga Dam to the east. This is in addition to a number of tubewells, which contribute less than 10% of Delhi's water supply. By the end of the year 2004, only 640 MGD was being treated at the water treatment plants of Delhi Jal Board. A new plant has been constructed at Sonia Vihar with a capacity of 140 MGD and raw water for this plant is to come from the Tehri dam in Uttaranchal via the Upper Ganga Canal (till Muradnagar in UP and then through a 30-km 3,250 mm diameter conduit). Noteworthy feature of this scheme is that the contract for building and operating this plant for 10 years has been given to a private company.

The entire water supply of Mumbai is dependent on a series of dams such as the Vaitarana, Tansa and Bhatsa. In fact, there is now a proposal to construct another dam on the Vaitarana to meet the increasing needs of Bombay's water supply. The water demand of Pune town in Maharashtra is met by the Panshet and Khadakwasla dams. Hyderabad is mainly dependent for its water supply on the Manjira and Singur Dams. The acute scarcity of water supply in Madras is well known. Apart from the storage available in Poondi Reservoir, there was no other possibility of further augmenting the water supply. Therefore, the states of Andhra Pradesh, Maharashtra and Karnataka agreed to part with their allocated share of Krishna water to provide 430 million m^3 of water supply to Madras city, which will be made available through the stored waters at Srisailam Reservoir in Andhra Pradesh and carried through 430 km long canal to reach Madras city. A major dam at Bisalpur is now being constructed mainly to provide water supply to Ajmer city in Rajasthan. Warangal town in Andhra Pradesh depends for its water supply on Sriramsagar Dam.

In India, water charges are paid either in the form of property tax, or based on water consumption, or a combination of the two. For consumption also, the charges are either on the basis of flat rates or depending on the quantity is consumed. In the later case also, two variations are found: the uniform rate or the rates increasing as the consumption increases. However, in many cases, the meters are faulty or broken. Often the water prices are very low and even a fraction of operation and maintenance expenditures are not recovered. Regarding pricing for domestic water supply, the average monthly water bill for consuming 25 kilo litres is Rs. 31.50 in Delhi, Rs. 67.50 in Chandigarh, Rs. 75.00 in Mumbai, Rs. 100.00 in NOIDA, Rs. 140.00 in Bangalore, and Rs. 225.00 in Chennai (The Times of India, 2004). Clearly, among these cities, the consumers in Delhi pay the least while the consumers of

Table 3. Cost of unreliable supply and poor quality water

Cost item	Amount (per annum)
Purification cost	Rs. 648.91
Booster pump	Rs. 1,147.74
Borewell motor	Rs. 644.63
Additional water	Rs. 844.53
Sickness cost	Rs. 516.34
Storage cost	Rs. 688.27
Total cost	Rs. 4,490.42

Chennai pay nearly eight times more than that. However, the charges in Delhi were recently increased so that the consumers will be paying about 1.5 to 3 ties what they were paying earlier. For instance, those consuming up to 10 kilo litres and living in posh area will pay Rs. 162 per month. But even with the increased charges, the authorities are not able to recover full operation and maintenance charges. Here it is pertinent to note that the bottled water is sold at rates varying from Rs. 12 to 15 per litre.

Attempts have been made to determine the cost of unreliable supply and of poor quality of water borne by people in planned settlements. Table 3 gives the results of a survey conducted by Dutta and Tiwari (2005) for New Delhi. According to this study, people residing in planned settlements are paying about Rs. 4,490 per annum or Rs. 374 per month on account of unreliable and poor quality water.

17.4. IRRIGATION

Agriculture is the most fundamental form of human activity. The word ‘agriculture’ has its roots in Latin word ‘agri’ meaning soil and ‘culture’ meaning cultivation. It is said that agriculture sustains the society while irrigation sustains agriculture.

The gross irrigated area in India was only 22.6 M-ha in 1950–51. The net irrigated area was 20.9 M ha, comprising of 9.71 M ha (8.3 M ha net) by major and medium projects, 6.4 M ha by minor surface water schemes and 6.5 M ha by groundwater. Since the food production was much less than the requirement, sincere efforts were made by the Govt. of India for expansion of irrigation through surface and ground water projects. An ambitious target for large-scale water resources development was set and efforts initiated to achieve it through five-year plans.

In India, irrigation projects are classified as major (having culturable command area or CCA more than 10,000 ha), medium (CCA between 2,000 ha to 10,000 ha), and minor (CCA below 2,000 ha). The source of water in major and medium schemes is surface water while the dominant source in minor schemes is ground water. The ultimate total irrigation potential from major, medium, and minor irrigation schemes has been estimated as 140 M-ha. Out of this, 76 M-ha would come from surface water and 64 M-ha from ground water sources. The ultimate potential from major and medium projects is about 59 M-ha and it is 81 M-ha from minor projects.

In India, progress of irrigation development is reported through three sets of figures: irrigation potential created; potential utilized; and irrigation by source (surface or ground water). The potential created by the year 1997 was about 91 M-ha comprising of 45 M-ha by surface water and 46 M-ha by ground water sources. Thus about 65% of the potential had been developed by 1997. The actual utilization was about 81 M-ha; the share of major and medium projects was 28 M-ha and 53 M-ha was the contribution of minor projects. Source-wise utilization was 39 M-ha from surface and 42 M-ha from ground water.

At the time of independence, canals provided irrigation to large commands over the country. Major systems included Sirhind canal (providing irrigation to 0.6 M ha), Upper Bari Doab canal (0.33 M ha), Sone canal (1.35 M ha), Yamuna canals (0.68 M ha), and Upper Ganga canal (0.7 M ha). According to Vaidyanathan (1999), total irrigation from diversion based systems was about 80% of the irrigation from major and medium projects. Proponents of diversion schemes argue that adverse social and environment impacts of these structures are much less than those of big dam projects. But it may be noted that as far as command area is concerned, both have the same impact and storages are often needed to improve reliability of diversion schemes.

The cumulative irrigation potential created and utilized for each five-year plan in the country is shown in Table 4. Further, Table 5 lists statewise irrigation potential, potential created and gap from surface water and ground water.

The data about cumulative irrigation potential created under the various plans as given in Table 4 (above) is shown in Figure 1 for a better visualization. It can be readily appreciated from the figure that the addition to the potential is very small from 1997 onwards. From 1969 to 2000, the potential created by surface water nearly doubled from 24.6 (18.1 + 6.5) to 48.12 (35.10 + 13.02) M-ha but

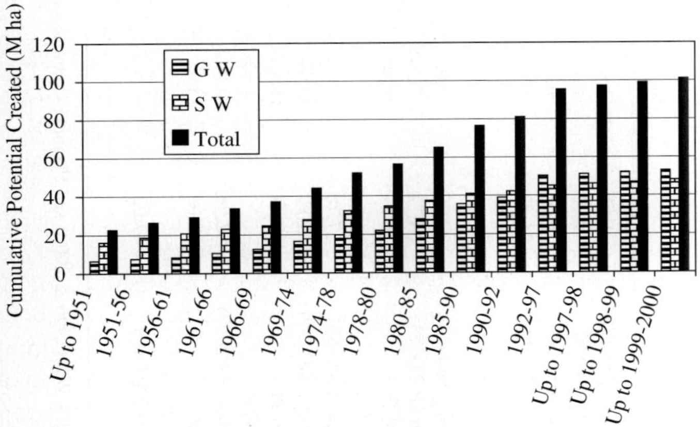

Figure 1. Cumulative irrigation potential created under various plans

Table 4. Planwise Cumulative Irrigation Potential Created and Utilized (in M-ha)

Plan	Potential created					Potential utilized				
	Major & medium	Minor			Total	Major & Medium	Minor			Total
		Surface Water	Ground Water	Total			Surface Water	Ground Water	Total	
Pre-Plan up to 1951	9.70	6.40	6.50	12.90	22.60	9.70	6.40	6.50	12.90	22.60
I Plan (1951–56)	12.20	6.43	7.63	14.06	26.26	10.78	6.43	7.63	14.06	25.04
II Plan (1956–61)	14.33	6.45	8.30	14.75	27.08	13.05	6.45	8.30	14.75	27.80
III Plan (1961–66)	16.57	6.48	10.52	17.00	33.57	15.17	6.48	10.52	17.00	32.17
Annual Plan (1966–69)	18.10	6.50	12.50	19.00	37.10	16.75	6.50	12.50	19.00	35.75
IV Plan (1969–74)	20.70	7.00	16.50	23.50	44.20	18.69	7.00	16.50	23.00	42.19
V Plan (1974–78)	24.72	7.50	19.80	27.30	52.02	21.16	7.50	19.80	27.30	48.46
Annual Plan (1978–80)	26.61	8.00	22.00	30.00	56.61	22.64	8.00	22.00	30.00	52.64
VI Plan (1980–85)	27.70	9.70	27.82	37.52	62.22	23.57	9.01	26.24	35.25	58.82
VII Plan (1985–90)	29.92	10.99	35.62	46.61	76.53	25.47	9.97	33.15	43.12	68.59
Annual Plan (1990–92)	30.74	11.46	38.89	50.35	81.09	26.32	10.29	36.25	46.54	72.86
VIII plan (1992–97)	32.96	12.09	50.31	53.30	86.26	28.44	7.73	39.58	48.80	77.24
IX Plan (1997–2002)	37.08	13.02	53.01	56.90	93.98	31.03	8.27	41.80	49.05	86.80
X plan (2002–07) Target	47.02			63.71	110.73					

Source: CWC (1996 & 2002) and Planning Commission (2006).

Table 5. Statewise Details of Irrigation Potential created and Gap (all in 1,000 ha)

State/UTs	Ultimate irrigation potential		Potential created up to 1994–95		Gap in Irrigation Potential		
	S.W.	G.W.	S.W.	G.W.	S.W.	G.W.	Total
Andhra Pradesh	7, 300	3, 960	4, 422	1, 635	2, 878	2, 325	5, 203
Arunachal Pradesh	150	18	73	2	77	16	93
Assam	1, 970	900	622	190	1, 348	710	2, 058
Bihar	8, 400	4, 947	4, 203	3, 965	4, 197	982	5, 179
Goa	25	29	32	2	–7	27	20
Gujarat	3, 347	2, 756	1, 532	1, 751	1, 815	1, 005	2, 820
Haryana	3, 050	1, 462	2, 112	1, 524	938	–62	876
Himachal Pradesh	285	68	140	15	145	53	198
Jammu & Kashmir	650	708	534	9	116	699	815
Karnataka	3, 400	2, 574	2, 317	758	1, 083	1, 816	2, 899
Kerala	1, 800	879	948	130	852	749	1, 601
Madhya Pradesh	8, 200	9, 732	3, 302	1, 508	4, 898	8, 224	13, 122
Maharashtra	5, 300	3, 652	3, 140	1, 608	2, 160	2, 044	4, 204
Manipur	235	359	126	1	109	358	467
Meghalaya	105	63	41	9	64	54	118
Mizoram	70	0	12	0	58	0	58
Nagaland	85	0	66	1	19	–1	18
Orissa	4, 600	4, 203	2, 143	665	2, 457	3, 538	5, 995
Punjab	3, 050	2, 917	2, 477	3, 349	573	–432	141
Rajasthan	3, 350	1, 778	2, 627	2, 019	723	–241	482
Sikkim	70	0	24	356	46	0	46
Tamil Nadu	2, 700	2, 832	2, 414	1, 289	286	1, 543	1, 829
Tripura	200	81	78	21	122	60	182
Uttar Pradesh	13, 700	16, 799	8, 029	20, 644	5, 671	–3,845	1, 826
West Bengal	3, 610	3, 318	2, 744	1711	866	1, 607	2, 473
UTs	201	5	40	62	161	–57	104
Total	75, 853	64, 040	44, 198	42, 868	3, 1655	21, 172	52, 827

Source: MOWR (1999).

during the same period, the potential created by ground water increased from 12.5 to 53.01 M-ha, more than four-fold increase.

It is noted from this table that there is large gap between ultimate irrigation potential and potential created in Madhya Pradesh, Orissa, Andhra Pradesh, Bihar, and Maharashtra. In Madhya Pradesh alone, of the estimated potential of 17,932 M-ha, only about 4,810 M-ha or only 27% of the potential has been created. One can also notice the instance of mining of ground water in some states, e.g. Haryana, Punjab, Rajasthan, Uttar Pradesh, and the Union Territories.

The state wise irrigation potential in India from groundwater for the year 1996–97 has been presented in Table 6. The full development of ultimate irrigation potential by construction of major, medium and minor irrigation projects by 2025 would be necessary to meet the food requirement of the projected population. Ground water irrigation is mainly through electrical and diesel pump sets.

The gross figures of irrigation potential created and irrigation potential utilized have been presented in Table 7. Note that there are differences among the figures given by different institutions and this is due to different definitions and methods of data collection.

An analysis of past data shows that the net sown area in the country has marginally increased over past 30 years from 140 M-ha to 143 M-ha. It is likely that the net sown area would hover around 145 M-ha between the years 2000 to 2050. To increase the net sown area, multi-prong effects are needed, which include conservation of rainwater, developing irrigation potential, increasing water use efficiency, efficient management of water resources and improvement in agriculture practices, etc.

Recently, Water Resource Consolidation Project (WRCP) has been taken up in the States of Haryana, Orissa and Tamil Nadu. WRCP envisages the completion of some major and medium irrigation projects and strengthening of institutions through Participatory Irrigation Management (PIM). This effort should result in efficient water use in irrigation sector but unless the efforts are sustained, the systems are likely to degrade.

Foodgrains production in the year 1950–51 was 50.82 million tons; it was about 200 million tons in 1996–97, and about 210 million tons in 2004. Thus, the production has not seen major changes over last 7–8 years. Lack of any significant breakthrough in seed technology and major improvements in agricultural practices are cited as the main reasons for the slow growth. Large variations in land yields per unit area across the country show that significant increase in grain production can be achieved if infrastructure is improved and scientific practices are followed in the areas where current yields are low.

Besides agricultural production, India has a variegated nature of horticultural potential. This can be judged by the organized upland tea and coffee plantations, the extensive coastal strips of coconut trees, and the subterranean tuber and root crops. India is famous for its high quality tea and coffee. Darjeeling tea is to be the best among the tea lovers in the world. Recent years have also witnessed high growth in the production of flowers and these are now being exported to many countries. Export of Indian fruits has also registered impassive growth in recent

Table 6. Statewise Irrigation potential from ground water (as on 1996–97)

S N	States	Net utilizable Ground Water Resource for Irrigation (M-ham/year)	Weighted Average Delta (m)	Utilizable Irrigation Potential (M-ha)	Potential Created (M-ha)	Potential Utilized (M-ha)	Percentage Development of Irrigation Potential (6/5*100)	Balance Irrigation Potential for Development (5–6) (M-ha.)
1	2	3	4	5	6	7	8	9
1	Andhra Pradesh	2.69981	0.047–1.472	3.96008	1.77420	1.73910	44.80	2.18588
2	Arunachal Pradesh	0.11005	–	0.01800	0.00240	0.00240	13.33	0.01560
3	Assam	1.89102	1.283	0.90000	0.20680	0.15180	22.98	0.69320
4	Bihar	2.56439	0.40–0.65	4.94763	4.29180	3.81590	86.74	0.65583
5	Goa	0.01670	0.57	0.02928	0.00190	0.00170	6.49	0.02738
6	Gujarat	1.55881	0.45–0.714	2.75590	1.77890	1.69370	64.55	0.97700
7	Haryana	0.65236	0.385–0.600	1.46170	1.54490	1.49930	105.69	−0.08320
8	Himachal Pradesh	0.02637	0.385	0.06850	0.01570	0.01150	22.92	0.05280
9	Jammu & Kashmir	0.33858	0.385–0.600	0.70795	0.01160	0.01100	1.64	0.69635
10	Karnataka	1.23821	0.18–0.74	2.57281	0.78010	0.76410	30.32	1.79271
11	Kerala	0.59281	0.53–0.83	0.87925	0.14060	0.12420	15.99	0.73865
12	Madhya Pradesh	3.89298	0.4	9.73249	1.62290	1.50630	16.68	8.10959
13	Maharashtra	2.29231	0.43–1.28	3.65197	1.63630	1.58840	44.81	2.01567
14	Manipur	0.24129	0.65	0.36900	0.00060	0.00050	0.16	0.36840
15	Meghalaya	0.04128	0.65	0.06351	0.01020	0.01000	16.06	0.05331
16	Mizoram	Not assessed						
17	Nagaland	0.05535	negligible					

(Continued)

Table 6. (Continued)

S N	States	Net utilizable Ground Water Resource for Irrigation (M-ham/year)	Weighted Average Delta (m)	Utilizable Irrigation Potential (M-ha)	Potential Created (M-ha)	Potential Utilized (M-ha)	Percentage Development of Irrigation Potential (6/5*100)	Balance Irrigation Potential for Development (5–6) (M-ha.)
1	2	3	4	5	6	7	8	9
18	Orissa	1.53009	0.34–0.44	4.20258	0.71720	0.60070	17.07	3.48538
19	Punjab	1.51109	0.518	2.91715	3.41200	3.35300	116.96	–0.49485
20	Rajasthan	0.96418	0.457–0.600	1.77783	2.04840	2.01250	115.22	–0.27057
21	Sikkim	Not assessed						
22	Tamil Nadu	2.01892	0.37–0.93	2.83205	1.31450	1.31190	46.42	1.51755
23	Tripura	0.05076	0.63	0.08056	0.02120	0.02120	26.32	0.05936
24	Uttar Pradesh	6.41233	0.20–0.50	16.79896	22.63400	20.35800	134.73	–5.83504
25	West Bengal	1.76653	0.33–0.75	3.31794	1.85520	1.41310	55.91	1.46274
	Total States	32.46622	–	64.04514	45.82140	41.99030	71.55	18.22374
	Total UTs	0.00642	–	0.00504+	0.06240	0.06180		
	Grand Total	32.47264	–	64.05018	45.88380	42.05210	71.64	18.16638

Note: +– Relates to Dadra & Nagar Haveli only

Source: Ministry of Water Resources http://wrmin.nic.in/resource/default3.htm

Table 7. Irrigation development from different sources (in thousand ha)

Data	Surface water	Ground water	Total
Irrigation potential created	43,271 (140%)	41,506 (110%	84,777(124%)
Irrigation potential utilized	37,534 (124%)	38,579 (103%)	76,113 (111%)
Irrigation as per land use	30,869 (100%)	37,498 (100%	68,367 (100%

Source: CWC (1996).

year. Technology and infrastructure support for handling, packing, processing and preservation can save large losses of fruits and vegetables. Compared to developed nations, immense possibilities of value addition in agricultural products exist in India.

17.4.1. Rainfed Agriculture

As large areas of the country do not have assured means of irrigation, rainfed agriculture is widely practiced in the country. Even with large scale development of water resources, the contribution of rainfed farming in Indian agriculture are immense. Estimates show that about 68.0 of the cropped area is under rainfed agriculture and this area is responsible for nearly 44% of total production. Further, rainfed farming supports 40% of the human population and 60% of the live-stock population. Of course, productivity in rainfed situation is quite low. While the average productivity ranges from 2.0 to 2.5 ton/ha in irrigated areas, it is in the range of 0.7–0.8 ton/ha in rainfed areas.

Table 8 gives some irrigation related projections for the years 2010, 2025, and 2050. As shown here, cropping intensity is likely to marginally increase to over 150% by the year 2050. This table also shows that maximum 64% of the gross cropped area is likely to be irrigated by the year 2050 although it will be difficult to attain this target. It is important to realize that about 36% of the gross cropped area is likely to remain rainfed even by the year 2050. Further, by the year 2050, little over half of the total irrigation will be from surface sources.

Table 8. Irrigation related projections for three years

Particulars	Year 2010	Year 2025	Year 2050
Net sown area (M-ha)	143	144	145
Cropping intensity (%)	135	140–142	150–160
Ratio of irrigated to gross cropped area (%)	40–41	45–48	52–64
Ratio of rainfed foodcrop area to gross irrigated area (%)	70	70	70
Ratio of rainfed foodcrop area to gross rainfed area (%)	66	66	66
Ratio of surface water irrigation to total irrigation (%)	47	49–51	54.3

Source: MOWR (1999) Perspective of Water Requirements.

Table 9. Area sown and irrigated from 1950–51 to 1997–98

Year	Area sown (in M ha)		Irrigated area (in M ha)		Cropping intensity (percentage)	Irrigation Intensity (percentage)
	Net	Gross	Net	Gross		
1950–51	118.8	131.9	20.9	22.6	111.0	108.1
1955–56	129.2	147.3	22.8	25.7	114.0	112.7
1960–61	133.2	152.8	24.7	28.0	114.7	113.4
1965–66	136.2	155.3	26.3	30.9	114.0	117.5
1970–71	140.3	165.8	31.1	38.2	118.2	122.8
1975–76	141.6	171.3	34.6	43.4	121.0	125.4
1980–81	140.0	172.6	38.7	49.8	123.3	128.7
1985–86	140.9	178.5	41.9	54.3	126.7	129.6
1990–91	143.0	185.7	47.8	62.5	129.9	130.8
1993–94	142.1	186.4	51.5	68.4	131.2	132.8
1994–95	143.0	188.0	53.0	70.7	131.5	133.4
1995–96	142.2	187.5	53.4	71.4	131.9	133.7
1996–97	142.8	189.6	55.1	73.3	132.8	133.0
1997–98	142.0	190.8	54.6	72.8	134.4	133.3

Source: CWC (2002).

The area sown and irrigated in India from 1950–51 to 1997–98 is given in Table 9. It can be noted that the gross sown area in the country was 190.8 M ha in 1997–98 whereas irrigated area was 72.8 M ha.

The crop wise irrigated area under principal crops in 1997–98 is shown in Table 10. Note that while the gross irrigated area of wheat and rice is of the same order, nearly 85% of wheat crop is irrigated while only 48.6% of rice crop is irrigated. Even in case of rice which is a water intensive crop, more than half of the cropped area is under rainfed cultivation. This is because rainfed rice is quite extensively cultivated in many southern states. Some crops like Sorghum are almost exclusively grown in rainfed areas as the Table 10 shows.

Despite large growth in irrigation and use of fertilizers etc., crop productivity remains quite low in India although the production has grown many folds in the past 50 years. Ambast (2005) has complied data about productivity of some crops in India and the same is shown in Table 11. Note that as compared to some advanced countries, the productivity is very small in India and there is considerable scope of improvement.

Crop productivity in the country shows large regional variations. For two principal crops, rice and wheat, productivity for some states in the country is shown in Table 12. The productivity is very high in Punjab and Haryana and it is 1.5 to 2 times as compared to the other states. Here crop water productivity (CWP) is defined as follows:

$$\mathrm{CWP}(\mathrm{kg/m^3}) = \mathrm{Yield}(\mathrm{kg/ha})/[\text{Water consumed in ET} + \mathrm{Losses}(\mathrm{m^3/ha})] \quad (1)$$

Table 10. Area irrigated under principal crops during the year 1997–98

Name/Type of the crop	Gross irrigated area (M ha)	Share of the irrigated area in the total cropped area for the given crop (percent)
Wheat	21.4	85.1
Rice	20.7	48.6
Rapeseed and Mustard	3.7	59.4
Sugarcane	3.0	89.0
Cotton	2.5	34.3
Groundnut	1.6	19.0
Gram	1.5	23.9
Maize	1.4	22.6
Jowar	0.8	6.2
Bajra	0.6	6.6
Oilseeds	–	24
Pulses	–	20
Pearl millet	–	6.0
Sorghum	–	5.0

Source: MOA (1997) and others.

Table 11. Productivity of selected crops in India

Crop	Production (Million Ton)			Average Yield (t/ha)	Frontline demonstration yield (t/ha)	Yield in other countries (t/ha)
	1950–51	1996–97	2003–04			
Rice	20.6	81.31	87.0	2.09	3.53	8.879 (Egypt)
Wheat	6.5	69.27	72.1	2.77	3.79	6.347 (Egypt)
Maize	1.7	10.61	14.7	2.02	3.99	7.974 (Canada)
Millet	13.7	–	23.0	0.88 (For Pearl millet)	1.99	
Pulses	8.4	14.46	15.2	0.85 (For chickpea)	1.57	2.255 (Egypt)
Jowar	–	11.09	–	0.958		
Gram	–	5.75	–	0.81		
Groundnut	–	9.02	–	1.155		
Cotton	–	14.25	–	0.266		
Sugarcane	–	277.25	–	66.510		

Many factors constrain productivity in rainfed areas; some are natural and technological and some are social. The major such factors are: climatic, soil, technological, and socio-economic. These are explained in Table 13.

17.4.2. Water Requirement to Satisfy Food Grain Demand

Based on the net sown area, cropping pattern, yields etc. for a particular crop, the water requirements for irrigation from surface and ground waters were estimated

Table 12. Crop water productivity in some states of India

Region/Crops	Land Productivity (kg/m^2)		Crop water Productivity (Experimental) in (kg/m^3)
	Avg.	Exp.	
		Rice	
Punjab	0.35	0.66	0.34
Haryana	0.27	0.64	0.44
Uttar Pradesh	0.21	0.46	0.38
Chhattisgarh	0.14	0.70	0.46
Orissa	0.16	0.17	0.21
West Bengal	0.25	0.42	0.36
Karnataka	0.22	–	0.61
		Wheat	
Punjab	0.45	0.54	1.40
Haryana	0.41	0.49	1.44
Uttaranchal	0.19	0.50	1.00
Uttar Pradesh	0.28	0.43	1.11
West Bengal	0.22	0.30	1.15

Source: CWC (2002), Ambast (2005), and others.

Table 13. Constraints in rainfed areas

Factor	Description
1. Climatic	Limited rainfall with erratic distribution, extreme temperatures, higher wind speed.
2. Soil	Shallow, low organic matter, low fertility, sloppy.
3. Technological	Non availability of drought resistant varieties, lack of adoption of production practice, good data base not available.
4. Socio-Economic	Smaller holding size, fragmented holdings, higher population growth rate and illiteracy, problems of marketing and distribution.

by NCIWRD (1999) and are presented in Table 14. The requirements have been computed for low and high demand scenarios and the gross requirements under these two scenarios are likely to be 628 and 807 km^2, respectively.

Although irrigation is the major water user, due to more pressing and competing demands from other sectors, its share in the total demand is bound to decrease from the present 83%. It is also expected that the price of water for agriculture will rise in future so as to recover a part of expenditure in operation and maintenance of irrigation works and this will result in higher water use efficiency. Estimates show that a 10% increase in the efficiency in irrigation will yield enough water to irrigate an additional 14 M-ha land. This much improvement in efficiency will require very moderate investment and user education. Consequently, the share of agriculture in total water demand by the year 2025 is expected to be about 75 percent.

Table 14. Water requirement for irrigation

Particulars	Unit	Year 2010		Year 2025		Year 2050	
		Low Demand	High Demand	Low Demand	High Demand	Low Demand	High Demand
Food grain demand	10^6 tones	245	247	308	320	420	494
Net cultivable area	10^6 hectares	143	143	144	144	145	145
Cropping intensity	%	135	135	140	142	150	160
% of irrigated to gross cropped area	%	40	41	45	48	52	63
Total cropped area	10^6 hectares	193	193	202	204	218	232
Total irrigated cropped area	10^6 hectares	77	79	91	98	113	146
Total unirrigated cropped area	10^6 hectares	116	114	111	106	104	86
Food crop area as % of irrigated area	%	70	70	70	70	70	70
Food crop area as % of unirrigated area	%	66	66	66	66	66	66
Food crop area -irrigated	10^6 hectares	54.1	55.4	63.5	68.7	79.2	102.3
Food crop area -unirrigated	10^6 hectares	76.4	75.2	73.2	70.2	68.9	56.7
Average yield-irrigated foodcrop	Tonne/ ha	3	3	3.4	3.4	4	4
Average yield-unirrigated foodcrop	Tonne/ ha	1.1	1.1	1.25	1.25	1.5	1.5
Foodgrain production from irrigated area	10^6 tones	162	166	216	234	317	409
Foodgrain production from unirrigated area	10^6 tones	84	83	91	88	103	85
Total surrogate food production	10^6 tones	246	249	307	321	420	494
Assumed % of total irrigation potential to potential from surface water	%	47	47	49	51	54.3	54.3
Irrigated area from surface water	10^6 hectares	36.3	37.2	44.5	50.1	61.4	75.9
Irrigated area from ground water	10^6 hectares	40.9	41.9	46.3	48.1	51.7	70.3
Assumed "delta" for surface water	Metre	0.91	0.91	0.73	0.73	0.61	0.61
Assumed "delta" for ground water	Metre	0.52	0.52	0.51	0.51	0.49	0.49

(Continued)

Table 14. continued

Particulars	Unit	Year 2010		Year 2025		Year 2050	
		Low Demand	High Demand	Low Demand	High Demand	Low Demand	High Demand
Surface water requirement for irrigation	km^3	330	339	325	365	375	463
Ground water requirement for irrigation	km^3	213	218	236	245	253	344
Total water required for irrigation	km^3	543	557	560	611	628	807

Source: WG (1999).

17.4.3. Water Distribution Practices in India

Water distribution or conveyance of water from the source to the field head is the most important link between the supplier and the farmers. Rotational methods of distribution of water are generally followed in India using supplier controlled schedules (frequency) with either constant or variable amount of water application. The rotation may be between: (i) sections of main/ branch canal (ii) distributaries and minors, (iii) outlets, and (iv) individual farmers or groups of farmers. Different types of rotation methods of water distribution followed in the country are discussed here.

17.4.4. Warabandi or Osrabandi

The term *warabandi* is composed of two words: *wara* (turn) and *bandi* (fixation). Thus warabandi means fixation of turn for supply of water to different fields. The word *osrabandi* is also synonymously used in place of warabandi. Warabandi is widely practiced in northern states of Punjab, Haryana, Rajasthan, Uttar Pradesh, Bihar, etc. for more than a century. According to this practice, the available volume of water is equitably allocated to all the users in the command irrespective of location of the holdings of farmers. The share of a farmer is in the ratio of the size of his field and is allocated in terms of time interval as a part of total hours (168) in a week.

In a typical surface irrigation system, the main canal feeds two or more branches which operate by rotation and may or may not be able to run full. This constitutes the primary distribution system which runs through out the irrigation season. Branch canals supply water to a number of distributaries – the secondary distribution system – which must run at full supply level by rotation. Distributaries feed water courses called guls (the tertiary system) through ungated, fixed discharge outlets which are generally semi-modular type (Malhotra, 1982). Water courses run at full supply level when the distributary is running and the water is allocated between the farmers on a water course by a time roster (Sahni, 2000).

Two types of warabandi are followed: Khatawar (area-wise) and Nakawar (location-wise). In Khatawar warabandi, the rotation period is divided amongst the applicants in the ratio of the area of their holdings in the command. In the Nakawar warabandi system, the filling and the depletion time of the field channels are also considered while fixing the turns. Besides the date and time of water delivery, the beneficiary is also informed about the exact *naka* (location) from where the water is to be withdrawn. The nakas are allocated keeping in view the position of the farmer's fields. If the turns are continuous, they occur at regular intervals irrespective of water availability. Alternatively, the turns are considered in force only when water is available.

The osrabandi method that is prevalent in U.P., distributes water among the farmers at the outlet command thokwise, chakwise, and village-wise. In the first system, the farmers of the outlet are aggregated to form *thok*s or groups and the share of different groups is proportional to the areas of their holdings. The leader of a thok or 'Thokdar' is given a list of irrigators belongings to his Thok and the time entitlement noted against each. He is responsible to distribute water among all the members of his thok. Thokwari system often fails down due to discriminatory practice by Thokdars, poor maintenance of field channels, and defiance by powerful farmers. In the chakwise system, water allocation is on the basis of *chak* (farm). In many parts of India, scattered land holdings of farmers have been consolidated under chakbandi (consolidation of holdings) and this has helped in overcoming the problems associated with fragmentation of land. This system is now preferred.

In Warabandi system, the infrastructure is sized so that the capacity of successive channels is proportional to the area served. Each unit of culturable command is allocated a certain rate of flow termed as water allowance which is the basis of designing the carrying capacity of distributaries and water courses. The outlets to water courses are so constructed that all of them on a distributary draw their authorized share concurrently. The value of the water allowance at water course head is (0.68 cumec) per 404.69 ha (1,000 acres) of culturable command area (CCA).

Management

The distribution of water in warabandi is a two tier operation. Up to outlet it is managed by the government agency. The distributaries are always run full in 8-day periods. The distribution of water coming out of an outlet ('mogha') is managed by cultivators. The distribution water is done on a 7-days rotation basis with the help of an approved roster which divides 168 hours (i.e. 7 days) in proportion to size of the holdings. The outlets have no gates and it is illegal to keep any of them closed when the parent distributary is in operation. The distributary runs for 8-days so that all water courses can get water at least for 7 full days.

Water distribution proceeds from head to tail along the water course. Each cultivator is entitled to receive the entire water in a water course only on a specific day of the week at a specified time. There is no provision in this system to compensate any individual farmer, who fails to receive his share of water for any reason, including those for which he is not responsible such as fault in operation or breaches in canal etc. Also this system does not compensate for the losses in the water course. No distributary operates for the entire period of the crop(s). The ratio of the operating period of a distributary to the total crop period called its Capacity factor is a compromise between demand and supply.

Roster of turns

When the water is delivered in the water course, it takes some time to each delivery point called nakka. This time of filling is called Bharai. The total time of filling is debited to the common pool time of 168 hours and credited to the individual account of each farmer. At the end of the rotation, the whole length of the water course which has been filled with common pool time, is utilized by the last farmer. Hence, part of total bharai time considering sudden reduction of supply is debited to the account of the last (tail-end) farmer. This is called the Jharai time which is credited to the common pool. Thus,

$$\text{Net running time(hours/unit\_CCA)} = \frac{168 - \text{Total Bharai} + \text{Total Jharai}}{\text{Total Area}} \tag{2}$$

and,

$$\begin{aligned}\text{Total time for a farmer (hr)} &= \text{Net time per unit area} \times \text{area} \\ &\quad + \text{his Bharai} - \text{his Jharai}\end{aligned} \tag{3}$$

where Jharai is zero for all farmers except for the last.

Although, each cultivator's right to share water is guaranteed under section 68 of the North India Canal and Drainage Act (1873) and canal officer can enforce this right if he receives a complaint, farmers themselves are vigilant in their own interest. This is one of the major strengths of the system. A second major strength is the equity. In the event that river flows fall below the combined capacity of the system, a schedule is adopted which rotates priorities between groups of branch canals and distributaries, the number of groups being determined by the variability of supply. Inspired by the achievements and success of warabandi in Punjab and Haryana, initiatives have recently been made to implement this practice in irrigation schemes elsewhere in the country.

17.4.5. Shejpali, Block and Satta Systems

The Shejpali and block systems are followed in Gujarat, Maharashtra and some parts of Karnataka. In these systems, the supplier of the water or the government enters in an agreement with the farmers for supply of water. In the Shejpali system, water is distributed according to a pre-determined schedule in each rotation. Based on expected water availability, a preliminary delivery schedule is prepared each year. Farmers give indent for water indicating the crops they plan to grow and the respective areas. This forms the basis of allocation of water. If the total demand of water exceeds the availability, the irrigated areas proposed by the farmers are proportionately reduced. Now for each rotation, a schedule known as Shejpali is prepared which fixes the turns of applicant farmers. In this system, the farmers at the tail end of the command get water first and those at the head are served last. The actual delivery of water to the crops also depends upon the nature of the crop. Some crops require frequent irrigation and are irrigated in each rotation while less water demanding crops may be irrigated on alternative rotation. After the final schedule is prepared, it is notified so that each farmer knows when his turn will come.

Most of the irrigation projects in these states are storage scheme with direct canals from the storage or from a pick-up weir downstream of a storage. Distribution of water essentially follows rotation-cum-demand approach and the word 'demand' is interpreted as follows. A farmer in the command area has to apply for the area and crops to be irrigated in each season except for perennial crops for which long term sanction (6 to 12 years basis) is given. Demand given by the farmer is sanctioned as per the cropping pattern of the project and availability of water. Water is supplied in each rotation according to crop requirement.

At the end of the Kharif season, the amount of water available in the reservoir is known. Based on this, crops to be grown and areas for each crop are decided before the start of Rabi season (15th October). Preliminary Irrigation Programme (PIP) is prepared by the in-charge of the project for proper utilization of the stored water, considering the duties of the various crops, river gains, evaporation and seepage losses in reservoir, losses in conveyance system and commitments for non-irrigation purposes. This is essentially water budgeting.

Public notices are then issued inviting application for water from farmers for season wise sanctions. If the area demanded for irrigation is more, suitable cut is applied, guaranteeing minimum area of up to 2 acres to small farmers. The cultivators are informed in advance about the area sanctioned by issuing passes to them.

Rotation program i.e., days 'off' and 'on' is prepared and published. Water is supplied to the crops at pre-decided rotation period of 14 to 21 days. Having fixed the rotation periods and the area to be permitted during the season, the quantity required to be released from the storage in each rotation is calculated on the basis of observed AI/DC or ET_c and losses. Generally in Rabi season, AI/DC of 4 is taken at distributory head and 6 at outlet head.

After scrutinizing and sanctioning the indent by the Executive Engineer, water is released and supplied from tail to head reach of the canal. On each outlet, the canal inspector prepares a list of sanction holders, their sanctioned area of calculates the total quantity of water in terms of day cusecs and the total time for which a particular outlet is required to run. A time table known as Pali-patrak giving the date and time for which water will be supplied to each sanction holder in his beat is prepared by canal inspector for each outlet and is communicated to irrigators by calling a meeting.

Supply is rotated tail to head among farmers on each outlet, among different distributaries and in different sections of canal. At the end of each rotation AI/DC statement are prepared and after completion of each season Completed Irrigation Report (CIR) is prepared which includes abstract of water utilized for irrigation and non-irrigation uses, rotation-wise discharge utilized by each sub-division, crop-wise area irrigated in each rotation, duty of water utilized on canal and distributary, conveyance losses, and graph showing planned and actual withdrawal from the storage. This CIR is compared with PIP to evaluate the performance of the system and management staff (WALMI, 1987).

Outlets and water supply in field channels

All outlets in irrigation commands in Maharashtra are non-modular type and of standard size generally 30 to 40 cm diameter fixed at the bed level of the parent channel and are provided with lift type gates to control the water supply. A constant discharge of 30 litre/second (about 1 cusec) is maintained at the outlet head. The time allotted to the farmers at the tail end is required to be adjusted to account for the conveyance losses in the field channel. When the time allotted to each irrigator is rigidly prescribed, the distribution system is called the Rigid Shejpali or Rotational Water Supply (RWS).

Under the block system, a long-term arrangement for water supply is made but irrigation from season to season proceeds in the same manner as in the Shejpali system. The blocks are sanctioned for a period ranging from six to twelve years. Since this system ensures supply of water for a considerably long period, the farmers can plan over a long-time horizon.

The Satta (means agreement) system has many features which are similar to the Shejpali or the block system. The Satta system is practiced in the Sone Command area in Bihar. This is amongst the oldest irrigation systems of India.

In Madhya Pradesh, different water distribution methods are practiced. While the Punjab or U.P. model is followed in the northern part of the state, the Shejpali system of Maharashtra is followed in the southern part. This situation is the result of merger of units from adjoining states after state reorganization. In Vindhya Pradesh and Bhopal region where Rabi irrigation is comparatively more than Kharif, the responsibility of irrigation beyond the outlets rests with Irrigation Department. Mixed cropping pattern prevails in this region. On the other hand, in the old Mahakoshal region which is mostly rice growing area, the distribution of water beyond the outlet is the responsibility of Irrigation Panchayats (IPs). In M.P., the formation of IPs is compulsory under the M.P. Irrigation Act. Each village has its own IP which has a small committee consisting of a Sarpanch (Head of Panchayat) and two or more members elected once in 3 years and serves on an average 250 ha (Raju, 1992).

In Tamil Nadu the Varavaram rotational water supply method is practiced. Paddy is the main crop both in Kharif and Rabi seasons. During Kharif season, in the Cauvery system all channels run continuously without any control till the maturity of the paddy crop. Distribution below the outlet is the responsibility of farmers who practice field to field irrigation. In Lower Bhawani Project, the distributaries run in rotation in alternate years supplying water to two crops in a year -paddy and an irrigated dry crop for the full area covered by these distributaries which run continuously throughout the crop seasons (January to April and August to December). During the Rabi season, rotational system of water distribution is used. Water to each cultivator is supplied at an interval of 1.5 to 4 days for paddy and larger interval for other crops.

17.4.6. Improvement in Water Use Efficiency

Most old canal systems in India were unlined and did not have adequate and appropriate structures to control flow of water. This results in large transmission losses. In view of this, a number of schemes have been taken up in the recent past to modernize such systems. The basic premise here is that the efficiency of a canal can be improved by providing impermeable lining. Some important canals modernization projects include the Haryana Canal modernization project, Punjab canal modernization project, and the Upper Ganga Canal modernization project. Work has also been taken up to improve some old canals such as the Khadakwasla right bank canal, Tungabhadra canal system and the Chambal canal system. Another factor is that the older systems were designed keeping in view the demands and their projections. As the demands have increased significantly over the years due to several reasons the older systems are unable to function well.

Inadequate maintenance of the canals is another cause of inefficient utilization of irrigation water. At present, many canal systems are in poor shape due to negligence and inadequate finances for their maintenance. A major chunk of the allocation funds is usually spent on the establishment costs and thus a very small amount is actually spent on maintenance work. To overcome this, it is necessary that prices of water are fixed such that wastage is reduced and the major portion of operation and maintenance expenditure is recovered. This will make available adequate funds for maintenance works.

Zero tillage practice

This is a technique that has come in news during the last decade. In this technique, after harvesting the crop, the new crop is planted with tilling the ground. Avoidance of tillage leads to conservation of moisture as well as nutrients. Proponents of this technique call it 'conservation farming' which helps in large reduction of input costs, saving in water to the tune of 25%, and about 10% saving in seeds. Estimates show that zero tillage practice is being followed in about 8 lakh ha in UP and abut 10 lakh ha in Haryana.

Systems of rice intensification (SRI)

SRI is drawing attention world-wide as a viable paddy cultivation practice that boosts yield while reducing water use and cost of cultivation. This technique was developed after experiments conducted in Madagascar where the conditions are quite similar to those observed in India. SRI promises significant increase in rice yields even without the introduction of high yield variety (HYV) seeds, increase in application of chemical fertilizers and, most importantly, with much lees water. In Madagascar, average paddy yields rose from 2 tons/ha to 8–10 tons/ha with up to 50% reduction in water consumption. SRI holds out a big promise for India where rice cultivation is extensively carried out, even in water deficient regions and where rice cultivation is frequently blamed for declining water tables in many areas.

Recent developments in bio-technology and genetically modified seeds

Recent advances in GM and bio-technology could drastically change crop water and fertilizer requirements in the future. Introduction of this technology has been slow in India so far and much of the efforts and debate has been concentrated around BT cotton rather than food crops. However, next decades are probably going to witness large changes in the way food is grown in many parts of the world. Some experts predict that poor countries might see dumping of cheaply produced food grains by advanced countries using GM technologies. While this may alleviate food shortages, it will have serious repercussions on the third world economies and will be a big challenge for the planners and policy makers.

17.4.7. Sprinkler and Drip Irrigation

In this method, water is carried through a network of pipes under medium or high pressure and is spread like light rain/drizzle. The system consists of pumping unit, main, sub-main and lateral pipes, couplers, risers and sprinkler heads. With careful selection of nozzles of the sprinkler, operating pressures, lateral spacing, and sprinkler spacing on lateral, the irrigation water required to refill the crop root zone can be applied nearly uniformly to suit the soil intake rates, thereby obtaining efficient irrigation. Both the rotating head and perforated pipe sprinkler systems are used in India. Rain gun sprinkler systems are also becoming popular with Indian farmers. Further details regarding planning, design, operation and maintenance of sprinkler system in Indian conditions can be found in Sahni (1996b).

The sprinkler irrigation was first introduced mainly in hills for plantation crops like tea, coffee, etc. in the Western Ghats and in the North Eastern States. Subsequently, when water scarcity, waterlogging, and salinity problems in command areas started manifesting on a big scale, the government induced the farmers (often through subsides) to adopt sprinkler irrigation. Sprinklers are being extensively used in Haryana, Madhya Pradesh, Rajasthan, etc. These are very well suited to all closely spaced crops except rice, i.e. cereals, oil seeds, pulses, and other cash crops.

Application of these techniques has resulted in efficiencies as high as 90% in some cases and, therefore, the use of these methods on large scale should be encouraged. The NCIWRD (1999) estimated the area under the sprinkler irrigation to be about 700,000 ha at that time. The commission also estimated the cost of sprinkler irrigation to be about Rs. 10,000 to Rs. 20,000/- per ha and the cost of micro-irrigation to be from Rs. 20,000 to Rs. 50,000/- per ha.

Studies conducted at various institutions in the country and at farmers' fields have proved that there is about 30 to 50% saving in water and yield increase 10 to 40% for various crops with sprinkler irrigation compared with surface irrigation. Results from some field studies are summarised in Table 15. The data collected from 22 farmers in Haryana have indicated that by introducing sprinkler method area of irrigation has increased by 38%, crop intensity is increased from 66 to 121%, saving in human labour and above all net income of the farmers. The benefit-cost ratios for plastic sprinkler and metallic sprinkler have been worked out to 4.65:1 and 3.42: 1 respectively (Shivanappan, 1998).

Micro irrigation

Micro irrigation is well suited to all row crops and especially for wide spaced high value crops. The required quantity of water to meet the ET requirement is provided to each plant daily at the root zone. Hence very little water is lost by evaporation. The use of drip irrigation on experimental basis began in India in 1970s. The area under micro irrigation has increased in India from 1,000 ha in 1985 to 60,000 ha in 1993 and further to 170,000 ha in 1997 covering about 30 different crops (Shivanappan, 1998). Maharashtra is the leading state with more than 60,000 ha area under micro irrigation. NCIWRD (1999) estimates place the area under drip and micro-jet system etc. to be 200,000 ha at that time.

Table 15. Summary of field studies on sprinkler – water saving and productivity

Crops and station	Delta in cm		Saving of Water (%)	Yield Q/ha		Increase in Yield%
	Flow Irrigation	Sprinkler Irrigation		Flow Irrigation	Sprinkler Irrigation	
Wheat (HD-2,189) Vadgaon	64.0	44.1	31.0	21.75	23.25	7.0
Wheat (HD-2,189) Phaltan	57.3	42.6	26.0	22.00	25.60	16.4
Wheat (HD-2,189) Akola	91.2	48.0	47.0	20.70	21.70	5.0
Rabi Jowar; Phaltan	64.7	53.7	17.0	15.40	20.00	30.00
Rabi Gram; Phaltan	59.3	54.4	8.3	14.30	17.80	24.50
HW Groundnut Vadgaon	82.0	56.0	32.0	11.00	12.50	14.00
HW Groundnut Phaltan	86.6	54.9	36.6	12.90	18.30	42.00
Kharif Ground nut; Akola	25.5	19.1	25.0	4.10	4.70	15.0
Kharif Jowar; Phaltan	33.8	29.5	13.0	31.50	34.50	9.5
Kharif Bajara; Phaltan	26.8	21.0	22.0	22.00	23.00	5.0
Sugarcane (Co740); Vadgaon	279.3	190.40	32.0	135*	151*	12.0

* Sugarcane Yield in tonnes.
Source: Sahni (2000).

The potential for drip irrigation in India is estimated to be about 10.5 M ha. Govt. of India provides subsidies to encourage wider adoption of drip irrigation. Experiments show that use of drip irrigation can lead to up to 60% saving in water use and to increase crop yields up to 50% (Postel, 1999). By following this method, water productivity can increase up to 250% for cotton. The drip system has a potential to use saline water. These systems are in use in Madhya Pradesh, Haryana, Tamil Nadu, Karnataka, West Bengal and Maharashtra. The micro irrigation systems has several advantages such as large crop yields, higher efficiency of fertilizer use, low labour requirement and reduced weed. Use of these systems is expected to grow significantly with time.

The drip system essentially consists of a pump or overhead tank, main, sub-main, laterals, drippers or emitters, filters, fertiliser tanks or venture pumps, and other accessories. Filters are essential for drip system to prevent clogging of drippers. Biwall system is used for closely spaced crops like sugarcane, vegetables, cotton etc. Micro sprinklers and micro sprayers are essentially a combination of sprinkler and drip irrigation. Water is sprinkled or sprayed around the root zone of the plants with a small sprinkler working under low pressure. This method is very suitable for tree or orchards crops. Details of selection, planning, design, operation and maintenance for Indian conditions can be found in Sahni (1996a).

The cost of micro-irrigation system depends on the type of crops grown, spacing, water requirement, location of water source, etc. It varies from Rs. 15,000 per ha for wide spaced crops like coconut, mango, etc. to Rs. 40,000 per ha for closely spaced row crops like sugarcane, vegetables and cotton. Studies by various institutions have revealed that water saving by drip irrigation compared to surface irrigation is 40–50% and increase in yield is up to 100% (Table 16).

Table 16. Water used and yield for various crops in micro and conventional methods

Crop	Yield (Quintal/Ha)			Water Supplied (cm)		
	Conventional	Drip	Increase in Yield (%)	Conventional	Drip	Water saving (%)
Banana	575.00	875.00	52	176.00	97.0	45
Grapes	264.00	325.00	23	53.20	27.8	48
Mosambi	100.00	150.00	50	166.00	64.0	61
Pomegranate	55.00	109.00	98	144.00	78.50	45
Sugarcane	1, 280.00	1, 700.00	33	215.00	94.0	56
Tomato	320.00	480.00	50	30.00	18.40	39
Cotton	23.00	29.50	27	89.53	42.00	53
Brinjal	280.00	320.00	14	90.00	42.00	53
Papaya	134.00	234.80	75	226.00	73.30	68
Chillies	42.33	60.88	44	109.71	41.77	62
Sweet Potato	42.44	58.88	39	63.14	25.20	60

Source: NCPA (1990).

The B.C. ratio for drip system in Maharashtra and Tamil Nadu is found to range 1.31 to 2.60 for various crops (Sahni 2000) except grapes including the proportion of water saving. In case of grapes, it is about 13.35. If water is also taken into account the ratio goes up from 2.78 to 11.05 for various crops and 30.0 for grapes. This explains why grape growers in Maharashtra have adopted drip irrigation in a big way.

Use of sewage water for irrigation

Sewage water is being used for irrigation in some parts of the country. In Punjab, 34 million litres of industrial effluents and municipal wastewater is used for irrigation. However, since most of it is untreated, toxic elements enters the food chain through plants (Singh 1997). In Gujarat (around Baroda and Ahmedabad) and Tamil Nadu (around Chennai) also, industrial and municipal effluent water is used for agriculture with or without treatment, with mixed results. Domestic sewage is also used to irrigate vegetables due to high content of organic waste. The presence of chemicals in such water on the soils, crops and groundwater will certainly be harmful. Many diseases, especially among children, are said to have been caused due to consumption of fruits or vegetables grown with contaminated water.

17.4.8. Traditional Methods of Irrigation in India

In India, all sources other than canals, wells and tubewells are considered as traditional sources. Small and medium irrigation works continue to play an important role in developing irrigation in the country. These works have smaller capital outlay, short gestation period, benefits from them are spread over a wider area, they do not face opposition from environmentalists, and give employment opportunities to local people.

Traditional methods in mountainous region

Traditional Methods of Irrigation in hilly region are characterized by in – situ water harvesting techniques using locally available skills and material. Terrace cultivation of crops which additionally conserves soil and water is a characteristic feature of the mountainous regions of India. The terraces are cut along the contours and are leaned up against the stone retaining walls to check soil erosion. Channels or bamboo pipes are placed to divert water from the streams or rivulets to irrigate these terraces.

Angami system, a form of terrace cultivation, is practiced in the north – eastern part of India (CSE, 1997). A unique means of terrace cultivation practiced in Nagaland is called *zabo* system. In fact, Zabo combines forest protection, soil conservation, water resources development, and protection of environment. In Meghalaya, natural spring water is diverted through Bamboo pipes to irrigate the plantation crops. These bamboo pipes are widely prevalent and are so fabricated that 18–20 litres of water entering the pipe gets reduced to 20–80 drops per minute at the site of plantation (CSE 1997). The irrigation system practised in Himachal Pradesh is called as *Kul* (diversion channel) irrigation which transports water from the glaciers to the fields. *Haveli* system is a traditional harvesting method prevalent in upper part of Narmada valley in the MP. Rainwater is stored in fields by constructing earthen embankments. After the passage of monsoon, farmers drain off the water and then plant crops which grow using available soil moisture. In Uttaranchal hills also, people have evolved various techniques for water conservation and utilization.

17.4.9. Tanks for Irrigation

Tanks are being built in India for many centuries for multiple objectives including social works. Some authors date the commencement of these works to as far back as 5th century. Most of the tanks were built by the rulers or local communities but maintenance was mainly by the community. In earlier times, tanks were part of social culture, customs, and rituals, and some of these also had religious connotations. The beginning of the 20th century saw gradual stoppage of construction of new tanks.

According to Mishra (1993), there were about 1.1 to 1.2 million tanks of various sizes in the country in the beginning of the 19th century. In 1800, there were more than 39,000 tanks in Mysore. Thakkar (1999) reports that a census of minor irrigation schemes carried out in 1986–87 showed that 507,200 minor irrigation tanks were in use in the country (no census done in Rajasthan). Four peninsular states (Andhra Pradesh, Tamil Nadu, Karnataka, and Kerala) accounted for about 60% of the irrigated area under tanks in the country. These states together with MP, Maharashtra, UP and West Bengal accounted for 97% of the total tanks of the country.

Anicuts (small weirs) are unique devices to divert water from rivers into the canals dug on their bank(s). Network of channels and tanks are typically connected with these *anicuts* which are dominant in Southern India. These *anicuts* range from small earthen barriers across small streams to that of concrete structures like the Grand Anicut on Cauvery River. The water from these tanks is utilized for irrigation through an outlet or sometimes irrigated through natural seepage. Typically, these tanks irrigate from less than 1 ha to about 1,000 ha and above.

Besides irrigation, the other benefits of these structures are recharge of aquifers, arresting soil erosion, and reducing the flood peaks. In Tamil Nadu, tanks are classified as individual tanks, systems tanks, and tank cascades (Saravanan, 1994) based on the hydrological characteristics. In Tamil Nadu, the *Panchayat* maintains tanks with command area of less than 40 ha that are rainfed or are fed by non – perennial streams while the PWD maintains both rainfed and canal fed tanks with command area of more than 40 ha.

Although the area under irrigation by the traditional sources was only 14.5% in 1987–88, Thakkar (1999) notes that these are practically the only sources of irrigation in the Himalayan states of Himachal Pradesh, Sikkim, Assam, Arunachal Pradesh, Nagaland, Mizoram, Meghalaya, Tripura and Manipur. Further, in Jammu and Kashmir and Kerala, they account for over half the net irrigated area. In Bihar,

Karnataka, Tamil Nadu & West Bengal, they account for over a quarter of net irrigated areas. In MP and AP, they account for over 20% of net irrigated area. In UP (including Uttaranchal), these sources irrigate over 0.4 M ha of net irrigated area, most of which is in Uttaranchal where other sources of irrigation are scant.

Rehabilitation of tanks and water bodies

It is common to hear that the traditional systems were largely neglected in the last century because the society has been gradually sidelined in their management and enough attention was not paid for their maintenance. The area irrigated by the traditional means is witnessing decline, especially in the last fifty years. In 1960–61, traditional water sources irrigated 8.2 M ha area. This area declined to 6.24 M ha by 1987–88 indicating that these sources were neglected over the years (Sengupta 1995). Thakkar (1999) noted that the area under tank irrigation reached an all time high of 4.78 M ha in 1962–63, which came down to 3.071 M ha in 1985–86.

The deterioration of tank irrigation in Tamil Nadu can be noted from the fact that the tanks were irrigating an area of 0.565 M ha (about in 1950–51 reached a peak of 0.936 M ha in 1960–61, and steadily reducing thereafter. According to Saravanan (1994), tanks contributed 60% of the irrigated area in 1950's which declined to less than 30% in 1989. A study by Indian Institute of Management, Bangalore, found that more than half of the estimated 43,000 tanks in Karnataka have been more or less silted up, causing loss of irrigation to an area of 0.23 M ha (Dhawan, 1993). Many tanks and dug wells in and around urban areas have been deliberately filled-up over the past 50 years due to pressure on land as a result of rapid urbanization.

Though the need to restore these systems was realized since 19th Century, the importance of community-based restoration (or currently termed as rehabilitation) has only been realized in the 1990's due to importance of participatory governance and impending water crisis. In recent years rehabilitation of tanks is gaining prominence. Recently, Govt. of India has formulated a massive scheme of rehabilitation of water bodies in collaboration with concerned state governments.

17.4.10. Under-Use of Irrigation Potential

The fact that larger potential created remains utilized is a cause of concern for water resources planners and decision makers. Two important reasons behind this are:

a. Poor maintenance

In many cases, the earlier developed facilities have become degraded and suffer from lack of proper maintenance. As a result, such facilities are unable to deliver the desired objectives. Reasons for such a scenario are poor recoveries, relatively higher operation costs, and relatively low allocation of funds for maintenance. Un-accountability is another reason for sub-optimal performance.

b. Under utilization and over utilization of resources

Faulty operation plans etc. also contribute to inefficient resource utilization. Many a times, the optimal benefits are not derived from the system because of lack of on-farm development. Seepage losses from minors and sub minors may be considerable at times. There are also issues related to over-utilization by one stakeholder at the cost of others. It is also seen that the areas or regions where water resources have been developed, there is a tendency to maximize the benefits by switching to more crops or cash crops, thus increasing the pressure on water usage.

Another cause of concern with regard to irrigation development is that a significant variation between potential created and ultimate potential exists in various regions. The stark variation is clear from the following:

- Northern Region States indicate variation between 41% (J&K) to 100% (Punjab).
- In Eastern States, the variation is between 37% (Orissa) to 79% (West Bengal).
- In Southern States, the variation is from 47% (Kerala) to 66%) (Tamil Nadu).
- Western and Central States have variation from 21% (MP) to 69% (Maharashtra).
- North-Eastern States indicate very low potential as regards major and medium schemes but under minor irrigation, the variation is from 16% (Manipur) to 100% (Nagaland).

Such large disparities will lead to regional imbalances in growth and development. Clearly, efforts are needed to ensure sustainability of the created facilities so that they can deliver the planned benefits. The conventional approach of maintenance of the system and its operation by designated units without involvement of stakeholders is not able to: (a) cope up with increasing demands for water, and (b) check the deterioration of created facilities. A holistic management approach involving the users and aimed at sustainability is essential.

17.4.11. Command Area Development Programme

Although very high priority was given to irrigation after the independence so as to increase agricultural production, it was noticed in 1960s that the utilization of potential created was increasingly lagging behind. While looking into the problems faced by major and medium irrigation projects, Irrigation Commission (1972) felt that utilization of created potential was impeded by deficiencies in implementation of projects, such as completion of reservoirs before canal networks, lack of field channels to carry water from the government outlets to the individual farms, etc. Systematic development of command area was recommended to solve this and other related problem like:

- equitable distribution of water,
- drainage,
- deficiencies in planning, design, construction, operation and maintenance of the projects,
- provision of inputs like seeds,
- agricultural support services such as credit and extension,
- lack of farmer's involvement in system operation and maintenance, and

- lack of coordination between various agencies involved in the development irrigation command, and
- dissemination of advance technology among the farmers,

Accordingly, the command area development program (CAD) was launched in 1974–75 chiefly to achieve speedier utilization of irrigation potential and efficient use of water. Initially, it covered 60 major and medium projects with a cultivable command area (CCA) of 13 M ha. By 1991, 54 CADAs were functioning, covering 131 commands and representing a total command area of 18.5 M ha. Gradually, a number of projects were brought under this scheme. Till the year 2000, total 236 projects were covered under the program with a culturable command area of 23 M ha.

The CAD programme is mainly being administered by the state government through Command Area Development Authorities (CADAs). CADAs are funded by the central and state government. The major activities that are taken by CADAs are:

- Development of field channels and drains,
- Enforcement of water deliveries practices such as the Warabandi system,
- Land leveling in the commands,
- Realignment of field boundaries and consolidation of holdings (chakbandi),
- Reclamation of waterlogged area.
- Introduction of suitable cropping pattern, and
- Development, maintenance and effective operation of irrigation systems.

To appreciate the activities initiated under the CAD programme, the achievements as of 1996–97 have been given in Table 17. Targets for a year are also listed in Table 17 so that the reader can appreciate as to how much work is practically possible in the given circumstances. Up to the end of 2000–01, Rs. 7,097 crore were spent on CADP.

Farmer's involvement in system management was identified as part of the CAD program when it was started in 1974. Ideally, the measures being taken up under CAD should be part of the project right from the planning stage which is not happening. Evaluation of selected projects has shown that the CADP has had a positive impact by better utilization of created potential, increase in irrigation intensity, improvement in water use efficiency, use of better variety of seeds, increase in agricultural production, more use of fertilizers, improvement in farm

Table 17. Cumulative achievements of CAD programme as of 2000 and targets for a year

Item	Cumulative achievements (million ha)	Target 1997–98
Field Channel	15.72	0.360
Warabandi	0.57	0.410
Field Drainage	0.68	0.030
Land leveling and shaping	2.19	0.015

Source: WG (1999a) and Planning Commission (2006).

income and reduction in water logging, soil salinity, etc. At the same time, some shortcomings of the programme were also noted. Some of these are:

(i) Progress of development of field channels was slow,
(ii) Slow progress in realignment of field boundaries and consolidation of holdings,
(iii) High water use crops like paddy, sugarcane had increased in head reach areas,
(iv) Conjunctive use did not pick up due to various constraints,
(v) Maintenance and upkeep of the canal system above outlet was lacking, and
(vi) Due to the neglect of intermediate and main drains, field drains were not effective in preventing waterlogging.

The CADA personnel are drawn from the Irrigation, Agriculture, Cooperatives, and Revenue Department. The essential pre-requisites for a multidisciplinary team to function as an effective and sustainable interdisciplinary team are the three C's: commitment, cooperation and communication. Adequate administrative coordination with personnel from the irrigation departments could be typically achieved only when CADA happened to be placed under the direct control of the state Irrigation Department itself, as in the state of Maharashtra (Sahni 2000). In states like Gujarat and Maharashtra, where decentralised democratic administrative system has been operating, the CADAs receive agricultural extension assistance from the local self government institutions. At the same time, there are states like Tamil Nadu, Punjab and Haryana where the production achievements are among the best in spite of these states not having CADA.

To give boost to wider implementation of the CAD programme in various states, many positive steps have been taken. There was a persistent feeling in the past that there is a lack of dialogue between the Union government and CAD departments in the State Governments. With this background, the state level interactions are being organized. Further, in 1997 the Government of India constituted an Inter-ministerial Coordination Committee for the implementation of the Centrally Sponsored CAD programme and Management of Waterlogged, Saline and Alkaline land in the commands of irrigation projects.

17.4.12. Participatory Irrigation Management (PIM)

The present irrigation management approach is top down wherein the prescriptions are handed down from the top and irrigation is managed more according to standard administrative procedures than by local needs. This decoupling between the water users and the suppliers is one of the major reasons behind the poor performance of most irrigation projects. Participation of stakeholders and users in joint management is being increasingly viewed as essential for effective management of irrigation projects. This is being sought through the Participatory Irrigation Management (PIM). When properly carried out, PIM provides benefits not only to the farmers but also to the government through equity in distribution, reduced costs, and improved agricultural production. The operational objective of PIM is to transfer the management of irrigation facilities at appropriate level from government to WUAs. In most cases, management transfer will focus at the level of a minor or distributary for large systems or the entire system in case of small schemes. The objectives of PIM are to:

1. Improve service deliveries through better operation and maintenance which in turn will result in improved efficiency and equity as well as higher reliability.
2. Ensure the physical sustainability of irrigation infrastructure. As water users associations (WUAs) assume management of the secondary levels of the system (e.g. a minor), the irrigation agency can focus its limited resources on the main system.
3. Promote a sense of partnership between the farmers and the irrigation agency which will ensure better co-operation leading to mutual benefits.

The essential functions of a typical WUA are:

1. Management of allotted water to WUA among the members and non-members on principles of equity to be decided by WUA,
2. Repairs and maintenance of the system within their jurisdiction,
3. Assessment and collection of water charges from farmers and remittance to the Irrigation Agency (IA) at rates fixed in the memorandum of understanding between WUA and IA, and
4. To ensure measured quantity of water supply from the IA.

It is to be mentioned here that the concept of PIM is not new. Up to the nineteenth century, most irrigation schemes were managed by farmers themselves as illustrated by the following examples:

- In Vijayanagar Empire (13th – 16th Century AD) which is now part of Karnataka State, a series of diversion weirs and canals were built by rulers in which farmers participated voluntarily. Resolution of conflicts and routine maintenance was done by the farmers themselves.
- In Tamil Nadu, the Chola King Karikala built an anicut on Cauvery River. Certain parts of the irrigation system known as 'Sarkari' were maintained by the government and the lower parts known as 'Kudimarammat' (maintenance by donated labour) were maintained by the farmers.
- Himachal Pradesh had an ancient system called 'Kuhl' in which irrigation was managed by the community. Similarly in the hilly regions of Uttar Pradesh, there was a tradition of farmers managing very small diversion schemes.
- In Maharashtra, the 'Phad' system of irrigation that evolved in Nashik and Dhule in the 17th century is a small-scale community irrigation which is entirely managed by irrigators.

Thus, the concept of PIM has been practiced in this country for a long time in some form or the other. However, PIM in its present context and scope of modern irrigation systems is quite recent in the country.

All the five – year plans since the sixth five year plan, various committees made recommendations for PIM. But planned and systematic effort on PIM started under the CAD Programme in 1985. At that time, the Government of India requested the state governments participating in the CAD Programme to take up farmers participation on pilot basis in at least one minor (comprising about 1,000–2,000 ha area) in each CAD Project. The aim was to involve farmers in water management and maintenance of water courses/field channels with the objective that eventually the minor level system could be handed over to WUAs. Around 1996 when the Ninth Five year Plan formulation started, Planning Commission constituted a working group on PIM to review and suggest the strategies for the Ninth Plan. This working group had made detailed recommendations to promote PIM and many of these were implemented. PIM concept is based on the widely tested premise that the involvement of users in management of irrigation systems creates a win-win situation – it is beneficial for the users as well as for the government. To promote the concept of PIM, several steps have been initiated to involve the farmers in irrigation management. These steps include organizations of conferences at national/ state level to discuss the associated issues, to give legal status to Water User's Associations (WUAs), to conduct training programs, etc. Under this scheme, the states have elected a large number of WUA's. The details of the number of WUAs formed and the command area covered is given in Table 18.

Two overarching issues that are addressed through PIM are: a) optimal use of water and associated resources, and b) equitable distribution of water among users as well as uses. These objectives can be realized in different ways and through participation by all stakeholders at different levels. For example, at the micro level, a distributary of an irrigation system can be managed exclusively by the stakeholders who can form a Water User's Association (WUA), Pani Panchayat, or similar bodies. Likewise, policies/guidelines can be evolved for regulation, in consultation

Table 18. The number of water user's associations formed and the command area covered by 1998

States	Number of water user's associations (Hydraulic level)	Approximate area (thousand ha)
Andhra Pradesh	10,292 (Minor)	4, 700.00
Assam	2 (Minor)	1.00
Bihar	1 (Distributary)	12.20
Goa	39 (Minor)	4.59
Gujarat	71 (Minor) 405 (Lift irrigation schemes)	19.00
Haryana	554 (Outlet)	110.80
Himachal Pradesh	875(Micro schemes)	35.00
Karnataka	193 (Minor)	138.39
Kerala	3,712 (Outlet)	148.48
Madhya Pradesh	65 (Minor)	26.80
Maharashtra	142 (Minor)	55.80
Manipur	62 (Minor)	49.27
Orissa	53 (Minor)	27.60
Rajasthan	35 (Minor)	15.63
Tamil Nadu	328 (Minor)	426.40
Uttar Pradesh	1 (Minor)	0.25
West Bengal	10,000 (Tubewell)	37.90
Total		5, 809.11

Source: WG (1999a).

with stakeholders and these policies can be implemented by the project authorities. At the basin level where there may be multiple stakeholders including the state governments etc., the involvement of all the stakeholders needs to be ensured right from the planning stage.

Some Non-governmental Organisations (NGOs) have also made significant efforts in formation of WUAs in the country. Efforts have also been made for formation of WUAs, as a part of the conditions under externally aided projects such as Water Resources Management and Training Project (WRMTP) of USAID and National Water Management Project (NWMP) and Water Resources Consolidation Project (WRCP) of World Bank. Efforts made by various Water and Land Management Institutes (WALMIs) and State Training Institutes (STIs) in Water Management under the Action Research Programme undertaken by them have also been instrumental in formation of some WUAs.

In AP, the government had formed some 10,800 WUAs and 172 distributary committees by Nov. 1997 through legislation. Some other states like Gujarat, Tamil Nadu and Haryana have also made progress but the number of successful cases is limited. Of course, while one frequently gets to read about the successful cases like Mohini distributary in Ukai command, there are relatively fewer write ups about the large number of failures. Even in case of Mohini distributary also there are some skewed developments. In 85% of the command of this distributary, water intensive crop like sugarcane are being grown while the maximum stipulated area under these crops was 15%.

Hydraulic unit has been accepted as the basis for formation of WUAs. At the outlet level, the WUA is considered too small to be effective. But WUAs at minor level having CCA of 500 to 750 ha are considered to be ideal as basic unit. Small irrigation schemes having CCA up to 1,000 ha can have a single tier. But for medium and major projects, multi-tiered WUA and a federation of WUAs at Distributary/Branch Canal level has been proposed. The successful formation of WUA and its sustainability depends largely on the motivation both on the part of the farmers and the irrigation bureaucracy. This motivation can largely come from the understanding of how they stand to gain from such an organization.

It may be emphasized that public involvement is beneficial only when the public is properly informed. To ensure desired level of involvement of stakeholders, infrastructure needs to be created and mechanisms have to be put in place. Detailed guidelines in this respect are available, for example as discussed by Jain and Singh (2003).

17.4.13. Accelerated Irrigation Benefits Programme (AIBP)

Time and cost over-runs are two major malaises associated with WRD projects in India. A large number of river valley projects have spilled over from one five-year plan to the next, mainly because of financial constraints being faced by the State Governments. Prolonged litigations are another cause of delay. Whatever be the cause, despite a huge investment having already been made on these projects, the country has not been able to derive the desired benefits. At the end of the VIII Plan, there were 171 major, 259 medium and 72 ERM (Extension, Renovation and Modernisation) on-going irrigation projects in the country at various stages of construction with spillover cost of Rs 75,690 crore. Indeed, this is a cause of great concern which is amplified by the fact that the things don't seem to be improving with time.

With this background the Government of India launched the Accelerated Irrigation Benefits Programme (AIBP) during 1996-97. As the name suggests, the aim of AIBP was to speed up implementation of ongoing WRD projects on which substantial progress has already been made. Thus the twin objectives of AIBP are: (i) to accelerate ongoing irrigation projects and (ii) to realize full benefits from completed irrigation projects at the earliest. Priority is given to a) those projects for which progress is hampered due to resource shortage at State Governments level, and b) those which are in advanced stage of construction with substantial investment and could yield benefits in near future.

Assistance under AIBP is given to only those projects which have been cleared by the Planning Commission. Large projects are given assistance for their phased completion so that benefits could start flowing early with comparatively smaller investments. Projects benefiting tribal/drought prone areas are given due preference provided they are otherwise eligible. Priority is also given to inter-state projects. Projects with larger irrigated area per unit of additional investment are preferred. For the projects supported under AIBP, Central Government monitors the progress. Sometime ago, AIBP has been linked with economic reforms in the irrigation sector. But with al the efforts, AIBP has not been very much successful. Out of 181 projects in the AIBP, only 32 have been completed and only 2.66 M-ha of potential has been created while the scope was to create 9 M-ha (Planning Commission 2006).

At present, the priority of projects under AIBP is fixed on the basis of likely completion date and no further projects are financed till the pending projects are complete. In this way, the selected projects preempt resources at the cost of other projects. Planning commission has suggested another strategy in which the projects can be prioritized based on likely additional irrigation from a given investment in a given time frame.

17.4.14. Economic Criteria of Irrigation Projects

At present, all the major and medium sector, are entirely funded by the Government. Institutional finance through financial institutions is available for the minor irrigation sector. National Bank for Agriculture and Rural Development (NABARD) extends assistance to minor and sometimes major & medium projects as well. The unit cost of irrigation development has steadily increased. It was Rs. 1,500/ha in the first plan which grew to nearly Rs. 2.5 lakh per ha of CCA in the tenth plan.

Before any major and medium project is taken up for execution, techno-economic appraisal is carried out. The criteria for establishing viability of a project have changed with time and with philosophy of development. Before independence, the economic return was the main criteria used for sanctioning projects. After independence, the minimum acceptable rate of return on investment was fixed at 3.75% since irrigation was considered to be a responsibility of the state and an essential part of social welfare. Later, this rate was raised to 4.5% and then to 5%. In 1960s, it was felt that a minimum acceptable rate of return on capital criterion for sanctioning irrigation projects was inappropriate from a social welfare point of view. After deducting charges for land levelling, interest on capital, depreciation and administrative expenses, projects having a benefit-cost (BC) ratio of 1.5 and above were considered project viable. The Irrigation Commission (1972) decided that the BC ratio of one should be considered acceptable for drought prone areas.

Later, there was a feeling that benefit-cost analysis alone is not sufficient to properly rank the projects. Accordingly, since late 1980s, besides the BC ratio, the internal rate of return (IRR) of the projects on a discounted cash flow basis is also being computed. A committee constituted by the Planning Commission under the chairmanship of Dr Nitin Desai had recommended that the minimum cut-off point of IRR should be 9% for general cases. Further, for drought-prone, chronically flood-prone, hilly areas, and in areas where 75% of the dependable flows of the basin had already been tapped, a cut-off IRR of 7% was suggested. Most irrigation projects in India have IRR varying from 12 to 16%. This is an indication of large economic benefits of the major and medium irrigation projects. But there are allegations that in these computations, many related indirect costs and benefits (which, at times are substantial) are not accounted for.

17.4.15. Working Expenses and Receipts of Irrigation Projects

In many countries, canals are operated by following "on demand supply system". Canals are operated in India by following schedule delivery, an upstream operational concept and downstream depth method. This system has evolved as a practical method to satisfy irrigation needs using the traditional structures by following a supply based operation. This system has several weaknesses. Notable among them is that there may be large differences between the demand and actual delivery of flows. In case the available water is inadequate, the users located at the tail end of the canal may not get adequate water. Modern technology is being employed to overcome the problems associated with present systems. For instance, the main canal of the Sardar Sarovar Project with a capacity of 1,100 cumec will be a computer controlled, centrally operated canal system.

Data of working expenses on canal systems and revenue generated from 1975–76 to 1991–92 is shown in Table 19. The table shows that the losses have tended to gradually mount and the gross receipts from irrigation fall short of working expenses by 2,500 crore by the end of 1991–92. One alternative to fill this gap will be to suitably increase the water rates. However, any significant adjustment of water price seems unlikely in view of current political situation. In fact, provision of free electricity to farmers to withdraw groundwater is quite common despite its obvious harmful consequences. The state wise working expenses, gross receipts and ranges of water rates for the year 1991–92 has also been presented in Table 20.

Table 19. Financial results of WRD projects (All India) (in crore rupees)

Year	Capital Outlay		Gross receipts (G.R.)	Working expenses (W.E.)	Interest on capital outlay at the end of year (I)	Profit (GR-WE+I)	% recovery of WE through GR (GR/WE* 100)
	During the year	At the end of year (cumulative)					
1974–75	403.6	3, 847.8	60.7	94.6	137.9	-171.8	64
1975–76	509.3	4, 454.3	86.9	95.4	157.7	-166.2	91
1976–77	680.5	5, 137.8	104.7	112.8	174.9	-183.0	93
1977–78	863.2	5, 996.3	96.9	127.2	215.5	-245.8	76
1978–79	970.2	6, 966.4	108.1	155.2	255.5	-302.6	70
1979–80	1, 123.7	8, 090.0	100.7	140.5	292.3	-332.1	72
1980–81	1, 256.7	9, 346.7	103.4	225.7	301.5	-423.8	46
1981–82	1, 438.6	10, 790.5	120.2	265.3	415.6	-560.7	45
1982–83	1, 551.8	12, 342.3	117.1	237.7	872.7	-993.3	49
1983–84	1, 720.3	14, 042.1	165.1	273.9	562.8	-671.6	60
1984–85	1, 867.6	15, 929.7	129.7	334.0	635.7	-840.0	39
1985–86	2, 042.1	17, 971.2	223.8	486.9	681.7	-944.8	46
1986–87	2, 260.1	20, 234.3	166.7	489.6	867.3	-1,190.2	34
1987–88	2, 274.8	22, 311.2	138.7	1, 400.3		-1,261.6	10
1988–89	2, 491.0	24, 608.0	16.4	2, 128.0		-2,016.2	1
1989–90	2, 481.0	27, 088.8	207.6	2, 223.8		-2,111.6	9
1990–91	2, 589.6	27, 292.3	222.4	2, 418.4		-2,196.0	9
1991–92	2, 951.7	30, 253.5	215.8	2, 716.5		-2,500.7	8

Source: WG (1999a).

Table 20. Irrigation project – working expenses, gross receipts, and range of water rates in 1991–92

States	Working expenses (per ha)	Gross receipts (per ha)	Range of water rates (Rupees/ha)
Andhra Pradesh	1, 377	48	99–222
Bihar	NA	NA	30–158
Gujarat	3, 605	231	25–830
Haryana	792	88	17–99
Jammu & Kashmir	529	15	6–289
Karnataka	1, 639	252	37–556
Kerala	596	46	37–99
Madhya Pradesh	748	182	15–297
Maharashtra	5, 627	206	65–1,000
Orissa	189	40	6–185
Punjab	412	65	14–81
Rajasthan	852	99	20–143
Tamil Nadu	579	15	6–65
Uttar Pradesh	808	64	7–237
West Bengal	514	16	74–593
All India	1,032	82	NA

Source: WG (1999a).

Table 20 shows that the revenue collection in Gujarat, Karnataka, Madhya Pradesh and Maharashtra is much higher than the other states. But in these states also, the revenue collection woefully inadequate in comparison to the working expenses. One may also notice that there is huge disparity in water rates across the states and the rates are very low in certain states.

17.4.16. Economic Development Through Irrigation

Irrigation projects are known to be a catalyst in socio-economic development of the region. A number of benefits flow to the society which can be categorized as direct benefits, indirect benefits, and tertiary benefits. Several attempts have also been made to quantify the benefits of irrigation projects.

Direct benefits

Studies of the gross income and cost of inputs indicate that net income per hectare of irrigated areas with food crops rises by about Rs 6,000 to Rs 8,000 per ha at present prices. With irrigated cash-crops, the figures would be Rs 20,000 to Rs 50,000 per ha. A detailed study of a major irrigation project carried out by the National Council of Applied Economic Research some years back indicated that gross receipts per man-day in irrigation areas have increased by more than 100% in all economic activities. Irrigated farming also results in an increase in permanent employment which is of the order of 30 to 50 man-days per ha, or about 60 to 100 million man-days per year at the present growth rate. World Bank (2005) notes that the biggest gainers from the green revolution were the landless whose income increased by 125% as a result of the large increase in demand for their labour. Figure 2 demonstrates the impact of Bhakra dam on crop production in Punjab. After the construction of dam, wheat production increased by 8.5 times and rice production by 29 times.

Studies also indicate that the manpower used per ha of cropped area is about three times higher in irrigated agriculture as compared with unirrigated. The obvious reason is the elimination of risk in crop cultivation due to availability of irrigation

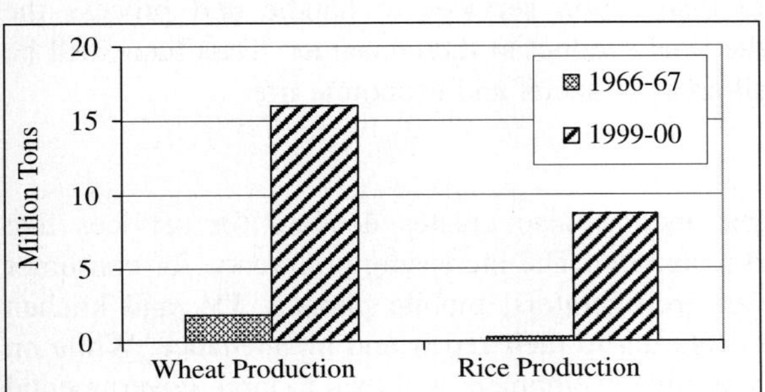

Figure 2. Impact of Bhakra dam on crop production in Punjab

resulting in higher yields and consequently more returns from agriculture. Advent of irrigation in an area results in higher and more stable employment and the poor are the major beneficiaries. Prevalence of poverty in irrigated districts was found to be about one-third of that in unirrigated districts (World Bank 2005). The increase in employment alone is a very significant aspect in the context of large-scale unemployment/ under-employment in India at present.

Indirect benefits

An examination of overall economic development in the states of Punjab and Haryana in the last four decades reveals that there is a close relationship between water resources development and socio-economic transformation of rural areas. With the availability of adequate water and power, and with the adoption of new agricultural technology, and of course with hard work of the farmers, one can see remarkable improvement in the quality of life. The interdependence of production processes arises from the fact that each such activity demands input that are supplied by other similar activities. One sector is linked with another sector which supplies inputs to it and also with those sectors which use output as their own inputs.

Regarding impacts on manufacturing industry, introduction of irrigation creates demand for inputs such as chemical fertilizers, pesticides, irrigation pumps, motors, new tools, and other agricultural machinery. This growing demand in turn, encourages setting up of industries to manufacture these products as well as paraphernalia to service and repair machines and tools. All these activities help generating additional employment opportunities in the ancillary sector.

Another off shoot of growth of agro-based industries as well as other small industries by attracting the surplus income generated. The development of capital-intensive industry as well as increased surplus agricultural production also requires an adequate and extensive transport system to move raw materials and finished goods, road and communication network, storage godowns, markets for wholesale and retail trade, and other services and utilities. The increase in crop output provides adequate food and fodder for livestock and thus accelerates animal husbandry activity. Besides, the water spread also promotes fisheries and forestry. All these factors would obviously result in additional income which would create a demand for additional production and distribution services to handle and process the increased output and deliver the final product to the consumer. Thus there will be various multiplier effects on all areas of social and economic life.

Impact on tertiary sector

Increase in personal disposable income also creates demand for services like transport, communications and recreation. The increasing preference for consumer durables like motor cycles, cars, refrigerators, mobile phones, TV, and kitchen gadgets will also need services catering to their repair and maintenance. While on the one hand the process of economic development will lead to large governmental activities in the form of regulation and administration, on the other hand it will bring about a rapid change in the fields of housing, education, entertainment, health and tourism. Such an overall change in the socio-economic scene has already been experienced in Punjab and Haryana on a very wide scale and in some of the other irrigation projects such as the Rajasthan Canal, etc., on a lesser scale.

Income, consumption and asset formation

Average receipts for households with irrigation are found to be 2.2 times higher than households without irrigation. In fact the receipts from crop farming alone are found to be almost six times higher for households with irrigation.

It is an established fact that, as income rises, a relatively larger part of it is spent on productive activities and less on consumption. From the analysis it has been seen that consumption expenditure in the case of non-irrigated areas is 73% of total expenditure. On the other hand, in irrigated areas it is only 57%, the balance accounting for productive activities. Further examination of the consumption pattern also brought out that people with a higher income spend relatively less on food and more on non-food items as compared with those on lower incomes. The analysis further indicated that, with increase in income, capacity to save and invest also increased. It is estimated that the investment made by households with irrigation was 57% higher than those without. It was also observed that those with irrigation invest in farm equipment in a big way to increase their income still further.

This clearly demonstrates the higher standard of living of those in irrigated areas. It further proves that in irrigated areas the people have been able to satisfy their demand for food and they are using their increased income for non-food items. This itself sets in motion the forces of economic development.

17.5. HYDROPOWER

Hydropower generation follows a simple concept: water falling under gravity turns the blades of a turbine, which is connected to a generator. The rotating generator produces electricity. The pre-historic man was aware of the energy contained in falling water. One of the earliest devices to utilize this energy was the water wheel. Romans used the energy of falling water to do many useful things. They had constructed paddle wheels that turned with the riverflow and lifted water to troughs built higher than river level. The Egyptians and Greeks harnessed the power of river currents to turn wheels and grind grain before 2,000 B.C. In the Middle Ages, more efficient water wheels were built for milling grain.

In the nineteenth century, the water turbine gradually replaced the water wheel, and dams were built to control the flow of water. Since then, the hydroelectric potential of rivers continued to be developed. The first modern turbine design was developed in 1849 by James Francis. According to estimates, the hydroelectric power production has now risen to a whopping 2,000 billion kilowatt hours worldwide.

The principle and the technique to generate electricity from water remain the same regardless of the size of the project. A plant may serve a small community or a country. For example, many communities in remote area of Nepal are not connected to the national grid and get electricity from mini/micro hydropower (size < 2 MW) plants. The largest hydroelectric complex in the world now is on the Parana River, between Paraguay and Brazil. It is known as the Itaipu Dam and its 18 turbines produce 12,600 megawatts (MW) of electricity. But this plant will be soon surpassed by the hydro-power plant at the Three Gorges dam (under construction) across river Changjiang (China) whose installed capacity (IC) will be about 18,000 MW.

Hydropower generation offers many distinct advantages:

(1) Water is a renewable source of energy and hence no fuel cost is involved;
(2) Hydro plants are ideally suited to supply peak energy during the day, thereby saving large amounts of capital and running cost for the installation of alternative thermal capacity;
(3) Operation and maintenance costs are very low, being of the order of 1% of the capital cost against 2.5% (excluding the cost of fuel) in the case of coal-based thermal plants;
(4) Auxiliary consumption is only 0.5% to 1% of the total energy generated against 8 to 10% in the case of coal-based thermal plants;
(5) Start-up and shutdown of hydro units can be achieved in a couple of minutes and therefore, these are ideally suited for meeting peak demands;
(6) Hydro plants do not create any problem of air and surface pollution;
(7) The cost of generation is relatively low, ranging from Rs 0.50 to Rs 1.00 per kWh on average, depending on the type of scheme, i.e. run-of-the river or storage-based, and size of the installation, and is not subjected to inflationary pressures; and
(8) Generation of one kWh of thermal energy requires about 0.65 kg of coal. If full hydro potential is developed, the country could save over 250 million tons of coal every year. Apart from conservation of coal reserves, the relief which the transport sector could obtain would be immense. The consumptive use of water is practically negligible except by way of evaporation in the storage-based schemes.

The amount of power generated is a function of discharge and the hydraulic head. It can be computed as

$$P = 9.817 Q H \eta \tag{4}$$

where P is the electric power in kW, Q is the discharge through power plant in m^3/s, H is the net head in m, and η is the overall efficiency of the power plant expressed as a ratio (usually about 0.85). The overall efficiency of the power plant is obtained by multiplying the turbine efficiency with the generator efficiency. The hydroelectric power generation depends on the volume of water passing through turbines and the effective head. Thus, the same amount of power can be produced by releasing more water at a low head or less water at a high head.

The demand for electrical energy is known as load. The ratio of the average power demand to peak power demand for the time period under consideration is known as a load factor and this is computed on a daily, weekly, monthly or annual basis. Thus,

$$\text{Load factor} = [\text{Average power demand}]/[\text{Peak power demand}] \quad (5)$$

an appropriate time unit is chosen in this equation. The amount of power generated over a time, or energy, is expressed in kilowatt-hour (kW-hr). It can be computed as:

$$\text{KWHR} = 9.817\text{QHT}\eta \quad (6)$$

in which KWHR is the hydropower generated during the period in kW-hr and T is the number of hours in the period.

With respect to types of site development, there are four major classifications of hydroelectric projects: storage, barrages, run-of-river, and pumped storage. Storage projects usually have heads in the medium to high range (>25 m) and can store relatively large volumes of water during periods of high streamflow to generation energy during periods of low streamflow. The power house is commonly located at the toe of the dam, although in some cases it might be away from the dam.

Peaking operation is frequently associated with storage projects and this requires large and sometimes rapid fluctuations in releases of water through the generating units. It is often necessary to provide facilities to even out the fluctuations in the discharge if rapid changes of discharge below the project are not desired. Such an arrangement exists in the Bhakra Nangal project where a small barrage (Nangal barrage) has been constructed at some distance downstream of the Bhakra dam.

A barrage, also known as pondage, has a very small storage capacity. It can regulate the flow only up to minor extent. Hence, the tail water fluctuations are usually quite large, particularly in peaking operations. Run-of-river plants have little or no storage and, therefore, generate power from streamflow as it occurs with little or no benefit from at-site regulation. These projects generally have productive heads in the low to medium range (5 to 30 m). For a base-load run-of-river project to be feasible, the stream must have a relatively high baseflow. Sometimes, the falls in irrigation canals are also used to generate energy, e.g., on the Upper Ganga Canal.

It is sometimes argued that hydropower can be exploited by a series of small dams or diversion structures rather than by constructing major dams. This argument is wrong for two seasons. First, the quantum of hydropower which can be generated by small dams will be limited. Second, studies have indicated that the cost of one kW of IC increases appreciably in small dams compared to large dams.

17.5.1. Quantitative Hydropower Facts

It can be seen from Table 21 that coal accounts for nearly 40% of the electricity generation in the world. Hydropower is world's cleanest source of renewable energy. In the year 2003, hydropower accounted for nearly 16.3% of the world's electricity production.

Table 21. World Electricity Generation by Source (2003)

Production from	Production (TWh)	Percent of total
Coal	6,681.339	39.91
Gas	3,224.699	19.26
Water	2,725.824	16.28
Nuclear	2,635.349	15.74
Oil	1,151.729	6.88
Biomass	138.207	0.83
Waste	62.493	0.37
Geothermal	53.735	0.32
Other Sources	67.406	0.41
Total	16,741.884	100.00

Source: International Energy Agency (www.iea.org), 2006.

Canada produces more than 13% of the global output of hydropower and is the world's second largest exporter of electricity after France. Europe and North America have developed more than 60% of their hydropower (see Table 22); most of it was developed during the twentieth century. In contrast, many countries in Asia, South America, and Africa currently utilize only a small portion of their potential hydropower and large hydropower potential still remains un-exploited.

Canada is the world's large producer of hydropower, generating 346 TWh/year (nearly 62% of the country's total electricity production). The electricity supplied by hydropower far exceeds the capacity of any other renewable energy resource. Norway meets virtually entire (99.6%) electricity demand by hydropower. Twenty-five countries world-wide depend on hydropower for more than 90% of their electricity needs. Though there is dominance of fossil fuels for electricity generation, worldwide more than 60 countries currently use hydropower for half or more of their electricity needs. Most of the installed hydroelectric capacity resides in North America, Brazil, Russia, China, and Europe. Hydropower generation by major producers is shown in Table 23.

Table 22. Continent-wise percentage of developed hydropower potential

Continent	Percentage of developed hydropower potential	Percentage of electricity generated by hydropower
Africa	7	2
Asia	20	39
Australia	40	2
Europe	65	13
N. America	61	26
S. America	19	18

Table 23. Hydropower generation by major producers of the world

Country	Hydro-power produced (TWh)	% of world total
Canada	346	13.0
United States	319	12.0
Brazil	293	11.0
China	204	7.7
Russia	161	6.1
Norway	122	4.6
Japan	96	3.6
India	81	3.0
France	77	2.9
Sweden	72	2.7
Rest of the world	888	33.4
Total World	2, 659	100.0

Source: IEA (2001).

Table 24. Hydro Composition Vs Total Electricity Generation

Source	Share in %		
	1975	1996	India (2000)
Hydro	21.0	18.4	24.8
Nuclear	3.3	17.7	2.1
Gas	12.1	18.4	71.8
Oil	24.6	9.3	
Coal	38.3	38.4	
Others*	0.7	1.4	1.2
Total Electricity (TWh)	6,118	13,652	242

* Other includes geothermal, solar, wind, combustible, renewable & waste.
Source: IEA (2001).

Out of world's total primary energy supply of about 9,376 millions of tons oil equivalent, about 2.3% comes out of hydro sources. Out of a total 13,652 TWh of electricity production, only 18.4% of electricity is generated through hydro sources, as evident in Table 24.

Table 25 is a compilation of quantity of hydropower produced by large and small hydropower projects in different regions of the world.

17.5.2. Electric Energy Scenario in India

India is currently facing large deficit of electric energy, particularly during peak hours and the IC would have to be substantially increased to ensure that the availability of electric energy does not become a handicap in economic development.

Table 25. Capacity and Production of Global Regional Hydropower

Region	Large hydro projects		Small hydro projects		Percentage of small w.r.t. large hydro	
	Capacity (GW)	Production (TWh)	Capacity (GW)	Production (TWh)	Capacity	Production
North America	133.7	576.8	4.302	19.738	3.20	3.42
Latin America	94.0	390.0	1.113	4.607	1.18	1.18
Western Europe	136.7	405.3	7.231	30.239	5.29	7.46
E Europe and CIS	82.3	260.2	2.296	9.438	2.79	3.63
Mid East and N. Africa	13.1	40.2	0.045	0.118	0.34	0.29
Sub-Saharan Africa	16.5	45.1	0.181	0.476	1.10	1.05
Pacific	12.1	38.7	0.102	0.407	0.84	1.05
China (2002)	79.35	243	24.58	80	32	30%
Asia	100.7	397.4	0.343	1.353	0.34	0.34
Total	627.0	2,281.2	19.503	81.709	3.10	3.6%
India (2002)	25.5	74.5	1.45	N.A.	5.68	N.A

At the same time, it calls for accelerated efforts for development of hydropower projects.

India has a per capita electricity consumption of about 300 kW/hour/year. Thermal and conventional hydropower contributes about 96% of the total IC. India has a total non-conventional renewable energy potential of about 126,000 MW (wind 20,000 MW, micro-hydro 10,000 MW, biomass/bio-energy 17,000 MW, ocean thermal power 50,000 MW, tidal power 9,000 MW, and sea wave power 20,000 MW). Of the total energy consumption in the country, almost 60% is met by conventional energy sources and the rest is met by non-conventional and renewable energy sources.

By the end of the year 2005, the IC of thermal plants was 81,681 MW (66%), hydropower plants 33,570 MW (27.29%), nuclear 3,310 MW (2.69%), rest 4,453 (3.62%), giving total 123,014 MW. It has been estimated that to meet the demand for power by 2012, an additional 100,000 MW of IC would be required. Table 26 gives the envisaged additions to IC during two five-year plans.

The estimated energy and peak demands for the recent times are presented in Table 27. According to Planning Commission, by the year 2002 the installed

Table 26. Hydropower capacity additions during 10th and 11th plans

	Hydropower Capacity Addition (MW)	
	10th Plan (2002–07)	11th Plan (2007–12) – Likely
Total	14, 393	16, 872
Central sector	8, 742	11, 691
State sector	4, 481	3, 589
Private sector	1, 170	1, 592

Table 27. Estimated energy and peak load requirement

Particulars	1995–96	1998–99	2001–02	2004–05
Energy demand (Billion KWh)	386.97	481.17	594.52	726.10
Peaking demand (MW)	68,541	84,960	104,641	127,401
Annual load factor (%)	64.45	64.45	64.86	65.06

hydropower capacity in the country was 26,269 MW (which increased to 33,570 MW by Sept. 2005), installed thermal power capacity was 76,057 MW, and the total IC was 105,046 MW. By the end of 10th plan (March 2007), installed hydropower capacity is likely to be 37,069 MW, installed thermal power capacity will be 95,247 MW and the total IC in the country is likely to be 136,336 MW.

17.5.3. Water Requirement for Generation of Electricity

Water is required for generation of electricity by all means. The consumptive water requirement for hydropower stations is not significant because the water coming out of power plant is fully available for other uses. In India, evaporation loss from storage projects is considered as the water required for this purpose.

According to the projections, the needed installed generation capacities by the year 2025 will be 531,000 MW for low demand scenario and 587,000 MW for high demand scenario. Further the requirement for the same by the year 2050 will be 2,278,000 MW for low demand scenario and 2,518,000 MW for high demand scenario. The total water requirement for the year 2025 is likely to be 31.1 km^3 for low demand scenario and 33.1 km^3 for high demand scenario. For year 2050, the annual requirements are expected to be 62.6 km^3 for low scenario and 69.8 km^3 for high scenario.

The water requirements for thermal and other means of power generation depend upon a number of factors such as the size of unit, technologies used for heat transfer and cooling, source of energy etc. In case of coal-fired power stations using cooling towers, the consumptive requirement is about 1 to 4 m^3/hr/MW. With the advancement in cooling technology and by efficient operation of plants, there is immense possibility of reduction in consumptive use of water in thermal power plants. Estimates show that about 15% to 40% reduction in consumptive water uses may be feasible by the year 2025 and 2050 respectively. Accordingly, consumptive use of water in the years 2025 and 2050 is likely to be 2.1 Mm^3/year/100 MW of IC and 1.5 Mm^3/year/100 MW of IC respectively.

17.5.4. India's Renewable Energy Sources

The importance of increasing the use of renewable energy sources was recognized in India in the early 1970s. During the past quarter century, a significant effort has gone into the development and induction of a variety of renewable energy technologies. Presently, India has one of the world's largest programs for renewable energy. The activities cover all major renewable energy sources of interest, such as biogas, biomass, solar energy, wind energy, small hydropower and other emerging technologies. The Ministry of Non-Conventional Energy Sources (MNES) looks after all matters relating to non-conventional/renewable energy. In India, the total estimated renewable energy availability is about 47,000 MW from commercially exploitable sources. It is estimated that small hydro plants can generate about 10,000 MW per year. MNES Web site can be accessed at *http://mnes.nic.in/*.

17.5.5. Hydropower Potential in India

The Central Electricity Authority (CEA) undertook reassessment of the hydropower resources of the country in 1980s. In this survey, theoretical and the economic hydro potential of the rivers was worked out. The potential was assessed by identifying specific suitable sites and water availability corresponding to a 90% dependable year. CEA had identified 845 economically feasible schemes in various river basin of the country. The firm energy was converted into hydroelectric potential by applying a suitable system load factor. On this basis, the total theoretical potential at 60% load factor was assessed as 301,117 MW and the economic potential at 84,044 MW as detailed in Table 28. Figure 3 diagrammatically shows the hydropower potential of Indian river basins.

Table 28. Basin-Wise of Hydroelectric Potential of Indian River System

S. N.	Name of River	Number of Schemes	Firm Potential (MW)	Potential (MW) at 60% LF		Needed IC (MW)
				Economic	Theoretical	
1.	Great Indus	190 (including 23 storage)	11,992.8	19,988	50,172	33,832
2.	Great Brahmaputra	226 (including 76 storage)	23,951.9	34,920	146,170	66,065
3.	Ganga	142 (including 35 storage)	3,409.0	10,715		20,711
5.	West flowing river of South India	94	3,689.4	6,149	9,437	9,430
6.	East flowing rivers of South India	140	5,719.0	9,532	26,972	14,511
7.	Central Indian river system	53	1,664.2	2,740	14,888	4,152
Total		845	50,426.3	84,044	301,117	148,701

Source: CEA and others.

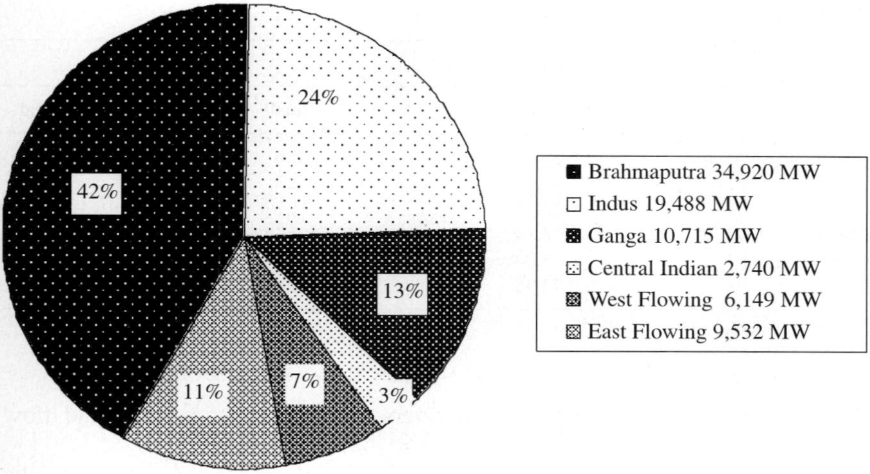

Figure 3. Basin wise hydroelectric potential of Indian rivers 60% load factor

On account of wide fluctuations in the discharges in the Indian rivers, large storage capacity is often required to increase the firm power of hydroelectric plants. Run-of-the-river type hydroelectric projects yielding large firm power are feasible only on the Himalayan rivers which carry substantial discharges during the lean months. Even small storage capacities created on Himalayan rivers help increase the firm power generation appreciably.

It is pertinent to mention that the assessment of hydro potential by CEA was based on limited data on hydrology, irrigation and other consumptive uses, etc. Computerized geographical databases were not used in this assessment. The economics of power development has substantially changed since then. Many sites which were not considered feasible earlier due to economic parameters may now be found to be economically exploitable. Further, the availability of water at many sites may also be substantially different now. Therefore, re-assessment of the hydropower potential of the country using latest tools and geographic/hydrological databases is necessary.

17.5.6. Hydropower Potential of Indian Rivers in Other Countries

Many Indian rivers have their source in neighbouring countries such as Napal. Karnali (10,800 MW), Pancheshwar (5,600 MW), and Saptkoshi (3,300 MW) in Nepal; Tala (1,020 MW), Kurichu, and Manas (2,800 MW) in Bhutan; and Tamanthi (1,200 MW) in Myanmar are some of the hydro projects which have been identified for development under mutual cooperation with these countries. Out of the above schemes, Tala and Kurichu are under construction and other schemes are at various stages of planning/investigation/discussions. Table 29 gives

Table 29. Hydro electrical potential of Nepal

River basins	Power potential (MW)		
	Major	Small	Total
Kosi	18.75	3.6	22.35
Gandak	17.95	2.7	20.65
Karnali	28.84	3.17	32.01
Mahakali	3.84	0.32	4.16
Southern rivers	3.07	1.04	4.11
Total	72.45	10.83	83.28

Source: WG (1999a).

hydropower potential of selected major rivers that originate in Nepal and flow through India.

17.5.7. Hydropower Development in India

A project with capacity of 130 kW installed at Sidrapong (Darjeeling) in the year 1897 was the first hydropower installation in India. A few old installations, e.g., Shiva Samundram in Mysore (2,000 kW), Chamba (40 kW) in 1902, Gagoi in Mussoorie (3,000 kW) in 1907, Jubbal (50 kW) in 1911, Chhaba (1,750 kW) in Shimla in 1913, are the known hydro-power stations that are still working. Energy from these plants was primarily used for domestic needs in adjacent towns.

The Gagoi Power House that Lt. Col. W.W. Bell had built near Mussoorie was to provide electricity and to pump water upward to Mussoorie town. It was for the first time that water was pumped to a height of 516 m (the highest in Asia at that time). The water was pumped from Murray springs below Mussoorie town. In the early 20th century, the princely State of Mysore established a major hydro development – the Shiva Samudaram powerhouse – on Cauvery River with initial capacity of 7.92 MW. The capacity of this powerhouse was increased to 47 MW by 1938. Initially power was supplied to Kolar Gold fields for mining development and operations and later to Bangalore and Mysore cities too. Mysore's second development was in 1940, the Shimsapura hydro power station (2×8.6 MW) again on Cauvery River.

In northern India, 4 MW Mohora hydropower station on Jhelum River in the year in 1905 was the first major hydro development in the then princely state of Jammu and Kashmir. A major contribution to hydropower development was made by the Tata industrial group who successfully established three major hydropower plants in the Western Ghats in Maharashtra: 40 MW Khopoli (5×8 MW) in 1915, 48 MW Bhivpuri (4×12 MW) during 1922–25, and 90 MW Bhira (5×18 MW) in 1927.

In 1932, two major hydropower projects, namely 48 MW Joginder Nagar station (Himachal Pradesh), and 14 MW Pykara hydro plants were taken up and completed by the then provincial governments of Punjab and Madras, respectively. Further notable developments in Madras state were Mettur dam hydro station (40 MW) in 1937 and Papanasam (14 MW initial) in 1944. Travancore-Cochin princely state (now in Kerala) and United Provinces (now Uttaranchal and Uttar Pradesh) also carried out some significant hydro developments namely the Pallivassal hydro station (with 15 MW initial capacity), and a series of hydro stations on Ganga canal, respectively. With the development of high voltage transmission lines in the early twentieth century, a shift occurred from small hydro plants serving local electricity markets to large plants feeding into distribution grids.

At the time of independence (1947), the IC of hydropower projects was 508 MW, which was about 37% of the total IC at that time. But with the taking up of five-year plans, work began on many multi-purpose river valley projects, the so called 'temples of modern India'. Bhakra dam was the notable showcase for a long time to come. At the end of 1998, the installed hydropower capacity was about 22,000 MW which was 24.85% (lowest % so far) of the total IC of 88,543 MW. In the year 1962–63, the hydro: thermal ratio was the maximum at 50.62 However, over the years, the share of hydropower has continuously come down. It may be noted that a thermal-hydro mix in the ratio of 60:40 is considered as ideal. Figure 4 shows the growth of IC of hydropower projects in India. Table 30 lists those hydropower projects that are in operation (IC of 50 MW or more) and Table 31 lists those hydropower projects that are under construction (IC of 50 MW or more).

At the time of independence, total IC in the country was very small. The first 5-year plan for comprehensive and overall development of the country including power sector was launched in 1951. Many medium- and large-scale hydropower projects were taken up – some individually, and others as part of multipurpose projects (irrigation, flood control and power etc.).

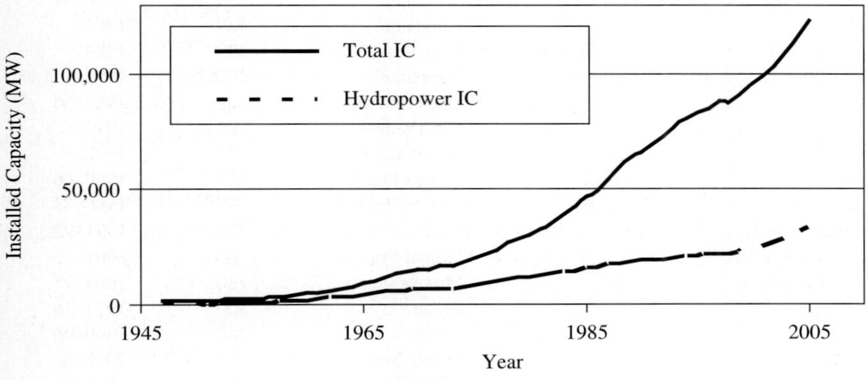

Figure 4. Growth of installed capacity of hydropower generation in India

Table 30. Hydropower stations in operation (IC of 50 MW and above)

Name of Station	Name of River	State	IC (MW)	Year Commissioned
Lower Jhelum	Jhelum	Jammu & Kashmir	105	1978–79
Salal	Chenab	Jammu & Kashmir	690	1987–95
Uri	Jhelum	Jammu & Kashmir	480	1996–97
Upper Sindh II	Sindh nallah	Jammu & Kashmir	105	2000–01
Shanan	Uhi	Himachal Pradesh	110	1932–82
Bhakra LB	Sutlej	Himachal Pradesh	540	1960–61
Bhakra RB	Sutlej	Himachal Pradesh	660	1966–68
Bassi	Uhi	Himachal Pradesh	60	1970–81
Dehar	Beas	Himachal Pradesh	990	1977–78
Giri	Giri	Himachal Pradesh	60	1978
Pong	Beas	Himachal Pradesh	360	1978–83
Baira Siul	Baira/Siul	Himachal Pradesh	180	1980–81
Sanjay Bhaba	Bhabha Khad	Himachal Pradesh	120	1989
Chamera I	Ravi	Himachal Pradesh	540	1994
Malana	Malana	Himachal Pradesh	86	2001
Baspa II	Baspa	Himachal Pradesh	300	2003
Nathpa Jhakri	Sutlej	Himachal Pradesh	1,500	2003–04
Ganguwal & Kotla	Sutlej	Punjab	155	1955–62
UBDC	Ravi	Punjab	91	1971–91
Mukerian	Beas	Punjab	207	1983–89
Anandpur Sahib	Sutlej	Punjab	134	1985
Ranjit Sagar	Ravi	Punjab	600	2000
Rana Pratap Sagar	Chambal	Rajasthan	172	1958–69
Jawahar Sagar	Chambal	Rajasthan	99	1972–73
Mahi I	Mahi	Rajasthan	50	1986
Mahi II	Mahi	Rajasthan	90	1989
Dhalipur	Yamuna	Uttaranchal	51	1965–70
Chibro	Tons	Uttaranchal	240	1975–76
Ramganga	Ramganga	Uttaranchal	198	1975–77
Chilla	Ganga	Uttaranchal	144	1980–81
Khodri	Tons	Uttaranchal	120	1984
Maneri Bhali I	Bhagirathi	Uttaranchal	90	1984
Tanakpur	Sarda	Uttaranchal	120	1992
Rihand	Rihand	Uttar Pradesh	300	1962–66
Obra	Rihand	Uttar Pradesh	99	1970
Khara	Ashan	Uttar Pradesh	72	1992
Gandhisagar	Chambal	Madhya Pradesh	115	1960–66
Bargi	Narmada	Madhya Pradesh	90	1986–88
Bansagar Tons I	Tons	Madhya Pradesh	315	1991–92
Bansagar Tons III	Sone	Madhya Pradesh	60	2001
Hasdeo Bango	Hasdeo	Chhatisgarh	120	1994–95
Ukai	Tapi	Gujarat	300	1974–76
Kadana	Mahi	Gujarat	240	1990–96
Khopoli	Ganga	Maharashtra	72	2001–03
Bhivpuri	Andhra	Maharashtra	72	1998–99
Bhira	Mula & Neela	Maharashtra	150	1927–51

Koyna I & II	Koyna	Maharashtra	560	1962–67
Koyna III	Koyna	Maharashtra	320	1975–78
Vaitarna	Vaitarna & Aliwandi	Maharashtra	60	1976
Pench	Pench	Maharashtra	160	1986
Tillari	Tillari	Maharashtra	60	1987
Bhira Tail race	Bhira & Kundlika	Maharashtra	80	1987–88
Bhira PS	Mula & Nila	Maharashtra	150	1997
Koyna IV	Koyna	Maharashtra	1,000	1999–2000
Upper Sileru	Sileru	Andhra Pradesh	240	1967–95
Lower Sileru	Sileru	Andhra Pradesh	460	1976–78
Nagarjunsagar	Krishna	Andhra Pradesh	815	1978–85
Srisailam	Krishna	Andhra Pradesh	770	1982–87
Nagarjunsagar RC	Krishna	Andhra Pradesh	91	1983–90
Nagarjunsagar CH	Krishna	Andhra Pradesh	61	1992
Srisailam LB	Krishna	Andhra Pradesh	900	2001–03
Mahatma Gandhi (Jog Falls)	Sharavathy	Karnataka	120	1947–52
Sharavathy	Sharavathy	Karnataka	891	1964–77
Sharavathy TR	Sharavathy	Karnataka	240	2001–02
Linganamakki	Sharavathy	Karnataka	55	1979–80
Kalinadi	Kalinadi	Karnataka	810	1979–84
Supa	Kalinadi	Karnataka	100	1985
Varahi	Varahi	Karnataka	230	1989–90
Kadra	Kalinadi	Karnataka	150	1997–99
Kodasalli	Kalinadi	Karnataka	120	1998–99
Pykara Singara	Pykara & Mukuthy	Tamil Nadu	70	1932–54
Periyar	Periyar	Tamil Nadu	140	1958–65
Kundah I	Avalanche & Emerald	Tamil Nadu	60	1960–64
Kundah II	Kundah	Tamil Nadu	175	1960–64
Kundah III	Pegumbahallah & Nirallapallam	Tamil Nadu	180	1965–78
Kundah IV	Bhavani	Tamil Nadu	100	1966–78
Mettur Tunnel	Cauvery	Tamil Nadu	200	1965–66
Aliyar	Aliyar	Tamil Nadu	60	1970
Kodayar I	Kodayar	Tamil Nadu	60	1970
Sholayar I	Sholayar	Tamil Nadu	70	1971
Kadamparai	Kadamparai	Tamil Nadu	400	1987–88
Lower Mettur I/ II/ III/ IV	Cauvery	Tamil Nadu	120	1988–89
Sholayar	Sholayar	Kerala	54	1966–68
Sabarigiri	Pamba & Kakki	Kerala	300	1966–67
Kuttiyadi	Kuttiyadi	Kerala	75	1972
Idukki	Periyar & Killivally	Kerala	780	1976–86
Idamalayar	Idamalayar	Kerala	75	1987
Lower Periyar	Periyar	Kerala	180	1997
Kakkad	Moozhiyar Veluthode	Kerala	50	1999
Panchet Hill	Damodar	Bihar	80	1959–91
Subarnarekha I & II	Subarnarekha	Jharkhand	130	1977–80
Machkund	Machkund	Orissa	120	1955–59
Hirakud PH I	Mahanadi	Orissa	236	1956–90
Hirakud II	Mahanadi	Orissa	72	1962–64

(Continued)

Table 30. (Continued)

Name of Station	Name of River	State	IC (MW)	Year Commissioned
Balimela	Sileru	Orissa	360	1973–77
Rengali	Brahmni	Orissa	250	1985–92
Upper Kolab	Kolab	Orissa	320	1988–93
Upper Indravati	Indravati	Orissa	600	1998–2000
Maithon	Barakar	West Bengal	60	1957–58
Rammam II	Rammam	West Bengal	51	1995–96
Teesta canal I/II/III	Mahananda	West Bengal	67	1997–99
Rangit III	Greater Rangit	Sikkim	60	2000
Khandong	Kopili	Sikkim	50	1984
Kopili	Umrong nallah	Assam	200	1998–97
Umiam III	Umiam	Meghalaya	60	1979
Umiam IV	Umtru	Meghalaya	60	1992
Ranganadi	Ranganadi	Arunachal Pradesh	405	2002
Doyang	Doyang	Nagaland	75	2000–01
Loktak	Loktak	Manipur	105	1983
Pykara Ult.	Moyar	Tamil Nadu	150	2003–04
Chamera II	Ravi	Himachal Pradesh	300	2004–05
Tehri I	Bhagirathi	Uttaranchal	1,000	2005–06
Dhauliganga	Dhauliganga	Uttaranchal	280	2005–06
Almati	Krishna	Karnataka	290	2004–06
Bhavani Barrage I/ II/ III	Bhavani/Cauveri	Tamil Nadu	90	2004–05

The status of hydropower development in major basins is quite unsatisfactory. For example, the hydropower potential of Brahmaputra basin is about 34,920 MW which is 41.5% of the total potential of the country. Only 1.86% of this was developed by the year 2005. The percentage development in Northern rivers was about 20.6%. The major hydroelectric projects (above 750 MW) in India are: Bhakra, Dehar, Koyna, Nagarjunasagar, Srisailam, Sharavarthy, Kalinadi and Idukki. Among the projects that are nearing completion, the prominent are Tehri and Sardar Sarovar. The region wise status of hydro power development in India has given in Table 32. Major hydropower installations have been discussed in the chapters of individual rivers.

Table 33 shows statewise hydropower potential Status of Hydro Electric Potential Development as on 31.08.1998.

17.5.8. Small Hydropower Projects (SHP)

In India, small hydro schemes are classified by the Central Electricity Authority (CEA) according to the norms given in Table 34.

The small hydroelectric development in Himalayan region of India has so far been in those areas only where alternative sources of electricity were not available.

Table 31. Hydropower stations under construction (IC range – 50 MW and above)

Name of Station	Name of River	State	IC (MW)	Commissioning Target
Dulhasti	Chenab	Jammu & Kashmir	390	2006–07
Sewa II	Sewa Ravi	Jammu & Kashmir	120	2006–07
Baglihar	Chenab	Jammu & Kashmir	450	2006–07
Sawalkot	Chenab	Jammu & Kashmir	600	NA
Larji	Beas	Himachal Pradesh	126	2006–07
Dhamai Sunda	Pabar	Himachal Pradesh	70	2006–07
Rampur	Sutlej	Himachal Pradesh	400	2006–07
Keshang I	Sutlej	Himachal Pradesh	66	2006–07
Koldam	Sutlej	Himachal Pradesh	800	2008–09
Parbati II	Parbati/Beas	Himachal Pradesh	800	2009–10
Uhi III	Uhi/Beas	Himachal Pradesh	100	NA
Allian Duhangan	NA	Himachal Pradesh	192	2008–09
Shapurkandi I/II/III		Punjab	168	2006–07
Tehri II (PS)	Bhagirathi	Uttaranchal	1,000	2006–07
Koteshwar	Bhagirathi	Uttaranchal	400	2005–06
Maneri Bhali II	Bhagirathi	Uttaranchal	304	2005–06
Lakhvar Vyasi I/II	Yamuna	Uttaranchal	420	2007–12
Vishnu Prayag	Alaknanda	Uttaranchal	400	2006–07
Srinagar	Alaknanda	Uttaranchal	330	2007–08
Indira Sagar	Narmada	Madhya Pradesh	1,000	2014
Omkareshwar	Narmada	Madhya Pradesh	520	2006–07
Maheshwar	Narmada	Madhya Pradesh	400	2012
Sardar Sarovar	Narmada	Madhya Pradesh	1,200	2006–07
Sardar Sarovar Canal Head	Narmada	Madhya Pradesh	250	2003–07
Ghatgar (PS)	Pravara	Maharashtra	250	2004–07
Priyadarshini Jurala	Krishna	Andhra Pradesh	235	2006–08
Kutiadi Augm	Kutiadi	Kerala	100	2006–07
Balimela	Sileru Balimela	Orissa	210	2005–08
Purulia PS	Kishtobazar nallah	West Bengal	900	2006–07
Teesta L Dam III	Teesta	West Bengal	132	2006–07
Teesta L Dam IV	Teesta	West Bengal	168	2006–07
Teesta V	Teesta	Sikkim	510	2006–07
Karbi Langpi	Borpani	Assam	100	2006–07
Myntdu	Myntdu	Meghalaya	84	2006–07
Loktak d/s	Loktak	Manipur	90	2008–09
Tuirial	Tuirial	Mizoram	60	2006–07
Bairabi	Dhaleshwari	Mizoram	80	2006–07

The development was confined to the small hilly streams in regions such as Jammu & Kashmir, Himachal Pradesh, Uttaranchal, Uttar Pradesh, West Bengal and Arunachal Pradesh etc. During 1960's, a few small hydro stations were constructed in the hilly regions mainly to serve isolated settlements. Some plants were installed on falls of the Upper Ganga Canal during 1930–50. These plants typically utilized relatively larger discharges and smaller head on the canal falls. Some installations

Table 32. Status of Hydro Electric Potential Development (region wise) by 2005

Region	Potential (MW)			Balance (%)
	Assessed (MW)	Developed (MW)	Developed (%)	
Northern	53, 395	11, 000	20.61	79.39
North Eastern	58, 971	1095	1.86	98.14
Eastern	10, 949	2466	22.53	77.47
Western	8, 928	6, 276	70.30	29.70
Southern	16, 458	11, 026	67.00	33.00
All India	148, 701	31, 865	21.42	78.58

were also constructed in 1970s on Kosi and Gandak canals. Currently, small hydropower accounts for about 1,450 MW of installed power.

Assessment of small hydro potential

A detailed study to assess small hydropower potential on the basis of information/data from various States was taken up in 1988–1989 by the CEA. This study arrived at an aggregate potential of about 6,000 MW from about 1,512 identified schemes/ sites (including the ones already in operation or under construction at the time). Another more exhaustive study for a comprehensive and more precise estimate of the SHP potential covering schemes with IC up to 25 MW (now defined as the upper limit for SHP) was completed by MNES in 2002–03. This study determined a potential of about 10,280 MW from about 4,215 identified schemes. An estimate by planning commission places the small hydropower potential as 15,000 MW. Table 35 gives state-wise small hydropower potential of the country.

The small hydro capacity (considering stations with IC up to 15 MW) developed so far aggregates to about 850 MW comprising (a) 360 micro mini stations (IC up to 3 MW) with total 315 MW and (b) 68 small stations (IC higher than 3 MW up to 15 MW) with total 535 MW). Considering stations with IC up to 25 MW under SHP, the aggregate small hydro capacity totals to about 1,045 MW from 454 stations in operation. Small hydro capacity under construction (stations with IC up to 25 MW) presently is 520 MW from about 201 schemes (Kumar 2004). Considering the plants up to 25 MW capacities, small hydropower development by March 2004 was 1,705 MW.

17.5.9. Pumped Storage Schemes in India

Pumped storage schemes are commonly used in hydropower projects to meet the peak power demands. Pumped storage projects consist of a high level forebay where inflow or pumped water is stored until it is needed for power generation and a low level afterbay where the power releases are stored. These projects depend on pumped water as a partial or total source for generating electric energy. The pumping and generation are done by units composed of reversible pump turbines

Table 33. Status of Hydro Electric Potential Development (State Wise) As on 31.08.1998

State	Potential at 60% load factor			Potential under-development at 60% load factor		% of Potential developed under-development	CEA cleared schemes Potential at 60% load factor (MW)	% of total % of Potential developed under-development + CEA cleared
	Assessed (MW)	Developed (MW)	% developed	In MW	In %			
Jammu & Kashmir	7, 487	480.17	6.41	407.17	5.44	11.85	503.17	18.57
Himachal Pradesh	11, 647	2, 007.00	17.23	549.17	4.72	21.95	583.25	26.95
Punjab	922	454.67	49.31	375.00	40.67	89.99	39.33	94.25
Haryana	64	51.67	80.73	11.67	18.23	98.96	0.00	98.96
Rajasthan	291	192.67	66.21	8.00	2.75	68.96	0.00	68.96
Uttar Pradesh	9, 744	1127.00	11.57	1, 117.67	11.47	23.04	405.67	27.20
Madhya Pradesh	2, 774	579.50	20.89	1, 211.05	43.66	64.55	234.45	73.00
Gujarat	409	138.67	33.90	110.67	27.06	60.96	0.00	60.96
Maharashtra	2, 460	1, 108.00	45.04	197.67	8.04	53.08	0.00	53.08
Goa	36	0.00	0.00	0.00	0.00	0.00	0.00	0.00
Andhra Pradesh	2, 909	1, 392.92	47.88	43.70	1.50	49.39	36.23	50.63
Karnataka	4, 347	2, 139.50	49.22	490.33	11.28	60.50	4.83	60.61
Kerala	2, 301	1, 068.67	46.44	276.13	12.00	58.44	41.50	60.25
Tamil Nadu	1, 206	944.67	78.33	69.33	5.75	84.08	11.33	85.02
Bihar	538	119.95	22.30	211.00	39.22	61.51	0.00	61.51
Orissa	1, 983	722.17	36.42	387.28	19.53	55.95	0.00	55.95
West Bengal	1, 786	91.33	5.11	9.83	0.55	5.66	101.67	11.36
Sikkim	1, 283	28.83	2.25	33.67	2.62	4.87	355.67	32.59
Meghalaya	1, 070	121.67	11.37	0.00	0.00	11.37	0.00	11.37
Tripura	9	8.50	94.44	0.00	0.00	94.44	0.00	94.44
Manipur	1, 176	73.17	6.22	5.33	0.45	6.68	42.50	10.29
Assam	351	111.67	31.81	90.83	25.88	57.69	0.00	57.69
Nagaland	1, 040	0.00	0.00	81.88	7.87	7.87	0.00	7.87
Arunachal Pradesh	26, 756	16.50	0.06	108.33	0.40	0.47	251.67	1.41
Mizoram	1, 455	1.00	0.07	36.83	2.53	2.60	168.17	14.16
Total	84, 044	12, 979.93	15.44	5, 832.55	6.94	22.38	2,779.43	25.69

Table 34. Classification of Micro, Mini & Small Hydel Schemes in India

Type	Station Capacity	Unit rating
Micro	Up to 100 kW	Up to 100 kW
Mini	101 kW to 2,000 kW	101 kW to 1,000 kW
Small	2,001 kW to 25,000 kW	1,001 kW to 5,000 kW

Table 35. Identified Small Hydro Projects (Up to 25 MW Capacity)

S. N.	State	Total	
		Number	Capacity (MW)
1	Andhra Pradesh	377	250.50
2	Arunachal Pradesh	452	1, 243.47
3	Assam	40	119.54
4	Bihar	74	149.35
5	Chhattisgarh	47	57.90
6	Goa	4	4.60
7	Gujarat	287	186.37
8	Haryana	23	36.55
9	Himachal Pradesh	288	1, 418.68
10	Jammu & Kashmir	208	1, 294.43
11	Jharkhand	89	170.05
12	Karnataka	221	534.76
13	Kerala	207	455.53
14	Madhya Pradesh	85	336.33
15	Maharashtra	221	484.50
16	Manipur	99	91.75
17	Meghalaya	90	197.32
18	Mizoram	53	135.93
19	Nagaland	84	149.31
20	Orissa	206	217.99
21	Punjab	122	124.22
22	Rajasthan	55	27.82
23	Sikkim	70	214.33
24	Tamil Nadu	155	373.46
25	Tripura	10	30.85
26	UT (A & N Islands)	5	1.15
27	Uttar Pradesh	211	267.06
28	Uttaranchal	354	1, 478.24
29	West Bengal	141	213.52
	TOTAL	4278.00	10, 265.45

and generator motors connecting the forebay and afterbay. The water is pumped from the afterbay to the forebay when the normal power demand is low and released from the forebay to the afterbay to generate power when the demand is high. Such projects derive their usefulness from the fact that the demand for power is generally low at night and on weekends and therefore, pumping energy at a very low cost will be available from idle generating facilities. The feasibility of pumped storage developments arises from the need for relatively large amounts of peaking capacity, the availability of pumping energy at a cheap rate and a load with an off-peak period long enough to permit the required amount of pumping.

There are three types of pumped storage development: diversion, off-channel, and in-channel. The diversion type of development usually consists of pumping in one basin to a forebay on or near the divide between that basin and an adjacent basin and it does not recirculate the water between the forebay and afterbay. The water is released through generating units into an afterbay located in the adjacent basin. The off-channel type of pumped storage development is most suitable when a forebay site exists on a hill above a stream where an afterbay can be constructed. The head differential should be large and the forebay site should be close to the afterbay to avoid head loss and reduce construction costs. The water requirement to support this type of development is not large after the initial supply has been provided. Since the system primarily recirculates water, it is necessary to provide water only to replace losses due to evaporation and leakage.

In the in-channel type of pumped storage development, the reservoir of a conventional power project is used as a forebay. The afterbay could be a reservoir from a downstream project or a reservoir provided solely to serve as afterbay. This type of development is more attractive if the cost of the afterbay is shared with other purposes. In an in-channel pumped storage project, the maximum possible amount of water is pumped back from the afterbay to the forebay during low flow period. During the less severe dry periods, only a part of the water that is used to generate power is pumped back into the forebay. During periods of high streamflow, none of the water is pumped back (Jain and Singh, 2003).

In India, the installed capacities of various pump storage schemes are shown in Table 36.

The Kadamparai site was initially proposed to be developed for a conventional hydro power station only (1 × 35 MW) but considering usefulness of the site (high head, short WCS, lower reservoir having been already in existence as a head reservoir feeding a downstream conventional hydro station (Aliyar), it was thought advisable to develop the site as a pumped storage station with a large peaking capacity (400 MW). The other two stations also had the advantage of both the reservoirs being in existence for other conventional stations.

Pumped storage type generating units and other associated equipments for the purpose were installed at Kadana Powerhouse (4 × 60 MW), Nagarjunasagar Dam Powerhouse (7 × 100 MW reversible units and 1 × 110 MW conventional unit 1), Paithan Powerhouse (1 × 12 MW), Ujani Powerhouse (1*12 MW), and Panchet Hill Powerhouse (1*40 MW). These powerhouses however were not to be operated as pumped storage initially. The waters available for flow down the five irrigation dams for the stations mentioned above were to be initially ample but were to diminish gradually over a period of time as a result of development of up-stream-utilization-schemes including high level canals taking off from the dams. Initially these were to operate as conventional hydro station their operation was to be switched over to pumped storage type when the waters diminished. As there was no

Table 36. Pumped storage development in India

Scheme	IC (MW)	
	No. of units * Unit size	Total
	Schemes in Operation	
Kadana stage I & II – (Gujarat)	$2 \times 60 + 2 \times 60$	240
Paithon – Maharashtra	1×12	12
Nagarjunsagar – A.P.	7×100	700
Kadamparai – T. N.	4×100	400
Panchet Hill – D.V.C.	1×40	40
Ujani – Maharashtra	1×12	12
Bhira – Maharashtra	1×150	150
Total		1, 554
	Schemes under construction	
Sardar Sarovar – Gujarat	6×200	1,200
Ghatgar – Maharashtra	2×125	250
Srisailam – A.P.	6×150	900
Purulia – West Bengal	4×225	900
Koyana Stage IV – Maharashtra	4×250	1,000
Bhivpuri – Maharashtra	1×90	90
Total		4,340
	Schemes approved by CEA	
Tehri stage II – U.A.	4×250	1,000
Total		6,894

need for the construction of lower tail pool dams the same was deferred in most of these cases. For Ujjaini it was built early after commissioning and for Paithan a little later after commissioning. These stations have been rendering pumped storage service for some years now. The tail pool dams at Nagarjunsagar, Kadana and Panchet Hill were taken up much later. The Ujjaini, Paithan and Panchet Hill being of too small capacity for pumped storage are not much significant. When Nagarjunsagar and Kadana become operational as pumped storage this will be another significant addition to pumped storage capacity in India which will get raised to 2,454 MW. (About 3,350 MW pumped storage capacity comprising Tehri II (1,000 MW) Uttaranchal, Sardar Sarovar (1,200 MW) Gujarat, Ghatgar (250 MW) Maharastra and Purulia (900 MW) West Bengal is under active construction scheduled for completion within 1 to 4 years (as of June 2003). With all this capacity coming into operation as pumped storage-the pumped storage capacity in the country would become quite substantial-5,804 MW.

There are a quite a number of favourable sites in India for development of pumped storage station. A study to identify the sites by CEA identified 56 such sites with a total installation possibility of 94,000 MW.

17.5.10. Hydro Power Policy of India

Government of India announced its new policy on hydropower development in August 1998. The following objectives for accelerating the pace of hydropower development have been set:

(i) Ensuring targeted capacity addition during the 9th Plan.
(ii) Exploitation of vast hydro electric potential at a faster pace.
(iii) Promoting small and mini hydroelectric projects.
(iv) Strengthening the role of PSUs/ SEBs for taking up new hydroelectric projects.
(v) Increasing private investment.

There has been a revival of interest in hydropower schemes in recent years in the context of harnessing all sources of energy to meet the rapidly growing energy needs of the nation. In many states, the private sector has been invited to tap the hydropower resources for captive use as well as for commercial purposes. Among the states, Himachal Pradesh, Uttar Pradesh, Uttaranchal, Kerala, Karnataka and Andhra Pradesh have taken a lead by advertising and allotting sites to the private sector.

To systematic accelerate the development of hydropower potential in the country, a 50,000 MW hydropower initiative was launched in May 2003. This initiative includes 162 hydroelectric projects with an aggregate capacity of over 48,000 MW (Planning Commission 2006). The proposed schemes include 17,000 MW of storage scheme and 31,000 MW of run-of-the-river projects. Note that around half of this capacity lies in Arunachal Pradesh alone.

17.6. INDUSTRIAL USES

The estimates for future water requirements for industries in India are highly uncertain due to several reasons: Chief among these is that there is lack of requisite data-base on the present water use by industries in India. The second factor to compound uncertainty is non-availability of data on future developments in industries. The last decade of the 21st century has witnessed large changes in the GDP growth from one year to another. Further, there has been tremendous acceleration in economic growth and it now appears that India will witness GDP growth at about 8% for several years.

When the industries are looked product-wise, water use by paper, petrochemicals, mining, fertilizer, chemical and steel industries is much higher than the other industries. Rough estimates indicate that the present water use in the industrial sector is of the order of 15 km^3. The water use by thermal and nuclear power plants with IC of 40,000 MW and 1,500 MW (1990 figures), respectively, is estimated to be about 19 km^3.

With the rapid pace of industrialization in India, the availability of adequate quantity of water for industrial use is becoming a problem. In some instances, water shortage has restricted the development of new projects and smooth functioning of existing projects. On the one hand, adequate surface water in some areas is not available for industrial use due to priority of domestic and irrigation supply while on the other hand, due to over-exploitation, there is a big problem of ground water in several parts of the country. Many industries have realized the importance of water and are adopting advanced technology to optimize use of water. Bodies

Table 37. Water consumption by industries (year 1996–97)

S. N.	Industry	Mm^3/Year
1	Large & medium industries	15,282.92
2	Small scale industries	6,882.15
	Total	22,165.07

Source: WG (1999a)

such as, Federation of Indian Chambers of Commerce and Industries (FICCI) and Confederation of Indian Industries (CII) have also initiated studies and programs in water sector. Water auditing of industries is also being initiated under their auspices.

Industrial requirements of water constitute only a small percentage of the total consumption of water and some of these are being met by the construction of storage dams. For example, for the needs of the Bokaro Steel Plant, a dam at Tenughat in the Damodar Basin was constructed. Similarly, a cluster of thermal and super-thermal stations in Uttar Pradesh are entirely dependent on the water stored at Rihand Dam on a tributary of the Sone River. Similarly, for the Vishakapatnam Steel Plant, water is being supplied from the Yeleru Reservoir.

The water consumption by industries in 1996–97 has shown in Table 37. The shortage of water is likely to force the industries to switch over to water efficient technologies. The projections for the future have assumed the development of water saving technologies and sliding scale of water requirement. If the present rate of water use is continued, the water requirement for industries in the year 2050 would be 103 km^3 and if water saving technologies are adopted on a large scale, the requirement is likely to be nearly 81 km^3.

The total water requirement for the industries was around 11 km^3/year for the base year 1996–97. It was observed that the pulp and paper, integrated iron & steel and textile sub sectors account for approximately 60% of the water requirement of the industry. For future projections, water requirement for each sub sector has been separately calculated. The overall water requirements for the year 2010, 2025 and 2050 have been estimated as 37, 67 and 81 km^3 respectively. The production figures and water requirements in the year 2010, 2025, and 2050 for the different industries have been presented in Table 38. It needs to be stated here that till 1990s, Indian economy was on a low growth path.

17.7. RE-CYCLING AND REUSE OF WATER

Re-cycling is internal use of waste water by the original user before disposal. In this process, the wastewater is recovered, fully or partially treated and then reused by the same user. The term reuse is applied to utilization waste water by a user other than the discharger. Waste water potentially available for reuse including discharges from municipalities, industries, and the agriculture irrigation. The water

Table 38. Water requirements for different industries for 2010, 2025, 2050

Category of Industry	Production		Water requirement per unit of production (m^3) 1997–2010	Water requirement km^3		
	Year 1997 (1,000 Tons)	Growth rate %		2010	2025	2050
Integrated iron & steel	119, 390	9	22	5.8377	5.7393	10.941
Smelters	174	5.8	82.5	0.0241395	0.03179792	0.043008
Petrochemicals & Refinery	1, 255	3	17	0.03061003	0.035536	0.049065
Chemicals-Caustic soda	1, 459.30	21.40	5.5	0.010200025	0.01036575	0.01213695
Textile & Jute	24, 730.20	11.57	200	19.01878	36.517893	35.192625
Cement	76, 000	11	5.5	1.2045	1.3825	1.8725
Fertilizer	11, 155.30	8.00	16.7	0.6309594	1.025686615	1.19274795
Leather Products	912.5	5.8	40	0.08765	0.0899725	0.1428975
Rubber	403.24	5.75	6.6	0.0042999	0.005634468	0.006357824
Food Processing	96, 450	14	11	5.56765	8.043034	8.3194496
Inorganic chemicals	2, 200	7	200	1.6	3.346	3.0076
Sugar	14, 500	12	2.2	0.07117	0.3344	0.31845
Pharmaceuticals	4, 190	12	22	0.18414	0.243012	0.3434
Distillery	1, 790.80	4.20	22	0.06731098	0.0980012	0.116788
Pesticides	86	7.3	6.5	0.001989975	0.004081275	0.0057996
Paper & Pulp	3, 100	4	280	2.898	10.18888	18.9053
General Engineering	5, 779	12	2.2	0.0236588	0.0278157	0.0556314
	Average	9.07	Total	37.26276	61.12383	80.52476

Source: WG (1999).

is generally reused for agricultural irrigation, for cooling in power plants and other industrial uses.

Since the demand for water use in the industrial sector is likely to increase many folds, recycling and reuse of the wastewater can reduce the net water demand. The reuse of wastewater can also reduce the cascading impact on downstream water users. Reuse of wastewater within the industry would help in minimization of freshwater requirement as well as reduction in wastewater volume. The amount of wastewater generated in various industries and the percentage of effluent that can be recycled is presented in Table 39.

17.8. NAVIGATION AND OTHER USES

Transport is one of the vital infrastructure components and has a key role in economic development of the country. Inland navigation historical means of

Table 39. Waste water generation from different types of industries and achievable reuse

Industry	Average Volume of Waste water per unit of product	Percentage reuse achievable
Thermal Power Plant	155×10^3 lit/hr/MW	98
Pulp & Paper	250×10^3 lit/ton	50
Iron & Steel	150×10^3 lit/ton	40
Pharmaceutical	4.5×10^3 lit/kg	40
Distillery	1.5 lit/lit of alcohol	25
Textile	250 lit/kg. Cloth	15
Tannery	34 lit/kg. Of raw hides	12

Source: Tyagi et al. (1994).

transport especially in Ganga and Brahmaputra rivers. Inland water transport is the cheapest mode of transport compared with the other modes of surface transport, namely, road and rail, especially when the two terminal points fall on a river. In the wake of the current energy crisis and the need to conserve energy, this mode of transport is even more attractive.

The navigable waterways in India including rivers, canals, backwaters, creeks etc. extend to about 14,500 km. Although India has a large network of rivers and canals, most of these waterways are not suitable for navigation because of shallow depth of flow, narrow widths, unstable beds and banks, and inadequate vertical clearances at road bridges. Due to these problems, presently about 5,200 km of major rivers and 485 km of canals are suitable for navigation in India. Consequently, the cargo movement through inland waterways is about 1 million tonne-km per annum compared to 900 billion tonne-km per annum by other surface means. Cargo transportation in an organized manner is confined to Goa, West Bengal, Assam and Kerala.

The consumptive use of water for navigation is not substantial as the wastage is only at the terminal points of the waterways. Water requirement for navigation is computed keeping in view the two-way movement of containers of required capacity and size. According to the guidelines of National Transport Policy Committee, minimum width of 45 m and minimum depth of 1.5 meter is necessary for a reach to be declared as a National Waterway. In most cases, flow requirement in waterways can be met by seasonal flow. However, some water may have to be released from upstream reservoirs, particularly in the lean flow season. It has been estimated that volume of water required for this purpose will be 10 km^3 for the year 2025 and 15 km^3 for the year 2050. After studies, ten important waterways have been identified and declared as national waterways. These are:

1. The Ganga-Bhagirathi-Hoogli system
2. The Brahmaputra river
3. The West Coast Canal
4. The Mandavi Zuari river and the Cumbarjua canal in Goa

5. The Mahanadi river
6. The Godavari river
7. The Narmada river
8. The Sunderbans area
9. The Krishna river, and
10. The Tapi River.

Three of these, viz., Ganga – Bhagirathi – Hoogli River system, the Brahmaputra River, and the West Coast canal have been declared as national water ways. The details of these national waterways are given below.

The ganga-bhagirathi-hoogly river system (National waterway no. 1)

The Ganga-Bhagirathi-Hoogly river system from Allahabad to Haldia has been declared as first national waterway which became operational in October 1986. The length of this waterway is 1,620 km. From Haldia to Nabadwip the waterway is available in Hooghly River. From Nabadwip to Jangipur & Farraka barrage the water way is formed by Bhagirathi River. Loading/unloading/terminal facilities are available at Haldia, Calcutta, Pakur, Farakka, Bhagalpur, Munger, Patna, Balia and Varanasi.

The brahmaputra river system (National waterway no. 2)

The waterway in river Brahmaputra from Dhubri to Sadiya was declared as National waterway no. 2. The length of this water way is 891 km. The important stations along the river bank in this waterway are Dhubri, Jogighopa, Guwahati, Tezpur, Neamati, Dibrugarh, Sadia and Saikhowa. Loading/ unloading/ terminal facilities are available at Dhubri, Jogighopa, Pandu/Guwahati, Tezpur, Neamati.

The west coastal canal system (National waterway no. 3)

The west coasty canal from Kottapuram to Kollam was declared as National waterway no. 3. The total length of the canal system from Kottapuram to Kollam including Champakara and Udyogmandal canals system is 205 km. The waterway comprises of natural lakes, river sections, back water and man-made canal system. The industrial centers are available at Ambalamugal and Udyaogmandal with the Kochi Port.

Sethusamudram project

Sethusamudram Project is a shipping canal, like the Panama or Suez Canal. Presently the ships going from India's east coast to west coast, or even from Chennai to Tuticorin, have to go around Sri Lanka, because the depth of water in the stretch of the ocean between India and Sri Lanka is not enough to allow passage of ships. Sethusamudram canal will allow the ships to pass through this stretch directly, reducing the shipping distance, time, and fuel costs. In addition to the benefits for mercantile shipping, it will enable navigation between India's east coast and west coast entirely through India's territorial waters, something that has immense strategic significance for India.

17.8.1. Water for Environment

Each river has developed a well-established ecosystem in its course having different habitats and seasonality. All the biological processes are highly timed and spaced. To accomplish these processes, a minimum quantity of water is required. Water quantity has ecological impact in a number of ways. Flood flows flush out spawning areas, leaving clean new gravels, sand washed out of the hills. Controlling flows by dams prevents both cleaning and renewal. High flow rates sweep debris from river channels and wash down new gravels and sand needed for spawning of many fish. In the past and even now, dilution was considered to be an acceptable "solution to pollution" and self-purifying capacity of a stream. This has been included in most of the effluent standards (Minimum National Standards, MINAS) notified under Environment (Protection) Act, 1986. It is assumed that at least ten times dilution is available in a stream where the effluent is going to be discharged. Because all deleterious material is not removed in wastewater treatment, the role of dilution is very significant in protecting the health of a river. In our country, the need for fresh water is growing at a fast rate. Thus, the focus is laid on utilisation of every drop of water. This has resulted in drastic reduction in flow conditions of many rivers in the country. Reduced flow followed by increased waste load rendered many rivers almost ecologically dead.

Environment is increasingly being considered a legitimate water user. Many aquatic species require a certain flow regime to live and breed. The environmental or instream flow requirement (EFR) for a river is the minimum flow required to maintain aquatic life. A range of methods are available to assess EFR. The decision on which method to use is dependent on the type of river (e.g. perennial, seasonal, flashy), perceived environmental importance, complexity of the decision to be made, and difficulty of collecting large amounts of information. In India, the estimates put a requirement of $10\,\text{km}^3$ for the year 2025 and $20\,\text{km}^3$ for the year 2050 for EFR purpose.

17.8.2. Fisheries

Water bodies are frequently utilized for fish production in many parts of our country. India is the seventh largest producer of fish in the world. As per the data from Ministry of Agriculture, the inland fish production increased substantially from 2 lakh tonnes in 1950–51 to about 24 lakh tonnes during 1997–98. Data of statewise inland fish production during 1990–98 shows that West Bengal continues to occupy the foremost position among fish producing States, accounting for about one third of the country's total fish production. Bihar, Andhra Pradesh, Assam and Uttar Pradesh are the other major fish producing states.

17.9. TOTAL WATER REQUIREMENTS

Based on the various studies, total annual requirement of fresh water for the country from various sectors including irrigation, domestic, industrial, hydropower and other uses is estimated to be about 694 to 710, 784 to 850 and 973 to 1, 180 km^3 by the years 2010, 2025 and 2050 respectively depending on the low and high demand scenarios. Estimates show that this demand will be met by harnessing 700 km^3 of surface water and 350 km^3 of ground water. Sector wise and source wise break up of these demands is shown in Table 40.

17.9.1. Basinwise and Statewise Total Water Requirements

There is a vast difference in availability of water from one basin to other. Similar situation exists between different states also. The NCIWRD has estimated the basin wise and state wise water requirements for different periods. Basinwise water requirement for the year 2050 are presented in Table 41. Similarly, statewise net water requirement for the years 2010, 2025 and 2050 for low and high demand scenarios have been presented in Table 42.

It is clear from Table 43 that irrigation sector is the main consumer of water. Therefore, future water resources planning as a whole will be greatly affected by the needs of this sector. Predictions (Table 43) show that in future, percentage of water use by agriculture sector will come down from the present 83% to about 68% while the usage by domestic, industries, and others will increase.

17.10. VIRTUAL WATER TRADE AND WATER FOOTPRINTS OF INDIA

The concept of 'virtual water' was introduced by Allan (1993). Water used to produce a commodity is called the 'virtual water' contained in the product; it is the water embedded in a product in a virtual sense. If 1,000 kg of water is needed to produce 1 kg of grain and if 1,000 kg of this grain is exported from one country to another, one million kg of water is also being exported with this product. To compute virtual water content of products, a distinction is made between primary products (e.g., vegetables), processed products (say sugar), and transformed products (say cheese). Table 44 gives virtual water contents of a few selected products. Some processes may yield multiple products and in this case, total quantity of water used is apportioned amongst these. Clearly, if a product does not require water for its production, its virtual water content is nil. The virtual water content of a product is an indicator of the environmental impact of consuming the product.

There can be two different ways to look at the concept of virtual water. From the view point of the producer, virtual water content of a product is the quantity of water that is consumed to produce that commodity. This quantity depends upon the technology and conditions of production. There can be considerable difference in the quantity consumed depending upon these factors. Considerable saving of water is possible if water efficient technology is employed to produce, say, steel. Further, more water is needed to produce each unit of a crop in arid climates as compared to humid areas. This view point is helpful when a country or region is involved in

Table 40. Annual Water requirement for different uses (in km^3)

Uses	Year 1997–98	Year 2010			Year 2025			Year 2050		
		Low	High	%	Low	High	%	Low	High	%
Surface Water										
Irrigation	318	330	339	48	325	366	43	375	463	39
Domestic	17	23	24	3	30	36	5	48	65	6
Industries	21	26	26	4	47	47	6	57	57	5
Power	7	14	15	2	25	26	3	50	56	5
Inland Navigation		7	7	1	10	10	1	15	15	1
Flood Control		––	––	0	––	––	0	––	––	0
Environment (1) Afforestation		––	––	0	––	––	0	––	––	0
Environment (2) Ecology		5	5	1	10	10	1	20	20	2
Evaporation Losses	36	42	42	6	50	50	6	76	76	6
Total	399	447	458	65	497	545	65	641	752	64
Ground Water										
Irrigation	206	213	218	31	236	245	29	253	344	29
Domestic	13	19	19	2	25	26	3	42	46	4
Industries	9	11	11	1	20	20	2	24	24	2
Power	2	4	4	1	6	7	1	13	14	1
Total	230	247	252	35	287	298	35	332	428	36
Total Water Use										
Irrigation	524	543	557	78	561	611	72	628	807	68
Domestic	30	42	43	6	55	62	7	90	111	9
Industries	30	37	37	5	67	67	8	81	81	7
Power	9	18	19	3	31	33	4	63	70	6
Inland Navigation	0	7	7	1	10	10	1	15	15	1
Flood Control	0	0	0	0	0	0	0	0	0	0
Environment (1) Afforestation	0	0	0	0	0	0	0	0	0	0
Environment (2) Ecology	0	5	5	1	10	10	1	20	20	2
Evaporation Losses	36	42	42	6	50	50	6	76	76	7
Total	629	694	710	100	784	843	100	973	1, 180	100

Source: WG (1999).

Table 41. Summary of Basin-wise total water requirement for 2050

Basin name	Catchment area in India km^2	Population year 1991 million	Total water requirement km^3		
			Surface water	Ground water	Total
Indus	321,289	41.9	47.24	29.88	77.12
Ganga	862,769	356.8	311.96	182.11	494.08
Brahmaputra	197,316	29.1	28.46	27.37	55.83
Meghna	41,157	6.2	3.76	8.57	12.33
Subernarekha	29,196	9.5	7.43	2.62	10.05
Brahmani-Baitarni	51,822	9.8	17.53	3.59	21.12
Mahanadi	141,589	26.6	36.5	24.46	60.96
Godavari	312,812	54	56.45	42.33	98.78
Krishna	258,948	60.8	6.88	30.64	91.53
Pennar	55,213	9.7	9.93	3.92	13.84
Cauvery	87,900	29.3	20.08	15.10	35.19
Tapi	65,145	14.8	13.31	4.88	18.19
Narmada	98,796	14.7	23.81	6.90	30.71
Mahi	34,842	10.5	7.18	3.0	10.19
Sabarmati	21,674	10.6	5.77	2.89	8.66
West flowing rivers of Kachchh & Saurashtra including Luni	334,390	29.2	16.98	11.75	28.73
West flowing rivers between Tapi and Kanyakumari	113,057	58.4	40.73	10.35	51.08
East Flowing rivers between Mahanadi & Pennar	86,508	23.6	20.99	6.42	27.41
East flowing rivers between Pennar and Kanyakumari	99,777	45.0	19.28	11.16	30.44
Minor rivers draining into Bangladesh & Myanmar	36,302	2.1	2.54	1.21	3.75
Area of North Ladakh not draining into Indus	28,478				
Drainage area of Andaman, Nicobar & Lakshdweep islands	8,280				
Total	3, 287, 260	842.6	750.8	429.0	1, 180.0

Source: MOWR (1999).

large-scale exports of a commodity. If the country in question is facing shortage of water, it may be worthwhile to review the export policy.

Table 42. State wise net water requirement in 2010, 2025 & 2050

States/UTs	Net water requirement (km^3)					
	Year 2010		Year 2025		Year 2050	
	Low	High	Low	High	Low	High
Andhra Pradesh	45.7	46.9	53.3	58.5	70.6	85.6
Arunachal Pradesh	0.9	0.9	1.3	1.4	11.4	11.7
Assam	12.1	12.5	15.5	17.2	28.5	38.5
Bihar	32.8	33.3	43.5	46.9	58.7	81.0
Goa	0.4	0.4	0.5	0.6	0.6	0.7
Gujarat	25.8	26.4	32.8	35.0	39.0	44.3
Haryana	22.3	22.6	23.2	23.5	24.4	24.6
Himachal Pradesh	4.8	4.9	4.7	4.9	5.4	5.5
Jammu & Kashmir	4.5	4.5	5.9	6.2	9.1	12.0
Karnataka	25.3	26.0	29.6	31.9	36.3	46.4
Kerala	7.6	7.8	10.4	11.3	19.4	23.8
Madhya Pradesh	35.6	36.7	46.9	50.7	64.6	89.9
Maharashtra	39.1	40.6	50.5	55.4	65.5	78.9
Manipur	1.0	1.0	1.1	1.2	1.8	4.0
Meghalaya	0.7	0.8	0.9	1.1	1.5	1.7
Mizoram	0.2	0.2	0.4	0.4	0.8	0.9
Nagaland	0.8	0.8	1.1	1.2	5.6	5.7
Orissa	16.9	17.3	19.6	24.2	32.5	38.6
Punjab	36.8	36.9	36.9	37.2	37.4	37.6
Rajasthan	37.8	39.9	40.2	40.9	44.4	45.8
Sikkim	0.3	0.3	0.3	0.3	0.5	0.6
Tamil Nadu	30.6	31.0	35.7	37.9	40.0	47.5
Tripura	1.1	1.1	1.3	1.4	5.9	6.2
Uttar Pradesh	84.9	86.2	96.0	102.5	112.3	133.1
West Bengal	26.7	27.4	31.0	33.5	40.7	51.6
UTs	1.1	1.1	1.5	1.6	2.3	2.6
Total States	496	508	584	627	759	919

Source: MOWR (1999).

From the view point of the consumer, virtual water content of a product is the amount of water that would be needed if the commodity is to be produced at the place of consumption. This quantity will also depend upon the technology and

Table 43. Water use – Changing trends of the future

Present scenario		Likely scenario for 2050	
Use	Percentage	Use	Percentage
Agriculture	83%	Agriculture	68%
Industry	5%	Industry	9%
Domestic	5%	Domestic	7%
Others	7%	Others	16%

Table 44. Virtual water contents of a few selected products

Product	Virtual water content (litres)	Product	Virtual water content (litres)
A glass of milk (250 ml)	200	A glass of apple juice (200 ml)	190
A cup of coffee (125 ml)	140	A glass of orange juice (200 ml)	170
A cup of tea (125 ml)	18	Potato (1 kg)	250
A loaf of bread (300 g)	400	One egg (40 g)	135
Potato chips (200 g)	185	Tomato (1 kg)	185
Apple (1 kg)	700	Orange (1 kg)	500
Cotton shirt (500 g)	4, 100	Leather shoes (1 pair)	8, 000
A4 size paper (100 sheets, 80 g/m^2)	1, 000	Microchip (2 g)	32

conditions of production. This view point can be helpful in taking a decision if the country is facing a shortage of water and plans to import certain goods that require large quantity of water.

Water footprints

Related to the concept of virtual water is notion of water footprint. Water footprint of a nation is defined as the total volume of freshwater used to produce the goods and services used by that nation. In a similar way, this concept can be defined for an individual.

Some countries of the world do not have adequate water to meet their current and projected water needs while in some others, surplus water may be available. Further, in big countries, there are regions of water surplus or deficit. A possible approach to overcome this spatial mis-match is to transport water from surplus regions to deficient regions. Due to the involvement of large distances and associated infrastructure and other costs, transportation of real water between water-rich and water-poor countries may be very difficult. Therefore, a viable option for water-scarce countries could be to import water-intensive products rather than produce them domestically. At the same time, water-rich countries could reap benefits from their abundant water resources by exporting products that consume large quantities of water. Of course, in reality things are not so simple and additional questions of food security, energy security, employment, etc. enter in the picture. Table 45 lists top ten virtual water importing and exporting countries.

For India, a number of issues are relevant when discussing about the virtual water trade. Some important issues are listed below.

1. Are investments in development of infrastructure (mainly irrigation) necessary and justified for food self-sufficiency. Could these investments be more gainfully used for other sectors while importing the required food? But since a large population of India is traditionally engaged in agriculture and related industries, investment in agriculture is a must for rural and national growth.

Table 45. Top ten virtual water importing and exporting countries

Rankwise countries with net export				Rankwise countries with net import			
Country	Virtual water flow (Gm^3/year)			Country	Virtual water flow (Gm^3/year)		
	Export	Import	Net export		Export	Import	Net export
Australia	73	9	64	Japan	98	7	92
Canada	95	35	60	Italy	89	38	51
USA	229	176	53	UK	64	18	47
Argentina	51	6	45	Germany	106	70	35
Brazil	68	23	45	South Korea	39	7	32
Ivory Coast	35	2	33	Mexico	50	21	29
Thailand	43	15	28	Hongkong	28	1	27
India	43	17	25	Iran	19	5	15
Ghana	20	2	18	Spain	45	31	14
Ukraine	21	4	17	Saudi Arabia	14	1	13

2. How reliable are global producers, the international food market, the access to this market, and food prices particularly when a big country like India has to make large purchases?
3. There is large virtual water flow within the country itself from one region to another. There are regions where water tables are falling rapidly but even then the farmers are cultivating water intensive crops. Pricing could be a mechanism to control water use but given the political set-up, it is difficult to raise water charges.
4. Large-scale imports of agriculture products will result in severe depletion of foreign exchange reserves.
5. For the sake of food security and to claim the rightful place in the international arena, India has to have a well-developed agriculture production and distribution system.
6. There are risks associated with genetically modified food and the exporting country may interfere in the policies of importing countries.

CHAPTER 18

PROBLEMS RELATED TO WATER RESOURCES MANAGEMENT IN INDIA

To safeguard India's economic and social prosperity, it is imperative that enough freshwater is available to meet the requirements of agriculture, industries, and the domestic sector in the coming years. Unfortunately, inadequate water planning, lack of water awareness, and lack of implementation of desired measures have created a difficult-to-manage situation. As a result, an alarming scenario of freshwater scarcity is gradually unfolding in India. The scarcity of water is already evident in many parts of India, varying in scale and intensity at different times of the year. This situation is the result of natural factors and human actions. Intense competition among water users — agriculture, industry and domestic sector — is pushing the groundwater table deeper and deeper. Widespread pollution of surface water and groundwater is degrading the quality of freshwater resources. Attempts to introduce and enforce legislation have, by-and-large, failed. The requirement for freshwater is increasingly taking the center stage on the economic and political agenda, as more and more disputes between and within states, regions, and even at the community level are rising.

At the time of independence, when the population of India stood at around 350 million, the per capita annual water availability was more than 5,000 m^3. Now, with the population at more than one billion, it stands at 1,950 m^3 and is likely to fall further to less than 1,000 m^3 in the next decade. The shortages will be more at local levels and are bound to spread to regional levels as the population continues to grow. The situation of plentiful water resources in the past is rapidly becoming one of water scarcity. The problems of pollution from municipal sewage, industrial effluents, agrochemicals, and pesticides have further threatened the availability of good quality water. In short, the country's fragile resources are stressed and depleting fast in both quality and quantity.

In certain areas between Pennar and Kannyakumari, the availability of water is as low as 400 cubic metres per person per year. Further, the proliferation of tube wells from 360,000 in 1947 to 6 million had disturbed the ground water situation in Punjab and Haryana. In 12 districts of Punjab and 3 districts of Haryana, exploitation of ground water exceeds recharge.

18.1. CAUSES OF WATER RELATED PROBLEMS IN INDIA

The root causes of water related problems in India can be stated as:

i. Highly uneven distribution of water availability, often leading to floods and droughts.
ii. Rampant pollution of freshwater resources mainly by agricultural, industrial and municipal activities.
iii. Uncontrolled use of the bore-wells that has allowed extraction of groundwater at very high rates, often exceeding recharge.
iv. Inadequate attention to water conservation, efficiency in water use, water re-use, groundwater recharge, and eco-system sustainability.
v. Very low water prices, which do not discourage wastage.
vi. Prevalent system of water rights which gives unlimited ownership of groundwater to the landowner, despite the fact that groundwater is a shared resource from common pool aquifers.
vii. Dis-association of communities in water resources management.

The main water-related problems in India are elaborated in what follows. Problems related with reservoirs and lakes are discussed in Chapter 19 and those related with water quality in Chapter 20.

18.2. UNEVEN DISTRIBUTION OF WATER AVAILABILITY

Water availability in India has a large variation – both spatial and temporal. The basin wise per-capita water availability varies between 14,100 m^3/year for the Brahmaputra-Barak basin to about 300 m^3/year for the Sabarmati basin. As per international norms, if the water availability is less than 1,700 m^3 per capita/year then the country is categorized as water stressed and if it is less than 1,000 m^3 per capita/year then the country is classified as water scarce. Growing water scarcity in India can be gauged from the fact that the available water per capita per year has decreased from 6,008 m^3 in 1947 to 2,384 m^3 in 2000. Although India is above the water stressed category, the real situation of per capita water availability is more disturbing than what is depicted by the average values. In Figure 1, the per capita water availability in selected river basins of the country is depicted, highlighting huge differences among them.

Rainfall pattern and distribution in a region is a good index of its water resources. South-West monsoon, North-East monsoon, cyclonic depressions and local storms contribute to rainfall in different degrees in various regions of the country. As noted previously, India receives nearly 75–80% of annual precipitation during the four monsoon months. Of the remaining amount, a large fraction is received during the winter monsoon. Further, out of 8,760 hours in a year, most of the precipitation is received in about 100 hours. Instances where 10% of annual rain falls in just 3 hours are not uncommon. Such a high concentration of precipitation and streamflows makes it imperative to regulate rivers. Moreover, the uneven distribution of rainfall across the country at different times of the year makes several parts of India fall under the water stressed, water scarcity and absolute water scarcity category.

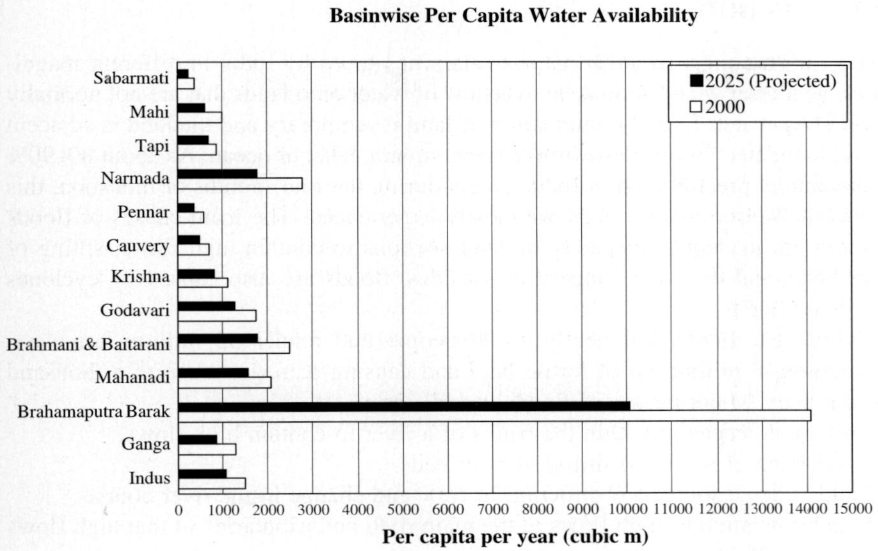

Figure 1. Basin-wise per-capita water availability

Analysis of the long series of meteorological records in India shows that quantity wise, Indian summer monsoon is reasonably stable. Of course, floods and droughts are common features which occur year after year. Notably, the regions with low seasonal rainfall also experience high variability and this makes them drought prone. Southern Oscillation Index (SOI) has been correlated with Indian monsoon – weak monsoons in India are associated with a large negative SOI and occurrence of *El Niño*. Further, large positive SOIs and absence of *El Niño* events have strong correlation with large monsoon rainfalls.

Several global and regional parameters have been found to be related with monsoons and are being used in forecasting. Analysis of monsoon rainfall data of India shows an absence of any trend and the variation seems to be random in nature over a long period of time. However, in some pockets, significant long-term changes in rainfall have been noticed. Monsoon seasonal rainfall has been found to be increasing in the areas along the West Coast, north Andhra Pradesh and north-west India while it has been found to be decreasing over east Madhya Pradesh and adjoining areas, north-east India and parts of Gujarat and Kerala.

To overcome problems due to this highly uneven distribution of water resources over the country, it is necessary to create storage space so that streamflows can be beneficially regulated. This matter has been discussed in detail in a previous chapter. Highly skewed and water resources distribution result in seasonal abundance and devastating floods in some areas while large tracts in other regions are chronically drought affected.

18.3. FLOODS

Floods are the most frequent natural calamities faced by India in different magnitudes, year after year. Flood is an overflow of water onto lands that are not normally covered by water. Usually, inundation of land is temporary and the land is adjacent to and inundated by overflow from a river, stream, lake, or ocean. As about 80–90% of the annual precipitation in India occurs during the four months of monsoon, this is also the season when floods are mostly experienced. The main causes of floods in India are inadequate capacity of river sections to contain high flows, silting of river beds, and drainage congestion. Besides, floods are also caused by cyclones and cloud bursts.

Every year floods kill nearly 1,600 people and render 33 million homeless, submerging 8 million ha of fertile land and causing damages close to a thousand crore rupees. Major causes of floods in India are:

- Inadequate capacity within the banks of a river to contain high flows.
- River bank erosion and silting of river beds.
- Land slides leading to obstruction of flow and change in the river course.
- Synchronization of high flows in the main river and tributaries so that high flows occur in all rivers at the same time.
- Retardation of flow due to tidal and backwater effects, resulting in stagnation of water and inundation of adjoining areas.
- Poor natural drainage in the area.
- Cyclone and associated heavy rainstorm/ cloud bursts.

A review of floods in India during the period from 1910 to 1985 was carried out by Ramaswamy (1985). He also studied the meteorological aspects of severe floods in India up to 1980 (Ramaswamy, 1987).

18.3.1. Flood Prone Areas in India

In Indian practice, any area that has been subject to flooding at any time is considered as a flood prone area unless it has been effectively protected. The Rashtriya Barh Ayog (RBA) made an assessment of the extent of the country exposed to the possibility of flooding. The maximum area affected due to floods in a state in any one of the years is taken as the area liable to flooding in that State. The total of such maxima of the various states is considered to be the area liable to flooding in the country.

On the basis of available data for the period from 1953 to 1978, RBA (1980) assessed the flood prone area in the country as 40 M-ha (see Table 1) which is about $1/8^{th}$ of the geographical area of the country. According to the estimates, the average area annually affected by floods is 7.52 M-ha out of which the agricultural area is 3.52 M-ha. The variability of submerged areas ranges from 1.46 million ha (1965) to 17.5 million ha (1978). The flood prone areas of India are shown in Figure 2. According to NCIWRD (1999), during the second half of 20th century, on an average, 1,515 lives were lost and 95,285 heads of cattle were lost every

Table 1. Flood prone and protected area in India

Flood Prone Area	40 mha
Area which can be given reasonable protection	32 mha
Area protected prior to 1954	3 mha
Area Protected till 2004	15.8 mha

Source: RBA (1980) and others.

year. Heavy flood damage had occurred in the country during the monsoon of 1955, 1971, 1973, 1977, 1978, 1980, 1984, 1988, 1989, 1998 and 2005. Further, floods have affected about 33 million persons between 1953–2000, and the number of people affected each year may increase due to population growth.

The area affected by the floods each year for the second half of the last century is shown in Figure 3. It can be seen that this area does not show much increase with time and the trend line is almost horizontal. It may be added that there is

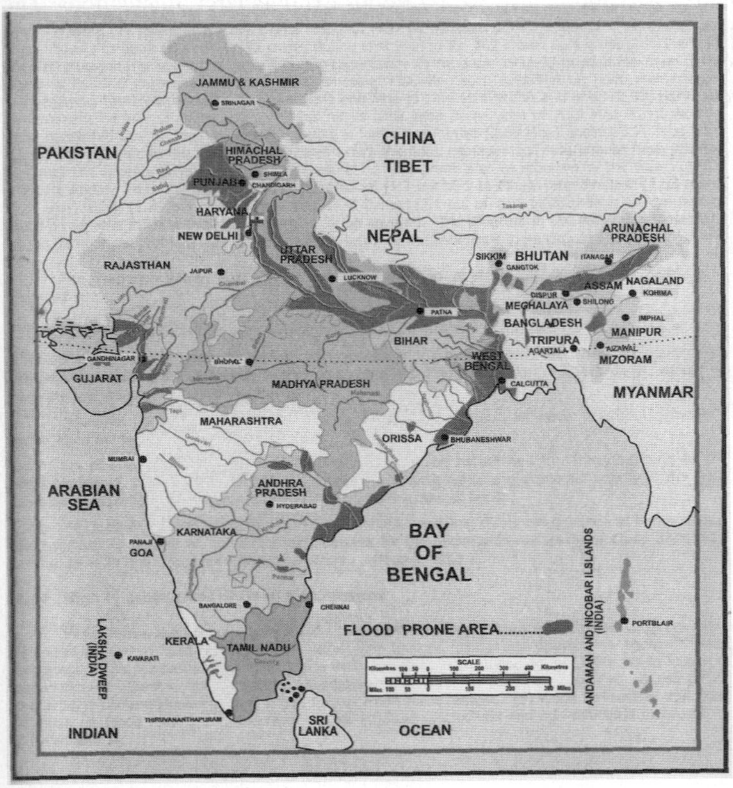

Figure 2. Flood prone areas of India

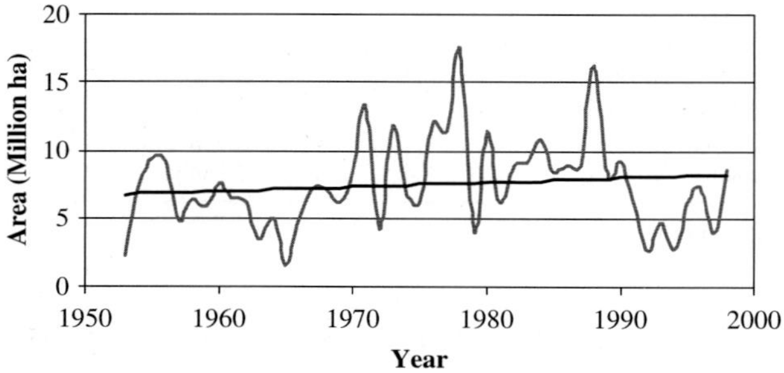

Figure 3. Area affected by floods in India

a feeling that the assessment of flood prone area is not being done scientifically. Often higher figures are reported and the actual area submerged may reduce if a careful scientific assessment is carried out.

For the same period, the population affected by floods is shown in Figure 4. This figure shows that the population affected by floods exhibits large fluctuations. The overall trend is increasing with a slope of 0.77 million/year. Analysis of data on population affected by floods, adjusted for population growth with a representative growth of 2% per annum, also exhibits a small rising trend with a slope of 0.11 million/year.

The pattern of crop damage increase (Figure 5) at the 1993–94 prices, both unadjusted and adjusted for productivity, is almost similar to that of the population affected. The trend is, however, increasing without productivity adjustment but is almost static with productivity adjusted.

Finally, Figure 6 shows the human lives affected by floods. There was unprecedented loss of lives in the year 1977 when about 11,300 persons lost their life due to floods.

There is a general perception that notwithstanding large amounts of money being spent for various flood management works in the country, the flood damage is

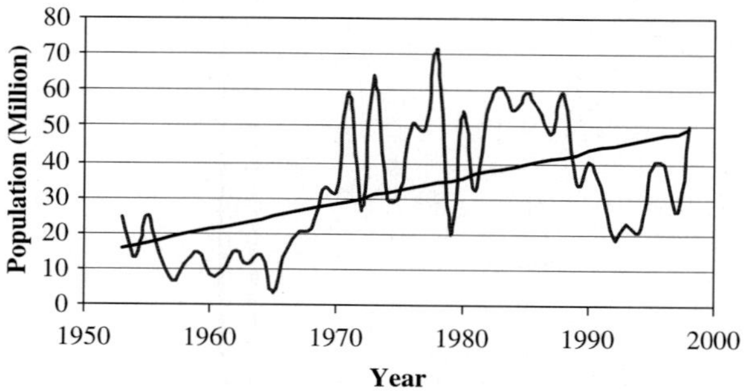

Figure 4. Population affected by floods in India

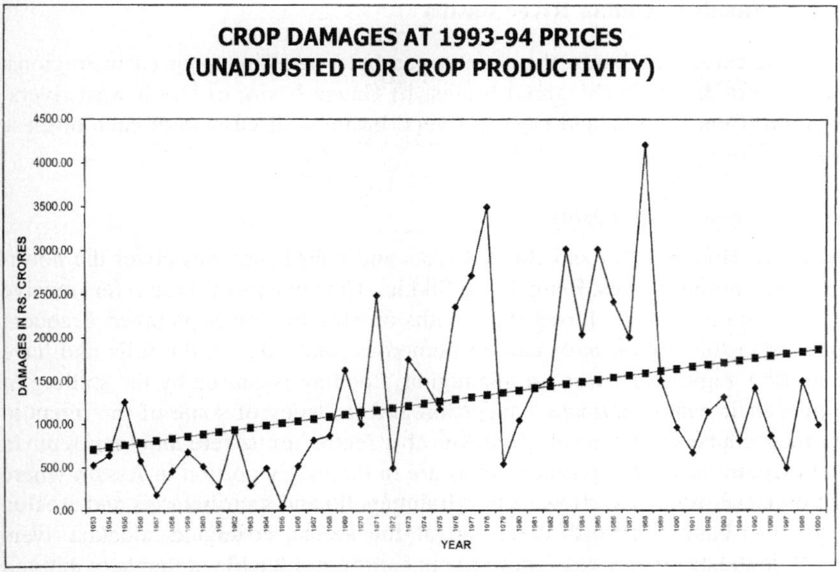

Figure 5. Crop damages by floods in India

increasing. In case of damage data also, it is likely that inflated figures being reported for seeking higher relief. It is seen that at constant prices, neither the crop damage nor the total damage figures show a definite pattern or an overall trend. Another important observation is that the yearly area affected by floods is more or less static. Assam, UP and Bihar are among the most flood prone states in the country. The maximum damage of Rs. 58,460 million due to floods was reported during the year 1998.

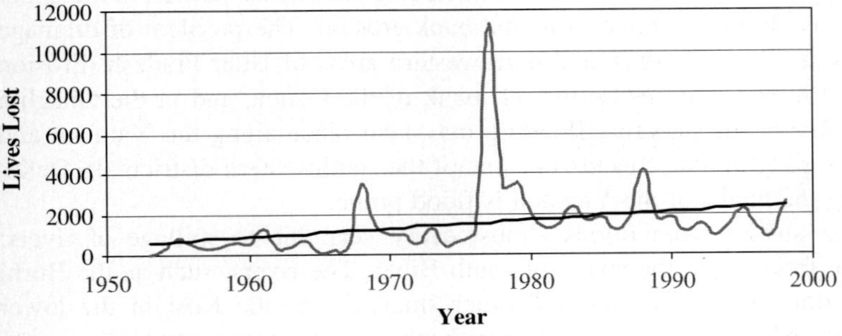

Figure 6. Human lives affected by floods in India

18.3.2. Floods in Indian River Basins

Based on the causes of floods, the country can be divided into four basins/regions: a) Brahmaputra & Barak (Meghna) basins; b) Ganga basin; c) North west rivers' basin; and d) Central India and Deccan rivers' basin. A discussion of each of these follows.

Brahmaputra and Barak Basins

The basins of Brahmaputra and Barak Rivers and their tributaries cover the north-eastern states, northern West Bengal, and Sikkim. Catchments of these rivers receive very heavy rainfall mostly during the months of May/June to September. Frequent earthquakes in this region have caused numerous landslides in the hills and have upset the flow regime of rivers. In this region, flooding is caused by the spilling of rivers over their banks, drainage congestion and tendency of some of the rivers to change their courses. As a result, floods in this region are severe and recurrent. In the Brahmaputra basin, the problem areas are in the valley portion in Assam where inundation is caused by overflow of the Brahmaputra and its tributaries and erosion along the river banks. In North West Bengal, the Teesta, Torsa and Jaldakha rivers frequently inundate large areas. These rivers carry considerable amounts of silt and have a tendency to change their courses. The rivers in Manipur frequently spill over their banks.

Ganga Basin

In the Ganga basin, the flooding problem is mainly confined to the middle and lower parts of the basin. In general, the problem becomes more severe as one moves from west to east and from south to north. In the Ganga basin, the flood problem is mostly in the areas on the northern bank of the Ganga River and damage is mainly caused by the northern tributaries of Ganga which spill over their banks and keep changing their courses. Ganga is a mighty river carrying huge discharges of 57,000 to $85,000\,m^3$/sec (2 to 3 million cusecs). Uttar Pradesh, Bihar, and West Bengal are the worst affected states in the Ganga basin. In eastern Uttar Pradesh, the rivers that cause flooding include the Sarada, the Ghagra, the Rapti, and the Gandak besides the main Ganga River. The major causes of flooding are intense rainfall in the upper catchment areas, drainage congestion, and bank erosion. The problem of drainage congestion exists in the western and northwestern areas of Uttar Pradesh. Erosion is experienced in some places on the left bank of the Ganga, and in the Ghaghra and Gandak Rivers. In Haryana, flooding may take place along the Yamuna and the problem of poor drainage exists in some of the southwestern districts. In Delhi, the area along the banks of the Yamuna is flood prone.

North Bihar suffers from floods almost every year due to spillage of rivers; floods are not frequent in the rivers of south Bihar. The rivers, such as the Burhi Gandak, the Bagmati, the Kamla and other small rivers, the Kosi in the lower reaches, and the Mahananda, spill over their banks causing considerable damage to crops and dislocation of life. High floods occur in the Ganga in some years, causing considerable inundation in Bihar. Figure 7 shows the flood prone area of Bihar. During the monsoon of 1998, large scale devastation took place in Uttar Pradesh and Bihar due to high floods in the Ghaghra, the Rapti, the Gandak, the Kosi, the Mahananda, the Bagmati & Adhwara Group leading to loss of lives, dwellings, properties, installations, communication, and infrastructure facilities.

In West Bengal, floods are caused due to drainage problems as well as tidal effects. In south and central West Bengal, the Mahananda, the Bhagirathi, the Ajay, the Damodar, etc., cause flooding due to the inadequate capacity of river channels and tidal effect. In Rajasthan and Madhya Pradesh, the problem is not serious but these states have also experienced some isolated incidents of heavy floods in recent years.

North West Rivers Basin

Flooding is not a very severe problem in the north-west part of India. Whatever floods are caused here arise due to the inadequate carrying capacity of the channels. The flood problem is not very severe in Sutlej, Ravi, Beas, Jhelum and Ghaggar rivers. The major problem in the states of Haryana and Punjab is that of inadequate surface drainage, which causes inundation and waterlogging over vast areas. In

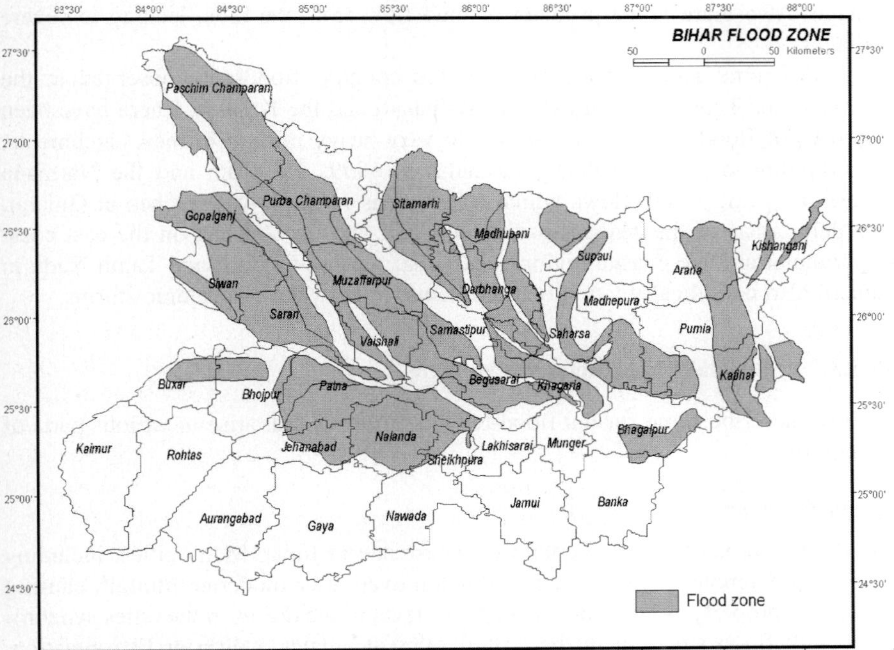

Figure 7. Flood prone area of Bihar

Rajasthan, floods were rare in the past but now isolated instances of floods are noticed. In Haryana, topography is the prime cause of flooding.

Ghaggar River carries a huge quantum of flow during the monsoon period; about 150 MCM of this can be utilized for recharge. Presently this water gets collected and retained in 19 natural depressions causing water logging problems in the adjoining low lying areas like Baropal, where water levels have risen from 50 m below ground level to less than 5 m below ground level. Judicious management of Ghaggar flood water would provide significant quantity of water for recharge.

Central India and Deccan Rivers Basin

This region covers all the southern states, namely, Andhra Pradesh, Chhattisgarh, Karnataka, Tamil Nadu and Kerala and the states of Orissa, Maharashtra, Gujarat and parts of Madhya Pradesh. Major rivers in this region, the Narmada, the Tapi, the Mahanadi, the Godavari, the Krishna and the Cauvery have mostly well defined and stable courses but occasionally there are flood problems in their lower reaches and delta areas. The lower reaches of major east coast rivers have been embanked, thus largely eliminating the flood problem.

In Orissa, damage due to floods is caused by the Mahanadi, the Brahmani and the Baitarni which have a common delta. When these rivers area in spate simultaneously, they cause considerable havoc. The deltaic districts are densely populated. The small rivers of Kerala, when in high floods, occasionally cause damage. Also there is the problem of mud flow from the hills, leading to severe damages.

In the central and southern parts of the country, floods are observed in the Narmada, the Tapi, the Godavari, the Mahanadi and the Krishna. There have been instances of floods in these rivers due to very heavy rainfall in their catchments, e.g., flooding of Hoshangabad (Narmada) in 1999. The Tapi and the Narmada occasionally carry high flows which affect areas in the lower reaches in Gujarat. The delta areas of the Mahanadi, the Godavari and the Krishna on the east coast in particular and the coastal regions of Orissa, Andhra Pradesh and Tamil Nadu in general also periodically face flood problems in the wake of cyclonic storms.

18.3.3. Special Flood Related Problems

Here we describe some special flood-related problems that arise in various parts of the country.

Urban Drainage

Flooding of urban areas is a common occurrence in India. Many cities, including Delhi and Mumbai, witness such a situation even after moderate rainfall, causing damage to property. When this rise in water level in the drains in the cities synchronizes with flood situation in the outfall river, it creates a disaster like situation. A flood in the Hyderabad city and adjoining areas recently is one such example. In July 27,2005, the meteorological observatory at Santa Cruz (north Mumbai) recorded 94.4 cm of rain fall. At the same time, rainfall at Vihar Lake was 105 cm and at Colaba observatory in South Mumbai, it was only 7.3 cm. Intense rainfall in north Mumbai led to severe flooding and disrupting, road, rail, air communication and killing 108 people. Although this was the maximum rainfall over the last 100 years, it shows that the city infrastructure is not prepared to handle such extreme events. Drainage congestion was identified as the main reason for flooding. Figure 8 shows the extent of waterlogging that was witnessed in Mumbai.

Urban drainage problems also lead to unhygienic conditions leading to spread of epidemics. Absence of any planned urban drainage system and inadequacy of the existing drains make the situation worse. Often people living in low-lying areas belong to poorer sections of the society and are the worst affected. It is necessary that urban drainage be given due importance. Some of the suggestions to improve the situation are:

(i) Along with master plan for an urban area, there should also be a master plan for drainage of the area,
(ii) This drainage plan should take into account the topographic conditions, existing drainage system, future development pattern of the city etc.,
(iii) The land use policies and future developments must follow the drainage plan of the city,
(iv) Existing lakes and depressions which form parts of drainage plan in the city should not be allowed to be filled up, and
(v) Adequate attention should be given to proper maintenance of the drainage system every year.

Figure 8. Inundation in Mumbai city in July 2005 when nearly 95 cm of rain fell in 24 hours

Flash Floods

Flash floods are characterized by very fast rise and recession of flow of small volume and high discharge that cause high damages because of suddenness. In case of flash floods, there is virtually no time to issue warnings and very little time is available for evacuation. These types of floods occur in hilly regions and sloping lands where heavy rainfall and thunder storm or cloud burst are common. Severe flash floods in Arunachal Pradesh during the year 2000 are an example of floods from such a situation. Punjab also experiences such floods occasionally. Large reservoirs with sufficient vacant space upstream of areas prone to flash floods can absorb the flood wave, protecting the downstream regions.

In Himachal Pradesh, a flash flood was experienced on the night of 31st July and 1st August, 2000. The discharge in Sutlej River at the Rampur site rose sharply from 1,500 cumecs to 5,000 cumecs in a period of 2.5 hours beginning 03:00 hrs. The discharge at Suni, downstream of the Rampur site, rose from 1,800 cumecs to 6,000 cumecs over a period of 2 hours only. Thereafter, the water level receded. The flash flood caused widespread damage to life and property in Kinnour district, washing away many bridges and villages. The Rampur town also suffered heavy damage.

In Arunachal Pradesh, the Siang River (Brahmaputra) experienced flash floods in June 2000. There was no heavy rainfall before the flood. Suddenly, flood water entered the Gelling area from its source Tsang-po River. The rate of rise of water level was very rapid – there was a rise of 3.21 m at Passighat between 12:00 hrs to 13:00 hrs on 11.6.2000 and the total rise was 5.3 m between 09:00 hrs to 18:00 hrs. The water level receded quickly thereafter. Vast areas in 4 districts on both sides of the river were inundated in the Passighat township, causing considerable losses. Though no specific reason could be ascribed for the flood, breach in a natural dam caused by landslides in Tibet or blockade in the upstream catchment in Tibet was suspected to be the reason.

Snow & Glacial Melt Generated Floods

Sudden release of water from glaciers are known as glacier floods and may represent the outflow either of water which has been held within the ice body or on surface due to ice dam. Snowmelt is a slow process and usually does not result in major floods in India. Generally, glacier melt is slower than snowmelt but sometimes glaciers hold large quantities of water which may be suddenly released with melting of ice dam which may be holding this water. This results in glacier outburst floods. This hydrological phenomenon is common in Karakoram Himalayas where 30 glaciers have potential to create floods in the upper Indus River.

Many rivers originating from Himalayas in North India also receive melt water from glaciers. Glacial outburst is one of the suspected reasons for flash floods in some Himalayan rivers. The outburst flood of the year 1929 by Chong Khundam glacier had a volume of 1,500 MCM. Some $300,000\,m^3$ of ice were carried with the flood. It was completely drained in 48 hours with flood peak discharge of $22,650\,m^3/sec$ in Indus at Attock.

Another example is the glacier flood in Aug. 1959 in Kashmir, which caused a rise in the water level of more than 30 m at a distance of over 40 km from the point of outburst. Major floods occur in some rivers when river ice breaks up into large blocks during the early stages of spring melt and piles up to form ice jam. After a jam develops, the water level behind it rises, often rapidly, until eventually the obstruction gives way releasing water and debris into the channel. Therefore, the flood hydrograph close to the ice jam is often extremely sharp peaked until the flood wave moves down the channel.

Cyclones and Storm Surges in Coastal Areas

Floods in coastal areas are also caused by rainstorms which are generally associated with low pressure systems like tropical cyclones. During the last century, over 1,000 tropical cyclones and depressions originating in the Bay of Bengal and Arabian Sea moved across the Indian subcontinent. Passage of such storms in quick succession over a river basin invariably leads to severe floods. Parts of the country, mainly coastal areas of Andhra Pradesh, Assam, Orissa, Tamil Nadu and West Bengal, experience such cyclones leading to extensive flooding. The flood due to super cyclone combined with heavy rainfall during October 1999 in the coastal belt of Orissa is an example of this type of floods.

Cloud Bursts

Some parts of the country are prone to sudden unprecedented heavy rains, known as cloud bursts. During the year 2000, coastal districts of Andhra Pradesh experienced severe flooding caused by such bursts. Cyclonic circulations during monsoon conditions too sometimes lead to cloud bursts leading to floods. In July 1981 such a condition developed over eastern Rajasthan when stations around Jaipur recorded a 3-day rainfall which was more than the annual normal. Annual rainfall at these places varies from 550 mm to 700 mm. But during the period, the daily rainfall was 250–590 mm, two-day rainfall of 430–800 and 3-day rainfall of 450–900 mm, which caused severe floods. High rainfall over Mumbai in July 2005 could also be termed as an instance of a cloud burst.

Human Activities

Human activities also sometimes result in floods. Sudden breaches in dams/reservoirs may cause devastating floods. The Machhu-II, constructed in Gujarat in 1972, failed during 1979 due to overtopping, causing severe damages in Morbi town in the downstream reach. Around 2,000 people were killed in the accident. However, the record of India in regard to dam safety has been very good.

The Ukai dam is the terminal dam on the Tapi River harnessing a catchment area of about 62,000 sq. km. Due to the large storage space reserved for flood control, the Ukai dam is able to moderate the peaks of incoming floods and has largely freed the town of Surat from flooding. However, recently the city of Surat is experiencing a flood problem whenever there is a large release of water from Ukai dam because of: (i) considerable reduction in safe carrying capacity of the river due to the encroachment of flood plains, (ii) reduction in carrying capacity of the river due to silting, and (iii) the effect of tides, etc. In the 1970s a discharge of 24,000 cumec (about 8.5 lakh cusecs) was considered safe for Tapi River in the Surat city area. But due to encroachments in the areas around Surat, the safe carrying capacity is now about half of the previous value. Thus, whenever Ukai dam managers are forced to make large releases due to high inflows, there is inundation of low-lying areas of Surat.

Tsunamis

Tsunami or the harbour waves were not considered to be a problem for India, not until 2004 when giant tsunami waves triggered by an earthquake near the Sumatra Island, Indonesia, caused widespread damage in more than 25 countries bordering the Indian Ocean. More than 180,000 people are believed to have perished in one of the worst natural disasters of Asia. In India, these waves led to widespread flooding and damages in states on the eastern coast of India, viz., Tamil Nadu, Andhra Pradesh, Pondicherry, and Andaman and Nicobar Islands.

In view of the devastation caused by the tsunamis, the Government of India is setting up a comprehensive warning system for tsunamis.

18.3.4. Flood Management Measures

Flood management activities can be broadly classified into two major groups: structural measures, and Non-structural measures. Broadly, all physical measures taken up to modify the flood are termed as *Structural Measures*, while those under other activities are grouped as *Non-structural Measures*: a) attempts to modify the susceptibility to flood damage, b) attempts to modify the loss burden, and c) bearing the loss.

Attempts to modify the "damage susceptibility" involve actions designed to reduce the vulnerability of property and other developmental activities in flood plains to flood hazard. Attempts to "modify the loss burden" consist of actions to modify the incidence of losses, by spreading them over a large segment of the community. Bearing the loss means living with floods.

Structural Measures

These measures attempt to prevent flood waters from reaching potential damage centres. These consist of embankments, flood walls, reservoirs, natural detention basins, channel improvement, drainage improvement, and diversion of flood waters. Each of these measures aims to protect an area; normally these may involve high capital costs.

Embankments restrict the flow of a river in its defined course. Generally these are constructed with available earth in the nearby area if spills are of small depth. Where adequate space is not available for an earthen embankment or land is very expensive, concrete or masonry flood walls are constructed. In the 19th century many embankments were constructed on some rivers which were causing recurrent damages. These measures were to protect some areas in northern India and the deltaic tracts of east flowing rivers in Orissa, Andhra Pradesh and Tamil Nadu.

Reservoirs store floodwaters and the stored water can be gradually released when the flood has receded. In a multipurpose project, besides moderating flood peaks, the stored water can be used for purposes such as irrigation, power generation, industrial and domestic uses, etc. From the flood control point of view, storage space is specifically allocated in the reservoir although some incidental flood control benefit is available from every reservoir. Many reservoirs in India have storage spaces specifically allocated for flood control, e.g., Maithon, Tilaiya (Damodar), Hirakud (Mahanadi), Bhakra (Sutlej), Rengali (Brahmani) and Ukai (Tapi).

Channel improvement involves increasing the area or velocity of flow or both so that larger flow can be safely carried. Channel improvement has not been widely resorted to in India, mainly because it is costly. Diversion of flood water is just like a road bypass for a congested city. Flow that may cause damage can be diverted into a natural or artificially constructed channel.

Non-Structural Measures

Non-structural measures strive to keep the people away from flood waters, bearing in mind that flood plains belong to the river. These contemplate judicious use of flood plains for beneficial purposes and vacating the same for use of the river whenever she wants it. In India, considerable thrust is on non-structural flood management measures. The non-structural measures are broadly grouped as: Flood plain zoning, Flood proofing, Flood forecasting and warning, Disaster preparedness, Disaster relief, Flood fighting, and Flood insurance. Figure 9 shows the concept of flood plain zoning.

Of all the non-structural flood management measures, the one which is gaining increased attention is flood forecasting and warning. Flood forecasting involves estimation of river water level and discharge at future times. A nationwide flood forecasting and warning system has been established by CWC. The system under CWC is largely on major interstate rivers and states often supplement these by their own efforts at other stations. With reliable advance information/warning about impending floods, loss of human lives and moveable properties and human miseries can be considerably reduced. People and cattle can be shifted to safer places. Similarly, valuable moveable properties can be removed to safer places.

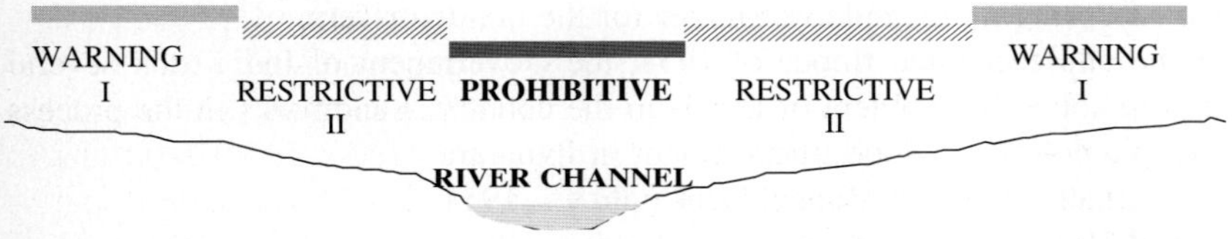

Figure 9. The concept of flood plain zoning

Disaster Preparedness and Response Planning

The impact of flood disaster can be mitigated by timely rescue, relief and rehabilitation operations. Responsibility for relief and rehabilitation is shared at two levels: State and Central. The role of the Central Government in the event of a natural calamity is at policy and administrative level while the state governments deal with the actual ground work. Advanced plans for disaster mitigation involve co-ordination among central, state and local authorities. The Ministry of Agriculture, Government of India, has prepared a contingency action plan for natural calamities, including floods. Most of the states have Relief Commissioners who are in charge of relief measures. Besides, there is a National Crisis Management Committee (NCMC) with Cabinet Secretary as Chairman and a Crisis Management Group (CMG). Voluntary agencies, Panchayats and co-operative societies can play a very vital and constructive role in creating awareness among people towards disaster management.

Flood Fighting covers building temporary dykes along the river, dowel bunds on the banks, immediate closing of breaches, attending to scour, evacuating movable property and people out of the flooded zone, protecting equipment, etc. At the time of floods, the facilities for water supply and sewerage get disrupted affecting the health of the population. Public health operations to ensure the availability of supplies and equipment, co-ordination with other organizations engaged in disaster relief and procedure for immediate mobilization of personnel to eliminate health hazards is an important part of relief operation. Flood fighting measures normally involve evacuation of stranded people, air dropping of food packets, supply of fodder and other essential commodities, restoration of road/rail links, pumping of water from inundated areas and restoration of public utilities, such as roads, bridges, electric supply, sewerage and water supply schemes, and drains, etc.

18.3.5. Flood Management in India

We often hear that "the flood plains are the playgrounds of rivers and should be left as such, without any human interference". Although it is a wise statement, this philosophy has not been found to be practical in overpopulated developing countries like India. Firstly, the limits of these "playgrounds" cannot be defined precisely. Most alluvial rivers are of meandering or braided form covering wide areas. The flow is confined to narrow channel(s) during fair weather. Typically, the width of these "playgrounds" could be about ten times the width of such a waterway width for the meandering rivers and even larger for the braided rivers.

After the unprecedented floods of 1954, the Government of India took several initiatives to solve the problem of floods in the country. Landmarks in the process of evolving a policy for flood management in India are:

- Announcement of Flood Management Policy – 1954
- Report of High Level Committee on Floods – 1957
- Report of Rashtriya Barh Ayog (RBA) – 1980

- Report of National Commission for Integrated Water Resources Development Plan – 1999
- National Water Policy – 2002

A commission, known as the Rashtriya Barh Ayog (RBA or National Flood Commission), was set up by the Government of India to cover the entire gamut of flood problem in the country. The commission submitted its comprehensive report in 1980 which contained 207 recommendations. This report formed the basis for formulating flood management programs. Subsequently, the National Commission for Integrated Water Resources Development Plan also, among other things, studied the flood problem and made many useful recommendations in its report published in 1999. The National Water Policy (2002) of India also has made several recommendations concerning flood control and management. Among other things, the National Water Policy (2002) has recommended preparation of basin-wide master plans for flood management in each flood prone basin, watershed management and catchment area treatment plans and measures like flood forecasting and flood plain zoning for damage minimization. The National Commission for Integrated Water Resources Development also studied the flood management problem and gave its recommendations (NCIWRD 1999). The commission rightly pointed out that there is no possibility for complete protection against floods. Therefore, it suggested that emphasis should be on better management of flood plains, flood forecasting, flood preparedness, and flood insurance.

In the policy statements listed above, the underlying objective has been to get rid of the menace of floods by managing them. Emphasis has been laid on a flexible approach to the problem of floods as part of an integrated approach for the utilization of land and water resources. A summary of the important recommendations follows:

1. It is not possible to provide complete protection against floods and the country has to focus attention on efficient flood management involving flood forecasting, disaster preparedness, wise use of flood plains, flood proofing, flood fighting, and flood insurance;
2. Basinwise master plan for flood management in each flood prone basin should be prepared;
3. To the extent feasible flood control schemes should fit in with other water related plans;
4. Future multi-purpose projects should simultaneously consider flood control aspects;
5. Properly designed, executed and maintained embankments are satisfactory means of flood protection. A suitable combination of embankments with other methods such as storage dams, detention basins, etc. is usually more efficient and should be adopted as resources permit;
6. Performance of embankments has to be evaluated and suitable changes be made in their design, construction and maintenance for better results;
7. Priorities for soil conservation work relating to flood control should be as under:
   (a) Catchment areas of multi-purpose dams,
   (b) Himalayas with their foothills,

(c) Indo-Gangetic plain, and
(d) Deccan plateau;

8. Works relating to watershed management be prioritized. Work commenced in a catchment should not be left incomplete to take up work in other catchments;
9. Data be collected to provide information on long-term performance of flood management measures and their impact on various socio-economic factors;
10. Legislation and enforcement by states is necessary to prevent unauthorized river bed cultivation and encroachments;
11. States to enact legislation amending section 17 (II) of land acquisition act to make the existing provisions for emergent situations applicable for flood control works;
12. Priority be assigned for measures to modify the susceptibility of life and property to flood damage;
13. Priority be given for completion of continuing schemes and adequate funds be allocated for maintenance;
14. Adequate flood-cushion should be provided in water storage projects, wherever feasible; and
15. The network of flood forecasting and warning is to be extended to remaining flood prone areas.

During the past 50 years, different methods of flood protection, both long term and short term, have been adopted, depending upon the nature of the problem and local conditions. In this period, 34,400 km of new embankments and 51,318 km of drainage channels have been constructed. In addition, 2,400 town protection works have been completed and 4,760 villages were raised above flood levels mainly in Uttar Pradesh, West Bengal, North Bihar, and Assam. The major embankment projects that have been completed are on rivers Kosi and Gandak (Bihar), Krishna and Godavari (Andhra Pradesh), Mahanadi (Orissa), Tapi (Gujarat), Damodar (West Bengal), and Brahmaputra (Assam). Needless to say, these embankments have provided reasonable protection to adjoining areas. While there have been occasional breaches in embankments, these have given reasonable protection to an area of about 15.07 M-ha. In addition an area of about 3 M-ha had been protected in some of the states by the structures which existed prior to 1954. A flood spill channel of Jhelum has been built to protect the Srinagar city.

A flood plain zoning bill was mooted by the Government of India. However, the response of the state governments towards enactment of flood plain zoning bill is not encouraging. Manipur had enacted flood plain zoning legislation in 1978 but demarcation of flood zones is yet to be taken up. The State Government of Rajasthan has also enacted legislation for flood plain management. One of the main requirements to implement flood plain zoning measure is availability of maps on a suitable scale to demarcate flood prone areas. For this purpose, maps on a scale 1:15,000 or larger preferably with contours at an interval of 0.5 m for plain areas and 1 to 2 m for hilly areas are needed. Such maps for the states of Assam, Bihar, Haryana, J & K, Punjab, U.P., West Bengal and Delhi covering 54,700 sq. km were prepared but the scheme remains unimplemented.

Planwise expenditure on flood management works up to the eighth plan period ending 1997 was Rs. 4,856.68 crore. For the ninth plan, the anticipated expenditure was Rs. 2,629.23 crore and the area likely to be benefited is 1.14 M-ha. The recommended outlay for the Tenth Plan is Rs. 10,631.84 crore. Of this, Rs. 7,624 crore is for the state sector and Rs. 3,007.91 crore is for the central sector. This is expected to benefit 2.781 M-ha. Among other activities, the works to be taken up involve strengthening of the flood forecasting network of the Central Water Commission (CWC), flood plain management through zoning and people's participation in the maintenance of embankments.

Organizations involved in flood management

The primary responsibility for the management of floods rests with the respective state governments. Schemes for flood management are investigated, planned and executed by state governments. Most states have Departments of Irrigation, Water Resources, or PWD which also deal with flood control. The Ganga Flood Control Commission was set up in 1972 for preparation of comprehensive plans for flood management, monitoring the execution of important flood control schemes, techno-economic examination of major and medium flood control, drainage and anti erosion schemes of Ganga basin. Brahmaputra Board was constituted in 1982 to plan and integrate implementation of measures for control of floods and bank erosion in the Brahmaputra Basin. Likewise, one of the prime objectives of the Damodar Valley Corporation is flood management in the Damodar basin.

In India, flood forecasting and warning commenced in 1958 with the establishment of a unit in CWC, New Delhi, for flood forecasting for the Yamuna River at Delhi. This set-up has now grown to cover most of the flood prone interstate river basins in the country. Presently, flood forecasts are issued at 173 stations: 145 stations for river stage forecast and 28 for reservoir inflow forecast. These stations are located on 62 of the country's most flood-prone river basins. Most of the forecasting stations are situated in 12 northern and north-eastern states. Eighty stations are in the Ganga basin and 27 in the Brahmaputra basin.

During a flood event, engineering departments and local administration work closely to fight the flood menace. In extreme situations, help is also sought from army, air force, and navy to evacuate people from the affected areas and to provide relief.

Capacity Building for Better Flood Management

For effective flood management, a larger capacity is needed in the country for understanding the problem and planning mitigation measures. This strategy should consist of: a) Building a scientific database, b) Building a large R & D capacity, and c) Human resources development.

Currently, no evacuation plans and no inundation maps showing the areas likely to be affected by dam burst floods exist for most Indian dams. Clearly, these should be prepared. In the recent past, dambreak studies have been conducted for some dams in India using mathematical models.

18.3.6. Problem of Inundation in Mokama Tal Area Bihar

Mokama Group of Tals in Middle Ganga Plains (Bihar) is well known for its typical flood problem and due to this reason, it is discussed here. Water accumulates and remains stagnant in large areas up to the end of September and the area becomes dry without adequate irrigation facilities after September/October. This prevents Kharif cultivation and delays Rabi crops. Attempts made to drain out the area for Kharif crops through various drainage schemes and embankments have proved to give no satisfactory results. A solution is possible, if the problem area is used for suitable purposes, such as very low areas for ponding and fishing, shallow areas for agricultural purposes using compatible cropping patterns and high level areas for other supporting infrastructure activities. Introducing new crop production technology; minikit seed distribution; training demo to farmers; ensuring easy availability of agricultural implements like pumping sets, tractors, etc.; ensuring irrigation during dry season privately or by government or both; easy distribution of high yielding varieties of seed; introducing suitable plant protection measures to check damage to standing crops providing supporting infrastructure to the above activities, etc. are some of the measures that may improve agricultural production and thus the economy of the area. Using the chronically inundation prone Tal areas as flood moderation detention basins to moderate peak floods of Ganga is also an option.

The Mokama group of Tals lies in Kiul-Harohar River basin and extends over parts of Patna, Nalanda, and Munger districts. It has a total area of about 1,062 sq. km and is a saucer shaped depression, extending from Fatuha in the west to Lakhisarai in the east. The length of the depression is about 100 km and its width varies from 6 to 17 km. It runs close and almost parallel to the right bank of Ganga River (See Figure 10). General slope of the Tal area is mainly from west to east and the Harohar River is the master drain of the Tals.

The Mokama Tal area is a fertile and low-lying land. It has been divided into several sub-areas by south-north embankments, rails, roads or other natural divides. The land between the railway line and Ganga is rather high and consequently natural drainage across this land is not possible. In the south of the Tal area, the terrain is very flat. In the Tal area the average rainfall depth is about 101 cm; about 85% of it occurs during the monsoon months. The tragedy of the area is its topography which renders drainage difficult. Consequently, the area remains water-logged for the whole of the monsoon period.

Various rivers that rise from south Bihar and flow in northward direction enter the Tal from the south and form one channel named Dhowa (later Harohar), flowing from west to east in the middle of the Tal. The Tal area acts as a delta for right bank tributaries of Ganga River in between Mohane and Nata valleys. The main central river flows through the Tal area from west to east. Finally the Harohar River, which is the main outlet channel for the Tal area, flows eastward and drains into the Kiul River near Rahuaghat. From this confluence the combined Kiul-Harohar flows in the north east direction and ultimately falls in the Ganga River near Surajgarh. The Tal and all these rivers lie in Kiul-Harohar basin. The total drainage area of

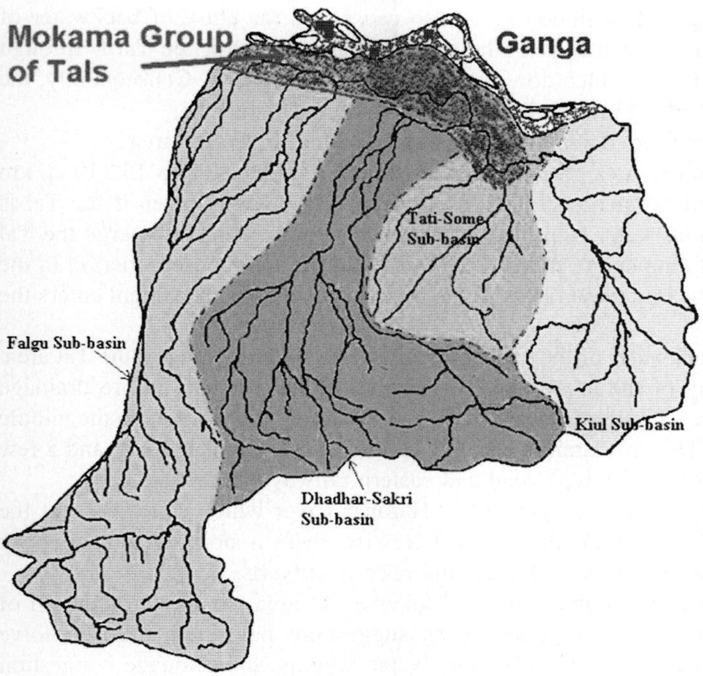

Figure 10. Mokama group of Tals

the Harohar River above its confluence with Kiul River which directly feeds these Tals is about 13,340 sq. km. The drainage areas of the major rivers are: Tilaiya (181 sq. km), Dhadhar (240 sq. km), Lilagan (646 sq. km), Mohane (990 sq. km), Job (43 sq. km), and Morhar (256 sq. km).

The slope of the Tal varies from 0.057 m/km to 0.095 m/km from west to east. The underlying soil for Tal area is impervious clay and silt having low moisture yielding capacity. The core of the Tal area has a surface cover of dense, poorly drained, unoxidised and humus-rich clayey soil. The Tal area is essentially built up by the major contribution from the upland drainage system originating from South Chotanagpur highland which transported silty or fine sand into this backswamp. Sediment brought by the spill-over water of Ganga also gets deposited here.

Cropping Pattern in the Tal Area: Approximately three-fourth of the total Tal area has Rabi cultivation, 17% is paddy and double cropland. In parts of Patna and Munger districts, the Kharif intensity is more than Rabi. The main crops in the Tal area are pulses, oilseeds, and wheat. Land remains under water for about two months during Kharif but is intensively cropped during Rabi.

Water Logging Problem: In general, the flood problem in the Kiul-Harohar River system is limited to the problem of the Mokama Tal area. In addition to discharge from the streams draining into the Tal, this area gets spills from the Punpun in the West and from Ganga. Flooding is also caused by the entry of backwater of Ganga through the Harohar River. The only drainage outlet for the Tal is through the Harohar-Kiul River, which does not have adequate capacity. Consequently, the land remains submerged practically up to the middle of November.

The following are the major causes of water logging in the Tal area.

1. The total catchment area of various rivers draining into the Tal is 13,340 sq. km against the total submergence area of 1,062 sq. km. Clearly, even if the Tal is dry, the incoming water from the catchment can appreciably submerge the Tal area even with a moderate runoff. Inundation during the monsoon period in the Tal area is caused by these rivers as the runoff of this large catchment enters the Tal area.
2. The entry of backwater of the Ganga River causes water logging in the Tal area.
3. The topography of the area does not permit any provision to ensure drainage until the Ganga water level starts receding. Sometimes, this goes up to the middle of November. The only drainage of the Tal is through Harohar River and a few culverts in the Patna-Munger road and eastern railway embankments.
4. There is no regulating arrangement in Harohar River which could prevent the entry of the backwater of the Ganga. Likewise, there is no provision to check the entry of backwater through road and railway culverts.

Solution: Water-logging problem in the Mokama Tal area has drawn attention of decision makers for a long time and many suggestions have been made to solve it. Broadly, the suggestions to solve this water logging and drainage congestion problem are:

1. Construction of storage reservoir and embankment on the right side of Punpun River to check flow into Tal.
2. Construction of an anti-flood sluice on Harohar River.
3. Constructing river embankments on both banks of Harohar and irrigation outlets for sluicing water for irrigation.
4. Use of mathematical modeling and system analysis techniques to derive a long-term sustainable solution of the Mokama Tal problem.

Despite studies and reports by several committees, the problem of Mokama Tal remains unsolved. A detailed description of the area and problem is given in NIH (1996).

18.4. DROUGHTS

Drought is a phenomenon which occurs in different parts of the Indian sub-continent with unpredictable frequency. Normally, the term drought is used to express shortage of precipitation or water at a place when it is expected. In India, droughts mainly occur due to: i) delay in the onset or failure of monsoons, ii) large variability of monsoon rainfall, and iii) long break in monsoon.

As water is used for various life supporting uses, drought can also be defined as a situation where water is expected nut not available. For example, in Bali, six consecutive rainless days would be termed as drought while in some parts of North

Africa, drought is said to have taken place if no rain is received for more than two consecutive years. Another contrasting picture can be quoted from Karnataka where a year in which rainfall is 1,000 mm will be called a drought year in Sirsi but a wet year in Chitradurga (Prasad 1990). In Egypt, which depends upon the flow of Nile for most of its water needs, drought depends not on rainfall but on the flow of Nile.

Meteorological drought is said to occur, when there is more than 25% decline in rainfall from the normal value. According to the norms of IMD, a meteorological subdivision is considered to be affected by drought if its total seasonal rainfall is less than 75% of the normal. If meteorological drought is prolonged over an extensive area, it triggers hydrological drought which is the condition of marked depletion of water resources in rivers, reservoirs, lakes, and springs and fall in groundwater levels. A marked depletion of soil moisture and precipitation leads to Agricultural Drought (Figure 11).

To identify a drought prone area, CWC (1982) has adopted the same criteria as followed by the Irrigation Commission (1972). Accordingly, an area is considered as drought prone if:

- The annual rainfall is less than 75% of the normal in 20% of the years examined.
- Less than 30% of the cultivated area is irrigated.

Depending upon the percent departure of runoff volumes, drought is classified as "severe" if percent departure is more than 50% and moderate if the percent departure is 25–50%.

The most important factor in understanding drought is that it is basically a situation of water deficit for a given use due to below-normal natural water availability. Basically, hydrological drought means a deficit of water supply involving duration, areal extent, severity (intensity), probability of occurrence, initiation, and termination.

Besides the amount of rainfall and its variability, the problem of drought also depends on the extent to which irrigation has been developed. In India, the areas that suffer from chronic drought are confined to West Rajasthan, particularly the districts of Jaisalmer, Barmer, Jodhpur and Bikaner; and Kutch in Gujarat. Providing

Figure 11. A field during drought

irrigation and water harvesting are the major steps in relieving the drought-affected areas from scarcity conditions. For example, the Indira Gandhi Canal in Rajasthan has transformed parts of Ganganagar, one of the intensely arid districts in the state, into a prosperous agricultural area.

The methods of drought analysis include deterministic, stochastic, and statistical. Low flow analysis is frequently employed to identify hydrological drought. The magnitude of low flow determines the amount of water available for various uses and the severity of drought. Drought indices are used to describe features of drought. A commonly used index is the depth of precipitation or runoff for a given duration viz., week, fortnight, month, with the long-term mean value for the corresponding duration. The numerical value of this index gives the severity of drought.

Characteristics of Drought

Droughts are cyclical and regional climatic phenomenon. They differ from other natural calamities in several ways:

- Drought is a 'creeping phenomenon', making its onset and end difficult to determine. The effects of drought accumulate slowly over considerable period of time and may prolong beyond a period of year(s) for major events.
- The absence of a precise and universally accepted definition of drought adds to the confusion about whether or not a drought exists. In the literature, the concept of drought definition is conflicting. it varies according to subject of interest of researchers and among regions of differing climates. This ultimately effects in drought management decisions.
- The societal impacts of drought are less obvious and extend over the larger geographical area than damages that result from other natural calamities. Drought seldom results in structural damage. For these reasons the quantification of impacts and the provision of disaster relief is a far more arduous task than it is for other natural hazards.

Droughts are recurring natural phenomena; therefore, they cannot be prevented. However, coping with droughts is possible through proper prediction of regional drought characteristics and planning. Droughts are characterized by their intensity, duration, frequency, and severity. These characteristics form a basis in the planning of management strategies to cope with drought catastrophe in a given place or region. To reduce the impact of drought hardships, it is necessary understand its characteristics, i.e., its possible duration (How long will it last?), its intensity (How severe will it be?), and its frequency (How often will it recur?). Once these characteristics are known for a given place/climatic region, they can be used as a management tool for drought mitigation.

Drought intensity (I): Drought intensity (I) refers to the magnitude to which actual precipitation/stream/soil moisture are lesser than the mean or a given threshold value. (i.e., the precipitation deficit/ streamflow deficit/ soil-moisture deficit).

Drought intensity is nearly independent of the duration and this fact is very well discussed in the literature (e.g., Woo and Tarhule, 1994).

Table 2. Social, environmental, and economic impacts of droughts

Aspect	Direct	Indirect
Social	Drinking water	Conflicts water users, Unemployment, Famine, Poverty, Health (heat stress and respiratory), Migration, Deaths
Environmental	Soil moisture, Groundwater levels, Groundwater discharge, Spring yield, Streamflow, Lake levels, Reservoir levels, Water velocity, Water depth	Water quality, Plant growth, Habitats, Endangered species, Dust storms, Forest fires
Economic	Groundwater abstraction, Surface water abstraction, Reservoir outflow, Crop yield	Irrigation water, Domestic water, Crop failure, Farm animal death, Navigability, Hydropower, Higher food and fodder prices, Economic growth

Drought duration (D): Drought duration (D) is the period of time when there is a deficiency of precipitation/ stream/soil-moisture preceded and followed by periods when there is no deficiency.

A drought event is a series of one or more consecutive drought months/seasons or years. A drought can have duration of one or more months/seasons or even years. Drought persistence is the tendency of a drought event to last more than one, season or year. For instance, a 4-year drought is a very persistent drought.

Drought Impacts: Drought is always viewed as having a negative effect meaning that magnitude of variable is lower, smaller or reduced, except for the indirect social effects. Its impacts are classified as direct or indirect. Table 2 lists the social, environmental, and economic impacts of droughts.

18.4.1. Historical Drought Occurrences

Drought is a disaster that frequently hits parts of India. Since the year 1800, there have been around 40 droughts in the country with varying degrees of severity. Some past severe droughts have taken place in 1965, 1966, 1969, 1973, 1974, 1976, 1980, 1985, 1986, 1987, and 1991. The drought of 1987 was one of the severest in the recent past and it had affected about half of the country.

- There are evidences of a terrible famine in India, which lasted for twelve consecutive years (310 to 298 BC).
- Severe drought occurred in Kashmir in 1917–1918 when Jhelum River completely dried up.
- After Independence, India faced droughts in 1966–67, 1972–73, 1979–80 and 1986–87. In each case food production fell below the national average. There were large-scale losses through starvation, depletion of assets and livestock and high mortality. The 1987 drought, which was one of the worst droughts of the

Table 3. Drought years in the past centuries

Period	Drought years	Number of years
1801–1850	1801, 04, 06, 12, 19, 25, 32, 33, 37	9
1851–1900	1853, 60, 62, 66, 68, 73, 77**, 91, 99**	9
1901–1950	1901*, 04, 05*, 07, 11, 13, 15, 18**, 20, 25, 39, 41*	12
1951–2003	1951, 65*, 66, 68, 72**, 74, 79*, 82, 85, 87**, 2002*	11

* Indicates severe drought (> 39.5% area affected), ** Phenomenal drought year (> 47.7% area).
Source: Ali (2003).

20th century with overall rainfall deficiency of 19%, affected 58–60% of cropped area and a population of 285 million.

The droughts that have occurred in India over the last 200 years are listed in Table 3.

Further, Table 4 lists droughts that have taken place since independence, the percent area affected by each, and their categories.

Most of the drought prone areas fall either in semi-arid or arid regions. The probabilities of occurrence of drought in different meteorological sub-divisions are shown in the Table 5. The probabilities are high in the arid region of the western Rajasthan compared to other sub-divisions. Communities like marginal farmers that are dependent on rainfed agriculture are most affected by drought. Drought not only affects the socio-economic condition of millions of people in the affected area, plants, and livestock but also damages the economy.

Table 4. Details of droughts since independence

S. N.	Year	Percentage of affected area in India	Category
1	1951	33.2	Moderate
2	1952	25.8	Slight
3	1965	42.9	Moderate
4	1966	32.3	Moderate
5	1968	20.6	Slight
6	1969	19.9	Slight
7	1971	13.3	Slight
8	1972	44.4	Severe
9	1974	29.3	Moderate
10	1979	39.4	Moderate
11	1982	33.1	Moderate
12	1985	30.1	Moderate
13	1986	19.0	Slight
14	1987	49.2	Severe
15	2002	Areas in 14 States	Severe

Source: Ali (2003).

Table 5. Probability of occurrence of drought in different meteorological sub division

Meteorological sub division	Frequency of deficient rainfall (75% of normal or less)
Assam	Very rare, once in 15 years
West Bengal, Madhya Pradesh, Konkan, Bihar and Orissa	Once in 5 years
South Interior Karnataka, Eastern Uttar Pradesh and Vidarbha	Once in 4 years
East Rajasthan, Gujarat and Western Uttar Pradesh	Once in 3 years
West Rajasthan, Tamil Nadu, Jammu & Kashmir and Telangana	Once in 2.5 years

Source: Ali (2003).

18.4.2. Drought Prone Areas in India

Out of the total geographical area of India, almost one-sixth area with 12% of the population is drought prone; the areas that receive an annual rainfall up to 60 cm are the most prone. The Irrigation Commission (1972) had identified 67 districts as drought prone. These comprise 326 talukas located in 8 states, covering an area of 49.73 m ha. Subsequently, the National Commission on Agriculture (MOA 1976) identified a few more drought prone areas with slightly different criteria. Later, based on detailed studies, 74 districts of the country have been identified as drought prone as given in Table 6. Figure 12 shows drought prone areas of the country.

Most of the drought-prone areas are found in arid, semi-arid, and sub-humid regions of the country, which experience less than average annual rainfall. Broadly, the drought-affected areas in India can be divided into two tracts. The first tract comprising the desert and the semi-arid regions covers an area of 0.6 million sq. km. It is rectangle shaped area whose one side extends from Ahmedabad to Kanpur and the other from Kanpur to Jullundur. In this region, rainfall is less than 750 mm and at some places it is even less than 400 mm. The second tract comprises the regions east of the Western Ghats up to a distance of about 300 km from coast. Known as the rain shadow area of the Western Ghats, rainfall in this region is less than 750 mm and is highly erratic. This region is thickly populated and periodic droughts cause considerable suffering and distress.

Besides these two tracts of scarcity, there are many pockets of drought in India. Some of these are: (i) Tirunelveli district, south of Vaigai River in Tamil Nadu, (ii) Coimbatore area in Kerala, (iii) Saurashtra and Kutch regions in Gujarat,

Table 6. Drought Prone Area of India

Drought Prone Area	51.1 mha
Affected Districts	74
Affected States	13

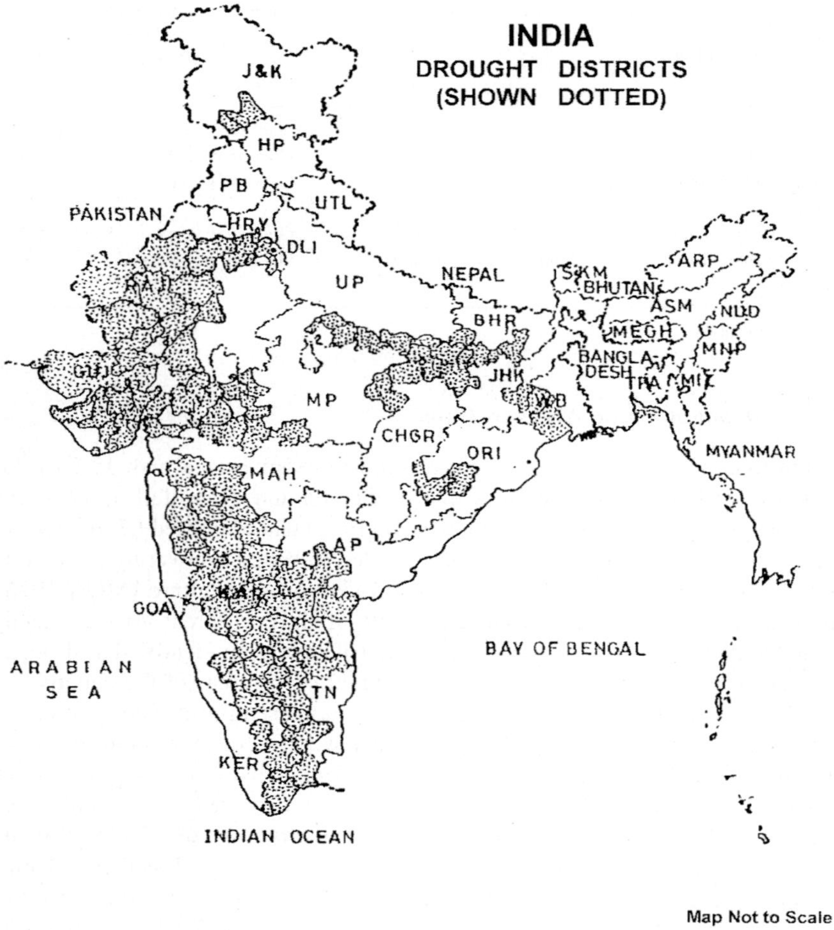

Figure 12. Drought prone area of the country

(iv) Mirzapur plateau and Palamu regions in Uttar Pradesh, (v) Purulia district of West Bengal, and (vi) Kalahandi region of Orissa. Together, these scattered pockets occupy an area of 0.1 million sq. km. Drought is a recurrent phenomenon in Andhra Pradesh where no district is entirely free of droughts. Table 7 gives the percentage occurrence of drought of class moderate and above in the Kharif season in some meteorological sub-divisions of India.

Rajasthan is one of the most drought prone areas of India. Eleven districts of the state are in arid regions including Jaisalmer as the driest district. No perennial river flows in Jaisalmer. Groundwater level in the district is 125–250 ft deep and at some

Table 7. Percentage occurrence of drought of class moderate and above in the Kharif season

S. N.	Sub-division	June	July	Aug	Sept	Oct	Nov
1	Telangana	22	17	19	22	30	30
2	Rayalseema	24	28	30	27	27	27
3	Tamil Nadu	20	22	23	23	21	17
4	Interior Mysore South	20	15	19	15	23	20
5	Interior Mysore North	25	23	29	26	30	29
6	Madhya Maharashtra	20	27	25	23	31	30
7	Vidarbha	20	20	24	21	25	24

places 400 ft deep. The rainfall in the district is extremely low at 164 mm. Out of 365 days of a year, on an average 355 days are dry.

The total area which receives inadequate rainfall is just over one million sq. km. The regions with rainfall less than 400 mm occupy 12% of the total geographical area, and the area below 750 mm rainfall is 35% or a little over a third of the country. Thus out of the total gross cultivated area of the country, 56 million ha is subject to inadequate and highly variable rainfall.

Large areas in the four states that utilize Narmada water falls in arid and semi-arid regions. As shown in Table 8, nearly 57% of Rajasthan and 32% of Gujarat falls in arid zone. Also, nearly 61% of Maharashtra and 46% of the area of Gujarat is semi-arid. This shows how important it is for these states to properly use available water.

Natural Disasters are huge economic burdens on developing economies, such as India. Droughts greatly and adversely affect vegetation and local biodiversity prevailing in that region. Every year huge amounts of resources are spent on rescue, relief and rehabilitation works following droughts. The Central Government plays a major role as far as mobilization of financial resources is concerned. A scheme called Calamity Relief Fund (CRF) has been constituted for each state with contribution from the Central and State Governments to undertake relief and rehabilitation measures. In addition to CRF, a National Fund for

Table 8. Statewise area of arid and semi–arid zones

State	Area (km^2)	
	Arid	Semi-arid
Gujarat	62,180 (31.72)	90,520 (46.18)
Madhya Pradesh	–	59,470 (13.41)
Maharashtra	1,290 (0.42)	189,580 (61.61)
Rajasthan	196,150 (57.31)	121,020 (35.396)

Note: figures inside brackets represent the % of the area of the state under that category.

Calamity Relief (NFCR) has been created to deal with hazards of rare severity managed by a National Calamity Relief Committee (NCRC).

18.4.3. Drought Indices

In a drought situation streamflows are low or even zero. Hence, indices characterizing the low flow regime of a river or time series of low flow values can be said to characterize drought behaviour of a river. Similarly, indices or time series of low groundwater recharge, levels and discharge can be said to characterize the drought behaviour of groundwater. These time series or indices characterize droughts by considering only the flows over pre-defined duration. They do not provide a complete characterization of the drought such as the start, the end, the total duration, and the severity of the drought. To quantify the latter characteristics, it is necessary to define a threshold level below which the flow or groundwater is regarded as being in a drought situation.

Drought events can be defined by introducing a threshold below which the flow is regarded as being in a deficit. Each event has a beginning and an end. The deficits can be described by different characteristics like duration, severity, time of occurrence and spatial extent. An index is a single number characterizing an aspect of drought or low flow behaviour at a site or in a region. A drought event definition implies the use of a method to select drought events from a time series. Whether the drought will have adverse effects or not will depend on the vulnerability of the region.

Low flow characteristics

The variation in climate combined with differences in physiographical catchment properties gives a wide variety of river flow regimes which might require the usage of several different methods. Most methods to derive low flow characteristics have been developed and used for perennial rivers, and applying them to intermittent and ephemeral streams should be done with caution. Many indices are derived by involving the whole spectrum of flows. These include the flow duration curve, from which the low flow percentiles are selected.

Mean annual minimum flow

One of the most frequently applied low flow indices is derived from a series of the annual minima of the n-day average flow, AM(n-day). This could be the mean annual 1-day flow,1-week flow, etc. For $n > 1$, the method consists of deriving a hydrograph whose values are not simply daily flows but are average flows over the previous n-days or alternatively the previous and the n/2 days. The derived data can thus be regarded as the outcome of passing a moving average filter of n-day duration through the daily data. Based on the 'filtered' hydrographs mean annual minimum n-day indices, MAM(n-day), can be derived.

The Base Flow Index (BFI) gives the ratio of base flow to total flow. BFI is a measure of the river's runoff that derives from stored sources. Values of the index range from above 0.9 for a permeable catchment with a very stable flow regime to 0.15–0.2 for an impermeable catchment with a flashy flow regime.

Complex indices are based on several variables, often including many elements of the hydrological cycle, for example a combination of precipitation, streamflow, and evapotranspiration. Several complex drought indices have been developed. The most frequently applied index, especially in the US, is the Palmer Drought Severity Index. Although the PDSI is sometimes used as an indicator of hydrological drought, it is more correctly referred to as a meteorological drought indicator.

Palmer Drought Severity Index

The Palmer Drought Severity Index (PDSI) was developed by Wayne Palmer in the 1960s and uses temperature and rainfall data to determine dryness. Palmer defined a drought period as "an interval of time, generally of the order of months or years in duration during which the actual moisture supply at a given place rather consistently falls short of the climatically expected or climatically appropriate moisture supply". PDSI is based on the supply and demand concept of the water balance equation taking into consideration monthly mean precipitation and temperature as well as the local available water content of the soil. It is most effective measuring impacts on sectors sensitive to soil moisture conditions, such as agriculture (Willeke et al. 1994). This index generally ranges from −6 to +6, with negative values denoting dry conditions and positive values denoting wet conditions.

PDSI has also been useful as a drought-monitoring tool and has been used to trigger actions associated with drought contingency plans (Willeke et al. 1994). It has typically been calculated on a monthly basis. The water balance model that forms the basis of the PDSI also allows the development of other indices that have different rates of response to changes in supply and demand for moisture and therefore represent different types of drought.

The PDSI is based around a supply and demand model of soil moisture at a location. The supply is the amount of moisture in the soil plus the amount that is absorbed into the soil from rainfall. The demand or the amount of water lost from the soil depends on several factors, such as temperature and the amount of moisture in the soil. The basis of soil modeling is the calculation of the potential evapotranspiration (PET) and evapotranspiration (ET). Monthly PET depends on that month's average temperature, and the latitude of the weather station. Besides PET, there is also potential recharge (PR), potential runoff (PRO), and potential loss (PL).

The moisture departure, d, is the deficit or surplus of moisture, adjusted for seasonal changes in climate. However, the moisture departure does not give any information about how severe that deficit or surplus is relative to the local climate. In order to do that, the moisture departure is adjusted again to create the Moisture Anomaly, Z, which represents how wet or dry it is with respect to the current season and the local climate.

With the moisture anomaly calculated, the PDSI itself can now be calculated. There are actually three intermediate indices, $X1$ is severity of a wet spell that may or may not be developing, $X2$ is the severity of a dry spell that may or may not be developing, and $X3$ is the severity of the current, "established" spell. The actual PDSI value is actually determined by picking one of the three indices according to a set of rules. The whole reasoning behind the three different indices is that a single wet month in the middle of a drought does not end the drought. However, several wet months in a row may be enough to end a severe drought. The three indices allow the flexibility to determine when a drought has ended using information from more than a single month. The detailed computational procedure developed by Palmer can be found in George et al. (1972), and Sikka (1986); WMO (1994) describes procedures for low-flow analysis.

The objective of development of PDSI was to provide measurements of moisture conditions that were standardized so that comparisons using the index could be made between locations and between months. PDSI is most effective in determining long term drought (of the order of several months) and is not as good with short-term forecasts (a matter of weeks). It uses 0 as normal, and drought is shown in terms of negative numbers. PDSI can also reflect excess rain using a corresponding level reflected by plus figures; i.e., 0 is normal, +2 is moderate rainfall, etc. Table 9 gives a drought classification based on PDSI.

Some authors have reported that PDSI failed to explain the well-known 1981 drought in the country. The Food and Agricultural Organization (FAO) has developed an algorithm for the generation of the time series of Moisture Adequacy Index to characterize agricultural droughts.

18.4.4. Drought Prediction and Mitigation

Drought is a major problem demanding a more systematic approach to its tackling. Improved drought management calls for effective drought prediction and assessment procedures. National Remote Sensing Agency in collaboration with the Government of Andhra Pradesh and IMD has undertaken a project on drought monitoring and identify the parameters for the areas that have been repeatedly declared as drought affected and to evaluate drought occurrence. Though the State government identified as much as 18 parameters on the causes and impact of drought only percentage rainfall deviations from the normal rainfall are expressed quantitatively. The drought impact is in general explained with quantitative figures with reference to area not

Table 9. Drought classification based on PDSI

PDSI	Drought class
−0.50 to −0.99	Incipient
−1.00 to −1.99	Mild
−2.00 to −2.99	Moderate
−3.00 to −3.99	Severe
≤ −4.00	Extreme

shown and extent of crop damage. However even those figures are not consistent between space and time (Thiruvengadachari, 1989).

Analysis of SST of the northern and southern hemispheric oceans for a 120-year period from 1871 to 1990 was carried out by IITM. It was observed that the two periods which witnessed frequent droughts – from 1901 to 1920 and from 1965 to 1990 – broadly corresponded with warm SST Index. During the earlier warm SST Index periods from 1901 to 1920, and 1965 to 1987, droughts occurred as frequently as every two or three years. In the last spell, India recorded droughts in 1965, 1966, 1968, 1972, 1974, 1979, 1982 and 1987. The reverse trend of the SST Index cooling started from 1990 and was likely to continue for one or two decades, bringing in good rainfall for the country. This could be one reason why India has recorded a series of normal monsoons from 1992 onwards in spite of El – Nino phenomenon in other regions.

Like those of other natural hazards, the impacts of drought can be reduced through mitigation and preparedness. Drought preparedness should be an integral part of water resources management. Drought risk is a product of a region's or community's exposure to the natural hazard and its vulnerability to extended periods of water shortage. If nations, regions, and communities are to reduce the serious consequences of drought, they must improve their understanding of the hazard and the factors that influence vulnerability. An integrated early warning system can provide timely and reliable information to decision makers and aid in reducing the impacts of drought.

18.4.5. Actions to Minimize Adverse Impacts of Drought

It is the responsibility of every government to implement measures to mitigate the effect of natural disaster on its citizens, particularly on the highly vulnerable communities. Of course, help should be provided such that people organize themselves and are not entirely dependent on the government. Government assistance should be targeted such that it increases the resilience of vulnerable communities without encouraging long-term dependence. After a disaster, it is imperative that authorities make every effort to restore the affected communities to pre-disaster status as early as possible.

A number of actions based on either increasing available water supply or reducing demand or to minimize impacts can be and are initiated to mitigate drought consequences. The drought mitigation measures can be classified into three groups:

- Water-supply measures: intended to increase the available water supply during drought and to develop new supplies,
- Water-demand measures: intended to reduce the water demand during drought,
- Impact-minimization measures: intended to minimize drought impacts even with increased water supply and reduced demands.

Supply-oriented measures include:

- Maximizing surface water storage capacity,

- Water conservation measures including control of evaporation, and rainwater harvesting,
- Soil and water conservation practices,
- Control of snow and ice melt,
- Relaxation of water quality standards,
- Increasing ground-water use,
- Water transportation by trains, tankers, ships, or boats,
- Water transfers between river basins or within the same river basin (long-term), and
- Desalination of salt or brackish waters.

Demand reduction oriented measures include:

- Issue appeals to consumers to reduce water use and create awareness,
- Encourage use of sprinkler and drip irrigation systems through subsidies,
- Enforce restrictions on water use and encourage water reuse,
- Induce change in patterns of water consumption, and
- Create economic incentives to reduce water consumption.

Measures to minimize impacts of drought include:

- Drought forecast and warnings,
- Drought insurance, and
- Disaster aid programs.

All these measures have varying degrees of effectiveness related to the circumstances of each drought and a judicious combination of measures can yield an ideal solution. Different kinds of reactions are observed from various individuals and groups on the forecast and warnings issued regarding various natural events like floods or droughts. When forecast and warnings are of creeping phenomenon such is the case with drought, an indefinite response is much more complex than with the other extreme events. The response of people to drought forecast and warnings is influenced by human assessment of the warning reliability. As people become more convinced to the forecast their reaction may become more prompt to the situation. The knowledge of response to forecast and warnings, and to adjustments and measures is useful for involving more information and management of drought control. Proper forecast of drought with reliabilities will make the common man understand the likely impacts of drought which might be there and may allow them for the adjustment of life styles to prevailing situations of water scarcity. Present drought and information which is presently in vogue is not comprehensive in nature and has limited usefulness. Efforts are required to be made.to provide to the farmers a list of the effect of drought for a particular state of evaluation and agro-climatic regions.

Actions are initiated by the Government of India from time-to-time to minimize the adverse impacts of droughts. Important of these are:

- Higher thrust is being given for watershed development.
- Dryland farming using available moisture is being developed.
- Low cost structures for soil and moisture conservation are being encouraged which can be constructed with local skills.

- Dovetailing crop production into watershed projects along with soil conservation activities.
- Over-exploitation of groundwater resources is being checked.
- Use of sprinkler and drip irrigation systems is being encouraged.
- Construction of suitable water harvesting structures for conservation and optimal use of surface water and recharge of underground aquifers.
- Renovation and restoration of old tanks/farm ponds in the villages is being taken up.
- Afforestation and pasture development.
- Animal husbandry and fodder development are being taken up systematically.
- People's participation in drought proofing is being sought.

The Government of India operates a command area development programme (CADP) to strengthen water management capabilities and enhance the effectiveness of irrigation water application. The Drought Prone Areas Programme (DPAP), the Desert Development Programme (DDP), and the National Watershed Development Project for Rainfed Areas (NWDPRA) have been devised and are under implementation. These programmes include development measures for all the spatial components of watersheds, i.e., arable land, non-arable and drainage lines as one organic geo-hydrological entity. The objective is to achieve conservation of rainwater, control of soil erosion, regeneration of green cover and promotion of dry land farming systems, including horticulture, agro-forestry and pasture development and livestock management as well as household production systems.

The DPAP was launched in 1973 in arid and semi-arid areas with poor natural resource endowments to promote more productive dryland agriculture by better soil and moisture conservation, more scientific use of water resources, afforestation, and livestock development through development of fodder and pasture resource, and in the long run to restore the ecological balance. The DPAP covers 615 blocks of 91 districts in 13 States. The DDP which was initiated in 1977–78 covers both the hot desert regions of Gujarat, Rajasthan, and Haryana and the cold desert areas in Jammu, Kashmir, and Himachal Pradesh. It is functional in 131 blocks of 21 districts in 5 States covering an area of about 0.362 million km^2 and a population of 15 million. The objectives of the programme include controlling the process of desertification, mitigating the effects of drought, restoring the ecological balance, and raising the productivity of land, water, livestock, and human resources.

Natural disasters, particularly droughts, create unemployment and under-employment problems in rural areas. The Jawahar Rozgar Yojana (JRY) is the largest programme in the country aimed at generating additional gainful employment for the unemployed and under-employed men and women in rural areas.

18.4.6. Irrigation for Drought Proofing

Irrigation is the most effective drought proofing strategy and is the single biggest factor in bringing stability in agricultural production. The statewise irrigation status of the drought prone areas and the other states is given in Table 10. It can be seen

Table 10. Statewise Irrigation Status of Drought Prone and other states (in ha)

S. N.	State	Total Cultivable Area (CA)	Gross Sown Area	Gross Irrigated Area (GIA)	Potential Created (PC)	Ultimate Irrigation Potential (UIP)	Indicators (%)	
							PC to CA	GIA to UIP
				Drought prone states				
1	A.P.	15,856	12,135	5,158	8,134	11,260	51.30	45.81
2	Bihar	10,859	10,012	4,664	7,360	13,347	67.78	34.94
3	Gujarat	12,361	10,609	3,643	4,067	6,103	32.90	59.69
4	Haryana	3,821	6,143	4,829	4,408	4,512	115.36	107.03
5	J & K	1,050	1,083	446	542	1,358	51.62	32.84
6	Karnataka	12,897	11,696	2,912	3,424	5,974	26.55	48.74
7	M.P.	22,899	26,070	6,527	7,930	17,932	34.63	36.40
8	Maharashtra	21,104	21,740	3,149	7,218	8,952	34.20	35.18
9	Orissa	7,975	8,645	2,318	3,198	8,803	40.10	26.33
10	Rajasthan	25,692	22,325	6,676	6,675	5,128	25.98	130.19
11	Tamil Nadu	8,301	6,558	3,519	5,208	5,532	62.74	63.61
12	U.P.	20,739	26,522	17,467	22,531	30,499	108.64	57.27
13	West Bengal	5,836	9,208	2,491	4,635	6,918	79.42	36.01
				Other states				
14	Arunachal Pradesh	293	250	36	66	168	22.53	21.43
15	Assam	3,257	3,994	572	708	2,870	21.74	19.93
16	Goa	198	169	36	36	116	18.18	31.03
17	H.P.	805	972	176	214	353	26.58	49.86
18	Kerala	2,444	2,969	417	1,049	2,679	42.92	15.57
19	Manipur	164	207	75	104	604	63.41	12.42
20	Meghalaya	1,082	251	54	82	168	7.58	32.14
21	Mizoram	446	113	10	10	75	2.24	13.33
22	Nagaland	636	260	70	68	90	10.69	77.78
23	Punjab	4,250	8,042	7,377	9,417	5,967	221.58	123.63
24	Sikkim	114	142	16	26	70	22.81	22.86
25	Tripura	310	456	60	49	281	15.81	21.35
	All India	183,593	190,762	72,784	97,266	140,008	52.98	51.99

Source: CWC (2002).

that substantial potential has been created in the drone prone states which varies from 25% to 115%. Only in five of these states the potential is below 40%.

Villagers in Rajasthan have built nearly 4,500 dams to make 90 villages drought-proof. About 500 to 700 dams are made every year. These dams have been made of mud and stone and cost only about Rs. 7 lakh. These dams prevent rain water from being washed away, carrying rich top soil, and also help increase recharge. The area that was declared a "dark zone" by the government with low ground water levels where no wells were to be drilled has five rivers – Ruparel, Bhagani, Sarsa, Jahajwali and Arvari – which have been made perennial by these dams.

Figure 13 shows the actual production of foodgrains in India since 1975–76 and the dips in the production line depicts the years of droughts – as can be seen in the years 1979–80, 1982–83, 1986–88.

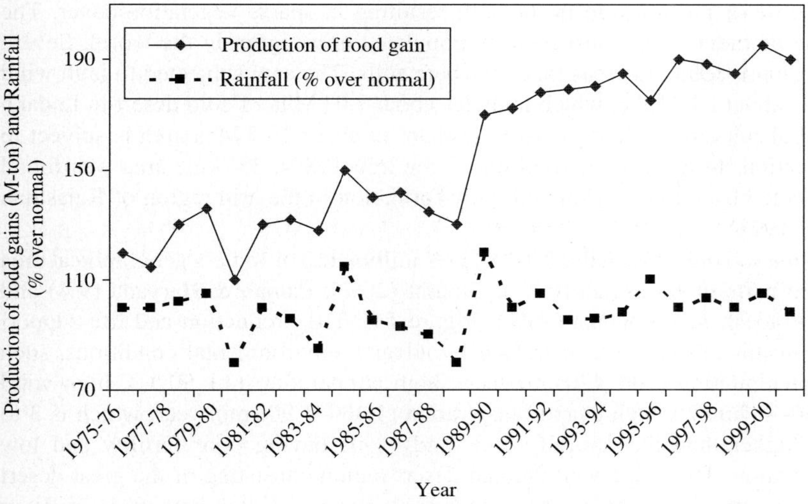

Figure 13. Actual production of food grains in India since 1975–76 and the dips in the production

The bold straight line gives the long-term trend of food grain production in India. The trend line has a positive slope, implying that the production of food grain is increasing. The rise in production is a consequence of increase in land under irrigation, use of high yielding seeds, etc. Creation of more irrigation has given enhanced protection to crops against monsoon failure and the possibility of more than one crop. During drought years, there is significant downward divergence from the trend line. Therefore, the rise in productivity has resulted in the increasing trendline, which also implies that the output levels of the troughs have also been increasing over time. Thus, from one drought to another, the floor of output has risen implying that any year of drought would have a lesser impact on the foodgrain production level than the previous one.

18.4.7. Special Problems of Arid Zones

Low and erratic rainfall, frequent droughts, high summer temperatures and high wind velocity resulting in high evapotranspiration characterize the Indian arid zone which covers 12% of the country's geographical area. Besides poor soils, undulating topography, limited availability of good quality ground water and increasing biotic pressure further aggravate the magnitude of the problems faced by the desert dwellers.

Out of the total geographical area of the country, about 54% falls in arid and semi-arid regions. In these regions, rains are erratic and often come in a few heavy storms of short duration. Rather than properly meeting crop demands or replenishing ground water, these storms result in high run-off. For the most part of the year, there is little or no moisture in the soil, resulting in sparse vegetation cover. The arid zone of India is the most densely populated arid region in the world. Severe wind erosion is found in areas that have bare soils. The area subjected to high wind erosion is about 59.2 Mha, which includes about 7.03 Mha of cold desert in Ladakh and the Lahaul valleys. In western Rajasthan, in about 13.3 Mha area is subject to desertification. In a survey carried out in the mid 1980s, 355 km^2 area was found to be affected by water erosion and gully formation in the arid region of Rajasthan (CAZRI, 1997).

The hot arid zone covers about 10% (31.7 million ha) of India's geographical area located in parts of Rajasthan (61%), Gujarat (20%), Punjab & Haryana (9%) and Andhra Pradesh & Karnataka (10%) (Figure 14). The production and life support systems in this region are constrained by adverse environmental conditions, such as low precipitation (100–420 mm/year), high temperature (45–50 °C), high wind speed (30–40 km/hr), high evapotranspiration (1,400–1,900 mm/year, which is 3 to 5 times higher than the rainfall), and sandy soils having poor fertility and low water retention. There is a well defined desert region consisting of the great desert and little desert. The great desert extends from Rann of Kutch beyond Luni River northwards. The little desert is located between Jodhpur and Jaisalmer and the two are divided by a zone of sterile rocky land cut up by limestone ridges. The annual

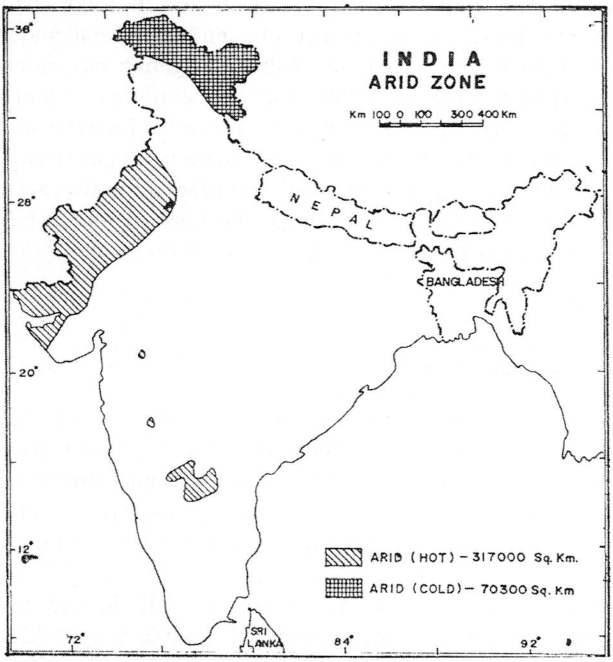

Figure 14. Arid zone in India

rainfall in western Rajasthan varies from less than 150 mm in west of the Jaisalmer district to about 500 mm in the Jalar district. Arid areas in Haryana, Punjab, Gujarat, Maharashtra, Karnataka and Andhra Pradesh receive rainfall in the range of 250 to 700 mm. New problems are cropping up due to ever-increasing human and livestock population.

India also has a cold arid region in Ladakh which receives only 90–120 mm of rainfall in a year. This region occupies nearly 2% area of the country. ICAR has identified the following agro-climatic zones in India (Table 11), which broadly represent the arid to extreme climatic conditions.

Two-thirds of the Indian desert is made up of Aeolian sands — these are low in clay and silt with predominance of fine sand. The dune soils are also highly sandy (75 to 98% sand) and are mainly concentrated to the west of 250 mm rainfall zone (where rainfall is still lower). In the region with slightly higher rainfall (250 to 450 mm), light brown sandy soils are extensively found. These are characterized by sandy, occasionally loamy fine sand. In the arid parts of Andhra Pradesh and Karnataka, one finds red and black soils.

With increasing population, low and erratic rainfall and absence of any other source of water, over-exploitation of groundwater is rising. The water table is falling at a rate of 0.20 to 0.40 m per year in the Jhunjhunu, Sikar, Nagaur and Pali districts. If the present trend of water use continues, it is feared that by 2020 AD a major part of Thar Desert will run out of economically viable groundwater resource. Nearly 65% area in the Thar Desert has saline groundwater. Water harvesting in the form of village ponds (nadis) and tankas is common for meeting drinking water demand.

The Indira Gandhi Nahar Pariyojana (IGNP), earlier known as the Rajasthan Canal, an ambitious irrigation power project was launched in the 1950's to bring Himalayan water to the Thar Desert through a 649 km long network of canals. The north-western part of Rajasthan desert is undergoing a great agro-ecological transformation due to the construction of IGNP. A great extent of canal command area

Table 11. Arid agro-climatic zones of India

Name of the region	Rainfall (mm)	Major soil
Arid western plains of Rajasthan	100–300	Desert (Rhegosolic)
Irrigated north Rajasthan	100–350	Desert/grey brown
Transitional plains of Rajasthan	300–350	Desert/grey brown
Transitional plain of Luni basin	300–350	Desert/grey brown
Western zone of Haryana	<500	Calcareous/sierozemic
Western zone of Punjab	<400	Old alluvial
North-west Gujarat	250–500	Grey brown/deltaic alluvium/red sandy/ medium black
Scanty zone of Maharashtra	500–700	Medium/deep black
Scarce rainfall zone of AP	500–700	Red loamy/medium black
Central dry zone of Karnataka	455–717	Red sandy/deep black medium black
Cold arid zone of Jammu & Kashmir	83	Skeletal (mountane meadow soils, tarai soils)

is being promoted for crop based agriculture. This project began to pay dividends in the late 1980's. But despite the availability of canal water, more than 80% of the region is expected to remain under rainfed farming. Irrigation through IGNP has also led to water logging and salinity problems over large tracts. Whereas water table is falling in many parts of Thar Desert, it is rising @ 0.16 to 0.33 m per year in the IGNP Phase I area. The area degraded by irrigation with sodic waters and waterlogging is estimated to be 2,000 sq. km by 2020 AD. Unless the degradation is effectively checked, this area is likely to increase to 3,000 sq. km area by 2020 AD. A view of Indira Gandhi canal has been shown previously as Figure 9.

The human population in arid Rajasthan was 17.44 million in 1991 which is projected to increase to 28.45 million by 2020 AD. Due to the rising population, the average size of land holding has been decreasing. It was 17.77 ha in 1951, 14.69 ha in 1961, 10.20 in 1971 and is only 6.0 ha in 1995. If the trend continues, it is likely to be less than 4.0 ha by 2020. The land distribution is uneven, with nearly 11.2% households owning 50% of the total land, whereas 47% hold only 10% of the total land.

Owing to the reduction in the average size of land holding, the practice of fallow farming is reducing and cultivation of marginal lands is rising. For example, fallow lands in Barmer district have reduced from 73% to 38% in the last four decades. In the Jaisalmer district, the net sown area increased from 11,000 ha in 1952 to 234,000 ha in 1988. Another consequence of the population hike is the increase in overall cropping area, being 32% in 1952, 50% in 1991 and is likely to be around 55% by 2020 AD.

As per the present trends in arid Rajasthan, the production of cereals, pulses and oils will be 26.27, 10.71 and 1.52 lakh tones by 2020. Despite this, there is likely to be a decrease in per capita availability. The value of agricultural and livestock production in Rajasthan was Rs. 900 million in 1960 and the same is expected to be more than Rs.1.3 lakh million by 2020 AD. However, this is not expected to lead to proportional rise in per capita income because the rural population is likely to increase from 6.28 million (in 1960) to 22.35 million by 2020 AD.

18.5. EXCESSIVE GROUNDWATER EXPLOITATION

Groundwater has contributed vastly to agricultural development in India, particularly during the last four decades. It has helped attain food security through green revolution and commercialization of farming practices. However, its exploitation, which has grown exponentially over the years and is a matter of concern, is largely in the hands of private individuals. In 1950, there were 3.86 million dug-wells and 3,000 deep tube-wells. In a span of four decades, as many as 10.2 million dug-wells, 5.4 million private tube-wells and 60,000 deep tube-wells became operational in the country (Nagaraj et al., 1999). Large-scale extraction of groundwater has led to overdraft and a drastic fall in water table in some basins. This in turn has created a chaotic situation, especially in the water scarce hard-rock regions of southern India, where assured sources of surface irrigation are rare and rainfall is non-uniform.

Currently, about 32% of the annual utilizable groundwater potential of 432 km^3 is actually exploited, and only 8% of the groundwater sources have been exploited above 85% of their potential. However, in states like Punjab, Rajasthan and Tamil Nadu, large areas fall under the dark category. Table 12 shows ten states where the percentage of dark areas has increased considerably between 1984–85 and 1997–98. In coastal regions, e.g., in Tamil Nadu and Gujarat, regional declines in water table have resulted in saltwater encroachment in aquifer systems.

Groundwater sources have been classified in three categories depending upon the extent of exploitation. In the 1st category (termed 'white'), the level of exploitation is below 65% of the annual utilizable potential. The 2nd category (termed 'gray') includes areas and sources in the range of 65% to 85% exploitation levels and the third and the worst category (termed 'dark') has the level of exploitation exceeding 85%. About 673 assessment units (Blocks/Talukas/Watershed) out of 7,928 units come under the category of over-exploited blocks where the ground water table has gone down considerably. Further, 425 assessment units have fallen under the category of dark/critical units where utilization has reached a level beyond which further development will exceed the natural replenishment of ground water.

In Gujarat, the water table in over 90% of all the wells monitored by Central Groundwater Board (CGWB) has dropped by 0.5 meter to 9.5 meters. In Haryana, the average depth has fallen by 1 to 33 cm annually in different parts of the state during the 1980's (TERI, 1998). Figure 15 shows the districts in various states of India where the water table has fallen more than 4 m during 1981–2000.

The adverse effects of over-utilization of ground water are lowering of the ground water table, depleting of the resource and deterioration of the quality of water. In coastal areas, this has also led to sea-water intrusion. Interestingly, in many areas ground water resources have not been adequately tapped.

Table 12. Blocks with intensive exploitation of groundwater

State	Number of Dark Blocks		
	1984–1985	1992–1993	1997–1998
Andhra Pradesh	0	30	26
Bihar	14	1	11
Gujarat	6	26	28
Haryana	31	51	41
Karnataka	3	18	16
Madhya Pradesh	0	3	3
Punjab	64	70	83
Rajasthan	21	56	94
Tamil Nadu	61	97	103
Uttar Pradesh	53	31	40
Total	253	383	445

Source: Chaddha (2002).

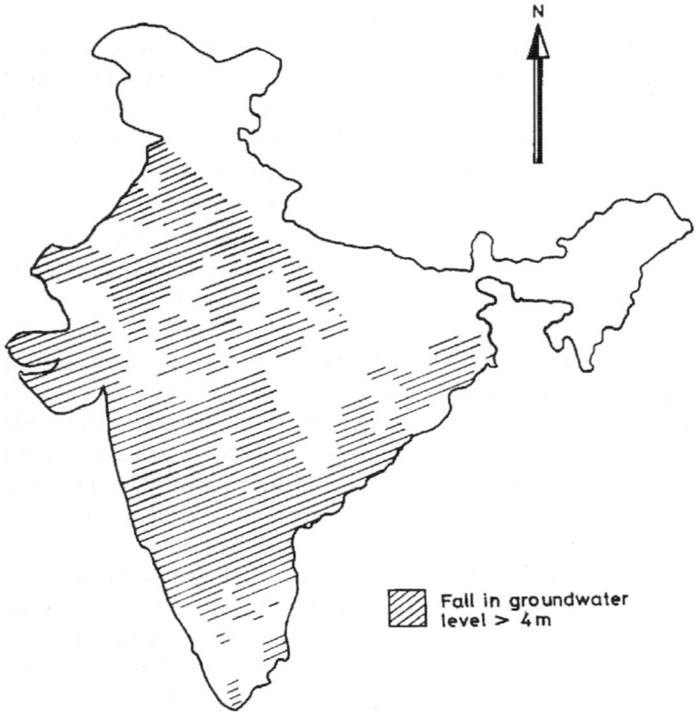

Figure 15. Over-exploitation of groundwater in India. The hatched portions show areas where, due to extraction of groundwater, especially for irrigation, the groundwater levels have in general fallen by more than 4 m (@ > 20 cm/yr) during 1981–2000

The water table in Haryana has been declining for the last two decades. The average water table in the state has declined by 1.90 m, whereas in fresh belt areas of the state its average fall is 4.38m. It varies from 0.94m in the Yamunanagar district to 12.09 m in the Mahendragarh district. However, on the other hand, the water table in saline areas has risen during the period ranging from 0.41m in Rohtak to 6.02m in the Sirsa district.

There is reason for concern in Rajasthan, as the number of blocks in the state categorized as "dark zones" has gone up from 33 in 1984 to 109 in 1995. Area-wise the dark zones now cover over 92,285 sq km compared to 18,000 sq. km in 1984. As such there is no perennial river system in the state other than the Chambal and the Mahi. The climate is arid and semi-arid with scanty and unpredictable rainfall contributing to the slow groundwater recharge.

According to the Rajasthan Ground Water Department, Jodhpur, the depletion of water level has been noticed in 67 per cent area of the 2.13 lakh sq. km land identified as groundwater worthy. Of this, in 39 per cent area, depletion of more than 3 metres has been recorded. Though a rise in water level was observed in mainly the command areas of the Indira Gandhi Nahar Project, despite normal monsoons a declining trend has been noticed in all other zones. The trends over the past 12 years indicate that as many as 13 districts in eastern and central parts of the state, such as Alwar, Bharatpur, Jhunjhunu, Jaipur, Dausa, Ajmer, Sawai Modhopur, Sikar, Nagaur, Jodhpur, Chittorgarh, Rajasamand and Jalore, come in the "danger category" in a classification of three. Several areas in these districts have problems of groundwater salinity and high concentration of hazardous constituents like fluoride and nitrate.

18.6. DRINKING WATER PROBLEMS

A recent report by the Washington-based Worldwatch Institute has concluded, "One of the first and the foremost basic changes that lengthen life-expectancy is the supply of clean water". It adds that this perhaps is the reason why the average Indian's life-expectancy today has increased by about 20 years during the last 45 years. For example, for the people of Kerala, the availability of safe drinking water has been responsible for a much higher average life-expectancy which is 73 years for men and 67.5 years for women.

The provision of safe drinking water is primarily the responsibility of the state governments and the Central Government supplements their efforts in providing drinking water systems like hand pumps, tube wells and regional piped water schemes. There is an acute shortage of drinking water in various parts of the country and different reasons are cited. In many places, the shortage of water is a temporary affair with water sources drying up during summer months. But almost all over the country, sources often go dry owing to a fall in the ground water table. Over the past few decades, the expansion of the water supply and sanitation infrastructure in Indian cities has not kept pace with growing population. Consequently, running water is available for only a few hours in the morning and evening each day. In some cities, people get up in early hours just to fill up their in-house storage tanks/vessels. During summers or droughts, running water may be available for only a few hours every week and that two at very low pressure. Municipal supply is only a small fraction of the demand in many major cities. Scenes like that shown in Figure 16 are quite common in the country. Another area of concern is the falling quality of drinking water in many areas. About 217,000 rural habitations did not have good quality drinking water till April 1999.

It may be noted that only 5% of the total water resources available in the country are required for drinking and domestic use. The total 'Not-Covered' (NC) habitations with regard to potable water supply stood at 5,759 as on Dec. 31, 1997. According to assessments, the states where NC habitations are fairly large are: Assam, Rajasthan, Punjab, Orissa, Bihar, Gujarat, Himachal Pradesh, Jammu and Kashmir, Karnataka, Kerala, and Meghalaya. The government norm for providing drinking water to villages is at least 40 litres per capita per day (LPCD) of safe drinking water for human beings, which is supposed to take care of all their needs like ablutions, drinking, washing and cooking.

Figure 16. People queuing for water for domestic use

Those habitations which get less than 40 LPCD are termed as 'Partially Covered' (PC) habitations. The country had 354,305 habitations, which were partially covered as of Dec. 31, 1997. Andhra Pradesh (29,071) Assam (23,698), Bihar (17,112), Himachal Pradesh (41,133) Karnataka (15,973), Madhya Pradesh (21,057) UP (53,725) West Bengal (25,039), Tamil Nadu (27,543), Rajasthan (41,672), Orissa (16,749), Maharashtra (37,436) and Meghalaya (1,692) are some states with large PC habitations. As on Dec. 31, 1997, the country had 1,430,663 habitations of which 49,374 were not covered, 354,305 were partially covered, and 1,026,984 were fully covered.

At the beginning of 10th plan (April 2002), out of 1,422,664 habitations in the country, 15,798 habitations wee not covered by any drinking water schemes while 133,305 habitations were partially covered. Thus, about 1,273,561 (89.5%) habitations were fully covered by the drinking water supply schemes in the country. Only in 10 states and Union Territories all habitations had been covered partially or fully under the drinking water supply schemes. The highest number of 2,974 habitations which had remained uncovered in the drinking water supply schemes by March 2004 was in Rajasthan, followed by Maharashtra, and Jammu and Kashmir. By March 2004, 75,607 habitations remained to be covered including 5,759 in the non-covered category (Planning Commission 2006).

The availability of safe drinking water in some cities (Pangare et al. 2006) of the country (in litres per capita per day) is: Ahmedabad 116, Allhabad 111, Bangalore 100, Bhopal 130, Chennai 65, Delhi 155, Hyderabad 90, Jaipur 97, Kanpur 118, Kolkata 100, Lucknow 264, Nagpur 130, Patna 107, Pune 220, and Vijaywada 137.

A centrally sponsored scheme, known as the Accelerated Rural Water Supply Programme (ARWSP) was initiated in 1972–73 to supply safe drinking water to all the rural areas. Priority is given to areas that are either 'not covered' or 'partially covered'. Funding is provided by central as well as the concerned state government. Under this programme, 3.5 million hand pumps and over 0ne lakh water supply schemes have been installed (Pangare et al. 2006). About 85% habitations are fully covered and 14% are partially covered. Swajaldhara is another programmer which is implemented through Panchayats and in which the users also bear about 10% of the expenses. Nearly 2,500 project proposals from 10 states and union territories have been approved under this programme.

An innovative project, the Swajal project, was launched in 1996 to empower rural communities to plan, construct, and manage their own water supply and sanitation systems. This six year project was implemented in the Bundelkhand area of UP, covering 1,200 villages and a population of 12 lakh.

Two important schemes were launched in the past for urban area. The Accelerated Urban Water Supply Programme (AUWSP) was aimed to provide safe water supply to urban area with population below 20,000 (1991 census). Nearly 240 schemes have been completed under this project. A mega city scheme catering to Bangalore, Chennai, Hyderabad, Kolkata, and Mumbai was initiated in 1994 for development of water supply and sanitation facilities.

The Rakesh Mohan Committee has estimated that as much as Rs. 27,000 crore to Rs. 28,000 crore is needed each year to provide safe drinking and sanitation in all urban areas of the country while the plan allocation is only to the tune of Rs. 5,000 crore every year. What is worse, around 35 to 45% of the water supplied is never paid for – either it leaks through the pipes or gets stolen. There are several models across the world that the Indian states can examine. Four different models are frequently discussed: Argentina, a developing country, has successfully achieved privatization through a concessional agreement of water supply; the corporatized system of Sydney in Australia has been a success; the lease agreement of Guinea (only partially successful), and the British model. Three states – Gujarat, Karnataka and Tamil Nadu – have made small beginnings on private participation.

18.6.1. Water Supply for Metropolitan Cities

Almost all the metropolitan areas in the country are facing shortage of water. As against the demand of 4,000 million litres/day in Mumbai in the year 2001, the supply was only 1,030 million litres/day. The situation in some other cities were (demand/supply in million litres/day): Bangalore 840/135, Bhopal 335/70, Chennai 1,894/105, Delhi 3,830/880, Hyderabad 956/186, and Kolkata 2,258/690.

Bangalore gets water from Cauvery and Arkavathi Rivers with the Cauvery contributing 432 million litre/ day and Arkavathi's share being 130 million litres, while this caters to 4.5 million people of the city with an average per capita of 100 litres, the projected per capita consumption is 140 to 200 litres.

In the case of the National Capital Territory (NCT) Delhi, water supply is not only deficient but also suffers from the problems of uneven distribution, unreliability, large scale undetected leakages, poor and inefficient maintenance of system, non-sustainable pricing of water and ignorance of the people about value of water. Long term solutions lie not only in construction of new infrastructure but also in demand management and conservation practices, in development of appropriate strategies and regulatory framework, and in transforming of existing institutions to become increasingly more efficient. Surface water contributes major share in the Delhi's total water supply. The allocation of water to Delhi is governed by interstate agreements. As per different interstate agreements, the allocation of water for municipal purpose from various sources is shown in Table 13.

From Table 13, it is clear that out of availability of 1,149.20 MCM, about 295.2 MCM of surface water still remains unutilized. Delhi gets a major part of its water supply from the Yamuna. The water demand in the capital city has been estimated at 800 MGD (million gallons per day). Against this, the supply is only 650 MGD, resulting in a shortage of 150 MGD. Some houses also have borewells to supplement the Delhi Jal Board (DJB) supply. Further, as many as 40% of the Capital's population are without adequate piped water supply. The demand for raw water during the next 20 years is expected to increase by 75%.

In some parts of Delhi, like West Delhi, ground water is the main source of water supply. The underground water reserves of Delhi are 1,900 MCM. As there is over-withdrawal in some places, it might lead to alterations in the city's immediate sub-strata. Places like Mexico City have experienced subsidence and collapse of buildings because of unplanned and excessive withdrawal of ground water. According to reports, there has been steep decline in the level of groundwater in certain pockets of Delhi during last 15 to 20 years. In 1977, the water table was almost within 6 m in most parts of Delhi, the deepest level being 23 m in Mehrauli. The level declined to 10 m by 1982 and was in the range of 10–20 m in1995 with the deepest level being 35 m at Gadiapur, Chapparpur basin.

Based on a norm of 70 gallon per capita per day (GPCD) as per Delhi Master Plan – 2002 Guidelines, the water requirement for NCT Delhi for 2001–02 stands at 965 MGD. As per CPHEEO manual, the per capita per day water requirement shall be 60 GPCD. Based on norm of 60 lpcd, the water requirement of Delhi in March 2001 comes to 827 MGD against the supply capacity of 650 MGD of DJB. The DDA's working group on physical infrastructure has estimated Delhi's water requirement as 1,500 MGD by the year 2021 based on 50 GPCD norms for projected population of 30 million.

Delhi's total water demand stands at 750 MGD, while the supply is only 600 MGD. Added to this the 17 percent loss due to leakages and faulty pipes, the deficit adds up to a substantial 270 MGD. Delhi's water demand is met through Bhakra dam (680 million litres), Yamuna River (1,000 million litres), Ramganga dam (450 million litres) and some from groundwater. However, one-third of Delhi's population does not get drinking water from the Delhi Jal Board. Moreover, 30

Table 13. Annual availability and withdrawal of water for drinking purpose in Delhi

River System	Annual Availability (in MCM)	Annual Withdrawal (in MCM)
Yamuna	724.00	442
Ganga	178.80	166
Bhakra system	246.40	246
Total	1,149.20	854

percent of the population living in the 1,071 unauthorized colonies, 600 slum clusters and a large number of villages depend on hand pumps and tubewells and the rest on rivers and canals which are not fit for consumption.

DJB has already made some investment for the proposed Renuka Dam in Himachal Pradesh, which will provide the capital city with 275 MGD of water. DJB also plans to make investments in the Lakhwar Vyasi dam in Uttaranchal to get nearly 100 MGD of water. Delhi will get an assured supply of 270 cusec from the Tehri dam for which a water treatment plant has been constructed at Sonia Vihar. Water will be brought here from the Upper Ganga Canal through close conduits.

18.7. WATERLOGGING

When the water table rises to such an extent that soil pores in the root zone of a crop become saturated, resulting in restriction of the normal circulation of air, decline in the level of oxygen and increase in the level of carbon dioxide, the area is said to be waterlogged. Waterlogging problem is predominant in those areas which have saline and brackish groundwater and canal irrigation is widely practiced. In India, large areas have become waterlogged because of overuse of water for irrigation. Such areas witness a rise in water table, leading to salinization and poor crop productivity. The abundant supply of cheap irrigation water tempts the farmer to use more water than the requirement of the crop. Many farmers harbour the notion that the crop yield increases as more and more irrigation water is supplied. But in fact, the excess water supplied to the field percolates into the soil. Due to percolation of large quantities of water, the groundwater table may rise so much that it might completely saturate the root zone of the crop. Further, in due course of time such land turns saline or alkaline and ultimately becomes unfit for cultivation.

Waterlogging is very common in the arid tracts of the Indo-Gangetic plains of north India, the arid tracts of Rajasthan and Gujarat, the semi-arid Deccan plateau and coastal tracts. It has adversely affected the productivity in these areas. Overuse of the available water mostly in head reaches of canals results in less availability to tail-enders.

18.7.1. Norms for Categorization of Waterlogged Areas

The depth of the water table at which it tends to make the land waterlogged and start harming the crops depends on the height of the capillary fringe and the type of crop. The height of the capillary fringe is the height to which water will rise above the water table due to capillary action. It is more in fined grained soils and less in coarse-grained soils. For most of the agricultural soils, the height of the capillary fringe varies from 0.9 to 1.5 m. Consider a crop whose yield is adversely affected when soil up to 0.6 m below the surface gets saturated. This crop will start deteriorating due to waterlogging when the water table is within 1.5 m $(0.6 + 0.9)$ to 2.1 m $(0.6 + 1.5)$ below the ground surface.

Table 14. Norms for Categorization of Waterlogged Areas in Some States

State/ Agency	Category	Water table depth below ground level (m)
Haryana	i) Critically Waterlogged	< 1.5
	ii) Moderately Waterlogged	1.5–3.0
Karnataka	Waterlogged	< 2.0
Maharashtra	i) Fully Waterlogged	Water at surface
	ii) Waterlogged	< 1.2
Uttar Pradesh	i) Worst Zone	< 1.0
	ii) Bad Zone	1.0–2.0
	iii) Alarming Zone	2.0–3.0
	iv) Safe Zone	> 3.0
As per MOWR Working Group 1991	i) Waterlogged area	< 2.0
	ii) Potential waterlogged area	2.0–3.0
	iii) Safe area	> 3.0

Source: WG (1999).

Different states have adopted different norms to categorize waterlogged areas depending upon prevailing soil types, crops grown, etc. Table 14 shows the norms used by Uttar Pradesh, Haryana, Maharashtra and Karnataka.

The Government of India and the states generally follow the norms adopted by MOWR-WG as guidelines. However, as the agro-climatic and subsoil conditions in different river basins are quite diverse, different states continue to adopt their own norms.

18.7.2. Causes of Waterlogging

Being a continent size country, India has widely varying agro-climatic conditions (see Chapter 1). Consequently, the causes, extent and occurrence of waterlogging are also different. Rainfall induced waterlogging is found in inadequately drained land throughout the country during the monsoon period. However, this is purely a temporary phenomenon which gradually disappears as the monsoon withdraws.

With the objective of extensive and speedy development of irrigation in India and to project an attractive benefit-cost ratio of the projects, provision of drainage is rarely made during planning and execution of irrigation projects. Irrigation induced waterlogging and salinity is mainly caused by application of irrigation water in excess of crop needs. It is found in command areas of many irrigation schemes throughout the country mostly in the form of waterlogging in the fields or as waterlogging and soil salinity caused by it.

Following are the basic reasons behind waterlogging:

(i) Seepage from Canals: In a system of unlined canals, water percolates from the channels and joins ground water. This causes a gradual rise of water table in the region. For example, with the advent of the Ganga Canal, the water table in the Ganga-Yamuna Doab rose from a depth of 12.2 m to about 4–6

meters below ground level in 100 years. Similarly, in the area commanded by the Western Yamuna Canal, the water table rose at an annual rate of 16 cm. Further, the rate of water table rise is faster in the Chambal canal system.

(ii) *Over-Irrigation of Field:* When large quantities of irrigation water are applied to the fields, the excess water percolates deep into the ground, which results in augmenting the groundwater storage and rising up of the water table.

(iii) *Nature of Soil:* Waterlogging depends upon the nature of the soil. A soil having low permeability, such as a black soil, is prone to waterlogging due to over-irrigation or flooding because of slow rate of movement of water through it.

Inadequate drainage is the reason behind the formation of sodic salinized soils crust in command areas in UP. Eventually this resulted in a highly salinised hard pan and the land became unsuitable for cultivation. Gradual rise in water table and related problems of water logging and soil salinity/alkalinity have surfaced mainly because of the lack of drainage, improper waste management, inadequate maintenance, etc. Improper planning of development works such as roads and railways obstructs the natural flow of water and can cause waterlogging.

18.7.3. Extent of Waterlogging in India

The first signs of irrigation induced waterlogging and salinisation were reported in the 1920s and the problem began to become widespread in certain districts of Punjab and Haryana from 1950s. Yields were affected and some areas could not be cultivated.

Several different estimates of waterlogged area in India are available. According to the Working Group of the Ministry of Water Resources (1996), 2.46 M ha area was water logged (see Table 15) in India at that time. CGWB has assessed the area for which water table is within 2 m of the land surface at 3.1 M ha for the year 1998. Another assessment of waterlogged area in canal command was reported by CSSRI (1997) according to which, the statewise figures of affected area vary from those in Table 15 but the total affected area is reported to be 2.19 M ha. Chaddha (2006) reports the water logged area (water table 0 to 2 m) to be 1.85 M-ha in May 1999 and at the same time, area prone to water logging at 10 M-ha. Waterlogging is surfacing in both old (e.g. Chambal, Tungabhadra) and new (e.g. IGNP) projects.

According to Sandra Postel (Bio Science, Aug. 1998), about 20% of the command areas in India are likely to be affected by salinisation and that crop yields have reduced in at least 10% of the commands. But this is likely to be over-estimate.

Introduction of canal irrigation through the Bhakra System since 1953 in the arid to semi-arid region of the western part of Haryana brought about a substantial change in the groundwater regime both in respect of quality and quantity. More than 60% of the area in the state has been classified as marginal (30%) saline land. Rapid and constant rise in the water levels have been registered due to applications of canal irrigation. About 0.286 M-ha area in Punjab has a water table depth of less than 1.5 m even in the dry and hot month of June. The water table further rises by

Table 15. Extent of Waterlogged Area in India

State	Affected area (1,000 ha)
Andhra Pradesh	266.40
Assam	Not available
Bihar	616.70
Gujarat	172.59
Haryana	249.00
Jammu & Kashmir	1.50
Himachal Pradesh	0.20
Karnataka	24.54
Kerala	11.61
Madhya Pradesh	73.12
Maharashtra	15.35
Orissa	196.26
Punjab	200.00
Rajasthan	179.50
Tamil Nadu	16.19
Uttar Pradesh	430.00
West Bengal	Not available
Union Territories	Not available
Total	2,455.96 thousand ha or 2.46 M ha

Source: MOWR Working Group (1996).

0.5 to 1.2 m during monsoon. These areas are normally subjected to waterlogging; the degree depends upon the local topography of the area.

The Gangetic plain is the cradle of agro-based civilization in India. But currently, almost 40% of the Gangetic plain suffers from acute salinity due to water logging, the use of chemical fertilizer and inadequate drainage. Analysis of data has shown that out of a total of 160 million ha of porous land in the country, about 100 million ha is slowly becoming unproductive. India's agricultural productivity is quite low at 1.6 tons/ha against a global average of 2.6 tons/ha and over 5 tons/ha in developed countries. About 1.2 M ha in UP is facing problems because of high concentration of exchangeable sodium. Such land is said to be having alkaline conditions (World Bank, 1998). In Tamil Nadu saline and alkaline soils are spread over in 0.41 M ha and acidic soils occur in 0.12 M ha (World Bank, 1995). In Orissa, water logging is a major problem of the irrigated delta of the coastal zones: about 85,000 Ha of the Mahanadi delta is affected (World Bank, 1995a).

The rise in ground water level has become a cause of worry in Rajasthan since 1997 after the commissioning of the Indira Gandhi Canal water to Jodhpur through the Kailana and Takht Sagar lakes. The Kailana Lake is located about 12 km from the city and has waters 12–16.7 m deep compared to the earlier times when it used to be 3–12 m in depth.

In the Brahmaputra valley, shallow aquifers have witnessed a rise in the water table during the last decade; the rise has been around 0.08 to 3.27 m. If the rate of rise is not controlled, large parts of the basin may become unsuitable for cultivation.

In the valley on the northern bank, the rise is 0.13 to 2.0 m and on the southern bank, the rising trend is 0.08 to 3.27 m. These waterlogged areas in Assam covers an area of 23,500 ha.

Studies have indicated that there has been a considerable rise in the water table in a number of irrigation command areas all over the country due to improper water management. According to Ministry of Water Resources, the Krishna River delta in AP has experienced a 2 to 4.4 m rise in the water table. In Karnataka, the Chitradurga area has experienced 2 to 6.8 m increases. In MP, the rise has been 2 to 9.3 m in certain areas. In Faridkot, Punjab, the rise has been 2 to 11.2 m (Vaidyanathan, 1999).

18.7.4. Control of Waterlogging

Over the years, the problem of land degradation due to waterlogging has begun to surface in varying magnitude and intensity in various parts of the country. In view of the growing menace, there is now considerable awareness and many programs have been launched to arrest degradation of lands. Several irrigation project authorities are undertaking ameliorative measures in their commands.

Waterlogging problem can be tackled by providing adequate drainage. By means of drainage, excess water which is harmful for crop growth is removed. The drainage can be on the surface or below the surface. In large parts of Punjab, Haryana and Rajasthan, where poor natural drainage is coupled with poor groundwater quality, artificial drainage of agricultural lands is a must. States like Rajasthan and Haryana, have taken up drainage works in about 19,000 hectares of land and sub-surface pipe drainage system is being provided. The Indo-Canadian Rajasthan Agricultural Drainage Research Project (RAJAD) is the largest project in progress in the country where the feasibility of subsurface drainage in reclaiming waterlogged saline soils of the Chambal Command is being researched in all its aspects and aimed is at providing subsurface drainage in about 20,000 ha. The successful efforts made in recent years in problem areas of Indo-Gangetic alluvial soils of Haryana and in black soil regions in Gujarat, Maharashtra and Rajasthan have shown the usefulness and feasibility of sub surface drainage system in India.

Conjunctive use of ground and surface water is an effective way of tackling the problem of waterlogging. Besides, overcoming the waterlogging problem, it has many other advantages such as saving in cost and the optimum use of available water. Integrated development and management of surface and ground water is widely recognized as the most suitable strategy for irrigation development in alluvial plains.

Bio-drainage which involves planting trees which help in removal of excess water by evapotranspiration has also been attempted in a few cases. Bio-drainage is being tried in IGNP command areas in Rajasthan. A study carried out in the command of IGNP has indicated that eucalyptus and other plantations are affective in lowering ground water table. Research on agricultural drainage is being carried out by CSSRI, Karnal, by ICAR, Agricultural universities, WALMIs of state irrigation departments, and academic institutions. A three- pronged strategy appears to be workable for the country. Prevention of the problem in newly irrigated areas, remedial measures for areas where the problem has manifested in recent times, and reclamation of degraded older areas. Due importance is now being given for reclamation of waterlogged areas and a target of 60,000 ha was kept for reclamation during the Ninth 5-Year Plan.

18.8. SOIL SALINITY

Along with waterlogging, soil salinity is also associated with irrigation in many parts of the world, India being no exception. In arid and semi-arid regions, soil salinity is caused due to many reasons, the main being salt deposits during the formation of the soil, saline shallow water table, and dissolved salts in water used for growing crops. As a result of evaporation of soil moisture, salts are left behind and their concentration in the root zone keeps on increasing. At some stage, the soil has such a high concentration of salts that it interferes with the physiological activities of most of the crops. A saline soil mostly has fluoride and sulphate of sodium, calcium, and magnesium, while alkaline soils contain bicarbonates, carbonates and silicates of sodium.

18.8.1. Classification of Salt-Affected Soils

The nature and intensity of salinity varies, depending upon the causes that are responsible for the formation of saline soils. Soil salinity in coastal belt is caused by sea-water intrusion and inundation, whereas the inland salinity is mostly encountered in areas with high ground water tables. Expectedly, these areas mostly fall in canal commands. The criteria laid down by the US Salinity Laboratory Riverside (California), was adopted by the MOWR-WG to classify salt-affected soils as indicated in Table 16.

In Table 17, ECe = electrical conductivity of the saturated soil paste extract taken from the root zone of plants (usually to a depth of 1.2 m to 1.5 m) as averaged over time and depth in deci-Siemen per meter (dS/m), and ESP = exchangeable sodium percentage expressed in percentage.

In India, saline soils are mostly found in Uttar Pradesh, Gujarat, West Bengal, Rajasthan, Punjab, Haryana, Maharashtra, Kerala and Andhra Pradesh. The states of Haryana, Punjab and Uttar Pradesh have the problem of alkali soils also. Saline

Table 16. Criteria adopted to classify salt-affected soils

Class	ECe in dS/m	ESP (%)	pH
Saline soils	> 4	< 15	< 8.5
Saline alkali soils	> 4	> 15	Variable
Alkali	< 4	> 15	> 8.5

Source: WG (1999).

Table 17. State-Wise Salt Affected Area in India (Thousand ha)

S.N.	State	Total (saline/alkaline)
1.	Andhra Pradesh	28
2.	Bihar	224
3.	Gujarat	911
4.	Haryana	197
5.	Karnataka	51
6.	Madhya Pradesh	36
7.	Maharashtra	5
8.	Punjab	490
9.	Rajasthan	70
10.	Tamil Nadu	140
11.	Uttar Pradesh	1,150
Total		3,302

Source: MOWR Working Group (1996).

sandy areas are mostly found in marine states. About 3.3 million ha area is reported to be salt affected. The statewise distribution of salt-affected area is given in Table 17.

Salt-affected areas in irrigation commands of major and medium schemes of 11 states was assessed by MOWR-WG (1991). It was reported that more than 97% salt-affected area was located in seven states, viz., Uttar Pradesh (35%), Gujarat (28%), Punjab (15%), Bihar (7%), Haryana (6%), Tamil Nadu (4%) and Rajasthan (2%). The remaining 3% salt-affected area lies in Karnataka, Madhya Pradesh, Andhra Pradesh and Maharashtra.

As detailed in Table 18, the Working Group of the Ministry of Water Resources (1991) reported 3.06 M ha as saline and 0.24 M ha as alkaline areas in the commands of major and medium irrigation schemes in the country.

18.8.2. Measures to Prevent Soil-Salininty

The following measures are generally adopted to control water logging and soil salinity:

a) Undertaking construction of appropriate surface drainage networks along with irrigation conveyance networks in new schemes.
b) Lining of distribution networks at all vulnerable reaches to reduce seepage losses.
c) Carrying out repairs of leaking and deficient structures causing unintended seepage to minimize waterlogging.
d) Undertaking site-specific on-farm water management measures, such as irrigation scheduling for water distribution and application, adoption of improved surface irrigation lay outs, and use of sprinkler and drip system etc.
e) Training of irrigation personnel in scientific irrigation water management and training of farmers to adopt improved irrigation practices.

Table 18. The number of resettlers from some major projects

Project	Country	Number of resettlers
Three Gorges	China	1,250,000
Sanmenxia	China	319,000
Upper Krishna II	India	220,000
MCIP III Irrigation	India	168,000
Andhra Pradesh Irrigation II	India	150,000
Gujarat Med. Irrigation II	India	140,000
Sardar Sarovar	India	127,000
Tehri	India	105,000
Subarnarekha Group	India	100,000
Srisalam	India	100,000
Mahaweli I-IV	Sri Lanka	60,000
Kariba	Zambia and Zimbabwe	67,000

18.9. PROBLEMS IN CONSTRUCTION OF NEW PROJECTS

In recent times, new issues have emerged that have to be dealt with before taking up a new WRD project. Due to these, there has been considerable delay in some projects. On its part, the Government of India has recently simplified the procedure for clearance of the major/medium and multipurpose WRD projects. The new procedure entails a two-stage clearance. In the first stage, the soundness of the proposal is established and in principle approval for preparation of DPRs is given. The second stage requires necessary forest and environment clearance as also clearance from the ministry of Social Justice and Empowerment/Tribal Affairs (if it involves displacement of SC/STs).

In this section, we discuss the problems that are being faced in taking up new WRD projects in India.

18.9.1. Rehabilitation and Resettlement

The issue of rehabilitation and resettlement (R & R) is of critical importance and it affects all major reservoir projects. Since the population affected generally belongs to the poorest strata of society, including tribal population, the R&R should be taken as a challenge to improve the living conditions of these people by providing them with better housing, better employment, better civic amenities and opportunities for education, health, etc. irrespective of the cost involved. The various schemes of the government can be combined to evolve a satisfactory R & R package in consultation with the affected people.

In the densely populated foothill areas of India, any construction of large dams is sure to inundate human settlements and fertile agricultural lands. In the case of the proposed Tehri Dam, direct involuntary displacement will occur for about 80,000 people. Indirectly, a large number of villages would lose parts of their land or access road to the nearest towns in the plains. The national R&R policy in India stipulates that the "living standards of those displaced should be maintained at least at the same level, if not improved, to what they were prior to their involuntary displacement". The richer and more influential residents of the town of Tehri as well as the government offices have been rehabilitated in the newly constructed New Tehri township. There are many problems, however, with the displaced people of rural origin. Satisfactory rehabilitation is a term which is difficult to evaluate. With the present level of marginalization of the upland people, whose habitat is threatened by submersion, adequate and satisfactory compensation and rehabilitation may remain difficult for some time to come. There is a need to redefine the ethical and procedural dimensions of the process, on the basis of which new legislation could be enacted for the purpose. There is no doubt that in an era characterized by growth of a free market and privatization, compensation and rehabilitation, official policy for land acquisition should be based on market values. Such a step, however, is sure to push up the cost element of the projects. Its acceptance would also provide the upstream people in the hills and mountains with good bargaining power for agreeing to allow water projects for the benefit of downstream economies. The disregard for ensuring proper rehabilitation and lack of human concern towards displacement may not be possible under that the changed situation.

In view of the density of the present population in the lower parts of the Ganga basin and scarcity of forest lands that could be used for rehabilitation, this basin presents a difficult situation in terms of compensation to the involuntarily displaced people. This is going to be increasingly important with growing awareness of the upland people of their continued marginalization. The issue of displacement and compensation should be sorted out through satisfactory and negotiated agreements. If the mountain waters are economically essential for the residents of plains, the people in headwaters should be adequately paid for by the consumers. Any effort at making the upstream people sacrifice their economic interests will only enhance the delay in project execution and lead to far more economic losses. The delay in the case of the Tehri and Sardar Sarovar dams have already shown how costly this can be. Lessons from these and many other projects show that if the water potential of the Indian rivers has to be economically utilized, the development of effective economic instruments to adequately compensate and rehabilitate the displaced people should be an essential component of the project.

Displacement of population is one of the main criticisms against major dams. Any development activity involves some displacement. A study carried out in respect of a number of major dams indicates that the population displaced on account of construction of dams is about 2–4% of the population benefited by irrigation facilities which again is a very small percentage. It is a fact that in the past adequate care was not taken of the displaced population. However, this defect is now being rectified and it is being ensured within the project cost that the displaced population is rehabilitated in such a way that they get a better standard of living than in the existing/ pre-dam conditions. It will be desirable to have a National Rehabilitation Policy to obviate this recurring controversy.

Sometimes, it is also argued that tribal people should not be displaced at all, as they cannot live in a different environment. This extreme approach is also faulty and will result in perpetuation of their backwardness. In fact, the tribal population has not yet joined the mainstream of economic development, as no efforts have been made to improve their lot. Experience shows that the tribal people definitely want to enhance their economic prosperity.

R&R of the population displaced or affected by a WRD project has been a hotly debated topic of recent times. A major reason for opposition to WRD projects in recent years is the displacement of population due to these projects. It is now a concern because of the realization that displacement, by its very nature, results in the breakdown of family and community networks, and causes social and economic distress.

The problem of resettlement attains gigantic proportions in densely populated or resource starved nations. In recent times, a number of dams have been constructed or are being constructed in developing countries. Table 18 gives the number of people that were or will be subject to displacement due to large water resources projects. The Three Gorges Dam in China is likely to result in displacement of over a million people. The Srisailam dam in South India has displaced about 100,000 people and the Sardar Sarovar project on Narmada will affect even more number of people. In many cases, the actual displacement is much more than estimated. The feasibility report of the Kiambere dam in Kenya estimated that some 1,000 people would be displaced but during construction, this number turned out to be 6,000.

The requirement for funds for resettlement also increases day by day. In countries where population densities are higher (e.g., India), this amount will be considerably more. The finances required for resettlement of people affected by the Sardar Sarovar Project in India are quite high.

R&R of tribals or hill-area people whose life styles and culture are radically different from the people of plains needs careful attention. This problem is frequently faced in India. The tribals are used to living in hills and forests form an integral part of their life. They practice their own way of agriculture. Their relocation even in command areas of nearby canals can inflict a cultural shock on them and could be a cause of avoidable social conflict. It would be desirable to relocate them within the forests on the fringes of the reservoir. Side-by-side, measures for their health and economic uplift should also be initiated so that they are brought in the national mainstream while retaining their own identity.

Importantly, R&R is not an administrative problem; it is a human problem. Since R&R deals with people, pragmatic policy decisions with involvement of the affected people at appropriate levels and stages is necessary. Many people, particularly the old, who may have been born and brought up on that land, consider it as their ancestral property and have an emotional and nostalgic attachment. Such cases should be dealt with care, respect, and patience that they deserve. The use of force should always be avoided and people should be motivated to move through giving them a higher standard of living and sustainable means to lead a dignified life. It is also necessary to ensure that the displaced people are made partners in prosperity due to the project and do not turn into adversaries.

18.9.2. Submergence of Forest Areas

Every major dam project results in submergence of some forest area. Major dams were responsible for about 2% of the forest land losses during the period 1951 to 1985 in India. While considerable protest is being raised about this loss, it is intriguing that not much is being said or done for the 98% loss of forest area from other causes. In addition, it must be realized that construction of storage means submergence of some areas. Obviously, in view of the limited cultivable area, the country can ill afford the loss of that area. Since most of the dams are located in the upper reaches of a river basin, the submergence has to involve waste land or forest land.

The submergence of forest land has also to be viewed in the context of the benefits of irrigation over a larger area. Analysis of data from a number of major projects indicates that the total area submerged by construction of dams varies from 3 to 10% of the area to be irrigated by these projects. The submergence of forest area accounts for only about 1–2% of the area irrigated. It is thus obvious that the loss of forest land is an insignificant percentage of the economic benefits to a large area of land.

As per the current guidelines, the forest land submerged by major dams is to be replaced by compensatory afforestation at project cost. Therefore, at present and in future, there will not be any net loss of forests as a result of major dams.

18.9.3. Environmental Issues

Impact assessment is a pointer to the environmental compatibility of the projects in terms of their location, suitability of technology, efficiency in resource utilization, recycling and so on. EIA was introduced in India in 1978 and now covers water resources development projects also. These days, Environmental Impact Assessment is statutory for 29 categories of developmental projects including irrigation and hydropower schemes. Typical adverse environmental impacts of water resources development projects include submergence of forested areas, water logging, soil salinity, landslides, threat to biodiversity, flora and fauna. Relevant aspects to this topic are discussed in Chapter 20.

18.9.4. Financial Aspects

Presently, the Government of India is the major source of funding in the irrigation sector, particularly for major and medium projects. Public investment in irrigation and flood control sector as a percent of overall plan outlays has gone down very much from 22.5% in the 1st Plan to about 6.5% in X Plan.

Despite reduced allocation for the irrigation sector, the state governments have taken up a number of new projects. This has resulted in thin spreading of resources leading to time and cost overruns in most of the projects. As a result, benefits are not commensurate with the investment. To promote early completion of projects and derive benefits, the Government of India has taken up Accelerated Irrigation Benefits Program (AIBP) in 1996–97 under which Central Loan Assistance is provided to state governments.

Private investments in major/medium river valley projects have not been forthcoming. This can be attributed to heavy investment, long gestation period, political risk, and involvement of multi stakeholders. Recently, some companies have begun investing money in the hydropower sector.

So far as minor schemes are concerned, adequate funds are available due to substantial contribution from Rural Infrastructure Development Fund created by NABARD. Such agencies have also provided funds for some medium schemes. Very large parts of the investment on dug wells, bore wells, tube wells and lift irrigation schemes as well as on pumps has been made by the private sector. Water bonds are also one of the viable options for raising funds for water resources projects. The Krishna Bhagya Jal Nigam (Karnataka), Maharashtra Krishna Valley Development Corporation, etc., have adopted this approach to generate resources.

External Assistance in the water resources is currently being provided by the World Bank, the Asian Development Bank, the European Union (EU), etc. Some schemes are also being funded by the Japanese Bank for International Cooperation (JBIC), Kreditanstalt Fur Wiederausban (Kfw), Germany and France on bilateral basis.

World Bank has provided support for water resource projects since 1970. Nearly 77% of the World Bank assistance has been for the implementation of irrigation projects, while 11% is towards consolidation of existing water resources infrastructure. The funding for multipurpose and other projects is about 9% and 3%, respectively. In recent years, the World Bank funding is generally focused on consolidation projects.

18.9.5. Problems Related with Command Area Development

Feedback from many major and medium irrigation projects reveal that utilization of irrigation potential created was inadequate and the efficiency of the system was poor. In 1974–75, the command area development program (CADP) was initiated to ensure equitable and timely supply of irrigation water to hold agricultural holdings within the command area of a project. Currently, CADP covers more than 200 projects in 22 states and 2 union territories, covering a culturable command area of about 21 billion hectares. As large areas of the country suffer from soil sediment and water logging, the focus of CADP is now being re-oriented for the improvement of land and development of drainage facilities. It will be important to formulate water management plans using data of soil survey and land use capability.

18.10. CLIMATE CHANGE AND ITS IMPACT ON WATER RESOURCES

Although climate change has concerned scientists for many years, it became a major international issue only in the late 1980s. The report of the World Commission on Environment and Development (1987) of the United Nations focused international attention on the threat climate change to the Earth's environment. This led the United Nations Environment Program (UNEP) and the World Meteorological Organization (WMO) to establish the Intergovernmental Panel on Climate Change (IPCC) to bring the world's leading scientists together to assess climate change and devise remedial steps.

India had signed the United Nations Framework Convention on Climate Change (UNFCCC) in 1993 and has initiated efforts to comply with the relevant obligations. To fulfill the commitments and requirements of the UNFCCC, many research programs have been initiated on (1) understanding climate change, (2) impacts of climate change on social and economic development, and (3) response strategies.

India largely depends on fossil fuels to meet its energy requirements. Coal accounts for about 60% of fossil fuel use in calorific terms followed by petroleum products (30%) and the remaining by natural gas. The relative emission of carbon dioxide (CO_2) for 1989–90 from coal, petroleum products and natural gas were estimated at 328.4 Tg/yr (65%), 162.7 (32%) and 17.5 (3%), respectively *(ALGAS, India, 1998*). Electricity generation in India accounts for the largest share of coal consumption as out of the total installed electricity generation capacity 70% is coal-based. Further, nearly 72% of the Indian population lives in rural areas and is dependent on agricultural and related activities. They use biomass resources, like wood, agricultural crop residues, dung, etc., for energy. In India, the amount of biomass burnt annually is estimated to be about 426 Tg or about 6.3% of the global level of 6,800 Tg/yr. India also has a large cattle population that contributes around 40% to total methane emissions from the country *(ALGAS, India, 1998)*. Industrial growth and increasing urbanization in India have led to associated environmental changes. A variety of industrial processes are responsible for emissions of various greenhouse gases, like CO_2, CH_4 and N_2O.

Preliminary studies on the impact of a rise in sea level indicate that 1 mm rise of sea level will inundate 0.41% of India's coastal area and will put 7.1 million persons at risk. Studies also suggest that as a result of global warming, cyclones will become more frequent and destructive, making islands, such as Andamans and Nicobar, Lakshadweep, highly vulnerable. The danger of frequent storms which generally originate in the Bay of Bengal is, however, higher in the Andamans and Nicobar than Lakshadsweep. Also, if sea level rises, fresh water aquifers of these islands as well as the mainland will be subjected to saline intrusion.

The NATCOM (India's National Communication, *www.natcomindia.org*) process comprises comprehensive scientific and technical exercises for estimating GHG emissions from different sectors, reduce uncertainties in current estimations, develop sector- and technology-specific emission coefficients pertinent to India, and assess the adverse impacts of climate change and strategies for adapting to these impacts.

NATCOM will also provide the general description of steps taken or envisaged to implement the convention. NATCOM will lead to developing a reliable database and capacity that will help fulfill commitments under the Convention. The process is also expected to initiate efforts to identify areas of Targeted Research on climate change according to sustainable development plans of the country.

Himalayan glaciers feed seven of Asia's greatest rivers – Ganga, Indus, Brahmaputra, Salween, Mekong, Yangtze, and Huange He. According to experts, the rapid melting of Himalayan glaciers will first increase the volume of water in rivers, causing widespread flooding. But after a few decades, the situation will change and the water levels in rivers will decline, meaning massive economic and environmental problems for people in northern India, western China and Nepal. As glacier water flows dwindle, the energy potential of hydroelectric power will decrease, causing problems for industry as well as in irrigation. Nepal's snow-fed rivers have shown declining discharge and its annual average temperature rise of 0.06 degree Celsius has been noticed. Gangotri in India is receding at an average rate of 23 metres a year.

18.10.1. Retreat of Glaciers

The Himalayas have one of the largest concentrations of glaciers outside the Polar regions – there are nearly 1,500 glaciers and it is estimated that these cover an area of about 33,000 km^2. These glaciers provide water during dry season so that these rivers run throughout the year. A useful feature of snow and glacial-melt runoff is the fact that the glaciers release more water in a warm year and less water in a wet (flood) year thus evening out variations in the flows. World Glacier Monitoring Service publishes reports on fluctuations of glaciers. Recently IUGG (CCS) have published this report covering the period 1995–2000 but no Indian glacier has been listed in this report.

Fears have been expressed that a significant number of mountain glaciers are shrinking due to climatic variations. The rate of retreat of several important glaciers in the Himalaya is presented in Table 19. Almost 67% of the glaciers in the Himalayan mountain ranges have retreated in the past decade. The mean equilibrium line altitude at which snow accumulation is equal to snow ablation for glacier is estimated to be about 50–80 m higher relative to the altitude during the first half of the 19th Century.

In the recent years, some important changes have been noticed in the glaciers and the snow covered areas in the Himalayas. Many prominent glaciers, such as the Gangotri glacier are retreating. The rate of retreat of the snout of Gangotri glacier has demonstrated a sharp rise in the first half of the 20th century. This trend continued up to around the 1970s, and subsequently there has been a gradual decline in its rate of retreat. Some of the possible causes of this retreat are increased tourist activity leading to rise in the ambient temperature near the snout area and atmospheric warming. The diminishing rate of retreat of the snout of the Gangotri glacier could be a consequence of the diminishing rate of rise in the temperatures.

Table 19. Retreat of selected Himalayan glaciers

Name of Glacier	Name of Basin	Glacier Area (sq. km)	Year of Observation	Retreat (m)	
				Total	Rate/year
Bilare Bange	Satluj	2.8	1962–1997	90	2.6
Shaune Garang	Satluj	8.8	1962–1997	923	26.4
Janapa Garang	Satluj	12.3	1962–1997	696	19.9
Milan	Ganga		1966–1997	940	30.3
Dokariani Bamak	Ganga	5.8	1962–1997	585	16.7
Gangotri	Ganga	143	1977–1996	535	28.1
Chipa	Dhauliganga	5.0	1961–2000	1, 050	26.92
Meola	Dhauliganga	14.0	1961–2000	1, 350	34.62
Jhulang	Dhauliganga	3.3	1962–2000	400	10.53
Miyar	Chenab	87.8	1961–1996	575	16.43
Meru Bamak	Ganga	4.7	1977–2000	395	17.20
Parbati	Parbati	188.0	1990–2001	578	52.00

Although the warming processes continue unabated, the rate of rise in temperatures in the Gangotri glacier area has nevertheless demonstrated a marked gradual decline since the last quarter of the past century.

Parbati glacier is one of the largest glaciers in the Parbati river basin, Kullu district, Himachal Pradesh. This glacier is fed by nearly 36 glaciers, covering an areal extent of 188 sq. km. The glacier had retreated 578 m between 1990 and 2001 which amounts to about 52 m per year. This high rate of retreat of the glacier could be because it is located in the lower altitude range. About 90% of the glacier is located in the altitude range lower than 5,200 m which is about the average altitude of the snow line at the end of the ablation season. The specific mass balance of the glacier, estimated using accumulation area ratio method for the year 2001 is −86 cm. A high rate of retreat of the Parbati glacier will affect the availability of flow in the river.

Global warming is likely to increase the melting of snow and glacier far more rapidly than the accumulation. Glacial melt is expected to increase under changed climate conditions, which would lead to increased summer flows in some river systems for a few decades, followed by a reduction in flow as the glaciers disappear (IPCC, 1998). If India wants to exploit the increased flow for beneficial use (hydropower generation, irrigation) and to make sure that this high flow does not cause damages, infrastructure in the form of reservoirs and power projects will have to be put in place.

18.11. HYDROLOGICAL PROBLEMS OF HARD ROCK REGION

Even though there is no universally accepted definition of a hard rock, it is generally understood that hard rocks are crystalline, (igneous and metamorphic) rocks. Some hydrogeologists define “hard rocks” as igneous and metamorphic, non-volcanic and non-carbonate rocks. But other rock types (especially well cemented sedimentary rocks often occurring in areas built also by crystalline rocks) may be characterized by the same hydrogeological environment as crystalline rocks themselves. It is often impossible to define exact geological boundary between "hard rocks" and some other rock types. Therefore, in many hydrogeological studies, the term "hard rock" is used in a wider sense, rather vaguely and not exactly defined. From a groundwater exploration point of view, the term "hard rock" might include all rocks without sufficient primary porosity and conductivity for feasible groundwater extraction.

Since a large part of India is covered by hard rocks, typical problems of this region are being discussed separately here.

18.11.1. Hydrology of Hard Rocks

The hard rock region of India comprises Maharashtra, Karnataka, Goa, Parts of Andhra Pradesh, Tamil Nadu & Kerala. Major portion of the Deccan basalt of hard rock region in India is semiarid. Due to insufficient surface water, groundwater has been an essential element in meeting domestic, agricultural and industrial demands. In general, the groundwater potential of hard rocks is poor, though relatively high yields may be obtained in restricted locations under a favorable combination of topography and rainfall. The size and the frequency of openings in fractured rocks are normally restricted to shallow depths, resulting in low void ratio and hydraulic conductivity. The drainage developed in individual lava flows during intertrappean periods gives rise to productive zones under favorable conditions of topography.

Western Ghats, from which most of the southern rivers originate, was formed by hard volcanic rocks (Deccan traps), which have withstood the rigours of sun and rain for ages. The rivers of the peninsula are of great antiquity, compared to the youthful rivers of the north and their channels have reached the base level of erosion.

Two water-bearing zones can be generally identified (Narasimhan, 1990; Briz-Kishor, 1983) in the hard rock area: the composed or weathered zone and water-bearing joints and fractures. In the weathered and decomposed part of bedrock, groundwater occupies the intergranular spaces of the formation material. The yield of this zone is often limited and is seasonal in character. The groundwater flow systems are of local type, where each local system has its recharge area at a topographic high and its discharge area at a topographic low, which are adjacent to each other. The intermediate and regional groundwater flow systems do not exist because of negligible hydraulic conductivity with depth. The crystalline rocks generally don't possess original or primary openings, and fresh crystalline rocks have less than 1% porosity and negligible hydraulic conductivity. The ability of crystalline rocks to store and transmit water is dependent on the development of secondary openings, which were formed by fracturing and weathering. The weathered part of these crystalline rocks is of particular importance both as storage zone for groundwater and as aquifer for open wells and shallow tube wells.

The occurrence and movement of groundwater in the Deccan basalt of hard rock region is controlled by the fracture pattern. The fracture porosity forms the main criteria in defining the ground water flow systems. The fractured zone generally constitutes the potential aquifers and therefore the geometry of the fractured aquifer system assumes considerable importance while exploring the groundwater in hard rock terrain. Hence, the exploration, development and management of groundwater in the hard rock basaltic terrain are very important. Groundwater in these regions occurs under unconfined and semi-confined conditions. The presence of vesicles, fractures, zeolites, intertrappean red-boles and tuffaceous formations give rise to varying porosity in the basalt. The aquifer transmissivity ranges from 2 to 140 sq. m/day, higher values of which were noticed over fractured and jointed basalt. Aquifer resistivity ranges from 10–30 ohm m over alluvium and 20–60 ohm m over weathered basalt (Narayanpethakar et al., 1997).

The hydrology and groundwater resources in Deccan Traps have been studied by many hydrologists. The black trap is hard, compact and is traversed by joints to a shallow depth only up to 10 to 15 m. Beyond this depth, the rock becomes more and more compact and the fresh presence of such a massive variety of trap is noticed approximately from 630 m contour and below. Weathering extends hardly to 0.5 to 1.0 m depth. The depth of water in the wells varies from 2 to 10 m. In comparison, the pink trap appears to be a better aquifer. They are weathered to an average depth of 12 to 15 m and have more blowholes and amygdoloidal structures which are filled by secondary minerals like zeolites and silica. More of fractures and fissures are noticed which help retain water percolated after rainfall. Pink traps are seen at an approximate altitude of 660 to 675 m and extend approximately up to 630 m contour. The depth of water table in such formations varies from 6 to 12 m depending upon the topography.

In the Deccan trap, ground water occurs under water table conditions in weathered and jointed traps, and under confined conditions in the zeolitic and vesicular traps wherever they are overlain by hard traps. The depth of weathering, in general varies from 2 to 18 m. Wells range in depth from 3.7 to 17.8 m below ground level (bgl) and the depth to water table ranges from 1.1 to 16.2 m bgl. The yield of dug wells ranges from 20 m^3/day to 250 m^3/day for the pumping period of 2 to 8 hours. Wells in valleys nearer to rivers and in zeolitic traps give better yield. The transmissivity figures obtained by the Papadopulos and Cooper method range from 21.5 m^2/day to 150 m^2/day. The specific capacity of the wells ranges from 2.42 to 19.13 m^3/h/m and the unit specific capacity is in the range of 0.039 to 0.1995 m^3/h/m. The Deccan trap does not contribute appreciably to tube well yields and the contained water can be tapped only by constructing large-diameter wells.

The saturated fractures and joints found in the relatively unweathered bedrock at greater depths are capable of yielding substantial quantities of water. The fractures and joints are mostly horizontal and interconnected with a network of joints and fissures. The yield from these zones is not readily affected by seasonal changes. In the granite and gneiss of south India, such saturated zones are normally encountered at depths ranging from 10 to 50 m. In tectonically disturbed areas, they may even occur at a greater depth of 100 m or more. These saturated zones are usually weathered and have a small vertical extent. The normal yield of a tube well tapping such zones is around $5.5\,m^3$/hour. Very low yields of about 450–900 liters/hour are frequent, whereas quite large yields up to 90,000 liters/hour have been reported from a few isolated tube wells. In the consolidated or fissured formation, the occurrence of groundwater is restricted to weathered residue and fracture zones having secondary porosity, and the yield is above $20\,m^3/h$ in the Mesozoic and Paleozoic formations, while it goes down to $5–20\,m^3/h$ and even below $5\,m^3/h$ in the Precambrian and Archean formations.

Intensive exploratory drilling in igneous and other hard rocks in parts of peninsular India have shown that the openings at a greater depth become less pronounced and less abundant and in some cases they are not favorable for the movement of ground water. Relatively higher yields from hard rocks are obtained within 40 to 50 m depth from the surface. The optimum depth beyond which drilling is normally not warranted is about 100 m, while rock type is commonly of secondary importance to the control of weathering and structure. The geometry of the fracture or joint sets is determined by the types of the rocks and the stress to which they have been subjected, besides the effect of weathering and relief which makes the void space constituting the system progressively larger on approaching the surface. The topographic conditions and the rainfall regime maintain a high level of saturation in hard rocks. Thus topographic lows and high rainfall will offer a better advantage.

18.11.2. Typical Hydrological Problems of Hard Rock Region

Variations in the geomorphological features as well as geohydrological differences give rise to a variety of hydrological problems in this heterogeneous physiological unit of peninsula. A discussion of major hydrological problems of the hard rock region follows.

a. Groundwater Related Problems

Hard rock region is witnessing a growing demand for groundwater resources, as adequate surface water sources are not available. Hence, there is an urgency to quantify the availability of groundwater. The water balance technique has been extensively used to make quantitative estimates of water resources and the impact of anthropogenic activities on surface and groundwater.

Failure of open and borewells in many parts of the hard rock region are a common phenomenon. This problem usually arises either because of over-abstraction in existing wells or due to the failure in identifying the exact water bearing zones. In the hard rock region, water exists mainly in fractures and joints but locating such zones and predicting the flow processes is difficult. The problem is quite common in northern districts of Karnataka and in many parts of Tamil Nadu.

b. Ground Water Quality Problems

Groundwater quality poses a serious health hazards in various parts of hard rock region. The major contaminants founds in this region are high concentration of fluoride, chloride, sulphate, sodium and bicarbonate. Table 20 shows the ground water pollution in the hard rock region of India.

Excessive salinity in drinking water is undesirable because of objectionable tastes and the laxative effect associated with sulphate. Groundwater generally contains higher dissolved solids concentrations compared to surface water of the same locality. Most of the minerals present in greater amounts are those which contribute to hardness (Calcium and Magnesium) and alkalinity (Bicarbonate, Carbonate and Hydroxide). This is largely due to the increased amounts of carbon dioxide in groundwater.

The quality of groundwater varies from place to place as well as from strata to strata. It may also vary with seasons. The water drawn from strata at a particular time of the year may be unsuitable, whereas it may be good enough at other times of the year. The groundwater quality problem can be detected only by regular monitoring of the quality of water. Status of research on groundwater quality in hard rock regions is discussed below.

The highest value of fluoride was recorded in Bankapatti, where concentrations up to 17–20 mg/litre have been found. Fluorite ($Ca\ F_2$) is the most common fluoride bearing mineral and occurs either as an igneous, metamorphic or detrital mineral. The occurrence and distribution of fluoride in the shallow aquifer in the upper reaches of Pennar basin has also been observed. Further, high concentrations (1.5 mg/l) of fluoride have been found in groundwater at a number of network stations in the Karnataka state comprising districts of Gulberga, Bijapur, Raichur, Bellary, Chitradurga, Tumkur, Kolar, Shimoga, Dharwar and Belgaum.

Table 20. Ground water pollution in Hard Rock Region

SN	Pollutant	State	Districts of occurrences
1.	Fluoride	Andhra Pradesh	Krishna, Anatpur, Nellor, Chittoor, Cuddapah, Guntur and Nalgonda
		Tamil Nadu	Chengalpett, Madurai
2.	Salinity (Inland)	Maharashtra	Amravati, Akola
3.	Nitrate	Andhra Pradesh	Vishakapanam, E. Godavari, Krishna, Prakasam, Nellure, Chittoor, Anatpur, Cuddapah, Kurnool, Khamman and Nalgonda
		Karnataka	Bidar, Gulbarga, Bijapur
		Maharashtra	Jalna, Beed, Nanded, Latur, Osmanabad, Solapur, Satara, Sangali, and Kolhapur
		Tamil Nadu	Coimbatore, Periyar and Salem
4.	Chloride	Karnataka	Dharwar, Belgaum
		Maharashtra	Solapur, Satara, Amaravati, Akola and Buldana
5.	Zinc	Andhra Pradesh	Hyderabad

c. Tank studies

Minor irrigation tanks are widely distributed in Karnataka, Tamil Nadu, Andhra Pradesh and Maharashtra to meet water requirements for agriculture and community use. However, in many cases, there is no inventory of these tanks, their storage capacities are not known and there is no information about inflow and outflow from them. Some tanks have gone into disuse, because they were not properly maintained. For proper management of tanks, it is necessary to prepare an inventory. Therefore, it is essential to take up the studies to estimate the yield and other related hydrological parameters, which are typical for hard rock regions. A coordinated effort needs to be made to rejuvenate the tanks that are in poor state and these should be used in an integrated manner.

d. Drainage problems in black soil areas

The drainage problem, particularly in irrigated heavy soils, is caused by over-irrigation or by surface runoff resulting from excess rainfall. This kind of problem has been reported in the black soil areas of Maharashtra and Karnataka. The solution depends mainly upon the ratio of precipitation to the rate of downward flow through the soil system consisting of poorly pervious layers and the presence or absence of a pervious sub-soil. Therefore, it is necessary to develop a suitable model for estimating field drainage requirement and install appropriate type of drains.

18.12. THREAT TO BIODIVERSITY AND WETLANDS

About 6.5% and 12.5% of the world's animal and plant species, respectively, are found in India. Out of these almost 7,000 are endemic to the subcontinent. Unfortunately, habitat destruction in both freshwater and coastal areas has endangered many endemic species. Most vulnerable are the freshwater fish, since they are more susceptible to water pollution and environmental change. Other endangered species include freshwater aquatic animals like the Gangetic dolphin and several species of aquatic birds, amphibians, reptiles and insects.

Wetlands in India cover a land area of about 4.1 M-ha. Most of these have become degraded due to pollution and development pressures, like conversion of wetlands for agriculture. This is threatening not only the local fauna but also the livelihood of the residents dependent on the wetland ecosystem. In coastal areas, industrial and domestic pollution has severely degraded estuarine and coastal environments. Estimates show that over 20,000 million liters per day of mostly untreated domestic sewage reaches India's coastal areas (Development Alternatives, 2001).

Mangroves are one of the most biologically productive environments, but their economic value has not been realized properly. Mangrove forests cover a significant portion of the coastal zone in India. Increasing development and destruction arising from fodder and fuel-wood collection is threatening their existence, which implies significant losses in terms of the overall biodiversity and the ecosystem health.

CHAPTER 19

RESERVOIRS AND LAKES

A reference exists [see Rangachari et al. (2000)] to an ancient enquiry by Rishi Narada of King Yudhistira (circa 3,150 BC): "Are the farmers sturdy and prosperous? Are there dams full of water and big enough and distributed in different parts of the kingdom and does agriculture not depend on rains only ?". According to some myths, dam engineering appears to have originated in India, for it is evidenced by the development of embankment on sloping grounds around 1,500 BC. A rock inscription near Junagarh in Gujarat describes a dam named Sudarashan, which created a big reservoir near Mount Girnar. This dam was constructed during the reign of Chandragupta Maurya and renovated during the reign of Ashoka. Rigveda (2000–1500 BC), an ancient sacred book on wisdom, contains 1,028 hymns which advocate building dams for irrigation purposes and caution against their breaches, especially during war. A similar mention was made in Arthashastra, written in 330 BC by Kautilya, which advises kings to build dams and encourages people to help in their construction by providing sites, approach roads, and free materials. It considered breaching of a dam a serious offence and advised imposing severe punishment against it. Rangachari et al. (2000) have stated that Magasthenes, the Greek ambassador to the court of Emperor Chandragupta (circa 300 BC), recorded that the district officers used to inspect the sluices by which water is distributed into the branch canals (watercourses) so that every one may enjoy his fair share of the benefit.

The Grand Anicut on the Cauvery was one of the earliest canal systems built, dating back probably to the 2nd century (Sengupta, 1993). It is functional even after nearly 2,000 years. Many earthen dams of moderate height were built in south India from the very early days and there are presently over 39,400 such reservoirs in Tamil Nadu alone (Rangachari et al. (2000). Many of these were built by the different ruling dynasties between 500 AD and 1,500 AD. Advances in technology coupled with greater confidence arising out of experience in undertaking such measures boosted dam construction. The Ramappa Lake (1,213) in Andhra, Karla (1,514) and Vihar (1,860) in Maharashtra, Rajsamand (1,671), Jai Samand (1,730) in Rajasthan and the Barwasagar (1,500) are instances of such works of the past. The Periyar Project which was completed in 1,895 involved the construction of a 48 m high and 378 m long concrete dam on the west-flowing Periyar River in Western Ghats.

Dam engineering received a big boost especially after independence with the construction of many reservoirs. Dams like Bhakra, Tungabhadara, Nagarjunasagar, Hirakud, Idukki are a few examples among many others. Considering the growth in population with consequent escalation in water demands for food grain production, the country needs many more dams for harnessing the water resources. However, the construction of new projects has slowed down during the last decade due to a number of reasons, opposition on environmental ground being one of them. Mighty water resources projects that are currently under construction are Sardar Sarovar Project, Tehri Hydro Electric Project, and Almatti dam.

19.1. NEED FOR STORAGE RESERVOIRS

A dam contains a number of structural features other than the main wall itself. Spillways are usually operated to discharge water when the reservoir level becomes dangerously high. Dams built across broad plains may include long lengths of ancillary dams and dykes. Weirs and barrages are constructed to divert river flow; they do not have significant storage and cannot effectively regulate flows. A weir is normally a low masonry or concrete wall. A barrage is a bigger (usually metallic) structure across a wide river.

There are many reasons why storage projects are needed in India. The principal function of a reservoir is regulation of natural streamflow by storing surplus water in the high flow season to control floods and releasing the stored water in the dry season to meet various demands. Generally, the major part of the annual streamflow is available during a few months of rainy season. But the demands for water arise all year round and therefore it is necessary to store the excess water in the rainy months so that it can be used when the natural streamflow is not sufficient to meet the demand. The water stored in a reservoir may be diverted by means of pipes or canals to far away places where it is needed; this diversion results in spatial changes of water availability. The water may also be kept in the reservoir and released later for beneficial uses resulting in temporal changes. In short, the aim of a reservoir is to match the temporal and spatial availability of water with demands. Depending on the magnitude of natural inflows and demands at a particular time, the reservoir storage is either built up or water is supplied from the storage.

Reservoirs also help transform the available resource into utilizable. Many river basins of the country are water short basins. Despite nearly full development of water resources of these basins, less than 30% of the cultivable area in many basins is able to receive assured irrigation. Even now, large tracts of cultivable land on the eastern coast of peninsular India depend on the vagaries of monsoons, making them highly vulnerable. In fact, delta areas of major rivers systems have been protected by the old irrigation systems. Most of these old projects have been rejuvenated and modernized.

As a result of storing water, a reservoir provides a water head which can be used for generation of electric power. The reservoir also provides an empty storage space for moderating inflow peaks. A reservoir also provides a pool for navigation to negotiate rapids, habitat for aquatic life and facilities for recreation and sports. It enhances scenic beauty, promotes afforestation, and supports wild life. Thus, it is of paramount importance for India to create as much storage space as possible so that adequate quantity of water can be stored and be converted to utilizable resource.

The need for reservoirs can be appreciated by the fact that a large part of water supply of Mumbai, Pune, Hyderabad, Bhopal, and Warangal cities is dependent on reservoirs like Vaitarana, Tansa, Bhatsa, Khadakwasla, Panchet, Majira, Singur, Kolar, and Sriramsagar. Numerous dams including Bhakra, Chambal valley projects, Ujjani, Tungbhadra, Almatii, in the country are providing water to irrigate crops and are essential for food security. Bhakra, Pong, Srisailam, and Balimela are generating electricity at very small price. The super thermal power stations in Uttar Pradesh are entirely dependent on the storage of Rihand reservoir for their water supply.

It is not rational to oppose the philosophy behind dams without going into merits and demerits of the individual project. Without dams, it would be impossible to ensure food security and supply of water for energy and industrial sectors in the monsoon climate of India. As the precipitation over the country becomes more erratic due to climate change, reservoirs would increasingly reduce our dependence on the vagaries of monsoons. A rational approach would be to examine all the options available to us and make the best use of our resources.

19.1.1. Classification of Dams

The Indian code classifies dams as small, intermediate, and large with respect to the storage in the reservoir and hydraulic head. These are briefly described in Table 1. Note that the design flood of a dam depends upon its type.

The International Commission on Large Dams (ICOLD) defines a dam as a large dam if it has a damwall greater than 15 m in height (from the lowest general foundation to the crest). However, dams between 10–15 m in height are also classified as large dams if they satisfy at least any one of the following criteria:

- Crest length > 500 m,
- Reservoir capacity > 1 MCM,
- Maximum flood discharge > 2,000 m3/s,
- Dam has difficult foundation problem, and
- Dam is of unusual design.

Table 1. Classification of dams

Type of Dam	Storage Capacity (million m^3)	Hydraulic head (m)	Inflow design flood
Small	0.5–10	7.5–12	100 year
Intermediate	10–60	12–30	SPF
Large	>60	>30	PMF

19.2. DAMS IN INDIA

Dams in India are constructed mainly for irrigation, hydroelectric generation, flood control, and water supply; a few are constructed for other purposes. At the turn of twentieth century (1900), there were 42 dams in India. During 1901 to 1950, about 250 dams were added. That is, at the time of the beginning of first five-year plan period (1950–51), there were a total of about 300 dams. During the next twenty years, there was a spurt in dam construction activity in which 695 dams were added, bringing the total number of dams to nearly 1,000 up to the year 1970. The dam building activity intensified during the next two decades and at the end of 1990, the total number of Indian dams stood at 3,244 (not accounting for 236 dams for which the year of construction is not available). Due to slow down in economy, opposition on socio-economic grounds, and other reasons, only 116 dams could be added between 1990 and 2000. India had 4,291 large dams and about 250 large barrages by 2003, including 695 dams under construction. Distribution of large dams in India according to their age is given in Table 2.

For better visualization, the age of dams is plotted in Figure 1. This figure shows that our dams are getting old. Nearly 50 dams are more than 100 years old. An implication of aging is that some of these dams would require replacement and maintenance cost of this infrastructure will increase with time. Since replacement of dams is proving to be very difficult, ways and means have to be found to prolong the useful life of the created infrastructure.

Table 3 details live storage position of completed, ongoing and proposed projects in various river basins of India. Table 3 shows that the live storage capacity is the largest in the Ganga basin. Of course, this capacity comes from the dams on tributaries while there is no major completed project on the main river. Next to Ganga, Krishna River has the highest amount of storage. Upon completion of on going projects, Narmada will be a highly regulated basin.

Table 4 summarizes the storage capacities created and projects under consideration in various states of India. As can be seen, Andhra Pradesh has created the

Table 2. Distribution of large dams according to age

Year of completion	Number of dams
Up to 1900	42
1901–1950	251
1951–1960	234
1961–1970	461
1971–1980	1,190
1981–1990	1,066
1991–2000	116
Year of construction not available	236
Under construction	695
Total	4,291

Source: CWC (1994).

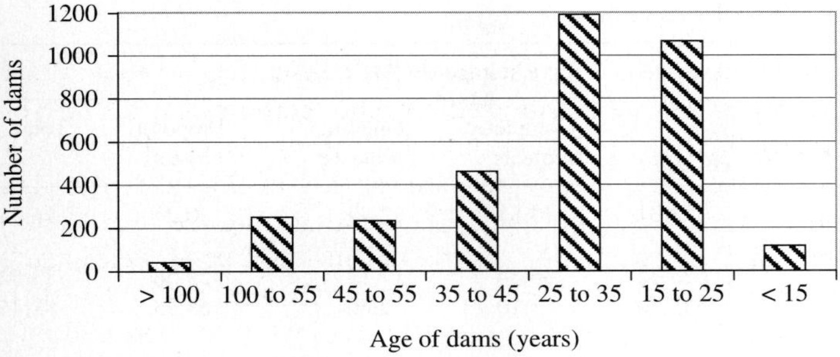

Figure 1. Distribution of dams and their age

maximum storage capacity among all the states and accounts for about 14% of the storage capacity created in the country. Maharashtra is very close to it, accounting for 12.7% of the capacity created. If, however, the projects under construction are also considered, the first place is taken by Madhya Pradesh where the on-going projects will add 21.63 km^3 of storage capacity. With this inclusion, Madhya Pradesh will account for 16% of the total storage space of nearly 249 km^3.

Table 5 summarizes the distribution of dams in various states of India. Here the first place is claimed by Maharashtra which has 1,229 dams followed by Madhya Pradesh with 946 dams. Note that even after considering the on-going projects, Maharashtra will continue to retain the first place since nearly 300 dams are under construction and this will take the total number of dams to 1,529. This number is more than the number of dams in the group 'other states'.

Table 6 summarizes the state wise major large dams of India. Further, the dams are also listed basin wise in each state.

There are a large number of tanks in the country whose water serves a variety of uses such as societal needs and irrigation etc. Comparatively more tanks are present in South India. Table 7 shows the distribution of small reservoirs and tanks in the states of India.

Table 8 shows that number of medium and large storages and total number of storages. Tamil Nadu has the maximum number of such structures whereas Karnataka comes a distant second. Note that the scheme of classification of dams in these tables is not the same as given in Table 1.

The analysis further reveals that more than 92% of the dams built in India mainly serve the purpose of irrigation, 2.2% hydroelectric power generation, less than 1% water supply and less than 35% serve a combination of purpose of irrigation, water supply and hydropower generation. Table 9 gives the details of 50 highest dams of India. As can be seen, 20 dams have height 100 m and above. Nearly 20 Indian dams have height exceeding 100 m. Further 34 reservoirs have live storage capacity more than 1 billion cubic m (BCM). Further, 10 dams have gross storage capacity more than 5 BCM.

Table 3. Storage projects in river basins of India

S. N.	Name of the River Basin	Average annual flow (BCM)	Live storage (BCM)			
			Completed projects	Ongoing projects	Proposed projects	Total
1.	Indus (Area in Indian Territory)	73.31	13.83	2.45	0.27	16.55
2.	a) Ganga	525.02	36.84	17.12	29.56	83.52
	b) Brahmaputra, Barak, and others	585.60	1.10	2.40	63.35	88.45
3.	Godavari	110.54	12.51	10.65	8.28	31.44
4.	Krishna	78.12	34.48	7.78	0.13	42.39
5.	Cauvery	21.36	7.43	0.39	0.34	8.16
6.	Pennar	6.32	0.38	2.13	–	2.51
7.	EFR and rivers from Mahanadi to Godavari and Krishna to Pennar	22.52	1.63	1.45	0.86	3.94
8.	EFR between Pennar and Kanyakumari	16.46	1.42	0.02	–	1.44
9.	Mahanadi	66.88	8.49	5.39	10.96	24.84
10.	Brahmani & Baitarani	28.48	4.76	0.24	8.72	13.72
11.	Subarnarekha	12.37	0.66	1.65	1.59	3.90
12.	Sabarmati	3.81	1.35	0.12	0.09	1.56
13.	Mahi	11.02	4.75	0.36	0.02	5.13
14.	WFR of Kutch & Saurashtra including Luni	15.10	4.31	0.58	3.14	8.03
15.	Narmada	45.64	6.60	16.72	0.46	23.78
16.	Tapi	14.88	8.53	1.00	1.99	11.52
17.	WFR from Tapi to Tadri	87.41	7.10	2.66	0.84	10.60
18.	WFR from Tadri to Kanyakumari	113.53	10.25	2.31	1.70	14.26
19.	Minor rivers draining to Myanmar (Burma) & Bangladesh	31.00	0.31	–	–	0.31
	Total	1, 869.35	173.73	75.42	132.30	381.45

Source: CWC (2002).

Currently, the constructed dams have created a total live storage capacity of $177\,km^3$ by the year 1995 as against $15.60\,km^3$ in the pre-plan period. Besides, dams to create an additional live storage capacity of $75\,km^3$ are under various stages of construction. Also, dams under formulation/consideration will create an additional live storage of $132\ km^3$ when all the dams under construction

Table 4. Storage capacities (km^3) in different states

State	Completed	Under Construction	Total
Andhra Pradesh	24.85	7.12	31.97
Assam	–	1.05	1.05
Bihar	4.66	4.35	9.01
Goa	0.04	0.67	0.72
Gujarat	14.92	7.25	22.17
Himachal Pradesh	13.81	0.11	13.92
Jammu and Kashmir	–	–	–
Karnataka	21.56	3.01	24.57
Kerala	4.62	1.62	6.23
Madhya Pradesh	18.56	21.63	40.21
Maharashtra	22.10	12.92	35.01
Manipur	0.40	0.12	0.52
Meghalaya	0.70	–	0.70
Nagaland	–	1.22	1.22
Orissa	14.29	3.30	17.59
Punjab	0.02	2.34	2.37
Rajasthan	8.32	1.59	9.91
Tamil Nadu	6.72	0.04	6.76
Tripura	0.31	–	0.31
Uttar Pradesh	16.29	7.06	23.40
West Bengal	1.48		1.48
Pondicherry	0.01	–	0.01
Total	173.73	75.42	249.15

Note: Storages having capacity 10 MCM and above are considered.
Source: CWC (2002), pp23.

Table 5. Distribution of dams in India

State	Completed	Under Construction	Total
Maharashtra	1,229 (34%)	300 (43%)	1,529 (36%)
Madhya Pradesh	946 (26%)	147 (21%)	1,093 (25%)
Gujarat	466 (13%)	71 (10%)	537 (13%)
Other States	955 (27%)	177 (26%)	1,132 (26%)
Total	3,596	695	4,291

Note: Figure in the bracket is percentage of total under the column.
Source: CWC (2002).

and those under consideration are completed, India will have a total live storage of 384 km^3, which is about 56% of assessed utilizable surface water resources of 690 km^3.

Figure 2 illustrates another important point. It shows the number of days whose average flows can be stored in various reservoirs in India. In this respect, the storage capacity available in Tapi and Krishna is the highest in India, 230 and 220 days respectively. As compared to this, the Colorado dam in the USA has enough

Table 6. Statewise and riverwise distribution of large dams

State	Name of river	Name of dam
Andhra Pradesh	Godavari	Sriram Sagar dam
	Kaddam (tributary of Godavari)	Kaddam dam
	Krishna	Nagarjuna Sagar dam
	Krishna	Srisailam hydroelectric project
	Machkund	Jalaput dam
	Sileru	Forebay dam
Bihar	Badua	Badua reservoir
	Barakar	Maithon dam
	Barakar	Tilaiya dam
	Chandan	Chandan reservoir
	Damodar	Panchet hill dam
	Damodar	Tenughat dam
	Konar	Konar dam
	Subarnarekha	Getalsud dam
Gujarat	Banas	Dantiwada dam
	Machhundri	Machhundri irrigation scheme
	Mahi	Kadana reservoir
	Raval	Raval irrigation scheme
	Sabarmati	Dharoi dam
	Sakra	Tapar dam
	Shetrunji	Shetrunji irrigation scheme
	Tapi	Ukai dam
Himachal Pradesh	Beas	Beas dam at Pong
	Beas	Pandhoh dam
	Sutlej	Bhakra dam
Karnataka	Arkavally and Kumudwathy	Chamarajasagar dam
	Bhadra	Bhadra reservoir
	Ghataprabha	Hidkal dam
	Harangi	Harangi reservoir
	Kabini	Kabini dam
	Krishna	Narayanpur dam
	Main Cauvery	Krishnarajasagar dam
	Malaprabha	Indira Gandhi dam
	Sharavathy	Linganamakki dam
	Talakalale	Talakalale dam
	Tungabhadra	Tungabhadra dam
	Vedavati	Vani Vilasa Sagar dam
Kerala	Ayalar	Pothundy dam
	Karuvannur	Peechi dam
	Malampuzha	Malampuzha dam
	Neyyar	Neyyar dam
	Periyar	Idukki dam
	Wadakkancherry	Vazhani dam

Madhya Pradesh	Barna	Barna dam
	Chambal	Gandhi Sagar dam
	Mahanadi	Mahanadi reservoir project
	Tawa	Tawa dam
Maharashtra	Ambi	Tanaji Sagar dam
	Aner	Aner dam
	Bagh	Sirpur dam
	Bhogawati	Radhanagari dam
	Boladwadi Stream	Kolkewadi dam
	Garvi	Itiadoh dam
	Godavari	Paithan dam
	Kadwa	Karanjwan dam
	Katepurna	Katepurna dam
	Koyna	Koyna dam
	Krishna	Dhom dam
	Mula	Mula dam
	Mutha	Khadakwasla dam
	Nira	Vir dam
	Nirguna	Nirguna dam
	Pawna	Pawna dam
	Pench	Kamthikhairy dam
	Pravara	Wilson dam
	Purna	Sidheshwar dam
	Purna	Yeldari dam
	Pus	Pus dam
	Waghadi	Waghadi dam
	Wuna	Kanholi dam
	Yelwandi	Bhatghar dam
Orissa	Kolab	Upper Kolab dam
	Machkund	Balimela dam
	Mahanadi	Hirakud dam
Rajasthan	Chambal	Jawahar Sagar dam
	Chambal	Rana Pratap Sagar dam
	Gomti	Jaisamand tank
Tamil Nadu	Aliyar	Upper Aliyar dam
	Amaravati	Amaravathi dam
	Avalanche Stream	Avalanche dam
	Bhavani	Lower Bhavani dam
	Bhavani	Upper Bhavani dam
	Cauvery	Mettur (Stanley) dam
	Emerald	Emerald dam
	Gatanandi	Gatana dam
	Karuppanadhi	Karuppanadhi dam
	Kodayar	Kodayar dam I
	Kodayar	Kodayar dam II
	Kodayar	Peechiparai dam
	Kundah	Kundapalam dam
	Manimuthar	Manimuthar dam
	Mukurthi	Mukurthi dam
	Nirar	Lower Nirar dam
	Palar	Thirumurthi dam

(Continued)

Table 6. (Continued)

State	Name of river	Name of dam
	Palar-Porandalar	Palar Porandalar dam
	Paralayar	Perunchani dam
	Parambikulam	Parambikulam dam
	Parappalar	Parappalar dam
	Parson's Valley Stream	Parson's valley dam
	Pegumbahalla	Pegumbahalla dam
	Periyar	Periyar dam
	Ponnaiyar	Sathanur dam
	Porthimund Stream	Porthimund dam
	Ramanadhi	Ramanadhi dam
	Sandy Nullah Stream	Sandy Nullah dam
	Sholayar	Sholayar dam
	Thambraparani	Thambraparani dam
	Tributary of Karampuzha	Western catchment no.2 dam
	Vaigai	Vaigai dam
	Varahapallam West	West Varahapallam dam
Uttaranchal	Bhagirathi	Maneri Bhali hydroelectric project (stage 1)
	Ramganga	Ramganga dam
	Ganga	Tehri dam
Uttar Pradesh	Rihand	Obra dam
	Rihand	Rihand dam
	Tons	Ichari dam
	Betwa	Matatila dam
West Bengal	Kangsabati and Kumari	Kangsabati-Kumari dam
	Mayurakshi	Massanjore dam

Source: CBIP (1987) and others.

capacity to store mean flow of about 900 days or about 2.5 years. Further, Mahanadi, Ganga and Brahmaputra are the rivers which carry large flow volumes but the storage capacity in their basins is abysmally low at 80, 40, and 2 days, respectively. Obviously, we need to create more storage space in India so that our rivers can be regulated in a more meaningful way. This argument gets further buttressed when one recalls that precipitation in India has very high temporal skewness due to monsoon type climate.

Per capita storages in India and various countries are shown in Figure 3. It can be immediately noticed that India has quite small storage space as compared to other countries of similar characteristics. Further, precipitation is far more evenly distributed in many other countries but despite that, they have attained a much higher degree of regulation to beneficially use their water resources.

Table 7. Distribution of small reservoirs and irrigation tanks in India

States	Small reservoirs		Irrigation tanks		Total	
	Number	Area (ha)	Number	Area (ha)	Number	Area (ha)
Tamil Nadu	58	15, 663	8, 837	300, 278	8,895	315,941
Karnataka	46	15, 253	4, 605	213, 404	4,651	228,657
Andhra Pradesh	98	24, 178	2, 800	177, 749	2,898	201,927
Orissa	1, 433	66, 047	–	–	1,433	66,047
Gujarat	115	40, 099	561	44, 025	676	84,124
Rajasthan	389	54, 231	–	–	389	54,231
Bihar	112	12, 461	–	–	112	12,461
Uttar Pradesh	40	20, 845	–	197, 806	40	218,651
Kerala	21	7, 975	–	–	21	7,975
Madhya Pradesh	6	172, 575	–	–	6	172,575
West Bengal	4	732	–	–	4	732
Haryana	4	282	–	–	4	282
North East	4	1, 639	–	600	4	2,239
Maharashtra	–	–	–	–	–	119,515
Himachal Pradesh	1	200	–	–	1	200
Total	2, 331	551, 695	16, 803	933, 862	19,134	1,485,557

Source: http://www.fao.org.

Table 8. Distribution of medium, large and total reservoirs in India

States	Medium		Large		Total Storages*	
	Number	Area (ha)	Number	Area (ha)	Number	Area (ha)
Tamil Nadu	9	19,577	2	23,222	8,906	358,740
Karnataka	16	29,078	12	179,556	4,679	437,291
Andhra Pradesh	32	66,429	7	190,151	2,937	458,507
Orissa	6	12,748	3	119,403	1,442	198,198
Gujarat	28	57,748	7	144,358	711	286,230
Rajasthan	30	49,827	4	49,386	423	153,444
Bihar	5	12,523	8	71,711	125	96,695
Uttar Pradesh	22	44,993	4	71,196	66	334,840
Madhya Pradesh	21	169,502	5	118,307	32	460,384
Kerala	8	15,500	1	6,160	30	29,635
West Bengal	1	4,600	1	10,400	6	15,732
Northeast	2	5,835	–	–	6	8,074
Himachal Pradesh	–	–	2	41,364	3	41,564
Maharashtra	–	39,181	–	115,054	–	273,750
Haryana	–	–	–	–	4	282
Total	180	527,541	56	1,140,268	19,370	3,153,366

* includes data of Table 7.

Table 9. Fifty highest Indian dams

Rank as per height	Name of Dam	Year of completion	State	Dam's feature (m)		Storage capacity (MCM)		Rank by Live storage capacity
				Height	Length	Gross	Live	
1	Tehri	U/C	Uttaranchal	261	575	3, 540	2, 615	19
2	Kishau	U/C	Uttaranchal	236	680	1, 810	1, 330	30
3	Bhakra	1963	Himachal Pradesh	225	518	9, 621	7, 191	3
4	Lakhwar	U/C	Uttar Pradesh	204	2, 800	580	330	46
5	Idukki	1974	Kerala	169	366	1, 996	1, 460	28
6	Sardar Sarovar	U/C	Gujarat	163	1, 244	9, 500	5, 800	9
7	Srisailam	1984	Andhra Pradesh	145	512	8, 722	7, 080	5
8	Jamrani	1990	Uttar Pradesh	140	765	206.6	206.6	48
9	Chamera	1994	Himachal Pradesh	140	295	391.3	110	50
10	Cheruthoni	1976	Kerala	138	650	1, 996	1, 460	27
11	Pong	1974	Himachal Pradesh	133	1, 956	8, 570	7, 290	2
12	Ramganga	1974	Uttar Pradesh	128	715	2, 450	2, 196	21
13	Nagarjuna Sagar	1974	Andhra Pradesh	125	4, 865	11, 555	6, 940	6
14	Salal (Rockfill)	1960	J&K	118	630	285	285	47
15	Kakki	1966	Kerala	114	336	455	447	45
16	Sholayar	1971	Tamil Nadu	105	1, 244	152.7	143.1	49
17	Koyna	1964	Maharashtra	103	805	2, 797	2, 652	17
18	Supa	1987	Karnataka	101	322	4, 150	3, 710	11
19	Idamalayar	1985	Kerala	100	375	1, 153	1, 018	33
20	Lower Karjan	U/C	Gujarat	100	903	630	581	44
21	Rihand	1962	Uttar Pradesh	90	934.2	10, 608.3	8, 979.94	1
22	Ukai	1972	Gujarat	81	5, 065	8, 511	7, 100	4
23	Bhadra	1963	Karnataka	72	440	2, 025	1, 785	25
24	Rengali		Orissa	71	1, 040	5, 150	3, 432	12
25	Balimela	1977	Orissa	70	372	3, 610	2, 676	16
26	Mettur	1934	Tamil Nadu	70	1, 615	2, 708	2, 647	18
27	Bargi	1980	Madhya Pradesh	69	5, 337	3, 920	3, 180	14
28	Kadana	1979	Gujarat	66	575	1, 542	1, 203	32
29	Gandhisagar	1960	Madhya Pradesh	65	514	7, 746	6, 910	7
30	Tenughat	1981	Jharkhand	63	408	1, 023	814	40

31	Lower Bhavani	1955	Tamil Nadu	62	8,797	928.8	907.8	37
33	Hirakud	1957	Orissa	61	1,248	8,136	5,818	8
32	Linganamakki	1965	Karnataka	61	2,749	4,418	4,276	10
34	Machkund	NA	Orissa	61	*NA*	*NA*	893	38
35	Ghatprabha	1979	Karnataka	60	10,035	1,448.70	1,387	29
36	Tawa	1974	Madhya Pradesh	58	1,815	2,310	2,050	23
37	Rana Pratap Sagar	1967	Rajasthan	58	1,143	2,900	1,567	26
38	Malprabha	1973	Karnataka	56	69	1,068.39	866	39
39	Upper Kolab	1990	Orissa	55	631	1,215	935	35
40	Yeldari	1968	Maharashtra	51	350	934	809	41
41	Vanivilassagar	1901	Karnataka	50	405.4	850.3	802	42
42	Dharoi	1976	Gujarat	50	6,847	907.88	775.89	43
43	Almatti	NA	Karnataka	49	1,565	3,485	2,986	15
44	Hemavathy	NA	Karnataka	45	4,692	1,050.83	1012.8	34
45	Sriramsagar	1983	Andhra Pradesh	43	14,600	3,172	2,322	20
46	Mahi Bajaj Sagar	1986	Rajasthan	43	3,062	*NA*	2,070	22
47	Kangsabati	1975	West Bengal	41	10,400	1,052	914	36
48	Krishnarajsagar	1932	Karnataka	40	2,620	1,368.8	1,244	31
49	Somasila	NA	Andhra Pradesh	38	*NA*	2,210	1,994	24
50	Tungabhadra	NA	Karnataka	36	2,449	3,737.82	3,308	13

Source: CBIP (1987), CBIP (1996), CWC (2006), and others.

As of now, the storage capacity created (including projects under construction) is about 50% of that ultimately possible. It is doubtful whether even the storage capacity of 400 km^3 can actually be created because of objections on environmental and other grounds. It is interesting to note that the USA, which has almost the same surface water potential, has already a storage capacity of about 700 km^3, i.e. about the same as India's ultimate storage capacity. Storage capacity in the former USSR is of the order of 1,100 km^3.

19.3. PROBLEMS IN DECISION MAKING FOR DAMS

Reservoir projects have immensely contributed in irrigation development in the country and in generating cheap hydropower. Impressive progress in creating irrigation facilities along with the other input has contributed in tremendous increase in agricultural products and making India self-sufficient in food grains. In fact, in recent past, India has even exported considerable quantities of food grains and other agricultural products. Contributions in other fields such as flood control and water supply are also immense.

In this section, we discuss some problems related with dam and reservoir development.

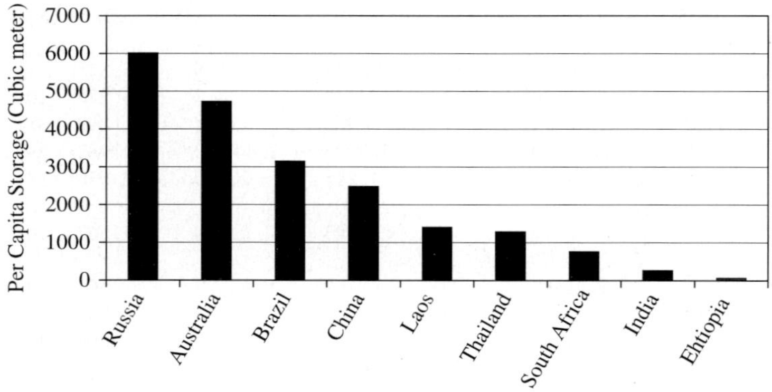

Figure 2. Number of days of average flows that can be stored in reservoirs in various river basins in India. For comparison, the data of Colorado River (USA) has also been plotted

19.3.1. Relative Economics of Projects

Sometimes, the opponents of major dams quote the capital cost of minor irrigation projects to show that they are cheaper than major dams. Undoubtedly, the capital cost of a minor dam will be low. However, minor dams normally do not have any spillway to regulate flood waters and breaches are common. Experience with tanks (small dams) in the states of Tamil Nadu, Karnataka and Rajasthan has indicated that, in years of high rainfall, breaches are frequent which require costly repairs.

Similarly, it is often advocated that the cost of groundwater development is very cheap compared to that of surface water through major and medium irrigation. This again is not true, as only the capital cost in both cases is taken into account; the recurring annual maintenance cost in the case of groundwater is comparatively high. Besides, normally the life of tubewells is about 20 years, while the useful

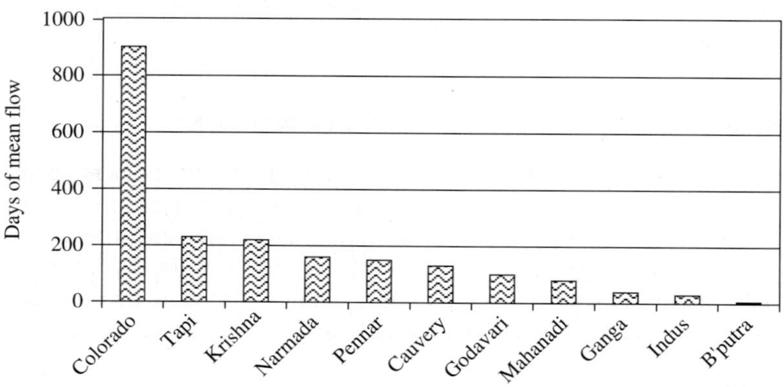

Figure 3. Per capita reservoir storage space in selected countries of the world

life of pumps is only ten years, at the end of which period complete replacement is required. Therefore, if the annual costs over a period of 100 years – which is taken as the economic life of a major project – are compared, taking into account both the capital and the annual maintenance and replacement costs, the picture changes. Even after considering the present cost of major and medium irrigation, the annual cost becomes less compared with minor tanks and groundwater. At the same time, to meet the needs of the country for agricultural production, irrigation has to be provided through all the means, namely major and medium projects, minor surface projects and also ground water.

19.3.2. Small Versus Big Dams

An oft-repeated discussion related to the dams is whether small dams are more suitable than big dams. Basically, there are three places to store water: soil profile, surface storages, and aquifers. The storage in soil profile is very important for agriculture but only small quantities of water can be stored for a short period. A comparative analysis of advantages, limitations, and key issues associated with groundwater, a small reservoir, and a large surface reservoir are given in Table 10.

The main argument against big reservoirs is that they submerge large areas compared to many small reservoirs. Any dam, big or small, needs a suitable site.

Table 10. Comparative advantages, limitations, and key issues associated with groundwater, small surface reservoir, and a large surface reservoir

	Groundwater storage	Small surface reservoirs	Large surface reservoirs
Advantages	Negligible evaporation loss	Ease of operation	Multipurpose
	Ubiquitous distribution	Multiple use	Large, reliable yield
	Operational efficiency	Groundwater recharge	Carryover storage
	Available on demand		Low cost per m^3 water stored
	Good water quality		Groundwater recharge
Limitations	Slow recharge rate	High evaporation loss fraction	Complexity of operations
	Groundwater contamination	Relatively high unit cost	Sedimentation
	Cost of extraction	Absence of over-year storage	High initial investment
	Recoverable fraction		Large gestation period
			Requires good sites
Key Issues	Declining water levels	Sedimentation	Social and environmental impacts
	Rising water levels	Population displacement	Rehabilitation and resettlement
	Regulation of use	Submergence	
	Groundwater pollution	Environmental impacts	

One just cannot build a dam wherever one wants. If a major dam is to be replaced by a number of small dams, there must be a number of suitable sites on the same river. Keller et al. (2,000) argued that it is very difficult to construct safe small dams. Of course, one will also have to build dams in lower reaches which will mean more submergence (often of good agricultural land) due to flat slopes and more population displacement due to higher population density.

It is useful now to examine some real cases. Consider Britain's largest reservoir Quoich on Quoich River near Fort Augustus in the Scottish Highlands. The lake behind 38 m high earth-rockfill dam has a storage capacity of 3,828 million m^3. According to Robbroeck (1996), if all British reservoirs are arranged in an ascending order of size and their volumes and surface areas aggregated, the total volume of the 327 smallest reservoirs would be needed to replace the volume of the largest, and that the total submerged area would be 6,705 ha, 3.5 times the area submerged by Quoich. A similar analysis for South Africa shows that 433 small reservoirs would be needed to replace the volume (5,246 million m^3) of the Gariep reservoir, with an aggregate area 222 times larger. A comparison was also made by Shah (1993) in India between the proposed Girna dam in the Mahanadi basin in Orissa and a smaller Girna dam plus 8 satellite storages making up the same volume. In case of smaller dams, the cost would be 150% higher, 60% more land would be submerged, considerably less energy would be generated, and evaporation will be 50–60% higher.

Although social and environmental problems are probably not in direct proportion to the area submerged, it can be safely deduced that a large number of small reservoirs will be far less acceptable from that point of view. Economics would also be much worse: loss of advantage of scale, more site establishment, more spillways, and diversion and outlet works. Silt accumulation is also substantially less, as the United States Department of Agriculture figures show: reservoirs smaller than 10 acre feet (ac-ft) silted up at an average rate of 3.5%/year, smaller than 100 ac-ft at 2.7%/year and smaller than 1,000,000 ac-ft at 0.16%/year. This alone is a powerful argument against a large number of small reservoirs. A comparative study of some key characteristics of three sizes of structures is made in Table 11 to illustrate the argument.

While dealing with hydropower projects, comparisons must look at impacts per unit of output. The impacts of a single large hydro project must be compared with the cumulative impacts of several small projects yielding the same power and level of service. The most fundamental determinant of the nature and magnitude of impacts of hydropower projects are the specific site conditions and not the scale of the project (www.hydropower.org).

Clearly, the degree of regulation and reliability that is provided by a large dam is not possible with a small tank. Small tanks tend to dry up fast during droughts as the surface area is large. In such periods, major and medium projects are the mainstay of water supply. One has to consider this aspect also when planning projects in drought-prone and arid regions. To conclude, one must go for the optimal size of the project rather than getting bogged down in small vs. big controversy.

Table 11. Contrast of characteristics of the High Aswan Dam, Dharoi reservoir, and a minor tank in Sri Lanka

Characteristic	High Aswan Dam	Dharoi reservoir	Typical minor tank
Storage capacity	168.9 km^3	1.321 km^3	4.1 ha-m
Surface area	6, 500 km^2	138 km^2	5.0 ha
Net irrigated area	2.648 million ha	36,827 ha	5.0 ha
Storage fraction of area times depth	0.29	NA	0.4
Annual evaporation loss	14 km^3	0.15 km^3	2.0 ha-m
Annual evaporation depth	2.7 m	2.458	1.0 m
Dam height	111 m	45.87 m	2 m
Crest length	3.830 m	1207 m	170 m
Embankment volume	44.3 million m^3	NA	2, 600 m^3
Command area	3.4 million irrigated ha	773,778 ha	< 10 ha

19.3.3. Regional Disparities in Dam Construction

Besides the large regional variations in utilizable flows and good sites for dams, there are large regional disparities in development of storage dams in the country. In basins such as Satluj, Beas, Krishna, Pennar Cauvery, Mahi, Tapi, and Sabarmati, more than 80% of the potential has already been developed. On the other hand, on the tributaries of Brahmaputra and Barak, less than 5% of storage potential has been developed. On the tributaries of Ganga, Yamuna, Mahanadi, and Godavari also very less storage potential has been developed. Finally, there are no large reservoirs on Ghagra, Gandak, and Kosi rivers which carry large flows. Of course, regional disparities in economic and social development are noted across the country.

Among the states, Bihar, Orissa, Madhya Pradesh, and the North Eastern States have least progress in creation of large reservoirs. Notably, the states with limited financial resources have smallest development of LSD during the past 20 years. As the reservoirs are catalysts in economic development, this regional disparity is not a welcome sign. To overcome this problem, a financially strong public sector undertaking may be assigned the responsibility to develop LSD having hydropower component. There are a few such organizations in India in the field of hydropower.

Not the absolute number but the water storage capacity and hydraulic head (energy generation capacity) of the reservoirs is of critical importance. Although the largest number of dams have been built in Madhya Pradesh and Maharashtra, these states could not harness as much benefits as Punjab, Haryana, and Western Rajasthan where only three projects, namely Bhakra, Pong, and Pandoh with huge storage capacities have provided assured irrigation to millions of hectares of land and cheap hydropower for ground water extraction. Of course, besides the WRD projects, the entrepreneurship and labour of the people also has a role to play in this prosperity. The optimum development of water resources potential of Satluj, Beas, and Ravi rivers through a system of large reservoirs, diversions, long distance water transfer and extensive canal network and the resulting socio-economic prosperity of the region is an excellent model for similar development in other river basins of India.

Although there is an extensive canal system in the Ganga basin, this is not backed up by storage reservoirs in the headwater areas. This has hampered assured irrigation through canals in the basin. The Tehri project is nearing completion now and the benefits should start flowing soon. In many other rivers, such as Mahanadi and Godavari, bulk of the monsoon runoff just flows to the unutilized and may also cause damage on its way. Unfortunately, more than 60% of the cultivated area in India is rainfed or dependent on the vagaries of monsoons. One can easily imagine how the agricultural yields will increase if a substantial part of this area can be provided with assured irrigation. As shown in many studies, besides increasing productivity, this can lead to almost magical socio-economic transformation of the country by removing poverty, unemployment, and promoting agro-based industries.

19.3.4. Design of Dam

There are many problems related with decision making for dams in India. According to World Bank (1991), these are basic weaknesses in data, detailed engineering design, and project planning and appraisal, resulting in inappropriate projects. Nine projects were assessed by them in 1989, out of which seven had unacceptable economic rate of returns (ERRs), all below opportunity cost of capital in India. Projects are frequently plagued with cost and time overruns, poor implementation, reduction in irrigated areas, reduction in yields, downward adjustments in cropping patterns, etc. Further, optimistic design assumptions on water availability and irrigation efficiency are made which are not supported by facts.

World Bank (1991) also noted that sensitivity analysis is rarely performed in India in irrigation project planning. The benefit/cost ratios are calculated at the end of project preparation, instead of a decision making aid during project identification and when evaluating design alternatives. The scrutiny of key technical assumptions and economic and financial evaluation are often weak. Although this evaluation by the World Bank is quite old, things don't seem to have improved much over time.

Due to increase of irrigation demands in excess of what can be accommodated with available resources, too many projects were started by the state governments in (particularly in 1970s and 1980s) without proper investigation, and spreading available resources thinly. Another weakness is that the management of irrigation sector in India is highly centralized. Users have very little participation in management of irrigation systems. While taking a decision, political interference is common; there is little accountability and very little performance evaluation.

The available information on rainfall, stream flow, and climate are often inadequate or unreliable. Lack of adequate staff with desired qualifications for analysis and design along with laxity in the technical and economic scrutiny of projects before approval have contributed to a lowering in the quality of project reports. Thakkar (1999) notes that many projects have been initiated by states without prior mandatory approvals, and changes in the scope, design, and cost estimates are large and frequent. Unfortunately, there are no rigorous and detailed analyses of the factors responsible for revision in cost estimates or the justification for the large divergence between projected and actual costs. The underestimation of costs leads to higher final costs for the WRD projects. A study by Public Accounts Committee of 32 projects showed cost escalation of 500% and beyond.

The entire project evaluation process is strictly internal to the government agencies. Unlike many countries, public hearings are not held nor information is disseminated to the general public. Due to this lack of transparency, mistakes often propagate undetected. However, things seem to be changing recently. Feasibility reports of some ILR proposals have been recently put in public domain on Internet. The Right to Information act has been passed and implemented recently but it as not yet made any impact on decision making in water sector.

In several cases, the projects have been started much before the mandatory approval. Thakkar (1999) has cited many examples to support this statement. For example, construction of Nagarjunasagar project in AP was started in 1955, whereas the approval came in 1960; construction of Kosi project in Bihar was started in 1955, three years before getting the approval; construction on Malprabha project in Karnataka started three years before the approval came in 1963; clearance for the Tawa project in MP and Kangasabati in West Bengal came four to five years after the commencement of construction work. In case of Sardar Sarovar Project (SSP), even though final sanctions came only in 1988, the work on the project was started much earlier in 1978, the United Nations Development Programme (UNDP) funded studies started in 1980 and the World Bank loan was approved in 1985.

Possibly the underestimation of the costs were deliberate and not unintended. Moreover, the large dam projects have invariably taken very long to be completed. It is noted that very few projects have been completed within the stipulated target date since independence. Singh (1997) has noted that average delay in the large dam projects studied have been 160%.

Unsatisfactory quality of construction work in many cases due to neglect or corruption is quite common. Paradoxically, there have been many notable developments in project planning and execution in infrastructure sector India but none in water sector.

19.3.5. Construction of New Reservoirs

Construction of new WRD projects is becoming more difficult with the passage of time. Future water resources development will become more complex and capital intensive. Poor planning, lack of coordination, and inept management have been the bane of many WRD projects that have been recently taken up in India.

Completion time of most medium and major projects is significantly higher than the scheduled completion time. These days, completion of a multipurpose project takes around 15 years or more; a major project takes about 10–12 years. There are many reasons behind time and cost overrun of water resources projects and the chief among them are listed below.

a) Delays due to interstate and trans-boundary Issues: Most major Indian rivers flow through more than one state and in addition, major northern rivers are international. Construction of projects in their catchments requires agreements among the involved entities which usually leads to excessive delay.
b) Delays due to environment, forest and rehabilitation and resettlement issues: Stiff opposition is faced while constructing new projects on the plea that these damage environment, submerge forest and agriculture lands, and require rehabilitation and resettlement of large number of people. Construction of projects has been stalled or even cancelled on these grounds. Consider the example of the Sardar Sarovar Project on the Narmada River in Gujarat. At the time of Tribunal Award in 1978, the number of project affected families was estimated around 8,000; currently it is more than 40,000.
c) Insurgency: Adverse law and order has been responsible in delay or some recent projects and survey and investigation work in some parts of the country suffers on account of this factor.
d) Funding: Lack of funding has also resulted in delay in many WRD projects in India.
e) Suitable sites for large and medium storages are limited and most of them have already been exploited. Naturally, the remaining sites are going to present more difficult problems which increase construction time.
f) Institutional mechanisms need to be created or strengthened at State, grass-roots and basin levels, particularly enhancing non-governmental stake holder participation as a core to all institutional initiatives.

Basu and Joshi (2000) identified the following construction related reasons for slippage in respect of time and cost over-run of WRD projects: (1) Insufficient funding, (2) Delay in land acquisition, (3) Incomplete survey and investigation, (4) Construction schedules were too optimistic, (5) Changes in design were made during execution, (6) Cost and time estimates were inaccurate, (7) Underbidding by contractors, and (8) Delays due to contractor related factors. The delay occurring between submission of a proposal and sanction is the most significant reason for delay among different phases of work.

Analysis shows that the present approach of project execution followed by many government departments is inadequate and incomplete; the present practice of scheduling does not meet the requirements; and the present practice of project monitoring has not stood the test of time. Also inadequate are the present practice of inventory management, manpower planning, and general estimating principles. Appropriate level of details and proper documentation are not given in project reports. Moreover, the level and commitment for quality control has significantly dropped and there is an urgent need for a quality assurance system.

For the new and complex projects of current times, the simple technique of a bar chart for planning, scheduling and monitoring are not enough. A conventional approach which is based on experience and common sense is often not helpful when making the decision to speed up project progress by incurring extra expenditure.

Programme Evaluation and Review Technique (PERT) and Critical Path Method (CPM) are extremely beneficial and effective in efficient management of projects and should be followed. Of course, these techniques do not relieve the management of the responsibility for making quick and right decisions.

The construction environment is prone to dispute and in the event of a dispute, the progress of work suffers and slows down. Every contract has clause for arbitration and dispute settlement but frequent resort to such measures is a costly and lengthy process. Proper coordination and trust between the concerned department and contractor is must for successful and timely implementation of a project. Both have to work closely as partners to attain the desired goals. Cost escalation frequently creates problems. It may be helpful to follow a construction cost index which will result in realistic assessment of escalation of construction cost.

The present approach towards environmental consideration in project planning and the execution phase is insufficient and inappropriate. All environmental issues should be properly addressed in the project planning phase. Before starting a construction project it is essential to assess the present environment without the project, and the likely impact of the project on the environment when it is completed. Taking into account certain environmental considerations during the construction phase may help in the environmentally sound construction of projects. While large dams are inevitable in the future development of India, all efforts should be made to minimize submergence of land.

It will be better if project estimates are prepared by making suitable provision for future escalation in price during the period of construction and uncertainty in project cost and time forecast should be incorporated in project estimates and schedules. Project execution on a turn-key basis may help to reduce time and cost over-runs.

19.4. RESERVOIR OPERATION

After creation of storage facilities, the benefits that can be reaped from them largely depend upon how efficiently these are managed. Water stored behind the dams is a precious resource collected at a large cost and it is necessary that it is used in an optimal manner. Scientifically developed operation procedures are not available for a large number of reservoirs in India. While operation manuals have been prepared for some major storage projects, for most Indian reservoirs, well-defined and scientific operation policies do not exist. To that end, it is necessary that optimum operation policies should be developed at least for all the important reservoirs in the country. Expertise is available in the country and to begin with, this exercise could be taken up for all the reservoirs in Table 9 (Ht $>$ 100 m). Later, operation policies can be prepared for the other reservoirs.

Another related issue is that even where such manuals exist, these might have been prepared long ago and the procedures may no longer be optimal as the basin conditions might have undergone large changes over time (say, upstream utilization, changes in downstream reaches) and the demands (purpose of dam, magnitude, and pattern of demands) may have changed. Moreover, the hydrology of the basin might have undergone changes as a result of the so called climate change and the other factors and this also calls that the design and operation practices of storage projects should be periodically reviewed.

Studies have shown that improvement in operation of reservoirs by a few percentage points translates into large sum of money. Systems analysis is an area which has seen enormous growth in applications to water resources management problems. The techniques that are commonly employed include simulation, optimization, and multi-objective analysis. The current trend is to take a holistic view of the river basin and take decisions after taking into consideration engineering, financial, environmental, legal, and social aspects. Recent development in this field includes genetic algorithms and fuzzy approaches, etc. Indigenously developed softwares as well as expertise is available in the country and hence this aspect should be given due importance.

Wherever a network of dams is present in a basin, these should be ideally operated as an integrated system. This is what leads to synergy and is advocated by professionals. Many countries are following this practice to their benefit. However, in India this is rarely done. In case the projects are located in different states, there is hardly any sharing of information and in some cases, sharing is discouraged. As a result, individual reservoirs in many systems are operated in isolation. At times, this leads to undesirable situations in which some reservoirs of the system may be nearly full and some nearly empty. Although the national water policy advocates basin planning and management, this is rarely practiced in the country and it is high time that we create infrastructure to implement this concept.

Reservoir projects provide numerous benefits to the society. In the Indian context, water is stored behind the dams is a precious resource which no operating engineer can let off without expecting re-filling over the next few days/weeks. Further in most projects, reservoirs are gradually filled during the monsoon season. Incidental flood moderation is achieved and sometimes reservoirs are depleted if high flood is anticipated. Such constraints create a situation when flood moderation storage is exclusively unavailable when a flood wave impinges. In many of our projects, we have gated spillways with large discharge capacity at the full reservoir level (FRL). But arrangement like this places gates with high discharge capacity in the hands of operating personnel. However, such a control has the potential to create intense flooding downstream either due to human error or misjudgment. This fact makes it necessary that very clear operation policies are communicated to the reservoir staff and during the flood season, advance information on incoming flood rate and volumes is available for judicious operation of spillway gates. Another necessary facility is a dependable system for operators to fallback upon in case the normal system break downs.

It has also been seen that construction of storages often reduces the occurrence of relatively moderate floods and this tends to attract human activity to the downstream flood plains. Occupations like growing melons, vegetables, and other crops or cattle grazing or fishing begin. These people suffer damage when a big flood comes.

Such situations demand that no economic activity be permitted in the flood plains downstream of a reservoir and whenever this is un-avoidable, reservoir operating staff should warn people downstream before making big release to avoid loss of human or cattle lives and other damages. At all major reservoirs particularly those that are operated for flood control, a real-time forecast network and information about condition in downstream is very essential.

19.5. HYDROLOGICAL ASPECTS OF DAM SAFETY

Safety of dams is vital as a failure can be catastrophic and huge investment and time is needed to construct a dam. Aging of dams, high replacement cost, and increasing difficulties in constructing new dams are the major forces to increasing emphasis on dam safety and operation and maintenance. In India, some catastrophic failures in past have led to focusing attention on hydrological aspects of dam safety. This consists of the need of more critically monitoring the integrity of dams, in evaluation of hazards due to failure of dam operation of spillways, review of design flood, preparation of emergency action plan, etc.

The hydrological aspect is one of the major factors affecting safety of a dam. International organization such as the World Bank have stressed on developing hydrologic criteria to ensure a more specific direction for design flood estimation. The hydrological aspects, which are of interest here, need a regular review of dam safety with the availability of more and more hydro-meteorological data. For all important dams under planning, dam-break study is a must and this study is also being carried out now for major existing dams. A dam-break study attempts to simulate the flooding conditions in the downstream areas should a dam fail. The results help planners understand the risks and consequences of dam failure and in design of necessary follow-up measures.

19.5.1. Flood Control Through Reservoirs

Flood management basically aims to reduce the frequency and magnitude of flood damage caused by overflowing water. The structural measures keep water away from flood prone areas. In the Indian context, water is impounded behind dams at great cost and are a precious resource which should not be let off unless there is a good possibility of re-filling the reservoir.

Many spillways of Indian dams have considerable discharge capacity at the full reservoir level (FRL). This capacity may range from 100 year return period flood to the PMF depending on storage available between FRL and maximum water level in the reservoir. Thus gates with high discharge capacity are handled by operating personnel and slip either due to human error or misjudgment can cause intense flooding downstream. This fact makes it incumbent that both policies of reservoir operation as well as advance information on incoming flood should be available for judicious operation of gates. There should be an emergency operation system for operators. Unfortunately, scientifically derived operation procedures are available for very few dams and many reservoirs are regulated based on ad-hoc rules. Many operators are neither interested nor one trained to improve the performance of their system.

Construction of storages often reduces the occurrence of relatively moderate floods downstream of dams and tends to spread human activity to the downstream flood plains. Occupations like growing melons, vegetables and other crops or cattle grazing or fishing are resorted to. Such situations warrant that reservoir operating staff has to warn people downstream even for normal release to avoid loss of human or cattle lives. Further in the coastal areas the swell in the sea is important as any releases at the time of swell can aggravate drainage congestion and this has to be taken into account in the timing of reservoir releases.

One of the methods to control floods is to store the excess flood water in reservoirs behind dams. For instance, the Hirakud Dam (Mahanadi River) has considerably mitigated the problem of floods in the fertile Mahanadi delta. Likewise, the series of dams constructed in the Damodar Basin, with one of the principal objectives of flood control, have again substantially controlled the recurring flood menace in the Damodar Basin. The Bhakra Dam has more or less solved the problem of recurring floods in the Satluj River, where a large extent of the wide bed of the Satluj below the Bhakra Dam is now put to intensive agricultural use. Surat, a rich and prosperous town in Gujarat, which used to suffer serious flood damage, is now largely protected by the Ukai Dam. Similarly, the problem of floods in quite a few rivers has been mitigated by the moderation effected by the storage space available in major reservoirs although of these were not specifically designed for flood control.

As no major dams have so far been constructed in the Ganga (except Tehri which is nearing completion) and Brahmaputra Basin, the problem of floods is faced year after year. Of course, there is little scope for construction of dams on the major northern tributaries of the Ganga in India as suitable sites for construction of major dams are not available within Indian territory. This is a limitation in solving the problem of floods in lower Ganga basin.

19.5.2. Failure and Safety of Dams

Dams may fail due to one or a combination of the following: i) foundation failure, ii) inadequacy of spillway capacity, iii) poor construction, iv) uneven settlement, v) improper operation, etc. Nearly 40% of large dam failures were reported due to foundation failure, 23% due to inadequate spillway, 12% due to poor construction, 10% due to uneven settlement, and nearly 15 may be attributed to others. The most common causes of failure are overtopping and internal erosion of dam structure. Penman (1996) indicated that out of 2,700, dams 17 were reported to have failed during first filling, 35 during operation caused by internal erosion, and 45 failed by overtopping. Penman also concluded that maximum failures of newer dams have been due to human error. The major concern for the older dams is the estimated design floods, derived using the methods that were available before the time of construction and were not well advanced. These dams have endured long periods of operation. There have been about 200 dam failure cases all over the world and nearly 8,000 people are reported to have lost their lives. Among the dam failures, notably are Valont of Italy (1,963–2,600 lives); South Fork and Teton of USA of 1889 and 1976 with 2,209 and 11 casualties, respectively; Tailing of Italy, Malpasset of France, and Kantle of Sri Lanka in 1985, 1959, 1986 with 200, 421 and 100 lives, respectively. The Table 12 (ICOLD, 1995) indicates the number of major dam failures up to 1965.

In India, 32 m high Waghad embankment dam failed in 1883 during construction. Sometimes, small dams also cause major disaster. For example, 25 m high Tigra dam failure in 1917 claimed 1,000 lives. The major disaster of Machhu II dam failure in Gujarat (26 m, Masonry and Earth-fill dam) occurred in 1979 and led to more than 2,000 casualties. Kaddam (22.5 m), Panshet (53 m), Khadakwasala (20 m), Chikkhole (36.7 m), constructed respectively in 1957, 1961, 1875, and 1968 failed in the years 1958, 1961, and 1972.

Safety of dams is the responsibility of the agency that constructs and operates them. In India, several unfortunate incidents in the last three decades have forced to focus attention on the need to more critically monitor the integrity of dams. The aspects of dam safety include flood warning system, instrumentation, inspection and maintenance, seismic aspects, extreme precipitation events, seepage aspects etc. The aspects related with design flood estimation have been covered in Chapter 5. With the presence of over 4,000 large dams, India has a very good record in dam safety. Detailed procedures of dam inspection are available in the country and are carefully followed by the operating agencies.

19.5.3. Review of Historic Flood Events

Floods are chiefly caused by intensive rainfall. The Machhu II Dam disaster of 1979 is reported to have been caused due to local phenomenon of cloud burst. The unprecedented flood of 1968 in Tapi River caused devastating downstream inundation up to the city of Surat. The flood recorded of Ukai was 42,475 m^3/s from the catchment of 62,000 km^2. Narmada flood of 1970 was 69,500 m^3/s from a drainage area of 87,000 km^2 and is considered to be the highest recorded. In 1973, the Mahi, Sabarmati and Banas river basins experienced large floods, and so did

Table 12. Number of major dam failures up to 1965

Year	No. of failures	Year	No. of failures
Before 1900	38	1940–1949	11
1900–1909	15	1950–1959	30
1910–1919	25	1960–1965	10
1920–1929	33	Unknown	25
1930–1939	15		

Table 13. Design floods and observed highest floods for a few projects

Projects	Catchment area (km^2)	Spillway design flood (m^3/s)	Highest observed flood (m^3/s)	Revised spillway design flood (m^3/s)
Ukai	62, 225	40, 917	42, 475	–
Kadana	25, 206	23, 100	33, 000	49, 544
Dharoi	5, 476	11, 212	14, 150	21, 662
Dantiwada	2, 862	6, 654	11, 950	18, 123
Machhu (II)	1, 929	5, 663	16, 307	20, 925
Damanganga	1, 813	11, 100	12, 900	12, 854
Machhu (I)	735	3, 313	9, 340	5, 947

Damanganga River in 1976. In many parts of the world during last two decades, the data have shown hydro-meteorological uncertainty and sharp deviations both in frequency and intensity and extent of storms. Design floods and observed highest floods for a few projects in Gujarat are given in Table 13.

It may be pointed out that the occurrence of a flood higher than the design flood is not a failure of design engineers. Design flood is worked out by analysis of past data and in future, extreme weather conditions which surpass the past values can develop and may eventually result in higher floods. What is of paramount importance is that such events are carefully analyzed and the practices and policies are updated.

19.6. RESERVOIR SEDIMENTATION

Owing to the geological and climatic peculiarities, rivers in India particularly the Himalayan rivers carry some of the highest sediment loads in the world. A large number of river valley projects have been constructed to serve various purposes. A major input in the assessment of the economic performance and efficiency of dams is the rate of sedimentation, both suspended material and bedload. One of the principal factors, which threaten the longevity of such projects is the accumulation of sediments in the reservoirs. Sedimentation reduces the storage capacity of reservoirs and hence their ability to conserve water for various intended purposes. Sedimentation also reduces the survival of aquatic species and restricts the use of water for multiple purposes. It further increases evaporation due to the increase in the exposure area of water.

Soil erosion and consequent transportation of silt by streams is a natural phenomenon which is a function of soil and or rock type, slope, valley shape, land use, and land cover of the catchment. The reservoirs by themselves do not accelerate the sedimentation rate and their impact on soil erosion in the catchment, if at all, is marginal.

19.6.1. Consequences of Reservoir Sedimentation

All rivers transport water as well as sediment and dam construction impacts the transport of both substances but with important differences. As reservoirs age, the impacts of sedimentation become severe and are attended with more care. Consequences of reservoir sedimentation can be classified in two groups: upstream consequences and downstream consequences.

a) Upstream Consequences

A wide range of sediment-related problems can occur upstream of dams as a result of sediment trapping.

Storage loss: Sediment deposition in the reservoir pool gradually reduces the storage capacity of the reservoir. After certain stage, the reservoir is unable to provide useful service.

Delta deposition: The coarser portion of the inflowing sediment load is deposited where rivers enter reservoirs, forming delta deposits, which not only deplete reservoir storage but can also cause channel aggradations extending many kilometers upstream from the reservoir. Channel aggradations can increase flooding of infrastructure, communities, and agricultural lands on floodplains; increase groundwater levels, creating waterlogging and soil salinization; reduce navigational clearance beneath bridges; and submerge upstream intakes.

Navigation and Recreation: Navigation can be severely impaired by sediment accumulation, especially in delta areas and in the vicinity of locks. Recreational access can be impaired when sediment accumulates at the periphery and boat ramps.

Abrasion: In hydropower facilities, sediments coarser than 0.1 mm greatly accelerate the erosion of turbine runners and Pelton wheel nozzles. In case of high-head operation, damage can be caused by even smaller particles.

Energy loss: Loss of storage reduces the ability to capture high flows for energy generation.

Intakes and outlets: Sediments can block intakes and low-level outlet at dams and can damage gates not designed for sediment pass.

Earthquake hazard: Sediment deposits have a greater mass than water. Therefore, the presence of sediment against the dam can significantly increase the force of earthquake shaking against the structure.

Ecology: Changes in sediment loading and sediment accumulation within the pool can dramatically alter reservoir ecology, affecting species composition and fishing.

b) Downstream Consequences

River reaches downstream of dams suffer large environmental impacts due to flow reduction and alteration of hydroperiod, reduction of sediment load, and altered nutrient dynamics. Stream morphology downstream of dams can be dramatically impacted by the reduction in the supply of bed material sediment. Clear water in the river channel downstream of the dam will tend to scour the streambed. Channel degradation can increase bank height and bank erosion rates, increase scour at downstream bridges, lower water levels at intakes, and reduce navigational depth at critical locations. Sediment trapping by reservoirs reduces the suspended solids concentration downstream, which may have many beneficial effects.

19.6.2. Management of Sedimentation

Engineering measures aim to reduce or prevent sheet erosion and gully erosion and restore degraded agricultural and non-agricultural lands. They provide a necessary gestation period and help build up a desired moisture regime to carry out other measures of land stabilization, such as afforestation. Important principles to be kept in view while planning engineering control measures are:

1. Increase the time of concentration and allow more runoff to be absorbed and held by the soil,
2. Break a long slope into several short ones to reduce the flow velocity below critical limits, and
3. Prevent excessive soil and water loss.

In the earlier times, it was assumed that sediment would settle only in the dead storage. But this is not supported by actual observations. Actually, sedimentation takes place throughout the reservoir. The sediment inflow rates need to be checked up through reservoir resurveys. A practice in India is that the elevation capacity table after 50 years of sediment deposition in the reservoir is used in simulation studies for the project. IS: 12,182 (1987) "Guidelines for Determination of Effects of Sedimentation in Planning and Performance of Reservoirs" is the national practices on this topic. This standard emphasizes the need for periodic resurveys and has the following main features:

a) The sedimentation rate is to be decided on the basis of observations of sediment inflow and reservoir surveys.
b) The live storage is to be so planned that the benefits do not reduce for a period of 50 years (full service time) for irrigation projects or 25 years for hydropower projects connected to a grid on account of sedimentation.
c) The live storage is to be planned so that sedimentation beyond the outlet, causing operational problems, would not occur for 100 years for irrigation projects and 75 years for hydropower projects.
d) If sedimentation is not serious, the conditions obtained at the end of full service period are to be used throughout the simulation period. If the problem is serious, studies are to be done by redistributing sediment and recomputing trap efficiency in 10-year blocks. The extent of studies to be done is linked to the seriousness of the problem, as assessed in a preliminary study. For this purpose, the problem is categorized in three classes:

Insignificant – If the annual loss of capacity is less then 0.1 percent, the problem is insignificant. No check on full service time needs to be made. The availability of adequate feasible service time, however, has to be ensured.

Significant – If the annual loss of capacity is between 0.1 percent to 0.5 percent, simulation or working table studies may be done for the reservoir geometry as obtained at the end of the full service time. This would simplify the simulation

study, and would also ensure that the planned benefits are available for this period. The availability of adequate feasible service time is also to be checked.
Serious – If the annual loss of capacity is beyond 0.5 percent, the recompilation of trapping efficiency and reservoir geometry for every 10 years in the simulation studies is preferred.

In the approach incorporated in the Indian Standard IS: 12,182 (1987), the end of Phase-I will depict the end of the period in which the reservoir is capable of yielding the full planned benefits. The Phase-II would depict a period when the operation of the reservoir is also trouble free in regard to sedimentation, although the efficiency of the reservoir gradually reduces. The Phase-III would be a period of troubled operation, and unless some new engineering solutions are implemented, the project may have to be given up in phase-IV or phase-V.

Some people suggest comprehensive soil conservation measures as panacea to arrest and control soil erosion and thereby sedimentation of reservoirs. This is neither possible nor practical. Firstly the cost of treatment would be prohibitive (the rate of catchment area treatment for Indus, Ganga and Brahmaputra basins works out to more than Rs. 25,000/- per ha which for Tehri project would mean an investment of over Rs. 900 crore at 2,000 price level and would take about 10 year time). Secondly the occurrence of big slides in these valleys is natural and not necessarily caused by construction of any water resources project. There are regions in the country whereon major project has been constructed yet instances of soil erosion are observed.

Thus, a practical and constructive approach in this respect would be to make a comprehensive catchment survey in respect of soil erosion and target those zones which are likely to witness severe soil erosion.

19.6.3. Reservoir Sedimentation in India

On average, about 16.75 ton/ha/year of soil is lost through erosion every year in India. This means that more than 5,000 million tons of topsoil is eroded annually. Further, almost 173.64 M ha area (slightly half of the country), is threatened by various types of degradation, such as salinity, alkalinity, water logged areas, ravinous and gullied lands, areas under ravages of shifting cultivation, desertification, etc. About 800 ha of arable land are being lost annually due to ingress of ravines. Frequent occurrences of floods and droughts in different parts of the country are, to some extent, evidence of improper land use in the catchments and inadequate conservation of rainwater.

An important consideration in planning of reservoirs in India is gradual reduction in the storage capacity due to deposition of the sediment brought by rivers. This is particularly important in view of the fact that good storage sites are limited and hence it is imperative to find ways and means to prolong the life of reservoirs and to ensure that realistic assumptions about the rate of sedimentation are made at the planning stage itself.

In the 20 big and medium rivers of the country, the rate of land erosion is between 10 MT and 799 MT per sq. km. During the last few years, due to the deposition of sediments, the capacity of Nizam Sagar Dam has gone down to almost half. Every year about 330 MT of sediments are deposited in the Bhakra Nangal Dam and this can be highly dangerous. In July 1970, the initial 12 km of upper Ganga canal near Haridwar was thoroughly filled up with sediments which could only be cleaned in six months at a considerable expense.

Although sedimentation surveys of reservoirs in India date back to early as 1870, systematic surveys started only in 1958 when the Central Board of Irrigation and Power undertook a coordinated scheme of reservoir sedimentation and entrusted this task to several research stations in the country, viz., Karnataka Engineering Research Station; Directorate of Irrigation Research, Bhopal; Maharashtra Engineering Research Station; UP Irrigation Research Institute; Andhra Pradesh Engineering Research Laboratory, etc. Under the scheme 28 major reservoirs were surveyed. During the VIII five-year plan period, Central Water Commission also formulated a scheme for carrying out capacity survey of thirty important reservoirs in the country through consultants available in the field.

Details of surveys in respect of 144 reservoirs of the country have been collected and analyzed by Central Water Commission (2001). An abstract showing the average rate of sedimentation, percentage loss of storage and other useful information are given in Table 14. The percentage of loss of capacity up to the last survey has been given in the table and it varies from 0.65% to 60.47%. Out of 144 reservoirs, 46 reservoirs have lost less than 10% of their capacity, 34 reservoirs have lost 10–20% of their capacity, 31 reservoirs have lost 20–30% of their capacity, and 33 reservoirs have lost more than 30% of their capacity up to the last survey. In all 47 reservoirs had lost more than 25% of their capacity up to the last survey. Twenty-nine reservoirs have served more than 50 years of their useful life. The sedimentation problem is high in Salal dam and is discussed in what follows.

Data of sixty-seven reservoirs with known design rate of sedimentation were analyzed by CWC (2001). The results are given in Table 15. It is observed that the actual rate of sedimentation is more than the design rate of sedimentation in most of the reservoirs. In 16 reservoirs, the actual rate is more than 5 times the design rate.

CWC (2001) reported analysis of sedimentation data of 144 reservoirs. The annual percentage loss in gross storage (APL-GS) was worked out as

$$APL-GS=\frac{\text{Total annual loss in gross storage of 144 reservoirs}}{\text{Total gross storage of 144 reservoirs}}*100 \tag{1}$$

The other indicators were also computed in a similar manner. Table 16 contains the results. The weighted average annual loss in gross storage due to siltation is 0.44%. Thus, assuming the current total gross storage at 217 km^3, the likely annual loss due to siltation is 0.95 km^3. Based on the data of 42 reservoirs, the annual

Table 14. Sedimentation of Reservoirs in India (10% of storage and above)

Name of Reservoir	Name of River	Year of first impoundment	C.A in sq. km	Storage capacity in MCM	Designed rate of siltation th. Cu. m/sq. km/yr	Number of surveys (Year of last survey)	Observed rata of siltation th. Cu. m/sq. km/ yr	Total storage loss till last survey (MCM)	Percentage annual loss of capacity
				Andhra Pradesh					
Himayatsagar	Issa	1927	1, 307.94	107.79	*NA*	1(1976)	0.447	28.63	0.54
Kaddam	Kaddam	1958	2, 656.25	124.43	*NA*	1(1977)	0.916	46.251	1.96
Manjira	Manjira	1966	16, 770.2	50.94	*NA*	1(1977)	0.102	18.74	3.34
Nizamsagar	Manjira	1930	21, 694.0	841.18	0.238	3(1992)	0.378	508.68	0.975
Sriramsagar	Godavari	1970	91, 751.0	3, 171.94	0.357	2(1994)	0.28	616.46	0.81
				Bihar					
Konar	Konar	1955	997.15	281.23	*NA*	1(1996)	1.750	71.58	0.62
Maithon	Barakar	1955	6, 294.0	1, 348.80	0.905	6(1994)	1.075	264.04	0.50
Panchet Hill	Damodar	1956	10, 878.0	1, 581.00	0.667	6(1996)	0.510	222.91	0.35
Tilaiya	Barakar	1953	984.20	335.83	*NA*	1(1997)	2.857	2.812	0.84
				Gujarat					
Bhadar	Bhadar	1983	406.00	46.72	0.357	1(1995)	2.260	11.01	1.96
Bramani	Bramani	1953	699.27	74.95	0.719	1(1986)	0.72	16.62	0.67
Dantiwada	Banas	1965	2, 860.90	464.4	0.361	3(1994)	1.748	145.08	1.07
Dharoi	Dharoi	1976	5, 540.00	907.88	0.286	5(2000)	0.763	105.63	0.46
Dhatarwadi	Dhatarwadi	1975	429.94	32.73	0.190	1(1986)	1.254	5.93	1.65
Karjan	Karjan	1984	1, 404.00	657.72	0.476	2(1998)	3.945	77.51	0.84
Khodiyar	Shetrunji	1967	383.30	40.35	0.357	1(1987)	1.360	10.41	1.29
Limbdi Bhogavo	Limbdi Bhogavo	1960	331.50	30.15	*NA*	2(1986)	0.890	7.66	0.98
Machhu-11	Machhu	1972	1, 928.00	100.55	0.476	1(1997)	0.827	39.86	1.59
Meshwo	Meshwo	1968	258.96	82.12	0.0857	1(1997)	3.78	28.49	1.18
Moj	Bhadar(s)	1955	440.30	53.01	0.357	2(1986)	1.031	14.07	0.86
Ranghola	Kalubhar	1952	370.37	44.52	0.143	2(1986)	0.623	7.84	0.52

(Continued)

Table 14. (Continued)

Name of Reservoir	Name of River	Year of first impoundment	C.A in sq. km	Storage capacity in MCM	Designed rate of siltation th. Cu. m/sq. km/yr	Number of surveys (Year of last survey)	Observed rata of siltation th. Cu. m/sq. km/ yr	Total storage loss till last survey (MCM)	Percentage annual loss of capacity
Sasoi	Sasoi	1954	562.03	51.02	*NA*	1(1986)	0.670	12.05	0.74
Shetrunji	Shetrunji	1959	4, 317.00	415.41	*NA*	2(1996)	0.630	106.72	0.60
Ukai	Tapi	1972	62, 224.0	8, 510.00	0.149	3(1992)	0.813	1, 013.1	0.595
				Karnataka					
Basavasagara	Krishna	1970	47, 850.0	1067.37	0.0184	1(1990)	0.151	144.91	0.68
Tungabhadra	Tungabhadra	1953	28, 180.0	3, 751.17	0.429	6(1993)	0.527	593.64	0.40
				Kerala					
Anayirankal	Panniar	1964	65.68	49.84	*NA*	1(1997)	7.11	15.41	0.94
Kuttiyadi	Kuttiyadi	1972	39.00	38.40	*NA*	1(1989)	16.77	11.12	1.70
Madupetty	Periyar	1967	104.90	55.22	*NA*	1(1995)	1.614	6.433	0.31
Peechi	Manali	1957	107.10	110.43	*NA*	2(1995)	7.570	30.82	0.90
Ponmudi	Panniyar	1962	220.52	51.54	*NA*	1(1992)	1.674	11.08	0.72
Poringalkuttu	Chalakudi	1957	512.00	31.99	*NA*	1(1993)	0.443	8.157	0.71
				Maharashtra					
Asolamendha	Jadam	1918	246.00	92.96	0.994	2(1994)	1.602	29.97	0.42
Ekruk	Adelanalla	1871	412.00	94.30	0.154	2(1991)	0.534	26.436	0.234
Gangapur	Godavari	1965	357.40	212.51	0.335	1(1997)	2.30	48.89	0.72
Khadakwasla	Mutha	1870	507.00	110.00	*NA*	1(1940)	0.674	23.92	0.31
Mangi	Kanala	1955	304.00	33.839	0.050	3(1995)	0.283	3.435	0.253
Manar	Manar	1969	1, 585.08	138.35	0.167	1(1999)	0.394	18.726	0.451
Mhaswad	Man	1888	1, 243.20	86.94	0.176	2(1990)	0.357	45.24	0.51
Nalganga	Nalganga	1963	315.98	76.201	0.190	1(1985)	0.624	4.338	0.26
Ramsagar	Sur	1914	212.35	117.18	0.206	1(1987)	0.953	14.78	0.17
ShivajiSager	Koyna	1961	891.80	2, 797.45	0.667	1(1986)	0.810	18.05	0.03
Visapur	Hanga	1937	412.00	42.76	0.357	3(1988)	0.835	17.55	0.80

				Meghalaya					
Yeldari	Puma	1963	7, 362.00	934.45	0.357.	1(1983)	0.577	84.95	0.45
Umium	Umium	1965	2, 21.50	181.42	*NA*	1(1990)	2.61	14.44	0.32
				Orissa					
Balimela	Machkund	1972	4, 908	3, 610.53	10.46	1(1999)	2.131	282.55	0.29
Hirakud	Mahanadi	1957	83, 395.0	8, 105.00	0.25	4(1994)	0.635	1, 959.3	0.653
				Punjab					
Bhakra	Satluj	1958	56, 980.0	9, 868.00	0.429	26(1998)	0.61	1, 390.4	0.35
				Tamil Nadu					
Aliyar	Aliyar	1962	195.00	109.40	*NA*	2(1981)	0.848	3.14	0.15
Amaravathy	Amaravathy	1958	839.16	117.16	*NA*	1(1986)	0.717	16.84	0.51
Emerald	Emerald	1961	58.534	156.75	*NA*	2(2000)	2.83	6.626	0.106
Avlanche	Avalanche								
Kaveripakkam	Palar	1902	31.08	41.73	*NA*	2(1996)	1.644	4.803	0.12
Krishnagiri	Pannalaru	1957	543.00	68.25	*NA*	1(1976)	0.172	17.78	1.37
Lower Bhavani	Bhavani	1953	4, 200.00	932.78	*NA*	4(1983)	0.30	37.752	0.14
Manimutharu	Kilmani Muthar	1958	161.62	159.734	*NA*	1(1980)	1.220	4.324	0.12
Mettur Stanley	Cauvery	1934	42, 200.0	2, 708.76	*NA*	2(1984)	0.253	533.33	0.39
Mukurthy	Pykara	1938	25.25	50.976	*NA*	1(1993)	11.80	16.391	0.58
Pechipparai	Kodaiyar	1907	172.00	150.27	*NA*	1(1992)	1.790	6.46	0.20
Perunchani	Paralaiyar	1953	159.49	65.03	*NA*	2(1995)	1.154	4.787	0.23
Pillur	Bhavani	1966	943.54	44.40	*NA*	1(1982)	1.157	17.27	2.43
Sathanur	Ponniar	1957	10, 826.0	243.828	*NA*	2(1982)	0.102	27.526	0.47
Thirumurthy	Palar	1966	80.29	54.80	*NA*	2(1994)	2.490	5.39	0.36
Upper Bhawani	Kundah	1965	33.59	101.148	*NA*	1(1985)	5.460	3.6676	0.18
Vaigai	Vaigai	1958	2, 253.30	194.79	*NA*	3(1983)	0.40	22.405	0.46
Veeranam	Cauvery	1923	427.35	40.805	*NA*	1(1991)	0.449	13.061	0.47
Wellington	Vellar	1924	129.50	71.46	*NA*	1(1985)	2.046	16.14	0.37
				Uttar Pradesh					
Dhukwan	Batwa	1907	21, 340.0	106.45	0.042	5(1980)	0.030	47.42	0.61
Matatila	Betwa	1956	20, 720.0	1, 132.70	0.133	9(1994)	0.469	369.19	0.86
Nanak Sagar	Deoha	1962	570.00	209.80	*NA*	1(1992)	1.212	20.73	0.33

Source: CWC (2001).

Table 15. Comparison between actual and design rate of sedimentation in some Indian reservoirs

Actual rate of sedimentation	Number of reservoirs
Less than design rate	8
1–2 times the design rate	13
2–3 times the design rate	14
3–4 times the design rate	8
4–5 times the design rate	8
More than 5 times the design rate	16

loss in live storage came out as 0.31% which implies that the likely annual loss in the total live storage is 0.55 km^3. Assuming the average density of sediments to be 1.137 ton/m^3 based on observed data, the weight of the total sediment deposited in all the Indian reservoirs is 1,080 M ton/year.

19.6.4. Empirical Equations for Indian Reservoirs

Various researchers have developed empirical equations to determine sedimentation rates in Indian reservoirs, based on actual sedimentation data and sediment carrying capacity of inflowing rivers. Some important equations are given next.

Dr. Khosla's findings (1953)

The earlier designs of reservoirs in India were based on the silting rate as suggested by Dr. A.N. Khosla which, on an average was, 3.6 ha-m per 100 km^2 of catchment area. Later surveys on some Indian reservoirs revealed that the actual silting experienced by them has been much higher. Dr. Khosla proposed an enveloping curve of the form:

$$y = 32/A^{0.72} \tag{2}$$

where y = Annual silting rate in ha-m/100 km^2 and A = catchment area at the site in km^2.

Table 16. Annual percentage loss in gross, live and dead storages

Indicator	Minimum value	Maximum value	Weighted average	Remarks
Annual percentage loss of gross storage	0.04	3.34	0.44	Based on weighted average data of 144 reservoirs
Annual percentage loss of live storage	0.11	5.00	1.40	Based on weighted average data of 42 reservoirs
Annual percentage loss of dead storage	0.03	1.62	0.31	Based on weighted average data of 42 reservoirs

Source: CWC (2001).

Murthy's equation

Murthy (1980) suggested the following equation to determine silt deposit in dead storage zone:

$$S = KC^n \tag{3}$$

where S = Percent of total sediment in the dead storage, C = Percent of dead storage capacity to total capacity, and K and n are constants depending on the type of reservoir. Their values are given in Table 17.

Garde and Kothyari equation

An analysis of sediment yield from Indian catchments was carried out by Garde and Kothyari (1985, 1986). Using data from 50 small and large catchments in India, they developed the following relationship

$$y = 0.02p^{0.60}F_e^{1.70}\overline{S}^{0.25}D_d^{0.10}\left(\frac{P_{\max}}{P}\right)^{0.19} \tag{4}$$

where y = sediment erosion rate in cm, P_{max} = average maximum monthly rainfall in cm, $\overline{S}$ = average catchment slope, F_e = erosion factor. Further,

$$S = \frac{\sum A_i S_i}{A} \tag{5}$$

$$F_e = \frac{1}{\sum A_i}(0.8A_A + 0.6A_G + 0.3A_F + 0.1A_W) \tag{6}$$

where, A_A = arable area (km^2), A_G = scrub and grass area (km^2), A_F = protected forest area (km^2), and A_W = waste area (km^2). Kothyari et al. (1986) derived the following equation to predict annual sediment yield:

$$y = 0.02F_e 1.7\overline{S}^{0.25}D_d^{0.10}\left(\frac{P_{\max}}{P}\right)^{0.19} P_a^m \tag{7}$$

y_a = annual sediment yield in cm, P_a = annual rainfall in cm, m = coefficient depending on coefficient of variation, 0.6 to 0.607 for C_v changing from 0.1 to 0.7.

Table 17. Values of K and n in Murthy's Equation

Reservoir Type	Values of	
	K	n
I (Lake)	3.30	0.78
II (Flood plain in foot hill)	9.33	0.56
III (Hill)	25.12	0.35
IV (Gorge)	32.36	0.30

19.6.5. Trap Efficiency of Some Indian Reservoirs

The major reservoirs constructed in the country have capacity-inflow ratio (a measure of detention time) ranging from 0.2 to 1.3. As per Brune's curve, such reservoirs are almost 100% efficient in trapping the incoming sediment. Relatively few data have been obtained on the actual trap efficiency of existing reservoirs. In order to arrive at the trap efficiency of some of the reservoirs, observations on the sediment load entering into the reservoir from different sources at its head reaches and the sediment load passing through the spillway and other outlet through the dam and power house are needed. Varshney (1997) presented the results of observations at Matatila Reservoir (UP), Hirakud Reservoir (Orissa), Govind Sagar (HP) and Gandhi Sagar (MP) on sediment inflow and outflow. Briefly, the findings are given below.

Matatila Reservoir

The trap efficiency for sand particles is 100% while that for the silt particles ranges from 70 to 90% and for clay particles, it ranges from 60 to 90%. The overall trap efficiency of the reservoir for the period 1962 to 1972 was found to vary from 67 to 90%. According to Brune's trap efficiency curve, this reservoir has a capacity-inflow ratio 0.2 and a trap efficiency of 90%.

Hirakud Reservoir

The overall trap efficiency of the reservoir was found to vary from 65 to 90%. The capacity-inflow ratio of Hirakud is 0.2 and according to Brune's curve a trap efficiency of about 92% is expected.

Gandhi Sagar

This reservoir has a capacity inflow ratio of 1.3 and the trap efficiency according to Brune's curve is almost 100%. Actual observations also support this.

Bhakra Reservoir

According to observations, the trap efficiency of the reservoir is almost 99%. The Bhakra Reservoir has a capacity-inflow ratio of 0.7 and according to Brune's curve the trap efficiency works out to 95%.

It has been observed that in reservoirs having small sluicing capacity with respect to normal floods and no reservoirs above them, the siltation rate is comparatively high in the first 15–20 years of their operation and thereafter, it falls off. This is because the obstruction by the dam causes the dips and the flanks of the storage basin to fill up with silt in early years. A stage comes when the river section is adjusted to carry the normal discharge and disposal of sediment load in the area of reservoir is harmonized with the condition of flow. Besides the progressive development of deltas above reservoir helps in trapping of some of the silt load. Shrinkage and settlement of deposited silt also takes place due to superimposed

Table 18. Reduction of Sedimentation Rate with Time – Indian Reservoirs

Reservoir	Sedimentation rate (ha-m/100 km/yr)	
	At first survey	At last survey
Maithon	12.53	9.06
Panchet	12.13	3.36
Matatila	11.82	5.29
Bhakra	6.53	5.66

loads of additional silt deposits. This results in reduction of silt volume thereby reducing the computed sedimentation rate.

CWC studied data of four reservoirs in detail to determine the trends of siltation rate with passage of time. The reservoirs chosen were Maithon and Panchet (both in Bihar), Matatila (UP) and Bhakra (in HP). The conclusion drawn by CWC was: "The enveloping curves of these reservoirs show that the silting rate has been comparatively higher during first 15–20 years and thereafter has fallen significantly". The data in respect of these four reservoirs are given in Table 18.

Use of remote sensing technique for reservoir sedimentation assessment is becoming popular as this technique offers many advantages compared to the other methods. Many organizations in the country are using remote sensing technique for this purpose. Ministry of Water Resources had launched a scheme sometime ago involving its concerned organizations to assess sedimentation rates in about 125 reservoirs. Results of this work are being published by the concerned organizations. As an outcome of this effort, useful database has been created and the latest status in important Indian reservoirs is now available.

19.7. LAKES

Under favorable conditions, depressions are formed on the earth's surface. Large depressions that are filled with water are described as lakes. Lakes receive water as precipitation on their surface, from surface inflow and from ground water. A lake loses water in the form of evaporation from its surface, streams flowing out, contribution to ground water, etc. They may disappear through a process of natural eutrophication involving the filling up with nutrient containing sediments. The process of cultural or man made eutrophication is much faster than the slow rate of natural eutrophication.

There are about 500,000 lakes on Earth, storing volume of water equaling 103,000 cub. km. Most of the world's water lakes are found in North America (25%), Africa (30%) and Asia (20%). The world's lakes contain about four times more fresh water than its rivers. Lakes are used extensively in many countries as the natural centers of civilization. A lake plays a significant role in shaping the hydrological, ecological, environmental, socio-economical balance of the region.

Lake is also a place for sanctuary for migrating birds, development of flora and fauna and an excellent spot for habitation of aquatic biota.

Lakes exist in all sizes and it is obvious that a small pond of 0.01 km^2 is totally different in most ways than lake of 50 km^2. For example, one of the Great Lakes in North America with surface area of around 50,000 km^2. Climate, geology and anthropogenic influences are other factors that make each lake unique. Due to the presence of a large number of lakes, Udaipur City in India is known as the city of lakes. Figure 4 shows one of the lakes of Udaipur.

19.7.1. Differences Between Lakes and Reservoirs

Although lakes and man-made reservoirs appear to be similar on a casual look, there are many fundamental differences (Table 19) between them and these also influence the chemical and biological processes. The differences in capacity-inflow ratio and water level fluctuations in lakes and reservoirs are largely responsible for this difference in behavior. A small residence time means that many species do not have an opportunity to reproduce in most reservoirs.

Both natural lakes and artificial reservoirs receive a wide variety of sediments and nutrients as inputs and these get accumulated and deposited. There are large variations in deposition rates depending on the catchment properties as well as the properties of the water body (including regulation in case of reservoirs). In general, the rate of sedimentation in man-made reservoirs is much higher compared to natural lakes. The process of sedimentation leads to a gradual reduction in the storage capacity and over time, the lake or reservoir may ultimately reduce to a marshy land.

Geometrically, lakes and reservoir are quite different and the reason lies in their origin. Natural lakes are typically circular or oval shaped. The central portion is deepest in a lake. In contrast, the site for a reservoir is decided based on many

Figure 4. A picturesque lake of Udaipur

Table 19. Important differences between natural lakes and man-made reservoirs

Feature	Natural lake	Man-made Reservoir
Age	1000s of years	100s of years
Capacity-inflow ratio	Large	Wide range
Water level fluctuations	Small	Usually large and seasonal
Location of maximum depth	Commonly near the center	Close to the dam
Source of inflows	Surface and subsurface	Predominantly surface
Outlet of water	Surface and subsurface	Almost totally surface
Catchment : water surface area	Small	Large
Shape	Mostly oval or circular	Usually linear or dendritic
Sediment and nutrient loading	Low	High
Bio-diversity	Higher	Lower
Primary productivity	Lower	Higher
Water quality gradients	Concentric	Longitudinal

factors; a deeper reservoir will have less surface area and less loss of water due to evaporation. Such a reservoir is created where river slopes are high. The upstream end of a reservoir has shallow depth and the maximum depth is near the dam. Since most of the flow enters at the farthest upstream end and leaves at the downstream end, there is a strong longitudinal flow and density currents may also be present.

19.7.2. Hydrology of Lakes

Hydrologic characteristics of lakes vary considerably because of difference in depth, breadth, width, surface area, basin material, surrounding ground cover, reservoir, prevailing winds, climate, surface inflows and outflows and other factors. Lakes may have some common features but often exhibit different performance characteristics. This individuality has environmental value and as such it presents the problem of understanding both, the general nature of the system and variations due to local conditions.

Lakes influence the local hydrology either directly or indirectly. In hydrologic cycle, the water passing through the lake will be filtered both with respect to its physical properties (including velocity) and its chemical properties. The water balance of lakes is different than that of the rest of the drainage basin. Hence water balance of a drainage basin having high percentage of lakes will be different than low percentage dittos. Because of storage of large mass of water, a lake moderates flood and climate factor in the region. Smaller the lake, the more responsive it is to changes in energy inputs. Small lakes behave as a single system, but almost all large lakes respond as a complex of sub-basins, each of which may be significantly different in size, form and depth. The large free surface area of the lake provides a great contact area between water and air and thus evaporation is enhanced. It is an accepted fact that large lakes possess specific system characteristics that distinguish them from small lakes.

Heat storage in lakes helps stabilize the air temperature, minimizing and lagging variation in adjacent region in winters and summers. Lakes also exhibit thermal stratification like reservoirs. In order to evaluate the lake problems and more realistic approach for their management, heat budget, nutrient budget and water budget of lakes are needed.

19.7.3. Eutrophication

The word "eutrophic" is generally taken to mean "nutrient rich" and is used sometimes a contrast to dystrophic (ill-nourished). Eutrophication is the biological response of a water body to excess nutrient inputs. As a result of nutrient availability, production of biomass and its death and decay results in a number of effects. Individually and collectively these result in impaired water use. The most important of these effects are decreased dissolved oxygen levels, release of odorous compounds (e.g. H_2S) and siltation. Many important lakes in India, e.g., Hussein Sagar (Hyderabad), Nainital (Uttar Pradesh), Khajjiar (HP), and Dal (Jammu and Kashmir) have reportedly progressed to advanced eutrophication levels.

Eutrophication has dramatically increased since the sixties, mostly in industrial countries with intensified agriculture, due to excessive anthropogenic input loads of plant nutrient, phosphorous and nitrogen. Therefore as contrast to natural eutrophication, the recent problem is anthropogenic eutrophication. This is usually observed as the excessive growth of phytoplankton that turns standing water and sluggish streams into green, often termed as "algae bloom". It is frequently associated by the increased growth of attached algae or macrophytes.

19.8. LAKES OF INDIA

Depending upon the depth of water and uses, the lakes are known by different names in different parts of the country like, Jheels, Bheels, Marshes, Talab and Tank. In India, the lakes have been extensively utilized, often resulting in overexploitation. A national inventory of lakes entitled "The All-India Wetland Survey" was completed in the mid 1980s.

Some lakes in India have significance from religious point-of-view. The city of Amritsar in Punjab which is famous for the Golden Temple got this name (Amritsar means the lake of nectar, the drink of Gods. He who drinks nectar becomes immortal) because of the holy lake (see Figure 5). A bath in the lake water is considered to be a privilege.

Similarly, a big artificial lake is present in the Kurushetra city (Haryana) which was the venue of the famous Mahabharata war. Known as Brahma Sarovar (Figure 6), this lake is fed by the water of Bhakra canals. A bath in this lake on a full moon day as well as on the occasion of solar eclipse is considered to be spiritually highly rewarding.

In the following, some important lakes of India are briefly described. After that, summary information of some lakes will be given.

Figure 5. The holy lake of Amritsar with the Golden Temple

Figure 6. The Brahma Sarovar, in Kurushetra city (Haryana)

19.8.1. Dal Lake

The Kashmir valley is gifted with exotic natural scenic beauty of its landscape and water bodies, which have given it the sobriquet "paradise on the earth". These water bodies are of great ecological and socio-economic significance. The Dal Lake with its multi-faceted eco-system and grandeur has been inviting the attention of national and international tourists.

Dal Lake is one of the most beautiful lakes of India and is the second largest in the J&K state. Being located in the heart of the Srinagar City (latitude 34°18′N, longitude 74°91′E, average altitude of 1,583 m), Dal can be considered to be an urban lake. Srinagar is the summer capital of J&K state. During winters when the temperature may fall to as low as −11 °C, the top crust of the lake freezes. Early spring and summers are the wet periods when large rainfall occurs; the average annual rainfall at this place is 655 mm. It is in this season that the snow melts in the higher catchment which results in high discharge in Dachigam and Dara Nallah which feed water into the lake. The maximum depth of the lake is reported to be in the range of 6 to 9 m. The maximum area of the Dal Lake has been estimated to be 24 sq. km out of which open water area is around 15.42 sq. km.

The normal temperature of the Srinagar varies from 11 °C to 1 °C in winters and 12 °C to 30 °C in summers. Mountains from three sides surround the lake. A large number of gardens and orchards have been developed along the shores of the lake. Another unique feature of the Dal Lake is hundreds of houseboats or floating houses which provide an opportunity to tourists to reside on the lake in an atmosphere of peace and tranquility. Of concern is the fact that many of these houseboats dump their waste directly in the lake.

As stated earlier, Dal is a typical urban lake which is mainly used for tourism and recreation. Fishery and harvesting of economically important water plants are of secondary importance. During winter months the Dal water is also used for drinking water purposes.

The Dal Lake basin can be classified into five basins, namely, 1. Nehru Park Basin, 2. Nishat Basin, 3. Hazratbal Basin, 4. Nagin Basin, and 5. Brari Nambal Basin. All the basins are interconnected with navigation routes in the shape of intertwined waterways.

The water quality of Dal Lake has deteriorated considerably in the last three decades. Dal Lake underwent far-reaching changes close to its banks and also in the catchment. Two additional islands were built which further obstructed water movements. A road was laid along the southwest part to improve communication, which separated a large part of the lake creating marshy areas along the fringes of Shankaracharya and Zabarwan mountains. This marshy land had since been reclaimed and developed into huge residential and commercial complexes. With the increase in the tourist influx, a large number of residential buildings, restaurants and hotels have come up along the lakefront. The number of houseboats has also increased. As a result of rapid and unplanned urbanization, a large quantity of sewage finds its way into the lake water and at some locations, the lake water has turned into a cesspool of filth.

Inflow and Outflow: A perennial inflow channel known as Telbal Nallah enters the lake from the north and supplies 80% of the water from a high attitude lake called Marsar Lake. Within the lake basin itself, there are number of springs, which act as permanent water source to the lake.

Towards the southwest side, an outflow channel drains the lake water into a link channel connected to the Jhelum River. Parallel to this exit, a stone lined canal serves as a navigation lock used for movement of boats in and out of the lake. Flow of water in the channel is regulated by two lock gates, which are generally open when the level of the lake is higher than that of the river. The gates are closed during floods in the Jhelum River to prevent the river water from entering into the lake and resulting in inundation of the floating gardens and human settlements. A small canal connects the Nagin basin of the Dal Lake with Anchar Lake and acts as an additional outflow channel called Nallah Ameer Khan.

Nagin Lake

Nagin Lake is known as the 'Jewel in the Ring'. It is considered to be the most beautiful of the Dal Lake region. This lake gets its name (Nagin means female snake) from a large number of trees, which encircle the small, deep blue lake. The waters are edged by willow and poplar trees whose reflection is mirrored in the lake. The lake water is pleasantly cool from mid-May to mid-September. Nagin Lake, which is farther from the Srinagar town provides a magnificent view of the mountains. Nagin is separated from the larger Dal Lakes by a narrow causeway. It usually has a number of houseboats floating around its periphery.

19.8.2. Surinsar Lake

The Surinsar Lake is situated about 40 km to the north east of Jammu city at an elevation of 605 m above mean sea level and lies at 75°02′30′ east longitude and 32°46′30 north latitude. It is a fine picturesque sweet water lake with a circumference of 2.496 km. The maximum length, breadth and depth of the lake are 888 m, 444 m and 24.05 m respectively. The water spread of the lake varies from 27.92 to 29.14 hectares. The water level of the lake oscillates by about 1.20 m and touches its peak during August. The excess water flows towards the western side of the lake and goes into a channel lying by its side. The lake does not have any permanent inlet. The main source of water is monsoon rain which is aided by certain natural springs. The water usually comes from the surface drainage and runoff from the watershed surrounding the lake. There are agriculture fields towards eastern side of the lake and fine covered hills are situated towards the northern and western side.

The Surinsar Lake is nearly oval shaped with the deepest part towards its north west. A small island is located inside the lake, more towards its north east and is rich in terrestrial flora and fauna. The littoral zone of the lake has a thick vegetation cover of emergent floating, submerged plants providing suitable cover for roosting of migratory and nesting of the resident water fowl (known as Murgabi and Jal Moorgi in the local language) in the lake.

The Surinsar Lake is situated in the lower Shiwalik ranges of the Western Himalayas. These outer hills are formed entirely of the younger tertiary rocks, the geological studies of the area mainly comprises of anticlines. The rocks surrounding the lake are mainly composed of sand stone, silt stones and shales of maroon to buff colour.

The climate of the area is subtropical. Monsoon rains are received from July to September. The average annual rainfall is 150 cm. Winters are mostly dry with occasional rains during the month of January. The temperature, the atmosphere and the water remain well above the freezing point. The air temperature during summer ranges from 35 to 40 °C. The temperature of the water surface ranges from 17 to 29 °C. The wind speed ranges from 1.5 to 5 km/ hour throughout the year.

The catchment area of the Surinsar Lake is badly degraded due to large-scale deforestation and denudation which, among other things, has caused a lot of siltation into the lake. The precipitation received in the form of rain and snow cannot be trapped in absence of vegetative cover. The top loose soil gets eroded and finds its way into the lake as surface run off. The silt deposits have reduced water spread area of the lake.

Water Quality of Lake

- With average pH values of the lake water varying from 8.27 to 3.43, the water in the late is alkaline in nature.
- The average concentrations of major cations including calcium, magnesium, sodium and potassium vary from 19.55 to 38.10 mg/1,6.80 to 13.37 mg/1,5.51 to 10.15 mg/1 and 2.64 to 4.01 mg/1 respectively.
- The average concentrations of major anions such as chloride, sulphate, bicarbonate, nitrate and phosphate vary from 3.75 to 9.75 mg/1,6.48 to 11.28 mg/1, 81.50 to 126.50 mg/1,3.14 to 5.05 mg/1 and 0.015 to 0.05 mg/1, respectively.
- Typical range of DO values is 7.39 mg/1 against prescribed minimum value (6.0 mg/1) for Class-A drinking water (Indian Standards). The maximum value of BOD is about 1.63 mg/1 during peak winter.
- The average values of calcium and magnesium hardness vary from 48.75 to 95.0 mg/1 and 28.0 to 55.0 mg/1, respectively.
- The lake water is good for irrigation purposes.

19.8.3. Mansar Lake

The Mansar Lake is located 55 km east of Jammu at a latitude of 32°41′N and longitude of 75°05′E, 666 m. above mean sea level. The basin area of the lake is 1.67 km^2. The maximum length and width of the lake is 1,204 m and 645 m respectively. The annual average rainfall in the catchment area of the lake is 1,500 mm. The total volume of lake water is 12.37 Mm^3. The maximum depth of water in the lake is 38.25 m. The slope of the lake in between 0.0–5.75 m depth is

0.21 m/m, 5.57–10.75 m depth is 0.30 m/m (maximum) and 35.75–38.25 m depth is 0.04 m/m The Lake is fed by rainfall and groundwater.

Apart from the hydrological aspects, Mansar Lake has the religious importance too. On the eastern bank of the lake there is a shrine of Sheshnag, a mythological snake with six heads. Two ancient temples of Umapati Mahadev & Narsimha as also a temple of Durga are situated in the vicinity of the Mansar Lake. People take a holy dip in the water of the lake on festive occasions.

With all religions belief and heritage behind the Mansar Lake, it is popular tourist destination also picking up its fame among the tourists with all its flora & fauna. There is a wildlife sanctuary housing animals like Spotted Deer, Neelgai besides water birds such as Cranes, Ducks, etc. Mansar Lake has a unique distinction: probably it is the only sweet water lake in the world which has a huge population of turtles. Turtles are known to normally prefer salt water bodies.

NIH has conducted study for bathymetry, sedimentation and water quality of this lake. Mean rate of sedimentation of the lake is $0.23 \pm 0.02\,\text{cm/year}$ and life of the lake is $9,110 \pm 790$ years.

19.8.4. Khajjiar Lake

Khajjiar Lake and its surroundings are one of the most picturesque saucer shaped plateaus and a tourist attraction. The watershed is located at a height of 1,940 m in a valley between Dhauladhar and Pir Panjal ranges of the Himalayas. The watershed lies at latitude of 32.5 °N and longitude of 76.1 °E. The size of the lake is not large and is more or less like a pond of about 60 to 80 m radius. The green pastures surrounding the lake are approximately 1.5 km long and 1.0 km wide. The lake region experiences a moist temperature climate. Precipitation is in the form of snow during January to March and rain during the south-west monsoon. The annual average precipitation is 1.2 m. The soils are sandy loam in the meadow part, while under the forest they are highly organic with humus and litter. The soil thickness varies from 30 to 40 cm.

A thick forest of Deodar and Spruce Fir surrounds the Khajjiar Lake and the meadow. The lake has a small floating island in the center. Currently the lake is in a state of neglect with grass and weeds all over. The lake as such has plenty of slush, weeds and decaying organic matter, resulting in the lake becoming more or less like a swamp. However, the state government is planning to clear the lake of weeds, divert water coming from rivulets feeding the lake, control the silt and develop greener pastures around the lake.

19.8.5. Nainital Lake

The Nainital Lake is situated at an altitude of 1,937 m above sea level and is 1,433 m long and 423 m broad at its widest. The total surface area is about 4.65 ha and the volume is approximately 8.33 MCM. The catchment area of the lake is 3.6 km^2 ranging in height from 1,937 m to 2,600 m. The mean hillslope of the area is 19° where large part being confined to the slope group of 20° to 25° and the maximum slope reaching 47° to 49°. The average slope of the snow view Sherka Danda ridge is 18°, varying between 5° to 35°. At many places, the slopes exhibit convex bulges resulting from continuing creep movement. The slopes are locally broken by scarps and fringed at the base by a succession of debris cones and fans. The bathymetric study reveals that the lake consists of two V-shaped basins, one in the north and the other in the south with maximum depths of 27.3 m and 25.5 m respectively. The depth of lake at the dividing ridge is 8.5 m. An outlet for draining out excess water is situated at the north-eastern end. The shoreline is steep, except at a few places where the drains have deposited silt and debris.

The annual rainfall is high in the catchment area of the lake and varies between 2,245 mm to 2,480 mm. The average monthly rainfall is 189 mm with a maximum of about 624 mm in August and a minimum of about 2.4 mm in March. Besides rainfall, there are occasional snowfalls in and around the lake catchment during the winter, varying between 200 mm to 600 mm in recent years.

Deforestation of the surrounding hills is causing land slides due to which heavy load of boulders and sediment get dumped in the lake from time to time. The lake receives 30% of its total inflow as surface runoff from the catchment and 16% as input from the direct rainfall. Total inflow from perennial drain such as Naina Devi is 15% of the input. The rest of input is from smaller drains. Outflow from the lake is in the form of springs.

The mean rate of sedimentation in the lake is $3{,}462\,m^3/year$ and life of the lake is $2{,}160 \pm 80$ years. However useful life of the lake is only 82 to 380 years. Trophic state index is 74 indicating hypereutrophic conditions.

19.8.6. Harike Lake

A barrage was constructed in 1952 at Harike, at the confluence of Beas and Sutlej rivers at about 60 km from Ferozpur, Punjab. Harike Lake came into being as a result of the construction of this barrage. The lake is located at latitude of 31°10′ N and longitude of 74°56′ E, 210 m above mean sea level. In the beginning, the lake had water spread area of 41 sq. km. It is one of the six most important wetlands in the country. Over the time, this lake began attracting migratory birds and became famous. However, later on, the fast spreading hyacinth plants have reduced the open water area to a mere 28 sq. km, leaving little space for migratory birds. Growth of hyacinths in the Harike wetland has pushed it to the brink of an ecological disaster. About 70% of the lake water surface is covered with water hyacinth. The major problems, which are facing by the lake is acute soil erosion and silting. However, a turn around began with the launch of "Operation Sahyog", initiated jointly by the Punjab Government and the Indian Army. World Wildlife Fund provided its expertise to restore the lake to its former glory.

During the monsoon period, enormous quantity of water flows out of Harike barrage. In the decade (1992–2002), the quantum of outflow water from Harike barrage varied from 765 MCM to 11,735 MCM.

19.8.7. Loktak Lake

Loktak Lake is situated 38 km south of Imphal city, the capital of Manipur State. The Manipur State has two distinct river basins, namely the Barak basin and the Manipur basin. The Manipur River arises in the north at Karong. Its tributaries are the Iril River, the Thoubal River, the Sekmai River, and the Khuga River. Manipur River has been regulated by two barrages for irrigation and hydropower. The Imphal barrage downstream of Lilong regulates the flow to irrigate about 6,000 ha area. The second barrage at Ithai diverts the river flow into the Loktak Lake for lift irrigation and a hydropower project.

Manipur is dotted with several shallow lakes, locally called pats. The important pats are Lamphel Pat between the Nambul and Imphal Rivers; Waithou Pat between the Iril and Thoubal Rivers; Ikoppat, Karung Pat and Lousi Pat between Thoubal and Sekmai Rivers; and Khoidum, Lamjao and Pumlen Pat south of Sekmai River. The Loktak Lake comprises about 20 small and large pats including Loktak, Takmu, Ungamen, Laphupat, Thammumacha, Khulak, Vena and Tharopokpi are fairly large (more than 80 ha). Two other pats just north of Loktak Lake are Sana pat and Utra pat. During the rainy season, most of these pats become continuous and merge under large sheet of water but these can be distinguished during the dry season at a water level of about 766 m above msl.

Several small rivers and streams rising in the surrounding hills drain into these pats on both sides of the Manipur River. The Nambul River and Nambol River are two major streams from the north flowing into Loktak basin. Besides these, there are over 34 small streams draining from the hills on the west into the Loktak basin. Loktak Lake is connected to Manipur River by a small channel (about 15 m wide and 3 m deep) called Khordak cut which drains the lake into the river during the dry period. However, when the river is in spate, it flows back into the lake through the channel. Similarly, the Pumlen pat is also connected to the Manipur River by a narrow channel, the Marambakhong, which also allows flow both ways. Thus, during the period of high flood, a large part of the valley on both sides of the Manipur River becomes a large lake.

Loktak Lake had vast area of 2,000 km^2 in 1950 that reduced to 495 km^2 in 1971 and 289 km^2 in 1995. The maximum depth of lake has reduced from 29 m in 1950 to about 20 m in 1980. The lake is located between longitudes 93°46′ & 93°55′ E and latitudes 24°25′ &, 24°42′ N. It is a shallow water lake, the depth of which during dry season ranges between 0.5 m to 1.5 m. Main water body of the lake is surrounded by shallow water stagnating over marshy/swampy land on 11 sides. About 40% of the lake surface area is covered by different types of weeds both floating and submerged. Southern portion of Loktak Lake (south of Thanga, Ithing and Sendra islands/hills) forms the Keibul Lamjao National Park and is the only floating wildlife sanctuary in the world. It is composed of a continuous mass of floating *phumdi* occupying an area of 40 sq. km. The park is the only natural habitat of the most endangered mammal, the brow-antlered deer *(Cervus eldi eldi)*.

Loktak Lake basin has a direct catchment area of 980 sq. km and indirect catchment of 7,157 sq. km. Out of the direct catchment area, 430 sq. km is under paddy cultivation, 150 sq. km under habitation/ settlement, and 400 sq. km under forest. The direct catchment area in the hills covers 96 hill villages. The elevation varies from 780 m at the foothills adjoining the central valley to about 2,068 m above mean sea level at peak. A number of streams originate from the hill ranges immediately to the west of the lake and these streams flow directly into Loktak Lake.

The Loktak Development Authority has been created for integrated and sustainable management of the resources of Loktak Lake. Its office is located at Imphal.

19.8.8. Chilika Lake

Chilika is the largest brackish water lagoon that sprawls along the east coast of India in the Mahanadi delta. It is a tidal lagoon created by a beach barrier berm that developed by the accretion of the coastal sediments following the stabilization of sea levels about 3,000 to 4,000 years ago. The pear shaped lagoon has a maximum length of 64 km and an average width of 20 km. The water depth generally fluctuates between 50 cm and 3.7 m. The water-spread area of the Lagoon varies between 906 and 1,105 sq. km. A 35 km long, narrow outer channel connects the main lagoon to the Bay of Bengal, near the village Motto. The mouth connecting the channel to the sea is close to the northeastern end of the Lagoon. High tide near this inlet mouth drives in salt water through the channel during the dry months from December to June. With the onset of the rains, the rivers falling into the northern zone bring in fresh water currents that gradually push the seawater out. As a result of these, the inlet mouth constantly changes its position. The inlet channel is connected with Chilika at Magarmukh. The other connection with the Bay of Bengal is through Palur Canal on the southeastern side. Several islands are located in the Lagoon covering an area of 223 sq. km, which include hills situated both inside and around the lagoon.

The total catchment area of the lagoon is 4,300 sq. km out of which 3,212 sq. km (74%) lies in Eastern Ghats and 1,088 sq. km (26%) in Mahanadi River system. About 30% of the catchment area is under degraded forest. The total annual fresh water inflow into the lagoon by surface has been estimated at 1,760 MCM. Direct precipitation to the Lagoon contributes 870 MCM of water. The total evaporation loss from the lagoon has been estimated at 1,286 MCM. Chilika Development Authority has estimated that 53 rivulets that drain in Chilika deposit about 365,000 tons of silt; the maximum contribution (89%) being from Mahanadi river system.

In general, lagoon water is alkaline with pH ranging from 7.1 to 9.6. The dissolved oxygen values have been recorded between 3.3–18.9 mg/l. The salinity concentration levels show remarkable variations, both temporally and spatially. A complex combination of freshwater discharge, evaporation, wind conditions, and tidal inflow of seawater govern the seasonal changes in salinity levels.

Biodiversity

The Chilika lagoon is identified as one of the hotspots of biodiversity in India. The spatial and temporal variations in salinity in combination with nutrient rich shallow warm water of the lagoon makes it a unique site rich in biodiversity. It is the largest wintering ground far migratory waterfowl in the country. Based on its unique biodiversity and socio-economic importance, Chilika Lagoon was designated as a *Wetland of International Importance* under the Ramsar Convention of 1981.

Chilika supports some of the largest congregations of aquatic birds in the country, particularly during the winter. Flocks of migratory waterfowl arrive from as far as the Caspian Sea, Baikal Aral Sea, remote parts of Russia, Kirghiz steppes of Mongolia, Central and South East Asia, Ladakh and the Himalayas. In 1989–90 an estimated population of two million birds visited the lagoon. The lagoon hosts over 160 species of birds during the peak migratory season. Much of the Lagoon bottom is covered by aquatic weeds.

During the last few decades, the lagoon has been showing signs of deterioration mainly due to siltation and over-exploitation of its resources. The gradual transformation of brackish water characteristics of the lagoon into fresh water, increase in freshwater weeds, reduction in fish productivity and changes in faunal diversity are quite apparent.

The denudation of the catchment area, construction of dams, barrages and other hydraulic structures and increasing anthropogenic pressures coupled with natural processes have led to degradation of the Chilika Lagoon. Siltation has seriously affected the Lagoon area, which has shrunk from 824 sq. km in 1972–73 to 750 sq. km in 1995. The cross section of the outer channel has also significantly reduced due to shoal formation, which leads to considerable hydraulic head loss and poor flushing. The summer depth has also reduced to 0.3 m at Magarmukh. Salinity in the lagoon has decreased significantly during the past few decades while weed coverage has increased. CWPRS, Pune, carried out a model study of the Chilika Lagoon to restore salinity level and increase its productivity.

Chilika Lake Development Authority looks after the maintenance of the lake. Its office is located at BJB Nagar, Bhubneshwar, Orissa.

19.8.9. Pushkar Lake

Pushkar is a famous pilgrimage center of Rajasthan. It is one of the few places on the Earth where Lord Brahma (the creator of the universe) is worshipped. The city has lent its name to the lake that lies inside it. In ancient times, the lake had a waterspread of over 71 bighas. However, of late, the Pushkar Lake is dying due to a variely of reasons. On an average the lake attracts 5,000 pilgrims daily. People from all over the country converge here to wash off their sins and immerse the ashes of their dead. The depth of water has plunged to just 1.5 m from a depth of 9 m observed in the late nineteen eighties. The condition of the depth lake is so bad that it can no longer sustain life. In the recent past, fish, weighing five to twenty kilograms, have died in the viscous brown depths of the lake due to the lack of oxygen.

Besides the authtopogenic causes, successive droughts have severely depleted the natural water sources. The water depth in the lake is so shallow that it rakes up the mud from the bed of the lake and forms a foul-smelling mixture. Over the past 10 years the water level has been receding continuously, partly because of the numerous tubewells dug by farmers living in the adjoining areas. The problem is aggravated by the stealthy advance of the desert into this area. To maintain water level in the lake, tubewells have been installed near its periphery. At times, these tubewells are operated round-the-clock to withdraw water from aquifers and put it in the lake.

19.8.10. Kolleru Lake

The scenic Kolleru Lake, situated in the Krishna and West Godavari districts of Andhra Pradesh, 50 km east of Vijaywada at latitude of 16°30′ N and longitude of 81°15′ E , 0–5 m above mean sea level, is spread over an area of more than 900 sq. km. The lake area is about 1,090 km^2 at the time of maximum flooding. The lake has a maximum depth of more than 3.5 m when full during the rainy season and a minimum depth of 1 m during summer. Kolleru Lake is one of the largest freshwater inland lakes in the world and is one of the largest bird sanctuaries, home to nearly 188 species of birds. It is also the world's largest natural freshwater fish producer – the produce is about 30,000 tonnes per year. The sedimentation in the lake is high. The lake bed is rising at 2.5 cm/year. Unfortunately, the lake is fast drying up due to indiscriminate exploitation and encroachment. High use of inorganic fertilisers in the catchment area and in pisciculture has resulted in increased nutrient load and reduction in storage capacity of the lake. A high growth of aquatic weeds and eutrophication is rendering the lake unfit for sustenance of aquatic animals like fish and freshwater otter. At the time of flooding in river Krishna and Godavari, sea water gets trapped in middle of the lake which makes the lake water saline.

19.8.11. Udhagamandalam and Kodaikanal Lakes

Due to growing human activity in the hilly regions and associated functions, such as intensive agriculture, animal husbandry and opening of commercial establishments, the Udhagamandalam (Ooty) and the Kodaikanal (Kodai) Lakes have come under enormous pressure. The Ooty Lake is an artificial lake which was constructed by John Sullivan, the first collector of Ooty. Originally it was intended to be an irrigation tank but during the years 1823–1825, it was dredged and enlarged. This lake is a big tourist attraction and supports water-based recreation. Over the years, increasing siltation and eutrophication have inflicted enormous damage to the lakes.

The National Lake Conservation Programme (NLCP) broadly envisages pollution abatement of the water bodies. The Ooty and Kodaikanal Lakes were chosen under this programme to improve and restore the health of these lakes from the viewpoint of ecology and tourism. The project components are conservation, protection and restoration of forest cover, erosion control measures on agricultural and vacant lands by permanent vegetative cover/engineering structures, desilting, and improvement of treatment system for sewage and solid waste, including vermi-compost disposal. Pollution in the Udhagamandalam Lake is mainly due to discharge of domestic sewage in the lake. The situation in Kodaikanal Lake is less severe compared to the Ooty Lake.

19.8.12. Other Lakes in India

Besides the lake described above, there are many small and big lakes in the country. Some other important lakes are described in Table 20.

In addition to this, there are numerous lakes in Uttaranchal, which are listed in Table 21. It can be seen from the table that the Nainital district has the largest area under the lakes.

19.9. PROBLEMS OF INDIAN LAKES

During recent years, due to alteration of landscape in India by denuding forests, urbanization, increasing tourist activities, and waste discharge in the lakes, sedimentation and eutrophication of lake water have increased. Even many high altitude lakes in Kashmir and in Garhwal Himalayas, which have remained clean and without eutrophication for centuries, are showing signs of deterioration. The famous Dal Lake of Kashmir, which covered about 40 km^2 in the beginning of nineteenth century, presently covers only about 20 km^2. Almost half of Renuka Lake (water spread of 670 ha), the biggest lake of Himachal Pradesh in the lesser Siwaliks, is slowly filling up with sediments. Things are much worse in the plains or in peninsular India. Upper Lake in Bhopal, Poondi, Red Hills in Madras and Osmansagar in Hyderabad, sources of drinking water for their cities have shrunk considerably in the recent past causing great hardship to the city dwellers. Not only that, in many cases lakes of smaller sizes located in the urban areas are used as a dumping spot of wastes, both solid and liquid, due to which the temperature structure of the lake changes. As a result, many lakes in India are experiencing the problem of eutrophication. Occurrence of inorganic nutrients in water and the resulting increase in plant productivity has become a serious water quality consideration all over the country.

Studies have been carried out in India mainly with ecological, environmental, socio-economic, and limnological aspects of lakes. However, by far only a few lakes have been studied and it is necessary to initiate studies of many more important lakes. Besides, it is of utmost importance that the recommendations of these studies are implemented in the right earnest.

In India, natural and/or manmade changes in storage of lakes, both in quality and quantity of water, have altered not only the stream flow regime but also the water balance of the region. Alteration of flow regime and quality of lake water, a common

Table 20. Salient features of some of the important lakes in India

Name	Location	Dimensions	Storage	Inflow	Outflow	Use of lake water	Other information
Lakes in Rajasthan							
Fatehsagar Lake (Rajasthan)	North-west of Udaipur city	L: 2.6 km, B: 1.8 km. Mean Depth: 5.4 m, Max Depth: 13.4 m, A: 4.0 km^2	Gross, live, and dead storage capacity: 12.1, 7.0, and 5.1 Mm^3 respectively.		Connected to Swaroopsagar Lake through a canal.	CCA: 680 ha; also used for drinking purposes	
Udaisagar Lake (Rajasthan)	13 km from the Udaipur city	L: 4.0 km, B: 2.5 km, D: 10.7 m, A: 10.4 km^2	Gross, live and dead storage capacity: 31.15, 27.6 and 3.5 Mm^3 respectively.	Ahar River; Lake also receives industrial effluents and domestic waste.	Berach River originates from the lake	CCA: 6,450 ha; also used for fishing, industrial, & drinking purpose.	
Sardarsamand lake (Rajasthan)	Pali, 55 km south of Jodhpur. at confluence of Sukri and Guhiya Rivers	Max. depth 7.6 m, A: 36 km^2	Capacity is 88 Mm^3			CCA: 10,336 ha.	
Rajsamand Lake (Rajasthan)	Kankroli in Rajsamand district at a distance of about 65 km from Udaipur city. Latitude: 25°04′ N; longitude: 73°48′ E	L: 7.0 km, B: 2.5 km	Gross, live and dead storage capacity: 107.2, 98.7 and 8.5 Mm^3 respectively.	River Gomati, a tributary of Banas	More than 4,000 m^3 of water/ day, withdrawn through tubewells by mines and other processing units.		

Ramgarh Tal (Rajasthan)	About 35 km north-east of Jaipur. Latitude and Longitude are 27°03′ N and 76°04′ E	Mean depth: 4.6 m; max depth: 18.0 m, CA: 760 km^2	Gross and live storage capacity: 59.0, 58.98 Mm^3 respectively.	Banganga River		CCA: 9,990 ha. Also used for drinking supply to Jaipur	
Jaismond Lake (Rajasthan)	About 51 km from Udaipur; Latitude: 25°05′ N and Longitude: 73°28′ E. Altitude from msl: 720 m.	Max depth: 20 m; A: 30 km^2;		Average Rainfall: 800 mm; No stream connected			Due to rapid defor-estation sediment load in the lake is increasing.
Sambhar Lake (Saline lake, Rajasthan)	About 64 km to the west of Jaipur; Latitude: 27°00′ N, Longitude 75°00′ E; Altitude from msl: 360 m.	L: 32.0 km, D: 22.0 m, CA: 233.0 km^2		By seasonal rivers Bandi, Medha, and several other streams		78 km^2 of the eastern arm of Sambhar Lake is devoted for salt production.	Tropical monsoon climate; Lake often dries in early summer
Phulera lake (Saline lake, Rajasthan)	Lies at 65 km to the north west of Sambhar; Latitude: 27°12′ N, Longitude: 74°34′ E	A: 2 km^2					Water level fluctuates from few cm to a maximum of 2 m

(Continued)

Table 20. (Continued)

Name	Location	Dimensions	Storage	Inflow	Outflow	Use of lake water	Other infor-mation
Lakes in Rajasthan							
Diwana lake (Saline lake, Rajasthan)	Lies at 60 km to the north west of Sambhar; Ltitude: 26°52′ N, Longitude: 75°11′ E.	A: 2 km^2				20% of Didwana Lake is used as salt pans.	Water level fluctuates from few cm. to a maximum of 2 m.
Pichola Lake, (Rajasthan)	Located in the western part of Udaipur city; latitude: 24°34′ N, longitude: 73°41′ E. altitude: 587 m above msl	Max L: 3.6 km, Max B: 2.6 km. Min B: 1.93 km, Mean Depth: 4.5 m, Max Depth: 10.5 m, A: 10.8 km^2	Gross, storage capacity: 13.67 Mm^3	Average rainfall: 700 mm: Sisarama River, a tributary of Kotra River is the chief source of water	Bechar River outflows from the lake	Municipal supply to Udaipur City, irrigation, bathing, fishing, recreation and socio-cultural activities	Due to rapid defor-estation sediment load is increasing year by year
Rangasagar Lake (Rajasthan)	Middle of the chain of lakes of Udaipur	L: 1.03 km, B: 0.25 km. Max Depth: 7.0 m, A: 0.136 km^2		Pichola lake to its southern side	Swaroopsagar lake on its northern eastern part		
Swaroopsagar lake (Rajasthan)	Lying to the south of Rangasagar lake		Gross, live and dead storage capacity of Pichola-Swaroopsagar	Rangasagar lake	Fatehsagar lake through a canal	CCA of the complex: 580 ha	

			lake complex: 13.7, 9.0 and 4.67 Mm^3 respectively.				
Nalsarwar Lake (Gujarat)	60 km south west of Ahmedabad; latitude: 22°47′ N, longitude: 72°03′ E, altitude: 11.5 m. above msl	CA: 115 km^2		Gets water only from monsoon runoff		Nalsarwar Bird Sanctuary on the sides of the lake, famous for migrating birds, even from Siberia	Large, shallow, fresh water lake; dries completely before onset of the monsoon
Pangon Tso Lake (Sikkim)	100 km south east of Leh-Ladakh in Indus basin; latitude: 33°50′ N, longitude 78°53′ E, altitude: 4,218 m above msl	Max Depth: 100 m, CA: 650 km^2		Input to the lake is only from precipitation		Has protected high altitude cold desert National Park	Narrow brackish lake, gets frozen in winter
Wullar Lake (J & K)	Lies at a distance of 60 km from Srinagar, between Bandipore and Sopore	L: 24 km, B: 10 km.		Excess water from the Jhelum,			Largest fresh-water lake in India
Mirgund Lake (J & K)	Located at a distance of 15 km north west of Srinagar; latitude: 34°08′ N, longitude: 74°38′ E, altitude: 1,580 m above msl.	Max. depth: 1.05 m, A: 3 km^2		Fed by precipitation only	No visible outflow from the lake		Shallow fresh water lake with reed beds, highly eutrophic

(Continued)

Table 20. (Continued)

Name	Location	Dimensions	Storage	Inflow	Outflow	Use of lake water	Other infor-mation
Lakes in Rajasthan							
Renuka Lake (H.P.)	Located in Nahan, at a distance of 173 km south west of Shimla; Latitude: 30°36′ N, longitude: 77°27′ E, altitude: 700 m above msl.	L: 1.5 km, B: 0.3 km. Min Depth: 5.0 m, Max Depth: 10.0 m, A: 0.5 km^2		Fed by monsoon runoff from the catchment area and perennial underground source		Destination for more than 600 migratory birds, 443 species of fauna, mammals, birds, reptiles and amphibians	Largest, narrow, oblong shaped natural water body, a pilgrim center of Renukaji (mother of Hanuman)
Naukuchiatal Lake	At 15 km from Nainital; latitude: 29°32′ N, longitude: 79°21′ E, altitude: 1,320 m above msl.	Mean Depth: 21.89 m, Max Depth: 41.25 m, maximum aerial extent of the lake is 0.90 km^2			A small artificial outlet leading to Bhimtal Gadhera		The deepest and most voluminous of the Kumaun lakes
Bhimtal Lake	5 km from Nainital; latitude: 29°21′ N, longitude: 79°34′ E, altitude: 1,345 m above msl.	Min Depth: 12 m, Max Depth: 24.75 m, Volume of lake: $4,245 \times 10^3\ m^3$		Perennial lake, receives water from direct rainfall and catchment runoff	Outflow from the lake goes to nanda drain		Largest amongst Kumaun lakes, irregular in shape, shore line index: 1.67, development index: 1.74.

Pullicat Lake (Tamil Nadu)	45 km north of Chennai; latitude: 13°25′ to 13°55′ N, longitude: 80°05′ to 80°19′ E, altitude: 0–10 m above msl.	L: 60 km, B: 5 to 15 km. Mean Depth: 2.0 m, A: 720 km^2		Fed by two rivers Separated from Bay of Bengal by Sri Harikota Island		Drinking, fishing, and navigation; Lake has also a bird-sanctuary	
Upper Bhopal Lake (Madhya Pradesh)	Situated in Bhopal city; latitude: 23°25′ N, longitude: 77°15′ E, altitude: 508 m above msl.	Min Depth: 3 m, Max Area: 10 m, Original A: 195 km^2: Area reduced in 2002: 30 km^2		The lake is fed by river Kolans.	Spillway from the lake is discharged by 11 gates and joins Kalisor River which finally meets in Betwa River	Encroached for agriculture, socio-economic activity, urban devel-opment transport facilities, recre-ational activities, etc.	1,100 years old man made fresh water lake, polluted by non point source pollution like domestic and industrial wastes
Chandpatta Lake (Madhya Pradesh)	5 km east of Shivpuri in Madhya National Park; latitude: 25°26′ N, longitude: 77°42′ E, altitude: 450 m above msl.	Max depth: 12 m, A: 2 km^2					Lake is eutrophic due to organic pollution and sedimen-tation.

(Continued)

Table 20. (Continued)

Name	Location	Dimensions	Storage	Inflow	Outflow	Use of lake water	Other information
Lakes in Rajasthan							
Sagar Lake (Madhya Pradesh)	Situated in Sagar city; latitude: 23°50′ N, longitude: 78°45′ E, altitude: 517 m above msl.	Max L: 1.247 km, B: 1.207 km. Periphery: 5.23 km; Max depth: 5.38 m, Mean depth: 2.48 m, A: 1.07 km^2; Volume: 4×10^6 m^3		Feeder Kanera Canal			Total annual quantity of silt deposition about 4, 500 m^3
Lonar Lake (Maharashtra)	Situated 140 km. from Aurangabad; latitude: 20°10′ N, longitude: 71°45′ E, altitude: 350 m above msl.	Circular Dia at Rim: 1.8 km, diameter at water surface: 1.2 km. Mean depth: 100 m		Underground feeding of the lake bringing pollutant into the lake			Around 50,000 years old, largest and oldest natural lake of meteor origin in the world
Rudrasagar Lake, Tripura	Located in the Agratala City of Tripura; latitude: 20°15′ N, longitude: 92°10′ E, altitude: 40 m above msl.	Max depth has been reduced from 20 m to 6 m in 1950; Water spread area of the lake has reduced to about 6 km^2 from 12 km^2 in 1950		Input to the lake is monsoon runoff from the catchment area	Spillway from the lake is discharge to River Gomti which flows downstream of the lake	The lake water is used for fishing, recreation, tourism and irrigation.	A palace, "Neer Mahal" was constructed inside the lake. Night soils and sewage from Agartala is dumped into lake.

Deepar Beel (Assam)	Located 10 km south west of city of Guwahati; latitude: 26°10′ N, longitude: 91°45′ E, altitude: 53 m above msl.	Max depth: 4 m; A: 40 km^2		The outflow from the lake goes to river Brahmaputra depending upon the relative water level of the lake.	Half lake dries in winter and converted to rice fields. continuous encroachment by the farmers resulting in shrinkage of the lake	Natural fresh water lake, famous for Jatia Hills on the southern side where migrating birds collide with the hills and die
Pakhal Lake (Karnataka)	Located 40 km east of Warangal; latitude: 17°57′ N, longitude: 80°00′ E, altitude: 85–90 m above msl.	A: 15 km^2	Numerous ephemeral and semi perennial streams feed the lake			
Krishnaraja Sagar lake (Karnataka)	Situated in the Mysore city; latitude: 18°45′ N, longitude: 80°10′ E, altitude: 100 m above msl.	Mean depth: 30 m, Max depth: 100 m, A: 75 km^2			Several varieties of flora and fauna are in abundance in the lake and its catchment.	Man made fresh water lake, constructed during 1895–1940 by King Krishnaraja Wodeyar IV.
Sultanpur Jheels (Haryana)	Located 15 km west of Gurgaon; latitude: 28°28′ N, longitude: 76°55′ E, altitude: 220–230 m above msl.	A: 137 km^2		The lakes flood during monsoon and heavy rainfall.	Excavation of sand from bed of the lake for lime industry.	Lake is eutrophic due to organic and urban pollution.

Note: L=Length; B= Breadth: D; Depth; A= Lake Area; CA= catchment Area.

Table 21. Lakes in the Uttaranchal State

Name of the District	No. of lakes	Total Area (in ha)	Altitude range (m)	Names of Lakes
Uttarkashi	5	8.00	2,330 to 3,352	Dyodital, Nachiketatal, Barsutal, Syabatal, Jakholtal
Chamoli	9	14.20	2,230 to 3,900	Brahmtal, Jha Tal, Bheekaltal, Sukhatal, Dewtal, Loktal, Basukital, Trijugi Narayan Dhikharital
Tehri	6	3.95	1,700 to 4,450	Masartal, Marintal, Jaraltal, Chandyaratal, Banghanital, Nekhrital
Rudraprayag	1	2.00	2,230	Dewaria tal
Nainital	9	267.04	1,061 to 1,600	Bhital, Naukuchiatal, Sattal, Nainital, Khurpatal, Garur tal, Lohakham tal, Hrish tal, Nal Damyanti tal
Champawat	1	1.50	1,500	Shyamtital
Total	31	296.69		

problem faced everywhere, arising from increase in demand and developmental activities, are resulting in the hampering of activities supported by these lakes. There are also side effects of ecological imbalances in the region. Impact of degradation is less in the initial stage but the cumulative impact in longer period proves to be deleterious. In many cases, if the state of the problems and their remedial measures is not looked into, it would be difficult to save the lakes and therefore, the adjoining area from economical, hydrological, and environmental disasters.

Many lakes in India are under threat due to a variety of somewhat inter-related reasons. So severe is the problem in some cases that the very existence of the lake is under threat and in fact several lakes have either largely filled up or are close to death. With a national perspective, the most important problems connected with lakes are:

i. Heavy pollution and deterioration of the quality of water in the lakes,
ii. Loss of storage capacity due to siltation, dumping of garbage or intentional filling-up,
iii. Reduction in inflow into the lakes due to upstream diversions, extractions, and over-use of lake water, and
iv. Eutrophication and deterioration of lake water quality due to inflow of pollutants from industries, and municipal areas.

Undoubtedly, concerted efforts are needed to main the health of our lakes and to restore the status of those lakes that the deteriorating due to reasons stated here. Such efforts are urgently needed because of the important role that these water bodies play for the benefit of nature and the society.

CHAPTER 20

WATER QUALITY AND RELATED ASPECTS

The study of relationship between natural resources (water, air and land), human beings, living creatures, plants and microorganism is known as ecology. Due to tremendous increase in population and human activities, there is rampant degradation of environment as well as natural resources. Environmental degradation is caused by deforestation, soil erosion, unplanned industrialization, indiscriminate use of fertilizers and pesticides in agriculture, water logging, and water pollution, etc. Protection of water resources, abatement of water pollution, and restoration of environment are necessary if the society wishes to continue to reap uninterrupted benefits from the use of natural resources.

Water has a key role in sustaining ecological balance. The widespread scarcity, gradual destruction and aggravated pollution of the water resources also lead to degradation of ecology. Water quality issues are gaining recognition as river waters are getting heavily polluted in many places and ground water quality at many places is beginning to deteriorate. In view of this, the Parliament of India had passed *The Prevention and Control of Pollution Act*. This was the first such enactment and thereafter many acts/rules have come into force to protect the environment and to safeguard against disposal of toxic industrial and hazardous wastes. *The Environment Protection Act* 1986 is an umbrella act which covers all aspects of environment including air, water and land. Under this act, Govt. of India is empowered to take all measures, required for the protection and improvement of the quality of environment and abating environmental pollution.

Water quality deterioration has serious implications for the supply of water for drinking, irrigation, industrial use, and is important determinant of public health. The level of natural contaminants such as fluoride and arsenic, and chemical pollutants such as pesticides and insecticides, is high and rising at several places. Although several studies and pilot programs sponsored by the Government and external funding agencies are under way, by far these have had mixed success. New risks are emerging from the rise in wastewater production and its inappropriate disposal that accompanies the increased coverage and service levels.

Water pollution is acquiring a serious dimension in India as almost 70% of its surface water resources and a large proportion of groundwater reserves are already contaminated by biological, toxic organic and inorganic pollutants. In many cases, the available water has been rendered unsafe for domestic consumption, irrigation, and industrial needs. Degradation of quality in turn leads to water scarcity as it limits water availability for human use.

20.1. UN MILLENNIUM DEVELOPMENT GOALS

During the UN Millennium Summit in September 2000, a Millennium Declaration was adopted by 189 nations. Eight goals drawn from the actions and targets contained in the declaration were identified and it was agreed that these are to be achieved by the year 2015. These goals are known as *The Millennium Development Goals* (MDGs) and these are supposed to respond to the world's main development challenges. The 8 MDGs break down into 18 quantifiable targets that are measured by 48 indicators. The goals are:

Goal 1: Eradicate extreme poverty and hunger,
Goal 2: Achieve universal primary education,
Goal 3: Promote gender equality and empower women,
Goal 4: Reduce child mortality,
Goal 5: Improve maternal health,
Goal 6: Combat HIV/AIDS, malaria and other diseases,
Goal 7: Ensure environmental sustainability, and
Goal 8: Develop a Global Partnership for Development.

Two targets under the Goal 7 are of relevance to water sector:

Target 1: Integrate the principles of sustainable development into country policies and programmes; reverse loss of environmental resources,
Target 2: Reduce by half the proportion of people without sustainable access to safe drinking water.

20.2. WATER QUALITY REQUIREMENT FOR DIFFERENT USES

For any water body to function adequately in satisfying the desired use, it must have corresponding degree of purity. Drinking water should be of highest purity. As the magnitude of demand for water is fast approaching the available supply, the concept of management of the quality of water is becoming as important as its quantity.

Each water use has specific quality need. Therefore, to set the standard for the desire quality of a water body, it is essential to identify the uses of water in that water body. In India, the Central Pollution Control Board (CPCB) has developed a concept of *designated best use*. According to this, out of the several uses of water of a particular body, the use which demands highest quality is termed its *designated best use*. Five *designated best uses* have been identified, as presented in Table 1. This classification helps the water quality managers and planners to set water quality targets and design suitable restoration programs for various water bodies.

Table 1. Designated best uses of water

Designated best use	Class	Criteria
Drinking Water Source without conventional treatment but after disinfection	A	1.Total Coliforms Organism MPN/100ml shall be 50 or less 2. pH between 6.5 and 8.5 3. Dissolved Oxygen 6mg/l or more 4. Biochemical Oxygen Demand 5 days 20 °C, 2mg/l or less
Outdoor bathing (Organised)	B	1.Total Coliforms Organism MPN/100ml shall be 500 or less 2. pH between 6.5 and 8.5 3. Dissolved Oxygen 5mg/l or more 4. Biochemical Oxygen Demand 5 days 20 °C, 3mg/l or less
Drinking water source after conventional treatment and disinfection	C	1. Total Coliforms Organism MPN/100ml shall be 5000 or less 2. pH between 6 and 9 3. Dissolved Oxygen 4mg/l or more 4. Biochemical Oxygen Demand 5 days 20 °C, 3mg/l or less
Propagation of Wild life and Fisheries	D	1. pH between 6.5 and 8.5 2. Dissolved Oxygen 4mg/l or more 3. Free Ammonia (as N) 4. Biochemical Oxygen Demand 5 days 20 °C, 2mg/l or less
Irrigation, Industrial Cooling, Controlled Waste disposal	E	1. pH between 6.0 and 8.5 2. Electrical Conductivity at 25 °C micro mhos/cm, maximum 2250 3. Sodium absorption Ratio Max. 26 4. Boron Max. 2mg/l
	Below-E	Not meeting any of the A, B, C, D & E criteria

Source: CPCB (2005).

A colour coding is frequently used to depict the quality of water on maps:

Blue water – This water can be directly used for drinking, industrial use, etc.,

Green water – Water contained in soil and plants is termed as green water,

White water – Atmospheric moisture is white water,

Brown or grey water – Various grades of wastewater are shown by brown or grey colour.

In India, CPCB has identified water quality requirements in terms of a few chemical characteristics, known as primary water quality criteria. Further, Bureau of Indian Standards has also recommended water quality parameters for different uses in the standard IS 2,296:1992. Important water quality standards of IS code are shown in Table 2.

Table 2. Water Quality standards in India (Source IS 2296:1992)

Characteristics	Designated best use				
	A	B	C	D	E
Dissolved Oxygen (DO)mg/l, min	6	5	4	4	–
Biochemical Oxygen demand (BOD)mg/l, max	2	3	3	–	–
Total coliform organisms MPN/100ml, max	50	500	5,000	–	–
pH value	6.5–8.5	6.5–8.5	6.0–9.0	6.5–8.5	6.0–8.5
Colour, Hazen units, max.	10	300	300	–	–
Odour	Un-objectionable			–	–
Taste	Tasteless	–	–	–	–
Total dissolved solids, mg/l, max.	500	–	1,500	–	2,100
Total hardness (as $CaCO_3$), mg/l, max.	200	–	–	–	–
Calcium hardness (as $CaCO_3$), mg/l, max.	200	–	–	–	–
Magnesium hardness (as $CaCO_3$), mg/l, max.	200	–	–	–	–
Copper (as Cu), mg/l, max.	1.5	–	1.5	–	–
Iron (as Fe), mg/l, max.	0.3	–	0.5	–	–
Manganese (as Mn), mg/l, max.	0.5	–	–	–	–
Cholorides (as Cu), mg/l, max.	250	–	600	–	600
Sulphates (as SO_4), mg/l, max.	400	–	400	–	1,000
Nitrates (as NO_3), mg/l, max.	20	–	50	–	–
Fluorides (as F), mg/l, max.	1.5	1.5	1.5	–	–
Phenolic compounds (as C_2H_5OH), mg/l, max.	0.002	0.005	0.005	–	–
Mercury (as Hg), mg/l, max.	0.001	–	–	–	–
Cadmium (as Cd), mg/l, max.	0.01	–	0.01	–	–
Salenium (as Se), mg/l, max.	0.01	–	0.05	–	–
Arsenic (as As), mg/l, max.	0.05	0.2	0.2	–	–
Cyanide (as Pb), mg/l, max.	0.05	0.05	0.05	–	–
Lead (as Pb), mg/l, max.	0.1	–	0.1	–	–
Zinc (as Zn), mg/l, max.	15	–	15	–	–
Chromium (as Cr^{6+}), mg/l, max.	0.05	–	0.05	–	–
Anionic detergents (as MBAS), mg/l, max.	0.2	1	1	–	–
Barium (as Ba), mg/l, max.	1	–	–	–	–
Free Ammonia (as N), mg/l, max	–	–	–	1.2	–
Electrical conductivity, micromhos/cm, max	–	–	–	–	2,250
Sodium absorption ratio, max	–	–	–	–	26
Boron, mg/l, max	–	–	–	–	2

Guidelines are available to evaluate quality of water for irrigation. For irrigation, water can be classified in five classes depending upon its chemical properties. The criteria for the same are given in Table 3.

Indian Standards code IS 10,500:1991 prescribes specifications for drinking water. These are given in Table 4.

Table 3. Guidelines for evaluation of irrigation water quality

Water class	Sodium (Na) %	Electrical conductivity (μS/cm)	SAR	RSC meq/l
Excellent	<20	<250	<10	<1.25
Good	20–40	250–750	10–18	1.25–2.0
Medium	40–60	750–2,250	18–26	2.0–2.5
Bad	60–80	2,250–4,000	>26	2.5–3.0
Very bad	>80	>4,000	>26	>3.0

Table 4. Drinking water specifications (IS 10,500:1991)

Characteristics	Desirable limit	Permissible limit
	Essential Characteristics	
Colour, Hazen Units, Max	5	25
Odour	Unobjectionable	–
Taste	Agreeable	–
Turbidity, NTU, Max	5	10
PH value	6.5 to 8.5	–
Total Hardness (as $CaCo_3$), mg/l, Max	300	600
Iron (as Fe), mg/l, Max	0.3	1.0
Chlorides (as Cl), mg/l, Max	250	1,000
Residual free chlorine, mg/l, Max	0.2	–
	Desirable Characteristics	
Dissolved solids, mg/l, Max	500	2,000
Calcium as (Ca), mg/l, Max	75	200
Magnesium (as Mg), mg/l, Max	30	75
Copper (as Cu), mg/l, Max	0.05	1.5
Manganese (as Mn), mg/l, Max	0.1	0.3
Sulphate (as So_4), mg/l, Max	200	400
Nitrate (as No_3), mg/l, Max	45	100
Flouride (as F0, mg/l, Max	1.0	1.5
Phenolic compounds (as C_6H_5OH), mg/l, Max	0.001	0.002
Mercury (as Hg), mg/l, Max	0.001	–
Cadmium (as Cd), mg/l, Max	0.01	–
Selenium (as Se), mg/l, Max	0.01	–
Arsenic (as As), mg/l, Max	0.05	–
Cyanide (as CN), mg/l, Max	0.05	
Lead (as Pb), mg/l, Max	0.05	–
Anionic detergents (as MBAS), mg/l, Max	0.02	1.0
Chromium (as Cr^{6+}), mg/l, Max	0.05	–
PAH, mg/l, Max	–	–
Mineral oil, mg/l, Max	0.01	0.03
Pesticides, mg/l, MAX	Absent	0.001
Alkalinity, mg/l, Max	200	600
Aluminum (as Al), mg/l, Max	0.03	0.2
Boron, mg/l, Max	1	5

After discussion on specifications for various uses of water, the attention is now focused on various sources and activities that pollute water.

20.3. SOURCES OF POLLUTION

Sources of pollution can be divided basically into two groups, natural and anthropogenic. The source can be further classified as either point or diffuse (non-point) sources. Point sources enter the pollution transport routes as discrete, identifiable locations and can be measured directly or otherwise quantified; and their impact can be evaluated directly. Major point sources include effluent from industrial and sewage treatment plants, and effluent from farm buildings or solid waste disposal sites. Table 5 describes categories of major sources of ground water contamination.

Pollution from diffuse sources can be related to weathering of minerals, erosion of virgin lands and forest including residues of natural vegetation, or artificial or semi artificial sources. The last can be related directly to human activities such

Table 5. Categories of major sources of ground water contamination

Source	Category		
	Point	Line	Diffuse
Municipal			
Sewer leakage	x	x	
Sewage effluent	x	x	X
Sewer sludge	x		X
Urban runoff	x	x	X
Solid wastes	x		
Lawn fertilizers			X
Agricultural			
Evapotranspiration and leaching (return flow)			x
Fertilizers			x
Soil amendments			x
Pesticides and herbicides			x
Animal wastes (Feedlots and dairies)	x		x
Stock piles	x		
Industrial			
Cooling water	x		x
Process water	x		
Storm run-off	x		x
Boiler blow down	x		
Water treatment plant effluent	x		
Hydrocarbons	x		
Tank and pipeline leakage	x	x	
Mining/oil field wastes			
Hydrocarbons	x		
Mining wastes	x	x	x

return flows from agricultural farming areas, wash-overs from animal feedlots, construction sites, transportation, strip mining and others.

Causes of water pollution are diverse: untreated sewage, industrial discharge, leaching from municipal waste, and drainage from the residues of agricultural fertilizers and pesticides. This section provides a discussion on the major sources.

20.3.1. Domestic Sources of Pollution

With burgeoning cities and increasing industrialization, the quantum of waste dumped into rivers has also increased. Tables 6 and 7 show the increasing urbanization and the growth of metro cities in India.

During the 50 year period from 1951 to 2001, urban population increased more than four fold. By the year 2001, more than one-fourth of the total population was living in urban agglomerations. From 1981 to 2001, the number of urban agglomerations with population in excess of 1 million increased three times as shown in Table 4. Out of 35 urban agglomerations with population exceeding 1 million, three had population exceeding 10 million in the year 2001: Mumbai (16.38 million), Kolkata (13.22 million), and Delhi (12.8 million).

Rapidly increasing urbanization has given rise to a number of water-related problems: water supply, wastewater generation and its collection, treatment and disposal. Water for domestic and industrial use in urban areas is drawn from rivers, streams, wells and lakes. Approximately 80% of the water supplied for domestic use is turned into wastewater and in India, most of it is let out untreated. Consequently, it either percolates into the ground and contaminates the aquifers or is dumped into the nearby river causing pollution in downstream areas. Consequently, river stretches near big cities are at their worst, e.g., Ganga at Kanpur, Yamuna at Delhi, Gomti at Lucknow, Sabarmati at Ahmedabad, Musi at Hyderabad, and Adyar at Chennai. Coliform count in Yamuna at Delhi reaches up to 10 crore per 100 ml.

Table 6. Increasing urbanization in India

Year →	1951	1991	2001	2021 (projected)
Number of Urban Agglomerations/towns	2,795	3,768	4,378	–
Urban population (million)	62.0	217.0	285.0	550.0
Percentage of total population	17.3%	25.72%	27.8%	41%

Table 7. Growth of metro cities in India

Year →	1981	1991	2001
Number of Urban Agglomerations (population over 1 million)	12	23	35
Population (million)	42	70	108
Percentage of urban population	26	32	39

The domestic sector is responsible for the majority of the wastewater generation in India. About 50 million m^3 of untreated sewage discharged into rivers per year significantly contribute towards pollution of India's fourteen major river systems. The 22 largest cities in the country produce over 7,300 million liters of domestic wastewater per day and only about 80% of it is collected for treatment (Development Alternatives, 2001).

The number of Class-1 (population > 1 lakh) cities increased from 142 in 1978 to 299 in 1995, their combined population also nearly doubled from 60.3 million to 128.1 million. In the year 1995, the volume of water supply to these cities was 8,638 mld and the waster water generated by 299 Class-1 cities was 16,660 mld which was approximately 81% of the water supplied. Only 74% of the total wastewater generated is collected and only 24% of the wastewater is treated. Out of 299 Class-1 cities, 160 cities had sewerage cover for more than 75% of the population and in 92 cities, between 50 and 75% of population had sewerage coverage. On the whole, 70% of the population of Class-1 cities was provided with the sewerage facility. Figure 1 shows the number, population, water supply, and wastewater – growth in Class-1 cities for three past years.

From 1978 to 1995, the number of Class-2 cities (population 0.5 lakh to 1 lakh) increased from 190 to 345; in this period their combined population increased from 12.8 million to 23.6 million. In the year 1995, the volume of water supply to these cities was 1,936 mld and the waster water generated by 345 Class-2 cities was 1,650 mld which was approximately 85% of the water supplied. Less than 4% of the wastewater was treated in 1995. Figure 2 shows the number, population, water supply, and wastewater – growth in Class-2 cities for three past years.

Currently, there are more than 900 towns with population exceeding 50,000 and these generate 26,000 mld of waste. Out of this, only about 7,000 mld is treated before letting out and the rest, i.e., 19,000 mld is disposed off untreated. The gravity of situation can be judged from the fact that in spite of sewage treatment plants, only 1,400 mld of about 3,600 mld of sewage generated in Delhi is treated; the rest

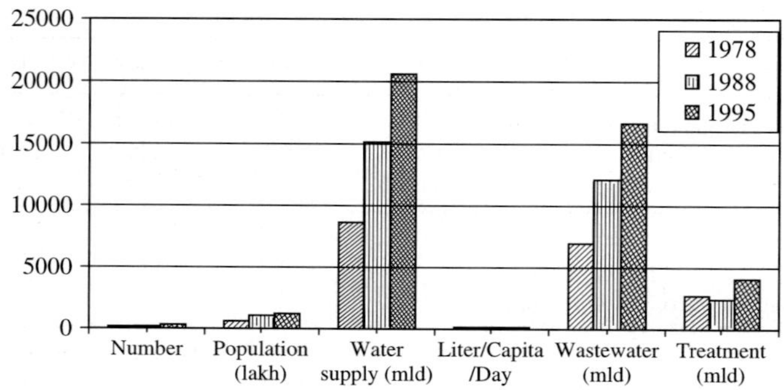

Figure 1. Number, population, water supply, and wastewater – growth in Class-1 cities for three past years. Note that the variables represent different features and the Y-axis scale should be viewed accordingly

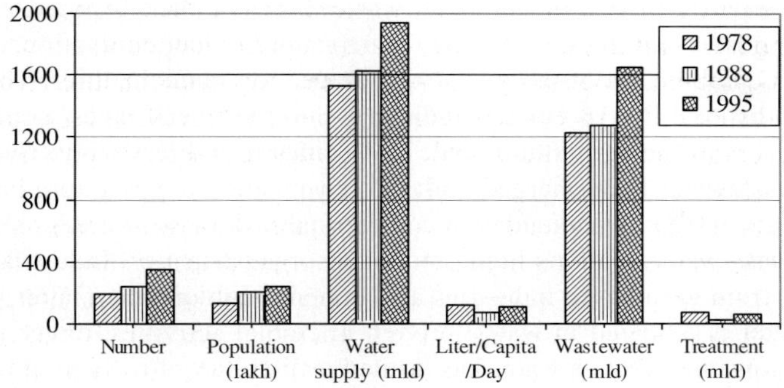

Figure 2. Number, population, water supply, and wastewater – growth in Class-2 cities for three past years. Note that the variables represent different features and the Y-axis scale should be viewed accordingly

of the untreated sewage is discharged into the Yamuna River. Twenty-seven cities have only primary treatment facilities and forty-nine have primary and secondary treatment facilities. The level of treatment available in cities with existing treatment plant varies from 2.5% to 89% of the sewage generated.

Comparative data of water supply, wastewater generation, collection and treatment in 1978, 1988 and 1995 for Class-1 and Class-2 cities is given in Table 8. Population of Class-1 cities is nearly 5 times that of Class-2 cities but the water supply in Class-1 cities is 10.6 times that of Class-2 cities and wastewater generation is nearly 10.1 times more in Class-1 cities. Treatment of wastewater in Class-1 cities is a lot larger, nearly 65 times that of Class-2 cities. Comparison of Indian cities with those of the other countries shows that water supply and sanitation facilities are woefully inadequate even in Class-1 cities in India. Under the circumstances, it is easy to image the kind of unhealthy conditions that are prevalent in Class-2 cities.

Table 8. Water supply and sanitation status in Class-1 and Class-2 cities in three years

S. N.	Item	Class-1 cities			Class-2 cities		
		1978	1988	1995	1978	1988	1995
1.	Number	142	212	299	190	241	345
2.	Population (million)	60	102	128	12.8	20.7	23.6
3.	Water Supply (mld)	8,638	15,191	20,607	1,533	1,622	1,936
4.	Per capita water supply (LPCD)	143	147	183	125	78	103
5.	Wastewater generated (mld)	7,007	12,145	16,662	1,226	1,280	1,650
6.	Wastewater treated (mld)	2,756	2,485	4,037	67	27	62
7.	Untreated disposed wastewater (%)	61%	79.5%	76%	94.56%	97.88%	96.27%

Treated or partly treated or untreated wastewater is either disposed into natural drains joining rivers or lakes, or is used for irrigation/ fodder cultivation, or diverted to the sea; a combination of these is also followed the municipalities. The mode of wastewater disposal in 118 cities is indirectly into the rivers/ lakes/ ponds/ creeks; in 63 cities it is to the agriculture land; in 41 cities it is directly into rivers, and in 44 cities, wastewater is discharged both into rivers and on agriculture land.

In India, about 0.37 kg/capita/day of commercial & domestic wastes are produced. Domestic wastewater contains high amount of organic matter. Industrial effluents particularly from agro-based industries also contain high organic matter. When this organic matter is oxidized in water through microbial activities, oxygen dissolved in water is consumed. Since water has limited availability of oxygen, this results in depletion of dissolved oxygen and the survival of aquatic life is endangered.

In many water bodies, massive input of organic matter sets off a chain of chemical and biological events. This results in high bacterial population, cloudy appearance, high BOD and strong disagreeable odour; all these symptoms indicate general depletion of oxygen. During monsoon, the sludge deposited in such stretches is flushed by flood water and stays in suspension, causing rise in oxygen consumption. Heavy fish mortality occurs every year due to this sudden oxygen depletion, during first flushing after the onset of monsoon.

20.3.2. Unsewered Domestic Waste

Most of the small towns and villages in India do not have running water supply or sewered sanitation. Consequently, most residents in these places use open field for defecation, with the remaining use pit-latrines or septic-tanks. Similar is the situation in numerous jhuggy-jhopady (shanty town) that have sprung in the vicinity of big cities and industrial complexes. Much of the bathing and washing (clothes, utensils etc.) in these habitations is in or near a water body or near a hand pump/municipal tap. This causes in-situ diffuse pollution or the waste water just flows and gradually infiltrates or evaporates. Generation of liquid effluents is very small in villages but most wastewater flows in the nearby depressions. Investigations show significant quantities of pollutants including salts, nutrients, organisms, and micro-organisms from such hamlets and rural areas reaching ground or surface water bodies through leachate seepages and as washings in the storm run-offs. CPCB (1995) estimated that on an average, 15 g of BOD_{5-20} per capita per day of the rural population reaches the major rivers draining the basin; the amounts of pollutants percolating to ground waters may be much larger.

Over the past four decades, the use of soaps and detergents has considerable increased in towns as well as villages. Washing machines have also become common in Indian households. Combined together, all these factors have led to significant increase in generation of pollution loads of salts, nutrients, organics, micro-organisms, soaps, detergents and other chemicals per capita per day. Unfortunately, provision of sewerage has lagged far behind water supply due to many reasons.

A large number of the cities/towns either do not have any sewerage system or the system is overloaded or defunct. Even in metropolitan cities like Delhi, the entire sewage is not treated. This results in large amount of uncollected and untreated wastewater. Most of the pollution generated flows as sullage to streams and other water-bodies through storm-drains or other channels. The bulk of pollution gets retained on land to percolate, leach or get washed-off to streams or ground water as in the case of pollutants from habitations of category (i) and (ii). Sewer lines are also known to be break or leak due to: a) poor construction material, b) leakage from corroded casting, and c) poor design of filter pack.

According to CPCB (1995) only about 40–50% of the populations of the major Indian metro-cities of Delhi, Mumbai, Calcutta, Madras and Bangalore are served by sewer systems. The percentage of sewered population is nearly negligible in most of the rural areas and is quite meager (0 to 50%) in most medium and small towns. Even where sewers exist, they often leak or overflow, releasing their contents. With time, these join storm-water or surface drains, or percolate in the soil to reach ground-water, or enter into streams.

Although during percolation, sewage loses part of bacteria while passing through the soil and geological materials (natural purification), many cases of bacterial pollution of ground water have been reported. This pollution may be due to direct contamination of well water, due to cesspools near the well casing, or through gaps between the casing and surrounding wells, resulting in infiltration of contaminated water; many incidences of gastro-enteritis can be linked to sewage pollution of ground water. When the polluted water is used for drinking purposes, it results in the spread of this disease.

Methods of wastewater disposal include infiltration ponds, spreading or spraying onto the ground surface and discharge to stream or dry stream beds, which if not carefully regulated may provide a rapid pollution pathway to underlying, shallow aquifers. In some areas, deep soakaways or abandoned wells are used for the disposal of liquid domestic, industrial or farming wastes into aquifers. Lack of monitoring, supervision or management adds to the problem.

Even if the intention is to dispose the waste at depth, improper sealing or corrosion of well linings often produces leaks and subsequent pollution of the shallow ground water which is used for water supplies. In urban areas covered by sewerage systems, an economical and common method of partial treatment of sewage is wastewater stabilization by retention in shallow oxidation lagoons before subsequent discharge into rivers or onto land for irrigation systems. These lagoons are often unlined and, if constructed over coarse-textured soils, may have high rates of seepage loss. Further, the use of such effluent in irrigation may also lead to similar problems. Most of the unplanned industrial complexes and industrial units scattered in urban areas and the agricultural based industries in remote villages are examples of this.

20.3.3. Industrial Pollution

Whereas the advanced technology is being widely applied in production processes by industries in India, proper treatment of industrial effluents has lagged far behind. If a river, canal or any other surface body is present nearby, the industries in most cases use this 'opportunity' to discharge their effluents in it without proper treatment. Alternately, the effluents are discharged through unlined channels on land or depressions near the plant. Although some of constituents present in effluents may be absorbed by soil and unsaturated zone, this capacity is limited. With the passage of time, toxic constituents percolate down to ground water and may damage the entire aquifer system. In case of industrial units, effluent in most of the cases are discharged into pits, open ground, or open unlined drains near the factories, thus allowing it to move to low lying depressions resulting ground water pollution.

Many large industries often release wash-over from storage yards, waste dumping, ash-ponds, sludge-pits etc. Partly-treated effluents, containing some pollutants, get leached or washed to streams as diffuse pollution. A number of examples could be cited, e.g., oily wastes present in the storm-water channel along an oil refinery, ammonia pollution in ground water around a urea factory, pollution of ground water near sugar factories in UP, etc.

The number of industries that can cause damage to ground water quality is substantial. However, groundwater in the vicinity of the following industries should be monitored regularly to check the adverse effects of effluents: 1) food processing, (2) breweries and distilleries, (3) soaps and detergents, (4) dairy and meat, (5) fertilizers and agricultural chemicals, (6) pharmaceuticals (7) textile and rayon, (8) paper and allied products, (9) chemicals, (10) pharmaceuticals and refineries, (11) paints, varnishes and enamels, (12) steel and foundries, (13) electroplating, (14) explosives, (15) mining, (16) wood preservation, (17) photofilms, (18) nuclear fuel processing, and (19) rubber and tires.

The industries generating chemical pollution can be divided in two categories: i) those which generate high TDS bearing wastes and (ii) those which generate toxic wastes. The major contributors of pollution in terms of organic load are distilleries followed by paper mills. Distilleries generate very concentrated wastewater which is hard to treat. The paper and board mills also generate heavy organic pollution load. A large number of paper mills are in small-scale sector and do not have adequate arrangement for wastewater treatment. Thus, they create heavy pollution in many areas. The other significant contributors of organic load are sugar and engineering industries. Although there are guidelines regarding discharge of waste by the industries, the implementation is tardy resulting in heavy pollution of water bodies by these industries.

Major contributors of suspended solid load are thermal power plants that dispose a significant quantity of cooling water, followed by paper mills and tanneries. Thermal power plants generate large quantities of fly ash but this is not usually dumped in water bodies. Thus, the major contributors of suspended solids in treated effluent are paper mills followed by tanneries. Fertilizer industry and steel industry generate toxic wastes consisting of cyanide and arsenic. Besides, steel plants and oil refineries are the major contributors of phenol. Engineering industries contribute large quantities of effluents containing oil and grease followed by oil refineries and edible oil (vanaspati) industry. Major pollution in terms of ammonia load is contributed by fertilizer plants (nitrogen) followed by steel plants.

Tanneries generate higher concentration of chloride in hide processing whereas fertilizer industry (phosphate) and oil refineries respectively generate Fluoride and Sulphide. Mercury is generated by caustic soda industries which employ mercury cell process.

Studies indicate extensive damages to the ground water quality from effluents of electroplating wastes. High concentrations of cyanide (to the extent of 2 mg/l) and hexavalent chromium (to the extent of 12.8 mg/l) have been observed in ground water of Ludhiana, Punjab.

Available estimates of industrial waste production differ widely. Although the industrial sector accounts for about 4% of the annual water withdrawals, its contribution to water pollution, particularly in urban areas, is significant. The wastewater generation from this sector has been estimated at 55,000 million m^3 per day, out of which 68.5 million m^3 is discharged into river streams (MoWR, 2003). Of the total pollution load, 40%–45% is contributed by the processing of industrial chemicals, while nearly 40% of the total organic pollution, arises from the food industries followed by industrial chemicals, and the pulp and paper industry.

Encouragement of cottage and small-scale industries through policies and subsidies has been an important component of stated economic development programme of Govt. of India. There were about 3 million such units in the country at the turn of the last century. Majority of these units do not have expertise about pollutant disposal systems. In past, such systems were not considered necessary for these small units on the plea that these industries can not afford to spend money on these paraphernalia which may even jeopardize their economic viability. In fact, some of these industries have adopted highly polluting production technologies such as chrome-tanning of leather, use of azo-dyes in fabrics, use of cadmium in ornaments and silver-ware, electroplating with cyanide baths, production of dye-intermediates and other toxic chemicals, etc. Many industrial areas and peri-urban centres in the country such as Pali, Balotra, Patancheru, Jajmau, Faridabad, Ghaziabad, etc. have many such units. Wastewater and sludge from such congregations gets collected in unlined pits, stagnates in depressions to percolate, flows to streams through storm-drains, or leaches or get washed-off during the rainy season. Ultimately, such activities result in significant diffuse water pollution.

20.3.4. Agricultural Practices

Agricultural land use and cultivation practices exert major influences on ground water quality. Under certain circumstances, serious ground water pollution can be caused by agricultural activities the influence of that may be very important because of the large areas of aquifer affected. Of particular concern, in India, is the leaching of fertiliser chemicals (e.g., nitrate) and pesticides from regular intensive cultivation of crops. The impact of cultivation practices on ground water quality is greatest, as are most anthropogenic effects, where relatively shallow, unconfined aquifers are used for potable supply.

With the advent of green revolution, fertilizer consumption in the country has grown several times. The consumption of N, P_2O_5 and K_2O fertilizers in 1951–52 was 45,000, 11,000, 8,000 tones, respectively, which in the year 1982–83, rose to 4,362, 1,420, 735 thousand tones, respectively. Large numbers of cattle in rural areas also generate wastes. Agricultural activities in rural areas which include application of barn yard manure to soils, leakage from composting pits, increase in the use of fertilizers and insecticide and unscientific management of crops, fertilizers and soil water may result in ground water pollution. The efficiency of N-fertilizes in India is about 30% for paddy (partly being lost by volatilization-microbial decomposition) and about 50% for wheat. For potassium and phosphatic fertilizers, the efficiency is around 50% and 15–20%, respectively (Handa, 1975).

In 1950, the use of fertilizers in India was 0.55 kg/ha. However, by 2001–02 this figure increased to around 90.12 kg/ha (Kolanu and Kumar, 2004). The consumption of pesticides, which rose from less than 1 million tones in 1948 to 66.36 million tones during 1994–95, was around 43.59 million tones during 2001–02 (MoCF, 2002). Pesticides are often applied in quantities that are more than necessary in the belief that without these, the crops will die. It has been reported that the largest amount of pesticide is used by cotton growers. The chemical analyses results of ground water at shallow levels indicate high concentrations of nitrate, potassium at several places in the country. For high concentrations of nitrate, animal and human wastes are also likely to be the contributing factors, besides the leaching of nitrogenous fertilizers.

Some of the chemicals in these fertilizers and pesticides, which enter water bodies through runoff, irrigation return flow, and leaching, are considered hazardous by the World Health Organization (WHO). Such chemicals are banned or under strict control in many countries. Water quality studies on the Ganga River indicate the presence of chemicals, such as HCH, DDT, dimethoate, endosulfan and malathion in quantities exceeding standards set by international organizations. Some of these chemicals persist in the environment over long periods and some others are known to accumulate in certain organisms. Consumption of such organisms increases risk to human health.

A high level of fertilizer use has been associated with increased incidences of eutrophication in several important water bodies of India, such as the Hussain Sagar in Hyderabad and the Nainital Lake in Uttar Pradesh (MoWR, 2003). Eutrophication of water bodies is caused by nutrient enrichment. The resultant spurt in the production of biomass and its death and decay causes a number of adverse effects, including decreased dissolved oxygen levels, release of odorous compounds, and siltation.

In India, a high proportion of the rural population in agricultural areas obtain their domestic water supplies from shallow, private bore holes, which suffer the impact of nitrate pollution to a much greater extent than the deeper, public supply aquifers utilised for urban water supply. These deeper aquifers can also be affected by nitrate contamination although this pollution often takes much more time to percolate to these depths.

Increasing salinity resulting from the effects of irrigated agriculture is one or the oldest and most widespread forms of ground water pollution. It is caused by the dissolved salts in irrigation water being deposited following evaporation of the water. The addition of further excess irrigation water merely leaches salts from the soil and transfers the problem to the underlying ground water. Drainage waters from irrigated agricultural land are always high in salts, since they also have to carry the salts originally contained in the evaporated fraction of the irrigation water. In economically developed areas this was appreciated long ago and arrangements were made to drain away such saline water without causing undue damage to freshwater bodies. In developing countries, while irrigation has been expanding exponentially, little is done to tackle the problem of the high salinity return-waters.

20.3.5. Leachates and Wash-Overs

The most common method of disposal of solid municipal waste in India is by deposition in landfills. Improper disposal of municipal solid waste is another cause of surface and groundwater pollution. Most of the municipal waste collected is dumped in low-lying areas outside the city/town limits, and many of the existing landfills to dispose the waste are neither well-equipped nor managed efficiently (TERI, 1998). Leaching from landfills and garbage pits transports toxic substances and heavy metals to the water table. Runoff from garbage dumps and city streets carries litter, particulate matter and chemicals to nearby streams and canals. The annual production of total solid waste in India by the turn of the century was estimated at around 2,000 million tones. This figure will undoubtedly continue to increase with growing population.

Less attention was given in the past to the leaching of pesticides from agricultural land to the underlying ground water in spite of the dramatic increase in the use of pesticide formulations over the years. These days, the problem is receiving attention but not much is being done to control it. There are currently few laboratories with the capability of analyzing pesticides.

The principal threat to ground water comes from inadequately controlled landfills where leachate generated from the fill material is allowed to escape to the surrounding and underlying ground. The chemical composition of such leachate depends on the nature and age of the landfill and the leaching rate. Most leachates emanating from municipal solid wastes are not only high in organic content but also contain some toxic material. Leachates from solid wastes of industrial origin, however, often contain a much higher proportion of toxic constituents, such as metals and organic pollutants. To minimize the impact of such landfills on ground water quality and environment, it is necessary to properly design and build these facilities to prevent pollution and put in place strict management controls to ensure they are operated correctly. Unfortunately this is rarely done as few towns and industries in the country make the requisite effort to ensure that their solid waste is treated or disposed off properly.

Under certain hydrogeological conditions, unsewered domestic waste can cause severe ground water contamination by pathogenic bacteria, nitrate and other pollutants. Unsewered waste normally means septic tanks or pit latrines of the ventilated, dry or pour-flush types. There are important differences between the two in relation to the risk of ground water contamination. Septic tank soakaways discharge at a higher levels in the soil profile than pit latrines and such conditions are preferable as far as the elimination of bacteria is concerned. Pit latrines are often deep excavations (to allow a long useful life) and the soil may be entirely removed thus offering less opportunity for bacterial death. Further the loading from septic tanks is likely to be less compared to some of the pit latrine types. Septic tanks are lined and their solid effluent of high nitrogen content is periodically removed, whereas most pit latrines are unlined and the solid material remains in the ground.

20.3.6. Effluents, Wash-Over from Cattle-Farms

Cattle-farms rarely have adequate arrangements to collect, treat, and properly dispose their solid and liquid wastes. In villages, almost every home has a small cattle-farm and the waste from these farms just flows over the ground to either infiltrate or get accumulated in small depressions from where, they infiltrate or water is evaporated. However, waste from these sources frequently does not contain much harmful chemicals.

20.3.7. Leakages and Accidents

With increasing storage, handling and transport of various chemicals, including those highly toxic and/ or hazardous, the contribution of these has been growing. Accidents involving mineral oils, acids, chlorine, ammonia etc. are too well known. A sizable part of chemicals spilled in such accidents travels to water bodies.

20.3.8. Mining Activities

A range of ground water pollution problems can be associated with mining activities. The nature of the pollution depends on the materials being excavated and extracted. Both surface and underground mines usually extend below the water table and often dewatering is required to allow mining to proceed. The water pumped either directly from the mine or from specially constructed bore holes, may be highly mineralized and its usual characteristics include low pH and high levels of iron, aluminum, and sulphate. Disposal of this mine drainage effluent to surface water or ground water can cause serious impacts on water quality for all uses. Pollution of ground water can also result from the leaching of mine tailings and from settling ponds and can, therefore, be associated with both present and past mining activity.

20.3.9. Radioactive Substances

Ground water frequently contains trace levels of naturally occurring radioactive substances or their by-products. The types and levels vary with area to depending principally on subsurface materials. In addition to natural radii radioactive substances in ground water can result from human activities. With increase in nuclear plants for power generation, contamination of ground water by radioactive sources may take place. Operations like mining and milling of radioactive ores, chemical reprocessing and radioactive waste disposal can source of radioactive pollution.

20.4. RIVER WATER QUALITY

Increase in population, urbanization and industrialization is causing an ever-increasing threat to the quality of waters in rivers, lakes and ponds in India. Each time water is used for some activity, its quality is degraded. Unfortunately, the sanctity attached to rivers in the country does not ensure that the rivers are clean. Pollution problems of various rivers have been discussed in the chapters on individual rivers. The Ganga Action Plan and the Yamuna Action Plans have also been described in details in the relevant chapter. Therefore, this section provides a brief discussion on the topic.

Table 9 gives average chemical composition of some Indian rivers.

Water pollution varies in severity from one region to the other, depending on the density of urban development, agricultural and industrial practices, and the systems for collecting and treating wastewater. The Central Pollution Control Board (CPCB), India, has identified some of the polluted river stretches and possible sources of pollution (Table 10). Most of the polluted stretches exist in and around large urban areas.

Table 9. Average chemical composition (ppm) of some Indian Rivers

River	HCO_3	Cl	SO_4	SiO_2	Ca	Mg	Na	K	TDS
Brahmaputra	56	11	4	7	14	5	7	3	107
Cauvery	135	20	13	23	21	9	43	4	272
Ganga	128	10	11	18	25	8	11	3	241
Godavari	105	17	8	10	22	5	12	3	181
Gomti	274	9	15	15	30	19	27	5	394
Indus	64	5	23	5	54	12	10	0.3	173
Krishna	178	38	49	24	29	8	30	2	360
Mahanadi	122	23	3	17	24	13	14	8	224
Narmada	225	20	5	9	14	20	27	2	322
Tapi	150	65	1	16	19	22	48	3	322

Source: Subramanian (2004) and others.

Table 10. Some polluted river stretches in India

River	Polluted stretch	Desired class	Existing class	Critical parameters	Possible source of pollution
Chambal	Downstream of Nagda and downstream of Kota	C	D/E	BOD, DO	Domestic and industrial waste from Nagda and Kota
Damodar	Downstream of Dhanbad	C	D/E	BOD, Toxicity	Industrial wastes from Dhanbad, Durgapur, Asansol, Haldia and Burnpur
Godavari	Downstream of Nasik and Nanded	C	D/E	BOD	Wastes from sugar industries, distilleries and food processing industries
Gomti	Lucknow to confluence with Ganga	C	D/E	DO, BOD, Coliform	Industrial wastes from distilleries and domestic wastes from Lucknow
Hindon	Saharanpur to confluence with Yamuna	C	D	DO, BOD, Toxicity	Industrial and domestic wastes from Saharanpur and Ghaziabad
Kali	Downstream of Modinagar to confluence with Ganga	C	D/E	BOD, Coliform	Industrial and domestic wastes from Modinagar
Krishna	Karad to Sangli	C	D/E	BOD	Wastes from sugar industries and distilleries
Sabarmati	Immediate upstream of Ahmedabad up to Sabarmati Ashram	B	E	DO, BOD, Coliform	Domestic and industrial waste from Ahmedabad
	Sabarmati Ashram to Vautha	D	E	DO, BOD, Coliform	Domestic and industrial waste from Ahmedabad
Satluj	Downstream of Ludhiana to Harike	C	D/E	DO, BOD	Industrial wastes from hosieries, tanneries, electro-plating and engineering industries and domestic waste from Ludhiana and Jalandhar
	Downstream of Nangal	C	D/E	Ammonia	Wastes from fertilizer and chloralkali mills from Nangal
Subarnarekha	Hatia dam to Bharagora	C	D/E	-do-	Domestic and industrial waste from Ranchi and Jamshedpur
Yamuna	Delhi to confluence with Chambal	C	D/E	DO, BOD, Coliform	Domestic and industrial wastes from Delhi, Mathura and Agra
	In the city limits of Delhi, Mathura and Agra	B	D/E	DO, BOD, Coliform	Domestic and industrial wastes from Delhi, Mathura and Agra

20.5. GROUND WATER POLLUTION

In ground water evaluation, the quality is of nearly same importance as quantity – physical, chemical, and biological characteristics of ground water determine its usefulness for various uses, viz., irrigation, domestic, and industrial. Variation in ground water quality occurs from place to place and at place from season to season. The ground water quality data give important clue to the geological history of rocks and indications of ground water recharge, discharge, storage and movement.

The most troublesome factor about ground water quality is that after pollution has entered the sub-surface environment, it may remain hidden for many years, is dispersed over wide areas, and renders ground water unsuitable for many uses. If an aquifer is contaminated, it may have to be abandoned and this may require development of costly alternate water supplies. Prevention is the best way to protect ground water quality. Slow movement of ground water is a favorable factor when the contaminants are degradable (biological, bacteriological or radioactive) with time. In these cases, long underground detention times may more or less completely remove undesired substances. On the other hand, slow movement causes long contact with the subsurface minerals, most of which are soluble in water and may cause natural degradation of subsurface water.

The ground water quality zones, viz., fresh (EC up to 2.0 ms/cm), marginal (EC 2.0 to 6.0 ms/cm) and saline (EC $>$ 6.0 ms/cm) have been demarcated in India on the basis of electrical conductivity of water. The ground water in most of the areas of the country is fresh. Brackish ground water occurs in the arid zones of Rajasthan, close to coastal tracts in Saurashtra and Kutch, and in some zones in the east coast and certain parts in Punjab, Haryana, Western Uttar Pradesh etc., which are under extensive surface water irrigation. The fluoride levels in ground water are considerably higher than the permissible limit in vast areas of Andhra Pradesh, Haryana and Rajasthan and in some places in Punjab, Uttar Pradesh, Karnataka and Tamil Nadu. In northeastern regions, ground water with iron contents above the desirable limit occurs widely. Pollution due to human and animal wastes and fertilizer application has resulted in high levels of nitrate and potassium in ground water in some parts of the country.

Based on physiographic and geo-hydrological considerations, attempts have been made to delineate different ground water provinces of India (Handa 1964). Depending upon the chemical characteristics of ground water, Handa (1964) first made an attempt to divide the country into six broad hydro-chemical provinces. These are described in Table 11.

With this background, we described the common ground water contaminants.

20.5.1. Common Ground Water Contaminants

Common contaminants that are found in ground water and their effects are listed in Table 12.

Ground water quality problems can be broadly divided in two categories: those due to natural causes and those due to anthropogenic causes. The natural

Table 11. Ground water provinces based on hydro-chemical quality

A. Bicarbonate type	Ground water falling in the category is characterized by relatively high bicarbonate contents. The Cl and SO_4 concentrations are normally below 2.0 and 1.0 mell respectively; the Cl : SO_4 ratio is normally more than unity. Coverage of this category is confined mainly to the humid and sub-humid areas in India, viz., northeast (Assam, Meghalaya, Nagaland, Mizoram, Manipur and Tripura); eastern India (West Bengal, excluding Calcutta and the coastal tracts); north India (Jammu and Kashmir, Himachal Pradesh, northern Punjab, north eastern Haryana, major parts of Uttar Pradesh excluding some parts of southwestern Uttar Pradesh, Bihar and parts of Rajasthan and Gujarat). It also includes the whole of Madhya Pradesh, Kerala (excluding the coastal fringe) and parts of Maharashtra, Andhra Pradesh, Karnataka and Tamil Nadu.
B. Bicarbonate-Chloride type	In this category, the dominance of bicarbonate is much less than as compared to the bicarbonate type ground water; the bicarbonates ions being 35 to 50 % of the total anions present. The concentration of chloride and sodium ions is quite significant. The EC of these waters is generally below 2,500 microsiemens/cm, although in a few cases, ground water with EC up to 5,000 microsiemens/cm have been encountered.
C. Chloride-Bicarbonate type	This category has chloride ions as the most prominent ions, followed by bicarbonate ions. The EC of these waters is generally above 5,000 microsiemens/ cm. The ground water falling in this category is normally super-saturated with respect to calcite and dolomite, but under-saturated with respect to gypsum and celestite.
D. Sulphate-Chloride type	Sulphate and chloride ions are the predominant anions, while among the cations, the sodium ions are the most important. Examples from this category occur are found in Tiruchirapalli district in Tamil Nadu and isolated patches in southern Haryana, southern Punjab and western Rajasthan.
E. Choride-Sulphate type	This category has chloride ions as the predominant ions followed by sulphate ions. Among the cations, the anions are most important.
F. Chloride type	The ground water belonging to this category is normally quite high in dissolved mineral content and has Cl and Na as the dominant ions. The EC of these waters is normally more than 15,000 microsiemens/cm, although in some cases ground water having lower EC values can also be of the chloride type. Northeast India and parts of Haryana come under this category.

causes are: arsenic pollution, occurrence of high salinity, high concentrations of fluoride, and high concentrations of iron, etc. The anthropogenic causes are over-exploitation, domestic pollution, industrial pollution, agricultural pollution, etc. Some problems, such as seawater intrusion, could be due to both, natural as well as anthropogenic causes.

20.5.2. Ground Water Quality Problems due to Natural Causes

Arsenic pollution

Arsenic pollution of ground water in West Bengal was first reported in the early eighties. The occurrence of arsenic is mainly due to two reasons: natural and

Table 12. Common ground water contaminants and their effects

Contaminant	Source	Effects
Dissolved nitrate	Sewage, fertilizers, air pollution, landfills & industries	Blue baby disease (Methamoglobinamia) in children, formation of carcinogens & acceleration of eutrophication in surface waters
Pathogens (bacteria & viruses)	Sewage, landfills, septic tanks & livestock's	Water borne diseases such as typhoid, cholera, dysentery, polio, and hepatitis
Trace metals (arsenic, lead, mercury, cadmium, copper, chromium & nickel)	Natural, industrial & mine discharges, fly ash from thermal power plants	These are toxic and carcinogenic and can cause various types of diseases
Organic compounds (volatile & semi-volatile organic compounds like petroleum derivatives, and pesticides)	Agricultural activities, street drainages, sewage landfills, industrial discharges, spills, vehicular emissions fall out etc.	Various ailments of stomach, breathing, etc.

anthropogenic. Arsenic is widely distributed in nature and principally occurs in the form of inorganic or organic compounds. Inorganic compounds consist of arsenite, the most toxic form and arsenate the less toxic form. The main ores of Arsenic are arsenopyrite, arpiment, realgar and arsenopalledenite. It is present in nature as iron arsenate, iron sulphate and in calcareous soil as calcareous arsenolite. In flood deposits, it is found as arsenite. The main anthropogenic sources are industrial waste, phosphate, fertilizers, coal, oil, cement, mine tailing, smelting, ore processing, metal extraction, metal purification, chemicals, glass, leather, textiles, alkali, petroleum refineries, acid mines, alloys, pigments, insecticides, herbicides and catalysts.

The problem of arsenic contamination in ground water from the vast tract of alluvial aquifers in Bengal, Bihar and UP is known to have affected a population of about 50 million in different districts of India and an equal number in Bangladesh. About 63 lakh people in West Bengal State live in the arsenic belt; 69 blocks are arsenic-affected, while two are affected by fluoride. Arsenic in ground water have been reported in a range (0.05–3.2) mg/l in shallow aquifers from 61 block in 8 districts of West Bengal namely Malda, Mushirbad, Nadia, North and South 24 Pargana, Bardhaman, Howrah, and Hugli.

In West Bengal, there is increasing concern of arsenic induced diseases due to exposure of high concentration of arsenic in the Natural Geochemical environment. In this area the source of arsenic is geogenic and associated with iron pyrites in arsenic rich layers occurring in the alluvium along the Ganga River. The availability of arsenic is possible due to excessive use of ground water irrigation (e.g. up to 80% of the annual replenishable recharge in north 24 Parganas for multiple cropping which causes dropping of water levels resulting exposure of the arsenic rich beds to air – oxidation of the pyrite and soubilisation of arsenic).

Arsenic concentration in ground water has been found to be in excess of permissible limit of 0.05 mg/l in a number of localized patches in Murshidabad and North 24-Pargana Districts in West Bengal. Population in the area is reported to be suffering from "Arsenic Dermatosis" by drinking arsenic rich ground water. In Ramnagar and Domkal blocks of Murshidabad district, Arsenic levels range form 0.06 mg/l to 1.90 mg/l, while in North 24-Parganas, it ranges from 0.66 to 0.9 mg/l.

Consumption of ground water with elevated arsenic levels (up to $3,700\,\mu g/l$ in certain wells) over a prolonged period of time has resulted in serious health hazards, especially among the rural and semi-urban population in the region. Symptoms of arsenic toxicity are manifested as skin lesions, hyperkeratosis, melanosis, cancer in different organs and several other health disorders, which in some cases have proved to be lethal. Need of water for domestic as well as irrigation purposes had triggered rapid development of ground water resources in the region during the last two decades. Overdraft of ground water in an indiscriminate manner is one of the key factors responsible for the spreading of arsenic epidemic in this region. A large number of government organizations and NGO are working on this problem and to find a lasting solution.

High salinity

There are large areas in the semi-arid and arid belt consisting parts of Rajasthan, southern Punjab, southern Haryana and Gujarat where salinity of ground water is moderate to fairly high. In several places in Rajasthan and southern Haryana, the EC values of ground water are greater than 10,000 micro siemens/cm, making water unpalatable (Handa 1979). In some of the areas, such as Kumher and Dig (Rajasthan), Farruknagar and Sultanpur (Haryana), and Gujarat, salinity of ground water is so high that salt is being manufactured by solar evaporation. For drinking purposes, high salinity is one of the major ground water quality problems in these areas. In parts of southern Punjab, UP, and AP, there are pockets where ground water has EC values exceeding 5,000 microsiemens/cm. This topic has been discussed in Chapter 6 also.

High fluoride

Fluoride is often called a two-edge sword – in small dosages, it has remarkable influence on the dental system by inhibiting dental carries, while in higher dosages causes dental and skeletal fluorosis. When present in concentration of 0.8–1.0 mg/L, fluoride is beneficial for calcification of dental enamel especially for the children below 8 years of age. At higher concentrations (1.5–2.0 mg F/L), fluoride effects adversely and leads to dental fluorosis. At still higher concentration, (3–6 mg F/L) skeletal fluorosis occurs. The disease affects the bone and ligaments. Intakes of 20–40 mg F/day over long period have resulted in crippling skeletal fluorosis.

High concentrations of fluoride in ground water are common in some of the semi-arid areas of Rajasthan, southern Punjab, Gujarat, Karnataka, Tamil Nadu, Madhya Pradesh, and southern Haryana. Several areas of Andhra Pradesh have high concentrations of fluoride in ground water (exceeding 5 mg/l). There are a number of cases of dental and skeletal fluorosis in these areas. In several parts of Karnataka, Tamil Nadu, U.P., and other areas, fluoride concentrations of groundwater are more than the permissible level of 1.5 mg/l. An extensive survey of the community water supplies has shown that around 25 million people in rural areas consume water with fluoride content more than this limit. Various authorities have given permissible limits for fluoride content in drinking water as shown in Table 13. Concentrations of fluoride in drinking water in different parts of the country varies from 0.5 to 50 mg/L.

Fluoride contaminated ground water is creating health problems in India. Nearly 90 million people including 6 million children in the country in 200 districts in 15 states are affected with dental, skeletal and/or non-skeletal fluorosis. The extent of fluoride contamination in ground water varies from 1.0 to 48 mg/l. The fluoride affected states are: Andhra Pradesh, Bihar, Delhi, Gujarat, Haryana, Jammu and Kashmir, Karnataka, Kerala, Madhya Pradesh, Maharashtra, Orissa, Punjab, Rajasthan, Tamil Nadu, and Uttar Pradesh.

In Rajasthan, in the absence of perennial rivers, surface and canal systems, ground water remains the main source of drinking water for about 95% population. Ground water of 18 districts contains high fluoride (2 to 20 mg/l) affecting approx. 3 million people. Studies have revealed that three million people in the state are consuming water with excess fluoride. Rajasthan has to depend on ground water resources to a greater extent and in arid in semi-arid areas, the ground water is the only water resource for drinking as well as agricultural purposes. According to the survey of states for drinking water supply in rural habitation conducted by the Rajiv Gandhi Drinking Water Mission, 9,741 villages and 6,819 habitations have fluoride content more than 1.5 mg/l in ground water.

In Tamil Nadu, fluoride monitoring is carried out at 1286 observation wells by TWAD Board throughout the State twice in a year – during pre-monsoon and post monsoon periods. About 121 blocks in 19 districts are found to have high fluoride concentration in ground water.

High iron content

High concentrations of iron in ground water have been found in several areas of high rainfall particularly in north-eastern states. In Assam, some wells have an iron content as high as 20 mg/l. In the Nadia district of West Bengal, the iron

Table 13. Permissible limit of fluoride in drinking water prescribed by various organizations

Name of the organization	Desirable limit (mg/L)
Bureau of Indian Standards (BIS)	0.6–1.2
Indian Council of Medical Research (ICMR)	1.0
The Committee on Public Health Engineering Manual and Code of Practice, Government of India	1.0
World Health Organization (International Standards for Drinking Water)	1.5

concentration of 2–3 mg/l has been reported in ground water at several places. In West Bengal, high concentrations of iron in ground water may be due to lowering of redox potential which is because of the presence of organic matter. Also, at a greater depth, lignite is found which can be associated with high iron content.

Nitrate pollution

The problem of nitrate pollution in ground water is also severe in the country. The state of Maharashtra alone showed severe problem of nitrate pollution. The studies carried out by Central Ground Water Board in the state of Maharashtra revealed that out of 688 samples, 518 (75.23%) samples had nitrates below desirable limit (45 mg/l) while 125 (14.08%) samples had nitrate concentration above 100 mg/l. The districts like Bhandara, Nagpur, and Pune had nitrate concentration above 45 mg/l. About 10.70% of total wells had nitrate concentration more than 45 mg/l but in the samples from rest of the districts, nitrate concentration was between both 45 and 100 mg/l.

Under certain conditions, nitrate consumed by babies is reduced to nitrite which reacts with haemoglobin present in the blood, thereby reducing the oxygen carrying capacity of the blood and the skin turns blue. This disease is known as methemoglobinemia or 'blue baby' syndrome in babies. Nitrogen based fertilizers are widely used in the country. The use has rapidly increased, partly because of subsidies offered by the government. Fertilizer consumption in India was nearly 175 tonnes in 2003–04 (Pangare et al. 2006).

20.5.3. Ground Water Pollution due to Anthropogenic Causes

Four main anthropogenic causes of ground water pollution can be readily identified: over-exploitation, industrial, domestic, and agricultural. The first category also includes areas of seawater intrusion. Ground water quality is being increasing threatened by agricultural, urban & industrial wastes, which leach or are deliberately injected into aquifers. Detailed discussion on these has already been provided in Section 20.3.

Ground water in certain geological formations may not be of desired quality for specific uses. Naturally occurring fluorides, arsenic and salinity are known to adversely affect the quality of water for drinking water supplies.

20.5.4. Problems due to Over-Exploitation

The ground water supply is neither unlimited nor it is always available in good quality. Due to uneven distribution of rainfall in time and space and ever-increasing demand, water resources are over-exploited in many parts of the country. This is resulting in shrinking of many water bodies and many ponds get dried up and remain so for considerable period in a year. Maintaining desired quality in such water bodies is difficult. In many cases, the abstraction of excessive quantities of ground water has resulted in the drying up of wells, salt-water intrusion & drying up of rivers that receive their base flows from ground water.

Most ground water quality problems caused by over-exploitation are difficult to detect. The solutions are usually very expensive, time consuming and not always effective. An alarming picture is emerging in many parts of our country. Ground water quality is slowly but surely declining everywhere. Usually the contamination is not detected until noxious substances actually appear in the water used. Unfortunately by this time the pollution may already have dispersed over a large area.

Although the green revolution in Punjab and Haryana has brought prosperity, it also brought the problems of soil and water degradation. While the irrigated area in these states doubled during 1965–1995, increased groundwater pumping has resulted in more than five meter fall in groundwater level in some places during the last 25 years. This has forced farmers to lower the pumps and further deepen the wells, increasing the cost of pumping and energy use and thus decreasing profitability and efficiency of agriculture. As per the Indian Council of Agricultural Research, the quality of deep groundwater in most parts of Haryana is marginal. Pumping of highly saline water from greater depth could result in the use of saline water for irrigation which will have serious unwanted consequences. In Delhi, the fall has been more than 10 m at some places. Similarly, fall in water table has been steep in some parts of Gujarat such as Mehsana.

A total of 106,019 sq. km area (about 31%) of Rajasthan comes under saline ground water. Of this 88,675 sq. km area falls in Western Rajasthan districts of Ganganagar, Barmer, Bikaner, Churu and Jaisalmer. The electrical conductivity of ground water in Western Rajasthan is over 8 ds/cm and in Eastern Rajasthan, it is over 6 ds/cm. Vast low lying alluvial tract from North-Western part of Banaskantha district through the western part of Mehsana and Ahmedabad districts, western and north-eastern parts of Surendranagar district, Southern part of Ahmedabad and South-Western part of Kheda district in Gujarat are underlain by saline ground water (EC 3.46 ds/cm). Ground water in Sangrur, Bhatinda, Ferozpur and Faridkot districts in Punjab has high (about 11.30 ds/cm) salinity. Ground water is saline in almost whole lift canal system of South-Western part of Haryana. About 3,766 sq. km. area in Haryana is underlain by saline ground water.

The Central Ground Water Authority (CGWA) formed under Environment Protection Act (1986) has already taken initiatives to regulate withdrawal of ground water. Similar restrictions on water abstraction from different water bodies and meeting environmental flow requirement for rivers need to be assessed and ensured.

20.5.5. Land Use and Ground Water Pollution

Large scale, point sources of pollution such as industrial discharges, landfills & subsurface injection of chemicals & hazardous wastes readily pollute ground water. These sources are easy to detect and regulate. However, the difficult problem is associated with non-point sources of pollution like leaching of agrochemicals and animal wastes, subsurface discharges from septic tanks and infiltration of polluted urban run-off and sewage. Non-point sources can affect a large swath of aquifers and the remediation is difficult. The best way to handle non-point sources of pollution is integrated land use and water management and control the use of fertilizers and chemicals. Table 14 presents land-use activities and their potential threat to ground water quality.

20.5.6. Ground Water Pollution Scenario in India

Leachates from compost pits, animal refuse of garbage dumping grounds nutrient enriched return irrigation flows seepage from septic tanks, seepage of sewage etc. has adversely affected the ground water quality in several parts of India. Solid waste disposal is also adding to ground water pollution problem. With increase of human and livestock population the quantum of waste produced has increased tremendously. The estimated annual waste production from these sources is around 2,000 million tons. Studies on chemical composition of ground water in phreatic zone have revealed that in many cases anomalously high concentration of Nitrate, Potassium and even phosphate (total phosphate) are present in contrast to their virtual absence or low concentration (NO_3 and K less than 10 mg/l) in semi-confined and aquifers. Unsystematic use of synthetic fertilizers coupled with improper water management practices have resulted in deterioration of ground water quality in many parts of the country.

Based on studies by CGWB, CPCB, and State Pollution Control Boards, state wise details of contamination of ground water in some areas of the districts due to various pollutants are given in Table 15.

Table 14. Land-use activities & their potential threat to ground water quality

Land use	Activities that cause to ground water pollution
Residential	Unsewered sanitation
	Discharge of sewage on land and in stream
	Infiltration from sewage or oxidation ponds
	Leakage from sewer, solid waste disposal sites, landfills
	Run-off from roads and urban areas
Industrial and Commercial	Process water, effluent lagoon
	Discharge of effluent on land and in stream
	Leakage from tanks and pipeline, and accidental spills
	Disposal of effluents through wells
	Disposal of solid and hazardous wastes through landfill
	Spillage and leakages of chemicals during handling
Mining	Waste discharge from mines
	Oilfield spillage at group gathering stations
Agriculture	Application of agrochemicals, fertilizers, pesticides, insecticides
	Irrigation with wastewater
	Livestock rearing

Table 15. State wise details of contamination of ground water in some areas of the districts due to various contaminants

Pollutant	State	Place of occurrences
Salinity (Inland)	Bihar	Begusarai
	Delhi	Najafgarh, Kanjhawala, and Mehrauli Blocks
	Haryana	Karnal, Sonepat, Rohtak, Hissar, Sirsa, Faridabad, Jind, Gurgaon, Bhiwani, Mahendragarh
	Madhya Pradesh	Gwalior, Bhind, Morena, Jhabua, Khargaon, Dhar, Shivpuri, Shajapur, Guna, Mandsor, Ujjain
	Maharashtra	Amravati, Akola
	Punjab	Bhatinda, Sangrur, Faridkot, Firozpur
	Rajasthan	Barmer, Jaisalmer, Bharatpur, Jaipur, Nagaur, Jalore, Sirohi, Jodhpur
	Uttar Pradesh	Agra, Mathura, Mainpuri, Banda
Salinity (Coastal)	Andhra Pradesh	Vishakapatnam, East Godavari, West Godavari, Krishna, Guntur, Prakasam
	Gujarat	Junagarh, Kachch, Varahi, Banskanta, Surat, Bharauch, Mehsana, Ahmedabad, Surendranagar, Kheda, Jamnagar
	Karnataka	Bijapur, Raichur, Bellary, Dharwar
	Kerala	Ernakulam, Trichur, Alleppey
	Pondicherry	Pondicherry
	Orissa	Puri, Cuttak, Balasore
	Tamil Nadu	Karaikal, Nagapattanam, Pudukottai, Ramananthapuram, North Arcot – Ambedkar, Dharampuri, Salem, Trichy, Coimbatore
	West Bengal	Haldia and 24 Parganas
Flouride	Andhra Pradesh	Cuddapah, Guntur, Nalgonda, Prakasam, Nellore, Anantapur, Rangareddy, Adilabad
	Bihar	Giridih, Jamui, Dhanbad
	Gujarat	Banskanta, Kachch, Amreli, Surendranagar, Rajkot, Ahmedabad, Mehsana, Sabarkantha
	Karnataka	Tumkur, Kolar, Bangalore, Gulbarga, Bellary, Raichur
	Haryana	Hissar, Kaithal, Gurgaon, Rohtak, Jind, Bhiwani, Mahendragarh, Faridabad
	Kerala	Palaghat, Ananipur, Nellore, Chittoor.
	Madhya Pradesh	Bhind, Moerana, Guna, Jhabua, Chhindwara, Seoni, Mandla, Raipur, Vidisha
	Maharashtra	Bhandara, Chandrapur, Nanded, Aurangabad
	Orissa	Bolangir, Bijapur, Bhubaneshwar and Kalahandi
	Punjab	Jalandhar, Amritsar, Bhatinda, Faridkot, Ludhiana & Sangrur
	Rajasthan	Barmer, Ganganagar, Jalore, Nagaur, Pali, Sirohi, Ajmer & Bikaner
	Tamil Nadu	Dharampuri, Salem, North Arcot, Viluppuram, Tiruchirapalli, Pudukottai, Chengalput, Madurai
	U.P.	Bulandshahar, Unnao, Agra, Aligarh, Mathura, Ghaziabad, Meerut, and Rai Bareilly
	West Bengal	Birbhum
Sulphide	Orissa	Balasore, Cuttak, and Puri
Iron	U.P.	Mirjapur
	Assam	Darrang, Jorhat, Kamrup, Northern Bank of Brahmaputra
	Orissa	Parts of Coastal Orissa, Bhubneshwar

(Continued)

Table 15. (Continued)

Pollutant	State	Place of occurrences
	Bihar	E. Champaran, Muzaffarpur, Gaya, Mungher, Deoghar & Madubani, Patna, Palamau, Nalanda, Nawada, Banka
	Rajasthan	Bikaner, Alwar, Bharatpur, Dungarpur
	Tripura	Dharmnagar, Amarpur, Agartala, Kamalpur, Khowai, and Parts of Agartala Valley
	West Bengal	Midnapur, Howrah, Hoogly and Bankura,
Manganese	Orissa	Bhubaneshwar
	U.P	Moradabad, Basti, Rampur, Unnao
Arsenic	West Bengal	Malda, Murshidabad, Nadia, Malda, South-24 Paraganas, Hoogly, Bardhaman, Howrah
Nitrate	Bihar	Patna, East Champaran, Gaya, Nalanda, Nawada, Banka, and Bhagalpur.
	Andhra Pradesh	Vishakapatnam, East Godvari, Krishna, Prakasam, Nellore, Chittoor, Anantapur, Cuddapah, Kurnool, Nalgonda, Mehboobnagar, Rangareddy, Medak, Adilabad, and Khammam.
	Delhi	Naraina, Shahadara (Blocks)
	Haryana	Ambala, Sonepat, Jind, Gurgaon, Faridabad, Hissar, Sirsa, Karnal, Kurukshetra, Rohtak, Bhiwani, Mahendragarh
	Himachal Pradesh	Kulu, Solan, Una
	Uttar Pradesh	Orai, Jhansi, Lalitpur, Faizabad, Sultanpur, Maharajganj, Gorakhapur, Deoria
	Karnataka	Bidar, Gulbarga, and Bijapur
	Madhya Pradesh	Sehore, Bhopal, and West & Central part of state
	Maharashtra	Thane, Jalna, Beed, Nanded, Latur, Osmanabad, Solapur, Satara, Sangli, Kolhapur, Dhule, Jalgaon, Aurangabad, Ahmednagar, Pune, Buldana, Amravati, Akola, Nagpur, Wardha, Bhandara, Chandrapur, Gadchiroli
	Punjab	Patiala, Faridkot, Firozpur, Sangrur & Bhatinda
	Rajasthan	Jaipur, Churu, Ganganagar, Bikaner, Jalore, Barmer, Bundi, and Sawaimadhopur
	Tamil Nadu	Coimbatore, Periyar, and Salem
	West Bengal	Midnapur, Howrah, Uttar Dinajpur, Malda, Birbhum, Murshidabad, Nadia, Bankura and Purulia
	Delhi	Shahadara and Mehrauli Blocks
Chloride	Karnataka	Dharwad, Belgaum
	Madhya Pradesh	Bhind, Shajapur and Sehore
	Maharashtra	Solapur, Satara, Amravati, Akola, and Buldana
	Rajasthan	Barmer, Jaisalmer, Jodhpur, and Jalore
	West Bengal	Digha, Haldia
Zinc	Andhra Pradesh	Hyderabad, Osmania University campus
	Delhi	R.K. Puram
	Rajasthan	Udaipur
Chromium	Punjab	Ludhiana
Heavy Metals	Andhra Pradesh	Anantapur, Mehboobnagar, Prakasam, Visakhapatnam, Cuddapah, Nalgonda
	Assam	Digboi
	Bihar	Dhanbad, Muzaffarpur, Begusarai
	Haryana	Faridabad
	Himachal Pradesh	Purwanoo, Kalaamb

Karnataka	Bhadrawati
Madhya Pradesh	Bastar, Korba, Ratlam, Nagda
Orissa	Angul, Talcher
Punjab	Ludhiana, Mandi, and Gobindgarh
Rajasthan	Pali, Udaipur, Khetri.
Tamil Nadu	Manali, North Arcot.
Uttar Pradesh	Singrauli, Basti, Kanpur, Jaunpur, Allahabad, Saharanpur, Aligarh
West Bengal	Durgapur, Howrah, Murshidabad, Nadia.
Delhi	Alipur, Kanjhawala, Najafgarh, Mehrauli, and Shahdara Blocks

The Central Pollution Control Board undertook a major groundwater quality survey and the report published in 1995 identified about 20 locations in various states of India as critical sites of ground water pollution. CPCB found that industrial effluents are the primary reason for ground water pollution. In the industrial and urban fringe zones of cities, sub-soil water has already been polluted by industries which release highly toxic substances. The wells in many residential areas are contaminated with nitrate and detergents. Summary results of these studies are given in Table 16.

20.6. GROUND WATER QUALITY

In the state of Haryana, the pesticides use was 5,100 ton in 1995–96. About 85 to 90% of this quantity dissipated into the environment. In 1994–95, 54% of the pesticides used in agriculture and 94% used in public health in India were banned or severely curtailed in the West. The eight Haryana districts through which the Yamuna flows used 327,000 tones of chemical fertilizers. In addition, the towns of Yamunanagar, Karnal, Sonepat and Panipat dump their own industrial and domestic effluents into the river. Ten firms in Yamunanagar and Panipat alone dump 72 million litres per day of wastewater into the Western Yamuna Canal which brings raw water to the Haiderpur water works in Delhi.

The National Environmental Engineering Research Institute (NEERI) has established that the seeds of the Nirmali tree cleanse muddy water by coagulating the suspended impurities. Similarly, Cardamon leaves, drumstick (Moringa olifera) seeds and many other herbs and roots, which are locally available and have been traditionally in use for ages, also possess ability to clean polluted water.

20.7. POLLUTION IN COASTAL AREAS

Rivers contribute to coastal pollution by transporting a wide range of pollutants through land drainage. About 135 billion metric tones (MT) of sediment and 32 billion MT of soluble matter enters the ocean through various Indian rivers each year. This constitutes 90 per cent of the total waste going into the ocean; the rest 10% of the waste matter going into the ocean is constituted by wind, rain and earthquakes. It may be noted that the water flowing through all Indian rivers is 5%

Table 16. Results of chemical analysis carried by CPCB

Area	Industrial activities	Ground water quality problem
Dhanbad (Bihar)	Fertilizers, Chemicals, Soft drink plants, Cement explosive factory and Ancillary units	Low pH, high NO_3, Al, Ca, TDS, Fe, Mn, Cr, Zn, Cu, and metals like Hg & Cd, Pesticides and micro-organism
Digboi (Assam)	Oil refinery	Fe and Mn were more than permissible limit. Ni, Zn, Cd, Cr, Pb were also reported. Total Coliform were also present.
Durgapur (West Bengal)	Coal field	Heavy metals except Cu exceeded the desirable limit. Hg was also reported as high as 9.5 mg/l. Phenolic compounds were in traces. Total pesticides levels have violated the desirable limit.
Howrah (West Bengal)	Foundries, Electroplating & other mechanical type	Heavy metals viz, Pb, Cd, Cr were very high and Zn, Cu, were within limit. Hg was also present Fe & Mn were also very high, CN & Phenolic compounds in traces. EC, Cl, TDS were some time very high. Pesticides were also on high side.
Botharam Patncheru (AP)	Pesticides, Pharmaceuticals	Phosphates, Hg & As were beyond limit, Cd, Fe, Mn & Pb also exceeded. Pesticides were also found to be present. The Coliform were also on higher side. TDS, Ca exceeded the desirable limit.
Greater Cochin (Kerala)	Fertilizer, Pesticides, Chemicals	Low acidity. Heavy metals, Pesticides etc. were present in low concentrations. The presence of Total Coliform was the single major concern. The Coliform of fecal origin was also high.
North Arcot (TN)	Tanneries, dyeing Units	Hg, N, Cd, Pb & As were in traces. Zn, Cu, Cr, Fe & Mn exceeded the limit at several locations. Total Coliform was generally high and feacal Coliform were also on higher side.
Bhadravathi (Karnataka) Steel,	Steel, Paper Mills	Zn, Fe & Mn were reported to be high. Pesticides like Aldrin, Dieldrin, Lindane & DDT were present in very high concentrations. Pathogenic organisms were also on higher side.
Ratlam, Nagda (MP)	Distillery Dye (intermediates) Pharmaceutical (intermediates)	TDS, TH, Hg, Pb were on higher side considerable amounts of pesticides were also reported. Fecal Coliforms were also present particularly at Nagda.
Vapi (Gujarat)	Dyes, pesticides, paper & pulp mills, organic & inorganic chemicals.	Phenolic compounds, cyanide & heavy metals were present within limit as per drinking standards.
Chembur (Maharashtra)	Petrolieum, refineries, fertilizer & petrochemical, thermal	TDS, Alkalinity, TH were

	power plant	higher. Heavy metals, pesticides, phenolic compounds were present in concentrations, but not very significant. Coliform were on higher side.
Korba (MP)	Thermal power plants, ancillary units, Aluminum industries, mining	The presence of Cd, Fe, Cr & Cu has exceeded the presented limits, Pesticides were also present. Coliform, F, TDS, B & phenolic compounds also exceeded the standards.
Singrauli (UP)	Thermal Power Plants, aluminum Plants, organic chemicals industries and other subsidiary units with Carbon plants, caustic soda & pesticides.	Fe, Cr & Cu were present in predominance. Presence of high Aldrine, Dieldrin & Lindane levels were also observed. Beside this F, Ca, Mg, B, Coliforms, Phenols exceeded the prescribed limit.
Mandi Gobindgarh (Punjab)	Wooden, chemicals, electro-plating units and other steel metals units.	Pb, Cu, Cd, exceeded the desirable limit of drinking water, Phenolic compounds & Cyanide was also present on higher side.
Parwanoo (HP)	Ancillary, general industries, fruit proceeding plant, airopesticides.	Presence of Cd, Pb, Fe, Mn were observed on higher side. Traces of Pesticides were also present. Phenolic were above the toxic limit.
Kala Amb (HP)	Paper mills	Phenolic compound were higher than the toxic limit; Coliforms were also present. Heavy metals like Cd, Pb & Mn and also pesticides had polluted the ground water.
Pali (Rajasthan)	Textile, dyes	Lead and Zinc were found higher. Fluoride concentration was noticed to be high. TDS, and Cl were in concentrates exceeding standards. Ground waters at some places were rich in organic method.
Jodhpur (Rajasthan)	Textile, steel, engineering foundry, chemicals, minerals dye plastic, oil, pulses and rubber.	Heavy metals such as Fe, Cr, Mn, were higher. Pb was also present NO_3, were also on higher side. Na, TDS exceeded the limit.
Najafgarh (Delhi)	Insecticides, Caustic Soda, Vanaspati, Electroplating etc.	EC, TDS both exceeded tin drinking water standards. Coliform, F, NO_3 were found significantly in higher values. Fe & Cr were in levels exceeding standards limit.
Angul Talcher	Thermal power station, fertilizers, chemicals, mining activities & aluminum	Cr, Fe, Cd, Pb & F all were found in concentration level exceeding standards limits. NO_3 values were high.

Source: Central Pollution Control Board.

of the water flowing through all the rivers of the world but Indian rivers carry 35% of the sediments that go to all the oceans in the world.

Concentrations of lead and cadmium (820 microgram per litre (mgpl) and 336 mgpl, respectively) have been observed in the Thane creek, Mumbai. The mercury concentration is 778 mgpl in 12 the Karnataka towns that are situated along the 300 km long coast, discharging untreated wastewater or about 30 tones of organic load directly into coastal waters daily. Three major industries besides a host of medium and small units along the coast dump 30,000 cubic metres of industrial effluents into the sea everyday.

Along the Kerala coast, 14 municipal bodies discharge about 650 tones of organic matter per day into coastal waters. Another 300,000 cubic m of industrial effluents, emanating from 200 large and medium and 2,000 small scale industries are directly discharged into the sea. The heavy traffic of marine vessels for fishing, port and defence-related operations makes the coastal waters vulnerable to water quality degradation. In the 1,050 km long Tamil Nadu and Pondicherry coasts, the major polluting industries are textiles, chemicals, fertilizers, pulp, paper, cement, sugar, mining and mineral processing. In Andhra Pradesh, which has a coastline of 980 km, pollution is caused not just by industrial units but by port related operations at Vishakhapatnam and Kakinada where intensive mechanized fishing leads to heavy traffic of marine vessels. Similar problems exist in Orissa and West Bengal. The water quality has deteriorated because of similar operations at the Haldia and Calcutta ports. Besides, a large quantity of effluents from the Calcutta and other municipal areas find their way into the coastal waters mainly through the Hooghly estuary.

20.7.1. Sea Water Intrusion

The coastal tract of India extends over 5,400 km on mainland and 7,500 km including the islands. Mainland coastal tract is densely populated. Anthropogenic activities have increased the demand of water and stressed the limited fresh groundwater resources. Although the coastal tract has enormous ground water potential, the availability of fresh ground water along the coastal stripe has always been a constraint vis-à-vis the demands. At many places, because of increasing abstraction and consequent disturbed hydrodynamic equilibrium, the seawater has intruded inland contaminating the fresh groundwater.

In coastal areas, ground water salinity problems are widely present with varying degrees. These problems are more extensive in the east coast in comparison to west coast, due to greater ground water development. The coastal alluvial tracts, including deltaic tracts of major rivers, such as the Ganga, the Mahanadi, the Krishna, the Godavari, the Cauvery, have varying degrees of salinity. The major causes of salinity of ground water in coastal areas can be attributed to: tidal water ingress in upper aquifers, sea water ingress in lower aquifers, over-exploitation of ground water, and less natural recharge of ground water.

Seawater intrusion is one of the most common forms of groundwater contamination in coastal areas. This contamination is of major concern because even a small proportion of seawater (about 2–3%) renders freshwater un-potable and may lead to abandonment of aquifers in extreme cases. Sea Water intrusion occurs when withdrawal lowers the hydraulic potential in a coastal aquifer, allowing seawater to migrate inland.

Reports from coastal districts of Gujarat, Maharashtra and Andhra Pradesh highlight the origin of seawater intrusion in response to increased demands for domestic, industrial and irrigation purposes. The major factors for seawater ingress in Gujarat are low natural recharge, overexploitation of groundwater by farmers and industrial growth. In Thane district of Maharashtra, it is the intensive pumping of groundwater associated with urban and industrial development projects that has caused the problem. In coastal Andhra Pradesh, it is the lack of canal irrigation facilities which necessitated heavy dependence on groundwater resulting in seawater intrusion.

The south Chennai aquifer that holds substantial quantity of groundwater, meets 20% of the city's water requirement. Due to constant pumping and improper management, this aquifer is facing a severe threat of being contaminated. In drought years, ground water is the sole safeguarding arrangement to meet the drinking water requirement. Possible options and strategies for addressing the problems in each of the above cases include site-specific technical solutions, legislative measures, and social awareness.

In general the situations are encountered in coastal areas; saline water overlying fresh water aquifer or fresh water overlying saline water or alternating sequence of fresh and saline water aquifers. Some of the locations of sea water intrusion in India are Coastal Saurashtra, Karnataka, Kerala, Tamil Nadu, Andhra Pradesh, Coastal Orissa and Coastal West Bengal.

20.8. WATER RELATED DISEASES

Estimates suggest that nearly half of the world's population suffers from diseases that are associated with insufficient or contaminated water. Although, provision of safe drinking water and sanitation has been one of the guiding principles of many national and international programs (recall the MDGs discussed in the beginning of this chapter), a significant amount of population, mostly poor people, are yet to receive assured water supply. In 1981, International Drinking Water Supply and Sanitation Decade was launched with the objective to give people access to drinking water in quantity and of a quality equal to their basic needs. In India, the existing water supply norms for rural population is 40 litres of drinking water per capita per day (lpcd) and a public stand post or a hand pump for 250 persons. Further, the sources of water supply should be within 1.6 km (1 mile) in plains, or within 100 metres elevation distance in hills. For urban population, the norm is 125 lpcd piped water supply with sewerage system, 70 lpcd without sewerage and 40 lpcd in towns with spot sources.

Water borne diseases are the most important water quality problems affecting the society in India. These diseases are mainly due to inadequate arrangements for transport, treatment, and disposal of wastes into natural water bodies, resulting in contamination of both surface and ground waters. Moreover, contribution of pathogens through diffuse sources is also quite significant. Thus, most of the surface water bodies and many aquifers are contaminated. A large population of the country still uses water directly for drinking and other uses without any treatment. Thus, they are exposed to water borne diseases resulting in high incidence of such diseases.

Several states in India have been identified as endemic to fluorosis due to abundance in natural occurring fluoride bearing minerals. There are nearly half million people in India suffering from ailment due to excess of fluoride in drinking water. In some villages of Rajasthan and Gujarat level of the fluoride goes up to 11.0 mg/lit.

Although iron content in drinking water may not affect the human system as a simple dietary overload, but in the long run, prolonged accumulation of Iron in the body may result in homochromatosis, where the tissues are damaged. In some districts of Assam and Orissa, ground water have high iron content ranging from 1 to 10 mg/l. Iron levels are also high in the north-eastern and eastern parts of the country (affecting 58,000 habitations or 29 million persons).

Inadequate treatment of human and animal wastes adds to the high incidence of water-related diseases. Only about 14% of rural and 49% of urban inhabitants (1997 estimates by Shukla, 1999) have access to adequate sanitation facilities, and hence the water contaminated by human waste is frequently discharged directly into watercourses or seeps into the groundwater from faulty septic tanks. As a result, the level of fecal coliform bacteria in many rivers often exceeds WHO standards which causes a number of gastrointestinal ailments among the population.

Water borne diseases are single most important factor, responsible for high human mortality in India. Children are worst affected, especially in rural areas and urban slums. In the past three decades, there has been considerable reduction in death rates due to water related diseases such as cholera, typhoid and diarrhea diseases. For example, cholera has been largely eliminated from West Bengal and Rajasthan while cases of malaria are very few now. Important water-borne and water related diseases, and the causative organisms are given in Table 17. Dengue fever still occurs in many parts of the country during post monsoon season and there are several deaths each year due to this. Some other water borne diseases are Hepatitis A, polio and Amoebiasis. Amoebiasis is an infection of large intestine and is closely related with poor sanitation. Scheistosomiasis is a water related disease which is associated with perennial irrigation. Fortunately, this disease is not common in India.

One of the main objectives of Government should be to improve health of population and environment, notably air and water. Providing adequate nutrition, drinking water and sanitation facilities will go a long way in overall development. To attain this, it is necessary to establish protected areas for sources of drinking water supply, suitable means of disposal of waste and develop appropriate mechanism and

Table 17. Water related diseases and Causative factors

	Name of the disease	Causative organism
Water-borne diseases	Typhoid	Salmonella typhi
	Cholera	Vibrio cholerae
	Paratyphoid	Slmonella parayphi
	Gastroenteritis	Enterotoxigenic
	Bacterial dysentery	Escherichia coli
	Infectious hepatitis	Hepatitis-A virus
	Poliomycetis	Polio-virus
	Diarrhoeal diseases	Rota-virus, Norwalk agent, other virus
	Other symptoms of enteric diseases	Echono-virus, Coxsackie-viru
	Amoebic dysentery	Entamoeba hystolitica
Water-washed diseases	Scabies	Various skin fungus species
	Trachoma	Trachoma infecting eyes
	Bacillary dysentery	E. coli
Water-based diseases	Schistosomiasis	Schistosoma sp.
	Guinea worm	Guinea worm
Infection through water related insect vectors	Sleeping sickness	Trapanosoma through tsetse fly
	Malaria	Plasmodium through Anaphelis
Infections primarily due to defective sanitation	Hookworm	Hook worm, Ascaris

regulations to reduce pollution. Public awareness campaign needs to be initiated to encourage participation of community in protecting and improving environmental quality.

20.9. WATER QUALITY MONITORING

The Central Water Commission (CWC) has a large network of stations for hydrological observation. In addition to observation of river flow, CWC is also monitoring water quality, covering all the major river basins of India; the details have been given in Chapter 5. At present CWC is maintaining a three-tier laboratory system for analysis of the water samples to determine quality indicators. The Level-I laboratories are located at 319 field water quality monitoring stations on major rivers of India where physical parameters such as temperature, colour, odour, specific conductivity, total dissolved solids, pH and Dissolved Oxygen (DO) of river water are observed. There are 21 Level-II laboratories located at selected divisional headquarters to analyses the physico-chemical characteristics and biochemical oxygen demand of river water. Three Level-III laboratories are functioning at Varanasi, Delhi and Hyderabad which are equipped with sophisticated equipments to analyse various, toxic elements, polyaromatic hydrocarbons, insecticides and microbiological inhabitants of river water. The data generated is computerized in data base system and disseminated in the form of hydrological yearbook, status reports & bulletins.

To monitor the quality of ground water, Central Ground Water Board (CGWB) uses a network of 14,995 stations. The Central Pollution Control Board (CPCB) has set up a network of 480 stations to monitor water quality in collaboration with various State Pollution Control Boards (SPCBs). This network operates in a 3-tier system. Tier-1 caters to the need of Global Environment Monitoring System (GEMS). It consists of 50 stations, which monitor 22 parameters on a monthly basis. Nearly 430 stations form Tier-2 network under a programme known as 'Monitoring of Indian National Aquatic Resources' (MINARS) on a monthly/quarterly basis. Stations operated by SPCBs carry Tier-3 level monitoring. Overall, water quality monitoring network covers 14 major, 12 medium, 9 minor river basins, 16 other small rivers, 35 lakes, 24 aquifers, 3 creeks, 2 tanks, and 1 pond.

CPCB regularly monitors water quality of various Indian rivers at several places. Summary results of this monitoring have been listed in the chapters of respective rivers. Their web site may be referred to for further details and updated information.

20.10. WATER QUALITY AND ENVIRONMENTAL LAWS IN INDIA

Many laws have been enacted in India to protect water and environment. A list of relevant laws is given in Table 18. Some laws have also been enacted by the Sate Governments.

A number of environment related laws have been passed and implemented in India to protect environment. Some important laws are:

- The Water (Prevention & Control of Pollution) Act, 1974 as amended in 1988.
- The Water (Prevention & Control of Pollution) Cess, Act, 1974 as amended in 1991.
- The Air (Prevention & Control of Pollution) Act, 1981 as amended in 1987.
- The Environment (Protection) Act, 1986.

Table 18. Relevant laws to protect water and environment in India

Central Enactment	
i	The River Board Act, 1956
ii.	The Merchant Shipping (Amendment) Act, 1970
iii.	The water (Prevention and Control of Pollution) Act, 1974
iv.	The Water (Prevention and Control of Pollution) Cess Act, 1977
v.	The Water Act, 1981
vi.	The Environment (Protection) Act, 1986
vii.	Environmental Impact Assessment Notification, 1994
State Enactment	
i.	Orissa River Pollution Act, 1953
ii.	Maharashtra Prevention of Water Pollution Act, 1969

- The National Environmental Tribunal Act, 1995.
- The National Environment Appellate Authority Act, 1997.

Further, many projects now require environmental clearance from the Central Government under the Environmental Impact Assessment Notification, 1994. These are the projects that may cause serious harm the natural environment. A partial list of the projects that are related to water sector and that require environmental clearance is given below.

1. River valley projects including hydroelectric power, major irrigation projects and their combination including flood control,
2. Ports, harbours, airports (except minor ports and harbours),
3. All tourism projects between 2,000 m – 500 m of High Water Line and at locations with an elevation of more than 1,000 meters with investment of more than Rs. 5 crore.
4. Thermal power plants.
5. Mining projects (major minerals) with leases more than 5 ha.
6. Tarred roads in Himalayan and/or forest areas.

20.11. EPILOGUE

Water is the most essential natural resource for life next to air and is likely to become a critical scarce resource in many region of the world in the coming decades. In order, for any water body to function adequately, in satisfying any one of the beneficial uses, it must have corresponding degree of purity. For example drinking water needs highest purity of water, whereas disposal of wastes can be done in any quality of water. In recent years, as the demand for water has nearly approached in magnitude to the available supply. The concept of management of the quality of water is quite as important and obligatory as its quantity has since been widely recognized and strongly advocated throughout the world.

CHAPTER 21

CONSTITUTIONAL PROVISIONS, INTER-STATE WATER DISPUTES AND TREATIES

Broadly speaking, the utilization of water in India comes under the jurisdiction of states. But it is important to note that even for a river which flows entirely in one state, it is possible that one state's intervention might produce detrimental environmental or social consequences (for example submergence, water logging, etc.) in another. Water withdrawal in a state may also have an impact on groundwater aquifers in the adjacent state. Moreover, operation of a dam may cause inundation in areas beyond the boundaries of the state or nation. All these aspects emphasize the need for coordinated approach towards water resources utilization in the country.

21.1. CONSTITUTIONAL PROVISIONS REGARDING WATER

Under the Indian constitution, a state government has the power to make laws in respect of water resources of that state. The Parliament has the power to legislate the regulation and development of interstate rivers. Thus the authority of the state Government over water can be exercised, subject to certain limitations that may be imposed by the Parliament.

Here it is important to make clear distinction between an inter-state river and an intra-state river. A river that lies within one state from its source to its mouth is an intra-state river and any river which flows in the territory of two or more states is an inter-state river.

The legislative framework of the constitution related to water is based on Entry 17 of the State List, Entry 56 in the Union List, and Article 262 of the Constitution. These are:

a) Entry 17 in List II (State List) in Schedule VII

Water, that is to say, water supplies, irrigation and canals, drainage and embankments, water storage and water power subject to the provisions of Entry 56 of List-I.

It can be seen immediately that it is not an unqualified entry. Although water is in the State List, this is subject to the provisions of Entry 56 in the Union List, which reads as:

b) Entry 56 of List I (Union List):

Regulation and development of inter-state rivers and river valleys to the extent to which such regulation and development under the control of the Union is declared by Parliament by law to be expedient in the public interest.

c) Also relevant here are the provisions of Article 262:

(1) Parliament may by law provide for the adjudication of any dispute or complaint with respect to the use, distribution or control of the waters of, or in, any Inter-State river or river valley.

(2) Notwithstanding anything in this Constitution, Parliament may by law provide that neither the Supreme Court nor any other court shall exercise jurisdiction in respect of any such dispute or complaint as is referred to in clause (1).

Some other articles and entries may also have a bearing on the matter.

Article 248 (Residuary powers of legislation):

Parliament has exclusive power to make any law with respect to any matter not enumerated in the Concurrent List or State List.

Article 254: Inconsistency between laws made by Parliament and laws made by the Legislatures of States

1. If any provision of a law made by the Legislature of a State is repugnant to any provision of a law-made by Parliament which Parliament is competent to enact, or to any provision of any existing law with respect to one of the matters enumerated in the Concurrent List, then, subject to the provisions of clause (2), the law made by Parliament, whether passed before or after the law made by the Legislature of such State, or, as the case may be, the existing law, shall prevail and the law made by the Legislature of the state shall, to the extent of the repugnancy, be void.

   Provided that nothing in this clause shall, prevent Parliament from enacting at any time any law with respect to the same matter including a law adding to, amending, varying or repealing the law so made by the Legislature of the State.

So far the Parliament has not made much use of the powers vested in it by the Entry 56 of the Union List. In view of the supremacy of the parliament, it is not totally right to say that water is a state subject. Moreover, most of India's important rivers are inter-state. Hence, water is potentially as much a union subject as a state subject. Further, the legislative power of a state under Entry 17 has to be exercised in such a manner that the interests of other states do not suffer adversely and a dispute is created.

The power and role given to the Union Government with regard to the management of inter-state rivers is very important and necessary. In this context, the provisions of Entry 20 in the Concurrent List, namely, economic and social planning, are also quite relevant. By virtue of this provision, the major and medium irrigation, hydropower, flood control and multi-purpose projects are required to seek clearance of the central government for inclusion in the national plan.

21.1.1. Water Rights

Traditionally water has been considered a social commodity which is a "basic requirement" for life. In India in ancient times, a fundamental premise was that nobody who is thirsty should be denied water, whatever be his income, purchasing power, and status. Even these days, the first thing that is offered to a guest is water to drink. This privileged social status of water was based on the doctrine of 'essential service' or 'public service'. Of late, due to various reasons among which scarcity being the chief, water and related services are becoming more and more an economic asset with production and conservation costs, utilization values, opportunity costs, and demands that vary with price.

In traditional societies like India, water was perceived as the common property of society rather than belonging to any individual. The term 'water rights' refers to the right to use water. The riparian system considers that anybody who owns land on the banks of a stream has a right to reasonable use of water for his needs. At the same time, he has a duty to let the water go beyond his land without unduly diminishing its quality and quantity.

The need to consider water rights arises when water resources are scarce and rigid attitude of users requires clear definition of rights and entitlements. According to the Indian Easements Act (1882) the government has the sole right to regulate the collection, detention and distribution of the waters of rivers and streams flowing in natural channels and/or natural lakes or ponds or the water flowing, collected, detained or distributed in/or by any channel constructed at the public expense for irrigation. Many experts have argued that the state may claim such a right for the major and medium rivers and the communities should have a right to utilize waters of small rivers and minor streams. It is argued that this approach would lead to more sustainable use of water. However, this might results in disputes due to encroachments in each other's jurisdiction.

The Indian Easements Act (1882) confers on the owner of the land, the right to collect and dispose, within his own limits, all water under the land which does not pass in a defined channel. This enables the owner full control of the water beneath his property and he is free to withdraw and use it as he feels appropriate. However, this has led to a situation where a resourceful farmer can dig deeper tube-wells and pump large quantities of water and thereby deprive nearby land owners from their legitimate rights. This has also resulted in mining of ground water in many places. Besides quantity, the quality of ground water has deteriorated due to uncontrolled disposal of industrial wastes, sewage and use of chemical fertilizers and pesticides. The Government of India formulated a model bill to regulate groundwater quantity and quality and has created a groundwater authority to enforce use. According to the model ground water control bill, the state governments will have powers to restrict construction of structures for ground water abstraction for all uses, exempting small and marginal farmers provided these structures are for their exclusive use. No permission is necessary if water is to be withdrawn with manual devices. So far groundwater legislation has been enacted only in a few states.

The Central Ground Water Authority has been constituted to check indiscriminate withdrawal of ground water and to issue directions to protect ground water resources of the country. Note that the constitution of this authority implies a change from the unrestricted use of ground water to a controlled withdrawal regime by way of licensing. The state governments are supposed to establish state ground water authorities to implement the rules and monitor.

21.2. RIVER BOARDS ACT

With the objective of promoting integrated development of the waters of inter-state rivers, the Parliament of India had enacted the River Boards Act (RBA), 1956, under Entry 56 of List I. This act contemplates the constitution of river boards by the Governments of India in consultation with State Governments. These boards are supposed to advise and promote integrated development of water resources of inter-state rivers for beneficial purposes and for flood control. The highlights of the River Boards Act (1956) are (Gosain and Singh 2005):

- RBA provides for the establishment of River Boards for regulation and development of interstate rivers either on the request of a riparian state or even otherwise.
- Different boards may be established for different interstate rivers.
- The board is to consist of the Chairman and such other members as the Central Government deems fit to appoint. They must be persons having special knowledge and experience in irrigation, electrical engineering, flood control, navigation, water conservation, soil conservation, administration or finance.
- Functions of the board are set out in detail in Section 13 of the Act. Subject wise, they are very wide, covering conservation of the water resources of the interstate river, schemes for irrigation and drainage, development of hydroelectric power, schemes for flood control, promotion of navigation, control of soil erosion and prevention of pollution.

Although the national water policy supports basin wise planning and management and this appears to be a rational and optimal way, there is no real river basin authority in India. The Betwa River Board was set up by a separate enactment specifically to oversee a particular project. The Brahmaputra Board was also established by a parliamentary enactment and was vested with powers of execution of projects. However, its role has been confined largely to the preparation of a Master Plan for the Brahmaputra basin. Likewise, the Bansagar Control Board was set up to supervise specific projects; the Narmada Control Authority came into being with well-defined functions under the orders of the Narmada Waters Dispute Tribunal; and the role of Ganga Flood Control Commission is limited to preparation of master plans for flood control in the Ganga basin. Clearly, there is a need to establish appropriate organizations for the planned development and management of a river basin.

21.3. INTER-STATE WATER DISPUTES (ISWD)

Most big rivers in India have catchments in more than one state or are of inter-state in character. Disputes do arise amongst the basin states regarding utilization, distribution or control of water in these inter-state rivers. It is also common to see disputes in the interpretation of the terms governing the use or control of water resources while implementing such agreements.

Presently, there is no well-defined procedure to allocate water of inter-state rivers and tribunals are doing this task when there is a dispute. At international level, guidelines such as "the Helsinki Rules" have been formulated for resolving inter-state water disputes all over the world and these were taken into account by tribunals while resolving the water disputes. The guidelines have been codified by the U.N. Convention on the law of the non-navigational use of international watercourses and adopted by the U.N. General Assembly in 1997.

Broadly, ISWD are settled by either of two processes: negotiations or adjudication. There is also the possibility of mediation by a senior political leader or statesman. Of course, such processes do not foreclose arbitration or adjudication. Conflict resolution through a dialog is a workable approach to solve ISWD and it should always be tried first. But there have been instances when this process could not yield the desired results.

Adjudication involves study and solution of a dispute or controversy. There is a view that adjudication is not the best means of settling ISWD. Another argument against the adjudication of ISWD by judicial tribunals is that such disputes do not involve questions of law but questions technical nature involving hydrology, engineering, agriculture, climate, meteorology, sociology, and so on, and should be dealt with scientific spirit.

The provisions of the Interstate Water Disputes Act, 1956 state that a State Government which has a water dispute with another State Government may request the Central Government to refer the dispute to a Tribunal for adjudication. The Central Government, if it is of opinion that the dispute cannot be settled by negotiation shall refer the dispute to a Tribunal:

- The Tribunal should consist of a Chairman and two other members, nominated by the Chief Justice of India from among persons who, at the time of such nomination, are Judges of the Supreme Court.
- The Tribunal can appoint assessors to advise it in the proceedings before it.
- On the reference being made by the Central Government, the Tribunal adjudicates the matter and gives its award. The award is published and is final and binding on the parties.
- Jurisdiction of the Supreme Court and other courts in respect of the dispute referred to the Tribunal is barred.
- The Central Government may frame a scheme, providing for all matters necessary to give effect to the decision of the Tribunal. The scheme may, inter-alia, provide for establishing an authority for implementing the award.

The ISWD Act of 1956 provides that the Union Government shall publish the decision of the Tribunal in the Official Gazette and the decision shall be final and binding on the parties to the dispute and shall be implemented by them. To facilitate their functioning, water tribunals have court-equivalent powers for gathering of information, requiring witnesses to testify and recovering the costs of the tribunal. But the central government cannot enforce the tribunal's award if a state government refuses to implement the award. Thus the tribunals lack the most important power – the litigants know that the tribunal cannot enforce its decisions. The ISWD Act states:

> Notwithstanding anything contained in any other law, neither the Supreme Court nor any other court shall have or exercise jurisdiction in respect of any water dispute which may be referred to a Tribunal under this Act.

This provision could be interpreted as vesting water tribunals with a status equivalent to the Supreme Court meaning thereby that the decisions of the tribunals have the same force as those of the Supreme Court. However, the tribunal's decisions have been questioned in the courts. In general, the resolution of water disputes is complicated by various aspects of centre-state relations.

The wide scope for discretion, the extensive bargaining and the multiplicity of potential vetoes work in tandem to undermine the clarity and transparency needed for speedy dispute resolution. In May 2002, the ISWD Act was amended by placing time limits on tribunals.

The existing processes and institutions for resolving inter-state river disputes are not sufficiently well defined. All too often this causes delays in settling a dispute. It is, therefore, better that the dispute resolution mechanism is clearly defined. Long delays can be too very costly in utilizing water resources in India and project funding.

There are four reasons behind delays in settling water disputes under the ISWD Act.

1. Delays in constituting tribunals. Under the ISWD Act (Section 21.4), the Union government is required to set up a tribunal only when it is satisfied that the dispute can not be settled by negotiations. Theoretically, the Union Government centre can withhold the decision to set up a tribunal for a long time till it is satisfied that negotiations had failed. In past, the Godavari and Krishna disputes started in the mid 1950s and the states began formal requests for reference in the early 1960s. After a long delay, the disputes were referred to tribunals in 1969. Similarly, in the Cauvery dispute, two of the basin states, i.e., Tamil Nadu and Kerala, asked for reference to set up a tribunal in the 1970s but the tribunal was constituted only in 1990. The 2002 amendment to the ISWD Act places a limit of one year on the central government to constitute a tribunal which should shorten such delays.
2. Delays in reaching a decision. The Krishna dispute tribunal took four years to arrive at a decision while the Godavari Tribunal took 10 years. Some other tribunals have not reached final decisions even after longer periods. Of course, at times there are delays in assembling facts and hearing arguments. Attempts at a political level to reach solutions may also delay the functioning of tribunals.

The new amendment to the ISWD Act places time limits, but still these are too long (up to six years, with possibilities for extension).

3. Delays and uncertainties in enforcement, due to delays in notifying the orders of tribunals in the official gazette of the Government of India. Although this step of adjudication should be quick, the process took three years in the case of the Krishna award and one year in the case of the Godavari award. These delays go against the spirit behind the dispute settlement process.
4. Finally, some tribunal decisions have been difficult to enforce. State governments have sometimes rejected tribunal awards; the Ravi-Beas Tribunal award was not accepted by the Punjab government and has not been implemented so far. More recently the Punjab assembly has terminated all previous water related agreements. Due to some reason, the central government has avoided notifying the award. In the case of the Cauvery dispute, the Karnataka government sought to nullify the tribunal's interim order through an ordinance. Though the Supreme Court pronounced that the ordinance was unconstitutional, the Karnataka government was unwilling to implement the tribunal's interim order, until a compromise was reached through political negotiations.

Such instances go against the underlying philosophy that the awards of the tribunal are final and binding. This also tends to make the dispute resolving mechanism an endless process, at time hoping that the divine intervention will set the things right.

ISWD are serious national problems. Many disputes are lingering on for quite some time because there is lack of will and determination on the part of the concerned parties to find a permanent solution to the problem. Earlier, the problem was confined to river major systems but now it seems to be spreading even to minor sources. Frequent and severe droughts have further aggravated the situation, causing problems like poverty, migration, etc. Due to the disputes, large quantities of water remain unutilized. These problems are not insurmountable however. An amicable solution can be found, given enough will and desire of the concerned parties to reach an agreement.

According to Iyer (1994), while an inter-state water dispute may not involve questions of formal law, it does involve questions of principles relating to water-sharing and water rights. Successive tribunals have discussed issues of riparian rights and the various theories on the basis of which the states concerned argue their case for a certain share in the flows of a river: the territorial sovereignty theory, the 'natural flows' theory, the prior appropriation or prescriptive rights theory, the theory of equitable apportionment for beneficial uses, and so on. Some people believe that a negotiated settlement is the more appropriate course. The central government has in the past shown preference for a negotiated settlement and the adjudication process at different times. In general, a negotiated settlement is preferable to arbitration and voluntary adoption for arbitration is preferable to adjudication. However due to its nature any process of adjudication requires a considerable amount of time and this results in delays in settling the dispute.

Although the River Boards Act was passed in 1956 after ISWD Act, it came into force only in 1957, much later than ISWD. However, no river boards have been established under RBA so far. The fact is that the various governments which have come at the Central level in the country have directly resorted to adjudication in case the negotiations fail, without going in for the intermediate step of arbitration as provided in the RBA. The result has been an overuse of the ISWD act which has led to a lot of wastage of time as well as resources of the nation as a whole (Gosain and Singh 2005). In place of this practice, if the Central government had gone in for setting up of a river board for each and every interstate river in the country, the problems would have been resolved long ago.

A comparison of the ISWD Act and the RBA (Gosain and Singh 2005) reveals the following differences in their provisions:

- The ISWD Act falls under the purview of judicial functions of the government whereas the RBA is an expression of the welfare and developmental functions of the government.
- RBA provides for a suo moto action on the part of the Central Government whereas the ISWD Act provides for the action of the Central Government in only those cases in which it is approached by the State Governments of the riparian states concerned.
- RBA is a comprehensive act that provides for the overall development of the river basin as a whole whereas the ISWD Act is limited to resolving disputes over the shared water resources.
- Under Section 21.8 of the ISWD Act, any matter that can be referred to arbitration under the RBA cannot be brought before any Tribunal under the ISWD Act. This makes it clear that the intention of the framers of the two laws was to encourage the application of the RBA while the ISWD Act was to be used only sparingly and that too as a last resort.
- The Tribunal created under the ISWD Act ceases to function after its decision is made whereas the River Boards created under the RBA are permanent bodies which are involved in all the aspects of river basin planning, development and management.

Failure of the adjudication process to provide timely resolution of ISWD has led to several attempts to improve the situation. In 1983, the National Water Resources Council (NWRC) was created by a central government resolution. With the Prime Minister as Chairman, its membership includes Chief Ministers of States, Lieutenant Governors of Union Territories and several union ministers. The NWRC met first in October 1985 and adopted the National Water Policy in 1987. This policy emphasized an integrated and environmentally sound basis for developing national water resources. NWRC, however, provided no specific recommendations for institutions to achieve this and has been an effective body.

So far in India, the following five tribunals have been set up to adjudicate water disputes:

i. Godavari Water Disputes Tribunal (in April, 1969),
ii. Krishna Water Disputes Tribunal (in April, 1969),
iii. Narmada Water Disputes Tribunal (in October, 1969),
iv. Ravi and Beas Waters Tribunal (in April, 1986), and
v. Cauvery Water Disputes Tribunal (in June, 1990).

Three Tribunals have already given their final awards and the remaining two Tribunals (Cauvery and Ravi-Beas) are still adjudicating the issues referred to them. The Government of India has constituted the Second Krishna Water Dispute Tribunal in 2005.The details of these tribunals are given in what follows.

21.4. GODAVARI WATER DISPUTE TRIBUNAL

The Godavari Water Disputes Tribunal (GWDT) headed by Justice Bachawat was constituted by the Government of India in April 1969. After considering the matter, the Tribunal gave its award in July, 1980. In fact, this award referred to a series of agreements among the party states. Brief details of such agreements are given here. According to the award:

1. The agreement on Polavaram Project provides for diversion of $2,266\,\text{Mm}^3$ of the Godavari Water from Polavaram Project to Krishna River upstream of Vijayawada Anicut. The water thus diverted in the Krishna will be shared as given in Table 1 among Andhra Pradesh, Karnataka, and Maharashtra.
2. The Inchampalli Multipurpose Project will be a joint venture of Madhya Pradesh, Maharashtra and Andhra Pradesh and will be executed and operated under the directions of a Tripartite Inter-State Control Board. The cost of storage, power and benefits will be shared by these States in agreed proportions. Andhra Pradesh is allowed to divert $2,407\,\text{Mm}^3$ of waters from Inchampalli Reservoir for its use. The remaining available water is to be used for power generation at Inchampalli Power House. After hydropower generation, Andhra Pradesh can use water in any manner.
3. As per the award any alteration, amendment or modification to any of the provision of the Tribunal can be made by agreement between the party States or by legislation of Parliament.

21.4.1. Discussions of 6.10.1975

A meeting between the chief ministers of Maharashtra and Andhra Pradesh was held in Hyderabad on the 6th October 1975. The discussions related to the clearance of the projects and the use of waters of Godavari River and its tributaries. After discussions, the following points were agreed to:

1. Maharashtra can use for their beneficial use all waters up to the Paithan dam site on the Godavari and up to Siddheswar dam site on the Purna.

Table 1. Allocation of Godavari water by GWDT among three states

State	Flow	
	In TMC	In MCM
Andhra Pradesh	45	1, 274.4
Karnataka & Maharashtra	35	991.2
Total	80	2, 265.6

2. (i) From the waters in the area of Godavari basin below the Paithan dam site on the Godavari and below the Siddeswar dam site on the Purna and below the Nizamsagar dam site on the Manjra and up to Pochampad dam site on the Godavari, Maharashtra can utilize waters not exceeding $1,700\,\text{Mm}^3$ for new projects including any additional use over and above the present sanctioned or cleared utilization, as the case may be.
   (ii) Andhra Pradesh can go ahead with building its Pochampad Project with FRL 1,091 and MWL 1,093 and is free to utilize all remaining waters up to the Pochampad dam site in any manner it chooses for its beneficial use.
3. (i) In the Manjira sub-basin above the Nizamsagar dam site, Maharashtra can utilize waters not exceeding $623\,\text{Mm}^3$ for new projects including any additional use over and above the present sanctioned or cleared utilisation as the case may be.
   (ii) Andhra Pradesh can withdraw $113\,\text{Mm}^3$ for drinking water supply to Hyderabad city from their proposed Singur project on the Manjira.
   (iii) Andhra Pradesh can construct Singur project with a storage capacity of $850\,\text{Mm}^3$. Andhra Pradesh can also use $1,443\,\text{Mm}^3$ under Nizamsagar project.
4. Maharashtra concurs with the agreement arrived at between the States of Andhra Pradesh and Karnataka in regard to the use proposed by Karnataka in the Manjira sub-basin upstream of the Nizamsagar dam site.
5. Maharashtra and Andhra Pradesh will be free to use an additional quantity of $8,496\,\text{Mm}^3$ of water each below the Pochampad dam site for new projects.
6. Maharashtra and Andhra Pradesh agree in principle to the taking up of the Inchampalli Project with FRL as commonly agreed to by the interested States, viz., Maharashtra, Andhra Pradesh and Madhya Pradesh.
7. Maharashtra and Andhra Pradesh agree to take up the joint projects at the appropriate time with agreed utilization: Lendi Project, Lower Penganga Project, and Pranhit Project, and to set up joint committees for this purpose.
8. The States of Maharashtra and Andhra Pradesh agree that this agreement will be furnished to the Government of India and a report will be submitted before the Godavari Water Disputes Tribunal at the appropriate time.

21.4.2. Discussions of 7.11.1975

In this meeting between the chief ministers of the Madhya Pradesh and Andhra Pradesh held at New Delhi, discussions related to the clearance of the projects and use of the water of Godavari River and its tributaries. After discussions, the following points were agreed.

1. Madhya Pradesh and Andhra Pradesh will be free to use an additional gross quantity of $8,496\,\text{Mm}^3$ each, out of the water in the Godavari River and its tributaries below the Pochampad Dam site for new projects.
2. Madhya Pradesh concurs generally with the agreement between Andhra Pradesh and Maharashtra on 6-10-1975. The quantity of the $8,496\,\text{Mm}^3$ mentioned in

clause I above will not be in addition to 8,496 Mm^3 already agreed between Andhra Pradesh and Maharashtra.

3. In agreeing to 8,496 Mm^3 referred to in clauses I and II above, for Andhra Pradesh, Madhya Pradesh on its part, has taken into account the estimated requirements within the basin only.
4. Madhya Pradesh and Andhra Pradesh agree in principle to the taking up of the Inchampalli project with F.R.L., as commonly agreed to by the interested states, viz., Maharashtra, Andhra Pradesh and Madhya Pradesh.
5. It is also agreed that Madhya Pradesh and Andhra Pradesh will consider the feasibility of taking up the Inchampalli project as a joint project with costs and benefits equitably shared amongst the above 3 States in accordance with a common agreement.
6. Madhya Pradesh agrees to the taking up of Taliperu project by Andhra Pradesh involving the use of 142 Mm^3 (Gross) of water out of the 8,496 Mm^3 agreed to in Clause I and to the submersion of river bed only in Madhya Pradesh. Andhra Pradesh agrees to put up at its cost such protective measures as would be necessary in consultation with Madhya Pradesh to prevent submersion of other areas in Madhya Pradesh.
7. The States of Madhya Pradesh and Andhra Pradesh agree that nothing in this agreement will be treated as a concession by either state in respect of any of their contentions in any other water dispute with any other state or with respect to the dispute regarding the sharing of the balance of water in Godavari and its tributaries.
8. The States of Madhya Pradesh and Andhra Pradesh agree that this agreement will be furnished to the Government of India and they would be requested to expedite the clearance of the projects. This Agreement will also be jointly filed before the GWDT at the appropriate time.

21.4.3. Discussions of 9.12.1975

In the meeting between the chief ministers of Orissa and Madhya Pradesh held at New Delhi, the discussions related to the use of waters of Godavari River and the clearance of the projects of Orissa and Madhya Pradesh. After discussions the following agreement was reached:

I. Pending final allocation of the Godavari water, Madhya Pradesh and Orissa will be free to use additional gross quantity of 8,496 Mm^3 and 5,664 Mm^3 respectively, out of the water of the Goadvari basin below the Pochampad Dam site for new projects in such manner as they deem fit.

II. In agreeing to 5,664 Mm^3 referred to in Clause I for Orissa, Madhya Pradesh on its part has taken into account the estimated requirements within the basin only. All the utilization by Orissa and Madhya Pradesh contemplated in the various Clauses shall be only as a part of the 5,664 Mm^3 and 8,496 Mm^3, respectively, agreed to in Clause I above. The States of Orissa and Madhya

Pradesh will not be entitled on the basis of the subsequent Clauses to utilize in any way more than $5,664\,\mathrm{Mm}^3$ and $8,496\,\mathrm{Mm}^3$, respectively.

III. Below the dam sites of the Upper Indravati Project, as proposed by Orissa, there is a catchment area of about 4,800 sq. km in the Indravati sub-basin up to the Orissa border with Madhya Pradesh. From this catchment there is some natural flow across the Jaura Nallah to Sabari (Kolab) River. It was agreed that Orissa will ensure at its border with Madhya Pradesh a flow of $1,274\,\mathrm{Mm}^3$ in the Indravati and its tributaries at 75% dependability for use by Madhya Pradesh. In the years of shortage, the shortage will be shared proportionately between the two states and the assurance of flow in the Indravati and its tributaries, referred to above, will stand proportionately reduced. Both the states agree to joint gauging at suitable points to ascertain the yield data and to ensure the flow of $1,274\,\mathrm{Mm}^3$ at 75% dependability of the proportionately reduced flow in years of shortage that has to flow below the common border. The figure of $1,274\,\mathrm{Mm}^3$ is on the assumption of total yield of $5,777\,\mathrm{Mm}^3$ from the Indravati sub-basin in Orissa and $2,577\,\mathrm{Mm}^3$ utilization for the Upper Indravati Project. If the assessment of $5,777\,\mathrm{Mm}^3$ is found to be high and the correct figure is lower than $5,777\,\mathrm{Mm}^3$ and the utilization for the Upper Indravati Project gets reduced from the figure of $2,577\,\mathrm{Mm}^3$ then the figure of $1,274\,\mathrm{Mm}^3$ will get reduced in the same proportion.

IV. In view of the agreement incorporated in the above clauses, Madhya Pradesh agrees to the clearance and execution of Upper Indravati Project, as proposed and submitted by Orissa to the Government of India. Orissa also agrees to the clearance and execution of Bodhghat Project, as may be modified by Madhya Pradesh taking into account the water availability specified in Clause III.

21.4.4. Discussions of 19.12.1975

On the basis of a series of discussions held between the representatives of the States of Maharashtra, Madhya Pradesh, and Andhra Pradesh, an agreement was arrived at regarding the sub-basin wise allocation of the waters of the Godavari and its tributaries downstream of the Pochampad Dam in the States of Andhra Pradesh, the projects therein and other allied matters taking into consideration the allocations already agreed to.

21.5. KRISHNA WATER DISPUTE TRIBUNAL

The Krishna Water Dispute Tribunal (KWDT), headed by Justice R.S. Bachawat, was constituted in April 1969 for adjudication of inter-state water dispute regarding the sharing of Krishna waters. The KWDT gave its award in 1973, and it was published in 1976. As per this award, 75% dependable flow of the Krishna water at Vijaywada was assessed as 2,060 TMC ($58,339\,\mathrm{Mm}^3$) which was allocated as given in Table 2.

Table 2. Allocation of Krishna water by KWDT among three states

State	Allocation in TMC	Allocation in Mm^3
Maharashtra	560	15,859
Karnataka	700	19,824
Andhra Pradesh	800	22,656
Total	2,060	58,339

The KWDT addressed three issues: (1) the extent to which the existing uses should be protected as opposed to future or contemplated uses; (2) diversion of water to another watershed; and (3) rules governing the preferential uses of water. The tribunal relied on the principle of equitable apportionment for allocation of water.

About the first issue, KWDT decided that projects that were in operation or under consideration as in September 1960 should be preferred to contemplated uses and should be protected. It also directed that except by special consent of the parties, a project committed after 1960 should not be entitled to any priority over contemplated uses. Regarding the second issue, the tribunal decided that diversion of the Krishna water was legal when water was diverted to areas outside the river basin but within the political boundaries of the riparian states. It was silent regarding the diversion of water to areas of non-riparian states. As for the third issue, KWDT concluded that all existing uses based on diversion of water outside the basin would receive protection.

After deliberating over the issue, KWDT passed the following order:

- That the States of Maharashtra, Karnataka and Andhra Pradesh will be free to use ground water within their respective sites in the Krishna basin. The use of underground water by any State shall not be reckoned as use of the water of the Krishna River.
- That the waters of the Krishna River be allocated to the three States of Maharashtra, Karnataka and Andhra Pradesh for their beneficial use to the extent provided in Table 2.
- The states are not permitted to use in any water year more than the specified quantity of water of the Krishna River.

In addition, liberty was given to Andhra Pradesh to use in any water year any excess flows that may be available without conferring any right whatsoever in the matter. The Tribunal allowed the States to utilize their allocated share of water for any project as per their plans.

- Beneficial use shall include any use of the water of the Krishna River by any State for domestic, municipal, irrigation, industrial, hydropower generation, navigation, pisciculture, wildlife protection and recreation purposes.
- Evaporation losses from reservoirs of projects using 85 Mm^3 or more annually shall be excluded in computing the 10% figure of the average annual utilizations mentioned in selected sub-clauses of the order.

- The depletion of the waters of the Krishna River in any manner whatsoever, including losses of water by evaporation and other natural causes from man-made reservoirs and other works. Without deducting in the case of use for irrigation the quantity of water may return after such use to the river.

The uses of water shall be measured in the manner indicated below:
Domestic and municipal water supply: 20% of the quantity of water diverted or lifted from the river or any of its tributaries or from any reservoir, storage or canal.
Industrial Use: 2.5% of the quantity of water diverted or lifted from the river or any of its tributaries or from any reservoir, storage or canal.

Recently, the Government of India has constituted the Second Krishna Water Dispute Tribunal.

21.5.1. The Almatti Dam Dispute

The Almatti dam is part of the Upper Krishna Project, a joint venture of the states of Karnataka, Andhra Pradesh, and Maharashtra. Karnataka planned the Upper Krishna project (consisting of reservoirs at Almatti and Narayanpur) to use 4,897 Mm^3 (173 TMC) of water in two stages: 3,369 Mm^3 (119 TMC) in the first stage and 1,528 Mm^3 (54 TMC) in the second stage. Narayanpur reservoir has been completed but the construction of the Almatti dam was under dispute. The Bacchawat Tribunal (the Krishna Water Dispute Tribunal) stated that Karnataka had proposed to complete the Almatti dam in the second stage of the project, without specifying the exact full height. The Central Water Commission had given clearance to build the Almatti dam to a height of 519 m in the first stage. However the Karnataka government interpreted the Bacchawat award to mean the full height of the project as it had proposed. This envisaged a final full reservoir level of 524.25 m. The problem of Almatti pertains to the storage of what Andhra Pradesh believes to be an excessive quantity of water into the Srisailam and Nagarjunasagar reservoirs and the Krishna barrage at Vijayawada which sustains the main rice bowl of Andhra Pradesh.

In the case of Almatti, the height of the dam or its storage capacity has been the issue of dispute. The proposed 524.25 m high Almatti dam can hold 6,429 Mm^3 of water whereas the allocation for the entire Upper Krishna Project was 4,899 Mm^3. The argument of Karnataka that it will use the extra 54 TMC feet (1,529 Mm^3) of water to generate power is not convicting to Andhra Pradesh because this water can irrigate 540,000 acres of land. Andhra Pradesh feels that Karnataka will delay flows to Andhra Pradesh in a normal year by detaining more water at the dam.

In 1996, the then Prime Minister appointed a committee of four ministers to resolve the dispute. They constituted an expert committee in 1997 who opined that the stage had not reached for Karnataka to go in for a higher storage capacity and that the full reservoir level be kept at 518.7 m in the second stage. The group suggested that it was not necessary for Karnataka to build a bigger storage. In July 1998, the case was referred to a five-judge bench of the Supreme Court. Here, the major issues were the interpretation of Article 262 of the constitution and Section 21.11 of the Inter State Water Disputes Act-1956. Karnataka sought the apex court's directions in regard to distribution and allocation of surplus water of the Krishna River. But Andhra Pradesh demanded to restrain Karnataka from raising the height of the Almatti dam from 519 m to 524 m and certain other relief.

21.6. NARMADA WATER DISPUTE TRIBUNAL

Under the Inter-State Water Disputes Act, 1956, the Central Government constituted Narmada Water Disputes Tribunal (NWDT) on 6th Oct. 1969 to adjudicate upon the sharing of Narmada waters and for Narmada River valley development. Justice V. Ramaswami was the chairman of NWDT. The Tribunal gave its Award on 7th Dec., 1979. The Award specified a quantum of utilizable waters at 75% dependability to be shared by the four States of Gujarat, Madhya Pradesh, Maharashtra and Rajasthan as stated in Table 3.

The Tribunal determined that the height of the Sardar Sarovar Dam should be fixed for Full Reservoir Level (FRL) of 138.68 m (455 ft) and also directed the Government of Gujarat (GOG) to take up and complete the construction of the dam accordingly. According to Clause-16 of the final orders of the Tribunal, the parameters of shares of utilizable waters by the States, FRL of the reservoir and Full Supply Level (FSL) of Navagam Canal are made subject to review at any time after a period of 45 years from the date of publication of the Award of the Tribunal in the official gazette.

The Tribunal has assessed the utilizable quantity of water at 75% dependability on the Narmada at Sardar Sarovar dam site as 28 MAF (34,537.45 MCM) considering the inflow at 75% dependability as 27.01 MAF (33,316.0 MCM) and bringing this up to 28 MAF (34,537.45 MCM) considering regeneration, carryover and evaporation loss. As per the NWDT report, the evaporation loss is 4 MAF, regeneration or return flow (+)2 MAF (2,467 MCM) and carry over (+)3 MAF (3,700 MCM).

21.6.1. Decisions of the Narmada Water Disputes Tribunal

The important decisions of the Tribunal are:

(i) The Full Reservoir Level of Sardar Sarovar Dam was fixed at 138.68 m.

(ii) The power benefits from the project are to be shared as: Madhya Pradesh 57%, Maharashtra 27%, and Gujarat 16%. The cost of power component of

Table 3. Allocation of Narmada water by NWDT among four states

State	Allocation in MAF	Allocation in Mm^3
Gujarat	9.00	11, 101.32
Madhya Pradesh	18.25	22, 511.01
Maharashtra	0.25	308.37
Rajasthan	0.50	616.74
Total	28.00	34,537.44

the project is to be shared by Gujarat, M.P. and Maharashtra in the ratio of their benefits in power.

(iii) M.P. has to make a uniform release of 8.12 MAF (10,015.86 MCM) ex-Maheshwar to meet the requirements of Gujarat and Rajasthan from Sardar Sarovar Dam in a normal year having 28 MAF (34,537.45 MCM) flow.

(iv) The Indira Sagar project is to be taken up by MP and completed with a FRL of 262.13 m (860 ft) either concurrently or earlier than the construction of Sardar Sarovar.

(v) Gujarat is required to credit to MP each year 17.63% of the expenditure on Indira Sagar dam.

(vi) The allocation of Unit I – Dam and Appurtenant works cost between irrigation & power was done as 43.9% and 56.1%. The irrigation component of the project is to be shared by Gujarat and Rajasthan in the ratio of water allocation for the dam and canal.

(vii) An organization, viz., Narmada Control Authority (NCA) was set up for implementation of the decision of the Tribunal. A Review Committee was also set up under Union Ministry of Water Resources to review decisions of the Authority.

(viii) For efficient, economical and early execution of Unit I & III of Sardar Sarovar Project, a construction advisory committee was set up. The Review Committee was empowered to decide on issues on which there is disagreement in the construction committee.

(ix) The decisions of the Tribunal are subject to review after a period of 45 years from the date of publication of the final award in the Gazette of Govt. of India.

21.6.2. Rules of Regulation

As per the NWDT award, NCA shall frame the rules of regulation and water accounting within the ambit of given guidelines. The procedure for preparing this account and rules of regulation are spelled out by the Tribunal. The Authority shall also ensure implementation of the orders of the Tribunal in respect of the quantum and pattern of regulated releases by M.P.

The water year is to be reckoned from 1st July to 30th June of the next calendar year. The Authority has to determine the volume of water flowing in the Narmada River and its tributaries in a water year. The utilizable flow in excess or falling short of 28 MAF is to be shared by the four party States in the same ratio as allocation of utilizable quantum of Narmada waters at 75% dependability, i.e., 28 MAF.

The requirements at Sardar Sarovar have to be met by releases by MP ex-Maheshwar and by inflows from the intermediate catchment between Narmadasagar and SSP surplus to the requirements of MP and Maharashtra. The releases ex-Maheshwar works out to be 8.12 MAF and uniform monthly releases of 0.677 MAF (835.06 MCM) would make up this amount. Although it would be fully known only in October whether the year is normal, surplus or deficit, the releases by MP in the filling period would have to be more or less on the basis of a normal year. As the months of July and early part of August are crucial for Kharif sowing, it is important that regulatory arrangements are made to ensure due shares of various parties.

The Authority was directed to review the ten-day releases made by M.P. at least once a month and even often made considered necessary for directing any change in the releases. NCA would also ensure by directing the releases by MP that there is sufficient utilizable water in Sardar Sarovar at all times to meet the requirements of the next ten days subject to water being available in the storage of M.P. after taking into account the proportionate requirements of M.P. For this purpose, Gujarat and Rajasthan are required to intimate their requirements of the 10-day period well in advance.

The Narmada Control Authority was directed by the Tribunal to determine, from time to time, the volume of water stored by each state in reservoirs and other storages and it may, for that purpose, adopt any device or method. Further, the water available in the live storage of the various reservoirs on 30th June shall be reckoned as an inflow to be shared in the next water year.

Note that the apportionment relates to actual withdrawals and not consumptive use. The available utilizable waters on any date will be inclusive of return flows and exclusive of evaporation losses in various reservoirs. The Tribunal did not specify how regeneration is to be worked out but as per the NWDT report, regeneration including return flow should be taken as 10% of irrigation use in upstream major, medium and minor projects in any month with a lag of one month. Further, 60% of the water used for domestic and industrial purposes within the Narmada basin may be taken as return flow uniformly available throughout the year.

The surplus water shall first be utilized for filling up the reservoirs to their capacity and further extra surplus water should be utilized for irrigation and other purposes only after that has been ensured. After meeting the storage requirements and withdrawals, the surplus waters in the filling period which would go waste to sea even without generating power can be allowed to be utilized by party States to the extent they can. Gujarat is required to inform NCA and the designated representative of all the concerned states whenever water starts going waste to sea as also when flows cease. During the period of such flows, the party States, whose reservoirs are spilling and the spill water cannot be stored elsewhere, may utilize such flows from the said reservoirs as they like and such utilization will not count either towards allotment of supplies to them nor will it establish prescriptive rights.

21.7. RAVI AND BEAS WATER TRIBUNAL

Punjab and Haryana, the main parties involved in this dispute, are both agricultural surplus states, also termed as the ‘granary of India’. Both states produce large quantities of grains. In view of the scarcity and uncertainty of rainfall in this arid area, irrigation is the mainstay of agriculture and is responsible for its tremendous progress. With the introduction and widespread adoption of high-yielding varieties of food grains by forward looking farmers of these states, irrigation became increasingly important from the late 1960s onwards. Through an inter-state meeting, an initial agreement on the sharing of the waters of the Ravi and Beas after partition of India was reached in 1955.

With the reorganization of Punjab in November 1966, Punjab and Haryana were carved out as successor states. Thereafter, the present dispute between Punjab and Haryana about Ravi-Beas water started. Four perennial rivers, Ravi, Beas, Satluj and Yamuna, flow through these states.

As a result of protests by Punjab against the 1976 agreement allocating water from Ravi-Beas, further discussions were conducted (now including Rajasthan as well), and a new agreement was accepted in 1981. This agreement faced opposition and a series of events led to the constitution of a tribunal to examine the Ravi-Beas issue in 1986. Both states sought clarifications of aspects of the award by this tribunal, but the centre has not provided these. Hence, the original award has not been notified (listed in the government gazette to give it final force), and does not have the status yet of a final, binding decision.

On the basis of the flow data available from 1921–45, the waters of the Ravi and the Beas (mean flows) were estimated to be 15.85 MAF (19,550.66 MCM) over and above the actual pre-partition use. In an inter-state meeting convened by the Central Government in January 1955, these waters were allocated to various states as: Rajasthan 8.00 MAF (9,767.8 MCM); Punjab 7.20 MAF (8,881 MCM); and Jammu and Kashmir 0.65 MAF (801.76 MCM).

The Indus Waters Treaty was signed in 1960 whereby waters of the three eastern rivers (the Ravi, the Beas and the Satluj) were reserved for exclusive use by India. The state of Punjab was reorganized with effect from 1 November 1966 and this raised the question of Haryana's share in the waters allocated to Punjab under the 1955 agreement.

In exercise of the power conferred by the Punjab Reorganization Act, Central Government issued a notification in March 1976 allocating 3.5 MAF (4,317 MCM) to Haryana out of the 7.2 MAF (8,881 MCM) earlier allocated to the composite Punjab state. However, Punjab was not happy with this decision. Therefore, after a number of discussions, a fresh agreement was accepted between the party states in December 1981 whereby the available supplies were estimated to be 17.17 MAF (21,178.86 MCM) on the basis of the 1921 to 1960 flow series. These were allocated to different states as: Punjab 4.22 MAF (5,205 MCM); Haryana 3.50 MAF (4,317 MCM); Rajasthan 8.60 MAF (10,608 MCM); Jammu and Kashmir 0.65 MAF (801.76 MCM); and Delhi Water Supply 0.20 MAF (246.7 MCM).

Even after this agreement, sharing of water resources continued to be a politically live issue in Punjab. An accord called *The Punjab Settlement* was signed between the then Prime Minister of India and Sant Longowal on 24th July 1985. In the accord, the clauses relevant to water were as follows:

(a) Quantum of usage of waters from the Ravi-Beas System by different states as on 1st July 1985 should be verified by a tribunal.

(b) The claim of Punjab and Haryana regarding their shares in the remaining waters was to be referred to for adjudication to a tribunal the decision of which would be binding on both parties.

Accordingly, the tribunal was constituted on 2nd April 1986. It was to give its award within a period of six months, which was extended thereafter. Since the main dispute was over the allocation of waters between the Punjab and Haryana the tribunal examined various factors in the Punjab and Haryana, such as the geographical area, basin area, cultivable area, water requirements, population density, extent of arid areas, rainfall, etc. It found that the ratios of these parameters between Punjab and Haryana varied from 1.05 to 1.30. Finally, it allocated the waters between Punjab and Haryana in the ratio of 1.3:1. Apart from accepting the 1981 assessment of the available water by the Central Government as 17.17 MAF (21,113.5 MCM), the tribunal also considered the surplus water available below the base stations, of which 40% (amounting to 2,882 MCM) was considered utilizable. Out of these utilizable supplies, only 60% (i.e., 1,369 MCM) was considered for allocation between Punjab and Haryana. Thus, the final allocation among the various states as given by the tribunal was: Rajasthan 8.60 MAF (10,608 MCM); Jammu and Kashmir 0.65 MAF (801.76 MCM); Delhi Water Supply 0.20 MAF (246.7 MCM); Punjab 5.00 MAF (6,167.4 MCM); Haryana 3.83 MAF (4,724.2 MCM): total 18.28 MAF (22,548 MCM).

21.8. CAUVERY WATER DISPUTE

The Cauvery basin covers large parts of Karnataka and Tamil Nadu and a small part of Kerala and the Union Territory of Pondicherry. Being riparian to the river, the states of Karnataka, Tamil Nadu, Kerala, and the Union Territory of Pondicherry, are entitled to the Cauvery water in a reasonable and beneficial manner.

The dispute over the sharing of Cauvery water goes back to the 19th century. An agreement was made between the then states of Mysore and Madras in 1924 under which Mysore became entitled to construct a dam known as Krishnarajasagar across the Cauvery River. The Mysore Government was at liberty to carry out future extensions of irrigation in Mysore under the Cauvery and its tributaries to an extent fixed at 110,000 acres in addition to the area of irrigation fixed under the Rules and Regulations. The Madras Government was at liberty to construct any new storage reservoirs on the Bhavani, Amaravathi or Noyil rivers. The agreement was open to reconsideration at the expiry of fifty years from the execution date. After expiry of the agreement in 1974, water allocation between Karnataka and Tamil Nadu required a fresh solution. Although Kerala and Pondicherry were not parties to the 1924 agreement, these are involved in the present dispute.

The Cauvery dispute is between a downstream state (Tamil Nadu) that has a long history of irrigated agriculture by the use of Cauvery water and an upstream state (Karnataka) that was a late starter in irrigation development but has made rapid progress and has the advantage of being an upper riparian with greater control over the flows. Clearly, the two major parties to the dispute are Karnataka and Tamil Nadu, although the original issues dated back to colonial times. Between 1968 and

1990, more than 25 meetings were held at the ministerial level but no agreement could be reached. Hence, the Cauvery Water Dispute Tribunal was constituted under the ISWD Act 1956.

Presently, the grievance of Karnataka is that its late start on irrigation development should not curtail its right to make the maximum use of Cauvery water for agricultural and other development. Tamil Nadu feels threatened because its long established agriculture using the Cauvery water now critically depends on diminishing flows as a result of upstream development.

Intermittent talks between Karnataka and Tamil Nadu went on for over two decades from the 1970s onwards but did not produce any result. On their part, the Government of India made unsuccessful efforts to bring about an agreement. While the central government favoured a negotiated settlement, on a petition by some Tamil Nadu farmers to get assured irrigation water from the Cauvery, the Supreme Court of India ordered the Central Government to immediately establish a tribunal. Accordingly, the Government of India established the Cauvery Water Dispute Tribunal (CWDT) on June 02, 1990.

The share of water claimed by basin states in the Tribunal is as under:

a) Karnataka – $13,169\,\text{Mm}^3$.
b) Tamil Nadu – Flows in accordance with the agreements of 1892 and 1924.
c) Kerala – $2,826\,\text{Mm}^3$.
d) Union Territory of Pondicherry – $263\,\text{Mm}^3$.

The Tamil Nadu State filed a petition before CWDT praying that the state of Karnataka be directed not to impound or utilize water of the Cauvery River beyond the extent impounded or used by them as on 31.5.1972, as agreed to by the Chief Ministers of Cauvery basin states and the Union Minister for Irrigation and Power. It further sought passing of an order restraining the state of Karnataka from undertaking any new projects, dams, reservoirs, canals, etc. The Union Territory of Pondicherry sought an interim order from the Tribunal directing the states of Karnataka and Tamil Nadu to release the water already agreed to ($2,650\,\text{Mm}^3$) during the months from September to March.

There was a basic difference between Tamil Nadu on the one hand and the central government and Karnataka on the other in their approach towards the sharing of the Cauvery water. The Tamil Nadu government argued that since Karnataka was constructing dams on the Cauvery River and was expanding ayacuts (irrigation works), Karnataka was unilaterally diminishing the supply of waters to Tamil Nadu, and adversely affecting the prescriptive rights of the already acquired and existing ayacuts. The government of Tamil Nadu also maintained that the Karnataka government had failed to implement the terms of 1892 and 1924 agreements relating to the use, distribution and control of the Cauvery waters. It asserted that the entitlements of the 1924 agreement were permanent, and that only those clauses that dealt with utilization of surplus water for further extension of irrigation in Karnataka and Tamil Nadu, beyond what was contemplated in the 1924 agreement could be changed. In contrast, Karnataka questioned the applicability of the 1924 agreement, emphasizing equity and regional balance in future sharing arrangements.

There were several reasons why the negotiations of 1968–90 failed to find a consensus. There was a divergence of interest between Karnataka and Tamil Nadu on the question of pursuing negotiations. Karnataka was interested in prolonging the negotiations and avoiding the reference to a tribunal to gain time to complete its new projects.

While the Cauvery Tribunal passed an interim order in 1991, it could not be implemented. In the end, the process of political bargaining led to the implementation of the interim order, and the formation of a Cauvery River Authority and Monitoring Committee in 1998. These bodies continue to meet, and the final outcome of the tribunal is awaited.

21.9. YAMUNA WATER DISPUTE

Yamuna is a tributary of Ganga whose waters in the head reaches have been largely utilized. Out of a total available flow of about 7.6 MAF (9,374.5 MCM) at Tajewala and about 9.6 MAF (11,841 MCM) at Okhla, almost 85% to 90% of the waters are already being utilized mainly for irrigation in Uttar Pradesh and Haryana, mostly from Tajewala Barrage and additionally from Okhla Barrage. A small portion is also being utilized for Delhi Water Supply and some irrigation in Rajasthan.

No major storage reservoir has been constructed in the Yamuna basin to regulate river flow. Barring a few days in the monsoon season, all the available flows at Tajewala are diverted for irrigation in Uttar Pradesh and Haryana through the Eastern and Western Yamuna Canals. Consequently, the Yamuna is virtually dry downstream of the Tajewala Barrage. Even the capital city of Delhi, situated on the banks of the river, gets only 30% of its present drinking and municipal water requirements from the Yamuna. Due to negligible flow in the river in the lean season, there is also a problem of Yamuna pollution in Delhi.

In the head reaches of Yamuna, two sites have been identified for construction of dams – at Kishau in Uttaranchal and at Renuka in Himachal Pradesh. These two will have storage capacities of about 1,603.5 MCM (about 13% of annual flows). However, neither of these dams have been sanctioned and taken up for construction till date on account of a dispute among the states on sharing of the Yamuna waters. A number of technical and political inter-state meetings have failed to resolve the dispute and come to an agreement. The claims of all the basin states aggregate to about 26 MAF (32,070.5 MCM), while the available water is far less, only about 10 MAF (12,334.8 MCM).

Although a consensus was supposed to have reached in 1989, no agreement could be signed among the party states. In the absence of an agreement on sharing of waters of Yamuna, the Kishau dam project could not be taken up. As a result of delay, among the sufferers is the city of Delhi since all the feasible sources to meet its domestic water supply have been exhausted, and its supply critically hinges on the Kishau and Tehri Dams.

21.9.1. Satluj-Yamuna Link Canal Dispute

In 1965 the Punjab Government appointed two committees: an expert committee of engineers called the Food Committee of Land and Water Use in Punjab and a committee known as the Haryana Development Committee. These committees were appointed to review the order of the Punjab Government of 1961. After a scientific study, the first committee recommended the allocation of 4.56 MAF (5,624.7 MCM) of water to areas now called Haryana out of 7 MAF (8,634.4 MCM) available for Punjab according to an agreement reached on January 29, 1955. The second committee recommended the allocation of Ravi-Beas waters for Haryana based on the principle of equitable apportionment between Haryana and Punjab.

A high level committee of independent experts was appointed by the Central Government in 1970. In its report of 1971, the committee recommended allocation of 3.78 MAF (4,662.58 MCM) water to Haryana. In another recommendation, Mr. D.P. Dhar, the then Deputy Chairman of the Planning Commission who was also asked to examine the question, suggested in 1973 that 3.74 MAF (4,613.22 MCM) water be allocated to Haryana.

In 1976, the Central Government allocated 3.5 MAF (4,317.2 MCM) of water to Haryana and 3.5 MAF to Punjab on the completion of further conservation works on the Ravi. It was decided that the share of Punjab would not exceed 3.5 MAF (4,317.2 MCM). If the availability of water in the Beas at Mandi increased, Haryana would get a pro-rata share of erstwhile Punjab according to the 1955 agreement.

Based on the analysis of flow series of 1921–45, the surplus water in Ravi-Beas Rivers was estimated as 15.85 MAF (19,551 MCM). However, on the basis of the 1921–60 flow series, the net surplus of these rivers was estimated as 17.17 MAF (21,179 MCM), an increase of 1.32 MAF (1,628.2 MCM). This increase was distributed between Punjab and Rajasthan: Punjab was allotted 0.72 MAF (888.11 MCM) and Rajasthan 0.60 MAF (740.1 MCM). Punjab was, therefore, allocated total 4.22 MAF (5,205 MCM). Punjab was also allowed to use all the water that Rajasthan was unable to use out of its share of 8.60 MAF (10,608 MCM).

Subsequently, there a demand from various quarters that issue of water allocation be referred to a tribunal. This demand was accepted and included in the Rajiv-Longowal accord. On April 2, 1988, the Ravi and Beas Waters Tribunal, headed by Justice V. Balakrishna Eradi, was appointed. The tribunal submitted its report on January 30, 1987. The Eradi Tribunal allocated the water of Satluj Yamuna link between Rajasthan, Punjab Haryana, J & K and Delhi as given in Table 4.

The SYL canal project was initiated to transfer water from Ravi and Beas Rivers and as per the agreement between the two states and the Central Government, the allocation of water of these rivers was made among Punjab, Haryana and Rajasthan. The Satluj-Yamuna Link (SYL) Canal has been a major political issue in Punjab and Haryana in the recent times. Punjab has opposed the construction of the canal on the plea that the state has the sole right over its river waters by applying the riparian principle.

Table 4. Allocation of water of Satluj-Yamuna link by Eradi tribunal in 1987

Name of state	Allocation of water	
	In Million acre feet	In Million m^3
Punjab	5.00	6, 167.40
Haryana	3.83	4, 724.22
Rajasthan	8.60	10, 607.93
J & K	0.65	801.76
Delhi	0.20	246.7

Source: Satluj-Yamuna Link: Troubled Waters, The Times of India, New Delhi, August 5, 2004.

The construction work on the canal was stopped after militants gunned down two senior engineers connected with the project in 1990. By then, only a few km of the total 121 km of the canal portion on the Punjab side had been constructed. A large sum of money has been spent by the State and the Central governments on the SYL in the Punjab portion alone. Haryana has already constructed the entire 90 km stretch of the canal in anticipation of getting river water from Punjab. On a writ petition filed by the Haryana Government, the Supreme Court gave a directive to central government in 2002 to take up the construction work in the Punjab portion of the Canal and to complete it within a year.

In July 2004, the Punjab Assembly passed the Punjab Termination of Agreements Act. According to this act, Punjab government has abrogated water accords prior to 1991 with all other states. The matter is now under consideration.

21.10. INTERNATIONAL TREATIES

The settlement of international water disputes through treaties is the most effective and popular medium of resolving such disputes. Of course, this mode is equally suitable for intra-national conflict resolution.

There are three main water related international treaties relevant to India:

- The Indus water Treaty between India and Pakistan,
- The Mahakali Treaty between India and Nepal, and
- The Ganga Treaty between India and Bangladesh.

In the following, we cover very briefly the background, nature of dispute, important features of the treaties and the manner in which they have been operating.

21.10.1. Indus Water Treaty (1960)

After the Partition of India and Pakistan in 1947, an understanding on the sharing of water of Indus River between the two countries became necessary to facilitate the development of water resources of this basin. After prolonged talks between the two Governments, the Indus Water Treaty was signed in September 1960.

According to this Treaty, waters of the three western rivers (the Jhelum, the Chenab, and the Indus itself) were allocated to Pakistan, and those of the three eastern rivers (the Ravi, the Beas and the Sutlej) were allocated to India. Certain restrictions about water utilization were placed on India which is the upper riparian country. The highlights of the Treaty are:

- India has not been allowed to build storages on the rivers where water has been allocated to Pakistan.
- Restrictions have also been imposed on the extension of irrigation development in India. Less significant restrictions have been placed on Pakistan, being the lower riparian.
- The Treaty contains provisions regarding the exchange of data on project operation, extent of irrigated agriculture, etc.
- The Treaty provides for certain institutional arrangements. A permanent Indus Commission consisting of a Commissioner each for India and for Pakistan has been set up and there are periodical meetings and exchanges of visits between the two sides.
- Detailed guidelines were included in the Treaty for conflict-resolution. The Indus Commission is the first step for conflict resolution. If an agreement cannot be reached at the Commission level, the dispute is to be referred to the two governments. If the governments too fail to reach an agreement, the Treaty provides an arbitration mechanism.

The Indus Water Treaty is a successful instance of conflict-resolution. It has been working reasonably well despite the strained political relationship between India and Pakistan. Importantly, it continued to be honoured even during wars between the two countries. Undoubtedly differences do arise from time to time, but these usually get resolved within the framework of the Treaty. Minor differences are settled within the Commission, and major disputes go to the two Governments. One of the important differences that arose during the 1970s was connected with the Salal Hydro-Electric Project in Jammu and Kashmir. This conflict could not be settled at the level of commissioners and was referred to the two governments. After lengthy negotiations, the issue was eventually resolved.

Another major unresolved dispute is regarding the Tulbul Navigation Project (or the Wular Barrage Project) on the Jhelum River. Pakistan has raised objections to this project on the ground that it involves the creation of storage on a river that is allocated to Pakistan and this constitutes a violation of the Treaty. India's view is that no storage has been created and the proposed barrage will merely provide a head of water temporarily to extend the period during which navigation is possible. Further, such regulation will also benefit Pakistan. The intergovernmental talks on the subject have not been successful.

Another long-pending dispute is regarding the Baglihar project which is a run-of-the-river scheme. In this case, the two countries have significant differences: India feels that there is no violation of agreement while Pakistan feels that there is. This dispute could not be settled at the commissioner and government levels. Ultimately, Pakistan approached the World Bank who had brokered the Treaty. But the World Bank has refused to get involved and the dispute remains unresolved. The matter has gone for arbitration by a third party. The World Bank has appointed Swiss expert Prof. Raymond Lafitte to mediate the dispute.

21.10.2. Mahakali Treaty

The Mahakali Treaty, signed in February 1996 between India and Nepal, pertains to sharing water of a river by the same name. Now the Treaty is in force and is in the process of implementation, although there have been ups and downs in its implementation.

The Mahakali Treaty basically aims at an integrated development of water resources in the Mahakali River and has been finalized on the basis of equal partnership. The Mahakali originates in Nepal and forms the border between the two countries for a considerable distance. The scope of the Treaty covers the Sarada Barrage, the Tanakpur Barrage and the proposed Pancheswar project. From the Sarada Barrage, the Treaty gives Nepal 28.3 cumec (1000 cusec) of water in the wet season and 4.25 cumec (150 cusec) in the dry season. This quantity is to be supplied from the Tanakpur Barrage if the Sarada Barrage turns non-functional. Further, the Treaty also directs that not less than 9.91 cumec (350 cusec) should flow downstream of the barrage to maintain and preserve the ecosystem of the river. On Tanakpur, the Treaty reaffirmed the Nepalese sovereignty over the land (2.9 ha) needed for building the eastern afflux bund, as well as the 9 ha of pondage area. In lieu of the eastern afflux bund, the Treaty gave Nepal the right to 28.3 cumec (1,000 cusec) of water in the wet season and 8.5 cumec (300 cusec) in the dry season; and 70 million kilowatt-hours (kWh) of electricity (as against the earlier agreed figure of 20 million kWh).

When the Pancheswar Project comes into being and augments the availability of water in dry season at Tanakpur, Nepal would be provided with additional water and additional energy. Nepal will bear a portion of the cost of generation of incremental energy. The Pancheswar Project, which was to be located on the Indo-Nepal boundary and was to be a joint project. For this project, some general principles applicable to border rivers (an important one being "equal entitlement in the utilization of the waters of the Mahakali River without prejudice to their respective existing consumptive uses of the waters") were laid down. These were further elaborated in a document exchanged by the two Prime Ministers. The detailed project report (DPR) was to be jointly prepared in six months; the energy, irrigation and flood control benefits to the two countries were to be assessed, and the capital cost shared accordingly; the power benefit was to be assessed on the basis of savings in costs as compared with the relevant alternatives available and so on. There was to be a bi-national Mahakali River Commission, guided by the principles of equality, mutual benefit and no harm to either party. There would also be a specific joint entity to develop, execute and operate the Pancheswar Project. There were other provisions relating to the life of the Treaty (75 years), review after 10 years, arbitration in the event of disputes, etc.

Although, the Treaty is formally in operation but the progress in its implementation has been tardy. The DPR which was to have been prepared in six months got stalled partly because of certain technical differences.

21.10.3. Ganga Water Sharing

The dispute between India and Bangladesh (East Pakistan till 1971) over the sharing of the Ganga water arose when in 1960s the Indian government decided to construct a barrage at Farakka, close to Indo-Bangladesh border. The objective behind the construction of the Farakka Barrage was to increase the lean period flow of the Bhagirathi-Hooghly branch of Ganga to increase the water depth at the Kolkota port which was threatened by siltation. As irrigation withdrawals increased in Bangladesh, dispute arose between India and Bangladesh over the sharing of the lean season flow at Farakka. The inadequacy of water during the lean season to meet the assessed demands in the two countries is the root cause of the conflict. The Bangladesh government feels that the reduction in flow caused damage to agriculture, industry and ecology in the basin in Bangladesh. The government of India feels that such misgivings are misplaced. Because of the inability of the concerned governments to come to any lasting agreement over the last few decades on sharing the river water, this problem has grown and now it is also viewed as a case of upstream-downstream dispute.

In the relationship between India and Bangladesh, the dispute over Ganga water was an important component for over two decades. In 1977, the two countries reached a five-year agreement on water sharing which was signed in November 1977. However, the basic issues remained un-solved and hence, the agreement was not renewed and lapsed in 1982. Later, a proposal was mooted by India to augment the flow of Ganga at Farakka by constructing a barrage across the Brahmaputra at Jogighopa in India and transfer water to Farakka through a canal. A proposal from Bangladesh was to construct a series of reservoirs on the tributaries of Ganga in Nepal. But this would have brought Nepal also in the picture. None of these proposals could be implemented.

According to the Bangladesh view of this dispute, there was a "unilateral diversion" of the waters of the Ganga by India at Farakka to the detriment of Bangladesh and the resulting reduction in flows had severe adverse effects on Bangladesh. It was also projected that this was a case of a large and more powerful country disregarding the interests of a small and weaker neighbour. This has been one topic on which there has been complete unanimity in Bangladesh. Sentiments grew to such an extent that India was projected as being responsible for all water related disasters, whether drought or floods and the Farakka project was supposed to be the cause behind all water related problems.

The perceptions on the Indian side were entirely different. Ganga River is closely intertwined with the religious and social life of the people. Also important are the legitimate needs of the Kolkota Port in the light of siltation and growth of future traffic.

After the liberation of Bangladesh, the general perception was that two countries should strive to establish stable and friendly relations. Thus, the position of Bangladesh was unbelievable to many Indians and the general feeling was that Bangladesh had taken an un-necessarily rigid and unreasonable stand on this issue. It was felt that Bangladesh had greatly overstated its water needs and its claim was disproportionate to its fair share by any logic. Regarding diversion of water for the Kolkota port through the Farakka Barrage, the view was that Bangladesh had magnified the adverse effects due to reduced flows and it was unfair on their part to blame India for flood problem. A further complication was the feeling on the parts of the concerned State Governments in India that their interests had not been properly addressed by the Central Government during bilateral talks with Bangladesh. For several years after 1990, there was a virtual stalemate between the two governments on this issue.

Finally, an agreement was reached and a Treaty on the sharing of Ganga water between India and Bangladesh was signed on 12 December 1996. The Treaty is essentially regarding the sharing of lean-season flows, although there is an article which recognizes the need for cooperation in finding a solution to the long-term problem of augmentation. The sharing formula agreed in the Treaty is related to actual flows at various levels and not to 75% dependable flows as in past agreements. The basic formula is that of equal sharing of the lean-season flows by the two countries. This applies to a range of flows, with two modifications at the upper and lower extremes. At the upper end, there is a slight increase in India's share to enable it to divert 1,132.66 cumec (40,000 cusec) (the full diversion capacity of the Farakka feeder canal) at a flow level of 2,123.74 cumec (75,000 cusec) instead of 2,265.33 cumec (80,000 cusec). Above 2,123.74 cumec (75,000 cusec), India's share is held at 1,132.66 cumec and the balance goes to Bangladesh.

Some people view the Treaty as a political rather than a technical settlement. Although the issue now stands resolved, some people feel that it has not been resolved for ever. The Treaty is a 30-year Treaty (renewable further), with a provision for a review at the end of five years, or even at the end of two years, if either party wants it.

21.11. COOPERATION WITH NEIGHBORING COUNTRIES FOR JOINT WATER RESOURCES PROJECTS

Since most Himalayan rivers originate either in Nepal or Bhutan, International collaboration is necessary between India and these upstream countries.

There is a huge demand for electricity and water in densely populated India and Bangladesh which are downstream countries. The two upstream countries, Nepal and Bhutan, have an enormous hydropower potential that is largely untapped. Selling of hydropower will be an effective instrument for economic growth in these countries. On the other hand, sorting out a mutually acceptable sharing and augmentation arrangement will only help the growth of the economies of Eastern India and Bangladesh. While there is no doubt that, in the interest of the largely poor population living in the GBB, the water flowing in the rivers of the basin could be a tremendous economic resource, collaborations among the riparian countries on joint projects is seriously handicapped by many obstacles. The removal of these obstacles requires a very high level of political stature and leadership in the region. Here, we first discuss the cooperation between India and Bhutan and then between India and Nepal.

21.11.1. Bhutan-India Collaboration

Although the geographical area of the mountain kingdom of Bhutan is only 45,000 sq. km, it receives copious precipitation which is drained by several major rivers, such as Wangchu (Raidak), Mochu (Sankosh), and Dangmechu (Manas). Originating in Bhutan, these rivers enter India. It is estimated that the total hydropower potential of Bhutan at 30% load factor is 20,000 MW. Although there is limited demand for electricity in Bhutan, there is a large demand for power from the areas of India across the border. Naturally, collaboration between the two countries is a win-win situation for both. This idea led to the Bhutan-India accord in 1974 for the construction of the 336 MW run-of-river Chukha hydropower project on the river Wangchu. This plant was designed, financed and constructed through a grant and loan agreement between Bhutan and India. An autonomous Indo-Bhutan Chukha Project Authority which had staff from Bhutan and India constructed the project. The Chukha project is a fine example of Indo-Bhutan friendship and cooperation. The availability of power from this project increased electricity consumption in Bhutan, sparked some industrial growth, contributed significantly to Bhutan's revenue, and provided the eastern region of India with much needed electricity. However, the project has also generated some controversies and differences between Bhutan and India.

According to Verghese & Iyer (1992), there are several potential collaborative projects between India and Bhutan, full development of which, within a decade, could raise Bhutan's installed capacity to about 2,400 MW. If mere market forces of supply and demand for peak power were decisive, the Chukha project would have been followed up with several others. A wider and deeper examination of the issues, however, leads to different viewpoints which need to be appreciated to have a realistic understanding of the impediments to effective collaboration between the two countries. Verghese & Iyer (1992) stated that "the Royal Government of Bhutan has been disinclined to pursue any such initiative [of collaboration] for the moment on socio-cultural and other considerations". Surely, mere economic growth is not the guiding factor in this and many other cases.

The controversies and differences related to the Chukha project identify several important obstacles to effective upstream-downstream cooperation. These controversies, are less environmental or economic and more cultural and political. There were some problems in selling electricity produced by Bhutan, a technically less equipped and small upstream country, to India, a technically superior, larger economy, and a monopoly buyer. Undoubtedly, for a larger economy such as India, the quantitative importance of the hydroelectric projects in Bhutan is not very high. But, because of its smaller economy, the same projects can make a huge impact on economic transformation in Bhutan. Thus, Bhutan had to negotiate in a buyer's market while the decisions on the project were taken. More recently, increasing contact of Bhutan with the rest of the world, easy availability of international aid from global sources and enhancement of her own technical competence have made Bhutan much more sensitive towards her bargaining position in the agreements between the two countries. Such sensitivities have substantially reduced early prospects of further cooperation between the two countries. The difficulties in reaching a simple sales agreement for Chukha electricity between the governments of Bhutan and India is a good indicator of the degree of the difficulties associated with developing new collaborations on joint water development projects.

21.11.2. Nepal-India Collaboration

Although the nature of the upstream-downstream linkage of both Bhutan and Nepal with India is largely similar, cooperative development of water resources between India and Nepal has a longer history and wider dimension. Among the first formal agreement for cooperation between India and Nepal was made as far back as 1927 for the construction of a barrage near Tanakpur across the Sharda River (Mahakali) which forms the western border between the two countries. The two countries have deep cultural ties, friendly links and extensive people-to-people interaction in the region. Nepal's immense hydropotential of 45,824 MW cannot be economically utilized within the country and the growing industrial economy of India will surely be willing to buy the peak power produced in Nepal. Collaboration on water projects to harness the water resources and control floods are in the joint interest of both countries. In the case of Nepal-India collaboration, both hydropower and irrigation have received significant attention, while flood control is also expected to be an important element in future large dam projects. However, the problems associated with collaboration between a technically superior and economically stronger India, as noted in the case of Bhutan, are also present in the case of Nepal. The Sharda barrage, instead of paving the way for more and larger cooperative projects, has also generated mistrust.

In 1950s, two project agreements materialized; a barrage on the Kosi River in 1954 and another barrage on Gandak River in 1959. The Trisuli project which supplies electricity to the capital city of Nepal, Kathmandu, was constructed by India.

In the case of India-Nepal cooperation, political differences have also been an obstacle for progress in larger collaborative projects such as on the Karnali at Chisapani, with hydropower potential of more than 12,000 MW. Analysing the background for the Nepali attitude behind this complex situation, Gyawali (1991) concluded that in the last three decades, Nepali people have begun to realize the value of their flowing water. Another cause behind growing distrust between India and Nepal was the feeling in Nepal that India had used its influence to the detriment of Nepal. A proof of the sensitivity of Nepali people is the provision in the Nepali constitution that any treaty related with natural resources must be ratified by a two-thirds majority in the national legislature.

Large storage dams on Mahakali, Karnali, Gandak and Kosi are necessary for the development of large quantities of hydropower and this can harbinger progress and prosperity in Nepal and India. In addition to hydropower, these projects will also moderate floods and augment flows during dry periods. Efforts should be made to reach agreements that are fair and beneficial to both the parties because that is the only way for long-term sustainable development of resources and to establish friendly relations with neighbours.

CHAPTER 22

INTER-BASIN WATER TRANSFER

22.1. INTRODUCTION

India has 16% of the world's population, 4% of the world's freshwater resources and 2% of world's land area. In India, precipitation distribution is highly variable – both temporally and spatially. The rainfall over the country is primarily associated with tropical depressions originating in the Arabian Sea and the Bay of Bengal. Nearly 80% of the annual precipitation is received during 4 months of summer monsoon season. Spatially, precipitation varies from less than 100 mm in Western Rajasthan to more than 10,000 mm in parts of Meghalaya. Large parts of the country are not only in deficit in rainfall but also subject to large variations, resulting in frequent droughts and causing immense hardship to the population and enormous loss to the nation. The uncertainty of occurrence of rainfall coupled with prolonged dry spells and large fluctuations in seasonal and annual rainfall is a serious problem in the country. The water availability even for drinking purposes becomes critical, particularly in the summer months as the rivers dry up and the ground water recedes.

Regional variations in precipitation lead to situations when some parts of the country do not have enough water even for raising a single crop while intense rainfall in some other parts creates havoc due to floods. Floods are a recurring feature, particularly in Brahmaputra and Ganga Rivers, which carry close to 60% of the river flows of the country. Table 1 shows the water resources potential and per capita water availability in the river basins of India. Due to topographical and other constraints, only 690 BCM of surface water out of 1,869 BCM and 432 BCM of ground water can be put to beneficial use. The reasons behind lesser quantities of utilizable flows are the skewed spatial distribution of these flows and that potentially good storage sites are not available in basins with plenty of water resources.

As can be seen from Table 1, nearly 60% of the potential is available in the Ganga-Brahmaputra-Barak River system in the north. About 11% is available in the high rainfall region of the Western Ghats which flows down through many small west flowing rivers. Hardly 16% of the potential is available in all the rivers – including mighty rivers like Mahanadi, Godavari, Krishna and Cauvery – of the peninsular India flowing towards east.

To illustrate disparities in the water availability, the per capita water availability for the year 2000 and projected data for the year 2025 for major basins

Table 1. Water Resources Potential of River Basins of India

S. N.	Name of the River Basin	Average annual potential as per NCIWRDP	Estimated utilizable surface water (km^3)	Per capita water availability (m^3)	Ground water potential (km^3)
1.	Indus (Area in Indian Territory)	73.31	46.00	1,757	26.5
2.	a) Ganga	525.02	250.00	1,260	171.57
	b) Brahmaputra, Barak, and others	677.41	24.00	14,616	35.07
3.	Godavari	110.54	76.30	2,026	40.65
4.	Krishna	69.81	58.00	1,058.0	26.41
5.	Cauvery	21.36	19.00	750.0	12.3
6.	Pennar	6.32	6.86	648.0	4.93
7.	East flowing and rivers from Mahanadi to Godavari and Krishna to Pennar	22.52	13.11	650.0	18.8
8.	East flowing rivers between Pennar and Kanyakumari	16.6	16.73		18.2
9.	Mahanadi	66.88	49.99	2, 546	16.46
10.	Brahmani & Baitarani	28.48	18.30	2, 696	4.05
11.	Subarnarekha	12.37	6.81	1, 392	1.82
12.	Sabarmati	3.81	1.93	182	3.2
13.	Mahi	11.02	3.10	1, 057	4.0
14.	West flowing rivers of Kutch & Saurastra including Luni	15.10	14.98	631	11.23
15.	Narmada	45.64	34.50	2, 855	10.83
16.	Tapi	14.88	14.50	1, 091	8.27
17.	West flowing rivers from Tapi to Tadri	87.41	11.94	3,366	17.69
18.	West flowing rivers from Tadri to Kanyakumari	113.53	24.27		
19.	Area of inland drainage in Rajasthan desert	Negligible		Negligible	–
20.	Minor rivers draining to Myanmar (Burma) & Bangladesh	31.00		14, 616	–
	Total	1, 952.87	690.00		431.98

in India is shown in Figure 1. It can be seen here that per capita utilizable water in Sabarmati basin is 182 m^3/year (1991 census), 2,050 m^3/year in Mahanadi, 2,900 m^3/year in Narmada basin and more than 14,000 m^3/year in the Brahmaputra basin. The utilizable water per ha of culturable area varies from 1,244 m^3/ha in the Sabarmati basin to 8,320 m^3/ha in the Mahanadi. Table 2 further illustrates the

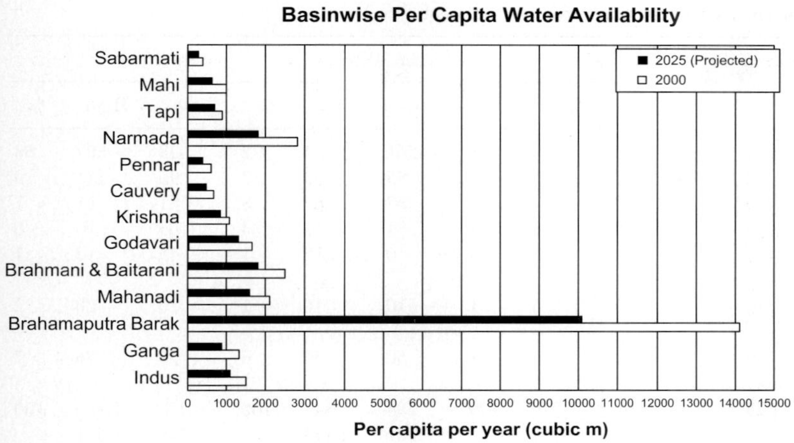

Figure 1. Per capita per year water availability for selected Indian basins

skewed distribution of water availability. Such a large variation in the availability of water has created a flood-drought-flood syndrome in the country, with some areas suffering from flood damages, while some other areas facing acute water shortage in many years.

In addition to the water availability, there are large variations in the demand of water. The reason is that population, industries, and agricultural areas are not always located at places where water is abundant. Consequently, some river basins have excess water compared to the present and projected needs whereas others just do not have enough water even to meet their present needs. Further, the water-short areas may not be part of the same basin as the surplus areas. Since water is a prime input in many economic activities, its shortage seriously downgrades the quality of life as well as economic development.

The total annual requirement of freshwater from various sectors as estimated for the year 2010, 2025 and 2050 is shown in Table 3 for two scenarios of demands: low

Table 2. Skewed distribution of water resources in India

Basin	Area M-ha	Water resource km^3	Utilizable surface water km^3
Ganga- Brahmaputra-Barak	110.13 (33.5%)	1,202 (62%)	274 (40%)
West flowing rivers south of Tapi	11.31 (3.5%)	201 (10%)	36 (5%)
Other basins	207.29 (63%)	550 (28%)	380 (55%)
Total	328.73	1953	690

Table 3. Annual water requirement (km^3) for different uses

Uses	Year 1997–98	Year 2010			Year 2025			Year 2050		
		Low	High	%	Low	High	%	Low	High	%
Irrigation	524	543	557	78	561	611	72	628	807	68
Domestic	30	42	43	6	55	62	7	90	111	9
Industries	30	37	37	5	67	67	8	81	81	7
Power	9	18	19	3	31	33	4	63	70	6
Inland Navigation	0	7	7	1	10	10	1	15	15	1
Environment–Ecology	0	5	5	1	10	10	1	20	20	2
Evaporation Losses	36	42	42	6	50	50	6	76	76	7
Total	629	694	710	100	784	843	100	973	1,180	100
Population (million)		1,146	1,157		1,286	1,333		1,346	1,581	

Source: NCIWRD (1999).

growth and high growth. Note that the available water will be just enough to meet the projected requirements and if the population grows at higher rates, there might be a shortage of water.

The river basin is a fundamental unit for water resources planning and management. This concept has received wide support from engineers and planners. But in view of the shortage of water in some basins, a question that can be asked is: should we not examine the possibility of a nationwide sharing of resources? A positive answer is the main reason for looking for transfer of water from a surplus basin to a deficit basin. Creation of storage and inter-basin transfer of water from surplus to deficit regions is a rational option to overcome the problems caused by the mismatch of supply and demands. Inter-basin transfer of water over long distances has been mooted as a long-term strategy for India for meeting the increasing water demand in water short areas.

22.2. INTER-BASIN WATER TRANSFER

In the event of a shortage of water in an area, two options are available: supply management and demand management. In supply management, steps are taken to increase the availability of water and the means that can be adopted include interbasin transfer of water, artificial recharge, desalinization of water, etc.

The term *water transfer* refers to the transport of water through engineering structures, usually across river basins for some beneficial purpose. Interbasin Water Transfer (IBWT) is one of the possible solutions of water deficiency and is somewhat similar to other alternatives, such as dams, desalination, groundwater extraction, etc. IBWT involves transportation of surplus water from a basin to another basin, which is deficient in water. The starting point of IBWT is an unsustainable situation in the receiving basin in the sense that it suffers from recurrent water shortages. If the surplus and deficient basins are not near to each other, which quite often is the case, this will involve transfer of water over long distances, sometimes of the order of thousands of kilometers. IBWT is an ancient approach and under certain conditions, it is a rational measure. In fact, most water resources development projects involve some kind of water transfer, may be over only short distances. Diversion of water by IBWT increases the resilience of the water system and decreases the risk of shortages.

The most common purpose of IBWT projects is water demand of agricultural areas or mega cities. As the human settlements are not always near the places of water occurrence or the available water may be inadequate to meet all demands at a place, waters have been transferred from one basin to another since time immemorial.

Typically long distance water transfer is carried out to improve national/regional economy, self-sufficiency in national/regional agricultural outputs and to remove regional disparities in development activities. The special attributes of long distance water transfer are:

1. Large amounts of water is transferred,
2. Water is transferred over large distances,
3. Infrastructure costs are large,
4. There is a possibility of extensive and irreversible environmental consequences, and
5. Such projects have significant influence on the economy of the receiving area.

Most commonly, canals carry water from one basin to another; tunnels and pipelines are used to negotiate ridges. However, the final selection depends on the topography of the area, climate, soil properties en-route, and the volume of water involved. Many IBWT projects involve pumping of water in some stretches when a mountain is to be crossed and construction of a tunnel is not feasible. To the extent possible, a gravity flow system is preferred over a system that involves pumping, even if a slightly longer route is to be employed. The running cost of a system with pumping is significantly higher and this places an additional burden on the infrastructure particularly in those countries, which are deficient in electric energy. Additional maintenance problems arise if water to be pumped contains sediments because these cause much wear and tear.

22.2.1. The Need of Inter-Basin Water Transfer in India

An overview of the water availability in various regions of India shows that except for the northern part, the north-east part and the Western Ghats areas, there is existing/expected scarcity of water elsewhere. To meet the water demands in the water scarce parts of the country, one option is to transfer water from surplus areas to deficit areas. With this scenario, the possibilities of transferring water from one river basin to the others were thought of.

IBWT from surplus to deficit regions is one of the options for augmenting utilizable water resources. IBWT in India is expected to augment irrigation potential and will also mitigate recurring floods and droughts. By creating storage reservoirs on surplus rivers, viz., Brahmaputra and Ganga (particularly their northern tributaries), Mahanadi, Godavari and west flowing rivers originating from the Western Ghats and connecting them to other parts of the country, regional imbalances could be reduced significantly and large benefits by way of additional irrigation, domestic and industrial water supply, hydropower generation, navigational facilities, etc. would accrue.

The growing water stress faced in some of the regions and resulting restlessness and political fallouts amongst states (e.g., Cauvery) is well known. Some of the peninsular basins in India like Vaigai; Cauvery, Krishna, Pennar, Sabarmati, etc., have virtually exploited their intra-basin resources fully. Gradually, more and more river basins will become water scarce in the coming decades. This is essentially due to the increasing population and resulting needs for the society, which is on the increase continually. Measures to stabilize the population are likely to result in success by around mid 21st century but by that time, Indian population might reach 1.6 to 1.8 billion. Thus there will be a pressure on water for the demands in respect of food production, people sector needs (drinking, industrial and other uses) besides environmental requirements and ecological needs (aquatic systems and the Nature Sector). A long range planning for the country's overarching self reliant economy indicates the need for augmenting supplies in deficit basins by other means. If ILR could help face these requirements, should this program not be given a serious consideration?

India uses nearly 70% of its water for agriculture. Population projection and per capita food grain requirements indicate that by the year 2050, India should produce around 550 million tones of food grains for the projected population of around 1,600 millions. Development of irrigation coupled with high yielding varieties of crops may be the only strategy available to achieve the required production.

The ultimate irrigation potential from in-basin development has been estimated to be 113 million hectares (M-ha). In-basin water resources development alone cannot increase the irrigated area beyond a certain limit. But to achieve a food production level of 550 million tones, it is imperative that the irrigation potential is enhanced. A major strategy to achieve an increase in irrigation potential is IBWT.

Many large towns and mega cities, particularly those situated in water deficit basins, are facing problems in meeting their municipal and industrial water demands. In 1991, the urban agglomerations were 3,768 with a combined population of 217 millions (26% of the total population). Metropolitan cities with more than a million people were 23 in 1991 accounting for nearly one-third of the total population. By the year 2050, urban population in India is expected at 820 millions which would be nearly 50% of the total population. As the economic status of the people improves, per capita water demand will also grow. Meeting the water requirement of large cities will be a challenging task. In many large cities like Mumbai, Delhi, Hyderabad and Chennai; water demands are already being met through IBWT.

Apart from these, rural areas are also facing problems in domestic water supply, particularly during years of less rainfall. During drought, ground water recharge is less and the sources for meeting domestic requirements get exhausted well before the onset of the next monsoon, resulting in an acute shortage of water, even for drinking. IBWT may be a lasting solution to meet the water requirements of such chronically water short areas.

22.3. PLANNING FOR IBWT PROJECTS

IBWT has been an option in water management for a long time in many parts of the world and will remain so in the future. Before any large-scale water transfer project is taken up, it would be helpful to mull over the following questions:

- Is water transfer the only option to overcome the present and likely problems?
- Is water transfer the most efficient alternative?
- What are the tradeoffs involved in the water transfer?
- Is the requisite institutional and infrastructural support available?

As per the practice being followed in India, if the water available in a basin is more than the demands that are likely to arise in future, then this basin is considered as a water surplus basin. The volume of water over and above the projected demands is labeled as surplus for that basin and this can be made available for transfer to other deficient basins.

An important issue in IBWT is the sharing of water resources between the donor and receiving basins. Sustainable development in both basins should be practiced through shared economical and social benefits from the project. In the donor basin, water transfer must not have negative impacts on the sustainability of water use. Water transfer agreements should take care of monitoring and periodical assessment, with the possibility of adjustments of mutual obligations.

IBWT projects are generally cost-effective solutions and technical problems are seldom the limiting factor except for those projects that involve long distances of transfer. The necessity or otherwise of an IBWT project can be evaluated using the following criteria:

1. The recipient basin must have a substantial deficit in meeting the present or projected future water demands after considering alternative water supply sources and all reasonable measures for reducing the water demand.
2. The future development of the donor basin must not be substantially constrained by the water scarcity.
3. An IBWT project should be taken up after comprehensive environmental impact assessment (EIA) and should indicate that it will not substantially degrade environmental quality within the area of origin or area of delivery.
4. A comprehensive assessment of socio-cultural impacts must indicate a reasonable degree of certainty that the project will not cause socio-cultural problems in the donor or recipient basins.
5. The net benefits from transfer must be shared equitably between the donor and recipient basins.

EIA is a necessary step in the evaluation of any major IBWT project. Due to serious and long-term environmental implications, EIA should be an important part, and not just a formality for project clearance. Importantly, EIA should not be viewed as an obstacle to the project. Due to water transfer, the ecological balance in the recipient basin may improve, and the transfer of water can help sustain cultural and emotional values that are associated with a water body. Jain and Singh (2003) have described the various aspects that should be considered while formulating an IBWT project.

22.4. IBWT PROJECTS IN OTHER COUNTRIES

IBWT concept has been in practice for a long time. In arid and semi-arid regions, IBWT is sometimes crucial to alleviate acute water shortages and to strengthen the resilience of water systems in case of droughts. In many countries, interbasin water transfer projects are taken up because they often offer the most attractive solution to a given water problem. These projects are often necessary because in many instances, the regions where water resources are abundant are not necessarily the regions where most of the population resides or where the industrial or agricultural activities are concentrated. For example, about 60% of water in Canada flows towards the north, while 90% of the population and majority of industries are concentrated within 300 km of its border with the United States (Sewell, 1985). Some existing and under construction IBWT projects in Canada include Kemano, Churchill Diversion, Welland Canal, James Bay, and Churchill Falls. Table 4 gives an overview of IBWT schemes world wide.

The Lingua Canal (completed in 214 BC) and the Grand Canal (completed in 605 AD) are two examples of IBWT from ancient China. Recently completed projects in China include Biliuha-Dalian interbasin water supply system and trans-basin transfer of water of Luhana River to the Tiajian and the Tengshan. The southern part of China is abundant in water resources whereas the northern part is water deficient. The basin of Huang He, Huai He, and Hai He Rivers suffer from water deficiency. The Chang Jiang River basin is abundant in water resources and this will be the main basin to supply water to north China. Hence, the south-to-north water transfer was conceived as a long-term solution to this problem. In fact, this project includes three components: the west, middle and east route with respective serving areas. The east route component was to be taken up first. In the middle route component, a large amount of water from Han Jiang and Chang Jiang (middle reaches) rivers will be transferred. Diversion of Quiantang River water and diversion of Yellow River surpluses are other ambitious proposed projects.

The major water transfer projects that have been proposed for the United States include the North American Water & Power Alliance, the Texas Water Plan, and the California State Water Project. The Texas Water Plan envisages redistribution of water in Texas and New Mexico to meet the needs up to the year 2020. The California State Water Project, the first phase of which was completed in 1973,

Table 4. Overview of IBWT schemes world wide

S. N.	Country	Number of completed schemes	Number of proposed schemes	Water transfer by completed schemes (BCM/Year)	Water transfer by proposed schemes (BCM/Year)
	Americas				
1	Bolivia	–	1	–	0.2
2	Brazil	3	2	5	2.5
3	Canada	38	10	262	464
4	Chile	2	–	3.15	–
5	USA	19	7	38	382
	Total Americas	62	20	308	849
	Europe				
6	Czech Rep	4	2	15	2.2
7	Finland	1	–	0.09	–
8	France	5	–	2.35	–
9	Germany	2	–	0.47	–
10	Portugal	1	–	0.01	NA
11	Romania	3	NA	NA	NA
12	Russia	5	2	60	47
13	Spain	3	1	1.3	1
	Total Europe	24	5	79	51
	Asia/Australia				
14	Australia	1	–	1.5	–
15	China	6	3	NA	45
16	India	27	30	56	150
17	Iraq	6	–	16	–
18	Japan	1	–	NA	–
19	Malaysia	1	1	NA	0.14
20	Pakistan	8	–	100	–
	Total Asia/Australia	50	34	173	196
	Africa				
21	Morocco	1	–	1.5	–
22	South Africa	24	–	2.5	–
23	Sudan	1	–	7.3	–
	Total Africa	26	–	11	–
	Grand Total	162	59	571	1,096

Source: Thatte (2005).

provides for diversion of 4 cubic km of flow from better watered northern California to the drier central and southern parts of the state. The conveyance system comprises a 715 km California Aqueduct, a complex system of lined and unlined canals, pumping stations, siphons and tunnels. The lift involved is nearly 1,000 m (IWRS, 1996). Similarly the waters of the Colorado River (an international river between U. S. A. and Mexico) are being diverted outside the basin to the Imperial Valley in California. In Mexico, the project for water supply of Mexico City through the transfer of ground waters from the Lerma basin was completed in 1958. The Mahaveli–Ganga project of Sri Lanka includes several IBWT links.

A notable IBWT scheme executed in the former USSR is the Irtysh Karganda scheme in the central Kazakhistan. The link canal is about 450 km long with a maximum capacity of 75 m^3/s and the lift involved is 14 to 22 m. There is another plan to transfer 90,000 MCM from the north flowing river to the area in south. Other proposals include partial redistribution of water resources of northern rivers and lakes of European part to the Caspian Sea basin involving 2 M-ha-m of water. More examples include IBWT for urban drinking water supply in Spain, France, and Germany; for irrigation, drinking water and hydropower in India, navigation and environmental aspects in Europe, and for environmental improvement in U. S. A. and Australia. Glubev and Biswas (1985) and IWRA (1986) contain many case studies of IBWT projects.

22.5. PAST INDIAN EXPERIENCES ON INTER-BASIN WATER TRANSFER

Interbasin transfer of water is not a new concept; it has been in practice in India for over five centuries. The Western Yamuna Canal and the Agra Canal for carrying water from the Himalayas to the distant parts of Punjab, Uttar Pradesh and Rajasthan were built in the Mughal times and carried water from the Himalayas to the distant parts of Punjab, U.P. and Rajasthan. The Kurnool–Cuddapah Canal (1860–1870) and Periyar-Vaigai Project (1896) were built in the 19th Century. In recent times, the Indira Gandhi Canal diverts waters to the deserts of Rajasthan. The Sardar Sarovar project will help transfer water from Narmada basin to the Sabarmati and Saurashtra areas of Gujarat. The Telugu Ganga project is another classic example of IBWT.

Dr. K. L. Rao, the then Irrigation Minister of India mooted the first ever large-scale long distance transfer of water in India in 1975. He proposed the National Water Grid of India in which the famous Ganga–Cauvery Link was one of the components (see Figure 2). The Ganga–Cauvery link was planned for diversion of the surplus discharge in the Ganga near Patna to Pennar and Cauvery basins. The total lift involved was of the order of 565 m and the net power requirement was 5,000 to 7,000 MW for 150 days. The irrigation benefit was assessed to be about 4 M-ha but the project had no flood control benefits (IWRS 1996). Though this proposal generated considerable interest and excitement, a closer scrutiny showed that the proposal was extremely expensive. Another proposal that received considerable publicity during the 1970s was the Garland canal by Captain Dastur. He proposed a 4,200 km long Himalayan canal and a 9,300 km long southern Garland canal with a connection between the two canals. Figure 3 shows the Dastur proposal. However, the proposal was found to be technically infeasible and economically not viable and was not given further consideration.

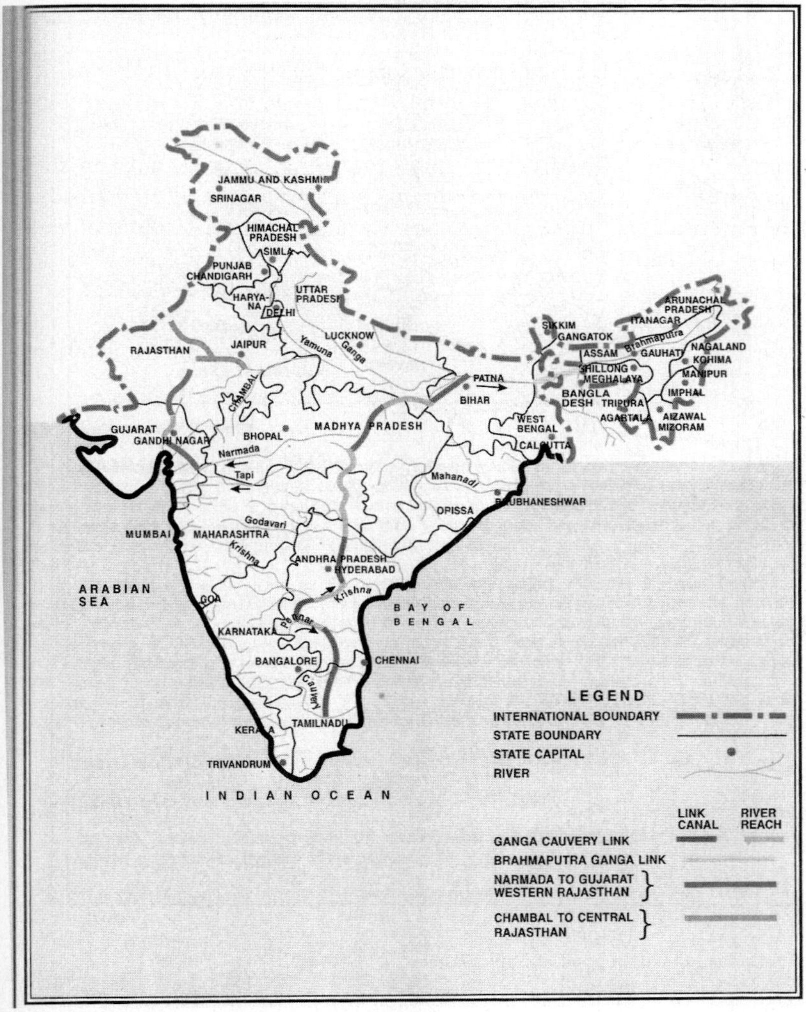

Figure 2. National water grid proposed by Dr. K. L. Rao

The increasing requirement of water in many parts of the world to meet varied demands especially in arid and semi-arid regions has given a boost to large inter-basin transfers in the last few decades. Accordingly, many schemes of large-scale water transfer projects have been planned and some of them implemented and are standing as a landmark for over all development of respective regions.

ILR has been in vogue in India even prior to the 20th century. The Telugu Ganga Project is a fine example of not only of hydraulic engineering but also of Inter-State cooperation which has been recently implemented primarily to meet the

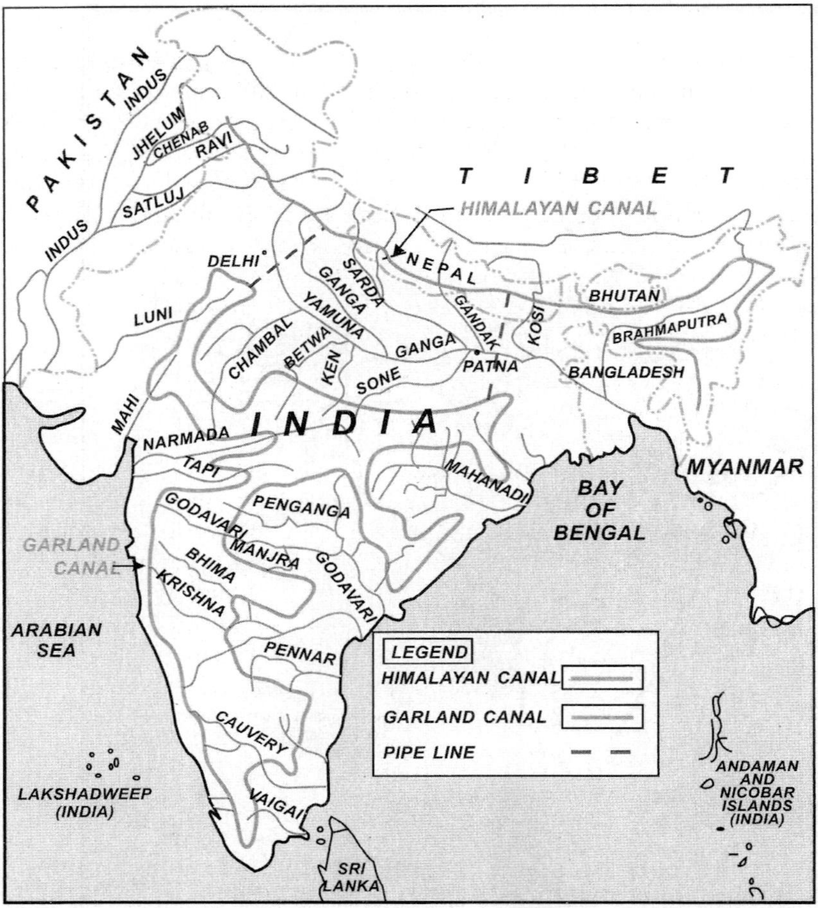

Figure 3. The Garland Canal proposal of Capt. Dastur

pressing need of water supply to the Chennai metropolitan area. The Beas-Satluj link in combination with the Indira Gandhi Nahar Project (IGNP) is an example of how large inter basin transfers have brought about all round socio-economic growth with overall enhancement in the ecology and environment of the region. The transfer of surplus waters of Ravi, Beas and Satluj to Rajasthan right up to Jaisalmer and Barmer through the IGNP has eliminated drought conditions by transforming desert wasteland into an agriculturally productive area by bringing irrigation and vegetation. The recently commissioned Sardar Sarovar Canal has linked more than 8 rivers to the North of Narmada River and is on the threshold of dramatically transforming the scarcity regions of Gujarat and Rajasthan.

Some landmark projects are described in what follows.

22.5.1. Indus Basin Project

In North India, inter sub-basin transfers in the Indus basin and Rajasthan canal project are some of the projects executed in the 19th & 20th centuries. In the Indus Basin a diversion dam, Pandoh, 140 km upstream of Pong Dam on the Beas, enables diversion of water from the Beas to the Bhakra reservoir on the Satluj River and generates 165 MW of power on the way. The Beas–Satluj link is 37.25 km long. It includes a 25 km long tunnel through difficult rock formations. The Rajasthan Canal diverts water from the Himalayas to the deserts of Rajasthan. The project comprises a huge multi-purpose project constructed across the Beas River at Pong, a barrage at Harike and a grand canal system. There are other examples also like the Sarda Sahayak (1960) and the Ramganga-Ganga Link (1978).

22.5.2. Periyar Project

The Periyar dam (47.28 m high) is a unique structure built during 1887–95 in the Western Ghats on the Periyar River. The dam was built to create a reservoir and divert the stored waters to the east tunneling through the Western Ghats. The hilly ridge between the Periyar and Vaigai basins was crossed by a tunnel (1,739 m long, discharge capacity of 40.75 cumec) with an open cutting in the foreshore of the Periyar Lake. The diverted water would augment the flow in the east flowing Vaigai River in the then Madras Presidency for irrigating the drought prone regions of the Madurai and Ramanathapuram districts. Perhaps this was the earliest attempt of transbasin diversion of water of such a large magnitude in the world.

Initially the project provided irrigation to 0.58 lakh ha, which has since been extended to 0.81 lakh ha. With the formation of lake, there was a big boost to the flora and fauna. No habitation that existed at the dam site or the water spread had to be shifted and no affected person required to be rehabilitated in this project. Practically no loss has been sustained, whereas the benefits have flowed in abundant measure from the time of completion of the project in 1895.

This project transformed the entire area from drought prone to a flourishing green belt and the standard of living improved beyond expectation. The project originally envisaged benefits for 36,423 ha of first crop paddy and 24,282 ha of second crop paddy. The hydropower scheme was completed in 1965 with four units of 35 MW totaling to 140 MW. Now the total ayacut is more than 81,000 ha. The secondary benefits like setting up of several industries in Madurai and other areas, tourism and trade have developed. The lake is also serving for fresh water fishing and several tons of fish are harvested daily. The monetary value of benefits that flow each year itself have far exceeded the total amount spent on the project.

22.5.3. Parambikulam Aliyar Project

The project is a complex multi-basin multi-purpose project of seven streams, five flowing towards the west and two towards the east. These have been dammed and their reservoirs interlinked by tunnels. The water is ultimately delivered to drought prone areas in the Coimbatore district of Tamil Nadu and the Chittur area of Kerala state. The command area for irrigation is presently about 162,000 ha. There is a 185 MW of power generation capacity at four power houses. This project was built during the second and third five-year plans.

22.5.4. Kurnool–Cudappah Canal

The famous Kurnool–Cudappah (KC) Canal was originally constructed by the Madras Irrigation Company in 1866 during the British rule of India. The scheme was taken over by the government in 1882. A 8.23 m high anicut was built on the Tungabhadra River upstream of Kurnool town. A 304 km long canal with a capacity of 84.9 cumec at its head extends from the Krishna basin to the Pennar basin and irrigates 52,746 ha. The scheme serves the irrigation and water supply needs of a part of the Kurnool and Cuddapah districts in the Pennar basin with the waters diverted from the Tungabadhra River of the Krishna Basin. The Pennar basin, as is well known, is a zone of poor and uncertain rainfall and has witnessed many famines in the past. After completion of the KC canal, the area served by it has become free from famine.

22.5.5. Telugu Ganga Project

This project has been recently implemented primarily to meet the pressing need of water supply to the Chennai metropolitan area. It brings the Krishna water from the Srisailam reservoir through an open canal, first to the Somasila reservoir in the Pennar valley. This involves rock cuts up to 35 m deep. From Somasila, the water is taken through a 45 km canal to Kandaleru and then to the Poondi reservoir in Tamil Nadu through another 200 km long canal. By mutual agreement, 3,394 MCM (12 TMC) of water will be delivered to Tamil Nadu at the border from the Krishna basin. The canal also irrigates 2.33 lakh ha in Andhra Pradesh en-route. The project was made possible by Maharashtra, Karnataka and Andhra Pradesh voluntarily foregoing 1,415.85 MCM (5 TMC) each from their entitlements. This project is a fine example of not only of hydraulic engineering but also of Inter-State cooperation.

22.5.6. Diversion of Ravi–Beas–Satluj Water Through Indira Gandhi Nahar

The Beas–Satluj link in combination with the Indira Gandhi Nahar Project (IGNP) is an example of how IBWT has brought about socio-economic growth with overall enhancement in the ecology and environment of the region. Under the Indus Water Treaty, the water of three eastern rivers, viz., Satluj, Beas and Ravi, were allocated to India. As the land to be benefited in India lies mostly to the east and south of these rivers, it was planned to interlink river and convey water to canal systems for serving vast tracts in India. The main storage on the Satluj is at Bhakra, while that on the Beas is at Pong. The Bhakra system provides irrigation to 26.3 lakh ha of new area besides the stabilization of existing irrigation on 9 lakh ha. The aggregate generation capacity of power on the Bhakra Nangal Project is 1,354 MW. A diversion dam, Pondoh, 140 km upstream of Pong on the Beas, enables diversion of water from the Beas to the Bhakra reservoir and generates 165 MW of power.

The Beas–Satluj link is 37.25 km long of which 25.45 km is tunnel through difficult rock formations. The capacity of the tunnel is 254.70 cumec. Another dam on the Ravi namely, the Ranjit Sagar dam, provides additional water to the Beas and also generates a large amount of power. A large quantity of water of this system is diverted to the Indira Gandhi Canal for irrigation in Rajasthan state. Among the major benefits of the IGNP are mitigation of drought conditions in Jaisalmer and Barmer, and generation of hydropower. As a result of assured irrigation, about a 2 million ha area has been transformed from the desert wasteland to an agriculturally productive area and one can see a lot of vegetation at places which were barren. Additional agricultural production alone has been estimated to be worth nearly Rs. 2,000 crore annually. The water of IGNP is also being used to meet domestic needs as well as the needs of army posts near the border. To sum up, the project has dramatically changed the living standard and socio-economic conditions of the people in the region.

As per the Indus water treaty of 1960, the waters of eastern flowing rivers of India, like the Ravi, the Beas and the Satluj, have been fully allocated to the unrestricted use by India. The average flow of water from eastern rivers was estimated as 43,153 MCM. Among the states in India, water allocation was made in 1981 as given in Table 5.

Kutch is an extremely water scarce arid region. The Government of Gujarat has examined the possibility of bringing the Indus water to Kutchh by the Indira Gandhi Canal by extending it from Rajasthan State.

22.5.7. Tungabhadra Project

Agreements signed among the Governments of Madras, Hyderabad and Mysore between 1944 and 1946 facilitated construction of the Tungabhadra project. This project is essentially a dam across the Tungabhadra River from which the left

Table 5. Allocation of water of Indus basin among the Indian states

State	Water share (MCM)	Water share (MAF)
Punjab	5,203.00	4.22
Haryana	4,315.28	3.5
Rajasthan	10,603.26	8.6
Delhi	246.6	0.2
Jammu & Kashmir	801.4	0.65
Total	21,169.5	17.17

and the right bank canal systems originate and carry water to far away places. Construction of the project commenced in 1945 and it was completed in phases from 1953 onwards. This scheme is an important inter basin scheme transferring the Tungabhadra water to drought prone areas of the country.

22.5.8. Cauvery Mettur Project

The Cauvery Mettur Project was an important work completed during the British rule of India. It largely serves the delta of the Cauvery River but also irrigates areas outside the basin. This scheme was first contemplated by Arthur Cotton in 1834. Col. Ellis reformulated this scheme in 1910 to secure adequate supplies for the delta irrigation for double and single crops, as also for extension of benefits to new areas. The project continued to serve well till the 1970s. Thereafter, some problems came up because of the differences among the basin states. Presently, the matter is under consideration of the Cauvery Water Disputes Tribunal.

22.5.9. Inter Basin Transfer of Water Through Narmada Main Canal

Work on the Sardar Sarovar project in the Narmada Basin is nearing completion and water has begun to flow in the Narmada main canal. This canal will carry the Narmada water to far-flung areas in Saurashtra across many river basins. The canal will feed water en-route to rivers from 13 escapes on crossings. There are plans to fill the rivers and dams in Saurashtra and North Gujarat with a 1,232.9 MCM (1 MAF) flood waters of Narmada River allocated to each region. In the Saurashtra region, filling of 17 rivers and dams through the Saurashtra branch of the Narmada main canal has been proposed in phase I. Plans have also been made for storage of flood waters of Narmada in 25 dams in different river basins in the drought prone region of Saurashtra. For the drought prone North Gujarat Region, plans are to fill the dams, such as Mazam, Meshwo, Dharoi, Hathmati, Guhai, Mukteshwar, Sipu and Dantiwada, by Narmada floodwater through pipelines.

The benefits of Narmada canal have already begun to flow. Large areas in Gujarat that were facing severe water shortage year after year are no longer cursed by drought. Similarly, many rivers which used to remain dry except the monsoon months now have flow for longer periods. Canal water is also being used to recharge aquifers at many places.

22.6. OBJECTIVES OF INTER-BASIN TRANSFERS OF WATER

While planning IBWT schemes, the Government of India has articulated the following broad objectives:

1. Equitable distribution of available water resources within the nation or regions,
2. Increased economic and water use efficiency,
3. Self-sufficiency in basic water related outputs such as food,

4. Providing in-situ livelihood and employment opportunities in various parts of the nation. This is necessary so that the migration of population, seasonal or permanent, short distance or long distance, is avoided through a balanced regional economic development.

Besides economic development, long distance water transfer proposals should also be planned as tools for employment generation, environmental improvement, and poverty alleviation.

22.6.1. Options to Inter-Basin Transfers

Before discussing IBWT any further, it is a good idea to examine other options to overcome water deficit. The possible options that need serious consideration are:

1. To meet growing food demands, evolve new crops which give higher yields while consuming less water,
2. Evolve cropping patterns and crop varieties suitable for local agro-climatic conditions,
3. Ensure efficient use of available water through measures such as
    - Watershed management, water harvesting and micro irrigation,
    - Improved recharge and conjunctive use of surface and ground waters,
    - Water saving in distribution and field application,
    - Reuse and recycling of water,
4. Rationalize prices to reflect the scarcity value of water,
5. Target subsidies in tariffs for water and power keeping in view the paying capacity while avoiding wastage.

All these options have to be tested meticulously both at policy and field levels and need to be simultaneously adopted, if necessary. While the technological improvements in agriculture and water management will continue unabated, they are not likely to solve the massive water shortage faced in many regions in India. In that situation, it is important to consider IBWT through interlinking of rivers.

22.7. NATIONAL PERSPECTIVE PLAN

The then Union Ministry of Irrigation (now the Ministry of Water Resources) came up with the National Perspective Plan (NPP) for Water Resources Development in the year 1980. The national perspective plan has two components: The Himalayan Rivers development and the Peninsular Rivers development. The Peninsular rivers development component comprises 16 link canals and the Himalayan rivers development component comprises 14 link canals. The Himalayan Rivers component envisages construction of reservoirs on the Ganga and Brahmaputra Rivers and their major tributaries as well as canals to transfer surplus flows from the eastern region to the western region. Detailed feasibility studies for this component are in progress. While the second component will be an inter-state venture, the first will involve neighbouring countries too.

A distinctive feature of NPP is that the transfer of water is essentially by gravity and only in small reaches by lifts (not exceeding 120 m). The total length of links in the scheme will be about 11,000 km. On completion, these would provide additional irrigation benefits of 35 M-ha including 10 M-ha by increased use of ground water. This is over and above the ultimate irrigation potential of 140 M-ha from major, medium and minor projects. Further, the scheme will generate 34,000 MW of power, apart from benefits of flood control, navigation, water supply, fisheries, salinity and pollution control, etc. The initial estimates had put the cost of the whole project at 5.6 lakh crore. The links of the scheme are shown in Figure 4. The scheme was prima-facie found to be technically feasible and economically viable (IWRS 1996).

When NPP is implemented completely, it would give additional irrigation benefits of 25 million ha from surface waters and about 10 million ha by higher use of ground water. The additional generation of hydropower will be on the order of 34,000 MW. Besides, the country would also get large benefits from domestic water supply, flood control, drought mitigation, navigation, fisheries, development of infrastructure, control of pollution, and improvement of environment. The link

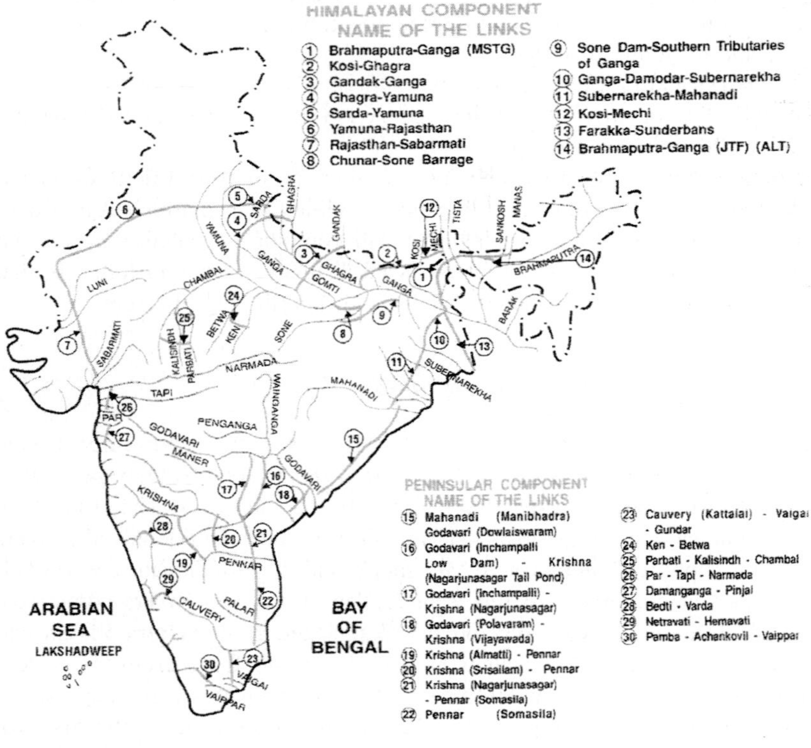

Figure 4. Overview of links of river linking scheme

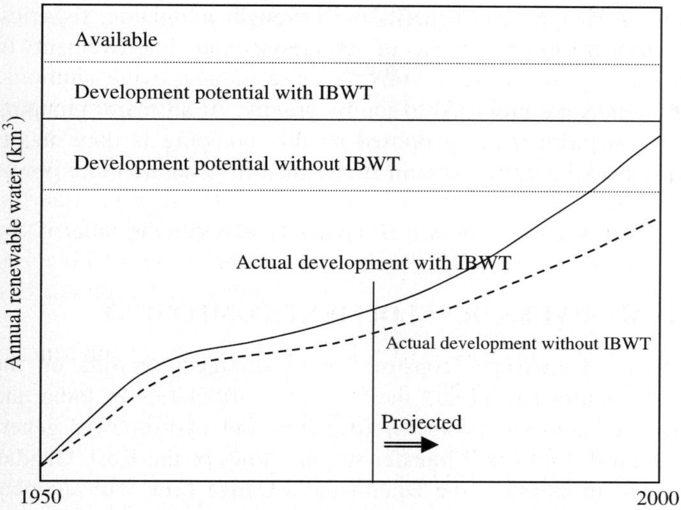

Figure 5. Diagram showing the effect of IBWT on utilization (Curve for the future depict expected scenario) of water resources. Adapted from Thatte (2003)

system is conceived on the basis of "substitution and exchange". Figure 5 is a conceptual representation showing how IBWT will enhance water utilization.

Some state governments have suggested some changes in the alignment and design of the links proposed by NWDA. These are under consideration and some of these changes may be incorporated in the final design.

22.7.1. Highlights of NPP

Among the major highlights of the NPP to inter-link rivers are the following:

- This scheme will permit simultaneous solution of two critical problems of water management in India – floods and droughts. The construction of new projects will, to a large extent, help in attainment of these twin objectives.
- Many cities will get additional water (total about 12 km^3) to meet municipal water supply demands.
- It will provide assured irrigation to additional 35 M-ha of land (25 million ha from surface water and 10 million ha by ground water). This area is about 25% of the ultimate irrigation potential of the country.
- Peak flood discharge at many locations will be reduced by about 30% due to regulation of reservoirs.
- Additional electricity to the tune of 34,000 MW can be generated.
- The rivers and canals can be used to transport goods and traffic at less cost and in an environmentally friendly manner.

- There are possibilities of additional benefits from drought mitigation, fisheries, development of infrastructure, control of pollution, and improvement of environment.

To the extent possible, links are proposed to run by gravity. At all major pumping stations, reversible pump turbines are proposed so that pumping is done in off peak hours and during peak hours the system is operated to generate more power for the grid.

At this stage, a detailed discussion of two components of NPP is in order.

22.8. HIMALAYAN RIVERS DEVELOPMENT COMPONENT

The Himalayan component envisages construction of storage reservoirs on the main Ganga and Brahmaputra Rivers and their principal tributaries in India and Nepal so as to conserve monsoon flows for irrigation and hydro-power generation, besides flood control. Links will transfer surplus flows of the Kosi, Gandak and Ghagra to the west. In addition, the Brahmaputra-Ganga Link will augment dry-weather flow of the Ganga. Surplus flows that will become available on account of inter-linking of the Ganga and the Yamuna are proposed to be transferred to the drought prone areas of Haryana, Rajasthan and Gujarat. With this proposal about 14 Mha-m of additional water would be available from these river systems for irrigating an estimated 22 M-ha in the Ganga-Brahmaputra basin apart from Haryana, Punjab, Rajasthan and Gujarat. It would also provide 1,120 cumec to Calcutta Port and would provide navigation facility across the country. It will also provide flood moderation in the Ganga-Brahmaputra system. The Himalayan component will benefit not only India but also Nepal and Bangladesh.

Fourteen links are proposed in the Himalayan component. These are shown in Figure 6 and are listed in Table 6.

In view of the ongoing dispute on the sharing of the Ganga water with Bangladesh, little details of this component are available. In broader terms, storages and links of the Himalayan component are of mammoth size. Due to size, topography and other reasons, construction and environmental problems might be enormous. Further, there appears to be some anomalies in the planning of this component. Since no additional storages are proposed on the Ganga and the Yamuna, their monsoon flows will continue to go to the Bay of Bengal while huge funds are to be spent to transfer water of Kosi, Ghagra, Gandak and Sarda to the West. The Satluj Yamuna link (which has not been made operational due to inter-state dispute) would transfer water from west to east, the proposed Sarda-Yamuna link towards east will flow in the opposite direction. Similarly, the Narmada canal transfers Narmada waters across Sabarmati towards North-West, the proposed Rajasthan-Sabarmati link will flow in the opposite direction towards South-east.

Highlights of some important links of the Himalayan Component are discussed in the following.

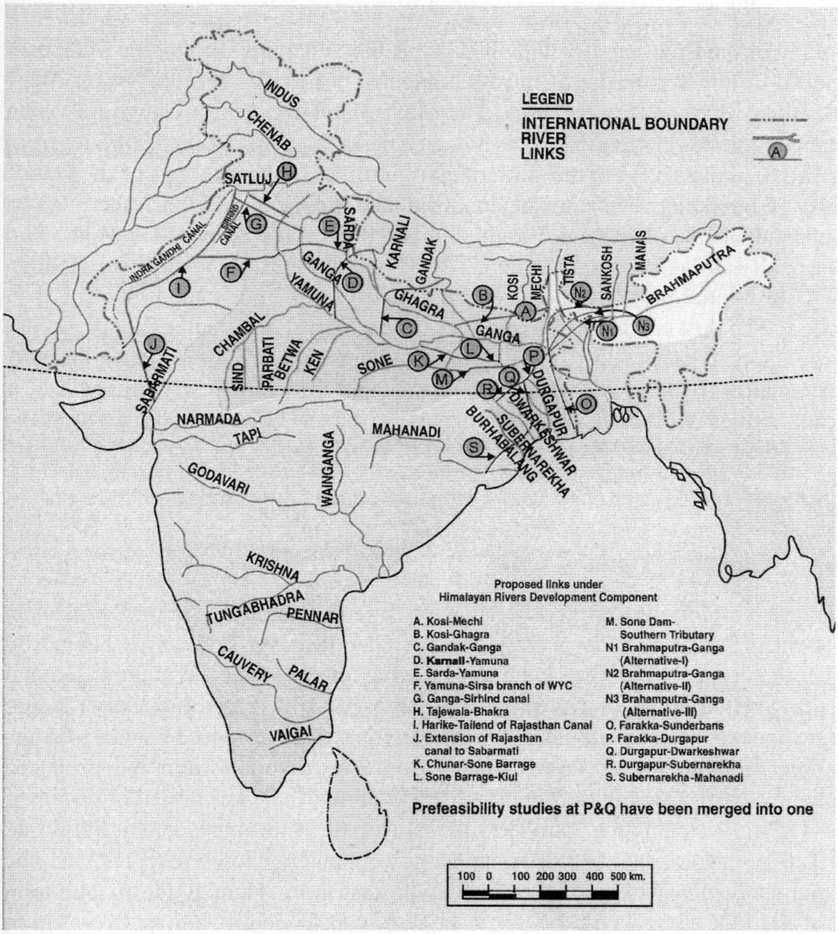

Figure 6. Himalayan component of the ILR scheme

Table 6. Proposed Fourteen Links in the Himalayan Component

1	Kosi–Mechi	2	Kosi–Ghagra
3	Gandak–Ganga	4	Ghagra–Yamuna
5	Sarda–Yamuna	6	Yamuna–Rajasthan
7	Rajasthan–Sabarmati	8	Chunar–Sone Barrage
9	Sone Dam–South Tributaries of Ganga	10	Brahmaputra–Ganga (MSTG)
11	Brahmaputra–Ganga (JTF)(ALT)	12	Farakka–Sunderbans
13	Ganga–Damodar-Subernarekha	14	Subernarekha–Mahanadi

22.8.1. Manas–Sankosh–Tista–Ganga (MSTG) Link

Interlinking of the Brahmaputra with the Ganga, the Subernarekha and the Mahanadi is proposed to transfer waters of the Brahmaputra to benefit areas in Assam, West Bengal, Bihar, Jharkhand and Orissa. The Manas-Sankosh-Teesta-Ganga link is an important link in this component. This link envisages diversion of surplus water from Manas and Sankosh rivers in the Brahmaputra basin to augment flows of the Ganga upstream of Farakka. A link to the Peninsular component through Subernarekha and Mahanadi is also envisaged. For this link high dams are proposed at Manas and Sankosh with storage capacities of 8.75 BCM and 4.93 BCM, respectively (Singh 2002). A substantial part of the cost of these dams will be allocated to hydropower generation. The 114 km long link canal between Manas and Sankosh will have a discharge capacity of 3,725 m^3/s. Beyond Sankosh and up to the Teesta barrage, the link canal is 137 km long with a capacity of 1,092 m^3/s. Clearly, this will be a huge canal which will cross major drainages. The MSTG link passes through the narrow chicken neck in West Bengal (north of Bangladesh) and may have security aspects.

22.8.2. Ghaghra–Yamuna Link

The Ghagra-Yamuna link project is an inter-dependent link under the Himalayan Component of NPP. A study reveals that the Ghagra River (known as Karnali in Nepal) at the proposed the Chisapani dam site has surplus water. It is proposed that the existing requirement of water for the Sarda Sahayak Pariyojna, Saryu Nahar Pariyojna and various pump canals would be met from the proposed Gandak–Ganga link project and the water saved thereby could be diverted from the proposed Chisapani reservoir through the Ghagra–Yamuna link canal. The height of proposed dam is 175 m. A regulating dam downstream of the Chisapani dam is proposed with a full reservoir level of 200 m and a minimum drawdown level 193 m. The link canal shall join Yamuna River in Etawah district of Uttar Pradesh. The total length of the link canal would be about 417 km with its depth varying from 8 m in the head reach to 5 m in the tail reach and the width varying from 85.5 m in the head reach to 18 m towards the tail end.

The status of existing and additional irrigation facilities proposed under the Ghagra–Yamuna link project is given in Table 7.

The link canal project will have a direct positive impact on the farmers of central region of U.P. because of additional production and job opportunities would improve their socio-economic conditions.

22.8.3. Sarda–Yamuna–Rajasthan–Sabarmati Link Canal

This is a continuous link having a combination of three links, viz., the Sarda-Yamuna link, the Yamuna-Rajasthan link, and the Rajasthan-Sabarmati link. This link canal is planned to divert 17,906 MCM (14.52 MAF) water of Himalayan rivers.

Table 7. Status of proposed existing and additional irrigation facilities under the Ghagra – Yamuna link project

S.N.	Name of en-route command	CCA en-route of the link (ha)	Annual irrigation in the existing command taken over by the link canal (ha)	Additional irrigation in new areas & intensification in existing areas from the link canal (ha)	Proposed intensity of irrigation (%)
1	Un-irrigated areas (western side including delta area)	1, 05, 630	––	2, 53, 512	240
2	Part of Sarda Canal Command	12,20, 675	6, 04, 234	6, 16, 441	100
3	Part of Lower Ganga Canal including PLGC	9,16,236	7, 55, 895	1, 60, 341	100
4	Un-irrigated areas of Uttar Pradesh	3,93,755	––	3, 93, 755	100
	Total	26, 36, 296	13, 60, 129	14, 24, 049	

Its length will be 1,835 km out of which 75 km will be in Gujarat State. A total of 4 states, Uttar Pradesh, Haryana, Rajasthan and Gujarat, are to be benefited by this link. About 1,627 MCM (1.32 MAF) water has been allocated to North Gujarat which is only 9% of the total divertible water at the canal head. A total 7.38 lakh ha area is to be irrigated by the Rajasthan-Sabarmati link, out of which 5.35 lakh ha in Rajasthan and 2.03 lakh ha in Gujarat.

22.8.4. Yamuna–Rajasthan Link Canal Project

The Yamuna-Rajasthan link proposal is an extension of the proposed Sarda–Yamuna Link beyond the Yamuna to provide irrigation to the drought prone areas of Haryana and Rajasthan. It envisages diversion of 8, 657 Mm^3 of water from the Sarda basin at Purnagiri. The Yamuna–Rajasthan link is to take off from the right bank of proposed Yamuna barrage and passes through the Karnal, Sonipat, Jind, Hisar and Bhiwani districts of Haryana and Churu, Hanumangarh, Ganganagar, Bikaner, Jodhpur and Jaisalmer districts of Rajasthan and ends on the Jaisalmer-Hamira-Shri Mohangarh Road at a distance of 4.5 km from village Kanod towards Jaisalmer. The length of the link canal is 786 km, out of which 196 km lies in Haryana and the rest 590 km in Rajasthan. The design discharge at head and tail are 572 cumec and 344 cumec, respectively. The longitudinal slope of the canal is 1:20,000. The full supply depth and bed width of the canal at head are 7 m and 53 m, respectively. The Yamuna–Rajasthan link will provide an annual irrigation of 244,200 ha in the districts of Ganganagar, Bikaner, Jodhpur and Jaisalmer of Rajasthan.

22.8.5. Rajasthan–Sabarmati Link Project

The Rajasthan-Sabarmati link canal is an extension of the proposed Yamuna–Rajasthan Link. The link envisages a transfer of 5,924 Mm^3 water available at the tail end of the Yamuna-Rajasthan link for drought prone areas of Rajasthan and Gujarat. The length of the canal is about 725 km out of which 650 km lies in Rajasthan and the rest 75 km in Gujarat. The design discharge at the head and the tail are 344 cumec and 60 cumec, respectively. The full supply depth and bed width of the canal at its head are 6 m and 39 m, respectively. The link canal on its way will cross the Luni River & its tributaries and the Banas River.

The Rajasthan-Sabarmati link will provide an annual irrigation of 535,000 ha in the districts of Jaisalmer, Barmer and Jalor of Rajasthan. The total annual irrigation thus envisaged in Rajasthan State through the above two interbasin water transfer links works out to be 779,200 ha.

Further, interlinking the Gandak, the Ghagra, the Sarda and the Yamuna, all tributaries of the Ganga, on to Rajasthan and the Sabarmati aims at transferring the waters of Gandak and Ghagra Rivers to benefit areas in Uttar Pradesh, Uttaranchal, Haryana, Rajasthan, Gujarat, Bihar and Jharkhand. Other important links proposed in the Himalayan component are the Kosi-Ghagra, Gandak–Ganga, Ghagra–Yamuna and Sarda–Yamuna links to supplement the supplies of the Ganga and the Yamuna and for further transfer of water towards the west to Rajasthan and Gujarat. A large canal parallel and to the east of the existing Rajasthan canal is proposed which will be extended beyond the tail of the present Rajasthan canal and be linked to the Sabarmati.

22.9. PENINSULAR RIVERS DEVELOPMENT COMPONENT

The main component of Peninsular Rivers Development is the "Southern Water Grid" which is envisaged to link Mahanadi, Godavari, Krishna, Pennar, and Cauvery rivers. The peninsular scheme was envisaged to provide additional irrigation benefits of over 13 million ha. The Peninsular component comprises the following four parts:

1. Diversion of surplus flows of Mahanadi and Godavari to Krishna, Pennar, Cauvery and Vaigai.
2. Diversion of west-flowing rivers of Kerala and Karnataka to the east.
3. Inter-linking small rivers flowing along the west coast, north of Mumbai and south of Tapi.
4. Inter-linking the southern tributaries of Yamuna.

The peninsular component of ILR has 13 major water storage/diversion structures situated in four basins. Three non-storage structures, viz., Dowlaiswaram barrage, Prakasam barrage, and Grand Anicut and storage node (Narayanpur) cater to only irrigation, while six storage nodes, viz., Inchampalli, Almatti, Nagarjunasagar, Pulichintala, Krishnarajasagar, and Mettur will serve both irrigation and power needs. One storage node, viz., Somasila is operated to meet domestic and irrigation needs and two storage nodes, viz., Polavaram and Srisailam are multi-purpose projects serving domestic, irrigation, and hydropower demands.

Among these, the interlinking of Mahanadi, Godavari–Krishna–Cauvery rivers will require the construction of a number of large dams and big canals. This system will be one of the largest and ambitious water transfer projects. The system will require huge financial outlays and will have immense influence on economic, social and environmental growth of the region. Logically, therefore, it would be necessary to closely examine the various components and arrive at the best solution. It is pertinent to note that water need not be transferred from a surplus basin, just because it is available. Before adopting such transfers, it would be necessary that all the resources of the recipient basin are put to the optimum use.

Sixteen links are proposed in the Peninsular Component. These are shown in Figure 7 and are listed in Table 8. Salient features of the links are given in Table 9.

In the following, highlights of some important links of the Peninsular Component are described.

22.9.1. Mahanadi (Manibhadra)–Godavari (Dowlaiswaram) Link

This link has been proposed between the Manibhadra reservoir on Mahanadi River to the Dowlaiswaram barrage on the Godavari. It will divert 11,176 Mm^3 of water out of which 3,854 Mm^3 is proposed to be used for irrigation of en-route command area and 6,500 Mm^3 would be delivered at the Dowlaiswaram barrage. The Manibhadra reservoir has gross and live storages of 9,375 Mm^3 and 6,000 Mm^3, respectively. The total length of the link canal is about 932 km. The design discharge of the link canal is 627 cumec as its head. The full supply levels at the head and tail are 74.00 m and 13.81 m, respectively.

22.9.2. Godavari (Inchampalli)–Krishna (Nagarjunsagar) Link

This link canal is proposed to divert 16,426 Mm^3 from the Inchampalli dam on Godavari River. Out of this, 14,200 Mm^3 will be transferred to the Nagarjunsagar reservoir on the Krishna River. The total length of the link canal will be about 298.7 km, including a 9 km long tunnel. The FSL at the head and tail are 142.00 m and 182.765, respectively, with a design discharge of 1,219 cumec. The link would involve a total lift of 116 m in four stages. For this purpose, power needed would be 1,705 MW.

22.9.3. Inchampalli–Pulichintala Link

This link has been proposed to divert 3,901 Mm^3 of surplus water from the Godavari and 470 Mm^3 of Inchampalli Right Bank Canal. Each year, the link would provide 1,382 Mm^3 of water in the existing Nagarjunasagar Left Bank Canal command, 746 Mm^3 in the proposed new area by extension of the Nagarjunasagar Left Bank

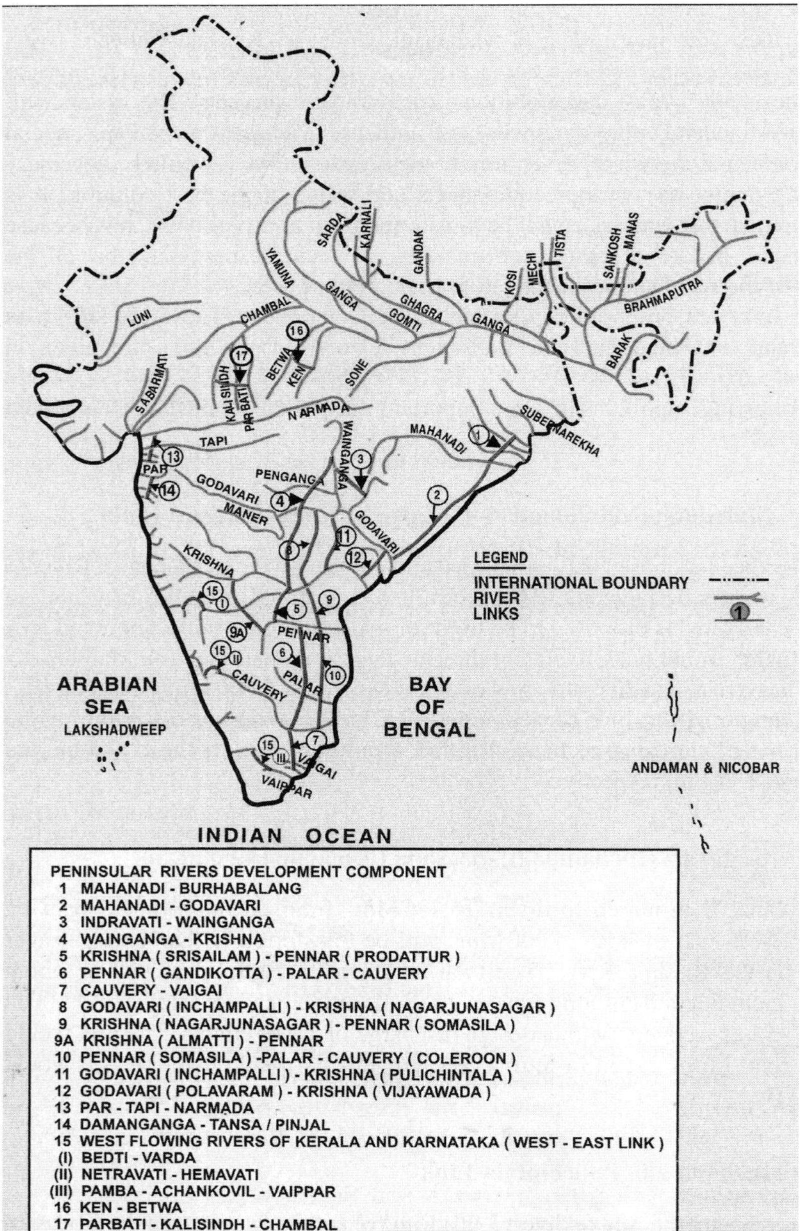

Figure 7. Peninsular component of the ILR scheme

Table 8. Proposed links in the Peninsular Component

1	Mahanadi(Manibhadra)–Godavari (d/s)	2	Godavari (Inchampalli)–Krishna (Nagarjunsagar)
3	Godavari (Inchampalli Low Dam)–Krishna (Nagarjunsagar Tail Pond)	4	Godavari (Polavaram)–Krishna (Vijaywada)
5	Krishna (Almatti)–Pennar	6	Krishna (Srisailam)–Pennar
7	Krishna (Nagarjunsagar)–Pennar (Somasila)	8	Pennar (Somasila)–Cauvery (Grand Anicut)
9	Cauvery (Kattalai)–Vaigai–Gundar	10	Ken–Betwa
11	Parbati–Kalisindh–Chambal	12	Par–Tapi–Narmada
13	Damanganga–Pinjal	14	Bedti–Varda
15	Netravati-Hemavati	16	Pamba–Achankovil-Vaippar

Canal command, 1,623 Mm^3 in the existing Nagarjunasagar Right Bank Canal command through the proposed Pulichintala Right Bank Canal and 470 Mm^3 in the command of the Inchampalli Right Bank Canal. In addition to the dam at the Inchampalli, a dam at Pulichintala on the Krishna River has been proposed. The total length of the lift channel will be 270 km including a 25.5 km long tunnel. The FSL at the head and tail will be 106.68 m and 69.68 m, respectively. The link is proposed to be operated for only 240 days in a year with a head discharge of 263 m^3/sec.

22.9.4. Godavari (Polavaram)–Krishna (Vijaywada) Link

This link canal has been proposed to divert 4,903 Mm^3 which include 1, 448 Mm^3 for Polavaram RBC command, 2,265 Mm^3 for the Krishna delta as committed under the Godavari Water Dispute Tribunal award and 1,190 Mm^3 for existing ayacut in the Krishna Delta. The proposed Polavaram Barrage will be used to divert the Godavari water to the existing Prakasam Barrage of the Krishna River at Vijayawada. The total length of the link canal will be 174 km and head discharge will be 361 cumec. The canal will operate round the year. The FSL at the head and tail are 40.23 m and 27.96 m, respectively.

22.9.5. Krishna (Srisailam)–Pennar Link

The link has been proposed to divert 2,310 Mm^3 of water from the Srisailam reservoir to Adinimmayapalli Anicut. The water would mostly flow through natural rivers and it is expected that about 2,095 M m^3 would reach the Somasila reservoir. This water is in exchange for surplus waters of the Mahanadi transferred from the Godavari to the Nagarjunasagar. The total length of the channel would be 171.30 km and design discharge will be 186 cumec. This channel would run for 180 days in a year.

Table 9. Salient features of the Peninsular Components

S. N.	Name of the Link	FSL at Head/ (m) Tail (m)	Length km	Discharge m^3/sec	Annual volume/ Transfer Mm^3	En-route use		Losses Mm^3
						Irrigation Vol. Mm^3/Area ha.	M & I Mm^3	
1	Manibhadra (Mahanadi) to Dowlaiswaram (Godavari)	74/13.81	932	627	11,176/ 6,500	3,854/ 4,54,229	802	822
2	Inchampalli (Godavari) to Nagarjunasagar (Krishna)	142/ 182.77	299	1,219	16,426/ 14,200	1,850/ 3,19,708	237	376
3	Inchampalli (Godavari) to Pulichintala (Krishna)	106.68/ 69.68	270	263	4,371	4,221/ 6,94,882	–	150
4	Polavaram to (Godavari) Vijaywada (Krishna)	40.232/ 27.965	174	361	4,903/ 3,305	1,448/ 1,48,418	162	150
5	Almatti (Krishna) to Pennar (Pennar)	510/ 434.4	564	208.12	1,980	1,778/ 2,34,589	56	202
6	Srisailam (Krishna) to Pennar (Pennar)	268.15/ 156.51	171.3	186	2,310/ 2,095	–	–	215
7	Nagarjunasagar (Krishna)- Somasila (Pennar)	151.67/ 102.63	394	555	12,146/ 8,648	3,166/ 5,60,606	124	332
8	Somasila (Pennar) to Grand Anicut (Cauvery)	91.96/ 59.7	538	616.4	8,565/ 3,855	3,170/ 4,91,200	1,105	385
9	Kattalai Regulator (Cauvery) to Vaigai (Vaigai)	100.75/ 78.865	250	174.14	2,252	2,007/ 3,53,337	185	136

Note : (i) In column 9, the upper figure indicates the gross diversion while the lower gives the quantity reaching the receipt river. The difference is accounted for the en-route irrigation and losses. (ii) In column 10, the upper figures are volume used en-route and the lower figures are area irrigated en-route. Source: NCIWRD (1999), NWDA, and others.

22.9.6. Krishna (Nagarjunasagar)–Pennar (Somasila) Link

This proposed link would divert 12,146 Mm^3 of water from the Nagarjunasagar reservoir to Pennar River at Somasila. Out of this quantity, 2,356 Mm^3 will be utilized to irrigate part of the command of the Nagarjunasagar RBC, about 810 Mm^3

will be used for en-route irrigation and 8,648 Mm^3 will be transferred to the Somasila reservoir. It is important to note that in most interlinking canals, provision has been made for en-route irrigation. Without this, farmers in the en-route area are likely to oppose water transfer and this might create many problems. The total length of the canal is 394 km and its design discharge is 555 cumec. The canal will be operated for 240 days in a year.

22.9.7. Pennar (Somasila)–Cauvery (Grand Anicut) Link

The aim of this link is to transfer 8,565 Mm^3 of water from the Pennar to the Cauvery. Of this quantity, 3,170 Mm^3 would be used for en-route irrigation, 279 Mm^3 for en-route domestic and industrial uses, 876 Mm^3 for the Chennai city water supply and 3,855 Mm^3 would be transferred to the Cauvery River at Grand Anicut. About 385 Mm^3 water is likely to be lost during transmission. The total length of the canal will be 538 km and its design discharge will be 616.38 cumec. The canal will be operated for 365 days in a year.

22.9.8. Cauvery (Kattalai Regulator)–Vaigai-Gundar Link

The link has been proposed to transfer 2, 252 Mm^3 of water from the Cauvery River to the Vaigai River to provide irrigation to 353,337 ha annually. The FSLs of the 250 km long link canal at the head and the tail will be 100.75 m and 78.865 m, respectively. This will be a lined canal which would be operated round the year.

22.9.9. Krishna (Almatti)–Pennar Link

The canal linking Krishna (Almatti) with Pennar (587 km long) will take off from right bank of the Almatti dam across the Krishna River in Karnataka with FSL of 510.00 m. The canal will run through Karnataka and Andhra Pradesh before joining Maddileru, a tributary of the Pennar near the Malakavemula village. A balancing reservoir is also proposed at Kalavapalli in Anantapur district. The canal will also supplement the Bukkapatnam tank across Chitravathi River. The link canal will carry about 1,980 Mm^3 water during the Kharif season and irrigate about 70,000 ha in Karnataka and 190,000 ha in the Anantapur district of AP. Allocation of 56 Mm^3 has also been made for domestic and industrial uses. There is a possibility of additional ground water recharge around the Kalavapalli reservoir and the Bukkapatnam tank.

22.9.10. Ken–Betwa Link

The Ken-Betwa and the Parbati–Kalisindh–Chambal links of the ILR project are the links on which urgent attention is being focused by the Government. The feasibility report of the Ken–Betwa link is available in the public domain at *www.riverlinks.nic.in*.

The Ken-Betwa link envisages diversion of surplus waters of Ken basin to water deficit Betwa basin. This link canal will provide irrigation to water short areas of upper Betwa basin of MP and also to en-route areas of MP & UP. It is proposed to transfer 1,020 Mm^3 of water from Ken basin to provide irrigation in Madhya Pradesh and Uttar Pradesh. Apart from drinking water facility and en-route irrigation of 47,000 ha in Chhatarpur & Tikamgarh districts of Madhya Pradesh and Hamirpur & Jhansi districts of UP, provision for downstream commitments of 1,375 Mm^3 for MP and 850 Mm^3 of water for UP has also been kept.

A dam is proposed on Ken River at Daudhan, 2.5 km upstream of existing Gangau weir. The 75% dependable yield of Ken at Daudhan site has been assessed as 6,188 Mm^3. The net water availability at dam site after accounting all the upstream requirements is 3,291 Mm^3. The downstream commitments from Ken at Daudhan are 2,225 Mm^3. Out of which, 850 Mm^3 is provided to UP and 1,375 Mm^3 to MP as per Interstate agreement (1981) on Ken River. The surplus water for diversion at Daudhan is 1, 020 Mm^3. Out of which, 659 Mm^3 will be transferred to Betwa River upstream of Parichha weir and 312 Mm^3 will be utilized in the en-route command.

The dam proposed at Daudhan is an earthen dam with two power houses (installed capacities of 3×20 MW and 2×6 MW). One Power House will be a pumped storage scheme. The design discharge of the link canal at its head is 72 cumec. The link canal after traversing about 230 km will outfall in existing Barwa Sagar reservoir from where the diverted water will join Betwa river through a natural stream in the upstream of Parichha weir. An area of 1.27 lakh ha in the Raisen and Vidisha districts of Madhya Pradesh will be benefited by water from this link. This link will also provide annual irrigation to 47,000 ha area en-route in the drought prone Chhatarpur and Tikamgarh districts of MP and Hamirpur and Jhansi districts of UP. The link will also provide 11.75 Mm^3 water for domestic uses in the en-route villages of Chhatarpur and Tikamgarh districts of MP and Hamirpur and Jhansi districts of UP.

The total cost of the link project has been estimated as Rs. 1,989 crore (1994–95 price level). The annual benefits accrued from the project are estimated to Rs. 450 crore (1994–95 price level). The Benefit-Cost ratio for the Ken-Betwa link project as a whole comes to 1.87. The internal rate of return of the project works out to 13.00%.

The Consensus Group of the MoWR has held a series of meeting to resolve the differences among the part states. Finally, a MOU was signed between UP and MP in August 2005. This has paved the way for preparation of the Detailed Project Report (DPR) for the link.

22.9.11. Par–Tapi–Narmada Link Canal

This link canal was proposed to divert surplus water of the rivers, like Par, Auranga, Ambica, Purna and Tapi, up to Vadodara branch of Narmada Command. About 1,350 MCM surplus water is proposed to be diverted by the Par-Tapi link canal up to Ukai Dam and 2,904 MCM surplus water is proposed to be diverted by the Tapi-Narmada link canal (including 1,554 MCM surplus water of Tapi at Ukai). The total length of the Par–Tapi–Narmada link canal is 402 km–the length of the Par-Tapi link will be 177 km and the Tapi–Narmada link will be 225 km. Seven reservoirs are proposed in the upstream catchment area of 2,573 sq. km. The link canal passes through dense forest and hilly region.

22.9.12. Damanganga–Pinjal Link Canal

The proposed Damanganga-Pinjal Link Project envisages the construction of reservoirs at Bhugad and Khargihill. The gross storage of these two reservoirs will be 426.39 & 460.79 million cubic meters (MCM) and live storage will be 400 & 420.56 MCM, respectively. The FRL will be 163.87 m and 154.52 m, respectively. The reservoirs will be connected by 16.85 km long pressure tunnel of 5.00 m diameter. Another 25.70 km long and 5.25 m diameter tunnel will connect Khargihill and Pinjal reservoirs. The surplus water from Bhugad and Khargihill reservoirs will be transferred through pressure tunnels to Pinjal reservoir for onward transmission to Greater Mumbai. This link canal is proposed to supply 909 MCM water annually to Mumbai City to improve the existing inadequate availability of domestic and industrial water.

The project lies partly in the Valsad district of Gujarat and partly in Nasik and Thane districts of Maharashtra. The Bhugad dam site on Damanganga River will intercept 141 km^2 catchment area of Gujarat State. It will be located near Bhugad village in the Nasik district and Modushi village in Valsad district. The Khargihill dam will be constructed on the Vagh River near Behapada village in Thane. The Bhugad-Khargihill and Khargihill-Pinjal tunnels lie entirely in Maharashtra.

22.9.13. Pamba–Achankovil–Vaippar Link Project

The proposed Pamba-Achankovil-Vaippar Link project has three storage reservoirs, two tunnels, necessary canal system and a few power generating units. The Punnamedu reservoir (reservoir-2) is located on river Pamba Kal Ar in Pamba basin in Kerala state, which serves a part/full of its downstream mandatory requirements and supplies surplus water to reservoir-1 through tunnel-2. The Achankovil Kal Ar reservoir (reservoir-1) located on the Achankovil Kal Ar River in the Achankovil river basin of Kerala state, supplies water for irrigation purposes to the state of Tamil Nadu, through tunnel-1 to the main canal. The water from the main canal is then distributed to the command area of Vaippar basin in Tamil Nadu state. Besides this, reservoir-1 releases 10 MCM of water daily during six hours of peak load period for power generation. The Achankovil reservoir (reservoir-3), which is located on Achankovil River in the Achankovil river basin of Kerala state, besides acting as a pumped storage scheme accommodating the water drawn from the upstream reservoir-1, also serves the purpose of releasing water downstream to meet its downstream mandatory demands. The 10 MCM of water drawn to the downstream reservoir-3 from reservoir-1 for power generation is pumped back to reservoir-1 in a 16 hours period. Also, if there is deficit at reservoir-1, the surplus water of reservoir-3 can be pumped back to reservoir-l.

22.10. PRESENT STATUS OF ILR

While planning activities for these schemes was proceeding at their own pace, a public interest litigation (PIL) was filed in the Supreme Court of India against inordinate delays in interlinking of rivers. After hearing the arguments, the Supreme Court directed the Government of India to initiate interlinking of rivers in a period of 10 years in a time bound framework.

This decision has resulted in a flurry of activities in this important sector. A Task Force (TF) was constituted by the Government of India to work out the modalities for implementing the scheme as per the time frame that indicates completion of all the feasibility reports by the year 2005 and the detailed project reports by 2006 and implementation by 2016. The TF had formulated action plans to achieve the targets. Terms of reference for preparation of detailed project reports (DPR) have been finalized and DPRs for a few schemes are being initiated. The task force was wound up in 2004. Subsequently, a special cell has been constituted under MoWR to take up further follow-up action on the ILR program. Currently, preparation of detailed project reports is in progress. The consensus building process among the states for taking up the preparation of DPR is in progress. After an agreement was signed between UP and MP for preparation of DPR for the Ken-Betwa link, the work has commenced. MoWR has also constituted a committee with officers from the Central Government, representatives of concerned State Governments, environmentalists, social scientists and other experts to make ILR a mutually consultative process. Besides this, dialogue with the States at the highest political level is also needed to expedite the process of consensus.

ILR projects would require undertaking a mammoth construction program. Nearly 12,500 km of canals would be constructed to transfer/distribute water from water-rich zones to water-short areas. These links would move about 173 billion m^3 of water, equivalent to one-fourth of the water in the Brahmaputra from the water surplus river basins to the deficient ones. Besides dams and canals, navigation locks, bridges and other structures like barrages for diversion, tunnels, would form part of the envisaged program. Some interesting relevant details are given in Table 10.

Steps in implementation

The steps involved in the implementation of link schemes after completion of the feasibility reports are:

- Negotiations and agreements amongst the concerned States regarding sharing of surplus water and other project benefits/outcomes,
- Preparation of Detailed Project Reports (DPRs),
- Technical, environmental, economic approvals and investment clearance by the Planning Commission,

Table 10. Some broad details of works involved in ILR

Item	Size / quantity based on rough calculations
Number of dams to be constructed	32
Length of the link canals (lined)	9,629 km
Length of distribution network up to minors (lined)	12,468 km
Size of link canals	Width varying from 3 m to 155 m and depth varying from 1.5 m to 10 m
Lining involved	737 million square meters
Number of canal structures	4,291
Drinking water supply	101 districts and Greater Mumbai, NCR of Delhi, Chennai, Kanpur and Lucknow cities
Employment generation	58 lakh man years

- Financing agreements,
- Construction, and
- Operation.

At present, most links are at the first and second stage.

22.10.1. Financing of the Project

As indicated earlier the total cost of the project has been roughly estimated to be of the order of Rs. 560,000 crore (Rs. 5,600 billion, 1 crore $= 1^*10^7$) or US $ 114 billion. This includes the cost of the power generation component. It may be noted that this is an approximate estimation. When DPRs are ready and actual costs of different projects are worked out, the final figure may be different than these numbers. Also, this cost does not include yearly inflation, costs of ecology, environment, wildlife, resettlement and rehabilitation of displaced people. It is pertinent to note that the GNP of India for 1998–99 was Rs. 1,110,000 crore. The estimated expenditure is more than twice the entire irrigation budget of India since 1950. It is the feeling of the opponents of the project that this huge expenditure is bound to push the country into more debt and might lead to privatization of water resources, denying basic water access to the poor.

Financing of the project would require careful consideration. Much of the finance would have to come from the public finance or from international funding agencies through soft loans. Special instruments may also have to be launched for this purpose. It may be possible to attract private investment on build-own-operate-transfer (BOOT) basis over a long period. The cost recovery also would pose serious challenges.

The sizeable investment required for implementation of the project can be classified into cost of developing the infrastructure, social, environmental and other costs. The ILR program can be developed through pure public investment and/or public-private partnerships including stakeholder involvement. The development of the irrigation infrastructure would require government support. There is a distinct possibility of generating large quantum of investible funds in the system. The savings level in India has been increasing over the last 30 years from 14.6% of GDP in 1970–71 to 24.0% of GDP in 2001–02. If the national savings level grows at this rate, it is expected to reach the level of other Asian countries such as Thailand (31%) in the next 12 years. This increase in savings can generate additional funds required.

A detailed mechanism to channel available funds into this program is a challenge. As far as hydropower is concerned, there is encouraging scope for implementation through public-private partnerships or Independent Power Producer agreements. Based on the independent revenue generation potential, these projects can be structured appropriately to attract financing by the private sector. The public-private partnership could be in various forms including traditional public contracts, build-operate-transfer (BOT) contracts, concessions or joint ventures. Considering the increases in aggregate assets of all the banks and financial institutions, the yearly investment requirement for these hydropower projects may be feasible subject to projects being structured in an appropriate manner. The detailed approach to structuring and funding such projects would also need to be evolved.

22.10.2. Challenges and Prospects

Some interlinking proposals are gigantic in nature. Some of the ideas mooted for the long-distance transfer of water may be techno-economically feasible but may require investment funds of a magnitude which cannot easily be found. They need to be examined from the point of view of environmental impact or social costs. Any proposal for large-scale inter-basin transfers needs to be approached with caution. Apart from the cost aspect, the implementation of the scheme is a very challenging task involving a large number of issues, both technical and others.

It is certain that in the ultimate scenario of development, some basins will be surplus in water while several other basins will face an acute shortage of water. Intra-state river transfers in India fall entirely under the purview of the concerned state but the inter-state water transfers fall under the purview of the central government. To begin with, a good strategy would be to concentrate on two or three ILR projects that are most promising candidates and where there is absolutely no doubt about the need and feasibility of IBWT. These projects could be taken up as a demonstration to showcase the techniques of analysis. A promising case is the Ken-Betwa link in central India.

A broad-based national perspective is essential for optimum development. In the following, various issues associated with ILR are briefly discussed.

22.10.3. Change in the Basin Level Planning Concept

It is widely accepted that a river basin (or a sub-basin in the case of very large river basins) should be considered a unit for planning and management of water resources in the framework of integrated water resources management (IWRM).

Here integration should be across many dimensions: all the sources of water (rainwater, surface water, and groundwater); all over the catchment from head waters down to the coastal area; use of water in various sectors (domestic use, irrigation, industrial use, maintenance of quality and ecosystems, equity in distribution of water between head-end users and tail-enders in the basin, etc.).

The IWRM concept is generally followed in India. Inter-State Water Dispute Tribunals adjudicate the disputes and allocate waters of a river basin among the co-basin states by following IWRM strategy.. The water dispute tribunal awards do not discuss about IBWT unless agreed upon by the co-basin states for any specific proposal. The ILR proposals call for IBWT from river basins like the Godavari and the Krishna for which water dispute tribunals have already given their awards. The ILR proposals go beyond the tribunal awards and adjustments to the awards are necessary to take into account the changed scenario of course, have to be made with the cooperation and involvement these if co-basin states. In fact, long-term planning by individual states has to be reoriented to take into consideration the inter-basin transfer schemes. While examining the economic soundness of interbasin water transfer, the criteria and methodology of economic analysis as applicable to the other water resources projects should normally be applicable.

The National Water Policy supports the use of non-conventional methods, such as IBWT, after meeting the legitimate requirements of the donor basin. The question of diversion of waters from a basin outside the basin boundary was examined in detail by the Krishna water dispute tribunal. The tribunal had decided that the diversion of water outside the river basin by a state is legally valid and the need for diversion outside the basin is relevant to equitable allocation. They however suggested that more weight should be given to the inter-basin use. Note that the supply of Narmada waters to the state of Rajasthan (which is not a basin state) is as per the agreement signed among the party states.

Most of the links of NPP work on the innovative principle of substitution and exchange. Substitution envisages that the surplus water is delivered at the downstream use points in the basins facing water deficit, substituting for the existing committed releases from the upstream locations. In exchange, water will be drawn from the upper reservoir to cover the needy upland areas in other basins, wherever feasible. Thus the command areas being served by existing projects are taken over by the inter-basin water transfer links and the water thus saved is tapped as exchange from a higher level in the headwater zone of the river. This way the upland areas of the basin can be benefited without resorting to high pumping.

The concept of substitution and exchange calls for a careful and integrated operation of the whole system. In some cases it may alter the basic planning of the projects proposed by the State Governments.

22.10.4. Water Rights and Ownership of Water

This is a very pertinent question in the context of IBWT. In India, all surface water is fully controlled by the Government. Water users do not have any well-defined rights in the form of water licenses etc. The allocation of the waters of a river or a river basin to the various parts of the state falls under the jurisdiction of the government. However, the actual ownership of water which is an element of the natural environment remains questionable. If one accepts 'ownership' of the concerned state government, the following observations require attention:

- Water is a property of basin States, and can be transferred to non-basin States only by basin States.
- Only the basin State and not the Union Government decides whether there is surplus water in a basin.
- The basin State is free to use or dispose off the surplus water in the manner that they deem fit.

The other popular concept of ownership of water is that of a negative community. In this concept, water is 'common' for all but owned by none, including the government. Accordingly, there can be no property rights, only 'usufruct' rights exist. Usufruct right is the right to use and enjoy the profits and advantages of something belonging to another as long as that thing is not damaged or altered in any way. The usufruct rights of the State are related to needs and uses, and are incapable of being sold. Delegation of these rights to the actual users within the state is possible.

In the international context, the present conventions (the international conventions for non-navigable use of international rivers) limit allocations of an international river to the co-basin nation states though this is not explicitly stated. In semi federal setup as India, the position is different. The States in India are not sovereign and the parliament has supreme power to enact legislation on any matter.

Considering both the issues of ownership and allocations, in Indian context, Mohile et al. (2005) suggested to firmly adopt the conceptual framework of treating water as a 'negative community'. Side-by-side, we need to frame appropriate laws regarding allocation of waters of interstate basins to non-co-basin states. Such laws will definitely serve vital national interests and will avoid many potential conflicts.

Many inter-state issues arise out of different interpretations of the ownership rights in respect of water. By interlinking of rivers water is transferred from a location (in a state) in a water rich river basin to water deficit basin(s). A question that is usually asked is whether the state from which water is transferred owns this water? Is it justified that the state should be compensated for agreeing to transfer water outside the state? Some experts strongly disagree with the suggestion about water royalty arguing that water cannot be compared with other natural resources such as oil and minerals. The question of water right is compounded by the fact that water is primarily a state subject in the Constitution of India. Another relevant question is whether only the state from where water is transferred to another state has to agree for the transfer of water or all the co-basin states have to agree for it? A classic example of the Telugu Ganga Project for supplying drinking water to Chennai city from Krishna can be quoted here.

The sensitivities about water transfer is leading to a situation that the States from which water is to be transferred to other States are challenging the studies establishing that surplus water is available in the basin for transfer to outside the basin. NWDA studies have established that all the waters of some water rich basins cannot be fully utilized within the basins at present and in near future and could be fruitfully used in water deficit basins. The States in general do not agree with the studies and try to act as the protectors of the water 'rights' of the States. This issue needs to be resolved before attempting to implement the interlinking of rivers scheme.

22.10.5. Defining Surplus and Deficit Basins

Besides identifying surplus and deficit basins, it is equally important to quantify the amount of surplus water in a basin. It is very much likely that there will be disagreements and the finally agreed quantity will be a 'compromise solution'. Experience from some recent disputes (the on-going Cauvery water dispute, dispute between Delhi and UP on Ganga water, between Delhi and Haryana on Yamuna water, etc.) shows that this might indeed be the toughest part of negotiations.

Unfortunately, so far we do not have any standard methodology for this purpose. Poor quality of data or absence of requisite data further complicates the picture since widely different assumptions can be made. Also, there is no consensus as to how ground water should be accounted for. Conjunctive use of surface and ground water is being emphasized in the country but still it is not a part of the feasibility reports. A realistic estimation of projected water demands is another dimension which plays a crucial role in estimation of surplus or deficit. Irrigation requirements account for about 70% of future water requirements and different experts can arrive at widely different estimates and yet be in a position to justify their estimates.

This aspect needs greater attention while quantifying the availability of surplus water in a basin. This also necessitates formulating an acceptable procedure with the participation of academicians, researchers, professionals, planners, and decision makers as it involves many gray areas depending upon an individual's viewpoint.

22.10.6. Environmental Concerns, Resettlement and Rehabilitation

With monsoonic type of climate in India, it is necessary to construct dams to store the water and use it throughout the year. Construction of dams naturally involves submergence of lands and habitation leading to environmental, ecological and resettlement and rehabilitation problems. About 4 to 5 lakh people may get affected or displaced due to creation of reservoirs and canals in the ILR projects. The displaced persons are to be provided with proper Rehabilitation and Resettlement (R&R) packages to ensure that their socio-economic condition improves after resettlement at new places and good models are available for the purpose. The 'National R&R Policy' is also in place, and it is necessary that it is implemented with due care and attention.

Due to construction of storages, about 79,000 ha of forest land is likely to come under submergence. Sufficient provisions are to be made for the protection of forests by means of raising compensatory afforestation, catchment area treatment and other land and water management programs. Bio-diversity likely to be affected particularly in Himalayas due to the construction of reservoirs. Preservation of the precious stock would have to be taken care of through gene-banks. It could also be used for new growth, where necessary. Environmental impact assessment and suitable mitigation measures for adverse effects should form an integral part of the program. However, no specific unmanageable adverse impact is likely to result from ILR.

22.10.7. Alternatives to Interlinking of Rivers

The huge cost and the social and environmental concerns involved in the interlinking of rivers have aroused opinions against the interlinking of rivers in general emphasizing the necessity for considering available other options. Other options are mainly oriented towards more efficient use of the available waters in a river basin through watershed development and water harvesting, improving the recharge of ground water, conjunctive use of ground water with surface water, water saving in various uses of water and reuse, and recycling of water.

Of course, the development of local water resources through watershed management, water harvesting, etc. needs to be done simultaneously with the implementation of any scheme for interlinking of rivers. However, watershed management and water harvesting are unlikely to solve the massive problems, since the intense rainfall during storms cannot be conserved and in semi-arid and arid regions the rainfall is much less than the requirements. Similarly the other measures advocated have only limited scope; they will no have any significant impact on the water requirement that will be confronting the people of India in the future. Although demand management is necessary, this has its own limits.

22.10.8. System Study of Inter-Basin Water Transfer

Systems analysis provides a framework for consideration of various alternatives and options and their analysis in an interdisciplinary approach; it can draw on the expertise in various disciplines and use the various state of the art tools of modeling, simulation, integrated ecological and economic cost-benefit analysis, and assessments of impacts. In addition to using these tools, it is important to use the concepts of integrated water resource management, integrating the policies of land use, water and energy with appropriate focus on the linkages among them, river basin as a unit for planning and management, and the role of recycling and reuse of water.

Inter-basin transfer of water is a complex program of water management. The NPP proposals we firmed up, based on sub-basin wise water balance studies. These studies have not analyzed the performance reliability of the system over a long period of operation. Further, the effect of carry-over capacities and operation policies of the major reservoirs in the system in making up the deficit or augmenting the surplus has also not been investigated. Jain et al. (2005) have carried out a system study of the peninsular component of the scheme. Detailed system studies are necessary for all the projects.

22.10.9. Risks Associated with ILR Scheme

Since the proposed interbasin water transfer project will be the biggest such exercise in terms of investment and geographical extent in India and indeed in any other country of the world, it will have a profound influence on national economy. On the other hand, any major setback on such a large project will have a devastating influence on national economy. Therefore, it is necessary to examine the associated risks before undertaking such a gigantic task. However, in view of the fact that the implementation of the project has received tremendous push due to the directive from the Supreme Court of India, the associated technical, political, and social aspects may not be examined as thoroughly and as meticulously as required. But in the interest of national development, it is necessary that all genuine concerns are carefully examined before a final commitment is made. Besides, working in haste, some important aspects may be overlooked and these may cause trouble later.

It is necessary to highlight the risks associated with the scheme so that these can be suitably accounted for in planning and execution of the project. This will, inter alia, help identify the steps that require more attention.

22.10.10. Technology of Interlinking

The envisaged ILR involves the transfer of large volumes of water across distant geomorphologic entities through link channels. As far as the question of technological competence is concerned, enough experience is available in the country to construct the civil engineering structures. However, considerations governing planning, design, implementation and operation of each of the subsystems associated with the task of ILR will have to be treated in an integrated manner.

It is important to take a holistic view of water resources development and management in river basins rather than just confining the task to interlinking of rivers for inter-basin transfers of water. Second, a truly interdisciplinary and system-analytic approach should be adopted. Experts from many disciplines should be involved while taking decisions. Third, all views, including those who are critical of the whole concept of ILR, should be taken seriously. Decisions should be taken on the basis of facts, analysis and rationality rather than passion and sentiment. Whenever there are issues on which people hold extreme and opposite views, reconciliation should be always attempted to reach the acceptable ground. The interests of the vulnerable and disadvantaged sections of the population should be considered with great sensitivity and empathy.

Energy will be required for lifting water to negotiate adverse gradients of interlinking. Hence, provision of energy will be an essential element of the design and operation of some link channels. For example, for transferring water for Inchampalli dam on Godavari to Krishna, a lift of about 116 m is involved. Most link channels will cover long distances; some of them will carry large discharges. The link channels are bound to cross several natural rivers. This will require numerous and costly cross-drainage works and will affect natural drainage of the en-route area. Acquisition of land may lead to massive problems of ecological disturbance, rehabilitation, and relief. In fact this aspect may cause discontent and resistance amongst the general public.

22.10.11. Political Consensus

Achieving political consensus between water-surplus and water-deficit basin states for IBWT is going to be difficult. For the water-surplus basin states it is a politically sensitive subject and the opposition parties may try to exploit the issue so as to derive political mileage. In some cases, the political response depends on whether the same political party is ruling at the center and the state or not. It is difficult but not impossible to achieve political consensus among the states and develop win-win solutions by negotiations. At the same time, there should be legislative support which could be adopted if negotiations tend to become endless.

While the receiving States readily agree to the link proposals, the donor States have generally opposed the idea. As expected, the donor States have reservations about the surpluses that are worked out in various river basins. Some States feel that the existing tribunal awards will get disturbed and hence no water can be taken out of the basin. Further, the link proposals may have adverse effect on existing irrigation and power generation. Also, the reliability and adequacy of the long distance water transfers are being questioned. Similarly, acquisition of land for the storage structures and long and wide link canals could pose a serious problem.

The pre-feasibility studies for the various links in the Himalayan Component have taken into consideration the existing, ongoing and proposed dams on the common river systems in Nepal, Bhutan and India. These studies need to be further detailed, surveyed and investigated in cooperation with Nepal and Bhutan. All relevant issues will have to be sorted out through bilateral discussions.

22.10.12. Institutional Aspects

Attention should be paid not only to the planning, financing and construction aspects and to create appropriate institutional set ups for management. Some re-engineering of the existing institutions and agencies in the water sector and establishment of new institutions may be necessary. A body for regulation of inter-state rivers, for example, Central Water Regulatory Board, might be necessary.

Strong and competent institutions will be necessary for sustenance of the system and its efficient working.

22.10.13. Transparency in Working

The organizations involved in the ILR works should maintain transparency and openness in all activities. Recently, many pre-feasibility and feasibility reports have been put in public domain to enable the experts and public to understand the basic philosophy. This will increase credibility of the government and inspire trust in the public and the experts who are generally wary of bureaucratic functioning. Thereafter, they may contribute to improving planning, designs, and management and decisions that have the support of the majority can be arrived at.

22.11. OPPOSITION TO THE ILR SCHEME

It is understandable that there will be opposition to divert water from a basin by the local people who may feel deprived of something which they consider their own. The various agencies and NGO's related to environmental issues are also expected to oppose such proposals. There is a distinct possibility that these issues might gain more importance than the technical and financial aspects. Clearly, the proposals require careful examination. According to preliminary estimates, the Peninsular component will submerge 2.5 lakh ha area and will require rehabilitation and resettlement of nearly 4.5 lakh persons. Besides, all the concerned state governments will have to arrive at an agreement for these projects.

While activating the government machinery, the judgment by the Supreme Court of India has also aroused considerable interest and curiosity about this scheme. Expectedly, there has been a lot of opposition from various quarters (for example, see Patkar, 2004). The main arguments of the opponents are that against the claims of the government, the project will neither solve the flood control nor the drought problem of India. Further, the introduction of irrigated agriculture in arid areas of the kind appropriate to wet areas may be unwise. A project of such gigantic nature will have serious environmental, economic, and social impacts which have not been studied as meticulously as needed. Another concern is that the entire planning work is being done under secrecy and the government should make available all the relevant data and reports in the public domain so that the citizens are aware of what is happening and can provide and make informed inputs.

It will be important that the demands and concerns of all sections of the society are carefully considered so that the final solution is in the larger interest of the nation. Key questions posed by the opponents of ILR are enumerated below.

Drought alleviation

Questioning the claim that ILR will help in drought alleviation, the opponents state that most of the uplands and dry lands of India are distant from rivers, and at elevations of 300 m to 1,000 m above mean sea level. Therefore, ILR will serve very few such areas. According to them, the primary answer to drought has to be local; it is only thereafter, and in some very unpromising places where rainwater-harvesting may not be feasible or may yield meager results, that the bringing in of some external water may need to be considered.

Irrigation

The question pertaining to water for irrigation is fundamental in nature. The augment runs as follows (Iyer, 2005). In irrigated areas, the question is whether large demands for additional irrigation water should be unquestioningly accepted and met through supply-side solutions, such as large dams or IBWT. Should serious attempts not be made to improve water-use efficiency in irrigated agriculture, get more value out of a given quantum of water, reduce water demand, and minimize the need for supply-side projects? In the context of the prevailing low efficiency of water-conveyance in canal systems and water-use in irrigated agriculture, bringing in more water from another basin would really amount to the provision of more water for being wasted. It would also mean that there would be no motivation at all for changing cropping patterns and shifting from water-intensive crops to crops that need less water; on the contrary, the tendency to grow water-consuming crops would receive strong encouragement. In arid or drought-prone areas, the introduction of irrigated agriculture of a kind that is appropriate to wet areas may be unwise. The opponents even argue that the Rajasthan Canal project was a misconceived project.

Gigantism/altering nature

This will be a massive intervention in nature, a severe case of technological hubris, of a kind that (one thought) had been discredited and was a thing of the past. Apprehensions about gigantism have been sought to be set at rest with the explanation that the flows will be largely by gravity with lifts (not exceeding 120 meters) at a few selected points, and that the need for a transfer of water through natural barriers will be obviated. The critics have raised two questions. First, such an approach may be possible in some cases, but its feasibility in some thirty projects seems prima facie doubtful. This, like the claim that the project will be a net generator of large quantities of electricity, needs to be looked at very carefully. Secondly, if indeed the links are to be largely by gravity with a few modest lifts, will such an approach not limit the extent of water transfer and the scope of the project? Can large claims still be made for the project?

Impacts and consequences

The adverse impacts and consequences of large-scale interventions have been observed in many projects and will need to be studied carefully in the case of each of the proposed links. Leaving aside submergence, displacement, impacts on flora and fauna, and so on, one must take note of one point, namely, that before diverting water and reducing downstream flows we must make sure that serious consequences will not follow. With the 'linking of rivers' project one may be headed for other unforeseen disasters and may discover this too late. A degree of caution seems warranted before the Government embarks on this enterprise.

Generating new conflicts?

We have not so far been able to persuade sates within a basin to share river water (e.g., the Cauvery Dispute). Instead of resolving such intra-basin disputes through better, more economical and more cooperative management of the resources of the basin, should we try to bring water from another and more distant basin? Further, even if we assume that the conflict within a water-short river-basin will be eased by importing external water, such an effort may initiate new conflicts between basins. Several State Governments are opposed to, or lukewarm about ILR. Bihar, West Bengal, Punjab, Maharashtra and Kerala are examples. Kerala is stoutly opposed to the proposed Pamba-Achankovil-Vaippar link. There is also some opposition on the part of Karnataka to the idea of the eastward diversion of west-flowing rivers. It may be argued that we should not allow ourselves to be deterred by such political difficulties, but is it really necessary to generate several new inter-state conflicts?

The proposals include a Brahmaputra-Teesta (or alternatively, a Brahmaputra Ganga) link, but one wonders whether major diversions from the Brahmaputra are likely. In any case, perhaps we need not discuss this further except to say that the sensitivities of the Northeastern States must be kept in mind.

Turning to international aspects, there are serious apprehensions about this project in Nepal and Bangladesh. These may be justified or misconceived, but criticisms of the kind that one heard in earlier years from Nepal in relation to the Kosi and Gandak Projects and from Bangladesh in relation to the Farakka Barrage are being voiced again. It is very necessary for the Government of India to explain matters and minimize if not eliminate misunderstandings.

Need for examination of individual links

If we proceed from the 'concept' or umbrella project to the individual links, it is clear enough that in each case there will have to be an examination of the calculation of surplus, and deficit, the need for the transfer/link, whether it represents the best out of the available options, its techno-economic soundness and viability, the projected benefits, the environmental/ecological implications and remedial measures, the human aspects (displacement, resettlement and rehabilitation, and the rights and risks of the affected people), the international dimensions (if any), and in the light of all these, a comprehensive techno-socio-economic-ecological analysis of the balance of costs and benefits (including the non-quantifiable aspects) and an appraisal of the project for an investment decision. We do not know what the outcome of that process will be: all projects may pass the test; all may fail; or some may survive a stringent scrutiny while others may not.

Pre-empting of resources

If this mammoth project, not reckoned earlier in the planning processes, is now to be inserted into the 5-year plan, there may be a pre-empting of resources, and a distraction of attention from the things that need to be done. Plan outlays are barely adequate even for the completion of projects already undertaken. From the Sixth Plan onwards the stress has been on consolidation rather than on new starts. Against that background, it seems strange to embark on a major river-linking undertaking. The rough figure mentioned in the Supreme Court in this context was Rs. 560,000 crore. That figure is likely go up in the course of actual implementation, but even if we ignore that point, the pre-empting of resources of this magnitude for this project will render the whole planning process meaningless. Equally important is effective demand management through improved efficiency and economy in water use, whether in agriculture or in industry or in domestic and municipal uses. A preoccupation with the gigantic river-linking project may prevent adequate attention to them.

22.12. CONCLUSION

Obviously, there are many questions that need to be answered and many answers that need to be questioned. For some reason, the information that is necessary to examine the proposals is not in public domain and is not accessible. Pre-feasibility and feasibility reports are said to have been prepared but these are also not available. Only one feasibility study report of the Ken-Betwa Link is in public domain. The fact that the information is not being made available has generated considerable skepticism and distrust. If this information is shared, it will only lead to better decision making since if some one is working in secrecy, there is no room for new ideas to flourish and mistakes may remain undetected.

Basin-wise studies indicate that in spite of the technical constraints and apprehensions of the states in assessing the water balance of river basins, some basins will be surplus in water, while several other basins will face acute shortages of water. To meet the ever-growing demands for increased food grain production and domestic and industrial water requirements, water resources development at state/basin level alone may not be adequate. A broad based national perspective is essential for optimum utilization of the water resources. The NPP is likely to facilitate meeting the demands of water-short regions and needs to be pursued.

We have to move progressively beyond the limited vision of satisfying the interest of a river basin. Harnessing water of surplus basins by inter-linking rivers is rational. Once the feasibility of inter-linking of rivers is established, discussions and negotiations between the states concerned with Central Government's support should be held for preparation of detailed project reports and the execution of the projects.

The National Commission on Integrated Water Resources Development Plan (1999), while agreeing that inter-basin transfer of water is a very important and extraordinarily large complex programme of water management, has emphasized that the basic problems are not techno-economic, but those of political, social and emotional character. As the Commission has aptly put it, the country's political system and civil society have to evolve institutions and techniques to deal with them and arrive at solutions in the best interest of the people.

Finally, interlinking of rivers is not panacea. Alone, it will not solve all water-related problems of the country. Concepts, such as water conservation, optimal regulation of existing facilities, rainwater harvesting, watershed management, water reuse, etc. will continue to be highly relevant and this grand scheme will be an important supplement to these. A holistic view of the scenario is always important and necessary.

CHAPTER 23

INSTITUTIONS IN THE FIELD OF HYDROLOGY AND WATER RESOURCES

The Government of India (GOI) conducts its business through various ministries. Under each ministry, there may be departments, academic and research institutes, and public sector undertakings. Each ministry has thus a large hierarchical structure. The apex website giving extensive information about the Government of India and links to the various ministries and organizations under it is *http://www.nic.in*. This website is maintained by the National Informatics Center. Information on this site is carefully arranged and a powerful search engine is also provided. This site is considered an authentic source about GOI and some of the information given in this chapter has originated from this site.

23.1. MAIN ORGANIZATIONS DEALING WITH WATER RESOURCES IN INDIA

A large number of organizations in India are associated with planning, development, management, research, and education in the field of water resources. This chapter provides an overview of the major organizations and their work. Table 1 displays the roles of various types of organizations in different tasks associated with water resources development and management.

For the purpose of subsequent discussion, the organizations have been classified into the following broad categories: central government organizations, state government organizations, research and academic organizations, river basin organizations, non-governmental organizations, and international organizations. We begin our discussion with the central government organizations.

23.2. MINISTRIES INVOLVED IN WATER SECTOR

Seven different ministries are associated with water sector under different subject domain as shown below.

Besides, some other ministries of the Govt. of India are involved in water sector. A discussion about these follows.

Table 1. Responsibilities for water sector functions by organizations

Task	Central and State Governments	Private sector	NGOs	Professional bodies
Policy formulation	M		S	S
Planning, analysis, and design	M	S	S	S
Construction	M	M		
Management	M	M	S	S
Financing	M	M		
Research	M		S	M
Capacity building	M		S	M

Note: M = major role, S = supporting role.

23.2.1. Ministry of Water Resources (MOWR)

In India, the Ministry of Water Resources is the nodal ministry responsible for the development and utilization of water resources. However, due to inter- and multi-disciplinary nature of development, several other ministries, e.g., Ministries of Agriculture, Power, Environment and Forest, Surface Transport, Urban Affairs, Rural Development, and Science and Technology, are also involved in the water sector.

The Ministry of Water Resources is headed by a cabinet rank minister. Usually, a Minister of State is also appointed in the ministry but there have been instances in the past when the Minister of State had the independent charge of the ministry. The Secretary (water resources) is the topmost civil servant (officer) in the ministry and he is assisted by a number of officers. A highly hierarchical system of decision-making is followed in the ministries in India and one can notice long movement of files which often slows down decision making.

S. N.	Ministry/Department	Subject domain
i.	Water Resources	Irrigation, command area development, flood control, ground water.
ii.	Rural Development	Rural drinking water supply and rural sanitation, watershed programmes.
iii.	Urban Development	Urban drinking water and urban sanitation.
iv.	Environment	Pollution control
v.	Power	Hydropower
vi.	Shipping	Inland navigation
vii.	Planning Commission	Allocation of Plan funds for various sectors and investment clearance.

The main functions of the Ministry of Water Resources are:

(a) Overall planning, policy formulation, coordination and guidance in the water sector.

(b) Coordination, mediation and facilitation in regard to the resolution of differences or disputes relating to inter-state rivers and overseeing of the implementation of inter-state projects.

(c) Overall planning for the development of water resources, assessment of utilizable resources and formulation of policies for exploitation, overseeing of and support to the State level activities.

(d) Regulation of ground water and creating mass awareness for artificial recharge.

(e) Formulation of the national water development perspective and determination of the water balance of different basins/ sub-basins.

(f) Negotiations with the neighboring countries, such as Bangladesh, Bhutan, China, Nepal and Pakistan, in regard to river waters, water resources development projects, and the operation of international treaties relating to water.

(g) Technical guidance, scrutiny, clearance and monitoring of irrigation, flood control and multi-purpose projects (major/medium) of States.

(h) Overall policy formulation, planning and guidance in respect of minor irrigation and command area development, and the administration and monitoring of centrally sponsored schemes in these areas.

(i) Operation of the central network for flood forecasting and warning on important rivers, the provision of central assistance for some State schemes in special cases and preparation of flood control master plans for the Ganga and the Brahmaputra River systems.

(j) Providing special central financial assistance for specific projects and assistance in obtaining external assistance from the World Bank and other international agencies.

(k) Infrastructural, technical and research support for sectoral development at the State level.

The Central Water Commission (see section 23.3.1) is the technical wing of Ministry of Water Resources and deals with technical matters pertaining to surface water development. The Central Ground Water Development Board (section 23.3.2) is responsible for development, assessment and management of ground water resources. The other important organizations under the Ministry of Water Resources include the National Institute of Hydrology, the National Water Development Agency, the Central Water and Power Research Station, the Brahmaputra Board, and the Ganga Flood Control Commission. A large number of river basin authorities have also been set up for the development and management of water resources of a particular basin, e.g., the Bhakra Beas Management Board and the Damodar Valley Corporation.

The office of MoWR is located in Shram Shakti Bhavan at Rafi Marg in New Delhi – 110 001. The Homepage of the MoWR is: *www.wrmin.nic.in*. It contains many useful data and other information about water resources of India.

23.2.2. Ministry of Agriculture

This ministry looks after all matters related to agriculture. The work of Ministry of Agriculture is organized under three Departments, viz., Department of Agriculture and Cooperation, Department of Agricultural Research & Education/ICAR, and Department of Animal Husbandry & Dairying.

The Department of Agriculture and Cooperation is responsible for the formulation and implementation of national policies and programmes aimed at achieving rapid agricultural growth through optimum utilization of the country's land, water, soil and plant resources. The department is entrusted with the responsibility of collection and maintenance of database of a wide range of statistical and economic data. The department is also associated with management of natural calamities, e.g., flood, drought, cyclone, etc.

Agriculture in India is the most vulnerable activity to climate change, especially in the arid and semi-arid tropics. It is expected that 1 m rise in the sea-level will to inundate about 1,700 km^2 of agricultural land in Orissa and West Bengal alone. The increasing frequency and intensity of extreme weather events will also have a direct bearing on agriculture. To build capacity and deal with climate change issues related to agriculture, a Climate Change Cell has been set up in Ministry of Agriculture.

To strengthen research, education and human resources development in agriculture, the Ministry of Agriculture has launched the National Agriculture Technology Project, through its national grid comprising 46 institutes including universities, research centers and regional stations in the Ninth Plan Period (1997–2002). All of these form a large infrastructure for research and outreach activities. The office of the Ministry of Agriculture is located in Krishi Bhavan in New Delhi – 110 001. The Homepage of the Ministry of Agriculture is: http://agricoop.nic.in/about.htm.

23.2.3. Ministry of Power

The Ministry of Power is primarily responsible for the development of electrical energy in the country. Among the responsibilities of the Ministry are matters concerned with perspective planning, policy formulation, monitoring of the implementation of power projects, and the administration and enactment of legislation in regard to thermal and hydro power generation, transmission, and distribution. This Ministry looks after all matters relating to hydroelectric power (except small/mini/micro hydel projects of and below 25 MW capacity). It is also concerned with research, development and technical assistance relating to hydroelectric and thermal power and transmission system network. The Ministry of Power is the coordinating agency for matters relating to energy efficiency for all conventional energy sources.

Two organizations under this ministry, viz., the National Thermal Power Corporation Limited and the National Hydro-electric Power Corporation Limited are involved in the construction and management of power projects. So far, these two organizations have constructed many large projects. North Eastern Electric Power Corporation Limited is responsible for power projects in the north-eastern region of India. Recall that about one-third of hydropower potential of India is available in this region. Another organization under this ministry, the Tehri Hydro Development Corporation, is involved in hydropower development through the construction of the Tehri and other projects in the Uttaranchal state. Likewise, Satluj Jal Vidyut Nigam Limited is involved in the development of power potential of the Satluj River.

The office of Ministry of Power is located in Shram Shakti Bhavan at Rafi Marg in New Delhi – 110 001. The Homepage of the Ministry of Power is: *http://powermin.nic.in*.

23.2.4. Ministry of Environment & Forests

The Ministry of Environment & Forests is the nodal agency for the planning, promotion, co-ordination and overseeing of the implementation of environmental and forestry programmes. The Ministry is also the nodal agency in the country for the United Nations Environment Programme (UNEP). As elaborated on its website, the principal activities undertaken by Ministry of Environment & Forests consist of conservation and survey of flora, fauna, forests and wildlife, prevention and control of pollution (including, of course, water pollution), afforestation and regeneration of degraded areas, and protection of environment. The main tools utilized for this include surveys, impact assessment, control of pollution, regeneration programmes, support to organizations, research to solve problems, collection and dissemination of environmental information and creation of environmental awareness among all sectors of the country's population.

The MoEF is the nodal agency for the subject of climate change in India. It has created various mechanisms for increasing public awareness and enhancing research in climate change by giving grants for wide-ranging research programmes and creating centers of excellence.

The office of Ministry of Environment & Forests is located in Paryavaran Bhavan, CGO Complex, at Lodhi Road in New Delhi – 110 003. Their homepage is: http://envfor.nic.in.

Environmental information system (ENVIS)

Realizing the importance of environmental information, the Environmental Information System (ENVIS) was established by the MoEF in 1982 as a plan programme. ENVIS is a decentralized system with a network of distributed subject-oriented centers. The aims of the ENVIS are to ensure the integration of national efforts in environmental information collection, collation, storage, retrieval and dissemination to all concerned (*http://www.envis.nic.in*). To assess the environment for pollution control, toxic chemicals, central and offshore ecology, the centers of ENVIS have been set up in different organizations in India. ENVIS supports environmentally sound and appropriate technology, bio-degradation of wastes and environment management research.

Since environment is a multidisciplinary subject, it is necessary to built a comprehensive information system on the environment for the involvement and effective participation of a range of institutions and organizations working in different spheres of the environment. To that end, ENVIS network comprising of 85 ENVIS Nodes and 25 ENVIS Centers covering the diverse subject areas of environment have been established. The Ministry of Environment and forest has the focal point of ENVIS.

23.2.5. Ministry of Surface Transport

Transport is a critical infrastructure for the social and economic development of a country, influencing its pace, structure, and pattern. Within its fold, the Ministry of Surface Transport encompasses shipping and ports, national waterways, and inland water transport. The headquarters of this Ministry is located in the Parivahan Bhavan at Parliament Street in New Delhi – 110 001.

An autonomous body under this ministry, the Inland Waterways Authority of India (IWAI), is responsible for the regulation and development of inland waterways for the purposes of shipping and navigation. IWAI was set up in October 1986 under the Inland Waterways Authority of India Act, 1985. The three waterways that have been declared as National Waterways and taken up for development are:

- Allahabad-Haldia stretch of the Ganga Bhagirathi Hooghly River system (1620 km) as National Waterway 1,
- Saidiya Dhubri stretch of the Brahmaputra River system (891 km) as National Waterway 2, and
- Kollam Kottapuram stretch of West Coast Canal (168 km) along with Champakara Canal (14 km) and Udyogmandal Canal (23 km) as National Waterway 3.

The major waterways projects currently in progress are shown in Table 2

Table 2. Major waterways projects in progress

Project	Responsible organization
Kakinada-Marakkanam NW-4 Waterway Project	Inland Waterways Authority of India
Inland Waterways (Thiruvananthapuram-Kasargode) Project	Government of Kerala
Inland Waterways (Dakshina Kannada) Project	Government of Karnataka
Inland Waterways (Udupi) Project	Government of Karnataka
Honavar-Gerusoppa Inland Waterways Project	Government of Karnataka
Kodibag-Kadra Inland Waterways Project	Government of Karnataka
Kochi-Kollam Waterway Project	Inland Waterways Authority of India
Inland Water Transport Terminal (Gaighat)	Inland Waterways Authority of India

23.2.6. Ministry of Urban Affairs

The Central Public Health and Environmental Engineering Organisation (CPHEEO) is the technical wing of the Ministry of Urban Affairs and Employment, functioning as the nodal organization at the national level to deal with urban water supply and sanitation, including solid waste management. The department is working on the three schemes for supplying of fresh water: Pesticides in potable water – development of removal technology, waste water recycling and groundwater recharge by natural methods, and rapid evaluation of performance of waste water treatment by dip slide technique.

Their contact address is: Adviser (PHEE), CPHEEO, Ministry of Urban Affairs and Employment, Nirman Bhavan, New Delhi – 110 001.

23.2.7. Ministry of Rural Development

This Ministry's main objective is to alleviate rural poverty and ensure improved quality of life for the rural population, especially those below the poverty line. Two departments under this Ministry are concerned with water resources: the Department of Land Resources and the Department of Drinking Water Supply. The department of land resources implements schemes to increase the bio-mass production by developing wastelands in the country. It also undertakes development of desert areas and drought prone areas in the country. The major programmes concerning land resources are the Drought Prone Area Programme (DPAP), The Desert Development Programme (DDP), and the Integrated Wasteland Development Programme (IWDP). These aim at increasing the soil and moisture conservation and productivity of the wastelands of the country.

The provision of drinking water supply and extension of sanitation facilities to the rural poor are the main activities of the department of drinking water supply. The major programmes of this department are *The Swajaldhara*, the Accelerated Rural Water Supply Programme (ARWSP), and the Total Sanitation Programme (TSP). The initiative *Swajalhara* was launched in December 2002, empowering the Panchayats (village level administrative bodies) to formulate, implement, operate and maintain drinking water projects. A related scheme, known as *Hariyali* was initiated in January 2003, to strengthen and involve Panchayati Raj Institutions in the implementation of Watershed Development Programmes. The homepage of the Ministry of Rural Development is: *http://rural.nic.in*.

23.2.8. Planning Commission

The Planning Commission of India was set up in March 1950, among other things, to promote a rapid rise in the standard of living of the people by efficient exploitation of the resources of the country. It was entrusted with the responsibility of assessing all resources of the country, augmenting deficient resources, formulating plans for the most effective and balanced utilisation of resources and determining priorities.

The Prime Minister is the Chairman of the Planning Commission, which works under the overall guidance of the National Development Council.

Since 1951, the development of India has been pursued through 5-years plans, meaning that the planning time scale is 5 years. The first Five-year Plan was launched in 1951 and two subsequent five-year plans were formulated till 1965. Two successive years of drought, devaluation of the currency, a general rise in prices and erosion of resources disrupted the planning process and after three Annual Plans between 1966 and 1969, the fourth Five-year plan was started in 1969. The Eighth Plan could not take off in 1990 due to the fast changing political situation at the Center and the years 1990–91 and 1991–92 were treated as Annual Plans. The Eighth Plan was finally launched in 1992. Since the launch of the Ninth Plan in 1997, the emphasis on the public sector has become less pronounced and the current thinking on planning in the country, in general, is that it should increasingly be of an indicative nature. The tenth plan was launched in 2002. The office of Planning Commission is located in Yojna Bhavan at Parliament Street in New Delhi – 110 001. The Homepage of the Planning Commission is: http://planningcommission.nic.in.

23.2.9. Ministry of Science & Technology

The Ministry of Science and Technology (MST) came into being in 1985. Main departments under this ministry are the Department of Science and Technology, the Department of Scientific and Industrial Research, and the Department of Biotechnology. The homepage of Ministry of Science & Technology is: http://mst.nic.in.

Two organizations under MST have an important and useful role in water sector. These are the India Meteorological Department (IMD) and the National Center for Medium Range Weather Forecasting function under MST. More information about these is given in later sections.

The Technology Information, Forecasting and Assessment Council (TIFAC) is an autonomous organization established by the Department of Science and Technology. The main aims of TIFAC are to monitor global trends, to formulate preferred technology options for India, promote key technologies and undertake technology assessment and forecasting studies in selected areas of the national economy. Many Technology Assessment and Technomarket Survey Study reports have been prepared by TIFAC and are available on the homepage of TIFAC at www.tifac.org.in.

23.2.10. Department of Space

Government of India had set up Space Commission and Department of Space (DOS) in June 1972. Indian Space Research Organization (ISRO) under DOS executes the space programme. In the mid-1970s, meteorology and weather forecasting were selected among the thrust areas by the Indian Space Programme. ‘Bhaskara’, one of the earliest satellites, had a microwave payload to study the atmosphere and ocean.

The Indian National Satellite (INSAT) series was conceptualized as a multi-purpose geostationary satellite system for communications, meteorology, oceanography, and weather services. In late 1980s, the INSAT 1 series was launched, which carried a Very High Resolution Radiometer (VHRR) payload that operated in two spectral bands—visible (0.55– 0.75 mm) and thermal infrared (10.5–12.5 mm). The INSAT system is designed to provide the following services:

i. Round the clock surveillance of weather systems and operational parameters for weather forecasting,
ii. Collection and transmission of meteorological, hydrological and oceanographic data from remote/inaccessible areas, and
iii. Timely dissemination of warning of impending disasters dissemination of meteorological information.

Summary information on the development and deployment of satellites in India is presented in Table 3; an exhaustive list is available with ISRO.

The INSAT 2 series includes many satellites to ensure the continuity of services in an enhanced manner. INSAT facilities are also being used for real-time hydro-meteorological data collection in some basins.

Indian remote sensing programme

Under the Indian Remote Sensing Satellite (IRS) program, ISRO has launched a series of land observation satellites. The operational first generation RS satellites IRS-1A and IRS-1B were placed in near-polar, sun-synchronous orbit, with repetitive time of 22 days. The satellites had two Linear Imaging Scanning Sensors (LISS-I and LISS-II) for providing data in four spectral bands: Visible, Infra Red (IR) and Near Infra Red (NIR). Their ground resolutions were 72.5 m and 36.25 m, respectively. The second generation, operational, multi-sensor satellites IRS-1C and IRS-1D were launched in 1995 and 1997. These were placed in near-polar, sun-synchronous orbit with a repetitive time of 22 days. These satellites have three on-board cameras. The PANchromatic (PAN) camera operates in the panchromatic region of the EM spectrum and has a spatial resolution of 5.8 m. It can be steered up to 26° across-track, thus enabling generation of steoroscopic imagery and improved revisit capability. The Linear Imaging Self Scanner-III (LISS-III) camera operates in four spectral bands: three in Visible/ Near Infrared (VNIR) and one in Short Wave Infrared (SWIR) region. It has a resolution of 23.5 m in VNIR bands (swath 141 km) and 70 m in SWIR band. The Wide Field Sensor (WiFS) is a coarse resolution camera with spatial resolution of 188.3 m. A satellite for oceanographic studies, namely, IRS-P4 (OCEANSAT-1) having an ocean color monitor with 8 spectral bands and a multi-frequency scanning microwave radiometer operating in four frequencies has been recently launched to enable measurement of physical and biological ocean parameters.

In view of rapid developments taking place in space technology and increasing application of remote sensing and GIS techniques, ISRO has planned a number of future satellite IRS missions. These will provide cadastral level information up to 1:5,000 scale thematic applications, vegetation and multi-crop discrimination,

Table 3. Indicative list of development and deployment of Indian satellites

Satellite	Launch Date	Achievements
Aryabhata	19.04.1975	First Indian satellite. Provided technological experience in building and operating a satellite system.
Bhaskara-I	07.06.1979	First experimental remote-sensing satellite. Carried TV and microwave cameras.
Ariane Passenger Payload Experiment (APPLE)	19.06.1981	First experimental communication satellite. Provided experience in building and operating a three-axis stabilized communication satellite.
Rohini Technology Payload (RTP)	10.08.1979	Intended for measuring in-flight performance of first experimental flight of SLV-3, the first Indian launch vehicle. Could not be placed in orbit.
Rohini (RS-1)	18.07.1980	Used for measuring in-flight performance of second experimental launch of SLV-3.
Rohini (RS-D1)	31.05.1981	Used for conducting some remote-sensing technology studies using a landmark sensor payload.
Stretched Rohini Satellite Series (SROSS-1)	24.03.1987	Carried payload for launch vehicle performance monitoring and for Gamma Ray astronomy. Could not be placed in orbit.
Stretched Rohini Satellite Series (SROSS-2)	13.07.1988	Carried remote sensing payload of German space agency in addition to Gamma Ray astronomy payload. Could not be placed in orbit.
Indian National Satellite (INSAT-1A)	10.04.1982	First operational multi-purpose communication and meteorology satellite.
INSAT-1B	30.08.1983	Identical to INSAT-1A. Served for more than design life of seven years.
INSAT-1D	12.06.1990	Identical to INSAT-1A.
INSAT-2A	10.07.1992	First satellite in the second-generation Indian-built INSAT-2 series. Has enhanced capability than INSAT-1 series.
INSAT-2B	23.07.1993	Second satellite in INSAT-2 series. Identical to INSAT-2A.
INSAT-2C	07.12.1995	Has mobile satellite service, business communication and television outreach beyond Indian boundaries.
INSAT-2E	03.04.1999	Multipurpose communication and meteorological satellite
INSAT-3B	23.03.2000	Multipurpose communication and meteorological satellite
INSAT-3C	24.01.2002	Multipurpose communication and meteorological satellite
INSAT-3A	10.04.2003	Multipurpose communication and meteorological satellite
INSAT-3E	28.09.2003	Multipurpose communication and meteorological satellite
Indian Remote Sensing Satellite (IRS-1A)	17.03.1988	First operational remote-sensing satellite.

Indian Remote Sensing Satellite (IRS-1E)	20.09.1993	Carried remote-sensing payloads. Could not be placed in orbit.
Indian Remote Sensing Satellite (IRS-P2)	15.10.1994	Carried remote-sensing payload.
Indian Remote Sensing Satellite (IRS-1C)	28.12.1995	Carries advanced remote-sensing cameras.
Indian Remote Sensing Satellite (IRS-P3)	21.03.1996	Carries remote-sensing payload and an X-ray astronomy payload.
Indian Remote Sensing Satellite (IRS-1D)	29.09.1997	Same as IRS-1C.
Kalpana	2003	Exclusive meteorological satellite, VHRR.
IRS-P6 (RESOURCESAT-1)	17.10.2003	Carries 3 cameras: a high resolution LISS-4 operating in 3 spectral bands with 5.8 m spatial resolution and steerable; a medium resolution LISS-3 operating in 3 spectral bands with 23.5 m spatial resolution; and an Advanced Wide Field Sensor (AWiFS) operating in 3 spectral bands with 56 m spatial resolution.
CARTOSAT – 1	05.05.2005	Moves in sun-synchronous orbit, has 2 cameras on-board, resolution 2.5 m, can provide stereo view.
IRS - P4 (OCEANSAT)	2005	Primarily built for ocean applications. Carries on board an Ocean Colour Monitor (OCM) and a Multi-frequency Scanning Microwave Radiometer (MSMR).

Source: http://www.isro.org/sat.htm and others.

species level discrimination etc. Further details of satellites and data can be obtained from http://www.isro.org or http://www.nrsa.gov.in.

The headquarters of ISRO are at Antariksh Bhavan, New BEL Road, Bangalore 560 094. The home page can be accessed at *www.isro.org*.

23.3. CENTRAL GOVERNMENT ORGANIZATIONS AND UNDERTAKINGS

This section briefly describes important central government organizations that deal with water.

23.3.1. Central Water Commission

Central Water Commission (CWC) is a premier technical organization in the country in the field of water resources since 1945 and is presently functioning as an attached office of the Ministry of Water Resources. The Commission is charged with the general responsibility of initiating and coordinating schemes for control, and conservation and utilization of water resources throughout the country. Water resources development schemes that are submitted by the concerned State Governments are examined by CWC from technical and economic angles and if found suitable, are recommended for funding. CWC may also undertake investigations, construction and execution of a scheme, if required.

Under the existing organizational setup, the Central Water Commission is headed by a Chairman who also holds the status of Ex-officio Secretary to the Government of India. CWC has three technical wings, namely, the Designs and Research Wing, the Water Planning and Projects Wing, and the River Management Wing. Each of these wings is headed by an officer designated as Member. Members are assisted by officers of the rank of Chief Engineer, Director/Superintending Engineer, Deputy Director/Executive Engineer, Assistant Director/Assistant Executive Engineer and other engineering and non-engineering supporting staff. A Human Resources Management Unit headed by a Chief Engineer looks after administrative and accounts matters.

A training unit, known as the National Water Academy (NWA), has been set-up under the CWC in Pune. This academy is headed by a Chief Engineer. NWA conducts training courses related to water resources for the in-service engineers of Central and State Governments.

CWC has a large network of field offices all over the country. It has thirteen regional offices located in Bangalore, Bhopal, Bhubaneshwar, Chandigarh, Coimbatore/Chennai, Delhi, Hyderabad, Lucknow, Nagpur, Patna, Shillong, Siliguri and Vadodara. Each of these offices is headed by a Chief Engineer. Under these regional offices, there are circle offices, division offices, and so on. CWC operates a large network of stream gauging stations on important rivers of the country. Observed data from these stations are used, among other purposes, for assessment of water resources and flood forecasting. CWC offices maintain a close coordination with the concerned state government departments.

The contact address of CWC is: The Chairman, Central Water Commission (CWC), Sewa Bhawan, R.K. Puram, New Delhi – 110 066. Homepage is: *www.cwc.nic.in*

23.3.2. Central Ground Water Board

The Central Ground Water Board is entrusted with the responsibility of hydro-geological surveys, exploration, assessment, development, and management of ground water resources in the country. The Board is mainly responsible for the following activities:

- Systematic hydro-geological surveys and their reappraisal,
- Ground water exploration aided by drilling,
- Monitoring of national hydrograph network stations,
- Periodic assessment of ground water resources,
- Publication of maps and reports,
- Findings ground water source for drought affected states under the National Drinking Water Mission,

- Construction of deposit wells,
- Hydro-geological and hydro-meteorological, chemical and geophysical studies,
- Mathematical modelling studies including pollution aspects,
- Data storage and retrieval,
- Water balance and artificial recharge studies,
- Studies on conjunctive use of ground water and surface water,
- Training in activities related to ground water, and
- Reviewing regulation of ground water development.

Similar to CWC, CGWB also has a large network of stations all over the country. The contact address of CGWB is: The Chairman, Central Ground Water Board (CGWB), New CGO Complex, NH-IV, Faridabad 121 001, Haryana.

23.3.3. National Institute of Hydrology (NIH)

The National Institute of Hydrology, a Govt. of India Society under the Ministry of Water Resources is conducting basic, applied and strategic research in the field of hydrology and water resources development since December 1978. The Institute has carried out research in various areas of hydrology and successfully completed a number of sponsored/consultancy projects in the country. The Institute with its headquarter at Roorkee has six regional centers located at Belgaum, Jammu, Sagar, Kakinada, Patna and Guwahati.

The mandate of NIH is:

- to undertake, aid, promote and coordinate basic, applied and strategic research on all aspects of hydrology, contributing to sustainable water resources development in the country;
- to act as a repository of knowledge and information, and dissemination of the same in the country;
- to act as a center of excellence for transfer of technology, human resource development and institutional development in specialized areas of hydrology;
- to conduct user defined, demand-driven research through consultancy in the field of hydrology; and
- to cooperate and collaborate with relevant national and international organizations in achieving the above objectives.

The research and development activities in the Institute are being carried out under the following six scientific divisions: Agricultural Hydrology, Environmental Hydrology, Ground Water Hydrology, Hydrological Investigations, Surface Water Hydrology, and Water Resources Systems. Five well equipped laboratories with state-of-art instruments are available to provide the necessary support to field studies: Hydrological Instrumentation, Nuclear Hydrology, Remote Sensing & GIS, Soil Water, and Water Quality.

The contact address of NIH is: The Director, National Institute of Hydrology, Jal Vigyan Bhavan, Roorkee-247 667, Uttaranchal. Homepage: *www.nih.ernet.iniah*

23.3.4. Central Water and Power Research Station

The Central Water and Power Research Station (CWPRS) is engaged in R & D work in the areas of water and energy resources development and water borne transport. The main functions of this organization are:

- Planning, organizing and undertaking specific research studies to evaluate, alter, modify or redesign the proposals and/or to redefine the objectives therein relating to all phases of water resources development, including water-borne transport, environmental aspects;
- Carrying out basic or fundamental research necessary aimed at the furtherance of knowledge relevant to water resources and related sciences;
- Rendering consultancy and/or advisory services to the Central and State Governments as may be called upon;
- Disseminating research findings and building up of a technical database in water resources; and
- Promoting/ assisting research activities in State and other institutions concerned with water resources, and carrying out training for research manpower.

The contact address of CWPRS is: The Director, Central Water & Power Research Station, P.O. Khadakwasla, Pune-411 024. Homepage: *www.mah.nic.in/cwprs*.

23.3.5. National Water Development Agency (NWDA)

This Agency was set up in 1980 to promote scientific development for optimum utilisation of water resources in the country and for preparing feasibility reports for inter-basin transfer of water. The main objectives of NWDA are:

- To promote scientific development for optimum utilisation of water resources in the country.
- To carry out detailed surveys and investigations of the possible storage reservoir sites and interconnecting link in order to establish feasibility of the proposal of Peninsular Rivers Development and Himalayan Rivers Development Components of National Perspective for water resources development.
- To carry out detailed studies about the quantum of water in various Peninsular/ Himalayan River Systems and transfer of water to other basins/States after meeting reasonable needs of basin States in the foreseeable future.
- To prepare feasibility reports of various components of the scheme relating to Peninsular Rivers Development and Himalayan Rivers Development.

NWDA has many regional offices all over the country. The contact address of NWDA is: The Director General, National Water Development Agency (NWDA), 18–20 Community Center, Saket, New Delhi 110 017.

23.3.6. Water and Power Consultancy Services (India) Limited

The Water and Power Consultancy Services (India) Limited (WAPCOS) provides an integrated package of consultancy services in the water resources sector. It is an organization under the Ministry of Water Resources. The main objects for which the company was established are:

- To provide engineering and related technical and consultancy services for development of water resources, irrigation and drainage, electric power, flood control and water supply projects.
- To establish, provide, maintain and perform procurement, inspection, management of construction and related services in connection with the construction of water resources development projects, thermal power stations, and transmission and distribution systems.

The company carries out business relating to:

- pre-investment investigations, planning, design, supervision of construction, operation and maintenance of works involved in the development and utilisation of water resources, generation and utilisation of electric power; and
- topographic surveys, aerial photography, hydrological surveys, meteorological surveys, geological surveys, material surveys, underground resources investigations, soil surveys and land classification surveys.

WAPCOS has provided its services for a large number of projects within India and abroad. Its head office is located at: Kailash, 26 Kasturba Gandhi Marg, New Delhi. Their Homepage is: *www.wapcos.net*.

23.3.7. Narmada Control Authority

The Narmada Water Disputes Tribunal, constituted to adjudicate upon the water dispute pertaining to Narmada River, in its final order has provided for the setting up of a machinery for implementing its directions and decisions. Accordingly, the Central Government framed the Narmada Water Scheme, inter-alia, constituting the Narmada Control Authority (NCA) to give effect to the decisions of the Tribunal.

Accordingly, NCA was set up in 1980 mainly for the purpose of overseeing the implementation of the award of the Narmada Water Dispute Tribunal (1979) for planning and management of the river basin, including storage apportionment, regulation and control of Narmada waters, sharing of power benefits from Sardar Sarovar Project, etc. The Authority is headed by the Secretary, Union Ministry of Water Resources as Chairman. The Executive Member of the Authority is in-charge of administrative working. The funding of the Authority is from the States of Madhya Pradesh, Maharashtra, Gujarat, and Rajasthan in equal share. A Review Committee headed by the Union Minister of Water Resources and Chief Ministers of the States of Madhya Pradesh, Maharashtra, Gujarat and Rajasthan as members has been set-up to take final decisions on matters of any disagreement in the Executive Committee. The decisions of the Review Committee are final and binding on all co-basin states. The main functions of the Authority are:

- Overall co-ordination and direction of the implementation of all the projects in Narmada Basin, including the engineering works, the environmental protection measures and the rehabilitation programme and to ensure the faithful compliance of the terms and conditions stipulated by the Central Government at the time of clearance of the aforesaid projects.

- To do any or all things necessary, sufficient and expedient for the implementation of the order of the tribunal with respect to:
  - the storage, apportionment, regulation and control of the Narmada waters,
  - sharing of power benefits from Sardar Sarovar Project,
  - regulated releases by Madhya Pradesh, and
  - acquisition of lands and properties likely to be submerged by the concerned States for the Sardar Sarovar Project.

Besides the activities of implementation of environmental safeguards and Resettlement and Rehabilitation activity in Sardar Sarovar Project, the Authority is now implementing the first phase of the scheme for setting up a hydro-met network in the Narmada basin. The head office of NCA is in Indore, Madhya Pradesh.

A contract was signed by NCA with M/s ECIL in September, 1996 for implementation of Real Time Data Acquisition System (RTDAS) Network comprising 26 Remote Stations (RS) and a Master Control Center (MCC) located at Indore. Depending upon the configuration, the Remote stations, will automatically collect various hydro- meteorological data such as water level, rainfall, evaporation, radiation, wind speed and direction, relative humidity and ambient temperature and transmit the same, via INSAT-2B to the MCC. The work is in an advanced stage of implementation.

23.3.8. India Meteorological Department (IMD)

The India Meteorological Department (IMD) was established in 1875. It is the national meteorological service and the principal government agency in matters relating to meteorology, seismology and allied subjects. IMD has units all over the country which are engaged in collecting meteorological and seismological data besides providing various meteorological services. Its main objective is to provide meteorological information for weather sensitive activities, such as aviation, shipping, agriculture, irrigation, off-shore oil exploration, and industries. IMD also issues warnings against severe weather phenomena, such as cyclones, dust-storms, heavy rains, and cold and heat waves. Besides, it also provides climatological information, records earthquakes and promotes research in meteorology. The department maintains an extensive network of modern observatories all over the country.

IMD also carries out operation and coordination of 22 types of atmospheric monitoring networks. The list of these atmospheric monitoring networks is presented in Table 4. Meteorological/climatological, air pollution and other specialized observation of trace atmospheric constituents are included in this list. Meteorological observations began in India as early as 1793, after the establishment of the first observatory at Madras (now Chennai). IMD maintains 559 surface meteorological, about 35 radio-sonde and 64 pilot balloon stations for monitoring the upper atmosphere. IMD also made specialized observations for agro-meteorological purposes at 219 stations and monitor radiation parameters at 45 stations. Besides, IMD also has 70 observatories that monitor current weather conditions for aviation.

Table 4. Atmospheric monitoring networks

SN	Type of observatory	Number
1	Surface observatories	559
2	Pilot balloon observatories	65
	a. RS/RW observatories	34
	b. Only RS observatories	1
3	Aviation current weather observatories	71
4	Aviation forecasting offices at national and international airports	19
5	Regional area forecast center	1
6	Storm detecting radar stations	17
7	Cyclone detection radar stations	10
8	High-wind recording stations	4
9	Stations for receiving cloud pictures from satellites	
	a. Low-resolution cloud pictures	7
	b. High-resolution cloud pictures	1
	c. INSAT – 1B cloud pictures (SDUC stations)	20
	d. APT Stations in Antarctica	1
	e. AVHRR station	1
10	Data Collection Platforms through INSAT	100
11	Hydro-meteorological observatories	701
	a. Non-departmental rain gauge stations	
	i. Reporting	3, 540
	ii. Non-reporting	5, 039
	b. Non-departmental glaciological Observations (non-reporting)	
	i. Snow gauges	21
	ii. Ordinary rain gauges	10
	iii. Seasonal snow poles	6
12	Agro-meteorological observatories	219
13	Evaporation stations	222
14	Evapotranspiration stations	39
15	Seismological observatories	58
16	Ozone monitoring	
	a. Total ozone and Umkehr observatories	5
	b. Ozone-sonde observatories	3
	c. Surface ozone observatories	6
17	Radiation observatories	
	a. Surface	45
	b. Upper air	8
18	Atmospheric electricity observatories	4
19	(a) Background pollution observatories	10
	(b) Urban Climatological Units	2
	(c) Urban Climatological Observatories	13
20	Ships of the Indian voluntary observing fleet	203
21	Soil moisture recording stations	49
22	Dew-fall recording stations	80

Source: http://www.imd.ernet.in

Although, monitoring of extreme events is done at all the weather stations by the India Meteorological Department, the monitoring and forecasting of tropical cyclones is specially done through three Area Cyclone Warning Centers as well as three cyclone warning centers located at Mumbai, Chennai, and Kolkata, and Ahmedabad, Vishakhapatnam and Bhubaneshwar respectively. These warning centers issue warnings for tropical storms and other severe weather systems affecting Indian coasts. To observe and forewarn severe weather events, particularly tropical cyclones, storm and cyclone detections radars have been installed all along the coast and some key inland locations. The radar network upgraded work using Doppler Radars is under progress at many locations.

IMD also monitors climate to detect change, predict climate change, and determine the effects of climate change; and contributes to global observational efforts. A National Climate Center (NCC) has been established at IMD's research wing in Pune to undertake climate research, applications, data collection and management, and impact awareness studies. Data collected from the network of observations are archived at the Department's National Data Center in Pune. IMD has a dedicated meteorological telecommunication network with the central hub at New Delhi. The Center holds over 65 million records in its archives. The National Meteorological Telecommunication Center (NMTC) under IMD functions as a Regional Telecommunication Hub (RTH) under the WWW Global Telecommunication System.

The IMD homepage (*http://www.imd.ernet.in*) contains dynamically updated information on all-India weather and forecasts, special monsoon reports, satellite cloud pictures updated every three hours, Limited Area Model (LAM) generated products and prognostic charts, special weather warnings, tropical cyclone information and warnings, weekly and monthly rainfall distribution maps, earthquake reports, etc. It also contains a lot of static information, including temperature and rainfall normals over the country and an overview of the activities and services rendered by IMD.

In the field of earth and atmospheric sciences, a center for exchange of satellite data has been established at IMD New Delhi. IMD has installed about 100 meteorological data collection platforms (DCP) all over the country. IMD plays an important role in collaboration with the National Physical Laboratory *(www.npl-cgc.ernet.in)* for long-term data collection related to climate change at the Indian base in East Antarctica (Schirmacher Oasis) where a Station known as *Maitri* is operational since 1983. The head office of IMD is located in New Delhi. Their contact address is: India Meteorological Department, Mausam Bhawan, Lodhi Road, New Delhi.

23.3.9. Indian Council of Agriculture Research (ICAR)

ICAR is a R & D organization under the Ministry of Agriculture, which acts as a repository of information and provides consultancy on agriculture, horticulture, resource management, animal sciences, agricultural engineering, fisheries, agricultural extension, agricultural education, and agricultural communication. It coordinates agricultural research and development programmes and develops linkages at national and international level with related organizations for improvement in agricultural practices and to enhance the quality of life of the farming community.

The principal executive officer of ICAR is the Director-General who is also the Secretary to the Government of India in the Department of Agricultural Research and Education (DARE). The Director General is assisted by eight Deputy Directors-General - one each in-charge of Crop Sciences, Natural Resource Management, Animal Sciences, Agricultural Education, Agricultural Extension, Fisheries, Horticulture and Agricultural Engineering.

ICAR has established various research centers to meet the agricultural research and education needs of the country. It pursues human resource development in the field of agricultural sciences by setting up numerous agricultural universities spanning the entire country. The Technology Intervention Programmes also form an integral part of ICAR's agenda which establishes Krishi Vigyan Kendras (KVKs) responsible for training, research and demonstration of improved technologies. For further information, please log on to *http://www.icar.org.in/*

23.3.10. Water Technology Center for Eastern Region, Bhubaneswar, Orissa

It is a pioneer research institute in the area of sustainable water management under Indian Council of Agricultural Research (ICAR). The goal of WTCER is to develop and transfer improved water management technologies for sustainable agricultural production in the eastern region.

23.3.11. National Water Resources Council

The National Water Resources Council (NWRC) was set up by the Government of India in March 1983. The Prime Minister is the Chairman, Union Minister of Water Resources is the Vice-Chairman and Union Ministers of Finance, Agriculture, Rural Development, Planning, Urban Development, Energy, Surface Transport and Science and Technology, Chief Ministers of State and Administrators/Lt. Governors of the Union Territories are the Members. Secretary, Ministry of Water Resources is the Secretary of the Council. The functions of the Council are as follows:

- To lay down the national water policy and to review it from time to time.
- To consider and review water development plans submitted to it (including alternative plans) by the National Water Development Agency, the River Basin Commission, etc.
- To recommend acceptance of water plans with such modifications as may be considered appropriate and necessary.
- To give directions for carrying out such further studies as may be necessary for full consideration of the plans or components thereof.

- To advise on the modalities of resolving inter-state difference with regard to specific elements of water plans and such other issues that arise during planning to implementation of the projects.
- To advise practices and procedures, administrative arrangements and regulations for the fair distribution and utilisation of water resources by different beneficiaries keeping in view optimum development and the maximum benefits to the people.
- To make such other recommendations as would foster expeditious and environmentally sound and economical development of water resources in various regions.

The Council is supposed to meet at least once in a year but it has not played an effective role in India's water resources development.

23.3.12. National Water Board

The Government of India constituted the National Water Board (NWD) in September 1990 with the Secretary, Ministry of Water Resources, as its Chairman. The Secretaries of concerned Union Ministries, the Chairman (Central Water Commission), and the Chief Secretaries of States/Union Territories are its members. The functions of the Board are:

- To review the progress of the implementation of the National Water Policy and report to the Council.
- To recommend the setting up of appropriate organizations and institutions for the integrated development of water resources as envisaged under the National Water Policy.
- To make recommendations on the pattern of financing of the water development projects for speedy and systematic development of the water resources.
- To suggest guidelines for the development and training of personnel required for the water sector.
- To make suggestions for undertaking appropriate programmes and investment priorities in the water sector for achieving the objective of the National Water Policy.
- To consider any matter/problem associated with the development and management of the nation's water resources and as may be brought up before it.

The Board normally meets once a year.

23.3.13. Bureau of Indian Standards (BIS)

BIS is an autonomous body under the Ministry of Consumer Affairs, Govt. of India. It has been established under the Bureau of Indian Standards Act with the main functions as standardization, quality certification of products and services.

Water resources department is one of the departments of BIS. The water resources division council looks after the work of Water Resources Department. The work of this department is accomplished through sectional committees listed in Table 5.

Table 5. Sectional committees of the Water Resources Department of BIS

SN	Title of the sectional committee
1	WRD 1 Hydrometry
2	WRD 2 Terminology Related to River Valley Projects
3	WRD 3 Ground Water and Related Investigations
4	WRD 5 Geological Investigation and Subsurface Exploration
5	WRD 6 Water Resources Planning, Management And Evaluation
6	WRD 8 Foundation and Substructures
7	WRD 9 Dams and Reservoirs
8	WRD 10 Spillways including Energy Dissipaters
9	WRD 12 Hydraulic Gates And Valves
10	WRD 13 Canals and Cross Drainage Works
11	WRD 14 Water Conductor Systems
12	WRD 15 Hydroelectric Power House Structures
13	WRD 16 Hydraulic Structures Instrumentation
14	WRD 21 Safety in Construction, Operation and Maintenance of River Valley Projects
15	WRD 22 River Training and Diversion Works
16	WRD 23 Measurement of Works of River Valley Projects
17	WRD 24 Environmental Assessment and Management of Water Resource: Projects
18	WRD 25 Geosynthetics

A list of standards relevant to this book is given in Appendix x. The website of BIS can be found at *http://www.bis.org.in*. Its head office is at Manak Bhavan, 9-Bahadur Shah Zafar Marg, New Delhi.

23.3.14. Central Pollution Control Board

The Central Pollution Control Board (CPCB) is the national apex body for assessment, monitoring and control of water and air pollution. The CPCB in collaboration with the State Pollution Control Boards (SPCBs) monitors the quality of fresh water resources of the country through a network of 480 monitoring stations located all over the country. Based on such monitoring, 13 heavily polluted and 26 medium-polluted river stretches have been identified.

CPCB, a statutory organization, was constituted in September, 1974 under the Water (Prevention and Control of Pollution) Act, 1974. Further, CPCB was entrusted with the powers and functions under the Air (Prevention and Control of Pollution) Act, 1981. It serves as a field formation and also provides technical services to the Ministry of Environment and Forests of the provisions of the Environment (Protection) Act, 1986. Principal functions of the CPCB, as spelt out in the Water (Prevention and Control of Pollution) Act, 1974, and the Air (Prevention and Control of Pollution) Act 1981, and relevant to Environment (Protection) Act, 1986 are: (i) to promote cleanliness of streams and wells in different areas of the States by prevention, control and abatement of water pollution, and (ii) to improve the quality of air and to prevent, control or abate air pollution in the country.

The Parliament of India enacted the Water (Prevention and Control of Pollution) Act (1974) to maintain and restore wholesomeness of our water bodies. One of the mandates of CPCB is to collect, collate and disseminate technical and statistical data relating to water pollution.

The major functions of the Central Board at the national level are:

- Advise the Central Government on any matter concerning prevention and control of water and air pollution and improvement of the quality of air.
- Plan and cause to be executed a nation-wide programme for the prevention, control or abatement of water and air pollution;
- Co-ordinate the activities of the State Boards and resolve disputes among them;
- Provide technical assistance and guidance to the State Boards, carry out and sponsor investigation and research relating to problems of water and air pollution, and for their prevention, control or abatement;
- Plan and organize training of persons engaged in programme on the prevention, control or abatement of water and air pollution;
- Organize through mass media, a comprehensive mass awareness programme on the prevention, control or abatement of water and air pollution;
- Collect, compile and publish technical and statistical data relating to water and air pollution and the measures devised for their effective prevention, control or abatement;
- Prepare manuals, codes and guidelines relating to treatment and disposal of sewage and trade effluents as well as for stack gas cleaning devices, stacks and ducts;
- Disseminate information in respect of matters relating to water and air pollution and their prevention and control; and
- Lay down, modify or annul, in consultation with the State Governments concerned, the standards for stream or well, and lay down standards for the quality of air;

23.3.15. National Environment Engineering Research Institute

The National Environment Engineering Research Institute (NEERI), Nagpur, was established in the year 1958, as Central Public Health Engineering Research Institute (CPHERI) and renamed in 1974 as National Environmental Engineering Research Institute (NEERI). While fulfilling its commitment towards the societal missions, NEERI has made significant contributions in identified thrust areas such as:

- Environmental Monitoring
- Environmental Biotechnology
- Hazardous Waste Management
- Environmental Systems Design, Modeling & Optimization
- Environmental Impact & Risk Assessment and Audit
- Environmental Policy Analysis.

NEERI is engaged in providing solutions to environmental and natural resource problems. The office of NEERI is located at Nehru Marg, Nagpur – 440 020. The home page of the NEERI is http://neeri.nic.in/.

23.3.16. Central Arid Zone Research Institute (CAZRI)

Research activities pertaining to various aspects of arid zones are being conducted at the Central Arid Zone Research Institute (CAZRI), Jodhpur. This Institute was established with the prime objective to undertake research into sand-dune stabilization, afforestation of the arid saline land of Rann of Kutch, Aravalli Hills, and the Indira Gandhi Nahar Project (IGNP) command area.

A Desert Afforestation Station was established in 1952 at Jodhpur, which was later expanded into Desert Afforestation and Soil Conservation Station in 1957, and finally upgraded to Central Arid Zone Research Institute in 1959. CAZRI became an ICAR institute in 1968. Many of the developed technologies were transferred to the field through various developmental programmes, including Drought Prone Area Programme (DPAP) and Desert Development Programme (DDP) of Government of India, initiated in 1972 and 1978, respectively.

There are four regional stations of CAZRI at Jaisalmer, Bikaner and Pali in Rajasthan and Bhuj in Gujarat to carry out location specific research. Besides, there are four experimental field areas located in different agro-ecological sub-regions.

23.3.17. Indian Institute of Tropical Meteorology (IITM)

This institute was established under the UNDP's Special Fund Project as the Institute of Tropical Meteorology (ITM) at Pune in 1962 under the India Meteorological Department (IMD). In 1971, the Government of India gave it an autonomous status with a new name, i.e., the Indian Institute of Tropical Meteorology (IITM). The IITM functions under the Department of Science and Technology, Government of India.

IITM is a national center for basic and applied research in monsoon meteorology of the tropics in general with special reference to monsoon meteorology of India and neighbourhood. Its primary functions are to promote, guide and conduct research in the field of meteorology. The office of IITM is located at Dr. Homi Bhabha Road, Pune. The web site is *www.tropmet.res.in*.

IITM has made significant contributions in weather forecasting, climatology and global change, hydrometeorology, monsoon, climate modelling, weather modification, instrumentation for the observational studies, and studies relating to land-surface processes. Following are its significant achievements:

- Developed weather forecasting models on different scales by identifying various teleconnections and related predictors.
- Constructed long homogeneous series of monsoon rainfall and surface air temperature to study their spatial and temporal variability and regional climate change.

- Documented the tele-connections of the Indian rainfall with El Nino events.
- Developed a multilevel global spectral model for investigation of the dynamics and predictability of monsoon flow.
- Developed an ocean model to simulate surface climatology and to study the interannual variability of the North Indian Ocean.
- Published an Atlas containing generalised Probable Maximum Precipitation (PMP) charts for different parts of India.
- Developed instruments for measurements of various atmospheric electric parameters for the study of atmospheric electricity under different environmental conditions and of different micrometeorological parameters in the atmospheric boundary layer.
- Undertaken the study of monsoon variability on different time scales using different global climate models.

23.3.18. National Center for Medium Range Weather Forecasting

The National Center for Medium Range Weather Forecasting is the premier institution in India to provide medium range weather forecasts through deterministic methods and to render agro-meteorological advisory services to farmers. The center offers challenging research opportunities in numerical weather prediction, diagnostic studies, and crop weather modeling. Its objectives are: i) development of operational and regional scale numerical weather prediction (NWP) models for forecasting weather in medium range (3-10 days) time scale, ii) to inform and guide the farmers in advance to undertake various farming activities based on the expected weather, iii) to set-up 127 agrometeorological advisory service (AAS) units, each unit representing one of the 127 agroclimatic zones spread all over India, and iv) to set-up satellite based V-SAT Network for enabling stable/fast dedicated communication with AAS units.

At NCMRWF, R&D activities on various areas of meteorology are carried out to improve the quality of medium range weather forecasts. All the research efforts lead to the development and improvement of the Operational Weather Forecast System of NCMRWF. The contact address of NCMRWF is: Mausam Bhavan Complex, Lodi Road, New Delhi-110 003, Home page *www.ncmrwf.gov.in.*

23.3.19. Central Electricity Authority

The Central Electricity Authority (CEA) is a statutory organization, under Ministry of Power. The authority comprises a Chairman and six full time Members. Chairman is the head of the organization and the Members are over all in-charge of different functional wings, such as Planning, Thermal, Hydro, Grid Operation and Distribution, Economic & Commercial. The CEA website is at *www.cea.nic.in.* Its head office is: Central Electricity Authority, Ministry of Power, Sewa Bhawan,

R.K. Puram, New Delhi-110,066. The Authority has been entrusted with the following main functions:

- To develop a sound, adequate and uniform national power policy, formulate short-term and perspective plans for power development and co-ordinate the activities of planning agencies in relation to the control and utilization of national power resources.
- To act as arbitrators in matters arising between the State Government or the Electricity Board and a licensee or other person as provided in the Electric (Supply) Act.
- To collect and record data concerning generation, distribution and utilization of power and carry out studies relating to cost, efficiency, losses, benefits and similar matters.
- To make public from time to time information secured under the Act and to provide for the publication of reports and investigation;
- To advise any State Government Board, Generating Company or any other agency engaged in the generation or supply of electricity on such matters as will enable such Government, Board, generating company or other agency to operate and maintain the power system under the ownership or control in an improved manner and where necessary in co-ordination with any other Government, Board, generating company or other agency owing or having the control of another power system;

23.3.20. G. B. Pant Institute for Himalayan Environment & Development

The Government has established the G.B. Pant Institute for Himalayan Environment and Development in Almora, Uttaranchal, with the specific mandate for generating and strengthening the knowledge about the ecology and sustainable development of the Indian Himalayas. The Institute is also involved in integrating and collecting traditional knowledge for sustainable and integrated development of watersheds.

23.3.21. National River Conservation Directorate

The Central Ganga Authority (established in 1985) was concerned with policies for works to be taken up under the Ganga Action Plan. With the approval of the National River Conservation Plan (NRCP) in July 1995, the Central Ganga Authority has been renamed as the *National River Conservation Directorate* (NRCD). NRCD coordinates the implementation of the schemes under the Ganga and other Action Plans and has important role in improvement in river water quality and environment in India. Its office is located at Paryavaran Bhavan, CGO Complex, Lodi Road, New Delhi. Further details about NRCD are available at its Website: http://envfor.nic.in

23.3.22. National Hydroelectric Power Corporation (NHPC)

National Hydroelectric Power Corporation was incorporated under the Company's Act (1956) in 1975 as a Central Government Enterprise for hydro power development in Central Sector. Since then, NHPC has become the largest organization for hydropower development in India, with capabilities to undertake all the activities from conceptualization to commissioning in relation to setting up of hydro projects. NHPC's areas of expertise in hydropower projects include detailed project report preparation; geological & geo-technical studies; design & engineering; underground construction - tunnels, shafts, power houses, transformer galleries; construction equipment, planning & management; operation & maintenance; and renovation, modernization & uprating.

Major projects that have been constructed or are under construction by NHPC include Dulhasti (390 MW), Chamera –II (300 MW), Dhauliganga –I (280 MW), Teesta –V (510 MW), Indirasagar (JV) (1,000 MW), Sewa –II (120 MW), Purulia PSS (JV) (900 MW), Omkareshwar (JV) (520 MW), Teesta low dam –III (132 MW), and Parbati -II (400 MW). Thus the total installed capacity created by NHPC stands close to 4,550 MW. Homepage of NHPC at www.nhpcindia.com contains latest details about the organization.

23.3.23. National Thermal Power Corporation (NTPC)

National Thermal Power Corporation Limited (NTPC) is the largest thermal power generating company of India. NTPC is a public sector company incorporated in the year 1975 to accelerate power development in the country as a wholly owned company of the Government of India. Within a span of 30 years, NTPC has emerged as a national power company, with power generating facilities in all the major regions of the country.

NTPC's core business is engineering, construction and operation of power generating plants and also providing consultancy to power utilities in India and abroad. By the year 2006, the installed capacity of plants owned by NTPC was 23,750 MW. Recognizing its performance and potential, the Government of the India has identified NTPC as one of the *Navratnas* (jewels) of Public Sector. Recently, NTPC has also made forays in the field of hydropower generation. Homepage of NTPC at www.ntpc.co.in contains detailed information about the organization.

23.4. RIVER BASIN ORGANIZATIONS

This section describes some important organizations that have been set-up to deal with planning and management of specific river basins or to attend to some water resources related problem in a river basin. Three types of river basin organizations have been set-up in India:

- For execution or regulation of projects: Damodar Valley Corporation, Tungabhadra Board, Bhakra-Beas Management Board, Mahi Control Board,

Chambal River Board, Betwa River Board, Bansagar Control Board, and Upper Yamuna River Board.

- For preparation of master plan of a basin: Ganga Flood Control Commission, Brahmaputra Board, and Sone River Commission.
- For implementation of Tribunal award: Narmada Control Authority.

Please note that all the organizations included here may not precisely meet the requirements of a river basin authority. Some of these are described next.

23.4.1. Damodar Valley Corporation

After the catastrophic flood of 1943, the Governor of Bengal appointed the Damodar Flood Enquiry Committee to suggest remedial measures, suggested creation of an Authority similar to that of Tennessee Valley Authority (TVA) in the United States. Mr. W.L. Voorduin, a senior engineer of TVA, prepared the preliminary memorandum embodying the outline of a plan designed for achieving flood control, irrigation, power generation and navigation. As a result, the Damodar Valley Corporation (DVC) came into existence in 1948 for the development and management of the basin as a whole. The functions of DVC include promotion and operation of irrigation, water supply, drainage, hydro-electric and thermal power generation, flood control navigation, afforestation, control of soil erosion, public health, agricultural, industrial, economic and general well being in the Damodar Valley. The Corporation is headed by a Chairman with two members appointed by the Central Government.

The Damodar Valley Project, as envisaged by Mr. Voorduin, was a multipurpose project to afford flood control, to generate 200 MW power and to provide irrigation facilities to about 0.308 Mha in West Bengal by constructing eight dams and a barrage. However, it was later decided to have only four dams, namely, Tilaiya, Konar, Maithon and Panchet and a barrage at Durgapur under the control of DVC. Subsequently, Tenughat dam was constructed by the Government of Bihar mainly for industrial water supply. The flood absorption capacity of 1,867 MCM could only be created as against the originally planned figure of 3,580 MCM due to the reduced scope of work. At present, as against the broader scope envisaged originally for functioning of DVC, irrigation has been handed over to the Government of West Bengal, the Tenughat Project operates outside the purview of DVC and the navigation has hardly taken off. DVC now continues to be in existence for the management and operation of all the projects under its control, excluding water and power distribution to consumers.

23.4.2. Ganga Flood Control Board and Ganga Flood Control Commission

The Government of India had set up the Ganga Flood Control Board in 1972. It was also decided to set up the Ganga Flood Control Commission (GFCC) for attending to the specific works of Ganga basin and for assisting the Ganga Flood Control Board. The important functions assigned to GFCC are:

- To prepare a comprehensive plan of flood control for the Ganga basin. The field investigation and collection of data for the purpose will be carried out by the State Governments.
- To draw out a phased and coordinated programme of implementation of works in the basinwise plans.
- To advise the concerned states to follow guidelines in respect of quality control material specifications, and maintenance in order to ensure the implementation of works and the maintenance thereof to proper standards.
- To evaluate the performance of major flood control measures executed by the states, including all the inter-state flood control schemes.
- To make an assessment of the existing waterways under the road and rail bridge and to determine additional waterways to be provided for reducing the drainage congestion to reasonable limits.
- To monitor the execution of the important flood control schemes particularly those receiving central assistance or being executed under the central sector:
- Examination of flood control drainage, anti-water logging and anti erosion schemes of Ganga sub basin states except for the schemes of the States of Haryana, Uttar Pradesh, and NCT of Delhi

The Commission is headed by a Chairman. In addition, it has two full time Members and part time Members from the states of Bihar, Uttar Pradesh, West Bengal Madhya Pradesh, CWC, CWPRS, Ministry of Surface Transport Railway Board and Chief engineers of all co-basin States of Haryana, Himachal Pradesh, Rajasthan and NCT of Delhi. So far, the GFCC has completed the master plans of 23 river systems of Ganga basin.

23.4.3. Brahmaputra Board

The Brahmaputra Board was set up in 1980 to prepare a master plan for the control of floods in the Brahmaputra valley giving due regard to the overall development and utilisation of the water resources of the valley for irrigation, hydropower, navigation and other beneficial purposes. The Board is headed by a Chairman and has one member each from the States of Assam, Meghalaya, Nagaland, Manipur, Tripura, Mizoram and Arunachal Pradesh apart from members from the Central Water Commission, Central Electricity Authority, Geological survey of India and India Meteorological Department. The main functions of the Board are:

- Preparation of plan for flood control and utilisation of water resources for various uses;
- Preparation of detailed reports and estimates for the proposed projects; and
- Construction, maintenance and operation of multi-purpose projects with the approval of Central Government.

The Board has so far prepared a Master Plan Part-I for main stem of Brahmaputra River, Part-II for the Barak Sub-basin and Master Plan for nine tributaries of

Brahmaputra River and six rivers of Tripura under the Master Plan Part-III. North Eastern Hydraulic and Allied Research Institute (NEHARI) under Brahmaputra Board also carries out hydraulic model studies and soil material research.

The works under progress include surveys and investigation for preparation of Master Plans in addition to survey and investigation of specific drainage schemes, multi-purpose projects, viz., Pagladiya, Tipaimikh, Subansiri, Dihang, Lohit and Kulsi. The Board is also setting up North Eastern Hydraulic and Allied Research Institute at North Guwahati. The headquarters of the board are at Guwahati.

23.4.4. Farakka Barrage Project, Farakka

The Farakka Barrage Project is designed to serve the need of preservation and maintenance of the Calcutta Port by improving the regime and navigability of the Bhagirathi-Hoogly River system. The Farakka Barrage Project organization has been assigned the work of execution of the following principal components of the project:

- A 2,245 m long barrage across the Ganga River at Farakka with rail-cum-road bridge, necessary river training works and a head regulator on the right side.
- A 213 m long barrage across the Bhagirathi River at Jangipur.
- Feeder Canal of 1,133 cumec carrying capacity and 38.38 km long, taking off the head regulator on the right of the Farakka Barrage.
- Navigation works, such as locks, lock channels, shelter basins, navigation lights and other infrastructure.

23.4.5. Upper Yamuna River Board (UYRB)

The Upper Yamuna River Board (UYRB) is a subordinate office of the Union Ministry of Water Resources. This Board was created as a follow-up of a MoU that was signed in 1995 amongst the party basin states for the sharing of the waters of Yamuna River up to and including Okhla Barrage. The Board has been set up primarily for:

- regulation of the allocation of available flows amongst the beneficiary states and also for monitoring the return flows;
- monitoring conserving and upgrading the quality of surface and ground water;
- maintaining hydro-meteorological data for the basin;
- over-viewing plans for watershed management; and
- monitoring and reviewing the progress of all projects up to and including Okhla barrage.

The head office of the Board is in New Delhi.

23.4.6. Tungabhadra Board

The Tungabhadra Board was constituted for completion of the Tungabhadra Project and for its operation and maintenance. The Board is in charge of the common portion of the Tungabhadra Project. The Krishna Water Disputes Tribunal has made a specific provision in its Award for the use of Tungabhadra waters by the States of Karnataka and Andhra Pradesh. The responsibility for carrying out these specific provisions relating to the use of Tungabhadra waters has been entrusted to the Tungabhadra Board by the Tribunal. The Board is regulating water for irrigation, hydropower generation and other uses from the reservoir.

The Board consists of the Chairman, a member representing Government of India, and two Members, one each representing the States of Andhra Pradesh and Karnataka. The Government of Andhra Pradesh and the Government of Karnataka provide funds in an agreed proportion.

The working table for canal wise distribution of water to the States is prepared every year by the Tungabhadra Board in consultation with the State Governments, and is reviewed from time to time during the water year. The regulation of water is carried out in accordance with this working table. The quantum of water assessed during the water year 2000–2001 was 4,956 MCM. Two power houses are maintained by the Tungabhadra Board, with a total installed capacity of 72.00 MW. Nearly 200.00 million units of power is generation during a water year. The generated power is shared between the States of Karnataka and Andhra Pradesh in the ratio of 20:80.

23.4.7. Bhakra Beas Management Board

The Bhakra-Nangal and Beas Projects were originally the joint ventures of the States of erstwhile Punjab and Rajasthan. On reorganisation of Punjab in 1966, the Bhakra Management Board was constituted by the Government of India in 1967 for the administration, maintenance and operation of Bhakra-Nangal Project. On completion, the Beas Project works were transferred to Bhakra Management Board in 1976 and it was renamed as Bhakra Beas Management Board (BBMB).

The Board is under the Ministry of Power and has been assigned administration, maintenance and operation of Bhakra & Nangal projects on Sutlej River and Beas Project (Unit I & Unit II) on Beas River. The Board consists of a whole time Chairman, two whole time Members, and one representative each from the States of Punjab, Haryana, Rajasthan and Himachal Pradesh besides two representatives from Government of India (one each from Ministry of Power and Ministry of Water Resources). The functions of BBMB are:

a. to regulate the supply of Sutlej, Ravi and Beas waters to the States of Punjab, Haryana, Rajasthan, Delhi and Chandigarh (UT), and
b. to distribute power from Bhakra Nangal and Beas projects to the States of Punjab, Haryana, Rajasthan, Himachal Pradesh, Jammu & Kashmir and Chandigarh (UT).

The Bhakra Nangal Project comprises of Bhakra Dam (225 m high), Nangal Dam to feed Nangal Hydel Channel and Anandpur Sahib Hydel Channel, and Nangal Hydel Channel, 65 km long with head discharge of 354 cumec. Beas Project comprises of two units – Unit I, the Beas-Sutlej Link Project and Unit-II, the Beas Dam at Pong:

Unit I consists of Pandoh Dam (76 m high); Pandoh Baggi Tunnel (13.1 km long, 254.9 cumec capacity), Sundernagar Hydel Channel (11.8 km long, 240.7 cumec capacity), and Sundernagar Sutlej Tunnel (12.35 km long, 403.5 cumec capacity); and Balancing reservoir with a capacity of 3.7 MCM. Unit II, the Beas Project consists of Pong Dam (132.6 m high).

The office of BBMB is at: Bhakra Beas Management Board, 19-B, Madhya Marg, Chandigarh.

23.4.8. Bansagar Control Board

The Bansagar Control Board was set up by the Government of India in 1976. The Board functions from Rewa (Madhya Pradesh) accordance with an agreement reached between the Governments of Madhya Pradesh, Uttar Pradesh and Bihar in 1973 for sharing the waters of Sone River and the cost of the Bansagar Dam. The resolution was amended in 1990.

The objectives of the board are to ensure an efficient, economical and early execution of Bansagar dam, including all connected works in Madhya Pradesh, excluding the canal systems. The Union Minister of Water Resources is the Chairman of the Board and the Minister of State for Water Resources, Union Minister of Power, Chief Ministers, Minister-in-Charge of Irrigation and Finance of the three States and Minister-in-Charge of Electricity of Madhya Pradesh are its members. The Executive Committee set up under the Chairmanship of the Chairman, Central Water Commission, manages the day-to-day affairs of the Board.

23.4.9. Betwa River Board

In accordance with the inter-state agreement of July 1972 between Uttar Pradesh and Madhya Pradesh, a decision was taken to constitute a Control Board for the execution of the Rajghat Dam Project, a joint venture of Madhya Pradesh and Uttar Pradesh. Accordingly, the Betwa River Board was constituted by the Ministry of Water Resources for an efficient, economical and early execution of the project. The headquarters of the Board are in Jhansi (Uttar Pradesh).

The Union Minister of Water Resources is the Chairman of the Board. Union Minister of Power, Union Minister of State for Water Resources, Chief Ministers and Ministers-in-charge of Finance, Irrigation and Power of the two States are its members. An executive committee of the board headed by the Chairman, Central Water Commission, manages the activities of the Board.

23.5. STATE GOVERNMENT ORGANIZATIONS

23.5.1. Water and Land Management Institutes (WALMIS)

Recognizing the need for educating the canal managers and canal water users, Water & Land Management Institutes (WALMIs) were established in a number of states during late 1980s. In addition to train canal managers and in-service staff of irrigation departments, WALMIs also conduct training courses for farmers covering efficient water management in the field. The first WALMI was established in Maharashtra in Aurangabad in 1980, followed by a similar Institute in Gujarat at Anand. Subsequently, WALMI's were set up in Hyderabad, Patna, Bhopal, Bhubaneswar, Kota, Trichy, Dharwad and NOIDA. A list of WALMI's is given in Table 6.

The basic objectives of WALMIs (WG 1999) are:

1. To provide in-service training of multidisciplinary nature to the personnel engaged in irrigated agriculture activities,
2. To undertake applied/action research on land, water and crop management,
3. Activities which will promote optimization of water use and land resources, including consultancy services, publication of literature, holding seminars and workshop etc.
4. Organizing training programme at project level for field operators and farmers for transfer of technology to the grass root level.

The WALMIs were conceived as state-government-supported, autonomous, inter-disciplinary training institutes, with learning through action research. However, the performance of the WALMIs has been highly skewed; some are doing well and some not. Many WALMIs do not have adequate faculty.

23.5.2. Central Design Organizations

Most state governments have established design organizations under the irrigation/water resources ministries for the purpose of design of hydraulics structures. Most such organizations do routine design work but at a few places, good research

Table 6. List of WALMIs in India with years of establishment

Name of institution/ location	State	Year of Establishment
WALMI, Aurangabad	Maharashtra	1980
WALMI, Anand	Gujarat	1980
WALMI, Hyderabad	Andhra Pradesh	1983
WALMI, Patna	Bihar	1984
WALMI, Bhopal	Madhya Pradesh	1984
WALMI, Bhubneshwar	Orissa	1984
IMTI, Kota	Rajasthan	1984
IMTI, Trichy	Tamil Nadu	1984
WALMI, Dharwad	Karnataka	1985
WALMI, NOIDA	Uttar Pradesh	1985
Center for Water Resources Development & Management (CWRDM), Calicut	Kerala	1988

Source: WG (1999).

work has also been and is being carried out. Water being the state subject, the state governments have very important role to play and development and management of the country. In this context, these design organizations have a key role to play.

23.5.3. Up Irrigation Research Institute

Beginning with a small research cell under the Irrigation Department in Lucknow in 1927, U.P. Irrigation Research Institute (UPIRI) came into being in the present form in Roorkee in 1954. It is a research station engaged in R and D activities for various multipurpose river valley projects in India. This station is associated with all the major activities of project planning, execution and operation, namely: (i) Investigation, (ii) Design, (iii) Execution, and (iv) Operation and Monitoring. Headed by a Director, UPIRI has three main wings: (i) Hydrology, (ii) Sub-surface exploration, and (iii) Material testing. A field station has been setup at Bahadarabad (about 15 km away) on the banks of the Upper Ganga Canal. It gets water supply from canal at 14 cumec for testing hydraulic models.

UPIRI has been providing basic help to the project formulators in the following fields: a) Runoff measurements, b) Sediment transport and sedimentation studies, c) Evapotranspiration, d) Seepage from canal systems, e) Rainfall and irrigation recharge to ground water, f) Foundation investigations and in-situ tests, g) Soil and rock mechanics, h) Advising on most viable layout of hydraulic structures, and i) Suggesting river training and appurtenant works.

UPIRI has made contributions in evolving unconventional designs of intake fall for a number of hydropower projects. It has also developed energy dissipators on important irrigation systems. Some important achievements of the institute are: Sharda type fall, hopper type sediment excluders and ejectors for canals, design of spillway of Rihand and Lakhwar dams, and model studies for the Tehri dam. UPIRI has also played an important role in suggesting the alignment of the new head regulator for Parallel Lower Ganga Canal, (PLGC) taking off from existing Narora Barrage. Detailed information about UPIRI is available at: *www.uttranchalirrigation.com/iri/iri-home.htm.*

23.5.4. AP Engineering Research Laboratory

The Andhra Pradesh Engineering Research Laboratory (APERL) is a research and development organization under the Ministry of Irrigation and CAD, Government of Andhra Pradesh. It was established in 1945, and is located just downstream of Himayatsagar reservoir, near the city of Hyderabad. Apart from the main laboratories at Himayatsagar, Hyderabad, it has three regional soil testing laboratories. Further details about APERL are available at: http://www.ap.gov.in/apirrigation/organisations/aperl.htm.

23.5.5. Central Water Resources Development and Management (CWRDM)

The Central Water Resources Development and Management (CWRDM) Center, originally established at Thiruvananthapuram, was shifted to Calicut in 1979. The center initially had six scientific divisions, dealing with surface water, ground water, water management-agriculture, water quality and environment, education and extension, and library and documentation. After a decade, scientific divisions to deal with computer applications and isotope hydrology were added. To take care of the special R&D needs of different hydro-ecologic regions of Kerala, five regional centers are also in operation since 1990. The CWRDM is one among the Water and Land Management Institutes (WALMIs) started in 1990 by the Ministry of Water Resources. CWRDM also conducts many training programs for government officials and farmers. The center had academic and research linkages with many universities abroad.

The center has substantially contributed to the scientific studies and water management in the region. The hydrologic data generated by CWRDM are expected to be of immense use to the water managers of Kerala. CWRDM has played important role in supporting the Kerala Government in sorting out several water management issues.

23.5.6. North Eastern Regional Institute for Water and Land Management

North Eastern Regional Institute of Water and Land Management (NERIWALM), Tezpur, Assam an Autonomous Society under North Eastern Council, Government of India. This institute was set up at Tezpur in December, 1989 by the NEC basically to impart training to the in-service professionals, NGOs and the farmers of the North Eastern Region. It has also been further developed as a research and consultancy institute on all matters relating to water and land management.

The institute has been conducting short courses so far. Recently, NERIWALM has taken initiative to open P.G. Diploma Course in Water and Land Management to provide training to the in-service personnel (engineers and scientists) to be sponsored by irrigation and agriculture departments of the states of the North Eastern region.

23.6. ACADEMIC ORGANIZATIONS

Hydrology and water resources are an important field of teaching and research at many academic institutions in India. The major organizations are mentioned in what follows. Besides, there have been significant efforts at many other engineering colleges but these were basically due to personal initiative and efforts of a few individuals and ended when that person moved elsewhere. At a number of agricultural universities, quite naturally, water subjects are taught and research is conducted with focus on irrigation water management.

23.6.1. IITs and NITs

Indian Institutes of Technology (IITs) are the topmost institutes for under-graduate technical teaching in India. These are located in Roorkee (www.iitr.ernet.in), New Delhi (www.iitd.ernet.in), Kanpur (www.iitk.ac.in), Mumbai (www.iitb.ac.in), Chennai (www.iitm.ac.in), Kharagpur (www.iitkgp.ernet.in), and Guwahati (www.iitg.ernet.in). Comprehensive information about the IITs is available at their respective web-sites. The IITs and the Indian Institute of Science, Bangalore (www.iisc.ernet.in), have made admirable contributions to hydrology teaching and research in the country. At the Banaras Hindu University, Varanasi, the Institute of Technology has close linkages with IIT .

Currently, there are eighteen National Institutes of Technology (NITs). Earlier, these were known as regional engineering colleges which have been upgraded to play a crucial role in technical education in India. At many NITs, useful work is being carried out in the field of water resources. Other notable institutes in the field of water resources are: the JNU New Delhi; JNTU, Hyderabad; and .

23.6.2. Center for Water Resources (Anna University)

Anna University (***www.annauniv.edu***) is a prestigious institute in the field of engineering and technology. It was started as a survey school in 1794. The Center for Water Resources (CWR) was established at the Anna University in 1979 to pursue teaching, research, consultancy and impart training programmes in hydrology and water resources engineering. It is basically the upgraded hydraulics division of the Department of Civil Engineering, Guindy College of Engineering. The Center had offered a Postgraduate Diploma course of one-year duration in Hydrology and Water Resources Engineering, between 1983 and 1991. This course was sponsored by UNESCO. These days, CWR is engaged in a range of activities including teaching, research, and consultancy.

23.6.3. Water Technology Center at IARI, New Delhi

The Water Technology Center (WTC), established in 1969 at the Indian Agricultural Research Institute (IARI), New Delhi is an inter-disciplinary facility for research, teaching, training and extension in the area of Water Resources Management in agriculture. Today, the Center addresses a wide range of activities centering around water management at on-farm level, large irrigation commands and watersheds. It renders guidance to farmers, policy makers and administrators at state and national levels on various aspects related to water management. The Center offers consultancy services on water management in agriculture. The headquarters of the All India Coordinated Research Project on Agricultural Drainage are also located here.

With the initiation of the Command Area Development Programme in 1974, the research expanded to include development of scientific guidelines for water distribution and management in irrigation project areas. A post-graduate course on Water Science and Technology is running since 1996.

The objectives of the center are:

- To initiate and conduct basic and applied research at all India level in all aspects of crop water management leading to more efficient use of water and land resources,
- To provide instruction in all aspects of water management at different levels and locations,
- To collaborate with agricultural universities and other institutions and agencies in the organization and development of water management research, teaching and extension programme.
- To facilitate exchange of information on crop water management and related issues,
- To assist personnel of agricultural universities, irrigation departments, agricultural departments, and various institutions of Indian Council of Agricultural Research (ICAR), in improving their professional capabilities.

The home page of the center can be accessed at http://www.iari.res.in/divisions/wtc/.

23.7. PROFESSIONAL SOCIETIES

Several professional societies are working in India in the field of water resources. In the following, activities of three national level societies are described.

23.7.1. Indian Association of Hydrologists (IAH)

Indian Association of Hydrologists (IAH) was established in year 1977 under the Society Registration Act, 1860 of Government of India. The IAH headquarters is currently located at National Institute of Hydrology in Roorkee.

The objectives of IAH are to advance the study of hydrology and related subjects and promote their application, disseminate the knowledge and application in the field, to honor individuals for their pioneering and meritorious contribution in the field of hydrology by electing them as honorary life members or inviting them to be patrons of the association and to publish relevant literature. The association maintains a close liaison with the Indian Associations dealing with the hydrology and related subjects.

IAH has been publishing "Hydrology Journal" on a quarterly basis since 1980. The Journal has a wide circulation covering members from 15 countries. It is devoted to the publication of scientific papers of national and international significance in the field of hydrology and water resources. The contact address of IAH is: The Secretary, Indian Association of Hydrologists, National Institute of Hydrology, Jal Vigyan Bhawan, Roorkee – 247 667, Uttaranchal, India. Its website is: *www.nih.ernet.iniah.*

23.7.2. Indian Water Resources Society (IWRS)

The Indian Water Resources Society (IWRS) was founded in 1980 at the Water Resources Development Training Center (WRDTC), the then University of Roorkee. The aims and objectives of IWRS, inter-alia, include:

- The advancement of research, planning, development, management and education, as well as establishment of a common meeting group of physical, biological and social scientists, engineers, and other persons, concerned with water resources.
- To provide a focal point for the collection, organization and dissemination of ideas and information relating to the broad spectrum of water problems.
- To provide a forum for communication in water resources science and technology through its meetings and publication.
- To honour individuals for their pioneering and meritorious contribution in the field of water resources, by electing them as Honorary Life Members or inviting them to be Patrons of the society.

IWRS organizes lectures, meetings, group discussions and national level workshops and seminars on subjects of common interest. It has organized a number of workshops and seminars on various subjects. As an annual event, it organizes the curtain raiser function of Water Resources Day in Delhi on World Water Day. The society publishes a quarterly technical journal (published in January, April, July and October) which contains technical papers on all aspects of water resources. Another important annual event is Dr. Kanwar Sain Memorial Lecture which is delivered by an eminent expert on a topic relevant to water resources development.

The contact Address for IWRS is: The Secretary, Indian Water Resources Society, Department of Water Development and Management, Indian Institute of Technology, Roorkee – 247 667, Uttaranchal, India.

23.8. NON-GOVERNMENTAL ORGANIZATIONS (NGOs)

Many NGOs are active in the field of water resources in India. However, only a handful of them have made any worthwhile contribution. Besides, most NGOs are limited to social aspects of WRD projects. An introduction to selected NGOs and their activities is given in the following. By no means, the list is exhaustive.

23.8.1. Center for Science and Environment (CSE)

The Center for Science and Environment, started in 1980, is a leading environmental NGO specializing in sustainable natural resource management. Its strategy of knowledge-based activism has won it admiration for its quality of campaigns, research and publications. As noted from its web-site (http://*www.cseindia.org*), CSE promotes solutions for India's numerous environmental threats – of 'ecological poverty' and extensive land degradation on one hand, and rapidly growing toxic degradation of uncontrolled industrialization and economic growth on the other.

CSE's role is to help India create a sustainable society even in the worst of circumstances, marked by poverty, illiteracy, inequality, and population growth. CSE's organisational culture is built around an information-knowledge-wisdom chain that helps make people aware of the emerging problems and helps in building constituencies. CSE identifies solutions (intellectual leadership) and then pushes for change through lobbying with politicians and bureaucrats. CSE's educational and training programmes are knowledge investments that seek to build capacities of managers and future leaders. The Center's efforts are built around five broad activities: Communication for Awareness, Research and Advocacy, Education and Training, Documentation, and Pollution Monitoring.

The CSE has also published the *State of Global Environmental Negotiations* (GEN) reports, which uncovered the issues and politics involved in these negotiations. It has launched a campaign to establish an equitable framework for a system of global environmental governance for climate change negotiations, and has been playing an important role at several international environmental negotiations. The CSE is also involved in creating awareness about climate change through research and publications. It has proposed strategies to address ecology, economy, social justice and equity.

The office of CSE is located at: 41, Tughlakabad Institutional Area, New Delhi-110 062, India.

23.8.2. The Energy and Resources Institute (TERI)

TERI was established in 1974. In the initial period, the focus was mainly on documentation and information dissemination activities. Around 1982, research activities in the fields of energy, environment, and sustainable development were initiated. The motivation for these activities was the belief that efficient utilization of energy, sustainable use of natural resources, large-scale adoption of renewable energy technologies, and reduction of all forms of waste would drive the process of development towards sustainability.

TERI has seven research divisions: Regulatory Studies and Governance Division, Policy Analysis Division, Energy-Environment Technology Division, Bioresources and Biotechnology Division, Information Technology and Services Division, Action Programmes, and Sustainable Development Outreach. It has established regional centers in Bangalore, Goa, Guwahati, and Kolkata, and a presence in Japan and Malaysia. It has set up affiliate institutes: TERI– NA (Tata Energy and Resources Institute, North America) Washington DC, USA, and TERI–Europe, London, UK. TERI is also involved in training, capacity building, and education. In 1999, it set up the TERI School of Advanced Studies, recognized as a deemed university by the University Grants Commission, India.

TERI's maintains an informative website (www.teriin.org). It contains substantial updated information with particular reference to India. TERI publishes research journals, newsletters, and has produced many documentary films on topics ranging from rural resources to global warming.

23.8.3. Tarun Bharat Sangh (TBS)

Tarun Bharat Sangh is an NGO working on issues of natural resource management and watershed development in semi-arid region. The organization has done path-breaking work in helping communities to come together, mobilize their own resources and regenerate the traditional water harvesting structures known as *johads*. As a result of this work covering around 600 villages, aquifers and ground water sources have been recharged and two rivers (Arvari and Ruparel) which had been dry for a generation, have again become perennial streams. Community organizations (*Lok Samitis)* initiated by TBS are directly implementing the 'Pavdi' watershed initiative of the Government of Rajasthan. The TBS, which had been set up in 1975, has built 4,500 johads in 850 villages over more than 6,500 kilometers over the last 15 years. Four rivers – Aravali, Ruparel, Bhagwani, Sarsa and Jahajwali – which had been reduced to drains were flowing round the year. The primary beneficiaries of the regeneration of water resources have been women, who have experienced an improvement in the material conditions of their lives.

In recognition of the exemplary work done, Mr. Rajendra Singh, the founder of TBS was chosen for the prestigious Ramon Magsaysay Award for the year 2001. The office of TBS is located at Village Bhikampura, P.O. Kishori, Block Thanagazi, Alwar, Rajasthan.

TBS helps in building dams or Johads and also repairing the existing ones. Interestingly, they give help with some conditions. For example, in Nimbi TBS wanted 25% of the cost of the dam to be borne by the villagers, and each villager had to donate either money or labour. In addition, deforestation and alcoholism had to be stopped at once. Villagers were ready to take drastic measures to improve their lot and pitched in with all they had. It took just a few years for the water level to rise, wells to get recharged and sub-surface water levels to rise.

23.8.4. Narmada Bachao Andolan (NBA)

The construction of large dams on the River Narmada in central India and its impact on millions of people living in the project affected areas has become one of the most important social issues in contemporary India. The Narmada Bachao Andolan (NBA or Save Narmada Movement) is a people's movement which started in 1985. It is a non-governmental organization which is active in the field of environment in general and 'saving' Narmada River from destruction by mega river valley projects in particular. Leadership to the movement is provided by Ms. Medha Patkar. NBA employs tactics that are entirely non-violent; sit-ins, fasts, rallies, marches, and petitions in the court of law. While NBA has opposed every large WRD project, it has not provided any long-term master plan to tackle water related problems in India. Now that the Sardar Sarovar Project has been completed to a substantial extent, the activists of NBA are protesting every move to increase the height of the dam. NBA alleges that the Madhya Pradesh and Gujarat governments had colluded to raise the height of the Sardar Sarovar dam without proper measures to rehabilitate the affected people. NBA believes that thousands of the villagers would be rendered homeless if the height of the dam is raised further since the government has not made enough efforts for R&R. information about the activities of NBA are available at many web-sites, including www.narmada.org.

23.8.5. Development Alternatives

Development Alternatives (DA) is a non-profit research, development and consultancy organization established in 1983. Its work includes design, development and dissemination of appropriate technologies, environmental resource management methods, and effective institutional systems. DA's outreach activities seek to create awareness among various stakeholders, such as NGOs, government agencies, industries, financial institutions, and communities on climate change issues. Its Climate Change Center (CCC) has developed training modules on incorporating sustainable development concerns in climate change projects in India. The Urban Environment System Group has a nationwide programme called CLEAN—India, to raise awareness about energy and resource conservation, and mobilizing communities for response measures.

23.8.6. The Sathya Sai Trust

Although water is not the main focus of work of this Trust, it is included here in view of a major water related initiative that it has undertaken. Rayalaseema region of South India which includes districts of Bellary, Anantapur, Cuddapah, and Kurnool, is a drought-prone region. Well water, if at all available, unfortunately has too much fluoride in it. In 1994, the Sathya Sai Central Trust, headed by Sathya Sai Baba stepped in to undertake a water supply project to bring water to the drought-stricken area of Rayalaseema, in South India. Other partners in this project were M/S Larsen & Toubro Limited and the Government of Andhra Pradesh. Never has a project of this magnitude been executed within a short span of less than a year. The project covers 20,000 kilometers and includes more than 750 villages hitherto without water. Nearly 2,500 kilometers of pipelines were laid (approximately the distance from Bangalore to Delhi) and 268 overhead storage tanks were constructed. Some of the summer storage tanks that were constructed are over 100 acres in area.

Probably it was the first time that a public charitable trust had ever undertaken such a major infrastructure project anywhere in India.

23.9. INTERNATIONAL ORGANIZATIONS

In this section, we describe a few international organizations related with water whose headquarters are located in India.

23.9.1. International Commission on Irrigation and Drainage

The mission of ICID is to stimulate and promote the development and application of the arts, sciences and techniques of engineering, agriculture, economics, ecology and social sciences in managing water and land resources for irrigation, drainage, flood control and river training and for research in a more comprehensive manner adopting up-to-date techniques for sustainable agriculture in the world. ICID has by now about half-a-century of experience in the promotion and transfer of water management technology and in the handling of related issues. Building on its past experience, accomplishments, and the comprehensive water management framework, ICID strives to promote programs to accelerate and enhance sustainable growth of irrigated agriculture.

Any recognized country, independently administered by a sovereign government and having interest in the activities of the Commission, is eligible to become its member. The member countries participate in the activities of the Commission through their National Committees constituted by them to further the objects of the Commission in their countries. Starting with 11 founding member countries in June 1950, 87 countries have so far been admitted to the fold of ICID for sharing its efforts in the direction of sustainable irrigation. Currently 70 countries are actively participating.

The head office of ICID is located at 48 Nyaya Marg, Chanakyapuri, New Delhi. Their homepage is: www.icid.org.

23.9.2. International Crop Research Institute for Semi-Arid Tropics (ICRISAT)

ICRISAT is an international organization concerned with agricultural development in semi-arid tropics (SATs). A SAT is characterized by limited and erratic rainfall, nutrient-poor soil, and unpredictable weather. ICRISAT is one of the 16 centers of the Consultative Group on International Agricultural Research (CGIAR, homepage: www.cgiar.org). It has been entrusted with the conservation and management of the seeds of five crops that are crucial in the diets of the poor: sorghum, millet, groundnut, chickpea, and pigeon pea. ICRISAT maintains a large gene bank and a district level database of farming systems. Headquarters of ICRISAT are located at Patancheru in Andhra Pradesh. It has seven regional offices in Africa.

23.10. RESEARCH AND DEVELOPMENT SETUP IN WATER RESOURCES

In India, most research and development activities in water sector are funded by the Government and are undertaken in central or state government organizations and academic institutions. Recently, some NGOs have also taken up research projects but these are mostly related to social aspects of water resources development and management. Institutes under the Ministry of Water Resources, under ICAR in the Ministry of Agriculture, under Ministry of Science and Technology, and the institutes in the State Governments under ministries of Water Resources/Irrigation/Command Area Development are the major contributors to research in water resources.

A number of agricultural universities conduct research on water management using funds channeled mainly by the Indian Council of Agriculture Research. Engineering institutions also do considerable water related research. Notable among the academic institutions are: the IITs; the Indian Institute of Science, Bangalore; Anna University, Chennai; JNU New Delhi; Jadavpur University, Kolkata; M. S. University, Vadodara; JNTU & Osmania University, Hyderabad; BHU, Varanasi; National Institute of Technology, Surathkal; GB Pant University, Pantnagar; Punjab Agricultural University, Ludhiana; and many others. Among the laboratories under various ministries, those involved in research and development related to water are: the National Remote Sensing Agency, Hyderabad; Regional Remote Sensing Service Centers; Space Application Center, Ahmedabad; Physical Research Laboratory, Ahmedabad; the National Environment Engineering Research Institute, Nagpur; the Bhabha Atomic Research Center, Mumbai; Snow and Avalanche Study Establishment, Chandigarh; and many more.

In the state governments departments, water related research is done mainly through Irrigation Research Institutes/River Research Institutes. In particular, the U.P. Irrigation Research Institute, Roorkee; the Irrigation and Power Research Institute, Amritsar; the Gujarat Engineering Research Institute, Vadodara; Karnataka Engineering Research Institute, K.R. Sagar; Maharashtra Engineering Research Institute, Nasik; and Hydraulic Research Laboratory, Poondy (Tamil Nadu) are conducting noteworthy research. Apart from these, twelve Water and Land Management Institutes (WALMIs)/Irrigation Management and Training Institutes (IMTI) have also been set up. Among the existing WALMIs, those at Aurangabad (Maharashtra), Anand (Gujarat) and IMTI, Trichi (Tamil Nadu) and the Center for Water Resources Development and Management, Calicut, are active in conducting research on irrigation management.

Three organizations under the Ministry of Water Resources, viz., NIH, CWPRS, and CSMRS are fully devoted to R&D in the water sector. NIH is devoted to systematic and scientific studies in all aspects of hydrology with the objective of improving the present practices in planning, design and operation of water resources projects. CWPRS is the premier national institute for research in the area of hydraulics of water resources structure related to irrigation, hydropower, navigation, coastal works and related instrumentation. CSMRS is involved in

Table 7. R and D activities in the Ministry of Water Resources

S. N.	Activities	Committee
1	Preparation of overall policy and plans for R&D in the field of water resources development.	Science & Technology Advisory Committee (STAC) of Ministry of Water Resources.
2	Finalization of research programme of various research institutes of MoWR and its monitoring.	Technical Advisory Committees of the respective institutes.
3	Support to academic institutes and other research establishments for taking up research schemes.	Standing Advisory Committee (SAC).
4	Monitoring of research schemes of other organizations.	R & D implementation and monitoring committee.

research related to construction materials, concrete technology, geophysics, rock and soil mechanics. Other organizations of the Ministry of Water Resources, viz., Central Water Commission, Central Ground Water Board, Brahmaputra Board, and National Water development Agency are also actively involved in studies related with water and these are described in Section 23.3 & 23.4.

The MoWR provides financial assistance by way of grants to academicians/ experts in the recognized laboratories/institutes under a R&D programme to promote research in the field of water resources. Under the existing arrangements, the R&D activities in the MoWR are planned and monitored through the committees shown in Table 7

Table 8. Various Indian National Committees (INCs) and their subject domains

Committee	Subject domain
INC for Hydrology (INCOH)	Meteorology, surface water hydrology, evaporation control, ground water hydrology and management, instrumentation, real time systems, application of GIS and remote sensing.
INC for Hydraulics (INCH)	Managements of floods, hydraulic structures (including masonry and concrete structure), river and estuarine hydraulics, river morphology, ground water hydraulics, instrumentation for seismic and geophysical measurements, open channel flow, pipe flow, hydraulic machinery, city water supply, and ports and harbours.
INC for Irrigation and Drainage (INCID)	Irrigation, drainage, agronomy, water management, environmental impact and socio- economic aspects of water resources projects, plasticulture development, geo-textiles.
INC for Geo-technical Engineering (INCGE)	Rock mechanics & tunneling technology; soil mechanics & foundation engineering; and instrumentation and measurement techniques.
INC for Construction Materials and Structures (INCCMS)	Construction materials, concrete technology and structures.

Considering wide range of topics covered by water resources engineering, five Indian National Committees (INCs), have been constituted to coordinate the R&D program. The five INCs and their subject domain are described in Table 8.

The INCOH was constituted in 1982 as High Level Technical Committee on Hydrology. In addition to coordination of R&D in Hydrology, its main function is to coordinate effective participation by India in the International Hydrological Programme (IHP) of UNESCO, Operational Hydrology Programme of WMO, and other inter-governmental programmes of hydrology in which India may wish to participate. INCID, besides looking after the R&D work in irrigation and drainage is also functioning as a national committee for ICID and coordinates all activities related to ICID.

CHAPTER 24

CONCEPTS OF WATER GOVERNANCE FOR INDIA

Water is a state subject in India which loosely implies that the state governments have the sole authority and responsibility to manage water resources of that state and the role of the central government is mainly limited to interstate projects. But almost all major rivers in India flow through more than one state which brings the central government in picture in all major projects. The central government also has the sanctioning and monitoring role with respect to technical, social, and environmental aspects of all major and medium projects. Through the Planning Commission, the central government decides allocation of funds to state governments. The central government also makes decisions with respect to Command Area Development projects. Major research institutes are also under central government. Clearly, the central government has substantial influence on state policies and practices with respect to the water sector.

As per the National Water Policy (2002) of the Government of India, water allocation priorities in the planning and operation of systems should broadly be: (i) drinking water, (ii) irrigation, (iii) hydropower, (iv) ecology, (v) agro-industries and non-agricultural industries, and (vi) navigation. In view of the current status of freshwater in India and the problems that are likely to arise in future, a well thought-out and planned long-term strategy is needed for sustainable water resources management in India. Key components of such a strategy are proposed in this chapter. Needless to say, the strategy has to be multi-pronged and all the stakeholders, including the general public, will have to be involved in the decision making and implementation.

In this chapter, we discuss the major initiatives that need to be taken up for improved water governance. Good quality data is the key to all hydrologic analyses – all decisions and management actions are based on data. Therefore, the discussion begins with hydrological information system in India.

24.1. HYDROLOGICAL INFORMATION SYSTEM (HIS)

A necessary pre-requisite to wise water management is accurate, comprehensive and timely hydrologic data and data about the economic, social and environmental dimensions. Rainfall stations constitute the bulk of the hydro-meteorological network and are mainly owned by the state departments. India Meteorology Department (IMD) collects and archives data from selected stations. The climate stations are mainly maintained by IMD; very few states have good network of climatic stations. Many rainfall and climate stations are quite old (ranging from 20 to 100 years) and many of these have not been adequately maintained in the recent past. Further, the observation networks of central and state agencies often have many duplications and gaps.

Surface water gauging stations have primarily been set up to collect water level and discharge data. At some of these stations, sediment and water quality parameters are also observed. Stations belonging to the Central Water Commission (CWC) are located on major rivers, whereas those of the states are on smaller tributaries and sub-tributaries. CWC mainly gathers data for assessing overall water resources of the country, to resolve interstate water sharing issues, and for flood forecasting. The networks of state agencies cover the basins more intensely and provide data for planning & designing smaller water resources projects.

It may be noted that in many states, very rich hydrologic data are available:

- Andhra Pradesh: 300*35 station years (rainfall), 150*25 station years (reservoir data),
- Gujarat: 300*30 station years (rainfall), 150*20 station years (Gauge and Discharge data),
- Karnataka: 400*50 station years (rainfall),
- Maharashtra: 500*40 station years (rainfall), 200*25 station years (GD data), and
- CWC: 265*25 years of data (GD & sediment data).

Most ground water (GW) observations have been through open dug wells, tapping the upper unconfined aquifers. Usually, observations are taken four times a year: pre-monsoon, monsoon, post-monsoon, and winter season. Presumably, these represent the troughs and peaks of the water table hydrograph, but these are too sparse to yield reliable conclusions. Awareness about water quality data has grown in the last few decades, primarily due to huge deterioration in the quality of dwindling water resources. Whereas the river gauging authorities take samples at the gauging stations, the pollution control boards take observations for surveillance near industrial or urban centers. In the past, water quality laboratories were inadequate in numbers and capabilities. Insufficient finances have also marred operations.

24.1.1. Problems in the Availability and Use of Hydrologic Data

Despite immense efforts and money spent in collecting hydrological data, the actual use in India continues to be very limited. There are a number of reasons behind it.

- In most cases, no inventory of the data is available at a centralized place and hence the user has to visit the offices of the data collecting agencies at different locations to get the requisite data. If the concerned person in the office is not available (may be on leave), it may not be possible to locate the data.
- Most data is in the form of paper manuscripts.
- The manuscripts get deteriorated with time and at times precious data are lost.

- Many observers have received little or no training in data collection and are not aware of the utility and significance of the data that they are collecting. Frequently, this is the least priority job for them and there are instances when the data is recorded by the observer without actually visiting the site.
- The observation procedures are non-standard and non-uniform across the agencies. Hence, it is not possible to ascertain the quality of data.
- There are duplications in data collection efforts among the involved agencies as well as large areas where no agency has any station. This makes the spatial coverage highly uneven.
- Streamflow data of many river basins, such as the Ganga and the Brahmaputra, are treated as confidential and not available to general users. Likewise, many state governments do not share river gauge and discharge data observed by them on the pretext of interstate water disputes. As a result, flow data from more than one-third of the country, accounting for 40% of the utilizable surface water of the country, are not in public domain.

Ever reducing financial support, coupled with poor staffing due to restrictions on new recruitment, has rendered many river gauging and meteorological stations (except of CWC and a few states) non-operational. The observation processes on most stations are manual, although a few automatic equipment have been installed recently. Velocity is mainly measured by current (flow) meters; floats were in vogue in many states.

Unfortunately, the hydrologic information systems in many parts of the world are inadequate. Difficulties arise due to the lack of funds, non-standard procedures for data collection, quality assurance, data management, and dissemination. In India, the availability of meteorological data is very good, that of surface water satisfactory (although the quality of data is doubtful at many locations), and data on ground water and its quality is limited. Although the ground water availability maps have been prepared for certain locations, extraction rates are often not available.

24.1.2. Upgrading of his Through the Hydrology Project

To overcome most of the above deficiencies and to improve the HIS in India, a giant step was taken by completing an ambitious Hydrology Project Phase 1 (HP-1). HP-1 was aimed at developing and improving the existing HIS of various government agencies in nine peninsular states of India: Andhra Pradesh, Chhattisgarh, Gujarat, Madhya Pradesh, Maharashtra, Karnataka, Kerala, Orissa and Tamil Nadu. The states covered under the first phase of the Hydrology Project and location of 31 data centers established under the project are shown in Figure 1. Five central government organizations participated in the project: Central Water Commission, Central Groundwater Board, National Institute of Hydrology, India Meteorology Department, and Central Water & Power Research Station. This assisted in gathering reliable and spatially intensive data on water quantity and quality and storage of this data in computerized databases.

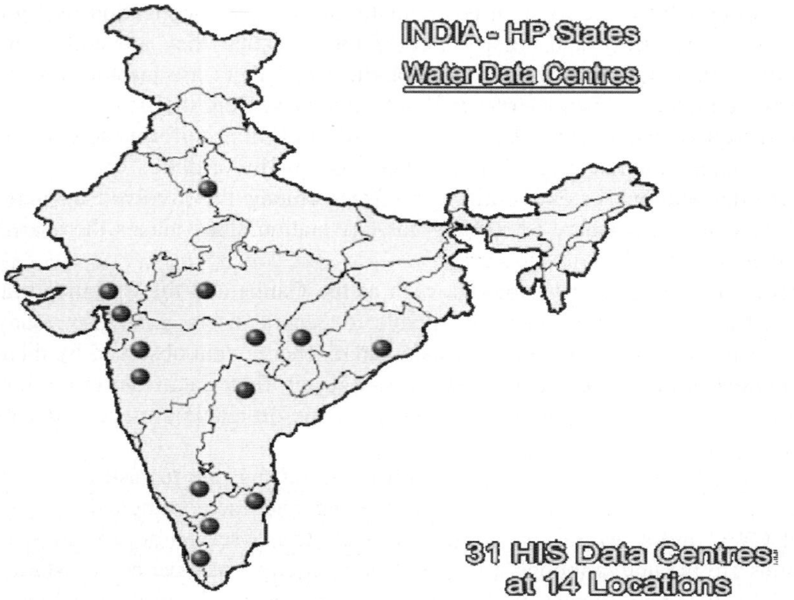

Figure 1. The states covered under Hydrology Project 1 (total area of 1.7 million sq. km) and location of data centers

HP-1 in India was a concerted effort to implement a state-of-the-art HIS for efficient dissemination of information. Augmentation and upgrading of various networks have provided fresh impetus to the (geo) hydrological monitoring in the country. Data processing with decentralised hierarchical structure ensures a completely participatory approach. The system will promote greater interaction between different HIS agencies and also ensure uniformity of tools and procedures. Improvements in infrastructure and institutional support are beginning to reflect in terms of the availability of organised hydrological databases and timely availability of better quality data to the user. It would only be fitting that the improved system further grows and the experience gained is utilised for similar improvements in all other states of India. Recent techniques, such as remote sensing and Geographic Information System (GIS) coupled with field based monitoring stations, may be utilized to monitor the data in real time and update the database. Special attention was paid to standardization of procedures for data observation and validation. Here, issues faced while upgrading the hydrological data acquisition and management system under HP are highlighted.

24.1.3. Salient Features of his Developed in HP-1

The primary objective of the HP was to develop a comprehensive, reliable, accessible and sustainable HIS in the concerned agencies, with emphasis on:

- collection of reliable and spatially intensive data on SW/GW quantity and quality,
- standardised procedures for various activities,
- dedicated hydrological data processing software,
- storage of data in well-defined databases,
- efficient dissemination of information, and
- institutional & human resources development

The overall objective of this HIS (see Figure 2) is to assist in implementing the Government of India's policies and strategies. Article 2 of India's National Water Policy stipulates the need for establishing a well-defined water information system. The new HIS was conceptualized and set-up through a number of steps as discussed in what follows.

Review of observation networks

The existing observational networks were thoroughly reviewed from three view points: (a) to establish new stations in the areas that were inadequately covered earlier or to replace non-representative stations with dedicated sites, (b) to avoid duplication of stations across agencies, and (c) to improve frequency and accuracy of observations through automated equipment and standard procedures. Comprehensive list of equipment and detailed specifications for each of these were drawn and equipment was procured by respective agencies. The old and defunct equipment was replaced with standard ones to reduce variability in observations on account of the use of non-standard equipment.

A major improvement in the meteorological network was by reactivating many old non-functional rainfall stations. Field inspections revealed that improperly

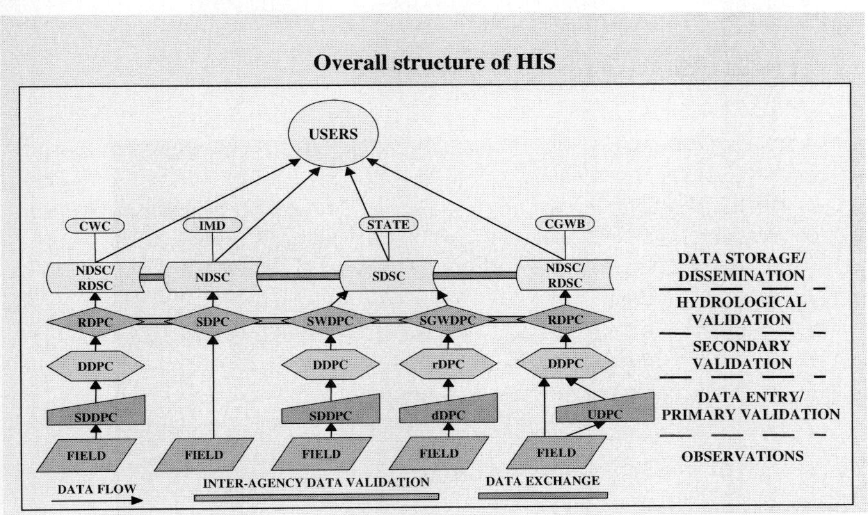

Figure 2. Overall structure of HIS and flow of data

located stations, ill maintained or defunct equipment and sub-standard observation practices were common. To revitalize the network about 460 new stations were added and another 1,530 stations upgraded. Many new full climatic stations were also set up. Figure 3 shows a typical hydro-met observatory set-up under HP-1.

The main improvements of river gauging stations were the introduction of digital recorders. Floats were replaced by current meters. In total, about 560 stations were upgraded and another 330 new stations were established. New techniques of discharge measurements, such as Acoustic Doppler Current Profilers (ADCPs), were introduced. Integrated bathymetric survey equipment for reservoir sedimentation surveying (Figure 4) were been introduced in all the participating states.

The objective of GW network is to provide improved understanding of quantity and quality of GW systems. Previously, there were about 27,000 observation wells in the project area, of which only about 6% were tubewells while others were open dug wells. These were not maintained adequately. Under the project, about 7,900 new

Figure 3. A typical hydro-met observatory set-up under HP-1

Figure 4. Measurement of discharge using moving boat method

piezometers were added. These are also being used for water quality monitoring. The network was optimized by integrating piezometers of different agencies. Nearly 6,400 piezometers were installed with Digital Water Level Recorders (DWLRs) to ensure measurement of head at the desired frequency (ranging from 1 to 6 hours).

An extensive network for monitoring SW quality at about 675 locations and GW at about 29,000 locations was established and stations categorized into "Baseline", "Trend" and "Flux"/"Surveillance" stations. Frequency of sampling and parameters to be analyzed for each category of stations were documented (Hydrology Project 2003a, 2003b). A comprehensive water quality laboratory development programme was completed by establishing or upgrading 290 laboratories that are capable of handling all water quality (WQ) analysis requirements. Analysis procedures for various parameters were standardized and documented (Hydrology Project 2003c).

Data processing, analysis and reporting

The existing system of manual or limited computerized data processing was replaced by fully computerized data processing using dedicated and user-friendly software. The raw data are passed through a series of operations, typically: data entry, validation, filling-in gaps, compilation in different forms, and obtaining the commonly required statistics, etc. Of particular importance is assuring the quality and reliability of the data through validation procedures. Reports bring out the salient characteristics of the hydrological regime of the region. Detailed procedures have been described in Hydrology Project (2003a, b, c).

Both surface water (SW) and GW agencies have employed dedicated software. HYMOS, a hydrological data processing software developed by WL | Delft Hydraulics of The Netherlands, was employed for all hydro-meteorological, SW quantity and quality data processing and a comprehensive GW data processing and resource estimation software GEMS has been developed by an Indian company. About 200 SW data processing centers (DPCs) and 180 GW data processing centers (DPCs) were established.

The primary modules for SW and GW are called Surface Water Data Entry System (SWDES) and Ground Water Data Entry System (GWDES) respectively. The first module is for easy data entry and preliminary validation while the second module carries out spatial consistency checks, data correction/completion, compilation & analysis. The higher level modules carry hydrological validation and reporting.

The Technical Assistance team of HP developed both these software. These have Microsoft Access database structure at the back-end and the front-end was built using Visual Basic for Application on Windows platform. The data entry screens of these systems look like the manuscripts used by observers for data recording. Comprehensive scrutiny of data is provided by graphical visualization. The software have become highly popular owing to user-friendly features. Another very useful work has been the digitization of about 10 themes (e.g., river drainage system, contours, geological setup, administrative units, etc.) on a 1:50,000 scale for the whole project area.

Huge volumes of historical data, mostly in manuscript or chart form, were available with many agencies. Some of this was even being lost due to gradual decay of manuscripts. A comprehensive program of historical data entry and processing had been formulated and completed. Such a mammoth organization of hydrological data was accomplished for the first time for a substantial part of the country. This would certainly unfold an excellent opportunity for all concerned to be able to easily access and use historical hydrological information.

Data storage and dissemination

All historical and current data were stored in well-defined computerized databases using industry standard SQL databases. For this, a data storage software WISDOM was developed by a private vendor. Both raw and processed data were stored to avoid loss of information. Features of data administration like data security, protection from data corruption, and controlled accessibility were implemented. An efficient query system aided with graphical visualization through maps for identifying the required data has been provided on the web. Hydrological data users can search and select the required information and make on-line requests to the data centers. Figure 5 conceptually shows the flow of data between HIS and the users.

Human resource development

An extensive training program was implemented to ensure skill building for all the personnel involved in HIS. A whole range of subjects, including SW, GW and WQ

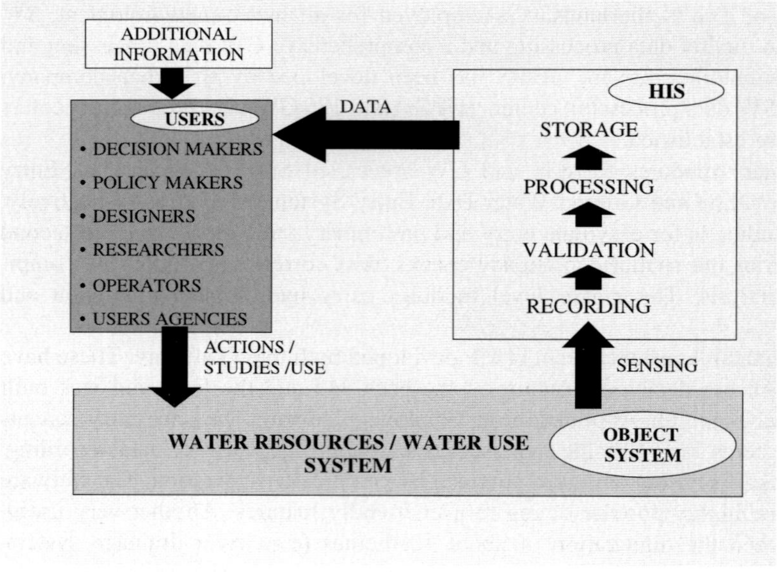

Figure 5. Flow of data between HIS and users

data acquisition, data entry and processing, were covered. Most training courses were institutionalized through the designated institutions called the Central Training Institutes (CTIs). A three-pronged approach was followed for imparting training to a large number (about 10,000) of trainees, using the concept of "training of trainers (ToT)". A core group of motivated officers of each state and/or CTIs were trained who subsequently trained the actual trainees. Comprehensive and well laid out training documents and presentation materials were prepared to ensure uniformity and standardization.

24.1.4. Important Issues in Upgrading of HIS

Various issues that require attention while upgrading a HIS in Indian circumstances are elaborated here.

Problems with data acquisition systems

Although most of the observation stations are operational, problems of non-functioning DWLRs were noticed at some sites. The working of some of the makes was not trouble-free in the rugged environment at many of the remote observation stations. A crucial factor for the low success rate of the DWLRs was the non-availability of the maintenance facility locally. Moreover, due to the lack of planning some agencies are not able to ensure timely availability of consumables. Vandalism and theft were also reported at some remote locations. Round the clock vigil of the equipment is not possible in many cases and it takes away the benefit of getting digital data without continuously manning the equipment. An indigenous maintenance support and manufacturing venture is essential for long-term sustainability and widespread use of such sophisticated equipment.

Development of dedicated data processing and management systems

Significant progress has been made in computerizing data processing and management. Huge amount of historical data has been organized in well-defined databases. Routine data processing has been institutionalized using dedicated software. A comprehensive training program has helped staff learn data management using these software. Although a good understanding has been developed by the personnel at most of the data processing centers, much is still desired in a few instances. The data entry and primary validation software (SWDES and GWDES) are well understood and used extensively but more efforts are needed for data processing and analysis modules (HYMOS & GEMS).

A good beginning has been made by developing a data management and dissemination software, WISDOM. Since the software was completed at the fag end of the project, sufficient regularity and completeness could not be achieved. Some more efforts are still required to make the web-based data dissemination regular, effective and efficient. Since this activity is basically to disseminate information to the data users, it would be important that the personnel involved in this activity have more service oriented outlook. Although a transparent data dissemination policy has been formulated, its implementation in the right earnest would increase the use of the hydrological data manifold. Importantly, upkeep of these databases in future should be meticulous and the software be upgraded on the basis of experience gained. It would be necessary to have close coordination between all state and central agencies to retain the characteristics of uniformity, standard tools and database formats.

Follow-up of HIS protocols and data sharing among agencies

HIS protocols defining all activities, the respective time frames and the person(s) responsible to carry out the same were clearly laid down. A few offices have internalized these protocols but a majority of offices have yet not fully implemented them. For the first time, various state and central agencies have started mutual exchange of data for validation, aimed at ensuring spatial consistency. However, there is scope for further improving the effectiveness of this activity. Cases of poor timeliness and inferior data quality can still be easily spotted. To arrest this slackness, a mechanism of HIS performance review is urgently required to be instituted at higher levels in the system. A national HIS coordination mechanism must be evolved to take up this responsibility. Positive attitude as exhibited by all the agencies in agreeing to follow the protocols and exchange data would be very helpful in consolidating and furthering this concept and practicing it with much more objectivity and regularity.

Periodic training requirement

Although the project has been successfully completed, there is a possibility of the created facilities withering away, the trained staff moving elsewhere and things gradually returning to the pre-project stage. New staff may join the department and will require adequate training before being able to contribute properly. Also, in this phase of reducing government expenditures there are very limited chances of new recruitment and therefore it would be quite practical to start training suitable existing staff for specialized jobs rather than waiting for the new staff. For all such reasons, efforts and provisions are required to make HIS training a continuous activity. The Central Training Institutions have sufficient strength to carry on these training programmes in future. Again, the important key for the success of such activity will lie in the initiatives taken by the departments in identifying their training needs and approaching the central training institutions with the same. Similarly, in the central training institutions should be showing due enthusiasm in taking upon themselves the responsibility in providing the desired training support.

24.1.5. Post-project Follow-up

Significant progress has been made under HP, as far as development of HIS is concerned. However, the assumption that everything has been done with the required efficacy and that all components of HIS are working with the desired efficiency would be a gross misjudgment. HIS, as developed under HP, must be viewed as a good start only and provisions have to be available for making it even more comprehensive, so as to be called a true water information system (WIS). Two important issues that should be taken up in the post-project period are as under.

With the completion of HP, all agencies in various states have set up updated observation networks and are collecting and processing data in a standardized manner. Expectedly, some agencies are more successful than others in operation of HIS. In view of insufficient time available for using, practicing, demonstrating and testing the effectiveness of the developed systems within the project period, it will be desirable that there is further consolidation of all HIS activities. Due consideration is to be given to: (a) all observation stations to have requisite material and staff, (b) further optimization of network and adjustment in monitoring frequencies wherever required, (c) bringing more water bodies (like reservoirs, tanks and lakes) under the purview of HIS (d) internalization of data validation, analysis and dissemination, (e) further promotion of data exchange among agencies (e.g., CPCB utilizing flow information collected by CWC while interpreting WQ data of its stations), and (f) objectively reaching out to the potential data users.

In the post HP-1 scenario, unfortunately the flow of data from a few states to the users has not at all improved. Due to interstate disputes about sharing of river waters, some states have stopped sharing gauge and discharge data observed by them. This goes against the basic spirit of HIS and negates the very purpose behind its creation.

Horizontal expansion and vertical expansion

HP was originally formulated for the whole country but was finally taken up only for the nine peninsular states. By all ac4counts, there has been a significant improvement in the HIS status. Implementing components such as those covered in HP in the remaining parts of the country is equally necessary and would be a logical follow up. It is hoped that such horizontal expansion covers the remaining parts of the country at the earliest.

Having strengthened the non-real time components, the next step is to ensure that the system is also able to respond to the real-time data needs. On the basis of available experience in setting up conventional flood forecasting systems and the lessons learnt during HP, the systems have to be upgraded and/or established in the identified basins/sub-basins. Another area that needs to be furthered is the Integrated Water Resource Management (IWRM). Fully integrated river basin management has seldom been practiced in India. Experience from working towards development of an IWRM will hugely benefit while setting up the River Basin Organizations (RBOs), as stipulated in the National Water Policy.

24.1.6. Hydrology Project Phase 2

HP-1 was concluded in the year 2003 and many valuable experiences were gained during this project. As a follow up of HP-1, the Government of India has signed another agreement with the World Bank for the second phase of the Hydrology Project. In this phase, besides the states that were part of HP-1 (Andhra

Pradesh, Chhattisgarh, Gujarat, Karnataka, Kerala, Madhya Pradesh, Maharashtra, Orissa, Tamil Nadu) four new states have been added to the list: Goa, Punjab, Himachal Pradesh, Pondicherry. Development of Decision Support Systems for water resources planning and real-time forecasting and development of hydrologic design aids are the main activities that have been planned under HP-2 by using the HIS of HP-1.

24.2. WATER CONSERVATION

Broadly speaking, water conservation implies improving the availability of water through augmentation by means of storage of water in surface reservoirs, tanks, soil, and geologic formations. It emphasizes the need to modify the space and time availability of water to meet demands. This concept also highlights the need for judicious use of water which is of utmost necessity because of rapidly dwindling water availability in the country. Per capita availability was about 5,200 m^3 in the year 1951 which fell to 2,200 m^3 in 1991, 1,820 m^3 in 2001. It is expected to further fall to 1,340 m^3 in 2025 and 1,140 m^3 in the year 2050 (See Figure 6).

If one looks at utilizable water resources in major river basins, these resources in the Indus, Ganga, Brahmaputra, and Godavari basins are 73.31, 525.02, 629.05 and 110.54 km^3 per year, respectively. The storages available in these basins, including projects under construction, are 16.28, 54, 3.5, and 30.16 km^3. Thus, only a small fraction of the available water is being regulated in these basins at present. These basins are subject to frequent flooding, making the argument for storage even stronger. Overall, out of 690 km^3 of utilizable surface water, the storage capacity created so far is only 177 km^3. Ongoing projects will add another 70 km^3 and those under planning 132 km^3. Table 19.3 gives basin wise storage position in the country. Note that even after completing the planned projects, 45% of the potential will

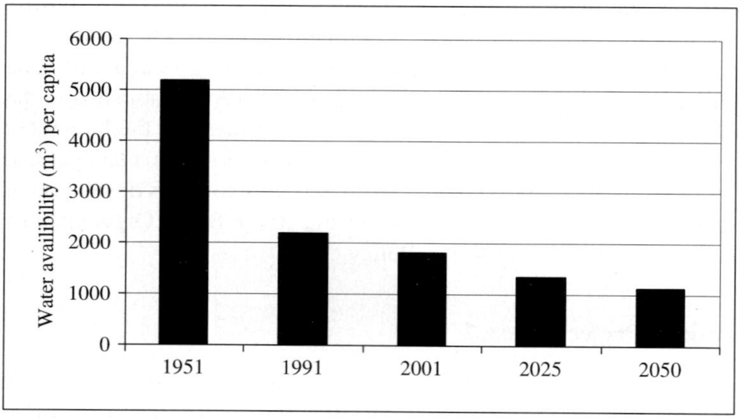

Figure 6. Per capita water availability in India for some past and future years. Note that the ordinates on x-axis are not even-spaced

remain unutilized. In view of rapidly rising population and demands for water, it will be necessary to conserve adequate quantity of water for later use.

No matter how freshwater is used—whether for agriculture, industry, or domestic purposes —there exists a huge potential for better conservation and management. On the demand side, a variety of economic, administrative and community-based measures can help conserve water. Side-by-side, it is necessary to control the growth of population, since India's large population is putting massive stress on all natural resources.

24.2.1. Water Conservation in Agriculture

Since agriculture accounts for about 83% of all water withdrawn, the greatest potential for water conservation lies in increasing irrigation efficiency. Just a 10% improvement in irrigation efficiency could conserve enough water to double the amount available for drinking. In India, sprinkler irrigation is being adopted in Haryana, Rajasthan, Uttar Pradesh, Karnataka, Gujarat and Maharashtra. The use of sprinkler irrigation saves about 56% of water for the fall crops of Bajra and Jawar, while for Cotton, the saving is about 30% as compared to the traditional gravity irrigation.

Drip irrigation systems that deliver water directly to the crop roots are in limited use in Tamil Nadu, Karnataka, Kerala and Maharashtra, mainly for irrigation of coconuts, coffee, grapes and vegetables. Experimental studies on sugarcane, banana and fruit crops have shown a very high profitability in addition to water conservation. Thus, there is an urgent need for large-scale adoption of sprinkler and drip irrigation in various parts of the country.

An important supplement to water conservation is to minimize its wastage. In urban water supply, for example, almost 30% of the water is wasted due to leakage, carelessness, etc., while most metro cities face deficit in supply of water. It is, therefore, imperative to prevent wastage. In industries also, there is potential for economy in the use of water. Prices of water for all uses should be fixed, keeping in mind its economic value, control of wastage, and the ability of users to pay. As water is becoming scarcer, pricing will be an important factor in avoiding wastage and ensuring optimal use.

24.2.2. Rainwater Harvesting

Rainwater harvesting (RWH) is the process to capture and store rainfall for its efficient utilization and conservation to control its runoff, evaporation and seepage. Some of the benefits of RWH are:

i. It increases water availability;
ii. It checks the declining water table;
iii. It is environmentally friendly;
iv. It improves the quality of groundwater through dilution, mainly of fluoride, nitrate, and salinity; and
v. It prevents soil erosion and flooding, especially in urban areas.

Even in ancient days, people were familiar with the methods of conservation of rainwater and had practiced them with success. Different methods of RWH were developed to suit the geographical and meteorological conditions of the region in various parts of the country. Traditional rainwater harvesting, which is still prevalent in rural areas, is done by using surface storage bodies, such as lakes, ponds, irrigation tanks, temple tanks, etc. For example, *Kul* (diversion channels) irrigation system which carries water from glaciers to villages is practiced in the Spiti area of Himachal Pradesh. In the arid regions of Rajasthan, RWH structures, locally known as *Kund* (a covered underground tank), are constructed near the house or a village to tackle the drinking water problem. In Meghalaya, *Bamboo Rainwater Harvesting* for tapping of stream and spring water through bamboo pipes to irrigate plantations is widely prevalent. The system is so perfected that about 18 to 20 litres of water entering the bamboo pipe system per minute is transported over several hundred meters.

There is a need to recharge aquifers and conserve rainwater through water harvesting structures. In urban areas, rainwater will have to be harvested using rooftops and open spaces, parks, etc. Harvesting rainwater not only reduces the possibility of flooding, but also decreases the community's dependence on groundwater for domestic use. These days RWH is being taken up on a massive scale in many states in India.

In the *Pani Roko Abhiyan* (a programme to stop the overland flow of water and allow it to infiltrate) the Kallipura Panchayat of the Jhabua block, the village community has constructed 613 check dams and three stop-dams. They have also completed the task of recharging a large number of hand pumps by constructing recharging pits. Similarly, the villagers of Bhagor and Manpura panchayats have built tanks to store water for their needs and to recharge ground water. Encouraged by the success of the *Pani Roko Abhiyan*, the Madhya Pradesh Government had decided in principle to institutionalize it in the State.

In earlier times, the Jhabua district in Western Madhya Pradesh was rich in forest and wildlife. But gradually the growing demand for timber and farmland ate up the forests. Slowly, wells dried up, perennial sources of water disappeared, and rains washed down the soil. This soil which clogged the channels, increasing the spread of water and the resultant flooding and devastation. People started migrating from Jhabua in search of work. In 1994, the Rajiv Gandhi Watershed Mission helped by the state government launched a programme of water harvesting by erecting trenches on the hill slopes to slow down the flow of water and through tree plantation. The project was a success – after a few years, one could see a rise in the water table in villages where work had completed to conserve water. Even during the drought of 2002, 88% pumps in the watershed villages were working and there was water in every well dug in such areas.

The low-lying area around the Tughlakabad Fort in Delhi suggests of the enormous amount of water it used to collect in earlier times. During the 12th century, the Khiljis had built a reservoir in the plains of Siri, the second city of Delhi, to harvest water from the Aravali hills. A large tank was built by Sultan Iltutmish inside Qila Rai Pithora of Mehrauli.

Limitations of Rain Water Harvesting

Revival of traditional systems and RWH are desirable and can be of help but they have their limitations. Of course, there are a few success stories like those of Relagaon Siddhi (Maharashtra), Alwar (Rajasthan), etc. mainly due to the dedication of those involved with these works. But there are also many failures in such attempts across the country. Some people have made exaggerated claims that RWH can meet a major part of current water demands. In the monsoon and (semi)arid climate of India, it will not be a good idea to store large quantities of water in small reservoirs. Lakhs of tanks would have to be constructed. These tanks will submerge millions of hectares of land, most of it fertile and cultivated. Loss of water by seepage will be large and more importantly, most of these tanks will be dry in summer months. Situation will be worse if the monsoon failed. Further, large scale storage of water in the headwater region and diversion affects water availability in the downstream areas. To sum up, rain water harvesting can have an important but a limited role in water conservation.

24.2.3. Recycle and Reuse of Water

Another way through which we can improve freshwater availability is by recycle and reuse of water. It is said that in the city of Frankfurt, Germany, every drop of water is recycled eight times. The use of water of lesser quality, such as reclaimed wastewater, for cooling and fire fighting is an attractive option for large and complex industries to reduce their water costs, increase production and decrease the consumption of energy. This conserves better quality waters for potable use. Currently, recycling of water is not practiced on a large scale in India and there is considerable scope and incentive to use this alternative. Gupta and Deshpande (2004) state that recyclable water ranges between 103 to 177 km^3/year for low and high population projections but these estimates appear to be on higher side.

24.2.4. Desalination of Water

Since 1970, there has been a significant commercial development using various desalination technologies, including distillation, reverse osmosis, and electrolysis. This technology is suitable for use in areas where freshwater is scarce, but saline water is available and energy is cheap. Compared to water recycling technologies, desalination presents fewer health risks.

Desalination, as currently practiced, mostly uses fossil fuels. Solar and wind energy are available in abundance in India and may be explored as alternative sources for this purpose in coastal states. Between the high capital and energy requirements, desalinated water costs several times more than water supplied by conventional means. But the costs are now coming down. Current production cost is about Rs. 50 (nearly one Euro) per m^3 (Times of India, New Delhi, dated 30 July 2004). Many facilities in coastal regions are using reverse osmosis for desalinization. For example, at Kalpakkam reactor, Tamil Nadu, 1.8 million litres water is being produced per day. It is expected that as the costs come down, desalinization would become commercially viable in the next 6 to 8 years.

The National Institute of Ocean Technology (NIOT, *www.niot.res.in*) has developed a Low Temperature Thermal Desalination (LTTD) plant. It makes use of the temperature difference which exists between the warm surface water and deep sea cold water. The technology was first demonstrated in a pilot project of 5,000 litres/day at Chennai and is now being used at a 100,000 litres/day plant at Kavaratti.

24.3. WATER QUALITY IMPROVEMENT AND ENVIRONMENT RESTORATION

Implementation of water pollution prevention strategies and restoration of ecological systems are integral components of all development plans. Somehow, this aspect has not been given adequate attention in the past in India. As a result, the quality of water in many of our rivers is extremely bad and many lakes and ponds have become dumping sites for waste products. To preserve our water and environment, we need to make systematic changes in the way we grow our food, manufacture the goods, and dispose off the waste (Lazaroff, 2000). For this purpose, society and individuals should have a greater knowledge and ability to bring about the required changes. Widely and readily available technical help about 'how to do this' will accelerate the process.

In India, agriculture is the biggest user and polluter of water. If pollution by agriculture is reduced, it would improve water quality and would also eliminate cost incurred for treatment of diseases. This would entail learning how to use less chemicals while boosting yields, e.g., eliminating the use of fungicides by planting more diverse varieties of grains and switching to organic farming so that a fewer chemicals are introduced on farms. Like all other inputs, there is an optimal quantity of fertilizer for given conditions and excess application does not improve crop yield. Pricing of fertilizers and pesticides as well as appropriate legislation to regulate their use will also go a long way in stopping indiscriminate use.

Industries need to carefully treat their waste discharges. Manufacturers may reduce water pollution by reusing materials and chemicals and switching over to less toxic alternatives. Industrial symbiosis, in which the unusable wastes from one product/firm become the input for another, is an attractive solution. Also, there is a need to encourage reductions or replacement of toxic chemicals, possibly through fiscal measures. Pollution taxes in the Netherlands, for example, have helped the country slash discharges of heavy metals, such as mercury and arsenic, into waterways by up to 99% between 1976 and the mid-1990s. Many countries discourage the use of equipment that contains harmful chemicals (such as thermometers that contain mercury). Such measures in India will also be helpful.

Environmental improvement and restoration should be planned and implemented such that the freshwater resources are protected and their quality is maintained and/or enhanced. A broad perspective is needed that unites social, economic, and environmental concerns in a landscape where upland forests and rangelands, wetlands, and agricultural and urban areas are integrated. An understanding of watershed linkages allows for long-term and sustainable solutions to a variety of natural resource problems. Model efforts in this direction include the capture, storage and safe release of water and the prevention of accelerated soil erosion through hydraulic structures and vegetation.

The Ministry of Environment and Forests serves is the focal point for environmental and water quality programmes. Other government departments in carrying out environmental protection activities are Central and State Pollution Control Boards, and the National River Conservation Authority (formerly Central Ganga Authority). A total of 1,532 grossly polluting industries in 24 States/Union Territories have been identified under the National River Action Plan. Comprehensive River Basin Documents for Ulhas, Brahmaputra, Pennar, Indus Part II, Rishkulya and Chaliyar Rivers are under preparation.

While utilizing water and land resources, their ability to serve other uses is often degraded either inadvertently or due to carelessness. Efforts should be made to restore landscapes and ecosystems to more efficiently protect water quality, aquatic and wildlife. On the legislative front, we require laws to check littering as well as to implement "polluter pays" principle. More importantly, these laws should be strictly enforced.

The National River Conservation Plan (NRCP) plans to tackle river pollution by building new sewage treatment plants where necessary, diverting raw sewage flowing in open drains to these plants, constructing sanitation facilities to check open defecation on river banks, setting up crematoria (electric or improved wood based), etc. Under NRCP, polluted stretches of major rivers have been identified for sewage collection and treatment. Currently, 153 towns have been included in NRCP. Out of these, 74 are located on the Ganga River, 21 on the Yamuna, 12 on the Damodar, 6 on the Godavari, 9 on the Cauvery, 4 each on the Tungabhadra and the Satluj, 3 each on the Subarnarekha, the Betwa, the Wainganga, the Brahmini, the Chambal and the Gomti, 2 on the Krishna, and 1 each on the Sabarmati, the Khan, the Kshipra, the Narmada and the Mahanadi. According to the Planning Commission, about 45% of the cleaning of the Ganga has been completed. However, the overall progress is poor because of delays in land acquisition and slow pace of work by municipal corporations.

The National Lake Conservation Plan (NLCP) was initiated in 1994 to clean important urban lakes with have high levels of silting and pollution. Ten lakes were initially identified – Ooty, Kodaikanal, Powai, Dal, Sukhna, Sagar, Nainital, Udaipur, Rabindra Sagar and Hussain Sagar. However, work has started on only one lake. The progress regarding other lakes is slow because of various reasons.

24.3.1. Environmental Flows (EF)

An environmental flow (EF) is the water regime provided within a river, wetland or coastal zone to maintain ecosystem and their benefits where there are competing water uses and where flows are regulated. Environmental flows provide critical contributions to river health, economic development and poverty alleviation. They ensure the continued availability of many benefits that healthy river and groundwater systems bring to society. Environmental flows normally include the flow requirements in rivers and estuaries for maintenance of riverine ecology. Some people view EF as wastage of water but clearly this is a narrow view.

Most Indian rivers have monsoon-driven hydrological regimes, where 70% to 80% of the annual flow occurs in 3 to 4 months. Such rivers fall into the category of highly variable flow regimes. The total environmental flow requirement (EFR) for most of the Indian rivers range between 20% to 27% of the renewable water resources. But these EFR estimates may be considered as preliminary. These need verification through comprehensive basin-specific assessments of EFR. At the same time, it is important to appreciate that the EFR allocations of less then about 20% of the mean annual flow are likely to degrade any river beyond the limits of possible re-habilitation. An additional factor, not yet considered in the assessment, is that a reduction in river flows decreases the ability of a river to cope with pollution loads. These loads are known to be massive in many Indian basins.

Un-utilizable portion of surface runoff in most Indian basins is adequate to meet EFR. Only in a few basins, namely Pennar, West flowing rivers in Kutch, Saurashtra and Luni, Cauvery and east flowing rivers between Pennar and Kanyakumari, EFR exceeds the un-utilizable runoff. In these basins, a part of the potentially utilizable water resources has to be earmarked for EFR.

24.4. INTER-BASIN WATER TRANSFER (IBWT)

The vast variation, both in space and time, in the availability of water in different regions of India has created a flood-drought-flood syndrome with some areas suffering from flood damages and other areas facing acute water shortages. The States of Karnataka, Tamil Nadu, Rajasthan, Gujarat, Andhra Pradesh and Maharashtra are the worst drought prone States. The States of Uttar Pradesh, Bihar, West Bengal, Orissa and Assam face severe flood problems. Inter-basin transfer of water in India is a long-term option to partly overcome the spatial and temporal imbalance of the availability and demand of water resources.

Recently, this topic has gained immense importance in India particularly in view of the proposed scheme of interlinking of rivers (ILR). Due to this reason, it is discussed in detail in Chapter 22. We may emphasize here that interbasin transfer of water at a mammoth scale may not be necessary in India. Nevertheless, IBWT is a rational option for solving water related problems.

24.5. GROUNDWATER MANAGEMENT

To protect aquifers from over-exploitation, an effective groundwater management policy oriented towards the promotion of efficiency, equity, and sustainability is required. Agricultural holdings in India are highly fragmented and the rural population density is large. The exploitation of groundwater resources should be regulated so as not to exceed the recharging possibilities, as well as to ensure social equity. The detrimental environmental consequences of over-exploitation of groundwater need to be effectively prevented by the central and state governments. Integrated and coordinated development of surface water and groundwater resources and their conjunctive use should be envisaged right from the project planning stage and should form an integral part of the project implementation. Over-exploitation of groundwater should be avoided, especially near the coast to prevent intrusion of seawater into freshwater aquifers (National Water Policy, 2002).

A joint management approach combining government administration with active people participation is a promising solution. The government can initiate a variety of programs and controls for recharge and discharge and implement regulatory measures, such as well spacing norms, control drilling of new wells by issuing permits, regulation of water intensive crops, and pricing of electricity for lifting groundwater. Elected committees through which the interests of all stakeholders can be voiced could govern these programs, especially in overexploited areas. These committees can allocate quota for utilizing water in such areas for every few years based on the current cropping pattern. It is essential that the quota be set corresponding to efficient irrigation techniques. To the extent possible, conjunctive use of surface water and groundwater should form an integral part of groundwater management policy.

In critically overexploited areas, bore-well drilling should be regulated till the water table attains the desired elevation. Artificial recharge measures need to be urgently implemented in these areas. Amongst the various recharge techniques, percolation tanks are least expensive in terms of initial construction costs. Many such tanks already exist but a vast majority of these structures have silted up. In such cases, cleaning of the bed of the tanks will make them reusable. Promotion of participatory action in rehabilitating tanks for recharging would go a long way in augmenting groundwater supply.

Due to declining water table, the cost of extraction of groundwater has been increasing over time and wells often go dry. This poses serious financial burden on farmers. Hence, special programs need to be designed to support these farmers. Finally, the role of government will have to switch from that of a controller of groundwater development to that of a facilitator of equitable and sustainable development.

24.5.1. Conjunctive Use of Water

Conjunctive use is the coordinated management of surface and groundwater resources, taking advantage of their complementary properties. The objectives of such coordination may be higher agricultural production, improved sustainability of the system and/or more acceptable socio-economic equity. The conjunctive use concept recognizes the unified nature of the surface water and ground water resources as a single natural resource and takes advantage of interactions between them in planning the use of water from the two resources. Interactions between surface water and groundwater resources include contribution to base flow from groundwater, recharge to groundwater from fields irrigated with surface water, artificial recharge of surface water to groundwater, indirect use of ground water through augmentation of tube wells into canals, and recharge from irrigation conveyance system to ground water. This topic has been covered in detail in Chapter 6.

24.5.2. Artificial Recharge of Ground Water

Artificial ground water recharge is resorted to in areas facing excessive lowering of water levels. These techniques are in vogue mainly in alluvial and other unconsolidated formations. However, on account of greater development of ground water in hard rock areas in recent years, attempts are also being made to adopt effective methods of recharge in such formations. The choice of a particular recharge method and its effectiveness depends upon the hydrogeological characteristics, viz., lateral and vertical continuity of aquifers, direction of ground water movement, interconnections between aquifers tapped by various structures, and hydraulic characteristics of aquifers.

Frequently used methods of artificial recharge in drought prone areas are percolation tanks, check dams, and percolation canals. Check dams are quite effective in limestones and other hard rock areas with steep slopes. Further, they serve the twin purpose of ground water recharge and soil conservation.

While preparing the National Perspective Plan for recharge to ground water by utilizing surplus monsoon run-off, the non-committed surplus monsoon run-off was assessed by CGWB vis-à-vis the available sub-surface space. This study was carried under different hydrogeological situations and assessment was made to saturate the vadose zone up to 3 m below ground level. The plan envisaged local conservation of surplus monsoon runoff through artificial ground water recharge. The surplus and feasible utilizable figures for various basins are given in Table 1. The surplus monsoon runoff for creation of subsurface ground water storage was estimated as 864 BCM. Saturating the vadose zone up to 3 m depth below ground level will create sub-surface storage potential of 591 BCM. Since the monsoon runoff is not uniform in all the basins, this will result in surplus and deficit of monsoon runoff vis-à-vis the water required to recharge the vadose zone. Hence, this potential needs modification keeping in view the availability of monsoon runoff and storage potential of vadose zone. Consequently, the retrievable storage potential would be 175 BCM. Substantial crop area can be irrigated by additional sub surface storage.

In India, the area which requires artificial recharge to ground water has been estimated to be 448,760 sq.km. This excludes the hilly terrain of Jammu and Kashmir (J&K), Himachal Pradesh (H.P.), Uttaranchal, North Eastern (N.E.) states, arid regions of Western India, and islands. In rural areas, artificial recharge by modifying natural movement of surface water through percolation tanks, check dams, nala

Table 1. Water availability and potential for enhanced recharge

Basin	Water available for recharge	Sub-surface storage potential for recharge	Feasible ground water storage	Retrievable ground water storage	Water availability to meet requirement of ground water storage	
					Excess	Deficit
Indus	31,326	143,903	31,326	21,665		112,577
Ganga	238,126	125,076	125,076	94,930	113,050	
Brahmaputra	307,752	19,177	19,177	15,883	288,575	
Barak	No Storage Potential					
Godavari	53,124	8,145	8,145	5,818	4,479	
Krishna	17,835	8,970	8,970	6,407	8,865	
Cauvery	8,132	4,746	4,746	3,406	3,386	
Pennar	2,742	1,940	1,940	1,386	802	
EFR between Krishna & Pennar, and Mahanadi & Godavari	8,562	832	832	594	7,730	
EFR between Pennar and Kanyakumari	10,882	5,731	5,731	4,145	5,151	
Mahanadi	26,187	2,055	2,055	1,468	24,132	
Brahmani & Baitarani	14,603	483	483	345	14,120	
Subarnarekha	4,406	868	868	630	3,538	
Sabarmati	1,489	8,134	1,489	1,116		6,645
Mahi	4,587	1,883	1,883	1,345	2,704	
WFR of Kachchh & Saurashtra including Luni	5,814	150,181	5,814	4,371		144,367
Narmada	10,244	77,356	10,244	7,687		67,112
Tapi	3,830	4,448	3,830	2,752		618
WFR of Tapi to Tadri	65,612	907	907	648	64,705	
WFR of Tadri to Kanya Kumari	49,537	518	518	370	49,019	
Areas of Inland drainage in Rajasthan	NIL	25,310	NIL	NIL		25,310
Minor rivers draining to Bangladesh & Myanmar	No Storage Potential					
Total	864,700	590,573	233,943	174,963	630,126	356,629

Note: EFR = East Flowing Rivers, WFR = West Flowing Rivers.

bunds, gully plugs, Gabion structures, sub-surface techniques of recharge shaft, well recharge, etc. have been recommended. Attempts to conserve ground water flow through ground water dams have also been made in different states. The Master Plan envisages construction of 2.25 lakh artificial recharge structures in rural areas. This number includes 37,000 percolation tanks, 110,000 check dams/nala bunds/gully plugs/anicuts, etc., 48,000 recharge shafts/dug well for recharge, revival of about 1,000 ponds, and 26,000 gully plugs and Gabion structures. In J&K, H.P., Uttaranchal, N.E. states, and Sikkim, emphasis is being placed on spring development and 2,700 springs have been identified for development.

A ground water recharge scheme has been formulated specially for Thar Desert of Rajasthan. About 70 BCM of water can be recharged to the deep water table areas of Thar Desert. IGNP water can be utilized during monsoon period to recharge the depleted aquifers through multipurpose lift canal system. There is potential to build up vast sub-surface reservoirs in Thar Desert.

Percolation tanks are very effective in recharging ground water aquifers and increase of yield of downstream wells. These have been used for artificial recharge in the hard rock formations in basalts, granites and other crystalline rocks in Tamil Nadu, Karnataka, AP, and MP. Romani and Sharma (1989) have given a detailed account of percolation tanks constructed in Jhabua district (M.P.) where the rocks are Deccan basalts and Bagh beds. Pardhasaradhi (1989) has described the effectiveness of percolation tanks from Siva and Man River basins in Maharashtra where the rock types are Deccan traps. The rate of seepage from these tanks was found to vary from 15 to 75 mm/day. A study of some tanks done in the state of Maharashtra indicates that for five tanks under study, 100 new wells have come up and increase in irrigated area is 355 ha (Saksena 2000). In zones with gentler slopes, percolation tanks generally perform better than check dams and are also cost effective.

Percolation by periodic release of water in canals and natural channels was tried in Anantpur District (AP). The area is underlain by Tadipatri shale of Cuddapah Group. This method of recharge was found to be effective and the ground water levels rose by 12 m.

24.5.3. Ground Water Dams

A ground water dam is an impervious subsurface structure which stops the lateral outflow of ground water, thereby augmenting water availability in an area. It may also serve as collecting structure that diverts ground water flow to recharge adjacent aquifers or to raise the ground water table in an aquifer with a limited flow of ground water, making it accessible for pumping. Damming ground water is a mean of bridging over seasonal dry periods. The purpose of the sub-surface dam is to arrest the movement of ground water out of a sub basin. The method of storing under ground water has received considerable attention during the last few years. Damming ground water for conservation purposes is not a new concept. Ground water dams were constructed on Islands of Sardinia in Roman times and structures in Tunisia show that damming of ground water was practiced by old civilizations in

North Africa. Most recently various small scale ground water damming techniques have been developed and adopted in many parts of the world, notably in India, southern and eastern Africa, Brazil and Japan.

Such structures have been constructed in the past in many countries. In India, the oldest example is perhaps from Ottapalam in Palghat district of Kerala constructed during 1962–64. According to Singhal (1992) this dam, constructed in 1962–64, is made up of a 4-inch plastered brick wall. It has a length of 155 m and has a depth of 5 to 9 m. The catchment area of the dam is about 10 ha and water is used for irrigation of 3.2 ha of paddy during October-December, and 1.2 ha of paddy during the dry season. In 1979, another subsurface dam was constructed by CGWB in Palghat, Kerala. It has a catchment area of 20 ha. The bedrock consists of gneisses and granites. A subsurface dam, made of plastic sheets, was built in Octacamund, Tamil Nadu, in 1981. CGWB has also constructed ground water dams at other places in Kerala. Sinha and Sharma (1990) and Saksena (2000) have given a detailed account of the design and criteria of site selection for these structures.

24.6. WATERSHED MANAGEMENT

For an equitable and sustainable management of shared water resources, flexible, holistic approach of Integrated Water Resources Management (IWRM) is required, which can cater to hydrological variations in time and space and changes in socio-economic needs along with societal values. Watershed is the unit of management in IWRM, where surface water and groundwater are inextricably linked and related to land use and management. Watershed management aims to establish a workable and efficient framework for the integrated use, regulation and development of land and water resources in a watershed for socio-economic growth. A conceptual framework for IWRM depicting the objectives, the inputs, and the tools is given in Fig. 7.

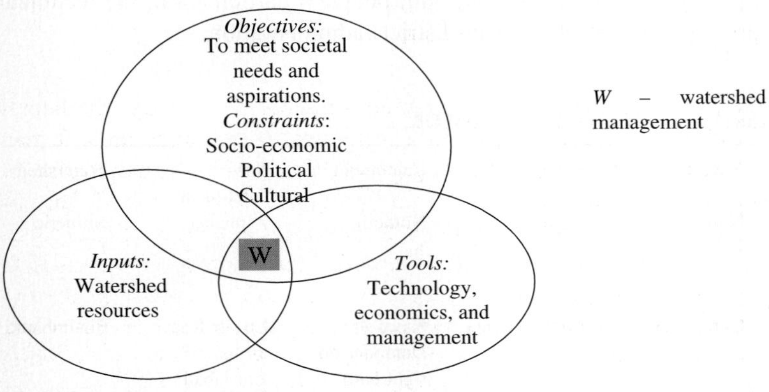

Figure 7. A conceptual framework for IWRM

Typical objectives of watershed development programs include:

(i). Raising the productivity of irrigated and rain-fed agriculture and non-arable lands,

(ii). Encouraging the sustainable management and optimal use of surface and groundwater,

(iii). Reducing soil erosion,

(iv). Conserving forests and other natural vegetation,

(v). Creating employment (both directly and indirectly), and

(vi). Promoting increased individual and collective responsibility for natural resources management and strengthening the social institutions.

Local communities play a central role in the planning, implementation and funding of activities within participatory watershed development programs. In these initiatives, people use their traditional knowledge, available resources, imagination and creativity to develop watershed and implement community-centered programs.

The All India Soil and Land Use Survey (AISLUS) has delineated watersheds at the national level and have delineated watersheds larger than 20,000 ha. They have adopted the scheme shown in Table 2 to identify watersheds.

Thus the code of the example watershed Bhairabanki in Table 2 would be 2A2A6.

Currently, many programs, campaigns and projects are underway in different parts of India to spread mass awareness and mobilize the general population in managing water resources. Some of these are being implemented by the central/state governments, while others have been taken up by various Non-Governmental Organizations (NGOs). The underlying watchword/motto for these programs is that "each one of us is responsible for the current state of water resources and environment in the country, and we cannot wait for someone else to solve it." A few water conservation efforts are briefly described in what follows.

Hariyali: Hariyali (meaning 'greenery') is a watershed management project, launched by the central government, which aims at enabling the rural population to conserve water for drinking, irrigation, fisheries and afforestation as well as generate employment opportunities. The project is being executed by the Gram Panchayats (village governing bodies) with people's participation; the technical support is provided by the block (sub-district) administration.

Table 2. Hierarchical coding of watersheds by AISLUS.

Details	Region	Basin	Catchment	Sub-catchment	Watershed
Type	Numeric	Alphabet	Numeric	Alphabet	Numeric
Characters	1	1	1	1	1
Example					
Code	2	A	2	A	6
Name	Ganga basin	Lower Ganga	Kasai to Damodar on right bank of Bhagirathi	Lower Kasai up to Kansabati	Bhairabanki

Neeru-Meeru: Another good example of water conservation efforts is the 'Neeru-Meeru' (Water and You) program launched in May 2000 by the Government of Andhra Pradesh. During the last three years, an additional storage space of more than 1,800 million m^3 has been created by constructing various water harvesting structures, such as percolation tanks, dugout ponds, check dams, etc., through people's participation. A water literacy drive 'Jalachaitanyam' (water awareness) has created a large-scale awareness about conservation measures and sustainable management of water resources in the state.

Water Harvesting: Tarun Bharat Sangh (Young India Association) or TBS is an NGO which promotes sustainable water management through rainwater harvesting in Rajasthan. Since 1986, TBS has helped with building or restoring nearly 10,000 water harvesting structures in Alwar and the neighboring districts in the Aravalli hills of northeastern Rajasthan. These structures not only hold surface water but recharge the groundwater as well. The central message of TBS is that good water management requires good land management. Emphasis is also put on protecting forests. The efforts of villagers are visible in the form of rising water table and regenerated forests on the rocky slopes of the Aravalli hills.

Despite the above success stories, so far there is no appreciable improvement on watershed resources utilization on national level. Undoubtedly, coordinated watershed development programs need to be encouraged and awareness about benefits of these programs must be created among the people. It is also pertinent to note that recently inflows to many lakes and reservoirs have decreased considerably. For instance, the Pichola Lake in Udaipur has almost dried on account of reduced inflows. Gandhisagar Reservoir in the Chambal basin is being able to fill up only partly due to meager inflows. A reason behind this may be large changes in the upstream catchment. It is, therefore, necessary that any watershed development activity should be holistically planned and carried out so that unintended and undesirable consequences can be avoided.

24.7. IMPACTS OF CLIMATE CHANGE

That the global warming is taking place is a fact and the consequences of the climate change require attention. The impact of climatic change may be more severe in developing countries, such as India, where agriculture supports more than half of the population and has a critical influence on economy. Despite the uncertainties about the magnitude and directions of climate change and the possible impacts, measures must be taken to minimize the anthropogenic causes of climate change and mitigate its adverse effects.

Surface temperature records reveal that the 1990s have been the warmest decade of the last millennium in the Northern Hemisphere and 1998 was the warmest year. Anthropogenic-induced CO_2 emissions are said to be the dominant factor causing the observed global warming. Substantial increases in greenhouse gases are likely in the future as a consequence of which global mean surface temperature is expected to increase scenarios and between 2.5° and 5.8 °C for high emission scenarios by 2100 with respect to 1990 (IPCC, 2001). Globally averaged precipitation is projected to increase and the global mean sea level is likely to rise by 0.14 to 0.80 m by the year 2100.

The summer monsoon is the predominant source of water for the country. Naturally, the interest is to know the likely impacts of climate change and global warming on the monsoons. Rainfall fluctuations over the last century have been largely random with no systematic annual or seasonal pattern. However, areas of increasing trend in the seasonal rainfall have been found along the West Coast, North Andhra Pradesh and Northwest India and those of decreasing trend over East Madhya Pradesh, Orissa and Northeast India during recent years. An analysis of air temperature data by Pant & Kumar (1997) has shown a warming trend of 0.57 °C per hundred years for India.

To predict the changes in temporal as well as spatial variability of the monsoon rainfall in response to increases in radiative forcing of the atmosphere, climate change scenarios over India under four emission possibilities were studied by Lal (2001). These scenarios cover a wide range of driving forces of future emissions. Among these, the scenario B1 projects the most conservative future emission of greenhouse gases while A2 scenario is characteristic of higher rates of emissions.

According to Lal (2001), the area-averaged annual mean surface temperature rise over India by 2080s is likely to range between 3.5 °C and 5.5 °C (least in B1 scenario and maximum in A2 scenario). During summer monsoon, the warming may range between 2.9 °C and 4.6 °C. The projected surface warming is more pronounced during winter than during summer monsoon season. A marginal increase of about 7 to 10% in area-averaged annual mean precipitation is projected over the Indian subcontinent by 2080s. A decline of between 5 to 25% in area-averaged winter precipitation is likely. During the monsoon season, an increase in area-averaged precipitation of about 10 to 15% over the land regions is projected and this will result in more frequent floods. A decrease of between 10 and 20% in winter precipitation over most parts of central India is expected for 2050s. Lal (2000) has concluded that the year-to-year variability in rainfall over central India during the monsoon season may not significantly change in the future. More intense rainfall spells are, however, expected over India in the future.

Some of the most pronounced year-to-year variability in climate features and the extreme weather events, such as droughts in many parts of Asia, have been linked to El Niño events. At least half the severe failures of the Indian summer monsoon since 1871 have occurred during El Niño years (Webster *et al.*, 1998). In the event of enhanced anomalous warming of the eastern equatorial Pacific Ocean, such as that observed during the 1997–98 El Niño, more frequent droughts in some parts of India are possible.

Formation of monsoon depressions and tropical cyclones over the Bay of Bengal and the Arabian Sea shows a falling trend since 1970. However, an increase in sea surface temperature due to climate change should lead to increase in cyclone intensity, higher storm surges, and an enhanced risk of disasters along the east coast of India. Climate change can also pose an additional threat to the mountainous region of India. Extreme precipitation events in the Himalayas may cause higher runoff, frequent slope failures, and more soil erosion and sedimentation in this region.

24.7.1. Coping with Climate Change and Adaptation

Climate change is just one of the many factors influencing the hydrologic cycle. One may assume that no special efforts or plans are required to protect against surprises or uncertainties. But this complacency by water managers may lead to severe impacts that could be prevented by simple actions taken now. In addition to climate change, population growth, land use changes, industrial growth, and increase in demands for various uses are also occurring simultaneously. A change in drought or flood risks due to climate change has the greatest implications for human well-being. Climate change is likely to produce, in many places over the country and over long-time horizons, hydrologic surprises (or shocks), variabilities, and extremes of a nature different from the one for which the current systems were designed and built. Relying solely on traditional methods and assuming that sufficient time and information will be available before the onset of large or irreversible change to permit managers to respond appropriately;

Many steps need to be taken to cope up with the climate change impacts on future water resources management in India:

(i) A well-planned national climate and water monitoring and research program should be developed as a collaborative effort between government departments and leading scientific organizations;
(ii) Decision makers should begin a systematic re-evaluation of engineering designs, operating rules, contingency plans, and water allocation policies for managing water resources in view of potential climate changes;
(iii) In view of regional variations, the state governments should begin assessing the likely climate impacts and the effectiveness of different operation and management options;
(iv) Improvements in the efficiency of end uses and the management of water demands must now be considered major tools for meeting future water needs, particularly in water-scarce regions. Water demand management and institutional adaptation are the primary components for increasing system flexibility to meet uncertainties of climate change;
(v) Assumptions about the probability, frequency, and severity of extreme events used for planning and their impact should be carefully re-evaluated;
(vi) Methods must be used that explicitly incorporate uncertainty into the decision process; and
(vii) New infrastructure with long-life should be designed considering a wider range of climate variability than shown by the historical record.

Records of past climate and hydrological conditions are no longer considered to be reliable guides to the future (Jain and Lall, 2001). Climate change will be imposed on the top of current and future non-climate stresses. In some cases, these changes will be larger than those expected from population growth, land-use changes, economic growth, and other non-climate factors. Another crucial fact is that the time lags between identifying the nature of the problems, understanding them, prescribing remedies, and implementing them are long. Hence, waiting for relative certainty about the nature of climate change before taking actions to reduce climate-change related risks might prove far more costly than taking certain pro-active management and planning steps now.

The areas for further research in this field include:

(a) Improvements in methods to downscale climate information to improve understanding of small-scale processes that affect water systems;
(b) Research on how summer monsoon will be affected due to climate change and its effect on water resources systems;
(c) Research on how the behaviour of cyclones and other extreme hydrologic events might change; and
(d) How climate change might affect recharge rates and response of aquifers.

24.8. WATER PRICING

Effective and efficient utilization of valuable water resources along with control of wastage and demands requires an appropriate pricing policy. There is a need for integrated pricing structure especially in groundwater extraction for different uses. The pricing policy should match not only the costs of supply (i.e., O & M and capital costs) but also opportunity costs, economic externality costs and environmental externality costs. Integrated pricing structure demands significant changes in current laws, regulatory mechanisms, operational transparency and facilities.

While drafting water pricing, water rights should be well defined without which the issue of water pricing and its enforcement could not be effective. Further, the water price should not only be determined on the basis of pure economic considerations alone; social considerations should as well be given rightful place. Also there is the need to target the subsidies to the needy groups whose social and economic status justifies such support. Another important point in this context is affordability of water by the users and its equity implications.

Ensuring an economic value to water without sacrificing social and environmental imperatives in a democratic system is a challenge for both the government and the society. If water prices are to be fixed so as to be able to control demands, it requires significant changes in regulating mechanism, operational procedures, accountability, and high quality of facilities. A steep rise in prices will invite opposition from the users. Hence, the water rates can be best increased in a phased manner. At the same time, frequent tinkering with rates causes confusion and distrust among the users.

24.9. PUBLIC INVOLVEMENT AND CAPACITY BUILDING

The principle objective of public involvement and capacity building should be to make various sections of the society aware of the issues and implications of decisions related to freshwater resources. This approach stresses the need to view freshwater resources in a holistic way and recognizes the fact that the socio-economic and hydro-environmental systems are mutually dependent. It is necessary to undertake a vigorous mass campaign to educate the various sections of the society about water conservation and efficient use. To that end, Water Resources Day is being observed every year on March 22 as part of a mass awareness programme.

When the communities begin to view water resources to be their own common property, they ensure that these are used in a sustainable manner. Capacity building should be perceived as the process whereby a community equips itself to become an active and well-informed partner in decision making. The process of capacity building must be aimed at both increasing access to freshwater resources and changing the power relationships between the stakeholders.

Capacity building is not only limited to officials and technicians but must also include the general awareness of the local population regarding their responsibilities in the sustainable management of the freshwater resources. Policy decisions in any water resources project should be directed to improve knowledge, attitude and practices about the linkages between health and hygiene, provide higher water supply service levels (quantity, quality, equity, reliability, coverage and access), and to improve environment through safe disposal of human waste. Sustainable management of freshwater requires decentralized decisions by giving authority, responsibility and financial support to communities to manage their natural resources and thereby protect the environment. So far, the involvement of communities in water resources planning and management in India is very little and is mostly confined to small irrigation areas. Efforts for dissemination of information from the government agencies are also minimal.

24.9.1. Use of Advanced Technology

Water resource development strategy, particularly for the surface water projects so far has focused on selection of best sites. The future development in water resources is going to be more challenging as most of the best sites have been developed. Selection of new sites will need advanced technical know-how for investigation, more elaborate planning and design as also to address the various constraints which will be associated with such developments. Due attention will have to be paid to social and environmental issues. This would mean relatively higher cost for the project. Technical know how for planning and design will have to be an important component in addressing many challenges. Important developments have taken place in information technology and instruments and these are now being gainfully employed in water sector in India. A few important fields are described in what follows.

Remote Sensing and GIS Techniques

Remote sensing and Geographic Information System (GIS) have an important role in water resources development and management. Space is the ideal vantage point to make primary observations from which plethora of useful information can be derived for input to various hydrological studies. The space born multi spectral data enable generating timely, reliable and cost effective information on various natural resources, namely surface water, ground water, landuse/cover, soil, forest cover and various environmental hazards, such as waterlogging, salinity and alkalinity, soil erosion by water etc. Multitude of spatially related data concerning topography, geomorphology, climatology, etc. along with satellite derived information can be appropriately integrated through GIS. India is among the pioneering countries that have launched remote sensing satellites.

Nuclear & Isotope Techniques

Isotope techniques have emerged as a viable tool to study various hydrological processes like, stream flow generation, groundwater flow and its salinisation, soil moisture movement through unsaturated zone and recharge to groundwater, and surface water and groundwater interaction, etc. Nowadays, these techniques are widely used in developed and developing countries. Every developed country has a national isotope database developed through a network of precipitation sample collection stations. Precipitation samples along with groundwater and major rivers water samples are collected and analysed for environmental stable and radioisotopes. The national isotope database provides the isotopic input functions to the various facets of hydrological cycle thus enabling carrying out important hydrological studies. In India, neither national isotope database nor any network of precipitation sample collection stations have been established for environmental isotopic analyses. Therefore, there is a need to establish sample collection stations covering all the major geographical regions in the country. National level programmes are being launched to make use of nuclear techniques in water sector.

Hydrological Instrumentation

In addition to the conventional monitoring system, country needs advanced observation systems for hydrological variables. Efficient and automatic sensing, data logging and communication features of the hydrological and hydro-meteorological observation systems enable faster collection of data and unattended operation of remote stations for long periods. The latest electronic sensors and data loggers consume less power than the electromechanical recorders that are presently used in the country. Many Automatic Weather Stations that can measure a number of hydro-meteorological variables have been recently installed. Robust digital water level recorders (DWLRs) are increasingly being used for river stage and ground water level observations providing near continuous records. However, the major hindrance in use of these equipment is their high cost and non-availability of facilities for cost effective maintenance and repair.

24.9.2. Gender Issues in Water Management

Women are involved in various ways in the management of irrigation systems and rural water supply. In most Indian homes, women have the major role in water management. In many villages, women have the responsibility to draw water for domestic consumption and carry it over long distance on foot, often in adverse and hot climate. Sub-groups of women assist village water and sanitation committees in many states. The Pani Panchayat in the Anuli irrigation project in Orissa is entirely managed by women (Planning Commission 2006). In Gujarat, the Self Employed Women's Association (SEWA) has trained women in the maintenance of hand pumps. Now that women have reservations in panchayats, their role in water management will also become more important. Therefore, if the involvement of women self-help groups in water management is increased, it is likely to result in improvement in the water use efficiency.

24.10. INITIATIVES NEEDED IN WATER SECTOR IN INDIA

While India has 16% of the world's population, its share in the world's fresh water is only 4% and it has only 2% of world's land. Precipitation is confined to a relatively short season over a year which makes irrigation imperative for reliable agricultural production and necessitates river regulation. Per capita availability of 1,700 m^3 is required to be free of water stress, while availability below 1,000 m^3 (this amount is made up of 200 m^3 for domestic/industrial use, 200 m^3 for environment and ecology, and 600 m^3 for food security) is termed as water scarcity. In India, the per capita availability in the year 2001 was 1,820 m^3 and per capita storage was 207 m^3. But this estimate is somewhat misleading, considering that about 630 BCM of water in the Ganga-Brahmaputra is unutilizable. Hence the actual per capita availability is only 1,200 m^3. Presently, 9 river basins with 20 crore population are already under water scarce conditions.

The falling per capita availability reflects the effect of rising population. The average availability is therefore likely to fall below the water-stress level in the near future and given the wide variations across the country the situation is likely to be worrisome. The per capita storage in the country of about 207 m^3 is much less than the storage created in many of the countries as described in Chapter 19. Even if all ongoing and potential storages are completed, the per capita storage will increase only up to 400 m^3. Rapidly increasing population and loss of storage due to sedimentation will further reduce the per capita storage.

The growing population and increasing water use in India are making freshwater scarce and polluted and posing a major threat to them. In the past, Indian scientists had ushered the green revolution by increasing food production. A new water revolution is needed to preserve, harness, develop and manage water resources keeping in view their quantity and quality. Undoubtedly, scientists will have a major role in the next blue revolution too but it would be necessary to involve all the stakeholders. It would be important to induce individuals and communities to appreciate their options, evaluate them, and then choose the one that is the most appropriate. Only when the empowered communities understand their rights and duties and assume their responsibilities, they can be true partners in the development of water resources.

24.10.1. Issues and Challenges in Water Sector

In the earlier chapters, details about water uses and problems have been given. The country thus seems to be on the threshold of a water crisis in the near future unless remedial steps are taken soon. To sum up, the main issues/problems in water sector in India are:

- Rapidly increasing demand for water accompanied by limited and variability;
- Extensive pollution of surface water and increasing pollution of ground water due to extensive agriculture, urbanization, industrialization, etc.;
- Provision of potable water to all villages, which is one of the monitorable targets of the Tenth Plan;
- Recurring problem of floods and droughts;
- Provision of assured supply of good quality water for domestic use in cities;
- Overuse of ground water and ineffectiveness of legislation to check it;
- A large number of on-going irrigation projects (more than 350 major and medium irrigation projects) with a huge balance cost exceeding Rs. 90,000 crore;
- A gap of nearly 14 M ha between created and utilized irrigation potential;
- Low water use efficiency in irrigation (25% to 35%) and in urban water supply (30% to 40% losses);
- Low tariffs for both irrigation and urban water supply, which do not cover even the operation and maintenance (O&M) costs;
- Water logging in canal command areas and drainage of surplus water;
- Inter- and intra-state water sharing disputes;
- Low coverage of urban and rural sanitation, leading to hygiene problems and water borne diseases;
- Growing problem of urban solid waste management and pollution from lechates; and
- Multiplicity of ministries/departments dealing with water.

24.10.2. Future Governance Strategy

The future strategies for water resources development and management have to take into consideration the above facts and issues. In view of restricted fund and other resources, relatively higher costs of the projects and the constraints, it is necessary to identify the strategic areas for focused attention. Similarly, management issues have to be given due emphasis and future planning will necessarily be a composite package of development and management. This can be achieved only with active participation of the stakeholders at all stages of development and management. But for that to happen, the government functioning will have to be more transparent, forthcoming, and participatory.

In future water resources management in India, the following aspects will have to be given due consideration:

- Most of the good project sites are already exploited and hence future developments will be much more challenging than the past;

- Besides technical aspects, social and environmental issues will dominate all future planning and it will be difficult to proceed with a project unless it has social support and acceptance;
- The imbalance resulting from over-utilization in one region and under-utilization in another will need to be addressed (equity issue);
- Efforts are required to optimally utilize the created potential in a systematic manner;
- Restricted fund availability calls for strict financial management;
- Co-ordination among the various agencies dealing with water sector;
- Problems of land acquisition and environment & forest clearance for the projects which often delay the work needs to be streamlined;
- Project-affected people need to be resettled and rehabilitated with care and respect;
- Delay in completion of the projects due to water disputes on sharing of waters between upper and lower Riparian states, control, operation and regulation of releases;
- Shift from project specific planning to integrated basin planning; and
- Improving water use efficiency through better water management.

Priority Areas

In view of the limited financial resources available and to appropriately address the needs of the society, it is necessary to prioritize the activities and support them accordingly. The Government of India has identified the following priority areas:

- Completion of on-going projects,
- Revival and restoration of existing water bodies,
- Command area development and PIM,
- Flood management and erosion control,
- Dam safety and rehabilitation, and
- Rural drinking water supply and sanitation.

The priority will, however, vary from region to region.

Based on the issues, challenges, and priority areas, the following initiatives are needed for the sake of good water governance:

- Identify critical projects as national projects and ensure their timely completion;
- Efforts to be made to achieve the Millennium Development Goals related to drinking water supply and sanitation;
- As is being done in other infrastructure projects, employ modern management tools for decision making in water sector projects and to monitor the progress;
- Water tariff for irrigation and other uses should be fixed to encourage saving as well as provide subsidized services to the needy;
- Anything that is given free including electricity for agriculture is likely to be wasted and this practice should be stopped;
- Review the issue of ownership rights on groundwater, as legislation to control over-exploitation has not been successful;

- Improve coordination among the agencies involved in water related matter at central and state government level; and
- Increase recycle and reuse of water and modernization of industrial process to reduce water demand

Important recommendations of Task Forces

Various Task Force/Task Groups have been constituted by the government to study and make recommendations on different aspects of water resources, such as floods, efficient utilization, environmental aspects, etc. Many recommendations of short-term and long-term have been made by them and the important recommendations are as follows:

1. Introduce schemes for system rehabilitation, ground water development and artificial recharge;
2. Promote cropping patterns suited to specific agro-climatic zones;
3. Include command area development works as part of the project;
4. Funds may be provided to states as additional central assistance to maintain embankments;
5. Community participation in maintenance of embankments should be encouraged;
6. The AIBP needs to focus not only on completing projects but also maximizing creation of potential at a given cost;
7. There should be full central funding of the flood component in storage dams;
8. State governments should be persuaded to implement flood plain zoning;
9. Increase the cost limit of major projects needing environment clearance from Rs. 100 crore to Rs. 250 crore;
10. Increase the cultural command area limit for major projects needing environment clearance from 10,000 ha to 25,000 ha; and
11. Exempt irrigation projects from the need to pay net present value of submerged forest area.

24.10.3. Future Research Directions

Indian researchers have done a commendable work in the field of hydraulics, hydrology, sediment transport, and irrigation management. The research and development work on design of unlined channels in alluvial soils by Kennedy, Lacey and several other irrigation engineers; and hydraulic analysis of barrages on permeable foundations conducted by Dr. Khosla and his team have found international acceptability. In recent times, one cannot fail to notice the large number of research findings that are published by Indian experts through various outlets, such as international journals, conferences, books, and technical reports and that have created a major impact in the field of water resources. Despite logistics and infrastructure limitations, leading institutes of India have provided their expert services to almost all major projects in the country and have fulfilled their role. Their contribution to water management is praiseworthy.

The country is facing a number of challenges in the water sector. In addition to the topics suggested elsewhere in the book, research on the following broad topics will be beneficial to the nation:

i) Methodology for integrated water resources assessment in a river basin under the various scenarios of land-use land-cover change and climate change;
ii) Optimal development of unutilized water resources;
iii) Improvement in the efficiency of present utilization and closing the gap between the demand and supply;
iv) Enhancement of the level of utilizable resource from the present estimate of 1,122 BCM to a higher potential;
v) Hydrological assessments and predictions in ungauged basins;
vi) Likely impacts of global warming and climate change on hydrologic systems in India and adaptation techniques;
vii) Optimal and sustainable methods suited for Indian conditions for water resources utilization while maintaining and improving the health of environment and eco-systems;
viii) Participatory decision making to resolve water related disputes using the modern technological advancements; and
ix) Low cost water treatment using locally available materials.

24.11. EPILOGUE

Water is a major factor in each of the three pillars of sustainable development – economic, social, and environmental. India has to initiate a series of measures to ensure that her people have access to clean water and sanitation, there is food security, and there are no water related conflicts. Water must meet the needs of the present population and those of future generations. To that end, sincere efforts will have to be continued, year-after-year, for the sake of peace, socio-economic growth and making the Earth a beautiful place to live.

REFERENCES

Abbi SDS, Gupta DK, Jain BC (1970a) A study of heavy rainstorms over North Bengal, India. J Meteorol Geophys 21(2): 195–210

Ahmad I (1997) Problems and planning aspects of Kharif Channel – a case study of east Ganga canal project. ME dissertation. Water Resources Development Training Centre, University of Roorkee, Roorkee, India

Ahmad M (2005) Irrigation benefits from Yamuna-Rajasthan and Rajasthan-Sabarmati link Proposals of NWDA to the western tracts of Rajasthan State. Proceeding of 11th National Water Convention, organized by National Water Development Agency at New Delhi

Alagh YK, Buch DT (1997) The Sardar Sarovar Project and sustainable development. In: Fisher WF (ed) Toward sustainable development: struggling over India's Narmada river. Rawat Publications, Jaipur

Alagh YK (2001) Water and food security in South Asia. Int J Water Resou Develop 17(1): 23–36

Ali S (2003) Drought indices in India-a review. Hydrol J 26(3): 31–40

Allan JA (1993) Fortunately there are substitutes for water otherwise our hydro-political futures would be impossible. In: Priorities for water resources allocation and management. ODA, London

Allen RG, Pereira LS, Raes D, Smith M (1998). Crop Evapotranspiration, irrigation and drainage. Paper no. 56. Food and Agriculture Organization, Rome, Italy

Ambast SK (2005) Benchmarking water productivity in India – policy issues and action plans. Paper presented at the XII World Water Congress "Water for Sustainable Development – Towards Innovative Solutions", New Delhi 22–25 November 2005

American Public Health Association (APHA) (1961) Methods for the examination of water and waste water. American Public Health Association, New York

Angstrom A (1924) Solar and terrestrial radiation. Q J Meteorol Soc 50: 121–125

Bahadur J (1992) Snow and glaciers and their contribution to India's water resources. In: Water science educational series, series no. 1. National Institute of Hydrology, Roorkee, India

Bandara WLHMT, Imbulana KAUS (1996) Hydrological safety of dams. Proceedings of the 2nd International Conference on Dam Safety Evaluation. Central Board of Irrigation and Power, Trivendrum, 26–30 Nov 1996

Bandyopadhyay J (1995) Water management in the Ganges-Brahmaputra basin: emerging challenges for the 21st century. Int J Water Resour Develop 11(4): 411–442

Basha MK (2003) River diversion planning for dam construction with special reference to Tehri dam. Dissertation of Master of Engineering in Water Resources Development, Water Resources Development Training Centre, Indian Institute of Technology, Roorkee, India

Basit M (2003) Flash floods on the Indus River in Pakistan. National Engineering Services of Pakistan (Pvt.) Ltd. Islamabad, Pakistan

Basnyat G (2000) Reappraisal of hydrology and water resources of Harbhangi irrigation project, Orissa, India. Dissertation of Master of Engineering in Hydrology, University of Roorkee, Roorkee

Basu PK, Joshi LS (2000) Study for the efficient planning, control and management of water resources development projects in India. Int J Water Resour Develop 16(4): 563–570

Bhar AK, Khobragade SD (1992–1993) Behaviour of different types of lakes and their effect and relationship with the catchment hydrology. Report No. TN-99. National Institute of Hydrology, Roorkee

Bhar AK (1993–1994) Water balances of lakes. Report No. SR-37. National Institute of Hydrology, Roorkee

Bhar AK (1994–1995) Groundwater-tank interaction in Jabalpur district, Madhya Pradesh. Report No. CS (AR) 165. National Institute of Hydrology, Roorkee

Bhar AK (1995–1996) Sedimentation in thermally stratified lakes of Kumaun region. Report No. CS(AR)-193. National Institute of Hydrology, Roorkee

Bhar AK (1996–1997) Wind erosion and lake sedimentation in desert area. Report No. TR(BR)-9/96–97. National Institute of Hydrology, Roorkee

Bhar AK (1997–1998) Systematic procedure for the components of water computation of the lakes – part I: evaporation. Report No. TR (BR)-15/97–98. National Institute of Hydrology, Roorkee

BIS (1985) Guidelines for fixing the spillway capacity. IS 11223:1985, Bureau of Indian Standards (BIS), New Delhi

Biswas AK (1969) Science in India. K.L. Mukhopadhyaya Publishers, Calcutta, India

Blaney RF and Criddle WD (1962). Determining consumptive use and irrigation water requirements, Technical bulletin 1275. U.S. Department of Agriculture, Washington, DC, 59 p

Bosen JF (1958). An approximate formula to compute relative humidity from dry bulb and dew point temperatures. Mon Weather Rev 86: 486

Briz-Kishor BH (1993) Assessment of yield characteristics of granite aquifer in southern India. Groundwater 31:921–928

CAZRI (1997) VISION 2020. Central Arid Zone Research Institute, Jodhpur

CBIP (1965) History of irrigation in India. Central Board of Irrigation & Power, New Delhi, India

CBIP (1987). Register of large dams in India. Publication No. 197. Central Board of Irrigation and Power, Malcha Marg, New Delhi, India

CBIP (1989) Workshop on unusual storm events and their relevance to dam safety to snow hydrology, Srinagar, July, 1989, p 82

CBIP (1989) Workshop on unusual storm events their relevance to dam safety, Bhubaneshwar, May, 1989, p 68

CBIP (1990) Workshop on unusual storm events and their relevance to dam safety, Nagarjunsagar, February, 1990, p 102

CBIP (1990) History of Ramganga project, vol. 1. Publication no. 214. Central Board of Irrigation & Power, New Delhi, India

CBIP (1996). National perspective plan for recharge to ground water by utilizing surplus monsoon run-off. Technical report no. 14. Central Board of Irrigation and Power, New Delhi, India

CBIP (1996) Proceedings of the 2nd International Conference on Dam Safety Evaluation. Central Board of Irrigation and Power, Trivendrum, 26–30 Nov 1996

CBIP (1998) Typical dams in India. Publication no. 272. Edited by Varma CVJ, Rao ARG, Sundaraiya E. Central Board of Irrigation and Power, New Delhi, India

CEA (1988) Hydro-electric power potential. Central Electricity Authority, Government of India, New Delhi

CGWB (1977) Water balance studies on upper Yamuna canal. Technical report no.18. Central Ground Water Board, Faridabad, India

CGWB (1995) Perspective plan for ground water resources in India. Central Ground Water Board, Ministry of Water Resources, Government of India, Faridabad, New Delhi, India

CGWB (1997). Report of the Ground Water Resources Estimation Committee. Ministry of water Resources, Government of India, New Delhi, India

CGWB (1999). Report of in storage fresh ground water resource of India (unpublished). Ministry of water Resources, Government of India, New Delhi, India

CGWB (2005) Perspective plan for ground water resources in India. Central Ground Water Board, Ministry of Water Resources, Government of India, Faridabad

Chachadi AG, Mishra GC, Singhal BBS (1991) Drawdown at a large diameter observation well. Journal of Hydrology, 127(1–4), 219–233

Chaddha DK (2002) State of art of artificial recharge applied in village level schemes in India – country report. NNC-IAH Publication No. 4. Proceedings of the Seminar on Management of Aquifer Recharge and Sub-surface Storage – Making Better Use of Our Largest Reservoir, Wageningen, Dec 2002, pp 19–24

Chaddha DK (2006) Development and management of ground water resources of India: an Overview. In: Ghosh NC, Sharma KD (eds) Groundwater modelling and management. Capital Publishing Company, New Delhi

Chandra S (1992) Key note lecture at the International Symposium on Hydrology of Mountainous Areas, Welcome address by Dr. Satish Chandra, Director, National Institute of Hydrology, Shimla, India 28–30 May, 1992

Chandra PC (2005) Geophysical investigations for ground water in coastal tracts. Jalvigyan Sameeksha (hydrology review) 20: 47–63

Chapman GP, Thompson M (1995) Global development and the environment series. Mansell Publishing Limited, England.

Chaturvedi MC, Rogers P (1985) Water resources planning – some case studies for India. Indian Academy of Sciences, Bangalore, India

Chitale MA (1992) Population and water resources of India. Umesh Communications, Pune, 452 p

Chitale MV (1997) The Narmada project. Water Res Dev 13(2):169–179

Chow VT (1964) Runoff section 14. Handbook of applied Hydrology. McGraw Hill Book Co. Inc., New York

Chowdhury A, Das HP, Singh SS (1993) Agro-climatic classification in India. Mausam 44(1): 53–60

Chowdhary H, Jain SK, Ogink HJM (2002) Upgrading a conventional hydrological information system – an Indian example. FRIEND 2002—regional hydrology: bridging the gap between research and practice. Proceedings of the fourth International FRIEND conference, March 2002, IAHS publication no. 274. Cape Town, South Africa, pp 35–42

CPCB (1995) Basin sub-basin inventory of water pollution. Central Pollution Control Board, New Delhi.

CPCB (1999) Water quality status and statistics 1996 & 1997. Central Pollution Control Board, New Delhi, India, 403 p

CPCB (2006) Web site of Central Pollution Control Board. http://www.cpcb.nic.in/cpcb/water/waternew/advance_serach/waterb.php. Cited January 2006

CSE (1997) Dying wisdom: rise, fall and potential of India's traditional water harvesting systems. Centre for Science and Environment, New Delhi

CSE (1999) The citizens' fifth report, vol I and II. Centre for Science and Environment, New Delhi

CSSRI (1997) Vision 2020 – CSSRI perspective plan. Central Soil Salinity Research Institute, Karnal

CWC (1969) Flood estimation by unit hydrograph method. In: CWC recommended procedure, chapter 5

CWC (1973–1989) Flood estimation reports of various climatic sub-zones. Directorate of Hydrology (Small Catchments), Central Water Commission, New Delhi

CWC (1973) Estimation of design flood peak. Report no.1/73. Flood Estimation Directorate, Central Water Commission, New Delhi, India

CWC (1982) Flood estimation report for Mahanadi subzone 3d – a method based unit hydrograph principle. Report no. M/5/1981. Hydrology Directorate, Central Water Commission, New Delhi, India

CWC (1984) Flood estimation report for upper Indo-Ganga Plains subzone-1e – a method based unit hydrograph principle. Hydrology Directorate, Central Water Commission, New Delhi, India

CWC (1987) Flood estimation report for Sone subzone-1 (d). Report no. S/15/1987. Directorate of Hydrology (small catchments), Central Water Commission, New Delhi, India

CWC (1990) Storages in river basins of India. Report no. 62/90. Central Water Commission, New Delhi, India

http://www.rajirrigation.com/3rivers4.htm. Accessed in February 2006

http://www.1upindia.com/states/andaman-nicobar/index.html. Accessed in February 2006

http://www.spiritualguides.net/Indiastate/LaksadweepR.htm. Accessed in February 2006

http://www.india-travel-tourism.com/Daman-Diu.htm. Accessed in February 2006

CWC (1995) Guidelines for planning conjunctive use of surface and ground waters in irrigation projects. Joint report prepared by the Central Water Commission and Indian National Committee on Irrigation and Drainage, New Delhi, India

CWC (1996) Water and related statistics. Water year book of Krishna basin. Central Water Commission, Government of India, New Delhi

CWC (2001) Compendium on silting of reservoirs in India. Water Planning and Projects Wing, Environment Management Organization, Watershed and Reservoir Sedimentation Directorate, Central Water Commission, New Delhi

CWC (2002) Water and related statistics. Water Planning Project Wing, Central Water Commission, Government of India, Bhubneshwar, New Delhi, India

CWC (1973–1989) Flood estimation reports of various climatic sub-zones. Directorate of Hydrology (small catchments). Central Water Commission, New Delhi, India

Das PK, Roy GK, Choudhary A (2004) Hydrogeological framework and groundwater development prospects in Orissa (unpublished report). Central Ground Water Board, Government of India, New Delhi, India

Dastur DJ (1974) This or else ...: A master plan for India's survival. Jaico Publishing House, Bombay

De RN (1972) Studies in the regulation and operation of the DVC reservoir system. Master of engineering thesis. Water Resources Development Training Centre, University of Roorkee, Roorkee, India

Development Alternatives (2004). Troubled waters: developing water sustaining livelihoods. www.devalt.org/water/WaterinIndia/issues.htm. Accessed in December 2004

Dey B, Goswami DC, Rango A (1983) Utilization of satellite snow-cover observations for seasonal stream flow estimates in the Western Himalayas. Nordic Hydrol 14: 257–266

Dey B, Goswami DC (1984) Evaluating a model of snow cover area versus runoff against a concurrent flow correlation model in the Western Himalayas. Nordic Hydrol 15: 103–110

Dhanju MS (1983) Studies of Himalayan snow cover from satellite. Hydrological applications of remote sensing and remote data transmission, IAHS publication no. 145, pp 401–409

Dhar ON, Bhattacharya BK (1975) A study of depth-area-duration statistics of the severe most storms over different meteorological divisions of North India. In: Proceedings of the national symposium on hydrology, Roorkee, India, G-4–11

Dhar DN, Nandargi S (1989) Analysis of severe rainstorm of contiguous Indian states. CBIP Regional Workshop, Bhubaneshwar

Dhar ON, Bhattacharya BK (1977) Relationship between central rainfall and its areal extend for severe most rainstorms of north Indian plains. J Irrigation Power 28(2): 245–250

Dhar DN, Kulkarni AK (1974) Estimation of probable maximum rainfall over plain areas of north India. Proceedings of the international tropical meteorological meeting, Nairobi, American Meteorological Society, pp 287–289

Dhar ON, Nandargi S (1993) Spatial disturbances of severe rainstorms over India and their associated areal rain depths. Mauasam 44(4): 373–380

Dhar ON, Rakhecha PR (1979) A review of rainfall relations based upon Indian data. Water Resourc J, ESCAP, Ser C-123: 16–25

Dhar ON, Rakhecha PR (1981) The effect of elevation on monsoon rainfall distribution in the Central Himalayas. In: Proceedings of the international symposium on monsoon dynamics, Cambridge University Press, Cambridge, pp 253–260

Dhar DN, Rakhecha PR, Kulkarni AK (1982) Estimation of extreme point rainfall for peninsular India. Proceedings of the international conference on rain water cistern systems, Honolulu, Hawaii

Dhar DN, Kulkarni AK, Mandal BN (1983) Estimation of probable maximum rainfall over plain areas of north India. Proceedings of the seminar on hydrology, Hyderabad, India

Dhawan BD (1993) Indian water resource development for irrigation: issues critiques and reviews. Commonwealth Publishers, New Delhi

Dhillon GS, Singh T, Paul D (1986) Technology for skimming canal seepage water in saline areas. Proceedings of 53rd R & D annual session of CBIP held at Bhubneshwar, Central Board of Irrigation & Power, New Delhi 8–10 May, pp 29–40

Dhillon GS The Pakistani experience. An article by former Chief Engineer (research) and Director (irrigation), Punjab

Doorenbos J, Pruit WO (1977). Crop water requirements. Irrigation and Drainage. Paper 24. Food and Agricultural Organization of the United Nations, Rome (revised), Italy

DOWR (1999) Basin planning report of Brahmani basin, second spiral study. Department of Water Resources, Government of Orissa, Bhubneshwar

Dutta V, Tiwari AP (2005) Challenges of sustainability: demand side analysis for urban water utility of Delhi, India. Paper presented at the XII World Water Congress "Water for Sustainable Development – Towards Innovative Solutions", New Delhi, 22–25 November 2005

Dyson M, Bergkamp G, Scanlon J (eds) (2003). Flow. The essentials of environmental flows. IUCN, IUCN Water & Nature Initiative, Switzerland

Ferguson HL, Znamensky VA (1981) Methods of computation of the water balance of large lakes and reservoirs. Studies and reports in hydrology #31. Unesco, Paris

Frater A (2002) Chasing the monsoon. Special Indian Edition,

Garg HK (1987) Water utilization model study of Rajghat multipurpose project and Betwa basin. Master of engineering thesis. Water Resources Development Training Centre, University of Roorkee, Roorkee, India

Garudkar AS (1991) Simulation for Mula irrigation project in Maharashtra. Dissertation of Master of Engineering in Hydrology, UNESCO Sponsored International Course in Hydrology, University of Roorkee, Roorkee

GEC (1984). Ground Water Resources Estimation Committee (1984). Ministry of Water Resources. Govt. of India.

GEC (1997). Ground water estimation methodoloy-1997. A report of the Ground Water Resources Estimation Committee. Ministry of Water Resources, Government of India, New Delhi

George CJ, Ramasastri KS, Rentala GS (1972) Incidence of droughts in Andhra Pradesh, Tamil Nadu and Mysore. Indian Meteorological Department Monograph, New Delhi

Ghosh NC, Sharma KD (ed) (2006). Ground water modelling and management. Capital Publishing Company, New Delhi, India.

GoAP (1984) Somasila Project Report, vol 1/4. Irrigation Department, Government of Andhra Pradesh, Hyderabad

http://waterresources.kar.nic.in/ongoing.htm. Accused in January 2006

http://waterresources.kar.nic.in/completed.htm. Accused in January 2006

http://waterresources.kar.nic.in/salient_features_krs.htm. Accused in January 2006

http://waterresources.kar.nic.in/projects.htm. Accused in January 2006

http://waterresources.kar.nic.in/river_systems.htm. Accused in January 2006

http://www.geol.lsu.edu/deltaweb/INDIARPT/CAUVERY/cauvery_delta.h

Gosain AK, Singh A (2005) Water rights in Indian transboundary watercourses. Jalvigyan Sameeksha (Hydrology Review) 19(1–2), INCOH, National Institute of Hydrology, Roorkee

Goswami DC (1985) Brahmaputra River, Assam, India: Physiography, basin denuadation, and channel aggradation. Water Resour Res 21(7): 959–978

Gowariker V, Thapliya V, Kulshrestha SM, Mandal GS, Roy NS, Sikka DR (1991) A power regression model for long range forecast of southwest monsoon rainfall over India. Mausam 42(2): 125–130

GSI (1978) Ground water observation in the wells in Bhopal district. Report of Geological Survey of India. Ground Water Survey Circle of Water Resources Department, Government of Madhya Pradesh, Bhopal

GSI (1988) Geology of Bhopal district. Report of Geological Survey of India. Ground Water Survey Circle of Water Resources Department, Government of Madhya Pradesh, Bhopal

Gujja B, Pangare G, Das S (2005) Role of the National Civil Society Committee on Interlinking of Rivers in India (NCSCILR) in steering a process of holistic water resource development: A case study of the Ken-Betwa link. Proceeding of 11th national water convention, organized by National Water Development Agency at New Delhi

Gupta SK, Deshpande RD (2004) Water for India 2050: first order assessment of available options. Curr Sci 86(9): 1216–1224

Gupta DB, Mitra S (2004) Sustaining Subarnarekha river basin. Int J Water Res Dev 20(3): 431–444

Gupta RK (2001) River basin management: A case study of Narmada valley development with special reference to the Sardar Sarovar Project in Gujarat, India. Int J Water Res Dev 17(1):55–78

Gyawali D (1991) Troubled politics of himalayan waters. Himalayan Journal 4(2)

http://www.water-mgmt.com/en/issues.htm

http://wrmin.nic.in/cooperation/disputes.htm

Handa BK (1964) Hydrochemical provinces in sedimentary basins of India, 22nd international geological congress, Part 12, pp 126–142

Handa BK (1975) Natural waters, their geochemistry, pollution, and treatment. Technical Manual No.2, Central Ground Water Board, New Delhi

Handa BK (1979) Saline water resources of India, Technical Manual No. 5, Central Ground Water Board, New Delhi

http://envfor.nic.in/nrcd/class.html

http://wrmin.nic.in/problems/pb_faced.htm

http://www.fao.org/DOCREP/003/V5930E/V5930E01.htm#ch1

http://mospi.nic.in/compenv2000_ch6.htm

http://edugreen.teri.res.in/explore/maps/biodivin.htm

Hande AP (1991) Multi-objective crop planning for Pench irrigation project, Maharashtra. Dissertation of Master of Engineering in Hydrology, UNESCO Sponsored International Course in Hydrology, University of Roorkee, Roorkee

Harihar Ayyer PS, Prasad B (1971) 1-hour rainfall of India for different return periods, Ind J Meterol Geophys 22(2): 219–222

Hasan Z (2005) Dams and water transfer. 11th National Water Convention. NWDA, New Delhi, India

http://www.indianchild.com/rivers_in_india.htm

http://www.water-mgmt.com/en/riverbasin.htm. Accessed in December 2005

http://www.water-mgmt.com/en/database_india11.htm. Accessed in December 2005

http://www.fao.org/DOCREP/003/V5930E/V5930E01.htm#ch1

http://en.wikipedia.org/wiki/Ganga

http://www.haryana-online.com/saraswati.htm

http://www.india-water.com

http://www.harappa.com and www.archaeolink.com

http://upgov.up.nic.in/irrigation/dams_reservoirs.htm. Accessed in January 2006

http://www.fao.org/DOCREP/003/V5930E/V5930E01.htm#ch1

http://doda.nic.in/others/hydromain.htm

http://earthtrends.wri.org/text/FRE/maps/355.htm. Accessed in January 2006

http://jammukashmir.nic.in/profile/bramula.htm#INTRODUCTION

http://udhampur.nic.in/tour/Lake.htm

http://wrmin.nic.in/riverbasin/indus.htm

http://wrmin.nic.in/cooperation/rvbstribunal.htm

http://www.hpseb.com/chenab_basin.htm

http://www.dams.org/images/maps/map_tarbela.htm

http://www.sanalist.org/kalabagh/a-4.htm

http://www.dal-naginlake.com/dallake.htm

http://www.dal-naginlake.com/phy-feature.htm

http://www.leh-ladakh.com/kashmir-india/dal-lake-nagin-lake.html

http://www.indiantravelportal.com/jammu-kashmir/lakes/nagin-lake.html

http://www.cpcb.nic.in/cpcb/water/waternew/advance_serach/waterb.php

Hasan Z (2005) Dams and water transfer. Paper presented at the 11th National Water Convention, NWDA, New Delhi 5–10 September 2005

http://envfor.nic.in/cpcb/rwq/99/nrwq_bai.html. Accessed in January 2006

http://www.orissawater.com. Accessed in January 2006

http://www.orissawater.com/basinmapping5.htm. Accessed in January 2006

http://wrmin.nic.in/riverbasin/mahanadi.htm. Accessed in January 2006

http://www.geol.lsu.edu/deltaweb/INDIARPT/KRISHNA/krishna_delta.htm. Accessed in February 2006

http://wrmin.nic.in/riverbasin/krishna.htm. Accessed in February 2006

http://waterresources.kar.nic.in/river_systems.htm. Accessed in February 2006

http://www.geol.lsu.edu/deltaweb/INDIARPT/india.htm. Accessed in February 2006

http://waterresources.kar.nic.in/ongoing.htm. Accessed in February 2006

http://waterresources.kar.nic.in/completed.htm. Accessed in February 2006

http://envfor.nic.in/cpcb/rwq/99/nrwq_krs.html. Accessed in February 2006

http://waterresources.kar.nic.in/projects.htm. Accessed in February 2006

http://www.ap.gov.in/apirrigation/MajorIrrig/moreprj/nsp.htm. Accessed in February 2006

Herschey RW (ed) (1978) Hydrometry principles and practices. John Wiley and Sons, New York

Hershfield DM, Wilson WT (1960) A comparison of extreme rainfall depths from tropical and non-tropical storms. J Geophys Res 65: 959–982

Hershfield DM (1961) Estimating the probable maximum precipitation. J Hydrol Div, Am Soc Civil Eng 87(5): 99–116

Hershfield DM (1965) Method for estimating probable maximum precipitation. J Am Water Works Assoc 57: 965–972

Higuchi K, Ageta PR, Yasunari T, Inoue J (1982) Characteristics of precipitation during monsoon season in high mountain areas of the Nepal Himalayas. Hydrological aspects of alpine and high mountain areas. IAHS publication no. 138, pp 21–30

Huq SM, Vijay Kumar, Sil SS (1982) Derivation of synthetic unit hydrographs for moderately steep catchments. Proceedings of international symposium on hydrological aspects of mountains watersheds. UOR, Roorkee, India, pp IV-27–IV-31

Hydrology Project (2003a) Guidelines for surface water quality network design, HIS (SW) design manual vol. 6 – water quality sampling. Ministry of Water Resources, Government of India, New Delhi

Hydrology Project (2003b) Protocol for water quality monitoring, HIS (SW) field manual vol. 6 – water quality sampling. Ministry of Water Resources, Government of India, New Delhi

Hydrology Project (2003c) "Guidelines on standard analytical procedures for water analysis, HIS (SW) operation manual vol. 7 – water quality analysis. Ministry of Water Resources, Government of India, New Delhi

Hydrology Project (2003d) Basic surface water data processing, HIS (SW) operation manual vol. 8 – data processing. Ministry of Water Resources, Government of India, New Delhi

ICID (2005) Homepage of ICID.
http://www.icid.org. Accessed in December 2005

ICOLD Bulletin 99 (1995) Dam failures-statistical analysis. International Commission on Large Dams, Paris

IEA (2001 and 2006) Key world energy statistics. International Energy Agency, Paris

IITM (1989) 1 day PMP atlas. Indian Institute of Tropical Meteorology, Pune

IMD (1978) Agro-climatic Atlas of India. India Meteorological Department, New Delhi, India

IMD (1970a) Krishna Godavari basin, hydrology monograph no. 1. India Meteorological Department, New Delhi

IMD (1970a) Narmada basin, hydrology monograph no. 2. India Meteorological Department, New Delhi

IMD (1971) Cauvery basin, hydrology monograph no. 4. India Meteorological Department, New Delhi

IMD (1971) Ganga basin, hydrology monograph no. 3 and 5. India Meteorological Department, New Delhi

IMD (1971) Mahanadi basin, hydrology monograph no. 6. India Meteorological Department, New Delhi

IMD (1972) Manual on hydrometeorology - part I. Climatological tables of observations in India. India Meteorological Department, New Delhi, pp 45–69

IMD (1977) Tapi basin, hydrology monograph no. 8. India Meteorological Department, New Delhi

IMD (1981). Climatological tables of observations in India. India Meteorological Department, New Delhi, India

Inamdar MS (1987) System study of Krishna irrigation project (Maharashtra). Dissertation of Master of Engineering in Water Resources Development, Water Resources Development Training Centre, University of Roorkee, Roorkee

Inder J (2005). Ground water resources of India-occurrence, utilization and management. Mittal Publications, New Delhi, India.

Indian Express (2004) Drying dam spells drought doom in Solapur. 2 Apr 2004

Interagency Advisory Committee of Water Data (1982) Guide-lines for determining flood wave frequency. Bull. 17B, US Department of the Interior, US GS, Office of Water Data Coordination, Reston

IPCC (1998) In: Watson RT, Zinyowera MC, Moss RH (eds) The regional impacts of climate change: an assessment of vulnerability. A Special Report of Working Group II and WMO/UNEP. Cambridge University Press, Cambridge

IPCC (2001) Climate change 2000: the science of climate change. In: Houghton JT et al. (eds), Intergovernmental Panel for Climate Change, Assessment Report of IPCC working group I and WMO/UNEP. Cambridge University Press, Cambridge

IRC (1970) General features of design-section-I. Indian Road Congress, Jamnagar House, New Delhi, India

Irrigation Commission (1972) Report of the volume III. Ministry of Irrigation & Power, Government of India, New Delhi

Irrigation Department (2005) http://upgov.up.nic.in /irrigation/dams_reservoirs.htm. Accessed in December 2005. Irrigation Department of UP, India

IS:12182:1987. Guidelines for determination of effects of sedimentation in planning and performance of reservoirs. Bureau of Indian Standards, Manak Bhawan, New Delhi, India

IS:11223-1985. Guidelines for fixing spillway capacity. Bureau of Indian Standards, Manak Bhawan, New Delhi, India

IS 9452: 1993. Code of practice for measurement of seepage losses from canals – part 1: ponding method (first revision). Bureau of Indian Standards, New Delhi, India.

IS:12752:1989. Guidelines for the selection of flow gauging structures. Bureau of Indian Standards, Manak Bhawan, New Delhi, India

IS:15119:2002 Measurement of liquid flow in open channels. Part 1: establishment and operation of a gauging station (superseding IS 2914:1964) and part 2: determination of the stage-discharge relation (superseding IS 2914:1964). Bureau of Indian Standards, Manak Bhawan, New Delhi, India

IS 6939 (1992). Methods for determination of evaporation from reservoirs (first revision). Bureau of Indian Standards, New Delhi, India

IS (4986:2002) Indian Standard Code of practice for installation of rain-gauge (non-recording type) and measurement of rain (second revision). Bureau of Indian Standards, New Delhi

IS (4987:1994) Indian Standard Code recommendations for establishing net work of raingauge stations. Bureau of Indian Standards, New Delhi

IS:5477-1969. Methods for fixing the capacities of reservoirs (parts I–IV). Indian Standards Institution, Manak Bhawan, New Delhi, India

IS (8389:2003) Indian Standard Code of practice for installation and use of raingauges, recording (2nd revision). Bureau of Indian Standards, New Delhi

IS:1192-1981. Velocity area methods for measurement of flow of water in open channels (first revision). Indian Standards Institution, Manak Bhawan, New Delhi, India

IS:4890-1968. Methods for measurement of suspended sediment in open channels. Bureau of Indian Standards, Manak Bhawan, New Delhi, India

IS:7784-1975. IS code of practice for design of cross drainage works (part I). Indian Standards Institution, Manak Bhawan, New Delhi, India

IS 9452: 1980. Code of practice for measurement of seepage losses from canals – part 2: inflow outflow method. Bureau of Indian Standards, New Delhi, India.

ISO (1982). Liquid flow measurement in open channels – part 2: determination of stage-discharge relation, ISO 1100/2-1982(E). International Standards Organization, Geneva, Switzerland

ISO (1983). Measurement of liquid flow in open channels. Handbook no. 16. International Standards Organization, Geneva, Switzerland

IUGG (CCS) – UNEP – UNESCO (2005) Fluctuations of glaciers 1995–2000, vol VIII. World Glacier Monitoring Service, Zurich

IWRS (1996) Inter basin transfers of water for national development – Problems and prospects. Theme paper. Indian Water Resources Society, Roorkee, India

IWRS (1998) Five decades of water resources development in India. Theme paper. Indian Water Resources Society, Roorkee, India

IWRS (1999) Water: vision 2050. Theme Paper, Indian Water Resources Society, Roorkee, India

IWRS (2001) Theme paper on management of floods and droughts. Water Resources Day, Indian Water Resources Society

Iyer RR (1994) Indian federalism and water resources. International Journal for Water Resources Development 10(2):191–202

Iyer RR (2005) River-linking project: A critique. Presented at 11th national water convention, organized by National Water Development Agency, 11 May 2005

Jain SK, Chalisgaonkar D (2000) Setting up stage discharge relations using ANN. J Hydrol Eng 5(4): 428–433

Jain S, Lall U (2001) Floods in a changing climate: does the past represent the future? Water Resour Res 37(12): 3193–3205

Jain SK, Singh VP (2003) Water resources systems planning and management. Developments in Water Science #51. Elsevier, Amsterdam, the Netherlands

Jain SK, Singh VP (2003) Qualitative risk analysis of the scheme of interlinking of rivers. Proceedings of international conference on water and environment: WE-2003. Held in Bhopal, December

Jain SK, Seth SM, Kumar R, Singh AK, Garg PK, Vijay T, Singh D, Jorgensen GH, Erlich M (1990) Field investigation in Kolar sub-basin of river Narmada. Report No. TR-81. National Institute of Hydrology, Roorkee

Jain SK, Storm B, Bathurst JC, Refsgaard JC, Singh RD (1992) Application of SHE to catchments in India Part 2 – field experiments and simulation studies with SHE on the Kolar sub-catchment of Narmada river. J Hydrol 140:25–47

Jain SK, Goel MK, Agarwal PK (1998) Reservoir operation studies for Sabarmati system, India. J Water Res Plann Manag, ASCE, 124(1):31–38

Jain SK, Reddy NSRK, Chaube UC (2005) Analysis of a large inter-basin water transfer system in India. Hydrological Sciences Journal, IAHS 50(1):125–138

Jain SK (1990) Application of SHE model to Kolar sub-basin of Narmada. Report No. CS-33. National Institute of Hydrology, Roorkee

Kane RP (1997a) On the relationship of ENSO with rainfall over different parts of Australia. Aust Meteorol Mag 46: 39–49

Kane RP (1997b) Relationship of El Nino-southern oscillation and Pacific sea surface temperature with rainfall in various regions of the globe. Mon Wea Rev 125: 1792–1800

Kane RP (1998) Extremes of the ENSO phenomenon and the Indian summer monsoon rainfall. Int J Climatol 18: 775–791

Kane RP (2004) Relation of El Nino characteristics and timings with rainfall extremes in India and Australia. Mausam 55(2): 257–268

Karunathilake AA (1997) Environmental safeguards and their impact in irrigated command areas of old and major canal systems in Gangetic plains of UP. Master of engineering thesis. Water Resources Development Training Centre, University of Roorkee, Roorkee, India

Kelkar P, Andey S, Nanoti M (2005) Management and mitigation measures for silt disposal problem of Beas-Satluj River link. Proceedings of 11th National Water Convention, 11 May 2005. National Water Development Agency, New Delhi, India

Keller A, Sakthivadivel R, Seckler D (2000) Water scarcity and the role of storage in development. Research Report 39. International Water Management Institute, Colombo

Kennedy EJ (1984) Techniques of water-resources investigations of the United States geological survey: discharge ratings at gauging stations. USGS, Washington, DC, Book 3, Chapter A10

KERS (1985) Reservoir sedimentation study report. Karnataka Engineering Research Station, Mysore

Keshari AK (2005) Decision making for conjunctive water use planning in inter-basin transfer framework. Jalvigyan Sameeksha (Hydrology Review), 19(1–2), INCOH, National Institute of Hydrology, Roorkee

Khanna RK (2005) Environmental aspects of river interlinking projects: Facts, fears and fallacies. Proceeding of 11th national water convention, organized by National Water Development Agency at New Delhi, on 11 May 2005

Khobragade, SD, Bhar AK (1992–1993) Classification of lakes and inventory of natural lakes in India. Report No. TN-98. National Institute of Hydrology, Roorkee

Khobragade SD (1995–1996) Major and important lakes of Rajasthan: status of hydrological research. Report No. SR-45. National Institute of Hydrology, Roorkee

Kimura K, Stephenson DG (1969). Solar radiation on cloudy days. Research paper 418. Div Bldge Res. Nat Counc, Ottawa, Canada

Koche CD, Chawla VK (2000) Dependability criteria in irrigation planning. J Ind Water Resour Soc 20(1): 1–15

Kolanu TR, Kumar S (2004) Greening agriculture in India – An overview of opportunities and constraints. www.fao.org/DOCREP/ARTICLE/AGRIPPA/658_en-04.htm. Accessed in November 2004

Krishna Murti CR, Bilgrami KS, Das TM, Mathur RP (1991) The Ganga – a scientific study. Northern Book Centre, New Delhi, India

Krishnan MS (1982) Geology of India and Burma. B. S. Publishers, Delhi, India

Kulkarni AV, Rathore BP, Mahajan S, Mathur P (2005) Alarming retreat of Parbati glacier, Beas basin, Himachal Pradesh. Curr Sci 88(11): 1844–1850

Kumar R, Chatterjee C (2005) Regional flood frequency analysis using L-moments for north Brahmaputra region of India. J Hydrol Eng 10(1): 1–7

Kumar R, Singh RD, Seth SM (1999) Regional flood formulas for seven subzones of zone 3 of India. J Hydrol Eng 4(3): 240–244

Kumar A (2004) Small hydro development in India. Lecture Notes, Alternate Hydro-energy Centre, Indian Institute of Technology, Roorkee

Kumaraswamy P (1978) Mathematical models of river basins in Tamilnadu. Institute of Hydraulics and Hydrology, Poondi

Kundra DN (1999) Geography today. Goyal Brothers Prakashan, New Delhi, India

Kusre BC, Patra SC (2005) Water for life in NE region: river systems, water demands, floods and water logging. Proceedings of 11th national water convention, 11 May 2005. National Water Development Agency, New Delhi, India

Lal M, Meehl GA, Arblaster JM (2000) Simulation of Indian summer monsoon rainfall and its intraseasonal variability. Reg Environ Change 1(3 and 4): 163–179

Lal M (2001) Climate change – implications for India's water resources. J Ind Water Resour Soc 21(3): 101–109

Lawrence P, Meigh J, Sullivan C (2003) The Water Poverty Index: international comparisons. http://www.nwl.ac.uk/research/WPI/images/wdpaper.pdf

Lazaroff C (2000) Hidden ground water pollution problem runs deep. Environment News Service, Worldwatch Institute, Washington, DC

Linarce ET, Hobbs JE (1977) The Australian Climatic Environment. John Wiley, Brisbane

Linsley RK, Kohler MA, Laulhus JLH (1989) Hydrology for engineers. McGraw Hill, New York

Lohani VK, Refsgaard JC, Clausen T, Erlich M, Storm B (1992) Application of SHE for irrigation command area studies in India. J Irrig Drain, ACSCE, 119(1):34–49

Mahto S, Ouesph (1996) Hydrological safety of Pechiparai dam (Tamil Nadu)-a case study, vol I. Proceedings of the 2nd International Conference on Dam Safety Evaluation. Central Board of Irrigation and Power, Trivendrum, 26–30 Nov 1996, pp 289–296

Mal Barna BD (2000) Sedimentation in Salal dam and its impact on environment. Hydrol J 23(1–4): 41–46

Male DH, Granger RJ (1981). Snow surface energy exchange. Water Resources Research 17: 609–627

Malhotra SP (1982) The warabandi and its infrastructure. Publication No. 157. Central Board of Irrigation and Power, New Delhi

Maps of India (2004) Forest. http://www.maps-india.com/overview/forest.htm. Cited July 2004

Mathur BS, Kumar V (1982) Modifications in synthetic unit hydrograph relations for Mahanadi subzone in India. Proceedings of International Symposium on Hydrological aspects of mountains watersheds, 4–6 Nov 1982. UOR, Roorkee, India, pp IV-11–IV-18

Mathur GN (ed) (2003) Hydroelectric power stations in operation in India. Publication no. 288. Central Board of Irrigation and Power, New Delhi, India

McBean EA, Al-Nassri S (1988) Uncertainty in suspended sediment transport curves. J Hydraul Div, ASCE, 114(1): 63–74

Mehta RD, Bhar AK (1997–1998) News regarding hydrological problems of the country. SR-3/ 1997–98. National Institute of Hydrology, Roorkee

Mehta RD, Chalisgaonkar D (2000) News regarding hydrological problems of the country. Report No. SR-4/1999–2000. National Institute of Hydrology, Roorkee

Mehta RD (1996) News regarding hydrological problems of the country. R.D. Mehta NIH Report No. 4/96–97. National Institute of Hydrology, Roorkee

Mehta RD (2001) Hydrological problems of India–a compilation of news items for the year 2000. SR-4/2000–01. National Institute of Hydrology, Roorkee

Ministry of Water Resources Physiographic conditions. http://wrmin.nic.in/resource/default3.htm. Accessed in October 2005

Mishra SK, Singh VP (2003) Soil conservation service curve number (SCS-CN) methodology. Water Science and Technology Library, Kluwer Academic Publishers, Dordrecht

Mishra GC (1992) Seepage and drainage. Hydrological developments in hydrology since Independence (a contribution to hydrological sciences) , National Institute of Hydrology, Roorkee, India

Mishra A (1993) Aaj bhi Khare Hain Talaab (in Hindi). Gandhi Peace Foundation, New Delhi

Mo CF (2002) Production and availability of pesticides. 37th Report. Standing Committee on Petroleum and Chemicals, Ministry of Chemicals and Fertilizers (Department of Chemicals & Petrochemicals), Government of India, New Delhi

MOA (1976) Report of National Commission on Agriculture. Ministry of Agriculture, Government of India, New Delhi

MOA (1997) Agriculture statistics at a glance. Department of Agriculture and Cooperation, Ministry of Agriculture, Government of India, New Delhi

MoCF (2002) Production and availability of pesticides. 37th report, Standing Committee on Petroleum and Chemicals, Ministry of Chemicals and Fertilizers (Department of Chemicals & Petrochemicals), Government of India, New Delhi

Mohile AD, Gupta LN (2003) ICID-CPSP Indian National Consultation. International Commission on Irrigation and Drainage, New Delhi

Mohile AD, Gupta LN, Saxena R (1984) Depth-area-duration analysis of 1983 Saurastra storm using short interval information. Hydrol J Indian Assoc Hydrol 7(1):

Mohile AD, Gupta LN, Gupta RK (2005) Rationalisation of water transfer projects in India. Proceedings of 12th world water congress, IWRA, New Delhi

Mohile AD (1996) Enhancing management capabilities of the water sector in India. Int J Water Resour Develop 12(4): 503–512

Monteith, JL (1965). The state and movement of water in living organisms. Proceedings. Evaporation and environment, XIXth symposium. Soc Exp Biol. Cambridge University Press, Swansea, New York, pp 205–234

Morris GL, Fan J (1997) Reservoir sedimentation handbook, design and maintenance of dams, reservoirs, and watersheds for sustainable use. McGraw-Hill, New York

Mosley MP, McKerchar AI (1993) Streamflow. In: Maidment DR (ed) Handbook of hydrology. McGraw-Hill Inc., New York

MOWR (1999) Report of the working group on perspective of water requirement. National Commission for Integrated Water Resources Development Plan, Ministry of Water Resources, Government of India, New Delhi September 1999

MOWR (2003). Fresh water for all. Issued by Ministry of Water Resources, Government of India, New Delhi

MoWR (2003) Problems facing the water/irrigation sector: Water quality issues. Ministry of Water Resources, Government of India, New Delhi (article available at MoWR website http://wrmin.nic.in/problems/default11.htm)

MoWR (2003) Annual Report for 2002–03. Ministry of Water Resources, Governemnt of India, Shram Shakti Bhawan, New Delhi

MoWR (2003) National Water Policy 1987 & 2003. Ministry of Water Resources, Government of India, New Delhi

MOWR (2004). Water Resources of India. http://wrmin.nic.in/resource/default3.htm. Website of Ministry of Water Resources, Government of India, New Delhi

MOWR (2005) Web site of Ministry of Water Resources. http://wrmin.nic.in/welcome.html

MOWR (2005) Water and Indian federalism: The "Rules of the Game". wrmin.nic.in/constitution/iswact.htm

MoWR (2005) Home Page of Ministry of Water Resources. http://www.wrmin.nic.in. Accessed in August 2005

MoWR (2005) Problems facing the water/irrigation sector: water quality issues. Ministry of Water Resources, Government of India, New Delhi. http://wrmin.nic.in/problems/default11.htm. Accessed in August 2005

Mukherjee K, Kaur S, Mehra AK (1991) Applicability of extreme value distribution for analysis of rainfall over India. Mausam, 42(1): 29–32

Municipal Corporation of Bhopal (2000) Planning for development of city of Bhopal. Report of Municipal Corporation, Bhopal

Murthi KVS, Ramakrishna M (1980). Structure and tectonics of Godavari-Krishna coastal sedimentaries basin. Bull ONGC 17(1)

Murthy BN (1980) Distribution of sediment in reservoirs. Technical Report No. 19. Central Board of Irrigation and Power, New Delhi

Nag BS, Kathpalia GN (1975) Water resources of India. Proceedings, second world congress on water resources, New Delhi

Nagaraj N, Marshall Frasier W, Sampath RK (1999) A comparative study of groundwater institutions in the western United States and Peninsular India for sustainable and equitable resource use. Department of Agricultural and Resource Economics, Colorado State University, Fort Collins

Nair R, Radhakrishna MD (2005) Interlinking of rivers for mitigation of droughts. Proceedings of the 11th National Water Convention, National Water Development Agency, New Delhi, 11 May 2005

Narasimhan TN (1990) Groundwater in the peninsular Indian shield. A frame work for rational assessment. J Geolog Soc India 36:353–363

Narayanpethakar AB, Gurunadha Rao VVS, Mallick K (1997) Groundwater in Deccan Basalts, India. National Seminar on Hydrogeology of Precambrian Terrains and Hard Rock Areas, Department of Studies in Geology, Karnataka University, Dharwad

NATCOM (2003) India's Initial National Communication (NATCOM) to the United Nations framework convention on climate change. Ministry of Environment and Forests, Government of India, New Delhi

National Commission on Agriculture (1976). Climate and agriculture. Final report of the National Commission on Agriculture, part IV. Ministry of Agriculture and Irrigation, Government of India, New Delhi

MOWR (2002) National Water Policy of India. Ministry of Water Resources, Government of India. New Delhi

Nayak RC (2005) Interlinking of rivers of India-the only possibility to reform irrigation water management sector in 2050. Proceedings of 11 National Water Convention, 11 May 2005. National Water Development Agency, New Delhi, India

NCA (2000) A working paper on formulation of a comprehensive integrated operation policy for reservoirs in Narmada basin. Working Group: Integrated Reservoir Operation, Narmada Control Authority, Indore

NCIWRD (1999) The report of National Commission for Integrated Water Resources Development Plan, vol 1. Government of India, Ministry of Water Resources, New Delhi

NCIWRD (1999) Report of working group on water management for domestic, industrial & other uses. National Commission for Integrated Water Resources Development. Govt. of India, Ministry of Water Resources, New Delhi

NCIWRDP (1999). Report of sub-group no. VIII on groundwater including conjunctive use, of the working group on water management for agriculture, hydropower, flood control and other allied sectors. National Commission for Integrated Water Resources Development Plan, Government of India, New Delhi, India

NCPA (1990) Status, potential and application for adoption of plastic in drip and sprinkler system. National Committee on the Use of Plastics in Agriculture (NCPA), Pune

NIH (1983–1984) Simulation of daily runoff of two sub basins of river Narmada using tank model. Report No. CS-5. National Institute of Hydrology, Roorkee

NIH (1985) Methodology for estimation of design storm. Report no. TR-12, National Institute of Hydrology, Roorkee

NIH (1985) Network Design of raingauges in Rajasthan. Report no. CS-19, National Institute of Hydrology, Roorkee

NIH (1986–1987) Land use mapping of upper Yamuna catchment using remotely sensed data. Report no. CS-14. National Institute of Hydrology, Roorkee, India

NIH (1986) Water availability studies for Mahanadi river basin at three sites, a joint study with NWDA. National Institute of Hydrology, Roorkee

NIH (1987–1988) Effect of additional surface irrigation supply on groundwater regime in upper Ganga canal command area-part I ground water balance. Report no. CS-10. National Institute of Hydrology, Roorkee, India

NIH (1987–1988) Watershed characteristics of ONG sub basin. Report No. TR-51. National Institute of Hydrology, Roorkee

NIH (1989–1990) Hydrological Studies carried out on Kolhai glacier. NIH report, TR-85, 1989–1990

NIH (1989–1990a) Application of SHE model to Sher sub basin of river Narmada. Report No. CS-31. National Institute of Hydrology, Roorkee

NIH (1989–1990b) Application of SHE model to Kolar sub basin of river Narmada. Report No. CS-33. National Institute of Hydrology, Roorkee

NIH (1989–1990c) Application of SHE model to Hiran sub basin of river Narmada. Report No. CS-30. National Institute of Hydrology, Roorkee

NIH (1989–1990d) Application of SHE model to the Ganjal sub basin of river Narmada. Report No. CS-28. National Institute of Hydrology, Roorkee

NIH (1989–1990e) Application of SHE model to the Barna sub basin of river Narmada. Report No. CS-32. National Institute of Hydrology, Roorkee

NIH (1990–1991) Application of HEC-1 to Hemavati (up to Sakleshpur) basin. Report no. CS-55. National Institute of Hydrology, Roorkee

NIH (1990) Hydrology in ancient India. National Institute of Hydrology, Roorkee, India

NIH (1991–1992) Water availability studies of river Tawi. Report no. CS-86. National Institute of Hydrology, Roorkee, India

NIH (1991–1992), Water availability studies-Ujh river basin. Report no. CS-87. National Institute of Hydrology, Roorkee, India

NIH (1991–1992) Hydrological land use mapping of Malprabha and Ghataprabha catchments of Krishna basin. Report No. CS-85. National Institute of Hydrology, Roorkee

NIH (1990–1991) Hydrological investigations of Chhota Shigri glacier. Report no. TR-106. National Institute of Hydrology, Roorkee, India

NIH (1992–1993) Evaluation of precipitation gauge density in Punpun catchment of Ganga River system. NIH report no. TR-180, 1992–1993.

NIH (1992–1993). Discharge measurement of River Teesta in Sikkim using tracer dilution method. Report no. CS-101. National Institute of Hydrology, Roorkee, India

NIH (1992–1993) Snow cover mapping for Baira catchment. Report no. CS-115. National Institute of Hydrology, Roorkee, India

NIH (1992–1993a) Changes in land use/land cover over Sarda river basin, A.P. NIH Report No. CS-124. National Institute of Hydrology, Roorkee

NIH (1992–1993b) Water balance studies in Suddagedda basin (part I)-status of network, data availability & instrumentation. NIH Report No. CS-0. National Institute of Hydrology, Roorkee

NIH (1992) Quantitative assessment of sediment distribution in the Tungabhadra reservoir using satellite imagery. Report No. CS-84/1991–92. National Institute of Hydrology, Roorkee

NIH (1993–1994) Assessment of water logged area in IGNP stage–I by remote sensing techniques. CS(AR)-138. National Institute of Hydrology, Roorkee, India

NIH (1993–1994) Hydrological studies for improvement of Khajjar lake (H.P.). Report no. TN 106. National Institute of Hydrology, Roorkee, India

NIH (1993–1994) Integrated regulation of a system of reservoirs for conservation purposes. Report No. CS(AR) 141. National Institute of Hydrology, Roorkee

NIH (1993–1994) Dam break study of Mitti dam. Report No. CS(AR) 126. National Institute of Hydrology, Roorkee

NIH (1993–1994). Infiltration studies in India. Report no. TN-103. National Institute of Hydrology, Roorkee, India

NIH (1994–1995) Hydrological study on Dokriani glacier in Garhwal Himalaya part II. NIH report, CS(AR)-160, 1994–1995

NIH (1994–1995) Water quality modeling of Kali river using QUAL2E. Report no. CS(AR)175. National Institute of Hydrology, Roorkee, India

NIH (1994–1995) Sedimentation studies in Massanjore reservoir of Mayurakshi Basin. Report no. CS-189. National Institute of Hydrology, Roorkee, India

NIH (1994–1995), Hydrological studies of lake Naini, District Nainital, Uttar Pradesh (part I). Report No. CS(AR) 184. National Institute of Hydrology, Roorkee, India

NIH (1994–1995) Water quality studies of Surinsar lake in Jammu region. Report no. CS(AR)-157. National Institute of Hydrology, Roorkee, India

NIH (1994–1995) Rainfall runoff modeling of Nagavali river upto Narayanapuram. NIH Report No. CS(AR)-163. National Institute of Hydrology, Roorkee

NIH (1995–1996) Assessment of water logging in Sriramsagar command area. Report No. CS(AR)-202. National Institute of Hydrology, Roorkee

NIH (1995–1996) Daily rainfall-runoff modeling of Sagileru river using HYSIM. Report No. CS(AR)-192. National Institute of Hydrology, Roorkee

NIH (1995–1996) Simulation of operation of Vellar system using HEC-5. Report No. CS(AR) 219. National Institute of Hydrology, Roorkee

NIH (1995–1996) Major and important lakes in Rajasthan: status of hydrological research. NIH Report No. SR-45. National Institute of Hydrology, Roorkee

NIH (1996–1997a) Dam break study of Barna dam. Report No. CS(AR)-20. National Institute of Hydrology, Roorkee

NIH (1996–1997b) Development of operation policy for Tawa dam. Report No. CS(AR)-18. National Institute of Hydrology, Roorkee

NIH (1996) Water logging and drainage congestion problem in Mokama Tal area Bihar, GPRC. Report No. CS (AR) 194. National Institute of Hydrology, Roorkee

NIH (1997–1998) Rainfall-runoff modeling for water availability study in Ken River basin using SCS-CN model and remote sensing approach. Report no. CS (AR) 12/97-98. National Institute of Hydrology, Roorkee, India

NIH (1997–1998) Daily rainfall-runof modelling of Rushikulya river, Orissa. Report No. CS/AR-16/97-98. National Institute of Hydrology, Roorkee

NIH (1998–1999) Hydrological inventory of River basins in eastern Uttar Pradesh. Report no. SR-1/98 - 99. National Institute of Hydrology, Roorkee, India

NIH (1998–1999) Hydrological inventory of south Bihar River basins. Report no. SR-2/98-99. National Institute of Hydrology, Roorkee, India

NIH (1998–1999) Rainfall-runoff modelling of Baitarani river basin up to Anandpur using HEC-1. Report no. TR/BR-13/98–99. National Institute of Hydrology, Roorkee

NIH (1999–2000) Application of artificial neural network in flood studies on Ajay River basin. Report No. TR/BR-5/1999-2000. National Institute of Hydrology, Roorkee, India

NIH (1999–2000) Inventory of water bodies in the western Himalayan region: part-I (Jammu & Kashmir). Report no. TR/BR-13/1999-2000. National Institute of Hydrology, Roorkee, India

NIH (1999) Comprehensive hydrological studies of Narmada basin. National Institute of Hydrology, Roorkee

NIH (2003) Monitoring and modeling of melt runoff from Gangorti Glacier, DST sponsored project report, National Institute of Hydrology, Roorkee, India

Nilsson C, Reidy CA, Dynesius M, Revenga C (2005) Fragmentation and flow regulation of the world's large river systems. Science 308: 405–408.

NVDA (2004) Indira Sagar Project (ISP). http://www.nvda.nic.in/indirasagar.htm. Accessed in December 2004

NVDA (2005) Web site of Narmada Valley Development Organization. http://www.nvda.nic.in/plan.htm. Accessed in November 2005

Paltridge GW (1974) Solar radiation statistics for Australia. Technical paper 23. Div Atmos Phys Common. Sci Ind Res Org, Melbourne, Australia

Pandharinath N (1987) A case study of unprecedented floods in the Godavari river during August 1986. Proceeding of the National Symposium on Hydrology, National Institute of Hydrology, Roorkee, 16–18 Dec 1987, pp 102–111

Pandit C (2005) Hydrology versus ideology. Proceedings of the 11th National Water Convention, National Water Development Agency, New Delhi, 11 May 2005

Pangare G, Pangare V, Das B (2006) Springs of life. Academic Foundation, New Delhi

Pant GB, Kumar KR (1997) Climates of South Asia. John Wiley & Sons Ltd., West Sussex, UK

Pardhasaradhi YJ (1989) Induced recharge by percolation tanks in parts of Maharashtra. Proceedings of international workshop on methodologies for development and management of ground water resources in dev. countries, vol. II. Hyderabad, India, pp 989–996

Patel DH, Patel IA (2003) Flood control operation of Ukai Multipurpose Reservoir –issues and need. Proceeding of the Brain Storming Session on Application of Systems Techniques for Water Resources Management in India-Current Trends and Future Directions, National Institute of Hydrology, Roorkee, pp 13–32 13th February 2003

Patel MS, Patel DH, Brahmbhatt VS, Ghadia LV (2005) Interlinking of rivers. Proceeding of 11th national water convention, organized by National Water Development Agency at New Delhi on 11 May 2005

Patil MD (2004) Optimal water utilization and intra-basin water transfers in Cauvery basin, India. Doctor of Philosophy Thesis, Water Resources Development Training Centre, Indian Institute of Technology, Roorkee

Patkar M (2004) River linking: a millennium folly? National Alliance of People's Movements, Mumbai

Penman RL (1948) Natural evaporation from open water, bare soil and grass, vol A193. Proceedings. Royal Society, London, pp 120–146

Penman ADM (1996) Dam safety. Proceedings of the 2nd International Conference on Dam Safety Evaluation. Central Board of Irrigation and Power, Trivendrum, 26–30 Nov 1996

Planning Commission (2006). Mid-term assessment of 10th plan. www.planningcommission.nic.in. Cited March 2006

Ponce VM (1989) Engineering hydrology: principles and practice. Prentice Hall, Englewood Cliffs, New Jersey

Postel S (1999) Pillar of sand: can the irrigation miracle last? W.W. Norton & Company, New York

Prasad R (1990) Drought analysis and management. Special lecture. National Workshop on Water Resources Project Management. Indian Institute of Technology, Madras 18–20 June 1990

Prasad T (2003) Interlinking of rivers in India: impacts and implications. In: Singh VP, Yadav RN (eds) Proceeding of the international conference on water and environment (WE-2003), held in Bhopal, India, December

Prasad RS (2005) Interbasin transfer of water – Issues and challenges. Proceeding of 11th national water convention, organized by National Water Development Agency at New Delhi on 11 May 2005

Quinn WH, Neal VT, Antunes de Mayolo SE (1987) El Nino occurrences over the past four and a half centuries. J Geophys Res 92: 14449–14461

Rahman A (1981) Construction planning including performance budgeting for Lakhwar project. Dissertation of master of engineering in water resources development. Water Resources Development Training Centre, University of Roorkee, Roorkee, India

Raju MB, Krupanidhi KVJR, Shyamprasad B (1982) Hydrogeological conditions in east Godavari district, AP (unpublished report). Central ground Water Board, Government of India, New Delhi, India

Raju KCB, Krupanidhi KVJR, Srinivasan KR, Ramana Rao V, Raju MB (1983) Ground water resources of deltaic and coastal aquifers in southern states with special reference to salinity problem (unpublished report). Central ground Water Board, Government of India, New Delhi, India

Raju KV (1992) Irrigation panchayats in Madhya Pradesh: a case study. ODI/IIMI Network Paper No. 21 Agriculture research and extension network, London 1–14

Rakhecha PR, Soman MK (1992) Hydrometeorological studies. Hydrological Development of India Since Independence, National Institute of Hydrology, Roorkee

Rakhecha PR, Kulakarni AK, Mandal BN, Deshpande NR (1990) Homogeneous zones of heavy rainfall of 1-day duration above over India. Theor Appl Climatolol 41: 213–219

Ram Babu, Tejwani KG, Agarwal MC, Bhusan LS (1979) Rainfall intensity-duration-return period equations and nomographs of India. Central Soil and Water Conservation Research and Training Institute, Dehradun

Ramachandra Rao A, Prasad T(1994) Water resources development of the Indo-Nepal region. Int J Water Resour Develop 10(2): 157–173

Ramasastri KS (1986) Depth-duration and depth-area duration analysis. Proceedings of the workshop on observation, processing and analysis of precipitation data, National Institute of Hydrology, Roorkee, 24–28 February 1986

Ramasastri KS (1987–88) Processing of precipitation data. Lecture notes of workshop on processing and analysis of precipitation data

Ramasastri KS (1987) Quality control procedures for precipitation data in an operational system. J Irrigation Power CBI&P 45(3): 173–184

Ramasastri KS (1987). Estimation of evaporation from free water surfaces. Proceedings of National Symposium on Hydrology, Organized by National Institute of Hydrology, Roorkee, India, pp II-16–II-27

Ramasastri KS (1992). Hydrology of mountain areas. Hydrological Development of India since independence. National Institute of Hydrology, Roorkee, India

Ramasastri KS (2002) Climatological data. Proceedings of training course on basic hydrology (surface water), 22 April–3 May 2002. National Institute of Hydrology, Roorkee, India

Ramasastry AA (1984) Weather and weather forecasting. Publications Division, Government of India

Ramaswamy C (1985) Review of floods in India during the past 75 years – a perspective. Indian National Science Academy, New Delhi

Ramaswamy C (1987) Meteorological aspects of severe floods in India, 1923–1979. India Meteorological Department, New Delhi

Rand McNally and Company (1998) Atlas of world geography. Rand McNally and Company, New York

Randhawa MS (1980) A History of agriculture in India (3 volumes). Indian Council for Agricultural Research, New Delhi, India

Rangachari R, Sengupta N, Iyer RR, Banerji P, Singh S (2000) Large dams: India's experience. Secretariat of the World Commission on Dams, Cape Town

Rangacharu R (2005) Water for life with special reference to interlinking of rivers-experience from some old schemes of south India. Proceeding of 11th national water convention, organized by National Water Development Agency at New Delhi

Ranganayakulum A, Naidu KS (2005) Rural renaissance through linking of rivers. Proceedings of the 11th National Water Convention, National Water Development Agency, New Delhi, 11 May 2005

Rani D, Srivastava DK, Gulati TR (2005) Derivation of intra-basin water transfer policy for PAV link project. Proceeding of 11th national water convention, organized by National Water Development Agency at New Delhi on 11 May 2005

Rantz SE (1982) Measurement and computation of streamflow: volume 2. Computation of discharge. U.S. Geological Survey, Water Supply Paper # 2175. Washington, DC

Rao KN, Ganesan HR (1972) Sunshine over India. Pre-published scientific report. India Meteorological Department, Poona, India

Rao KN, George CJ, Ramasastri KS (1971) Potential evapo-transpiration (PE) over India. I.M.D. Pre publication science report 136

Rao KN, George CJ, Ramasastri KS (1972) Agro-climatic classification of India. India Meteorological Department meteorological monograph, Agrimet no. 4, New Delhi, India

Rao KL (1973) India's water wealth. Orient Longman Limited, New Delhi, India

Rao PR (1975) Modelling of non-linear behaviour of drainage basins. PhD thesis, IIT, Delhi

Rao KL (1975) India's water wealth. Orient Longman Limited, New Delhi, India

Rao YP (1976) Monsoons. IMD meteorology monograph synoptic meteorology no. 1

Rao KL (1979) India's water wealth. Sujit Mukherjee, Orient Longman Limited, New Delhi, India

Rao PR (1990) Study of loss rates for application to severe rainfall storms. CBIP Regional Workshop, Nagarjuna Sagar Dam

Rao PCM (2001) A multiple-input single-output model for flow forecasting of Wardha river. Dissertation of Master of Engineering in Water Resources Development, Water Resources Development Training Centre, University of Roorkee, Roorkee

Rao PS (2003) Interlinking of rivers for inter-basin transfers of water: some issues. In: Singh VP, Yadav RN (eds) Proceeding of the international conference on water and environment (WE-2003), held in Bhopal, India, December

RBA (1980) Report of Rashtriya Barh Ayog (National Flood Commission). Ministry of Irrigation, Government of India, New Delhi

RBA (1990) Report of Rashtriya Barh Ayog (National Commission on Floods). Ministry of Energy and Irrigation (Department of Irrigation), Government of India, New Delhi

RDSO (1991). Estimation of design discharge based on regional flood frequency approach for subzones 3 (a), 3 (b), 3 (c) and 3 (e). Bridges and floods wing Report no. 20. Research Design and Standards Organization, Ministry of Railways, Lucknow, India

Reddy MB (1995) Sedimentation studies for lower Manair and Kaddam reservoirs in Andhra Pradesh. Dissertation of Master of Engineering in Water Resources Development, Water Resources Development Training Centre, University of Roorkee, Roorkee

Refsgaard JC, Seth SM, Bathurst JC, Erlich M, Storm B, Jorgensen GH, Chandra S (1992) Application of SHE to catchments in India part I – general results. J Hydrol 140:1–23

WG (1999) Perspective of water requirements. Report of working group, National Commission for Integrated Water Resources Development Plan, Ministry of Water Resources, Government of India, New Delhi

Richards A, Singh N (2002) Inter-state disputes in India: Institutions and policies. International Journal of Water Resources Development 18(4):611–625

Robbroeck, Theo van (1996) Reservoirs: bane or boon? Geoffrey Binnie Lecture. British Dam Society, London

Rogers P, Lydon P, Seckler D (1989) Eastern waters study. Report prepared for irrigation support project for Asisa and the near East, Arlington, Virginia, USA

Romani S, Sharma N (1989) Water conservation and artificial recharge techniques in hard rock areas. Proceedings of International Workshop on Methodologies for Development and Management of Ground Water Resources in Dev. Countries, vol. II. Hyderabad, India, pp 959–968

Romani S (2005) Linkage of surplus and deficient water areas – a succor to life. Proceedings of 11th National Water Convention, 11 May 2005. National Water Development Agency, New Delhi, India

Roy PK, Roy D, Mazumdar A (2004) An impact assessment of climate change and water resources availability of Damodar River basin. Hydrol J IAH, 27(3–4): 53–70

Roy RD (1992) In: Chatrath KJS (ed.) Case study of Loktak lake. Wetlands of India, Ashish Publishing House, Delhi

RRSSC (2004) Technical report on sedimentation assessment of Halali reservoir, MP through satellite Remote Sensing. Technical report No. RCJ/TR/2004/2, Oct 2004. Regional Remote Sensing Service Centre, Indian Space Research Organization, Jodhpur, India

RRSSC (2004a) Technical report on sedimentation assessment of Ramsagar reservoir, Rajasthan through satellite remote sensing. Technical report No. RCJ/TR/2004/6. Regional Remote Sensing Service Centre, Indian Space Research Organization, Jodhpur, India

RRSSC (2004a) Technical report on sedimentation assessment of Panam reservoir, Gujarat through satellite remote sensing. Technical Report No. RCJ/TR/2004/12. Regional Remote Sensing Service Centre, Indian Space Research Organization, Jodhpur

RRSSC (2004a) Technical report on sedimentation assessment of Dudhawa reservoir, Gujarat through satellite remote sensing. Technical Report No. RCJ/TR/2004/7. Regional Remote Sensing Service Centre, Indian Space Research Organization, Jodhpur

RRSSC (2004a) Technical report on sedimentation assessment of Damanganga reservoir, Gujarat through satellite remote sensing. Technical Report No. RCJ/TR/2004/16. Regional Remote Sensing Service Centre, Indian Space Research Organisation, Jodhpur

RRSSC (2004b). Technical report on sedimentation assessment of Parbati reservoir, Rajasthan through satellite remote sensing. Technical report No. RCJ/TR/2004/5. Regional Remote Sensing Service Centre, Indian Space Research Organization, Jodhpur, India

RRSSC (2004b) Technical report on sedimentation assessment of Jakham reservoir, Rajasthan through satellite remote sensing. Technical Report No. RCJ/TR/2004/4. Regional Remote Sensing Service Centre, Indian Space Research Organization, Jodhpur

RRSSC (2004b) Technical report on sedimentation assessment of Sondur reservoir, Gujarat through satellite remote sensing. Technical Report No. RCJ/TR/2004/8. Regional Remote Sensing Service Centre, Indian Space Research Organization, Jodhpur

RRSSC (2004b) Technical report on sedimentation assessment of Palitana reservoir, Gujarat through satellite remote sensing. Technical Report No. RCJ/TR/2004/3. Regional Remote Sensing Service Centre, Indian Space Research Organisation, Jodhpur

Sahni VS (1996a) Drip irrigation, 3rd edn. Water Management Technical Publication No. 34. WALMI, Aurangabad

Sahni VS (1996b) Sprinkler irrigation, 3rd edn. Water Management Technical Publication No. 38. WALMI, Aurangabad

Sahni BM (2000) Water management in command areas. Report no. INCOH/SAR-20/2000. National Institute of Hydrology, Roorkee

Sahu GC (2005) Two inter basin transfer projects in the drought prone areas of Orissa-a success story. Proceeding of 11th National Water Convention, National Water Development Agency, New Delhi, 11 May 2005

Saksena RS (2000) Conjunctive use of Surface and Ground Water. Report no. INCOH/SAR-23/2000. National Institute of Hydrology, Roorkee, India

Sampurno (2002) Comparative analysis of design practices for lined canals with reference to Indira Gandhi Main Canal. Dissertation of master of engineering in water resources development. Water Resources Development Training Centre, Indian Institute of Technology, Roorkee, India

Saravanan VS (1994). Selection methodology for rehabilitation of rainfed tank irrigation systems – a case study of Madurai district in Tamil Nadu. Masters Diploma Dissertation, Centre for Development Studies, University of Pune, Pune

Sarma V (1973). Evaporation over India. Ind J Meteorol Geophys 24
Sengupta N (1995) Environmental contributions of some traditional techniques. Paper presented at the Seventh International Conference of the International Rainwater Catchment Systems Association, Beijing, 19–25 June 1995
Seth SM, Rakesh K (1986). Rivers of India – ancient name. Bhagirath, vol. XXXIII. Central Water Commission, New Delhi
Seth SM, Singh RD (1985–1986) Regional unit hydrograph. Report no. RN-17. National Institute of Hydrology, Roorkee, India
Shah RB (1993) Role of major dams in the Indian Economy. Int J Water Resour Develop 9(3): 319–337
Shah RB (1994) Inter-state water disputes: A historical review. Water Resources Development 10(2):175–189
Shanmugham CR, Gurunathan A (2005) Water for life – A long range view. Proceeding of 11th national water convention, organized by National Water Development Agency at New Delhi
Shiklomanov IA (2000) Appraisal and assessment of world water resources. Water Int 25(1): 11–32
Shivanappan RK (1998) The status, scope and prospects of sprinkler irrigation in India. Proceedings of the Workshop on Micro Irrigation and Sprinkler Irrigation Systems, New Delhi, 28–30 April 1998
Shukla SR (1999) Management of urban water supply and sanitation – Challenges and strategies. Proc. Nat. workshop on challenges in the management of water resources and environment in the next millenium: Need for inter-institute collaboration, 8–9 October 1999. Delhi College of Engineering, Delhi, pp 309–324
Sikka AK (1986) Hydrological aspects of drought. Jalvigyan Sameeksha, I(1):89–110. National Institute of Hydrology, Roorkee
Singh P, Bengtsson L (2004). Hydrological sensitivity of a large Himalayan basin to climate change. Hydrol Process 18: 2363–2385
Singh P, Jain SK (2002) Snow and glacier contribution in the Satluj River at Bhakra dam in the western Himalayan region. Hydrol Sci J 47(1): 93–106
Singh P, Jain SK (2003) Modelling of streamflow and its components for a large Himalayan basin with predominat snow melt yields. Hydrol Sci J 48: 257–275
Singh P, Kumar N (1997) Effect of orography on precipitation in the western Himalayan region. J Hydrol 199: 183–206
Singh RD, Seth SM (1984–1985). Comparative study of unit hydrograph methods. Report no. CS-7. National Institute of Hydrology, Roorkee, India
Singh P and Singh VP (2001) Snow and glacier hydrology. Kluwer Publishers, Dordrecht, the Netherlands
Singh S, Thakural LN (1999–2000) Water balances of Sagar lake. Report no. TR (BR)-12. National Institute of Hydrology, Roorkee
Singh P, Ramasastri KS, Kumar N (1994) Study on snow distribution in Chenab basin. In: Proceedings of the international symposium on snow and its manifestations, SASE, Manali, Himachal Pradesh, India
Singh P, Ramasastri KS, Kumar N (1995) Topographical influence on precipitation distribution in different ranges of western Himalayas. Nordic Hydrol 26: 259–284
Singh RD, Mishra SK, Chowdhary H (2001) Regional flow-duration models for large number of ungauged Himalayan catchments for planning microhydro projects. J Hydrol Eng 6(4): 310–316
Singh VP, Sharma N, Ojha CSP (ed) (2004) The Brahmaputra basin water resources. Kluwer Academic Publishers, Dordrecht, the Netherlands
Singh VP (1982) Elementary hydrology. Prentice Hall, Englewood Cliffs, New Jersey
Singh IT (1983) A study on economics of canal lining with special reference to Salwa distributary. ME thesis, WRDTC. University of Roorkee, Roorkee, India
Singh RD (1985) Application of an efficient smoothed least square technique for unit hydrograph derivation. Hydrol J IAH 8(3): 21
Singh RL (1989) India a regional geography. National Geographical Society of India, Varanasi, Uttar Pradesh
Singh VP (1992) Elementary hydrology. Prentice Hall, Englewood Cliffs, New Jersey

Singh B (1995) Development of water resources in Ganga-Brahmaputra basin. 6th Kanwar Sain endowment lecture at Uttar Pradesh Irrigation Research Institute, organised by Indian Water Resources Society, WRDTC, University of Roorkee, Roorkee, India

Singh S (1997) Taming the waters: the political economy of large dams in India. Oxford University Press, New Delhi

Singh B (2002) Inter basin water transfer – Review of need and problems. Proceedings of all India seminar on water and environment-issues and challenges, IIT Roorkee, 12–13 October 2002, pp3–18

Singh KP (2005) Ghaghra-Yamuna link project-life line of farmers of central region of Uttar Pradesh. Proceeding of 11th national water convention, organized by National Water Development Agency at New Delhi

Singhal BBS, Singhal DC (1989) Evaluation of aquifer parameters and well characteristics in fractured rock formation of Karnataka, India, Abstracts, vol. 3. International Geol, Congress, Washington, pp 3-124–3-125.

Singhal BBS (1992) Hydrology in hard rocks. Hydrological developments in hydrology since Independence (a contribution to hydrological sciences). National Institute of Hydrology, Roorkee, India

Sinha SK, Jeyaseelan R (2005) Large scale surface water transfer-hydrological and other issues, Jalvigyan Sameeksha (Hydrology Review), 19(1–2), INCOH. National Institute of Hydrology, Roorkee

Sinha BPC, Sharma SK (1990) Ground water dams – concepts and case histories. Bhu-Jal News 5, pp. 3–13

Sinha SK, Saxena RP, Banerjee A, Pramanik K (2005) Interlinking of rivers-case studies. Proceeding of 11th national water convention, organized by National Water Development Agency at New Delhi on 11 May 2005

Sinha SK, Sinha AK, Chandra S (2005) Interlinking of rivers-issues and challenges. Proceeding of 11th national water convention, organized by National Water Development Agency at New Delhi

Sokolov AA, Chapman TG (eds) (1974) Methods for water balance computations. UNESCO Press, Paris

Soni B, Singh R (1992) Agricultural hydrology and drainage. Hydrological developments in hydrology since independence (a contribution to hydrological sciences). National Institute of Hydrology, Roorkee, India

Srivastava CS, Sinha AK (2003) Real-time data acquisition system in Narmada basin – necessity, configuration, and advantages. Brain Storming Session on Application of Systems Techniques for Water Resources Management in India – Current Trends and Future Directions, National Institute of Hydrology, Roorkee 13th February 2003

Subbarayudu M (1974) Study of siltation of reservoirs and its control with special reference to Nizamsagar project in Andhra Pradesh. Dissertation of Master of Engineering in Water Resources Development, Water Resources Development Training Centre, University of Roorkee, Roorkee

Subramanian S (2002) India assessment 2002 – water supply and sanitation. Planning Commission, Government of India, New Delhi

Subramanian V (2004) Water quality in South Asia. Asian Journal of Water, Environment and Pollution, 1(1&2): 41–54

Subramanya K (1984) Engineering hydrology. Tata McGraw-Hill Publishing Company Limited, New Delhi

Subudhi A (1979) Optimization of a multipurpose reservoir with special reference to upper Indravati project. Dissertation of Master of Engineering in Water Resources Development, Water Resources Development Training Centre, University of Roorkee, Roorkee

Talati J, Dinesh KM (2005) Quenching the thirst of Gujarat through Sardar Sarovar project. Paper presented at the XII World Water Congress “Water for Sustainable Development – Towards Innovative Solutions”, New Delhi, 22–25 November 2005

Tanwar BS (1997) Multi-benefits of augmentation tubewells along canals. Proceedings of the workshop on lift irrigation, 23–25 April 1997, CBIP, New Delhi, India

TERI (1998) India’s environment pollution and protection. Report no. 97ED52. TATA Energy Research Institute, New Delhi

Thakkar H (1999) Assessment of irrigation in India. Thematic Review IV.2. The World Commission on Dams.
http://www.dams.org/, Accessed in October 2005

Thambi DS, Chandra PC, Jayakumar B (2005) Fresh/saline groundwater interface movement in the coastal tracts of India (unpublished report). Central ground Water Board, Government of India, New Delhi, India

Thatte CD (2003) ICID-CPSP: Indian national consultation, International Commission on Irrigation and Drainage, New Delhi

Thatte CD (2005) Dams and inter basin water transfer for augmentation of water resources – A review of needs, plans, status and prospects. Keynote lecture presented at 12th world water congress, IWRA, New Delhi

The Times of India newspaper dated 15th February 2005, New Delhi, India

The Times of India (2004) The Times of India, New Delhi, 23 Nov 2004

The Times of India (2004) Satluj-Yamuna link: Troubled waters, New Delhi, August 5, 2004

Thiruvengadachari S (1986) Remote sensing application in drought monitoring. Jalvigyan Sameeksha, 3(1). National Institute of Hydrology, Roorkee

Thompson ES (1976) Computation of solar radiation from sky cover. Water Resour Res 12: 859–865

Thornthwaite CW (1948) An approach toward a rational classification of climate. Geogr Rev 38: 55–94

Tracks of storms, India Meteorological Department, 1877–1970

Tyagi P, Sengupta M, Chakraborty SP (1994) Making the Best Use of Waste Water. Proceedings of the Indo French Seminar on River Basin Management, New Delhi, 12–15 December 1994

UN (2003) Water for people, water for life. World Water Development Report. UNESCO Publication, Paris

UNESCO (1978) World water balance and water resources of the earth. United Nations Educational Scientific and Cultural Organization, Paris

Upadhyay DS, Bahadur J (1982) On some hydrometeorological aspects of precipitation in Himalayas. Proceedings of the international symposium on hydrological aspects of mountainous watersheds, 6 November, University of Roorkee, pp I:58–65

UPIRI (1989). Annual review, 1989. UP Irrigation Research Institute, Roorkee, India

Vaidyanathan A (1999) Water resources management: institutions and irrigation development in India. Oxford University Press, New Delhi

Valdiya KS (1991) The Uttarkashi eqrthquake of 20 October – implications and lessons. Curr Sci 61(12): 801–805

Varma CVJ, Rao ARG, Sundaraiya E (eds) (1998) Typical dams of India. Publication No. 272. Central Board of Irrigation and Power, New Delhi

Varshney RS (1997) Impact of siltation on the useful life of large reservoirs. Scientific Contribution No. INCOH/SAR-11/97. National Institute of Hydrology, Roorkee

Vasantha A (2003) Save the wetlands – a call from Ramsar. Paper from Press Information Bureau. Government of India, New Delhi

Verghese BG, Iyer R (1993). Harnessing the Eastern rivers – Regional cooperation in South Asia. Centre for Policy Research, Konark Publishers Pvt. Ltd., New Delhi

Vishwantaham R, Eashwaraiah K (2004) Report of capacity evaluation of Osmansagar reservoir by using remote sensing techniques satellite data. Remote Sensing Laboratory, Andhra Pradesh Engineering Research Laboratories, Government of Andhra Pradesh, Hyderabad
www.ap.gov.in/apirrigation/MajorIrrig/comp/kds.htm
www.ap.gov.in/apirrigation/MajorIrrig/moreprj/tgp.htm

Vyas JN (2001) Water and energy for development in Gujarat with special focus on the Sardar Sarovar Project. International Journal of Water Resources Development 17(1):37–54

WALMI (1987) Water distribution methods in Maharashtra. Publication No. 22. Water and Land Management Institute, p 67

Water Management Forum (2003) Inter-basin transfer of water in India – prospects and problems. Water Management Forum. The Institution of Engineers, New Delhi, India

Water Resources Council (1981) Estimating peak flow frequency for national ungauged watersheds-a proposed nation-wide test. US Government Printing Office, Washington, DC

Webster PJ, Magana VO, Palmer TN, Shukla J, Tomas RA, Yanagi M, Yasunari T (1998) Monsoons: processes, predictability and the prospects for prediction. J Geophys Res 103: 14451–14510

Wetlands.
http://www.water-mgmt.com/en/wetlands.htm. Cited date July 2004

WG (1999) Report of sub group no II on canal automation of the working group on water management for agriculture, HP, FC & other. National Commission for Integrated Water Resources Development. Ministry of Water Resources, Government of India, New Delhi

WG (1999) Report of sub group no IV on water logging and drainage of the working group on water management for agriculture, HP, FC & other. Ministry of Water Resources, Government of India, New Delhi

WG (1999) Report of sub-group 1 of the working group on water management for agriculture, HP, FC & other. Report of the National Commission for Integrated Water Resources Development. Ministry of Water Resources, Government of India, New Delhi.
http://www.annauniv.edu New Delhi

WG (1999a) Report of working group on water management for domestic, industrial & other uses. National Commission for Integrated Water Resources Development Plan, Government of India, New Delhi

WG (1999b) Report of the Working Group on Water Management for Agriculture, HP, FC & Others. National Commission for Integrated Water Resources Development Plan, Ministry of Water Resources, Government of India, New Delhi

Willeke GE, Guttman, NB, Hosking, JRM, Wallis, J. R. (1994), The National Drought Atlas (draft), IWR Report 94-NDS-4, Institute for Water Resources, U.S. Army Corps of Engineers, Fort Belvoir, VA, USA

WMO (1968) Quality control procedures for meteorological data. WMO/WWW report no. 26, World Meteorological Organization, Geneva

WMO (1969) Manual for depth-area duration analysis of storm precipitation. WMO no. 237, TP 129, World Meteorological Organization, Geneva

WMO (1982) Methods of correction for systematic error in point precipitation measurements for operational use. Operational hydrology report no. 21, WMO-589, World Meteorological Organization, Geneva

WMO (1983) Guide to meteorological instruments and method of observation. WMO no. 8, World Meteorological Organization, Geneva

WMO (1986) Advanced introduction to modelling, Pergamon Press Manual for estimation of probable maximum precipitation. Operational hydrology report no. 1, WMO no. 332, Geneva

WMO (1994) Guide to hydrological practices. WMO no. 168, World Meteorological Organization, Geneva

Woo M-K, Tarhule A (1994) Streamflow droughts of northern Nigerian rivers. Hydrol Sci J 39(1):19–34

Working Group Report (1980) Guidelines for preparation of detailed project reports of irrigation and multipurpose projects. Ministry of Irrigation, Government of India, New Delhi

Working Group (1996) Water logging, salt salinity and alkalinity. Report of the working group on Problem identification on irrigated areas with suggested remedial measures. Ministry of Water Resources, Government of India, New Delhi

Working Group (1999) The National Commission for Integrated Water Resources Development. Ministry of Water Resources, Government of India, New Delhi
www.hydropower.org. Accessed in October 2005

Workshop (2001) Proceedings on international workshop on fluoride in drinking water: Strategies, management, and mitigation, 22–24 January 2001

World Bank (1991) India irrigation sector review, vol I. The World Bank, Washington, DC

World Bank (1995a) Tamil Nadu water resources consolidation project: staff appraisal report. The World Bank, New Delhi

World Bank (1995b) Orissa water resources consolidation project: staff appraisal report. The World Bank, New Delhi

World Bank (1998) India – water resources management sector review: irrigation sector. The World Bank, New Delhi

World Bank (2005) India's water economy: bracing for a turbulent future. The World Bank, New Delhi

Yevjevich V (1972) New vistas for flood investigations. Academic Nationale Dei Lincei, Rama, Quaderno No. 169

Zope MY (1979) Assessment of water resources in Gomti Kalyani Doab. Dissertation of master of engineering in water resources development. Water Resources Development Training Centre, University of Roorkee, Roorkee, India

Websites

http://agricoop.nic.in/about.htm
http://dst.gov.in/scprog/ncfmrwf.htm
http://envfor.nic.in
http://planningcommission.nic.in
http://powermin.nic.in
http://mst.nic.in
http://necouncil.nic.in/power.htm
http://neeri.nic.in/
http://rural.nic.in
http://www.ap.gov.in/apirrigation/organisations/aperl.htm
http://www.bis.org.in
http://www.cseindia.org
http://www.envis.nic.in
http://www.iari.res.in/divisions/wtc/
http://www.icar.org.in/
http://www.imd.ernet.in
http://www.isro.org/sat.htm
http://www.kissankerala.net/kissan/kissancontents/research_others.jsp
http://www.nic.in
http://www.nrsa.gov.in
http://www.annauniv.edu
http://www.cea.nic.in
http://www.cwc.nic.in
http://www.icid.org
http://www.mah.nic.in/cwprs
http://www.mha.nic.in/nenl1004.htm
http://www.narmada.org
http://www.ncmrwf.gov.in
http://www.nih.ernet.in
http://www.npl-cgc.ernet.in
http://www.tifac.org.in
http://www.tropmet.res.in
http://www.uttranchalirrigation.com/iri/iri-home.htm
http://www.wrmin.nic.in
http://www.ncmrwf.gov.in
http://www.wapcos.net
http://www.isro.org
www.isro.org

Appendix A

ABBREVIATIONS

BBMB	:	Bhakra Beas Management Board
BIS	:	Bureau of Indian Standards
CAZRI	:	Central Arid Zone Research Institute
CDO	:	Central Design Organization
CGWB	:	Central Ground Water Board
CPCB	:	Central Pollution Control Board
CWC	:	Central Water Commission
CWPRS	:	Central Water and Power Research Station
DVC	:	Damodar Valley Corporation
ET	:	Evapotranspiration
GD or G & D	:	Gauge & Discharge
GOI	:	Government of India
HP	:	Hydropower
IAH	:	Indian Association of Hydrologists
ICAR	:	Indian Council of Agriculture Research
ICID	:	International Commission on Irrigation and Drainage
IITM	:	Indian Institute of Tropical Meteorology
IMD	:	India Meteorological Department
INCOH	:	Indian National Committee on Hydrology
IIT	:	Indian Institute of Technology
IRI	:	Irrigation Research Institute
IRS	:	Indian Remote Sensing Satellite
ISO	:	International Standards Organization
IWRS	:	Indian Water Resources Society
NBA	:	Narmada Bachao Aandolan
NCA	:	Narmada Control Authority
NIT	:	National Institute of Technology
NTPC	:	National Thermal Power Corporation
NHPC	:	National Hydroelectric Power Corporation
NIC	:	National Informatics centre
NIH	:	National Institute of Hydrology
NWDA	:	National Water Development Agency
MOEF	:	Ministry of Environment & Forest

MOWR	:	Ministry of Water Resources
MOA	:	Ministry of Agriculture
PMF	:	Probable Maximum Flood
SPF	:	Standard Project Flood
TERI	:	The Energy and Resources Institute
UH	:	Unit Hydrograph
UT	:	Union Territory
WALMI	:	Water & Land Management Institute
WRD	:	Water Resources Development
WQ	:	Water Quality

Appendix B

CONVERSION FACTORS

Conversion factors – length

	cm	m	km	inch	foot	mile
cm	1	0.01	1×10^{-5}	0.3937	0.032808	6.2137×10^{-6}
m	100	1	0.001	39.37	3.2808	6.2137×10^{-4}
km	10^5	1,000	1	39,370	3,280.8	0.62137
inch	2.54	0.0254	2.54×10^{-5}	1	0.083333	1.5783×10^{-5}
foot	30.48	0.3048	3.048×10^{-4}	12	1	1.8939×10^{-4}
mile	1.6093×10^5	1,609.3	1.6093	63,360	5,280	1

Example to use this table: 1 mile = 1.6093 km.

Conversion factors – area

	square meter	hectare	square km	square foot	acre	square mile
square meter	1	10^{-4}	10^{-6}	10.764	2.4711×10^{-4}	3.861×10^{-7}
hectare	10^4	1	0.01	107,639	2.4711	0.003861
square kilometer	10^6	100	1.0764×10^7	247.11	247.11	0.38610
square foot	0.092903	9.2903×10^{-6}	9.2903×10^{-4}	1	2.2957×10^{-5}	3.5870×10^{-8}
acre	4,046.9	0.40469	0.0040469	43.560	1	0.0015625
square mile	2.590×10^6	259.0	2.59	2.7878×10^7	640	1

Conversion factors – volume

	Liter	m^3	ha-m	ft^3	acre-ft	Thousand million cu ft (TMC)
liter	1	0.001	10^{-7}	0.035315	8.1071×10^{-7}	3.531×10^{-7}
m^3	1,000	1	10^{-4}	35.315	8.11071×10^{-4}	3.531×10^{-10}
hectare meter	10^7	10^4	1	353,147	8.1071	3.531×10^{-4}
ft^3	28.317	0.028317	2.8317×10^{-6}	1	2.2957×10^{-5}	1×10^{-9}
acre-foot	$1,233.5 \times 10^3$	1,233.5	0.12335	43,560	1	4.356×10^{-5}
Thousand million cu ft	28.317×10^9	28.317×10^6	2,831.7	1×10^9	4.356×10^{-5}	1

Indian Units

The units that are popularly used in India and their equivalents are described herein.

1 BCM = $1{*}10^9$ m^3

1 crore = 10 million = 100 lakh = 10,000,000

1 cumec(m^3/s) = 35.315 cusec(ft^3/s)

1 lakh or lac = 100,000 = 0.1 million

1 million = 10 lakh

1 MAF = 123.35 million m^3

1 MCM = $1{*}10^6$ m^3

1 TMC = 28.317 million m^3

Other Units

1 pound (lb) = 453.6 gm

1 ton (tonne) = 1,000 kg

1 horsepower = 745.7 watt

1 nautical mile = 1.151 miles = 1.852 km

Appendix C

NATIONAL WATER POLICY

NEED FOR A NATIONAL WATER POLICY

1.1 Water is a prime natural resource, a basic human need and a precious national asset. Planning, development and management of water resources need to be governed by national perspectives.

1.2 As per the latest assessment (1993), out of the total precipitation, including snowfall, of around 4,000 billion cubic metre in the country, the availability from surface water and replenishable ground water is put at 1,869 billion cubic metre. Because of topographical and other constraints, about 60% of this i.e. 690 billion cubic metre from surface water and 432 billion cubic metre from ground water, can be put to beneficial use. Availability of water is highly uneven in both space and time. Precipitation is confined to only about three or four months in a year and varies from 100 mm in the western parts of Rajasthan to over 10,000 mm at Cherrapunji in Meghalaya. Rivers and under ground aquifers often cut across state boundaries. Water, as a resource is one and indivisible: rainfall, river waters, surface ponds and lakes and ground water are all part of one system.

1.3 Water is part of a larger ecological system. Realising the importance and scarcity attached to the fresh water, it has to be treated as an essential environment for sustaining all life forms.

1.4 Water is a scarce and precious national resource to be planned, developed, conserved and managed as such, and on an integrated and environmentally sound basis, keeping in view the socio-economic aspects and needs of the States. It is one of the most crucial elements in developmental planning. As the country has entered the 21st century, efforts to develop, conserve, utilise and manage this important resource in a sustainable manner, have to be guided by the national perspective.

1.5 Floods and droughts affect vast areas of the country, transcending state boundaries. One-sixth area of the country is drought-prone. Out of 40 million hectare of the flood prone area in the country, on an average, floods affect an area of around 7.5 million hectare per year. Approach to management of droughts and floods has to be co-ordinated and guided at the national level.

1.6 Planning and implementation of water resources projects involve a number of socioeconomic aspects and issues such as environmental sustainability, appropriate resettlement and rehabilitation of project-affected people and livestock, public health concerns of water impoundment, dam safety etc. Common approaches and guidelines are necessary on these matters. Moreover, certain problems and weaknesses have affected a large number of water resources projects all over the country. There have been substantial time and cost overruns on projects. Problems of water logging and soil salinity have emerged in some irrigation commands, leading to the degradation of agricultural land. Complex issues of equity and social justice in regard to water distribution are required to be addressed. The development and overexploitation of groundwater resources in certain parts of the country have raised the concern and need for judicious and scientific resource management and conservation. All these concerns need to be addressed on the basis of common policies and strategies.

1.7 Growth process and the expansion of economic activities inevitably lead to increasing demands for water for diverse purposes: domestic, industrial, agricultural, hydropower, thermal power, navigation, recreation, etc. So far, the major consumptive use of water has been for irrigation. While the gross irrigation potential is estimated to have increased from 19.5 million hectare at the time of independence to about 95 million hectare by the end of the Year 1999–2000, further development of a substantial order is necessary if the food and fiber needs of our growing population are to be met with. The country's population which is over 1,027 million (2001 AD) at present is expected to reach a level of around 1,390 million by 2025 AD.

1.8 Production of food grains has increased from around 50 million tonnes in the fifties to about 208 million tonnes in the Year 1999–2000. This will have to be raised to around 350 million tonnes by the year 2025 AD. The drinking water needs of people and livestock have also to be met. Domestic and industrial water needs have largely been concentrated in or near major cities. However, the demand in rural areas is expected to increase sharply as the development programmes improve economic conditions of the rural masses. Demand for water for hydro and thermal power generation and for other industrial uses is also increasing substantially. As a result, water, which is already a scarce resource, will become even scarcer in future. This underscores the need for the utmost efficiency in water utilisation and a public awareness of the importance of its conservation.

1.9 Another important aspect is water quality. Improvements in existing strategies, innovation of new techniques resting on a strong science and technology base are needed to eliminate the pollution of surface and ground water resources, to improve water quality. Science and technology and training have to play important roles in water resources development and management in general.

1.10 National Water Policy was adopted in September, 1987. Since then, a number of issues and challenges have emerged in the development and management of the water resources. Therefore, the National Water Policy (1987) has been reviewed and updated.

INFORMATION SYSTEM

2.1 A well-developed information system, for water related data in its entirety, at the national/state level, is a prime requisite for resource planning. A standardised national information system should be established with a network of data banks and data bases, integrating and strengthening the existing Central and State level agencies and improving the quality of data and the processing capabilities.

2.2 Standards for coding, classification, processing of data and methods/ procedures for its collection should be adopted. Advances in information technology must be introduced to create a modern information system promoting free exchange of data among various agencies. Special efforts should be made to develop and continuously upgrade technological capability to collect, process and disseminate reliable data in the desired time frame.

2.3 Apart from the data regarding water availability and actual water use, the system should also include comprehensive and reliable projections of future demands of water for diverse purposes.

WATER RESOURCES PLANNING

3.1 Water resources available to the country should be brought within the category of utilisable resources to the maximum possible extent.

3.2 Non-conventional methods for utilisation of water such as through inter-basin transfers, artificial recharge of ground water and desalination of brackish or sea water as well as traditional water conservation practices like rainwater harvesting, including roof-top rainwater harvesting, need to be practiced to further increase the utilisable water resources. Promotion of frontier research and development, in a focused manner, for these techniques is necessary.

3.3 Water resources development and management will have to be planned for a hydrological unit such as drainage basin as a whole or for a sub-basin, multi-sectorally, taking into account surface and ground water for sustainable use incorporating quantity and quality aspects as well as environmental considerations. All individual developmental projects and proposals should be formulated and considered within the framework of such an overall plan keeping in view the existing agreements/awards for a basin or a subbasin so that the best possible combination of options can be selected and sustained.

3.4 Watershed management through extensive soil conservation, catchment-area treatment, preservation of forests and increasing the forest cover and the construction of check-dams should be promoted. Efforts shall be to conserve the water in the catchment.

3.5 Water should be made available to water short areas by transfer from other areas including transfers from one river basin to another, based on a national perspective, after taking into account the requirements of the areas / basins.

INSTITUTIONAL MECHANISM

4.1 With a view to give effect to the planning, development and management of the water resources on a hydrological unit basis, along with a multi-sectoral, multi-disciplinary and participatory approach as well as integrating quality, quantity and the environmental aspects, the existing institutions at various levels under the water resources sector will have to be appropriately reoriented/reorganised and even created, wherever necessary. As maintenance of water resource schemes is under non-plan budget, it is generally being neglected. The institutional arrangements should be such that this vital aspect is given importance equal or even more than that of new constructions.

4.2 Appropriate river basin organisations should be established for the planned development and management of a river basin as a whole or sub-basins, wherever necessary. Special multidisciplinary units should be set up to prepare comprehensive plans taking into account not only the needs of irrigation but also harmonising various other water uses, so that the available water resources are determined and put to optimum use having regard to existing agreements or awards of Tribunals under the relevant laws. The scope and powers of the river basin organisations shall be decided by the basin states themselves.

WATER ALLOCATION PRIORITIES

5 In the planning and operation of systems, water allocation priorities should be broadly as follows:

- Drinking water
- Irrigation
- Hydropower
- Ecology
- Agro-industries and non-agricultural industries
- Navigation and other uses.

However, the priorities could be modified or added if warranted by the area / region specific considerations.

PROJECT PLANNING

6.1 Water resource development projects should as far as possible be planned and developed as multipurpose projects. Provision for drinking water should be a primary consideration.

6.2 The study of the likely impact of a project during construction and later on human lives, settlements, occupations, socio-economic, environment and other aspects shall form an essential component of project planning.

6.3 In the planning, implementation and operation of a project, the preservation of the quality of environment and the ecological balance should be a primary consideration. The adverse impact on the environment, if any, should be minimised and should be offset by adequate compensatory measures. The project should, nevertheless, be sustainable.

6.4 There should be an integrated and multi-disciplinary approach to the planning, formulation, clearance and implementation of projects, including catchment area treatment and management, environmental and ecological aspects, the rehabilitation of affected people and command area development. The planning of projects in hilly areas should take into account the need to provide assured drinking water, possibilities of hydro-power development and the proper approach to irrigation in such areas, in the context of physical features and constraints of the basin such as steep slopes, rapid run-off and the incidence of soil erosion. The economic evaluation of projects in such areas should also take these factors into account.

6.5 Special efforts should be made to investigate and formulate projects either in, or for the benefit of, areas inhabited by tribal or other specially disadvantaged groups such as socially weak, scheduled castes and scheduled tribes. In other areas also, project planning should pay special attention to the needs of scheduled castes and scheduled tribes and other weaker sections of the society. The economic evaluation of projects benefiting such disadvantaged sections should also take these factors into account.

6.6 The drainage system should form an integral part of any irrigation project right from the planning stage.

6.7 Time and cost overruns and deficient realisation of benefits characterising most water related projects should be overcome by upgrading the quality of project preparation and management. The inadequate funding of projects should be obviated by an optimal allocation of resources on the basis of prioritisation, having regard to the early completion of on-going projects as well as the need to reduce regional imbalances.

6.8 The involvement and participation of beneficiaries and other stakeholders should be encouraged right from the project planning stage itself.

GROUND WATER DEVELOPMENT

7.1 There should be a periodical reassessment of the ground water potential on a scientific basis, taking into consideration the quality of the water available and economic viability of its extraction.

7.2 Exploitation of ground water resources should be so regulated as not to exceed the recharging possibilities, as also to ensure social equity. The detrimental environmental consequences of overexploitation of ground water need to be effectively prevented by the Central and State Governments. Ground water recharge projects should be developed and implemented for improving both the quality and availability of ground water resource.

7.3 Integrated and coordinated development of surface water and ground water resources and their conjunctive use, should be envisaged right from the project planning stage and should form an integral part of the project implementation.

7.4 Over exploitation of ground water should be avoided especially near the coast to prevent ingress of seawater into sweet water aquifers.

DRINKING WATER

8.1 Adequate safe drinking water facilities should be provided to the entire population both in urban and in rural areas. Irrigation and multipurpose projects should invariably include a drinking water component, wherever there is no alternative source of drinking water. Drinking water needs of human beings and animals should be the first charge on any available water.

IRRIGATION

9.1 Irrigation planning either in an individual project or in a basin as a whole should take into account the irrigability of land, cost-effective irrigation options possible from all available sources of water and appropriate irrigation techniques for optimising water use efficiency. Irrigation intensity should be such as to extend the benefits of irrigation to as large a number of farm families as possible, keeping in view the need to maximise production.

9.2 There should be a close integration of water-use and land-use policies.

9.3 Water allocation in an irrigation system should be done with due regard to equity and social justice. Disparities in the availability of water between head-reach and tail-end farms and between large and small farms should be obviated by adoption of a rotational water distribution system and supply of water on a volumetric basis subject to certain ceilings and rational pricing.

9.4 Concerted efforts should be made to ensure that the irrigation potential created is fully utilised. For this purpose, the command area development approach should be adopted in all irrigation projects.

9.5 Irrigation being the largest consumer of fresh water, the aim should be to get optimal productivity per unit of water. Scientific water management, farm practices and sprinkler and drip system of irrigation should be adopted wherever feasible.

9.6 Reclamation of water logged / saline affected land by scientific and cost-effective methods should form a part of command area development programme.

RESETTLEMENT AND REHABILITATION

10.1 Optimal use of water resources necessitates construction of storages and the consequent resettlement and rehabilitation of population. A skeletal national policy in this regard needs to be formulated so that the project affected persons share the benefits through proper rehabilitation. States should accordingly evolve their own detailed resettlement and rehabilitation policies for the sector, taking into account the local conditions. Careful planning is necessary to ensure that the construction and rehabilitation activities proceed simultaneously and smoothly.

FINANCIAL AND PHYSICAL SUSTAINABILITY

11.1 Besides creating additional water resources facilities for various uses, adequate emphasis needs to be given to the physical and financial sustainability of existing facilities. There is, therefore, a need to ensure that the water charges for various uses should be fixed in such a way that they cover at least the operation and maintenance charges of providing the service initially and a part of the capital costs subsequently. These rates should be linked directly to the quality of service provided. The subsidy on water rates to the disadvantaged and poorer sections of the society should be well targeted and transparent.

PARTICIPATORY APPROACH TO WATER RESOURCES MANAGEMENT

12.1 Management of the water resources for diverse uses should incorporate a participatory approach; by involving not only the various governmental agencies but also the users and other stakeholders, in an effective and decisive manner, in various aspects of planning, design, development and management of the water resources schemes. Necessary legal and institutional changes should be made at various levels for the purpose, duly ensuring appropriate role for women. Water Users' Associations and the local bodies such as municipalities and *gram panchayats* should particularly be involved in the operation, maintenance and management of water infrastructures / facilities at appropriate levels progressively, with a view to eventually transfer the management of such facilities to the user groups / local bodies.

PRIVATE SECTOR PARTICIPATION

13.1 Private sector participation should be encouraged in planning, development and management of water resources projects for diverse uses, wherever feasible. Private sector participation may help in introducing innovative ideas, generating financial resources and introducing corporate management and improving service efficiency and accountability to users. Depending upon the specific situations, various combinations of private sector participation, in building, owning, operating, leasing and transferring of water resources facilities, may be considered.

WATER QUALITY

14.1 Both surface water and ground water should be regularly monitored for quality. A phased programme should be undertaken for improvements in water quality.

14.2 Effluents should be treated to acceptable levels and standards before discharging them into natural streams.

14.3 Minimum flow should be ensured in the perennial streams for maintaining ecology and social considerations.

14.4 Principle of 'polluter pays' should be followed in management of polluted water.

14.5 Necessary legislation is to be made for preservation of existing water bodies by preventing encroachment and deterioration of water quality.

WATER ZONING

15.1 Economic development and activities including agricultural, industrial and urban development, should be planned with due regard to the constraints imposed by the configuration of water availability. There should be a water zoning of the country and the economic activities should be guided and regulated in accordance with such zoning.

CONSERVATION OF WATER

16.1 Efficiency of utilisation in all the diverse uses of water should be optimised and an awareness of water as a scarce resource should be fostered. Conservation consciousness should be promoted through education, regulation, incentives and disincentives.

16.2 The resources should be conserved and the availability augmented by maximising retention, eliminating pollution and minimising losses. For this, measures like selective linings in the conveyance system, modernisation and rehabilitation of existing systems including tanks, recycling and re-use of treated effluents and adoption of traditional techniques like mulching or pitcher irrigation and new techniques like drip and sprinkler may be promoted, wherever feasible.

FLOOD CONTROL AND MANAGEMENT

17.1 There should be a master plan for flood control and management for each flood prone basin.

17.2 Adequate flood-cushion should be provided in water storage projects, wherever feasible, to facilitate better flood management. In highly flood prone areas, flood control should be given overriding consideration in reservoir regulation policy even at the cost of sacrificing some irrigation/power benefits. (Ministry of Water Resources 7 April 1, 2002)

17.3 While physical flood protection works like embankments and dykes will continue to be necessary, increased emphasis should be laid on non-structural measures such as flood forecasting and warning, flood plain zoning and flood proofing for the minimisation of losses and to reduce the recurring expenditure on flood relief.

17.4 There should be strict regulation of settlements and economic activity in the flood plain zones along with flood proofing, to minimise the loss of life and property on account of floods.

17.5 The flood forecasting activities should be modernised, value added and extended to other uncovered areas. Inflow forecasting to reservoirs should be instituted for their effective regulation.

LAND EROSION BY SEA OR RIVER

18.1 The erosion of land, whether by the sea in coastal areas or by river waters inland, should be minimised by suitable cost-effective measures. The States and Union Territories should also undertake all requisite steps to ensure that indiscriminate occupation and exploitation of coastal strips of land are discouraged and that the location of economic activities in areas adjacent to the sea is regulated.

18.2 Each coastal State should prepare a comprehensive coastal land management plan, keeping in view the environmental and ecological impacts, and regulate the developmental activities accordingly.

DROUGHT-PRONE AREA DEVELOPMENT

19.1 Drought-prone areas should be made less vulnerable to drought-associated problems through soil moisture conservation measures, water harvesting practices, minimisation of evaporation losses, development of the ground water potential including recharging and the transfer of surface water from surplus areas where feasible and appropriate. Pastures, forestry or other modes of development which are relatively less water demanding should be encouraged. In planning water resource development projects, the needs of drought-prone areas should be given priority.

19.2 Relief works undertaken for providing employment to drought-stricken population should preferably be for drought proofing.

MONITORING OF PROJECTS

20.1 A close monitoring of projects to identify bottlenecks and to adopt timely measures to obviate time and cost overrun should form part of project planning and execution.

20.2 There should be a system to monitor and evaluate the performance and socio-economic impact of the project.

WATER SHARING/DISTRIBUTION AMONGST THE STATES

21.1 The water sharing/distribution amongst the states should be guided by a national perspective with due regard to water resources availability and needs within the river basin. Necessary guidelines, including for water short states even outside the basin, need to be evolved for facilitating future agreements amongst the basin states.

21.2 The Inter-State Water Disputes Act of 1956 may be suitably reviewed and amended for timely adjudication of water disputes referred to the Tribunal.

PERFORMANCE IMPROVEMENT

22.1 There is an urgent need of paradigm shift in the emphasis in the management of water resources sector. From the present emphasis on the creation and expansion of water resources infrastructures for diverse uses, there is now a need to give greater emphasis on the improvement of the performance of the existing water resources facilities. Therefore, allocation of funds under the water resources sector should be re-prioritised to ensure that the needs for development as well as operation and maintenance of the facilities are met.

MAINTENANCE AND MODERNISATION

23.1 Structures and systems created through massive investments should be properly maintained in good health. Appropriate annual provisions should be made for this purpose in the budgets.

23.2 There should be a regular monitoring of structures and systems and necessary rehabilitation and modernisation programmes should be undertaken.

23.3 Formation of Water Users' Association with authority and responsibility should be encouraged to facilitate the management including maintenance of irrigation system in a time bound manner.

SAFETY OF STRUCTURES

24.1 There should be proper organisational arrangements at the national and state levels for ensuring the safety of storage dams and other water-related structures consisting of specialists in investigation, design, construction, hydrology, geology, etc. A dam safety legislation may be enacted to ensure proper inspection, maintenance and surveillance of existing dams and also to ensure proper planning, investigation, design and construction for safety of new dams. The Guidelines on the subject should be periodically updated and reformulated. There should be a system of continuous surveillance and regular visits by experts.

SCIENCE AND TECHNOLOGY

25.1 For effective and economical management of our water resources, the frontiers of knowledge need to be pushed forward in several directions by intensifying research efforts in various areas, including the following:

- Hydrometeorology;
- Snow and lake hydrology;
- Surface and ground water hydrology;
- River morphology and hydraulics;
- Assessment of water resources;
- Water harvesting and ground water recharge;
- Water quality;
- Water conservation;
- Evaporation and seepage losses;
- Recycling and re-use;
- Better water management practices and improvements in operational technology;
- Crops and cropping systems;
- Soils and material research;

- New construction materials and technology (with particular reference to roller compacted concrete, fiber reinforced concrete, new methodologies in tunneling technologies, instrumentation, advanced numerical analysis in structures and back analysis);
- Seismology and seismic design of structures;
- The safety and longevity of water-related structures;
- Economical designs for water resource projects;
- Risk analysis and disaster management;
- Use of remote sensing techniques in development and management;
- Use of static ground water resource as a crisis management measure;
- Sedimentation of reservoirs;
- Use of sea water resources;
- Prevention of salinity ingress;
- Prevention of water logging and soil salinity;
- Reclamation of water logged and saline lands;
- Environmental impact;
- Regional equity.

TRAINING

26.1 A perspective plan for standardised training should be an integral part of water resource development. It should cover training in information systems, sectoral planning, project planning and formulation, project management, operation of projects and their physical structures and systems and the management of the water distribution systems. The training should extend to all the categories of personnel involved in these activities as also the farmers.

CONCLUSION

27.1 In view of the vital importance of water for human and animal life, for maintaining ecological balance and for economic and developmental activities of all kinds, and considering its increasing scarcity, the planning and management of this resource and its optimal, economical and equitable use has become a matter of the utmost urgency. Concerns of the community need to be taken into account for water resources development and management. The success of the National Water Policy will depend entirely on evolving and maintaining a national consensus and commitment to its underlying principles and objectives. To achieve the desired objectives, State Water Policy backed with an operational action plan shall be formulated in a time bound manner say in two years. National Water Policy may be revised periodically as and when need arises.

Source: Ministry of Water Resources, Govt. of India, 2002.

Appendix D

INDIAN STANDARDS RELATED TO HYDROLOGY AND WATER RESOURCES

Identifier	TITLE
	HYDROMETRY
IS 1191:2003	Hydrometric determination – Vocabulary and Symbols
IS 1192:1981	Velocity area Methods for measurement of flow of water in open channels (first revision)
IS 1194:1960	Forms for recording measurement of flow of water in open channels
IS 2912:1998	Liquid flow measurement in open channels – Slope-area method
IS 2951:1965	Recommendation for estimation of flow of liquids in closed conduits Part 1: Head loss in straight pipes due to friction resistance Part 2: Head loss in valves and fittings
IS 3910:1992	Requirements for rotating element current meters (cup type) for water flow measurement (first revision)
IS 3911:1994	Surface floats – Functional requirements (First Revision)
IS 3912:1993	Sounding rods – Functional requirements (first revision)
IS 3913:2005	Specification for suspended sediment load samplers (first revision)
IS 3917:1966	Specification for scoop type bed material samplers
IS 3918:1966	Code of practice for use of current meter (cup type) for water flow measurement
IS 4073:1967	Specification for fish weights
IS 4080:1994	Vertical staff gauges – Functional requirements (First Revision)
IS 4477:1975	Methods of measurement of fluid flow by means of venturi meters Part 2 Compressible fluids
IS 4858:1968	Specification for velocity rods
IS 4890:1968	Methods for measurement of suspended sediment in open channels
IS 4986:2002	Code of practice for installation of rain-gauge (non-recording type) and measurement of rain (second revision)
IS 4987:1994	Recommendations for establishing net work of raingauge stations
IS 5542:2003	Guide for storm analysis (first revision)
IS 6062:1971	Method of measurement of flow of water in open channels using standing wave flume-fall

(*Continued*)

Appendix D (*Continued*)

Identifier	TITLE
IS 6063:1971	Method of measurement of flow of water in open channels using standing wave flume
IS 6064:1971	Specification for sounding and suspension equipment
IS 6330:1971	Recommendation for liquid flow measurement in open channels by weirs and flumes-end depth method for estimation of flow in rectangular channels with a free overfall (approximate method)
IS 6339:1971	Methods of analysis of concentration, particle size distribution and specific gravity of sediment in streams and canals
IS 8389:2003	Code of practice for installation and use of raingauges, recording (second revision)
IS 9108:1979	Liquid flow measurement in open channels using thin plate weirs
IS 9115:2002	Method for estimation of incompressible fluid flow in closed conduits by bend meters (first revision)
IS 9116:2002	Specification for water stage recorder (float type) (first revision)
IS 9118:1979	Method for measurement of pressure by means of manometer
IS 9119:1979	Method for flow estimation by jet characteristics (approximate method)
IS 9163:1979	Dilution methods of measurement of steady flow Part 1: constant rate injection method
IS 9922:1981	Guide for selection of method for measuring flow in open channels
IS 12752:1989	Guidelines for the selection of flow gauging structures
IS 13083:1991	Liquid flow measurement in open channels- flat-V weirs
IS 13084:1991	Liquid flow measurement in open channels – round nose horizontal crest weirs
IS 13371:1992	Code of Practice for calibration (rating) of rotating element current meters in straight open tank
IS 14359:1996	Echo sounders for water depth measurements
IS 14371:1996	Measurement of liquid flow in open channels – Parshall and SANIIRI flumes
IS 14573:1998	Liquid flow measurement in open channels – Velocity area methods – collection and processing of data for determination of errors in measurement
IS 14574:1998	Measurement of liquid flow in open channels by weirs and flumes- end depth method for estimation of flow in non rectangular channels with a free overfall (approximate method)
IS 14615:1999	Measurement of fluid flow by means of pressure differential devices Part 1: Orifice plates, nozzles and venturi tubes inserted in circular cross-section conduits running full.
IS 14673:1999	Liquid flow measurement in open channels by weirs and flumes – Triangular profile weirs
IS 14869:2000	Liquid flow measurement in open channels Rectangular, trapezoidal and U-shape flumes
IS 14973:2001	Measurement of fluid flow in closed conduits Velocity area method using Pilot Static Tubes
IS 14974:2001	Liquid flow measurement in open channels by weirs and flumes – Rectangular broad-crested weirs
IS 14975:2001	Measurement of fluid flow in open channels – Stream lined triangular profile weirs
IS 15117:2002	Hydrometric determination- Cable way system for stream gauging
IS 15118:2002	Measurement of liquid flow in open channel – Water level measuring devices
IS 15119:2002	Measurement of liquid flow in open channels Part 1: Establishment and operation of a gauging station (superseding IS 2914:1964) Part 2: Determination of the stage-discharge relation (superseding IS 2914:1964)
IS 15122:2002	Measurement of liquid flow in open channels under tidal condition (superseding IS 2913:1964)

IS 15123:2002	Hydrometric determination – Flow measurement in open channels using structures-Trapezoidal broad crested weirs (superseding IS 4362:1992)
IS 15124:2002	Liquid flow measurement in open channels – Sampling and analysis of gravel-bed-material
IS 15352:2003	Liquid flow measurement in open channels – Position fixing equipment for hydrometric boats
IS 15353:2003	Liquid flow measurement in open channels by weirs and flumes – V- shaped broad crested weirs
IS 15358:2003	Liquid flow measurement in open channels – Flow measurement under ice conditions
IS 15359:2003	Liquid flow measurement in open channels Measurement of suspended sediment transport in tidal channels
IS 15360:2003	Measurement of liquid flow in open channels – Bed material sampling
IS 15362:2002	Liquid flow measurement in open channels – Flow measurements in open channels using structures-:compound gauging structure
IS 15454:2004	Liquid flow measurement in open channels – Velocity area method using a restricted number of verticals
IS 15527:2004	Measurement of liquid flow in open channels Measurement in meandering rivers and in streams with unstable boundaries
	TERMINOLOGY RELATED TO RIVER VALLEY PROJECTS
IS 4410:1991	Glossary of terms relating to river valley projects
	Part 1: Irrigation practice
IS 4410:1967	Part 2: Project planning
IS 4410:1988	Part 3: River and river training
IS 4410:1982	Part 4: Drawings
	Part 5: Canals
IS 4410:1983	Part 6: Reservoirs
IS 4410:1982	Part 7: Engineering geology
IS 4410:1992	Part 8: Dams and dam section
IS 4410:1982	Part 9: Spillways and syphons
IS 4410:1988	Part 10: Hydro-electric Power station including water conductor system
IS 4410:1972	Part 11: Hydrology Section
	Section 1 General terms
	Part 11/Section 2 Precipitation and run-off
IS 4410:1973	Part 11/Section 3 Infiltration and water losses
IS 4410:1973	Part 11/Section 4 Hydrographs
IS 4410:1977	Part 11/Section 5 Floods
IS 4410:1994	Part 11/Section 6 Ground water
IS 4410:1984	Part 11/Section 7 Quality of water
IS 4410:1993	Part 12 Diversion works
IS 4410:1985	Part 13 Operation, maintenance and repairs of river valley projects
IS 4410:1977	Part 14 Soil conservation and reclamation
	Section 1 Soil conservation
	Part 14/Section 2 Reclamation
IS 4410:1973	Canal structures
	Part 15/Section 1 General terms
	Part 15/Section 2 Transitions
	Part 15/Section 3 Flumes
	Part 15/Section 4 Regulating works
IS 4410:1992	Part 15/Section 5 Cross drainage works

(*Continued*)

Appendix D (*Continued*)

Identifier	TITLE
IS 4410:1999	Part 16 Gates and valves
	Section 1 Gates
IS 4410:1981	Part 16/Section 2 Valves
IS 4410:1977	Part 17 Water requirements of crops
IS 4410:1983	Part 18 Energy dissipator devices – Stilling basins
IS 4410:1996	Part 19 Grouting
IS 4410:1983	Part 20 Tunnels
IS 4410:1987	Part 21 Flood control
IS 4410:1994	Part 22 Barrages and weirs
IS 4410:1999	Part 23 Hoists, cranes and other related terms
	GROUND WATER RELATED INVESTIGATIONS
IS 13969:1994	Guidelines for sampling of groundwater
IS 14476:1998	Test pumping of water wells – Code of practice
	Part 1 General
	Part 2 Hydro-geological considerations
	Part 3 Pre-test planning
	Sec 1 General aspects
	Part 3 Sec 2 Test design
	Part 3 Sec 3 Observation wells
	Part 3 Sec 4 Test wells
	Part 4 Pre-test observations
	Part 5 Pumping test
	Part 6 Special tests
	Part 7 Post test observations and presentation of data
	Part 8 Water level and discharge measuring devices
	Part 9 Well development
	WATER RESOURCES PLANNING, MANAGEMENT AND EVALUATION
IS 4008:1985	Guide for presentation of project report for river valley projects (first revision)
IS 4186:1985	Guide for preparation of project report for river valley projects (first revision)
IS 4877:1968	Guide for preparation of estimate for river valley projects
IS 5510:1969	Guide for soil surveys for river valley projects
IS 6748:1973	Recommendations for watershed management relating to soil conservation
	Part 1 Agronomic aspects
IS 7560:1974	Guidelines for allocation of cost among different purposes of river valley projects
IS 10336:1983	Guidelines for preparation of completion reports of multi-purpose river valley projects
IS 13028:1991	Guidelines for overall planning of river basin
IS 13142:1991	Proforma for reporting progress of benefits created by river valley projects
IS 13218:1992	Proforma for reporting progress during construction for river valley projects
	Part 1 Irrigation works
IS 13218:1991	Part 2 Hydel works
IS 13218:1992	Part 3 Flood Control
IS 13668:1993	Guidelines for fixing intensity of irrigation
IS 13739:1993	Guidelines for estimation of flood damages
IS 14519:1998	Guidelines for fixing rates for irrigation water
IS 15087:2001	Guidelines for planning and design of drainage in irrigation projects

DAMS AND SPILLWAYS

SP 55:1993	Design aid for anchorages for spillway piers, training walls and divide walls.
IS 4997:1968	Criteria for design of hydraulic jump type stilling basins with horizontal and sloping apron
IS 5186:1994	Design of chute and side channel spillways – Criteria (first revision)
IS 6512:1984	Criteria for design of solid gravity dams (first revision)
IS 6934:1998	Recommendations for hydraulic design of high ogee over-flow spillways (first revision)
IS 7365:1985	Criteria for hydraulic design of bucket type energy dissipators (first revision)
IS 7894:1975	Code of practice for stability analysis of earth dams
IS 8605:1977	Code of practice for construction of masonry in dams
IS 8826:1978	Guidelines for design of large earth and rock fill dams
IS 9296:2001	Guidelines for inspection and maintenance of dam and appurtenant structures (first revision)
IS 9297:1979	Recommendations for lighting, ventilation and other facilities inside the dam
IS 9429:1999	Code of practice for drainage system for earth and rockfill dams(first revision)
IS 10135:1985	Code of practice for drainage system for gravity dams, their foundations and abutments (first revision)
IS 10137:1982	Guidelines for selection of spillways and energy dissipators
IS 10635:1993	Freeboard requirements in embankment dams Guidelines (first revision)
IS 11155:1994	Construction of spillways and similar overflow structures – Code of practice (First Revision)
IS 11216:1985	Code of practice for permeability test for: masonry (during and after construction)
IS 11223:1985	Guidelines for fixing spillway capacity
IS 11485:1985	Criteria for hydraulic design of sluices in concrete and masonry dams
IS 11527:1985	Criteria for structural design of energy dissipators for spillways
IS 11772:1986	Guidelines for design of drainage arrangements of energy dissipators and training walls of spillways
IS 12169:1987	Criteria for design of small embankment dams
IS 12200:2001	Code of practice for provision of water stops at transverse contraction joints in masonry and concrete dams (first revision)
IS 12720:1993	Criteria for structural design of spillway training walls and divide walls (second revision)
IS 12731:1989	Hydraulic design of impact type energy dissipators: Recommendations
IS 12804:1989	Criteria for estimation of aeration demand for spillways and outlet structures
IS 12966:1992	Code of practice for galleries and other openings in dams Part 1 General requirements
IS 12966:1990	Part 2 Structural design
IS 13048:1991	Recommendations for hydraulic design of duck bill spillways
IS 13144:1991	Recommendations for provision of facilities outside dams
IS 13195:1991	Preliminary design, operation and maintenance of protection works downstream of spillways – Guidelines
IS 13551:1992	Criteria for structural design of spillway pier and crest
IS 13645:1993	Guidelines for grouting the upstream face of masonry dams
IS 14591:1998	Guidelines for temperature control of mass concrete for dams
IS 14690:1999	Quality control during construction of earth and rockfill dams – Recommendations
IS 14954:2001	Distress and remedial measures in earth and rockfill dams-Guidelines
IS 15058:2002	Specification for PVC waterstops at transverse contraction joints in masonry and concrete dams

(*Continued*)

Appendix D (*Continued*)

Identifier	TITLE
	LAKES AND RESERVOIRS
IS 5477:1999	Methods for fixing the capacities of reservoirs Part 1 General requirements
IS 5477:1994	Part 2 Dead storage (first revision)
IS 5477:2002	Part 3 Live storage (first revision)
IS 5477:1971	Part 4 Flood storage
IS 6518:1992	Code of practice for control of sediment in reservoirs (first revision)
IS 6939:1992	Methods for determination of evaporation from reservoirs (first revision)
IS 7323:1994	Operation of reservoirs – Guidelines (first revision)
IS 8237:1985	Code of practice for protection of slope for reservoir embankments (first revision)
IS 12182:1987	Guidelines for determination of effects of sedimentation in planning and performance of reservoirs
IS 13665:1993	Sedimentation in reservoirs – Method of measurement
IS 14654:1999	Minimizing evaporation losses from reservoirs Guidelines
IS 15472:2004	Guidelines for planning and design of low level outlets for evacuating storage reservoirs
	HYDRAULIC GATES AND VALVES
IS 4622:2003	Recommendations for structural design of fixed wheel gates (third revision)
IS 4623:2000	Recommendations for structural design of radial gates (second revision)
IS 5620:1985	Recommendations for structural design criteria for low head slide gates (second revision)
IS 6938:1989	Code of practice for design of rope drum and chain hoists for hydraulic gates (first revision)
IS 7326:1992	Penstock and turbine inlet butterfly valves for hydropower stations and systems Part 1 Criteria for structural and hydraulic design (first revision)
IS 7326:1992	Part 2 Guidelines for design and selection of control equipment (first revision)
IS 7326:1976	Part 3 Recommendations for operations and maintenance
IS 7332:1991	Spherical valves for hydropower stations and systems Part 1 Criteria for structural and hydraulic design (first revision)
IS 7332:1993	Part 2 Guidelines for design and selection of control equipment (first revision)
IS 7332:1994	Part 3 Recommendations for operation and maintenance of spherical valves (first revision)
IS 7718:1991	Recommendations for inspection, testing and maintenance of fixed wheel and slide gates (first revision)
IS 9349:1986	Recommendations for structural design of medium and high head slide gates (first revision)
IS 10021:2000	Guidelines for de-icing systems for hydraulic installations (first revision)
IS 10096:1983	Recommendations for inspection, testing and maintenance of radial gates and their hoists Part 1: Inspection, testing and assembly at the manufacturing stage Section 1 Gates
IS 10096:1986	Part 1 Section 2 Rope Drum Hoists
IS 10096:1983	Part 2 Inspection, testing and assembly at the time of erection
IS 10096:2002	Part 3 After erection (first revision)
IS 10210:1993	Criteria for design of hydraulic hoists
IS 11228:1985	Recommendations for design of screw hoists for hydraulic gates
IS 11793:1986	Guidelines for design of float-driven hoisting mechanism for automatic gated control
IS 11855:2004	General requirements for rubber seals for hydraulic gates (first revision)

IS 13041:1991	Recommendation for inspection, testing and maintenance of hydraulic hoist (After erection)
IS 13591:1992	Criteria for design of lifting beams
IS 13623:1993	Criteria for choice of gates and hoists
IS 14177:1994	Guidelines for painting system for hydraulic gates and hoists
IS 15466:2004	Specification for rubber seals for hydraulic gates
	CANALS AND CROSS DRAINAGE WORKS
IS 3872:2002	Code of practice for lining of canals with burnt clay tiles (first revision)
IS 3873:1993	Laying cement concrete/stone slab lining on canals code of practice (second revision)
IS 4515:2003	Stone pitched lining for canals – Code of practice (second revision)
IS 4558:1995	Code of practice for under-drainage of lined canals (second revision)
IS 4701:1982	Code of practice for earthwork on canals
IS 4839:1992	Maintenance of canals – Code of practice
	Part 1 Unlined canals (second revision)
	Part 2 Lined canals (second revision)
	Part 3 Canal structures, drains, jungle clearance, plantation and regulation (second revision)
IS 5256:1992	Code of practice for sealing expansion joints in concrete lining on canals (first revision)
IS 5690:1982	Guide for laying combination lining for existing unlined canals (first revision)
IS 5968:1987	Guide for planning and layout of canal system for irrigation (first revision)
IS 6004:1980	Criteria for hydraulic design of sediment ejector for irrigation and power canals (first revision)
IS 6522:1972	Criteria for design of silt vanes for sediment control in off taking canals
IS 6936:1992	Guide for location, selection and hydraulic design of canal escapes (first revision)
IS 7112:2003	Criteria for design of cross section for unlined canals in alluvial soil (first revision)
IS 7113:2003	Code of practice for soil-cement lining for canals (first revision)
IS 7114:1973	Criteria for hydraulic design of cross regulators for canals
IS 7331:1981	Code of practice for inspection and maintenance of cross-drainage works (first revision)
IS 7495:1974	Criteria for hydraulic design of silt selective head regulator for sediment control in off taking canals
IS 7784:1993	Code of practice for design of cross drainage works
	Part 1 General features (first revision)
IS 7784:1995	Part 2 Specific requirement
	Section 1 Aqueducts
IS 7784:2000	Part 2/Section 2 Super passages (first revision)
IS 7784:1996	Part 2/Section 3 Canal siphons
IS 7784:1999	Part 2/Section 4 Level crossings
IS 7784:2000	Part 2/Section 5 Syphon aqueducts (first revision)
IS 7871:1975	Criteria for hydraulic design of groyne walls (curved wing) for sediment distribution at off take points in a canal
IS 7873:1975	Code of practice for lime concrete lining for canals
IS 7880:1975	Criteria for hydraulic design of skimming platform for sediment control in off taking canal
IS 7986:1976	Code of practice for canal outlets
IS 8835:1978	Guidelines for planning and design of surface drains
IS 9097:1979	Guide for laying lining of canals with hot bitumen or bituminous felts
IS 9447:1980	Guidelines for assessment of seepage losses from canals by analytical method

(*Continued*)

Appendix D (*Continued*)

Identifier	TITLE
IS 9451:1994	Guidelines for lining of canals in expansive soils (second revision)
IS 9452:1993	Code of practice for measurement of seepage losses from canals Part 1 Ponding method (first revision)
IS 9452:1980	Part 2 Inflow Outflow method
IS 9452:1988	Part 3 By seepage meter method
IS 9698:1995	Lining of canals with polyethylene film – Code of practice (first revision)
IS 9913:2000	Code of practice for construction of cross drainage works (first revision)
IS 10430:2000	Criteria for design of lined canals and guidelines for selection of type of lining (first revision)
IS 10646:1991	Canal linings – Cement concrete tiles – Specification (first revision)
IS 11809:1994	Lining for canals by stone masonry – Code of practice (first revision)
IS 12331:1988	General requirements for canal outlets
IS 12379:1988	Code of practice for lining water courses and field channels
IS 13143:1991	Joints in concrete lining of canals – sealing compound specification
	RIVER TRAINING AND DIVERSION WORKS
IS 6966:1989	Guidelines for hydraulic design of barrages and weirs Part 1 Alluvial Reaches (first revision)
IS 7349:1989	Guidelines for operation and maintenance of barrages and weirs (first revision)
IS 7720:1991	Criteria for investigation, planning and layout of barrages and weirs (first revision)
IS 8408:1994	Planning and design of groynes in alluvial river guidelines (first revision)
IS 9461:1980	Guidelines for data required for design of temporary river diversion works
IS 9795:1981	Guidelines for the choice of type of diversion works Part 1 Coffer dams
IS 10084:1982	Design of diversion works – criteria Part 1 Coffer dams
IS 10084:1994	Part 2 Diversion channels and open cut or conduit in the body of dam
IS 10751:1994	Planning and design of guide banks for alluvial river Guidelines (first revision)
IS 10788:1984	Code of practice for construction of diversion works Part 1 Cellular coffer dams
IS 11130:1984	Criteria for structural design of barrages and weirs
IS 11150:1993	Construction of concrete barrages – Code of practice (first revision)
IS 11532:1995	Construction and maintenance of river embankments (levees) Guidelines (first revision)
IS 12094:2000	Guidelines for planning and design of river embankment (levees) (first revision)
IS 12892:1989	Guidelines for the safety of barrage and weir structures
IS 12926:1995	Construction and maintenance of guide banks in alluvial rivers – Guidelines (first revision)
IS 13877:1993	Planning and design of fish pass- Guidelines
IS 13912:1993	Closure of diversion channel and open cut or conduit in the body of dam- Code of practice
IS 14262:1995	Planning and design of revetments – Guidelines
IS 14592:1998	Guidelines for planning and design of river powerhouses integrated with barrages Part 1 Investigation, planning and layout
IS 14815:2000	Design flood for river studies of barrages and weirs Guidelines
IS 14955:2001	Hydraulic model studies of barrages and weirs Guidelines

These standards can be obtained from the Bureau of Indian Standards, Manak Bhawan, Bahadur Shah Zafar Marg, New Delhi, or their regional offices.

INDEX

Water Science and Technology Library

1. A.S. Eikum and R.W. Seabloom (eds.): *Alternative Wastewater Treatment*. Low-Cost Small Systems, Research and Development. Proceedings of the Conference held in Oslo, Norway (7–10 September 1981). 1982 ISBN 90-277-1430-4
2. W. Brutsaert and G.H. Jirka (eds.): *Gas Transfer at Water Surfaces*. 1984 ISBN 90-277-1697-8
3. D.A. Kraijenhoff and J.R. Moll (eds.): *River Flow Modelling and Forecasting*. 1986 ISBN 90-277-2082-7
4. World Meteorological Organization (ed.): *Microprocessors in Operational Hydrology*. Proceedings of a Conference held in Geneva (4–5 September 1984). 1986 ISBN 90-277-2156-4
5. J. N.ĕmec: *Hydrological Forecasting*. Design and Operation of Hydrological Forecasting Systems. 1986 ISBN 90-277-2259-5
6. V.K. Gupta, I. Rodríguez-Iturbe and E.F. Wood (eds.): *Scale Problems in Hydrology*. Runoff Generation and Basin Response. 1986 ISBN 90-277-2258-7
7. D.C. Major and H.E. Schwarz: *Large-Scale Regional Water Resources Planning*. The North Atlantic Regional Study. 1990 ISBN 0-7923-0711-9
8. W.H. Hager: *Energy Dissipators and Hydraulic Jump*. 1992 ISBN 0-7923-1508-1
9. V.P. Singh and M. Fiorentino (eds.): *Entropy and Energy Dissipation in Water Resources*. 1992 ISBN 0-7923-1696-7
10. K.W. Hipel (ed.): *Stochastic and Statistical Methods in Hydrology and Environmental Engineering*. A Four Volume Work Resulting from the International Conference in Honour of Professor T. E. Unny (21–23 June 1993). 1994
    10/1: Extreme values: floods and droughts ISBN 0-7923-2756-X
    10/2: Stochastic and statistical modelling with groundwater and surface water applications ISBN 0-7923-2757-8
    10/3: Time series analysis in hydrology and environmental engineering ISBN 0-7923-2758-6
    10/4: Effective environmental management for sustainable development ISBN 0-7923-2759-4
    Set 10/1–10/4: ISBN 0-7923-2760-8
11. S.N. Rodionov: *Global and Regional Climate Interaction: The Caspian Sea Experience*. 1994 ISBN 0-7923-2784-5
12. A. Peters, G. Wittum, B. Herrling, U. Meissner, C.A. Brebbia, W.G. Gray and G.F. Pinder (eds.): *Computational Methods in Water Resources X*. 1994 Set 12/1–12/2: ISBN 0-7923-2937-6
13. C.B. Vreugdenhil: *Numerical Methods for Shallow-Water Flow*. 1994 ISBN 0-7923-3164-8
14. E. Cabrera and A.F. Vela (eds.): *Improving Efficiency and Reliability in Water Distribution Systems*. 1995 ISBN 0-7923-3536-8
15. V.P. Singh (ed.): *Environmental Hydrology*. 1995 ISBN 0-7923-3549-X
16. V.P. Singh and B. Kumar (eds.): *Proceedings of the International Conference on Hydrology and Water Resources* (New Delhi, 1993). 1996
    16/1: Surface-water hydrology ISBN 0-7923-3650-X
    16/2: Subsurface-water hydrology ISBN 0-7923-3651-8
    16/3: Water-quality hydrology ISBN 0-7923-3652-6

16/4: Water resources planning and management ISBN 0-7923-3653-4
Set 16/1–16/4 ISBN 0-7923-3654-2

17. V.P. Singh: *Dam Breach Modeling Technology.* 1996 ISBN 0-7923-3925-8
18. Z. Kaczmarek, K.M. Strzepek, L. Somlyódy and V. Priazhinskaya (eds.): *Water Resources Management in the Face of Climatic/Hydrologic Uncertainties.* 1996 ISBN 0-7923-3927-4
19. V.P. Singh and W.H. Hager (eds.): *Environmental Hydraulics.* 1996 ISBN 0-7923-3983-5
20. G.B. Engelen and F.H. Kloosterman: *Hydrological Systems Analysis.* Methods and Applications. 1996 ISBN 0-7923-3986-X
21. A.S. Issar and S.D. Resnick (eds.): *Runoff, Infiltration and Subsurface Flow of Water in Arid and Semi-Arid Regions.* 1996 ISBN 0-7923-4034-5
22. M.B. Abbott and J.C. Refsgaard (eds.): *Distributed Hydrological Modelling.* 1996 ISBN 0-7923-4042-6
23. J. Gottlieb and P. DuChateau (eds.): *Parameter Identification and Inverse Problems in Hydrology, Geology and Ecology.* 1996 ISBN 0-7923-4089-2
24. V.P. Singh (ed.): *Hydrology of Disasters.* 1996 ISBN 0-7923-4092-2
25. A. Gianguzza, E. Pelizzetti and S. Sammartano (eds.): *Marine Chemistry.* An Environmental Analytical Chemistry Approach. 1997 ISBN 0-7923-4622-X
26. V.P. Singh and M. Fiorentino (eds.): *Geographical Information Systems in Hydrology.* 1996 ISBN 0-7923-4226-7
27. N.B. Harmancioglu, V.P. Singh and M.N. Alpaslan (eds.): *Environmental Data Management.* 1998 ISBN 0-7923-4857-5
28. G. Gambolati (ed.): *CENAS. Coastline Evolution of the Upper Adriatic Sea Due to Sea Level Rise and Natural and Anthropogenic Land Subsidence.* 1998 ISBN 0-7923-5119-3
29. D. Stephenson: *Water Supply Management.* 1998 ISBN 0-7923-5136-3
30. V.P. Singh: *Entropy-Based Parameter Estimation in Hydrology.* 1998 ISBN 0-7923-5224-6
31. A.S. Issar and N. Brown (eds.): *Water, Environment and Society in Times of Climatic Change.* 1998 ISBN 0-7923-5282-3
32. E. Cabrera and J. García-Serra (eds.):*Drought Management Planning in Water Supply Systems.* 1999 ISBN 0-7923-5294-7
33. N.B. Harmancioglu, O. Fistikoglu, S.D. Ozkul, V.P. Singh and M.N. Alpaslan: *Water Quality Monitoring Network Design.* 1999 ISBN 0-7923-5506-7
34. I. Stober and K. Bucher (eds): *Hydrogeology of Crystalline Rocks.* 2000 ISBN 0-7923-6082-6
35. J.S. Whitmore: *Drought Management on Farmland.* 2000 ISBN 0-7923-5998-4
36. R.S. Govindaraju and A. Ramachandra Rao (eds.): *Artificial Neural Networks in Hydrology.* 2000 ISBN 0-7923-6226-8
37. P. Singh and V.P. Singh: *Snow and Glacier Hydrology.* 2001 ISBN 0-7923-6767-7
38. B.E. Vieux: *Distributed Hydrologic Modeling Using GIS.* 2001 ISBN 0-7923-7002-3
39. I.V. Nagy, K. Asante-Duah and I. Zsuffa: *Hydrological Dimensioning and Operation of Reservoirs.* Practical Design Concepts and Principles. 2002 ISBN 1-4020-0438-9
40. I. Stober and K. Bucher (eds.): *Water-Rock Interaction.* 2002 ISBN 1-4020-0497-4

Water Science and Technology Library

41. M. Shahin: *Hydrology and Water Resources of Africa*. 2002 ISBN 1-4020-0866-X
42. S.K. Mishra and V.P. Singh: *Soil Conservation Service Curve Number (SCS-CN) Methodology*. 2003 ISBN 1-4020-1132-6
43. C. Ray, G. Melin and R.B. Linsky (eds.): *Riverbank Filtration*. Improving Source-Water Quality. 2003 ISBN 1-4020-1133-4
44. G. Rossi, A. Cancelliere, L.S. Pereira, T. Oweis, M. Shatanawi and A. Zairi (eds.): *Tools for Drought Mitigation in Mediterranean Regions*. 2003 ISBN 1-4020-1140-7
45. A. Ramachandra Rao, K.H. Hamed and H.-L. Chen: *Nonstationarities in Hydrologic and Environmental Time Series*. 2003 ISBN 1-4020-1297-7
46. D.E. Agthe, R.B. Billings and N. Buras (eds.): *Managing Urban Water Supply*. 2003 ISBN 1-4020-1720-0
47. V.P. Singh, N. Sharma and C.S.P. Ojha (eds.): *The Brahmaputra Basin Water Resources*. 2004 ISBN 1-4020-1737-5
48. B.E. Vieux: *Distributed Hydrologic Modeling Using GIS. Second Edition*. 2004 ISBN 1-4020-2459-2
49. M. Monirul Qader Mirza (ed.): *The Ganges Water Diversion: Environmental Effects and Implications*. 2004 ISBN 1-4020-2479-7
50. Y. Rubin and S.S. Hubbard (eds.): *Hydrogeophysics*. 2005 ISBN 1-4020-3101-7
51. K.H. Johannesson (ed.): *Rare Earth Elements in Groundwater Flow Systems*. 2005 ISBN 1-4020-3233-1
52. R.S. Harmon (ed.): *The Río Chagres, Panama*. A Multidisciplinary Profile of a Tropical Watershed. 2005 ISBN 1-4020-3298-6
53. To be published
54. V. Badescu, R.B. Cathcart and R.D. Schuiling (eds.): *Macro-Engineering*. A Challenge for the Future. 2006 ISBN 1-4020-3739-2
55. To be published
56. G. Salvadori, C. De Michele, N.T. Kottegoda and R. Renzo: *Extremes in Nature*. An Approach to Using Copulas. 2007 ISBN 1-4020-4414-3
57. S.K. Jain, P.K. Agarwal and V.P. Singh: *Hydrology and Water Resources of India*. 2007 ISBN 1-4020-5179-4
58. To be published
59. M. Shahin: *Water Resources and Hydrometeorology of the Arab Region*. 2007 ISBN 1-4020-4577-8